Bio-Based Building Materials

RILEM Bookseries

Volume 45

RILEM, The International Union of Laboratories and Experts in Construction Materials, Systems and Structures, founded in 1947, is a non-governmental scientific association whose goal is to contribute to progress in the construction sciences, techniques and industries, essentially by means of the communication it fosters between research and practice. RILEM's focus is on construction materials and their use in building and civil engineering structures, covering all phases of the building process from manufacture to use and recycling of materials. More information on RILEM and its previous publications can be found on www.RILEM.net.

Indexed in SCOPUS, Google Scholar and SpringerLink.

Sofiane Amziane · Ildiko Merta · Jonathan Page
Editors

Bio-Based Building Materials

Proceedings of ICBBM 2023

Springer

Editors
Sofiane Amziane
Institut Pascal CNRS, INP Clermont Auvergne
Université Clermont Auvergne
Aubière, France

Jonathan Page
Laboratoire de Génie Civil et géo-Environnement (LGCgE)
Univ. Artois, IMT Nord Europe, Junia, Univ. Lille
Béthune, France

Ildiko Merta
Faculty of Civil and Environmental Engineering
TU Wien, Institute of Material Technology, Building Physics, and Building Ecology
Vienna, Austria

ISSN 2211-0844 ISSN 2211-0852 (electronic)
RILEM Bookseries
ISBN 978-3-031-33464-1 ISBN 978-3-031-33465-8 (eBook)
https://doi.org/10.1007/978-3-031-33465-8

This Springer imprint is published by the registered company Springer Nature Switzerland AG
The registered company address is: Gewerbestrasse 11, 6330 Cham, Switzerland

Preface ICBBM 2023

Dear colleagues,

After four excellent editions in Clermont-Ferrand in 2015 and 2017, in Belfast in 2019 and in Barcelona in 2021, the fifth edition will take place in the imperial city of Vienna. This capital is renowned for its high quality of life and its exceptional urban facilities.

This fifth ICBBM presents many innovations with the publication of this proceedings containing almost 100 papers that have undergone a selection and evaluation process. It represents the full diversity of the use of bio-based building materials and the great vitality of this scientific sector.

Yes, climate change is much more dramatic for humanity, and it is urgent for us to change our ways of building.

Yes, we have the immense responsibility to continue our march forward by taking maximum sanitary precautions.

ICBBM2023 is an international forum for information dissemination and exchange, discussions and debates on research and practice related to innovative bio-construction materials and technologies with objectives for sustainable development. The purpose of this international conference is to present the latest available scientific and technical information in the field of bio-based building materials. To dynamically address these, and to capitalize on real opportunities which they present for the future, the University Clermont Auvergne and the TU Wien bring together leading building- and civil engineering sector players and international experts, as well as other key stakeholders for what promises to be a landmark event in 2023's professional and business calendar. Working together in an international context is nowadays a basic condition for progress.

The conference attracted 200 participants including academics, scientists, researchers, students, designers, policymakers, and industrial actors from various fields of engineering, materials, sustainable architecture, ecological technologies, biomaterials, materials sciences, environmental engineering, government agencies and end-users. Participants have the opportunity to share ideas on the state-of-the-art innovations, state-of-the-practice and future trends of bio-based and sustainable building materials used in construction.

This proceedings include 78 original peer-reviewed papers part of more than 150 contributions presented at the ICBBM2023 conference. The papers cover a wide spectrum of topics related to the physical- and mechanical properties of bio-based building materials and constructions; their design, performance, testing and durability properties; their application in building constructions with opportunities and challenges as well as the sustainability, life-cycle assessment, modelling and recycling aspects of bio-based building materials.

On behalf of the Local Organizing Committee of ICBBM2023, we would like to take this opportunity to express our sincere thanks to all our contributors and participants for their carefully prepared, stimulating and thought-provoking manuscripts as well as to all the organisers of the conference for their dedicated task.

Thanks are extended to the members of the International Scientific Committee for supporting us in the review process of the papers. The proceedings could not have been published for distribution at the conference without their dedicated efforts. The cooperation of the authors in accepting reviewers' suggestions and revising their manuscripts accordingly is greatly appreciated.

The organisation of a conference of this scale could not have been possible without the support and contributions of many institutions, such as the TU Wien and the Association Universitaire de Génie Civil AUGC and GdR MBS Matériaux de construction biosourcés (CNRS). We also thank RILEM (the International Union of Laboratories and Experts in Construction Materials, Systems and Structures) for the support of the Young Members Evening in the frame of the conference along with the promotion of our conference.

Thanks go to all those individuals who have devoted their time and effort to the organization and realization of the conference, especially Assoc. Prof. Jonathan Page from the Artois University (France) supporting us in the preparation and publication of these proceedings. A special thank goes to the secretary, the staff and students of the research unit of Building Physics at the TU Wien-Institute of Material Technology, Building Physics, and Building Ecology (Austria) for their diligent work in bringing this ICBBM2023 to success!

June 2023

Organization

Chair of the Conference

Prof. Sofiane Amziane, Institut Pascal CNRS, INP Clermont Auvergne Université Clermont Auvergne Aubière, France

Co-Chair of the Conference

Prof. Ildiko Merta, Faculty of Civil and Environmental Engineering TU Wien, Institute of Material Technology, Building Physics, and Building Ecology Vienna, Austria

Local Organizing Committee-TU Wien

Ildiko Merta (chair)
Thomas Bednar
Jasmin Berger-Heda
Bojan Poletanovic
Sabine Sint
Manfred Grüner
Harald Hofbauer
Dalia Hammash

Local Scientific Committee

Ildiko Merta	TU Wien, Vienna
Thomas Bednar	TU Wien, Vienna
Andrea Rieger-Jandl	TU Wien, Vienna
Benjamin Kromoser	Boku, Vienna
Maximilian Neusser	TU Wien, Vienna

International Scientific Committee

K. Abahri	ENS Cachan, France
A. A. Akindahunsi	Obafemi Awolowo Univ., Nigeria
S. Amziane	Université Clermont Auvergne, France
A. Armada Bras	Liverpool J. Moores University, UK
T. Bednar	TU Wien, Austria
L. Bessette	Centre Technique Louis VICAT, France
P. Blanchet	University Laval, QC, Canada
J. Brouwers	University of Eindhoven, Netherlands
C. Buratti	Univ. of Perugia, Italia
T. Colinart	University of South Brittany, France
F. Collet	University of Rennes 1, France
D. Dodoo-Arhin	University of Ghana, Ghana
G. Escadeillas	University of Toulouse, France
P. Faria	Nova de Lisboa University, Portugal
R. Fangueiro	University of Minho, Portugal
F. Gauvin	University of Eindhoven, Netherlands
E. Ghorbel	Cergy-Pointoise, Paris, France
P. Glé	Cerema, Strasbourg, France
E. Gourlay	Cerema, Strasbourg, France

G, Habert	ETH Zurich, Suisse
Li. Junjie	Beijing Jiqotong University, China
S. R. Karade	CSIR-C. Building Research Inst., India
Said Kenai	University of Blida, Algeria
O. Kinnane	University College Dublin, Ireland
J. Khatib	UK/Beirut Arab Uni., Lebanon
Olonade Kolawole A	University of Lagos, Nigeria
R. N. Krishna	Contech, Chennai, India
G.Senthil Kumaran	The Copperbelt University, Zambia
C. Lanos	University of Rennes 1, France
E. Latif	Cardiff University, UK
M. Lawrence	University of Bath, UK
N. Lushnikova	Univ. of Water of Env., Ukraine
C. Magniont	University of Toulouse, France
S. Marceau	IFSTTAR, France
I. Merta	TU Wien, Austria
S. Mertens	BBRI, Belgium
L. Nunes	LNEC, Lisbon, Portugal
S. Nwaubani	Univ. of the Witwatersrand, South Africa
S. Ouldboukhitine	Université Clermont Auvergne
J. Page	Artois University, Béthune, France
M. Palumbo	Catalunya Polytech Uni, Spain
D. Panesar	University Toronto, Canada
S. Pavia	Trinity College Dublin, Ireland
A. Perrot	University Bretagne Sud, France
V. Picandet	University Bretagne Sud, France
S. Pretot	University Rennes 1, France
Akeem A. Raheem	Ladoke Akintola Univ., Nigeria
N. Radebe	Institute for Chemical Technology, Germany
H. Savastano	University of Sao Polo, Brazil
W. Schmidt	BAM, Germany
N. Sebaibi	ESITC Caen, France
M. Sonebi	Queen's University of Belfast, UK
P. Strandberg	Lund University, Sweden
F. Tittarelli	University Polytech Delle Marche, Italy
E. Toussaint	Université Clermont Auvergne, France
T. Tahenni	Djilali Bounaama University of Khemis Miliana, Algeria
RD. Toledo Filho	University Fed. -Rio de Janeiro, Brazil
Y. Xiao, ZJUI	Zhejiang University, China

List of the Reviewers

Akindahunsi, Dr. Akindehinde, Senior Lecturer
Amziane, Prof. Sofiane, Professor
Bednar, Prof. Thomas, Univ. Prof. DI, Dr.
Blanchet, Prof. Pierre, Professor
Colinart, Dr. Thibaut, Assistant Professor
Collet, Dr. Florence, Doctor
Faria, Prof. Paulina, Associate Professor
Ganesan, Prof. Senthil Kumaran, Associate Professor/Head of Department
Ghorbel, Prof. Elhem, Professor
Glé, Philippe, Researcher
Gourlay, Dr. Etienne, Researcher
Homoro, Dr. Omayma, Associate Professor
Karade, Prof. Sukhdeo, Chief Scientist
Keita, Dr. Emmanuel, Researcher
Kenai, Prof. Said, Professor
Lanos, Prof. Christophe, Professor
Lushnikova, Dr. Nataliya, Associate Professor
Magniont, Prof. Camille, Full Professor
Marceau, Dr. Sandrine, Researcher
Merta, Prof. Ildiko, Prof. Dr.
Nunes, Dr. Lina, Researcher
Nwaubani, Prof. Sunday, Professor
Page, Dr. Jonathan, Associate Professor
Palumbo, Mariana, Lecturer
Panesar, Prof. D., Professor
Pavia, Dr. Sara, Dr.
Perrot, Prof. Arnaud, Professor
Picandet, Vincent, Associate Professor
Poletanovic, Bojan, Dipl.-Ing.
Prétot, Dr. Sylvie, Dr.
Schmidt, Dr. Wolfram, Dr.-Dipl.-Ing.
Sonebi, Prof. Mohammed, Professor
Strandberg-de Bruijn, Dr. Paulien Brigitte, Researcher
Tahenni, Dr. Touhami, Dr.
Toledo Filho, Prof. Romildo Dias, Professor
Toussaint, Prof. Évelyne, Professor
Zea Escamilla, Dr. Edwin, Dr.sc.

Contents

Physical Properties of Bio-based Building Materials

Earth Constructions and Building Materials

Innovative Bio-Binders and Additives

Modeling and Digitalization of Bio-based Building Materials and Constructions

RILEM Publications

The following list is presenting the global offer of RILEM Publications, sorted by series. Each publication is available in printed version and/or in online version.

RILEM Proceedings (PRO)

PRO 1: Durability of High Performance Concrete (ISBN: 2-912143-03-9; e-ISBN: 2-351580-12-5; e-ISBN: 2351580125); *Ed. H. Sommer*

PRO 2: Chloride Penetration into Concrete (ISBN: 2-912143-00-04; e-ISBN: 2912143454); *Eds. L.-O. Nilsson and J.-P. Ollivier*

PRO 3: Evaluation and Strengthening of Existing Masonry Structures (ISBN: 2-912143-02-0; e-ISBN: 2351580141); *Eds. L. Binda and C. Modena*

PRO 4: Concrete: From Material to Structure (ISBN: 2-912143-04-7; e-ISBN: 2351580206); *Eds. J.-P. Bournazel and Y. Malier*

PRO 5: The Role of Admixtures in High Performance Concrete (ISBN: 2-912143-05-5; e-ISBN: 2351580214); *Eds. J. G. Cabrera and R. Rivera-Villarreal*

PRO 6: High Performance Fiber Reinforced Cement Composites - HPFRCC 3 (ISBN: 2-912143-06-3; e-ISBN: 2351580222); *Eds. H. W. Reinhardt and A. E. Naaman*

PRO 7: 1st International RILEM Symposium on Self-Compacting Concrete (ISBN: 2-912143-09-8; e-ISBN: 2912143721); *Eds. Å. Skarendahl and Ö. Petersson*

PRO 8: International RILEM Symposium on Timber Engineering (ISBN: 2-912143-10-1; e-ISBN: 2351580230); *Ed. L. Boström*

PRO 9: 2nd International RILEM Symposium on Adhesion between Polymers and Concrete ISAP '99 (ISBN: 2-912143-11-X; e-ISBN: 2351580249); *Eds. Y. Ohama and M. Puterman*

PRO 10: 3rd International RILEM Symposium on Durability of Building and Construction Sealants (ISBN: 2-912143-13-6; e-ISBN: 2351580257); *Ed. A. T. Wolf*

PRO 11: 4th International RILEM Conference on Reflective Cracking in Pavements (ISBN: 2-912143-14-4; e-ISBN: 2351580265); *Eds. A. O. Abd El Halim, D. A. Taylor and El H. H. Mohamed*

PRO 12: International RILEM Workshop on Historic Mortars: Characteristics and Tests (ISBN: 2-912143-15-2; e-ISBN: 2351580273); *Eds. P. Bartos, C. Groot and J. J. Hughes*

PRO 13: 2nd International RILEM Symposium on Hydration and Setting (ISBN: 2-912143-16-0; e-ISBN: 2351580281); *Ed. A. Nonat*

PRO 14: Integrated Life-Cycle Design of Materials and Structures - ILCDES 2000 (ISBN: 951-758-408-3; e-ISBN: 235158029X); (ISSN: 0356-9403); *Ed. S. Sarja*

PRO 15: Fifth RILEM Symposium on Fibre-Reinforced Concretes (FRC) - BEFIB'2000 (ISBN: 2-912143-18-7; e-ISBN: 291214373X); *Eds. P. Rossi and G. Chanvillard*

PRO 16: Life Prediction and Management of Concrete Structures (ISBN: 2-912143-19-5; e-ISBN: 2351580303); *Ed. D. Naus*

PRO 17: Shrinkage of Concrete – Shrinkage 2000 (ISBN: 2-912143-20-9; e-ISBN: 2351580311); *Eds. V. Baroghel-Bouny and P.-C. Aïtcin*

PRO 18: Measurement and Interpretation of the On-Site Corrosion Rate (ISBN: 2-912143-21-7; e-ISBN: 235158032X); *Eds. C. Andrade, C. Alonso, J. Fullea, J. Polimon and J. Rodriguez*

PRO 19: Testing and Modelling the Chloride Ingress into Concrete (ISBN: 2-912143-22-5; e-ISBN: 2351580338); *Eds. C. Andrade and J. Kropp*

PRO 20: 1st International RILEM Workshop on Microbial Impacts on Building Materials (CD 02) (e-ISBN 978-2-35158-013-4); *Ed. M. Ribas Silva*

PRO 21: International RILEM Symposium on Connections between Steel and Concrete (ISBN: 2-912143-25-X; e-ISBN: 2351580346); *Ed. R. Eligehausen*

PRO 22: International RILEM Symposium on Joints in Timber Structures (ISBN: 2-912143-28-4; e-ISBN: 2351580354); *Eds. S. Aicher and H.-W. Reinhardt*

PRO 23: International RILEM Conference on Early Age Cracking in Cementitious Systems (ISBN: 2-912143-29-2; e-ISBN: 2351580362); *Eds. K. Kovler and A. Bentur*

PRO 24: 2nd International RILEM Workshop on Frost Resistance of Concrete (ISBN: 2-912143-30-6; e-ISBN: 2351580370); *Eds. M. J. Setzer, R. Auberg and H.-J. Keck*

PRO 25: International RILEM Workshop on Frost Damage in Concrete (ISBN: 2-912143-31-4; e-ISBN: 2351580389); *Eds. D. J. Janssen, M. J. Setzer and M. B. Snyder*

PRO 26: International RILEM Workshop on On-Site Control and Evaluation of Masonry Structures (ISBN: 2-912143-34-9; e-ISBN: 2351580141); *Eds. L. Binda and R. C. de Vekey*

PRO 27: International RILEM Symposium on Building Joint Sealants (CD03; e-ISBN: 235158015X); *Ed. A. T. Wolf*

PRO 28: 6th International RILEM Symposium on Performance Testing and Evaluation of Bituminous Materials - PTEBM'03 (ISBN: 2-912143-35-7; e-ISBN: 978-2-912143-77-8); *Ed. M. N. Partl*

PRO 29: 2nd International RILEM Workshop on Life Prediction and Ageing Management of Concrete Structures (ISBN: 2-912143-36-5; e-ISBN: 2912143780); *Ed. D. J. Naus*

PRO 30: 4th International RILEM Workshop on High Performance Fiber Reinforced Cement Composites - HPFRCC 4 (ISBN: 2-912143-37-3; e-ISBN: 2912143799); *Eds. A. E. Naaman and H. W. Reinhardt*

PRO 31: International RILEM Workshop on Test and Design Methods for Steel Fibre Reinforced Concrete: Background and Experiences (ISBN: 2-912143-38-1; e-ISBN: 2351580168); *Eds. B. Schnütgen and L. Vandewalle*

PRO 32: International Conference on Advances in Concrete and Structures 2 vol. (ISBN (set): 2-912143-41-1; e-ISBN: 2351580176); *Eds. Ying-shu Yuan, Surendra P. Shah and Heng-lin Lü*

PRO 33: 3rd International Symposium on Self-Compacting Concrete (ISBN: 2-912143-42-X; e-ISBN: 2912143713); *Eds. Ó. Wallevik and I. Níelsson*

PRO 34: International RILEM Conference on Microbial Impact on Building Materials (ISBN: 2-912143-43-8; e-ISBN: 2351580184); *Ed. M. Ribas Silva*

PRO 35: International RILEM TC 186-ISA on Internal Sulfate Attack and Delayed Ettringite Formation (ISBN: 2-912143-44-6; e-ISBN: 2912143802); *Eds. K. Scrivener and J. Skalny*

PRO 36: International RILEM Symposium on Concrete Science and Engineering – A Tribute to Arnon Bentur (ISBN: 2-912143-46-2; e-ISBN: 2912143586); *Eds. K. Kovler, J. Marchand, S. Mindess and J. Weiss*

PRO 37: 5th International RILEM Conference on Cracking in Pavements – Mitigation, Risk Assessment and Prevention (ISBN: 2-912143-47-0; e-ISBN: 2912143764); *Eds. C. Petit, I. Al-Qadi and A. Millien*

PRO 38: 3rd International RILEM Workshop on Testing and Modelling the Chloride Ingress into Concrete (ISBN: 2-912143-48-9; e-ISBN: 2912143578); *Eds. C. Andrade and J. Kropp*

PRO 39: 6th International RILEM Symposium on Fibre-Reinforced Concretes - BEFIB 2004 (ISBN: 2-912143-51-9; e-ISBN: 2912143748); *Eds. M. Di Prisco, R. Felicetti and G. A. Plizzari*

PRO 40: International RILEM Conference on the Use of Recycled Materials in Buildings and Structures (ISBN: 2-912143-52-7; e-ISBN: 2912143756); *Eds. E. Vázquez, Ch. F. Hendriks and G. M. T. Janssen*

PRO 41: RILEM International Symposium on Environment-Conscious Materials and Systems for Sustainable Development (ISBN: 2-912143-55-1; e-ISBN: 2912143640); *Eds. N. Kashino and Y. Ohama*

PRO 42: SCC'2005 - China: 1st International Symposium on Design, Performance and Use of Self-Consolidating Concrete (ISBN: 2-912143-61-6; e-ISBN: 2912143624); *Eds. Zhiwu Yu, Caijun Shi, Kamal Henri Khayat and Youjun Xie*

PRO 43: International RILEM Workshop on Bonded Concrete Overlays (e-ISBN: 2-912143-83-7); *Eds. J. L. Granju and J. Silfwerbrand*

PRO 44: 2nd International RILEM Workshop on Microbial Impacts on Building Materials (CD11) (e-ISBN: 2-912143-84-5); *Ed. M. Ribas Silva*

PRO 45: 2nd International Symposium on Nanotechnology in Construction, Bilbao (ISBN: 2-912143-87-X; e-ISBN: 2912143888); *Eds. Peter J. M. Bartos, Yolanda de Miguel and Antonio Porro*

PRO 46: ConcreteLife'06 - International RILEM-JCI Seminar on Concrete Durability and Service Life Planning: Curing, Crack Control, Performance in Harsh Environments (ISBN: 2-912143-89-6; e-ISBN: 291214390X); *Ed. K. Kovler*

PRO 47: International RILEM Workshop on Performance Based Evaluation and Indicators for Concrete Durability (ISBN: 978-2-912143-95-2; e-ISBN: 9782912143969); *Eds. V. Baroghel-Bouny, C. Andrade, R. Torrent and K. Scrivener*

PRO 48: 1st International RILEM Symposium on Advances in Concrete through Science and Engineering (e-ISBN: 2-912143-92-6); *Eds. J. Weiss, K. Kovler, J. Marchand and S. Mindess*

PRO 49: International RILEM Workshop on High Performance Fiber Reinforced Cementitious Composites in Structural Applications (ISBN: 2-912143-93-4; e-ISBN: 2912143942); *Eds. G. Fischer and V. C. Li*

PRO 50: 1st International RILEM Symposium on Textile Reinforced Concrete (ISBN: 2-912143-97-7; e-ISBN: 2351580087); *Eds. Josef Hegger, Wolfgang Brameshuber and Norbert Will*

PRO 51: 2nd International Symposium on Advances in Concrete through Science and Engineering (ISBN: 2-35158-003-6; e-ISBN: 2-35158-002-8); *Eds. J. Marchand, B. Bissonnette, R. Gagné, M. Jolin and F. Paradis*

PRO 52: Volume Changes of Hardening Concrete: Testing and Mitigation (ISBN: 2-35158-004-4; e-ISBN: 2-35158-005-2); *Eds. O. M. Jensen, P. Lura and K. Kovler*

PRO 53: High Performance Fiber Reinforced Cement Composites - HPFRCC5 (ISBN: 978-2-35158-046-2; e-ISBN: 978-2-35158-089-9); *Eds. H. W. Reinhardt and A. E. Naaman*

PRO 54: 5th International RILEM Symposium on Self-Compacting Concrete (ISBN: 978-2-35158-047-9; e-ISBN: 978-2-35158-088-2); *Eds. G. De Schutter and V. Boel*

PRO 55: International RILEM Symposium Photocatalysis, Environment and Construction Materials (ISBN: 978-2-35158-056-1; e-ISBN: 978-2-35158-057-8); *Eds. P. Baglioni and L. Cassar*

PRO 56: International RILEM Workshop on Integral Service Life Modelling of Concrete Structures (ISBN 978-2-35158-058-5; e-ISBN: 978-2-35158-090-5); *Eds. R. M. Ferreira, J. Gulikers and C. Andrade*

PRO 57: RILEM Workshop on Performance of cement-based materials in aggressive aqueous environments (e-ISBN: 978-2-35158-059-2); *Ed. N. De Belie*

PRO 58: International RILEM Symposium on Concrete Modelling - CONMOD'08 (ISBN: 978-2-35158-060-8; e-ISBN: 978-2-35158-076-9); *Eds. E. Schlangen and G. De Schutter*

PRO 59: International RILEM Conference on On Site Assessment of Concrete, Masonry and Timber Structures - SACoMaTiS 2008 (ISBN set: 978-2-35158-061-5; e-ISBN: 978-2-35158-075-2); *Eds. L. Binda, M. di Prisco and R. Felicetti*

PRO 60: Seventh RILEM International Symposium on Fibre Reinforced Concrete: Design and Applications - BEFIB 2008 (ISBN: 978-2-35158-064-6; e-ISBN: 978-2-35158-086-8); *Ed. R. Gettu*

PRO 61: 1st International Conference on Microstructure Related Durability of Cementitious Composites 2 vol., (ISBN: 978-2-35158-065-3; e-ISBN: 978-2-35158-084-4); *Eds. W. Sun, K. van Breugel, C. Miao, G. Ye and H. Chen*

PRO 62: NSF/ RILEM Workshop: In-situ Evaluation of Historic Wood and Masonry Structures (e-ISBN: 978-2-35158-068-4); *Eds. B. Kasal, R. Anthony and M. Drdácký*

PRO 63: Concrete in Aggressive Aqueous Environments: Performance, Testing and Modelling, 2 vol., (ISBN: 978-2-35158-071-4; e-ISBN: 978-2-35158-082-0); *Eds. M. G. Alexander and A. Bertron*

PRO 64: Long Term Performance of Cementitious Barriers and Re inforced Concrete in Nuclear Power Plants and Waste Management - NUCPERF 2009 (ISBN: 978-2-35158-072-1; e-ISBN: 978-2-35158-087-5); *Eds. V. L'Hostis, R. Gens and C. Gallé*

PRO 65: Design Performance and Use of Self-consolidating Concrete - SCC'2009 (ISBN: 978-2-35158-073-8; e-ISBN: 978-2-35158-093-6); *Eds. C. Shi, Z. Yu, K. H. Khayat and P. Yan*

PRO 66: 2nd International RILEM Workshop on Concrete Durability and Service Life Planning - ConcreteLife'09 (ISBN: 978-2-35158-074-5; ISBN: 978-2-35158-074-5); *Ed. K. Kovler*

PRO 67: Repairs Mortars for Historic Masonry (e-ISBN: 978-2-35158-083-7); *Ed. C. Groot*

PRO 68: Proceedings of the 3rd International RILEM Symposium on 'Rheology of Cement Suspensions such as Fresh Concrete (ISBN 978-2-35158-091-2; e-ISBN: 978-2-35158-092-9); *Eds. O. H. Wallevik, S. Kubens and S. Oesterheld*

PRO 69: 3rd International PhD Student Workshop on 'Modelling the Durability of Reinforced Concrete (ISBN: 978-2-35158-095-0); *Eds. R. M. Ferreira, J. Gulikers and C. Andrade*

PRO 70: 2nd International Conference on 'Service Life Design for Infrastructure' (ISBN set: 978-2-35158-096-7, e-ISBN: 978-2-35158-097-4); *Eds. K. van Breugel, G. Ye and Y. Yuan*

PRO 71: Advances in Civil Engineering Materials - The 50-year Teaching Anniversary of Prof. Sun Wei' (ISBN: 978-2-35158-098-1; e-ISBN: 978-2-35158-099-8); *Eds. C. Miao, G. Ye and H. Chen*

PRO 72: First International Conference on 'Advances in Chemically-Activated Materials – CAM'2010' (2010), 264 pp, ISBN: 978-2-35158-101-8; e-ISBN: 978-2-35158-115-5, *Eds. Caijun Shi and Xiaodong Shen*

PRO 73: 2nd International Conference on 'Waste Engineering and Management - ICWEM 2010' (2010), 894 pp, ISBN: 978-2-35158-102-5; e-ISBN: 978-2-35158-103-2, *Eds. J. Zh. Xiao, Y. Zhang, M. S. Cheung and R. Chu*

PRO 74: International RILEM Conference on 'Use of Superabsorbent Polymers and Other New Addditives in Concrete' (2010) 374 pp., ISBN: 978-2-35158-104-9; e-ISBN: 978-2-35158-105-6; *Eds. O. M. Jensen, M. T. Hasholt and S. Laustsen*

PRO 75: International Conference on 'Material Science - 2nd ICTRC - Textile Reinforced Concrete - Theme 1' (2010) 436 pp., ISBN: 978-2-35158-106-3; e-ISBN: 978-2-35158-107-0; *Ed. W. Brameshuber*

PRO 76: International Conference on 'Material Science - HetMat - Modelling of Heterogeneous Materials - Theme 2' (2010) 255 pp., ISBN: 978-2-35158-108-7; e-ISBN: 978-2-35158-109-4; *Ed. W. Brameshuber*

PRO 77: International Conference on 'Material Science - AdIPoC - Additions Improving Properties of Concrete - Theme 3' (2010) 459 pp., ISBN: 978-2-35158-110-0; e-ISBN: 978-2-35158-111-7; *Ed. W. Brameshuber*

PRO 78: 2nd Historic Mortars Conference and RILEM TC 203-RHM Final Workshop – HMC2010 (2010) 1416 pp., e-ISBN: 978-2-35158-112-4; *Eds. J. Válek, C. Groot and J. J. Hughes*

PRO 79: International RILEM Conference on Advances in Construction Materials Through Science and Engineering (2011) 213 pp., ISBN: 978-2-35158-116-2, e-ISBN: 978-2-35158-117-9; *Eds. Christopher Leung and K. T. Wan*

PRO 80: 2nd International RILEM Conference on Concrete Spalling due to Fire Exposure (2011) 453 pp., ISBN: 978-2-35158-118-6, e-ISBN: 978-2-35158-119-3; *Eds. E. A. B. Koenders and F. Dehn*

PRO 81: 2nd International RILEM Conference on Strain Hardening Cementitious Composites (SHCC2-Rio) (2011) 451 pp., ISBN: 978-2-35158-120-9, e-ISBN: 978-2-35158-121-6; *Eds. R. D. Toledo Filho, F. A. Silva, E. A. B. Koenders and E. M. R. Fairbairn*

PRO 82: 2nd International RILEM Conference on Progress of Recycling in the Built Environment (2011) 507 pp., e-ISBN: 978-2-35158-122-3; *Eds. V. M. John, E. Vazquez, S. C. Angulo and C. Ulsen*

PRO 83: 2nd International Conference on Microstructural-related Durability of Cementitious Composites (2012) 250 pp., ISBN: 978-2-35158-129-2; e-ISBN: 978-2-35158-123-0; *Eds. G. Ye, K. van Breugel, W. Sun and C. Miao*

PRO 84: CONSEC13 - Seventh International Conference on Concrete under Severe Conditions – Environment and Loading (2013) 1930 pp., ISBN: 978-2-35158-124-7; e-ISBN: 978-2-35158-134-6; *Eds. Z. J. Li, W. Sun, C. W. Miao, K. Sakai, O. E. Gjorv and N. Banthia*

PRO 85: RILEM-JCI International Workshop on Crack Control of Mass Concrete and Related issues concerning Early-Age of Concrete Structures – ConCrack 3 – Control of Cracking in Concrete Structures 3 (2012) 237 pp., ISBN: 978-2-35158-125-4; e-ISBN: 978-2-35158-126-1; *Eds. F. Toutlemonde and J.-M. Torrenti*

PRO 86: International Symposium on Life Cycle Assessment and Construction (2012) 414 pp., ISBN: 978-2-35158-127-8, e-ISBN: 978-2-35158-128-5; *Eds. A. Ventura and C. de la Roche*

PRO 87: UHPFRC 2013 – RILEM-fib-AFGC International Symposium on Ultra-High Performance Fibre-Reinforced Concrete (2013), ISBN: 978-2-35158-130-8, e-ISBN: 978-2-35158-131-5; *Eds. F. Toutlemonde*

PRO 88: 8th RILEM International Symposium on Fibre Reinforced Concrete (2012) 344 pp., ISBN: 978-2-35158-132-2, e-ISBN: 978-2-35158-133-9; *Eds. Joaquim A. O. Barros*

PRO 89: RILEM International workshop on performance-based specification and control of concrete durability (2014) 678 pp, ISBN: 978-2-35158-135-3, e-ISBN: 978-2-35158-136-0; *Eds. D. Bjegović, H. Beushausen and M. Serdar*

PRO 90: 7th RILEM International Conference on Self-Compacting Concrete and of the 1st RILEM International Conference on Rheology and Processing of Construction Materials (2013) 396 pp, ISBN: 978-2-35158-137-7, e-ISBN: 978-2-35158-138-4; *Eds. Nicolas Roussel and Hela Bessaies-Bey*

PRO 91: CONMOD 2014 - RILEM International Symposium on Concrete Modelling (2014), ISBN: 978-2-35158-139-1; e-ISBN: 978-2-35158-140-7; *Eds. Kefei Li, Peiyu Yan and Rongwei Yang*

PRO 92: CAM 2014 - 2nd International Conference on advances in chemically-activated materials (2014) 392 pp., ISBN: 978-2-35158-141-4; e-ISBN: 978-2-35158-142-1; *Eds. Caijun Shi and Xiadong Shen*

PRO 93: SCC 2014 - 3rd International Symposium on Design, Performance and Use of Self-Consolidating Concrete (2014) 438 pp., ISBN: 978-2-35158-143-8; e-ISBN: 978-2-35158-144-5; *Eds. Caijun Shi, Zhihua Ou and Kamal H. Khayat*

PRO 94 (online version): HPFRCC-7 - 7th RILEM conference on High performance fiber reinforced cement composites (2015), e-ISBN: 978-2-35158-146-9; *Eds. H. W. Reinhardt, G. J. Parra-Montesinos and H. Garrecht*

PRO 95: International RILEM Conference on Application of superabsorbent polymers and other new admixtures in concrete construction (2014), ISBN: 978-2-35158-147-6; e-ISBN: 978-2-35158-148-3; *Eds. Viktor Mechtcherine and Christof Schroefl*

PRO 96 (online version): XIII DBMC: XIII International Conference on Durability of Building Materials and Components (2015), e-ISBN: 978-2-35158-149-0; *Eds. M. Quattrone and V. M. John*

PRO 97: SHCC3 – 3rd International RILEM Conference on Strain Hardening Cementitious Composites (2014), ISBN: 978-2-35158-150-6; e-ISBN: 978-2-35158-151-3; *Eds. E. Schlangen, M. G. Sierra Beltran, M. Lukovic and G. Ye*

PRO 98: FERRO-11 – 11th International Symposium on Ferrocement and 3rd ICTRC - International Conference on Textile Reinforced Concrete (2015), ISBN: 978-2-35158-152-0; e-ISBN: 978-2-35158-153-7; *Ed. W. Brameshuber*

PRO 99 (online version): ICBBM 2015 - 1st International Conference on Bio-Based Building Materials (2015), e-ISBN: 978-2-35158-154-4; *Eds. S. Amziane and M. Sonebi*

PRO 100: SCC16 - RILEM Self-Consolidating Concrete Conference (2016), ISBN: 978-2-35158-156-8; e-ISBN: 978-2-35158-157-5; *Ed. Kamal H. Kayat*

PRO 101 (online version): III Progress of Recycling in the Built Environment (2015), e-ISBN: 978-2-35158-158-2; *Eds. I. Martins, C. Ulsen and S. C. Angulo*

PRO 102 (online version): RILEM Conference on Microorganisms-Cementitious Materials Interactions (2016), e-ISBN: 978-2-35158-160-5; *Eds. Alexandra Bertron, Henk Jonkers and Virginie Wiktor*

PRO 103 (online version): ACESC'16 - Advances in Civil Engineering and Sustainable Construction (2016), e-ISBN: 978-2-35158-161-2; *Eds. T. Ch. Madhavi, G. Prabhakar, Santhosh Ram and P. M. Rameshwaran*

PRO 104 (online version): SSCS'2015 - Numerical Modeling - Strategies for Sustainable Concrete Structures (2015), e-ISBN: 978-2-35158-162-9

PRO 105: 1st International Conference on UHPC Materials and Structures (2016), ISBN: 978-2-35158-164-3, e-ISBN: 978-2-35158-165-0

PRO 106: AFGC-ACI-fib-RILEM International Conference on Ultra-High-Performance Fibre-Reinforced Concrete – UHPFRC 2017 (2017), ISBN: 978-2-35158-166-7, e-ISBN: 978-2-35158-167-4; *Eds. François Toutlemonde and Jacques Resplendino*

PRO 107 (online version): XIV DBMC – 14th International Conference on Durability of Building Materials and Components (2017), e-ISBN: 978-2-35158-159-9; *Eds. Geert De Schutter, Nele De Belie, Arnold Janssens and Nathan Van Den Bossche*

PRO 108: MSSCE 2016 - Innovation of Teaching in Materials and Structures (2016), ISBN: 978-2-35158-178-0, e-ISBN: 978-2-35158-179-7; *Ed. Per Goltermann*

PRO 109 (2 volumes): MSSCE 2016 - Service Life of Cement-Based Materials and Structures (2016), ISBN Vol. 1: 978-2-35158-170-4, Vol. 2: 978-2-35158-171-4, Set Vol. 1&2: 978-2-35158-172-8, e-ISBN: 978-2-35158-173-5; *Eds. Miguel Azenha, Ivan Gabrijel, Dirk Schlicke, Terje Kanstad and Ole Mejlhede Jensen*

PRO 110: MSSCE 2016 - Historical Masonry (2016), ISBN: 978-2-35158-178-0, e-ISBN: 978-2-35158-179-7; *Eds. Inge Rörig-Dalgaard and Ioannis Ioannou*

PRO 111: MSSCE 2016 - Electrochemistry in Civil Engineering (2016), ISBN: 978-2-35158-176-6, e-ISBN: 978-2-35158-177-3; *Ed. Lisbeth M. Ottosen*

PRO 112: MSSCE 2016 - Moisture in Materials and Structures (2016), ISBN: 978-2-35158-178-0, e-ISBN: 978-2-35158-179-7; *Eds. Kurt Kielsgaard Hansen, Carsten Rode and Lars-Olof Nilsson*

PRO 113: MSSCE 2016 - Concrete with Supplementary Cementitious Materials (2016), ISBN: 978-2-35158-178-0, e-ISBN: 978-2-35158-179-7; *Eds. Ole Mejlhede Jensen, Konstantin Kovler and Nele De Belie*

PRO 114: MSSCE 2016 - Frost Action in Concrete (2016), ISBN: 978-2-35158-182-7, e-ISBN: 978-2-35158-183-4; *Eds. Marianne Tange Hasholt, Katja Fridh and R. Doug Hooton*

PRO 115: MSSCE 2016 - Fresh Concrete (2016), ISBN: 978-2-35158-184-1, e-ISBN: 978-2-35158-185-8; *Eds. Lars N. Thrane, Claus Pade, Oldrich Svec and Nicolas Roussel*

PRO 116: BEFIB 2016 – 9th RILEM International Symposium on Fiber Reinforced Concrete (2016), ISBN: 978-2-35158-187-2, e-ISBN: 978-2-35158-186-5; *Eds. N. Banthia, M. di Prisco and S. Soleimani-Dashtaki*

PRO 117: 3rd International RILEM Conference on Microstructure Related Durability of Cementitious Composites (2016), ISBN: 978-2-35158-188-9, e-ISBN: 978-2-35158-189-6; *Eds. Changwen Miao, Wei Sun, Jiaping Liu, Huisu Chen, Guang Ye and Klaas van Breugel*

PRO 118 (4 volumes): International Conference on Advances in Construction Materials and Systems (2017), ISBN Set: 978-2-35158-190-2, Vol. 1: 978-2-35158-193-3, Vol. 2: 978-2-35158-194-0, Vol. 3: ISBN:978-2-35158-195-7, Vol. 4: ISBN:978-2-35158-196-4, e-ISBN: 978-2-35158-191-9; *Eds. Manu Santhanam, Ravindra Gettu, Radhakrishna G. Pillai and Sunitha K. Nayar*

PRO 119 (online version): ICBBM 2017 - Second International RILEM Conference on Bio-based Building Materials, (2017), e-ISBN: 978-2-35158-192-6; *Eds. Sofiane Amziane*

PRO 120 (2 volumes): EAC-02 - 2nd International RILEM/COST Conference on Early Age Cracking and Serviceability in Cement-based Materials and Structures, (2017), Vol. 1: 978-2-35158-199-5, Vol. 2: 978-2-35158-200-8, Set: 978-2-35158-197-1, e-ISBN: 978-2-35158-198-8; *Eds. Stéphanie Staquet and Dimitrios Aggelis*

PRO 121 (2 volumes): SynerCrete18: Interdisciplinary Approaches for Cement-based Materials and Structural Concrete: Synergizing Expertise and Bridging Scales of Space and Time, (2018), Set: 978-2-35158-202-2, Vol.1: 978-2-35158-211-4, Vol. 2: 978-2-35158-212-1, e-ISBN: 978-2-35158-203-9; *Eds. Miguel Azenha, Dirk Schlicke, Farid Benboudjema and Agnieszka Knoppik*

PRO 122: SCC'2018 China - Fourth International Symposium on Design, Performance and Use of Self-Consolidating Concrete, (2018), ISBN: 978-2-35158-204-6, e-ISBN: 978-2-35158-205-3; *Eds. C. Shi, Z. Zhang and K. H. Khayat*

PRO 123: Final Conference of RILEM TC 253-MCI: Microorganisms-Cementitious Materials Interactions (2018), Set: 978-2-35158-207-7, Vol.1: 978-2-35158-209-1, Vol.2: 978-2-35158-210-7, e-ISBN: 978-2-35158-206-0; *Ed. Alexandra Bertron*

PRO 124 (online version): Fourth International Conference Progress of Recycling in the Built Environment (2018), e-ISBN: 978-2-35158-208-4; *Eds. Isabel M. Martins, Carina Ulsen and Yury Villagran*

PRO 125 (online version): SLD4 - 4th International Conference on Service Life Design for Infrastructures (2018), e-ISBN: 978-2-35158-213-8; *Eds. Guang Ye, Yong Yuan, Claudia Romero Rodriguez, Hongzhi Zhang and Branko Savija*

PRO 126: Workshop on Concrete Modelling and Material Behaviour in honor of Professor Klaas van Breugel (2018), ISBN: 978-2-35158-214-5, e-ISBN: 978-2-35158-215-2; *Ed. Guang Ye*

PRO 127 (online version): CONMOD2018 - Symposium on Concrete Modelling (2018), e-ISBN: 978-2-35158-216-9; *Eds. Erik Schlangen, Geert de Schutter, Branko Savija, Hongzhi Zhang and Claudia Romero Rodriguez*

PRO 128: SMSS2019 - International Conference on Sustainable Materials, Systems and Structures (2019), ISBN: 978-2-35158-217-6, e-ISBN: 978-2-35158-218-3

PRO 129: 2nd International Conference on UHPC Materials and Structures (UHPC2018-China), ISBN: 978-2-35158-219-0, e-ISBN: 978-2-35158-220-6;

PRO 130: 5th Historic Mortars Conference (2019), ISBN: 978-2-35158-221-3, e-ISBN: 978-2-35158-222-0; *Eds. José Ignacio Álvarez, José María Fernández, Íñigo Navarro, Adrián Durán and Rafael Sirera*

PRO 131 (online version): 3rd International Conference on Bio-Based Building Materials (ICBBM2019), e-ISBN: 978-2-35158-229-9; *Eds. Mohammed Sonebi, Sofiane Amziane and Jonathan Page*

PRO 132: IRWRMC'18 - International RILEM Workshop on Rheological Measurements of Cement-based Materials (2018), ISBN: 978-2-35158-230-5, e-ISBN: 978-2-35158-231-2; *Eds. Chafika Djelal and Yannick Vanhove*

PRO 133 (online version): CO2STO2019 - International Workshop CO2 Storage in Concrete (2019), e-ISBN: 978-2-35158-232-9; *Eds. Assia Djerbi, Othman Omikrine-Metalssi and Teddy Fen-Chong*

PRO 134: 3rd ACF/HNU International Conference on UHPC Materials and Structures - UHPC'2020, ISBN: 978-2-35158-233-6, e-ISBN: 978-2-35158-234-3; *Eds. Caijun Shi and Jiaping Liu*

PRO 135: Fourth International Conference on Chemically Activated Materials (CAM2021), ISBN: 978-2-35158-235-0, e-ISBN: 978-2-35158-236-7; *Eds. Caijun Shi and Xiang Hu*

RILEM Reports (REP)

Report 19: Considerations for Use in Managing the Aging of Nuclear Power Plant Concrete Structures (ISBN: 2-912143-07-1); *Ed. D. J. Naus*

Report 20: Engineering and Transport Properties of the Interfacial Transition Zone in Cementitious Composites (ISBN: 2-912143-08-X); *Eds. M. G. Alexander, G. Arliguie, G. Ballivy, A. Bentur and J. Marchand*

Report 21: Durability of Building Sealants (ISBN: 2-912143-12-8); *Ed. A. T. Wolf*

Report 22: Sustainable Raw Materials - Construction and Demolition Waste (ISBN: 2-912143-17-9); *Eds. C. F. Hendriks and H. S. Pietersen*

Report 23: Self-Compacting Concrete state-of-the-art report (ISBN: 2-912143-23-3); *Eds. Å. Skarendahl and Ö. Petersson*

Report 24: Workability and Rheology of Fresh Concrete: Compendium of Tests (ISBN: 2-912143-32-2); *Eds. P. J. M. Bartos, M. Sonebi and A. K. Tamimi*

Report 25: Early Age Cracking in Cementitious Systems (ISBN: 2-912143-33-0); *Ed. A. Bentur*

Report 26: Towards Sustainable Roofing (Joint Committee CIB/RILEM) (CD 07) (e-ISBN 978-2-912143-65-5); *Eds. Thomas W. Hutchinson and Keith Roberts*

Report 27: Condition Assessment of Roofs (Joint Committee CIB/RILEM) (CD 08) (e-ISBN 978-2-912143-66-2); *Ed. CIB W 83/RILEM TC166-RMS*

Report 28: Final report of RILEM TC 167-COM 'Characterisation of Old Mortars with Respect to Their Repair (ISBN: 978-2-912143-56-3); *Eds. C. Groot, G. Ashall and J. Hughes*

Report 29: Pavement Performance Prediction and Evaluation (PPPE): Interlaboratory Tests (e-ISBN: 2-912143-68-3); *Eds. M. Partl and H. Piber*

Report 30: Final Report of RILEM TC 198-URM 'Use of Recycled Materials' (ISBN: 2-912143-82-9; e-ISBN: 2-912143-69-1); *Eds. Ch. F. Hendriks, G. M. T. Janssen and E. Vázquez*

Report 31: Final Report of RILEM TC 185-ATC 'Advanced testing of cement-based materials during setting and hardening' (ISBN: 2-912143-81-0; e-ISBN: 2-912143-70-5); *Eds. H. W. Reinhardt and C. U. Grosse*

Report 32: Probabilistic Assessment of Existing Structures. A JCSS publication (ISBN 2-912143-24-1); *Ed. D. Diamantidis*

Report 33: State-of-the-Art Report of RILEM Technical Committee TC 184-IFE 'Industrial Floors' (ISBN 2-35158-006-0); *Ed. P. Seidler*

Report 34: Report of RILEM Technical Committee TC 147-FMB 'Fracture mechanics applications to anchorage and bond' Tension of Reinforced Concrete Prisms – Round Robin Analysis and Tests on Bond (e-ISBN 2-912143-91-8); *Eds. L. Elfgren and K. Noghabai*

Report 35: Final Report of RILEM Technical Committee TC 188-CSC 'Casting of Self Compacting Concrete' (ISBN 2-35158-001-X; e-ISBN: 2-912143-98-5); *Eds. Å. Skarendahl and P. Billberg*

Report 36: State-of-the-Art Report of RILEM Technical Committee TC 201-TRC 'Textile Reinforced Concrete' (ISBN 2-912143-99-3); *Ed. W. Brameshuber*

Report 37: State-of-the-Art Report of RILEM Technical Committee TC 192-ECM 'Environment-conscious construction materials and systems' (ISBN: 978-2-35158-053-0); *Eds. N. Kashino, D. Van Gemert and K. Imamoto*

Report 38: State-of-the-Art Report of RILEM Technical Committee TC 205-DSC 'Durability of Self-Compacting Concrete' (ISBN: 978-2-35158-048-6); *Eds. G. De Schutter and K. Audenaert*

Report 39: Final Report of RILEM Technical Committee TC 187-SOC 'Experimental determination of the stress-crack opening curve for concrete in tension' (ISBN 978-2-35158-049-3); *Ed. J. Planas*

Report 40: State-of-the-Art Report of RILEM Technical Committee TC 189-NEC 'Non-Destructive Evaluation of the Penetrability and Thickness of the Concrete Cover' (ISBN 978-2-35158-054-7); *Eds. R. Torrent and L. Fernández Luco*

Report 41: State-of-the-Art Report of RILEM Technical Committee TC 196-ICC 'Internal Curing of Concrete' (ISBN 978-2-35158-009-7); *Eds. K. Kovler and O. M. Jensen*

Report 42: 'Acoustic Emission and Related Non-destructive Evaluation Techniques for Crack Detection and Damage Evaluation in Concrete' - Final Report of RILEM Technical Committee 212-ACD (e-ISBN: 978-2-35158-100-1); *Ed. M. Ohtsu*

Report 45: Repair Mortars for Historic Masonry - State-of-the-Art Report of RILEM Technical Committee TC 203-RHM (e-ISBN: 978-2-35158-163-6); *Eds. Paul Maurenbrecher and Caspar Groot*

Report 46: *S*urface delamination of concrete industrial floors and other durability related aspects guide - Report of RILEM Technical Committee TC 268-SIF (e-ISBN: 978-2-35158-201-5); *Ed. Valerie Pollet*

Mechanical Properties of Bio-based Building Materials

Assessing the Mechanical and Durability Properties of Recycled Polyethylene Terephthalate (PET) Plastic Soil

Théodore Gautier L. J. Bikoko[1(✉)], Jean Claude Tchamba[2], Ngomen Kouatchoua Fanny Gildas[3], and Sofiane Amziane[4]

[1] Department of Civil Engineering Science, University of Johannesburg, PO BOX 524, Auckland Park, Johannesburg, South Africa
lejeunegautier@rocketmail.com

[2] Civil Engineering Laboratory, ENSET, University of Douala, PO BOX 1872, Douala, Cameroon

[3] Department of Civil Engineering and Forestry Techniques, HTTTC, University of Bamenda, P.O Box 39, Bambili, NWR, Bamenda, Cameroon

[4] Department of Civil Engineering, Blaise Pascal University, Clermont Auvergne University, Polytech Clermont-Ferrand, PO BOX 20206, Clermont-Ferrand, France

Abstract. Non-biodegradable plastic wastes for example, polyethylene terephthalate (PET) bottles, wrapping etc. can be found in the street, farm, inside dustbin in major or main cities in Cameroon (Douala, Yaoundé); this may cause infertility of soil, contamination of groundwater and surface water and pollution of environment with consequences for both ecosystems and human health. It has been found that in Douala, the largest city and economical capital of Cameroon about 20 tons of plastic wastes are produced daily, also, plastic wastes account for about 10% of generated waste in Douala city. In a bid to put an end to plastic pollution in Douala city, Cameroon, there is a dire need to recycle and use them in a wide range of fields such as in Civil Engineering works, for example in the manufacturing of stabilized soil bricks. In this regard, the influences of PET plastic waste bottles on the strength and durability properties of stabilized soil brick were investigated in this study. To achieve this, we collected the soil at the depth of 70 cm from earth surface at EKOUDOUM around corners called "Carrefour de l'amitié", Yaoundé, Cameroon; the particle size distribution curve showed that the soil is composed of 5% gravel, 50% sand and 45% silts and clay. It optimal dry density is 1.61 g/cm^3 obtain when the water content is 21.4%. Atterberg limits tests have shown that the soil had liquid limit of 61.8%, plastic limit of 26.86% and the plasticity index of 34.95%. Liquid PET bottles were used as stabilizer agent /binder. The preparation of the soil-liquid PET bottles composites was as follows: Firstly, a mixture of soil and PET bottles were carefully mixed and then the composite was filled in the compaction mold and statically compacted. To evaluate the performance of the composites, such tests as compressive strength, flexural strength, resilience factor and water absorption were examined. The findings revealed that when we have increased the percentage of PET bottles, the compressive strength, flexural strength, water absorption have decreased while the resilience factor and Young's modulus have increased; the minimal compressive

S. Amziane et al. (Eds.): ICBBM 2023, RILEM Bookseries 45, pp. 3–13, 2023.
https://doi.org/10.1007/978-3-031-33465-8_1

strength, flexural strength, Young's modulus and resilience factor were 5.625 MPa, 2.11 MPa, 2.552 GPa and 77.55 J/cm^2, respectively while the maximum water absorption was 3.69%.

Keywords: Stabilizer · PET plastic waste bottles · mechanical properties · durability

1 Introduction

In Cameroon, non-biodegradable plastics can be found in all form such as wrapping, bottle etc. in the street, farm, inside dustbin etc. and are frequently and carelessly dumped either on unused land or in water ways with the risk of transportation to other waterbodies such as lakes or rivers; this may cause infertility of soil, contamination of groundwater and surface water and pollution of environment and it is known that increase in pollution correlates with an increase in the prevalence of malaria [1, 2]. In the same vein, Ngoran et al. [3] reported that water pollution is the main conduit for transmission of contagious diseases such as cholera, dysentery, polio, typhoid, and ascariasis. In Douala for example, the largest city and commercial center of Cameroon about 20 tons of plastic wastes are produced daily [2]. Besides, plastic wastes account for about 10 % of generated waste in Douala city [2]. In a bid to put an end to plastic pollution, it is necessary to recycle and use them in a wide range of fields such as in Civil Engineering works for example in the production of stabilized soil bricks. Previous researches have already used plastic wastes as a binder for soil stabilizers for road construction; for example, Nicoleta-Maria Ilies et al. [4] investigated the effect of polyethylene waste admixture or cement on the properties of soil. The authors concluded that using waste polyethylene material in the soil stabilization is an eco-friendly method. Rebecca Belay Kassa, et al. [5] conducted a study on the effect of waste plastic materials for reinforcing and stabilizing clayey soils. Gangwar and Tiwari [6] reported an increased in the UCS of soil upon addition of 0.5% of plastic waste into the soil-plastic waste composite. Hassan et al. [7] reported an increased in the UCS of soils upon addition of PE fibres. Ziegler et al. [8]; Babu and Chouksey [9]; Mondal [10]; Ahmadinia et al. [11]; Modarres and Hamedi [12]; Fauzi et al. [13]; Changizi and Haddad [14]; Rawat and Kumar [15]; Peddaiah et al. [16] and Salimi and Ghzavi [17] reported that using plastic waste materials for soil stabilization improve the unconfined compressive strength (UCS) of weak soil. Kabeta [18] mixed a weak clay soil with 0.2%, 0.3%, and 0.4% of plastic strips by weight of soil and found a significant improvement in the strength of weak soil stabilized with plastic waste strips. Abousninal et al. [19] stated that polyethylene water bottles fiber-stabilized soil improves the unconfined compressive strength (UCS) than tensile strength. Dhatrak et.al. [20] reported that plastic waste strips enhance the soil strength.

Due to its very low water absorption, the plastics have been used and continue to be used as stabilizer for soil brick.

To the best knowledge of the authors, no study was conducted on mixing the soil with plastic wastes PET bottles collected in the locality of Bambili in the North West region of Cameroon.

The main aim of this paper is to use liquid polyethylene terephthalate bottles commonly abbreviated PET as stabilizing agent of natural soil and determine the unconfined compressive strength, flexural strength, water absorption, Young's modulus and resilience factor of soil-liquid PET bottles composites.

2 Materials and Methods

2.1 Materials

Soil and polyethylene terephthalate (PET) plastic waste bottles were the materials used in this study. The materials are discussed as follows:

2.2 Soil

Soil sample was collected at Ekoumdoum suburb (Carrefour de l'amitié) in the Centre region of Cameroon. This soil was composed of 5% gravel; 50% sand and 45% silts and clay. Its optimal dry density was 1.61 g/cm^3 obtained when the water content is 21.4%. Its liquid limit was 61.8%, the plastic limit was 26.86% and the plasticity index was 34.95%. Prior to using, we dry it under the sun to remove some quantities of water and then we sieved the soil to obtain a grain with maximum diameter of 2 mm; after that we crushed the retained one and sieved it again and finally we put the sieving soil inside the oven for at least 24 h.

Figure 1 show the particle size distribution curve of soil sample.

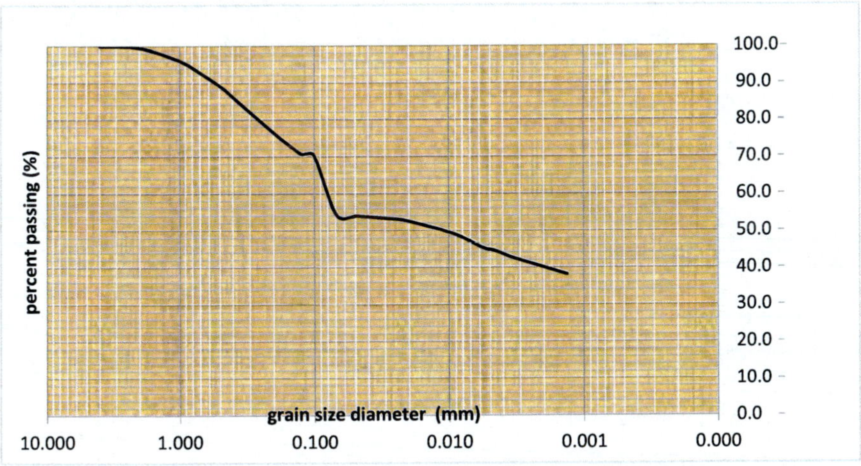

Fig. 1. Particle size distribution curve of soil sample

Figure 2 and Fig. 3 show the graph of water content vs dry density and the graph of moisture content for complete saturation vs water content, respectively. From Fig. 2, we can concluded that the optimum value of water content is 21.4% which correspond to maximum dry density of 1.61 g/cm^3, hence the quantity of water to be added inside a sample of block is equal to 21.4% of total mass of block. From Fig. 3, it can be observed that the minimum moisture content for complete saturation is equal to 25.2.

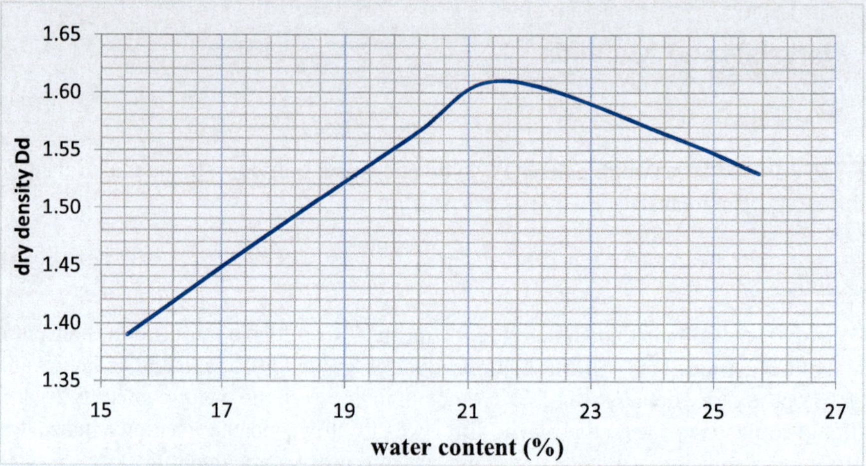

Fig. 2. Graph showing water content vs dry density

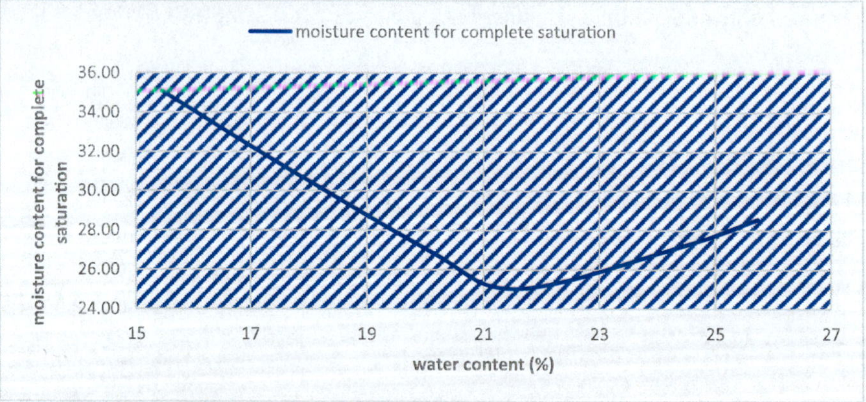

Fig. 3. Graph showing moisture content for complete saturation vs water content

2.3 Polyethylene Terephthalate (PET) Plastic Waste Bottles

The PET plastic waste bottles were collected in Bambili in the North West region of Cameroon. Prior to using, we washed them with clean water and soap and cut into small part with scissors. Figure 4 shows the collected PET plastic waste bottles. Its properties are shown in the Table 1.

Fig. 4. PET plastic waste bottles collected

Table 1. Characteristics of PET plastic waste bottles used

Density	1.38 g/cm^3 (20 °C), amorphous (transparent): 1.370 g/cm^3, single crystal: 1.455 g/cm^3
Melting point	>250 °C
Boiling point	>350 °C
Thermal conductivity	0.15 to 0.24 W/m.K
Refractive index	1.57–1.58
Young's modulus (E)	2800–3100 MPa
Tensile strength	55–75 MPa
Elastic limit	50–150%
Notch test	3.6 kJ/m^2
Water absorption	0.16

3 Methods

3.1 Compressive Strength Test

The compressive strength test was conducted to determine the load carrying capacity of bricks under compression. The test was carried out with the help of compression testing machine (hydraulic) as shown in Fig. 5 below. The samples at 28 days of maturity were mounted on to the universal compression testing machine at 1 mm/min crosshead speed (Emic DL10000) (see Fig. 5), which has a load capacity of 300 tons. The compressive strengths of the samples were calculated using Eq. (1).

$$\sigma = F/S \quad (1)$$

where: σ = compressive strength, F = maximum load applied before failure, S = cross sectional area of the specimen.

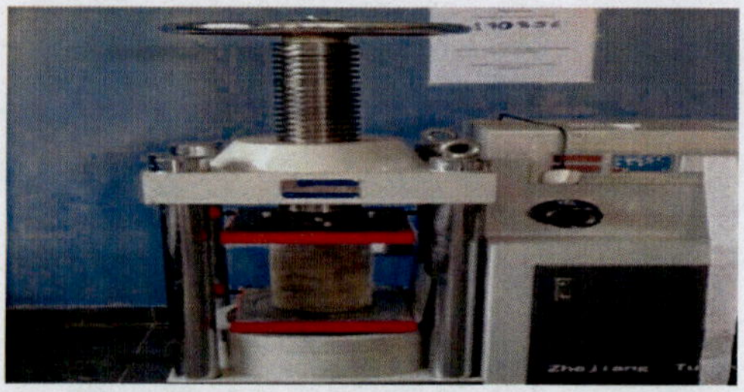

Fig. 5. Compressive strength testing machine

3.2 Flexural Strength Test

The aim of this test is to determine the maximum normal stress due to flexural plane and deduced the flexural strength.

3.3 Water Absorption Test

After soaking the samples in water for 24 h, they were dried in an oven. The water absorption (W) is given by the relationship:

$$\mathbf{W\%} = \frac{\boldsymbol{Ma - Ms}}{\boldsymbol{Ms}} \times 100 \tag{2}$$

where: Ms = mass of the dry sample after passing through an oven at 105 °C
Ma = mass of the sample soaked in water.

3.4 Young's Modulus Test

The Young's modulus of a sample was deduced from the compressive strength results above.

3.5 Resilience Factor

The resilience test is conducted to determine the capacity of the material to resist shock. This test was done using charpy impact testing machine with maximal supported energy of 500 J.

4 Results and Discussion

Figure 6 and Fig. 7 displays the compressive and flexural strengths for the control sample and the samples with 35, 40, 45 and 50% natural soil substituted with PET plastic waste bottles, respectively. We observe that when we added PET bottles intoa soil sample, the compressive and flexural strengths were increased. The control specimen has the compressive strengths smaller than that substituted by liquid PET plastic waste bottles. The compressive and flexural strengths decrease with the increase of the percentage of

PET bottles as replacement of natural soil into the soil-liquid PET bottles composite. This strength reduction is attributed to the reduction in friction angle and maximum dry density (MDD) [6]. The maximum compressive and flexural strengths were 17.1 MPa and 5.11 MPa, respectively obtained at 35% of PET bottles and 65% of soil. The compressive strength showed an increase of 14.5 MPa, 7.71 MPa, 4.28 MPa and 3.03 MPa at 35% of PET bottles + 65% of soil, 40% of PET bottles + 60% of soil, 45% of PET bottles + 55% of soil and 50% of PET bottles + 50% of soil, respectively comparing to the control sample. The most stable soil brick is obtained when the percentage of PET bottles is between 35 and 40% with the compressive strength ranging from 10.31 to 17.1 MPa. These are greater than the minimum compressive strength of 3.0 MPa required for masonry units or bricks as per the South African standard (SANS 1215) [21].

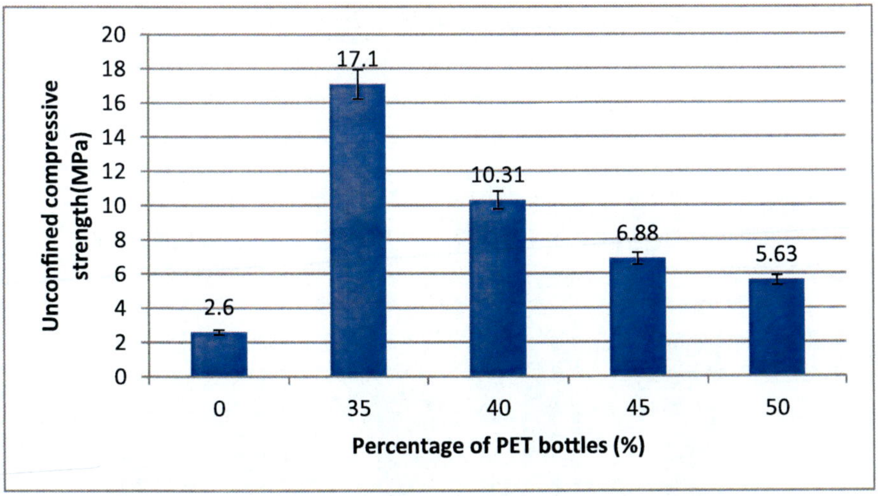

Fig. 6. Percentage of PET bottles vs Compressive strength

Figure 8 presents the water absorption of soil-PET bottles blends. From this Fig. 8, we observe that incorporating the PET bottles decrease the water absorptions of the composites except the 35% blend of PET bottle which is higher than the control sample i.e. the blend without PET bottle. The decrease of water absorption is due to the low water absorption of PET bottles, as the plastics have low water absorption, so increasing the PET bottle in the mixture lowering the water absorptions of the composites. The maximum water absorption is 3.69% obtained at the minimum percentage of PET bottles i.e. at 35%.

Figure 9 reports the Young's modulus of soil-PET bottles blends. From this Fig. 9, we observe that incorporating the PET bottles into the soil increase the Young's modulus of the composites. The Young's modulus increased with increasing the percentage of PET bottles in the mixtures. The average of Young's modulus is greater than the one obtained from 35% of PET bottles; this is but due to the discontinuity of sample at 40% of PET bottles. The optimum Young's modulus is 3212.473 MPa obtained at the maximum percentage of PET bottles i.e. at 50%.

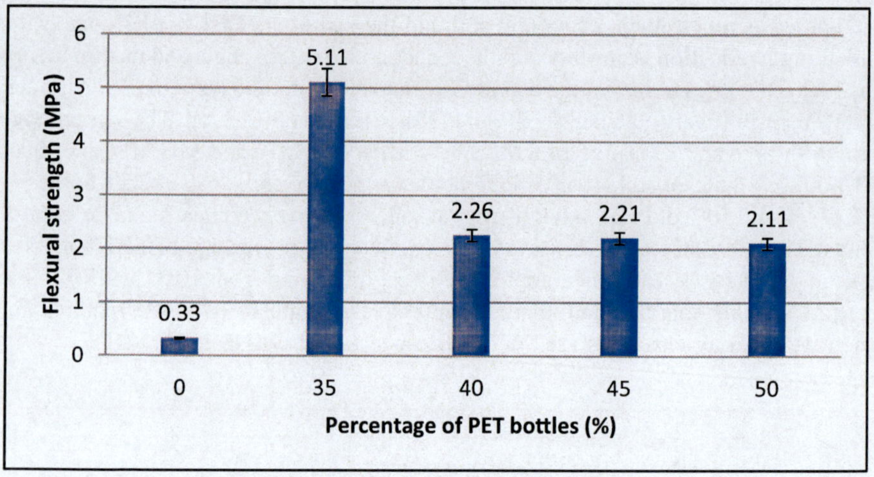

Fig. 7. Percentage of PET bottles vs Flexural strength

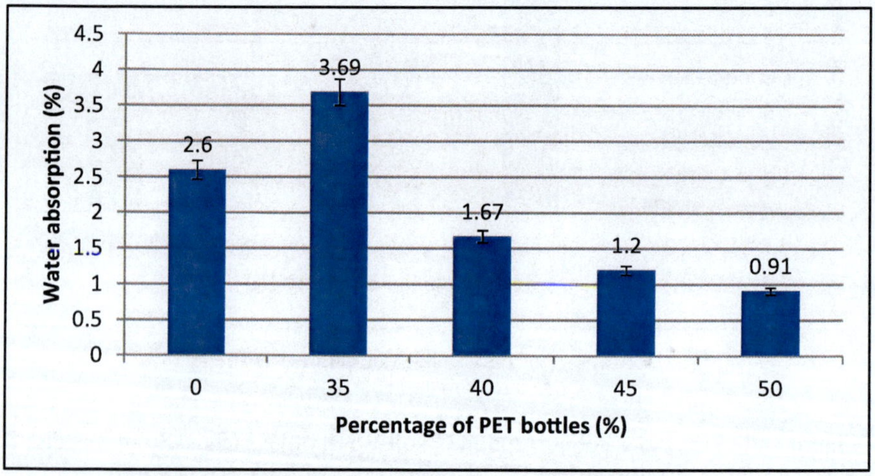

Fig. 8. Percentage of PET bottles vs Water absorption

Figure 10 displays the resilience factors of soil-PET bottles blends. It is observed that incorporating the PET bottles increase the resilience factors of the composites. But this resilience factor is slightly increased with increased content of PET bottles from 75.555 J/cm^2 to 85.555 J/cm^2, 85.555 J/cm^2 to 87.122 J/cm^2 and 87.122 J/cm^2 to 88.122 J/cm^2 at 35% to 40%, 40% to 45% and 45% to 50% of PET bottles, respectively. Such increase is due to the fact that the liquid plastic is thermoharden when it became cold.

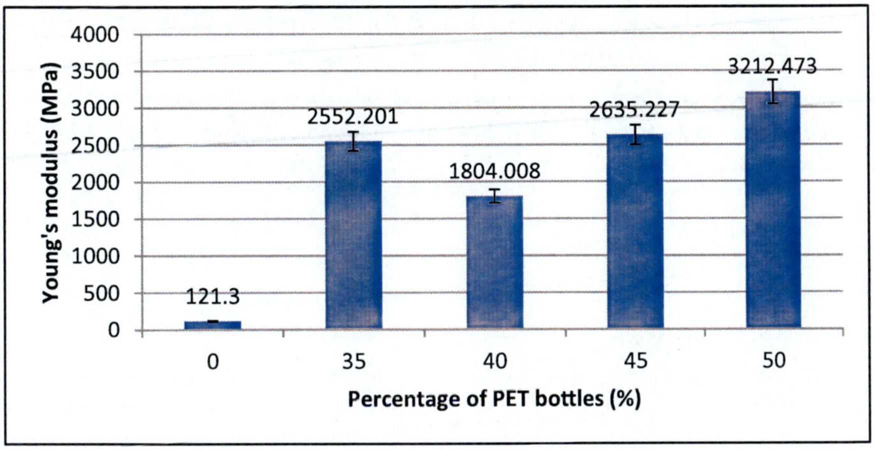

Fig. 9. Percentage of PET bottles vs Young's modulus

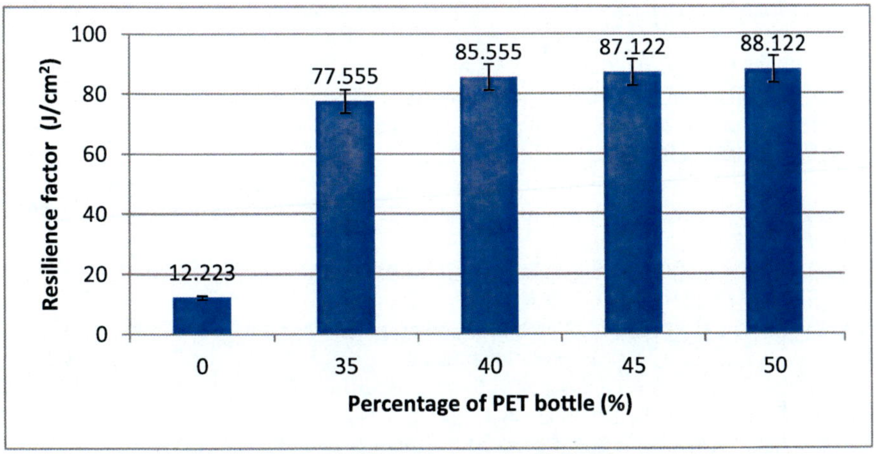

Fig. 10. Percentage of PET bottles vs Resilience factor

5 Conclusions

The goal of this paper was to assess the mechanical and durability properties of recycled polyethylene terephthalate (PET) plastic soil, based on test results, the following conclusions were made:

- When we increase the proportion of PET plastic waste bottles, the compressive and flexural strength decrease
- The most stable soil brick is obtained when the percentage of PET bottles is between 35 and 40% with the compressive strength ranging from 10.31 to 17.1 MPa.
- The water absorption of the composite i.e. blends of PET bottles and soils decrease with increasing amount of PET bottles into the soils

- The recycle PET bottles soils mixture meet the requirements for masonry units or bricks as per the South African standard (SANS 1215)

Overall, this paper investigates the use of PET plastic waste bottles in soil as a way of reducing the environmental impact from the pollution of PET plastic waste bottles. PET plastic waste bottles collected in Bambili in the North West region of Cameroon were used. General outcomes were that, compressive and flexural strengths of soil with PET plastic waste bottles were higher than soil without PET plastic waste bottles.

References

1. Keiser, J., Utzinger, J., De Castro, M.C., Smith, T.A., Tanner, M., Singer, B.H.: Urbanization in sub-Saharan Africa and implication for malaria control. The American journal of tropical medicine and hygiene **71**(2 suppl.), 118–127 (2004)
2. Christian, A., Hyeng, B., Yamb, E., Ewunkem, Akamu, J., Smith, A.: Impact of Globalization on Dilapidation, IJRRAS **36**(1), 1–5 (July 2018)
3. Ngoran, S.D., Xue, X., Ngoran, B.S.: The dynamism between urbanization, coastal water resources and human health: A case study of Douala, Cameroon. J. Econo. Sustain. Develop. **6**(3), 167–181 (2015)
4. Ilies, N.-M., Circu, A.-P., Nagy, A.-C., Ciubotaru, V-C., Zsombor, K.-B.: Comparitive Study on Soil Stabilization with Polyethylene Waste Materials and Binders. In: 10th International Conference Interdisciplinary in Engineering, INTER- ENG 2016, Science Direct Procedia Engineering **181**, 444-451 (2017)
5. Kassa, R.B., Workie, T., Abdela, A., Fekade, M., Saleh, M., Dejene, Y.: Soil Stabilization Using Waste Plastic Materials, scientific research publishing. Open J. Civil Eng. **10**(01), (2020) 55–68
6. Preeti, G., Sachin, T.: Stabilization of soil with waste plastic bottles. Materials Today: Proceedings **47**(2021), 3802–3806 (2021)
7. Aswad Hassan, H.J., Rasul, J., Samin, M.: Effects of plastic waste materials on geotechnical properties of clayey soil. Transportation Infrastructure Geotechnology **8**, 390–413 (2021)
8. Ziegler, S., Leshchinsky, D., Ling, H.I., Perry, E.B.: Effect of short polymericfibers on crack development in clays. Soils Found. **38**(1), 247–253 (1998)
9. Babu, S.G.L., Chouksey, S.K.: Stress–strainresponse of plastic waste mixed soil. Waste Manag. J. **31**, 481–488 (2011)
10. Mondal, P.K.: Behaviour of a clayey soil mixed with plastic waste. Jadavpur University Kolkata, Thesis of Civil Engineering Dept. In (2012)
11. Ahmadinia, E., Zargar, M., Karim, M., Abdelaziz, M., Ahmadinia, E.: Performance evaluation of utilization of waste polyethylene terephthalate (PET) in stone mastic asphalt. Constr. Build. Mater. **36**(2012), 984–989 (2012)
12. Modarres, A., Hamedi, H.: Effect of waste plastic bottles on the stiffness and fatigue properties of modified asphalt mixes. Mater. Des. **61**(2014), 8–15 (2014)
13. Fauzi, A., Djauhari, Z., Fauzi, U.J.: Soil engineering properties improvement by utilisation of cut waste plastic and crushed waste glass as additive. Int. J. Eng. Technol. (1), 8 (Jan 2015)
14. Changizi, F., Haddad, A.: Strength properties of soft clay treated with mixture of nano-SiO2 and recycled polyester fiber. Journal of Rock Mechanics and Geotechnical Engineering **7**(4), 367–378 (2015)
15. Rawat, P., Kumar, A.: Study of CBR behaviour of soil reinforced with HDPE strips. In: Indian Geotechnical Conference IGC2016. IIT Madras, Chennai, India (2016)

16. Peddaiah, S., Burman, A., Sreedeep, S.: Experimental study on effect of waste plastic bottle strips in soil improvement. Geotech. Geol. Eng. **36**(5), 2907–2920 (2018)
17. Salimi, K., Ghzavi, M.: Soil reinforcement and slope stabilisation using recycled waste plastic sheets (2019)
18. Kabeta, W.F.: Study on some of the strength properties of soft clay stabilized with plastic waste strips. Archives of Civil Engineering **LXVIII**(3), 385–395 (2022)
19. Abousnina, R., et al.: Effect of short fibres in the mechanical properties of geopolymer mortar containing oil-contaminated sand. Polymers **13**(17) (2021). https://doi.org/10.3390/polym13173008
20. Dhatrak, A.I., Konmare, S.D.: Performance of randomlyoriented plastic waste in flexible pavement. Int. J. of pure and applied research in Engineering and Technology **3**(9), 193–202 (2015)
21. South African National Standards: SANS 1215 - Concrete Masonry Units. Standards South Africa, Pretoria (2008)

Polyurethane Wood Adhesives from Microbrewery Spent Grains

Alex Mary[1], Pierre Blanchet[1], and Véronic Landry[1,2](✉)

[1] Department of Wood and Forest Sciences, NSERC Industrial Chair On Eco-Responsible Wood Construction (CIRCERB), QC G1V 0A6 Québec, Canada
alex.mary.1@ulaval.ca, veronic.landry@sbf.ulaval.ca

[2] Department of Wood and Forest Sciences, NSERC – Canlak Industrial Research Chair in Interior Wood-Products Finishes, Université Laval, Québec, QC G1V 0A6, Canada

Abstract. The building sector is responsible for nearly 40% of greenhouse gas emissions, which have a major impact on climate change. One of the strategies to alleviate this problem is to increase the use of wood in the construction of buildings. However, the adhesives used in the design of engineered wood products are synthetic adhesives that rely heavily on the use of materials of fossil origin and therefore non-renewable, such as formaldehyde. Common methods used to reduce formaldehyde emissions from wood panels are to use polyurethane adhesives, formaldehyde-free adhesives. In order to increase the biobased content, it is also common to add certain compounds such as proteins. Proteins are compounds present in appreciable quantities in plants, and can increase the adhesion strength of adhesives on different substrates, including wood. In this study, a protein concentrate was prepared from microbrewery spent grains. The nitrogen content, thermal behavior, molecular weight, and structure of these proteins were then evaluated to facilitate and understand their incorporation into a polyurethane adhesive system. The adhesives were formulated with different protein incorporation percentages: 5%, 10%, 15%, and 20% and compared to a petrochemical reference. This paper highlights the fact that the incorporation of proteins makes it possible to maintain, or even increase, the properties of the adhesives, particularly the mechanical strength. An increase in pot life was also observed.

Keywords: Wood · Engineered wood products · Building · Polyurethane adhesive · Microbrewery spent grains · Protein

1 Introduction

The building sector is responsible for nearly 40% of greenhouse gas emissions (UNEP, 2020). These emissions, which have an impact on climate change, could be significantly reduced by using existing strategies. One of the strategies to address this issue is to increase the use of wood in building construction. Wood is a renewable resource, sustainably harvested in Quebec, which allows for temporary carbon capture in buildings (Stevanovic and Perrin, 2009). However, the adhesives used in the design of wood structural elements (i.e. Cross Laminated Timber, GluLam, plywood, Laminated Veneer Lumber.

S. Amziane et al. (Eds.): ICBBM 2023, RILEM Bookseries 45, pp. 14–28, 2023.
https://doi.org/10.1007/978-3-031-33465-8_2

etc) are synthetic adhesives which, although interesting for wood structures, rely heavily on the use of fossil-based and therefore non-renewable materials, such as formaldehyde. Formaldehyde is classified by the World Health Organization as a human and animal carcinogen. Since 2021, regulations have been strengthened through the "Formaldehyde Emissions from Composite Wood Products Regulation", a regulation under the Canadian Environmental Protection Act. It reflects the critical need for biobased, formaldehyde-free adhesives for wood construction (Gui et al., 2016). The most compelling alternative to formaldehyde-emitting adhesives is polyurethane (PU) adhesives. PU adhesives can be used to bond many materials, are capable of forming hydrogen and covalent bonds with the wood substrate, and their small molecules allow the impregnation of porous substrates (Pizzi and Mittal 2005). They can be single-component, noted PUR 1K, or two-component, noted PUR 2K. PUR 1Ks are based on isocyanate-terminated polymers that can crosslink in the presence of moisture in the air, while the PUR 2Ks are composed of at least one prepolymer that contains an isocyanate, and a polyol. However, these adhesives are also petrochemical-based. Several studies have been conducted to increase the bio-based content of these adhesives, such as incorporating proteins into these adhesives. Proteins are biological macromolecules known to improve the adhesion of the adhesive to the wood substrate (Yang et al., 2006). Soy, cotton and milk proteins have been the most studied. Soy proteins increase the durability of adhesives but they also increase their viscosity and decrease their water resistance (Huang and Li, 2008; Lei et al., 2014; Thakur et al. 2017; Vnučec et al., 2017). Cotton proteins have proven to be a good alternative to soy proteins, as they increase the water resistance and tackiness of the adhesive (Cheng et al., 2013). Milk proteins-based adhesives, on the other hand, can create strong bonds with wood (Detlefsen, 1989). However, they are not able to withstand long-term exposure in humid environments without significant deterioration (Vick and Rowell, 1990). The proteins targeted for this study are derived from microbrewery spent grains, co-products from local resources. With the incorporation of proteins, the isocyanate will be able to react with the NH groups constituting the amino acids of the proteins, as well as with the OH groups of polyols. The proteins were extracted from the microbrewery grains and their nitrogen content, thermal behavior, molecular weight, and structure were evaluated to understand their incorporation into a polyurethane adhesive system. The incorporation of proteins was performed at different percentages and their impact on the adhesives properties has been studied.

2 Methodology

2.1 Materials

Microbrewery spent grains (MSG) are derived from grain crops and refer to microbrewery residues. These residues are generally thrown away or given as feed to livestock in limited quantities. The spent grains selected for this project come from the microbrewery Le Corsaire (Lévis, Canada), and have been used to produce a Pilsner made of 88% barley, 8% oats, and 4% wheat. Polymeric methylene diphenyl diisocyanate (pMDI) (mass equivalent amine = 32.5 Wt.%, viscosity = 129 mPa.s at 20 °C) and a polypropylene oxide-based triol (Multranol 8175) (acid value = 350–390 mg KOH/g sample, molecular weight = 450 kDa, viscosity = 232–412 at 25 °C) were purchased from Covestro (Pittsburgh, USA). All chemicals were used as received.

2.2 Protein Extraction from MSG

MSG protein concentrate (MSGPC) was prepared by alkaline extraction of MSG (17% w/v) with 0.1 M NaOH at 60 °C (Celus et al., 2009). After 60 min of extraction, the solution was centrifuged at 5000 RPM for 10 min at 20 °C and the filtrate was collected. Samples were then washed with 2 mL of distilled water to be centrifuged again and the filtrate was collected. Proteins in the filtrate were precipitated by acidification to pH 4.0 using 2.0 M citric acid and then placed at 4 °C for 3 h. The obtained protein precipitate was then centrifuged at 8000 RPM for 10 min at 4 °C. The filtrate was disposed of, and the precipitate was washed with 2 mL of 0.1M NaOH before being centrifuged. The protein precipitate obtained after centrifugation was finally freeze-dried to recover the samples in powder form and remove traces of water.

2.3 Protein Characterization

Nitrogen content. To determine the protein content of the samples, it is necessary to know the percentage of nitrogen present in MSGPC. The analysis was performed using the "carbon, nitrogen, and sulfur (CNS) in Plant Tissue" method at a 1350 °C temperature on a TruMAC CNS (LECO Corporation, Midland, Canada). Once the nitrogen content has been obtained, the protein content can be determined using a conversion factor specific to each raw material, which is equal to 6.25 for microbrewery grains (AOAC International, 2000). The analyses were performed in triplicate.

Thermal Stability. The thermal stability of MSGPC was determined using thermogravimetric analysis (TGA). TGA was performed on a TGA/DSC 3 + (Mettler Toledo, Columbia, USA). MSGPC samples of 4–10 mg were placed onto the TGA sample pan. The samples were heated from 25 to 800 °C with a heating rate of 20 °C/min under nitrogen flow. The temperature at which protein begins to degrade was considered the starting temperature for the second stage of weight loss, the first being attributed to moisture loss (Ricci et al., 2018). The analyses were performed in triplicate.

Molecular Weight and Protein Identification. *Samples solubilization, gel migration and protein digestion.* Protein digestion and mass spectrometry analyses were performed by the Proteomics Platform of Quebec City CHU Research Center (Quebec, Qc, Canada).

Approximatively 4 mg of MSGPC was solubilized in 2 mL of 50 mM ammonium bicarbonate, 1% sodium deoxycholate. 14 μL of the sample was transferred and the volume was adjusted to 20 μL in 1X gel sample buffer and 1X reducing agent (Biorad 161–0788 and 161–0792 respectively). Samples were heated at 95 °C for 5 min and spun at 16 000 g for 20 s. Denatured proteins were then deposited on a 4–12% Bis-Tris acrylamide gel and migrated in the stacking portion at 200 V for 8 min (Biorad Precast Gel Criterion XT 4–12% 3450123) (Fig. 1) (Bilgraer, 2014).

Bands of interest were extracted from gels and placed in 96-well plates and then washed with water. Proteins were reduced with 10mM DTT and alkylated with 55 mM iodoacetamide. Trypsin digestion was performed using 126 nM of modified porcine trypsin (Sequencing grade, Promega, Madison, WI) at 37 °C for 18h. Digestion products were extracted using 1% formic acid, 2% acetonitrile followed by 1% formic acid, 50% acetonitrile. The recovered extracts were pooled, vacuum centrifuge dried and then resuspended into 10 μl of 2% acetonitrile, 0.05% trifluoric acid and 5 μl were analyzed by mass spectrometry.

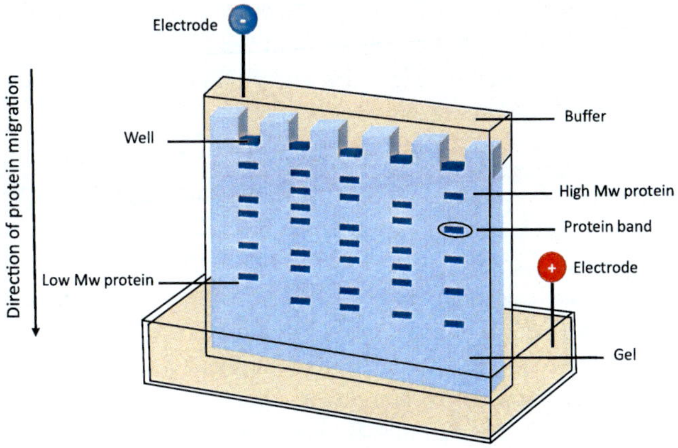

Fig. 1. Gel electrophoresis

Mass Spectrometry. Samples were analyzed by nano LC-MS/MS using a Dionex UltiMate 3000 nanoRSLC chromatography system (Thermo Fisher Scientific, San Jose, USA) connected to an Orbitrap Fusion mass spectrometer (Thermo Fisher Scientific, San Jose, USA). Peptides were trapped at 20 μl/min in loading solvent (2% acetonitrile, 0.05% TFA) on a 5 mm x 300 μm C18 pepmap cartridge pre-column (Thermo Fisher Scientific / Dionex Softron GmbH, Germering, Germany) for 5 min. Then, the pre-column was switched online with a Pepmap Acclaim column (Thermo Fisher Scientific, San Jose, USA) 50 cm x 75 μm internal diameter separation column and the peptides were eluted with a linear gradient from 5–40% solvent B (A: 0.1% formic acid, B: 80% acetonitrile, 0.1% formic acid) in 30 min, at 300 nL/min. Mass spectra were acquired using a data dependent acquisition mode using Thermo XCalibur software version 4.3.73.11. Full scan mass spectra (350 to 1800m/z) were acquired in the orbitrap using an automatic gain control target of 4e5, a maximum injection time of 50 ms and a resolution of 120 000. Internal calibration using lock mass on the m/z 445.12003 siloxane ion was used. Each scan was followed by MS/MS fragmentation of the most intense ions for a total cycle time of 3 s (top speed mode). The selected ions were isolated using the quadrupole analyzer in a window of 1.6 m/z and fragmented by Higher energy Collision-induced Dissociation with 35% of collision energy. The resulting fragments were detected by the linear ion trap in rapid scan rate with an automatic gain control target of 1e4 and a maximum injection time of 50 ms. Dynamic exclusion of previously fragmented peptides was set for a period of 20 s and a tolerance of 10 ppm.

Database Searching. Mascot generic format peak list files were created using Proteome Discoverer 2.3 software (Thermo Fisher Scientific, San Jose, USA). Mascot generic format sample files were then analyzed using Mascot (Matrix Science, London, UK; version 2.5.1). Mascot was set up to search a contaminant database using the following database: *Hordeum vulgare* (UP000011116, 35907 entries), assuming the digestion enzyme trypsin and with a fragment ion mass tolerance of 0.60 Da and a parent ion tolerance of 10.0 ppm. Carbamidomethylation of cysteine was specified in Mascot as a fixed modification. Deamidation of asparagine and glutamine, oxidation of methionine were specified as variable modifications.

Criteria for Protein Identification. Scaffold (version Scaffold_5.1.2, Proteome Software Inc., Portland, OR) was used to validate MS/MS based peptide and protein identifications. Peptide identifications were accepted if they could be established at greater than 89.0% probability to achieve an FDR less than 1.0% by the Scaffold Local FDR algorithm. Protein identifications were accepted if they could be established at greater than 99.0% probability to achieve an FDR less than 1.0% and contained at least 2 identified peptides. Protein probabilities were assigned by the Protein Prophet algorithm (Nesvizhskii et al., 2003). Proteins that contained similar peptides and could not be differentiated based on MS/MS analysis alone were grouped to satisfy the principles of parsimony.

2.4 Preparation of Polyurethane Adhesives

Polyurethane adhesive formulations were prepared with a ratio of isocyanate to hydroxyl functions (NCO/OH) of 1.13, to ensure a complete reaction between the polyol and the isocyanate (Meier-Westhues, 2019). The incorporation of proteins was done by substitution of the hydroxyl groups of the polyol with the amine groups of the proteins. Proteins were incorporated into the polyol and dispersed at 1000 rpm for 3 min with a Dispermat LC30 Dissolver (VMA-Getzmann, Reichshof, Germany) with a 45 mm flat turbine. The substitution was done at 5%, 10%, 15%, and 20%. The petrochemical reference, formulated with the same chemicals as the protein-based adhesives, is represented by the formulation containing 0% protein.

2.5 Adhesives Characterization

Viscosity. Viscosity is an important physical parameter that affects the behavior of the adhesive. Proper viscosity gives the adhesive good flowability and facilitates handling to achieve high bond strength of the bonded product (Luo et al., 2016). Viscosity measurements were performed, at 25 °C, on systems consisting of the polyols and proteins using a ViscoLab 4100 (Cambridge Applied Systems Inc., Boston, USA).

Optical microscopy. The polyol/protein systems were observed using a VHX-7000 digital microscope (Keyence Co. Ltd., Osaka, Japan).

Pot Life. Pot life is the maximum period during which a multi-component adhesive can be used after mixing the components (International Organization for Standardization,

2018). The pot life was determined using French Standards NF EN ISO 10364:2018 method 3. Twenty grams of adhesive were prepared by mixing isocyanate, polyol and proteins as described in Sect. 2.4 and temperature monitoring was performed. In this method, pot life is considered to be the time elapsed from the start of mixing until the critical temperature is reached. The critical temperature was determined from the petrochemical reference and was set at 30 °C. The analyses were performed in triplicate.

Kinetics and Conversion. The polymerization kinetics of the adhesive systems prepared were studied by real-time Fourier transform infrared spectroscopy (RT-FTIR). The instrument used is the INVENIO® R (Bruker Optics Inc., Billerica, USA). The spectra were recorded in the range 450–4000 cm^{-1} for a duration of 215 min, corresponding to 400 measurements of 32 scans with a resolution of 4 cm^{-1}. Details of the expected FTIR bands are shown in Table 1 (Maji and Bhowmick, 2009). The analyses were performed in duplicate.

Table 1. Principal Peak Assignments in the FTIR Spectra of the isocyanate, polyol and cured PU

Observed peaks (cm^{-1})	Peaks assignments
3510 - 3100	-NH, -OH stretching vibrations
2970 - 2870	-CH stretching vibration
2260	-N = C = O stretching vibration
1730 - 1710	-C = O- stretching vibration of urethane
1620	-NH stretching vibration

Activation Energy and Reaction Heat. The differential scanning calorimetry analysis was performed on DSC 822e (Mettler Toledo, Columbia, USA). PU adhesives of 4–6 mg were placed onto the DSC sample pan and heated under nitrogen flow from 30 to 220 °C at different heating rates: 5 °C/min, 10 °C/min, and 20 °C/min. Those heating rates were used to obtain sufficient information to calculate the activation energy and the reaction heat of each adhesive according to the ASTM E698:2011 method (Lépine, 2013). The reaction heat represents the amount of energy released during the polymerization reaction, and the activation energy is the energy required to initiate the polymerization reaction (Lépine, 2013).

Block Shear. Wood cutting, bonding, and testing were conducted according to the ASTM D905:2008 test method. Black spruce (*Picea mariana*, Mill.) wood specimens were cut into rectangular panels of 32 × 65 × 20 mm^3 and 3.75g of adhesive, prepared according to Sect. 2.4., was applied to the wood within 24 h after cutting. The two pieces, one with adhesive and one without, were placed together for the adhesives to be cured at room temperature with a pressure of 150 psi exerted onto the contact area for 24 h. After that, the glued elements were cut according to the ASTM D905:2008 method. Wood specimens were conditioned at 20 °C with a relative humidity of 65% for seven days. Block-shear strength tests were performed on the Alliance RT/50 (Frank Bacon Machinery Sales Co., Warren, USA). The load was applied with continuous movement of the moving head at a rate of 5 mm/min until failure. The analyses were performed on ten samples per adhesive.

3 Results and Discussion

3.1 Protein Characterization

Nitrogen content. Table 2 presents the protein content before and after the extraction described in Sect. 2.2. The protein content before extraction is higher than what can be found in the literature concerning barley, where it has been proven that it is between 10 and 15% (Yu et al., 2017). The increase in the protein content of the post-extraction samples, compared to the pre-extraction samples, suggests the viability of the extraction protocol used. In addition to obtaining a higher protein content, which should increase the water resistance of the adhesive according to Gui et al., extracted proteins under moderate alkaline conditions should enhance adhesive strengths compared with unextracted protein, as this extraction process makes polar and apolar groups of the proteins available (Gui et al., 2016; Hettiarachchy et al., 1995). The other components of barley grains are starch, cellulose and fat (Alijosius et al., 2016; Yu et al., 2017).

Table 2. Nitrogen and protein contents of microbrewery grains before and after extraction

Raw materials	Nitrogen content (%)	Protein content (%)
MG pre-extraction	3.7 ± 0.1	22.9 ± 0.6
MG post-extraction	9.6 ± 0.2	60.0 ± 2

TGA. TGA experiments were performed to study the thermal stability of MSGPC, which have been previously placed in an oven for 24 h at 50 °C (Fig. 2). The barley sample showed several peaks of mass loss following thermal events, including 71.50 °C and 315.33 °C. These losses were also observed by Borsato et al. (2019). The first mass loss (71.50 °C) is attributed to dehydration. The main phase of degradation, between 210 and 420 °C, presents a peak corresponding to the release of gases such as carbon monoxide, carbon dioxide, methane, and ethylene (Borsato et al., 2019). Since the adhesive system is formulated cold and samples degradation temperatures are higher than 300 °C, the proteins are not likely to degrade once incorporated.

Molecular Weight and Protein Identification. Protein separation on polyacrylamide gel electrophoresis allows the identification of a variety of proteins in MSGPC (Fig. 3). Liquid chromatography coupled with mass spectrometry ensured the identification and quantification of the various proteins present in raw materials. The most significant band present in the sample is the 40 kDa band associated with serpins, a superfamily of proteins commonly present in barley (Gettins, 2002). A second well-marked band is positioned around 23 kDa. This band has not been assigned, due to its low presence compared to the serpin protein, but can be assigned to a protein fragment. Having molecular weights

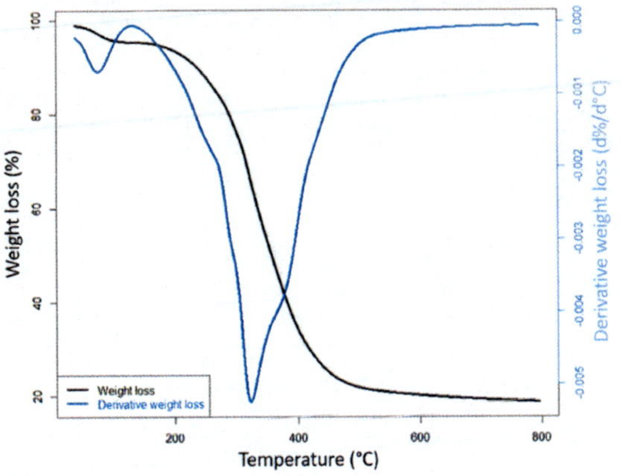

Fig. 2. TGA curves of MSGPC

around 40 kDa is not a barrier to adhesive formulation, as the literature has shown that protein-based adhesives up to 100 kDa can be obtained (Jenkins et al., 2013). In fact, having smaller proteins helps to promote protein incorporation into the adhesive and minimize the final viscosity of the system.

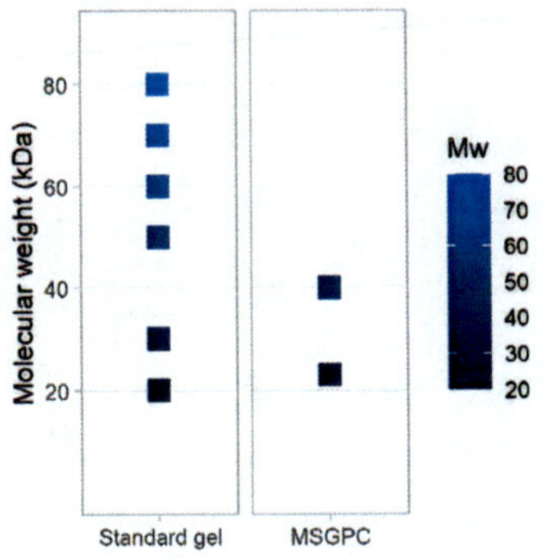

Fig. 3. Molecular weight of proteins contains in MSGPC

Viscosity. Viscosity measurements were performed at 25 °C on the systems consisting of polyols and proteins at different substitution rates (Fig. 4). The results show that protein incorporation causes an increase in viscosity. This increase is due to the protein molecules unfolding into the surrounding system once incorporated (Bacigalupe et al., 2015). These results are consistent with what can be found in the literature as it is well known that the addition of fillers to a polymer system leads to an increase in its viscosity (Markovičová, 2021; Schulze et al., 2003).

Optical Microscopy. The polyol/protein systems were observed using an optical microscope to better understand the behavior of the proteins once incorporated into the polyol (Fig. 5). Microscopic images were corrected with a contrast of 50%. The polyol is the

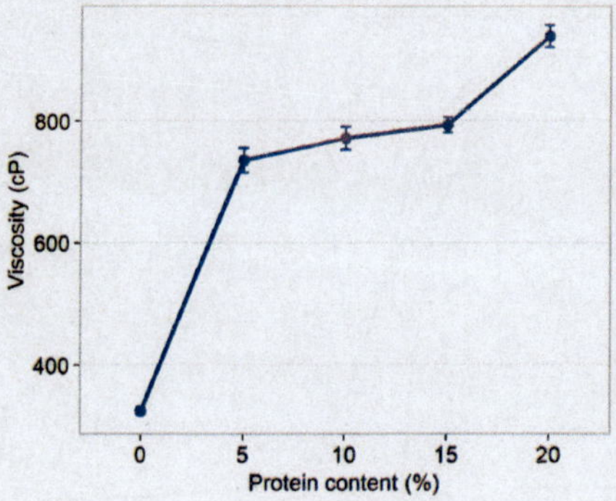

Fig. 4. Viscosity of polyol/protein systems at 25 °C

liquid matrix while the observable aggregates are the proteins. These analyses prove that when the percentage of proteins increases, protein clusters are formed and thus reduce the mobility of proteins in the polyol. At a protein incorporation of 20%, the protein clusters formed are prominent.

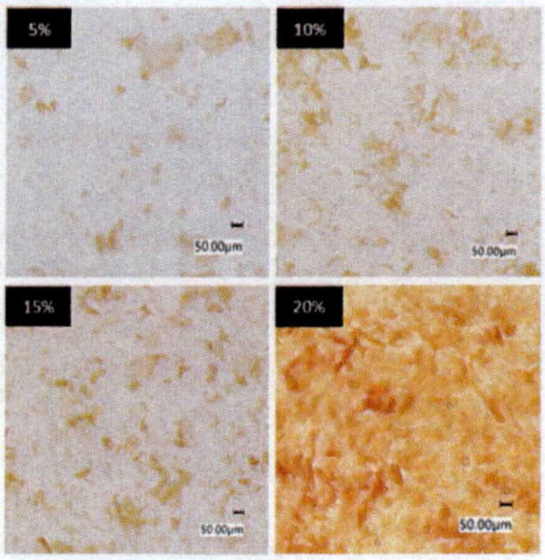

Fig. 5. Corrected microscope images of polyol/proteins systems at different protein content

Pot Life. Pot life, in this study, is considered to be the time from the start of mixing until the critical temperature, here 30 °C, is reached. The incorporation of proteins, regardless of the percentage, increases the pot life of the adhesive (Fig. 6). This increase means that the time before the adhesive significantly changes viscosity is lengthened, representing a definite advantage for the application of these adhesives. Since amine groups are more reactive than hydroxyl groups, it was expected that the pot life of protein-based adhesives would be shorter than the petrochemical reference. However, the opposite was found here. Several parameters must be taken into account to explain these results. First, part of the protein fraction is unreactive, making the probability of encounters between an isocyanate and the NH and OH groups lower. Second, the presence of cellulose and starch, which contain OH groups, may reduce the NCO/OH ratio set at 1.13 which could lead to a longer pot life. This increase in pot life could also be explained by an incomplete polymerization of the adhesives. The following kinetic and conversion of adhesive analyses will investigate this hypothesis.

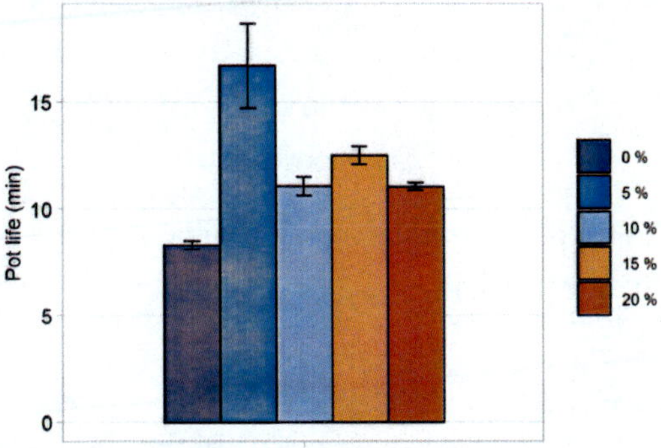

Fig. 6. Pot life of PU adhesives at different protein contents

Kinetic and Conversion of Adhesives. FTIR spectra of PU-MSGPC adhesives at different percentages of protein incorporation showed similarity in the absorption band. RT-FTIR spectra are shown in Fig. 7. A baseline correction was performed for all absorption peaks. The OH and NH stretching bands were recorded at 3335 cm^{-1}, the -CH stretching vibration band at 2920 cm^{-1}, the NCO stretching vibration band at 2260 cm^{-1}, the –C=O–Stretching vibration band of urethane at 1730 cm^{-1}, and the -NH stretching band at 1620 cm^{-1}.

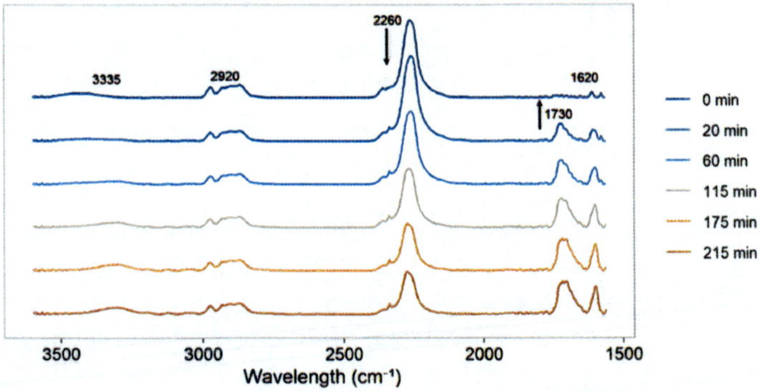

Fig. 7. FTIR spectra of PU adhesive with protein over time

The urethane formation can be monitored by the disappearance of the isocyanate's NCO vibration band at 2260 cm^{-1} and the appearance of the urethane's C=O vibration band at 1730 cm^{-1}. The isocyanate conversion (Fig. 8) can be used as the degree of curing reaction as follows (Eq. 1) assuming that there is an insignificant side reaction (Maji and Bhowmick, 2009).

$$\%\,NCO\ conversion = 100 * \left(1 - \frac{A_{NCOt}/A_{COt}}{A_{NCO0}/A_{CO0}}\right) \tag{1}$$

Isocyanate conversion

where, A_{NCO_0} is the 2260 cm^{-1} peak intensity at the initial time, A_{NCO_t} is the 2260 cm^{-1} peak intensity of absorbance at specified time during the curing, A_{CO_0} is the 1730 cm^{-1} peak intensity at the initial time, A_{CO_t} is the 1730 cm^{-1} peak intensity of absorbance at specified time during the curing. The curves representing the conversion rates being overlapped, it is possible to deduce that the incorporation of proteins has a few impacts on the conversion rate. Indeed, it is possible to see that the higher the protein level, the slower the reaction will start, which is consistent with the viscosity increase presented in Fig. 4.

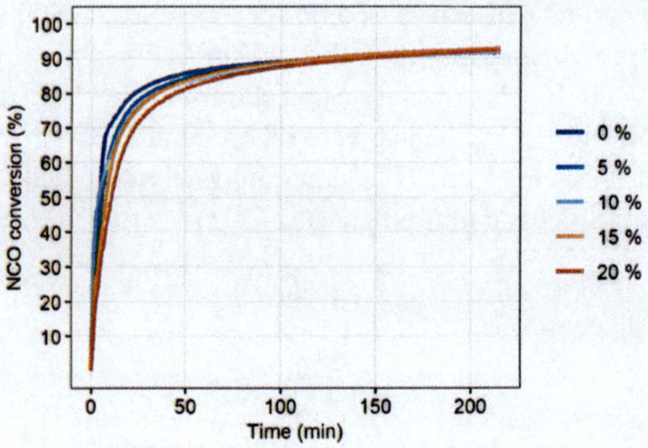

Fig. 8. Isocyanate band conversion rate versus time of PU adhesives at different protein contents

Activation Energy and Reaction Heat. Activation energy represents the energy required to initiate the polymerization reaction. The lower the energy, the easier it is for the resin to cure (Lépine, 2013). A 5% protein incorporation results (Fig. 9 left) in a decrease in activation energy, which can be explained by the fact that amine groups react faster with isocyanates than hydroxyl groups. The increase in activation energy at 10% and then a decrease at higher percentages can be explained by the increased viscosity of the adhesive. The heat of the reaction represents the amount of energy released during the polymerization reaction. Since the reaction is exothermic, high energy is characterized by greater chemical bond formation during the polymerization of the adhesive. The results obtained (Fig. 9 right) show that the incorporation of proteins has little effect on the formation of chemical bonds during polymerization.

Block Shear. This analysis covers the determination of comparative breaking point of adhesives used for bonding wood when tested in compression. The bar diagram obtained

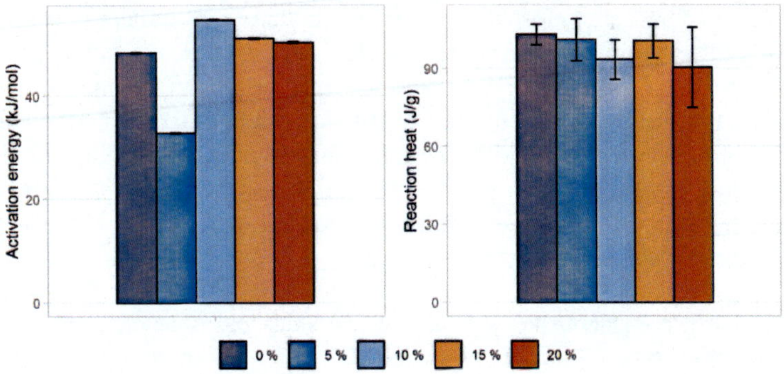

Fig. 9. Activation energy (left) and reaction heat (right) of PU adhesives at different protein contents

for comparing the maximum load at breaking point of the adhesives with an increasing percentage of protein is given in Fig. 10. The values obtained from the standard error analysis, which gives an indication of the sampling error, are also presented in Fig. 10. As it can be seen from the figure, the incorporation of protein appears to increase the maximum load at breaking point of the adhesive. This value increases with the percentage of protein. Parallels can be made with studies conducted on soy-based adhesives where increasing the percentage of protein in the adhesive was shown to increase the adhesion strength of the adhesive (Gui et al., 2016; Trinh, 2012).

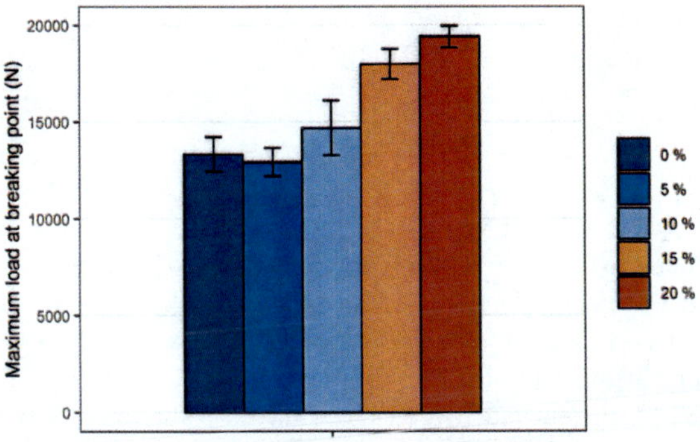

Fig. 10. Maximum shear force before the rupture of PU adhesives with different protein contents

4 Conclusion

Polyurethane adhesives are a commonly used alternative for reducing formaldehyde emissions from wood panels. Incorporating natural compounds, such as proteins, represents a step toward bio-based adhesives. The proteins selected for this study are from microbrewery spent grains and their structure as well as their chemical composition has been taken into account before their incorporation into polyurethane adhesives. The protein-based adhesives exhibited a longer pot life as well as better mechanical resistance than the petrochemical adhesive. It was also shown that, although the polymerization of the adhesives takes longer to start as the percentage of protein increases, the adhesives form with a similar degree of conversion as the petrochemical reference. Although the technical feasibility of protein incorporation into polyurethane adhesives has been positively demonstrated, the number and accessibility of OH and NH groups in the molecules remain to be determined.

Acknowledgments. The authors are grateful to Marie Soula and Aurélien Hermann for their valuable assistance. Thanks to Patrick Leclerc for the CNS analysis, Victor Fourcassié and Sylvie Bourassa for their expertise on proteins, and thanks to Yves Bédard, Daniel Bourgault, Luc Germain, Benoit St-Pierre and Jean Ouellet for their help with the various analyses. The authors would also like to thank the Renewable materials research center and the NSERC industrial chair on eco-responsible wood construction (CIRCERB) and its industrial partners.

Funding. The authors are grateful to Natural Sciences and Engineering Research Council of Canada for the financial support through its IRC and CRD programs (IRCPJ 461745- 18 and RDCPJ 524504–18) as well as the industrial partners of the NSERC industrial chair on eco-responsible wood construction (CIRCERB). The authors are also grateful to the Ministère de l'Économie et de l'Innovation du Québec through the PSO-I-009 project.

Conflicts of Interest. There are no conflicts to declare.

References

Alijosius, S., Kliseviciute, V., Sasyte, V.: The chemical composition of different barley varieties grown in Lithuania (2016). https://www.researchgate.net/publication/311434685

AOAC International: AOAC Official Method 997.09:1997 Nitrogen in Beer, Wort, and Brewing Grains Protein (Total) by Calculation. Combustion Methode (2000)

Bacigalupe, A., Poliszuk, A.K., Eisenberg, P., Escobar, M.M.: Rheological behavior and bonding performance of an alkaline soy protein suspension. Int. J. Adhes. Adhes. **62**, 1–6 (2015). https://doi.org/10.1016/j.ijadhadh.2015.06.004

Bilgraer, R.: Decipher the histone code : epigenetics and placental toxicology (2014). https://tel.archives-ouvertes.fr/tel-01195983

Borsato, V.M., Jorge, L.M.M., Mathias, A.L., Jorge, R.M.M.: Thermodynamic properties of barley hydration process and its thermostability. Journal of Food Process Engineering **42**(2) (2019). https://doi.org/10.1111/jfpe.12964

Celus, I., Brijs, K., Delcour, J.A.: Fractionation and characterization of brewers' spent grain protein hydrolysates. J. Agric. Food Chem. **57**(12), 5563–5570 (2009). https://doi.org/10.1021/jf900626j

Cheng, H.N., Dowd, M.K., He, Z.: Investigation of modified cottonseed protein adhesives for wood composites. Ind. Crops Prod. **46**, 399–403 (2013). https://doi.org/10.1016/j.indcrop.2013.02.021

Detlefsen, W.D.: Adhesive from Renewable Ressources, Chapter 31 : Blood and Casein Adhesives for Bonding Wood (1989). https://pubs.acs.org/sharingguidelines

Gettins, P.G.W.: Serpin Structure. Mechanism, and Function. (2002). https://doi.org/10.1021/cr010170

Gui, C., Zhu, J., Zhang, Z., Liu, X.: Research progress on formaldehyde-free wood adhesive derived from soy flour. In: Adhesives - Applications and Properties, pp. 187–200 (2016). InTech. https://doi.org/10.5772/65502

Hettiarachchy, A'.N.S., Kalapathy, U., Myers, D.J.: Alkali-modified soy protein with improved adhesive and hydrophobic properties. In: JAOCS, Vol. 72 (1995)

Huang, J., Li, K.: A new soy flour-based adhesive for making interior type II plywood. JAOCS, Journal of the American Oil Chemists' Society **85**(1), 63–70 (2008). https://doi.org/10.1007/s11746-007-1162-1

International Organization for Standardization: ISO:10364 Structural adhesives - Determination of pot life (service life) of adhesives (2018)

Jenkins, C.L., Meredith, H.J., Wilker, J.J.: Molecular weight effects upon the adhesive bonding of a mussel mimetic polymer. ACS Appl. Mater. Interfaces. **5**(11), 5091–5096 (2013). https://doi.org/10.1021/am4009538

Lei, H., Du, G., Wu, Z., Xi, X., Dong, Z.: Cross-linked soy-based wood adhesives for plywood. Int. J. Adhes. Adhes. **50**, 199–203 (2014). https://doi.org/10.1016/j.ijadhadh.2014.01.026

Lépine, E.: Synthèse d'adhésifs thermodurcissables à base de farine de soya et de furfural pour la fabrication de panneaux composites en bois (2013)

Luo, J., Luo, J., Li, X., Gao, Q., Li, J.: Effects of polyisocyanate on properties and pot life of epoxy resin cross-linked soybean meal-based bioadhesive. In: Journal of Applied Polymer Science, Vol. 133, Issue 17. John Wiley and Sons Inc. (2016). https://doi.org/10.1002/app.43362

Maji, P.K., Bhowmick, A.K.: Influence of number of functional groups of hyperbranched polyol on cure kinetics and physical properties of polyurethanes. J. Polym. Sci., Part A: Polym. Chem. **47**(3), 731–745 (2009). https://doi.org/10.1002/pola.23185

Markovičová, L.: The Effect of Filler Content on the Viscosity of Polymer Composites **3**, 293–298 (2021). https://doi.org/10.2478/cqpi-2021-0028

Meier-Westhues, U.: Polyurethanes: coatings, adhesives and sealants. European Coatings. (Vincentz Network). European Coatings Tech Files (2019)

Nesvizhskii, A.I., Keller, A., Kolker, E., Aebersold, R.: A statistical model for identifying proteins by tandem mass spectrometry. Anal. Chem. **75**(17), 4646–4658 (2003). https://doi.org/10.1021/ac0341261

Pizzi, A., Mittal, K.L.: Handbook of Adhesive Technology (2005)

Ricci, L., et al.: On the thermal behavior of protein isolated from different legumes investigated by DSC and TGA. J. Sci. Food Agric. **98**(14), 5368–5377 (2018). https://doi.org/10.1002/jsfa.9078

Schulze, K.A., Zaman, A.A., Söderholm, K.J.M.: Effect of filler fraction on strength, viscosity and porosity of experimental compomer materials. J. Dent. **31**(6), 373–382 (2003). https://doi.org/10.1016/S0300-5712(03)00091-5

Stevanovic, T., Perrin, D.: Chapitre 1 "Chimie et analyse des produits naturels." In : Chimie du bois, pp. 3–46. Presses polytechniques et universitaires romandes (2009)

Thakur, V.K., Thakur, M.K., Kessler, M.R.: Handbook of composites form renewable materials - Volume 1 : Structure and Chemistry (2017)

Trinh, E.H.: Formulation of green adhesives from mixtures of soy protein isolates and rosin ester resin (2012)

UNEP : Rapport 2020 sur l'écart entre les besoins et les perspectives en matière de réduction des émissions (2020)

Vick, C.B., Rowell, R.M.: Adhesive bonding of acetylated wood (1990)

Vnučec, D., Kutnar, A., Goršek, A.: Soy-based adhesives for wood-bonding–a review. In: Journal of Adhesion Science and Technology, Vol. 31, Issue 8, pp. 910–931. Taylor and Francis Ltd. (2017). https://doi.org/10.1080/01694243.2016.1237278

Yang, I., Kuo, M., Myers, D.J., Pu, A.: Comparison of protein-based adhesive resins for wood composites. J. Wood Sci. **52**(6), 503–508 (2006). https://doi.org/10.1007/s10086-006-0804-5

Yu, W., et al.: Relationships between protein content, starch molecular structure and grain size in barley. Carbohyd. Polym. **155**, 271–279 (2017). https://doi.org/10.1016/j.carbpol.2016.08.078

Correlation Between Length Change and Mechanical Properties of Mortar Containing Phragmites Australis Ash (PAA)

Jamal M. Khatib[1,2(✉)], Lelian W. ElKhatib[1], Mohammed Sonebi[3], and Adel Elkordi[1,4]

[1] Faculty of Engineering, Beirut Arab University, Beirut 11-5020, Lebanon
j.khatib@bau.edu.lb
[2] Faculty of Science and Engineering, University of Wolverhampton, Wolverhampton WV1 1LY, UK
[3] School of the Natural and Built Environment, Queens University of Belfast, Belfast, UK
[4] Department of Civil and Environmental Engineering, Faculty of Engineering, Alexandria University, Alexandria 21511, Egypt

Abstract. Due to the high increase in pollution levels, the use of biomass ashes become a fundamental need. These ashes are considered to be environmentally friendly since they emit less CO_2 to the atmosphere than other cementitious materials. This paper is a part of a wide investigation for the use of phragmites australis ashes (PAA) as a cementitious material in mortar mixes. It examines mechanical, durability and length change properties. Length change include drying, autogenous and expansion tests. The test results for total and capillary water absorption at 24 h, at all curing ages 1, 7, 28 and 90 days, show an increase as the percentage of replacing cement by PAA increases from 0 to 30% in increment of 10. Correlations between mechanical and length change properties are also examined.

Keywords: Bio-ash · Phragmites Australis Ash · Mortar · Sustainability · Cement Replacement

1 Introduction

Currently, concrete has become the most dominant material used in construction. Despite the unique properties concrete have such as its strength and the ability to form it in any shape, this material can be easily obtained [1]. However, climate change and the increase in the earth temperature are considered to be the most challenging and harmful problems facing the whole world. The global production of cement is evaluated to be around 4.6 billion tonnes. The combustion of fossil fuels and decomposition of calcium carbonates resulting from the process of manufacturing cement reveals a huge amount of carbon dioxide (CO2) into the atmosphere leading to the emission of around 7% of greenhouse gases all over the world [2–5]. Previous studies have shown that each ton of cement produced contributes to 800 kg of CO2 into the environment [6]. Due to the fact that

S. Amziane et al. (Eds.): ICBBM 2023, RILEM Bookseries 45, pp. 29–41, 2023.
https://doi.org/10.1007/978-3-031-33465-8_3

concrete has a long life time, excellence and unique properties, its use in construction has increased widely all over the world [7, 8]. So, in order to achieve sustainability, lessen the carbon footprint and the severe environmental damages occurring by the concrete industries, an environmentally friendly alternative should be found for some materials used in concrete mixtures. In the past centuries, vast alternatives for both cement and aggregates have been examined [9]. The use of some agricultural waste ashes as a substitute for cement could help in solving the management and disposal of these ashes in addition to providing a material that can take place instead of non-renewable raw materials that will be lost [10, 11]. Agricultural wastes are mainly shells, stems, leaves and husks which are burnt to obtain the ashes [12]. Many studies were conducted on agricultural wastes such as rice husk ash, pumpkin shell ash, coconut shell ash, sunflower shell ash, wheat straw ash, corn cob ash, pistachio shell ash and bamboo leaves ash [13, 14]. Different biomass ashes, when used as an alternative to cement can improve some properties of cementitious systems [15]. As an example, the use of rice husk ash as a partial replacement to cement in mortar mixtures increases the mechanical strength and lowers the permeability [16]. Although several researches and studies have been done on the use of agricultural waste ashes as supplementary cementitious material [1], there are null studies on the use of PAA in mortar, paste and concrete specimens. This paper examines the total and capillary water absorption of mortar specimens replacing cement with PAA by 0, 10, 20 and 30% and cured for 1, 7, 28 and 90 days of curing ages at 24 h exposure to water and correlation between length change and mechanical properties, flexural and compressive strength, at 1, 7 and 28 days of curing.

2 Materials

Portland cement Type I conforming EN 197–1 standards and sand passing through 4.75 mm and retaining on 200 μm was used. PAA was passed through the 300 μm sieve after burning and grinding. Tap water was also used.

2.1 Chemical Composition

The main chemical contents of cement and PAA are SiO_2, CaO and Al_2O_3. The percentages in cement were 18.1, 61.55 and 4.29 respectively whereas the percentages in the PAA were 76.15, 6.15 and 0.84 respectively.

2.2 Mix Proportions

Four different mortar mixes were cast. The cement:sand ratio for control mix was 1:3 for control mix. Cement in the other mixes was replaced by PAA with different percentages ranging between 10% to 30% in increment of 10. The water to binder ratio was kept constant for all mixes at 0.55. Details are shown in Table 1.

Table 1. Mix Proportions.

Quantities (kg/m^3)				
Mix Code	Cement	PAA	Sand	Water
0% PAA	475.0	0.0	1425	262
10% PAA	427.5	47.5	1425	262
20% PAA	380.0	95.0	1425	262
30% PAA	332.5	142.5	1425	262

3 Testing

3.1 Total Water Absorption

Mortar cubes of 100 × 100 × 50 mm size were used. After each curing period, specimens were removed from water and dried until reaching a constant mass at 80 °C in oven. Test were then conducted and total water absorption (TWA) percentage were calculated based on the following formula:

$$\mathrm{TWA} = \frac{\mathrm{M30} - \mathrm{Md}}{\mathrm{Md}} \times 100 \tag{1}$$

TWA: Total Water Absorption (%).
M30: Mass at time 30 min (kg).
Md: Mass for dry specimen (kg).

3.2 Capillary Water Absorption

For capillary water absorption (CWA) test, dimensions for specimens were 100 × 100 × 50 mm. After curing period, specimens were placed also in oven at 80 °C until reaching a constant mass. Then when removing them from oven, specimens were placed on 2 supports in a shallow tray where water was added to 2 mm level above the mortar specimen as shown in Fig. 1 and test was conducted. CWA were then calculated according to the following formula:

$$\mathrm{CWA} = \frac{\mathrm{Mt} - \mathrm{Md}}{\mathrm{A}} \tag{2}$$

CWA: Capillary Water Absorption (g/mm^2).
Mt: Mass at time t (kg).
Md: Mass for dry specimen (kg).
A: Area of specimen (mm^2).

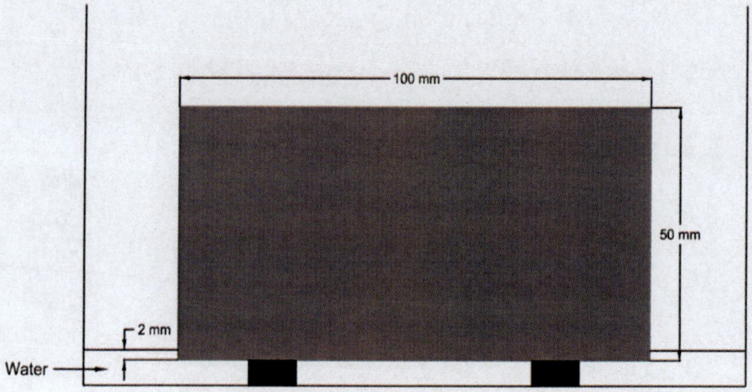

Fig. 1. Capillary Absorption Test

3.3 Length Change

Samples were casted according to a binder:sand ratio 1:3 and a constant w/c 0.55. All samples were cast with dimensions of 25 mm × 25 mm × 285 mm. After 24 h from casting, demoulding took place for all samples. Two demec points were placed on two sides of each sample at 200 mm distance apart and the expansion and shrinkage of mortar samples were measured by a dial gauge as shown in Fig. 2. For expansion, samples were tested according to ASTM C192 where all samples were placed in a water bath at room temperature 20 °C. However, drying shrinkage were tested according to ASTM C157 where samples were placed in a container in open air at room temperature. For autogenous shrinkage, test was performed according to ASTM C192, all samples were placed in a closed plastic bag to prevent any entrance of moisture. The length change was then recorded and the average was taken. Specimens length change was recorded up to 28 days after demoulding. The following equation was used for the relation between both slope (m) and intercept (n) with curing times.

$$y = mx + n \tag{3}$$

Fig. 2. Length Change Test

4 Results and Discussion

4.1 Total Water Absorption

The total water absorption percentage is presented in Fig. 3 as function of PAA replacement at 24 h for 1, 7, 28 and 90 days of curing. Results display that the control mix recorded the highest value. There is an increase in TWA percentage as PAA replacement percentage increases. However, as the curing age increases, the TWA percentage decreases for all different mortar mixes.

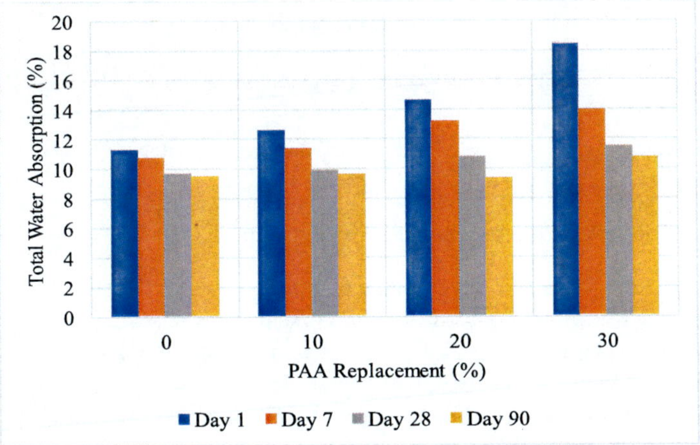

Fig. 3. Total Water Absorption at different PAA Replacements

4.2 Capillary Water Absorption

Figure 4 presents the capillary water absorption at 24 h at all different curing ages 1, 7, 28 and 90 days for the four different mortar specimens. Results show that the control mix and the mix with 10% PAA recorded approximately close results. As the PAA percentage increases from 10% to 30%, the CWA increases. As the curing time passes, the CWA decreases for all mixes.

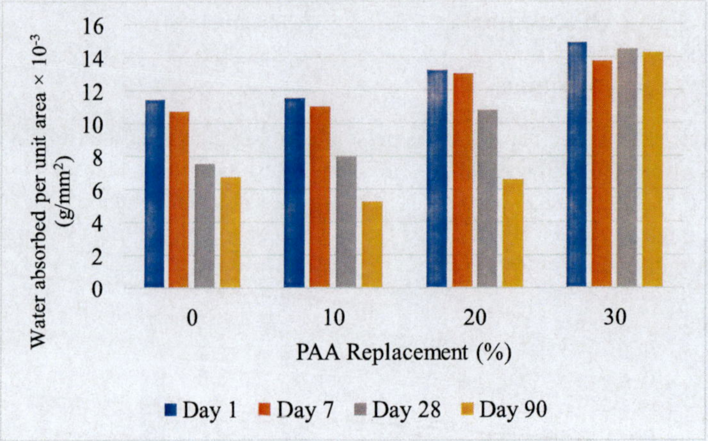

Fig. 4. Capillary Water Absorption at different PAA Replacements

4.3 Correlation Between Compressive Strength and Expansion

Figure 5 shows the relationship between the compressive strength and expansion for all mixes containing 0% to 30% PAA as partial cement replacement at 1, 7 and 28 days of curing. As the compressive strength increases, the expansion decreases. There is a linear relationship between compressive strength and expansion with high correlation coefficient. Figure 6 plots the relation between both slope (m) and intercept (n) with different ages of curing. Both the slope in absolute value and the intercept increases as curing time passes.

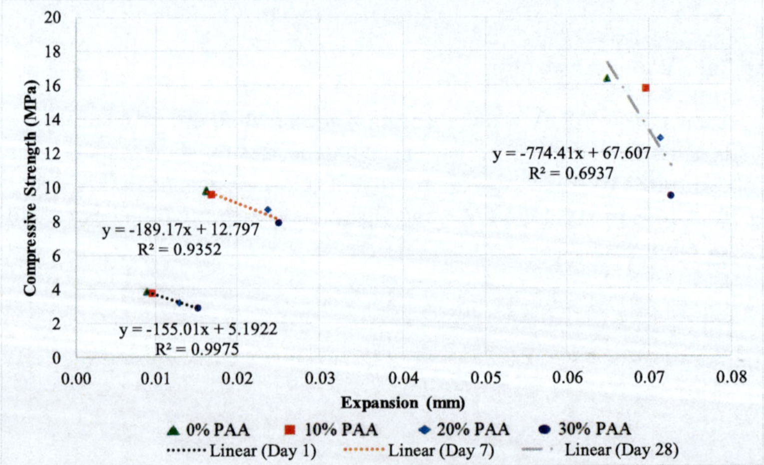

Fig. 5. Correlation Between Compressive Strength and Expansion

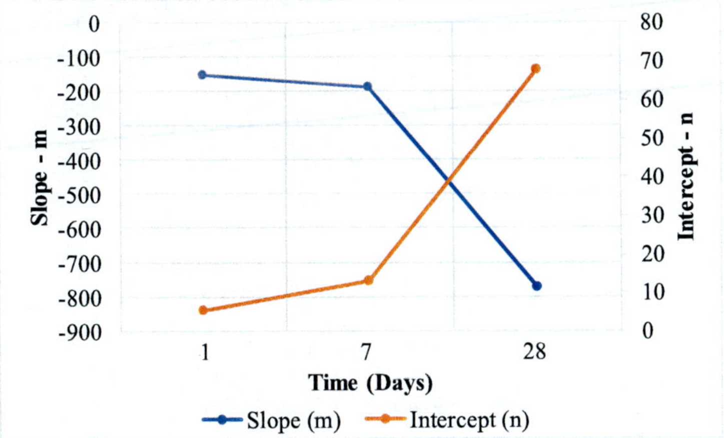

Fig. 6. Slope (m) and Intercept (n) for the linear relationship between compressive strength and expansion as a function of curing Time

4.4 Correlation Between Compressive Strength and Drying Shrinkage

Figure 7 represents the correlation between compressive strength and drying shrinkage for the four different mixes at 1, 7 and 28 days of curing. Results show a strong relationship between compressive strength and drying shrinkage. As the compressive strength increases between different mixes at same curing age, the drying shrinkage decreases. There appears a relationship between both slope (m) and intercept (n) with the age of curing where both increases as curing age increases as shown in Fig. 8.

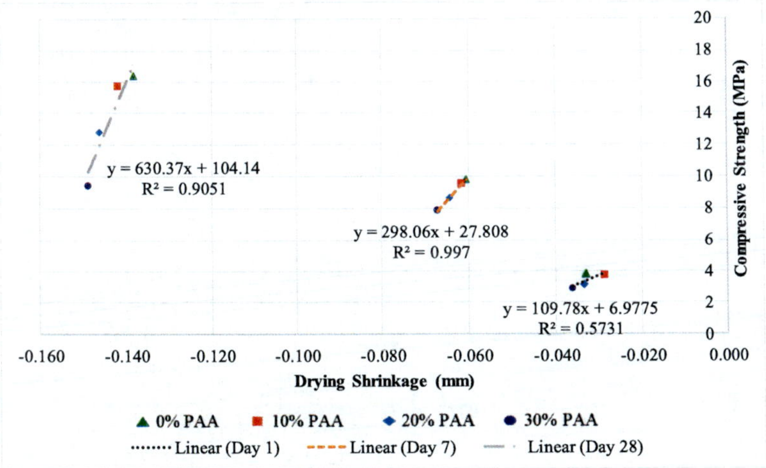

Fig. 7. Correlation Between Compressive Strength and Drying Shrinkage

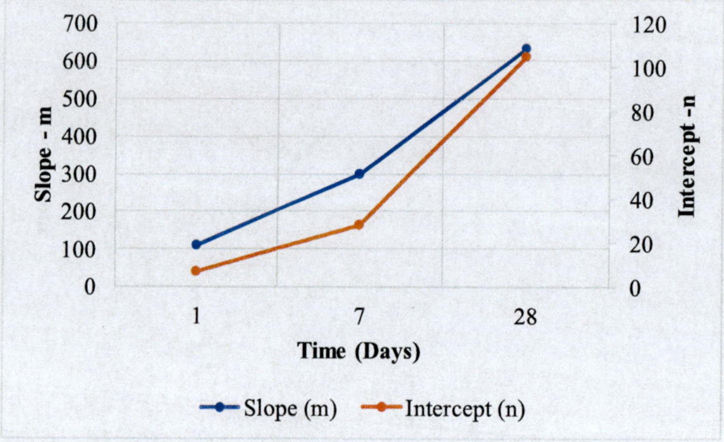

Fig. 8. Slope (m) and Intercept (n) for the linear relationship between compressive strength and drying shrinkage as a function of curing Time

4.5 Correlation Between Compressive Strength and Autogenous Shrinkage

Figure 9 displays the correlation between compressive strength and autogenous shrinkage for all mixes at different curing ages 1, 7 and 28 days. Results show a good relationship between these two properties with a high correlation coefficient R^2 above (0.75). As the compressive strength increases between different mixes at same curing age, the autogenous shrinkage decreases. Figure 10 shows the relationship of both slope (m) and intercept (n) with curing age.

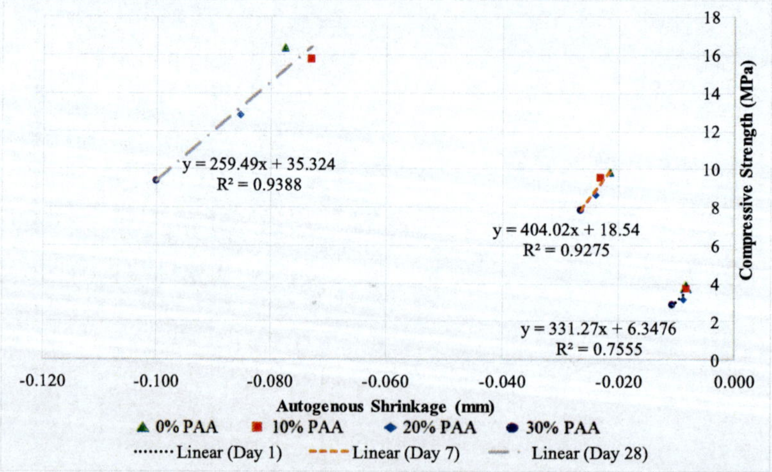

Fig. 9. Correlation Between Compressive Strength and Autogenous Shrinkage

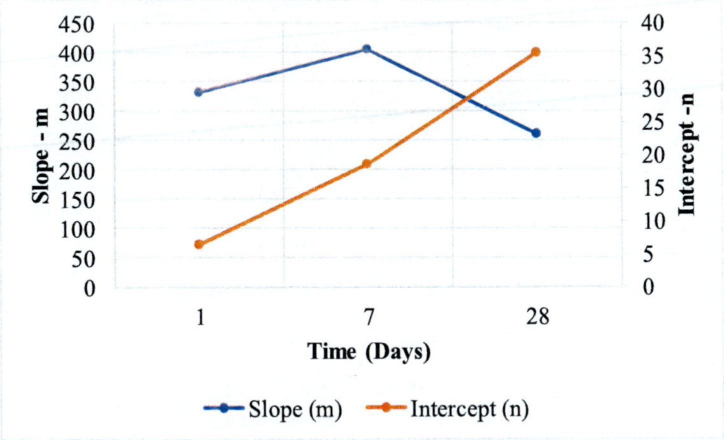

Fig. 10. Slope (m) and Intercept (n) for the linear relationship between compressive strength and autogenous shrinkage as a function of curing Time

4.6 Correlation Between Flexural Strength and Expansion

Figure 11 plots the correlation between flexural strength and expansion for all mixes containing 0% to 30% PAA as cement replacement at different curing times. Similar to Fig. 9, there appears to be a linear correlation between parameters where a high correlation coefficient was obtained. Figure 12 displays the relation between curing age and slope (m) and intercept (n). Both parameters increase in absolute as curing time passes.

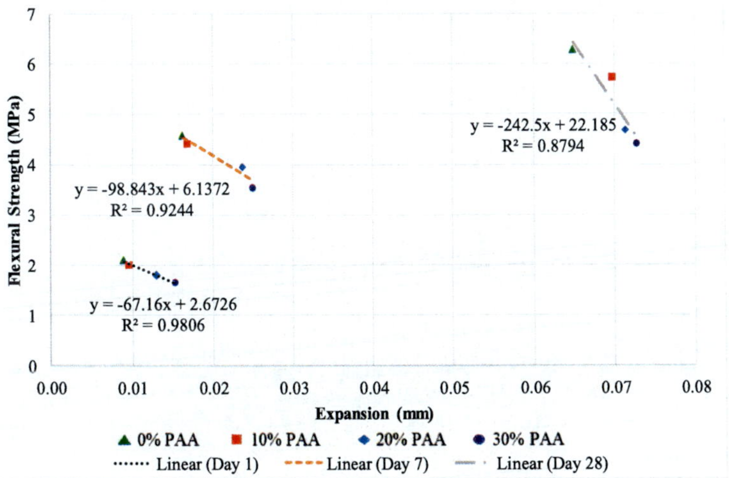

Fig. 11. Correlation Between Flexural Strength and Expansion

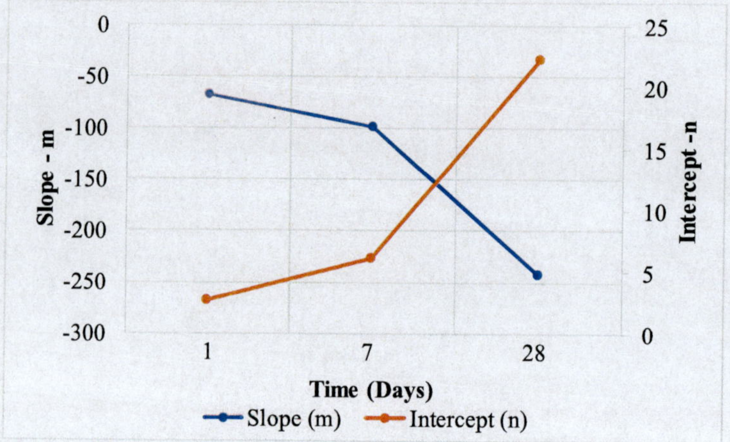

Fig. 12. Slope (m) and Intercept (n) for the linear relationship between flexural strength and expansion as a function of curing Time

4.7 Correlation Between Flexural Strength and Drying Shrinkage

The correlation between flexural strength and drying shrinkage is presented in Fig. 13. The results show that as the flexural strength increases as time passes, the drying shrinkage increase in all mixes. At same curing age, mixes with highest flexural strength records the lowest drying shrinkage where the control mix and mix with 10% PAA shows approximately same values. Figure 14 plots the relation between slope (m) and intercept (n) with curing time where both increase as time passes.

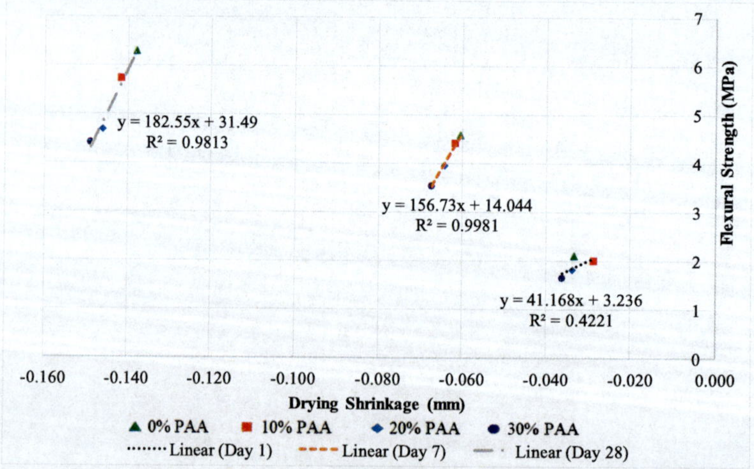

Fig. 13. Correlation Between Flexural Strength and Drying Shrinkage

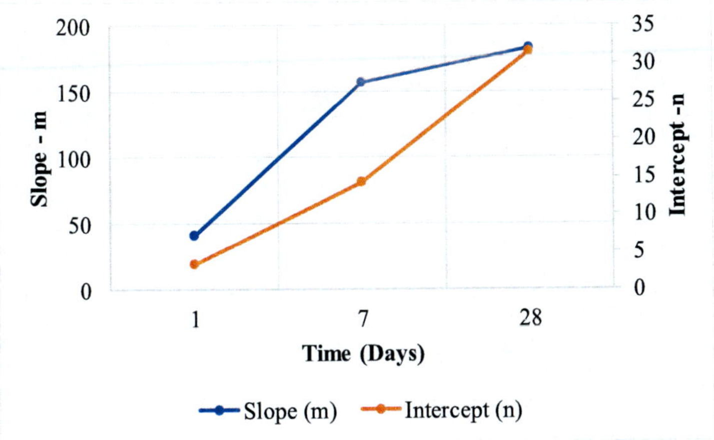

Fig. 14. Slope (m) and Intercept (n) for the linear relationship between flexural strength and drying shrinkage as a function of curing Time

4.8 Correlation Between Flexural Strength and Autogenous Shrinkage

The correlation between flexural strength and autogenous shrinkage is presented in Fig. 15 for mixes containing different PAA percentages at different curing ages. As flexural strength increases when time passes, the autogenous shrinkage increases in all mixes.

At same curing age, as PAA percentage increases causing the flexural strength to decrease, the autogenous shrinkage increases. Figure 16 shows the relation between curing time and slope and intercept.

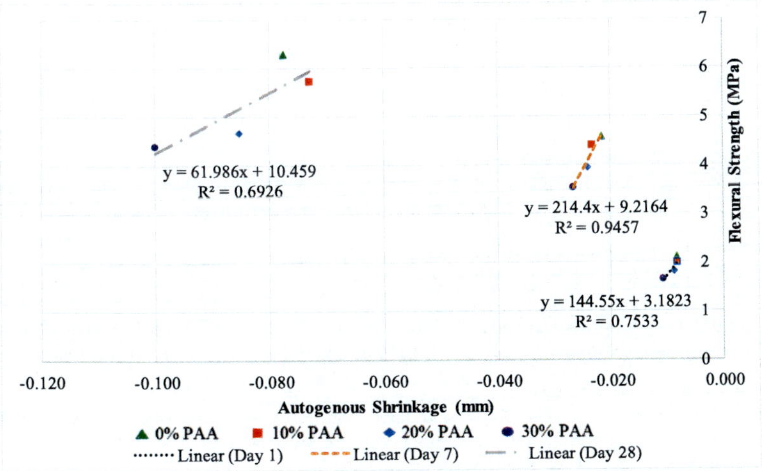

Fig. 15. Correlation Between Flexural Strength and Autogenous Shrinkage

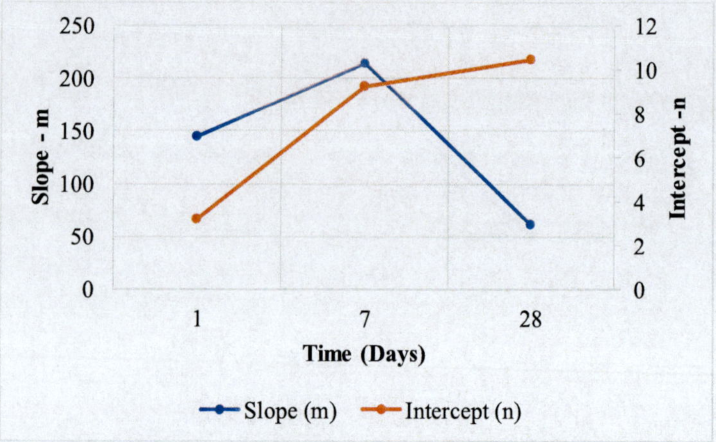

Fig. 16. Slope (m) and Intercept (n) for the linear relationship between flexural strength and autogenous shrinkage as a function of curing Time

5 Conclusion

There is a good relationship between mechanical properties such as compressive strength and flexural strength with length change properties. As the mechanical properties decreases as the percentage of PAA is increased, the length change increases for all curing days.

Using 10% PAA as cement replacement appears to be give similar results to the control mix. At 90 days of curing, the total water absorption of the control was 9.5% and 9.6% for mix containing 10% PAA.

Therefore, 10% of PAA can be used as partial replacement of cement. This can lead to a lower CO_2 emission due to the high energy required to produce cement.

References

1. ElKhatib, L., Al Aridi, F., ElKordi, A., Khatib, J.: Mechanical and durability properties of geopolymer concrete–a review. BAU Journal-Science and Technology. **3**(2), 8 (2022)
2. Syarif, M., et al.: Development and assessment of cement and concrete made of the burning of quinary by-product. J. Mater. Res. Technol. **15**, 3708–21 (2021 Nov 1)
3. Jannat, N., Al-Mufti, R.L., Hussien, A., Abdullah, B., Cotgrave, A.: Utilisation of nut shell wastes in brick, mortar and concrete: A review. Constr. Build. Mater. **26**(293), 123546 (2021). Jul
4. Berenguer, R.A., Capraro, A.P., de Medeiros, M.H., Carneiro, A.M., De Oliveira, R.A.: Sugar cane bagasse ash as a partial substitute of Portland cement: Effect on mechanical properties and emission of carbon dioxide. J. Environ. Chem. Eng. **8**(2), 103655 (2020). Apr 1
5. Thiedeitz, M., Ostermaier, B., Kränkel, T.: Rice husk ash as an additive in mortar–Contribution to microstructural, strength and durability performance. Resour. Conserv. Recycl. **1**(184), 106389 (2022). Sep
6. Khan, A., Sikandar, M.A., Bashir, M.T., Shah, S.A., Zamin, B., Rehman, K.: Assessment for utilization of tobacco stem ash as a potential supplementary cementitious material in cement-based composites. Journal of Building Engineering. **1**(53), 104531 (2022). Aug 1

7. ElKhatib, L.W., Al Aridi, F.K., ElKordi, A., Khatib, J.M.: Selected properties on geopolymer concrete – A review. III International Turkish World Engineering and Science Congress, 337–345 (2021)
8. Raheem, A.A., Ikotun, B.D.: Incorporation of agricultural residues as partial substitution for cement in concrete and mortar–A review. Journal of Building Engineering. **1**(31), 101428 (2020). Sep
9. Carević, I., Baričević, A., Štirmer, N., Bajto, J.Š: Correlation between physical and chemical properties of wood biomass ash and cement composites performances. Constr. Build. Mater. **30**(256), 119450 (2020). Sep
10. Adhikary, S.K., Ashish, D.K., Rudžionis, Ž: A review on sustainable use of agricultural straw and husk biomass ashes: Transitioning towards low carbon economy. Sci. Total Environ. **2**, 156407 (2022). Jun
11. Munshi, S., Sharma, R.P., Chatterjee, T.: Investigation on the mechanical properties of cement mortar with sustainable materials. Materials Today: Proceedings. **1**(47), 4833–4837 (2021). Jan 1
12. Roselló, J., Soriano, L., Santamarina, M.P., Akasaki, J.L., Monzó, J., Payá, J.: Rice straw ash: A potential pozzolanic supplementary material for cementing systems. Ind. Crops Prod. **1**(103), 39–50 (2017). Sep
13. Shahbazpanahi, S., Faraj, R.H.: Feasibility study on the use of shell sunflower ash and shell pumpkin ash as supplementary cementitious materials in concrete. Journal of Building Engineering. **1**(30), 101271 (2020). Jul
14. Rodier, L., Villar-Cociña, E., Ballesteros, J.M., Junior, H.S.: Potential use of sugarcane bagasse and bamboo leaf ashes for elaboration of green cementitious materials. J. Clean. Prod. **10**(231), 54–63 (2019). Sep
15. Patil, C., Manjunath, M., Hosamane, S., Bandekar, S., Athani, R.: Pozzolonic activity and strength activity index of bagasse ash and fly ash blended cement mortar. Materials Today: Proceedings. **1**(42), 1456–1461 (2021). Jan
16. Bonfim, W.B., de Paula, H.M.: Characterization of different biomass ashes as supplementary cementitious material to produce coating mortar. J. Clean. Prod. **1**(291), 125869 (2021). Apr 1

Structure and Properties of Portland-Limestone Cements Synthesized with Biologically Architected Calcium Carbonate

Madalyn C. Murphy, Danielle N. Beatty, and W. V. Srubar(✉)

University of Colorado Boulder, Boulder, CO 80303, USA
wsrubar@colorado.edu

Abstract. Portland cement is one of the most used materials on earth. Its annual production is responsible for approximately 7% of global carbon dioxide (CO_2) emissions. These emissions are primarily associated with (1) the burning of fossil fuels to heat cement kilns and (2) the release of CO_2 during limestone calcination. One proposed strategy for CO_2 reduction includes the use of functional limestone fillers, which reduce the amount of portland cement in concrete without compromising strength. This study investigated the effect of using renewable, CO_2-storing, biogenic $CaCO_3$ produced by *E. huxleyi* as limestone filler in portland limestone cements (PLCs). Biogenic $CaCO_3$ was used to synthesize PLCs with 0, 5, 15, and 35% limestone replacement of portland cement. The results substantiate that the particle sizes of the biogenic $CaCO_3$ were significantly smaller and the surface areas significantly larger than that of reagent grade $CaCO_3$. X-ray diffraction indicated no differences in mineralogy between reagent-grade and biogenic $CaCO_3$. The use of biogenic $CaCO_3$ as a limestone filler led to (i) increased water demand at the higher replacements, which was countered by using a superplasticizer, and (ii) enhanced nucleation during cement hydration, as measured by isothermal conduction calorimetry. The 7-day compressive strengths of the PLC pastes were measured using mechanical testing. Enhanced nucleation effects were observed for PLC samples containing biogenic $CaCO_3$. 7-day compressive strength of the PLCs produced using biogenic $CaCO_3$ were also enhanced compared to PLCs produced using reagent-grade $CaCO_3$ due to the nucleation effect. This study illustrates an opportunity for using CO_2-storing, biogenic $CaCO_3$ to enhance mechanical properties and CO_2 storage in PLCs containing biologically architected $CaCO_3$.

Keywords: *E. huxleyi* · calcium carbonate · portland limestone cement · nucleation effects · mechanical properties

1 Introduction

Functional fillers, such as limestone, and supplementary cementitious materials (SCMs), such as slag and fly ash, have been studied in the context of replacing cement clinker in cement and concrete production because of their beneficial effects on cost, mechanical properties, and sustainability. Functional limestone fillers (LFs) decrease the amount of

S. Amziane et al. (Eds.): ICBBM 2023, RILEM Bookseries 45, pp. 42–53, 2023.
https://doi.org/10.1007/978-3-031-33465-8_4

cement clinker used in cement or concrete, which in turn reduces energy consumption in cement production and CO_2 emissions [1, 2].

Multiple studies have detailed the various effects of LFs in cement paste and concrete, including the filler effect, nucleation effect, chemical effect, and dilution effect [3, 4]. Traditional LF fills space where cement grains would normally interact, augmenting the distance between interactions, thus improving cement paste workability. Due to this space filling, water consumption of PLCs can sometimes be reduced to achieve the same workability [5–7]. The nucleation effect has been observed with the incorporation of nano-limestone accelerating early age reactions, though hydration acceleration was not observed for coarser LF addition on the scale of 4–16 μm [8, 9]. Wang et al. posit that when LF is finer than cement particles, LF addition causes filler, nucleation, and dilution effects, while, when LF is coarser than cement particles, the dilution effect dominates. It has also been shown that fine LF enhances hydration, thereby offsetting the dilution effect [10]. Additionally, fine LF induces filler effects, where more space within cement paste is filled by limestone particles, thus increasing compressive strength and decreasing porosity and sorptivity [3]. LFs, while often considered chemically inert, actually induce chemical effects when added to cement paste. The chemical effect is mainly attributed to the reaction of LF with monosulfate and aluminate hydrate causing the suppression of early (16 h) C_3A hydration [4]. Overall, chemical and filler effects have little impact on setting time, and chemical and nucleation effects seem to have little impact on workability. Due to the dilution effect, compressive strength was reported to decrease with increased LF addition, particularly at replacement values exceeding 35% [11]. The effect on mechanical and compressive strength is trivial at replacement values up to 15% [12], and replacement of 35% LF in PLC may yield satisfactory mechanical and compressive strength for various applications. The American Society for Testing Materials (ASTM) currently sets the industrial limit at 15% replacement value of LF in PLC (ASTM C595), while the European Union (EU) has an increased limit of up to 35% (EN 197–1) [31, 32]. While the EU has been using PLC with LF replacement values up to 35% for some time, the United States (US) cement industry has begun to integrate PLC into industrial applications in lieu of OPC, taking advantage of its decreased cost and reduced CO_2 emissions [12, 13].

This study investigated the use of a biologically derived LF in PLCs, namely calcium carbonate ($CaCO_3$) biologically architected by the coccolithophore, *Emiliana Huxleyi*. *E. Huxleyi* is a species of microalgae ubiquitous in nearly every ocean on Earth. *E. Huxleyi* produces intricate $CaCO_3$ coccoliths through a photosynthetic process known as coccolithogenesis [14]. Major advantages of the coccolithogenesis process used by *E. Huxleyi* are (1) CO_2 is *consumed* during coccolith production and (2) very few nutrients are required for sustained growth of *E. Huxleyi*. The microstructures of $CaCO_3$ produced by coccolithophores are complex and uniform, with a particle size under 10 microns [15]. Due to this highly complex and intricate design (small particle size and high surface area), biogenic $CaCO_3$ from *E. Huxleyi* shows potential for increased nucleation of cement hydration products compared to industrial limestone when used as an LF in PLCs. In this study, we explore the suitability of biogenic $CaCO_3$ produced by *E. Huxleyi* as a filler for PLC, corresponding to both US (ASTM C595, up to 15% replacement) and EU (EN 197–1, up to 35% replacement) standards.

2 Materials and Methods

2.1 $CaCO_3$ Sources

Reagent grade $CaCO_3$ (>99%) was purchased from Sigma Aldrich. ASTM C150 Type I/II portland cement was purchased from Quikrete.

Lyophilized *E. huxleyi* biomass (containing both $CaCO_3$ and organic cell materials) was purchased from the Algal Resource Collective (ARC) at the University of North Carolina, Wilmington, USA. $CaCO_3$ was then purified to remove organics following a similar protocol to that used by Jakob et al., 2015 [16]. Approximately 0.15 g of freeze-dried biomass was added to 50 mL centrifuge tubes and suspended in MilliQ water to a volume of 40 mL, which then sedimented for at least 24 h. Tubes were then centrifuged for 17 min at 4696 x g and 4 °C. Supernatants were removed to the 10 mL marking, and contents were re-suspended via vortexing. Next, 3.3 mL of 12% NaOCl was added to each tube, shaken lightly, and allowed to sit for 15 min. Then 6 mM $NaHCO_3$ was added to the 40 mL marking and tubes were centrifuged for 6 min at 1500 x g and 4 °C. Supernatants were again removed to the 10 mL marking before vortexing to resuspend pellets. Washing with 6 mM $NaHCO_3$ and subsequent centrifugation was repeated four more times. Following the final wash, supernatant was removed to the $<$ 5 mL marking. Pellets were resuspended through gentle shaking and contents from 8 tubes were combined in a separate centrifuge tube before centrifugation for 6 min at 1500 x g and 4 °C. Maximum volume of supernatant was removed without disturbing the pellet, and purified $CaCO_3$ was dried in an oven at 80–90 °C for at least 24 h.

2.2 Limestone Characterization

Morphology and Particle Characteristics

Both $CaCO_3$ sources were imaged using a Hitachi SU3500 scanning electron microscope (SEM). Samples were coated in platinum prior to imaging to ensure sufficient conductivity. Images of both reagent grade and biogenic calcite were taken in secondary electron mode using 15 keV accelerating voltage, 2,500 x and 10,000 x magnifications, with $<$ 9 mm working distances.

Particle size distributions (PSD) of both $CaCO_3$ sources was measured using laser diffraction in a Malvern Panalytical Mastersizer3000. Particles were suspended in ultrapure MilliQ water, dispersed using ultrasonication prior to size analysis. Mean particle size and standard deviation were calculated using results from 5 replicate samples.

Mineralogy

Mineralogy of both $CaCO_3$ sources was confirmed using qualitative X-ray diffraction (XRD) with a Bruker D8 Advance X-ray diffractometer. Cu Kα X-ray radiation (wavelength 1.5406 Å) was used to scan from 5° to 90° 2θ with a step size of 0.02° and a dwell time of 1.5 s per step. The resulting patterns were analyzed with the Bruker DIFFRAC.EVA software that was equipped with the International Center for Diffraction Data (ICDD) PDF-4 AXIOM 2019 database [33] to identify phases.

2.3 Portland Limestone Cement Paste Studies

Hydration Kinetics

Cement pastes with a water-to-cement (w/c) ratio of 0.65 were mixed according to the mixture proportions given in Table 1. A w/c = 0.65 was chosen due to the anticipated workability concerns of high-$CaCO_3$ replacement percentage mixtures at lower w/c ratios, especially for samples containing biogenic $CaCO_3$. $CaCO_3$ was dry mixed with cement by hand for 5 min. Dry cement mixtures were then added to water and thoroughly mixed by hand for at least 2 min. A polycarboxylate-based superplasticizer (SP) was added dropwise to the 35% biogenic $CaCO_3$ PLC paste up to 0.18 g until a thick paste consistency was achieved. SP was only necessary for the 35% *E. huxleyi* cement paste, as acceptable workability was achieved for remaining pastes.

Table 1. Cement paste mix proportions.

	Cement (g)	Reagent-Grade $CaCO_3$ (g)	Biogenic $CaCO_3$ (g)	Water (g)	SP (g)
Control	100	0	0	65	0
5-R	95	5	0	65	0
15-R	85	15	0	65	0
35-R	65	35	0	65	0
5-B	95	0	5	65	0
15-B	85	0	15	65	0
35-B	65	0	35	65	1

Approximately 7 g of paste was added to each of two glass ampoules and sealed for ICC analysis across from a corresponding reference sample (siliceous sand). ICC was operated at 25 °C using an 8 channel Thermometric TAM Air calorimeter. For each $CaCO_3$ source, heat of hydration data were collected for a control mix (OPC), a 5%, 15%, and 35% $CaCO_3$ replacement mix. Heat of hydration and total heat data were collected for at least 72 h. Data were then normalized by weight of cement powder in each sample.

Compressive Strength Testing

Following mixing, 16.5 g of cement paste were poured into a silicone 1 cm^3 cube mold tray. A small metal rod was used to ensure even spreading of cement paste to the cube corners. The tray was then placed in a > 94% humidity chamber (created according to a modified ASTM E104 standard) and cured for 24 h. Cubes were then removed from the humidity chamber and placed in a supersaturated $Ca(OH)_2$ solution. Samples were tested at 7 days using an Instron Universal Testing machine with a 50 kN load cell and a 0.1 mm/sec compression rate. Samples were tested in triplicate.

3 Results

3.1 $CaCO_3$ Characterization

Particle Size and Scanning Electron Microscopy

Reagent grade $CaCO_3$ exhibited a significantly larger particle size than that of biogenic $CaCO_3$ grown from *E. huxleyi* (Fig. 1). Particle size analysis revealed that the median particle sizes (d_{50}) of the reagent-grade and biogenic $CaCO_3$ were 18.8 ± 0.274 μm and 1.77 ± 0.17 μm, respectively, which is an order of magnitude difference. The differences in particle size are evident in the SEM micrographs shown in Fig. 1.

Although some small, angular particles are seen for reagent grade $CaCO_3$ (Fig. 1b), the majority of particles are much larger than those seen for *E. huxleyi* $CaCO_3$ (Fig. 1c, d). The microstructures of biologically architected $CaCO_3$ appear much more intricate than the more granular and rigid microstructure of reagent grade $CaCO_3$. It can be inferred from these SEM micrographs that biogenic $CaCO_3$ also exhibits higher surface area than reagent grade $CaCO_3$ due to decreased particle sizes and their biologically architected, intricate shapes.

X-ray Diffraction

XRD results (Fig. 2) show nearly identical phases present for both reagent grade and biogenic $CaCO_3$. Both samples exhibit distinct peaks at expected angles characteristic of the calcite phase of $CaCO_3$, with minor peaks associated with additional, minor phases in each sample, specifically those shown near 27° 2θ.

3.2 Cement Paste Studies

Hydration Kinetics

Isothermal conduction calorimetry results are shown in Fig. 3. The rate of heat evolution and cumulative heat evolved were similar for the 5% and 15% PLC with reagent grade $CaCO_3$ filler, as compared to the OPC control. A slight increase and leftward shift of the main hydration peak was evident for the 35% PLC with reagent grade $CaCO_3$ filler. The cumulative heat also increased slightly with each increased addition of reagent grade $CaCO_3$ filler. PLC with 5% biogenic $CaCO_3$ exhibited a slight main hydration peak increase, as well as a slight cumulative heat increase, which was lower than the increase exhibited by 5% reagent grade $CaCO_3$ PLC (see Fig. 3). PLC with 15% biogenic $CaCO_3$ showed a significant heat increase and leftward shift of the main hydration peak. Notably, PLC with 35% biogenic $CaCO_3$ and SP showed a significantly delayed rate of heat evolution, reaching a maximum around 20 h, roughly 12 h later than other PLC pastes with lower amounts of biogenic $CaCO_3$. Both 15% and 35% biogenic $CaCO_3$ PLC exhibited increased cumulative heat at 72 h as compared to OPC and 5% biogenic $CaCO_3$ PLC paste, with 35% biogenic $CaCO_3$ PLC reaching approximately the same cumulative heat as 15% biogenic $CaCO_3$ PLC despite significantly delayed hydration. PLC with 35% biogenic $CaCO_3$ also shows a significant increase in the second hydration peak, associated with the reaction of C_3A and the formation of a calcium aluminate phase (ettringite).

Compressive Strength

The lowest 7-day compressive strength was observed for the 35% reagent grade $CaCO_3$

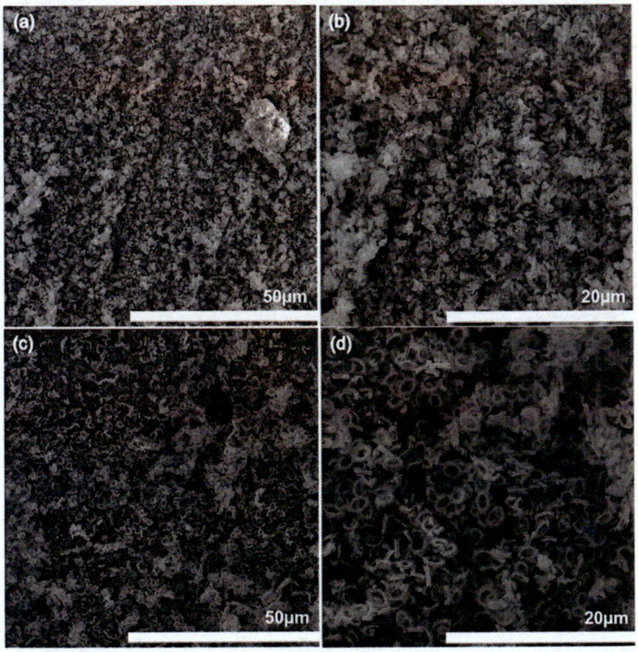

Fig. 1. Scanning electron micrographs of **(a-b)** reagent grade $CaCO_3$ compared to **(c-d)** biogenic $CaCO_3$ from *E. huxleyi.*

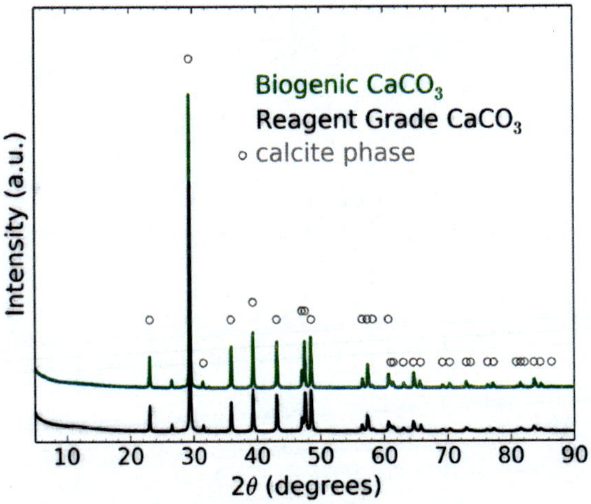

Fig. 2. X-ray diffraction results of reagent grade $CaCO_3$ compared with biogenic $CaCO_3$ from *E. huxleyi.* All significant peaks align with the calcite phase of $CaCO_3$ for both reagent grade and biogenic $CaCO_3$.

pastes. The 5% and 15% reagent grade $CaCO_3$ pastes displayed comparable strengths slightly lower than that of OPC. Contrarily, the highest compressive strength was observed for the 15% biogenic $CaCO_3$ PLC paste by a large, statistically significant margin. Both 5% and 35% biogenic $CaCO_3$ PLC pastes displayed similar strength, the latter of which is lower than that of OPC. At all tested replacement values of 5% or greater, PLC pastes with biogenic $CaCO_3$ showed increased compressive strength over their reagent grade $CaCO_3$ counterparts. Pastes with both types of LF at a 5% replacement value had comparable compressive strengths and did not show excessively decreased compressive strengths compared to OPC (Fig. 4).

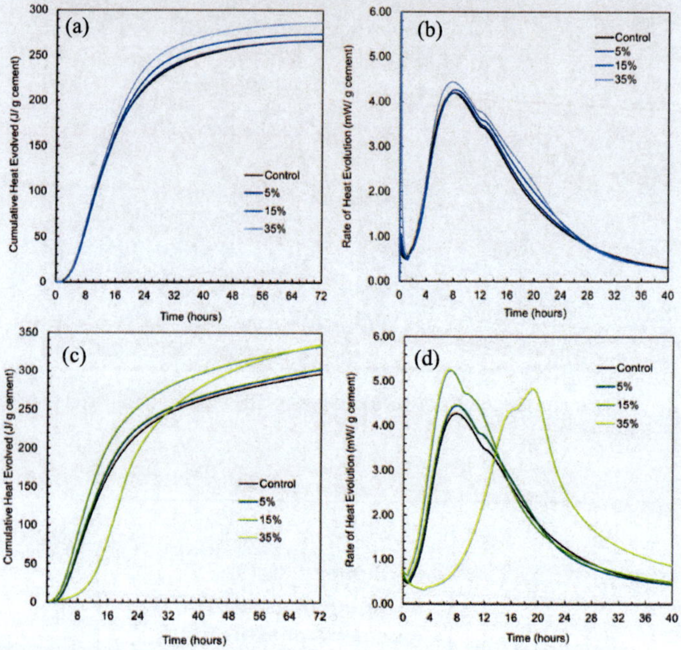

Fig. 3. Rate of heat evolution (mW/g cement) and cumulative heat evolved (J/g cement) for portland limestone cement mix designs with **(a-b)** reagent grade $CaCO_3$ limestone filler or **(c-d)** biogenic $CaCO_3$ limestone filler produced by *E. huxleyi*.

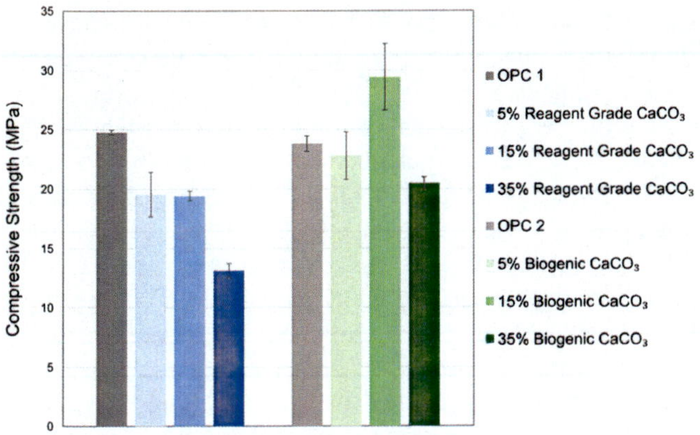

Fig. 4. A7-day compressive strength (in MPa) of PLC cement pastes (1 cm^3 cubes) containing 5%, 15%, or 35% reagent grade or biogenic $CaCO_3$ as limestone filler.

4 Discussion

ASTM C595 currently defines the LF requirement as having > 70% $CaCO_3$ by mass, similar to the EN 197–1 requirement of > 75% $CaCO_3$ by mass [31, 32]. EN 197–1 also dictates that the limestone shall contain no more than 0.50% total organic carbon (TOC) by mass. By these definitions, biogenic $CaCO_3$ produced by *E. huxleyi* as used in this study meets these requirements and is thus a suitable filler material for PLC according to current industrial standards. However, neither standard directly addresses particle size. Both PSD and SEM results reveal that biogenic $CaCO_3$ has a much smaller particle size than reagent grade $CaCO_3$, in addition to a much more complex structure. While reagent grade $CaCO_3$ is larger and more granular, biologically architected $CaCO_3$ has intricate ring-like structures. The difference in particle size, shape, and surface area contributed to nucleation effects during cement hydration, most clearly observed in pastes designed with 35% replacement with both biogenic and reagent grade $CaCO_3$.

The replacement of cement with 35% biogenic $CaCO_3$ LF yielded a comparable maximum rate of heat evolution and cumulative heat. However, maximum rate of heat evolution was delayed by roughly 12 h as compared to the remaining PLC pastes studied. The addition of SP to the 35% biogenic $CaCO_3$ paste most likely contributed to its significantly delayed hydration, a phenomenon well-documented in literature [17, 18]. The mechanism of hydration retardation by polycarboxylate-based SPs, such as the one used in this study, is largely based on sorption to solid phases, as well as steric and electrostatic dispersion mechanisms [19]. Javadi et al. found that high charge density in the form of carboxylate ions on the SP backbone led to higher SP adsorption onto cement grains and resulted in the longest retardation of hydration reactions [20]. Because the main hydration peak still occurred at 20 h mark, the presence of SP does not appear to significantly affect cumulative heat development at the 72-h timepoint.

The use of superplasticizers is not uncommon in the cement industry. SP addition reduces water demand in PLC, though simultaneously retards hydration significantly [17]. However, SP addition may have a positive effect on mechanical strength. Ghosh et al. found that SP addition enhanced mechanical strength while reducing water demand by 20% [21]. In this study, the mix design containing 35% biogenic $CaCO_3$ showed the most pronounced water demand (nucleation and filler effect) and required superplasticizer (SP) addition (0.05 g/g $CaCO_3$). These results are consistent with literature suggesting that LF finer than cement particles induces nucleation and filler effects [3, 8, 22]. While traditional LF typically increases workability and may decrease water demand, fine LF such as biogenic LF used here augments the nucleation effect and increases water demand. The 35% biogenic $CaCO_3$ PLC paste was the only paste to require SP addition, demonstrating the phenomenon that a smaller particle size and increased surface area of the LF leads to increased water adsorption onto LF surfaces, as well as increased nucleation of cement hydration products.

Compressive strength testing revealed that the 15% biogenic $CaCO_3$ PLC paste showed the highest compressive strength compared to all other mixtures, likely due to increased nucleation. Though the PLC paste containing 35% reagent grade $CaCO_3$ showed significantly decreased compressive strength compared to the 5% reagent grade $CaCO_3$ PLC paste, the 35% biogenic $CaCO_3$ PLC paste had (i) a higher compressive strength compared to its reagent grade counterpart due to enhanced nucleation and (ii) comparable compressive strength to the 5% biogenic $CaCO_3$ PLC paste. It is possible that both LF sources experienced some particle agglomeration, especially at high replacement percentages, which would reduce the propensity for nucleation [23, 24], and the inert filler effect is likely most responsible for decreased strength in both pastes with 35% replacement compared to OPC.

The advantages of using biogenic $CaCO_3$ as limestone filler are numerous. While traditional mined limestone takes centuries to regenerate, biogenic limestone is an unlimited resource on the human timescale. The mining of limestone in quarries causes dust emissions and erosion, impacts groundwater flow, contamination, and overall water quality and, in the majority of cases, increases CO_2 emissions due to operational energy demand and the need for product transportation [26, 27]. The cultivation of biogenic $CaCO_3$ has the potential to be conducted on-site at cement production plants, bypassing the need for transportation-related CO_2 emissions. Perhaps most significantly, biogenic $CaCO_3$ has a significantly reduced carbon footprint as compared to mined limestone, as it actively consumes CO_2 during its production and has the potential to be an opportunity for carbon storage in PLCs [28]. In order to optimize use of biogenic $CaCO_3$ as an LF and reduce CO_2 emissions, future studies should be conducted to test various microorganism culturing parameters. The optimal strain of *E. huxleyi* should be determined to maximize $CaCO_3$ production. Though some studies have explored scale-up practices for similar algal cultures, laboratory-scale classification and best culturing practice need to be determined before industrial scale-up can occur [29, 30].

5 Conclusions

To conclude, 5%, 15%, and 35% biogenic $CaCO_3$ PLC pastes showed enhanced hydration and increased strength compared to both OPC and their reagent grade $CaCO_3$ PLC counterparts. The 35% biogenic $CaCO_3$ exhibited the highest compressive strength (~29 MPa). All PLC pastes containing biogenic $CaCO_3$ exhibited 7-day compressive strengths > 20 MPa. Enhanced hydration and increased strength are attributable to the smaller particle size and higher surface area of the biogenic $CaCO_3$ compared to reagent-grade $CaCO_3$. While biogenic $CaCO_3$ shows promise as an LF for PLC production, biogenic $CaCO_3$ production at the lab scale must be optimized before scale-up to industrial levels and application can occur. Cost analyses should be conducted to compare cost savings of using biogenic $CaCO_3$ in lieu of traditional limestone. However, the advantages of biogenic $CaCO_3$ and its capability to greatly reduce CO_2 emissions during PLC production cannot be ignored. The potential of PLCs with biogenic $CaCO_3$ should be further examined to fully characterize and compare its mechanical properties and environmental impact to traditional PLCs.

Acknowledgements. This research was made possible by the Department of Civil, Environmental, and Architectural Engineering, the Materials Science and Engineering Program, the College of Engineering and Applied Sciences, and the Living Materials Lab at the University of Colorado, Boulder. This research was supported in part by the Colorado Shared Instrumentation in Nanofabrication and Characterization (COSINC): the COSINC-CHR (Characterization), College of Engineering & Applied Science, University of Colorado Boulder. The authors would like to acknowledge the support of the staff (Tomoko Borsa) and the facility that have made this work possible. This material is based upon work supported by the National Science Foundation under Grant No. CMMI-1943554, the Department of Energy Advanced Research Projects Agency-Energy (ARPA-E) Grant No. DE-AR0001629, and the National Science Foundation Graduate Research Fellowship under Grant No. DGE 2040434. Any opinions, findings, and conclusions or recommendations expressed in this material are those of the authors and do not necessarily reflect the views of the sponsors.

This material is based upon work supported by the National Science Foundation under Grant No. CMMI-1943554, the National Science Foundation Graduate Research Fellowship under Grant No. DGE 2040434, and the Advanced Research Projects Agency - Energy under Award No. DE-AR0001629. Any opinions, findings, and conclusions or recommendations expressed in this material are those of the authors and do not necessarily reflect the views of the National Science Foundation or the Advanced Research Projects Agency.

This material is based upon work supported by the Breakthrough Energy Foundation under Explorer Grant number AWD-22-05-0179.

Declaration of Competing Interest. W.V.S. is a listed coinventor on a patent application (PCT/US2020/020863) filed by the University of Colorado on April 3, 2020, related to biomineralized building materials. W.V.S. is a cofounder and shareholder of Prometheus Materials and Minus Materials Inc. And a member of their scientific advisory boards. D.N.B. is a cofounder, consultant to, and shareholder of Minus Materials Inc.

References

1. El-Hawary, M., Nouh, K.: Properties and sustainability of concrete containing fillers. Aust. J. Civ. Eng. **16**(2), 96–105 (2018)
2. Liew, J.Y.R., Xiong, M.-X., Lai, B.-L.: Design of steel-concrete composite structures using high-strength materials. Woodhead Publishing (2021)

3. Wang, D., Shi, C., Farzadnia, N., Shi, Z., Jia, H.: A review on effects of limestone powder on the properties of concrete. Constr. Build. Mater. **192**, 153–166 (2018)
4. Panesar, D.K., Zhang, R.: Performance comparison of cement replacing materials in concrete: Limestone fillers and supplementary cementing materials – A review. Constr. Build. Mater. **251**, 118866 (2020)
5. Irassar, E.F., Violini, D., Rahhal, V.F., Milanesi, C., Trezza, M.A., Bonavetti, V.L.: Influence of limestone content, gypsum content and fineness on early age properties of Portland limestone cement produced by inter-grinding. Cement Concr. Compos. **33**(2), 192–200 (2011)
6. Burgos-Montes, O., Alonso, M.M., Puertas, F.: Viscosity and water demand of limestone- and fly ash-blended cement pastes in the presence of superplasticisers. Constr. Build. Mater. **48**, 417–423 (2013)
7. Tsivilis, S., Chaniotakis, E., Kakali, G., Batis, G.: An analysis of the properties of Portland limestone cements and concrete. Cement Concr. Compos. **24**(3–4), 371–378 (2002)
8. Sato, T., Matoba, Y., Takahashi, N., Koike, H.: Interactive surface that can dynamically change the shape and touch sensation. In: Proceedings of the ACM International Conference on Interactive Tabletops and Surfaces - ITS '11, 1 (2011)
9. De Weerdt, K., Haha, M.B., Le Saout, G., Kjellsen, K.O., Justnes, H., Lothenbach, B.: Hydration mechanisms of ternary Portland cements containing limestone powder and fly ash. Cem. Concr. Res. **41**(3), 279–291 (2011)
10. Briki, Y., Zajac, M., Haha, M.B., Scrivener, K.: Impact of limestone fineness on cement hydration at early age. Cem. Concr. Res. **147**, 106515 (2021)
11. İnan Sezer, G.: Compressive strength and sulfate resistance of limestone and/or silica fume mortars. Constr. Build. Mater. **26**(1), 613–618 (2012)
12. Bentz, D.P., Irassar, E.F., Bucher, B.E., Wesis, W.J.: Limestone fillers conservecement, Part 1: An analysis based on Powers'model. Concr. Int. **31**, 41–46 (n.d.)
13. Ferrari, N.: A Milestone for Sustainable Infrastructure – 44 State Departments of Transportation Now Accept Use of Portland-Limestone Cement (2022, May 2). https://www.cement.org/newsroom/2022/05/05/a-milestone-for-sustainable-infrastructure-44-state-departments-of-transportation-now-accept-use-of-portland-limestone-cement
14. Müller, M.N.: On the Genesis and Function of Coccolithophore Calcification. Front. Mar. Sci. **6**, 49 (2019)
15. Jakob, I., Chairopoulou, M.A., Vučak, M., Posten, C., Teipel, U.: Biogenic calcite particles from microalgae-Coccoliths as a potential raw material. Eng. Life Sci. **17**(6), 605–612 (2017)
16. Tobias Rubner, I.H.: Investigation of Cell Growth and Chlorophyll a Content of the Coccolithophorid Alga Emiliania huxleyi by Using Simple Bench-Top Flow Cytometry. Journal of Bioprocessing & Biotechniques **05**(06) (2015)
17. El-Gamal, S.M.A., Al-Nowaiser, F.M., Al-Baity, A.O.: Effect of superplasticizers on the hydration kinetic and mechanical properties of Portland cement pastes. J. Adv. Res. **3**(2), 119–124 (2012)
18. Chiocchio, G., Paolini, A.E.: Optimum time for adding superplasticizer to Portland cement pastes. Cem. Concr. Res. **15**(5), 901–908 (1985)
19. Lothenbach, B., Winnefeld, F., Figi, R.: The influence of superplasticizers on the hydration of Portland cement. In: Proceedings ofthe 12th ICCC, Montreal, QC, Canada, 9–12 July 2007, pp. W1–W5.03
20. Javadi, A., Jamil, T., Abouzari-Lotf, E., Soucek, M.D., Heinz, H.: Working mechanisms and design principles of comb-like polycarboxylate ether superplasticizers in cement hydration: quantitative insights for a series of well-defined copolymers. ACS Sustainable Chemistry & Engineering **9**(25), 8354–8371 (2021)
21. Ghosh, R.S., Malhotra, V.M.: Use of Superplasticizers as Water Reducers, 2nd ed., Vol. 1. ASTM (1979). https://www.astm.org/cca10404j.html

22. Li, W., Huang, Z., Cao, F., Sun, Z., Shah, S.P.: Effects of nano-silica and nano-limestone on flowability and mechanical properties of ultra-high-performance concrete matrix. Constr. Build. Mater. **95**, 366–374 (2015)
23. Kumar, A., et al.: The filler effect: The influence of filler content and type on the hydration rate of tricalcium silicate. J. Am. Ceram. Soc. **100**(7), 3316–3328 (2017)
24. Knop, Y., Peled, A.: Setting behavior of blended cement with limestone: Influence of particle size and content. Mater. Struct. **49**(1–2), 439–452 (2016)
25. Pelletier-Chaignat, L., Winnefeld, F., Lothenbach, B., Müller, C.J.: Beneficial use of limestone filler with calcium sulphoaluminate cement. Constr. Build. Mater. **26**(1), 619–627 (2012)
26. Ganapathi, H., Phukan, M.: Environmental Hazards of Limestone Mining and Adaptive Practices for Environment Management Plan. In: Singh, R.M., Shukla, P., Singh, P. (eds.) Environmental Processes and Management. WSTL, vol. 91, pp. 121–134. Springer, Cham (2020). https://doi.org/10.1007/978-3-030-38152-3_8
27. Venkata Sudhakar, C., Umamaheswara Reddy, G., Usha Rani, N.: Delineation and evaluation of the captive limestone mining area change and its influence on the environment using multispectral satellite images for industrial long-term sustainability. Cleaner Engineering and Technology **10**, 100551 (2022)
28. Zondervan, I., Rost, B., Riebesell, U.: Effect of CO_2 concentration on the PIC/POC ratio in the coccolithophore Emiliania huxleyi grown under light-limiting conditions and different daylengths. J. Exp. Mar. Biol. Ecol. **272**(1), 55–70 (2002)
29. Hankamer, B., Lehr, F., Rupprecht, J., Mussgnug, J.H., Posten, C., Kruse, O.: Photosynthetic biomass and H_2 production by green algae: From bioengineering to bioreactor scale-up. Physiol. Plant. **131**(1), 10–21 (2007)
30. Gouveia, L., et al.: Microalgae biomass production using wastewater: Treatment and costs. Algal Res. **16**, 167–176 (2016)
31. C01 Committee: Specification for Blended Hydraulic Cements. ASTM International (n.d.)
32. European Committee for Standardization: Cement Part 1: Composition, specifications and conformity criteria for common cements. BSI Standards (2011). http://106.38.59.21:8080/userfiles/d46365fdde004ea0a5da5d9701142815/files/teckSolution/2019/10/EN%20197-1-2011_3750.pdf
33. International Centre on Diffraction Data: PDF-4/Axiom 2019 (2019). https://www.icdd.com/pdf-4-axiom/#1512062099603-612b7e4c-35a3

Bricks Geopolymer Based on Olive Waste Fly Ash: Mechanical Properties

I. Labaied[1,2](✉), O. Douzane[1], M. Lajili[2], and G. Promis[1]

[1] EMIR Laboratory, University of Monastir, 15 Avenue Ibn Eljazzar, 5019 Monastir, Tunisia
labaiedines49@gmail.com

[2] Innovative Technologies Laboratory, University of Picardie Jules Verne, Avenue Des Facultés, Le Bailly, 80025 Amiens, France

Abstract. Each year, billions of bricks are produced in the world, but their production is far from environmentally friendly. A growing field of research is focusing on geopolymer bricks constructed with lignocellulosic ashes and calcined clays as a long-lasting, durable, and sustainable binder. This study aims to develop geopolymer bricks utilizing clay from a burned brick factory's quarry that had been calcined at 700 °C for an hour and fly ash (10 to 30%) generated from biomass combustion (olive waste) as an aluminosilicate's sources. As an alkaline activator, a solution of 8 M NaOH was utilized. According to DRX, while 700 °C is inadequate to convert the crystalline phases of clay into more reactive amorphous phases and, as a result, to initiate the geoplymerization reaction, adding potassium and calcium-rich ashes results in stronger bricks.

Keywords: brick geopolymer · olive waste ash · mechanical properties · calcination temperature

Nomenclature

RC	Raw Clay
$RC_{/c}$	Calcined Raw Clay
OPFA	Olive Pomace Fly Ash
FWHM	Full Width at Half Maximum

1 Introduction

In the 1970s, Davidovits used the term "geopolymer" to describe research on how metakaolin reacts with alkaline conditions to form aluminosilicate polymers [1]. Geopolymers are created by activating an aluminosilicate source derived from natural minerals, calcined clays, or industrial by-products [2]. These alumiosilicates degrade rapidly in an extremely alkaline medium, releasing tetrahedral units of SiO_4 and AlO_4, which hastens the polycondensation process [3]. The most widely used activating agents are hydroxides (NaOH or KOH) and silicate solutions (Na_2SiO_3 or K_2SiO_3)

S. Amziane et al. (Eds.): ICBBM 2023, RILEM Bookseries 45, pp. 54–62, 2023.
https://doi.org/10.1007/978-3-031-33465-8_5

[4]. Low temperatures (less than 100 °C) are required to produce amorphous geopolymers, whereas high-pressure autoclaves can produce crystalline geopolymers at temperatures as high as 200 °C [5].The majority of research has been concentrated on creating metakaolin-based geopolymer materials, which are composed primarily of the clay mineral kaolinite, are silica and alumina-rich, and become amorphous when heated to temperatures between 500 °C and 700 °C [6]. Heat treatment converts crystalline phases into more reactive amorphous phases. These amorphous phases, which are active during the geopolymerization process, regulate the final characteristics of the geopolymers [7]. In fact, a geopolymer with high mechanical properties and a strength range of 20 to 50 MPa is produced when these amorphous phases interact with alkaline solutions. Research on using raw clay as a source of aluminosilicates is currently relatively restricted. In fact, Clay is a fairly complex mineralogical mixture of several distinct minerals, and the origin of the source rocks has a significant impact on this complexity. These substances, on the other hand, can be found in abundance and, when thermally activated, can exhibit some reactivity. Indeed, thermal activation of clay minerals generally results in dehydroxylation, which is associated with mass loss due to the departure of the mineral's crystalline hydroxyl groups (OH) as water and breakdown in a disordered amorphous state [8]. The ideal activation temperature is significantly influenced by the mineralogical composition of the clay. If the clay is primarily composed of kaolin, a temperature of 700 °C is frequently sufficient to produce a dehydroxylated and completely amorphous metakaolin; however, this temperature will not be sufficient to dehydroxylate a clayey material primarily composed of illite [9].

The principles of eco-sustainability now permit the use of waste materials from industrial processes as raw materials. The creation of geopolymeric materials from industrial wastes like fly ash and slag as a partial replacement for clay has received the majority of the study's attention [10–12]. However, the production of geopolymers from lignocellulosic biomass ashes is incredibly rare. These lignocellulosic ashes are very promising because they are rich in potassium oxide, which is the activation solution that forms geopolymerization. They are also rich in silica and alumina, which confirms their use in the geopolymerization reaction by alkaline activators [13]. Geopolymeric materials based on natural clay and lignocellulosic biomass ashes seem to be a practical and eco-friendly option for the preservation of the environment, the reduction of toxicity and landfill issues associated with these ashes, as well as the development of eco-friendly, economical, and sustainable binders that require less energy and have potent qualities like good mechanical properties, low liquid permeability, resistance to high temperatures, etc.

In the current study, calcined natural clay reinforced with olive pomace ash was used as an aluminosilicate source precursor. These substances were activated with an alkaline sodium hydroxide solution to produce geopolymeric materials. To demonstrate the impact of clay minerals' mineralogical behaviors on calcination temperature as a crucial factor in the creation of geopolymeric binders, natural clay and calcined clay were chemically compared.

2 Materials and Methods

Geopolymer bricks were made using raw clay (RC) that was recovered from a Tunisian brick factory and fly ash from olive pomace (OPFA) that was gathered from the "Zouila Tunisia" olive oil factory. The aforementioned "Zouila" factory uses olive pomace as fuel in gall ovens to generate heat that is then turned to electricity to power the dryers. Fly ash is extracted from the kiln's electro filter. The raw materials' chemical compositions and mineral phases were determined using a high-resolution SEM imaging approach with a low vacuum pressure setting of 0.5 "Tm" and an X-ray diffractometer utilizing the D8 Discover Diffractometer.

2.1 Sample Preparation

The raw clay was crushed and sieved through a 200 μm sieve to guarantee that the grains were all the same micrometer size. The clay was then calcined at 700 °C for an hour in order to produce more reactive aluminosilicate minerals.

Geopolymer bricks were made in two steps. First, the dry components were combined at low speed for 3 min (calcinated clay ($RC_{/c}$) and various percentages of OPFA (160 μm)). The water was then added and mixed for one minute on high speed with the dry ingredients. The created geopolymer materials were molded in 7 × 3.5 cm cylindrical molds and allowed to cure for 8 h at 80 °C (Table 1).

Table 1. Brick geopolymer families with varying percentages of $RC_{/C}$, distilled water, and OPFA.

SAMPLE	$RC_{/c}$ (% in mass)	OPFA (% in mass)	Water (% in mass)
0-FAOS	100	0	15
10-FAOS	90	10	15
15-FAOS	85	15	15
20-FAOS	80	20	15
25-FAOS	75	25	15
30-FAOS	70	30	15

A second batch of samples was created using NaOH (8 M) to examine how the addition of the alkaline solution affected the growth of the geopolymer bricks.

2.2 Sample Characterization

Mechanical testing was carried out on cylindrical specimens using a mechanical press of the Tinius Olsen H50KS type, which was linked to a 200 KN force transducer and three 50 mm displacement transducers (Fig. 1). The experiments were carried out at loading speeds of 1 mm/min. The crystalline properties of raw clay and calcined clay, including percent crystallinity (Cr%) and FWHM, were investigated using an X-ray diffraction pattern and High Score Plus software. The air peak method was used to calculate the percent crystallinity from X-ray diffraction spectra, according to Eq. (1) [14].

$$Cr\,(\%) = \frac{\text{Area of crystalline peaks}}{\text{area of all peaks(crystalline} + \text{amorphe)}} * 100 \quad (1)$$

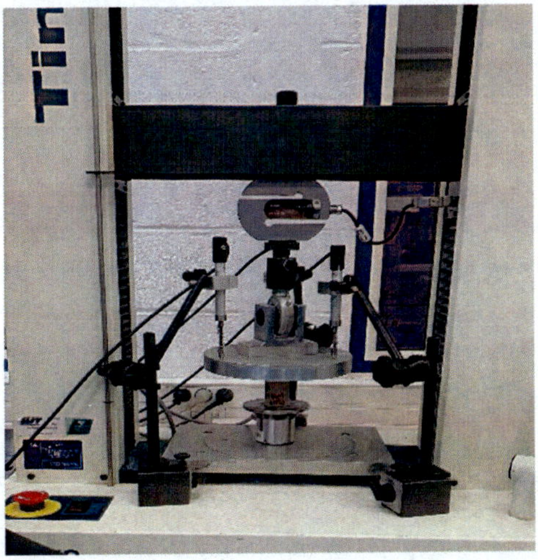

Fig. 1. Compression test of geopolymer materials

3 Results and Discussion

3.1 Initial Testing of By-Products

This section describes the results and conclusions reached from tests aimed at determining the chemical properties of the study's elements, raw clay and olive pomace fly ash. First, chemical analyses of various materials used in the design of geopolymeric materials were performed to determine the elemental chemical composition of each component (Figs. 2 and 3).

The raw clay used to make geopolymer bricks is composed of 35.5% crystalline SiO2 and Al2O3, which serve as precursor materials for the geopolymerization reaction (Fig. 2). However, amorphous aluminosilicate minerals are the most reactive minerals. The reason for this is that an hour-long calcination process at 700 °C was performed to convert the crystalline phase to more reactive amorphous aluminosilicate minerals.

The components of fly ash made from olive pomace are 2.15% SiO2, Al2O3, and 25% potassium (Fig. 3). The chemical composition is very promising because it is rich in silica and amorphous alumina, which confirms their use in the geopolymerization reaction, it also contains a significant amount of potassium oxide, which is responsible for the generation of geopolymerization.

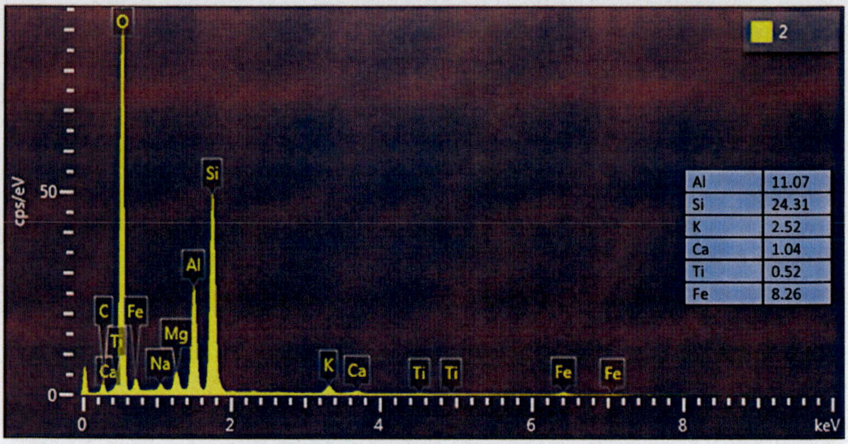

Fig. 2. SEM analysis – raw clay

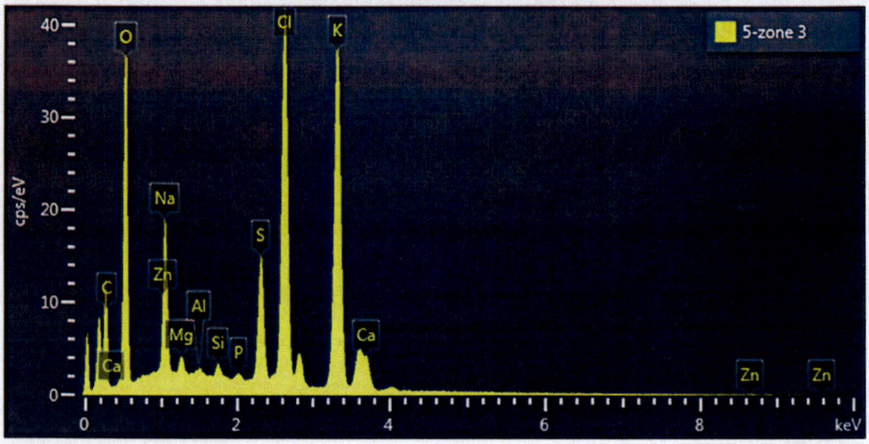

Fig. 3. SEM analysis – FAOS

XRD was used to determine the raw clay's mineralogical composition. Figure 4 depicts the diffractogram obtained on raw clay powder. X-ray diffraction (XRD) analysis reveals the presence of clay minerals and crystalline phases, primarily in the form of tectosilicate. The latter is reflected as a line of significant intensity relative to quartz observed at the angular position $2\theta = 26.62°$. Moreover, the reflection at 8.81° in 2θ, corresponds to the presence of illite. Other relatively more intense peaks are observed at 12.33° (7.16 Å) and 20.84° (4.25 Å). These peaks highlight the presence of kaolinite. The diffractogram also reveals the presence of diffraction lines corresponding to clay minerals present respectively at 29.03° and 30.84° which are attributed to calcite and dolomite. The determination of mineralogical compositions was made possible the clay's X-ray diffraction analysis results. The clay is mostly illito-kaolinite with impurities such as quartz, calcite, and dolomite.

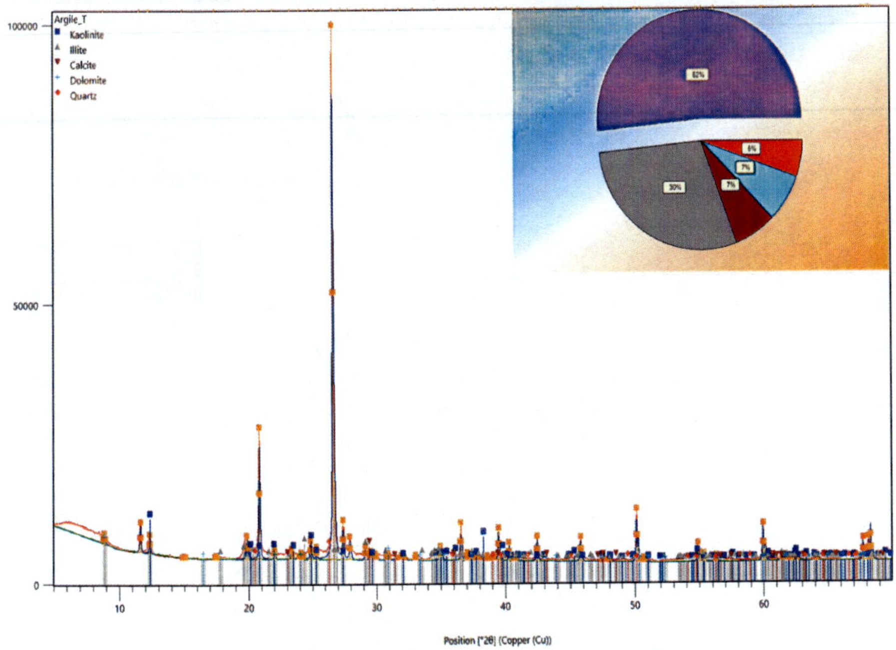

Fig. 4. X-ray diffraction (XRD) analysis of raw clay

3.2 Initial Testing of Geopolymer Materials

The sections that follow provide details on the tests' findings that are listed in the material characterization section. The findings of this study's tests provide preliminary conclusions that allow us to assess the feasibility of a binder made from calcined clay and olive pomace biomass fly ash.

X-ray Diffraction:
An X-ray diffraction analysis was carried out on the calcined clay at a temperature of 700 °C for one hour in order to confirm the viability of the calcination temperature used to produce reactive amorphous aluminosilicate minerals in the geopolymer reaction. Figure 5 compares the raw clay's diffraction pattern to that of the calcined clay.

Analysis of the XRD peak profiles of calcined clay and raw clay indicated the disappearance of peak observed at angular position $2\theta = 12.33°$ which highlighted the presence of kaolinite as well as an increase in the full width at half maximum (FWHM) of peak observed at angular position $2\theta = 20.84°$ (0.103 [°2θ] for calcined clay versus 0.082[°2θ] for raw clay) highlighting the beginning of kaolinite amorphization. Indeed, a higher FWHM means a less improved crystal quality due to the heat treatment process during the preparation of the calcined clay [14]. No change in the reflections of illite was found, due to its higher thermal stability. Indeed, the reflection at 8.81° in 2θ, corresponding to the presence of illite in the raw clay, is still present in the calcined clay and it still keeps the same FWHM (0.165 [°2θ]). Indeed, the raw clay had a crystallinity percentage of 77.63%, whereas the calcined clay had a crystallinity percentage of 63.85%, indicating that the chosen temperature was unable to convert the clay minerals from crystalline to amorphous phases.

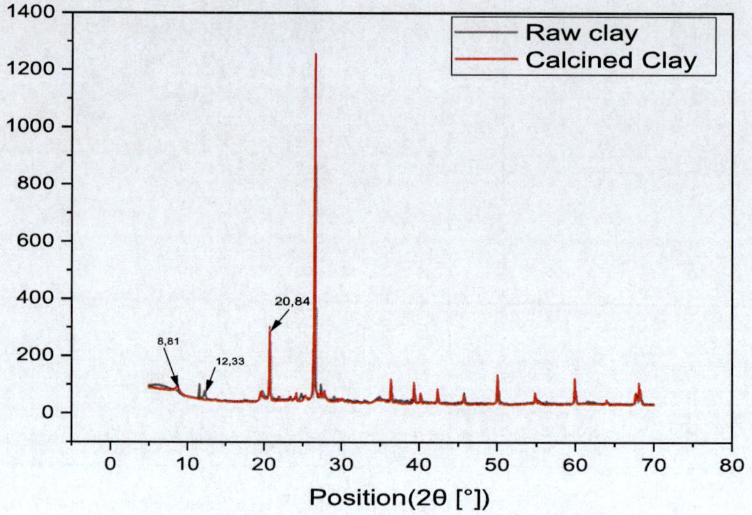

Fig. 5. Diffraction patterns of raw clay and calcined clay.

Table 2. Compressive strength of geopolymer specimens

SAMPLE	Water	NaOH (8M)
0-FAOS	1.6 MPa	2.7 MPa
10-FAOS	0.5 MPa	3.1 MPa
15-FAOS	0.4 MPa	4.3 MPa
20-FAOS	1.5 MPa	4.6 MPa
25-FAOS	1.5 MPa	6 MPa
30-FAOS	1.1 MPa	4.7 MPa

Mechanical Characterization:

The compressive strength of geopolymeric specimens showed that materials elaborated with water had a low strength (Table 2). In fact, geopolymerization takes place in the presence of an alkaline solution and an aluminosilicate precursor. The goal of this solution is to chemically attack the precursor, allowing for its complete dissolution, and releasing Si or Al elements that will be used during the geopolymerization process. However, clays rarely dissolve completely and quickly in an alkaline medium due to their crystalline structure. This explains why all specimens have such low mechanical resistance. Indeed, as was previously shown in the DRX analysis, neither the complete transformation of kaolinite into metakoalin nor the dehydroxylation of illite and its transformation into more reactive illite anydrydes were possible at the calcination temperature used. Furthermore, the use of water and fly ash did not result in geoplymeric bricks with good mechanical properties. In fact, the potassium concentration in the fly ash was insufficient to serve as an alkaline solution. In fact, the effect of fly ash addition on compressive strength values is only noticeable at significant percentages of OPFA (> 25%).

The elaborated materials' mechanical properties were enhanced by the addition of an alkaline sodium hydroxide solution and potassium from the olive pomace ashes. In fact, adding 25% olive pomace ash increased the developed specimens' mechanical strength by 122%. However, after 30% by weight of olive biomass fly ash incorporation, compressive strength decreases. The cause is that potash, at more than 30% of its weight, is excessive and does not interfere with the freezing of the geopolymer. Therefore, the geopolymerization process is hampered by a higher proportion of fly ash from olive biomass [15].

4 Conclusion

In this study, geopolymer bricks are produced using calcined clay and fly ash as aluminosilicate precursors and sodium hydroxide as an alkali activation solution.

The following results were obtained:

- The data presented showed that a calcination temperature of 700 °C for one hour was insufficient to produce reactive aluminosilicate minerals, resulting in the specimens' low mechanical strength. The presence of illite in the mineralogical structure of the clay used will necessitate more thermal activation energy associated with the dehydroxylation of the (OH) group. To determine the optimal calcination temperature for dehydroxylation, a thermogravimetric analysis (TGA) must be performed.
- The addition of a sodium hydroxide solution and olive pomace fly ash allowed for improved mechanical performances. Indeed, these ashes are high in potassium, present in the form of K2O, as well as alumina and reactive silica, which promotes hydraulic consolidation and the elimination of free water, resulting in higher values of compression resistance.

The most traditional method of creating a geopolymer with good mechanical properties and durability is through the alkaline activation of calcined kaolin, also known as metakaolin. However, this approach costs money and uses a lot of energy. Alkaline activation of clay soils at low temperatures to produce economical and environmentally friendly geopolymer bricks is becoming increasingly researched.

References

1. Davidovits, J.: Geopolymers: Inorganic polymeric new materials. J. Therm. Anal. **37**, 1633–1656 (1991)
2. Cong, P., Cheng, Y.: Advances in geopolymer materials: a comprehensive review. J. Traffic Transp. Eng. (English Edition) **8**(3), 283–314 (2021). https://doi.org/10.1016/j.jtte.2021.03.004
3. Davidovits, J.: Geopolymers: ceramic-like inorganic polymers. J. Ceram. Sci. Technol. **8**(3), 335–350 (2017). https://doi.org/10.4416/JCST2017-00038

4. Gómez-Casero, M.A., Moral-Moral, F.J., Pérez-Villarejo, L., Sánchez-Soto, P.J., Eliche-Quesada, D.: Synthesis of clay geopolymers using olive pomace fly ash as an alternative activator. Influence of the additional commercial alkaline activator used. J. Market. Res. **12**, 1762–1776 (2021). https://doi.org/10.1016/j.jmrt.2021.03.102
5. Amran, Y.H.M., Alyousef, R., Alabduljabbar, H., El-Zeadani, M.: Clean production and properties of geopolymer concrete; A review. J. Cleaner Prod. **251**, 119679 (2020). https://doi.org/10.1016/j.jclepro.2019.119679
6. Sore, S.O., Messan, A., Prud'Homme, E., Escadeillas, G., Tsobnang, F.: Comparative study on geopolymer binders based on two alkaline solutions (NaOH and KOH). J. Miner. Mater. Charact. Eng. **8**(6), 407–420 (2020). https://doi.org/10.4236/jmmce.2020.86026
7. Seiffarth, T., Hohmann, M., Posern, K., Kaps, Ch.: Effect of thermal pre-treatment conditions of common clays on the performance of clay-based geopolymeric binders. Appl. Clay Sci. **73**, 35–41 (2013)
8. Shvarzman, A., Kovler, K., Grader, G.S., Shter, G.E.: The effect of dehydroxylation/amorphization degree on pozzolanic activity of kaolinite. Cem. Concr. Res. **33**(3), 405–416 (2003). https://doi.org/10.1016/S0008-8846(02)00975-4
9. Gualtieri, A.F., Ferrari, S.: Kinetics of illite dehydroxylation. Phys. Chem. Miner. **33**(7), 490–501 (2006). https://doi.org/10.1007/s00269-006-0092-z
10. Ahmad, M., Rashid, K., Hameed, R., Haq, E.U., Farooq, H., Ju, M.: Physico-mechanical performance of fly ash based geopolymer brick: influence of pressure – temperature – time. J. Build. Eng. **50**, 104161 (2022). https://doi.org/10.1016/j.jobe.2022.104161
11. Qaidi, S.M.A., Tayeh, B.A., Isleem, H.F., de Azevedo, A.R.G., Ahmed, H.U., Emad, W.: Sustainable utilization of red mud waste (bauxite residue) and slag for the production of geopolymer composites: a review. Case Stud. Const. Mater. **16**, e00994 (2022). https://doi.org/10.1016/j.cscm.2022.e00994
12. Ferone, C., Colangelo, F., Cioffi, R., Montagnaro, F., Santoro, L.: Mechanical performances of weathered coal fly ash based geopolymer bricks. Procedia Eng. **21**, 745–752 (2011). https://doi.org/10.1016/j.proeng.2011.11.2073
13. Labaied, I., Douzane, O., Lajili, M., Promis, Geoffrey: Bricks using clay mixed with powder and ashes from lignocellulosic biomass: a review. Appl. Sci. **12**(20), 10669 (2022). https://doi.org/10.3390/app122010669
14. Segal, L., Creely, J.J., Martin, A.E., Conrad, C.M.: "Opportunity for new developments in all phases of textile manufacturing. 'Literature Cited An Empirical Method for Estimating the Degree of Crystallinity of Native Cellulose Using the X-Ray Diffractometer." (1952)
15. Carrillo-Beltran, R., Corpas-Iglesias, F.A., Terrones-Saeta, J.M., Bertoya-Sol, M.: New geopolymers from industrial by-products: olive biomass fly ash and chamotte as raw materials. Constr. Build. Mater. **272**, 121924 (2021). https://doi.org/10.1016/j.conbuildmat.2020.121924

Mechanical and Thermal Properties of an Innovative Bio Based Concrete

Maya Hajj Obeid[1](✉), Omar Douzane[1], Lorena Freitas Dutra[1], Geoffrey Promis[1], Boubker Laidoudi[2], and Thierry Langlet[1]

[1] University of Picardie Jules Verne, Amiens, France
maya.hajj.obeid@u-picardie.fr
[2] CODEM, 56 Rue André Durouchez, 80080 Amiens, France

Abstract. Generally, bio-based concrete is applied as non-load-bearing insulation. In this work, an innovative material, considered a load-bearing bio based concrete, was developed. Rapeseed concrete is primarily composed of rapeseed straw, with the addition of gravel, sand, and cement-based binders. The formulation of BIP requires the preparation of rapeseed straws, which is called the "inerting" of the straws. It consists of coating the straws with a layer of binders in order to protect the pores of the straws from the destruction of other, more rigid materials. Also protects the hydration of the binders, particularly cement, from the sugars and soluble matters that exist in the rapeseed straws and that are easily released upon contact with water. The mechanical and thermal performances of this novel material are characterized according to its density and by using two different sizes of rapeseed straw. According to the results, the thermal conductivity of these concretes ranges from 0.2 to 0.48 $W.m^{-1}.k^{-1}$. This concrete has compressive strengths ranging from 2 to 7 MPa and densities ranging from 1150 to 1500 kg/m^3, proving that BIP can be used as an insulating, load-bearing bio-based concrete.

Keywords: rapeseed concrete · mechanical strength · bio based concrete · thermal conductivity

1 Introduction

In France, the building and residential sector is responsible for 45% of energy consumption and 26% of greenhouse gas emissions [1]. Bio-based materials, such as those derived from rapeseed straw, are viewed as a promising solution in the construction industry due to their low thermal conductivity, relatively high specific capacity, and ability to regulate temperature and humidity in enclosed spaces, thereby decreasing the energy consumption of buildings [2, 3]. Their transportation and its effect on their carbon footprint is also a significant aspect [4]. This is why we are interested in rapeseed straw, as this bio-based material is readily available throughout France and in practically all regions. In addition, this plant is planted year, ensuring a consistent, local, and speedy supply. After processing, the plant can yield powders, fibers, or aggregates, which can be utilized as insulation or in the composition of bio-sourced concrete by combining the plant with

S. Amziane et al. (Eds.): ICBBM 2023, RILEM Bookseries 45, pp. 63–69, 2023.
https://doi.org/10.1007/978-3-031-33465-8_6

a binder, water, and occasionally additives. These concretes can also contribute to the improvement of the acoustic comfort of dwellings [5], as well as provide outstanding thermal insulation [6] and hygric performance [7]. However, their primary weakness is their low mechanical performance [8]. Certain elements play a significant influence in determining the qualities of bio-based materials, including the kind of binder, the size of aggregates, the compaction energy, the molding method, and the manufacturing process [9, 10]. Due to the plant aggregates, their high porosity, and, of course, their low density, bio-sourced concretes are often lightweight and, through their thermal efficiency, contribute to energy savings.

Rapeseed concretes are part of the family of bio-sourced concretes, which allow for the combination of agro-resources, renewable raw materials, and CO_2 storage, with a binder that has, if possible, a low environmental impact. In order to achieve this goal, load-bearing insulating concrete (BIP), a novel material that combines rapeseed straw with standard concrete components, was developed. Its goal is to increase the concrete's thermal resistance while preserving its good mechanical strength. These results enable the properties of rapeseed straw and its aggregates to be optimized in order to enhance the performance of rapeseed straw concrete.

2 Materials and Methodology

2.1 Rape Straws

The rapeseed straws used in the study were sourced from the French regions of Somme and Marne and harvested in 2015 and 2016. They were processed by external providers who crushed the straws into various sizes (0.25–20 mm and 0.25–10 mm) using the shearing method. This resulted in 6 different aggregate lots, each identified by the name and year of the harvest region (e.g. Marne 2015 or M15), and the size of the grind (e.g. S16–10 for Somme straw harvested in 2016 and crushed to 10 mm).

2.2 Rape Straw Concrete: BIP (Load-Bearing Insulating Concrete)

The formulation of the BIP is done in two phases. The first phase is the inerting of the rape straw, this step consists in coating the rape straw with a layer of binder in order to protect the pores of the rape straw from destruction during the mixing with other more rigid materials (gravel). The inerting prevents the problems of lack of hydration of the binders, especially the cement binders, due to the intervention of sugars and soluble materials that exist in the rape straw and are easily released when in contact with water. The inerted straws are stored for a minimum period of two weeks before their use for the manufacture of BIP. Table 1 shows the quantities of binder and water used to elaborate the inerting of 200 L of rapeseed straw.

Step 2 consists of the final production of the load-bearing insulating concrete. The inert straws are placed in a 20 L mixer, with the gravel, sand and pre-mixing with a premolding water for 2 min. After that, we add the binders and the final water and mixing lasts for 5 min. The samples are made with several different densities, this density is controlled by the initial settling during the casting of the samples. Table 2 shows the composition of the formulation of the insulating load-bearing concretes.

Table 1. Composition of the inerting

Rape straw	Pre-molding Water	Binder	Molding Water
200 L	5 L	12 kg	13 L

Table 2. Volume percentage of BIP composition.

Inerted rape straw	62 ± 5%
CEM I	16 ± 3%
Lime	5 ± 2%
Sand	5%
Gravel	11%
Water/binder	0,4–0,6

2.3 Mechanical Compression Test

The compressive strength of rapeseed concrete were studied to differentiate the effect of the type and size of rapeseed straw on the compressive strength, as well as the density of the samples. The compression testing machine (hydraulic servo drive) was used to perform the compressive strength test after 28 days of curing on 15 cm × 15 cm × 15 cm samples with a displacement speed of 0.02 mm/s. Compressive strength tests are performed in accordance with NF EN 12390–2 [11] and NF EN 196–1 [12]. The strength of the cubic samples was calculated by averaging the results of three samples.

Figure 1 shows the mechanical test samples of BIP made with rape straws.

Fig. 1. Mechanical sample of load-bearing insulating concrete made with rape straws (BIP).

2.4 Thermal Conductivity Test

The thermal conductivity of building materials is determined through thermal conductivity tests, which assess a material's ability to transfer heat under specific temperature and equilibrium conditions. The conductivity of rape-seed concrete was measured using the guarded hot plate method and conformed to standards NF EN 12667 [13] and NF EN 12664 [14]. Samples size is 25 cm × 25 cm × 8 cm, were cured for 30 days, pre-dried at 70 °C, and left in an air-conditioned room at 25 °C until they reached equilibrium. During testing, the hot and cold plates were set to average temperatures of 10 °C, and a thermocouple was placed between the overlapping sample surfaces to prevent any contact resistance.

3 Results

3.1 Mechanical Compression Behavior

Figure 2 shows the behavior of rapeseed bearing insulating concretes as a function of deformation, using two different sizes of rapeseed straws, 0.25–10 cm and 0.25–20 cm. Regarding bio-based materials, the size of the straw plays an important role on the compressive strength of concrete [8]. In a previous study, on a non-bearing insulating concrete made of rapeseed, it's shows that the size of the straws is directly related to the compressive strength, since the latter increases with the size of the straws [15].

In this study, the influence of aggregate size on mechanical performance is less significant than in previous studies, but using larger aggregates still results in better mechanical capacity. The use of 20 mm sized aggregates improves cohesion among different materials. The BIP M20 has a compressive strength of 4.3 MPa, the average density of the samples is 1479 kg/m^3. While the compressive strength of BIP S20 concrete is 3.5 MPa, the average density of the samples is 1360 kg/m^3. This great difference is due to the difference in the density of the samples. While the BIP made with the 10 mm straws have a compressive strength of about 3.2–3.7 MPa, and their density is 1282–1320 kg/m^3. Regarding to the results, the size of the straws affects the plastic part of the concrete (Fig. 2). It can be noticed that the concrete made with 20 mm straws has a higher plasticity behavior than the concrete made with 10 mm straws. In this study, the improved plasticity of the load-bearing insulating concrete with 20 mm sized rapeseed straw is attributed to the tight coating of the straw, forming a barrier against cracks in the binder matrix.

3.2 Thermal Conductivity

Table 3 shows the values of the BIP thermal conductivity as a function of the type of aggregate and its dry density. The thermal conductivities of these materials range from 0.208 W.m^{-1}.K^{-1} to 0.445 W.m^{-1}.K^{-1} depending on their apparent densities. BIP is a complex material due to two factors; firstly, the effect of the binder layer with a bond on the rape straws can decrease the thermal performance of these materials. And also the effect of integrating several raw materials makes this concrete very heterogeneous, hence the difficulty in understanding its microstructure and generalizing it.

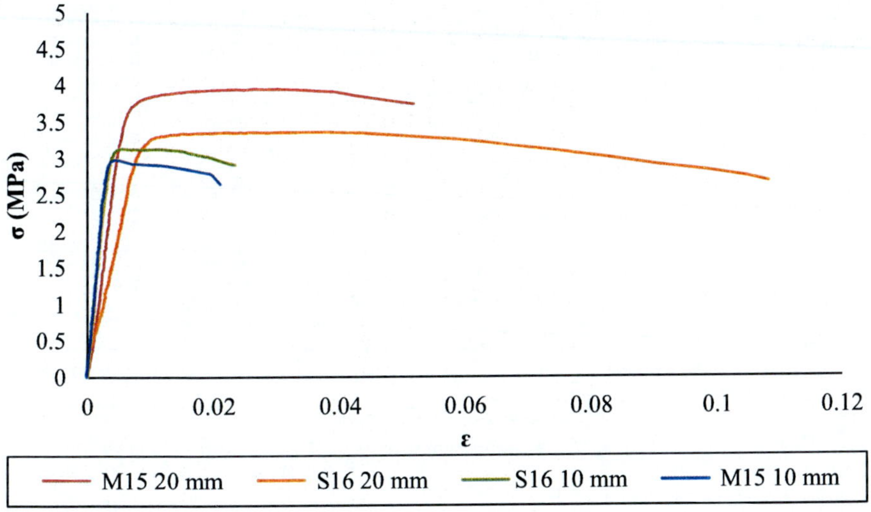

Fig. 2. Mechanical compressive strength as a function of deformation of the load-bearing insulating concrete using two sizes of straw.

Table 3. Thermal conductivity of different rape straw concrete in function of the dry density.

	Dry density (Kg/m^3)	Thermal conductivity ($W.m^{-1}.K^{-1}$)
M15 - 20	1796	0,363
S 16 - 20	1763	0,307
M15 - 10	1627	0,258
S 16 - 10	1683	0,258
M 15 - 20	1804	0,449
M15 - 10	1500	0,208
S16 - 20	1339	0,429
M15 - 20	1299	0,419

Figure 3 shows the thermal conductivity of BIP with 10 mm and 20 mm straws, it is clear that the aggregate size plays a very important role in the thermal conductivity of Insulating Concrete Beams. As explained in the previous study on non-bearing insulating concrete [15], this difference is due to the rearrangement of the microstructure of these materials, particularly the pores.

The Fig. 3 presents the thermal conductivity of the Insulating Concrete load-bearing (BIP) based on the dry apparent volume mass of the samples. The samples studied were made with the percentage in Table 2, and straws of size 0.5–20 mm from both Marne and Somme were used. The place of harvest does not influence the thermal conductivity. The thermal conductivity of the concrete varies between 0.265 and 0.45 $W.m^{-1}\ K^{-1}$;

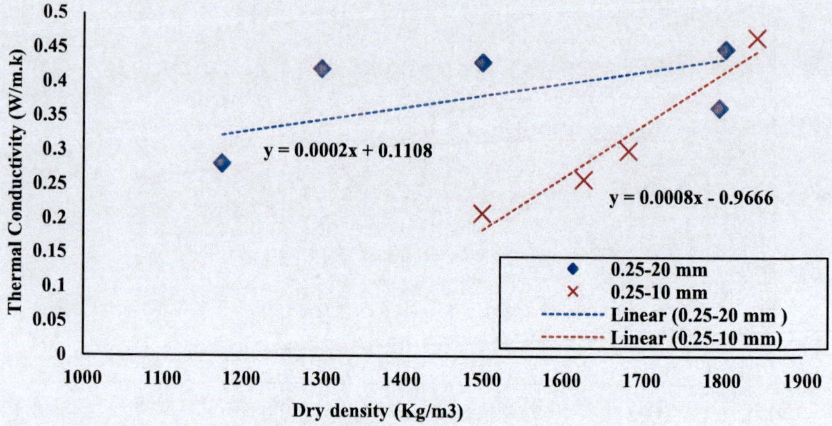

Fig. 3. Thermal conductivity of BIP as a function of bulk density of samples made with 20 mm and 10 mm straws.

the difference in this conductivity is directly related to the apparent volume mass and the size aggregate. The relationship between mass and concrete depends on each material. A literature review by Asadi [16] on the thermal conductivity of materials and its relationship with volume mass, explains that thermal conductivity is directly related to volume mass, particularly the pores or voids in the material, hence cellular concrete is considered a good insulator, as well as concrete based on plant aggregates. The equation of thermal conductivity with density found for BIP resembles to this found by Liu [17] on cellular concrete integrating geopolymers, as well as Cerezo [10] found a relationship between volume mass and thermal conductivity that is close to the equation found with the BIP samples made with the largest aggregates.

4 Conclusion

The study aimed to improve the mechanical and thermal properties of a new bio-sourced material known as load-bearing insulating concrete (BIP). It was noted that the origin and harvest year of the material did not have any effect on its mechanical and thermal behavior.

The compressive strength ranges between 3.2 MPa and 4.5 MPa, while thermal conductivity varies more greatly, due to the variation in density and aggregate size. These values range between 0.2 $W.m^{-1}.k^{-1}$ and 0.4 $W.m^{-1}.k^{-1}$ for a density between 1300 kg/m^3 and 1800 kg/m^3. The size of the aggregates and the density of the material were considered as variables in the optimization process. The results indicated that using 20 mm aggregates provided better mechanical performance and increased plasticity before breaking, and this is due to the good cohesion achieved with the 20 mm straws compared to the breaks in the binder. While using 10 mm aggregates offered better thermal resistance at lower densities. However, when the density increased, the thermal resistance of BIP made with both aggregate sizes was found to be similar. An increase in density carries the risk of damaging the material's pores.

Acknowledgement. This work is done within the framework of the BIP-Colza project, with the partnership of Codem-batlab. This work is financed by the Haut de France region FEDER and the environmental agency ADEME. Authors thank their financial support. The authors also express gratitude to CODEM technician Florent Bordet for his involvement in the production and thermal and mechanical testing.

References

1. Énergie Dans Les Bâtiments | Ministères Écologie Énergie Territoires Available online: https://www.ecologie.gouv.fr/energie-dans-batiments Accessed on 26 Oct 2022
2. Dennis Jones, C.B.: Performance of Bio-Based Building Materials. 1st ed. Woodhead Publishing (2017)
3. Buildings, M.C.-E.: Undefined Young's modulus and thermophysical performances of bio-sourced materials based on date palm fibers. Elsevier (2016)
4. La Filière Des Produits Biosourcés – Ademe
5. Delannoy, G., et al.: Aging of hemp shiv used for concrete. Mater. Des. **160**, 752–762 (2018). https://doi.org/10.1016/J.MATDES.2018.10.016
6. Collet, F., Pretot, S.: Thermal conductivity of hemp concretes: variation with formulation density water content. Constr. Build. Mater. **65**, 612–619 (2014). https://doi.org/10.1016/j.conbuildmat.2014.05.039
7. Mazhoud, B., Collet, F., Pretot, S., Chamoin, J.: Hygric and thermal properties of hemp-lime plasters. Build. Environ. **96**, 206–216 (2016). https://doi.org/10.1016/J.BUILDENV.2015.11.013
8. Arnaud, L., Gourlay, E.: Experimental study of parameters influencing mechanical properties of hemp concretes. Constr. Build. Mater. **28**, 50–56 (2012). https://doi.org/10.1016/j.conbuildmat.2011.07.052
9. Nguyen, T.T.: Contribution à l'étude de La Formulation et Du Procédé de Fabrication d'éléments de Construction En Béton de Chanvre (2010)
10. Cerezo, V.: Propriétés Mécaniques, Thermiques et Acoustiques d'un Matériau à Base de Particules Végétales: Approche Expérimentale et Modélisation Théorique Véronique CEREZO Soutenue Le 16 Juin 2005 Devant La Commission d'examen (2005)
11. Standard NF EN 12390–2 Available online: https://www.boutique.afnor.org/en-gb/standard/nf-en-123902/testing-hardened-concrete-part-2-making-and-curing-specimens-for-strength-t/fa190565/83458. Accessed on 28 Feb 2022
12. Standard NF EN 196–1 Available online: https://www.boutique.afnor.org/en-gb/standard/nf-en-1961/methods-of-testing-cement-part-1-determination-of-strength/fa184622/57803. Accessed on 1 Mar 2022
13. Norme NF EN 12667 Available online: https://www.boutique.afnor.org/fr-fr/norme/nf-en-12667/performance-thermique-des-materiaux-et-produits-pour-le-batiment-determinat/fa045167/18796. Accessed on 17 Oct 2021
14. Norme NF EN 12086 Available online: https://www.boutique.afnor.org/fr-fr/norme/nf-en-12086/produits-isolants-thermiques-destines-aux-applications-du-batiment-determin/fa177323/41201. Accessed on 16 Feb 2022
15. Obeid, M.H., et al.: Physical and mechanical properties of rapeseed straw concrete. Materials **15**(23), 8611 (2022). https://doi.org/10.3390/ma15238611
16. Asadi, I., Shafigh, P., Bin, Z.F., Hassan, A., Mahyuddin, N.B.: Thermal conductivity of concrete-a review. J. Build. Eng. **20**, 81–93 (2018). https://doi.org/10.1016/j.jobe.2018.07.002
17. Liu, M.Y.J., Alengaram, U.J., Jumaat, M.Z., Mo, K.H.: Evaluation of thermal conductivity, mechanical and transport properties of lightweight aggregate foamed geopolymer concrete. Energy Build. **72**, 238–245 (2014). https://doi.org/10.1016/J.ENBUILD.2013.12.029

Evaluation of the Potential of Plant Aggregates from Corn and Sunflower Stalks for the Design of Building Materials

Alina Avellaneda[1], Philippe Evon[2], Laia Haurie[1], Aurélie Laborel-Préneron[3], Méryl Lagouin[3], Camille Magniont[3](✉), Antonia Navarro[1], Mariana Palumbo[1], and Alba Torres[4]

[1] GICITED, Universitat Politècnica de Catalunya, Barcelona, Spain
[2] LCA, Université de Toulouse, INP, INRAe, ENSIACET, Toulouse, France
[3] LMDC, Université de Toulouse, INSA, UPS, Toulouse, France
camille.magniont@insa-toulouse.Fr
[4] SUSCAPE, Universitat Rovira i Virgili, Tarragona, Spain

Abstract. This work has been realized within the framework of Action No. 4 of the SAVASCO collaborative project (EFA353/19/SAVASCO) co-funded by INTERREG POCTEFA (INTERREG V-A Spain - France - Andorra 2014–2020). This action was carried out between June 2020 and May 2022, and is devoted to the development and characterization of a constructive solution based on corn and sunflower stalks as raw material.

At the end of Action No. 3 of the project, different fractions of plant aggregates from corn and sunflower stalks were produced. The objective of Action 4 is to formulate different construction materials and products from these raw materials. The present paper will focus on the mechanical and thermal behavior of samples made of lightweight earth and concrete implemented by vibrocompaction at laboratory scale, as a model for prefabricated construction products assembled in a dry state on site.

The influence of the nature of the aggregate on the properties of the composites was studied by working on the entire crushed stalks of corn, but also on sunflower pith and bark fractions produced from the separation process developed in the project. Finally, the influence of the particle size distribution of the sunflower bark particles was also evaluated. In the entire experimental campaign, hemp shives were used as the reference aggregate.

Considering the binders, a commercial formulated lime binder was selected as well as two alternative clay binders produced locally in the regions of Tarbes, France and Barcelona, Spain.

The first part of the paper summarises the main physical characteristics of the different aggregate fractions studied: bulk and particle densities, particle size distribution, water absorption, mechanical behavior and thermal conductivity. Some of these characteristics are used as formulation parameters for composites, others could be indicators to correlate with the performance of composites.

The second part aims to evaluate the thermal and mechanical performance of composites formulated from these fractions. The highly anisotropic character of these composites and the significant influence of the nature of the particles and

S. Amziane et al. (Eds.): ICBBM 2023, RILEM Bookseries 45, pp. 70–86, 2023.
https://doi.org/10.1007/978-3-031-33465-8_7

binders is highlighted. Several fractions, in particular from sunflower stalks, show a definite potential as an alternative plant aggregate to hemp shiv, in regions where the availability of the latter is limited.

Keywords: corn · sunflower · plant concrete · lightweight earth · mechanical behaviour · thermal conductivity

1 Introduction

Due to its negative environmental impacts, the building industry should reconsider its activities. Indeed, buildings are responsible for approximately 40% of energy consumption and 36% of greenhouse gas emissions in the European Union [1]. On the trans-Pyrenean territory, the construction sector is responsible for the production of more than 4 million tons of waste in Midi-Pyrénées [2] and 5,7 million tons in Catalonia [3], which exceeds the production of domestic waste. In 2018, the residential/tertiary sector was leading energy consumption item (45%) in the Occitanie region and accounts for 29% of the region's greenhouse gas emissions [4]. The reduction of the environmental impact is therefore a major challenge for this territory that possesses high level of activity in the construction/renovation sector. In 2020, with 9 billion euros, Occitanie ranked as the fifth region in France in terms of turnover related to construction activities [5]. Like Andalusia, Catalonia stands out as the region in Spain that have hosted most works (>5 000) in 2021 [6]. To meet these challenges, one needs to assess and reduce the environmental impacts generated by all building-related activities. To do so, circular economy and resource efficiency principles should be applied to buildings.

Bio-aggregate based building materials offer promising prospects to achieve these objectives thanks to the use of agricultural by-products, which have considerable environmental benefits. First, bio-aggregates are derived from an abundant, renewable vegetal resource, which is also a carbon sequestration material as agricultural products have the property of capturing carbon dioxide from the atmosphere during their growth and storing it in their organisms. Besides, the use of plant particles in building materials helps to economize valuable natural resources thanks to agricultural waste recovery [7].

Hemp concrete has been the most commonly investigated bio-aggregate based material in recent years. However, hemp is not widely produced in France nor in Spain and little is available in the trans-Pyrenean territory. To minimize transport distances and thus carbon emissions, the raw material has to be locally available. It is the case for corn and sunflower stems, two agricultural by-products produced in greater quantities in this region.

The Spanish POCTEFA areas account for more than a third of corn production in Spain [8]. According to France AgriMer figures [9], assuming that 52% of the stalks are returned to the soil to maintain its fertility, the potential deposit corn stalks in Occitanie would be of 154 560 t/year. In France, about one third of the national cultivated surface area of sunflower is located in Occitanie (207 505 ha in 2017) [10].

Moreover, corn and sunflower stalks exhibit strong potential for the development of insulating materials for construction. Indeed, previous works conducted by the project's partner laboratories demonstrated the feasibility of separating the two main stalks fractions (pith and bark) on an industrial scale. Those studies also highlighted the interest and potential of using by-product of corn or sunflower cultivation in order to develop insulating panels [11–14] or bio-based concretes [15–17].

The INTERREG V-A (POCTEFA 2014–2020) project entitled 'StructurAtion of a cross-border Value chAin for corn and Sunflower stems for COnstructions' (SAVASCO) helps to meet these challenges by substituting non-renewable resources of mineral or petroleum origin with renewable agricultural waste. SAVASCO aims to structure an innovative and eco-efficient trans-Pyrenean construction sector based on corn and sunflowers stalks. It relies on innovative technologies for the use of currently non-valued agricultural by-products to develop a low-cost process for collecting and processing stalks into aggregates with controlled characteristics.

This article focuses on the insulating construction products developed using those aggregates in association with a mineral binder. The first part of the paper summarises the basic physical characteristics of the different aggregate fractions studied: bulk and particle densities, particle size distribution and water absorption, mechanical. Some of these characteristics are used as formulation parameters for composites, others could be indicators to correlate with the performance of composites. The second part aims to evaluate the thermal and mechanical performance of composites formulated from these fractions.

2 Materials and Methods

2.1 Materials

Plant Particles. A multi-stage process for the collection, fractionation and separation of corn and sunflower stalks was previously develop in the frame of the SAVASCO project [18] to obtain bio-aggregates (Fig. 1).

Field collection was carried out in collaboration with the Federació de Cooperatives Agràries de Catalunya (FCAC) (a federation of agricultural cooperatives), and plant aggregates were obtained thanks to the use of industrial tools available in LCA laboratory. The combination of grinding, dedusting, sieving, separating and screening steps allowed obtaining different fractions of plant particles compatible with a use as plant aggregates for construction. The applied process is summarized in Fig. 1.

The 7 fractions written in grey frame in Fig. 1 were characterized and used as aggregates for the development of lime or clay concretes during the project. Among them, we will present, in the present article, the results concerning 4 SAVASCO aggregates (ground corn stem, sunflower pith and sunflower barks larger and smaller than 4 mm) as well as those obtained with hemp shiv also studied as a reference plant aggregate.

The physical characterization of aggregates was conducted according to the methodology proposed by Ratsimbazafy [16] for the determination of the densities and intra and interparticle porosities, the dry density of aggregates compacted in wet state and the real water absorption after compaction which have been proven to be good formulation parameters for composites. Those characteristics are presented in Sect. 3.1.

Mineral Binders. Among the 5 distinct mineral binders studied in SAVASCO, the results obtained with 2 of them will be presented and analysed in the present paper. The

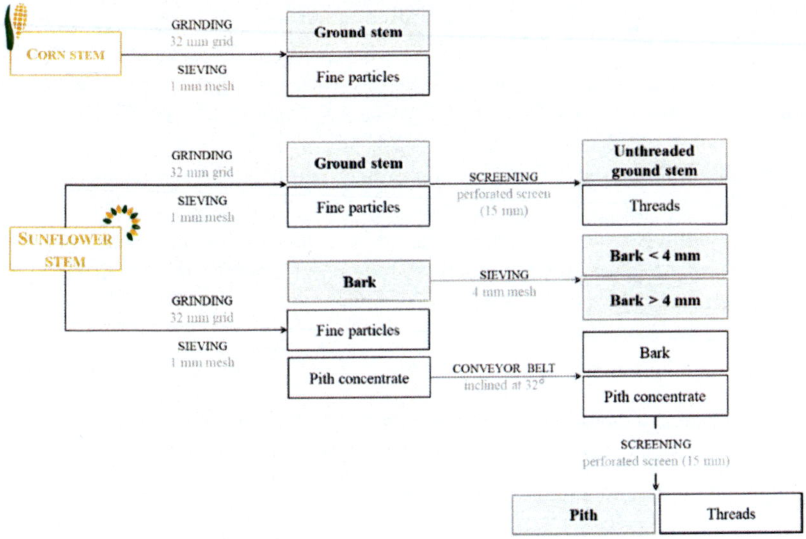

Fig. 1. Mechanical separating process of the two main parts of corn and sunflower stems.

first one is a preformulated commercial lime-based binder BATICHANVRE® produced by the company 'Chaux & Enduits de Saint-Astier'. This type of binder is highly recommended for the manufacture of hemp concrete [19]. According to the technical data from the manufacturer, BATICHANVRE® is composed of 70% natural lime and 30% hydraulic and pozzolanic binder plus specific additives. With the aim of reducing the environmental impact of composites and benefiting from the high hygroscopicity of the clay minerals, the second studied binder was a clayish binder composed of fines from the aggregate washing process of a quarry in south-west France.

2.2 Methods

Plant Particles Characterization

Bulk Density, Particles Size Distribution and Water Absorption. Measurements were carried out according to the recommendations of the RILEM TC 236-BBM [20].

Dry Density of Aggregates Compacted in Wet State, and real Water Absorption After Compaction. Measurements were carried out according to the methods detailed in [21].

Particle Density, Solid Density and Porosities. The method applied for particle density measurement is a pycnometric method using fine sand adapted by Ratsimbazafy [16] and previously mentioned by Monreal et al. [22] and Karaky et al. [23]. It consists in determining the real volume (particle volume) of a representative quantity of plant aggregates of known mass by "embedding" them in very fine sand capable of filling all inter-particle voids, without saturating the intra-particle porosity.

Solid density is deduced from the determination of:

- the parietal constituents (cellulose, hemicelluloses and lignins) via the gravimetric method known as ADF-NDF (ADF for Acid Detergent Fiber, NDF for Neutral Detergent Fiber) of Van Soest and Wine [24, 25]
- the total mineral content, determined by mineralization in an ash furnace according to ISO 749:1977 [26].
- the water solubles content by measuring the mass loss in aqueous solutions at pH 7.
- the mean densities of these elementary chemical components deducted from a literature review and reported in Table 1.

Table 1. Minimum, maximum and average density values of chemical components of lignocellulosic materials from the literature used for the determination of solid density.

Chemical components	Density (kg/m^3)			References
	Min	Max	Mean	
Cellulose	1450	1630	1543	[27–36]
Hemicelluloses	1457	1622	1516	
Lignins	1278	1397	1347	
Minerals	1500	1740	1595	
Watersolubles	1500	1560	1535	

Assuming that the dry solid part of the aggregates is formed only by cellulose, hemicelluloses, lignins, minerals and water-solubles extracted at pH 7, the density of the solid (ρ_s) is calculated from the Eq. 1.

$$\frac{1}{\rho_s} = \sum \frac{w_i}{\rho_i} \tag{1}$$

With:

w_i: the mass content of component i in proportion to the sum of the contents of these 5 components

ρ_i: the density of the component i.

The inter- and intra-particle porosities (n_{inter} and n_{intra}) can then be determined from the Eq. 2 and Eq. 3 respectively.

$$n_{inter} = \left(1 - \frac{\rho_b}{\rho_p}\right).100 \tag{2}$$

$$n_{intra} = \left(1 - \frac{\rho_p}{\rho_s}\right).\frac{\rho_b}{\rho_p}.100 \tag{3}$$

With:

ρ_b: the bulk density (kg/m^3)
ρ_p: the particle density (kg/m^3)
ρ_s: the solid density (kg/m^3).

Composite Characterization

Mechanical Behaviour. The compressive strength tests on the cubical specimens (15 × 15 × 15 cm^3) were performed after one year of curing, using a 100kN capacity hydraulic press. Three to four samples were tested for each formulation. The load was applied at a constant deflection rate of 3 mm/min while the unloading was carried out at 6 mm/min. The following 4-step loading cycle was applied to samples:

- loading up to 1% of strain - unloading back to 0%;
- loading up to 2% - unloading to 0%;
- loading up to 3% - unloading to 0%;
- loading up to 10% - unloading to 0%.

Thermal Conductivity. Cubical samples of 15 × 15 × 15 cm^3 were dried in an oven at a temperature of 60 °C until the sample mass change was less than 0.1% between three weighings at least 24 h apart. Thus, tests were conducted with a Neotim-FP2C hot wire apparatus that consists in placing a shock probe between two pieces of material so that a power supply can be broadcast and the rise of temperature within the material may be measured. The heat flow and heating time were chosen so that a temperature rise higher than 5 °C and a correlation coefficient between experimental data and theoretical behaviour higher than 0.999 could be reached. This transient method only allows local measurements. To overcome the question of accuracy and representativeness for heterogeneous materials, thermal conductivity is defined as the average of at least five measurements. For each formulation, the device was placed at several locations on different samples.

3 Results and Discussion

3.1 Plant Particles Properties

Particle Size Distribution. The particle size distributions of the 5 aggregates studied are shown in Fig. 2.

Two groups of aggregates can be identified in terms of size. The first group includes hemp particles, sunflower pith and the sunflower bark fraction passing the 4 mm sieve. The second is composed of the ground corn stalk and sunflower bark particles retained on the 4 mm sieve. Among the smaller aggregates (D50 between 4 and 4.5 mm), hemp shiv has the narrowest particle size, probably due to its calibration in the processing plant. In the pith and bark fractions, on the contrary, larger elements are identified.

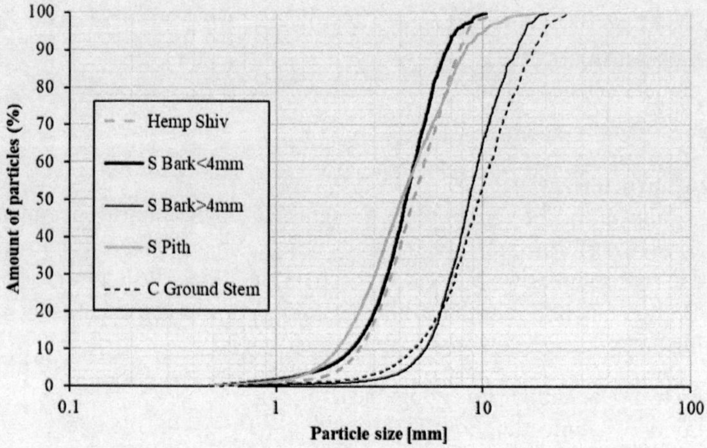

Fig. 2. Particle size distribution of the five studied plant aggregates.

As far as particle morphology is concerned, clear differences can also be observed in Fig. 3.

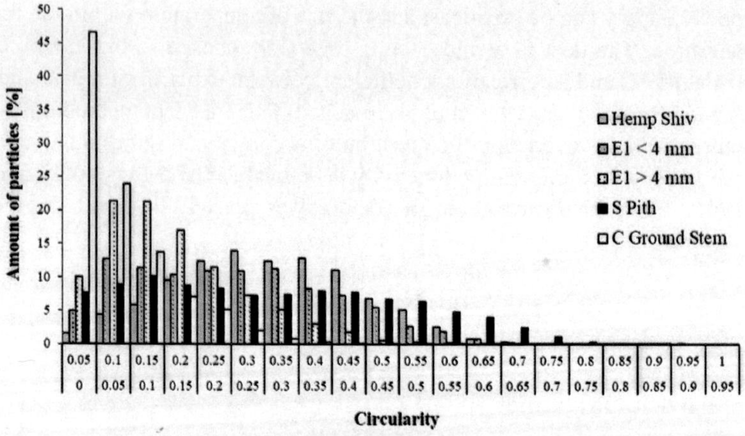

Fig. 3. Circularity of the five studied plant aggregates.

It can be seen that size reduction by grinding also tends to limit the elongation of the particles. The larger particles (ground corn stalk and sunflower bark particles constituting the 4 mm retain) are also those with the lowest circularity. The arrangement of ground corn stalk particles in particular contains a large proportion of very elongated particles (more than 45% of particles with a circularity of less than 0.05). Amongst the smaller particles, sunflower pith is characterized by the presence of very rounded particles (more than 18% of the particles with a circularity higher than 0.5). The hemp shiv and sunflower bark of small size present varied morphologies.

Densities and Porosities. In order to understand and predict the properties of composites including plant particles, it is necessary to characterize their porosity. The measurement of the bulk density of the particles gives a first indication but it is not sufficient. It does not provide information on the distribution between interparticle porosity (which will be reduced by rearrangement and partially filled by the binder at the time of mixing of the composite) and intraparticle porosity, which will remain in the composite or be reduced by compression of the particles. In order to access this data, a measurement of the density of the particles is carried out, the results obtained for the 5 aggregates considered are presented in Table 2 as well as the values of bulk density and solid density. From these measurements, it is then possible to calculate the different volume proportions of solid, interparticle voids and intraparticle voids in each arrangement. The results are shown in the graphs in Fig. 4.

Table 2. Bulk and particle densities of the five studied plant aggregates.

Plant Aggregate	Bulk density ρ_b (kg/m^3)	Particle density ρ_p (kg/m^3)	Solide density ρ_s (kg/m^3)
Hemp shiv	111 ± 1	434 ± 17	1520
S Bark < 4 mm	144 ± 3	496 ± 20	1453
S Bark > 4 mm	125 ± 5	489 ± 23	1444
S Pith	64 ± 3	190 ± 12	1569
C ground stem	43 ± 4	485 ± 31	1472

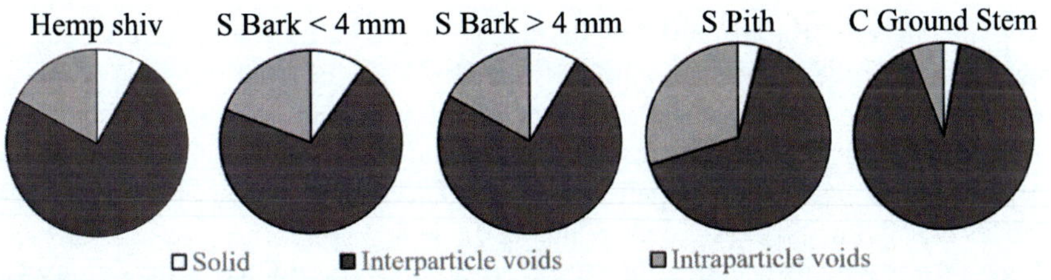

Fig. 4. Distribution of solid and pore volume fractions in the five studied plant aggregates.

Whatever the nature of the particles considered, the total porosity of the arrangements is very important, it varies between 90% for the small sunflower bark and 97% for the ground corn stalk.

The distribution of voids is close for hemp shiv and sunflower bark aggregates. On the contrary, sunflower pith has a much higher intraparticle porosity than the other particles, while the ground corn stalk arrangement has the highest interparticle porosity, caused by the interlocking of the large, elongated particles that mainly constitute this arrangement, as evidenced by the particle size analysis.

Water Absorption. The capacity of the particles to absorb water is an important parameter especially for the formulation of composites. In Fig. 5, different behaviours can be seen. Sunflower pith, due to its high particle porosity, absorbs the most water (almost 500% by mass after only 30 s of immersion). The three other aggregates from sunflower and maize stalks have lower absorption capacities than the reference aggregate, which seems rather favourable.

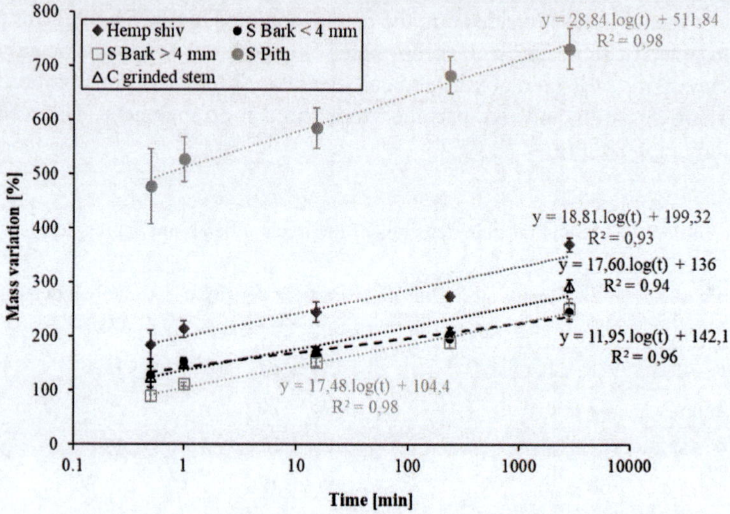

Fig. 5. Water absorption capacity of the five studied aggregates.

Design Parameters. Based on the methodology proposed by Ratsimbazafy [16], the characterization of the aggregates was completed by the determination of the properties of the aggregates compacted in the wet state, which is considered to be representative of the state of the aggregates in a vibrocompacted plant concrete or lightened earth. The dry density of aggregates compacted in wet state and the real water absorption of the aggregates during this process are presented in Table 3.

Regarding water absorption, the water absorption determined after compaction exceeds the absorption after one minute of immersion for all aggregates except the small sunflower bark aggregate. This parameter will be retained for the formulation of composites. A relative difference between the bulk density and the dry density of aggregates compacted in wet state is observed, which differs from one aggregate to another. Indeed, depending on their morphology and stiffness, the behavior of the arrangement of aggregates subjected to compaction is distinct. The phenomena of rearrangement and then compression of the different particles will tend to increase the density of the arrangement but, at the same time, the absorption of water by the aggregates will lead to a swelling and bulking of the particles after demolding which can compensate for this phenomenon as shown in Fig. 6.

Table 3. Formulation parameters of biocomposites.

Plant Aggregate	Bulk density ρ_b (kg/m3)	Dry density of aggregates compacted in wet state ρ_c (kg/m3)	Water absorption after compaction (mass %)	Water absorption after 1 min of immersion (mass %)
Hemp shiv	111 ± 1	86 ± 3	234 ± 15	214 ± 15
S Bark < 4 mm	144 ± 3	131 ± 5	147 ± 3	153 ± 7
S Bark > 4 mm	125 ± 5	103 ± 12	147 ± 6	111 ± 3
S Pith	64 ± 3	61 ± 4	600 ± 14	527 ± 41
C ground stem	43 ± 4	88 ± 1	186 ± 5	149 ± 10

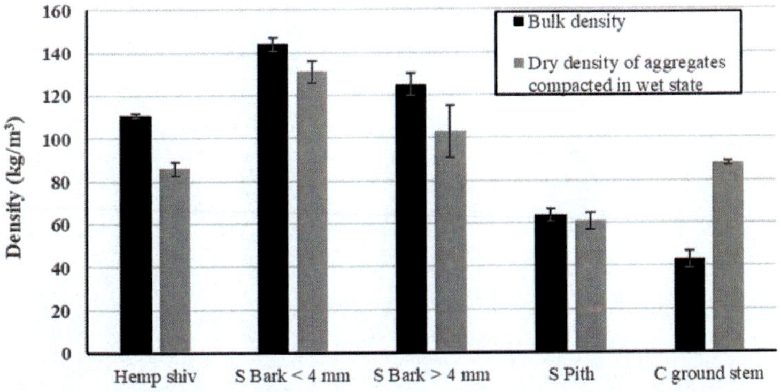

Fig. 6. Comparison of bulk and wet compacted densities of the five studied plant aggregates.

In the case of hemp shiv and sunflower bark, wet compaction leads to a delayed swelling of the particles and a decrease in the bulk density. In the case of sunflower pith, the two densities are equivalent. On the contrary, in the particle arrangement of ground corn stalks, the one with the highest interparticle porosity, compaction significantly reduces these voids and results in an increase in density.

3.2 Composites Design and Fabrication

Based on the formulation method developed by Ratsimbazafy et al. [16], plant-based concrete or lightweight earth were produced with a constant binder dosage in order to enable evaluation of the influence of the properties of the aggregates on the composites performances. The designation of the studied mixes is reported in Table 4. The proportions of the other components were determined from a reference formulation listed in the first row of Table 5. Based on this wall type hemp-lime mix design, the water requirement and the mass of plant aggregates to be introduced were adjusted according to the dry density of aggregates compacted in wet state ρ_c and to the real water absorption capacity after compaction, in order to take into account the specificities of each aggregate. The composition of each mix as well as their fresh and dry densities are presented in Table 5.

Table 4. Designation of the studied mixes.

Binder	Hemp shiv	S Bark < 4 mm	S Bark > 4 mm	S Pith	C ground stem
Lime-based preformulated binder Batichanvre®	BH	BB < 4	BB > 4	BP	BC
Clayish binder (Chis)	CH	CB < 4	CB > 4	CP	CC

Table 5. Composition and densities in the fresh and hardened state of the studied mixes.

Plant Aggregate	Binder dosage (B) (kg/m^3)	Aggregates dosage (A) (kg/m^3)	Water dosage (W) (kg/m^3)	A/B	W/B	Fresh state density ρ_f (kg/m^3)	Dry density ρ_d (kg/m^3)
BH	258.3	111.1	361.6	0.43	1.40	728.4 ± 5.2	352.3 ± 9.9
BB < 4	258.3	164.7	358.6	0.66	1.39	782.3 ± 3.0	402.9 ± 16.5
BB > 4	258.3	132.8	342.2	0.51	1.32	719.0 ± 7.2	387.4 ± 13.9
BP	258.3	79	425.5	0.31	1.65	746.3 ± 10	361.9 ± 14.0
BC	258.3	113.4	346.9	0.44	1.34	711.7 ± 3.8	366.1 ± 7.6
CH	258.3	111.1	361.6	0.43	1.40	728.9 ± 3.4	393.1 ± 10.1
CB > 4	258.3	132.8	342.2	0.51	1.32	729.5 ± 3.1	425.2 ± 19.7
CP	258.3	79	425.5	0.31	1.65	761.7 ± 1.8	348.1 ± 20.8
CC	258.3	113.4	346.9	0.44	1.34	714.8 ± 1.1	383.7 ± 37.2

3.3 Use Properties

Mechanical Behaviour. Typical stress-strain curves obtained from the cyclic compression tests performed on the composites are shown in Fig. 7.

In the direction parallel to compaction, a maximum stress is reached before failure for some aggregates (BP Para), while progressive compression of the particles is observed for others (BC Para).

The comparison of the compressive behaviour of the plant concrete in the two directions of solicitation (parallel (BP Para) and perpendicular (BP Per) to the direction of compaction) highlights their mechanical anisotropy linked to the structural anisotropy induced during compaction, which generates a preferential orientation of the elongated

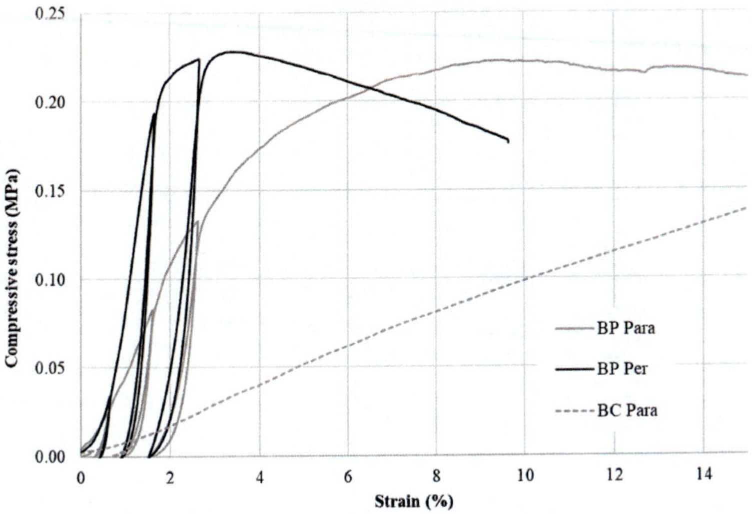

Fig. 7. Typical stress-strain curves for perpendicular and parallel solicitation

aggregates in the plane perpendicular to the compaction. In order to compare the mechanical performance of the different composites, the characteristic mechanical parameter chosen is the stress reached at a strain of 1.5%. The results are shown in Fig. 8.

Whatever the aggregate considered, the stress reached in the perpendicular direction exceeds that reached in the parallel direction (the ratio between the two stresses at 1.5% deformation is respectively 2.1, 2.3, 2.7 and 4.6 for the BB < 4, BP BH and BC mixtures). Indeed, during a stress perpendicular to the direction of compaction, i.e. parallel to the plane of preferential orientation of the plant particles, it is the stiffest element of the composite that controls its deformation.

Whatever the direction of loading or the type of binder, the best mechanical performance is achieved with the reference aggregate, hemp shiv. Sunflower pith comes next, followed by sunflower bark and then ground corn stalks. Considering the results obtained for parallel loading with lime binder, the increase in particle size of sunflower bark seems to have a beneficial effect on the mechanical performance. This result is in agreement with the majority of the few studies dealing with the subject in the literature [37–39] even if other authors observe an opposite tendency with a clay matrix [40].

The mixes formulated with a clay binder have significantly lower compressive stresses than those based on formulated lime. It is nevertheless important to underline that despite low mechanical compressive performances, these mixes could be used in the prototype built within the framework of action 5 of the SAVASCO project, either by spraying or by onsite casting for the construction of walls between 14.5 and 29 cm thick. This raises the question of the relevance of studying the behavior in compression of this type of composite, which is not very representative of its mechanical solicitation in the structure.

Figure 9 shows a more detailed comparison of the mechanical behavior of composite including the reference aggregate i.e. hemp shiv with the one including sunflower

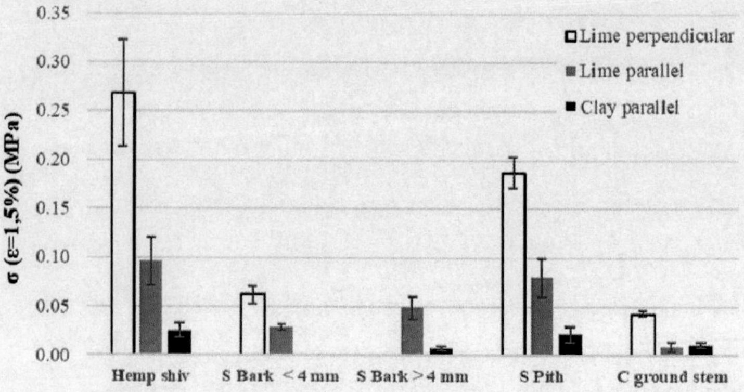

Fig. 8. Compressive stress at 1.5% strain for lime-based composite in perpendicular and parallel directions and for clay-based composites in parallel direction.

pith particles. For the hemp shiv and sunflower pith aggregates, whatever the nature of the binder and the direction of loading, the stress-strain curve reaches a maximum value before the sample fails. Three characteristic mechanical parameters were therefore retained: the stress reached at a strain of 1.5%, the maximum stress and the strain at this maximum stress.

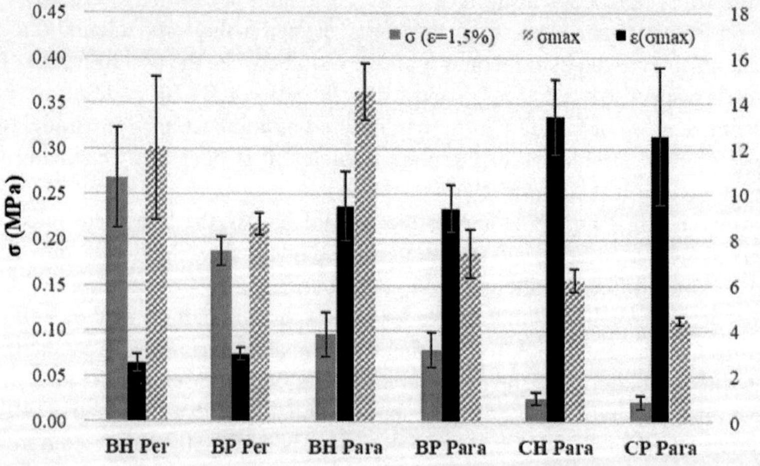

Fig. 9. Compressive stress at 1.5% strain, maximum stress and strain at maximum stress of biocomposites including hemp shiv (BH and CH) and sunflower pith particles (BP and CP)

The difference in performance between hemp shiv and sunflower pith aggregates is particularly marked for composites with a formulated lime matrix, loaded perpendicular to the direction of compaction. On the contrary, if we consider as a performance indicator the stress reached for a strain of 1.5%, the impact of the nature of the aggregates is not significant for clay-based mixes or lime-based mixes loaded parallel to the compaction direction. Finally, the nature of the aggregate has no influence on the deformation reached for the maximum stress.

Thermal Conductivity. The formulated composites will primarily function as thermal insulation in buildings. Figure 10 shows the results of dry thermal conductivity measurements.

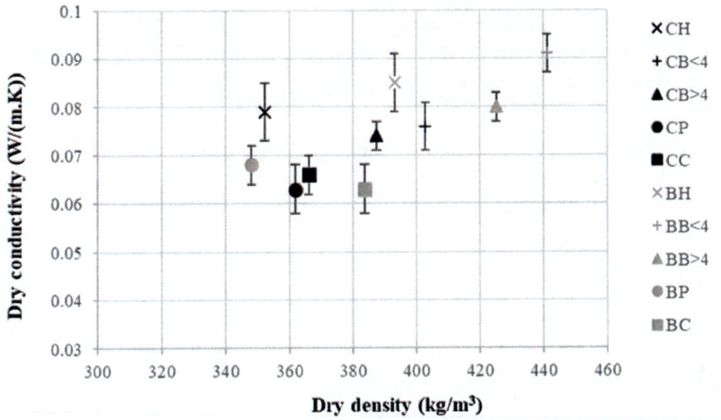

Fig. 10. Dry thermal conductivity of the biocomposites versus dry density

The composites have low thermal conductivities (between 0.063 and 0.091 W/(m.K)), consistent with their dry densities (348 to 441 kg/m^3) which place them in the range of the lightest mixtures for this type of material [41]. The results do not show any significant impact of the nature of the binder on the thermal performance. Regarding the nature of the plant particles, the mixes incorporating hemp shiv have slightly higher thermal conductivities at equivalent density. In contrast, mixtures based on sunflower pith and ground corn stalk particles achieve the lowest conductivities. These are the two aggregates with the highest intra- and inter-particle porosities respectively.

4 Conclusion and Perspectives

In the framework of the INTERREG POCTEFA SAVASCO project, different plant aggregates were produced from sunflower and corn stalks. 4 aggregates are studied in this paper and compared to hemp shiv, selected as reference aggregate.

The determination of the bulk, particle and solid densities shows a variable pore distribution within the different particle arrangements. Larger particles generally have lower circularity and higher inter-particle porosity. Sunflower pith is distinguished from other aggregates by high circularity and high intraparticle porosity.

Additional characterization parameters are proposed to assess the wet compaction ability of the aggregates and to serve as formulation parameters.

The different particles are combined with a commercial formulated lime binder and a clay binder to form composites with the same binder content. The mechanical and thermal characterization of these composites shows the high potential of sunflower pith aggregates which achieve the best mechanical performance among alternative aggregates and surpass the hemp shiv-based reference mix in terms of thermal conductivity.

Predicting the mechanical and thermal performance of composites from aggregate properties is a complex problem. Indeed, the size, the shape but also the stiffness of the aggregates condition their behavior during compaction and consequently the structuring of the pore network and of the binding matrix in the composites in the hardened state. Complementary modelling approaches using homogenization methods should be used to complete this experimental study. The morphology and size of the particles can then be optimized by additional steps of grinding, sieving, dedusting etc.

Acknowledgements. This research was co-funded by the European Regional Development Fund as part of the POCTEFA (Interreg V A Espagne/France/Andorre 2012–2020) program (project number EFA353/19/SAVASCO).

References

1. European Commission, 'Buildings - Energy', Energy (2018). /energy/en/topics/energy-efficiency/buildings. Accessed 30 May 2018
2. CRC Midi-Pyrénées. https://www.cercoccitanie.fr/IMG/pdf/midi-pyrenees_-_septembre_2015.pdf. Accessed 08 Aug 2022
3. Generalitat de Catalunya Departament d'Acció Climàtica, Alimentació i Agenda Rural and Agència de Residus de Catalunya. https://residus.gencat.cat/web/.content/home/lagencia/publicacions/estadistiques/estadistiques_2020.pdf. Accessed 08 Aug 2022
4. Fedene. https://www.fedene.fr/wp-content/uploads/sites/2/2022/06/FEDENE_Fiche-Region-2022-OCCITANIE.pdf. Accessed 08 Aug 2022
5. GIE Réseau des CERC. https://www.cercoccitanie.fr/IMG/pdf/cp_octobre_2021.pdf. Accessed 10 Aug 2022
6. Construnario.com. https://www.construnario.com/notiweb/55896/el-sector-de-la-construccion-crece-un-22-en-espana/. Accessed 10 Aug 2022
7. Peñaloza, D., Erlandsson, M., Falk, A.: Exploring the climate impact effects of increased use of bio-based materials in buildings. Constr. Build. Mater. **125**, 219–226 (2016). https://doi.org/10.1016/j.conbuildmat.2016.08.041
8. Encuesta sobre Superficies y Rendimientos Cultivos (ESYRCE). Encuesta de Marco de Áreas de España. https://www.mapa.gob.es/es/estadistica/temas/estadisticas-agrarias/agricultura/esyrce/default.aspx. Accessed 10 Aug 2022
9. FranceAgriMer: L'Observatoire National des Ressources en Biomasse Évaluation des ressources agricoles et agroalimentaires disponibles en France – édition 2020. FranceAgriMer (2020). https://www.franceagrimer.fr/content/download/66147/document/DON-ONRB-VF4.pdf. Accessed 08 Aug 2022
10. Agreste: Pratiques culturales en grandes cultures en 2017 – Occitanie. Agreste (2019). https://draaf.occitanie.agriculture.gouv.fr/IMG/pdf/premiers_resultats_pkgc_2017_cle8ff4e6.pdf. Accessed 08 Aug 2022
11. Laborel-Préneron, A., Ampe, C., Labonne, L., Magniont, C., Evon, P.: Thermal insulation blocks made of sunflower pith particles and polysaccharide-based binders: influence of binder type and content on their characteristics, pp. 43–50 (2022). https://doi.org/10.4028/www.scientific.net/CTA.1.43

12. Palumbo, M., Lacasta, A.M., Holcroft, N., Shea, A., Walker, P.: Determination of hygrothermal parameters of experimental and commercial bio-based insulation materials. Constr. Build. Mater. **124**, 269–275 (2016). https://doi.org/10.1016/j.conbuildmat.2016.07.106
13. Sabathier, V., Maaloum, A.A., Magniont, C., Evon, P., Labonne, L.: Développement d'un panneau isolant 100% biosourcé à base de moelle de tournesol. Auxerre, France (2017)
14. Verdier, T., Balthazard, L., Montibus, M., Magniont, C., Evon, P., Bertron, A.: Development of sunflower-based insulation materials coated with glycerol esters to prevent microbial growth. Acad. J. Civ. Eng. **37**(2), Art. no. 2, (2019). https://doi.org/10.26168/icbbm2019.7
15. Lagouin, M., Magniont, C., Sénéchal, P., Moonen, P., Aubert, J.-E., Laborel-préneron, A.: Influence of types of binder and plant aggregates on hygrothermal and mechanical properties of vegetal concretes. Constr. Build. Mater. **222**, 852–871 (2019). https://doi.org/10.1016/j.conbuildmat.2019.06.004
16. Ratsimbazafy, H.H.: Evaluation du potentiel de co-produits agricoles locaux valorisables dans le domaine des matériaux de construction. Université Toulouse III - Paul Sabatier, Tarbes (2022)
17. Magniont, C., Escadeillas, G., Coutand, M., Oms-Multon, C.: Use of plant aggregates in building ecomaterials. Eur. J. Environ. Civ. Eng. **16**(sup1), s17–s33 (2012). https://doi.org/10.1080/19648189.2012.682452
18. Evon, P., et al.: Development of a multi-stage process for the collection, fractionation and separation of corn and sunflower stalks to obtain bio-based construction materials. In: 4th International Conference on Bio-Based Building Materials, Barcelona, Spain, p. 1 (2021)
19. Pichon, Q., Naumovic, J.-M.: Guide des bonnes pratiques - Construire en chanvre (2016). https://www.construire-en-chanvre.fr/documents/pdf/documentation/CenC_Bonnes_Pratiques_Tome_2.pdf. Accessed 23 Nov 2020
20. Amziane, S., Collet, F., Lawrence, M., Magniont, C., Picandet, V., Sonebi, M.: Recommendation of the RILEM TC 236-BBM: characterisation testing of hemp shiv to determine the initial water content, water absorption, dry density, particle size distribution and thermal conductivity. Mater. Struct. **50**(3), 1–11 (2017). https://doi.org/10.1617/s11527-017-1029-3
21. Ratsimbazafy, H.H., Laborel-Préneron, A., Magniont, C., Evon, P.: Comprehensive characterization of agricultural by-products for bio-aggregate based concretes formulation. In: 4th International Conference on Bio-based Building Materials, Barcelona, Spain (2021)
22. Monreal, P., Mboumba-Mamboundou, L.B., Dheilly, R.M., Quéneudec, M.: Effects of aggregate coating on the hygral properties of lignocellulosic composites. Cem Concr Compos. **33**, 301–308 (2011)
23. Karaky, H., Maalouf, C., Bliard, C., Polidori, G.: Elaboration and physical characterization of an agro-material based on sugar beet pulp and potato starch. Acad. J. Civ. Eng. **35**(2), 606–612 (2017). https://doi.org/10.26168/icbbm2017.91
24. Van Soest, P.J., Wine, R.H.: Use of detergents in the analysis of fibrous feeds. Part IV. Determination of plant cell-wall constituents. J. Assoc. Off. Agric. Chem. **50**, 50–55 (1967)
25. Van Soest, P.J., Wine, R.H.: Determination of lignin and cellulose in acid-detergent fiber with permanganate. J. Assoc. Off. Agric. Chem. **51**, 780–785 (1968)
26. ISO. ISO 749:1977, Oilseed Residues - Determination of Total Ash; International Organization for Standardization: Geneva, Switzerland (1977)
27. Barnett, J., Jeronimidis, G.: Wood Quality and its Biological Basis. Wiley, Hoboken (2009)
28. Bouasker, M., Belayachi, N., Hoxha, D., Al-Mukhtar, M.: Physical characterization of natural straw fibers as aggregates for construction materials applications. Materials **7**, 3034–3048 (2014). https://doi.org/10.3390/ma7043034
29. Chabannes, M.: Formulation et étude des propriétés mécaniques d'agrobétons légers isolants à base de balles de riz et de chènevotte pour l'éco-construction (phdthesis). Université Montpellier (2015)

30. Ehrnrooth, E.M.L.: Change in pulp fibre density with acid-chlorite delignification. J. Wood Chem. Technol. **4**, 91–109 (1984). https://doi.org/10.1080/0277381840806228531
31. Nozahic, V., Amziane, S., Torrent, G., Saïdi, K., De Baynast, H.: Design of green concrete made of plant-derived aggregates and a pumice-lime binder. Cem Concr Compos. **34**, 231–241 (2012)
32. Qi, C., Hou, S., Lu, J., Xue, W., Sun, K.: Thermal characteristics of birch and its cellulose and hemicelluloses isolated by alkaline solution. Holzforschung **74**, 1099–1112 (2020). https://doi.org/10.1515/hf-2019-0285
33. Rowell, R.M.: Handbook of Wood Chemistry and Wood Composites. CRC Press (2012). https://doi.org/10.1201/b1248734
34. Salas, J., Alvarez, M., Veras, J.: Lightweight insulating concretes with rice husk. Int. J. Cem. Compos. Lightweight Concret. **8**, 171–180 (1986)
35. Sun, C.: True density of microcrystalline cellulose. J. Pharm. Sci. **94**, 2132–2134 (2005). https://doi.org/10.1002/jps.20459
36. Terashima, N., Kitano, K., Kojima, M., Yoshida, M., Yamamoto, H., Westermark, U.: Nanostructural assembly of cellulose, hemicellulose, and lignin in the middle layer of secondary wall of ginkgo tracheid. J. Wood Sci. **55**(6), 409–416 (2009). https://doi.org/10.1007/s10086-009-1049-x
37. Niyigena, C., Amziane, S., Chateauneuf, A.: Multicriteria analysis demonstrating the impact of shiv on the properties of hemp concrete. Constr. Build. Mater. **160**, 211–222 (2018)
38. Chabannes, M., Bénézet, J.C., Clerc, L., Garcia-Diaz, E.: Use of raw rice husk as natural aggregate in a lightweight insulating concrete: an innovative application. Constr. Build. Mater. **70**, 428–438 (2014)
39. Niyigena, C., et al.: Variability of the mechanical properties of hemp concrete. Mater Today Commun. **7**, 122–133 (2016)
40. Laborel-Préneron, A., Aubert, J.E., Magniont, C., Maillard, P., Poirier, C.: Effect of plant aggregates on mechanical properties of earth bricks. J Mater Civ Eng. **29**, 04017244 (2017)
41. Ratsimbazafy, H.H., Laborel-Préneron, A., Magniont, C., Evon, P.: A review of the multi-physical characteristics of plant aggregates and their effects on the properties of plant-based concrete. Recent Progr. Mater. **3**(2), 68 (2021). https://doi.org/10.21926/rpm.2102026

A Minimal Invasive Anchoring Technique for the Foundation of Technical Structures in Trees

Simon Loske[1(✉)], Ingo Muench[1], Panagiotis Spyridis[1], and Martin Zeller[2]

[1] Technical University Dortmund, 44227 Dortmund, Germany
simon.loske@tu-dortmund.de

[2] Technische Überprüfungsgesellschaft mbH, Eichstetten, Germany

Abstract. We present motivation, ideas, experiments, and numerical results for the anchoring of buildings or technical elements into trees. The task here is interdisciplinary since the tree with its biological tissues does not represent a building material in the classical sense but is used for building by the intended use. Therefore, as in medicine, a minimal invasive procedure must impart forces into the hard tissue layers via a technical connector, considering the metabolism as well as the growth processes of the tree. Our hypothesis is, that in this context only a two-bolt solution offers an acceptable anchoring technique to fulfill building laws. The consideration of biological aspects into the solution also has relevance in terms of the building law, as well as the durability and reliability of the connection depends on the vitality of the tree. However, an anchoring technique like that sets up new and ecologically oriented construction methods, which enable the preservation of tree populations in civil development areas.

First, by analysis of existing anchoring techniques different concepts are discussed. Then, the load capacity of a minimally invasive anchoring technique is investigated in a series of load tests including interpretation. Finally, we propose a solution based on two bolts with rigid coupling in order to increase significantly the load capacity of the connection.

Keywords: Construction with trees · fastening · anchoring · tree house building

1 Introduction

1.1 Motivation for Building with Trees

The importance of forested areas for mitigating global problems such as climate change, species extinction, soil erosion and flood events is of increasing importance for our society and is prompting the development of novel concepts such as "urban greening" in the building sector as described in [1], among others. In this context, the preservation and use of existing trees for this new generation of buildings is little discussed, even though such building development only requires a defined intervention and not a complete transformation of the ecosystem.

S. Amziane et al. (Eds.): ICBBM 2023, RILEM Bookseries 45, pp. 87–99, 2023.
https://doi.org/10.1007/978-3-031-33465-8_8

Yet neither in Germany nor in other countries, building laws or technical standards regulate the use of trees. Although wood (technically prepared) has always been important as building material, the use of living trees for the implementation of professional building projects is unusual. One reason for this is, that the development of civilization has had no need for tree-preserving construction methods up to date [11].

Another barrier to use trees for construction is the risk of strong wind events. Official guidelines rely primarily on the trees vitality or simple models but do not serve like engineering models as suggested in [2]. The same issue applies to the anchoring of technical loads into trees. To the best of the authors knowledge, neither technical guidelines nor documented deterministic models on this topic can be found so far.

1.2 Anatomy of Trees and Intervention by Fasteners

Anchoring techniques in trees can be summarized basically into **three categories:**

1. Form-locking force transfer via the stems surface into the wood matrix
2. Friction-based force transfer via the stems surface into the wood matrix
3. Form-locking force transfer directly into the wood matrix

Methods of category 1 & 2 do not remove or penetrate any layer as necessary in category 3. The basic structure of tissue layers in a tree are sketched in Fig. 1. The thin and soft vascular cambium positioned between the inner bark (phloem or cortex) and the sap wood (xylem) as described in [3] devide and diversify new cells to create vascular bundles of transporting tissues and increase stem girth [4, 5]. As a result, trees possess the distinctive and inherent characteristic of secondary thickness growth in radial direction r of the stems cross-section, which we define as r-φ-plane. The (primary) extension growth of trees in height is made principally from just below the topmost apical bud (growing point or "leader") [4]. The stem itself does not elongate in z-direction since it is solid wood.

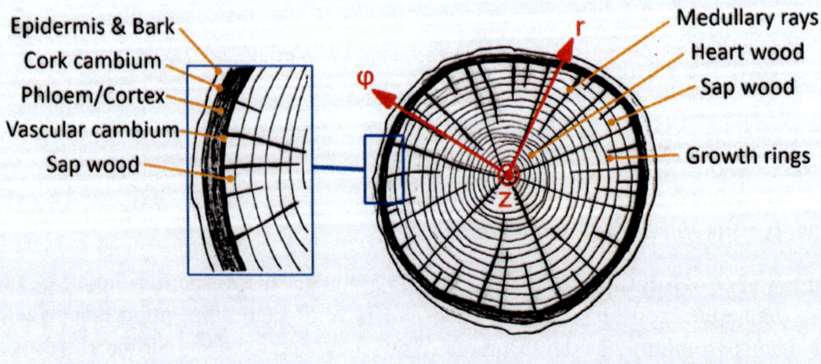

Fig. 1. Cross-section of a stem with layer structure and coordinate system.

The outer bark defines the surface of the stem and protects the soft tissues within [4]. It is more or less resilient to mechanical impact depending on the tree species. While oak trees (Quercus) in general exhibit a thick and robust outer bark, beech trees (Fagus) provide generally a thin and fragile outer bark, although both types belong to the same family (Fagaceae). It implies that any force transfer via the surface of a stem have to pass a series of tissue layers, which are

- different in thickness, strength and load capacity depending on the tree species
- thin and soft in relation to the heartwood.
- subject of the growth process.

Category 1: Methods of form-locking force transfer via the stems surface can be split up in a) anchoring techniques for the initiation of predominantly **vertical** forces and b) anchoring techniques for the initiation of predominantly **horizontal** forces.

a) Form-locking force transfer in predominantly vertical direction via the stems surface is often found in practice by laying textile slings over the shoulder of branch forks, see Fig. 2 a). Due to the natural rounding of the fork, for a wide range of force vectors a sling-position can be found such that the surface normal corresponds to the direction

Fig. 2. **a)** Form-locking force transfer through the stems surface under usage of textile slings. **b)** Result of form-locking force transfer under usage of textile slings when contact stress is greater than the cambium growth pressure. **c)** Result of form-locking force transfer under usage of (covered) steel cables when contact stress is greater than the cambium growth pressure.

of the force vector. Thus, the force transfer takes place almost without friction and thus without shear forces, provided that local effects due to the finite width of the sling as well as friction between the first contact of the sling to the surface and the actual pressure point in the branch fork are disregarded.

If the contact stress between the sling and the stems surface is less than the growth pressure of cambium and cork cambium, which is about 0,7 N/mm2, see [6], natural thickness growth can continue, lifting the sling by the newly grown layers. If the contact stress between the sling and the stems surface is greater than the growth pressure of the cambium, only the layers next to the sling can continue to grow. As a result, unnatural ingrowth and even death of tissue layers occurs, see Fig. 2b). In case of steel cables - even when they are covered with rubber hose as supposed protection of the tree- the contact stress between the sling and the stems surface in the most load cases is greater than the strength of cork or cortex or both, leading to immediate damage of layers, see Fig. 2c).

The mechanical strength of the wood matrix in the area of the branch fork is another complex issue, which is only predictable through exact knowledge of the local geometry as well as an estimation of the fiber course in the wood [10].

b) Form-locking force transfer in predominantly **horizontal** direction via the stems surface is often found in practice by wrapping techniques. To protect the tree in most cases wooden strips were placed between the steel cable and the outer bark of the tree, see Fig. 3a). In fact, the resulting forces are transferred radially into the tree directly through the wooden blocks.

Fig. 3. **a)** Form-locking force transfer through the stems surface using wrapping techniques. **b)** Different grades of failure under usage of wrapping techniques.

Depending on the contact stress between the wooden stripes and the outer bark the tree shows a wide range of reactions, that are from slight damage up to total failure, see Fig. 3b).

Category 2: Especially in adventure parcs established in forests the attachment of technical equipment to living trees via friction-based force transfer very common, see Fig. 4a).

Fig. 4. **a)** Examples of different methods of friction-based force transfer **b)** Failure mechanism of a friction-based force transfer.

For construction projects, friction-based force transfer across the stems surface into the wood matrix is critical for two reasons. First, the coefficient of friction must be provided on the surface. Since the surface is exposed to weathering and the bark can be attacked by fungi, this aspect is problematic. For solid friction (Coulomb friction), furthermore, the frictional force is proportional to the normal force, so that this must also be secured, for example when the trees move in the wind and actually requires clamping. Clamping, however, leads in principle to unnatural growth, damage to the growth tissue, waterlogging, etc., as described in [7–9].

Second, the low shear strength of the outer stem layers calls into question any **category 2** connection. In Fig. 4b) one can see how the cambium layer is peeled away from the wood core by shear stresses from vertical force components.

Category 3 anchorage methods bridge the outer tissue layers through stiff elements and thus the uncertainties and weaknesses listed above. However, tissue layers and wood matrix must be removed locally. Thus, minimall invasive elements are advantageous in order to spare the organism. For practical reasons, the assembly is prepared by drilling holes. Expansion anchors, through rods, and moderately long threaded pins, see Fig. 5 a)– c), are discussed below.

a) **Expansion anchors** achieve their anchoring effect by expansion, which compresses the material around the anchor. This deformation generates ring tensile stresses in the wood matrix above and below the drilled hole, which can cause fibers to delaminate locally in the longitudinal direction and lead to large-scale splitting of the wood matrix. If such splitting occurs, this leads to a loss of the anchoring effect of the expansion anchor, which speaks against its use.
b) **Through rods** require drilling through the complete stem to anchor on the opposite side of loading by means of compressive contact stress. Disadvantage of the usage of

Fig. 5. **a)** Expansion achor, **b)** Through rod, **c)** moderately long threaded pin. **d)** Through rods after mounting. **e)** and 3 years after mounting.

through rods is, that the accessibility and detachability of the connection is getting lost after some time and cannot be monitored any longer, see Fig. 5d) and e).

To make the drill hole as thin as possible, the bolt can be oriented in the direction of the load so that it is only subjected to tensile normal forces, allowing efficient small diameter design. However, the drilled hole will always create a continuous entry point for microorganisms and other subcellular pathogens, which can thus cause widespread decay in the wood. If the drilling is carried out professionally and in a sterile manner, the rod fills the drill hole completely sterile and airtight. If rod and anchor are made of non-corrosive steel, this type of anchoring technique can meet the requirements for safety and durability. But it cannot be called minimall invasive.

c) **Moderately long threaded bolts** are screwed into a blind hole and injure the tissue layers only at one side of the stem [12]. The bolts establish the form fit to the wood matrix via shaft and thread. In principle, the orientation of the bolt is selectable and can follow the loading direction in order to be free of shear force and bending moment. Thus, the angle between the bolt and the wood fiber direction is varied in our experiments in Sect. 2. We usually align the bolt in the r-z plane, see Fig. 6. But it can be inclined by an angle β with respect to the r-φ plane. If the load direction and the centerline of the bolt do not coincide, shear force must also be transferred into the stem via the bolt. This results in contact forces with the wood matrix especially in the outer regions (i.e. new annual growth rings).

The flanks of the thread yield form-locking to the wood matrix. It is recommended to ensure that the blind hole can absorb the displaced wood matrix through the flanks of the thread. Otherwise, undesirable ring pull forces will occur below and above the blind hole, as discussed for expansion anchors.

2 Anchoring with Moderately Long Threaded Bolts

2.1 Basic Structure

Anchoring devices with moderately long threaded bolts are available from various suppliers. They differ in material, thread type, diameter, shaft and other geometric features. The limited length of the thread is due to aspects of minimal invasive drilling and to the technically justifiable screw-in resistance. Generally, such anchors have the following areas and features:

The shaft and thread establish form-locking to the wood in longitudinal (via the thread) and transverse (mainly via the shaft) direction of the bolt. The pull-out resistance can be associated with the form-locking of the thread. In case of shear forces, complex mechanical processes occur, which are primarily based on bending and bedding of the bolt by the wood matrix.

The cause of tensile stress in ring direction of the stem is explained in Fig. 6. This kind of tensile stress can easily delaminate the vertical wood fiber structure below and above the bolt.

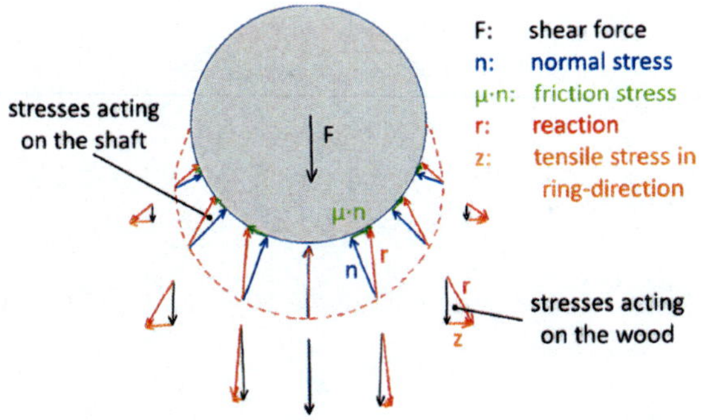

Fig. 6. Formation of ring tensile stresses by compression force

The shaft is followed by the so-called growth reserve section (GRS) of the anchor. Since this section is installed outside the bark or cambium, no or fewer tissue layers have to be removed. Thus, the diameter of the GRS is usually larger than the diameter of the shaft. Due to the secondary growth of the stem, this section becomes gradually enclosed by callus tissue and wound wood, see Fig. 7. Thus, the spatial inclusion of the GRS provides a significantly larger surface for the transfer of shear forces after appropriate time.

The **load connection point** is at the end of the GRS in order to couple construction components like ropes. For pure normal force loading, the distance between the shaft and the load connection point has no influence on the load-bearing capacity of the anchor. However, when shear forces act, this distance leads to a linear increase in the resulting bending moment.

2.2 Specification of the Test Body

In the following experiments threaded bolts with a total length of 220 mm are used. The shaft is 36.5 mm long and has 20 mm in diameter. The trapezoidal thread is 103.5 mm long with 20 mm outer and 16 mm core diameters. The GRS is 80 mm long with 50 mm in diameter. A movable swivel joint with eyelet is used as load connection point.

2.3 Specification of the Test Setup

The anchors are inserted into trunk sections of freshly cut copper beech (Fagus sylvatica). The bark and phloem is removed from the center of these sections by face milling and the hole is drilled for the shaft and thread.

For inclinations $\beta \neq 0°$, a wedge-shaped washer mediates between the face milling and the GRS, compare Fig. 7. We use a servo-hydraulic cylinder with 5 mm/min speed for displacement-controlled loading onto the movable swivel joint. The experiments vary the load application angle α in the interval [0°; 90°], and the inclination angle β in the interval [0°; 30°].

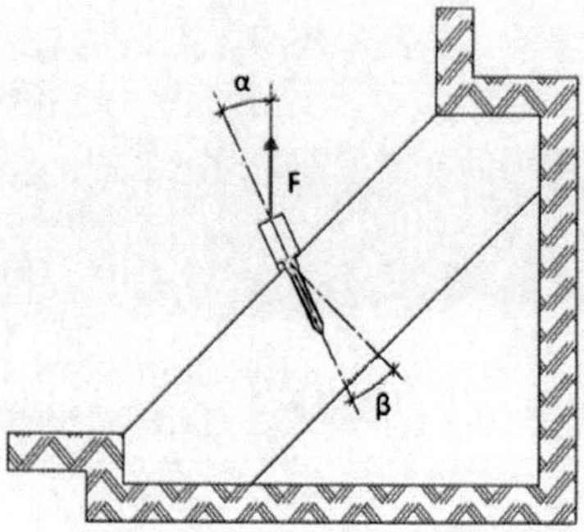

Fig. 7. Schematic representation of the experimental setup

2.4 Axial Pull-Out Test

The axial pull-out test for the anchor is obtained by the load application angle $\alpha = 0°$. We observe decreasing pull-out resistance for increasing inclination angles, see Table 1.

Table 1. Pull-out resistance of the anchor for different inclination angles β

Inclination angle β	0°	15°	30°
Pull-out resistance	58 kN	51 kN	47 kN

A typical force-displacement curve of this test can be seen in Fig. 8 and can be observed similarly for all inclination angles β After reaching the maximum resistance of approx. 58 kN the load curve drops abruptly to less than half (28 kN). This is explained by the loss of the form-locking due to shear-failure of fibers in the area of the thread, since the thread on the anchor presents itself undamaged after the test.

The slightly wavy course after failure has a wavelength of 4 mm, which corresponds to the thread pitch of the anchor. Since the amplitude of the wave is small and continues to decrease with increasing pullout, the residual stiffness of the connection can be attributed to friction between the thread and the wood matrix.

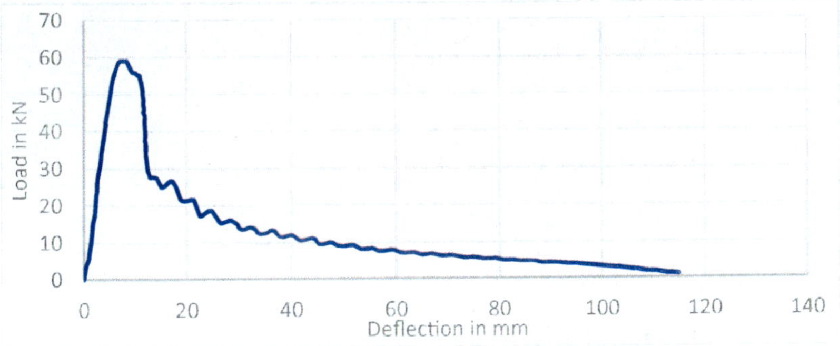

Fig. 8. Typical load deflection curve in the axial pull-out test.

2.5 Shear Force Test and Combined Loading

The pure shear force test requires the load application angle $\alpha = 90°$. Decreasing angles α yield combinations of tensile and shear loading.

Table 2. Shear load resistance of the anchor for different loading angles α

Loading angle α	90°	60°	45°
Load resistance *FE,k*	5 kN	6,5 kN	7,5 kN
Bending moment in the shaft *ME,k*	62,5 kNcm	58,8 kNcm	66,3 kNcm

Under pure shear loading, a system change in the mechanical behavior becomes apparent at approx. 5 kN, which already limits the ultimate load of the anchor, see. Fig. 9 This system change is identified by the attainment of the plastic cross-sectional load capacity of the shaft, which is correspondingly low for a diameter of 20 mm. Neglecting strain hardening effects in the yield zone the shafts plastic moment reads

$$M_{pl} = w_{pl} \cdot f_{y,k} = 1,33\left[cm^3\right] \cdot 35,5\left[kN/cm^2\right] = 47,2[kNcm] \tag{1}$$

The distance between the load connection point and the shaft can be used to determine its bending moment listed in Table 2:

$$M_{E,k} = F_{E,k} \cdot sin(\alpha) \cdot 12,5[cm] \tag{2}$$

Obviously, for all loading angles the shafts plastic load capacity is already exceeded at ultimate load $F_{F,k}$. This is in accordance to the significant loss of stiffness at the beginning of the force-displacement curve (less than 5 mm deflection).

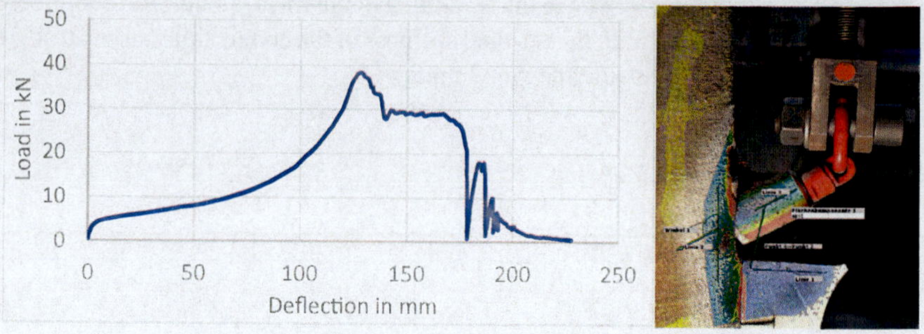

Fig. 9. Typical load deflection diagram (left) and image (right) of the GRS

In the sense of building regulations, the present system (cantilever) must not exceed the ultimate load from the elastic-plastic verification method. This confirms our hypothesis that at least a second anchor is required for practically efficient anchor systems.

Although the load deflection curve in Fig. 9 (left) considerable load increase as well as rotational capacity, such configurations are unacceptable for the state of use. For example, tissue layers close to the surface are damaged by the rotation of the GRS, Fig. 9 (right).

2.6 Basic Conceptual Design of the Coupling

The tests in Sect. 2.5 reveal the strong limitation of the load capacity of a single anchor. An increase of the shaft diameter is undesirable in terms of the minimally invasive anchoring technique. It would also increase the risk of greater ring tensile forces in the stem. However, the load capacity of the connection can be significantly increased by bending-resistant coupling of two anchors via an appropriate coupling element.

The rigid coupling is provided by a sleeve that clamps the GRS of the anchors. Thus, the coupling element is radially expandable and can follow secondary growth of the stem. Furthermore, the sleeves are rigidly connected to a square hollow section, see Fig. 10.

Up to the ultimate load, four yield hinges are necessary in the plastic-plastic verification method. The sequence in which the yield hinges occur depends primarily on the load application angle, the square hollow section and the bedding of the shaft by the wood matrix, which will be discussed in the next chapter using numerical simulations.

2.7 Specification of the Numerical Load Capacity Tests

Our simulations vary the load angle α within a simplified 2D model as suggested by the scheme in Fig. 10. The material properties of the coupling element and the anchors are

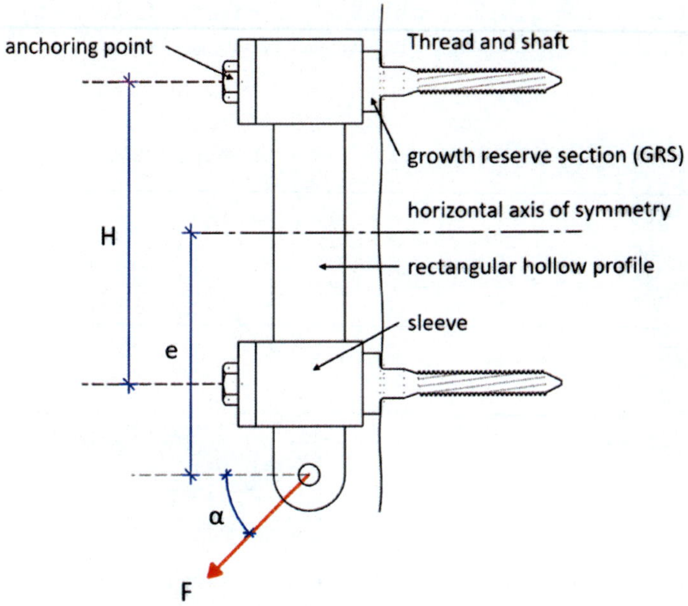

Fig. 10. Scheme of the coupling element

modeled as elastic-plastic material with linear strain hardening following the material data sheet *1.4301*. The material data of the green wood follow a bilinear Drucker-Prager model with values from Table 3. These values are in accordance with literature [13, 14] but partially determined from own experiments.

Table 3. Material parameters of the numerical model for beech

Modulus of elasticity E	7500 kN/cm^2
Shear modulus G	2586 kN/cm^2
Yield point (tension) fy,t	2,7 kN/cm^2
Yield point (compression) $f_{y,c}$	1,6 kN/cm^2
Modulus of hardening Ep	0,038 kN/cm^2

2.8 Calibration of the Model with Horizontal Loading

The model of the coupling element is calibrated with the axial pull-out resistance of a single anchor for purely horizontal loading with $\alpha = 0°$. We account the fact, that the load eccentricity e with respect to the horizontal axis of symmetry increases the pull-out force in the lower anchor. Thus, the horizontal limit load is $F_{max} = 47.3$ kN - generating the maximal pull-out force of 58 kN in the lower anchor.

The load eccentricity also causes a bending moment in the shaft of the lower anchor. However, this is not yet sufficient to close the yield zone, as shown in Fig. 11.

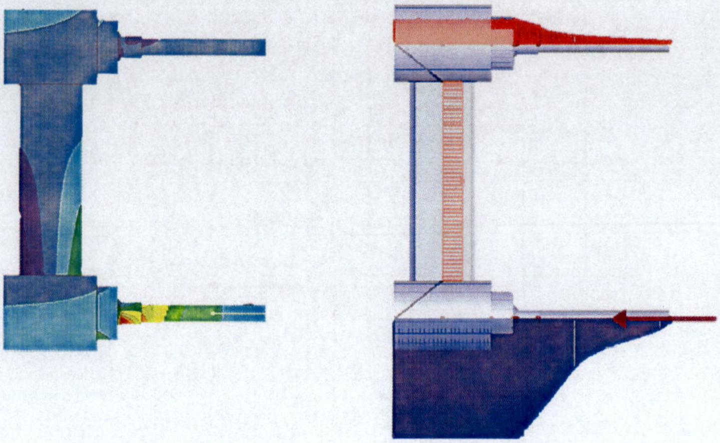

Fig. 11. Equivalent stress (left) and normal force distribution (right) for horizontal limit load.

2.9 Transverse Load

Under pure transverse load with $\alpha = 90°$ the system remains in the elastic state up to a load of 62 kN. The ultimate load is reached at $F = 85$ kN by closing four yield zones. The formation of the yield zones starts in the lower anchor.

The failure mechanism at the limit load is remarkable. The formation of yield zones in both – in the shaft and in the thread - leads to the kinematic sketched in Fig. 12.

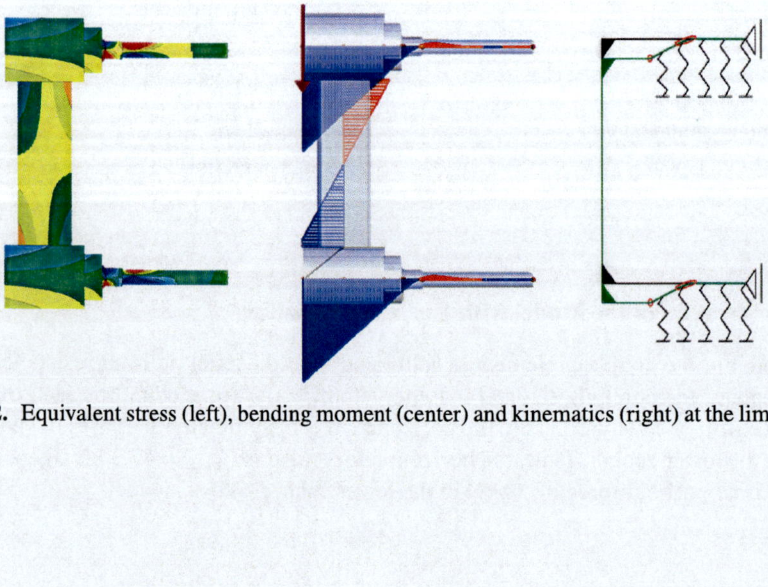

Fig. 12. Equivalent stress (left), bending moment (center) and kinematics (right) at the limit load.

3 Summary and Outlook

We motivate building in and with trees in accordance with building law and technical guidelines. On the one hand, this construction technique can reduce resource consumption and protect the used trees. On the other hand, the used trees can be damaged in the short or long term by improper anchoring techniques. Furthermore, safety-relevant aspects have to be considered, motivating our experiments on the load capacity. They confirm the working hypothesis that professional anchoring in trees requires at least two anchors per point to be safe. Both anchors should be coupled with a bending-resistant device to fully exploit an efficient and minimal invasive anchor system. This is supported by numerical simulations and analysis.

Further work aims load tests for the proposed anchor system in situ. It may result in further optimization of the simulation parameters as well as anchor geometries. Experiments should be performed in different tree species to extend the range of applications.

References

1. Shu, Q., Rötzer, T., Detter, A., Ludwig, F.: Tree information modeling: a data exchange platform for tree design and management. Forests **13**, 1955 (2022)
2. Loske, S., Muench, I.: Experiments and Modelling of the Load Capacity of Green Wood. Accepted for publication in Proc. Appl. Math. Mech. 20(1) (2022)
3. Shigo, A.L.: Tree Anatomy, Shigo & Trees Association (1994)
4. Brickell, Chr., Joyce, D.: Pruning & Training, Royal Horticultural Society (2017)
5. Shigo, A.L., Lang, J., Bernatzky, A.: Die neue Baumbiologie, Haymarket Media (1990)
6. Masselter, T., Speck, T.: Quantitative and qualitative changes in primary and secondary stem organization of aristolochia macrophylla during ontogeny: functional growth analysis and experiments. J. Exp. Bot. **59**, 2955–2967 (2008)
7. Jaffe, M.J.: Thigmomorphogenesis: The response of plant growth and development to mechanical stimulation. Ohio University, Department of Botany (1973)
8. Telewski, F.W.: A unified hypothesis of mechanoperception in plants. Am. J. Bot. 93(10), 1466–1476 (2006). Botanical Society of America
9. Markovic, D., Glinwood, R., Olsson, U., Ninkovic, V.: Plant response to touch affects the behaviour of aphids and ladybirds. Arthropod-Plant Interact. **8**(3), 171–181 (2014). https://doi.org/10.1007/s11829-014-9303-6
10. Mirabet, V., Das, P., Boudaoud, A., Hamant, O.: The role of mechanical forces in plant morphogenesis. Annu. Rev. Plant Biol. **62**, 365–385 (2011)
11. Ludwig, F., Schönle, D.: Wachsende Architektur – Einführung in die Baubotanik, Birkhäuser, Basel, 2023
12. Zeller, M., Muench, I.: Befestigung von Bauwerken in Bäumen mit Baumankern und doppelter Umreifung. Bautechnik **99**(S1), 13–22 (2022)
13. Niklas, K., Spatz, H.: Worldwide correlations of mechanical properties and green wood density. Am. J. Bot. **97**, 1587–1594 (2010)
14. Lavers, G.: The Strength Properties of Timber. Bulletin of the Forest Products Research Laboratory Princes Risborough, England (1969)

Towards Biobased Concretes with Tailored Mechanical Properties

Rafik Bardouh[1](✉), Evelyne Toussaint[1], Sofiane Amziane[1], Sandrine Marceau[2], and Nátalia Martinhão[1]

[1] Clermont Auvergne INP, Institut Pascal UMR 6602 – UCA/CNRS, 63178 Aubière, France
Rafik.bardouh@doctorant.uca.fr

[2] Universit´e Gustave Eiffel, MAST/CPDM, 77454 Marne-la-Vallée, Cedex 2, France

Abstract. In the context of the energy consumption and the climate change issue, a particular interest has been focused on the use of biobased materials in the construction industry because of their low embodied energy and carbon sequestration. The main problem resides in the prediction of the mechanical behavior of biobased concretes because of the Physico-chemical interactions between the plant aggregates and the mineral binders, which can lead to modifications of the hydration mechanisms of this latter. The objective of this work is to determine macroscopic mechanical properties of the plant concrete as well as the local mechanical information by the aim of a digital image correlation (DIC) method. The purpose is then to determine kinematic charts of displacement and surface deformation of the plant concrete specimens and to quantify the moduli and Poisson's coefficient of the plant concrete.

Keywords: Biobased Materials · DIC · Kinematic Charts · Physico-chemical interactions

1 Introduction

The building sector is a major emitter of greenhouse gases. In an attempt to reduce the impact of the construction materials on the environment, the research aims at exploring different alternatives to limit the loss of resources, energy consumption and the release of harmful compounds [1]. In this context, bio-based concretes are positioned as a serious alternative to traditional concrete, with a lower carbon footprint which can be described as environmentally friendly, long-term sustainable, and efficient multifunctional materials [2]. Another justified reason that led to the widespread of the plant concrete is the necessity to develop an efficient building material (thermal insulation, acoustic absorption, and acoustic insulation) with a lower impact on the environment [3].

In France, most of the plant particles that are used in construction are derived from plant stems such as hemp, flax, sunflower, rape, corn, cereals, miscanthus, reed, etc. They can be mixed with a binder (e.g. clay, plaster or lime) to make bio-based concrete or plaster [4]. Hemp shives are the most studied plant aggregates to be incorporated with a mineral binder to form a biobased concrete that is so called hemp concrete

S. Amziane et al. (Eds.): ICBBM 2023, RILEM Bookseries 45, pp. 100–108, 2023.
https://doi.org/10.1007/978-3-031-33465-8_9

[4]. Hemp shives stand out from the rest vegetable fibers due to their availability, low fertilization and irrigation requirements, continuous renewal, good moisture control and environmental sustainability [5].

Many factors have an influence on the mechanical response of green concrete such as the size and the morphology of the aggregates, mineral matrix properties, formulation process, casting process, and aggregates/matrix physicochemical interactions [6]. This latter is the main reason behind the heterogeneous state of green concrete that led to hard prediction of its mechanical behavior [7]. Digital image correlation has shown to be an ideal technique for the study of crack propagation and material deformation as it has demonstrated to be a cost effective, simple and accurate solution leading to a wide range of scientific applications [8]. Thus, this study came to evaluate the mechanical behavior of Hemp concrete by studying the deformation fields through a digital image correlation to detect the mechanical performance of this green concrete.

On a first level, this work aims at characterizing the hygrothermal properties of the hemp shives by the aim of a referred experimental procedure [9]. On a second level, thanks to the compression test, the mechanical properties of the vegetable concrete at the macroscopic scale will be determined. These latter's are then determined by the method of Digital Image Correlation (DIC) followed by a post treatment program in order to determine the local maps of deformation.

2 Material Properties and Mix Proportions

2.1 Materials

Hemp Shives

The hemp shives used in this study were conducted to a series of Granulometric, thermal, and water retention tests to characterize their physical and hygrothermal properties. The protocols of the tests were followed according to the recommendation RILEM TC 236-BBM (Amziane et al. 2017) [9]. The results of the tests applied on the hemp particles are listed in Table 1.

Table 1. Hygrothermal properties of Hemp shives

Bulk density (kg/m^3)	112
Water retention capacity (%)	225
Thermal conductivity (W/(m.K))	0.048

Portland Natural Cement (PNC)

The mineral binder used in the production of the green concrete was the Prompt natural cement (PNC). The mechanical properties of PNC were determined by Vicat according to a w/c ratio of 0.6. The compressive strength and the young's modulus were 10 MPa and 20 GPa respectively [10].

Hemp concrete was designed according to a mixing ratio of 2:1:2 in mass referred to PNC, hemp shives, and water respectively.

2.2 Specimens Preparation

The Hemp concrete specimens were casted in a cubic mold of 150 × 150 × 150 mm^3. At a first step, the hemp shives were Pre-wet by versing half of water amount weighted for the formulation process and hand mixed until they were well humidified. Then, the cement was placed in the same container followed by a manual mixing with the hemp particles. The rest amount of water was then versed and a continuous hand mixing was performed until the hemp particles were well adhered to the cement paste. The resulting mixture was filled in the molds according to three successive layers which were classified by compaction loads after each layer. The specimens were demolded after 10 min and placed in a climatized room under (20 °C and 50% RH) for 28 days before testing.

2.3 Experimental Methodology

Compression Tests

The mechanical characterization of the hemp concrete specimen was carried out using Zwick Roelle Machine of 50 KN load capacity. The force exerted on the specimen was registered by the Zwick software at an acquisition frequency of 10 Hz. The compressions tests realized were a cyclic test of charge - discharge with a global deformation of 1%, 1.5%, 2%, 2.5%, and 3%, which is then followed by recharge until the failure.

The failure mode was setup according to a drop of 80% of maximum load reached. The velocity of charge and discharge was 3 mm/min and 10 mm/min respectively. Figure 1 displays the orientation of the hemp shives as they appeared in perpendicular direction to the direction of compaction.

Digital Image Correlation (DIC)

Digital image correlation was used to determine the macroscopic mechanical properties of the Hemp concrete and to localize the mechanical information on the tested face. The principle of the digital image correlation is allowing non-contact measurements of the deformation of materials [11]. In order to setup DIC technique, a PCO 2000 digital camera with a CCD sensor of 2048 × 2048 $pixels^2$, coded on 14 bits and equipped with a 105 mm lens, was placed in front of the hemp concrete specimen. The images detected by the camera were registered by camware64 software at an acquisition frequency of 1 frames/second. The camera system was launched at the same time as the start of the compression machine. Furthermore, the surface was illuminated by a cold lighting device composed of a bar of LED lamps. The measurement pattern consists of covering the specimen with a black paint speckle in a random way on a white background. The white background is placed in order to have a good luminosity that will be a very necessary condition in the post-processing. SeptD software developed by Pierre Vacher at Symme laboratory was used to calculate the displacement and deformation fields between two images using the gray level distribution at the concerned points [12]. The two main standard parameters required to setup the correlation analysis are the subset

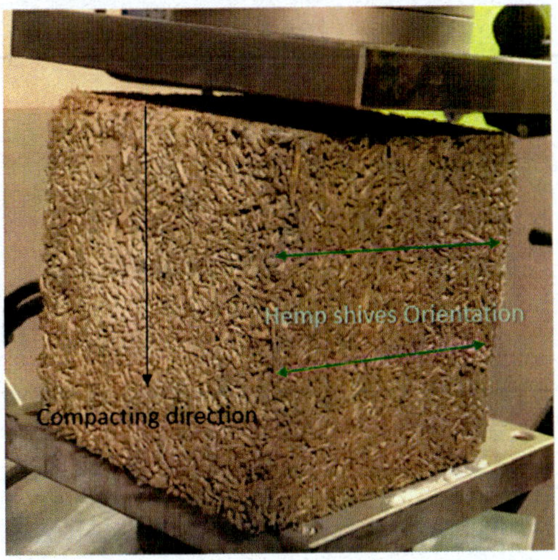

Fig. 1. Hemp concrete specimens showing the direction of load compaction and hemp shives orientation.

size and step size which represent the correlation pattern and the spatial resolution respectively. The value of subset size and step size was 10 pixels for each, i.e., the smallest distance between two independent measurement points. The resolution of the displacement method was evaluated by measuring the standard deviation obtained from the images of the specimen without any motion.

The results are in the order of 0.01 pixels, equivalent to 0,85 μm.

3 Results and Discussion

3.1 Mechanical Behavior of Hemp Concrete

Figure 2 shows the stress-strain compressive cycles with the 6 stages of charging and discharging. The methodology of calculating the young's modulus representing the slope at the zone of recharge as explored at the 5th stage on the curve. It must be noted that each image detected by the camera system during the test was referred to the corresponding amount of force exploited at the curve by an interpolation technique.

The values of the stress and the rigidity exploited from the curve were summarized in the Table 2.

Table 2. Mechanical strength and Rigidity for the different stages of the compression cycles.

	1st Stage	2nd Stage	3rd Stage	4th Stage	5th stage	6th Stage
Rigidity E (MPa)	19.07	23.00	26.77	28.15	28.19	–
σ_{max} (MPa)	0.08	0.14	0.19	0.24	0.28	0.51

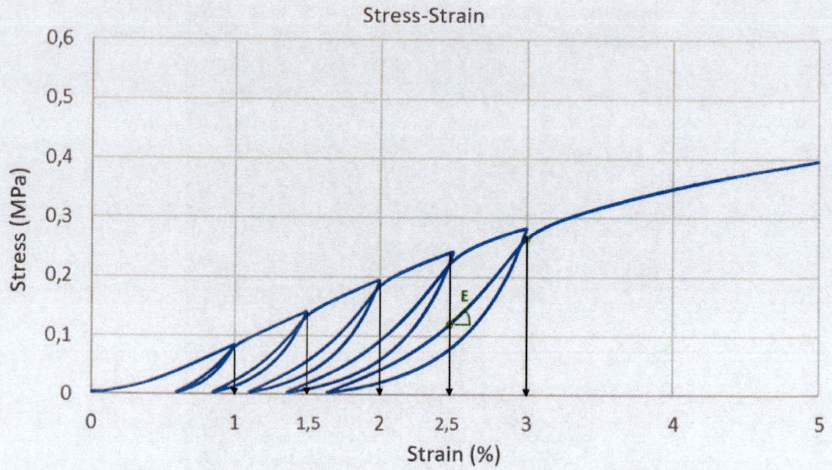

Fig. 2. Stress-Strain curve for the 6 charging stages showing the young's modulus exploitation methodology

Stiffening was the main mechanical behavior obtained from the stress-strain curve as the Young's modulus in the recharge area was increased after each cycle. The Young's modulus values calculated in each cycle are displayed in Table 2. The rigidity increased continuously from 19.07 MPa during the 1st cycle to 28.19 MPa during the 5th cycle. This increase was explained by the evolution of the stress as listed in Table 2 that has reached a gain of 250% between the first compressive cycle and the fifth compressive cycle.

The macroscopic mechanical properties were also determined by the digital image correlation. The mean values of the vertical displacement maps were calculated at each horizontal line along the map. These deformation values obtained were then distributed on a chart as a function of the actual gauge length of the specimen. The modulus of rigidity was determined according to Hooke's law by dividing the stress obtained over the mean values of the vertical deformation. The values of vertical deformation were similarly plotted with respect to the actual gauge length of the specimen and the determined rigidity was the slope between these points. On the other side, the same methodology was applied in the case of the horizontal displacement maps. The main difference is that the values of the horizontal displacement were calculated at each vertical line of the map. These values were similarly distributed on a chart as a function of the actual gauge length of the specimen. Hence, the Poisson's ratio represent the ratio of the mean deformation in x direction over the mean deformation in Y direction. The values of the rigidity and the Poisson's coefficient calculated were listed in Table 3.

The kinematic charts of displacement and deformation that were represented in the following section were determined between the two green curls at the 5th compressive cycle as mentioned in Stress-strain curve at Fig. 2. The following Figs. 3, 4, 5, and 6 were determined between two specific images that are numbered 268 and 283 which correspond to a specific amount of force of 3.25 KN and 6.30 KN, respectively.

Table 3. Rigidity and Poisson's coefficient determined by DIC

Cycles	1	2	3	4	5	6
Rigidité E (MPa)	45.73	45.42	50.19	47	44.32	–
Poisson's coefficient ∨	0.10	0.11	0.13	0.18	0.19	–

3.2 Displacement in the Vertical and Horizontal Directions

Figures 3 and 4 illustrate the displacement maps obtained for the hemp concrete specimen in the vertical and horizontal directions respectively. From a general perspective, large displacement was remarked on the upper portion of the hemp concrete specimens compared to the lower portion. This was due to stress concentration that the upper machine plate exerted on the contact surface with the specimens since it is the only movable plate in the test. A common homogeneous distribution of vertical displacement was seen at both edges of the specimen. This uniformly distribution was due to the flexibility of the hemp concrete to absorb load. However, the horizontal displacement was found to be mainly uniformly all along the hemp concrete specimen with a little more displacement at the right section of the specimen. This evidence can be explained by two main reasons, the first one was the ultimate capacity in horizontal deformation that the specimen has reached at the 5th cycle, and the second one was the orientation of the hemp shives. This latter was the responsible of vulnerability in displacement at a particular zone of the specimen.

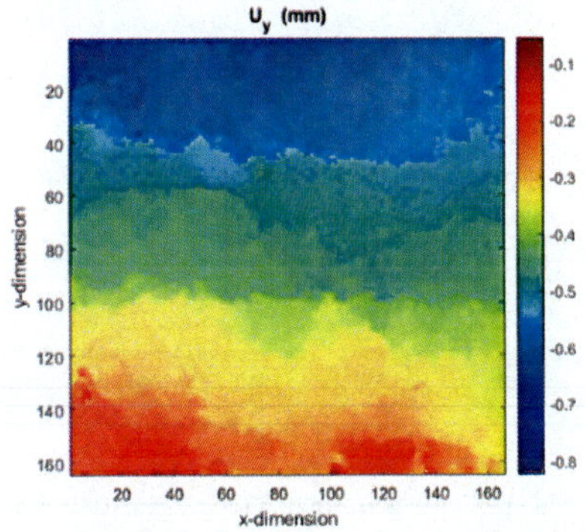

Fig. 3. Vertical displacement map

Figure 5 shows the distribution of the mean values by each line of the vertical displacement map obtained in Fig. 3. These points were then accumulated on a graph and by the aim of a Matlab function, the slope Uy that corresponds to the rigidity of the specimen was plotted. It can be noticed that a typical linear line trace the different vertical displacement points which reflects the behavioral law of elastic mechanical response of hemp concrete specimen. This elastic behavior was justified during the test execution, as the hemp concrete specimen returned to its original form at the discharge zones.

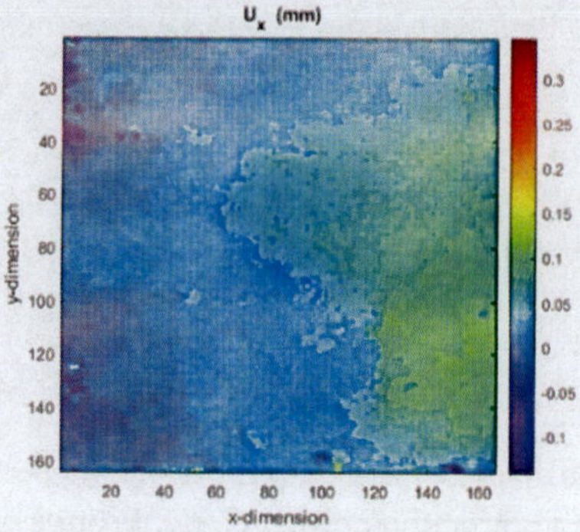

Fig. 4. Horizontal displacement map

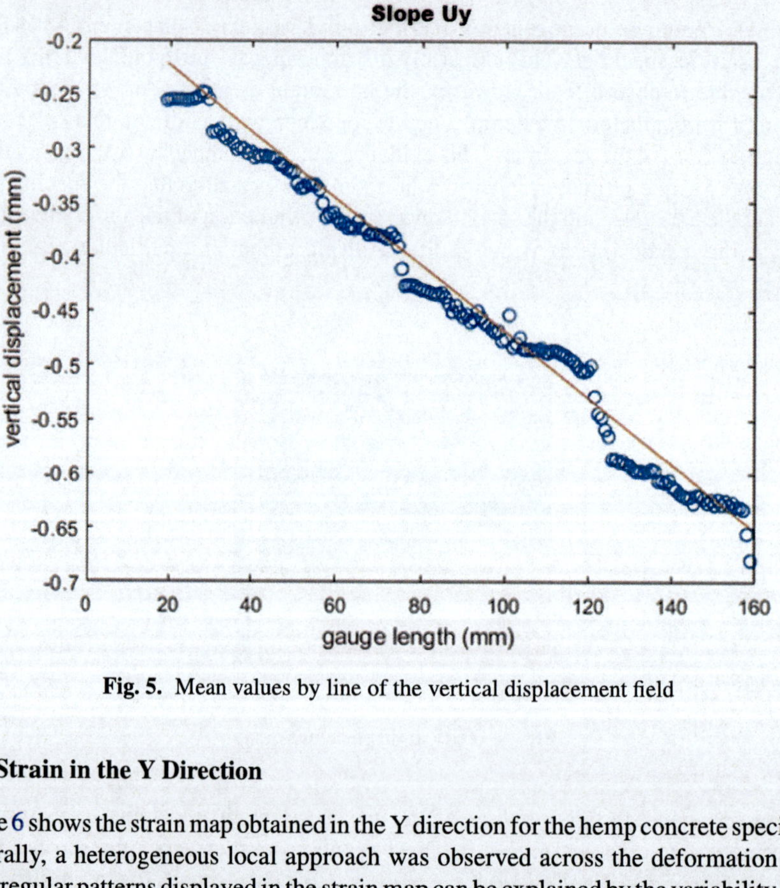

Fig. 5. Mean values by line of the vertical displacement field

3.3 Strain in the Y Direction

Figure 6 shows the strain map obtained in the Y direction for the hemp concrete specimen. Generally, a heterogeneous local approach was observed across the deformation map. The irregular patterns displayed in the strain map can be explained by the variability in the degree of participation of the hemp shives to the mechanical behavior of the specimen. A spotting lines covering certain zones at the deformation map were remarked. These lines refer to the 3 layers laid during the specimen preparation process. It can be noticed that the interface zones between the 3 different layers consists of a strain concentration zones in terms of vertical deformation. These particular zones explain the need of rubbing the end of each layer at specimen's edges after compacting forces to ensure a proper distribution of the hemp particles. This latter will then lead to an optimized mechanical properties. It must be noted that the negative sign shown on the map scale was attributed to the direction of the test execution which was downward and so opposite to the conventional Y-axis.

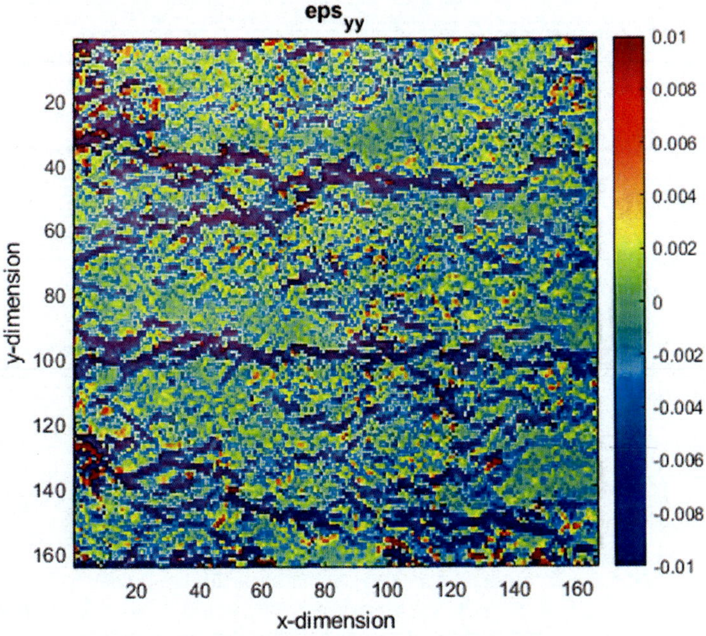

Fig. 6. Strain map in vertical direction

4 Conclusion

Based on the displacement and deformation maps obtained from the digital image correlation, the following points can be concluded:

1. A global heterogeneous response was observed in the deformation of hemp concrete.
2. The interfaces between the layers represent a concentration zone in the deformation of the specimen.
3. The mechanical stress-strain behavior of the specimen was characterized by a stiffening after each cycle.
4. As a result of the imperfect arrangement of hemp shives, these particles has a major influence on the mechanical response of hemp concrete based on their degree of participation.

5. New perspectives will be taken under investigation concerning the parallel orientation of hemp shives along the hemp concrete.

Acknowledgement. The work in this study was supported by the ANR BIO-UP project of the French Agence Nationale de la Recherche (ANR-21-CE22-0009).

References

1. Sandrine, U.B., Isabelle, V., Hoang, M.T., Chadi, M.: Influence of chemical modification on hemp–starch concrete. Constr. Build. Mater. **81**, 208–215 (2015)
2. Chinnu, S.N., et al.: Reuse of industrial and agricultural by-products as pozzolan and aggregates in lightweight concrete. Constr. Build. Mater. **302**, 124172 (2021)
3. Bourdot, A., et al.: Characterization of a hemp-based agro-material: influence of starch ratio and hemp shive size on physical, mechanical, and hygrothermal properties. Energy Build. **153**, 501–512 (2017)
4. Ratsimbazafy, H.H., et al.: A review of the multi-physical characteristics of plant aggregates and their effects on the properties of plant-based concrete. Recent Prog. Mater. **3**(2), 69 p. (2021)
5. Pantawee, S., et al.: Utilization of hemp concrete using hemp shiv as coarse aggregate with aluminium sulfate [Al2 (SO4) 3] and hydrated lime [Ca (OH) 2] treatment. Constr. Build. Mater. **156**, 435–442 (2017)
6. Niyigena, C., et al.: Variability of the mechanical properties of hemp concrete. Mater. Today Commun. **7**, 122–133 (2016)
7. Diquélou, Y., et al.: Impact of hemp shiv on cement setting and hardening: influence of the extracted components from the aggregates and study of the interfaces with the inorganic matrix. Cem. Concr. Compos. **55**, 112–121 (2015)
8. McCormick, N., Lord, J.: Digital image correlation. Mater. Today **13**(12), 52–54 (2010)
9. Amziane, S., Collet, F., Lawrence, M., Magniont, C., Picandet, V., Sonebi, M.: Recommendation of the RILEM TC 236-BBM: characterisation testing of hemp shiv to determine the initial water content, water absorption, dry density, particle size distribution and thermal conductivity. Mater. Struct. **50**(3), 1–11 (2017). https://doi.org/10.1617/s11527-017-1029-3
10. http://www.romanportland.net/files/doc/cahier_technique_cr_cnp_eng.pdf
11. Surrel, Y.: Les techniques optiques de mesure de champs: essai de classification. Instrumentation mesure et métrologie **4**, 11–42 (2004)
12. Cuynet, A., Toussaint, F., Roux, E., Scida, D., Ayad, R.: Apport des mesures de champ dans l'étude de composites renforcés en fibres de lin au cours d'essais de traction quasi-statiques. In: 1ère Conférence EuroMaghrébine des BioComposites (2016)

Physical Properties of Bio-based Building Materials

Optimisation of Production Parameters to Develop Innovative Eco-efficient Boards

Eleonora Cintura[1,2](✉), Paulina Faria[1], Luisa Molari[3], and Lina Nunes[2,4]

[1] CERIS, Department of Civil Engineering, NOVA School of Science and Technology, NOVA University of Lisbon, Caparica, Portugal
ecintura@fct.unl.pt

[2] Structures Department, National Laboratory for Civil Engineering, Lisbon, Portugal

[3] Department of Civil, Chemical, Environmental and Materials Engineering, DICAM, Alma Mater Studiorum, University of Bologna, Bologna, Italy

[4] CE3C, Centre for Ecology, Evolution and Environmental Changes and CHANGE, Global Change and Sustainability Institute, University of the Azores, Angra do Heroísmo, Portugal

Abstract. Laboratory tests were carried out to define production parameters of innovative eco-efficient composites made up of hazelnut shells as aggregate and a sodium silicate solution as adhesive. The aim was to maximize the content of bio-aggregates and minimize the amount of adhesive, guaranteeing the feasibility of producing samples. Therefore, after preliminary testing, the percentages of hazelnut shells and the sodium silicate solution were kept constant: 70% and 30% of the total volume, respectively. However, the characteristics of the considered composites did not allow the production of uniform samples. The sodium silicate solution was not rapidly absorbed by the bio-aggregates; during the drying process, it was deposited on the bottom side of the samples. The uniformity of the samples is required to guarantee a correct evaluation of their performance and future homogeneous panels. Hence, different production parameters were investigated, such as drying at T = 60 °C or T = 80 °C during different periods of time, and the addition of different percentages of sodium bicarbonate was also considered. The visual analysis during drying, the final uniformity, thus the distribution of the sodium silicate solution, the resistance and the crumbliness of the samples allowed defining the best production process. The most uniform sample was selected, and its production parameters can thus be applied to produce innovative composites.

Keywords: Bio-waste · Hazelnut shells · Sodium silicate

1 Introduction

The use of panels such as particleboards as building materials has several benefits. Besides the easy on-site installation, the economic advantages, and the several applications, there is the possibility of employing different materials as aggregates. Among them, recycled materials and by-products derived from agro-industrial practices [1, 2]. Considering the high environmental impact caused by both the construction sector and

S. Amziane et al. (Eds.): ICBBM 2023, RILEM Bookseries 45, pp. 111–122, 2023.
https://doi.org/10.1007/978-3-031-33465-8_10

agriculture [3], the production of agricultural waste-based materials seems to be an efficient solution that encourages a circular economy system [4]. Several studies addressed the feasibility of producing panels by using agro-industrial wastes, for example, cork, coffee chaff, rice husk and corn cob [5–7]. Many of the developed composites met the requirements of the standards, proving to be a good alternative to conventional materials.

In the development of these innovative composites, the origin of the considered agro-industrial wastes is extremely important, too. Using local by-products lowers, even more, the environmental impact, moderating the transport variable. Considering previous work [8] hazelnut shells were selected as agro-industrial waste. Indeed, according to FAO [9], the first three producers of hazelnut in 2020 were Turkey, Italy, and the United States. They produced 665,000 tonnes, 140,560 tonnes, and 64,410 tonnes, respectively.

Besides the production, availability was considered. The seasonality of agricultural products could be an obstacle to their supply. Nevertheless, hazelnuts are available all over the year, as reported by the industry. Indeed, they are harvested in August and September, and they are left to dry naturally or mechanically [10], a process that can require some months. The drying process guarantees several benefits, such as the improvement of the chemical and physical stability of the food and a greater resistance to mould growth [10, 11]. After drying, the hazelnuts are normally stored. The maximum storage period varies depending on the storage conditions and it could be even 48 months [10, 12]. Hence, there is a large quantity of hazelnut shells available, throughout the year, and not only during the harvesting season.

Another important parameter to produce eco-efficient composites, such as boards and panels, is the selected adhesive [13, 14]. A sodium silicate solution, also known as water glass, was used for its several benefits, such as harmlessness for human health (formaldehyde-free adhesive), high resistance to mould, and the prevention of chemical decomposition [15–17]. However, it has also some drawbacks, such as high hygroscopicity, and hence low moisture resistance [18, 19]. Furthermore, the sodium silicate solutions may be highly fluid and require much time for the hardening process. To improve the bonding and the setting time some strategies can be considered, such as heat or chemical treatments [20]. Several additives may accelerate the hardening process of the composites (e.g., sodium carbonate, sodium bicarbonate, aluminium sulphate [21, 22]), as well as drying at high temperatures.

Starting from this knowledge, the feasibility of producing boards made up of hazelnut shells as aggregates and sodium silicate solution as the adhesive was investigated. Practical tests were carried out to define the best production process. Different parameters were analysed, such as drying the samples at different temperatures for different durations, and the addition of sodium bicarbonate intended to accelerate the hardening process. The problem was the slow drying process of the sodium silicate solution and the slow absorption by the hazelnut shells. During the hardening phase, the adhesive was deposited on the bottom side of the samples making them non-uniform. Therefore, an experimental campaign was performed with the aim to produce uniform samples required to secure a correct evaluation of the composites' properties, and future production of panels.

2 Materials

2.1 Hazelnut Shells

The hazelnut shells were provided by Raccolti di Cin, Baldissero d'Alba (CN), Italy (Fig. 1a). Practical tests (not detailed for the sake of brevity) were carried out to evaluate the most suitable grain size to produce samples. The use of the hazelnut shells as they were provided did not allow cohesion between the aggregate and the binder. Hence, they were shredded by using a mechanical mill to have grain sizes mainly between 4 mm and 8 mm (Fig. 1b), similar to a previous study [23].

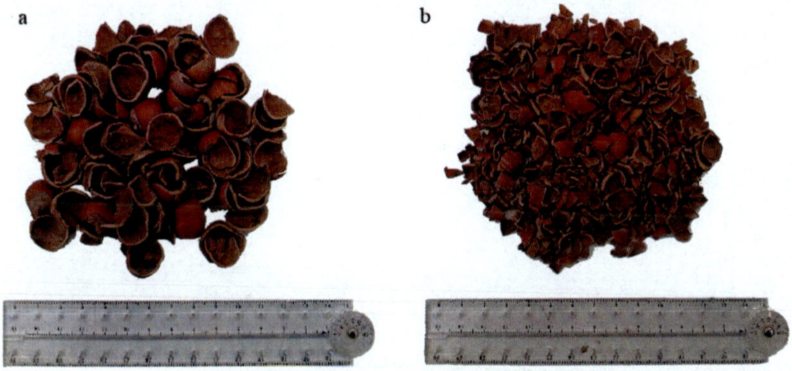

Fig. 1. Hazelnut shells used as aggregates: a) before shredding; b) after shredding.

Hazelnut shells were dried at T = 60 °C until constant mass (change in mass after 24 h less than 0.1%) and characterized according to the recommendation of RILEM Technical Committee 236-BBM "Bio-aggregate-based Building Materials" [24] and past work [23]. The initial water content was (6.5 ± 0.2)%; the loose bulk density was (469.3 ± 5.8) kg/m^3; the particle size distribution is reported in Fig. 2.

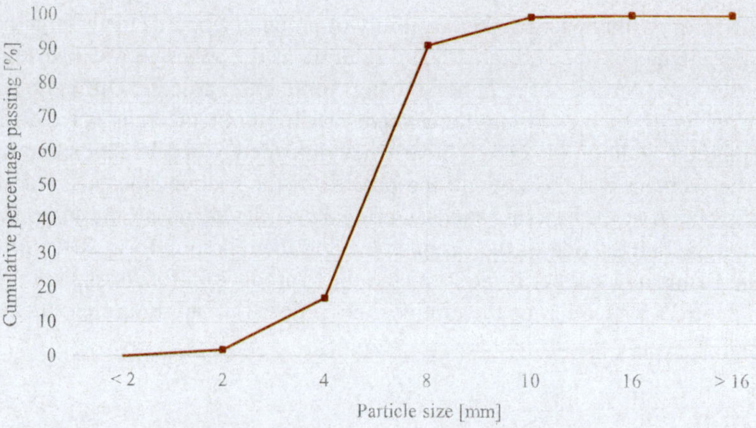

Fig. 2. Particle size distribution of the shredded hazelnut shells.

2.2 Sodium Silicate Solution and Sodium Bicarbonate

The sodium silicate solution was provided by Ingessil, Montorio (VR), Italy, which also carried out the chemical analysis. Table 1 reports the characteristics of the sodium silicate solution.

Table 1. Characteristic of the sodium silicate solution provided by Ingessil [25].

Property	Value
Weight ratio	2.4
Density [°Bè]	46.45
Molar ratio	2.48
Sodium silicate concentration [% p/p]	41.33
SiO_2 [% p/p]	29.17
Na_2O [% p/p]	12.16
Density [g/ml] at T = 20 °C	1.471
pH (T = 20 °C)	12.40

The sodium bicarbonate, $NaHCO_3$, was produced by Crastan S.p.A., Pontedera (PI), Italy [26]. This is a commonly used sodium bicarbonate. Its use will be justified and detailed in Sect. 3.

3 Production Parameters

As previously described, the mix design was selected trying to maximize the content of hazelnut shells and minimize the sodium silicate solution. The selected ratio was 70% of aggregates and 30% of adhesive (by total volume). Lower quantities of adhesive did not guarantee mechanical resistance.

The first samples were produced by mechanically mixing the hazelnut shells and the sodium silicate solution. Then, the mixture was put into moulds and left air-drying. As Fig. 3 reports, different moulds were tested: standaradized prismatic metallic (4 cm × 4 cm × 16 cm), quadrangular wooden home-made, and silicone ones (both with 10 cm × 10 cm × 4 cm high). The aim was to avoid the sample's bonding to the moulds.

Fig. 3. The considered moulds to produce the samples: metal, wooden and silicone.

The bonding of the samples was easily avoided by using the silicone mould. As for the other moulds, materials that guarantee the remotion of the samples were needed (e.g., resins or anti-glueing materials). The benefit of the wooden and the metal moulds was the possibility of opening them, demoulding the sample without moving it and leaving its sides drying. As a result, the silicone mould and the wooden mould covered by baking paper were selected as the best solutions.

Besides the type of mould, the uniformity of the samples was investigated. It is required for a correct evaluation of the samples' performance. Since the sodium silicate solution was not rapidly absorbed by the hazelnut shells, depositing on the bottom side of the samples, the hardening process had to be accelerated.

Different production parameters were investigated, such as drying at temperatures between T = 60 °C and T = 80 °C and the addition of sodium bicarbonate ($NaHCO_3$) as solid reactive. The mixtures of hazelnut shells and sodium silicate solution (and eventually sodium bicarbonate) were placed in the silicone moulds (8.8 cm × 5 cm ×

2.5 cm high) and 12 different production parameters were considered and compared (A to L). They are reported in Table 2, with the composites' designation. The samples were demoulded after 3 days.

Table 2. Sample's designation and description of the production parameters.

Composites' designation	Production parameters	Description
A	Reference	Dring at laboratory conditions without moving the sample
B	Rotation	Dring at laboratory conditions by rotating the sample each 3 h
C	Temperature	Dring at T = 60 °C for 2 h
D	Addition of reactive	Addition of 25% of $NaHCO_3$ (by vol. of sodium silicate solution); drying at laboratory conditions without moving the sample
E	Addition of reactive, rotation	Addition of 25% of $NaHCO_3$ (by vol. of sodium silicate solution); drying at laboratory conditions by rotating the sample each 3 h
F	Addition of reactive, temperature, rotation	Addition of 25% of $NaHCO_3$ (by vol. of sodium silicate solution); drying at T = 60 °C for 1 h by rotating the sample every 30 min
G	Temperature, rotation	Drying at T = 80 °C for 1.5 h, by rotating the sample each 30 min
H	Temperature	Drying at T = 80 °C for 1.5 h, without moving the sample
I	Addition of reactive, temperature	Addition of 10% of $NaHCO_3$ (by vol. of sodium silicate solution), drying at T = 60 °C for 1 h
J	Addition of reactive, temperature	Addition of 2% of $NaHCO_3$ (by vol. of sodium silicate solution), drying dry at T = 60 °C for 1 h
K	Addition of reactive, temperature, rotation	Addition of 2% of $NaHCO_3$ (by vol. of sodium silicate solution), drying at T = 60 °C for 30 min by rotating the sample every 15 min; then, drying at laboratory conditions by rotating the sample each 1 h
L	Addition of reactive, rotation	Addition of 2% of $NaHCO_3$ (by vol. of sodium silicate solution), drying at laboratory conditions by rotating the sample each 1 h

The visual analysis during drying allowed defining the most efficient production parameters. The final uniformity, hence, the distribution of the sodium silicate solution, the resistance and the crumbliness of the composites were evaluated. The production parameters that guarantee to have the most uniform samples were selected for future composites' production.

4 Results and Discussion

Figure 4 shows the samples of each composite after demoulding.

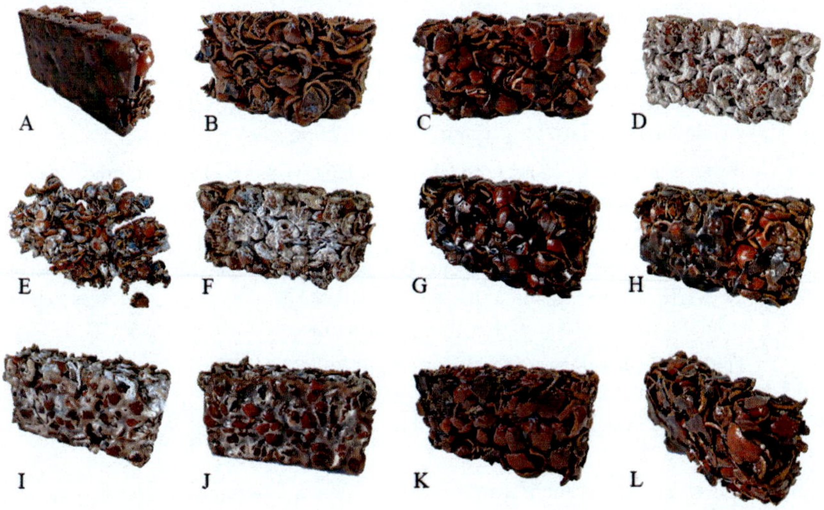

Fig. 4. Representative composite samples produced by using different production parameters after demoulding, from A to L.

Table 3 described the results of the visual analysis and the considerations during the drying process, after three days from the production and after demoulding.

Table 3. Results of the composite samples visual analysis.

Composite	Visual analysis
A	The sample was not uniform, and it was not dried after 3 days; the sodium silicate solution was mainly on the bottom side; six days were required to demould
B	The sample was quite uniform; in some parts, the sodium silicate solution was not completely uniformly distributed; overall, it was throughout the sample; demoulding was possible after 3 days; the sample was dried after 6 days
C	The sample was quite uniform; in a few parts, the silicate was not completely uniformly distributed (less than for sample B), but overall, it was throughout the sample; 3 days were required to demould; the sample was dried after 6 days
D	The sample seems to be uniform, even if the sodium silicate solution was not perfectly distributed; since after 2 days the sample was dried, the $NaHCO_3$ accelerated the drying process; however, after 6 days, the sample started to release the $NaHCO_3$, probably added in too high quantities; the $NaHCO_3$ was strongly visible on the surface of the sample as a white powder
E	It was broken after 1 day; probably the $NaHCO_3$ absorbed the sodium silicate solution decreasing its bonding performance; the $NaHCO_3$ was visible on the surface of the sample as a white powder
F	After 1 h the sodium silicate seemed to be completely dried but the quantities of the $NaHCO_3$ were too high; the sample did not seem resistant and it was highly crumbly; the $NaHCO_3$ was visible on the surface as a white powder
G	The sample was quite uniform; the sodium silicate solution was not completely uniformly distributed, but overall, it was throughout the sample; demoulding was possible after 3 days; after 6 days the sample was dried; it seemed similar to sample C
H	The sample was not uniform; the sodium silicate solution was deposited on the bottom surface, as happened for sample A; when the drying process finished, it was similar to sample C, even if less uniform
I	The sample did not seem uniform at the end of the drying process and after demoulding; the sodium silicate solution was on the bottom side of the sample; the $NaHCO_3$ was highly visible on the surface of the sample as a white powder
J	The sample was not uniform; the sodium silicate solution was on the bottom part of the sample; it seemed similar to sample I, but more resistant (probably due to the fewer quantities of $NaHCO_3$); due to the less employed quantities, the $NaHCO_3$ was less visible on the surface of the sample
K	The sample was not completely uniform after 3 days; however, the sodium silicate solution was throughout the sample and quite distributed; it seemed similar to sample C; due to the less employed quantities, the $NaHCO_3$ was less visible on the surface of the sample

(*continued*)

Table 3. *(continued)*

Composite	Visual analysis
L	The sample was not uniform; in many parts of the bottom side, there were high quantities of the sodium silicate solution; demoulding was possible after 3 days; the sample seemed similar to C; the $NaHCO_3$ was not highly visible on the surface of the sample

As shown in Fig. 4 and considering the descriptions reported in Table 3, the more relevant production parameters seemed to be the temperature and the rotation. The addition of sodium bicarbonate did not improve the hardening process. Furthermore, the sodium bicarbonate could determine a lower bonding capacity (e.g., sample E). This was in line with the results achieved by Lee and Thole [18]. These researchers analysed the properties of the sodium silicate used as a binder for particleboard production, modified by several additives, including sodium bicarbonate. The researchers concluded that the addition of sodium bicarbonate decreased the bonding performance.

The most uniform samples were C and G, hence a combination of the two production parameters was selected. To secure a lower environmental impact, the lower temperature, T = 60 °C, was chosen. The rotation of the sample was considered every 30 min for 3 h instead of only 2 h. These production parameters accelerated the hardening process of the sodium silicate, guaranteeing its distribution throughout the sample.

Thus, the production process was defined. First of all, the bio-aggregates and the sodium silicate solution were mechanically mixed for 10 min, until homogeneity. Then, the mixture was placed without compaction in silicone moulds or wooden moulds covered by baking paper. The mixture was levelled by using a spatula and the moulds were closed by a wooden top. After that, the samples were dried at T = 60 °C for 3 h, being

Fig. 5. Production process: a) mechanical mixing; b) wooden mould; c) levelling; d) closure of the wooden mould; e) silicone mould; f) closure of the silicone mould; g) drying at T = 60 °C; h) demoulding.

rotated every 30 min. Finally, they were dried at laboratory conditions, rotated every 30 min and demoulded after 2 days.

Figure 5 shows the production of two samples. Both wooden and silicone moulds are reported.

Finally, after 28 days of curing, the samples were put at T = 50 °C until reaching a constant mass (variation in mass after 24 h less than 0.5%) to secure a complete drying. This final procedure is defined considering past studies [4, 17].

5 Conclusions

Production parameters to produce boards made up of hazelnut shells as aggregates and sodium silicate solution as adhesive were tested. The drying at different temperatures for different periods, the rotation of the samples and the addition of sodium bicarbonate as an additive were assessed. The work aimed at producing uniform samples, required for a correct evaluation of their properties and future production.

The following conclusions were achieved:

- Both drying with thermal treatment (T = 60 °C and T = 80 °C) and the addition of sodium bicarbonate accelerated the hardening process of the sodium silicate solution.
- High quantities of sodium bicarbonate worsened the bonding properties of the sodium silicate solution; the produced samples were more crumbly and less resistant.
- Rotating the samples before the sodium silicate solution was completely dried allowed its distribution throughout the sample and increased the final uniformity.

None of the samples achieved perfect uniformity. This is probably due to the selected materials and the manually controlled production process (the impossibility to ensure a constant rotation of the samples during drying). Nevertheless, the combination of temperature and rotation seemed to be the best production parameters.

As a result, the production process to have uniform samples was defined and tested. It can be used to produce composites for future analysis of these innovative building products.

Acknowledgements. This research was funded by FCT, the Portuguese Foundation for Science and Technology, with Ph.D. grant PD/BD/150579/2020, as part of the Eco-Construction and Rehabilitation Program (EcoCoRe). The authors are grateful for the FCT support through funding UIDB/04625/2020 of the research unit CERIS and for the support of the project BIO-FIBRE funded by the Erasmus+ Programme of the European Union.

The authors acknowledge Lorenzo Coraglia (Raccolti di CIN) for giving information about the production and harvesting of hazelnut and for donating the hazelnut shells to carry out the research project. The authors acknowledge Dr. Mirko Braga (Ingessil) for his help and suggestion to correctly use the sodium silicate solution as adhesive and Prof. Stefania Liuzzi (Poliba) to provide suggestions to carry out the research project.

References

1. Gürü, M., Karabulut, A.F., Aydın, M.Y., Bilici, İ: Processing of fireproof and high temperature durable particleboard from rice husk. High Temp. Mater. Process. **34**(6), 599–604 (2015). https://doi.org/10.1515/htmp-2014-0092

2. Owodunni, A.A., et al.: Adhesive application on particleboard from natural fibers: a review. Polym. Compos. **41**(11), 4448–4460 (2020). https://doi.org/10.1002/pc.25749
3. Duque-Acevedo, M., Lancellotti, I., Andreola, F., Barbieri, L., Belmonte-Ureña, L.J., Camacho-Ferre, F.: Management of agricultural waste biomass as raw material for the construction sector: an analysis of sustainable and circular alternatives. Environ. Sci. Eur. **34**(1), 70 (2022). https://doi.org/10.1186/s12302-022-00655-7
4. Liuzzi, S., Rubino, C., Stefanizzi, P., Martellotta, F.: Performance characterization of broad band sustainable sound absorbers made of almond skins. Materials **13**(23), 5474 (2020). https://doi.org/10.3390/ma13235474
5. Ricciardi, P., Belloni, E., Merli, F., Buratti, C.: Sustainable panels made with industrial and agricultural waste: thermal and environmental critical analysis of the experimental results. Appl. Sci. **11**(2), 494 (2021). https://doi.org/10.3390/app11020494
6. Ramos, A., et al.: Thermal performance and life cycle assessment of corn cob particleboards. J. Build. Eng. **44**, 102998 (2021). https://doi.org/10.1016/j.jobe.2021.102998
7. António, J., Tadeu, A., Marques, B., Almeida, J.A.S., Pinto, V.: Application of rice husk in the development of new composite boards. Constr. Build. Mater. **176**, 432–439 (2018). https://doi.org/10.1016/j.conbuildmat.2018.05.028
8. Cintura, E., Nunes, L., Esteves, B., Faria, P.: Agro-industrial wastes as building insulation materials: a review and challenges for Euro-Mediterranean countries. Ind. Crops Prod. **171**, 113833 (2021). https://doi.org/10.1016/j.indcrop.2021.113833
9. FAO: FAOSTAT Crops. https://www.fao.org/faostat/en/#data/QCL. Accessed 30 Nov 2022
10. Ansell, M.P., et al.: Natural Plant-Based Aggregates and Bio-Composite Panels with Low Thermal Conductivity and High Hygrothermal Efficiency for Applications in Construction, Non-conventional and Vernacular Construction Materials, 2nd (edn.) Elsevier (2020). ISBN: 9780081027042. https://doi.org/10.1016/B978-0-08-102704-2.00010-X
11. Vrtodušić, R., Ivić, D., Jemrić, T., Vuković, M.: Hazelnut postharvest technology: a review. J. Cent. Eur. Agric. **23**(2), 423–454 (2022). https://doi.org/10.5513/JCEA01/23.2.3493
12. Markuszewski, B., Bieniek, A.A., Wachowska, U., Bieniek, A., Krzymińska, I.: Effect of biological treatment used before harvesting and storage methods on the quality, health and microbial characteristics of unripe hazelnut in the husk (*Corylus avellana* L.). PeerJ **10**, e12760 (2022). https://doi.org/10.7717/peerj.12760
13. Younesi-Kordkheili, H., Pizzi, A.: Improving the physical and mechanical properties of particleboards made from urea–glyoxal resin by addition of pMDI. Eur. J. Wood Wood Prod. **76**(3), 871–876 (2017). https://doi.org/10.1007/s00107-017-1242-3
14. Ansell, M.P., et al.: Natural Plant-Based Aggregates and Bio-Composite Panels with Low Thermal Conductivity and High Hygrothermal Efficiency for Applications in Construction, Non-conventional and Vernacular Construction Materials, 2nd edn., Elsevier (2020). ISBN 9780081027042
15. Nordström, J., et al.: Concentration- and pH-dependence of highly alkaline sodium silicate solutions. J. Colloid Interface Sci. **356**(1), 37–45 (2011). https://doi.org/10.1016/j.jcis.2010.12.085
16. Li, L., Tang, X., Ouyang, Q., Tao, N.: Combination of sodium dehydroacetate and sodium silicate reduces sour rot of citrus fruit. Postharvest Biol. Technol. **151**, 19–25 (2019). https://doi.org/10.1016/j.postharvbio.2019.01.006
17. Liuzzi, S., Rubino, C., Martellotta, F., Stefanizzi, P., Casavola, C., Pappalettera, G.: Characterization of biomass-based materials for building applications: the case of straw and olive tree waste. Ind. Crops Prod. **147**, 112229 (2020). https://doi.org/10.1016/j.indcrop.2020.112229
18. Lee, S.J., Thole, V.: Investigation of modified water glass as adhesive for wood and particleboard: mechanical, thermal and flame retardant properties. Eur. J. Wood Wood Prod. **76**(5), 1427–1434 (2018). https://doi.org/10.1007/s00107-018-1324-x

19. Song, L., Liu, W., Xin, F., Li, Y.: Study of adhesion properties and mechanism of sodium silicate binder reinforced with silicate fume. Int. J. Adhes. Adhes. **106**, 102820 (2021). https://doi.org/10.1016/j.ijadhadh.2021.102820
20. Fawer, M., Concannon, M., Rieber, W.: Life cycle inventories for the production of sodium silicates. Int. J. Life Cycle Assess. **4**(4), 207 (1999). https://doi.org/10.1007/BF02979498
21. Wang, Y., He, F., Wang, J., Hu, Q.: Comparison of effects of sodium bicarbonate and sodium carbonate on the hydration and properties of portland cement paste. Materials **12**(7), 1033 (2019). https://doi.org/10.3390/ma12071033
22. Liu, S., Shen, Y., Wang, Y., He, H., Luo, S., Huang, C.: Synergistic use of sodium bicarbonate and aluminum sulfate to enhance the hydration and hardening properties of portland cement paste. Constr. Build. Mater. **299**, 124248 (2021). https://doi.org/10.1016/j.conbuildmat.2021.124248
23. Cintura, E., Faria, P., Duarte, M., Nunes, L.: Bio-wastes as aggregates for eco-efficient boards and panels: screening tests of physical properties and bio-susceptibility. Infrastructures **7**(3), 1–26 (2022). https://doi.org/10.3390/infrastructures7030026
24. Amziane, S., Collet, F., Lawrence, M., Magniont, C., Picandet, V., Sonebi, M.: Recommendation of the RILEM TC 236-BBM: characterisation testing of hemp shiv to determine the initial water content, water absorption, dry density, particle size distribution and thermal conductivity. Mater. Struct. **50**(3), 1–11 (2017). https://doi.org/10.1617/s11527-017-1029-3
25. Ingessil: https://www.ingessil.com/it. Accessed 30 Nov 2022
26. Crastan S.p.A: https://www.crastan.it/. Accessed 30 Dec 2022

Biobased Façade Materials in Europe

Francisco Ortega Exposito[1](✉), Fred van der Burgh[2], and Willem Böttger[1]

[1] Centre of Expertise Biobased Economy, 63, 4818AJ Breda, The Netherlands
fa.ortegaexposito1@avans.nl, franorex@gmail.com
[2] Agrodome B.V., 42a, 6703BT Wageningen, The Netherlands

Abstract. On one hand, the majority of the construction materials utilized in mainstream applications require a considerable amount of energy to produce, resulting in a large carbon footprints and unsustainable processes. These issues arise from the use of virgin materials, the utilization of critical raw materials, and the generation of a substantial volume of waste. The construction sector needs a solution.

On the other hand, the amount of agricultural waste being produced in Europe is substantial, around 400 million tonnes of agricultural waste are produced each year, and is projected to increase to 488 million tonnes by 2050 [1]. Research intends to show the products of an inter-industrial relationship, showing how agricultural waste, can be upcycled in the construction industry, producing diverse biobased materials suitable for construction, and thereby solving two key problems. The facet of construction this research focuses on is the biobased materials in Europe for use in façades.

The research was divided into two main activities: desk research and taking up contact with country representatives on the subject of Biobased Building research throughout Europe. The research for façade materials in Europe commenced by obtaining information from previous research - Performance of Bio-Based Façades by Fred van der Burgh & Sissy Verspeek, to then use the findings and advice to orient the investigation [2]. These and other materials were further analyzed by obtaining manufacturers' and suppliers' information on the materials.

Compiling all this information allowed us to produce a list of the biobased façade materials available in Europe. To show the usefulness in construction, certain aspects of the materials were first-order investigated following criteria set in the case-study: biobased percentage, technical lifespan, fire resistance class, weather resistance, vandalism resistance, and form freedom. Information was obtained via technical sheets and interviews of the manufacturers.

The results of this investigation show that there are fifty European biobased façade material available. The results likewise show that most of these biobased materials have suitable attributes for use in façades. These results provide an indication of what are the next steps to optimize their use.

Based on these statements, it can be concluded that there is an increasing diversity of biobased façade materials in production in Europe, and that the characteristics of these biobased façades are suitable for the construction sector.

Keywords: Biobased material · Biobased construction · Biobased façade · Biobased Europe · Construction Europe

S. Amziane et al. (Eds.): ICBBM 2023, RILEM Bookseries 45, pp. 123–143, 2023.
https://doi.org/10.1007/978-3-031-33465-8_11

1 Introduction

The introduction chapter includes background information on biobased materials, the goal and aim for this project; information on the client and experts reviewing this paper.

1.1 Background Information

Biobased materials are growing in interest, the rate of new developments on these materials is increasing substantially, at par with its demand. The current global state requires clean and sustainable materials, especially in an industry which has an incredibly high impact [3]. Biobased materials can also provide a strong connection to nature, which is associated with restorative processes and nature-inclusiveness, which is rapidly becoming a significant aspect of construction [3].

Currently, there are several biobased materials on the market. They can be used in most construction aspects, including interior and exterior construction). Common biobased materials used in construction are materials produced with (not exclusive to): wood, wood fiber laminates, insulation materials like flax and hemp. For façades the products range from pure (wood), (bio-)composites, a combination of two or more (bio)materials), such as wood and flax composite, and/or can be processed such as thermally modified wood, dependent on the application. One of the possible applications of biobased materials in the construction industry is the façade [2].

1.2 Goal

Produce a list of biobased façade materials manufactured and/or available within Europe, with information relating to the requirements set by ProRail.

1.3 Aim

The research aims to raise awareness, understanding, and acceptance of the biobased façade materials market, while highlighting real clients' preferences in materials.

1.4 Client

To better relate the attributes to a real-life situation, and give an idea of what clients are looking for in today's market, this paper is oriented towards a case study regarding an investigation carried out by the Centre of Expertise on Biobased Economy (CoE BBE).

ProRail is a Dutch governmental organization responsible for the maintenance and extension of the national railway network infrastructure [4]. They ensure safe and efficient operations, while coordinating with train operators and promoting sustainable transport [4]. They approached the CoE BBE with the request of a list of biobased materials to replace the façade of their obsolete rail-road related buildings. The requirements set by ProRail are used as criteria for the materials included in the list.

1.5 Experts

This paper and investigation have been team-analyzed and co-authored by Willem Böttger and Fred van der Burgh, experts in the industry of biobased construction.

Fred van der Burgh has been dedicated to driving the transition towards ecological and biobased building materials in the construction industry for over 20 years. His work focuses on advancing knowledge development and dissemination, shaping policy and regulations, and addressing market issues. To support his mission, he has become a certified LCA specialist and an expert in biological building practices.

Willem Böttger's work has focused on developing sustainable materials and technologies that can help to reduce the environmental impact of building and construction while also promoting economic growth and innovation. Over 20 years of research into the technical and practical aspects of using sustainable composites in structural applications, in combination with exploration of the economic and environmental benefits of sustainable and biobased materials.

2 Case Study

ProRail is responsible for the maintenance and management of the main railway infrastructure in the Netherlands. This includes approximately 1300 rail-road related buildings containing systems and installations that support the functioning of the railway network.

One of those types concerns the substations in which transformers and rectifiers are located. A large part of these substations is technically obsolete and must be replaced in the coming years. ProRail intends to have a standard design and consists of a biobased and modular basic unit on which different types of facades can be placed, depending on the location. The aim is that sustainability, environment and technology come together in the design.

The Rail-bound Buildings Handbook expresses the ambition that rail-bound buildings are part of the local ecological system and contribute to biodiversity [5]. ProRail wants to gain insight into which biobased materials are most suitable for a nature-inclusive approach in facade construction. Figure 1 shows the façade design produced by Studio Marco Vermeulen for ProRail. The design features a 3D railway pattern and strategically placed openings to encourage biodiversity. The design additionally incorporates a living box, although outside the scope of this research, which is meant to serve as a habitat for species residing within the facade.

2.1 Requirements

The requirements for the case study were set by client ProRail in conjunction with Studio Marco Vermeulen. These requirements relate to the façade's design and application.

Biobased Material

The materials must be biobased. The material must have a certain biobased percentage; referring to products that mainly consist of a substance (or substances) derived from living matter (biomass), and either occur naturally (for example wood and fibres) or are

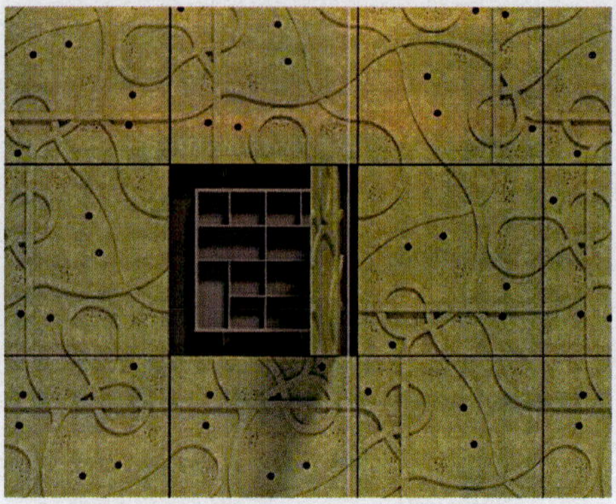

Fig. 1. ProRail's Façade design [4]

synthesized, or it may refer to products made by processes that use biomass (for example bio-composites) [6]. Typically, the term refers to modern materials that have undergone more extensive processing. Materials with low biobased percentage are also included in the list. This requirement is the key emphasis of this paper.

Fire Safety Classification

B-fire safety class based on EN-13501-1 standard is required for the materials. This safety classification is selected as benchmark due to most of the regulations in Europe requiring a minimum of a B fire safety class [7]. EN-13501-1 consists of four different experiments dependent on the fire behaviour of the material [7]. Materials of all fire safety classifications are included on the list, as the aim of this paper is to show the vast number of available biobased materials.

Weather Durable

Façades protect buildings against weather conditions [2]. Weather durability is determined by the ability of materials to resist UV light, temperature, humidity and rain [2]. The facades protect the buildings against the weather conditions. This implies that the materials in the facade should be able to cope with water, sand, salty conditions, extreme temperature changes, air pollution and frost conditions. All of the materials present in the list are believed or accredited to be weather durable.

Form-Freedom

Form-freedom refers to the ability of the façade material to be produced in any shapes and 3-D elements can be formed in the surface. The reason for this specific criteria is the design of the panels. Figure 2 shows the proposed design features of the new façade of ProRail's rail-road related buildings [4]. These panels are modular and can be placed in different orientations to produce different patterns.

Vandalism Resistant

This requirement focusses on the material's capability to resist vandalism in terms of impact, leaning and scratching. It is a requirement set for the material by Pro Rail, as their

Fig. 2. ProRail rail-road related building façade design [4]

current façades face several vandalism acts [4]. The list therefore only includes materials that are assumed to be resistant to vandalism based on their material composition.

Technical Readiness Level (TRL)

TRL is a measure of a material's development stage in terms of technical maturity [8]. It can relate directly to the amount of available certifications of the material for example for fire safety, and to the volumes of the materials. Figure 4 in the Appendix I illustrates the different levels of TRL. The client would like to place a high number of certified façades, therefore require a high technical readiness level. The list only includes materials with TRL > 8.

Lifespan

ProRail have requested for a material which has a technical lifespan higher than 40 years. The technical lifespan of the materials was, in some cases, assessed via their previous usage and composition. The list includes materials with any technical lifespan, as to reliably conserve a high number of material entries.

EPD Availability

This requirement assesses the availability of an environmental product declaration (EPD). It should be noted that the objective of this research is not to evaluate or quantify environmental impact. Rather, this requirement aims to demonstrate manufacturers' transparency and the number of materials that have an EPD available. If an EPD is available, it is hyperlinked for easy access.

To ensure compliance with their regular procurement rules, ProRail employs the environmental shadow costs of materials, which are determined based on the results presented in the EPD, as a crucial criterion for material selection. Thus, manufacturers who possess an EPD have a higher chance of securing contracts with ProRail. Furthermore, other large contractors like RWS also consider these shadow costs when making procurement decisions.

3 Methodology

The information from the materials was obtained via technical sheets (desk research), and interviews with manufacturers, practical research was not in the scope. Only materials manufactured and/or available within Europe are present in this list.

3.1 Desk Research

The desk research was conducted in three stages. The first stage being an orientation phase, which involved conducting a literature review to establish a starting point, gain familiarity with the biobased material market, and identify materials that have yet to be thoroughly researched. In the second stage, reliable sources of information for biobased materials, namely biobased material databases, were identified. Finally, in the last stage, relevant data was retrieved, according to the requirements of the case study. Figure 3 simplifies the phases of the desk research.

Fig. 3. Phases of the desk research.

The article produced for the 3rd edition of the ICBBM – Performance of biobased façades by Sissy Verspeek & Fred van der Burgh was selected for orientation [2]. This article became the starting point of the research, and considered the predecessor of this paper. The article consists on an overview of the properties of eight different biobased façades, and provides an advice on which biobased materials have potential and should be researched more thoroughly.

During phase 2, the identification of reliable sources of information for biobased materials concluded in the discovery of five databases were Biobased Consultancy (EU), Biobased Inkopen (Netherlands), Biobased Bouwen (Netherlands), Die Nachwachsende Produktwelt (Germany) and AgroBioBase (France) (Table 3 in the Appendix II).

In the final phase, data was retrieved from the five identified databases. In cases where the required information was not available through the databases, manufacturers' websites were examined to obtain the necessary data. The retrieved data was then matched with the requirements specified in the case study.

3.2 Interviews

The interviews conducted as part of the research helped improve both the quality and quantity of data gathered for the investigation. During the desk research, it was discovered that some materials lacked sufficient data, prompting contact with manufacturers to obtain further information. To ensure impartiality, this aspect of the research is kept confidential.

Furthermore, the interviews conducted as part of this study facilitated an expansion of the information horizon to encompass the entire European continent. Inquiry was made with research centers and experts from multiple European countries regarding the availability of biobased materials within their respective countries or networks. The successful contact interactions yielded the discovery of new materials and provided additional expertise on known materials. Table 4 in Appendix III lists the interviewed research centres and experts per country.

3.3 Data Analysis

Throughout the investigation, data regarding biobased materials was gathered using a combination of desk research and interviews. This information was directly pertinent to the requirements specified in the case study. The majority of the data was readily accessible through databases or communication with manufacturers. When data was unavailable, an alternative method was used to make informed assumptions, which involved examining the material type and production method.

4 Biobased Façade Materials

The list consists of 50 biobased materials available in Europe, seen in Table 1, starting on the page below. The established criteria are divided into identificatory (material, country, material type and short description) and qualitative (biobased %, technical lifespan, form-freedom, fire class (according to EN 13501-1) and EPD availability. This chapter presents the catalog of biobased materials.

Table 1 indicates that there are a high number of biobased materials currently in the market. The materials range includes natural and modified materials. All of the materials have some suitable properties, following the requirements of the case study. Only Nabasco 8010 complies with all the requirements in this case study, with Resysta complying with most of the requirements, except partial form-freedom.

Several other materials such as Accoya, Cellon and many others on the list comply with all but one requirement, in most cases form-freedom. Therefore, in a façade where form-freedom is not a requirement, these materials would be suitable. The most important requirement is fire safety class B, as it is a legislative requirement for façades in most European countries [7]. Around half of the materials included in the list do meet these requirements (in most cases using fire retardants).

Table 1. European Biobased Façade Material List

Material (*Country*)	Material Type	Short Description	Biobased %	Technical Lifespan	Form-freedom	Fire safety class (EN 13501-1)	EPD Available
Accoya (*NL*)	Modified wood	Acetylated wood (Radiate Pine from New Zealand) for external cladding	100%	60 years	Not form-free	{B-s1,d0}	✓
Accoya Shou Sugi Ban (*NL*)	Modified wood	Charred timber wood siding panels	100%	60 years	Not form-free	{B-s1,d0}	✓

(*continued*)

Table 1. (*continued*)

Material (*Country*)	Material Type	Short Description	Biobased %	Technical Lifespan	Form-freedom	Fire safety class (EN 13501-1)	EPD Available
Atmosphere 175 (*FR*)	Wood-plastic composite	Wood composite dust 67% & 33% (mass volume) polyethylene (PE) cladding boards made of different timbers	67%	>40 years*	Partially form-free (extrusion)	E	✓
Bamboo wall panelling (*DE*)	Other	Pre-oiled and processed stick glued bamboo façade which grows to harvest maturity within 5 years and is therefore a resource-saving material	90%	>50 years	Not form-free	D-C*	✗
BioBased tiles (*NL*)	Mineral composite	85% industrial waste stream aggregate (granite) and 15% BioCement, produced from bacteria	≈15%	>40 years*	Form-free	C-B*	✓
Brimstone façade (*UK*)	Modified wood	Thermally modified wood made from British-grown timber, superheated in an oxygen-depleted environment	100%	>30 years	Not form-free	{B-s1,d0}	✓
Cellon (*SZ*)	Biocomposite	Compact high-pressure panel (HPL): 70% cellulosic web and 30% phenolic resin	70%	>40 years*	Not form-free	B-s1,d0	✗
Cempanel (*UK*)	Mineral composite	69% cement and 31% fine wood particles with trace amounts of process additives and mineralizing agents,	≈31%	10–30 years*	Not form-free	B-s1,d0	✓

(*continued*)

Table 1. *(continued)*

Material (*Country*)	Material Type	Short Description	Biobased %	Technical Lifespan	Form-freedom	Fire safety class (EN 13501-1)	EPD Available
CLT Ekoflin (*NL*)	Wood	CLT consists of 99.4% sustainably produced wood and 0.6% of ecologically sound PU glue	100%	>50 years	Not form-free	D-C*	✗
Cross Trade Wood Facades (*DE*)	Wood	Ayous (African hardwood) timber cladding product	100%	>40 years*	Not form-free	D-C*	✗
DBP Siding (*NL*)	Wood-plastic composite	72% wood fibre from wastewood and 28% polypropylene	72%	>50 years	Partially form-free (extrusion)	D (improving to C),	✗
Emsiem wooden panels (*BG*)	Wood	Softwood harvested (Lamperia) from regulated places in the Rhodope Mountains, with molded elements	100%	>40 years*	Not form-free	D-C*	✗
Frake Noir (*NL*)	Modified wood	Thermally modified and coloured Movingui (from West Africa)	100%	>40 years	Not form-free	D-s2,d0	✗
Hasslacher Norica CLT (*AT*)	Wood	Cross-laminated timber (from birch)	100%	>50 years	Not form-free	B-s2,d0	✓
Hemp Fibre Corrugated Sheets (*UK*)	Biocomposite	Hemp non-woven fibre blended with farm bio-waste resin that mainly consists of corn cob, oat hulls, bagasse	100%	10–30 years*	Not form-free	C-s2,d0	✗
Hempcrete Block (*BE*)	Biocomposite	Hemp binded with concrete, very high thermal performance and a very low production energy content. Has been used as an external façade in one project	60%	10–30 years*	Form-free	B-s1,d0	✓

(continued)

Table 1. (*continued*)

Material (*Country*)	Material Type	Short Description	Biobased %	Technical Lifespan	Form-freedom	Fire safety class (EN 13501-1)	EPD Available
Holonite façade (*NL*)	Mineral composite	90% natural resources to 10% fossil resources. In total, 54% of the binding agent is waste from the vegetation and animal feed industry	≈5%	>40 years*	Not form-free	C-B*	✗
Hygrothermolytic wood (*NL*)	Modified wood	Hygrothermolytic process maintains an intermediate elevated steam pressure (4–7 bar), adapted to the applied steam temperature (160–180 °C), keeping wood moisture in equilibrium with the steam conditions	100%	>50 years	Not form-free	D-C*	✗
Imola Legno Laminated solid wood panels (*IT*)	Wood	Recycled plywood panels made up of mutually glued wooden boards	≈70%*	>50 years	Not form-free	D-C*	✗
Isokurk Expanded Insulation Corkboard (*PT/BE*)	Biocomposite	A Portuguese (from Amorim) and Belgian product, expanded cork insulation which can be used as a façade with different designs	100%	10–30 years*	Not form-free	E	✓
Isokurk spray cork (*BE*)	Biocomposite	The natural cork fibers (96%) are processed into an emulsion and are sprayed quickly and accurately on the surface by means of a textile spraying machine	96%	10–30 years*	Form-free (applied on another material, no mechanical properties by itself)	D-C*	✗

(*continued*)

Table 1. *(continued)*

Material (*Country*)	Material Type	Short Description	Biobased %	Technical Lifespan	Form-freedom	Fire safety class (EN 13501-1)	EPD Available
Kerloc (*NL*)	Biocomposite	Kerloc is a high-quality fibre-reinforced (Dutch pinewood) cold ceramic material with a natural appearance	≈50%*	>50 years	Form-free	A1-s1, d0	✓
Kerto® LVL (*FI*)	Modified wood	Structural laminated veneer lumber used in all types of construction projects. The veneers are bonded with weather- and boil-resistant phenol formaldehyde adhesive (7%) to form a continuous bille	93%	>40 years*	Not form-free	{B-s1,d0}	✓
Lunawood Thermowood (*FI*)	Modified wood	Thermally modified wood using heat and steam	100%	30 years	Not form-free	{B-s1,d0}	✓
MOSO Bamboo X-treme (*NL*)	Other	Large boards, planks or beams of thermally treated fibre bundles of Giant bamboo trunks harvested after 4–5 years	95%	>50 years	Not form-free	B-s1,d0	✓
Nabasco 8010 (*NL*)	Biocomposite	Natural fibers, filler from waste streams and a largely biobased resin	80%	>40 years*	Form-free	B-s1,d0	✗
Naturclad-W (*SP*)	Modified wood	A laminated wood board for exteriors that requires zero maintenance and can be installed as ventilated façade	100%	>40 years	Not form-free	B-s2,d0	✓

(continued)

Table 1. (*continued*)

Material (*Country*)	Material Type	Short Description	Biobased %	Technical Lifespan	Form-freedom	Fire safety class (EN 13501-1)	EPD Available
Neolife Cover (*FR*)	Biocomposite	Made of VESTA x.fire, which contains 85% of wood fibres, mineral resins, antioxidants and mineral pigments	85%	10–30 years*	Not form-free	B-s3,d0	✗
Nobelwood FRX (*NL*)	Modified wood	Timber fully impregnated with water soluble biopolymers made from bagasse in a high pressure treatment plant	100%	>50 years	Not form-free	{B-s1,d0}	✗
Platowood Façades (*NL*)	Modified wood	Platonised (hydrothermal modification) cladding produced from Fraké (West Africa), Spruce (Netherlands) & Poplar (Europe)	100%	>50 years	Not form-free	{B-s2,d0}	✗
Pura NFC (*NL*)	Biocomposite	Ventilated façade made of 57% cellulose and 43% phenol resin	57%	>40 years*	Not form-free	B-s2,d0	✓
Regge Hout Douglas Natural Cladding (*NL*)	Wood	Natural cladding from douglas wood	100%	>50 years	Not form free	{B-s1,d0}	✗
Resysta (*DE*)	Biocomposite	Resysta consists of approx. 60% rice husks, an agriculturally waste product. 40% PVC	60%	>40 years*	Partially form-free (extrusion)	{B-s1,d0}	✓
Robinia Natural Cladding (*NL*)	Wood	Consists of natural black locust cladding	100%	>50 years	Not form free	{B-s1,d0}	✗
Scotlarch panels (*UK*)	Wood	Timber cladding from Scottish Larch	100%	50–100 years	Not form-free	{B-s1,d0}	✓
Scottywood Thermowood (*UK*)	Modified wood	Thermally modified timber (Radiata pine, Nordic softwoods, Ayous, Frake and Ash)	100%	>40 years	Not form-free	{B-s1,d0}	✗

(*continued*)

Table 1. (*continued*)

Material (*Country*)	Material Type	Short Description	Biobased %	Technical Lifespan	Form-freedom	Fire safety class (EN 13501-1)	EPD Available
Sivalbp Woods (*FR*)	Wood	Several different natural wood products, these include woods from douglas fir, larch, nordic spruce, nordic pine and western red cedar. Has 4 different styles, new age, elegance (no nordic spruce), vintage (o*NL*y nordic pine), and authentic (only red douglas fir, larch and western red cedar)	100%	50 years (Nordic spruce 30 years)	Not form-free	D-C	✗
Stora Enso Thermowood (*FI*)	Modified wood	Heat-treatment involves heating wood material to a temperature of 180–220 °C while at the same time protecting it with steam. The steam protects the wood, but it also influences the chemical changes taking place in wood	100%	>50 years	Not form-free	D-s2,d0	✓
Techniclic (*LX*)	Wood	Different wood cladding for facades, with modular instalation. Modular mounting system is made of plastic	100%	>40 years*	Not form-free	D-C*	✗
Tholin Millboard Envello facade cladding (*NL*)	Wood	Oak and cedar natural wood cladding	100%	>50 years	Not form-free	{B-s1,d0}	✗

(*continued*)

Table 1. (*continued*)

Material (*Country*)	Material Type	Short Description	Biobased %	Technical Lifespan	Form-freedom	Fire safety class (EN 13501-1)	EPD Available
Trespa Meteon (*NL*)	Biocomposite	HPL consisting of layers of cellulose impregnated with thermosetting resins and surface layer(s) on one or both sides, with a transparent topcoat cured with electron beam curing	≈50%*	>40 years*	Not form-free	{B-s2,d0}	✓
Tricoya (*UK*)	Modified wood	Acetylated wood (accoya) with reduced swelling and shrinkage for façade MDF panelling	100%	60 years	Not form-free	{B-s1,d0}	✓
Vertical Grain Siberian Larch (*UK*)	Wood	Naturally durable and aesthetically refined timber cladding	100%	>40 years*	Not form-free	{B-s1,d0}	✓
Vingui Noir (*NL*)	Modified wood	Thermally modified and coloured Movingui (from Finland)	100%	>40 years	Not form-free	{B-s1,d0}	✗
Vivix (*UK*)	Biocomposite	Solid phenolic wood material produced from tough thermosetting resins reinforced with cellulose fibre for added strength and durability	≈30%	>40 years*	Not form-free	{B-s1,d0}	✓
Vuré Noir (*NL*)	Modified wood	Thermally modified and coloured Vuren (from Northern Europe)	100%	>40 years*	Not form-free	F	✗
Wind Hout Siberian Larch Natural Cladding (*NL*)	Wood	Natural wood cladding from siberian larch	100%	>50 years	Not form-free	{B-s1,d0}	✗

(*continued*)

Table 1. (*continued*)

Material (*Country*)	Material Type	Short Description	Biobased %	Technical Lifespan	Form-freedom	Fire safety class (EN 13501-1)	EPD Available
Zwarthout Yoroi (*NL*)	Other	Planks are made by compressing bamboo fibres (from China). The bamboo planks are individually furnace charred by han, creating an outside layer of char	92%	>50 years	Not form-free	B-s1,d0	✗

* Assumption; { } Using fire resistant
Notice that all the materials are hyperlinked to the manufacturer's webpage
EPDs are hyperlinked if they exist (✓), and (✗) if they do not exist or are unavailable
AT - *Austria.* BE - *Belgium.* BG - *Bulgaria.* DE - *Germany.* FI - *Finland.* FR - *France.*
IT - *Italy.* LT - *Lithuania.* LX - *Luxembourg.* NL - *Netherlands.* NO - *Norway.*
PT - *Portugal.* SP - *Spain.* SZ - *Switzerland.* UK - *United Kingdom.*

5 Discussion

Criteria such as vandalism resistance and form-freedom could, for the most part, not be retrieved directly from the databases, due to lack of direct testing standards. Vandalism resistance was determined by analyzing the material application and usage. Form-freedom was analyzed by assessing the material type and manufacturing of it. If the material is produced by molding, it is said to be form-free, being able to produce materials with several diverse shapes and 3-D elements dependent on the mold [6]. Materials made by extrusion are said to be partially form free, hence only being able to be formed in select shapes, without the ability to produce complex 3-D elements on the surface [7]. Material types such as wood are non-form free.

As evident in Table 1 some assumptions were made in the biobased percentage, fire safety class and lifespan, as the data gathering is limited to desk research, some of this information is unobtainable if not provided. The assumed biobased percentages were determined by analyzing the material composition and type. In the same way, assumed fire resistance classes were determined by examining the material type. If the material was a mineral composite, it was assumed to have a C-B fire resistance class, while any other material type was assumed be in the D-C resistance class. The assumed technical lifespan was determined based on the material composition and durability class, when applicable [9]. Practical testing in the laboratory would reduce the uncertainties in these three criteria.

In the Appendix IV, Table 5 shows non-biobased circular materials. When comparing some of the biobased materials to the non-biobased circular materials, it becomes apparent that the attributes (based on this case study) are very similar. KLP façades is the only material to comply with all the requirements (except being biobased), the rest of the materials lack form-freedom. For this case study, it can be concluded that biobased materials have similar, and in the case of Nabasco 8010, more suitable attributes than the non-biobased circular façade materials when using the client's requirements.

Overall, biobased materials have attributes to compete with non-biobased circular products and even with current non-biobased mainstream façade products, seen in Table 5. Based on this case study, biobased façade materials meet the current demands of the market, set by a real client. Consequently, these materials are as suitable as existing mainstream materials. Therefore, the next step for biobased façade materials is to gain greater acceptance within the architecture industry and society at large by advocating for legislation that is more favorable to the use of biobased materials in façades.

6 Conclusion

The list consisted of fifty biobased façade materials manufactured in Europe, with a high technical readiness level (TRL > 8). The materials can be collected into different material types (also present in Table 1) and are shown in Table 2. This gives an idea of what type of biobased materials are currently in the market. Table 2 only shows some examples of the different material types.

Table 2. Material types summary

Type of Material	Examples	
Wood	– Timber claddings – Mutually glued plywood – Cross-laminated Timber	
Modified Wood	– Thermally modified – Acetylated – Impregnated	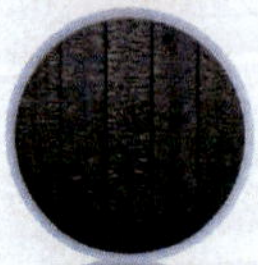
Wood-plastic Composites	– Wood-PE – Wood-PP – Wood-recycled plastic	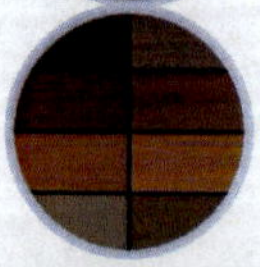
Bio-composites	– Cellulose – Flax – Hemp	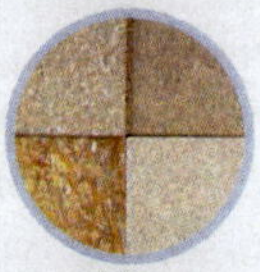
Other	– Bamboo – Cork – Pressed peat	

All of the materials have some suitable properties, following the requirements of the case study. Nabasco 8010 follows all the requirements in this case study, with Resysta complying with most of the requirements, except for being partly form-free. There are several materials that meet all but one requirement. As a result, these materials could be considered suitable for a facade where the specific attribute is not a requirement.

In conclusion, the comparison between biobased and non-biobased circular materials reveals remarkable similarities in their attributes, as evidenced by this case study. Biobased materials possess the capabilities to effectively compete with non-biobased circular products and even with existing non-biobased mainstream façade products. The investigation shows that biobased façade materials meet the present market demands established by a real client.

7 Follow-Up Research

While there has been significant progress in the development of biobased materials, as seen by the number of biobased façades are available currently in the market; there are several areas where data is scarce, hence where future work is needed to fully understand and realize the potential of these materials.

Firstly, research and development is the main area for further work. There is a need for continued research into the properties and characteristics of biobased materials. The list of biobased materials has several assumptions, reducing the reliability of the results. To reduce this uncertainty, practical testing of the materials should be carried out. The vandalism resistance and weather durability should be investigated, in addition to the fire resistance and biobased % being investigated further.

Via a scratch and compression test, the vandalism resistance can be measured and quantified. Weather durability can be tested by using diverse drying-wetting cycles in a climate chamber, also giving an indication of technical lifespan. Fire resistance could be researched by following procedures detailed in the EN-13501-1 internally, and biobased % composition could be determined by performing a chemical analysis on the amount of renewable carbon in each material, namely using the EN 16640 method,

An interesting research approach for the follow-up would be to purchase samples (of specific dimensions) of all the 50 biobased materials in the list, to carry out the tests mentioned above for all of them. This research would aid to obtain more information on the current state of biobased materials, and provide ideas and approaches which could eventually contribute to improving their quality, increase their durability, and making them more cost-effective.

Lastly, the manufacturing and processing should be investigated in depth. This would shed light on the environmental impact of biobased materials, which was not part of the scope of the case study. Even though these materials are biobased, it does not mean that they are environmentally sound. A material could be biobased and still have a high environmental impact in terms of carbon footprint and other emissions and impact types. The expected critical points are the resins used in composites, energy demand in the processing stage, origin of materials. Therefore, a dive into the raw materials, manufacturing and processing of the biobased materials via LCAs or EPDs of all the materials would inform about the environmental impact. Moreover, would derive on impact quantification and classification, optimization patterns and subsequent reduction of costs and environmental impact for biobased materials.

Appendix I

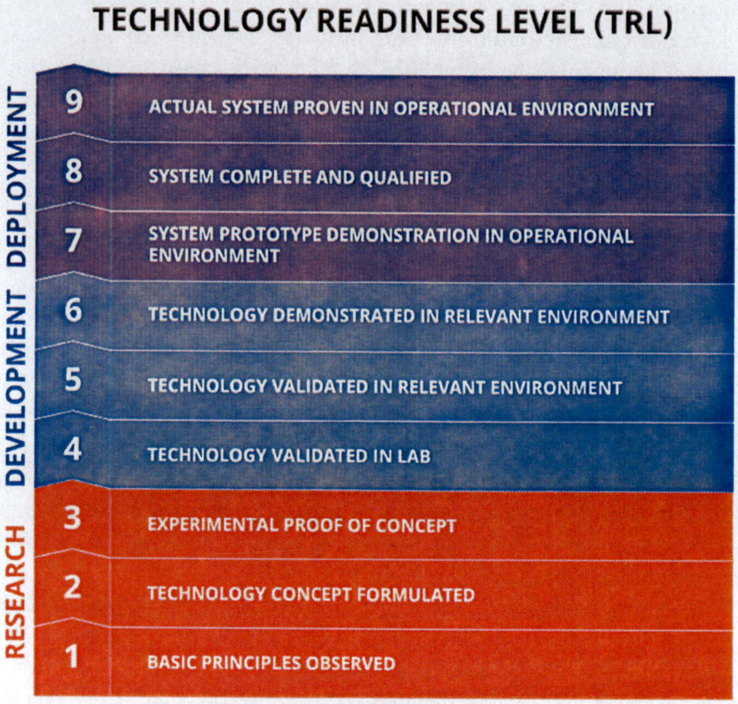

Fig. 4. Technological readiness level (TRL) [8]

Appendix II

Table 3. URL of databases

Database	URL
Biobased Consulancy	https://www.biobasedconsultancy.com/nl
Biobased Inkopen	https://biobasedinkopen.nl/
Biobased Bouwen	https://www.biobasedbouwen.nl/
Die Nachwaschende Produktwelt	https://www.die-nachwachsende-produktwelt.de/

(*continued*)

Table 3. (*continued*)

Database	URL
AgroBioBase	http://www.agrobiobase.com/

Appendix III

Table 4. Interviewed research centres & experts

Country	Research Centre/Expert
Netherlands	Agrodome
	CoE BBE
Germany	German Sustainable Building Council (DGNB)
	Nova Institute
	BioMat (Hanaa Dahy)
	Allthings.biopro
Belgium	KampC
	Vlaamse Confederatie Bouw (Embuild Vlaanderen)
	Eurabo
Austria	TUV Austria OK Biobased
Denmark	COWI
	Danish Materials Network (DMN)
Ireland	Biorbic
	Hempbuild
UK	Bath University - Centre for Innovative Construction Materials (CICM)
	BioVale
Spain	Tecnalia
	Aimplas
	EcoHUB Barcelona (centre for biobased products and green building)
Switzerland	Zurich University of Applied Science (ZHAW)
Slovenia	Innorenew Centre of Excellence (Innorenew CoE)
Hungary	BioEast
Bulgaria	
Estonia	

Appendix IV

Table 5. Non-biobased circular façade materials

Material (*Country*)	Material Type	Short Description	Biobased %	Technical Lifespan	Form-freedom	Fire safety class (EN 13501-1)
KLP Façade panels (*NL*)	Recycled plastic	Recycled plastic cladding	0%	>50 years	Form-free	B-s1,d0
Lamilux Facade panel (*DE*)	Recycled plastic	Made of glass-fibre-reinforced recycled plastic composite	0%	>40 years*	Not form-free	E-D (B2 in DIN 4102)
Pretty Plastic panel (*NL*)	Recycled plastic	Recycled PVC façades	0%	10–30 years*	Not form-free	B-s3,d0
Rockpanel Woods (*UK*)	Mineral composite	Compressed natural basalt bonded with organic binder (workability and aesthetic of wood), from waste of construction industry	0%	50 years	Not form-free	A2-s1, d0 (non-combustible)
STENI facades (colour, nature and vision) (*NO*)	Mineral composite	Fibreglass reinforced cured stone composite panel with core of crushed natural stone. Difference in between all types is the colouring (technique) and texture. Eco label. Due to production with renewable energy	0%	>60 years	Not form-free	B-s1,d0

References

1. European Commission. "Agricultural Waste Management in Europe," European Commission (2021). https://ec.europa.eu/environment/waste/agricultural_en.htm. Accessed 2023
2. van der Burgh, F., Verspeek, S.: Performance of bio-based facades. In: International Conference Biobased Building Materials (ICBBM), Belfast (2019)
3. Bardage, S.: Performance of buildings. In: Performance of Bio-based Building Materials, Stockholm, pp. 335–383. Science Direct (2017)
4. ProRail. "ProRail" (2023). https://www.prorail.nl/. Accessed 23 Jan 2023
5. Spoorbeeld door Bureau Spoorbouwmaster; ProRail, "Handboek Railgebonden Gebouwen" (2021). https://www.spoorbeeld.nl/sites/default/files/2021-06/Handboek%20Railgebonden%20Gebouwen.pdf. Accessed 1 Mar 2023
6. Environmental Policy Agency. Biobased Materials (2010). https://cfpub.epa.gov/si/si_public_record_report.cfm?Lab=NRMRL&dirEntryId=231873. Accessed 20 Jan 2023
7. Quintieri, E.: Fire Safe Europe (2017). https://firesafeeurope.eu/facades-fire-safety-accross-eu-countries/
8. TWI. What are technology readiness levels (TRL)? (2023). https://www.twi-global.com/technical-knowledge/faqs/technology-readiness-levels. Accessed 15 Jan 2023
9. Logic Bespoke Limited. Timber durability classes (n.d.). https://logic-bespoke.com/timber-durability-classes/. Accessed 1 Mar 2023

Moisture Buffering of Hemp-Lime with Biochar and Rape Straw-Lime as Surface Materials for a Stable Indoor Climate

Paulien Strandberg-de Bruijn[1](✉) and Kristin Balksten[2]

[1] Division of Building Materials, Department of Building and Environmental Technology, Box 118, 22100 Lund, Sweden
paulien.strandberg@byggtek.lth.se

[2] Uppsala University Campus Gotland, Visby, Sweden

Abstract. An appropriate and stable indoor climate in museums is crucial to guarantee an appropriate preservation of our cultural heritage. Depending on the collection, indoor temperature and relative humidity need to be kept within a certain range. Fluctuations in temperature and relative humidity could cause damage to museum artefacts and may require higher energy needs than necessary. Biochar is a material of which the use is relatively new in building materials. Previous studies have shown that biochar has unique moisture properties with a high surface area, high porosity and therefore high capability of moisture uptake. In Southern Sweden there are several biochar manufacturers that produce biochar from local biomasses such as seaweed, gardening wastes and residues from greenhouses.

The aim of this project was to investigate the impact of hygroscopic surface materials on the indoor climate of buildings, focusing on moisture buffering and hygrothermal properties. The building materials that were studied were hemp-lime (with and without biochar) and rape straw-lime. Passively influencing the indoor climate by choosing appropriate surface materials could contribute to lower energy needs and less need for mechanical ventilation in historic buildings and museums without the need for excessive HVAC solutions.

Keywords: Hemp · Rapeseed · Lime · Historic Buildings · Thermal Properties

1 Introduction

Traditional building materials used in the interior of historic buildings have different material properties than many conventional materials due to their hygroscopic nature. Also, historic artefacts often consist of different materials with different hygroscopic properties [1]. With moisture absorption the materials swell and shrink and as they change size objects will deteriorate faster. A stable climate without moisture fluctuation therefore would provide better conditions for preservation. If the building itself is made of hygroscopic materials with a high moisture buffering capacity they could contribute and help providing a more stable relative humidity (RH) in the air [2]. The aim of this study was to see if novel bio-based materials with high moisture buffering properties can

S. Amziane et al. (Eds.): ICBBM 2023, RILEM Bookseries 45, pp. 144–157, 2023.
https://doi.org/10.1007/978-3-031-33465-8_12

be an alternative as surface materials on plastered walls and roofs in historic buildings i.e. churches, castles and local museums. Many of those buildings have a low temperature in winter and the moisture levels can thus be higher than preferred. In this study the moisture buffering properties are investigated of hemp-lime, of hemp-lime with an addition of biochar as well as of rape straw-lime.

Previous research has shown that hemp-lime has an "excellent" moisture buffer value [3 4 5]. According to the classification in the Nordtest protocol, a material with "excellent" moisture buffering has a moisture buffer value greater than 2.0 kg/(m^2·% RH). Hemp-lime is a building material in which hemp shiv (the woody core parts of the hemp stem) are used in combination with building limes. Hemp-lime has been tested as additional insulation and is compatible with historic wooden houses as well as historic masonry [6, 7]. The material has both good thermal insulation and relatively high thermal inertia. Studies have also shown that it can contribute to a good indoor environment due to its acoustic properties as well as its good fire resistance [8–10]. The fact that the material also has a low climate impact makes it interesting for the transition towards more sustainable building constructions.

Hemp-lime is a hygroscopic material that can be used as a visible internal surface layer on walls and in roofs or, more commonly, as a base layer for a thin lime plaster surface finish in historic buildings.

1.1 Hemp-Lime and Rape Straw-Lime

Rapeseed (Brassica napus and Brassica rapa) and hemp (Cannabis sativa) are agricultural crops cultivated in Sweden. Statistics from the Swedish Board of Agriculture [11, 12] state that the total acreage cultivated in 2022 was 127,500 ha rapeseed (both Brassica napus and Brassica rapa), see Fig. 2. The same year 222 ha hemp was cultivated, see Fig. 3. Rapeseed is a significant oilseed crop in Sweden, while hemp cultivation is not as common. Hemp cultivation was prohibited in Sweden from the 1965 until 2003 [13], which caused knowledge on hemp cultivation techniques to vanish, as well as the commercial market for hemp seed, hemp fiber and hemp shiv. Since 2003 hemp cultivation has been allowed again in Sweden. In 2007 hemp cultivation peaked at 828 hectares. Since then, hemp cultivation has been approximately 200 ha annually. Today, hemp-lime is used for construction in Sweden on a modest scale, with a handful of hemp-lime houses built in southern Sweden. For hemp-lime the hemp shiv are used, which are the woody core parts of the hemp stem, see Fig. 1. When not used in construction, hemp shiv are commonly used for animal bedding. Rape straw-lime is a material that consists of chopped rape straw that is used in combination with building limes, in a similar fashion as the production of hemp-lime. For rape straw-lime (to our knowledge thus far not used in Sweden for construction) the straw of rapeseed was used, see Fig. 1.

Because of the greater abundance of rape seed in Sweden, residues from rapeseed cultivation would be interesting for use in building materials. Nowadays rape seed straw is usually ploughed under in the field to improve soil structure and add nutrients to the soil (so-called soil incorporation) or used as animal bedding [14].

The use of biomass from agriculture in building materials is advantageous in multiple ways. The building sector benefits from the production of building materials with a low environmental impact as increasing the biobased material content of a building generally

Fig. 1. Hemp shiv to the left, rape seed straw to the right.

results in a reduced climate impact if the biogenic carbon sequestration is accounted for in the life-cycle analysis [15]. Biobased residue materials from agriculture are renewable; new material can be cultivated each year. At the same time, the agricultural sector benefits from finding a commercial market for its residue materials.

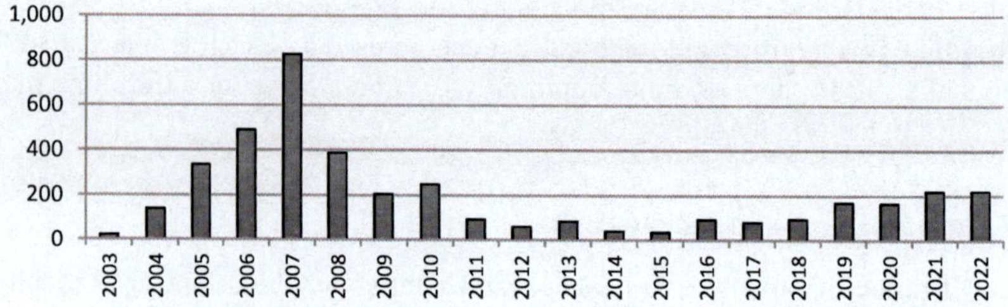

Fig. 2. Hemp cultivation (hectares) in Sweden 2003–2022 [11].

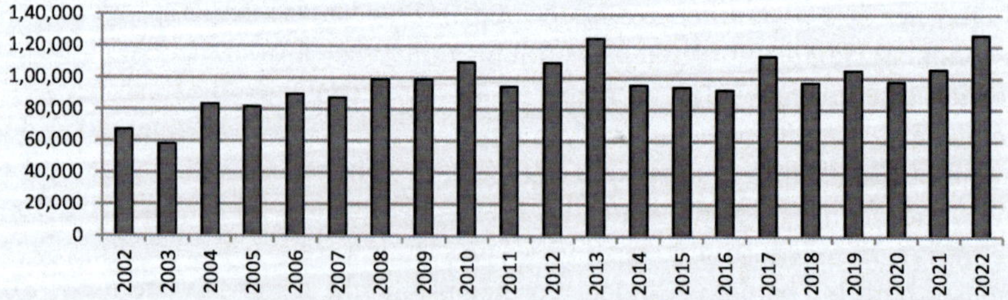

Fig. 3. Rape seed cultivation, Brassica napus and Brassica rapa, in Sweden 2002–2022 [12].

1.2 Biochar

Locally sourced seaweed (macroalgae) was used to produce the biochar used in this study. Seaweed from the Baltic Sea was collected at the shoreline of Trelleborg municipality, on the southern coast of Sweden. Every year problems arise with seaweed at the shoreline of Trelleborg municipality. These seaweed-related problems entail inconveniences for beach visitors such as difficulties to reach the beach because of beach-cast seaweed and an unpleasant smell, economic problems that arise when removing the seaweed, and loss of income from tourists that stay away. But mostly the seaweed accumulation leads to, and is amplified by, ecological problems such as eutrophication and biodiversity loss [16]. At the same time, seaweed is a material resource that is available in abundant amounts. Therefore, Trelleborg municipality is searching for ways to upcycle seaweed harvested from its shoreline and put it to use in a different function, such as biogas [16] or biochar [17]. The biochar in this study was produced by Skånefrö AB in Hammenhög, southern Sweden. The biochar was produced through pyrolyzing biomass (in this case seaweed) by subjecting it to high temperatures in the absence of oxygen, thereby causing thermal decomposition of the biomass [17].

Biochar is today mostly used as a fertilizer and to improve soil structure. However, there is also research ongoing exploring the use of biochar in concrete or in renders and mortars with a cement-based or lime-based binder [18–22]. Adding biochar to cement or lime could improve the material's environmental impact as biochar functions as a carbon sink [23]. Using biochar in building materials is effectively contributing to carbon capture in the material, thus decreasing the amount of carbon that is released into the atmosphere. In addition, the algae biomass is not returned to the sea thus reducing eutrophication in the Baltic Sea. According to [24] biochar plaster (biochar added to a lime-based binder) has a good moisture storage capacity and can function as a humidity regulator in environments that have occasional peaks in relative humidity. Therefore, biochar was added to the hemp-lime in this study to determine if adding biochar to the hemp-lime would improve its moisture buffering capacity.

Fig. 4. Left: Beach cast seaweed at the southern Swedish coast. Right: Biochar produced from local sea weed.

2 Materials and Methods

The following materials were studied in this paper;

- HL: Hemp-lime (Tradical® Thermo lime and Granngården hemp hurds),
- HL20: Hemp-lime with 20% biochar (20% of the lime binder replaced by seaweed-based biochar. Seaweed provided by Trelleborgs municipality, seaweed pyrolyzed by Skånefrö),
- HL50: Hemp-lime with 50% biochar (50% of the lime binder replaced by seaweed-based biochar),

- RL: Rape straw-lime (Tradical® Thermo lime, and rape straw provided by Halmeko Scandinavia).

The hemp-lime was produced in accordance with instructions provided by the manufacturer's product sheet for lining plaster application with a trowel. Rape straw-lime was produced in a similar way as the hemp-lime in this study, with Tradical® Thermo lime as a binder. For the Tradical® Thermo lime the material quantities for the application Lining with trowel were used. The hemp-lime in this study is thought to be used as an indoor lime plaster used for renovation purposes, in other words, the plaster would most likely be applied by a mason with a trowel on site. Samples were produced with a thickness of approx. 50 mm and a surface area of 150x150 mm. Exact measurements of the surface area as well as exact sample thickness were performed prior to each moisture buffer test (Table 1 and Fig. 5)

Table 1. Material proportions by weight.

Abbreviation	Lime [kg]	Shiv or straw [kg]	Water [kg]	Biochar [kg]
HL	2.70	1.0	4.0	0.0
HL20	2.16	1.0	4.0	0.54
HL50	1.35	1.0	4.0	1.35
RL	2.70	1.0	4.0	0.0

Fig. 5. Left: Hemp-lime, center: Hemp-lime with 20% biochar, right: Hemp-lime with 50% biochar.

Biochar was produced by Skånefrö AB of beach cast sea weed from the southern Swedish coastline near the city of Trelleborg. The biochar consisted of small lumps, see Fig. 4, and was therefore ground into smaller particles before it was added to the binder mix. Particle size distribution of the ground biochar is shown in Fig. 6.

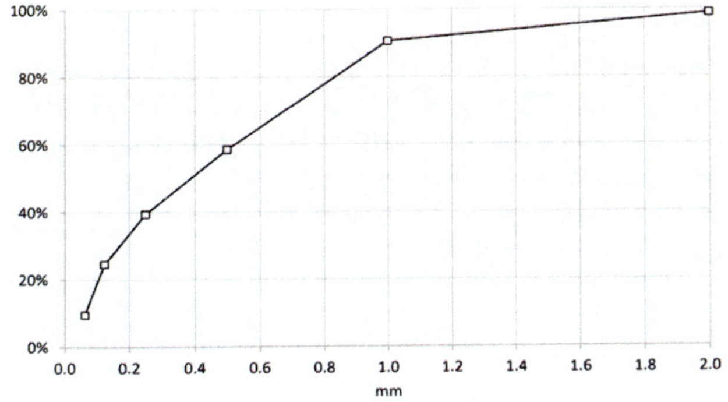

Fig. 6. Particle size distribution of the ground biochar.

2.1 Thermal Properties

Thermal conductivity, thermal diffusivity and specific heat capacity were measured by a Transient Plane Source (TPS) method using a Thermal Constants Analyzer (TPS1500, HotDisk®). After the moisture buffer measurements, specimens were dried in a heating cabinet at 40 °C until mass change was less than 0.01% between three consecutive weighing at 24-h time steps. Drying at this rather low temperature prior to testing thermal properties was done to perform the measurements on a sample that is sufficiently dry so that moisture flows within the material do not interfere with the measurement too much. Drying at a higher temperature might have altered material properties and would not have been representative to actual conditions in a building. The dried samples were placed in plastic bags, and the samples in their plastic bags were placed in plastic containers with a desiccant and a glass lid. Samples were transported to the TPS1500 HotDisk® device inside the plastic containers and remained there for 24 h to acclimatize to the surrounding temperature (21 °C). When performing the measurements, specimens were kept inside the plastic bags, with the sensor on top of the material surface. On top of the sensor, a rigid insulating material with known material characteristics was placed with a weight on top. At least three measurements were performed per sample, three samples per material. Measurements were performed at 2-h intervals with sensor Kapton 8563 (9.868 mm), measurement time 320 s and output power 40 mW.

After the measurements were completed, the samples were returned to the drying cabinet where they were kept at 40 °C. Once all measurements with the TPS1500 Hot-Disk® device were completed, the samples were dried at 105 °C and moisture content at 40 °C was determined, see Table 3.

2.2 Moisture Buffering

To determine the moisture buffering capacity of the materials in this study, an adaptation of the NORDTEST protocol [25] was used. In the NORDTEST protocol the moisture buffer performance of a room is described as"the ability of the materials within the room to moderate variations in the relative humidity" [25].

Usually, the sorption isotherm is used to define moisture absorption and moisture desorption of a building material. A sorption isotherm for a building material can be determined in a lab situation, waiting for the material to reach equilibrium with its surroundings. However, when a material is used in a building, the material will most likely not reach equilibrium because of fluctuations in temperature and relative humidity. Instead, the material will have a moisture content that fluctuates between its sorption and desorption isotherm. This is why the moisture buffer value of a material could be a more suitable indicator of the moisture absorption and desorption of a material in a shorter time span.

The practical moisture buffer value ($MBV_{practical}$) is defined by [25] as "the amount of water that is transported in or out of a material per open surface area, during a certain period of time, when it is subjected to variations in relative humidity of the surrounding air" [25].

The time period and variations in relative humidity proposed by the NORDTEST protocol are cyclic variations of 8 h of 75%RH followed by 16 h of 33%RH;

$$MBV_{practical} = \frac{\Delta m}{A \cdot \Delta RH} \quad [\mathrm{kg/(m^2\%\ RH)}] \quad [25]$$

Δm = mass, defined as the average of the weight gain moisture uptake (absorption), and the weight loss during moisture release (desorption) [kg].

ΔRH = the difference in relative humidity between the high RH (8 h of 75% RH) and low RH (16 h of 33%RH) [%].

A = area of the moisture buffering material exposed to the fluctuations in relative humidity [m^2].

As a direct cause of the Covid-19 pandemic access to the building lab was limited during a long period of time. This made it considerably more difficult to perform manual measurements. Therefore, instead a climate test cabinet (CTS C-20/1000) was used to create an environment with cyclic variations of 75%RH for 8 h followed by 33%RH for 16 h. Inside the climate test cabinet three scales (OHAUS Scout with an accuracy of two decimals) were placed that continuously measured the weight of the samples that were suspended underneath the scales.

Also, additional plastic containers were placed beneath the specimens to protect them from too strong air flows inside the climate test cabinet. Air velocity inside the climate test cabinet can influence the moisture buffer value of the material since air velocity has an influence on the surface transfer coefficient. The Nordtest protocol mentions an air velocity interval of 0.10 ± 0.05 m/s [25]. The air velocity inside the climate test cabinet was measured near the surface of each of the three samples and found to be 0.14 ± 0.10 m/s.

Prior to the moisture buffer test, samples were dried; 24 h at 40 °C followed by 24 h at 50 °C followed by drying at 60 °C until equilibrium was reached. Equilibrium was reached when variation in weight was less than 0.01% between three consecutive weighing at 24-h time steps. After drying, the samples were placed in a climate chamber with relative humidity 60 ± 5% and temperature 20 ± 0.5 °C until they reached equilibrium; variation in weight less than 0.01% between three consecutive weighing at 24-h time steps. Before the specimens were subjected to the moisture buffer test inside the climate test cabinet, aluminum tape was applied to all but one surface of the samples. The surface area was measured for each specimen. The surface that was left open would be able to absorb and desorb moisture once inside the climate test cabinet, see Fig. 7.

Fig. 7. Examples of specimens sealed with aluminum tape on all but one surface.

Each scale was placed inside a plastic container with a glass lid. Inside the plastic container, desiccant was placed to keep the relative humidity inside the container as stable as possible. However, it was found that the scales anyway were subjected to changes in moisture inside the climate test cabinet. To allow the suspended stirrup to move freely, a small hole was made in the bottom of the plastic container. Air could pass freely through this hole, changing the relative humidity in the scale environment somewhat. The scales were found to react to this change in relative humidity, affecting the measurements. Therefore, a calibration of the scales was performed by weighing steel weights while they underwent the same cyclic variations; 33%RH for 8 h followed by 75%RH for 16 h. A steel weight 500 grammes was weighed by each scale respectively, throughout a minimum of three cycles. These measurements were used to calibrate the data.

Between each cycle, 30 min was required for the climate test cabinet to change from 33%RH to 75%RH or from 75%RH to 33%RH. This time was not included in the 8-h period of 33%RH or in the 16 h period of 75%RH. Thus, a period of 8 h 33%RH was followed by 30 min changeover time. Then a period of 16 h 75%RH, again followed by 30 min to reach 33%RH. This led to a total phase time of 25 h (instead of 24 h as required by the NORDTEST protocol). In order to get a value for MBV that can be compared to other research results, it was important to have a moisture buffer value for 24-h cycles. Therefore, the weight change during these last 30 min of each relative humidity level was disregarded when analyzing the results, see Fig. 11.

Data analysis was performed in MATLAB R2020b. Measurement results were imported in MATLAB. The changes in weight that were registered for the steel weight throughout the cyclic variations were deducted from the measurement results. Measurements for all materials had been made every 15 s. A moving average was calculated for every 80 measurement values. The moving average was then used to calculate the moisture buffer value. Moisture buffer values were calculated both for absorption and desorption. The average moisture buffer value was calculated from a minimum of three relative humidity cycles per material for three specimens per material.

3 Results and Discussion

Dry apparent density of the materials was determined, three samples per material. When adding biochar to the material the overall apparent density decreased, see Table 2.

Table 2. Dry apparent density of material samples. Drying of samples at 60 °C.

Material	Dry apparent density [kg/m^3]	Standard deviation
HL	487.5	25.8
HL20	477.0	7.2
HL50	422.1	33.4
RL	468.0	37.6

3.1 Thermal Properties

Thermal conductivity, thermal diffusivity and specific heat capacity were determined by a Transient Plane Source (TPS) method. The moisture content of the materials at the time of the thermal testing was determined, see Table 3. The moisture content in the hemp-lime specimens with 50% biochar was the highest.

Table 3. Moisture content of material samples prior to thermal test. Dry weight at 105 °C.

Material	Moisture content	Standard deviation
HL	3.12%	0.04%
HL20	3.17%	0.05%
HL50	3.82%	0.09%
RL	3.06%	0.06%

The results from the thermal testing showed that the thermal conductivity of the materials decreased when biochar was added to the specimens. The hemp-lime specimens without biochar had the highest values for thermal conductivity. Rape straw-lime and hemp-lime with 50% biochar had similar values for thermal conductivity, see Fig. 8.

Thermal diffusivity also decreased for the material with biochar, with the specimens of hemp-lime with 50% biochar showing the lowest value for thermal diffusivity. This indicates that a hemp-lime with biochar would allow for a slower heat flow through the material, thus contributing to a more stable indoor climate. The specific heat capacity of the three hemp-lime materials was quite similar, see Fig. 9. Here the rape straw-lime material showed somewhat lower values for specific heat capacity, indicating that less heat is needed to change the temperature of the rape straw-lime, even though the standard deviation was quite high. Overall, the specific heat capacity of the tested materials can be compared to aerated concrete with values 0.4 MJ·m-^{3}K-1 for aerated concrete with density 400 kg·m^3 and 0.5 MJ·m-^{3}K-1 for aerated concrete with density 500 kg·m^3 [26].

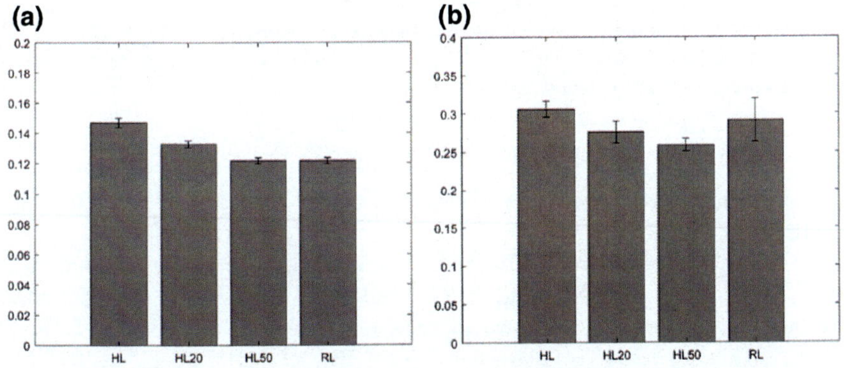

Fig. 8. (a) Thermal conductivity [$W \cdot m^{-1} \cdot K^{-1}$]. (b) Thermal diffusivity [$mm^2 \cdot s^{-1}$].

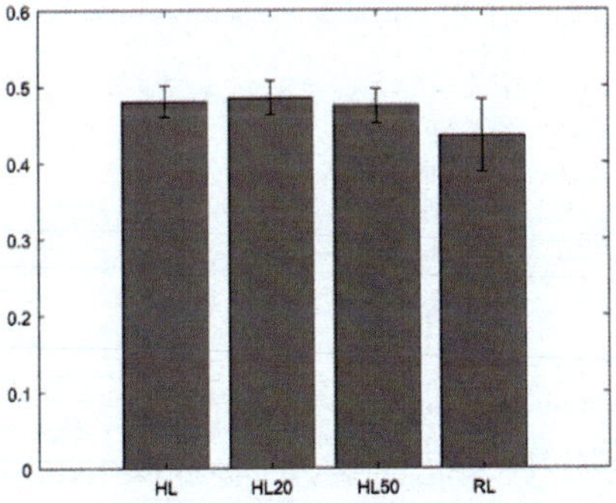

Fig. 9. Specific heat capacity [$MJ \cdot m^{-3}K^{-1}$]

3.2 Moisture Buffering

Data from the moisture buffer measurements had to be calibrated. Each of the three scales had a different calibration file. This data had to be deducted from the measured raw data.

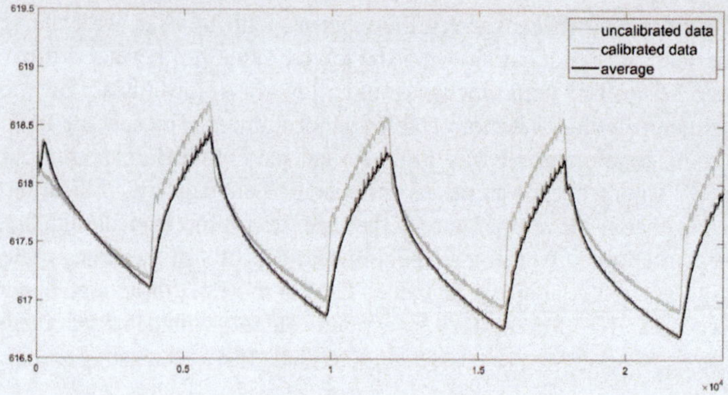

Fig. 10. Weight of one sample of hemp-lime undergoing the moisture buffering regime. Light grey line uncalibrated data, dark grey line calibrated data and black line average data for 80 data points (equivalent of 10 min). The data is shown before calibration and after calibration, taking into account the changes in the scale due to changes in RH.

After calibration, a moving average value was calculated, see the black continuous line in Fig. 10.

Also, half an hour of each relative humidity level had to be excluded as measurements had been made for 8.5 h 75%RH and 16.5 h 33%RH. To allow values to be compared to other $MBV_{practical}$ the data had to be adjusted to 8/16 h intervals, see Fig. 11.

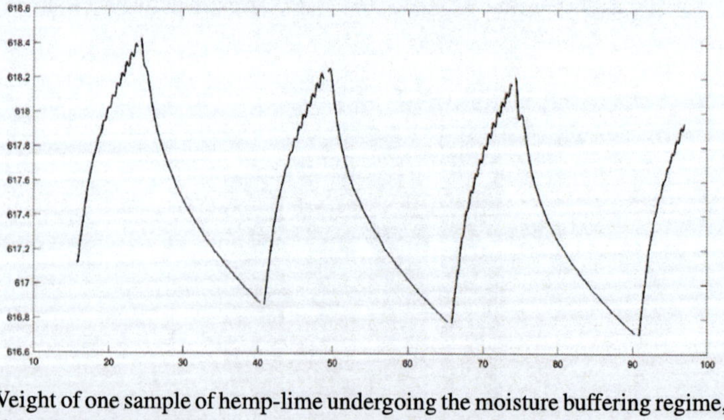

Fig. 11. Weight of one sample of hemp-lime undergoing the moisture buffering regime, excluding 30 min with each change in RH.

According to [22] the level of $MBV_{practical,}$ can be divided into five categories; negligible for values <0.2 g/m^2·%RH, limited 0.2–0.5 g/m^2·%RH, moderate 0.5–1.0 g/m^2·%RH, good 1.0–2.0 g/m^2·%RH and excellent >2.0 g/m^2·%RH.

Results from the moisture buffering measurements showed that the hemp-lime and rape-straw lime materials either have "good" or "excellent" practical moisture buffer values, see Fig. 12.

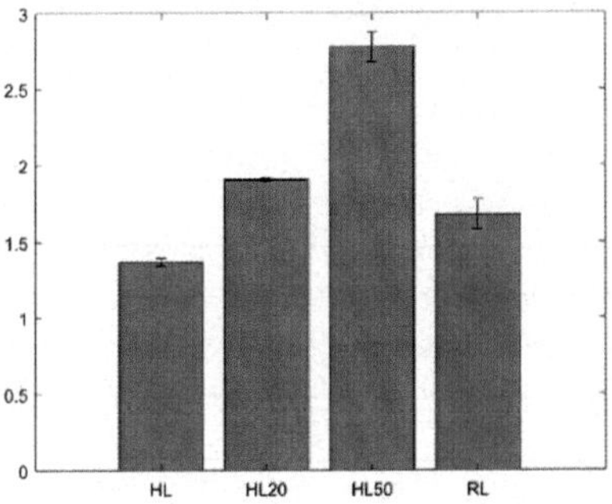

Fig. 12. Average moisture buffer value for all four materials.

Adding biochar to the hemp-lime material increased the moisture buffer value, with an average value of 2.78 $g/m^2 \cdot \%RH$ for hemp-lime with 50% biochar, thus providing *excellent* moisture buffer capacity.

4 Conclusions

Biochar is an interesting material for use in many different applications, one of which could be the use as an additive in building materials, for example in hemp-lime.

Before using biochar in building materials more research is needed. For example, the choice of biomass from which the biochar is produced will have an influence on the final material properties. Biochar based on seaweed will have some marine salts still present which could potentially negatively influence the material in which the biochar is used. Which salts and how much salt is still present in the biochar should be investigated prior to its use in building materials. Biomasses other than seaweed could be a potentially good source for biochar suitable for building.

Adding biochar to the hemp-lime gave a material with lower density, lower thermal conductivity, and lower thermal diffusivity. From a thermal insulation point-of-view it could therefore be beneficial to add biochar to the mix. Regarding moisture buffering, adding biochar increased the practical moisture buffer value, achieving a material with excellent moisture buffer capacity. Adding biochar to the surface finish of historic buildings could therefore contribute to achieving an indoor climate with stable temperature and stable relative humidity.

The rape straw-lime mix used in this study showed material properties similar to those of hemp-lime and could therefore be a valid alternative to hemp-lime in Sweden where hemp cultivation is scarce and rape seed is cultivated in abundance.

Acknowledgements. This research was funded by the Swedish Energy Agency through the *Spara och Bevara* research programme (grant number 50974–1) and the European Union through the project Greater Bio (Grant nr NYPS 20203414) funded by Interreg Öresund-Kattegat-Skagerak (ÖKS). The role of these funding sources was strictly financial.

The laboratory work in this study was performed in cooperation with research engineer Stefan Backe. His help is gratefully acknowledged.

References

1. Thomson, G.: The museum environment, 2. ed. London: Butterworth-Heinemann in association with the International Institute for Conservation of Historic and Artistic Works (1986)
2. Ferreira, C., de Freitas, V.P., Delgado, J.M.P.Q.: The influence of hygroscopic materials on the fluctuation of relative humidity in museums located in historical buildings. Stud. Conserv. **65**(3), 127–141 (2020). https://doi.org/10.1080/00393630.2019.1638666
3. Rahim, M., et al.: Characterization of flax lime and hemp lime concretes: hygric properties and moisture buffer capacity. Energy Build. **88**, 91–99 (2015). https://doi.org/10.1016/j.enbuild.2014.11.043
4. Collet, F., Chamoin, J., Pretot, S., Lanos, C.: Comparison of the hygric behaviour of three hemp concretes. Energy Build. **62**, 294–303 (2013). https://doi.org/10.1016/j.enbuild.2013.03.010
5. Latif, E., Lawrence, M., Shea, A., Walker, P.: Moisture buffer potential of experimental wall assemblies incorporating formulated hemp-lime. Build. Environ. **93**, 199–209 (2015). https://doi.org/10.1016/j.buildenv.2015.07.011
6. Strandberg-de Bruijn, P.B., Balksten, K.: Energy and moisture in historic masonry walls retrofitted with hemp-lime.In: IOP Conference Series: Materials Science and Engineering, vol. 660, p. 12070 (2019). https://doi.org/10.1088/1757-899x/660/1/012070
7. Strandberg-de Bruijn, P., Donarelli, A., Balksten, K.: Full-scale studies of improving energy performance by renovating historic Swedish timber buildings with hemp-lime. Appl. Sci. **9**(12), 2484 (2019). https://doi.org/10.3390/app9122484
8. Arnaud, L., Gourlay, E.: Experimental study of parameters influencing mechanical properties of hemp concretes. Constr. Build. Mater. **28**(1), 50–56 (2012)
9. Källgren, V.: Ljudmiljö i kulturmiljö - hampakalks akustiska egenskaper och tillämpbarhet i kulturhistoriskt värdefulla byggnader. Uppsala Universitet Campus Gotland (2017)
10. Persson, J., Edbladh, J.: Hampakalk -En jämförelsestudie med avseende på brandegenskaper. Lunds Tekniska Högskola, (2018)
11. Swedish Board of Agriculture, "Hemp cultivation," 2022
12. Swedish Board of Agriculture, "Jordbruksmarkens användning 2022. Slutlig statistik (2022). https://jordbruksverket.se/om-jordbruksverket/jordbruksverkets-officiella-statistik/jordbruksverkets-statistikrapporter/statistik/2022-10-20-jordbruksmarkens-anvandning-2022.-slutlig-statistik
13. Holstmark, K.: Hampa i ekologisk odling, Jönköping (2006)
14. Prade, T., Björnsson, L., Lantz, M., Ahlgren, S.: Can domestic production of iLUC-free feedstock from arable land supply Sweden's future demand for biofuels? J. Land Use Sci. **12**(6), 407–441 (2017). https://doi.org/10.1080/1747423X.2017.1398280
15. Peñaloza, D., Erlandsson, M., Falk, A.: Exploring the climate impact effects of increased use of bio-based materials in buildings. Constr. Build. Mater. **125**, 219–226 (2016). https://doi.org/10.1016/j.conbuildmat.2016.08.041
16. Risen, E., Tatarchenko, O., Grondahl, F., Malmstrom, M.E.: Harvesting of drifting filamentous macroalgae in the Baltic Sea: an energy assessment. J. Renew. Sustain. Energy, **6**(1), 013116 (2014).https://doi.org/10.1063/1.4862783
17. Paulsson, M.: Sammanfattande slutrapport för projektet Rest till Bäst (Steg 2) (2020). https://biokol.org/publikationer/pdf/sammanfattande-slutrapport-steg-2
18. Choi, W.C., Do Yun, H., Lee, J.Y.: Mechanical properties of mortar containing bio-char from pyrolysis." J. Korea Inst. Struct. Maint. Insp. **16**(3), 67–74 (2012)
19. Cuthbertson, D., Berardi, U., Briens, C., Berruti, F.: Biochar from residual biomass as a concrete filler for improved thermal and acoustic properties. Biomass Bioenerg. **120**, 77–83 (2019). https://doi.org/10.1016/j.biombioe.2018.11.007

20. Suarez-Riera, D., Restuccia, L., Ferro, G.A.: The use of Biochar to reduce the carbon footprint of cement-based materials. Procedia Struct. Integr. **26**, 199–210 (2020). https://doi.org/10.1016/j.prostr.2020.06.023
21. Tan, K., Pang, X., Qin, Y., Wang, J.: Properties of cement mortar containing pulverized biochar pyrolyzed at different temperatures. Constr. Build. Mater. **263**, 120616 (2020). https://doi.org/10.1016/j.conbuildmat.2020.120616
22. Maljaee, H., Madadi, R., Paiva, H., Tarelho, L., Ferreira, V.M.: Incorporation of biochar in cementitious materials: a roadmap of biochar selection. Constr. Build. Mater. **283**, 122757 (2021). https://doi.org/10.1016/j.conbuildmat.2021.122757
23. Gupta, S., Kua, H.W.: Factors determining the potential of biochar as a carbon capturing and sequestering construction material: critical review. J. Mater. Civ. Eng. **29**(9), 0401708 (2017). https://doi.org/10.1061/(ASCE)MT.1943-5533. 0001924
24. Abi Khaled, F., Enriquez, P.: Biochar addition to a lime-hemp insulating plaster, Politecnico di Torino, Italy (2020)
25. Rode, C., et al.: Moisture Buffering of Building Materials. Kongens Lyngby (2005)
26. Burström, P.G.: Byggnadsmaterial - Tillverkning, egenskaper och användning, _3rd ed. Lund: Studentlitteratur AB, Lund, Sweden (2021)

Effect of Silane on Physical and Mechanical Properties of Wood Bio-Concrete Exposed to Wetting/Drying Cycles

Amanda Lorena Dantas de Aguiar[1](✉), M'hamed Yassin Rajiv da Gloria[1], Nicole Pagan Hasparyk[2], and Romildo Dias Toledo Filho[1]

[1] Federal University of Rio de Janeiro, Rio de Janeiro, RJ 21945-970, Brazil
amanda.aguiar@numats.coc.ufrj.br
[2] Eletrobras Furnas, Rio de Janeiro, Brazil

Abstract. Wood bio-concrete is a promising bio-based building material due to its insulating properties and its contribution to increase energy efficiency and reduce environmental impacts. However, there is still a lack of knowledge concerning its durability over time. For that reason, this study investigates the influence of accelerated aging in laboratory on the properties of wood bio-concretes through a succession of wetting/drying cycles. In order to improve the durability, silane-based water repellent was used as a surface skin protection (SSP) of the bio-concretes and as an impregnation agent of bio-aggregates (IBA). The bio-concretes were produced with a volumetric fraction of 45% of wood shavings and a cementitious matrix composed by cement, rice husk ash and fly ash. The physical and mechanical properties of the samples before and after the wetting/drying cycles were evaluated. After accelerated aging, a decrease of compressive strength was observed for the control (without silane addition) and IBA samples. It was found that SSP was more effective in improving the capillarity water absorption resistance of wood bio-concrete in the early hours only, while for IBA samples, there was a decrease of 17% comparing to control samples at the end of testing. Analysis performed by scanning electron microscopy showed cracked transition zones of the bio-aggregate and cementitious matrix of bio-concrete, due to aging process.

Keywords: Wood bio-concrete · silane · durability · physical and mechanical properties

1 Introduction

The high environmental impact caused by the building sector has led to increased research on green or alternative materials such as bio-based construction materials using local vegetable resources as lightweight aggregates in combination with inorganic binders [1]. Named as bio-concrete, this material uses plant origin particles as aggregates, and for this reason is capable of sequestering CO_2 through photosynthesis during the plant growth [2] and has interesting thermal insulation properties combined with good hygric regulation performances [3].

S. Amziane et al. (Eds.): ICBBM 2023, RILEM Bookseries 45, pp. 158–170, 2023.
https://doi.org/10.1007/978-3-031-33465-8_13

Wood waste represents, nowadays, an important economic and environmental issue, and can be either used as energy recovery or reused as a building material [4]. Several authors have used wood as bio-aggregate to produce eco-friendly bio-concrete [4–6]. However, the durability of the bio-concrete is not yet itself sufficiently exploited and has been target of concern. Since the knowledge about the evolution of its long-term performance is scarce, it can limit the use of wood bio-concrete in the construction field.

Durability of a material depends on the conditions of its use and the environment of exposure [7]. Wood is a heterogeneous and anisotropic material, whose porosity is responsible for dimensional variations when exposed to outside climatic conditions, which consequently affects the physical and mechanical properties of the bio-concrete produced, despite of controlling thermal and acoustic dissipation phenomena. In addition, aging can modify the material microstructure with the hydration of the binder over time and generate dissolution/precipitation of different compounds [8]. Relative humidity, temperature, liquid water, and UV radiation cause changes in chemical and microstructural properties, modifying the functional properties of bio-based materials [7]. Water is a critical issue, since wood shavings are hydrophilic, and the presence of water leads to a swelling of the bio-aggregate which increases the stresses at the interface with the binder [9].

In this context, a hydrophobic product based on silane has been used as a protective agent for porous cement-based materials. Among commercial organic surface treatment materials, silane-based products, and especially the mixture of silane and siloxane, could be considered as favorable water repellents with good breathability [10]. Silanes belong to the pore liner category and are a group of silicones containing one silicon atom [11]. For surface impregnations, alkoxy and alkyl silanes are commonly used [12]. The alkoxy groups (CH_3O) attached to the silicate atom (Si) contain silicon-oxygen bonds that will link to silicates present in the concrete, while the organic alkylic (CH_3) group remaining will protrude from the pore structure and are responsible for the hydrophobic characteristics [13].

Silane can be used to coat exposed surface of concrete in order to protect it from attack by foreign agents. This type of treatment can suppress significantly capillary water absorption and therefore improve the durability of concrete [14]. Furthermore, silane can be used to impregnate bio-aggregates and create a hydrophobic surface capable of reducing the water absorption and improving durability issues. Alkoxy groups in silane can be hydrolyzed by reacting with water to form more reactive silanol groups, which can also attract each other and form -Si-O-Si- bonds, generating a siloxane network in the solution. Moreover, the silanol molecules attach to the surface of fibers and form hydrogen bonds with the hydroxyl groups of the cellulose in the step of adsorption, creating a monolayer of polysiloxane on the surface of the fibers and resulting in a coating characteristic [15]. However, the performance of silane in both cases depends on surface preparation, amount of silane used, time of curing or impregnation, cyclic wetting and drying and environmental conditions at the time of application.

The objective of this study was to evaluate the influence of silane-based water repellent on the durability of wood bio-concrete by exposing it to wetting and drying cycles. Two silane application methods were adopted: 1) a surface skin protection of the bio-concrete, and 2) impregnation of bio-aggregates before bio-concrete casting.

Their influence on some physical and mechanical properties of wood bio-concrete was investigated.

2 Materials and Methods

2.1 Materials

Wood Shavings

Wood waste obtained in shavings form from the city of Rio de Janeiro (Brazil) was used as bio-aggregates. The wood shavings were a mix of unknown fractions of four species: *Hymenolobium petraeum*, *Cedrela fissilis*, *Erisma uncinatum warm*, and *Manilkara salzmanni*. Wood particles were separated using a mechanical sieve and only the fraction retained in the sieve of opening 1.18 mm was used due to the high water absorption potential of the smaller particles, as suggest by Da Gloria and Toledo Filho [6]. In order to improve the chemical compatibility between the bio-aggregate and mineral binders and based on the work of Aguiar et al. [16], wood shavings were treated by immersion in calcium hydroxide solution ($Ca(OH_2)$) with concentration of 1.85 g per kg of water for 2 h. The main characteristics of the treated wood are presented in Table 1: apparent bulk density [17], water absorption, following the method proposed by Da Gloria and Toledo Filho [6], moisture content [18], and chemical composition [19].

Table 1. Physical parameters and chemical composition of wood shavings.

Physical parameters		Chemical composition (%)	
Apparent bulk density ($kg.m^{-3}$)	530	Cellulose	44.28
Water absorption (%)	70	Hemicellulose	26.43
Moisture content (%)	19	Lignin	26.84
		Extractive compounds	2.45

Supplementary Cementitious Materials

The Brazilian Portland Cement type CP II-F-40 (compound with filler) supplied by Holcim company was used as a binder. Rice husk ash provided by Silcca Nobre company and fly ash supplied by Pozofly were used as supplementary cementitious materials. The specific densities of the cement, rice husk ash and fly ash were 3050 kg/m^3, 2510 kg/m^3 and 1890 kg/m^3, respectively.

Water Repellent

The water repellent based on silane siloxane, supplied by Souza Filho Impermeabilizantes LTDA (São Paulo/Brazil), is non-flammable and has low emission of volatiles. Its density is 1.01 ± 0.02 kg/l.

2.2 Wood Bio-Concrete Manufacturing and Curing Conditions

Wood bio-concrete (WBC) was produced with a volumetric fraction of 45% of wood shavings (WS). The cementitious matrix fraction was partitioned, in mass, by 45% of Portland Cement (PC), 25% of rice husk ash (RHA) and 30% of fly ash (FA), which were defined based on previous works [20, 21] to reduce the cement consumption. The water-to-binder ratio was set at 0.30 and calcium chloride (CC) was used as setting accelerator at a content of 2% in relation to the mass of cementitious materials. Wood bio-aggregates absorb water during the mixing process and, for that reason, it must be considered a compensation water (Wc) to guarantee good workability of the bio-concretes. Thus, the total water (Wt) is the sum of the cement hydration water (Wh) and the compensation water. Table 2 shows the consumption of materials, in kg/m^3.

Table 2. Mix proportions of wood bio-concrete (kg/m^3).

WBC	WS	PC	RHA	FA	CC	Wh	Wc
WBC45	238.5	344.5	191.4	229.7	15.3	229.7	166.9

The production process started with the addition and mixing of cementitious materials and bio-aggregates in a 20-L mixer for 1 min. Afterward, the total water was added, previously mixed with CC, progressively over 1 min. Total mixing time was 4 min. Specimens were molded in three layers, mechanically vibrated (68 Hz) for 10 s each, in cylindrical molds of dimensions 50 × 100 mm (diameter × height). After 24 h, the samples were demolded and kept in a room at temperature of 22 ± 2 °C and relative humidity of 55 ± 5% until reaching 28 days of age.

2.3 Silane Treatment Procedures

Two treatment methods were evaluated using the water repellent based on silane. The first one consisted of the surface application of a water-based solution with 5% of silane on the surface of bio-concrete samples using a brush (Fig. 1a). This surface skin protection (SSP) was made in two application coats, with an interval of 10 min between applications, totaling a consumption of 13.7 mL/m^2. The second method was carried out based on the work of Al Abdallah et al. [22]. Silane emulsion was dissolved at 2 vol% in water, then wood shavings were immersed in the solution for 2 h at room temperature to couple the silane with the bio-aggregate (Fig. 1b). The ratio of wood-to-silane solution was fixed at 10 g of wood in 100 mL of solution. Drying was carried out in an oven at 60 °C for 48 h. Finally, the impregnated bio-aggregate (IBA) was used in the production of wood bio-concrete. The silane-impregnated bio-aggregate changes the characteristics of the mixture in the bio-concrete production process since it produces less bubbles and enhance density of bio-aggregate.

(a) (b)

Fig. 1. Silane treatment procedures (a) SSP, and (b) IBA.

2.4 Wetting/Drying Cycles Protocols

Assessment of the durability of materials can be done in situ, exposing them to natural weathering, or in laboratory controlled conditions by means of accelerated aging tests [23]. In this work, an accelerated aging test was chosen and it consisted of a succession of immersion cycles in water for 30 min followed by drying in a room at 40 °C for 2 h, for ten times. The ten cycles were divided into three blocks with two rest periods, in which the samples were placed in a room with T = 23 °C and RH = 70%, for 14 h. This type of aging was selected based on intense rains followed by hot weather in short periods of time, characteristic of the summer in Rio de Janeiro/Brazil, and can simulate the surrounding conditions of materials during their life cycle. At the end of the aging process, the samples were placed in a room (T = 22 ± 2 °C and RH = 55 ± 5%) for 7 days, where other control wood bio-concrete samples (non-weathered) were stored. This procedure was done in order to have the same hygric and thermal equilibrium at the moment of the tests.

2.5 Functional Properties

Compressive Strength

The compressive strength test was performed according to Brazilian standard NBR 5739 (2018) [24] using a Shimadzu - 1000 kN at a test velocity of 0.3 mm/min.

Capillarity Water Absorption

A method proposed by RILEM (protocol in progress – TC HDB 275) was followed to measure the capillary water absorption of the bio-concretes. The samples were dried in an oven at 60 °C for 72 h. To ensure one-dimensional water transport, the lateral surfaces were sealed with metallic tape. The bottom of the samples was immersed in water at a depth of 5 mm for the following times: 1, 3, 5, 10, 15, 30 min, 1, 2, 3, 4, 5, 6, 24, 48 and

72 h. The water absorbed by capillary suction was measured by weighting the samples at the indicated times.

Microstructure

To investigate the effects of the silane treatments and the accelerated aging on the microstructure of the bio-concrete, weathered samples were examined by a scanning electron microscope (SEM) using a Secondary Electron (SE) model Hitachi TM 3000 from the transcoding of the energy emitted by electrons. In addition, energy-dispersive X-ray spectroscopy was performed on wood particles with and without silane impregnation to verify their chemical composition.

3 Results and Discussion

3.1 Wetting/Drying Mass Control

A control of mass (gain and loss) was performed during the wetting (W)/drying (D) cycles and in the between cycles (BC) period. Figure 2 shows the mass gain of control (CONTROL), surface skin protection (SSP) and impregnation of bio-aggregates (IBA) bio-concrete samples, as an average of 5 specimens.

From Fig. 2a, it is noted that after the first wetting a mass gain of approximately 15% occured followed by a loss of just 4% after the drying for control samples. In the third cycle, the samples already reached their maximum absorption limit for 30 min of immersion and presented a similar behavior, with gains and losses of approximately 4%, until the 10th cycle.

For surface skin protection samples (Fig. 2b), the first wetting led to a mass gain of 1.7%, followed by a drying that caused a mass loss of 1.4%, practically returning to the initial sample mass and showing the potential of the silane to protect against water ingress in the first hours. In the second cycle, there was a mass gain of 5% and a loss of 2%. The third cycle followed the same trend. After the period between cycles (BC), a new wetting/drying block (cycles 4 to 7) behaved similarly to the two previous cycles (2 and 3), in which there is a mass loss of around 2.7%. The third block (cycles 8 to 10) showed the same trend. The use of silane as SSP was quite efficient in the first cycle, and despite the increase in water absorption in the other cycles it presented lower average mass gain (6.6%) compared to CONTROL samples (19.2%) under the study conditions. Christodoulou et al. [12] affirm that the Si-O bond that connects the hydrophobic layer and the substrate concrete can be destroyed by alkaline solution over time, UV radiation, high temperature and other factors, resulting in deteriorated protective performance.

IBA samples (Fig. 2c) showed a similar percentage of mass gain and loss from cycle to cycle (approximately 3% and 2%, respectively), which leads to a progressive mass gain over the cycles. In the 10th cycle, with absorption potential of 7.5%, it was not perceived a stabilization of an absorption plateau as for CONTROL and SSP samples. Despite the absence of a surface protection, IBA was able to reduce the ingress of water into the composite by creating a hydrophobic film on the wood shavings.

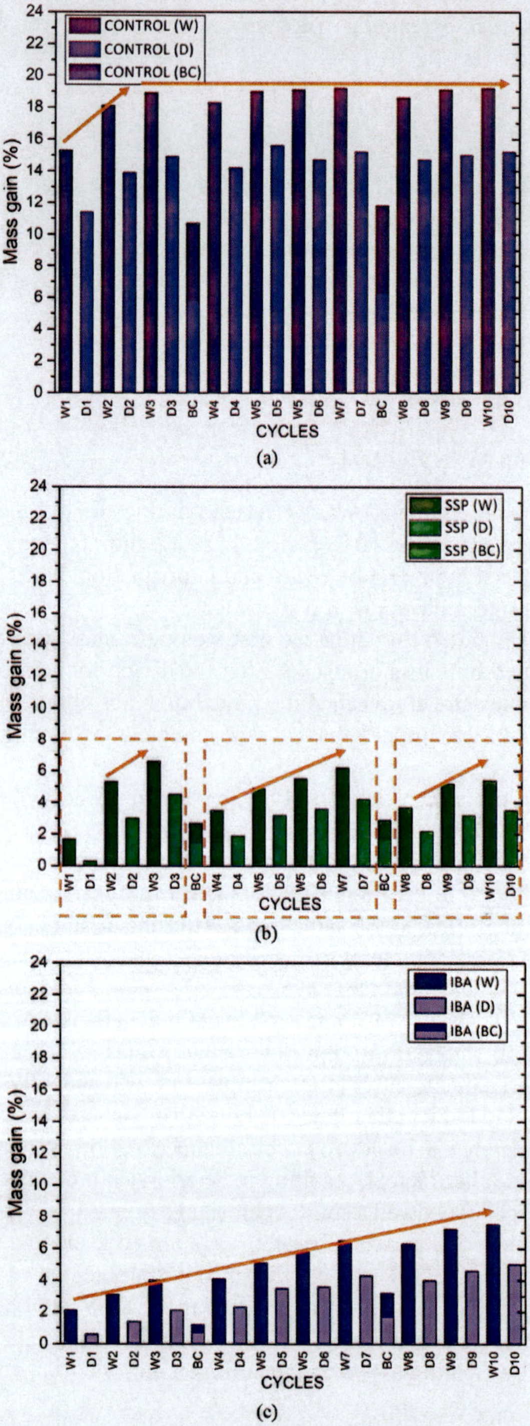

Fig. 2. Wetting/drying mass control of bio-concrete samples (a) CONT; (b) SSP and (c) IBA.

3.2 Functional Properties

Compressive Strength and Density

The compressive strength (fc_{28}) and density (ρ) values of the wood bio-concretes (weathered (W/D) and non-weathered) are shown in Fig. 3. Analyzing the results from Fig. 3a considering the non-weathered samples, an analysis of variance (ANOVA) was performed and there was no significant difference between CONTROL and SSP compressive strength, showing that the application of silane as surface treatment does not change this mechanical property of the bio-concrete. For IBA, a 23% increase in compressive strength was observed compared to CONTROL. This increase can be explained by the silane impregnation of the bio-aggregates, which changed the rheology of the mixture and made the bio-concrete denser, as can be seen in Fig. 3b.

The effect of wetting/drying cycles for CONTROL and IBA was similar, leading to a decrease of 13% and 9% of compressive strength, respectively. Behmahiddine et al. [23] explain a decrease in compressive strength after wetting/drying cycles of hemp concrete especially due to the degradation of the binder/bio-aggregate interface in addition to a possible increase in porosity, which weakens the mechanical behavior of the material. In addition, both showed greater water absorption capacity at the end of the wetting/drying cycles, compared to SSP. Despite being denser, CONTROL (W/D) has reduced mechanical strength due to the appearance of cracks in the transition zone between binder and bio-aggregate and in the bio-aggregate itself caused by the accelerated aging process. On the other hand, as the SSP samples showed a lower water absorption capacity until the last cycle, they did not present a substantial difference, according to ANOVA, in compressive strength compared to the non-weathered samples. Even after successive W/D cycles, the protective silane layer on the samples surface causes the bio-concrete to absorb less water in the wetting stage after the BC period, showing that the SSP samples were less damaged by the cycles than the others, which preserved their mechanical strength.

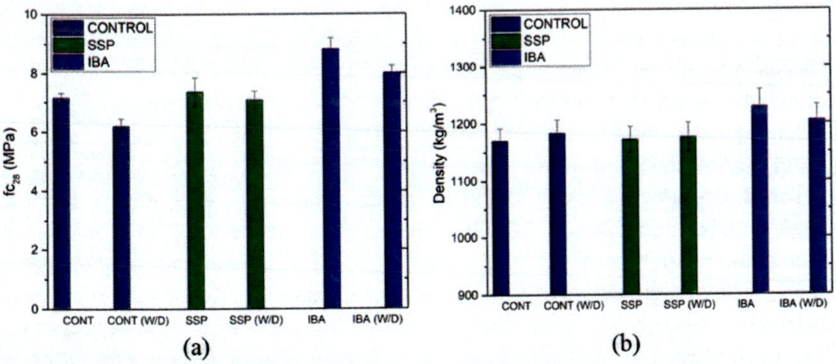

Fig. 3. Functional properties of bio-concrete samples before and after wetting/drying cycles (a) compressive strength, and (b) density.

Shen et al. [25] studied the efficiency of silane-based products as surface treatments for protecting degraded historic concrete and observed an increase in compressive strength of 16% using a water repellent based on silane/siloxane (50 wt. %) compared to untreated specimens, exhibiting a consolidation effect.

Capillarity Water Absorption

The water absorption curves of the silane treated bio-concretes are presented in Fig. 4. It can be seen that CONTROL samples presented a high potential to absorb water during the early hours, reaching about 20 kg/m^2 in 6 h of test and practically reaching saturation after 24 h of test (approximately 33 kg/m^2). Analyzing the difference of water absorption in silane application methods, SSP has a superficially hydrophobic characteristic capable of delaying the entrance of water into the sample during the first hours, with an absorption value of 9 kg/m^2 (55% lower than the CONTROL) in 6 h of test. On the other hand, this protection is temporary and the continuous contact of water with probably the sample disrupts the barrier of the silane film, meaning that after 48 h of testing SSP has similar absorption to CONTROL. IBA samples presented an intermediate behavior, because despite not having a surface protection, the water that entered into the cementitious paste did not have the same ease of penetration into the impregnated bio-aggregates. Thus, up to the 30 min of test, the absorption is low (3.5 kg/m^2), but it increased more than double after 1 h of test and the curve become closer to the CONTROL, reaching at 6 h of test a value of 17.3 kg/m^2 (13.5% less than CONTROL). The major difference is in the total water absorption capacity over time, which after 72 h of testing is 29.5 kg/m^2 (approximately 14% lower than CONTROL and SSP) and still does not show a level of stabilization. Zhu et al. [14], investigating the capillary water absorption coefficient of recycled aggregate concrete with surface silane treatment, found a decrease by 95% when compared to the control sample. This result can be related to the penetration depth of surface treated that depends on the type of hydrophobic agent applied, the porosity of the concrete, the initial moisture content, and the surface treatment of the concrete substrate [26].

By the capillarity water absorption tests, the bio-concrete (CONTROL) absorbed more water than the non-weathered ones. The wetting/drying cycles caused a tiny reduction in the absorbed water for CONTROL and IBA. For SSP, the behavior was inverse. The porous structure of bio-concrete deprived of any kind of treatment, even without the action of external agents, has a great potential for absorbing water, as expected, which consequently impairs the durability of this material. For SSP (W/D) samples, the wetting/drying cycles reduced the efficiency of the surface protection, and in the first hours. The absorption values of CONTROL after 1 and 6 h of testing were 150% and 56% superior when compared to SSP. However, in this same period, it presented a lower absorption value than control samples. After 24 h of testing, the values of SSP (W/D) are equal to SSP and show the same tendency for 48 h and 72 h. The behavior of the IBA (W/D) curve shows a greater absorption in the first minutes of the test, having a value 66% higher in 30 min of test compared to IBA. After 1 h of testing, the curves intersected and remained close until the end of the test. The rapid absorption in the first minutes may be related to the transition zone of the bio-aggregate and cementitious matrix that

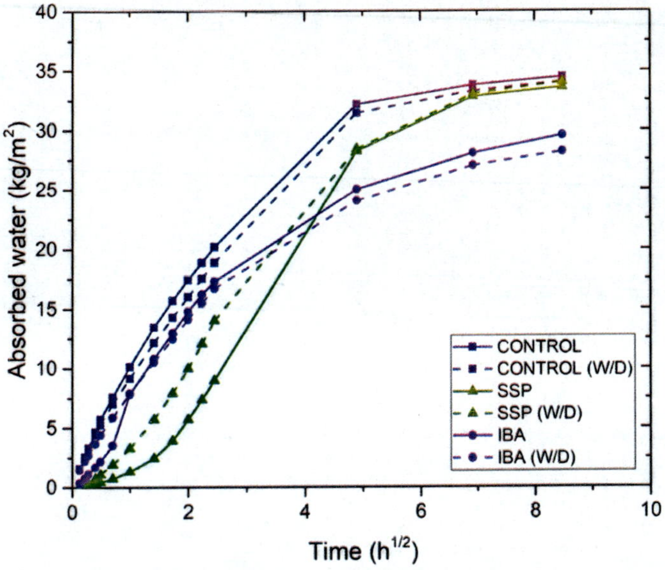

Fig. 4. Water absorption curves of the silane-modified bio-concretes non-weathered and after W/D cycles.

underwent changes during the wetting/drying cycles. But the final water absorption of IBA indicates the better performance when compared to SSP.

Microstructure

SEM images of internal bio-concretes samples before and after wetting/drying cycles can be observed in Fig. 5. For CONTROL (Fig. 5a) samples, the bio-aggregate (BA) is connected to the cementitious matrix (CM). CONTROL (Figs. 5a and 5b) and SSP (Fig. 5c) samples have a more porous structure due to a wood extractive that produces bubbles during the mixing process. These voids can further impair the durability of the material by an increase of concrete permeability, mainly under wetting/drying conditions (Figs. 5b and 5c). They create pathways for water transport and weaken the transition zone of the bio-aggregate and cementitious matrix. In addition, for CONTROL (W/D) samples, that promoted a faster entry of water, the drying process does not remove all the absorbed water, and the wood particle itself undergoes cracking. Figure 5d shows a sample of IBA, not as porous as others due to the impregnation process, but with the bio-aggregate-matrix interface cracked caused by wetting/drying cycles.

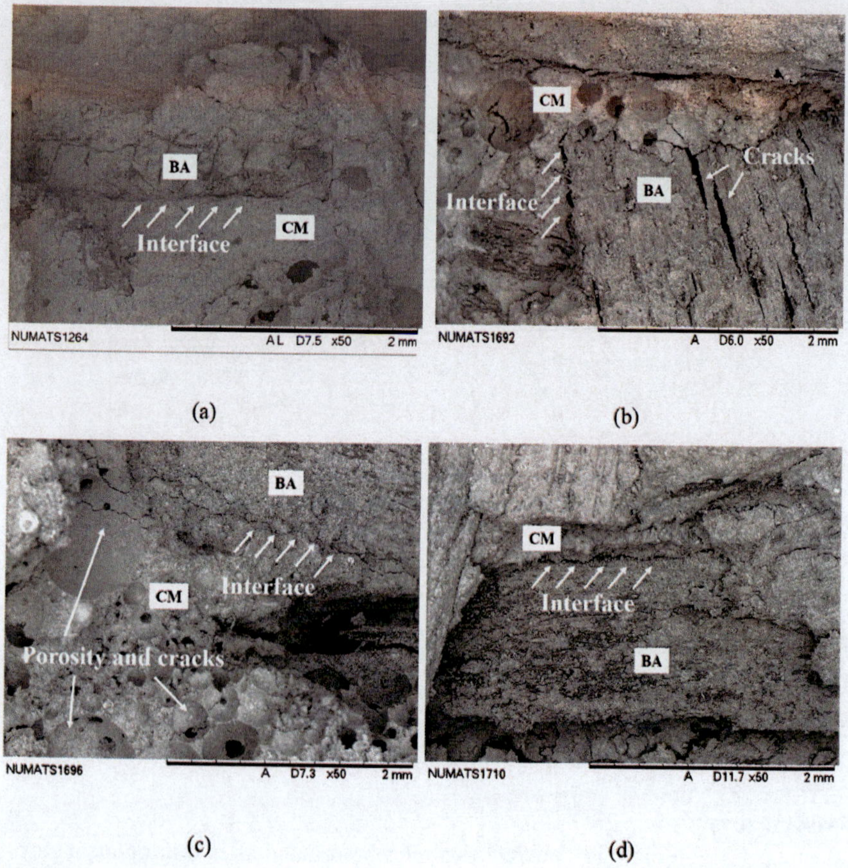

Fig. 5. SEM images of bio-concrete samples (a) CONTROL (b) CONTROL (W/D); (c) SSP (W/D) and (d) IBA (W/D).

4 Conclusions

The effect of silane application on the wood bio-concrete properties was evaluated. From the experimental results it is possible to draw the following conclusions:

- Both silane-based treatments studied proved to decrease the water absorption capacity of wood bio-concrete under immersion in the wetting stages;
- The wetting/drying cycles caused a decrease in compressive strength of 13% for CONTROL, 9% for IBA, and no change for SSP, showing that the surface treatment was more efficient for this mechanical property;
- According to capillary water absorption results, SSP samples had a temporary protection which delayed the ingress of water in the first hours, but up to 48 h of test, the water absorbed was the same with the CONTROL samples, while IBA has a 14% lower total water absorption capacity;
- Wetting/drying cycles negatively affected the performance of water absorption from the early hours for IBA and SSP samples;

- SEM images show the cracked transition zone of the bio-aggregate and cementitious matrix of bio-concrete, due to the aging process, which signals a reduction in the durability of the material.

Acknowledgments. This research was based upon a R&D Project from ANEEL - National Agency, "Uso de Bioconcretos e Bio-MMFs de Baixo Impacto Ambiental Visando o Aumento da Eficiência Energética de Prédios Públicos" – PD.0394-1719/2017, supported by Eletrobras Furnas with NUMATS/POLI/COPPE/UFRJ cooperation.

References

1. Amziane, S., Sonebi, M.: Overview on biobased building material made with plant aggregate. RILEM Tech. Lett. **1**, 31–38 (2016)
2. Caldas, L.R., Saraiva, A.B., Andreola, V.M., Toledo Filho, R.D.: Bamboo bio-concrete as an alternative for buildings' climate change mitigation and adaptation. Constr. Build. Mater. **263**, 120652 (2020)
3. Collet, F.: Hygric and thermal properties of bio-aggregate based building materials. In: Amziane, S., Collet, F. (eds.) Bio-aggregates Based Building Materials. RSR, vol. 23, pp. 125–147. Springer, Dordrecht (2017). https://doi.org/10.1007/978-94-024-1031-0_6
4. Berger, F., Gauvin, F., Brouwers, H.J.H.: The recycling potential of wood waste into wood-wool/cement composite. Constr. Build. Mater. **260**, 119786 (2020)
5. Bourzik, O., Akkouri, N., Baba, K., Nounah, A.: Study of the effect of wood waste powder on the properties of concrete. Mater. Today Proc. **58**, 1459–1463 (2022)
6. Da Gloria, M.H.Y.R., Toledo Filho, R.D.: Innovative sandwich panels made of wood bio-concrete and sisal fiber reinforced cement composites. Constr. Build. Mater. **272**, 121636 (2021)
7. Delannoy, G., et al.: Durability of hemp concretes exposed to accelerated environmental aging. Constr. Build. Mater. **252**, 119043 (2020)
8. Bennai, F., Issaadi, N., Abahri, K., Belarbi, R., Tahakourt, A.: Experimental characterization of thermal and hygric properties of hemp concrete with consideration of the material age evolution. Heat Mass Transf. **54**(4), 1189–1197 (2017). https://doi.org/10.1007/s00231-017-2221-2
9. Nozahic, V., Amziane, S., Torrent, G., Saïdi, K., De Baynast, H.: Design of green concrete made of plant-derived aggregates and a pumice–lime binder. Cem. Concr. Compos. **34**(2), 231–241 (2012)
10. Sun, H.Y., Yang, Z., Shan, G.L., Xu, N., Sun, G.X.: Current situation of research and application of silicone water repellent for protecting reinforced concrete. In: 7th International Conference on Bridge Maintenance, Safety and Management, pp. 7–11 (2014)
11. Concrete Society: Technical Report 50. Guide to surface treatments for protection and enhancement of concrete. Surrey, United Kingdom (1997)
12. Christodoulou, C., Goodier, C.I., Austin, S.A., Webb, J., Glass, G.K.: Long-term performance of surface impregnation of reinforced concrete structures with silane. Constr. Build. Mater. **48**, 708–716 (2013)
13. De Vries, J., Polder, R.B.: Hydrophobic treatment of concrete. Constr. Build. Mater. **11**(4), 259–265 (1997)
14. Zhu, Y.G., Kou, S.C., Poon, C.S., Dai, J.G., Li, Q.Y.: Influence of silane-based water repellent on the durability properties of recycled aggregate concrete. Cem. Concr. Compos. **35**(1), 32–38 (2013)

15. Xie, Y., Hill, C.A., Xiao, Z., Militz, H., Mai, C.: Silane coupling agents used for natural fiber/polymer composites: a review. Compos. A Appl. Sci. Manuf. **41**(7), 806–819 (2010)
16. Aguiar, A.L.D., Da Gloria, M.Y.R., Toledo Filho, R.D.: Influência do tratamento da serragem de madeira na resistência à compressão do bio-concreto de madeira. In: Congresso de Construção Civil, Brasília (2020)
17. Associação Brasileira de Normas Técnicas: Agregado miúdo – Determinação da densidade e da absorção de água. NBR 16916 (2021)
18. Associação Brasileira de Normas Técnicas: Agregado graúdo – Determinação do teor de umidade total – Método de ensaio. NBR 9939 (2011)
19. Abreu, H.D.S., et al.: Métodos de análise em química da madeira. Floresta e ambiente 20 (2006)
20. Aguiar, A.L.D., Da Gloria, M.Y.R., Hasparyk, N.P., Toledo Filho, R.D.: Influência do teor de bio-agregados na resistência à compressão e à tração de bioconcretos de madeira. In: Congresso Brasileiro de Concreto, Brasília (2022)
21. Rocha, M.A.F., Gomes, B.M.C., Aguiar, A.L.D., Landesmann, A., Hasparyk, N.P., Toledo Filho, R.D.: Determinação das propriedades de reação ao fogo de bioagregados e bioconcreto de bambu. In: Congresso Brasileiro de Concreto, Brasília (2022)
22. Al Abdallah, H., Abu-Jdayil, B., Iqbal, M.Z.: Improvement of mechanical properties and water resistance of bio-based thermal insulation material via silane treatment. J. Clean. Prod. **346**, 131242 (2022)
23. Benmahiddine, F., Bennai, F., Cherif, R., Belarbi, R., Tahakourt, A., Abahri, K.: Experimental investigation on the influence of immersion/drying cycles on the hygrothermal and mechanical properties of hemp concrete. J. Build. Eng. **32**, 101758 (2020)
24. Associação Brasileira de Normas Técnicas. Concreto-Ensaio de compressão de corpos de prova cilíndricos. NBR 5739 (2018)
25. Shen, L., Jiang, H., Wang, T., Chen, K., Zhang, H.: Performance of silane-based surface treatments for protecting degraded historic concrete. Prog. Org. Coat. **129**, 209–216 (2019)
26. Dai, J., Akira, Y., Yokota, H., Wittmann, F.H.: Surface impregnation of pre-conditioned concrete subjected to seawater immersion test. Restor. Build. Monum. **13**(4), 229–240 (2007)

Treatment Protocol Efficiency of Plant Aggregates to Their Influence on Swelling and Shrinkage

C. Achour(✉), S. Remond, and N. Belayachi

Univ. Orleans, Univ. Tours, INSA-CVL, LaMé – EA7494, 8 Rue Léonard De Vinci, 45072 Orléans, France
chafic.achour@univ-orleans.fr

Abstract. The use of plant resources for the development of insulation biocomposites, demands a solid understanding of their morphological, physico-chemical behaviour, and mechanical properties in relation to their conditions of use. The behavior of bio-composites and their durability are influenced by the high sensitivity of plant aggregates to hydro/hygrothermal conditions. This sensitivity of vegetable aggregates, due to water absorption, can be controlled by treatments. Several factors are important to assess the impact of the treatment on the properties of the aggregates. Among them, the used concentration, duration, and temperature. The objective of this research is to adjust the treatment procedure for vegetable aggregates to improve the behavior of the bio-composites. The study is interested in investigating the variations of the treatment duration and the rinsing time affects the aggregates.

Two salts are used in this work, sodium bicarbonate $NaHCO_3$ and sodium chloride NaCl at a concentration of 10% by weight to evaluate the swelling and shrinking phenomena. After completing the treatment, the aggregates are exposed to wetting and drying cycles to evaluate the treatment's efficiency on swelling and shrinking behavior. During the treatment, pH measurements are taken, and the dimensional variations of the aggregates are analyzed under the microscope.

The pH measurement during treatment shows an increase for $NaHCO_3$-treated aggregates compared to NaCL and untreated aggregates. At the same time, the pH measurement during rinsing decreases with the removal of salts after a 20-min rinse for $NaHCO_3$ versus 5 min for NaCl. Microscopic analysis confirms the effectiveness of $NaHCO_3$ treated aggregates in decreasing swelling and shrinkage behavior, as well as their durability, as they do not show a decline in effectiveness over time. Finally, it is effective to treat the aggregates with salt for 3 days, dry them at a moderate temperature, and then rinse them for 20 min to remove any remaining salt and impurities.

Keywords: Treatment · vegetable aggregates · swelling · shrinking

S. Amziane et al. (Eds.): ICBBM 2023, RILEM Bookseries 45, pp. 171–184, 2023.
https://doi.org/10.1007/978-3-031-33465-8_14

1 Introduction

Reducing the environmental impact of construction materials is a key concern in the building and construction industry. In this context, the interest in biobased material is growing for a few years. These materials are considered more sustainable and environmentally friendly than traditional materials [1]. The plant aggregates can guarantee not only the reduction of carbon emissions and energy consumption, but also provide thermal comfort [2]. Thanks to their interesting thermal, hydric and mechanical performances, they qualify as a good insulator for buildings [3].

However, as any building material, they are exposed to various environmental use conditions of long term such as humidity, temperature, or UV radiation. Moreover, studies on plant aggregates have revealed their hygroscopic behavior and high sensitivity to water vapor and contact with liquid water. The ability to absorb moisture help to control the humidity in each environment to maintain the freshness and quality of building comfort. Nevertheless, this affects their behavior by causing damage bio-composites. This consequence is highly related to the dimensional variation which is influenced by hygro and hydrothermal conditions [4]. When plant aggregates are exposed to environmental conditions such as humidity and temperature variations, they are likely to absorb moisture, resulting in swelling [5–7]. When subjected to repetitive immersion-drying cycles, the aggregates experience degradation resulting in a loss of swelling and shrinkage due to the combined effects of moisture absorption and material weakening [8].

To reduce the amount of absorbed water, there are a large number of treatments in the literature depending on the specific problem and the desired result. Treating the aggregates, weather by a chemical or physical treatment improve the mechanical, structural, or hygroscopic properties by modifying their chemical component. In this regard, studies on component modification using salt solutions show their benefit in decreasing their water absorption rate. These modifications are generally based on the use of hydroxyl functional groups capable of changing the composition of the aggregate structure [9]. In this way, the tendency to absorb moisture is reduced. The treatment principle consists in immersing the aggregates in a salt solution, for a studied time, and carry out the rinsing to release the quantity of salt. In their work, Fiore and al. [10, 11] soaks a concentration of 10% by weight of $NaHCO_3$ for a period of 5 days to treat the sisal fiber in order to improve their tensile and flexural properties. The same concentration was used by dos Santos et al. [12] for a duration of 96 h to treat coconut fibers to increase their mechanical properties. However, for Chaitanya et al. [13], a duration of 72 h presented optimal strength properties for aloe vera fibers. The second salt used to soak the aggregates in the literature is sodium chloride. Some attempts successfully proved the decrease in water absorption by treating the aggregates with NaCl [14] and subsequently their swelling and shrinkage. This is a commonly used salt for clay soils to decrease their liquid and plastic limit and plasticity index [15]. Nevertheless, the above-mentioned authors focused in their studies on the optimization of salt concentration and the effect of exposure time, and none of them studied the effect of rinsing and the duration of treatment and their response on the swelling and shrinking behavior.

For this purpose, an attempt is made to study the efficiency of the treatment protocol on vegetable aggregates. The objective is to propose an experimental protocol for the treatment of plant aggregates with emphasis on the effects of the treatment on their swelling and shrinkage behavior. This objective is achieved through two processes. The first one consists in studying the influence of the treatment time on the aggregates. The second part of the study is to focus on the influence of the rinsing time to remove salt to reach a stable pH after treatment [16] to avoid their later impact. Therefore, these two processes are monitored by measuring the pH of the salt solutions during the treatment of the aggregates in parallel with the microscopic analysis. The pH measurement allows the identification of interactions between the salt solution and the aggregates. The microscopic analysis confirms the influence of the treatments on the aggregates. Furthermore, the aggregates will be subjected to four cycles of wetting and drying after the end of each treatment to measure swelling and shrinkage to examine the effectiveness of the treatment on the aggregates.

2 Materials and Methods

2.1 Materials

Four types of aggregates are studied in this work, wheat straw, rapeseed straw, pith and sunflower bark and are all collected in the Centre Val de Loire region. These aggregates have been used for biocomposites developed for thermal insulation of buildings in previous work of the team [17, 18]. The aggregates chosen are first cleaned by a dry brushing to remove the dirt and debris from the aggregates. Then, they are cut to obtain a representative sample for microscopic analysis. Through this study, we tested the influence of two salt treatments, sodium chloride NaCl [19] and sodium bicarbonate $NaHCO_3$, in reducing the rate of water absorption and swelling and shrinkage of aggregates [20]. The salts sodium chloride (NaCl, M = 58,44 g/mol, $\rho = 2.17 g/cm^3$) and sodium hydrogen carbonate ($NaHCO_3$, M = 84.01 g/mol, $\rho = 2.21 g/cm^3$) are purchased from CARLROTH, a German company.

2.2 Aggregates Treatment

The concentration used to treat the aggregates is 10% for both salts as this is the ideal concentration used in other research cited previously [10, 12]. Portions of the aggregates are soaked in two different solutions of $NaHCO_3$ and NaCl of 10% concentration by weight. Figure 1 illustrates the protocol followed for the aggregate treatment process. This protocol is divided into two sequences. The first sequence is to study the soaking time of the aggregates in the salt solutions. The second process is to study the influence of rinsing time after each day of treatment. First, the salts are dissolved in preheated water at a temperature between 25 and 30 °C since this is the optimal temperature to dissolve the salts. Then, the prepared samples are added and stirred regularly. Portions of 10 samples per aggregate are taken every 24 h to reach a total duration of 120 h. The extracted samples are dried at 60 °C. These aggregates are then totally immersed for 5 min to measure the pH of the solution and to check the presence or disappearance of salt. Then, these same aggregates are put back in the oven for a new immersion cycle until the salt disappears. They will then be exposed to a wetting and drying cycle to test the efficiency of the treatment.

Microscopic analysis is performed at the beginning of the experiment in the dry condition and was continued through both sequences to assess the dimensional change of the vegetable aggregates. Ten samples are taken per aggregate per treatment day for analysis. These aggregates are measured before the treatment, after the treatment, during each immersion and drying cycle and finally after wetting and drying cycles.

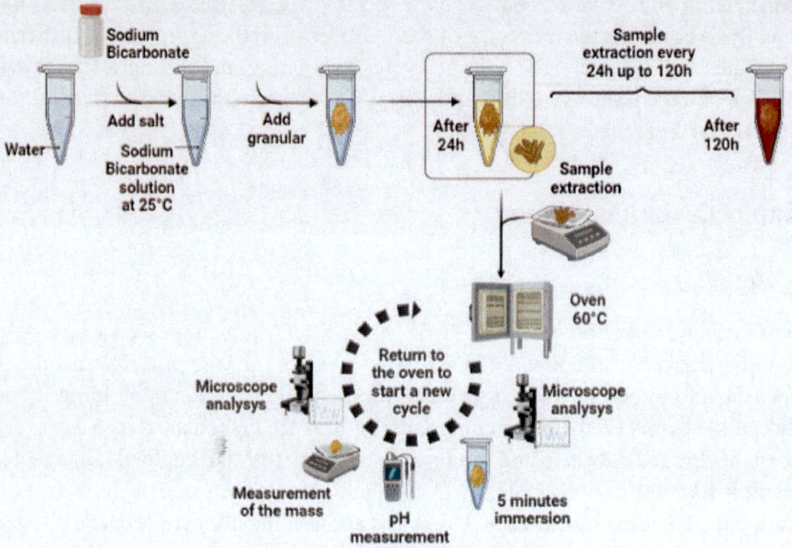

Fig. 1. Treatment protocol.

2.3 Efficiency of Treatment: Wetting-Drying Cycles

The swelling and shrinkage of aggregates due to wetting and drying cycles is observed under a microscope to assess whether the behavior remains constant after several cycles and whether the treatment maintains its effectiveness. This test is performed by subjecting the aggregates to wetting and drying cycles, during which they are wetted at 90% relative humidity at 23 °C and dried at 60 °C with 0% relative humidity. The steps of these cycles are shown in Fig. 2.

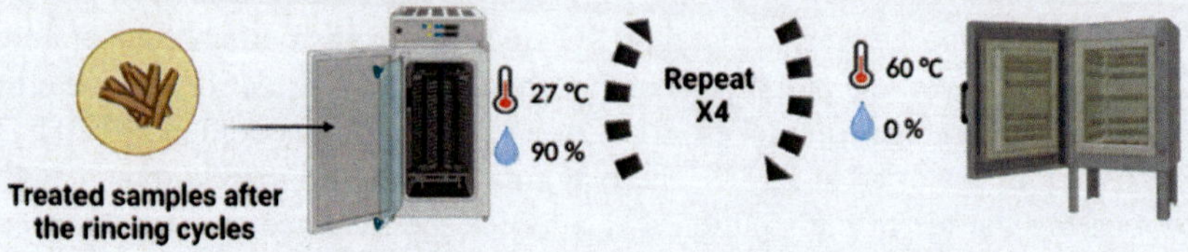

Fig. 2. Repetitive wetting and drying cycle principle

2.4 Microscopic Measurement

The influence of the variation of the treatment time on the morphology of the treated aggregates compared to the untreated aggregates is examined using an optical microscope. An image analysis of the aggregates is carried out using photos taken with the optical microscope "Leica Microsystems (Schweiz) AG". These images allow to examine the influence of the above processes on the morphology of the treated aggregates compared to the untreated ones. Figure 3 shows the microstructure of each type of aggregate and demonstrates the significant porosity that causes their highly hygroscopic behavior.

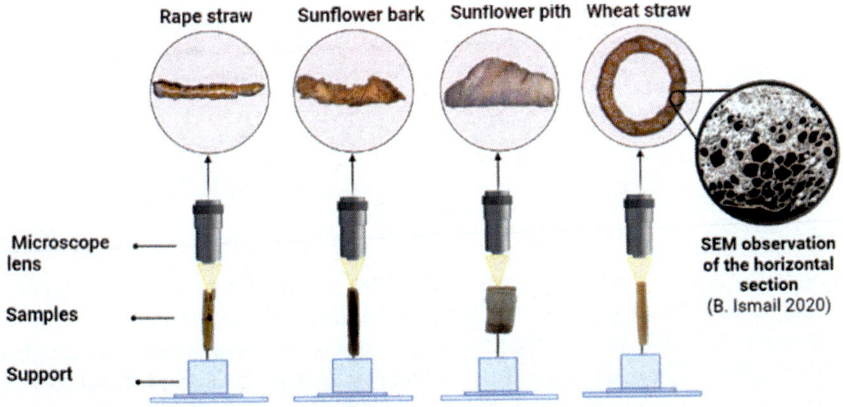

Fig. 3. Different types of aggregates studied as seen via the microscope a) wheat straw, b) rape straw, c) sunflower bark and d) sunflower pith

The microscopic measurement is based on surface measurements. It consists in delimiting the contour of the aggregate and measuring the surface of the aggregate by a region-based segmentation as presented in Fig. 4. This method consists in using the properties of the surface to group the pixels of similar appearance different from the black background. The percentage of variation between two states is defined by the ratio of the variation between the two surfaces to the initial surface. Swelling is thus defined by a positive percentage while a negative sign indicates shrinkage. The measurement error is limited by the distribution of the contour on the aggregate which can be result

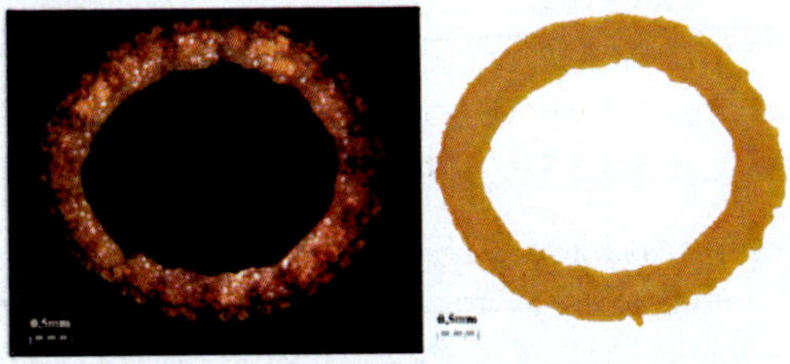

Fig. 4. Delimiting the contour of the aggregates for surface measurements.

from the number of pixels measured, the color and shape of the captured image with a margin of ±2% per measurement.

3 Results and Discussion

3.1 Impact of Treatment

The aqueous sodium bicarbonate solution produces a slightly alkaline solution due to the formation of sodium and hydroxide ions, as shown in Eq. (1) and (2) from which the basicity of the solution over time. The fiber containing hydroxyl groups and Na + ions react as Eq. (3).

$$NaHCO_3 + H_2O \rightarrow Na^+ + HCO_3^- \quad (1)$$

$$HCO_3^- + H_2O \rightarrow H_2CO_3^+ + OH^- \quad (2)$$

$$\text{Fiber-OH} + Na^+ + OH^- \rightarrow \text{Fiber-O}^-Na^+ + H_2O + \text{impurities} \quad (3)$$

The amount of Na + ions depend on the concentration and basicity of the solutions represented by the pH values. The histograms in Figs. 5, 6 and 7 presents the pH variations of the aggregates treated by $NaHCO_3$, NaCl and of untreated aggregates.

Record indicates an increase in basicity of the samples over time to reach a pH of 9 for the $NaHCO_3$ case. In contrast to the samples that are treated with NaCl, which showed no regular increase but a decrease over days. Nevertheless, untreated aggregates showed an initial variation defined by a decrease during the first 2 days followed by an increase of pH. One of the causes of the decrease in pH in aggregates when water is added can be attributed to the presence of microorganisms within the aggregates [21]. The measurements are repeated four to five times to reduce the error of the measurement to be less than ±4%.

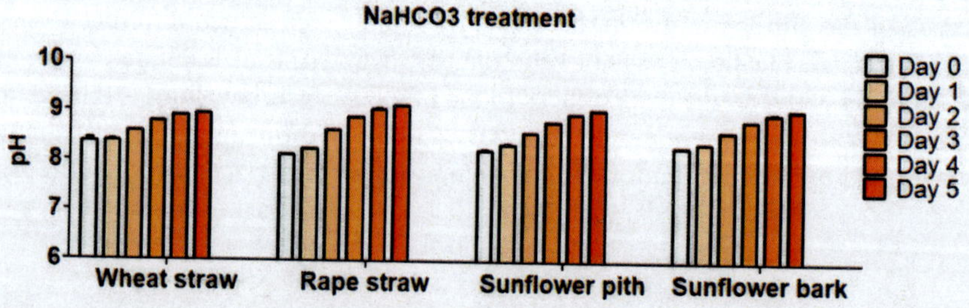

Fig. 5. pH measurement of different aggregates treated with $NaHCO_3$ solution over days

Treating the aggregates has also a consequence on water absorption since it can causes the change of the chemical composition. Water absorption behavior of wheat straw after treatment is highlighted in Fig. 8 allowing to further define the number of days of treatment. For a better understanding of the water absorption behavior, histograms

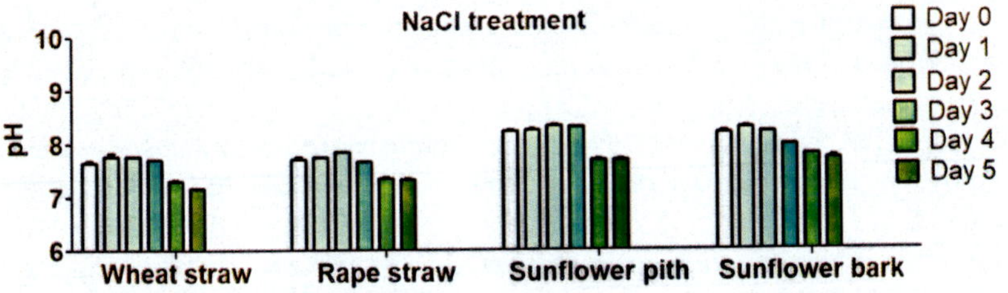

Fig. 6. pH measurement of different aggregates treated with NaCl solution over days

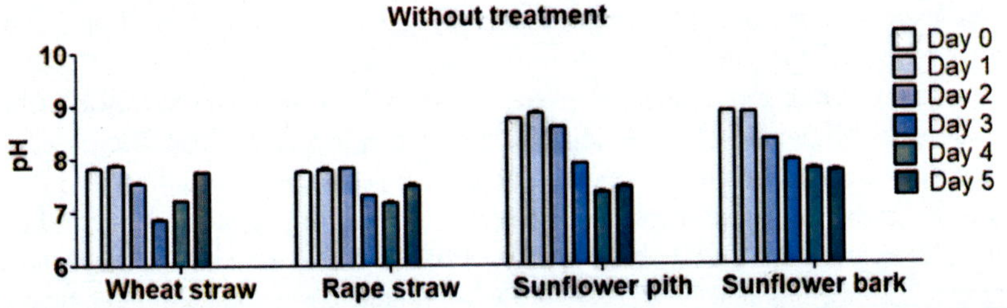

Fig. 7. pH measurement of different untreated aggregates over days.

are presented with the percentage of weight gain with respect to the treatment days. The decrease in water absorption of the treated aggregates is more significant for the aggregates treated with $NaHCO_3$ rather than for NaCl.

The treatment with $NaHCO_3$ reduces the water absorption of the aggregates per day of treatment, making them more resistant to moisture. This reduction is mainly due to the change in aggregate composition by the $NaHCO_3$ treatment. The treatment increases the cellulose content by the partial elimination of hemicellulose [22] which is highly hydrophilic controversy of lignin which is hydrophobic. From the third day of treatment onwards, the moisture absorption becomes 3% and decreases with the day of treatment up to 2.6% for 120 h. Meanwhile, NaCl treated wheat straw remains the same, around 5.5%.

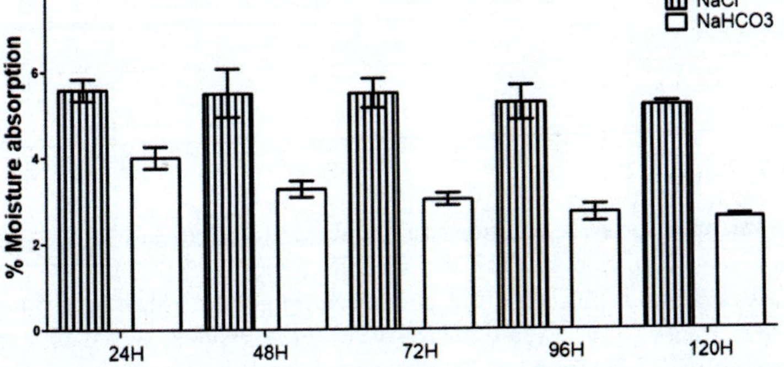

Fig. 8. Moisture absorption as a function treatment time for wheat straw aggregates

Figure 9 illustrates the variations in measurements of the dimensions of the extracted portions before and after treatment. Aggregates treated with $NaHCO_3$ show a reduction in surface of the treated aggregates from the first day of treatment. After 72 h of treatment, $NaHCO_3$ solution begins to have a high effect on dimensional variation. As the treatment goes on by day, the surface variation continues to decrease until it reaches a reduction in size of 9% at 120 h.

This agrees with the results of Raharjo et al. [23] who claimed that over the third day of treatment with sodium bicarbonate, the chemical composition of aggregates such as cellulose, lignin and hemicellulose, is modified. Meanwhile, according to Chaitanya et al., beyond 72 h of treatment, the tensile and flexural strength marginally decline. However, excessive removal of hemicellulose beyond 72 h of treatment time leads to a decrease in the load carrying capability of the aggregates.

According to Pejic et al. [24], lignin has an impact on the water sorption of hemp, and it has been shown that increasing the lignin content leads to a decrease in the water retention capacity of hemp fibers. Garat et al. [25] also found that higher levels of lignin are correlated with a lower percentage of swelling. This will strengthen the surface of the aggregates and make it rougher [26]. It can be concluded that after 72 h of treatment, the mechanical properties of the aggregates are enhanced, as indicated by other authors [11] and [9]. The aggregates exhibit a decrease in swelling and shrinkage behavior according to our results. However, NaCl treated aggregates does not prove any dimensional decrease, but a slight increase of for most of them around 2%.

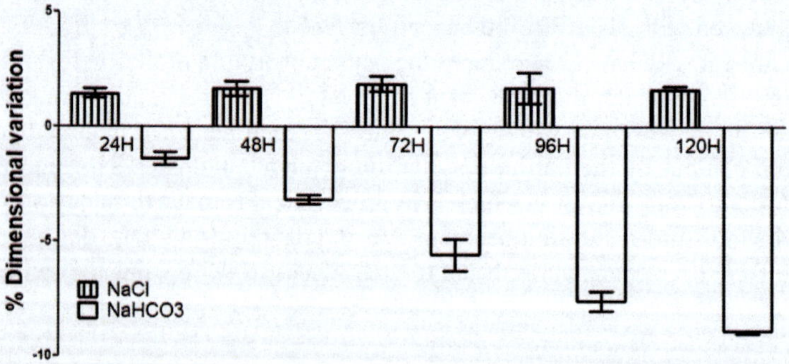

Fig. 9. Difference in measurement between the initial state before treatment and after treatment of the extracted portions per day.

3.2 Rinsing Impact

During the treatment process, samples are extracted and dried in the oven. Then, the rinsing and drying process begins. The aim is to determine the recommended rinsing time to remove excess salt from the aggregate, and to determine how time it takes for the pH to become stable. Figure 10 and 11, present the pH of the aggregates rinsed after extraction during a duration of 72 h of treatment. These aggregates are immersed and dried until they have a stable pH value. The pH of the aggregate treated with NaHCO3 was 9.2 during the treatment stage. After the aggregates are dried, the pH of the aggregates increases to 10.2 after the first rinsing. Then, the pH decreases with each subsequent rinsing until it reaches a stable value of 7.8. This variation is also observed in the extracted samples treated for 24 h, up to 120 h. Indeed, when the treated aggregates are dried in the oven at a temperature of 50 °C, the salt of the sodium bicarbonate decomposes under the effect of heat with loss of CO_2 and H_2O. In fact, according to [9], this occurs more normally with high basic pH. The authors [8] found the formation of amine bicarbonate salts in the drying process after treating their aggregate. It is believed that sodium bicarbonate with water creates carbonic acid and hydroxyl component which can react with hydroxyl functional group of flax fiber as similar to alkali treatment and thereby increase the tensile and interfacial properties of composites [27, 28].

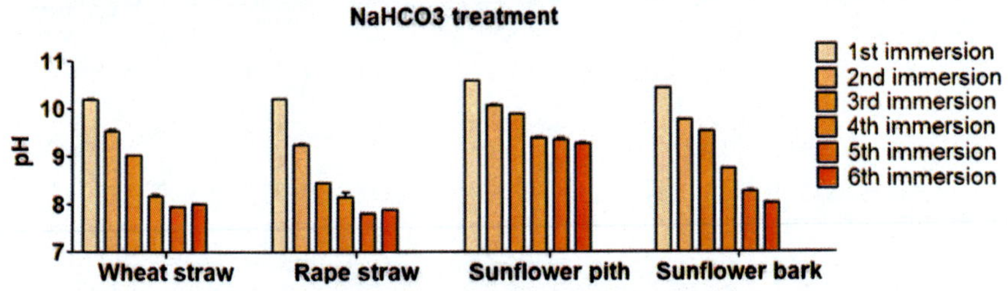

Fig. 10. pH measurement of $NaHCO_3$ treated aggregates after several rinses.

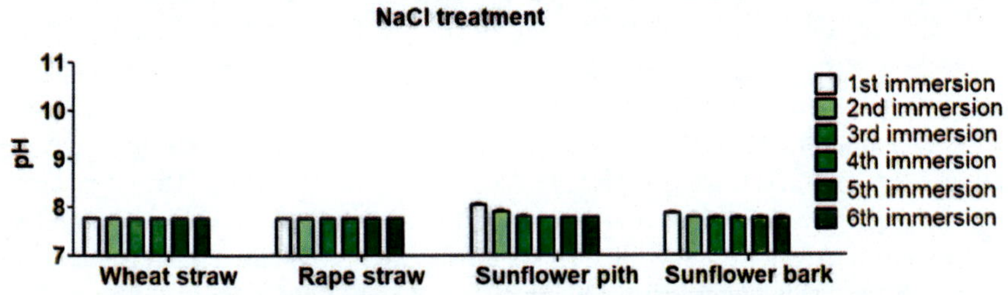

Fig. 11. pH measurement of NaCl treated aggregates after several rinses.

At the same time, the drying of the aggregates at moderate temperature creates a reaction of the amine groups of the sodium bicarbonate with the carbonyl groups of the cellulose to form schiff bases, which is responsible for the yellowing of the material [9]. Indeed, this is what the aggregates treated with sodium bicarbonate has experienced in Fig. 12. The yellowing marked much more the wheat straw, rapeseed straw, than the bark and pith of sunflower. This observation is different from those of treated with NaCl.

After the aggregates have been treated and dried in the oven, they are rinsed several times for 5 min each. The water in which the aggregates were rinsed became lighter and more transparent after the fourth rinse as shown in Fig. 13. The color change is visible when comparing the first rinse (on the left) with the water after the fifth rinse (on the right). Once the aggregates have been rinsed for a total of 20 min, the treatment process is complete, and the aggregates are ready to be used.

Fig. 12. Color change of wheat straw over time: untreated, NaCl-treated, NaHCO3-treated (left to right).

Fig. 13. The rinsing process and color variation of sunflower bark treated aggregates with $NaHCO_3$ (from left to right).

In summary, the rinsing test identifies the time required to rinse the treated aggregate to remove the amount of salt on the surface. Figure 14 shows the surface of $NaHCO_3$ treated aggregate after 4 rinses. The amount of salt is quite remarkable after 5 min of rinsing and gradually decreases after the fourth rinsing, for a total of 20 min.

According to the results, it took 20 min of rinsing to remove the salt from $NaHCO_3$ treated aggregate, while only 5 min of rinsing was needed for NaCl treated aggregate. Therefore, the treatment was repeated using 5 min of rinsing for NaCl treated aggregate and 20 min for $NaHCO_3$ treated aggregate. Microscopic images (Figs. 15a and c before rinsing the aggregates, 15b and d after rinsing the aggregates) show that the salt was completely removed after the specified rinsing times for NaCl treated aggregates, 5 min, and for $NaHCO_3$ treated aggregates, 20 min.

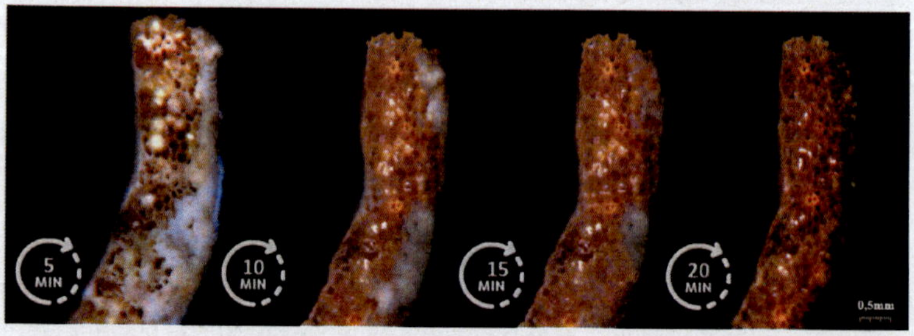

Fig. 14. Salt disappearance during time for nahco3 treated aggregates.

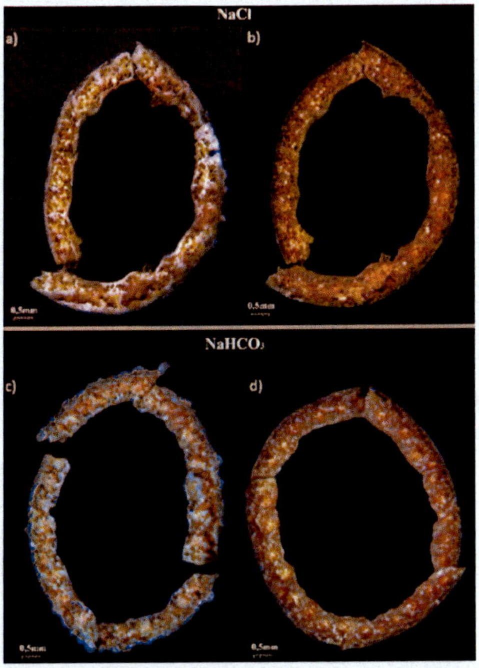

Fig. 15. Surfaces of treated aggregates: a) and c) before rinsing and b) after 5 min of rinsing for NaCl and d) after 20 min of rinsing for $NaHCO_3$.

3.3 Wetting and Drying Cycles

The percentage of dimensional change in the aggregates after wetting and drying cycles is shown in Fig. 16. Treatment with Sodium bicarbonate (NAHCO3) leads to a significant decrease in swelling by 60% ± 3%. This can be seen by comparing the swelling of untreated aggregates (21%) to that of aggregates treated with NAHCO3 (8%). Based on the data, the $NaHCO_3$ treatment is more effective and durable than NaCl treatment, as it does not show a decrease in effectiveness over time. The untreated samples began to become extremely soft, which impacted the swelling measurements.

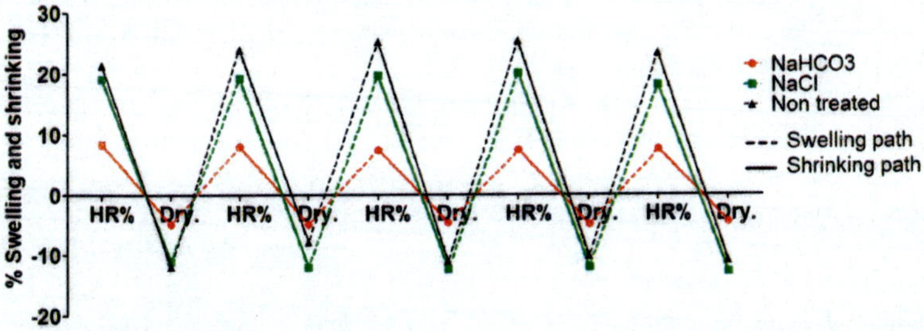

Fig. 16. Humidification and drying cycle for wheat straw for $NaHCO_3$ treated aggregates, NaCl treated aggregates and non-treated aggregates.

4 Conclusion

The study of the effectiveness of the treatment protocol of plant aggregates and the treatment effect on swelling and shrinkage is investigated. It is important to adapt a treatment protocol on vegetable aggregates that will have an influence on the bio-composite. According to the results obtained from pH measurements, microscopic analysis and weight analysis, the adaptation of the treatment for 3 days, followed by a drying at moderate temperature at 50 °C and then a rinsing of 20 min is recommended for $NaHCO_3$ treatment while only 5 min of rinsing for NaCl treatment.

For aggregates treated with 10% sodium carbonate, there is a notable reduction of swelling and shrinkage by 60 $\pm$ 3% in comparison to untreated aggregates. However, the NaCl-treated aggregates did not show a high enough rate of swelling and shrinkage prevention. A remarkable improvement of sodium bicarbonate treated aggregates to decrease their dimension variation is related to the removal of hemicellulose from the interfibrillar region according to the literature. It reduces water absorption up to 2.6% for 120 h and consequently swelling and shrinkage up to 10%.

Acknowledgments. The authors gratefully acknowledge the "REGION CENTRE VAL DE LOIRE" for the financial support of the research program MATBIO.

References

1. Yadav, M., Agarwal, M.: Biobased building materials for sustainable future: an overview. Mater. Today Proc. **43**, 2895–2902 (2021). https://doi.org/10.1016/j.matpr.2021.01.165
2. Asadi, E., da Silva, M.G., Antunes, C.H., Dias, L.: A multi-objective optimization model for building retrofit strategies using TRNSYS simulations. GenOpt. MATLAB. Build. Environ. **56**, 370–378 (2012). https://doi.org/10.1016/j.buildenv.2012.04.005
3. Ismail, B., Belayachi, N., Hoxha, D.: Hygric properties of wheat straw biocomposite containing natural additives intended for thermal insulation of buildings. Constr. Build. Mater. **317**, 126049 (2022). https://doi.org/10.1016/j.conbuildmat.2021.126049
4. Ismail, B.: Contribution au développement et optimisation d'un système composite biosourcé-enduit de protection pour l'isolation thermique de bâtiment. These de doctorat, Orléans (2020). http://www.theses.fr/2020ORLE3112. Accessed20 July 2022
5. Garat, W., Corn, S., Le Moigne, N., Beaugrand, J., Bergeret, A.: Analysis of the morphometric variations in natural fibres by automated laser scanning: towards an efficient and reliable assessment of the cross-sectional area. Compos. Part Appl. Sci. Manuf. **108**, 114–123 (2018). https://doi.org/10.1016/j.compositesa.2018.02.018
6. Rima, A., Abahri, K., Bennai, F., El Hachem, C., Bonnet, M.: Microscopic estimation of swelling and shrinkage of hemp concrete in response to relative humidity variations. J. Build. Eng. **43**, 102929 (2021). https://doi.org/10.1016/j.jobe.2021.102929
7. Achour, C., Remond, S., Belayachi, N.: Mesure du gonflement-retrait des granulats végétaux par analyse d'image. Acad. J. Civ. Eng. 104–107 (2022). https://doi.org/10.26168/AJCE.40.1.26
8. Sheridan, J., Sonebi, M., Taylor, S., Amziane, S.: The effect of long term weathering on hemp and rapeseed concrete. Cem. Concr. Res. **131**, 106014 (2020). https://doi.org/10.1016/j.cemconres.2020.106014

9. De La Orden, M.U., Matías, M.C., Martínez Urreaga, J.: Spectroscopic study of the modification of cellulose with polyethylenimines: cellulose modification with polyethylenimines. J. Appl. Polym. Sci. **92**(4), 2196–2202 (2004). https://doi.org/10.1002/app.20236
10. Fiore, V., Sanfilippo, C., Calabrese, L.: Influence of sodium bicarbonate treatment on the aging resistance of natural fiber reinforced polymer composites under marine environment. Polym. Test. **80**, 106100 (2019). https://doi.org/10.1016/j.polymertesting.2019.106100
11. Fiore, V., Scalici, T., Nicoletti, F., Vitale, G., Prestipino, M., Valenza, A.: A new eco-friendly chemical treatment of natural fibres: effect of sodium bicarbonate on properties of sisal fibre and its epoxy composites. Compos. Part B Eng. **85**, 150–160 (2016). https://doi.org/10.1016/j.compositesb.2015.09.028
12. dos Santos, J.C., Siqueira, R.L., Vieira, L.M.G., Freire, R.T.S., Mano, V., Panzera, T.H.: Effects of sodium carbonate on the performance of epoxy and polyester coir-reinforced composites. Polym. Test. **67**, 533–544 (2018). https://doi.org/10.1016/j.polymertesting.2018.03.043
13. Chaitanya, S., Singh, I.: Ecofriendly treatment of aloe vera fibers for PLA based green composites. Int. J. Precis. Eng. Manuf.-Green Technol. **5**(1), 143–150 (2018). https://doi.org/10.1007/s40684-018-0015-8
14. Day, L., Fayet, C., Homer, S.: Effect of NaCl on the thermal behaviour of wheat starch in excess and limited water. Carbohydr. Polym. **94**(1), 31–37 (2013). https://doi.org/10.1016/j.carbpol.2012.12.063
15. Afrin, H.: Stabilization of clayey soils using chloride components. Am. J. Civ. Eng. **5**(6), 365 (2017). https://doi.org/10.11648/j.ajce.20170506.18
16. Mukhtar, I., Leman, Z., Zainudin, E.S., Ishak, M.R.: Effectiveness of alkali and sodium bicarbonate treatments on sugar palm fiber: mechanical, thermal, and chemical investigations. J. Nat. Fibers **17**(6), 877–889 (2020). https://doi.org/10.1080/15440478.2018.1537872
17. Belayachi, N., Bouasker, M., Hoxha, D., Al-Mukhtar, M.: Thermo-mechanical behaviour of an innovant straw lime composite for thermal insulation applications. Appl. Mech. Mater. **390**, 542–546 (2013). https://doi.org/10.4028/www.scientific.net/AMM.390.542
18. Brouard, Y., Belayachi, N., Hoxha, D., Ranganathan, N., Méo, S.: Mechanical and hygrothermal behavior of clay – Sunflower (Helianthus annuus) and rape straw (Brassica napus) plaster bio-composites for building insulation. Constr. Build. Mater. **161**, 196–207 (2018). https://doi.org/10.1016/j.conbuildmat.2017.11.140
19. He, Y., Ye, W.-M., Chen, Y.-G., Zhang, K.-N., Wu, D.: Effects of NaCl solution on the swelling and shrinkage behavior of compacted bentonite under oneimensional conditions. Bull. Eng. Geol. Env. **79**(1), 399–410 (2019). https://doi.org/10.1007/s10064-019-01568-1
20. Mir, S.S., Nafsin, N., Hasan, M., Hasan, N., Hassan, A.: Improvement of physico-mechanical properties of coir-polypropylene biocomposites by fiber chemical treatment. Mater. Des. 1980–2015 **52**, 251–257 (2013). https://doi.org/10.1016/j.matdes.2013.05.062
21. Marceau, S., et al.: Impact de vieillissements accélérés sur les propriétés de bétons de chanvre. Acad. J. Civ. Eng. 712–719 (2020). https://doi.org/10.26168/AJCE.34.1.86
22. Sahu, P., Gupta, M.K.: Effect of ecofriendly coating and treatment on mechanical, thermal and morphological properties of sisal fibre. Indian J. Fibre Text. Res. IJFTR **44**(2), Art. n° 2 (2019). https://doi.org/10.56042/ijftr.v44i2.21100
23. Raharjo, P.W., Soenoko, R., Purnowidodo, A., Choiron, A.: Characterization of sodium-bicarbonate-treated zalacca fibers as composite reinforcements. Evergreen **6**(1), 29–38 (2019). https://doi.org/10.5109/2321001
24. Pejic, B., Kostic, M., Skundric, P., Praskalo, J.: The effects of hemicelluloses and lignin removal on water uptake behavior of hemp fibers. Bioresour. Technol. **99**, 7152–7159 (2008). https://doi.org/10.1016/j.biortech.2007.12.073

25. Garat, W., Le Moigne, N., Corn, S., Beaugrand, J., Bergeret, A.: Swelling of natural fibre bundles under hygro- and hydrothermal conditions: determination of hydric expansion coefficients by automated laser scanning. Compos. Part Appl. Sci. Manuf. **131**, 105803 (2020). https://doi.org/10.1016/j.compositesa.2020.105803
26. Hossain, S.I., Hasan, M., Hasan, M.N., Hassan, A.: Effect of chemical treatment on physical, mechanical and thermal properties of ladies finger natural fiber. Adv. Mater. Sci. Eng. **2013**, e824274 (2013). https://doi.org/10.1155/2013/824274
27. Sarker, F., Uddin, M.Z., Sowrov, K., Islam, M.S., Miah, A.: Improvement of mechanical and interfacial properties of hot water and sodium bicarbonate treated jute fibers for manufacturing high performance natural composites. Polym. Compos. **43**(3), 1330–1342 (2022). https://doi.org/10.1002/pc.26451
28. Zindani, D., Kumar, S., Maity, S.R., Bhowmik, S.: Mechanical characterization of bio-epoxy green composites derived from sodium bicarbonate treated punica granatum short fiber agro-waste. J. Polym. Environ. **29**(1), 143–155 (2020). https://doi.org/10.1007/s10924-020-01868-8

Hot-Lime-Mixed Hemp Concretes

Zifeng Wang and Sara Pavia(✉)

Department of Civil Engineering, Trinity College Dublin, Dublin, Ireland
paviast@tcd.ie

Abstract. Hemp-lime composites are sustainable, low-carbon, non-loadbearing materials with outstanding thermal properties and high vapour permeability, used in new construction or thermal upgrades. The abundant pores in the hemp shiv ($\geq$70%) are blamed for the low mechanical strength, and for a high water absorption that can result in long drying times and low early strength that delay building.

Hot-lime mixing involves using quicklime as a binder, so that the exothermic reaction of lime slaking takes place simultaneously to mixing. This paper investigates whether hot-lime mixing can improve the properties of hemp-lime concretes. It was hoped that the heat generated on lime slaking would reduce the water intake by the shiv, block shiv pores and promoting adhesion between the shiv and the lime matrix.

The properties of the hot-lime mixed concretes agree with former authors, with densities of 430–470 kg/m^3 and strengths of 0.32 MPa (compressive) and c.0.27 MPa (flexural) at 3 months. Their vapour diffusion resistance factor of 4.62, water sorption coefficient c.4.58 kg/m^{2}h$^{1/2}$, thermal conductivity 0.015–0.021 W/mK and specific heat capacity of c.1135 J/kgK are also typical of hemp concretes. The improvement in properties due to hot-lime mixing is very small.

The optimum proportions and method for hot-lime mixing are investigated. The mixing process needs to avoid water competition between the shiv and the lime slaking. Mixing the hemp with quicklime prior to slaking undermines the process lowering the slaking temperature. The best mixing method involved layering the hemp over the quicklime.

Keywords: Hot-Lime Mixing · Hemp-Lime Materials · Quicklime · Thermal Conductivity · Specific Heat Capacity · Water Vapour Permeability · Water Absorption Coefficient · Strength

1 Introduction

In today's environment, it is important to use low-carbon, sustainable materials for construction. Hemp-lime materials are a mix of a lime-based binder and hemp. As they are made with a renewable plant (hemp), they are a carbon sink, hence hemp materials are highly sustainable and have great environmental credentials. Hemp-lime materials have been investigated by many authors in the last two decades. Previous authors have demonstrated that they are materials with outstanding thermal properties and high

S. Amziane et al. (Eds.): ICBBM 2023, RILEM Bookseries 45, pp. 185–196, 2023.
https://doi.org/10.1007/978-3-031-33465-8_15

vapour permeability. As explained by many authors, they display ductile failure and low ultimate strengths under uniaxial compressive loads. Typical compressive strengths for 2:1 (binder: hemp by weight) range between 0.05–1.2 MPa, mainly depending on density and age, and the flexural strength is also low, ranging between 0.06 and 1.2 MPa (Amziane et al. 2017; Evrard 2003; Cerezo 2005; Elfordy et al. 2008; Arnaud and Gourlay 2012; Walker et al. 2014).

The abundant pores in the hemp shiv (over 70%) are blamed for the low mechanical strength (Nguyen 2010), and are responsible for the high porosity, water vapour permeability and water absorption of hemp concretes, which can result in long drying times and low early strength that delay building.

In hot-lime mixing, the exothermic reaction of lime slaking takes place either while mixing or quickly after. It was hoped that the heat released on slaking would help block hemp-shiv pores, speeding up drying and increasing strength. The slaking reaction is exothermic, and generates 1.14 MJ per kg CaO (JRC 2013). Equation 1 describes this reaction:

$$CaO + H_2O = Ca(OH)_2 + \text{energy (1.14 MJ per kg CaO)} \tag{1}$$

In hot-lime-mixed mortars, it is argued, that both the heat generated and the expansion of the quicklime on slaking produce early stiffening and fills voids which enhance the interface bond and improve microstructure (BLFI 2014 and 2016; Hunnisett 2016; Wiggins 2018; Artis 2018; Henry 2018; Copsey 2019). However, this is based on site observations has not yet been quantified. Furthermore, the composition of the quicklime and the quantity of water used for slaking control the resultant temperature, which in turn determines the surface area and particle size of the end hydrate, and this rules the properties of the resultant material (Pavia et al. 2023). The speed with which the temperature rises from 20 to 60 °C (on slaking under standardised conditions) indicates how reactive the lime is. The quicklime's reactivity depends on the burning temperature and time, the crystalline structure of the limestone, the impurities of the limestone and the kiln type and fuel. Soft burnt limes have high reactivity whereas hard burnt limes typically show low reactivity (JRC 2013).

This paper investigates whether hot-lime mixing with a pure quicklime of high reactivity (CL90Q) can reduce the water intake by the shiv and improve the properties of hemp concretes, promoting adhesion between the shiv and the lime matrix. It was hope that the exothermic reaction of lime slaking could seal the shiv pores and improve strength.

2 Materials and Methods

2.1 Materials

A quicklime of European designation CL90Q was used for hot-mixing. The quicklime was crushed and sieved into <1 mm powder. Some of the quicklime particles were unburnt, hence the limestone could hardly be separated from the quicklime. Therefore, the quicklime contains some unburnt limestone which lowers reactivity. Industrial hemp shiv was supplied by La Chanvrière De L'Aube in central France. The water content of the hemp was ~12% (at RH ~60%) prior to mixing. The hemp shiv particles vary between 5 and 30 mm in length and 0.3–2 mm width (Fig. 1).

Fig. 1. Quicklime prior to crushing and sieving (left) and hemp shiv.

2.2 Mechanical Properties

The compressive strength was measured on 100 × 100 × 100 mm cubes according to EN 459-2. The loading rate was 20N/s. The ultimate compressive strength was determined at the point where the behaviour departs from linear stress/strain. The flexural strength was measured with a three-point loading system, on 40 × 40 × 160 mm prisms, at 10 N/s loading rate, according with EN196-1. The results are the arithmetic mean of 6 tests. The flexural strength was calculated using Eq. 2.

$$Rf = 1.5 \times Ff \times l\,b^3 \quad (2)$$

where, Rf = Flexural strength (MPa)

Ff = Load applied to the middle of the prism (N).

b = Side of the squared cross-section of the prism (mm).

l = distance between the supports (mm).

2.3 Water Vapour Permeability

The water vapour permeability was measured with the method in EN 12806 using 4 × Ø 135 mm discs, 22 mm deep. The discs were sealed on top of containers with 75g calcium chloride inside as a desiccant. The containers were placed in a curing room at 18.5 ± 3 °C temperature and 50 ± 10% RH. Weight gain was monitored over five weeks. The water vapour permeability was calculated using Eq. 3.

$$\delta = G \times d\,A \times \Delta p \quad (3)$$

where, δ = Water vapour permeability

G = Weight gain (kg).

d = Thickness of the specimen (m).

A = Area of the specimen (m^2).

Δp = water vapour pressure difference.

The water vapour diffusion resistance factor was calculated with Eq. 4.

$$\mu = \delta \times \delta_{air} \quad (4)$$

where, μ = Water vapour diffusion resistance factor

δ_{air} = Water vapour permeability of air.

δ = Water vapour permeability of the material.

2.4 Water Absorption Coefficient by Capillary Action

The capillary suction was measured on 100 × 100 × 100 mm cubes (EN 1925). The cubes were placed on a wire grill and the weight gain measured once a day over five days. The water sorption coefficient was then calculated as:

$$A_w = m\,a\,\sqrt{t} \tag{5}$$

where, A_w = Water sorption coefficient (kg/m^2 hour$^{1/2}$)
m = Weight gain (kg).
a = Surface area (m^2).
t = Time (hour).

2.5 Thermal Conductivity

The thermal conductivity was measured with the hot-box method. The hot-lime mixes (25 mm thick) were applied on a hemp wall (750 mm high, 930 mm wide and 220 mm thick). The back of the hemp wall was close to the heating panel and the side of the wall was insulated with 50 mm thick PIR board, so that the external surface was covered with the hot-lime mixes. The heat flow through the wall was measured with a Hukseflux TRSYS01 system with two heat flux sensors attached to the external surface. Two thermocouples were also attached to measure the external surface temperature. The heat flux and surface temperature data were logged into TRSYS01 system every 10 min. The temperature of the heating panel was set as 60 °C and the test was running for 48 h until the temperature in the front surface was stabilized. The thermal transmittance (U-value) was calculated with Eq. 6. The test was then repeated with the hemp wall alone, and the U-value of the hot-lime mixes calculated.

$$U = \frac{1}{Tsi - Tse\underline{\qquad\qquad\qquad}Q + rint + rext} \tag{6}$$

where, U = U-value (W/m^{2}K)
T si = Internal surface temperature (K).
T se = External surface temperature (K).
Q = Heat flux (W/m^2).
r int = Internal surface resistance (standard 0.13 m^{2}K/W).
r ext = External surface resistance (standard 0.04 m^{2}K/W) (Fig. 2).

2.6 Specific Heat Capacity

Discs of 135 mm diameter (22 mm thick) were heated in an oven at 100 °C. Once taken from the oven, their temperatures were measured, and they were then directly transferred into insulated containers with water at 15 °C. The rise of the water temperature was monitored every 15 min over 1 h. The results are the arithmetic mean of 4 tests. The property was calculated with Eq. 7.

$$\Delta T_c \times m_c \times C_c = \Delta T_w \times m_w \times C_w \tag{7}$$

Fig. 2. Hot-box assemblage for thermal conductivity measurement.

where, C_c = Specific heat capacity (J / kgK)

ΔT_c = Temperature of the heated sample minus water temperature after immersion (°C).

m_c = Mass of sample (kg).

ΔT_W = Water temperature before immersion minus temperature after immersion (°C).

m_W = Mass of water in container (kg).

C_W = Specific heat capacity of water = 4187 J / kgK.

2.7 Microstructure

The microstructure and the cementing phases was investigated at 52–56 days with a Zeiss Ultra Scanning Electron Microscope -SEM- with an EDX (Energy Dispersive X-Ray analysis) attachment. SEM tests were run on both the QL mix at 56 days and the QL+GGBS mixes at age of 52 days.

3 Results

3.1 Reactivity of the Quicklime Used for Hot-Lime Mixing

The reactivity of the quicklime was studied with a slaking curve (EN459-2) where the measured temperature values were recorded as a function of time. According to the results, the lime has a high reactivity: t 60 = 30s to 1min; T′ max = 80–98 °C (100 g) and 130 °C (1 kg). It took between 30 s and 1 min for the quicklime to reach 60 °C depending on the size of the sample (100 g vs 1 kg). A large amount quicklime (1.5 kg) generated high heat and the reaction was violent. The highest temperature of the hot-mixing reaction was 130 °C. The slaking reaction of the lime is assumed to be 100% complete at the time when the maximum temperature (T′ max) is reached which is between 5 and 7 min for this lime. In general, the reaction begun after 15–20 s, and between 20s and 1 min the temperature raised rapidly to reach 50–60 °C. High temperature (>60 °C) continued over 5–10 min with 100 g quicklime. Between 1 and 5 min a violent reaction takes place releasing steam. Between 5 and 10 min the reaction is less violent with no steam. In all the tests, the temperature lowered after 10–15 min hence the reaction was finished (Table 1).

Table 1. Reactivity of the quicklime.

Mass of quicklime	t_{60} - seconds	T'_{max} - °C	t_{max} - min
100 g	30	80–98	5
1000 g	60	130	7

3.2 Mixing

The water content is especially important in hot-lime mixing because it strongly impacts the properties of the resultant material. The amount of water determines the slaking temperature which makes the resultant $Ca(OH)_2$ crystals vary from fairly large to extremely small, and this produces materials with different properties (Pavia et al. 2023). Therefore, the slaking temperature during hot-mixing was measured with thermocouples and a thermal image camera, and the optimum proportions were investigated by trial, based on cohesion and workability. Twelve mixes were tested with varying proportions. The proportions 1.43: 4.57: 1 (quicklime: hemp: water - specimen 12 in Table 2) displayed the best cohesion, workability and early strength at the lowest water content. Specimens were fabricated with this mix and their mechanical, hygric and thermal properties and microstructure investigated. This mix was blended with GGBS (at 70/30 - CL90Q/GGBS) to enhance the production of cementing hydrates (Walker et al. 2014). An amount of concrete was weighted to ensure a dry density of c.450 kg/m^3 (Fig. 3).

The thermocouples and a thermal image camera allowed to record the maximum temperatures of the exothermic reaction of lime slaking and the temperature change over time. It was evidenced that mixing the hemp with quicklime prior to slaking undermines the process lowering the slaking temperature (the highest temperatures using this mixing method were under 80 °C). Therefore, the hemp was layered over the quicklime so that the hemp shiv prevented heat loss during slaking. The heat generation continued over 15 min and high temperatures (over 50 °C) lasted for over 20 min assisted by the insulation provided by the hemp shiv. However, once the hemp was mixed with the binder, the temperature dropped very quickly (Fig. 4).

3.3 Properties of the Hot-Lime Mixed Hemp Concretes

The hot-mixed hemp concretes show the typical mechanical properties of hemp-lime concrete described by former authors, and the ultimate values are slightly higher than others previously reported (Cerezo 2005; Nguyen 2010; Elfordy et al. 2008; Amziane et al. 2017). Walker et al. (2014), obtained compressive strength between 0.02 and 0.04MPa at 5 days and 0.29 and 0.39MPa at 1 year, when investigating lime-hemp concretes with pozzolans comparable to those in this paper (Table 3).

Table 2. Trial mixes and selection. L-lime. H-hemp. W-water.

Mix No	L: H: W (mass)	CS (MPa)
1	2.00: 7.00: 1	0.14
2	2.00: 5.74: 1	0.23
3	~~1.00: 2.00: 2~~	~~Discarded~~
4	~~0.60: 3.00: 1~~	~~Discarded~~
5	2.00: 5.40: 1	0.30
6	1.43: 4.43: 1	0.20
7	1.25: 3.82: 1	0.20
8	1.43: 3.63: 1	0.31
9	1.30: 3.60: 1	0.26
10	1.43: 5.43: 1	0.17
11	1.43: 5.00: 1	0.18
	MIX QL = L: H: W	
12	1.43: 4.57: 1	0.33
	MIX QL+GGBS = L: GGBS: H: W	
13	1: 0.43: 3.60: 1	0.30-0.36

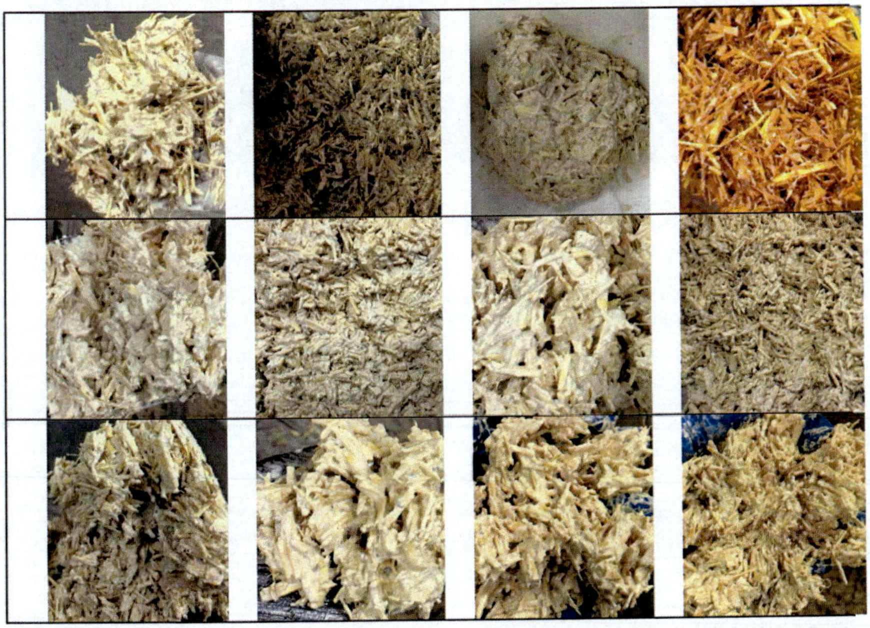

Fig. 3. Specimen trials in Table 2: specimen 1–12, from top left to bottom right.

The hygric properties fall within the range observed by previous authors. Hemp-lime concretes typically have high water vapour permeability. The results conform with the

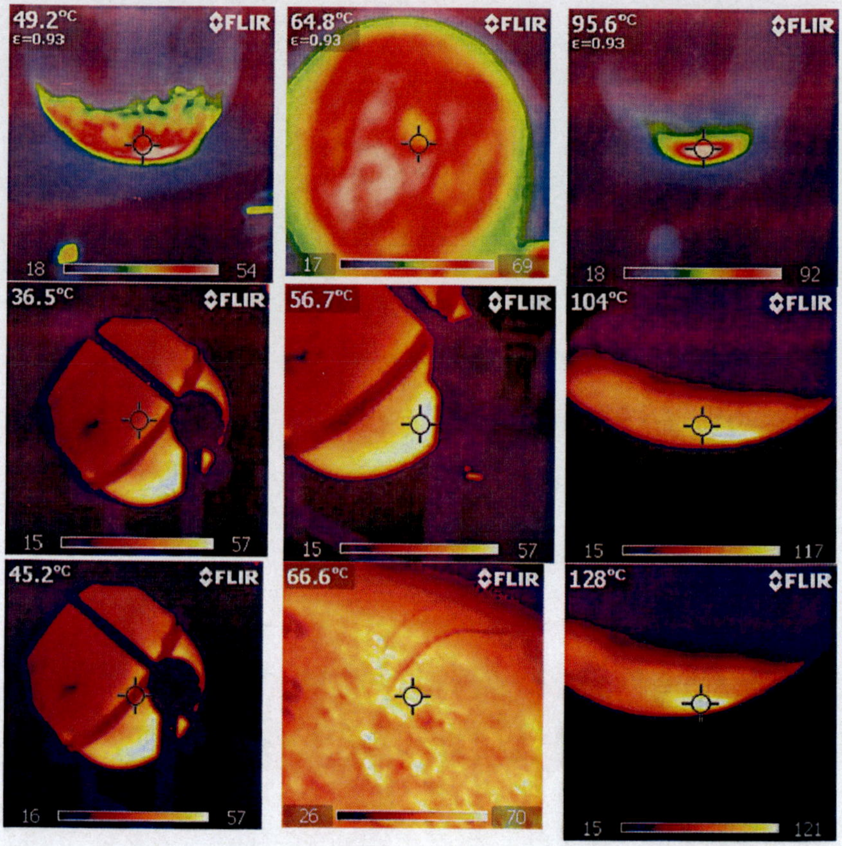

Fig. 4. Evolution of temperature during hot-lime mixing concretes 12 and 13-Table 2.

Table 3. Properties of the hot-lime mixed hemp concretes at 3 months. Density = 430–470 kg/m^3. λ = thermal conductivity. C_c = Specific heat capacity. δ = Water vapour permeability. μ = Water vapour diffusion resistance factor. A_w = Water sorption coefficient (kg/m^2 hour$^{1/2}$). COV: 0.6–9% for CS; 6.1–10.5 for FS; 18–32 for δ.

	CS MPa	FS MPa	μ	δ kg.m^{-1}.s^{-1}.Pa^{-1}	Aw kg/m^2 hour$^{1/2}$	R^2	λ W/(mK)	Cc J/kgK
QL	0.33	0.21	4.56	5.11×10^{-10}	4.58	*0.83*	0.021	1135
QL+GGBS	0.30–036	0.25	4.68	5.20×10^{-10}	4.24	*0.87*	0.015	1016

common industry figure of water vapour diffusion resistance factor (μ) of lime-hemp concrete is 4.85 $\pm$ 0.24 measured in accordance with EN12572 for samples with a binder:hemp:water ratio of 2:1:3 and a density of c.400 kg/m^3 (Evrard 2008; Evrard et al. 2006).

The water absorption values also agree with former authors such as Collet (2004, 2009), Amziane et al. (2017); and Evrard (2008) reporting a water absorption coefficient of 4.42 ± 0.27 $kg/m^2h^{1/2}$ (0.0736 ± 0.0045 $kg/m^2s^{1/2}$) for a 487kg/m^3 density concrete made with a proprietary binder (DIN52617). When compared with lime-pozzolan hemp concretes (Walker and Pavia 2014), the hot-lime mixed concretes show slightly lower vapour diffusion resistance factors (4.56–4.68 vs 5.42–5.71) hence a higher permeability; and a marginally greater capillary suction, as shown by the higher water absorption coefficient values – 4.24–4.58 vs 2.65–3.37).

As lightweight materials, the thermal conductivity of hemp-lime concretes is typically low, resulting in outstanding U-values and good insulation properties. It can be very low (0.04 W. $m^{-1}.K^{-1}$) with typical values ranging between 0.1 and 0.3 W. $m^{-1}.K^{-1}$) (Amziane et al. 2017). With respect to the specific heat capacity, hemp concrete has a high thermal mass when compared to other lightweight building materials. Previous research has identified a thermal heat capacity ranging between 1000 J/kgK for a concrete with a density of 413kg/m^3 and 1560 ± 30 J/kgK for a "wall mixture" with a density of 480 kg/m^3, and values ranging from 1240 ± 172 to1350 ± 279 (J/KgK).

The specific heat capacity of the hot-mixed hemp concretes (ranging from 1016–1135 J/kgK) is slightly lower than the 1240–1350 J/kgK reported for lime pozzolan hemp concretes (Walker and Pavia 2014) and similar to the values reached by LeTran et al. (2010) (1000 J/kgK) for hemp concretes of similar density to the ones in this paper.

3.4 Microstructure

As expected, the SEM analyses showed extensive carbonation, with abundant microcrystals of calcium carbonate resulting from the carbonation of the slaked quicklime. The calcium carbonate binder forms a continuous coating over the hemp particles that suggests a good adhesion at the interface which closely relates to strength and durability. However, the SEM analyses are qualitative, and hence no specific measurements can be made from the analyses. Both the QL and G+QL mixes show extensive carbonation, but some cementing hydrates were evidenced in the GGBS mixes. The presence of occasional hydrates does not seem to affect the physical properties of the concretes in a significant manner (Figs. 5 and 6).

Fig. 5. Hot lime mixed hemp concrete (QL mix) showing extensive carbonation, with abundant microcrystals of calcium carbonate (left images) coating hemp particles (right images) providing a good adhesion at the interface.

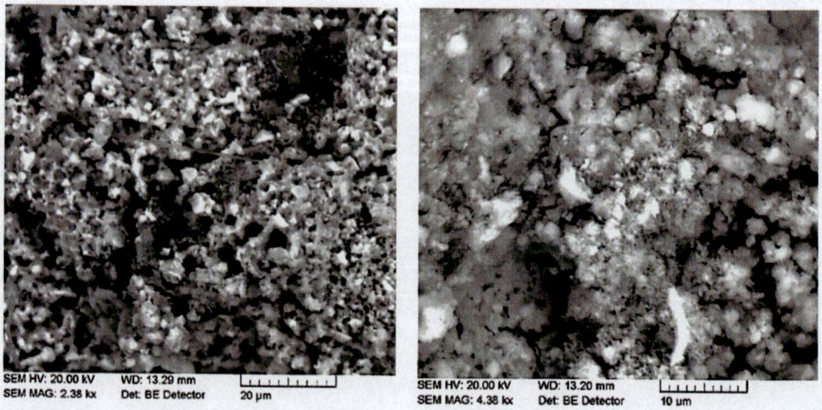

Fig. 6. Hot lime mixed hemp concrete (QL + GGBS mix) under the SEM showing extensive carbonation, with abundant microcrystals of calcium carbonate and some cementing hydrates (needles -left image) and low crystallinity gels (all over).

4 Conclusion

This paper investigates whether hot-lime mixing hemp concretes can improve the properties of these materials. In hot-lime mixing, the exothermic reaction of lime slaking takes place simultaneously to mixing. It was hoped that the heat generated on quicklime slaking would reduce the water intake by the shiv, block shiv pores and promoting adhesion between the shiv and the lime matrix.

It is concluded that the mixing process needs to avoid water competition between the shiv and the slaking process. Mixing the hemp with quicklime prior to slaking undermines the process lowering the slaking temperature. The best mixing method involved layering the hemp over the quicklime. The quicklime used in this research (CL90Q) is highly reactive, this results in a strong competition for water between the shiv and the slaking reaction: the shiv tends to absorb the water that the lime requires for slaking. A less reactive quicklime may render different results.

The properties of the resultant hot-lime-mixed, hemp materials slightly improved with respect to others in former literature, but they didn't improve substantially. The hot mixing method for hemp-lime composites provides a slightly superior mechanical strength than other binders at an early age (3 months), but the improvement is small.

As expected, the sem analyses showed extensive carbonation, with abundant microcrystals of calcium carbonate resulting from the carbonation of the slaked quicklime. The calcium carbonate binder forms a continuous coating over the hemp particles that suggests a good adhesion at the interface which closely relates to strength and durability.

References

Amziane, S., Collet, F. (eds.): Bio-aggregates based building materials. State-of-the-art report of the RILEM technical committee 236-BBM, vol. 23 (2017). ISBN 978-94-024-1030-3

Arnaud, L., Gourlay, E.: Experimental study of parameters influencing mechanical properties of hemp concretes. Constr. Build. Mater. **28**(1), 50–56 (2012)

Artis, R.: Historic Environment Scotland Technical Paper 28 Specifying Hot-Mixed Lime Mortars àrainneachd eachdraidheil alba, Edinburgh (2018)

Building Limes Forum Ireland BLFI: The Hot-lime Mortar (HLM) Project. Unpublished report (2014)

Building Limes Forum Ireland BLFI 2016 Hot-mix lime mortar guide. Published by the Building Limes Forum Ireland (2016)

Cerezo, V.: Propriétés mécaniques, thermiques et acoustiques d'un matériau à base de particules végétales: approche expérimentale et modélisation théorique. Ph.D. thesis, L'Institut National des Sciences Appliquées de Lyon, France (2005)

Collet, F.: Caractérisations hydrique et thermique de matériaux de génie civil à faibles impacts environnementaux. Ph.D. thesis, Institut National des Sciences Appliquées de Rennes, France (2004)

Collet, F.: Hydric characterization of sprayed hempcrete. In: Fourth International Building Physics Conference, Istanbul (2009)

Copsey, N.: Hot Mixed Lime and Traditional Mortars. The Crowood Press, Marlborough (2019)

Elfordy, S., Lucas, F., Tancret, F., Scudeller, Y., Goudet, L.: Mechanical and thermal properties of lime and hemp concrete ("hempcrete") manufactured by a projection process. Constr. Build. Mater. **22**(10), 2116–2123 (2008)

Evrard, A.: Hemp Concrete- A synthesis of physical properties. Construire en Chanvre, France (2003)

Evrard, A.: Transient hygothermal behaviour of lime-hemp material. Ph.D. thesis, Universite catholique de Louvain, Belgium (2008)

Evrard, A., de Herde, A., Minet, J.: Dynamical interactions between heat and mass flows in Lime-Hemp Concrete. In: Research in Building Physics and Building Engineering, Third International Building Physics Conference, Concordia University, Montreal, Canada (2006)

Henry, A.: Hot-mixed mortars: the new lime revival. Context **154**(5), 30–33 (2018)

Hunnisett Snow, J.: Hot-Mixed Lime Mortars. Historic Environment Scotland, 2nd edn. Àrainneachd Eachdraidheil Alba, Edinburgh (2016)

JRC– IPTS BREF. Best Available Techniques Reference Document for the Production of Cement, Lime and Magnesium Oxide, JRC - IPTS (Joint Research Centre - Institute for Prospective Technological Studies), Sevilla (2013)

Nguyen, T.: Contribution à l'étude de la formulation et du procédé de fabrication d'éléments de construction en béton de chanvre, Ph.D. thesis. Université de Bretagne Sud, France (2010)

Pavia, S., et al.: RILEM TC 277-LHS report: How hot are hot-lime-mixed mortars? A review. (MAAS-D-22-01517), Materials and Structures (2023)

Tran Le, A.D., Maalouf, C., Mai, T.H., Wurtz, E., Collet, F.: Transient hygrothermal behaviour of a hemp concrete building envelope. Energy Build. **42**(10), 1797–1806 (2010)

Walker, R., Pavía, S.: Moisture transfer and thermal properties of hemp–lime concretes. Constr. Build. Mater. **64**(2014), 270–276 (2014)

Walker, R., Pavía, S., Mitchell, R.: Mechanical properties and durability of hemp-lime concretes. Constr. Build. Mater. **61**(2014), 340–348 (2014)

Wiggins, D.: Historic environment Scotland technical paper 27 hot-mixed lime mortars microstructure and functional performance. Àrainneachd Eachdraidheil Alba, Edinburgh (2018)

Earth Constructions and Building Materials

Vibration as a Solution to Improve Mechanical Performance of Compressed Earth Blocks

M. Audren[1,2](✉), A. Perrot[2], S. Guihéneuf[2], D. Rangeard[1], and T. Leborgne[1]

[1] Laboratoire CBTP, Research & Development Department, Noyal-Sur-Vilaine, France
mathieu.audren@lcbtp.com

[2] IRDL, UMR CNRS 6027, Université de Bretagne Sud, BP 92116, 56321 Lorient Cedex, France

Abstract. Vernacular technics like earth construction has recovered much attention due to its low environmental impact and large availability. However, earth-based materials need to be standardized and industrialized to meet the global demand for affordable housing solutions.

Compressed earth blocks used for masonry construction seem to show potential for mass production. Nevertheless, its use remains marginal due to lack of dedicated compression devices. To solve this issue, factories must be designed. It is also worth noting that earthworks and concrete blocks production uses vibration to be more efficient. Therefore, vibration is expected to play a major role in the production of dense earth blocks.

In this study, samples composed of quarry waste (washing sludge originated from sand production) and coarse sand are compacted with and without additional vibration. The friction was measured during the test to evaluate the vibration impact on compaction behavior. It allows comparing friction generated by compaction under various vibration characteristics. It appears that the increase of the centrifugal force of the vibrators decreases the friction inducing a better densification of the earth. This effect is even more significant under a critical sample density, where the vibration could annihilate the friction. This vibration-induced reduction of the friction makes the compression more efficient to densify the granular skeleton of the earth blocks. Finally, this vibro-compression technique applied to earth-based materials allows decreasing the optimal moisture content of compaction and increasing the dry density and the compressive strength of the dried block.

Keywords: Vibration · Compressed Earth Blocks · Friction

1 Introduction

Humankind needs to answer the needs for building to satisfy social and economic demand [1]. This prospect must meet the reduction of CO_2 emissions, the scarcity of resources and the reduction of embodied energy which are the new problematics of modern construction [2, 3]. However, current construction methods hardly respond to these problems linked to climate change: for example cement production is responsible for 9 to 10% of

S. Amziane et al. (Eds.): ICBBM 2023, RILEM Bookseries 45, pp. 199–211, 2023.
https://doi.org/10.1007/978-3-031-33465-8_16

global emissions of carbon dioxide [4]. In this context, raw earth construction is regaining consideration due to its environment-friendly character [1]. This vernacular construction method has been replaced by fired-clay and concrete and has been considered as an obsolete and low performance solution [5]. However earth materials present a low environmental impact, are recyclable when not stabilized with hydraulic binder and can respond to crisis problematics [6]. Moreover, a large proportion of the fine-grained soils can be used to create earth construction, conferring to earth local, available and affordable characters [7]. Another advantage of this material is its thermo-hydric properties: indeed, its thermal and hydric capacity could efficiently regulate the indoor temperature and humidity [8, 9]. However, the lack of industrialization and standardization for these materials slows down the democratization of its use.

To answer the future demand for earth construction, high-scale production could be necessary. Concrete manufactured blocks could be adapted to make Vibro-Compressed Earth Blocks (VCEB) which consists in compressing earth under vibration at a low moisture content to reach a high dry density. The addition of vibration is likely to improve the compaction process, reducing friction between the material and the mold [10]. Due to the contact between grains, a force path headed to mold walls is created during compaction [11, 12]. The reduction of friction by vibration can be an efficient solution because friction dissipates energy and reduces the compaction force on the granular skeleton and finally harms the mechanical performances [13].

In this study an earth-based material made from quarry by-product is used. The friction between the material and the mold walls during compaction and under various vibration configurations is measured. The aim of this study is to understand how vibro-compaction could increase the dry density (and compressive strength) compared to conventional compaction. Modification of earth formulation to reach the best performance possible is also highlighted.

2 Scientific Background on Soils Vibration

A fine-grained soil could be described by a system composed of particles of dimensions less than 100 μm [14]. Common soil is partially composed of elements of a size less than 100 μm, such as clay and silt. It could be considered as a cohesive granular material due to the force present between the fine particles when water is added [15–17].

Friction in granular material and especially in earth is studied for earth reinforcement in public earthworks or civil engineering [18, 19]. When compacted, soils exhibit a pressure dependent behavior which can be described by the Coulomb model. The behavior of the material at the interface with a steel surface is the same: in a compaction mold, at a given position, the axial compaction load is proportional to the friction stress leading to an exponential stress profile along the height of the mold. Moreover, the increase of the compaction force also increases the stress at the interface and therefore to energy loss through friction.

Many parameters influence the amplitude of friction during compaction: packing fraction, surface roughness, level of axial load (pressure), particle size distribution, particle shape, compaction velocity, and moisture content [20–24]. Surface roughness is defined by the size of the asperities and directly interferes with the dimensions of the particles. If the surface is rough enough, the surface friction could reach the friction of the bulk material. On the opposite side, with a perfectly smooth surface (negligible roughness), the friction is insignificant. Some studies have already qualified the impact of the roughness on the friction behavior [25, 26].

During friction, the particle movement takes place in the interfacial zone [27] which has different definitions [28]. In this study, it could be considered as the area where the behavior of the material is elastic or elastoplastic. Different assumptions were made based on the median diameter of the particles to estimate the interface thickness [29–31]. When the surface moves, the behavior of the material depends upon the distance from the surface. Four areas can be described [24]:

- Transition area: within the thickness of the size of a couple of particles, its behavior depends on roughness. Particle movement is similar to surface movement.
- Area of volumetric steady-state behavior: the material has plastic deformation because the particles arrange themselves to reach steady-state property. In addition to the transition area, it represents the interface. The interface boundary could be defined by the distance to the surface where the granular speed has decreased by 90% [24, 32].
- Area of volumetric transitional behavior: transitional areas between those with a plastic deformation and an elastic deformation.
- Area of plastic behavior: the movement is too weak and the particles move between positions of equilibrium.

The mechanics of vibrated granular materials could be applied to earth vibration. Nevertheless, it is necessary to take into account the nature of earth [33]. The vibration used in soil application is regularly harmonic and can therefore be characterized by several parameters such as frequency, phase shift angle, amplitude, speed, acceleration and direction [34]. During vibration, granular materials could behave differently depending on the material and the vibrating system. The vibration energy is absorbed by induced soil deformations which decrease with the distance from the vibration source [35]. Moreover, the vibration damping increases as the material gets denser [36].

Depending on the deformation characteristics, the nature of the vibrated medium could be considered gaseous, liquid or solid [33]. At high speed, the pore pressure increases and the lack of contact between particles the soil leads to a fluid behavior [33, 35, 37]. In soil reinforcement, a 1.5 to 3 g acceleration could fluidize the medium [38]. Farther from the vibrator, the material is affected by plastic deformation making the soil consolidates or expands [39] depending on its initial state: when the vibration is sufficiently reduced, the particle movement is elastic.

To improve the impact of the vibration on soil reinforcement, the induced vibration has a frequency close to its resonance frequency which is approximately 15 Hz. It allows to achieve a fluidification behavior when the material is saturated with water [40]. At the interface between a soil and a wall, the behavior could be liquid and lead to granular convection [41–43]. Also, space may appear between the wall and the soil [44] due to the phase shift between the two materials.

The vibration could be described by the movement of a system around an equilibrium position: the works of Lamb on soil vibration [45] are pioneering on this topic. Today, vibrations are used for reinforced soil with a vibrating roller compactor [40]. Soil vibrations are also naturally present during earthquakes [46]. A previous study already shows the ability of vibro-compaction to increase the dry density of CEB and its compressive strength [47]. However, this study has not studied the effect of vibration parameters on the compaction process efficiency.

3 Materials

Quarries activity induces several by-products which cannot be sold because of their characteristics. In quarries production, aggregates such as standardized gravels are washed to be used in concrete application. This washing step produces Clay Muds (CM) with only fine particles of silt and clay which are kept and dry on site. These muds are considered currently as a waste but could be used as a clay binder in earth construction to design a clay-based concrete or in addition to natural earth exhibiting a low proportion of fine particles.

In this study, the CM is used in combination with a crushed sand to create an earth material dedicated to earth construction. Particle size distribution of particles smaller than 83 μm is measured by a laser particle size diameter [48, 49]. The studied CM exhibits a methylene blue value of 1.83. Its liquidity limit and plasticity index are respectively 64% and 24. Moreover, the proportion of elements smaller than 2 μm is 9.39% and the proportion of elements smaller than 63 μm is 91.6%. These characteristics suggest a high clay activity and a high proportion of clay and silt. The crushed sand presents a Methylene Blue Value of 0.1. Its proportion of particles smaller than 2 μm is 0.7%, and the proportion of elements smaller than 63 μm is 11.4%. The low Methylene Blue Value and fine particles content suggest a sandy material that cannot be used in earthen construction without additions (Fig. 1).

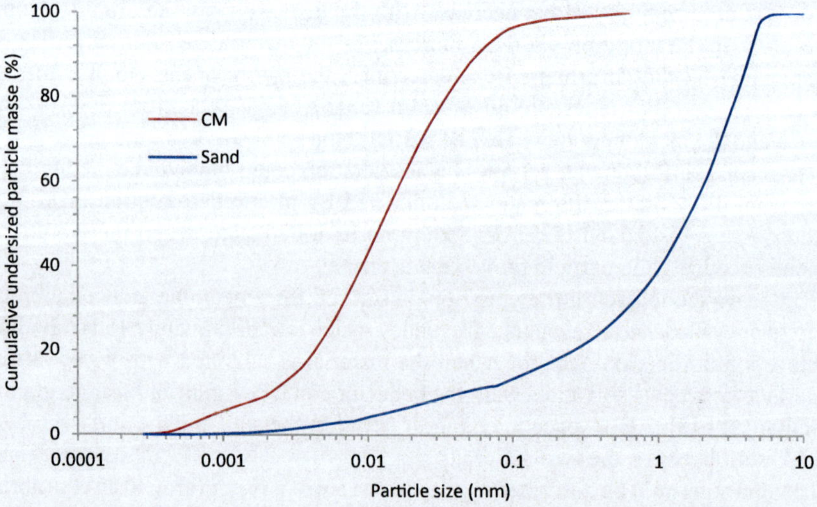

Fig. 1. Particle size distribution of the sand and the CM used in the frame of the study.

To determine the optimal proportion of CM, modified Proctor tests are performed with proportions of CM ranging between 0% and 50% with a step of 10% of CM between each test.

Compressive tests were performed on each sample produced during the modified proctor test: the highest density was reached with a proportion of 10% of mud at 2223 kg/m^3 but the highest compressive resistance was reached with a CM proportion of 30% and a dry density of 2135 kg/m^3. An optimum ratio of mud is then shown to be a trade-off between density and clay activity.

The material used in this study is composed of 30% of CM, 70% of sand and is prepared with a moisture content of 9.1%.

4 Methods

Measuring the vibration response of a system to vibrations requires complex analysis and knowledge of the material characteristics (stiffness and damping). In the case of earth compaction, this approach is complex because the material changes with density increasing. Also, during vibro-compaction, contact between the mold and the other elements, such as a piston, could create a non-linear response.

The vibration efficiency is estimated from the measurement of the friction during the vibro-compaction of earth. A schematic drawing of the system is shown in Fig. 2. Earth material is placed in the mold between perforated plates to drain air and water. The mold is then placed on a hollow support which contains a force sensor linked to the bottom of the sample using a piston. The force applied on the sample and the friction stress varies along the sample height but for the sake of simplicity local (top and bottom) or average values are considered in the frame of this study. For instance, an average friction stress has been defined and computed as the difference between the compressive force and the force measured at the base of the samples, divided by its lateral surface.

Fig. 2. Schematic drawing of the system used to compact the samples and measure friction force.

To induce vibration, 3 pneumatic vibrators are fixed on the mold surface. These turbine vibrators generate vibrations through the movement of a shifted mass. The movement is created by the flow of pressurized air. Three different models of vibrator are used: NCT 1, NCT 3 and NCT 5 with varying range of induced centrifugal force. A vibrator model could be used at different air pressures to reach different frequencies and centrifugal forces. These characteristics are described in the Table 1.

The earth samples have a slenderness of 1 and the compaction speed was set at 10 mm/min. The vibration began when the compacting stress reached 0.05 MPa. This value is sufficiently high to accurately measure the force, but sufficiently low to be negligible compared to the computed average friction stress estimated from the test data analysis.

Table 1. Characteristics of vibrators

Vibrator	Air pressure (bar)	Frequency (min^{-1})	Centrifugal force (N)
NCT1	2	29100	288
NCT1	4	38820	513
NCT1	6	45460	703
NCT3	2	26940	637
NCT3	4	34900	1069
NCT3	6	39700	1383
NCT5	2	22740	1389
NCT5	4	27840	2082
NCT5	6	30940	2572

To assess the impact of vibration, the work produced by friction force (W_τ) has been used and calculated along a piston displacement between two positions corresponding to samples height H_1 and H_2:

$$W_\tau = \pi D \int_{H_2}^{H_1} \tau \times H(dH) \quad (1)$$

with D the mold diameter. In this study, friction was performed for all the frequency modalities showing on Table 1. Also, samples have been compacted and vibro-compacted by the vibrator NCT 1 using 6 bars of air pressure, at a compaction stress of 10 MPa, with the same dry mass but with different moisture contents to show the impact of vibrations on the optimal moisture content.

Dry compressive strength tests were performed on samples with slenderness ratio equal to 1 (i.e. height equal to diameter) to evaluate the impact of vibration on this mechanical parameter. During these tests, a constant displacement rate of 1 mm/min was used, which allows to satisfy the NF EN 13286-41 standard [50]. The samples were dried in an oven at 60 °C until mass stabilization. The latter is considered to be stable when it varies by less than 0.1% of in 48 h.

5 Results and Discussions

5.1 Friction Behavior

The computed friction stress is plotted in Fig. 3 for compaction and vibro-compactions using each of the three vibrators at an air pressure of 6 bars. Whatever the vibration conditions, the friction increases to reach a value close to 1 MPa but the moment when this value is reached depends on the applied vibration. Without vibration, friction starts at the beginning of the compaction with a friction stress that grows exponentially as the sample height is reduced. The behavior of the earth under vibration is quite different and could be described by a two steps approach:

- A first step corresponding to granular rearrangement, influenced by vibration and leading to a friction free consolidation of the earth sample: without vibrations, the friction reaches a value of 0.05 MPa for a sample's height of 104.1 mm. With vibrators action, the height of the sample when friction reaches 0.05 MPa is decreased (for NTC1 at 3 and 5 to 6 bars of air pressure, the friction stress of 0.05 MPa is reached respectively at 99.3 mm, 94.8 mm and 86.8 mm of sample's height.)
- A second step corresponding to the compaction of a packing of rigid particles in contact. This kind of behavior induces friction that grows exponentially with the densification of the earth and tends toward a vertical asymptote as depicted in Fig. 3.

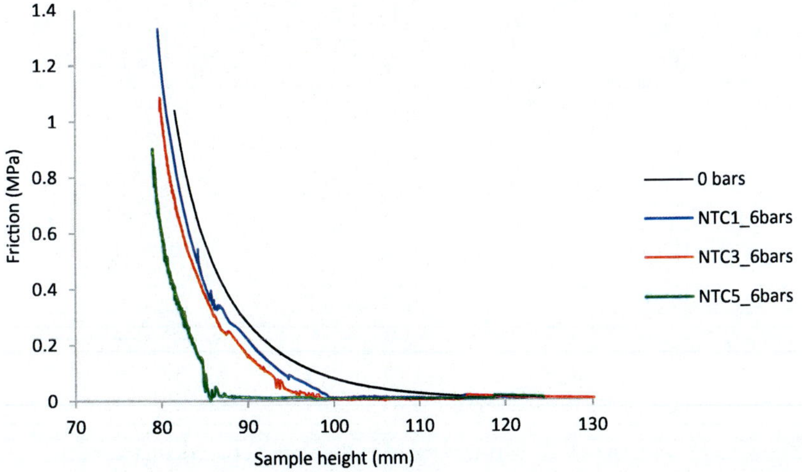

Fig. 3. Friction behavior during compaction test with and without application of vibrations (for the three tested vibrators running at a pressure of 6 bars).

Vibrations are efficient to reduce friction stress, especially at low density, where it allows for granular rearrangement and seems to lubricate the contacts and remove friction. At high density, the effect is less obvious as the friction stress always tends to infinite value. However, it is interesting to note that all friction curves are shifted to higher density (smaller samples) as the energy of the vibrator increases. This vibration-induced reduction of friction force allows to reach higher densities as it increases the compaction force on the granular skeleton. It is also worth noting that a constant friction stress, the sample height decreases with the increase of the vibration energy (for example at 85 mm).

The computed friction stress shows fluctuation that could result from a loss of contact between the sensors and the material or from non-linear effects.

To quantify the friction all along the entire compaction process, the work of the friction force is used. The Fig. 4 shows a decrease in friction work with the increasing centrifugal force. Vibrators NCT 1, 3 and 5 allow to reduce the friction work to 8%, 9% and 39% respectively. The impact on the work of friction comes from the longer period before a measurable value of friction force appears and from the reduction of friction at constant density after this 0-friction period. A linear decrease of the work of the friction force with the centrifugal force of the vibrators is observed.

As a consequence, an efficient solution to reduce friction when compacting CEB would be to use a vibrator with a high centrifugal force. It is worth noting that the centrifugal force received by the mold can differ from the one of the vibrators but the force of one vibrator and the energy received by the system are assumed to be proportionate.

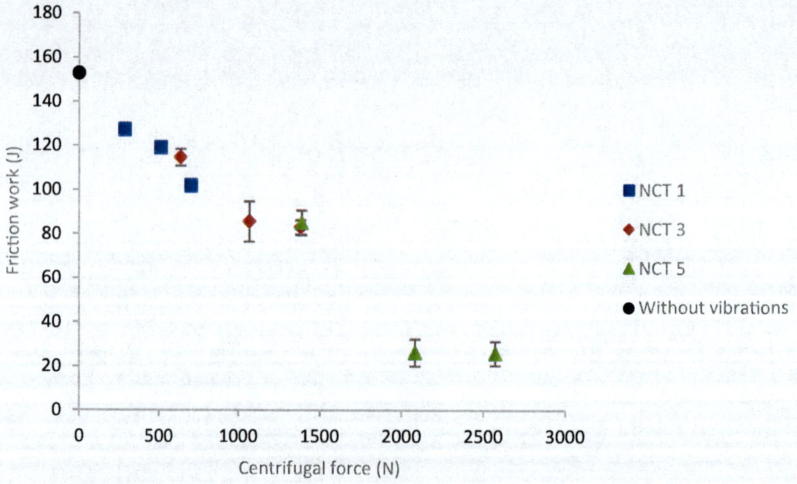

Fig. 4. Impact of the centrifugal force on the work of the friction force.

The compaction test has a good repeatability except for the test with the vibrator NCT 5 and the vibrator NCT 3 for 4 and 6bar of air pressure. For these modalities, the standard deviation is multiplied by factors between 2 and 5. This variation comes from the difficulty of maintaining the system stable with the highest centrifugal forces. The movement induced by the vibrators could rotate the mold and remove the damping layer use to limit shocks between the different parts of the system.

5.2 The Vibrations Characteristics on the Macroscopic Behavior

Samples compacted statically with no vibration at different moisture content show an optimum moisture content of approximately 9% to reach the highest dry density feasible with a compressive stress of 10 MPa. The Fig. 5 shows that the same optimal water content is found in the Proctor modified test. The small vibrations generated by the vibrator NCT 1 at 6 bar of air pressure show a 1% decrease in the optimum moisture content. The reduction of the moisture content allows a small increase in dry density of 0.01 t/m^3. However, this variation is too small to significantly affect the mechanical characteristics, but with higher vibration energy, the impact could be greater. Indeed, the reduction of the moisture content at this optimized state allows to increase the dry density and the compressive strength [51, 52]. The conventional vibration system used in the manufacture of concrete blocks may be relevant for CEB-making.

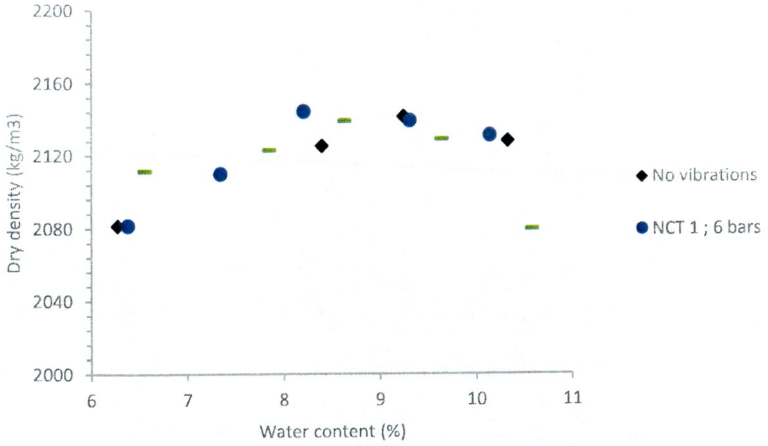

Fig. 5. Evolution of dry density with initial water content for different samples fabrication process. Process-oriented determination of the optimal water content for each type of fabrication method.

The relevance of using vibration during compaction can also be observed on the evolution of dry compressive strength of the material with centrifugal force as already shown [47]. Figure 6 shows that vibration allows to increase the compression strength of the materials in comparison with the reference non-vibrated samples. Moreover, the figure displays a slight increase in compressive strength with the increasing centrifugal force for a same vibrator. This behavior seems to depend on the type of vibrator used, suggesting that the centrifugal force is not the only parameter affecting this increase in compressive strength. Each vibrator induces a different effect on the strength due to varying compaction time between processing. The greater the vibration, the more the material will be packed and the shorter the compaction time will be. As a result, it is interesting to compute the efficiency of the process in terms of strength gain versus compaction energy lost by friction (Fig. 7). It can be seen that this ratio also increases with the centrifugal force showing the pertinence of using vibration to ease the production of strength-improved construction blocks.

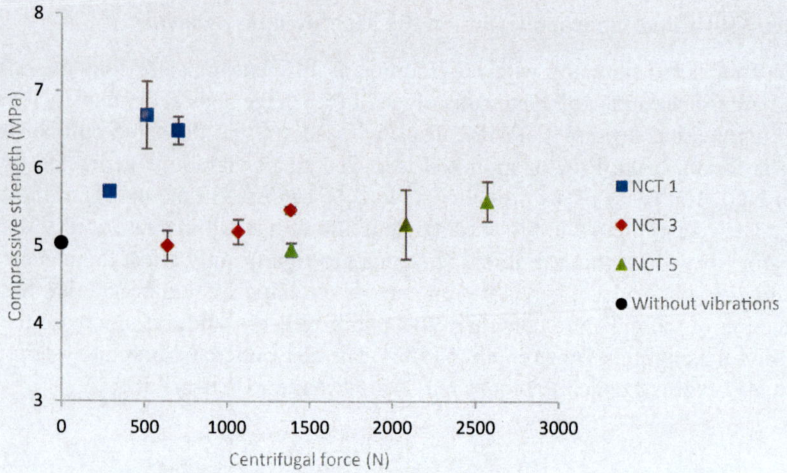

Fig. 6. Influence of the centrifugal force on the compressive strength.

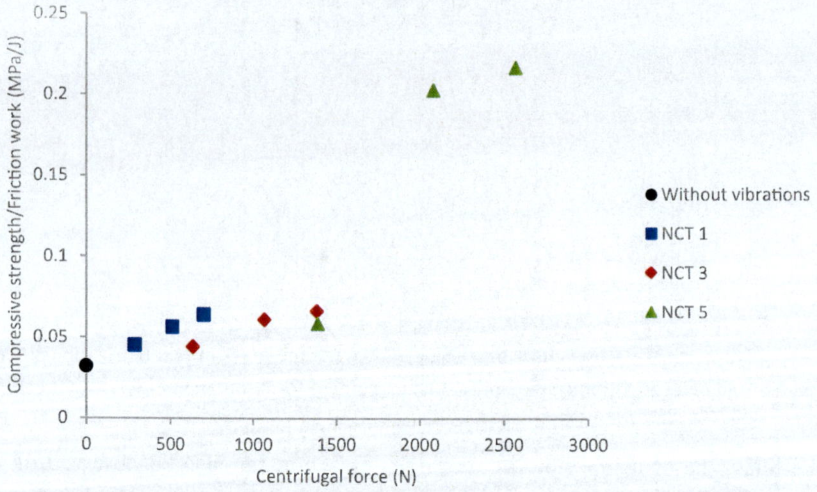

Fig. 7. Compressive strength to loss friction force in function of centrifugal force.

6 Conclusion

Compressed earth blocks represent an environment friendly solution to respond to the new challenges of the construction industry. This technique is currently used marginally and needs to be industrialized in order to become an affordable solution. Vibrations, already used in road compaction and in concrete block manufacture could improve the characteristics of compressed earth blocks.

When making blocks, friction with the mold reduces the compaction stress applied to the granular skeleton, reducing density and compressive strength. This study shows the relevance of vibration for reducing friction at low and high density. At low density, the vibrations delete friction stress during the earth compaction. At high density, the vibrations reduce the friction at a given density. The vibration efficiency seems to be related to the vibrator parameters and especially its centrifugal force. Moreover, the vibration brings a reduction in the optimal moisture content to optimize the compaction process. These different analyses show the relevance of vibrating systems commonly used in concrete blocks manufactures to improve CEB.

In another development, it could be interesting to evaluate the material and the system parameters leading to the reduction of friction, for a better understanding of this phenomenon in order to optimize the vibration solicitation. Also, tests need to be performed in block factories to assess how much vibration energy is required at this scale.

Acknowledgment. This study is funded by the ANR (Agence National de la Recherche) within the framework of the Labcom COLORE "Construction with local resources" ANR-21-LCV3-0008. The authors are grateful to ANR for the financial support.

References

1. Medineckiene, M., Turskis, Z., Zavadskas, E.K.: Sustainable construction taking into account the building impact on the environment. J. Environ. Eng. Landsc. Manag. **18**(2), 118–127 (2010). https://doi.org/10.3846/jeelm.2010.14
2. Zutshi, A., Creed, A.: An international review of environmental initiatives in the construction sector. J. Clean. Prod. **98**, 92–106 (2015). https://doi.org/10.1016/j.jclepro.2014.06.077
3. IPCC. IPCC Sixth Assessment Report; Climate Change 2022: Mitigation of Climate Change: Technical Summary (2022)
4. Taylor, M., Tam, C., Gielen, D.: Energy efficiency and CO_2 emissions from the global cement industry, p. 13 (2006)
5. Gallipoli, D., Bruno, A.W., Perlot, C., Mendes, J.: A geotechnical perspective of raw earth building. Acta Geotech. **12**(3), 463–478 (2017). https://doi.org/10.1007/s11440-016-0521-1
6. Van Damme, H., Houben, H.: Should raw earth be improved? An environmental assessment. In: Terra 2016, Lyon, France, pp. 180–182 (2016)
7. Hamard, E., Lemercier, B., Cazacliu, B., Razakamanantsoa, A., Morel, J.-C.: A new methodology to identify and quantify material resource at a large scale for earth construction – application to cob in Brittany. Constr. Build. Mater. **170**, 485–497 (2018). https://doi.org/10.1016/j.conbuildmat.2018.03.097
8. Giada, G., Caponetto, R., Nocera, F.: Hygrothermal properties of raw earth materials: a literature review. Sustainability **11**(19), 5342 (2019). https://doi.org/10.3390/su11195342
9. Guihéneuf, S., Rangeard, D., Perrot, A., Cusin, T., Collet, F., Prétot, S.: Effect of bio-stabilizers on capillary absorption and water vapour transfer into raw earth. Mater. Struct. **53**(6), 1–18 (2020). https://doi.org/10.1617/s11527-020-01571-z
10. Zhemchuzhnikov, A., Ghavami, K., dal Toé Casagrande, M.: Static compaction of soils with varying clay content. Key Eng. Mater. **668**, 238–246 (2015). https://doi.org/10.4028/www.scientific.net/KEM.668.238
11. Radjaï, F.: Features of force transmission in granular media. In: Powders and Grains 2001. CRC Press (2001)
12. Radjaï, F., Roux, S.: Features of the granular texture. In: Powders and Grains 2001. CRC Press (2001)

13. Mesbah, A., Morel, J.C., Olivier, M.: Comportement des sols fins argileux pendant un essai de compactage statique: détermination des paramètres pertinents. Mater. Struct. **32**(9), 687–694 (1999). https://doi.org/10.1007/BF02481707
14. Andreotti, B., Forterre, Y., Pouliquen, O.: Les milieux granulaires: Entre fluide et solide. EDP Sciences (2011). https://doi.org/10.1051/978-2-7598-0925-7
15. de Richter, S.K.: Étude de l'organisation des réarrangements d'un milieu granulaire sous sollicitations mécaniques. Ph.D. thesis, Université Européenne de Bretagne (2009)
16. Gelard, D., et al.: When physics revisit earth construction: recent advances in the understanding of the cohesion mechanisms of earthen materials. In: Proceedings of the International Symposium on Earthen Structures, IIS Bangalore, vol. 294302 (2007)
17. Van Damme, H., Houben, H.: Earth concrete. Stabilization revisited. Cem. Concr. Res. **114**, 90–102 (2018). https://doi.org/10.1016/j.cemconres.2017.02.035
18. Schlosser, F., Guilloux, A.: Le frottement dans le renforcement des sols. Rev. Fr. Géotechnique **16**, Article no. 16 (1981). https://doi.org/10.1051/geotech/1981016065
19. Sandeep, C.S., Senetakis, K.: Effect of young's modulus and surface roughness on the inter-particle friction of granular materials. Materials **11**(2), Article no. 2 (2018). https://doi.org/10.3390/ma11020217
20. Albatal, A., Stark, N., Castellanos, B.: Estimating in situ relative density and friction angle of nearshore sand from portable free-fall penetrometer tests. Can. Geotech. J. **57**(1), 17–31 (2020). https://doi.org/10.1139/cgj-2018-0267
21. Schöpfer, M.P.J., Abe, S., Childs, C., Walsh, J.J.: The impact of porosity and crack density on the elasticity, strength and friction of cohesive granular materials: insights from DEM modelling. Int. J. Rock Mech. Min. Sci. **46**(2), 250–261 (2009). https://doi.org/10.1016/j.ijrmms.2008.03.009
22. Lehane, B.M., Jardine, R.J., Bond, A.J., Frank, R.: Mechanisms of shaft friction in sand from instrumented pile tests. J. Geotech. Eng. **119**(1), 19–35 (1993). https://doi.org/10.1061/(ASCE)0733-9410(1993)119:1(19)
23. Reiffsteck, P., et al.: Mesures en laboratoire du comportement mecanique des sols heterogenes. Bull. Lab. Ponts Chaussees, no. 268–269 (2007)
24. Koval, G.: Comportement d'interface des matériaux granulaires (2008)
25. Uesugi, M., Kishida, H.: Influential factors of friction between steel and dry sands. 土質工学会論文報告集 **26**(2), 33–46 (1986). https://doi.org/10.3208/sandf1972.26.2_33
26. Unterreiner, P.: Contribution à l'étude et à la modélisation numérique des sols cloués: application au calcul en déformation des ouvrages de soutènement. Ph.D. thesis, Ecole Nationale des Ponts et Chaussées (1994)
27. Boulon, M., Plytas, C., Foray, P.: Comportement d'interface et prévision du frottement latéral le long des pieux et tirants d'ancrage. Rev. Fr. Géotechnique **35**, Article no. 35 (1986). https://doi.org/10.1051/geotech/1986035031
28. Desai, C.S., Rigby, D.B.: Modelling and testing of interfaces. In: Selvadurai, A.P.S., Boulon, M.J. (eds.) Studies in Applied Mechanics, vol. 42, pp. 107–125. Elsevier (1995). https://doi.org/10.1016/S0922-5382(06)80008-5
29. Fioravante, V.: On the shaft friction modelling of non-displacement piles in sand. 地盤工学会論文報告集 **42**(2), 23–33 (2002). https://doi.org/10.3208/sandf.42.2_23
30. Hoteit, N.: Contribution à l'étude du comportement d'interface sable-inclusion et application au frottement apparent. These de doctorat, Grenoble INPG (1990)
31. Yoshimi, Y., Kishida, T.: A ring torsion apparatus for evaluating friction between soil and metal surfaces (1981)
32. Frank, R.: Etude theorique du comportement des pieux sous charge verticale; introduction de la dilatance, no. 46 (1975)
33. Jaeger, H.M., Nagel, S.R., Behringer, R.P.: Granular solids, liquids, and gases. Rev. Mod. Phys. **68**(4), 1259–1273 (1996). https://doi.org/10.1103/RevModPhys.68.1259

34. Yang, X.-D., Zhang, W.: Nonlinear dynamics of axially moving beam with coupled longitudinal–transversal vibrations. Nonlinear Dyn. **78**(4), 2547–2556 (2014). https://doi.org/10.1007/s11071-014-1609-5
35. Triantafyllidis, Th., Kimmig, I.: A simplified model for vibro compaction of granular soils. Soil Dyn. Earthq. Eng. **122**, 261–273 (2019). https://doi.org/10.1016/j.soildyn.2018.12.008
36. Farzaneh, O.: Compactage par vibration des matériaux granulaires. Etude sur le comportement de l'ensemble sol-rouleau vibrant (1981)
37. Roger, A.A.: Vibrocompaction of cohesionless soils. Cementation Research Limited. Cementation Research Limited (1979)
38. Rodger, A.A.: Vibrocompaction of cohesionless soils. Cementation Research Limited, International Report **7**(97) (1997)
39. Reynolds, O.: LVII. On the dilatancy of media composed of rigid particles in contact. With experimental illustrations. Lond. Edinb. Dublin Philos. Mag. J. Sci. **20**(127), 469–481 (1885). https://doi.org/10.1080/14786448508627791
40. Massarsch, K.R.: Effects of Vibratory Compaction, p. 10 (2002)
41. Knight, J.B.: External boundaries and internal shear bands in granular convection. Phys. Rev. E **55**(5), 6016–6023 (1997). https://doi.org/10.1103/PhysRevE.55.6016
42. Möbius, M.E., Cheng, X., Eshuis, P., Karczmar, G.S., Nagel, S.R., Jaeger, H.M.: Effect of air on granular size separation in a vibrated granular bed. Phys. Rev. E **72**(1), 011304 (2005). https://doi.org/10.1103/PhysRevE.72.011304
43. Wassgren, C.R., Hunt, M.L., Freese, P.J., Palamara, J., Brennen, C.E.: Effects of vertical vibration on hopper flows of granular material. Phys. Fluids **14**(10), 3439–3448 (2002). https://doi.org/10.1063/1.1503354
44. Medved, M.: Connections between response modes in a horizontally driven granular material. Phys. Rev. E **65**(2), 021305 (2002). https://doi.org/10.1103/PhysRevE.65.021305
45. Lamb, H.: On the propagation of tremors over the surface of anelastic solid (1904)
46. Carlson, J.M., Langer, J.S., Shaw, B.E.: Dynamics of earthquake faults. Rev. Mod. Phys. **66**(2), 657–670 (1994). https://doi.org/10.1103/RevModPhys.66.657
47. Guihéneuf, S., Rangeard, D., Perrot, A.: Cast, compaction, vibro-compaction or extrusion: processing methods for optimizing the mechanical strength of raw earth-based materials. Acad. J. Civ. Eng. **37**(2), Article no. 2 (2019). https://doi.org/10.26168/icbbm2019.22
48. Konert, M., Vandenberghe, J.: Comparison of laser grain size analysis with pipette and sieve analysis: a solution for the underestimation of the clay fraction. Sedimentology **44**(3), 523–535 (1997). https://doi.org/10.1046/j.1365-3091.1997.d01-38.x
49. Beuselinck, L., Govers, G., Poesen, J., Degraer, G., Froyen, L.: Grain-size analysis by laser diffractometry: comparison with the sieve-pipette method. CATENA **32**(3–4), 193–208 (1998). https://doi.org/10.1016/S0341-8162(98)00051-4
50. AFNOR. NF EN 13286-41: Mélanges traités et mélanges non traités aux liants hydrauliques - Partie 41: méthode d'essai pour la détermination de la résistance à la compression des graves traitées aux liants hydrauliques (2021)
51. Morel, J.-C., Pkla, A., Walker, P.: Compressive strength testing of compressed earth blocks. Constr. Build. Mater. **21**(2), 303–309 (2007). https://doi.org/10.1016/j.conbuildmat.2005.08.021
52. Mansour, M.B.: Optimizing thermal and mechanical performance of compressed earth blocks (CEB). Constr. Build. Mater. **104**, 44–51 (2016). https://doi.org/10.1016/j.conbuildmat.2015.12.024

The State of the Art of Cob Construction: A Comprehensive Review of the Optimal Mixtures and Testing Methods

Kamal Haddad(✉), Eshrar Latif, and Simon Lannon

Cardiff University, Cardiff CF10 3NB, UK
{Haddadks,LatifE,lannon}@cardiff.ac.uk

Abstract. Earthen construction systems have potential hygrothermal, and environmental benefits over conventional building materials such as concrete. However, such systems are not yet fully optimised to be part of an energy-efficient building. Therefore, to further optimise the material, this review explores peer-reviewed research articles that relate to different materials used in cob mixes and the different testing methods used to assess the produced specimen's hygrothermal performance. For data collection, a systematic keyword search was carried out on ScienceDirect, Scopus, Google scholar search engines and relevant books. The filtering of journal articles was based on studying the abstracts followed by analysing their content within the scope of the review. The results show that the soil's constituents and the added fibre ratios critically affect the percentages of clay and water added to the mixture. Fibres' impact on the mix was experimented with by multiple researchers using distinct types of plant aggregates. The percentage of fibre addition ranged between 0.9% and 3% for structural specimens and reached 25% for non-structural specimens with optimised insulation properties. However, there is no consensus and robust collated data available about the ratios of the mixes concerning the hygrothermal performance of the specimens. Therefore, a matrix for mixes and testing methods was developed with the available data to aid the progression of any future optimisation effort.

Keywords: Earthen Construction · Mixing Ratio · Hygrothermal Testing · Cob

1 Introduction and Background

Conventional construction materials such as concrete and steel have gained momentum since the industrial revolution sparked due to their structural performance. Yet, global statistics have demonstrated that cement has a contribution of 8% to the worldwide total of CO2 emissions [1]. According to the International Energy Agency (IEA), such materials are responsible for 30% of total global final energy consumption and 27% of total energy sector emissions [2].

Earthen construction has been defined as construction that uses building materials in which clay is a binder [3]. Saxton, in one of his early publications on earthen buildings, has discussed that around one-third of the world's population lives in earth-constructed

S. Amziane et al. (Eds.): ICBBM 2023, RILEM Bookseries 45, pp. 212–231, 2023.
https://doi.org/10.1007/978-3-031-33465-8_17

domestic buildings [4]. Saxton has also elaborated that many developed countries, such as England, France, Germany, and China, in addition to a wide number of developing countries, have adopted this type of construction due to its affordability [4].

Even though the earth is an old construction and building material that has been used since early civilizations, recent demand for using the material has been on the rise due to the increased interest in sustainable and green building practices [5]. Besides, it is considered a natural concrete alternative with lower embodied carbon as well as having a lower need for energy for production and operation compared to other materials [6].

According to Hamard et al., earthen construction can be classified into two groups based on the construction method used: wet methods and dry methods. Dry earthen construction methods consist of masonry units like compressed earth blocks and monolithic walls that are implemented moistly, which is rammed earth construction. On the other hand, wet construction methods include four other systems of earthen construction, namely adobe, wattle and daub, earthen-based plaster, and cob as a monolithic wall [7].

The presented paper will explore cob as a potentially promising building material. Furthermore, it will review the current literature that relates to the different mixes of cob and the different ratios used in the mixes, as well as the mixing conditions and processes used in the production of the material. In parallel, the paper will discuss the various methods that have been discussed in the literature that evaluate the hygrothermal performance of cob.

2 Methodology

This paper focuses on presenting a comprehensive review of cob and its associated hygrothermal testing methods. In this research, data collection was fulfilled through a systematic keyword-based search using ScienceDirect, Scopus, and Google Scholar search engines and relevant books. The keywords in the search process were cob, earthen construction, cob mixtures, and hygrothermal tests of cob. Then, the papers' abstracts and their content were thoroughly analysed and studied, which resulted in the selection of 25 studies that studied cob in particular or as a partial study on earthen construction. The selected studies have discussed mix ratios and hygrothermal testing methods.

3 Cob as a Building Material

Generally, the term "cob" has been defined as a lump of rounded shape [8]. In the field of earthen construction, cob is defined as a sustainable material that consists of soil, water, and fibrous materials [8]. A cob wall is an ancient, traditional building material used primarily for structural purposes [9]. Meanwhile, recent research has started to focus on enhancing its hygrothermal performance [10–14].

Cob as a building material has unique characteristics when compared to other earthen materials. For instance, cob exhibits higher material ductility when compared to rammed earth and adobe [15]. Besides, as a wet-based construction material, cob gives further freedom in design during all phases of construction when compared to dry methods like rammed earth [16]. Furthermore, such a characteristic enhances the durability of the material since its maintenance will be smoother and more flexible.

4 Content of Cob Mixtures

Cob as a material consists of three main constituents: subsoil, fibres, and water. To enhance its structural stability, many researchers have added other materials such as clay, lime, or cement. Generally, it has been recommended for conventional construction of cob that the composition of the mixture averages 78% subsoil, 20% water, and 2% fibre (typically straw) by weight [16].

The general composition of earth that is used in cob mixtures has been suggested to be 15–25% clay to 75–85% aggregate/sand [15]. It is crucial to mention that in order to increase cob density, well-graded soils were preferred since they had good space-filling properties that improved cob strength. Topsoil has been deemed unsuitable for cob construction because it decomposes quickly after application and causes a mechanical weakening in earth walls. Therefore, it has been found that the most suitable soil for cob mixtures is just under the topsoil [7]. Literature has discussed that water content and the initial moisture level of a cob mixture have a significant effect on the strength of the material [15]. Furthermore, it has been concluded that mixtures that are reinforced with fibres tend to require more water content in the mixture [17].

Even though the tensile strength and bonding of straw help in the reduction of cracking [4], some studies have argued that adding large amounts would increase the strain at failure due to loading. Many studies investigate distinct types of fibres that are considered green and mostly biodegradable, such as cereal, straw, corn stalk, bagasse, rice straw, sunflower hulls and stalks, banana stalks, coconut coir, bamboo, durian peel, and palm leaves oil [8].

It has been discussed in the literature that fibre content has several advantages that contribute to the success of cob mixtures. For instance, it facilitates the mixing of cob, assists handling, accelerates the drying process, works on distributing shrinkage cracks throughout the wall mass, enhances cohesion and shear resistance of the wall, and helps improve weathering resistance. The effect of fibres on thermal insulation was discussed in some studies, whereas other studies suggested that fibres' effect on thermal conductivity would be noticeable when the content of fibres in the fabric is about 25% by mass [18].

As introduced in an exhaustive review on plant aggregates and fibres in earthen construction materials [19], distinct types of plant fibres and aggregates were studied in the earthen construction literature and classified into eight main categories, which can be classified as cereal straw, which includes wheat straw, barely straw, and oat straw; wood aggregates such as wood shavings and wood fibres; Bast fibres which includes hemp fibre, hemp hurds, jute fibre, kenaf fibre, and diss fibre; Palm tree fibres including coir, oil palm fibre, and date palm fibre; Waste and residues like cassava peel, millet residue, cotton residue, tea residue, tobacco residue, and grass; Leaf fibres, including sisal, banana fibres, and pineapple fibre; aquatic plants like Phragmites, Typha, and seaweed fibre; Wool (sheep wool).

5 Cob Mixes in Literature

Several researchers have started to study and assess the implications of using cob as a building material [3, 4, 7, 14, 16, 20]. As discussed, cob consists of subsoil, water, fibres, and a binder like clay. Most studies presented have focused on the optimisation of the structural performance of the material, while a few have focused on enhancing its hygrothermal performance. In this section, these studies will be explored at the constituent level. This section will discuss the different ratios of cob's constituents that are presented in literature. Table 1 demonstrates a comprehensive review of the mixing ratios that have been adopted in previous research.

5.1 Sub-soil and Binder Ratios and Granulometric Characterisation

The selection and creation of the soil that will be used in the mix have a significant impact on the materials' structural and hygrothermal performance [12, 13, 21–23]. The used soil needed to be obtained from a local source, which would help in the reduction of the embodied carbon emissions caused by transportation, as will be discussed later in the review.

One of the early papers on studying cob has been published by Saxton, who has concluded that typical soil contains 30% gravel, 35% sand, and 35% silt and clay [4]. Alassaad et al. have used two types of soil; the first's Unified Soil Classification (UCS) as low plasticity silt (ML) with a plasticity of 24% and a plasticity index of 3.6%, and the second's USC as silty sand with gravel (SM), which had a plasticity of 21% and a plasticity index of 2.7%. In the same study, both soils have quartz, mica, feldspar, iron oxyhydroxides, and limonite [21].

Quagliarini et al. have focused on performing a detailed analysis of a historic cob. After analysing the building's fabric, it is shown that it consists of 34% clay, 17% sand, and 49% silt, with a plastic index of 19% and a liquid index of 38% [24]. Meanwhile, the researchers have tried to make a mixture that has a similar character with 36% clay, 13.5% sand, and 50.5% silt with a plastic index and a liquid index of 21% and 42%, respectively, and used 3 kg of the created soil in the cob mix. Alhumayani et al. have used 80 kg of subsoil in their separate study [25].

Miccoli et al. have utilised a mixture that contained 18% gravel and sand, 61% silt, and 21% clay [22, 23, 26]. Ben-Alon et al. have used 257 kg of clay-rich soil with a 50% clay content for one square metre of cob and a thickness of 457 mm [3]. Medero et al. have experimented with soil that is characterised to contain 24.4% gravel, 19.7% coarse sand, and 32.5% fine sand [27]. In addition, Sangma and Tripura have worked on multiple studies to improve the structural performance of cob. For their mixtures, the soil's characterisation is 60.5% sand, 22.25% silt, and 14.25% clay, with a plastic index of 11.43% [17, 28].

Alqenaee and Memari have primarily focused on optimising a mixture that will be efficient and stable for 3D printing. The authors have created 36 different mixtures, nine of which are described in the paper and shown in Table 1. The clay content ranges from 38.53% to 52.63%, the sand content varies between 11.11% and 17.73%, and the lime content is between 7.08% and 11.63%. In their study, mix M30 and M31 have had enough cement to work as binders with values of 2.40% and 2.48%, respectively. The final mixture, M36 consists of 49% clay, 15.31% sand, and 10.00% lime [29]. For the same purpose of developing a 3D printable mixture, Gomaa et al. have used 73% of the soil that was found to contain 19–20% clay and 80–81% aggregate/sand of the total mixture [30].

Various standards for determining the soil properties and characterisation were used in studies and can be used for further research, such as ISO 13320:2009 [31] and IS 2720 Parts 4, 5, and 7 [32]. Furthermore, a study by Vinceslas et al. thoroughly discusses the methods to characterise the tested soil; Particle size distribution has been assessed using NF P 94-056 [33], the absorption capacity value has been determined using NF P 94-068 [34], the plastic limit, liquid limit, and plastic index have been determined using NF P 94-051 [35], the normal proportional water content and density of the main materials have been configured using NF P 94-093 [36], the specific gravity of the produced specimens using NF P 94-050 [37] and Soil characterisation has been tested by following ASTM D2487-11 [38].

5.2 Water Content in Mixes

The concentration of water in cob mixtures is controlled by other constituents, such as fibre content and type. For instance, Akinkurolere O.O. et al. have argued that the addition of fibre to a mixture with a high initial water content has beneficial effects on the strength of the mix as it will enhance the bonding and homogeneity between the components within the cob mixture [20].

Generally, mixtures that are designed for structural purposes have a significantly lower amount of water than those designed to be hygrothermally effective or insulation materials. Accordingly, the water content in studies that focused on evaluating the structural performance ranges between 19% and 40%, while the ones designed to enhance the thermal insulation varied between 62.1% and 131.3% as the fibre content was greatly increased in the mix [20].

Alassaad et al. have established the added water content as a ratio, where the water-to-soil ratio was equivalent to 0.3 [39]. Alhumayani et al. in their paper have used an amount of 20 kg of water, which have resulted in a ratio of 1:4 concerning the soil content in the mix [25] which aligns with Weismann and Bryce, who proposed a water-to-subsoil ratio of one part water to every four parts soil [40]. To determine the water content in the produced cob specimens, researchers like Vinceslas et al. have referred to the French standard NF P 94-050 [37] after drying the specimens at 105 °C [41].

As many of the selected papers in the review have worked on the traditional construction of cob methods, some of the studies have explored creating mixtures that are designed for 3D printing. Water content may be the main constraint for 3D printing cob as the mixture needs to be consistent and stable while being viscous enough to be extruded through the nozzle. In 2021, Gomaa et al. have published a paper that focuses on developing an extrusion system that can be used for cob's 3D printing. In their study, the authors have assessed different concentrations of water content in the mix (22%, 24%, 26% and 28%) where it has been concluded that the optimum water content was 25% [15]. Alqenaee and Memari have experimented with water contents between 24.19% and 34.25% [29]. It has been observed by Gomaa et al. that the moisture content of the final printed cob is slightly reduced by the 3DP extrusion process. This is caused by the pressurisation of the mixture inside the extrusion system, which leads to moisture release in the form of leakage around the cartridge connections [15, 16].

5.3 Fibres and Aggregates in Cob Mixtures

Fibre content ratios and fibre type are essential aspects that need to be considered while studying a material like cob. Fibres have a vital role that affects the structural, hygrothermal, and environmental performance under study. In current literature, many fibre types have been studied, like seaweed fibre [42], sheep wool [43], and tobacco residue grass [44]. In this review, different studies have used several types of straw, including hemp straw, flax straw, wheat straw, paddy straw, and rice straw, meanwhile, other studies have used other aggregates in their mixes, such as coconut coir [28, 45], hemp shives [10, 12–14], and reed [12, 14].

Generally, the use of fibres and added aggregates in cob mixtures has been extensively studied due to its effects on the mix's performance. The fibre content in the majority of studies ranged from 0.6% to 3%. Alassaad et al. have used 2.5% flax straw of the dry soil mass [39]. In parallel, Ben-Alon et al. in their research have used 10.1 kg of wheat straw that is added to 256 kg of clay-rich soil with a calculated soil/fibre ratio of 0.039 [3]. Goodhew et al., in a study that focuses on optimising cob for better insulation and thermal performance, have added a higher amount of fibre within different mixes [12].

Zeghari et al. have developed eight mixtures that are optimised for better structural performance using hemp straw, flax straw, wheat straw, and reed [14]. Simultaneously, the study has worked on developing two mixes using hemp shiv and reed for the insulation part of the dual-system wall.

Within the different studies, it was observed that the fibre lengths have ranged between 20 mm and 300 mm, as shorter fibres demonstrated better performance since they blended in the mix, which created a homogeneous cob mixture. In their study, Sangma and Tripura used a systematic comparative analysis to compare the structural performance of cob mixtures made with coconut coir and straw fibres, and they concluded that coconut coir performs better enhancing the structural performance [17, 28].

6 Assessing the Hygrothermal Performance of Cob

In general, earthen materials offer various economic advantages due to their thermal inertia. For example, cob walls with lower conductivity and lower density have a higher insulation value and are lighter in weight [30]. As for 3D-printed cob, a significant increase in thermal performance can be noticed in 3D-printed cob structures compared to their manually structured counterparts [30]. This section will go over the various tests that were carried out, such as porosity, bulk density, thermal conductivity, water vapour permeability, moisture buffering value, and moisture sorption isotherm.

6.1 Porosity Tests

Earth is a porous, unsaturated material. In addition to its effect on the hygrothermal performance of the material, it also has a significant effect on the structural performance of cob mixtures. For instance, the young modulus for raw earth material ranges between 1 GPa and 5.5 GPa, where the higher the porosity, the higher the young modulus value will be [49]. A study by Goodhew et al. discussed the correlation between adding fibres

Table 1. A matrix of mixes and ratios of studied cob mixtures.

Test		Soil				Total Soil content (%)	Additives			Water content (%)	Fibres		Ref.
#	Mix ID	Clay (%)	Silt (%)	Sand (%)	Gravel (%)		Cement (%)	PCM kg/m^3	Lime (%)				
1	Soil 0	15–25		75–85		78				20	2%		[7]
2	Soil 0	35		35	30					17–29	0%–3%		[4]
	Soil 1	4	76	17	3								[39]
	Soil 2	1	13	51	35								
3	Cob 0					1500 kg/m^3		0		375 kg/m^3	37.5 kg/m^3		
	Cob 2					1475 kg/m^3		30.2		368 kg/m^3	36.9 kg/m^3		
	Cob 5					1450 kg/m^3		74.3		362 kg/m3	36.3 kg/m^3		
	Cob 10					1390 kg/m^3		142.5		347 kg/m^3	34.8 kg/m^3		
4	Original soil	34	49	17									[21]
	Yellow soil	36	50.50	13.50		3 kg				28	0.02 kg		
5	Conventional Cob					78.0				20.0	2.0% Straw		[25]
	3DP Cob					73.0				25.0	2.0% Straw		
6	Soil 0	21	61	18							20–30kg/m^3		[23]
7	Soil 0					256 kg				185 kg	10.1 kg		[3]
8	Soil 0	20.60		52.2	24.40	79.20				1.55	1.25% Straw		[27]
9	All Mixes Average	14.25	22.25	60.50			3–10			31.7 - 40	3–10% Straw	3–10% Coir	[17]
10	UK3 Soil	12.83	68.93	17.80	0.44								[12]

(continued)

Table 1. (*continued*)

Test		Soil				Total Soil content (%)	Additives			Water content (%)	Fibres	Ref.
#	Mix ID	Clay (%)	Silt (%)	Sand (%)	Gravel (%)		Cement (%)	PCM kg/m^3	Lime (%)			
	UK4 Soil	5.59	58.64	16.74	19.03							
	FR3 Soil	12.85	65.43	12.36	9.36							
	Soil 0	UK3								65.60	Hemp shiv: 50%	
	Soil 1	UK3								107.30	Hemp shiv: 50%	
	Soil 2	UK3								107.3	Hemp shiv: 25%	
	Soil 3	UK3								107.30	Reed: 25%	
	Soil 4	FR3								131.30	Reed: 25%	
	Soil 5	FR3								131.30	Hemp shiv: 25%	
	Soil 6	UK4								62.10	Reed: 25%	
	Soil 7	UK4								62.10	Reed: 50%	
11	S1	Three different French soils were used								25.00	Hemp straw: 5%	[14]
	S2									28.00	Hemp straw: 5%	
	S3									28.00	Hemp straw: 2.5%	
	S4									28.00	Flax straw: 2.5%	
	S5									31.00	Flax straw: 2.5%	
	S6									31.00	Wheat straw: 2.5%	
	S7									31.00	Reed: 2.5%	
	S8									31.00	Wheat straw: 5%	
	I1									131.00	Reed: 25%	
	I2									131.00	Hemp shiv:25%	
12	Soil 0	3	18	42	37						0.9% Straw (by mass)	[46]
13	Soil 0									33 - 47	Rice straw that varies from 0.6%-3%	[20]

(*continued*)

Table 1. *(continued)*

Test		Soil				Total Soil content (%)	Additives			Water content (%)	Fibres	Ref.
#	Mix ID	Clay (%)	Silt (%)	Sand (%)	Gravel (%)		Cement (%)	PCM kg/m^3	Lime (%)			
14	S3	FR2 soil								28.50	2.5% hemp straw	[13]
	T1	UK 3 soil								107.30	25% hemp shiv	
15	WF	8	47	45		100 kg				19	0%	[10]
	FL-1%					100 kg				23	1% Flax yarn	
	HA-1%					100 kg				24	1% Hay stalk	
	HE-1%					100 kg				25	1% Hemp shiv	
	FL-3%					100 kg				24	3% Flax yarn	
	HA-3%					100 kg				23	3% Hay stalk	
	HE-3%					100 kg				23	3% Hemp shiv	
16	Soil 0	19–20		80–81		73				22, 24, 26, and 28	2% Straw	[15]
17	Soil 0									25		[16]
18	M23	50.51		11.11					10.10	28.28	-	[29]
	M25	52.63		11.58					7.37	28.42		
	M26	46.51		12.79					11.63	29.07		
	M27	49.78		14.60					7.08	28.54		
	M28	48.39		14.19					9.68	27.74		
	M30	38.53		17.12			2.40%		7.71	34.25		
	M31	39.89		17.73			2.48%		7.98	31.91		
	M34	51.23		16.01					7.47	25.29		
	M34 w/straws	50.69		15.84					7.39	25.03	1.06%	
	M36	49		15.31					10.00	24.19	1.50%	
19	Soil 0	21	61	18						24	1.7% Wheat straw	[26]

Table 2. A matrix of the hygrothermal tests of analysed research.

Test		Density (kg/m^3)	λ (W. m^{-1}.K^{-1})	Water Vapour Permeability (kg/m s Pa)	MBV (g m^{-2}%RH^{-1})	Moisture Sorption Isotherm (%) RH 12% - 90%		Ref
#	Mix ID					Adsorption	Desorption	
1	0	1200–2000	0.47–0.93					[47]
2	UK3 50% Shiv Dry (D)	398.73	0.12					[12]
	UK3 50% Shiv Wet (W)	426.82	0.13					
	UK3 25% Shiv (W)	702.78	0.2					
	UK3 25% Reed (W)	684.1	0.18					
	FR3 25% Reed (W)	637.92	0.16					
	FR3 25% Shiv (W)	654.54	0.18					
	UK4 25% Reed (W)	664.6	0.18					
	UK4 35% Reed (W)	542.87	0.14					
3	Hemp straw 5%	1520	0.57					[14]
	Hemp straw 2.5%	1530	0.74					
	Flax straw 2.5%	1460	0.42					

(continued)

Table 2. (*continued*)

Test		Density (kg/m^3)	λ (W. m^{-1}.K^{-1})	Water Vapour Permeability (kg/m s Pa)	MBV (g m^{-2}%RH^{-1})	Moisture Sorption Isotherm (%) RH 12% - 90%		Ref
#	Mix ID					Adsorption	Desorption	
	Reed 2.5%	1540	0.36					
	Wheat straw 2.5%	1320	0.32					
	Wheat straw 5%	1120	0.24					
	Reed 25%	780	0.2					
	HEMP SHIV 25%	955	0.22					
4	WF	1846	1.08	9.13×10^{-12}	1.06	0.62–2.17	0.66–2.79	[10]
	FL-1%	1841	1.06	8.69×10^{-12}		0.74–2.75		
	Ha-1%	1749	0.92	1.32×10^{-11}	1.73	0.71–2.78	0.81–3.52	
	He-1%	1709	0.91	1.14×10^{-11}	1.73	0.85–3.42	0.77–3.49	
	Fl-3%	1733	0.95	6.48×10^{-11}	1.68			
	Ha-3%	1471	0.7	1.05×10^{-11}	1.54	0.95–2.67	0.68–3.11	
	He-3%	1537	0.7	1.05×10^{-11}	1.74			
5	Straw			2.33×10^{-11}				[48]

to the mixture and its measured porosity, where the addition of fibre would increase the porosity since fibres are considered materials with high porosity [12].

Tchiotsop et al. have presented the differential distribution curve and the cumulative distribution curve of material for a number of fibres with different fibre contents. The modal porosity is the same at 2 μm for 1% fibered mixes, indicating that the earth matrix is consistently well-represented in mixtures. While Ha-3% samples have a uniform distribution and a variety of pore textures. This might result in a rise in the variety of attributes associated with fibre content [10].

6.2 Bulk Density of Specimens

Bulk density is one of the most important factors that need to be deeply studied and investigated, as it has an impact on the structural and hygrothermal performance of building materials. Alassaad et al. followed the French standard, NF ISO 5017 [50]. In his experiments, it was observed that the addition of PCMs will reduce the density of the samples as well as the mechanical strength of cob where the Unconfined Compressive Strength of cob decreased and became more ductile. The density of cob samples that had the Micronal DS 5038 X added was approximately 1500 kg/m^3 [39].

Zeghari et al.'s study resulted in a density range between 1107 kg/m^3 and 1583 kg/m^3 for structural walls, while the density of insulation walls was less than 700 kg/m^3 [14]. Tchiotsop et al. have studied the effect of plant add-ons on the hygric and thermal performance of cob buildings [10].

According to DIN 18945 (DIN 2013a) [51], Miccoli et al. have retrieved a bulk density of 1475 kg/m^3 after the testing specimens were dried [22]. Sangma and Tripura, in their study of the effect of stabilisers on the structural performance of cob, recorded a density of 1690 kg/m^3 for unstabilised mixtures. For cement stabilised samples, the densities varied between 1710 kg/m^3 for 3% cement content and 1780 kg/m^3 for 10% cement content. Samples with 3% added coir had a density of 1650 kg/m^3 and reached 1630 kg/m^3 when the added content was 10%. Straw samples recorded a density of 1640 kg/m^3 for 3% sand and reached 1610 kg/m^3 when 10% was added [28]. The chart below illustrates the densities variation with different fibre types and content in addition to the measured average moisture content. Miccoli et al. reported a bulk density of 1475 kg/m^3, which aligns with other studies and existing literature that showed density variation between 1200 kg/m^3 and 1700 kg/m^3 [22] (Fig. 1).

6.3 Thermal Conductivity

Thermal conductivity can be considered the most explored parameter that affects the hygrothermal performance of a material. This can be explained by the fact that the conductivity of cob has a significant effect on the thermal comfort of users, in addition to the heating and cooling loads that will be used to assure the users' comfort rather than minimising operational energy consumption [52].

Thermal conductivity λ can be defined as the amount of heat (in W) that goes through an area of one sq. meter/one-meter thickness when its interior and exterior faces differ in temperature by one Kelvin [52]. Besides, it is a physical parameter that characterises the ability of a substance to conduct heat and is measured by W/m.K. It is critical to

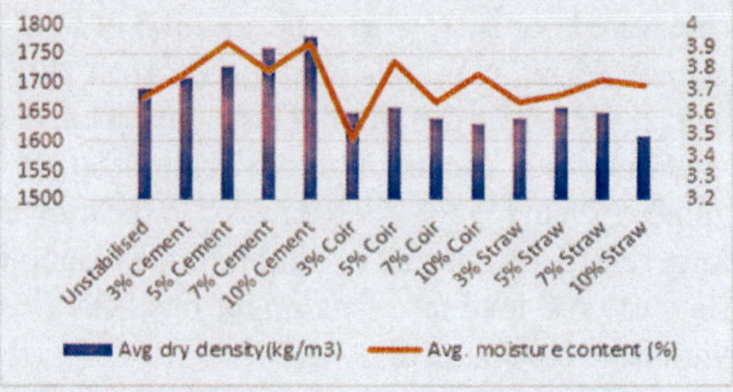

Fig. 1. The average dry density (left axis) and average moisture (right axis) content in cob mixtures with cement, coir, and straw as stabilisers [28].

mention that the lower the value of thermal conductivity, the better the insulation [52, 53].

A study by Laurent et al. have found that the thermal conductivity of cob was a mean of 0.45 W. $m^{-1}.K^{-1}$ while a study by Minke have resulted in a range from 0.47 W. $m^{-1}.K^{-1}$ and 0.93 W. $m^{-1}.K^{-1}$ [47] For the measuring of λ, a Netzsch HFM446 heat flow meter (HFM) was used by Goodhew et al. The researchers have used three different soils (UK3, UK4, and FR3) to make eight different mixes aiming to find a dual cob wall that performs structurally and hygrothermally, and the results are demonstrated in Fig. 2 [12].

Zeghari et al. have discussed that a mixture with a thermal conductivity of 0.4 $Wm^{-1}K^{-1}$ would have a low density, leading to better insulation [14]. In this study, the researchers have found the local thermal conductivity by evaluating the average thermal conductivity regarding the fibre distributions within the tested mixture. The test was carried out in a laboratory setting, with each specimen placed between the upper cooling brass plate and the lower heating brass plate. A highly conductive thermal paste was then used to ensure the success of the tests because the texture of the samples' surfaces was rough, which could have impacted the measurement's accuracy. For the structural samples in the dual wall system, the lowest thermal conductivity was recorded for samples with 5% wheat straw of 0.244 W. $m^{-1}.K^{-1}$ and the highest recorded result of 0.75 W. $m^{-1}.K^{-1}$ was for 2.5% of small fibre content, meanwhile, the highest value for the insulation section of the system was reported to be 0.19 W. $m^{-1}.K^{-1}$ [14].

Tchiotsop et al. have studied several specimens with different plant aggregates in the mixtures. The authors have used a hot disk device to measure and assess the thermal conductivity of the different mixtures. Mixtures without any added fibre have recorded a thermal conductivity of 0.062 W. $m^{-1}.K^{-1}$. The highest reported value of mixtures with added fibre was for 3% hemp shiv with 0.079%, while the lowest was for 3% flax yarn with a recorded result of 0.031 W. $m^{-1}.K^{-1}$ [10].

6.4 Water Vapour Permeability

The importance of vapour diffusion or water vapour permeability (k_m) in a building complies with allowing the moisture exchange between the indoor and outdoor envelopes of buildings, which significantly impacts the comfort of the users within that building

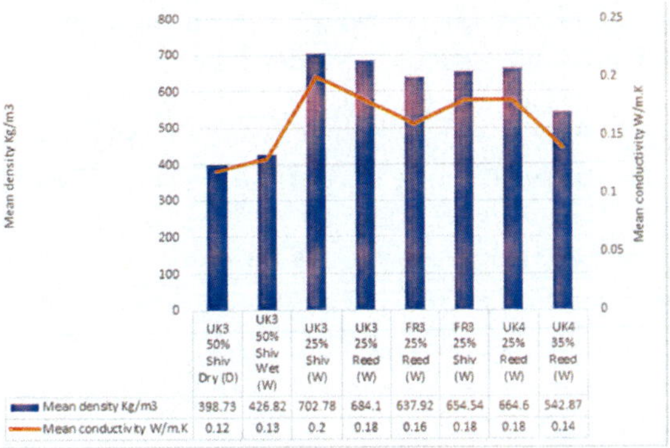

Fig. 2. The relationship between mean thermal conductivity and mean density for cob mixtures with several soil compositions and fibre content for the CobBauge project [12].

[52]. The higher the value of permeability, the better the exchange will be, which can be characterised using the factor of resistance to water vapour (μ). This factor is equal to the ratio between the permeabilities of air and the sample to water vapour, and the higher that factor value is, the harder it is for the moisture to be exchanged [52, 53]. According to EN ISO 1015-19 [54], permeability can be measured either using dry or wet cup methods, the saturated salt solution method (SSS), or by using a climate chamber.

Tchiotsop et al. have followed the dry cup test as stated in NF EN ISO 12572:2016 Standard [55] to evaluate k_m manually and automatically. Gravitest results have resulted in a global CV (coefficient of variation) of 11%, and the CV for specimens with no fibre was 12% higher than manual DCT. For composites having 1% hemp shiv and 1% flax yarn, manual DCT revealed CVs of 50% and 56%, while Gravitest has revealed CVs of 7% and 14%. Lower values (19% and 27% using manual DCT, respectively) have been investigated for He-3% and FL-3%. A modest 11% increase in CV with manual DCT is observed for hay stalk composites [10].

Alassaad et al. have also studied water vapour permeability, but by following the ISO 12572 standard [55] have preconditioned all samples to 23 °C and 50% relative humidity (RH). Therefore, a moisture gradient of around 0% RH to the outside with an approximate RH of 50% was achieved. Stazi et al., in their study that aimed to evaluate the effectiveness of the coatings in protecting the earthen walls against weathering, have experimentally determined the water vapour permeability to be 2.33E^{-11} kg/m/s/Pa [48]. Other researchers, such as Collet-Foucault discussed that the average permeabilities obtained at 23 °C, are between 1.5 E^{-11} kg/m/s/Pa and 1.7 E^{-11} kg/m/s/Pa [56].

6.5 Moisture Buffering Value (MBV)

Moisture buffering value is another key parameter that has a critical effect on the hygroscopic behaviour of a material as it describes the moisture uptake/release capacity of the material [52].

Tchiotsop et al. have used 110 × 40 mm PVC moulds to make specimens, which were kept at 20 °C and relative humidity of 50% before the tests according to the NORDTEST protocol, where specimens have been tested within a climate chamber at 23 °C and have been submitted to a daily relative humidity loading cycle of 75% RH during 8 h and 33% RH during 16 h. Specimens were then weighed regularly every 2 h during the adsorption cycle and twice during the desorption cycle. Afterwards, MBV has been calculated as the ratio of the mean mass variation of the last two cycles by the exchange surface of the specimen (A) and the gap in RH between cycles. MBV for non-fibred mixes was recorded at 1.06 g m^{-2}%RH^{-1}. The highest MBV was for mixes with 1% hemp shiv and 1% hey stalk of 1.73 g m^{-2}%RH^{-1}, while the lowest was for mixtures with 3% hay stalk with a value of 1.54 g m^{-2}%RH^{-1}[10].

6.6 Moisture Sorption Isotherm

The term "moisture sorption isotherm" refers to a material's capacity to absorb and expel moisture from its surroundings [52]. The moisture sorption isotherm has different values for each material and temperature. The adsorption isotherms and desorption isotherms values are the two main parameters used to determine sorption isotherms. Generally, adsorption and desorption differ primarily in that desorption refers to the release of an adsorbed substance from a surface, whereas adsorption refers to the process by which some substances keep the molecules of a gas, liquid, or solute in a thin layer [57].

In a study by Tchiotsop et al., 7 specimens of 10 × 10 × 10 mm^3 and 10 specimens of 50 × 50 × 20 mm^3 size have been tested for sorption isotherms with ProUmid and Saturated Salt Solutions box devices, respectively. For SSS, RH for the inner environment is at 12%, 33%, 55%, 65%, 76%, 86%, and 97%. The equilibrium moisture content was determined using a relative mass variation of less than 0.5% of the specimens. For specimens that were tested with ProUmid, the relative error of mass variation was set at 0.01% Instead, the specimens were dried at 50 °C and the specimen's water content was tested at different RH levels between 10% and 95%. For the majority of the imposed RH values, it was found that the water content values of the adsorption and desorption curves recorded using SSS were higher than those obtained with the ProUmid device [10]. Table 3 shows the normal values for the adsorption isotherms for the mixtures, and Table 4 demonstrates the desorption isotherms for the same mixtures.

Table 3. Distribution model parameters of adsorption isotherms of mixtures [10].

RH level (%)	Non-Fibred	Hay stalk 1%	Flask yarn 1%	Hemp shiv 1%	Hay stalk 3%
12	0.62	0.71	0.74	0.93	0.95
33	0.95	1.16	1.17	1.96	1.28
55	1.3	1.61	1.58	2.74	1.64
75	1.76	2.23	2.20	3.82	2.11
90	2.17	2.78	2.75	4.27	2.67

Alassaad et al. have conducted a different study using the dynamic vapour sorption (DVS) technique according to the ISO 12571 standard [58]. To assess the sorption/desorption values, a ProUmid SPSx-1μ sorption/desorption analyser was used with

Table 4. Distribution model parameters of desorption isotherms of mixtures [10].

RH level (%)	Non-Fibred	Hay stalk 1%	Flask yarn 1%	Hemp shiv 1%	Hay stalk 3%
12	0.66	0.81	0.77	0.93	0.68
33	1.31	1.67	1.58	1.96	1.45
55	1.52	2.24	2.20	2.74	1.45
75	2.23	2.83	2.97	3.82	2.40
90	2.79	3.52	3.49	4.27	3.11

a precision balance (±1 μg) and tight temperature and humidity control, allowing accurate measurements of sample mass and sorption kinetics [39].

As discussed, a large number of tests are needed to assess the hygrothermal performance of materials. Table 5 below summarises the various tests and the associated standards that can be followed to obtain a detailed view of the structural and hygrothermal performance of cob specimens.

Table 5. A list of the hygrothermal testing methods based on researched studies.

Test	Method	Ref
Bulk Density	DIN 18945 (DIN 2013a) [51]	[23, 26]
	NF ISO 5017 [50]	[39]
Water vapour permeability	Dry cup method following ISO 12572 [59]	[39]
Thermal conductivity	Transit hot wire method	[52]
	heat flow meter/Guarded hot plate	
	Flash method	
Sorption isotherms	Saturated salt solution (SSS)	[41]
	Dynamic vapour sorption (DVS)	
Thermal Diffusivity	Flash method	[52]
	Adiabatic calorimeter	
	Flash method	[52]
Specific heat capacity	The guarded hot plate method	[39]
	Differential Scanning Calorimetry (DSC) following ISO 11357-4 [60]	
Moisture buffering value	NORDTEST protocol	[10]

7 Conclusion

The paper presents a literature review on cob construction, its constituents, mixing ratios, and hygrothermal testing methods. It is observed that the different produced specimens vary in their mixing ratios and the use of fibres. Primarily, straw fibre has been the most used and explored throughout literature, followed by other fibres like coconut coir, flax, hemp, hay, and reed. Generally, only a few research articles have thoroughly studied and discussed the mix ratios. Furthermore, when discussing earthen materials such as cob, a fixed terminology may be required. For instance, some studies have referred to the added fibrous content as "aggregates," while other studies have used the term "aggregates" in reference to coarse sand and gravel within the subsoil.

The research has investigated studies that explored the hygrothermal performance of cob. The study presents results that support the significant hygrothermal performance advantage of cob when compared to other materials. On the other hand, it is noticed that no study has fully explored the structural and hygrothermal performances of a cob specimen.

According to the reviewed literature, the researchers believe that studying and developing a mixture that ensures adequate structural and hygrothermal performance is required for it to be part of an energy-efficient construction with lower embodied and operational carbon. Besides, the researchers encourage performing a comprehensive study that investigates the structural and hygrothermal performance of cob to fill the current gap within relevant research.

As this research has provided the available data on hygrothermal tastings and its assessment standards and methods, future work will undertake a detailed analysis of the effect of formulation on the hygrothermal performance of cob. Furthermore, the researchers will work on developing a comprehensive farmwork that provides a robust method to evaluate the hygroscopic and thermal properties of cob.

References

1. Chandler, D.: Researchers have created emissions-free cement. In: World Economic Forum (2019)
2. IEA. Global Energy Review: CO2 Emissions in 2021 (2022). https://www.iea.org/reports/global-energy-review-co2-emissions-in-2021-2. Accessed 12 Feb 2022
3. Ben-Alon, L., et al.: Cradle to site Life Cycle Assessment (LCA) of natural vs conventional building materials: a case study on cob earthen material. Build. Environ. **160**, 106150 (2019)
4. Saxton, R.: Performance of cob as a building material. Struct. Eng. **73**(7), 111–115 (1995)
5. MacDougall, C.: Natural building materials in mainstream construction: lessons from the UK. J. Green Build. **3**(3), 1–14 (2008)
6. Schroeder, H., Sustainable Building with Earth. Springer, Cham (2016). https://doi.org/10.1007/978-3-319-19491-2
7. Hamard, E., et al.: Cob, a vernacular earth construction process in the context of modern sustainable building. Build. Environ. **106**, 103–119 (2016)
8. Gounni, A., Louahlia, H.: Dynamic behavior and economic analysis of sustainable building integrating cob and phase change materials. Constr. Build. Mater. **262**, 120795 (2020)
9. Dente, A., Donahue, K.: Review of the Current State of Cob Structural Testing, Structural Design, the Drafting of Code Language, and Material Based Testing Challenges (2019)

10. Tchiotsop, J., et al.: Assessment of the natural variability of cob buildings hygric and thermal properties at material scale: influence of plants add-ons. Constr. Build. Mater. **342**, 127922 (2022)
11. Goodhew, S., Carfrae, J., Fox, M.: CobBauge - a hybrid walling technique combining mechanical and thermal performance (2020)
12. Goodhew, S., et al.: Improving the thermal performance of earthen walls to satisfy current building regulations. Energy Build. **240**, 110873 (2021)
13. Zeghari, K., et al., Comparison of the thermal performance between conventional and cob building. In: E3S Web of Conferences, vol. 111 (2019)
14. Zeghari, K., et al.: Novel dual walling cob building: dynamic thermal performance. Energies **14** (2021)
15. Gomaa, M., et al.: 3D printing system for earth-based construction: case study of cob. Autom. Constr. **124**, 103577 (2021)
16. Gomaa, M., et al.: Feasibility of 3DP cob walls under compression loads in low-rise construction. Constr. Build. Mater. **301**, 124079 (2021)
17. Sangma, S., Tripura, D.D.: Experimental study on shrinkage behaviour of earth walling materials with fibers and stabilizer for cob building. Constr. Build. Mater. **256** (2020)
18. Keefe, L.: Earth Building: Methods and Materials, Repair and Conservation. Routledge (2012)
19. Laborel-Préneron, A., et al.: Plant aggregates and fibers in earth construction materials: a review. Constr. Build. Mater. **111**, 719–734 (2016)
20. Akinkurolere, O., et al.: Engineering properties of cob as a building material. J. Appl. Sci. **6**(8), 1882–1885 (2006)
21. Alassaad, F., et al.: Improvement of cob thermal inertia by latent heat storage and its implication on energy consumption. Constr. Build. Mater. **329**, 127163 (2022)
22. Miccoli, L., et al.: Earth block masonry, rammed earth and cob: earthen components from different construction techniques and their structural performance. In: Proceedings of XIth International Conference on the Study and Conservation of Earthen Architectural Heritage, Lima, Peru (2012)
23. Miccoli, L., Müller, U., Fontana, P.: Mechanical behaviour of earthen materials: a comparison between earth block masonry, rammed earth and cob. Constr. Build. Mater. **61**, 327–339 (2014)
24. Quagliarini, E., et al.: Cob construction in Italy: some lessons from the past. Sustainability **2**(10), 3291–3308 (2010)
25. Alhumayani, H., et al.: Environmental assessment of large-scale 3D printing in construction: A comparative study between cob and concrete. J. Clean. Prod. **270**, 122463 (2020)
26. Miccoli, L., et al.: Static behavior of cob: experimental testing and finite-element modeling. J. Mater. Civ. Eng. **31**(4), 04019021 (2019)
27. Medero, G., et al.: Cob construction in Scotland: case study old school house at Cottown, Perthshire. In: Proceeding of the 1st Mediterranean Conference on Earth Architecture. UNESCO World Heritage Centre Cagliari, Italy (2009)
28. Sangma, S., Tripura, D.D.: Characteristic properties of unstabilized, stabilized and fibre-reinforced cob blocks. Struct. Eng. Int. **31**(1), 76–84 (2021)
29. Alqenaee, A., Memari, A.: Experimental study of 3D printable cob mixtures. Constr. Build. Mater. **324**, 126574 (2022)
30. Gomaa, M., et al.: Thermal performance exploration of 3D printed cob. Archit. Sci. Rev. **62**(3), 230–237 (2019)
31. ISO, ISO 13320: 2009 Particle Size Analysis Laser Diffraction Methods. International Organization for Standardization, Geneva (2009)
32. IS, Methods of test for soils: Grain size analysis-Part 4 (1985)
33. NF P 94-056 (1996). Standard test for soils investigation and testing-granulometric analysis-dry sieving method after washing (1996)

34. NF P 94–068 (1998). Standard test for soils investigation and testing-measuring of the methylene blue adsorption capacity of a rocky soil–determination of the methylene blue of a soil by means of the stain test (1998)
35. NF P, 94-051. Mars (1993)
36. NF P, 94-093 (1999). Standard test for soils investigation and testing-determination of the compaction characteristics of a soil–standard Proctor test and Modified Proctor test (1999)
37. NF P 94-050. Soils: investigation and testing–determination of moisture content–oven drying method. Association Française de Normalisation (1995)
38. Standard A D2487-11 (2011). Standard Practice for Classification of Soils for Engineering Purpose (Unified Soil Classification System) (2011)
39. Alassaad, F., Touati, K., Levacher, D., El Mendili, Y., Sebaibi, N.: Improvement of cob thermal inertia by latent heat storage and its implication on energy consumption. Constr. Build. Mater. **329** (2022)
40. Weismann, A., Bryce, K.: Building with Cob: A Step by Step Guide. Green Books, Devon (2006)
41. Colinart, T., et al.: Hygrothermal properties of light-earth building materials. J. Build. Eng. **29**, 101134 (2020)
42. Achenza, M., Fenu, L.: On earth stabilization with natural polymers for earth masonry construction. Mater. Struct. **39**(1), 21–27 (2006)
43. Aymerich, F., Fenu, L., Meloni, P.: Effect of reinforcing wool fibres on fracture and energy absorption properties of an earthen material. Constr. Build. Mater. **27**(1), 66–72 (2012)
44. Demir, I.: Effect of organic residues addition on the technological properties of clay bricks. Waste Manag. **28**(3), 622–627 (2008)
45. Sangma, S., Tripura, D.D.: Experimental study on shrinkage behaviour of earth walling materials with fibers and stabilizer for cob building. Constr. Build. Mater. **256**, 119449 (2020)
46. Vinceslas, T., et al.: Further development of a laboratory procedure to assess the mechanical performance of cob. Environ. Geotech. **7**(3), 200–207 (2018)
47. Minke, G.: Building with Earth: Design and Technology of a Sustainable Architecture Fourth and Revised Edition. Birkhäuser (2021)
48. Stazi, F., et al.: An experimental study on earth plasters for earthen building protection: the effects of different admixtures and surface treatments. J. Cult. Herit. **17**, 27–41 (2016)
49. Moevus, M.: State of the art of the raw earth characteristics. Final Report (2014)
50. Française, N.: NF-ISO 5017-Dense shaped refractory products. Determination of bulk density, apparent porosity and true porosity (1988)
51. German Institute for Standardisation Deutsches Institut für Normung: DIN 18945. Earth blocks—Terms and definitions, requirements, test methods (2013)
52. Giada, G., Caponetto, R., Nocera, F.: Hygrothermal properties of raw earth materials: a literature review. Sustainability **11**(19), 5342 (2019)
53. Moevus, M., Anger, R., Fontaine, L.: Hygro-thermo-mechanical properties of earthen materials for construction: a literature review. In: XIth International Conference on the Study and Conservation of Earthen Architectural Heritage Terra 2012 (2012)
54. EN, EN ISO 1015-19. Methods of Test for Mortar for Masonry. Determination of Water Vapour Permeability of Hardened Rendering and Plastering Mortars (1999)
55. ISO, Č., Hygrothermal performance of building materials and products-determination of water vapour transmission properties-cup method. ISO Geneva, Switzerland (2016)
56. Collet-Foucault, F.: Caractérisation hydrique et thermique de matériaux de génie civil à faibles impacts environnementaux. Rennes, INSA (2004)
57. Madhu: Difference Between Adsorption and Desorption. (2020). https://www.differencebetween.com/difference-between-adsorption-and-desorption/. Accessed 12 May 2022
58. ISO E 12571: Hygrothermal performance of building materials and products-determination of hygroscopic sorption properties. European Committee for Standardization, Brussels (2013)

59. ISO E 12572: Hygrothermal performance of building materials and products-determination of water vapour transmission properties (ISO 12572: 2001). CEN (European Committee for Standardization), Brussels (2001). 27 p
60. 11357-4 I: Plastics. Differential scanning calorimetry (DSC). Part 4: Determination of specific heat capacity. International Organization for Standardization (2014)

Influence of Aspect Ratio on the Properties of Compressed Earth Cylinders and Compressed Earth Blocks

Jack Andrew Cottrell(✉) and Muhammad Ali

School of Civil Engineering and Surveying, University of Portsmouth, Portland Building, Portland Street, Portsmouth PO1 3AH, UK
jack.cottrell@port.ac.uk

Abstract. A large proportion of the human population still resides in earthen structures all over the world. The benefits of earth construction are widely reported, but there is a lack of scientific understanding relating to standard production and test methods which has led to inconsistencies in the reporting of engineering parameters, such as compressive strength.

This study investigates the use of small-scale Compressed Earth Cylinders (CECs) to predict the compressive strength of equivalent full-scale Compressed Earth Blocks (CEBs). A full-scale manual CEB machine and a small-scale CEC moulding rig were utilised for the production of test specimens and the results obtained from both production methods were examined.

Two soil types with different engineering parameters were utilised in this investigation. It was found that a sample of un-stabilised CEB with an aspect ratio of 0.67 achieved a mean compressive strength of 6.73 N/mm^2 (Soil A) and 4.60 N/mm^2 (Soil B). A selection of CECs with an aspect ratio ranging from 0.50 to 2.00 were used to determine a relationship between the aspect ratio and compressive strength for each soil type. The theoretical relationship was used to predict the compressive strength of the equivalent CEBs within $\pm$ 3.0%. The theoretical relationship was also used to predict the unconfined compressive strength of the samples and enabled the determination of aspect ratio correction factors of Soil A and Soil B.

Findings from this study reveal that the conversion factors between cylinders and blocks are dependent on numerous variables including compaction pressure, aspect ratio, soil type and density.

Keywords: Compressed Earth Blocks · Compressed Earth Cylinders · Aspect Ratio · Conversion Factors

1 Introduction

Due to the increased demand for low-cost and sustainable building materials, there has been extensive research and development into Compressed Earth Blocks (CEBs) in recent years [1]. The ongoing experimental investigation into CEBs has highlighted the need for standard manufacturing and test procedures to ensure the performance of CEBs can be measured in a reliable and consistent manner [2].

S. Amziane et al. (Eds.): ICBBM 2023, RILEM Bookseries 45, pp. 232–248, 2023.
https://doi.org/10.1007/978-3-031-33465-8_18

Due to the load-bearing nature of masonry, compressive strength is regularly used as a basic measure of performance and quality, and is often empirically related to other engineering parameters. Currently, there are three main methods for testing the compressive strength of CEBs including the common direct (confined) compressive strength tests on single masonry units, the Réunion Internationale des Laboratoires et Experts des Matériaux, systèmes de construction et ouvrages (RILEM) method proposed by RILEM Technical Committee 164 [3], and indirect testing methods such as the 3-point bending test [4]. The compressive strength of a CEB is influenced by several factors including the material constituents, moisture content, density, and the magnitude of compaction pressure during manufacture, to name a few. However, assuming these parameters are all equal, the compressive strength of a CEB is known to be a function of the specimen size and geometry [2, 5–7].

The aspect ratio (height/width) of a test specimen may influence the Apparent Compressive Strength (ACS) of a sample, due to the interaction with the test apparatus. It is known that, as a vertical load is applied, the sample exhibits lateral expansion due to the Poisson's ratio effect [2]. Lateral expansion near the ends of the test specimen is restrained due to the friction developed between the test sample and the platens of the test machine. The restraint of the lateral expansion induces shear stresses which, when acting in addition to the uniaxial compression, causes a delay to the failure and results in an apparent increase in compressive strength [8].

Neville [8] stated that concrete test specimens with an aspect ratio greater than 1.7 would be free from the restraining effects of the platens, allowing the Unconfined Compressive Strength (UCS) to be determined. The UCS, also known as Uniaxial Compressive Strength, is the measure of maximum axial compressive stress that a sample of material can withstand under unconfined conditions, and is a fundamental engineering parameter used to assess the performance requirements of earthen construction [9–11]. The aspect ratio of most CEBs is less than 1.7, therefore the aspect ratio and the associated influence of platen restraint should be considered when undertaking compressive strength tests.

There are a few internationally recognised standard sizes of CEBs, with the most commonly reported to be 295 mm (L) × 140 mm (W) × 90 mm (H), with an aspect ratio (H/W) of 0.64 [2, 12]. Within existing literature, the size and aspect ratio of CEBs varies, as shown in Table 1.

Despite the range of size, geometry and aspect ratio found within the existing literature, there is often little or no regard for the influence of platen restraint on the results of compressive strength testing. This may be due to the lack of globally recognised standard manufacturing and testing procedures.

A recent publication by [21] outlines the current codes and standards for the manufacture and testing of earth construction. There are no specific British or European standards related to the manufacture or testing of CEBs, however, there is some technical guidance specific to un-stabilised compressed earth blocks, originating from Australia [9, 10], and New Zealand [11, 22, 23]. New Zealand Standard, NZS 4297:1998 [22] provides a theoretical relationship between the characteristic compressive strength of an individual specimen (f'_e), the unconfined compressive strength of an individual specimen (f'_{uc}) and

Table 1. Common Block Dimensions

Block Dimensions (mm)			Aspect Ratio (H/W)	Reference
Length (L)	Width (W)	Height (H)		
295	140	125	0.89	[13]
305	143	105	0.73	[14]
290	140	100	0.71	[15]
300	140	100	0.71	[16]
300	150	100	0.67	[17]
295	140	90	0.64	[2]
300	150	95	0.63	[6]
203	191	121	0.63	[18]
295	145	90	0.62	[16]
305	152	89	0.59	[19]
320	150	80	0.53	[20]

the compressive strength of wall construction (f_e), as shown in Eqs. (1) and (2).

$$f'_e \times k_a = f'_{uc} \tag{1}$$

$$0.5 \times f'_{uc} = f_e \tag{2}$$

New Zealand Standard, NZS 4298:1998 [11] contains a table of aspect ratio factors (k_a) used to adjust the characteristic compressive strength of earth bricks, ranging from an aspect ratio of 0.4 to 5.0. These factors are based on work carried out by Krefeld [24] and are generally the same factors used for fired clay masonry units.

Cylindrical samples are commonly used to assess the unconfined compressive strength of different materials, including undisturbed cohesive soils [25], soil-cement composites [26], soil-lime composites [27], rock core samples [28] and concrete [29–33]. To achieve a measure of UCS, test cylinders within a standard (minimum) aspect ratio (height/diameter) of 2.0 are commonly used, but sometimes specimens of other proportions are encountered. Several conversion factors have been published within existing literature to account for different aspect ratios for different materials, as shown in Table 2.

A limited number of existing studies have used cylindrical samples to test the compressive strength of compressed earth. One study [35] used cylindrical samples with an aspect ratio of 2.0 (80 mm diameter × 160 mm length) to test the compressive strength of soil reinforced with coconut and sisal fibres. The aforementioned study found that CECs with an aspect ratio of 1.2 achieved 23% greater compressive strength than CECs with an aspect ratio of 2.0. Another study [36] used CECs with an aspect ratio of 2.1 (57 mm diameter × 120 mm length) to test the unconfined compressive strength of nylon fibre-reinforced lime-stabilised soil. Silveira [37] established the relationship between

Table 2. Aspect Ratio Correction Factors for Cylindrical Test Samples

Material	Diameter (mm)	Aspect Ratio (H/D)	Correction Factor	Ref
Rammed Earth	150	2.00	-	[34]
	100	2.00	–	
	105	1.10	0.70	
Soil-Cement Cylinders	71.1	2.00	–	[26]
	101.6	1.15	0.91	
ASTM: Cylindrical Concrete Specimens	150	1.80	–	[33]
		1.75	0.98	
		1.50	0.96	
		1.25	0.93	
		1.00	0.87	
ASTM: Concrete Core Samples	≥94	1.80	–	[29]
		1.75	0.98	
		1.50	0.96	
		1.25	0.93	
		1.00	0.87	
British Standards Institution: Concrete Core Samples	75	2.00	–	[32]
		1.00	0.82	

the compressive strength of cubic specimens and the compressive strength of the cylinder, with a correction factor of about 0.94 for converting the cube strength to cylinder strength.

There are several practical advantages to the use of small-scale CEC test specimens, including a reduction in the time and materials required to manufacture test samples and the subsequent increase in the number of test specimens that can be made and tested using the same quantity of material, as shown in Fig. 1.

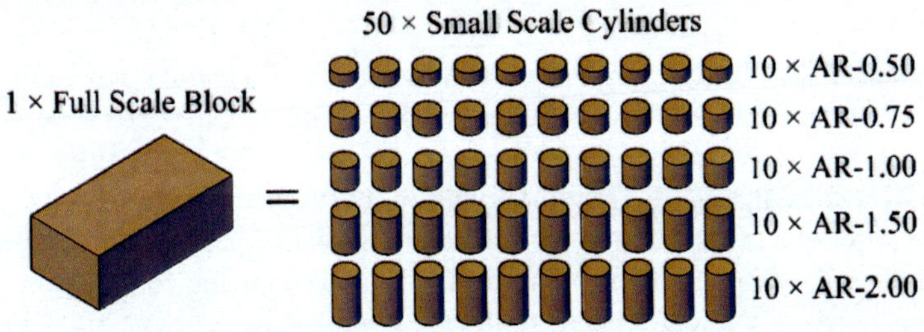

Fig. 1. Practical Advantages of Small-Scale CEC vs Full-Scale CEB

Furthermore, small-scale CECs can be manufactured using rudimentary cylindrical moulds, whereas the manufacture of full-scale CEBs requires access to full-scale CEB machinery. Access to such machinery is limited and the costs of procuring a block machine may discourage future research and development.

Several studies have investigated the influence of aspect ratio on compressed earth blocks [2, 5, 7, 37, 38] and the majority of which highlight the need for further investigation into the influence of geometric effects on compressive strength performance. This paper seeks to develop existing knowledge by investigating the use of small-scale un-stabilised CECs to estimate the ACS of full-scale un-stabilised CEBs and the UCS of the material.

2 Materials and Methods

2.1 Materials (Kent Brick Earth and British Standard - BS 8601 Subsoil)

Kent Brick Earth and BS 8601 Subsoil. Two soil types were used in this investigation. Soil A was a blend of Kent Brick Earth (KBE) obtained from Kent (south east of England) with 20% additional non-uniform particle size marine sand by weight. Soil A was used to enable direct comparison with existing research undertaken by the authors [6]. Soil B was a BS 8601 [39] compliant subsoil, obtained from Hampshire (south of England).

A particle size distribution test (sieve analysis) was performed as per BS1377-2:2022, Part 10 [40]. Further analysis of the clay and silt content was undertaken using a Mastersizer 3000 laser diffraction particle size analyser [41]. Other material properties including Optimum Moisture Content (OMC), Maximum Dry Density (MDD), Liquid Limit (LL), Plastic Limit (PL) and Plastic Index (PI) were determined as per BS1377-2:2022 [40], the results of which are presented in Table 3.

2.2 Methods

Summary of Test Specimens. A series of CEBs and CECs were manufactured and tested to determine the influence of aspect ratio, as shown in Table 4. A total of 25 unstablised CECs were manufactured (5 per test variable) from each soil type. The CECs were used to assess the influence of aspect ratio, ranging from 0.5 to 2.0. To provide a comparison, 3 full-scale CEBs were manufactured from each soil type with the same mix designs.

Manufacture of Full-Scale Compressed Earth Blocks. To facilitate the study, a University of Portsmouth Compressed Earth Block Machine (UoP-CEB machine) was designed and manufactured by the authors, as documented by Cottrell et al. (2021). Unlike any other manual CEB machine found in existing literature, the design incorporates a hydraulic ram positioned beneath the baseplate (see Fig. 2) to measure the amount of compaction pressure being applied to the block during compression.

A target pressure was determined following a qualitative assessment of the CEBs produced at different pressures. This assessment was performed through observation such as size of block, number of surface cracks and regularity of compaction. To ensure

Table 3. Properties of Kent Brick Earth and British Standard - BS8601 Subsoil

Properties	Soil Type	
	Soil A: Kent Brick Earth with Marine Sand	Soil B: BS 8601 Subsoil
Proctor Test		
Optimum Moisture Content (%)	12.5	17.5
Maximum Dry Density (Mg/m^3)	1.92	1.68
Atterberg Limits		
Liquid Limit LL (%)	26.3	33.7
Plastic Limit PL (%)	15.3	25.9
Plastic Index PI	10.9	7.8
Soil Classification		
Unified Soil Classification System	CL	ML
Particle Size Distribution		
Gravel (>2.0 mm) (%)	0.0	0.0
Sand (2.0–0.063 mm) (%)	38.8	68.0
Silt (0.063–0.002 mm) (%)	50.1	15.0
Clay (<0.002 mm) (%)	11.1	17.0

Table 4. Sample Dimensions

Sample Reference	Sample Dimensions (mm)				No. of Samples		Illustration
	Length	Width	Height	Diameter	Soil A	Soil B	
CEB	300	150	100	-	3	3	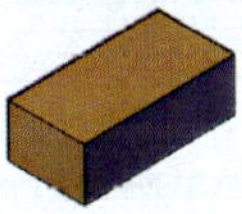
AR-0.50	-	-	21.8	43.6	5	5	
AR-0.75	-	-	32.7	43.6	5	5	
AR-1.00	-	-	43.6	43.6	5	5	
AR-1.50	-	-	65.4	43.6	5	5	
AR-2.00	-	-	87.2	43.6	5	5	

consistency in the manufacture of the blocks, a target pressure of 1.25 MPa was determined, with an acceptable tolerance of ±10%. At + 1.5 MPa, the physical limitation of the UoP-CEB machine was observed as the operator is unable to lower the lever arm without applying excessive force.

The dosage into the UoP-CEB Machine was calculated based on the maximum dry density of the material at optimum moisture content, obtained using the method defined BS 1377-2:2022, Part 11.3 [40]. Solid blocks, with dimensions 300 mm (L) × 150 mm (W) × 100 (H), were manufactured and tested to provide a comparison with the results obtained from the small-scale cylindrical samples.

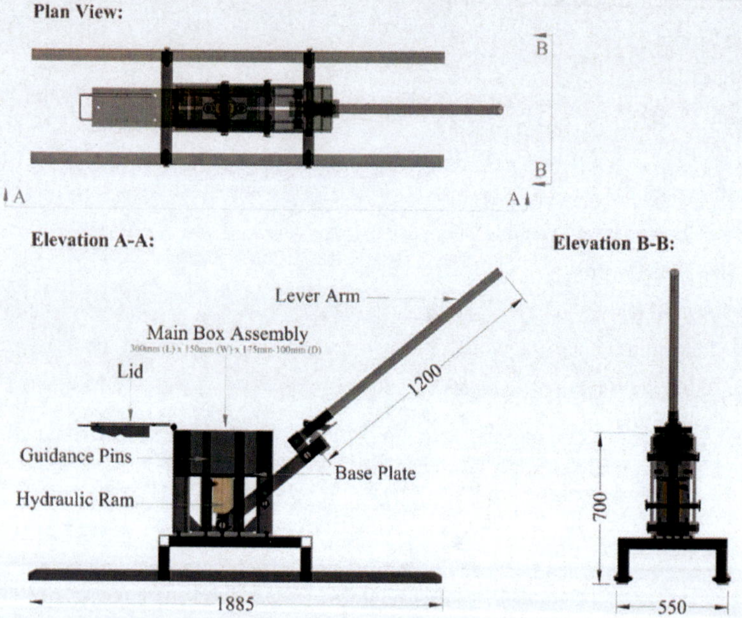

Fig. 2. Illustration of UoP-CEB Machine (50% Transparency to Show Hidden Critical Elements) Drawn in AutoCAD 2019 [6, 42]

Manufacture of Compressed Earth Cylinders. Cylindrical test samples were manufactured using a 200 mm long steel tube with a 43.6 mm internal diameter, as shown in Fig. 3. A 200 mm long steel rod, with an external diameter of 43.2 mm, was used to compress the soil. The pressure was applied using a Zwick/Roell Z250 250 kN Universal testing machine, at a rate of 20 mm/min. The cylindrical samples were ejected from the cylinder manually. The dosage of the CECs was calculated based on the maximum dry density of the material at optimum moisture content, obtained using the method defined BS 1377-2:2022, Part 11.3 [40]. To achieve the required size and density, the compaction pressure varied depending on the aspect ratio of the sample. The peak compaction pressure was recorded during the manufacturing process.

Sample Drying. Test samples were dried at 30 °C in a temperature controlled Genlab MINO/100 oven [43] to replicate sun-drying conditions. The cylinders and blocks were re-weighed throughout the week following the procedure detailed in BS 1377-2:2022

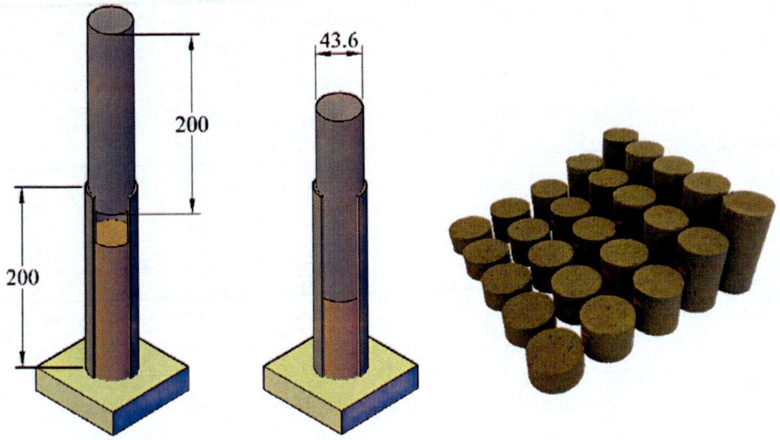

Fig. 3. Illustration of Apparatus used to Manufacture CECs (Left), CEC Samples (Right)

[40] until a consistent weight was observed ± 0.01 g indicating that the sample had reached hygrothermal equilibrium. Samples were found to reach hygrothermal equilibrium after 5 days. It is recognised that, due to the hygroscopic nature of the material, the samples may not be considered as fully dry. Prior to testing, the samples were re-weighed and re-measured at an ambient room temperature of 20 °C ± 0.2 °C and relative humidity of 58% ± 2%, to determine the amount of shrinkage during the drying process.

2.3 Mechanical Testing

There are currently no standards within the UK for the testing of CECs or CEBs. Furthermore, the commonly referred NZS 4298 [11] for the Materials and Workmanship For Earth Buildings does not cover the testing of cylindrical samples, other than the testing of mortar and grout. Due to this, British Standards and/or Eurocodes for the testing of Concrete and Masonry were adopted where appropriate. This approach has been taken by numerous existing studies [6, 15, 44–47].

Compression Test of Cylindrical Samples. A Zwick/Roell Z250 universal testing machine was used to perform the compression test of cylindrical samples. The tests were conducted in accordance with BS EN 12390-3:2019 [48]. The size and aspect ratio of the test samples we notable derivations from the aforementioned standard. The deformation was measured and recorded by gauges installed in the Zwick/Roell Z250. The load at which the sample failed was recorded and maximum compressive stress was calculated as per Section 8 of BS EN 12390-3:2019. The method of testing remained the sample for all cylindrical samples, therefore this test is deemed appropriate for determining a comparison between samples.

Compression Test Blocks Samples. A Losenhausen 1000 kN universal testing machine was used to perform the compression tests of the CEB samples. The tests were conducted in accordance with BS EN 772-1 [49]. A 10 mm steel plate was positioned on top and beneath the sample to ensure that the load was distributed evenly across the sample. The load was applied at a rate of 0.05 MPa/s, as per Table 2 of BS EN 772-1, until the block failed. The deformation was measured and recorded by gauges installed in the Losenhausen 1000 kN universal testing machine. The load at which the block failed was recorded and maximum compressive stress was calculated.

2.4 Statistical Analysis

To assess the statistical significance of the results obtained from physical testing, statistical analysis (ANOVA with Tukey pairwise comparison) was undertaken with Minitab 17.3.1 [50]. Within the results, statistically significant differences are indicated by different letters (A, B, C) at p-value < 0.05 (5%).

3 Results

3.1 Properties of CEC and CEB Samples

The dimensions of the samples were recorded throughout the manufacturing process, and the aspect ratio of each sample type was found to be consistent, as shown in Table 5. This is due to an accurate dosing method and controlled method of production.

The mean dry density of all samples manufactured from Soil A was 1930.1 kg/m^3 with a standard deviation (SD) of $\pm$ 7.4 kg/m^3. In contrast, the mean dry density of all samples manufactured from Soil B was 1742.7 kg/m^3 with a SD of $\pm$ 10.6 kg/m^3.

Table 5. Mean Aspect Ratio and Mean Dry Density of Test Samples

Material	Sample	Mean Aspect Ratio	Mean Dry Density (kg/m^3)
Soil A: Kent Brick Earth with Marine Sand	CEB	0.67 ± 0.01	1931.3 ± 4.1
	AR-0.50	0.50 ± 0.01	1922.5 ± 2.5
	AR-0.75	0.75 ± 0.01	1933.5 ± 4.1
	AR-1.00	1.00 ± 0.00	1934.0 ± 2.0
	AR-1.50	1.50 ± 0.01	1927.0 ± 9.0
	AR-2.00	2.00 ± 0.00	1933.4 ± 9.5
Soil B: BS 8601 Subsoil	CEB	0.67 ± 0.00	1742.3 ± 17.7
	AR-0.50	0.51 ± 0.01	1745.3 ± 5.5
	AR-0.75	0.76 ± 0.00	1738.0 ± 7.8
	AR-1.00	1.01 ± 0.00	1735.0 ± 6.0
	AR-1.50	1.51 ± 0.00	1743.8 ± 9.8
	AR-2.00	2.00 ± 0.00	1749.6 ± 1.2

Values represent mean ± standard deviation (σ).

3.2 Compaction Pressure

To achieve cylindrical samples of the correct size and density, the compaction pressure varied during the manufacturing process. The compaction pressure required to manufacture the CECs is presented in Fig. 4.

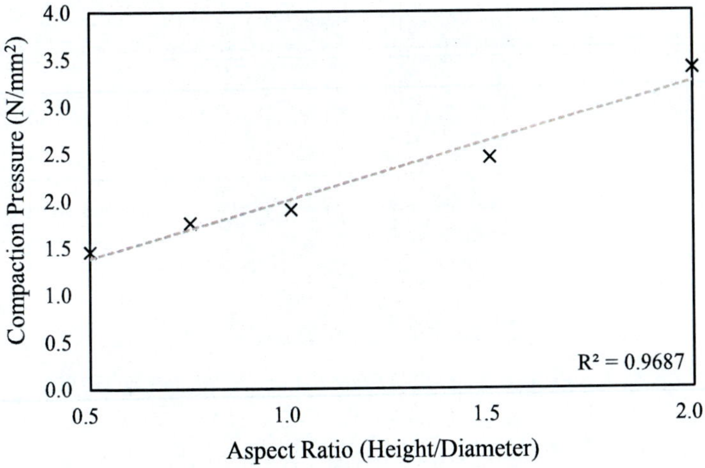

Fig. 4. Peak Compaction Pressure During Manufacture of CEC Samples vs Aspect Ratio (Soil A)

It can be seen from Fig. 4 that during the manufacture of cylindrical samples with equivalent density, greater compaction pressure is required to manufacture samples of a greater aspect ratio. A similar relationship was observed during the production of samples made from Soil B.

Using Eq. (3) derived from the linear trendline in Fig. 4, it is possible to interpolate to determine the compaction pressure required to manufacture a CEC with an aspect ratio equivalent to a CEB is 1.59 N/mm^2, whereby x = 0.67.

$$y = 1.2444x + 0.7598 \tag{3}$$

The CEBs were manufactured using the UoP-CEB machine at a consistent compaction pressure of 1.25 MPa, with an acceptable tolerance of $\pm$10%. The CECs required approximately 27% greater compaction pressure when compared to the CEBs to produce a sample with an equivalent aspect ratio and density. This may be due to friction and the boundary effects caused by the shape of the cylindrical versus cuboid mould.

The results in Fig. 4 also shows that a compaction pressure of approximately 1.38 N/mm^2 is required to manufacture a CEC with an aspect ratio of 0.5, whereas a compaction pressure of approximately 3.25 N/mm^2 is required to manufacture a CEC with an aspect ratio of 2.0, a difference of approximately 2.35 times. Figure 5 demonstrates the consequence of manufacturing a CEC AR-2.00 sample at a lower compaction pressure (2.0 N/mm^2 - 62% of the required compaction pressure). The subsequent sample did not achieve the correct height, resulting in a lower dry density. Furthermore, the sample did not achieve uniform compaction through its height, which will adversely influence its compressive strength and cause areas of weakness in the sample.

This is an important observation and demonstrates the importance of utilising the correct compaction pressure when manufacturing CECs of equivalent density and different aspect ratios.

Fig. 5. Insufficient Compaction (Left), Uniform Compaction (Right)

3.3 Aspect Ratio Factors for Un-Stabilised Compressed Earth

The compressive strength of un-stabilised CECs at different aspect ratios is shown in Fig. 6. As can be seen, samples manufactured from Soil A achieved a higher compressive strength when compared to those manufactured from Soil B. This is likely due to the different soil constituents and differences in dry density. Within each dataset, the results presented show that cylindrical samples with a low aspect ratio achieved a greater compressive strength than those with a higher aspect ratio. This observation is consistent with existing studies on the aspect ratio of compressed earth samples [2, 5, 7, 37, 38].

A second-order polynomial trendline has been applied to each data set in Fig. 6 with a coefficient of determination (R^2) of 0.9764 and 0.9882 for Soil A and Soil B respectively. This suggests that the trendline for each set of data is a close fit, and indicates that there is a non-linear relationship between compressive strength and aspect ratio.

Using Eqs. (4) and (5) derived from the polynomial trendlines in Fig. 6, it is possible to interpolate to predict the ACS of a CEB sample with an aspect ratio of 0.67. Note, the interpolation between data points may be considered but extrapolation beyond the dataset is not advised. The predicted values and the results from the physical testing of CEBs, can be found in Table 6.

$$\textbf{Soil A}: y = 1.8175x^2 - 6.5305x + 8.1209 \tag{4}$$

$$\textbf{Soil B}: y = 2.3768x^2 - 8.4797x + 11.531 \tag{5}$$

The relationship observed in Fig. 6 may also be used to predict the USC of the CEB, by considering the cylindrical test result at an aspect ratio of 2.0.

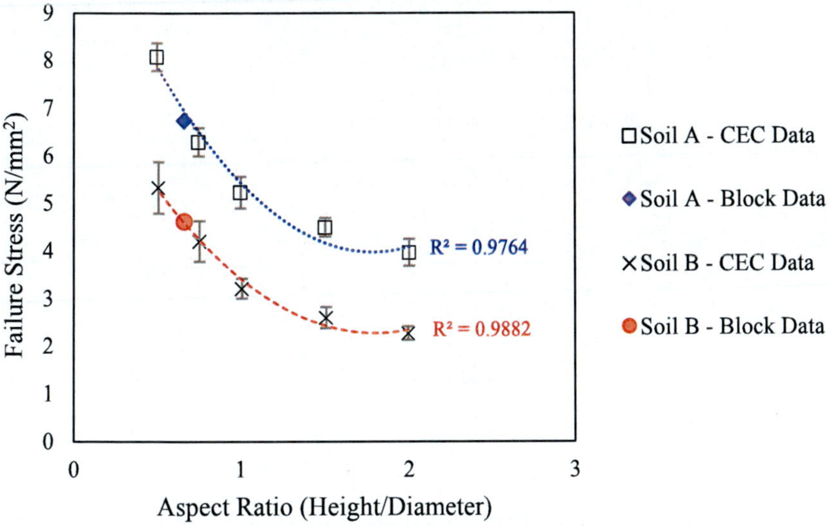

Fig. 6. Mean Compressive Strength of CECs and CEBS. Error Bars Represent the Standard Deviation.

Table 6. Predicted versus Actual Compressive Strength

Soil Type	Predicted Compressive Strength of CEB, based CEC at Aspect Ratio of 0.67 (N/mm^2)	Mean Compressive Strength of CEBs obtained from physical testing (N/mm^2)	Percentage Difference (%)
Soil A	6.92	6.73	2.78
Soil B	4.56	4.60	0.87

For Soil A, the CEB achieved an ACS of 6.73 N/mm^2, and the equivalent CEC at AR-2.00 achieved a compressive strength of 4.08 N/mm^2, suggesting a correction factor of approximately 0.61 is required to determine UCS. For Soil B, the CEB achieved an ACS of 4.60 N/mm^2, and the equivalent CEC at AR-2.00 achieved a compressive strength of 2.33 N/mm^2, suggesting a correction factor of approximately 0.51 is required to determine UCS.

This finding suggests that aspect ratio correction factors may be influenced by the relative strength of the samples. This theory is supported by Murdock and Kesler [51] who investigated the influence aspect ratio on concrete cylinders at different concrete strengths and found that lower-strength concrete is more affected by the height/diameter ratio of the specimen. The compressive strength of the CECs samples for Soil A and Soil B is shown in Fig. 7, whereby the compressive strength is presented relative to the strength of a sample with an aspect ratio of 2.0. A comparison is made with concrete cylinders, using data obtained from [8, 51].

Figure 8 contains aspect ratio correction factors, published with New Zealand Standards NZS 4298:1998 [11]. The standard allows linear interpolation between data points,

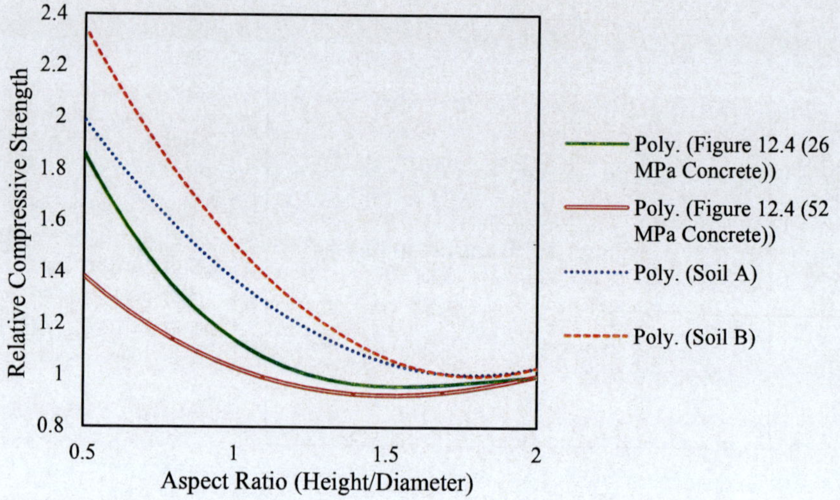

Fig. 7. Relative Compressive Strength of Samples versus Aspect Ratio. Data for 26 and 52 MPa concrete was obtained from [8, 51].

therefore, a correction factor of 0.59 can be determined for a CEB with an aspect ratio of 0.67. When compared to the results of this study, the application of a 0.59 correction factor would underestimate the UCS of Soil A by approximately 4%, which is deemed acceptable. However, a 0.59 correction factor would overestimate the UCS of Soil B by approximately 16%. The result of this investigation indicates that a single correction factor for all soil types and densities may not be appropriate, leading to inaccuracies.

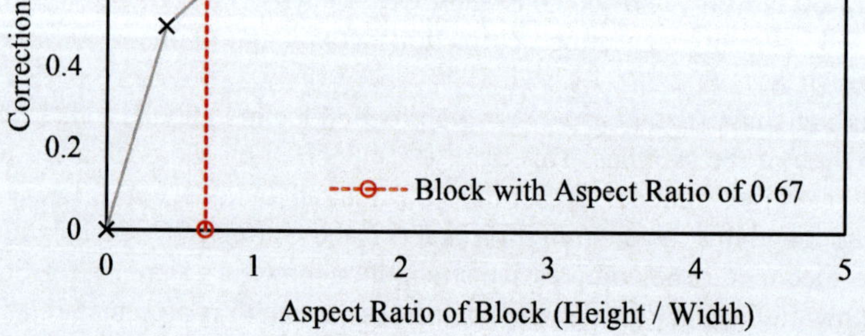

Fig. 8. Aspect Ratio Correction Factors, Published with New Zealand Standards NZS 4298:1998 [11]

It must be noted that the geometric features of a CEB, including the presence of vertical holes or frogs, also influence apparent compressive strength [6], therefore the aspect ratio factors presented in this study apply to solid CEBs only. Further correction factors should be investigated to correlate between cylindrical samples and CEBs with alternate geometric features.

Furthermore, the addition of fibre reinforcement highlights potential challenges associated with scaling and boundary effects. When fibres are incorporated into a soil mixture, the distribution and orientation of fibres is random and difficult to replicate. This often produces a wide range of results within an experimental investigation. Moreover, the addition of fibres will provide resistance to lateral expansion as the sample is subject to a vertical load. It is recommended that further testing should be carried out using fibre-reinforced CECs at aspect ratios of 0.50, 0.75, 1.00, 1.50 and 2.00 to assess and determine the relationship and influence of aspect ratio of fibre-reinforced CEBs.

4 Conclusion

The purpose of this study was to investigate the use of small-scale un-stabilised CECs to estimate the ACS of full-scale un-stabilised CEBs as well as the UCS of the material. The results obtained from the physical testing provide a greater understanding of the influence of aspect ratio on the compressive strength of un-stabilised CECs and CEBs. The important conclusions emerging from this study are:

- Small-scale un-stabilised CECs have been used to predict ACS of full-scale un-stabilised CEBs, with an accuracy of $\pm$ 3.0%.
- The use of small-scale un-stabilised CECs may be used to predict the UCS of the material.
- Results indicate that aspect ratio correction factors may depend on the relative strength, soil constituents and dry density of the compressed earth sample, and therefore care should be taken when applying generic correction factors.
- The results of this study demonstrate the importance of utilising the correct compaction pressure when manufacturing CECs of equivalent density and different aspect ratios.
- This study demonstrates several practical advantages to the use of small-scale CEC compared to full size test specimens, including a reduction in the time and materials required to manufacture test samples and the subsequent increase in the number of test specimens that can be made and tested.

In addition to the potential benefits of using small-scale CECs to predict the compressive strength of full-scale CEBs, this study also highlights the limitations of using small-scale CECs and highlights areas for future research and development to determine aspect ratio correction factors for non-homogenous, fibre-reinforced CEB mixes. Further investigation is also required to determine aspect ratio correction factors for CEBs with alternative geometric features, including the presence of vertical holes or frogs.

References

1. Aubert, J.-E., Faria, P., Maillard, P., Ouedraogo, K.A.J., Ouellet-Plamondon, C., Prud'homme, E.: Characterization of earth used in earth construction materials. In: Fabbri, A., Morel, J.-C., Aubert, J.-E., Bui, Q.-B., Gallipoli, D., Reddy, B.V.V. (eds.) Testing and Characterisation of Earth-based Building Materials and Elements. RSR, vol. 35, pp. 17–81. Springer, Cham (2022). https://doi.org/10.1007/978-3-030-83297-1_2
2. Morel, J.C., Pkla, A., Walker, P.: Compressive strength testing of compressed earth blocks. Constr. Build. Mater. **21**, 303–309 (2007). https://doi.org/10.1016/j.conbuildmat.2005.08.021
3. Olivier, M., Mesbah, A., El Gharbi, Z., Morel, J.C.: Mode opératoire pour la réalisation d'essais de résistance sur blocs de terre comprimée. Mater. Struct. **30**, 515–517 (1997). https://doi.org/10.1007/bf02486394
4. Morel, J., Pkla, A.: A model to measure compressive strength of compressed earth blocks. Constr. Build. Mater. **16**, 303–310 (2002)
5. Walker, P.J.: Strength and erosion characteristics of earth blocks and earth block masonry. J. Mater. Civ. Eng. **16**, 497–506 (2004). https://doi.org/10.1061/(ASCE)0899-1561(2004)16:5(414)
6. Cottrell, J.A., Ali, M., Tatari, A., Martinson, D.B.: An investigation into the influence of geometry on compressed earth building blocks using finite element analysis. Constr. Build. Mater. **273**, 121997 (2021). https://doi.org/10.1016/j.conbuildmat.2020.121997
7. Lan, G.Q., Wang, Y.H., Chao, S.S.: Influences of specimen geometry and loading rate on compressive strength of unstabilized compacted earth block. Adv. Mater. Sci. Eng. **2018** (2018). https://doi.org/10.1155/2018/5034256
8. Neville, A.M.: Properties of Concrete, 4th edn. Pearson, Essex (2004)
9. Middleton, G.F., Schneider, L.M.: Bulletin 5: Earth-wall Construction, 4th edn. CSIRO Division of Building, Construction and Engineering, Sydney (1987)
10. Walker, P.: Standards Australia, HB 195 - The Australian Earth Building Handbook. Standards Australia International Ltd, Sydney (2002)
11. Standards New Zealand, NZS 4298:1998. Materials and Workmanship for Earth Buildings, p. 91 (1998)
12. Rigassi, V.: Compressed Earth Blocks: Manual of Production, Deutsches Zentrum für Entwicklungstechnologien, Deutsche Gesellschaft für Technische Zusammenarbeit (1985)
13. Walker, P., Stace, T.: Properties of some cement stabilised compressed earth blocks and mortars. Mater. Struct. Constr. **30**, 545–551 (1997). https://doi.org/10.1007/bf02486398
14. Sekhar, D.C., Nayak, S.: Utilization of granulated blast furnace slag and cement in the manufacture of compressed stabilized earth blocks. Constr. Build. Mater. **166**, 531–536 (2018). https://doi.org/10.1016/j.conbuildmat.2018.01.125
15. Danso, H., Martinson, D.B., Ali, M., Williams, J.B.: Physical, mechanical and durability properties of soil building blocks reinforced with natural fibres. Constr. Build. Mater. **101**, 797–809 (2015). https://doi.org/10.1016/j.conbuildmat.2015.10.069
16. Delgado, M.C.J., Guerrero, I.C.: Earth building in Spain. Constr. Build. Mater. **20**, 679–690 (2006). https://doi.org/10.1016/j.conbuildmat.2005.02.006
17. Laursen, P.T., Herskedal, N.A., Jansen, D.C., Qu, B.: Out-of-plane structural response of interlocking compressed earth block walls. Mater. Struct. **48**(1–2), 321–336 (2013). https://doi.org/10.1617/s11527-013-0186-2
18. Donkor, P., Obonyo, E.: Earthen construction materials: assessing the feasibility of improving strength and deformability of compressed earth blocks using polypropylene fibers. Mater. Des. **83**, 813–819 (2015). https://doi.org/10.1016/j.matdes.2015.06.017
19. Sitton, J.D., Zeinali, Y., Heidarian, W.H., Story, B.A.: Effect of mix design on compressed earth block strength. Constr. Build. Mater. **158**, 124–131 (2018). https://doi.org/10.1016/j.conbuildmat.2017.10.005

20. Villamizar, M.C.N., Araque, V.S., Reyes, C.A.R., Silva, R.S.: Effect of the addition of coal-ash and cassava peels on the engineering properties of compressed earth blocks. Constr. Build. Mater. **36**, 276–286 (2012). https://doi.org/10.1016/j.conbuildmat.2012.04.056
21. Reddy, B.V.V., et al.: Codes and standards on earth construction. In: Fabbri, A., Morel, J.-C., Aubert, J.-E., Bui, Q.-B., Gallipoli, D., Reddy, B.V.V. (eds.) Testing and Characterisation of Earth-based Building Materials and Elements. RSR, vol. 35, pp. 243–259. Springer, Cham (2022). https://doi.org/10.1007/978-3-030-83297-1_7
22. Standards New Zealand, NZS 4297:1998. Engineering Design of Earth Buildings, p. 63 (1998)
23. Standards New Zealand, NZS 4299:1998, Earth Build. Not Requiring Specif. Des. Build. Code Compliance Doc. 4299, p. 131 (1998)
24. Krefeld, W.J.: Effect of shape of specimens on the apparent compressive strength of brick masonry. In: Proceedings of the American Society for Testing Materials, Philadelphia, pp. 363–369 (1938)
25. British Standards Institute, BS 1377-7:1990, Soils Civ. Eng. Purp.—Part 7 Shear Strength Tests Total Stress. (1990)
26. American Society for Testing and Materials, ASTM D1633 - 00, Standard Test Methods Compressive Strength Molded Soil-Cement Cylinder, pp. 1–15 (2007). https://www.astm.org/DATABASE.CART/HISTORICAL/D1633-00.htm
27. American Society for Testing and Materials, ASTM D5102 - 22, Standard Test Method Unconfined Compressive Strength Compact. Soil-Lime Mixtures, 04 August 2022, p. 7 (2022). https://doi.org/10.1520/D5102_D5102M-22
28. American Society for Testing and Materials, ASTM D7012 - 14, Standard Test Method Compressive Strength Elastic Moduli of Intact Rock Core Specimens under Varying States Stress Temperature, pp. 1–8 (2014). https://doi.org/10.1520/D7012-14.1.5.1.1
29. American Society for Testing and Materials, ASTM C39/C39M, Standard Test Method compressive strength of cylindrical concrete specimens, pp. 1–5 (2001)
30. BS EN 12390-1:2021, BSI Standards Publication Testing hardened concrete, p. 18. British Standard (2021)
31. Rangasamy, G., et al.: An extensive analysis of mechanical, thermal and physical properties of jute fiber composites with different fiber orientations. Case Stud. Therm. Eng. **28**, 101612 (2021). https://doi.org/10.1016/j.csite.2021.101612
32. C, BS EN 13791:2019, Assess. in-Situ Compressive Strength Struct. Precast Concr. Components (2019)
33. American Society for Testing and Materials, ATSM C42/C42M, Obtaining Test. Drill. Core Sawed Beams Concrete, p. 6 (2004)
34. Walker, P., Keable, R., Martin, J., Maniatids, V.: Rammed earth design and construction guidelines. In: BRE Bookshop, Watford (2019). https://doi.org/10.5040/9781350122741.1002057
35. Ghavami, K., Toledo Filho, R.D., Barbosa, N.P.: Behaviour of composite soil reinforced with natural fibres. Cem. Concr. Compos. **21**, 39–48 (1999). https://doi.org/10.1016/S0958-9465(98)00033-X
36. Jafari, M., Esna-ashari, M.: Effect of waste tire cord reinforcement on unconfined compressive strength of lime stabilized clayey soil under freeze-thaw condition. Cold Reg. Sci. Technol. **82**, 21–29 (2012). https://doi.org/10.1016/j.coldregions.2012.05.012
37. Silveira, D., Varum, H., Costa, A.: Influence of the testing procedures in the mechanical characterization of adobe bricks. Constr. Build. Mater. **40**, 719–728 (2013). https://doi.org/10.1016/j.conbuildmat.2012.11.058
38. Aubert, J.E., Maillard, P., Morel, J.C., Al Rafii, M.: Towards a simple compressive strength test for earth bricks? Mater. Struct. **49**(5), 1641–1654 (2015). https://doi.org/10.1617/s11527-015-0601-y

39. British Standards Institution, BS 8601:2013. Specification for Subsoil and Requirements for Use, p. 30 (2013)
40. British Standards Institution, BS 1377-2:2022, Methods Test Soils for Civil Engineering Purposes Part 2 Classification tests and Determination of Geotechnical Properties vol. 3, p. 158 (2022)
41. Malvern Panalytical Ltd, Mastersizer 3000 (2020). https://www.malvernpanalytical.com/en/products/product-range/mastersizer-range?creative=328687395069&keyword=mastersizer&matchtype=e&network=g&device=c&gclid=EAIaIQobChMImbDMvZvv6AIVhbTtCh1YeAhXEAAYASAAEgLaCPD_BwE. Accessed 1 Mar 2020
42. Autodesk Inc, AutoCAD 2019 (2019). https://www.autodesk.com/education/free-software/autocad
43. Genlab Ltd, Genlab Classic Ovens (2020). https://www.genlab.co.uk/57-genlab-classic-ovens. Accessed 10 June 2020
44. Danso, H., Martinson, D.B., Ali, M., Williams, J.B.: Effect of sugarcane bagasse fibre on the strength properties of soil blocks. In: 1st International Conference on Bio-Based Building Materials (2015)
45. Danso, H., Martinson, B., Ali, M., Mant, C.: Performance characteristics of enhanced soil blocks: a quantitative review. Build. Res. Inf. **43**, 253–262 (2015). https://doi.org/10.1080/09613218.2014.933293
46. Danso, H., Martinson, D.B., Ali, M., Williams, J.B.: Mechanisms by which the inclusion of natural fibres enhance the properties of soil blocks for construction. J. Compos. Mater. **51**, 3835–3845 (2017). https://doi.org/10.1177/0021998317693293
47. Danso, H.: Use of agricultural waste fibres as enhancement of soil blocks for low-cost housing in Ghana (2016)
48. British Standards Institute, BS EN 12390-3:2019, Testing Hardened Concrete Part 3 Compressive Strength Test Specimens (2019)
49. British Standards Institute, BS EN 772-1:2011+A1:2015. Methods Test Masonry Units. Determine the compressive strength (2011). https://bsol.bsigroup.com
50. Minitab Inc, Minitab 17.3.1, (2016). https://www.minitab.com/en-us/
51. Murdock, J.W., Kesler, C.E.: Effect of Length of Specimen on the Apparent Compressive Strength of Concrete, ASTM Committee. C9 Concrete and Concrete Aggregates (1955)

Lime Stabilized Rammed Earth – Influence of Curing and Drying Conditions

Adrian Sylwester Los, Marius Næraa-Nicolajsen(✉), and Ida M. G. Bertelsen

Technical University of Denmark, Brovej 118, 2800 Kongens Lyngby, Denmark
mariusnndk@gmail.com

Abstract. In this project, the use of excavation soil for the construction of noise barriers made of rammed earth (RE) is investigated. In its natural form, RE has poor durability when exposed to harsh Danish weather conditions, and one of the most common ways to improve the durability of earth-based construction materials is through stabilization with inorganic binders such as lime. Some of the precautionary measures one needs to take when stabilizing soil with lime is the suitability of the soil and the importance of proper curing before the material is exposed to drying. Unstabilized RE (USRE) specimens, as well as lime stabilized RE (LSRE) specimens with 5% lime are produced and exposed to different combinations of curing and drying before testing the unconfined compressive strength (UCS). The results show that the USRE specimens obtain the highest UCS in the dry condition but completely eroded in the wet condition. On the other hand, little to no erosion took place for the LSRE specimens and a UCS could be obtained in the wet condition. An additional curing time of 14–91 d prior to drying the RE specimens results in an increase in UCS, but the influence is highly dependent on the temperature. An increase in the temperature from 23 °C to 40 °C results in a higher increase in UCS than increasing the curing time from 0 d to 91 d for specimens cured and dried at 23 °C.

Keywords: Rammed earth · Lime stabilization · Curing and drying conditions · Unconfined compressive strength · Excavation soil

1 Introduction

Rammed earth (RE) construction is gaining increasing attention these years, due to its low carbon footprint and recyclability. Moreover, it is often possible to reuse excess soil from construction sites for earth construction materials. In Denmark, excavation soil is by far the largest waste fraction of 15 mio. tonnes per year [1]. Although not all the excavated material may be considered suitable for RE construction, most of Denmark is covered with till with adequate geotechnical properties for RE construction [2]. The transportation of soil to the deposits is also an economic and environmental burden, which can be reduced by reusing the soil locally. The motivation of this project is to use excavation soil from Vinge, Frederikssund, Denmark, as raw material in RE construction. The Vinge area is intended for a large city development project with an

S. Amziane et al. (Eds.): ICBBM 2023, RILEM Bookseries 45, pp. 249–258, 2023.
https://doi.org/10.1007/978-3-031-33465-8_19

expected amount of excavation soil of ~1.9 mio. tonnes [3]. The intended use of part of the excavation soil is to produce noise barriers of RE construction. As noise barriers are not shielded from weather exposure, it is considered necessary to stabilize the soil to improve its durability and reduce erosion. An effective way to improve the durability of RE construction is through lime stabilization (LS) [4–6]. However, adequate curing and drying conditions such as curing time, drying time, temperature, and relative humidity are essential for facilitating the formation of cementitious hydration products through pozzolanic reactions between lime and clay minerals [4, 7–9]. The importance of curing and drying conditions is valid for most types of inorganic binders used for soil stabilization [10] and also influences the mechanical performance of the material [4, 6]. The lime-soil interactions are generally well-understood and concern the immediate processes as cation exchange and flocculation/agglomeration of clay particles, and the slower pozzolanic reactions and carbonation [11, 12].

The aim of the present study is to investigate the influence of curing and drying procedures on the mechanical performance of unstabilized (USRE) and lime stabilized rammed earth (LSRE) specimens of excavation soil from Vinge. As it is important to make the production of the noise barriers as simple and fast as possible, it is essential to evaluate the necessity of curing and drying of the material after compaction.

2 Materials and Methods

2.1 Characterization of Raw Materials

The soil used in this study was collected from a large construction site in Vinge, Frederikssund, Denmark. Geotechnical drillings show that the top layer of the excavated soil is made up of humus, followed by at least three meters of till with an undrained shear strength in the range of 70–140 kN/m^2. The till was separated from the humus during excavation and stored in a large pile next to the construction site. The lime powder (CaO) used in this study was provided by Lhoist and had a purity of $95 \pm 2\%$ CaO and a grain density of 910 ± 50 kg/m^3 according to the datasheet.

The grain size distribution (GSD) of the soil was determined using the sieve analysis for particles >63 μm and the hydrometer method for particles <63 μm, following DS-CEN ISO/TS 17892-4 (2004). The Atterberg limits of the soil were determined according to CEN-ISO/TS 17892-12 (2004) using a 60 g/60° cone in the fall cone penetration test. The GSD and the plastic properties of the soil were compared with the recommendations described by Houben and Guillaud [13] to determine its suitability as raw material in RE construction. The optimum water content (OWC) for the soil was determined using the Proctor test based on DS/EN 13286-2 (2011) using an Automatic Digital Proctor (33-T8502) with Ø100 × 120 mm moulds. Please note that the machine was set up to the Standard Proctor test, but due to a mistake, a wrong rammer weight of 4.5 kg instead of 2.5 kg was used in the setup, which resulted in a higher compaction energy compared to the Standard Proctor test. The unstabilized soil mixes were hand-mixed, while the lime stabilized soil mixes were mixed in a Hobert-style mixer using a similar process as described in Sect. 2.2 for the RE specimen production.

2.2 Production and Testing of Ø60 RE Specimens

The soil was used to produce unstabilized and lime stabilized cylindrical Ø60 × ~120 mm specimens. Before production, the raw soil was prepared by drying at 50 °C for at least 72 h to evaporate all water without breaking clay minerals and afterwards grinded in a ball mill without balls to gently break down larger clay lumps. It was confirmed that the GSD was not affected by the process in the ball mill. The soil was slightly modified by dry sieving to < 8 mm to adjust the grain sizes to the selected mould size of Ø60 mm. Mixes to produce RE specimens were prepared with moulding water contents in accordance with the Proctor test ± 1–2%. Moulding water contents ranging from 9–11% were used for the USRE, while the LSRE specimens were produced by adding 5 wt% lime to the soil and moulding water contents of 14% and 16% (with respect to the total weight of the dry materials). The prepared soils were mixed in a Hobart-style mixer until homogenized. The mixing time for both lime stabilized and unstabilized mixes took a total of 9 min. The initial 3 min for both mixes consisted of mixing water and soil, while afterwards the next 3 min included the addition of lime for lime stabilized mixes and plain mixing for the unstabilized mix. The sides of the bowl were scrapped to ensure a homogenous mixture and the material was finally mixed for an additional 3 min.

RE specimens for unconfined compressive strength (UCS) tests were produced in steel moulds with the dimensions of Ø60 × 140 mm, using an electromechanical Universal Testing Machine from Instron, with a maximum force capacity of 10 kN. Each specimen was made up of three soil layers, weighing 220 g for LSRE and 225 g for USRE, resulting in heights of 119 ± 6 mm for LSRE specimens and 108 ± 2 mm for USRE specimens depending on the water content. A pre-programmed control system was used to operate the Instron machine, allowing each layer to be compacted with a force of 2 kN (0.71 MPa) five times before the piston head was removed and the soil for the next layer was added to the mould.

Table 1. Curing and drying procedures for Ø60 RE specimens for UCS testing. Dry = UCS test in dry conditions; Wet = UCS test in wet conditions after 2 h of water immersion.

Specimen type	ID	USRE	LSRE
7 d drying at 23 °C	Dr7-23C	Dry	Dry + wet
7 d drying at 40 °C	Dr7-40C	Dry	Dry
14 d curing at 23 °C + 7 d drying at 23 °C	Cu14-23C_Dr7-23C		Dry
14 d curing at 23 °C + 7 d drying at 40 °C	Cu14-23C_Dr7-40C		Dry
14 d curing at 40 °C + 7 d drying at 23 °C	Cu14-40C_Dr7-23C		Dry
14 d curing at 40 °C + 7 d drying at 40 °C	Cu14-40C_Dr7-40C		Dry + wet
91 d curing at 10 °C + 7 d drying at 23 °C	Cu91-10C_Dr7-23C		Dry + wet
91 d curing at 23 °C + 7 d drying at 23 °C	Cu91-23C_Dr7-23C		Dry + wet
91 d curing at 40 °C + 7 d drying at 23 °C	Cu91-40C_Dr7-23C		Dry + wet

After production, the specimens were demoulded and stored in different climate-controlled rooms, one with a temperature of 10 °C and a relative humidity (RH) of 65%; one with a temperature of 23 °C and a RH of 65%; or another one with a temperature of 40 °C and a RH of 12%. The specimens were exposed to different curing and drying procedures to evaluate their influence on the UCS, as shown in Table 1. A drying time of 7 d was selected for all specimens as it was sufficient for reaching a constant weight during that period. For the selected test series of LSRE specimens, an additional curing time of either 14 d or 91 d was applied before the drying period of 7 d. The specimens which were allowed to cure were tightly wrapped in plastic film directly after demoulding. It was attempted to avoid evaporation from the specimens during the curing period to maintain the free water to be available for the pozzolanic reactions by wrapping the specimens in several layers of plastic film. However, after the curing period, it was observed that some weight loss had taken place, especially for the specimens placed at the temperature of 40 °C and relative humidity of 12%. For future studies, the specimens should be kept at a high relative humidity or sealed better to avoid any drying during the curing period.

The naming system for the Ø60 specimens comprises several elements as given in Table 1, each providing information about lime content, water content, curing time, curing temperature, drying time, and drying temperature: L = lime content, W = water content, Cu = curing time + C for temperature, and Dr = drying time + temperature.

The UCS test was conducted using an Instron machine with a load capacity of 100 kN and a force-controlled loading rate of 0.11 kN/s (0.04 MPa/s). To ensure that irregularities on the specimen surface did not influence the results, a movable head was used in the setup. Five replicates were tested for each specimen type. The test specimens were all evaluated under dry conditions and selected test series were also tested after being immersed in tap water for 2 h prior to testing inspired by the procedure in [4].

3 Results and Discussion

3.1 Characterization of Raw Materials

The geotechnical properties are shown in Fig. 1. The GSD is slightly outside the range recommended by Houben and Guillaud [13] with a too high content of fine sand, while the plastic properties (PI of 16% and LL of 28%) is within the recommended range. According to Danish recommendations for soil stabilization [14], the soil falls within the recommended range for lime stabilization, but would also be suitable for stabilization with a mixture of lime and cement.

The OWC of the soil was determined from the Proctor test and the results are shown in Fig. 2. Both the results of the bulk density and the dry density from the proctor measurements are shown. Moreover, the bulk density of Ø60 RE specimens was measured directly after the production by electromechanical compaction as described in Sect. 2.2, were also included in Fig. 2 to compare the two production procedures. The OWC for the unstabilized soil (L0) was achieved at a water content of 9% resulting in a maximum dry density (MDD) of 2010 kg/m^3. The OWC for the LS soil (L5) was found to 16% with an MDD of 1830 kg/m^3. Thus, the curve for LS soil was shifted towards higher water contents and lower dry densities, which is in accordance with the findings in most previous studies [5, 15, 16]. The same effect of lime was valid for the bulk densities

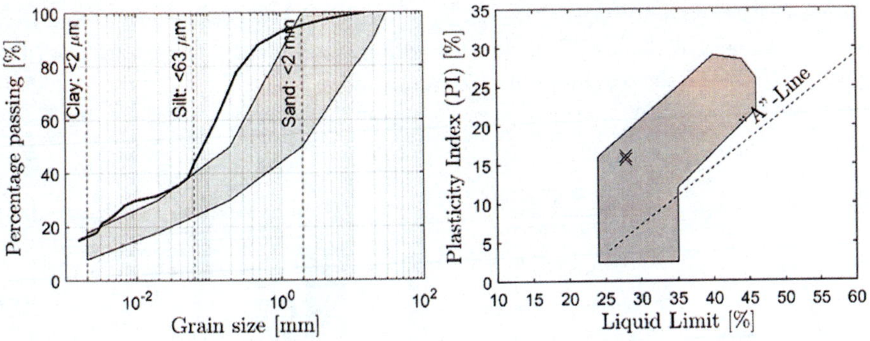

Fig. 1. Geotechnical properties of raw soil: a) Grain size distribution; b) Plastic properties. The grey areas are the recommendations according to [13].

independent on the production method (Proctor or Ø60 RE specimens by electromechanical compaction). The bulk densities for the two production methods revealed that the Proctor method produced specimens with slightly higher bulk densities compared to the Ø60 specimens, revealing that the Proctor compaction was more effective than the one used for RE specimen production.

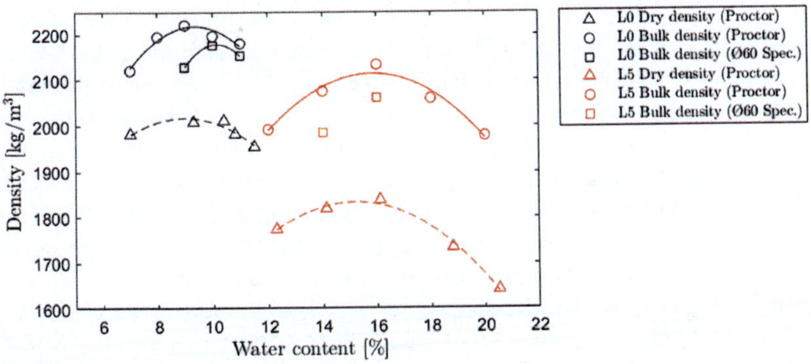

Fig. 2. Results from Proctor test (dry and bulk density) and bulk densities directly after production of Ø60 RE specimens

3.2 Unconfined Compressive Strength

Figure 3 shows the UCS of RE Ø60 specimens of USRE with moulding water contents of 9–11% and LSRE with moulding water contents of 14% and 16%. The specimens were subject to a drying period of 7 d right after production (Fig. 3a: 23 °C/65%; or Fig. 3b: 40 °C/12%). It is worth noticing that the specimens in Fig. 3 were not allowed to cure. The results show that the USRE specimens in all cases achieved higher UCS than the LSRE specimens independent of the drying temperature and relative humidity. This can be explained by the lack of curing time, which is important for LSRE specimens as the pozzolanic reactions occur over time leading to an increase in UCS [11, 15]. Moreover, the results show that the USRE specimens with a water content of 9% achieved the highest UCS, which agrees with the OWC determined by the Proctor test. Moulding water contents above OWC resulted in a decrease in the UCS for USRE.

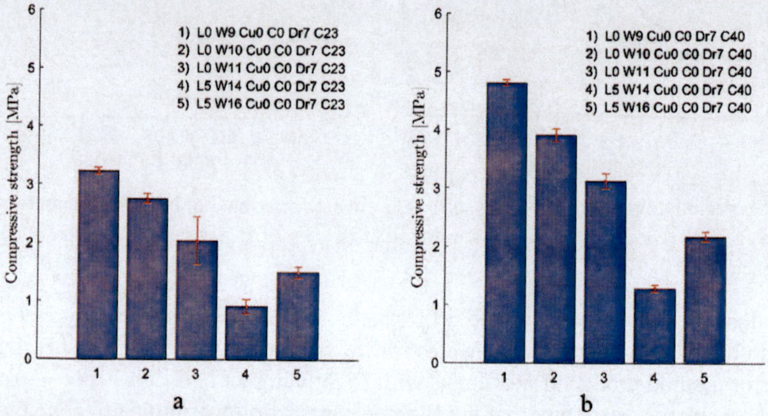

Fig. 3. UCS of USRE and LSRE Ø60 specimens dried for 7 d at 23 °C/65% (a); or at 40 °C/12% (b)

Regarding the LSRE specimens, it is well known that the addition of lime results in an increase in OWC and a decrease in MDD, which was also observed from the Proctor test [5, 15, 16]. The highest UCS was also found at the OWC for the LSRE specimens, as a moulding water content of 16% resulted in a higher UCS compared to that of 14% independent on the drying climate. The production of LSRE specimens with moulding water content >16% was not possible due to production limitations and difficulties with the demoulding process. However, there are contradictions in the literature regarding the moulding water content for lime stabilized soil and whether it should be below, at, or above the OWC [17, 18]. For the efficient formation of pozzolanic reaction products, the water content of the soil–lime mixture should be kept as high as possible [17]. Also, results have shown that rewetting after a curing period may lead to increased UCS [5], showing that additional water is beneficial for the pozzolanic reactions. Although no time was dedicated to the curing process between the production and drying period, some curing must have taken place in the LSRE specimens during the drying period with the available water as evaporation of free water from the specimens happened gradually during the drying period.

Concerning the drying climate, drying at higher a temperature and lower relative humidity (40 °C/12% in Fig. 3b) resulted in an increase in UCS for all specimens compared to their respective counterparts at lower temperatures and higher relative humidity (23 °C/65% in Fig. 3a). The overall increase in UCS for all specimens can be largely explained by the much lower relative humidity. The low relative humidity causes more water to evaporate from the specimens during the drying period which produces an increase in compressive strength [19, 20]. For the LSRE specimens, the increase in temperature could also accelerate the pozzolanic reaction [9].

Figure 4 shows the UCS of LSRE Ø60 specimens with 5% lime and moulding water contents of 14% (Fig. 4a) and 16% (Fig. 4b). The specimens were exposed to different curing and drying conditions to evaluate the influence of temperature and RH on the UCS. The results show that the additional curing period of 14 d prior to the drying period was beneficial for the specimens with a moulding water content of 14%, as an increase in UCS was detected between Dr7-C23 and Cu14-C23_Dr7-C23; and Dr7-C40 and Cu14-C23_Dr7-C40, respectively. However, no increase in UCS was observed for the specimens with a moulding water content of 16% (OWC). One possible explanation could be that the higher water content of 16% resulted in more free water for the pozzolanic reactions to occur and that more water was still available during part of the drying period for the specimens with 16% water. Temperatures of 23 °C does not accelerate the pozzolanic reactions to the same extend as temperatures of 40 °C [8].

In all cases, an increase in temperature from 23 °C to 40 °C, resulted in higher UCS, although it was not clear whether the increase in temperature is most important during the curing or the drying period. The highest UCS of 3.9 MPa was found for the specimens with 16% water cured and dried at 40 °C, which was more than twice the UCS for the respective counterpart cured and dried at 23 °C. Nshimiyimana et al. [8] also found that the increase in curing temperature from 20 °C to 40 °C of LS-CEB of kaolinite-rich soil resulted in a doubling of the UCS. However, the increase was not as great for LS-CEB of quartz-rich soil, which also shows that the effect of lime stabilization is highly dependent on the soil mineralogy and chemical composition [4, 21, 22].

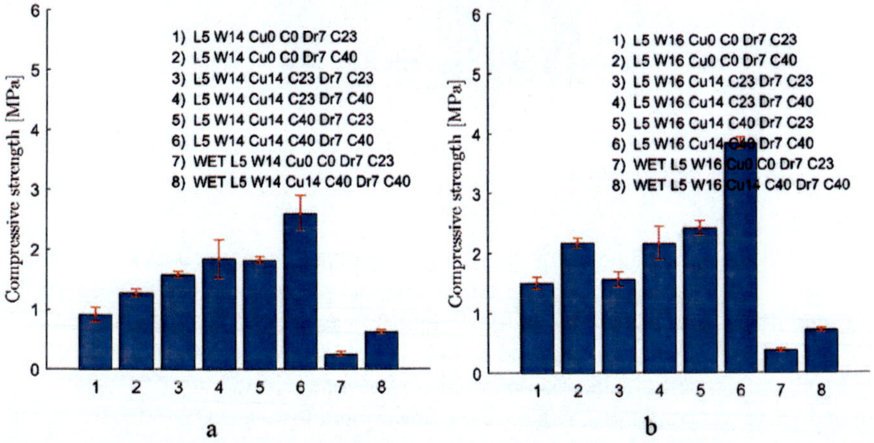

Fig. 4. UCS of LSRE Ø60 specimens (a: 14% water); (b: 16% water) cured for 14 d and dried for 7 d at 23 °C/65% or at 40 °C/12%.

Selected test series of LSRE specimens with 5% added lime and moulding water contents of 1%4 and 16% were immersed in a water bath for 2 h before the UCS test. The motivation behind these wet UCS tests was to determine whether the curing and drying process had any impact on the UCS of the specimens in the wet condition. This was done by comparing specimens WET Cu0_Dr7-C23 (7 d of drying at 23 °C/65% but no curing); with specimens WET Cu14-C40_Dr7-C40 (14 d of curing and 7 d of drying at

40 °C/12%). Thus, the selected test series were those that achieved the lowest and highest UCS in dry conditions, respectively. The results reveal a significant decrease in UCS in wet conditions compared to the same specimens in the dry conditions. A clear difference was also seen for specimens allowed to cure for 14 d and dry for 7 d at 40 °C compared to the specimens which were only allowed to dry for 7 d at 23 °C. The former UCS was more than twice that of the latter. It is important to note that different drying conditions were used for the specimens. Again, specimens with a moulding water content of 16% reached higher UCS than 14% water content. Ouedraogo et al. [15] found that earth bricks with 4% of lime needed a curing time of more than 7 d to obtain any compressive strength in wet conditions and a measurable strength was obtained after 21 d of curing. A similar water immersion procedure of 2 h before the UCS test was used in the study [15].

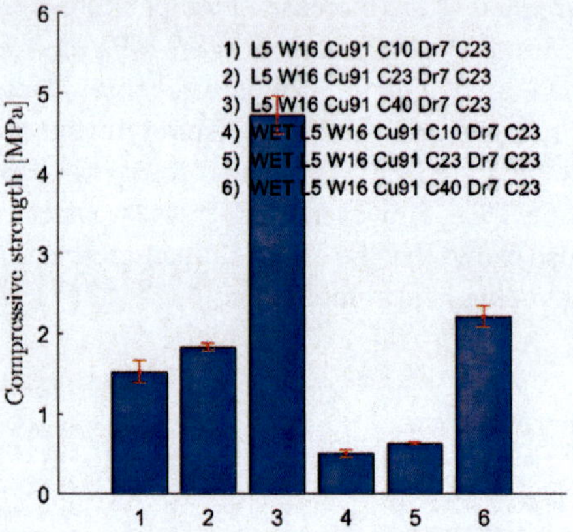

Fig. 5. UCS of LSRE Ø60 specimens with 91 d of curing at 10 °C, 23 °C or 40 °C and drying for 7 d at 23 °C in dry or wet condition

Figure 5 shows the influence of the curing temperatures for specimens with longer curing times of 91 d on the UCS. The three different curing temperatures were 10 °C, 23 °C and 40 °C were selected and the specimens were tested in dry condition (7 d of drying at 23 °C/65%) and after the additional immersion in water for 2 h (WET). Almost no difference in UCS was observed for specimens cured at 10 °C and 23 °C both in dry and wet conditions, while the UCS was more than doubled when increasing the temperature to 40 °C. In the wet condition, the relative strength increment was even higher when increasing the temperature to 40 °C, showing how important the curing temperature is for the pozzolanic reactions between the lime and the clay minerals.

4 Conclusion

The motivation of this study was to evaluate the effect of different curing and drying conditions for unstabilized rammed earth (USRE) and lime stabilized rammed earth (LSRE) on the unconfined compressive strength (UCS). Excavation soil from Vinge,

Denmark, was characterized with the aim of using the soil as raw material in RE construction. The UCS test revealed that RE specimens produced at the optimum water content (OWC) obtained the highest UCS for both USRE and LSRE. However, when the LSRE specimens were exposed to drying directly after production, low UCS's were obtained, although an increase in the drying temperature from 23 °C to 40 °C, increased the UCS, especially for the LSRE specimens produced with the OWC. According to the literature, it is important to allow LSRE to cure to enhance the pozzolanic reactions. By adding 14 d of curing at 23 °C prior to the drying period for LSRE, an increase in UCS was obtained for the specimens with a moulding water content of 14%, while those with a water content of 16% did not experience any beneficial effect. However, when the curing and drying was done at 40 °C, an increase in UCS was observed for specimens independent of the moulding water contents. A further increase in curing time to 91 d for specimens cured at 10, 23 or 40 °C resulted in a high increment in UCS for those cured at 40 °C compared to the lower curing temperatures. USRE specimens obtained higher UCS when tested in the dry conditions but had no strength in the wet condition. In this case, the effect of lime stabilization was evident, as the LSRE specimens did not erode and obtained a compressive strength in the wet conditions.

Acknowledgement. The Capital Region of Denmark is acknowledged for the financial support for the project "Udvikling og test af støjskærme ved anvendelse af stampet lerjord". The excavation soil was kindly provided by Remco A/S. Shankar Rimal is acknowledged for conducting the Proctor test for lime stabilized soil.

References

1. Nordpub: Survey of the emergence and use of naturally occuring materials (2021)
2. Holmboe, T.: Teglværksler i Danmark (2001)
3. Frederikssund kommune: Jordhåndteringsstrategi Vinge 2022 Strategi for bæredygtig jordhåndtering (2022)
4. Ouedraogo, A.K.J., Aubert, J.-E., Tribout, C., Escadeillas, G.: Is stabilization of earth bricks using low cement or lime contents relevant? **236** (2020). https://doi.org/10.1016/j.conbuildmat.2019.117578
5. Arrigoni, A., Pelosato, R., Dotelli, G., Beckett, C.T.S., Ciancio, D.: Weathering's beneficial effect on waste-stabilised rammed earth: a chemical and microstructural investigation. Constr. Build. Mater. **140**, 157–166 (2017). https://doi.org/10.1016/j.conbuildmat.2017.02.009
6. Guettala, A., Houari, H., Mezghiche, B., Chebili, R.: Durability of lime stabilized earth blocks. Courr. Du Savoir-N°02, 61–66 (2002)
7. Reddy, B.V.V.: Soil stabilisation. In: Reddy, B.V.V. (ed.) Compressed Earth Block & Rammed Earth Structures. STICEE, pp. 73–94. Springer, Singapore (2022). https://doi.org/10.1007/978-981-16-7877-6_3
8. Nshimiyimana, P., Moussa, H.S., Messan, A., Courard, L.: Effect of production and curing conditions on the performance of stabilized compressed earth blocks: Kaolinite vs quartz-rich earthen material. MRS Adv. **5**(25), 1277–1283 (2020). https://doi.org/10.1557/adv.2020.155
9. Al-Mukhtar, M., Lasledj, A., Alcover, J.F.: Lime consumption of different clayey soils. Appl. Clay Sci. **95**, 133–145 (2014). https://doi.org/10.1016/j.clay.2014.03.024
10. Al Haffar, N., Fabbri, A., McGregor, F.: Curing conditions impact on compressive strength development in cement stabilized compacted earth. Mater. Struct. **54**(3), 1–15 (2021). https://doi.org/10.1617/s11527-021-01702-0

11. Ciancio, D., Beckett, C.T.S., Carraro, J.A.H.: Optimum lime content identification for lime-stabilised rammed earth. Constr. Build. Mater. **53**, 59–65 (2014). https://doi.org/10.1016/j.conbuildmat.2013.11.077
12. George, S.Z., Ponniah, D.A., Little, J.A.: Effect of temperature on lime-soil stabilization. Constr. Build. Mater. **6**, 247–252 (1992). https://doi.org/10.1016/0950-0618(92)90050-9
13. Houben, G.: Earth Construction: A Comprehensive Guide (1994)
14. Vejregelgruppen for Jord grus og brolægning. Udbudsskrift for Jordstabilisering (2018)
15. Ouedraogo, K.A.J., Aubert, J.E., Tribout, C., Escadeillas, G.: Is stabilization of earth bricks using low cement or lime contents relevant? Constr. Build. Mater. **236**, 117578 (2020). https://doi.org/10.1016/j.conbuildmat.2019.117578
16. Ávila, F., Puertas, E., Gallego, R.: Mechanical characterization of lime-stabilized rammed earth: lime content and strength development. Constr. Build. Mater. **350**, 128871 (2022). https://doi.org/10.1016/j.conbuildmat.2022.128871
17. Ramesh, H.N.G., Sivapullaiah, P.V.: Role of moulding water content in lime stabilisation of soil. Proc. Inst. Civ. Eng. Gr. Improv. **164**, 15–19 (2010). https://doi.org/10.1680/grim.900040
18. Canivell, J., Martin-del-Rio, J.J., Alejandre, F.J., García-Heras, J., Jimenez-Aguilar, A.: Considerations on the physical and mechanical properties of lime-stabilized rammed earth walls and their evaluation by ultrasonic pulse velocity testing. Constr. Build. Mater. **191**, 826–836 (2018). https://doi.org/10.1016/j.conbuildmat.2018.09.207
19. Champiré, F., Fabbri, A., Morel, J.C., Wong, H., McGregor, F.: Impact of relative humidity on the mechanical behavior of compacted earth as a building material. Constr. Build. Mater. **110**, 70–78 (2016). https://doi.org/10.1016/j.conbuildmat.2016.01.027
20. Gerard, P., Mahdad, M., McCormack, A.R., François, B.: A unified failure criterion for unstabilized rammed earth materials upon varying relative humidity conditions. Constr. Build. Mater. **95**, 437–447 (2015). https://doi.org/10.1016/j.conbuildmat.2015.07.100
21. Bell, F.G.: Lime stabilization of clay minerals and soils. Eng. Geol. **42**, 223–237 (1996)
22. Aldaood, A., Bouasker, M., Al-Mukhtar, M.: Impact of wetting-drying cycles on the microstructure and mechanical properties of lime-stabilized gypseous soils. Eng. Geol. **174**, 11–21 (2014). https://doi.org/10.1016/j.enggeo.2014.03.002

The Effect of Natural Filler on Enhancing Strength for Unstabilized Rammed-Earth Walls

Hazem Abu-Orf(✉)

University of Palestine, Al-Zahra, Gaza Strip, Palestine
h.orof@up.edu.ps

Abstract. This research's experimental analyses have at their aim to improve the "unstabilized" rammed earth's mechanical properties while choosing sandy limestone as a particular soil under investigation. The parameters studied here are 1) natural infill materials coupled with bond substances as an independent variable and 2) a rise in compressive strength as a dependent variable. Existing research shows that the "unstabilized" technique not only inheres a relatively low strength test but also has been labeled with water "infiltration". Despite this technique's environmental advantages, moisture ingress remains yet unresolved. This research addresses this issue by interlocking the raw material particles that are induced by the compaction process while substantially relying on clay as a bond substance to achieve density. A noticeable rise in mechanical strength follows in this research's findings by the natural infill material physically interlocking the sandy limestone mixture grains through the "particle size distribution". Another finding suggests that the infill's effect contributes to the "unstabilized" rammed earth's material strength, however, in conjunction with the aid of bond substances to arguably help a chemical bonding to occur between the fine particles residing between the large grains; the gravels.

Keywords: Soil Stabilization · Granular Gradient · Infill Materials · Environment-friendly Additives · Material Strength

1 Introduction

The interest for this research to go for the "unstabilized" rammed earth (URE) lies in the latter adopting sustainable construction material inherent in a minimal embodied energy while enhancing the recycling potential [1, 2]. Yet, moisture ingress [3, 4] negatively affects compressive strength while causing walls' deformability due to the air's relative humidity that induces transient hygroscopic transfers into the compacted specimens [5]. The hygroscopic state is in command over suction [6] distribution which has an impact on the mechanical response of URE construction, particularly in association with stiffness and/or cracking failure. The importance of URE sensitivity to water grabs attention for investigation in this research because it exhibits a crucial issue that affects URE material's mechanical strength to the extent that its neglect results in URE material being coined with a relatively low strength test, as the next subsection of this article suggests.

S. Amziane et al. (Eds.): ICBBM 2023, RILEM Bookseries 45, pp. 259–270, 2023.
https://doi.org/10.1007/978-3-031-33465-8_20

1.1 Previous Research

To address the relatively-low strength test of the "unstabilized" technique, the literature shows two strategies. One is internal suction [7–10] by the latter reinforcing the contact to occur between the soil particles resulting in a dried-earthen material that in turn increases stiffness and/or strength. A particular attempt [11], has been initiated to quantify the structural behavior while considering hygroscopic conditions, which suggests excessive time necessary to establish earthen material's stability due to the latter's material slowly drying effect. Another technique is soil "particle size distribution" (PSD) [12, 13], or soil ingredient composition. PSD's importance outlines the rate at which humidity may well ingress into the compacted earthen material as a result of the capillary suction [14, 15]. PSD also affects "soil porosity and inter-particle friction/interlock" [15] by primarily relying upon dynamic compaction to maximize density, therefore a lower percentage of fine aggregates and clay content might be needed. Besides this strategy's physical stabilization, the literature studies the chemical bonding's effect [16] by holding on to the additives' strategy in an attempt to intervene in the "controlled modification of soil texture, structure, or physic-mechanical properties" [9]. For URE, clay (8–14%, by mass) remains the bonding substance [17]. A slight rise in URE's mechanical strength emerges, ranging between 1.27 MPa and 1.42 MPa [18] – common mechanical strengths for URE are 1–2.5 MPa [19, 20].

The relatively low compressive strength has encouraged many studies [21, 22] to investigate other bond elements; other than clay, particularly cement, or lime, to stem the pitfall in URE's strength. Their conclusions, albeit variations, agree on the excellent performance of the stabilized specimens, while the moisture content upon compaction and throughout the curing stage did not record a significant rise within the cement-stabilized walls. With an increase in the cement content from 2.5% to 7.5%, the wet-to-dry strength shows a rise in a strength ratio from 0.51 to 0.85 [23], while noting that the acceptable ratio starts from 0.35 [24]. Burroughs [25] has in particular studied the suitability of cement and lime for stabilizing varied soil types by investigating the soil's linear shrinkage and its plasticity index. Central to Burroughs's assessment is grading neatly associated with finished texture and friability. The literature concerned with grading bounds its analyses [9] with a correlation being investigated not only between type and %wet of stabilization but also between shrinkage and plasticity while holding a focus on soil's physical characteristics (or, grading) [27].

Both lime and/or cement nevertheless harm the environment through their materials' pH, largely because both materials are in nature industry-produced stabilizers that not only glue the earthen materials' grains through a chemical reaction [27] but also duplicate the "greenhouse gas emissions" [3]. Recent studies [15], therefore, have turned to environment-friendly additives. Despite the latter contributing to a rise in strength tests such altitude arguably remains limited. As this study explains, environment-friendly studies utilize ramming [28] at a moisture content to achieve workability rather than an optimum moisture content (OMC) for the maximum dry density (MDD). Indeed, Ciancio et al. [29] suggest the OMC being +1%–2%, while the New Zealand Standard NZS4298 [32] recommends that the moisture content upon compaction falls below 3% of the OMC; however, not exceeding 4% dry or 6% wet of the optimum. Furthermore, the concern in the previous studies (cited in [31]) has been either with the impact that rain has on the durability of earthen material or addressing how the wet conditions might influence the mechanical properties of earthen material specimens. A particular issue emerges in these studies which concerns a change in relative humidity on the internal suction. This issue has been well recognized in previous studies; nevertheless, less unaddressed, as this research suggests. A research gap emerges in this respect.

1.2 Research Question

To address the research gap identified in the previous subsection of this article – the gap is concerned with stemming a moisture ingress that occurs once completing the compaction process and goes throughout the curing stage, analyses in this article turn away from the strategies narrowed down by either investigating the wide-ranging choices of chemical bond substances or their effect while primarily relying on ramming to ensure workability – as the previous research suggests. Despite this article recognizing such strategies' vital importance, it instead adopts a two-fold strategy. Laboratory analyses undertaken for this article first seek the contribution of natural infill materials to physically interlock the compacted grains and, secondly, look into the effect of a combined, natural, and locally available bond substance. The strategy responds to the question for addressing in this research which concerns how a moisture ingress; arising as a result of internal humidity or rainfall, for example, and occurring upon completing the compaction process and thereafter, might be minimized.

2 Research Methodology

The methodology adopted here concerns laboratory research undertaken within a sandy limestone case study, chosen here as a particular material. The properties of URE specimens are examined to evaluate their suitability for URE construction in accordance with universal recommendations. The PSD of the limestone material has been carefully observed and described. Concerning the mechanical performance of the URE, compression and water absorption erosion tests have been undertaken. The next subsections of this article describe the stages of the methodology program employed here: 1) materials selection; 2) specimens' preparation; 3) specimens' manufacturing; and lastly 4) testing of the specimens. The methodology is described as follows:

2.1 Raw Material

The material is characterized by the means of expenditure (chemical analysis, drop test, and dry strength test) and the laboratory test (PSD analysis). The sandy limestone, or sedimentary, rocks have been transported from the superficial revelations that exist along the Gaza Strip's Mediterranean coastal line. Notably, the rocks' soil quarry appears to be unsuitable for cultivation due to being arguably located on the outer crust of a non-agricultural area. The choice of sandy limestone, therefore, endorses the argument that its use as a material for construction would not negatively affect our Earth's crust. Sandy limestone has further benefits, for example, it inheres to a greater consistency in mineralogy while depicting in its micrograph low-fine particles (especially, the sand ones), which might in turn interfere with cement hydration (or, "hydration water") [33].

2.2 Specimen's Preparation

The experiment tools include two pneumatic rammers; sieve trays (0.0150–4.75 mm), a sieve shaker, a wooden wrench, and five plastic piles. The limestone material necessary to manufacture the specimens falls within a man-made earthen soil mixture, which is predominantly composed of sand particles (55%, by mass). Initially, a realistic wooden formwork (90 cm × 45 cm in plan and 60 cm height) and three wooden cubes (10 cm × 10 cm in plan and 10 cm height) for testing have to be constructed. Three realistic sample walls and about 115 specimens have been constructed in this research. The specimens' preparation concerns setting the sandy limestone mixture; including the grain sizes and the latter's quantities.

At the start, the moistened sedimentary rocks brought from the coastal line have been laid exposed to the sun's heat to dry out for, at least, a week. Once the rocks have been dried, the specimens' preparation started with breaking down the sedimentary rocks using an iron hammer (1.689 kg) into small gravels (10–20 mm in diameter) while manually excluding the gravels <10 mm to avoid a reduction in a strength test, when the gravel size increases. Table 1 shows limestone particle quantities. Two notes worth noting in this table. Once concerns cockle shells/marble/natural stone that acts as infill materials. Another note is clay content of 5% combined with 3% burning residue of olive and orange branches which serves as an additive. Furthermore, the ranges, shown in Table 1, fall in agreement with the PSD envelope recommended by Houben and Gillard [32] and also with the material mixture adopted by most authors working within URE literature [15, 33–35].

Table 1. Limestone's mixture quantities.

Materials	% By mass	Particle size
Limestone gravel	20	4.75–2.36 mm (9%), 2.36–0.6 mm (11%)
Limestone sand	55	0.425–0.300 mm (13%), 0.150–0.300 mm (19%), and <0.150 mm (23%)
Cockle shells/marble or natural stone	10	<0.6 mm (Powder)
Ordinary lime	7	Dust
Ash (additive)	3	A burning residue of olive and orange branches (rich in potassium and magnesium)
Red clay (additive)	5	<0.0150 mm

Worth noting, that ordinary lime has not been perceived as a bond substance. Rather, it has been added to the mixture quantities shown in Table 1 at 7% (by mass) not only to compensate for the sandy limestone's natural loss of lime due to acidic rain but also to decrease the earthen material's sensibility to water while protecting the latter from an insect attack, or rot, for example. Chemical reactions in this vein occur due to the rainfall hitting the limey outer veneer of the sedimentary rocks due to the more reactive minerals that existed within the limestone than in the sandier beds [29].

The sieving stage has followed to set the grain sizes bypassing the small gravel <10 mm through the sieving nets to evict both the gravel (4.75–2.36 mm in diameter) and the sand grains in three levels (0.6–0.85mm, 0.150–0.6 mm, and <0.150 mm). The cockle shells have been fractured followed by sieving to produce the cockle shell particles (0.85–0.6 mm, in diameter). The latter's particles, waste marble powder, and natural stone have been ground to powder <0.6 mm using a manual grain mill.

2.3 Specimen's Manufacturing

This stage includes mixing, compaction and curing. The particle-size quantities have been thoroughly mixed dry four times for five minutes to enable a sufficient interlocking of the particles throughout the material mixture. A known amount of water w% has been added and the material mixture has been also mixed four times for a further five minutes. A "drop test" technique – recommended by rammed earth (RE) guidelines – ensures that the material mixture's "wet state" reaches the OMC. The "drop test" starts by dropping from a height of 1.5 m a moistened and compacted material mixture by hand onto a hard, and/or flat, surface. Once the material mixture is too dry the moistened ball fragments into many segments. By contrast, when the moisture content is nearing the OMC the ball fragments into a few segments. In case the moistened ball is too wet, the ball falls in one segment onto the surface.

On the site, the material mixture has been not moistened at once, but rather in several stages to release the energy embodied within the mixture which is necessary to initiate a chemical reaction, each time when water is sprayed over the dried mixture. When satisfactory, according to the "drop test", the moistened quantities have been poured loose into the sample formwork in layers that each has arisen about 20 cm in height. For the laboratory tests, the moistened mixture has been poured loose into the cubes (numbered 4B, 7, C, 5, 7a, and 5A) (10 cm × 10 cm × 10 cm) in five layers of 3 cm height; each layer has been compacted using both a wooden wrench and a 1.689 kg flat-based rammer at a height of 2 cm. The layers borrow the same sandy limestone-moistened mixture of its sample wall. The next stage concerns compacting each layer; namely being undertaken into three phases. The first uses an iron pestle to physically interlock the moistened mixture's particles. Secondly, ramming uses a timber pestle that enables a much tighter interlocking. The last phase uses a timber wrench that fine-tunes the edges, followed by repeating the second phase of compaction to work out the surface before proceeding. Each layer has been sacrificed before the start with the next layer to ensure a good bonding surface between the layers. The compaction phase has resulted in 4 layers (Fig. 1); each about 14 cm in height, and 5 layers for each cube. The sample walls and the cubes have been left to cure for at least 3 days before demolding (Fig. 1). Following molding, the sample walls and the cubes have been wrapped with a nylon film for 7 days to allow each sample and its cubes to hydrate (Fig. 1).

Once wrapped, the specimens have been cured or hydrated. The curing periods of 7, 28, and 90 days are adopted here following ACI (American Concrete Institute) standards; commonly used for curing the concrete, to monitor the development of strength recorded in association with curing time before undertaking the compressive strength tests. By

Fig. 1. Molding and wrapping the specimens with a nylon film.

adopting the same set of curing periods applied for concrete in this research, a comparison has been established between the hydration of cement and the sandy limestone mixture. The ACI in this vein recommends the minimum curing time that corresponds to concrete achieving 70% of its characteristic compressive strength. The 70% characteristic compressive strength would be accordingly achieved at 7 days upon casting the concrete. Therefore, a minimum curing period of 7 days has been set in this research. The curing period of 7 days has been sprayed with 7% water (by mass) early in the morning, namely to avoid spraying the cold water on the sun's heated surfaces. The other two curing periods follow this procedure, however, have been cured for 28- and 90-days and sprayed with water being set at 8% and 8.5%, respectively.

2.4 Testing

A rise in the sandy limestone mixture's unconfined compressive strength (UCS) has been sought in this research by not only intervening in the sandy limestone mixture's quantities but also altering the physical properties. Two laboratory tests have been conducted; one is a strength test while the other concerns porosity.

The first test measures the UCS: each specimen has been accordingly tested upon accomplishing its curing period and also being left for a day within the outdoor environment to dry out prior to testing (Fig. 2). The effect of age of curing on the UCS is shown in Fig. 8. Two notes are worth noting. One is a slight variation to be found between measured UCS of the specimens and even that existed within the same batch (Fig. 8). In comparison, as per Indian Standard (IS-456), concrete gains its strength rapidly in the initial days upon casting, namely a 65% compressive strength in 7 days and 99% in 28 days. It also gains 90% strength in 14 days and the remaining 9% strength is gained in 28 days. Hence, one may suggest that after 14 days of the curing period, concrete's strength slowly increases. It means that the process of hydration of cement sharply increases during the first 14 days upon casting, whilst it sharply drops down thereafter. It is therefore advisable that concrete should be cured for a minimum of 14 days upon casting to achieve durability-related properties.

Furthermore, the specimens' strength in this research, in contrast to concrete, slowly increases, and this has been clearly noticeable all through the set curing periods. This note suggests that the process of hydration remains ineffective primarily due to the weak chemical reaction occurring. An example is cube 4B (Fig. 3). The relative UCS obtained in this example, upon curing periods of 28 and 90 days, has arisen by a 14.08% and a 37.08%, respectively. The curing periods of 28 and 90 days have therefore arisen steadily in strength than that already recorded within 7 days of the curing period. Upon

Fig. 2. Specimen's fracture failure.

90 days, the UCS gained has arisen by a 20.01% in comparison with those reported within 28 days. The UCS's percentages (Fig. 3) witness an increase of 6.02% in 28 days; when multiplying the curing period of 7 days. These also show an increase of 17.07% in 90 days; when multiplying the curing period from about a month to almost two months. What the UCS's percentages further suggest is that the specimens require a much longer curing period than concrete specimens would do to attain durability-related properties – primarily blaming the slowly occurring hydration process. In essence, a three-time increase in the UCS has been scored; namely from 6.02% to 17.07%, by multiplying the curing period from nearly a month, in 28 days, to two months, in 90 days, of curing, instead of multiplying the initial curing period of 7 days in weeks, namely in 28 days of curing period. The rate of the UCS increment ranges from 0.3 to 1.8 (megapascal) MPa per every 7 days of the curing period.

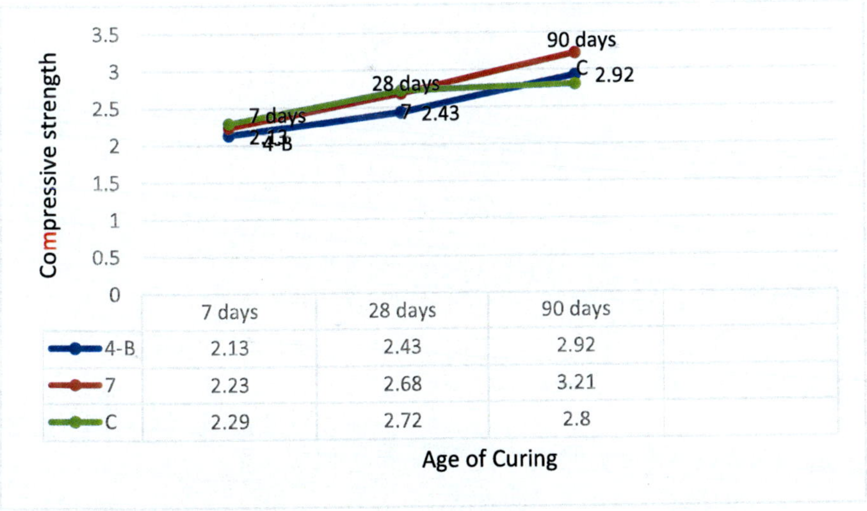

Fig. 3. UCS tests.

Maximum dry density (MDD), OMC, and water ingress: a dry density has been specified in this research by weighing the specimen dry and wet; that is, by weighing that occurs before and upon soaking the dried specimen into a water basin for 24 h (Fig. 4). Before testing, all specimens have been left within the outdoor environment for five days to dry properly. Specimens' porosity emerges as a result of subtracting the wet volume from the dry one, followed by dividing the outcome by the wet volume (Fig. 5). As is shown in Fig. 5, the moisture content decreases in all specimens tested here as the dry density increases. Or, it may be suggested that increasing the moisture content accompanies a decrease in the strength of compacted sandy sub-mixture.

Fig. 4. Porosity test.

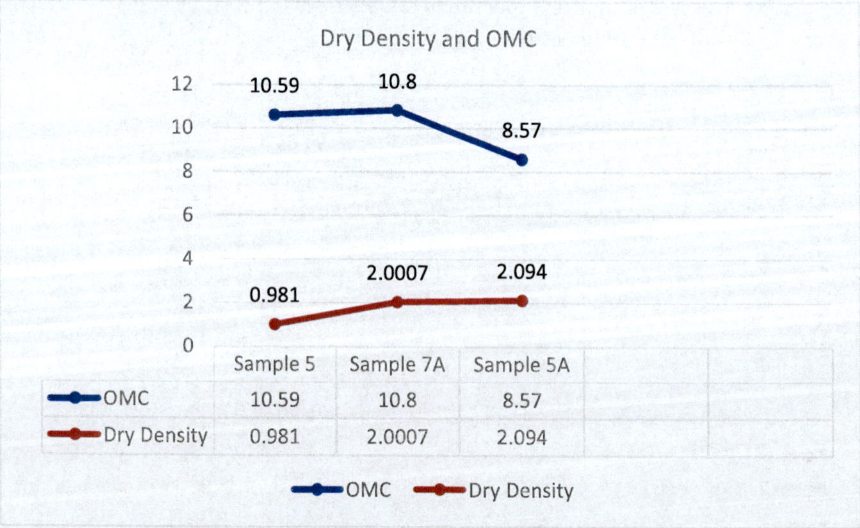

Fig. 5. Dry density to OMC.

A porous structure marks the sedimentary limestone rock suggesting a water capillary movement to easily sneak into its pore spaces. To address this issue, the strategy adopted here enhances the hydration process of the sandy limestone's compacted sub-mixture by leaning upon a filler effect, rather than an impact occurring through chemical bonding. It on the one hand uses in doing so sand particles (55%, by mass), less in size than the limestone's gravels, as infill material. This strategy benefits from the limestone's characteristic concerned with slowly dissolving in water; thereby the sand particles arguably hinder the spread of water to move into the pore spaces that exist between both the gravels and also within the gravels' porous.

On the other, the strategy turns to cockle shells considered here as an infill substance between the limestone's gravels which have been characterized by their greater surface area, or a "surface chemistry" [8, 32] that circumscribes a arise of moisture ingress [9, 27]. What their morphology structure suggests is a neatly bonded cockleshell's composition marked by a smaller size of the grain, and aggregates, offering "slender needle-like shapes" [35, p. 133] forming a morphology of structured mixture and small in size pores. The result is an increase in bending strengths. However, the sandy limestone sub-mixture's composition uses in this research as particle degrees, instead of ground cockle shells, due to the em manual working procedure. What remains yet unresolved are the pore spaces left unfilled between the limestone gravels. And this might explain why the specimens' moisture contents upon compaction remain relatively high (Fig. 4) than the porosity rate ranging between 5.5% to 10.2% (by mass).

3 Tests' Discussion

Two tests have been undertaken for this research which their analyses suggest being correlated. One concern measuring a rise in the UCS tests. URE, in this vein, has been labeled with low UCS tests, as it has been reported earlier in this article, namely not falling below 1.4 MPa while not exceeding 2.5 MPa. As shown in Fig. 2, the specimens have a UCS > 2 MPa – Uniform Building Code Standard 71-1 (UBS) specifies a satisfactory UCS starting from 2 MPa. An optimum UCS has been reached in this research upon 90 days of curing, namely a strength test of 2.92 MPa, which arguably lies within the highest ranges of UCS scores recorded for URE, and these range between 2–3 MPa [6]. To explain, a granular gradient strategy in this study contributes to such a UCS score by a "soil porosity and interparticle friction/interlock" [9]. The strategy employed in this research nevertheless does not rely upon a chemical reaction supported by sufficient compaction. Rather, it targets minimizing pore spaces by interlocking the mixture particles. Two methods are adopted in doing so. One concerns the particle degrees marked by a limestone density of 2.71 g/cm^3 to score a rise in the UCS tests. The second method echoes the effect of fine particles seen as an infill substance.

Yet, a granular gradient strategy has been arguably insufficient, and this has been witnessed in the relatively low score of 2.92 MPa, particularly when being compared to stabilized rammed earth's (SRE) UCS: 3 MPa–20 MPa. Clearly, a remarkable gap in the UCS tests remains. To overcome the difference between SRE's and URE's bending strengths, the UCS tests in this research increase when multiplying the curing time and, therefore, the argument put forward here advocates having a long curing period towards maximizing the mechanical strength. However, the argument may appear impractical. Explanation blames the mineral calcite of the limestone and the cockle shells, which their grains slowly dissolve in water (about 10–15%) as they are in need of acid and/or extensive heat in doing so to not only release Cao particles, necessary to chemically react with H_2O but also to produce the calcium hydroxide ($Ca(OH)_2$). In this vein, the samples have been in this research cured within a natural Mediterranean environment of about 30 Celsius throughout the summer period, thereby suggesting a slow process of calcination that requires temperatures ranging between 1070–1270 Celsius. The result has arguably been a low degree of bonding betweenthe particles, evident in a maximum UCS sore of 2,92 MPa. What remarkably remains a central theme for the specimens tested in this research, that has yet to be addressed, is the pore spaces.

Such a challenge becomes evident in the second test undertaken in this research which concerns measuring a moister content upon compaction. The literature in this respect suggests suction that endorses a decrease in the UCS tests when a moisture content upon compaction arises. And this is true in this research. The moisture content upon compaction remains relatively high (Fig. 4) despite the 90 days of curing; consequently, the specimens have scored a relatively low UCS test of 2,92 MPa. Critical, as analyses in this research suggest, has been the degree of saturation being measured through the lens of porosity which has not been yet minimized in this study, and this degree not surprisingly records a relatively high OMC (Fig. 4) despite the specimens showing an increase in the dry density. The problem, as this research further suggests, lies in particle sizes [24]. Cockle shells have been blamed because their particles have arguably resulted in an ineffective filler effect. This is because cockle shell particles have left infilled pore spaces between the limestone gravels suggesting an increase in suction, thereby a decrease in the UCS tests – as this research suggests. From the perspective of a capillary movement, pore spaces being left unfilled remain at stake.

Figure 4 shows a slight decrease in moisture content despite an increase in dry density. The pore structure of the limestone gravels and the cockle shells' particle sizes are arguably blamed for laying a template of network channels to be occupied by a moister content shown by a relatively high OMC when conducting the porosity test. What appears to suggest is the filler effect of cockle shells remains ineffective in this research. The cockle shell particle degrees pose a further problem by the limestone mixture composition necessitating a long curing period when a comparison is made with the ground cockle shells marked by a "surface chemistry". An explanation for a long time, as analyses in this research suggest, lies in a moisture content that sneaks into the pore spaces and, when evaporating, the particle size distribution points to the yet unfilled pore spaces, thereby scoring a low UCS test of 2,92 MPa.

4 Conclusions

The aim of this research is with scoring a rise in the UCS tests of URE. Key findings emerge in this research as follows:

- Granular gradient strategy has been effective, yet it remains insufficient on its own;
- In comparison with those suggested in previous research, a slight rise in the UCS tests of the URE has been achieved in this research, yet this rise remains relatively low in comparison with the stabilized soil mixture;

- Material strength has been arguably associated in this research with minimizing the pore spaces or creating less pore structure, necessary to penetrate the capillary movement of moisture content.
- A shortage concerned with the infill particles' effect has been evident in this research which arguably explains the relatively low rise in the UCS tests. The shortage blames the use of cockle shell degree particles, instead of employing powder, thereby not benefiting from the bonding effect of "surface chemistry";
- To obtain durability-related properties, this research's analyses suggest that URE does arguably need a long curing period in comparison with concrete specimens. And this appears to be impractical.
- This research's analyses endorse the contribution of an infill effect towards maximizing the material strength of URE, however, a chemical reaction to bond the fine particles residing between the gravels remains vital.

References

1. Jaquin, P., Augarde, C., Gallipoli, D., Toll, D.: The strength of unstabilized rammed earth materials. Géotechnique **59**, 487–490 (2009)
2. Taylor, P., Luther, M.: Evaluating rammed earth walls: a case study. Sol. Energy **76**(1–3), 79–84 (2004)
3. Paul, W., Taylor, P.: A comparison of occupants comfort and satisfaction between a green building. Build. Environ. **43**, 1858–1870 (2008)
4. Marco, M.: Energy analysis in buildings: a complementary approach to energy analysis. Ph.D. dissertation, KTH-Stockholm (2009)
5. Bui, Q.-B., Morel, J.-C., Venkatarama, R., Ghayad, W.: Durability of rammed earth walls exposed for 20 years to natural weathering. Build. Environ. **44**(5), 912–919 (2009)
6. Ciancio, D., Beckett, C., Carraro, J.: Optimum lime content identification for lime-stabilized rammed earth. Constr. Build. Mater. **53**, 59–65 (2014)
7. Gerad, P., Mahdad, M., McCormack, A., François, B.: A unified failure criterion for unstabilized rammed earth materials upon varying relative humidity conditions. Constr. Build. Mater. **95**, 437–447 (2015)
8. Gomes, M., Gonçalves, T., Faria, P.: Hydric behavior of earth materials and the effects of their stabilization with cement or lime: study on repair mortars for historical rammed earth structures. J. Mater. Civ. Eng. **28** (2016)
9. Bui, B., Morel, J., Hans, S., Walker, P.: Effect of moisture content on the mechanical characteristics of rammed earth. Constr. Build. Mater. **54**, 163–169 (2014)
10. Jaquin, P., Augarde, E., Gallipoli, D., Toll, G.: The strength of unstabilised rammed earth materials. Géotechnique **59**(5), 487–490 (2009)
11. Francoise, B., Palazon, L., Gerard, P.: Structural behaviour of unstabilized rammed earth constructions submitted to hygroscopic conditions. Constr. Build. Mater. **155**, 164–175 (2017)
12. Christopher, B., Daniela, C., Manzi, S., Bignozzi, C.: Strengthening mechanisms in cement-stabilized rammed earth. In: Ciancio, D., Beckett, C. (eds.) Rammed Earth Construction: Cutting-Edge Research on Traditional and Modern Rammed Earth, pp. 61–86. Taylor and Francis Group, London (2015)
13. Hall, M., Najim, K., Keikhaei, P.: Soil stabilisation and earth construction: materials, properties, and techniques. In: Modern Earth Buildings: Materials, Engineering, Constructions and Applications, pp. 222–255. Elsevier Ltd. (2012)

14. Hall, R., Djerbib, Y.: Rammed earth sample production: context, recommendations and consistency. Constr. Build. Mater. **18**, 281–286 (2004)
15. Hall, M., Djerbib, Y.: Moisture ingress in rammed earth: part I - the effect of soil particle-size distribution on the rate of capillary suction. Constr. Build. Mater. **18**, 269–280 (2004)
16. Bruno, A., Gallipoli, D., Perlot, C., Mendes, J.: Effect of stabilization on mechanical properties, moisture buffering, and water durability of hyper compacted earth. Constr. Build. Mater. **149**, 733–740 (2017)
17. Walker, P., Keable, R., Martin, J., Maniatidis, V.: Rammed Earth: Design and Construction Guidelines. BRE Bookshop, Watford (2005)
18. Otcovská, T., Mužíková, B., Padevět, P.: Mechanical properties of rammed earth with respect to clay mixture composition. Acta Polytechnica **59**(4), 372–383 (2019)
19. Ávila, F., Puertas, E., Gallego, R.: Characterization of the mechanical and physical properties of unstabilized rammed earth: a review. Constr. Build. Mater. **325**, 126693 (2022)
20. Walker, P., Keable, R., Martin, J., Maniatidis, V.: Rammed earth: design and construction guidelines. In: Watford, BRE Bookshop. BRE, Watford (2005)
21. Guettala, A., Abibsi, A., Houari, H.: Durability study of stabilized earth concrete under both laboratory and climatic conditions exposure. Constr. Build. Mater. **20**, 119–127 (2006)
22. Hall, M.: Assessing the environmental performance of stabilised rammed earth walls using a climatic simulation chamber. Build. Environ. **42**, 139–145 (2007)
23. Toufigh, V., Kianfar, E.: The effects of stabilizers on the thermal and the mechanical properties of rammed earth at various humidifies and their environmental impacts. Constr. Build. Mater. **200**, 616–629 (2019)
24. Heathcote, K.: Durability of earth wall buildings. Science **9**, 185–189 (2007)
25. Burroughs, S.: Soil property criteria for soil stabilization. J. Mater. Civ. Eng. **20**(3), 264–273 (2008)
26. Paul, J.: Analysis of historic rammed earth construction. Ph.D. dissertation, University of Durham (2008)
27. Van Stephan, B.: Soil property criteria for soil stabilization. J. Mater. Civ. Eng. **20**(3), 264–273 (2008)
28. Van Stephan, B.: Recommendations for the selection, stabilization, and compaction of soil for rammed earth wall construction. J. Green Build. **5**(1), 101–114 (2010)
29. Ciancio, D., Beckett, C., Carraro, A.: Optimum lime content identification for lime-stabilized rammed earth. Constr. Build. Mater. **53**(28), 59–65 (2014)
30. New Zealand Standard, NZS 4298:1998: Materials and workmanship for earth buildings (1998). https://www.standards.govt.nz/sponsored-standards/building-standards/nzs4298. Accessed 26 May 2022
31. Ghasemalizadeh, S., Toufigh, V.: Durability of rammed earth materials. Int. J. Geomech. **20**(11), 04020201 (2020)
32. Houben, H.: Guillaud, Earth Construction – A Comprehensive Guide, 2nd edn. Intermediate Technology Publications, London (1996)
33. Lertwattanaruk, P., Makul, N., Siripattarapravat, C.: Utilization of ground waste seashells in cement mortars for masonry and plastering. J. Environ. Manag. **111**, 133–141 (2012)
34. Hall, M., Allinson, D.: Assessing the effects of soil grading on the moisture content-dependent thermal conductivity of stabilised rammed earth materials. Appl. Therm. Eng. **29**, 740–747 (2009)
35. Hall, M., Djerbib, Y.: Moisture ingress in rammed earth: part 2 – the effect of particle-size distribution on the rate of static pressure-driven moisture ingress. Constr. Build. Mater. **20**, 374–383 (2005)

Evolutionary Approach Based on Thermoplastic Bio-Based Building Material for 3D Printing Applications: An Insight into a Mix of Clay and Wax

Yohan Jacquet(✉) and Arnaud Perrot

University of South Britany, UMR CNRS 6027, IRDL, 56100 Lorient, France
yohan.jacquet@univ-ubs.fr

Abstract. Digital manufacturing technique is becoming one of the most promising ways to automate construction. While applications of 3D printed concrete daily increase, those of bio-based or earth-based materials remain rather confidential at this time. The reduction of environmental footprint is one of the major 21th century issues for the construction industry. For this purpose, a now common strategy has progressively been adopted to design material mixes by reducing the amount of cement and sometime substitute this one by less impacting earth-based materials. Another strategy would consist in designing material with only bio-based and earth-based materials. Whereas the capability of earth-based material to sustain compressive load has been proven, two major challenges are remaining: the first is related to its water sensibility, while the second one deals with material drying kinetic. Common ways to build with earth-based material refer to vernacular techniques that leads to handling, insurability, and variability of materials issues. This study proposes to focus on a bio-earth-based material, mainly dedicated to 3D printing applications, that presents a thermo-dependant behaviour able to comply with industrial building rate, and to display a high potential to increase water-resistance whatever the raw resources variability, using a mix of clay and a low amount of vegetal wax. This paper aims to build the path towards thermoplastic material solutions for 3D printing in construction, targeting thus to automate, industrialize and insure 3D printed structures.

Keywords: 3D printing · thermoplastic · clay · wax · water-resistance

1 Introduction

Gradually forgotten and recently regaining interest, earth material represents today one of the most promising ways to reduce carbon footprint in the construction sector. Earth-based materials are often associated to mechanical poor properties and durability issues mainly because of raw material variable properties, the different vernacular existing techniques, and the high-water sensitivity. Stabilization of earth-based material is essential to enhance mechanical strength and water-resistance: a small addition of hydraulic blinder, such as cement, is a common way to stabilize this kind of material, increasing

S. Amziane et al. (Eds.): ICBBM 2023, RILEM Bookseries 45, pp. 271–279, 2023.
https://doi.org/10.1007/978-3-031-33465-8_21

at the same time its mechanical strength by up to four times compared to the non-treated material. However, using hydraulic binder in earthen material cannot be considered as an effective solution because of its huge impact on recycling possibilities and the associated carbon footprint [1]. Different kinds of natural additives can be extracted from plants, minerals or even animal and can be used to improve mechanical strength or water resistance of material [2–4]. Among the multiplicity of theses natural admixture, some of these can be encountered in traditional construction material recipes. It can be noted that these natural admixtures have an effectiveness highly dependent on the chemical composition of the media. The stabilization mechanisms can be correlated for instance to polymerization, ionic or hydrogen bonding. Making a focus on 3D printing applications, earth-based material differs by their flow behavior; while 3D printing of cement-based materials usually requires increasing the viscosity of the mixed material, earth-based one is usually too viscous to be pumped or extruded and need some mix adjustment to fulfil the requirement of this process. While cementitious materials often exhibit a frictional granular like behavior [5], the use of clay induces on the contrary a too sticky, and too viscous, behavior leading to high extrusion and pumping pressure. Clay can sometimes be advantageously added to mix design of 3D printed concrete mix design to ease extrusion process. One strategy to reduce viscosity of this clay-based mixture consists in an addition of non-clayey material to the mix design: in the same way than for concrete, the paste (exhibiting a clayey behavior) can be distinguished to the granular filler (lime, silt, sand, …). The amount of paste is thus related to its capability to create and mobilize viscosity. Another way is to control viscosity by controlling temperature of material during extrusion process. The viscosity of liquid always decreases when temperature increases: influence of this phenomenon can be balanced because the liquid phase in construction material is rather limited, reducing de facto the sensitivity of material to temperature change.

This study proposes to focus on phase-change thermoplastic material based on mix of clay, solid particles, and wax. Several kinds of vegetal wax are experimented in this study, but a same effect can be underlined: a very low content of wax allows to drastically decrease the stickiness induced by clay. This phase change behavior requires to fully reconsider the rheological test to evaluate the setting, or hardening, of cooling material. This paper proposes successively an overview on mix design possibilities, on rheological characterization, on mechanical assessment, and on the water-repellent aspect.

2 Materials

The phase-change thermoplastic material investigated in this study is based on mix of clay, sand, and wax. The choice of clay and sand is not the main purpose of this study, that's why only three kinds of mixes are considered. Three kinds of binders are used: the first one is mainly composed of kaolinite with a Methylene Blue Value (MBV) of 1.54, while the second one consists of a mix of clay and sea sediments, both from Brittany. The first one is envisioned to establish the proof of concept, while the second one aims to assess representativeness of the use of a given local resource and how a highly water-sensitive material can be upgraded in the construction field and thus become water-resistant by using very low amount of wax.

To go a step further, a last part of this paper investigates the influence of manufactured sand addition into the mix design regarding a particle size distribution between 0 and 1 mm, following various ratio between 0% and 70% in 10% increment. The goal is to maximize the amount of sand (or the amount of material unsensitive to water) to decrease the amount of sensitive material that must be protected (e.g. mud, sludge, clay, …), even without compromising the mechanical performances.

The origin of this wax can be mineral, animal or vegetal, but considering the amount of wax required and the potential negative impact on environment, special emphasis is given to the vegetal based ones (e.g. soy, rapeseed, sunflower, hemp, palm, or coconut), and more precisely to three of these: soy, rapeseed and coconut (see Table 1). Depending on its nature, the wax can exhibit different melting point usually from 30 °C to 90 °C.

Table 1. Description of melting points of three different common waxes

Wax	Coconut	Soy	Rapeseed
Melting point	30 to 35 °C	55 to 60 °C	80 to 85 °C

Vegetal waxes are obtained from hydrogenation reaction of a vegetal oil: the more oil is hydrogenated; the more stable the wax will be. Supposing that the provenance and the quality of raw material presents a low environmental impact, this process of hydrogenation can be responsible for the major carbon footprint of this mix design strategy. Amount of wax added into a mix design must be well-thought considering the context of use and the proportion of material in this mix effectively sensitive to water. In this way, it is important to recall that the addition of sand decreases de facto the amount of clay, highly sensitive to water, and allows thus to decrease the quantity of clay required to guarantee the wholeness of printed material. Different mass ratio of wax compared to the mass of clay are discussed in this study, but it is interesting at this point to note that a wax to clay ratio higher than 1% can improve the water-resistance of material effectively. For the sack of clarity, only the results obtained using soy wax will be presented in this study.

3 A Proof of Concept for Sustainable 3D Printing Applications

3.1 A Thermoplastic Behavior as a Way of Control Material Setting

Materials designed for building usually rely on setting mechanism such as for cement-based materials. The material printed in this study is based on a mix of clay and wax for which clay plays the rule of binder, and the wax is responsible of the thermoplastic behavior. Some aggregates can be mixed with this earth and bio-based paste to increase the building ability, to avoid shrinkage and to increase its compressive strength.

Considering the typical amount of wax used in this material (i.e. 2% ± 1%), the consequences on hardened material properties can be neglected. However, concerning the rheological behavior, influence of wax content must be assessed because of the huge changes that occurs. The melting point of wax being reached, this kind of material indeed presents an important slipping at the wall of the characterization device: a Couette geometry for example in this study.

Rheological methods have been adapted and deals with 22 mm in diameter and 40 mm in height vane geometry, combined with a customized 40 mm diameter cup within which 1 cm length fins are spread out the cup wall to avoid slippage after heating (see Fig. 1). The main concern is related to the difference between the heat capacity and thermal conductivity of clay and wax respectively, as well as their specific dependency upon water content. Material is heated 10° higher than the melting point of wax and mixed several times during temperature increase to ensure homogeneity inside the material. A thermocouple sensor is then introduced into material during the rise of temperature and until the beginning of shear test to control and manage thermal control of Peltier system. When the setpoint temperature is reached, shear test is started according to a shear rate of 0.1 s^{-1}.

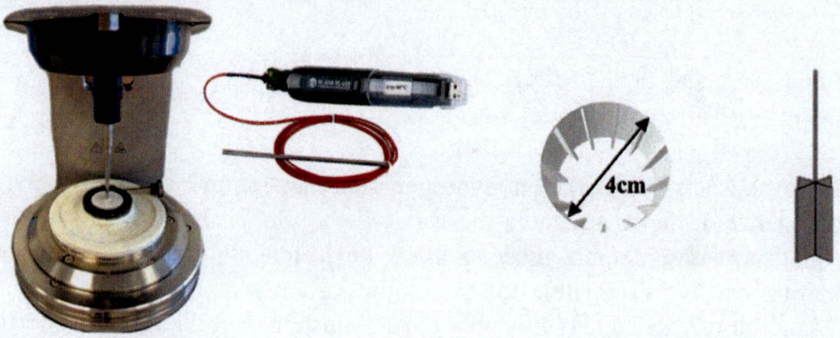

Fig. 1. Cup with ribbed wall surface, vane tool geometry, and temperature sensor

Assessment of cement-based materials are usually made in time to evaluate hydration rate influence of cement. In the same way, evolution in time of the thermal setting of the mix of clay and wax must be assessed. The cooling of this mix is mainly governed by its heat capacity and the thermal exchanges with the ambient environment, so from a temperature of 70 °C to and ambient one near to 25 °C.

The rheological method to characterize this thermal setting consists in a measurement of shear strength of material after different resting time. The correlation between this resting time and the temperature of the mix is represented in the background of the Fig. 2. Each measuring point plotted in this graph is obtained thanks to independent tests, or about ten points to analyze the influence of one hour of thermal setting. This test is performed on two different materials: the first with a water to clay mass ratio of 35%, while the second one is performed regarding a ratio 10% higher. The difference between thermal dissipation of these two mixes is negligible (lower than 1% in this case), that's why only one curve of the evolution of temperature in time is proposed.

After the first five minutes of cooling, the material shear strength is multiplied by three for the material with the lower solid volume fraction (i.e. water to clay ratio of 45%), while for the other one, the shear strength is multiplied by four to reach a value near to 2.6 kPa at 40 °C. All other things remaining equal, only the amount of water changes the thermal dependency of the material's behavior.

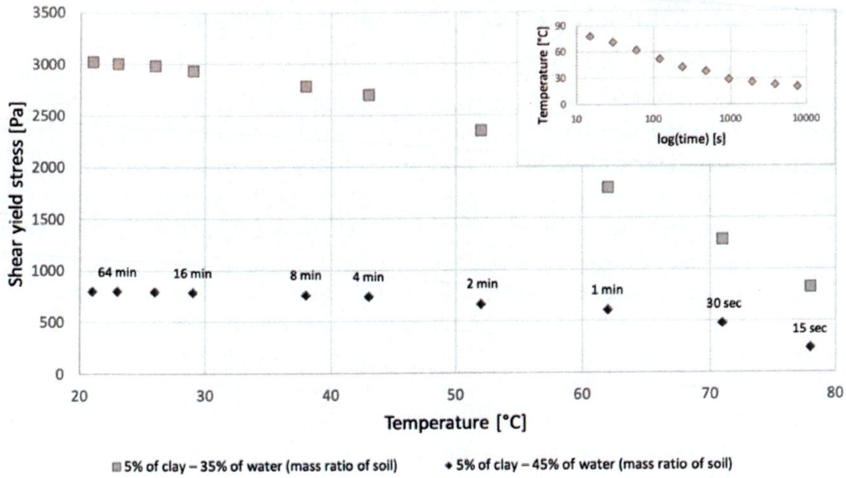

Fig. 2. Evolution of shear yield stress with the temperature of mix design

Over 5 min of cooling, the material that presents the highest solid volume fraction exhibits shear strength which is not progressing any longer. However, for the lowest one, the shear strength increases by 15% for about one hour of cooling. For 3D printing process, a short time of setting is preferred to enhance buildability. This setting time is not sufficient to ensure mechanical performances and only allows to increase the printing rate compared to common raw earth-based materials: drying of material still being a necessary condition to provide the intended final mechanical properties to the printed structure.

Finally, it has been shown the localization of wax on the walls when heating material makes difficult to perform rheological measurement without wall slipping, but, in return, allows to drastically decrease stickiness usually induced by clay. This sticky behavior is one of the major challenges for printing and/or pumping earth-based materials. By the same way, when this heated mix of clay and wax is extruded, shear stress decreases significantly, especially at the extruder die, improving thus the surface quality (see Fig. 3).

Fig. 3. Printed sample of clay and soy wax with 10 × 30 cm cross section extruder die

3.2 Assessment of Water-Resistance from Mechanical Performances

In addition to the thermo-dependent behavior, the major interest of this material is based on its ability to resist when exposed to water. We can distinguish two typical behaviors of materials: the water repellency and the water-resistance. For the first one, material can be exposed to water while rendering it impervious to water incursion for a limited period according to the natural water repellency power of raw earth material: in the field of earthen construction, this kind of behavior can be often encountered in vernacular technics. The second one consists in adding some bio-sourced or earth-based materials to enhance the water-resistance potential. Both technics are fully adapted to the earthen construction, but the respect of the standards of professional practices is crucial to limit direct exposure to water and ensure structure durability, while the other one can be favorably envisioned for different and more complex shapes when a physical protection against water cannot be totally guaranteed.

In this study, the material can be considered as highly water-resistant, especially for an earth-based one. A low dosage of wax leads indeed to a huge increase of water-resistance potential when samples are entirely immersed in water. For this purpose, a choice has been made to assess the material ability to resist water by measuring the degradation rate of different immersed samples.

This test is intended to be the simplest as possible and consists in immerging printed samples into tap water at 20° to estimate their performances when submitted to accidental water exposure. The compressive strength of these sample is measured depending on this time of direct exposure to water: in this study, tests are performed at 30 min and 12 h after the first contact of dried sample with water for each mix design respectively. Compressive tests are made on sample regarding a cross section near to 1600 mm^2 with a of 2:1 aspect ratio: depending on the water content, shrinkage can occur, and the cross section can be thus updated to increase computation accuracy of compressive strengths. Figure 4 shows the ratio between the compressive strength of a control sample printed and stored in controlled conditions (25 °C and between 5 and 10 h%), and the compressive strength of a sample immersed in water for 30 min and 12 h respectively. Half an hour after immersion, compressive strength is reduced by more than 90% without any wax in the mix and is destroyed 12 h after. Only 2% of soy wax into this mix design are sufficient to obtain a compressive strength decreased by 30%. Once wax is added into a mix design, the increase of immersion time always leads to an adverse effect on mechanical properties. The concept of accidental exposure to water ingress must be defined to push for a suitable dosage for human safety.

3.3 Mix Design Optimization and Outcomes

Two different strategies are envisioned in this study to reduce environmental impact of this material intended to be used for additive manufacturing purpose. The first one consists in optimizing mix design by adding manufactured sand (granular size distribution between 0 and 1 mm), while the second one aims to use and recover only local raw resources, regardless of their physical or chemical properties (water resistance, MBV, stickiness, viscosity, and so on).

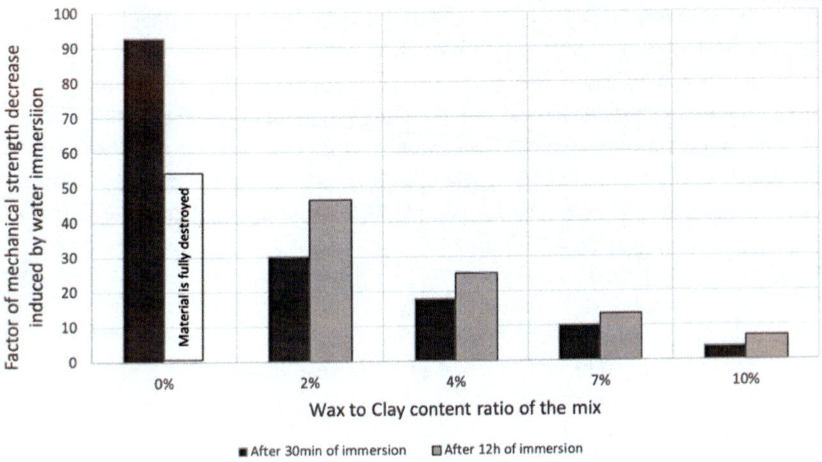

Fig. 4. Influence of immersion of samples on compressive strengths measurements

Based on the same mix design than the proof of concept exposed earlier in this study, a sand volume fraction is adjusted in 10% between 0% and 70% compared to the volume of binder (kaolinite, soy wax, and water). This protocol aims to investigate effect of addition of water-unsensitive solid particles on compressive strength. The material only composed of kaolinite following a water content of 45% exhibits a compressive strength near to 800 kPa. Figure 5 shows the evolution of this strength with addition of sand until a maximum value of 1.2 MPa for a sand ratio ranging between 50 and 60%. This means that only 50 to 40% of this mix design presents a poor water resistance and must be protected from water ingresses.

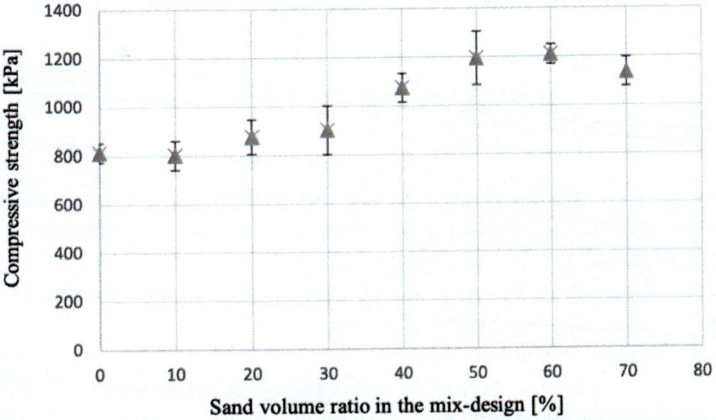

Fig. 5. Effect on compressive strength of the increase of sand volume fraction in a thermoplastic binder composed of clay, wax, and water.

The second strategy consists in a material based on a mix of local clay and sea sediments from Brittany. Both being sensitive to water ingresses, soy wax ratio of 1% is defined compared to the amount of clay and sediments. The first sample is only composed of clay (C) and presents a compressive strength of 2.35 MPa. With the addition of sediments comes an increase in mechanical strength until an optimal addition ratio of 20%, which could be explained by the optimization of granular packing. Over 70% of sediment, these mechanical properties decrease because of the poor particles bond properties of sediments compared to those of clay (see Fig. 6).

In any case, considering the water sensitivity of these both raw and by-products materials respectively, the wax content must be maintained at the same mass ratio value of at least 1%. Using by-product such as sediments sludge, can be a means to reduce environmental footprint of mix design and preserve resources.

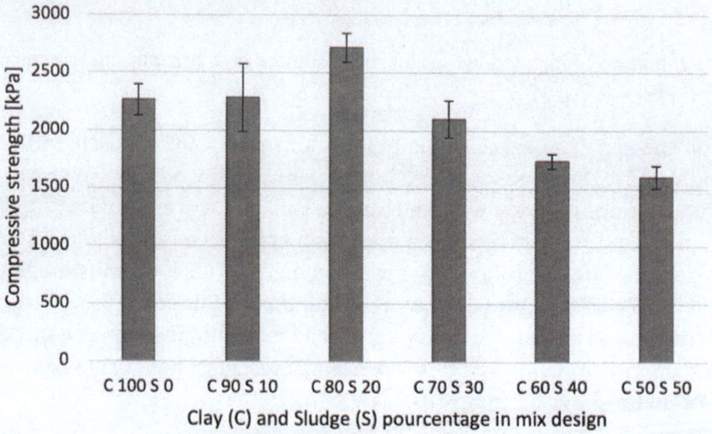

Fig. 6. Mixes of local resources regarding varying proportions (clay, sediments and soy wax)

4 Conclusion

This paper aims to propose a proof of concept of a new kind of earth-based material intended to be used for 3D printing applications in the field of construction industry. Insurability and buildability represent the two major challenges in the field of additive manufacturing of earth-based materials, especially because of their difficulties to reach an industrial building rate and their water sensitivity.

Firstly, the addition of a low amount of wax into a mix design of bio-reinforced earth materials allows to increase the acceptability of earth-based materials because of the high water-resistance properties of the dried material products. Secondly, thanks to various thermo-dependent properties of wax, thermal setting can be envisioned to pave the way toward industrial building rate during printing. The thermal setting provided by wax cannot be a substitute to drying and only aims to increase printability and buildability of printed earth-based structures.

In addition to these both main points, this thermo-dependent behavior results in two benefits; the first is related to the ability of heated wax to eliminate stickiness of clay regardless their initial viscosity, while the second provides a thermo-dependent behavior of a printed wall depending on the environmental conditions.

This last point allows to envision some outcomes dealing with thermos-hydric exchanges through materials porosity: depending on the ambient temperature and melting point of vegetal wax, pore size distribution can vary locally. High temperature can be responsible of a decrease pore surface tension, that therefor favors water vapor exchanges trough the walls and vice-versa. This thermo-dependent capillary network can be interesting both for housing and for solving urban heat island issues for instance.

Bibliography

1. Van Damme, H., Houben, H.: Earth concrete. Stabilization revisited. Cem. Concr. Res. **114**, 90–102 (2018). https://doi.org/10.1016/j.cemconres.2017.02.035
2. Guihéneuf, S., Rangeard, D., Perrot, A., Cusin, T., Collet, F., Prétot, S.: Effect of bio-stabilizers on capillary absorption and water vapour transfer into raw earth. Mater. Struct. **53**(6), 1–18 (2020). https://doi.org/10.1617/s11527-020-01571-z
3. Losini, A.E., Grillet, A.C., Bellotto, M., Woloszyn, M., Dotelli, G.: Natural additives and biopolymers for raw earth construction stabilization – a review. Constr. Build. Mater. **304**, 124507 (2021). https://doi.org/10.1016/j.conbuildmat.2021.124507
4. Achenza, M., Fenu, L.: On earth stabilization with natural polymers for earth masonry construction. Mater. Struct. **39**(1), 21–27 (2007). https://doi.org/10.1617/s11527-005-9000-0
5. Jacquet, Y., Perrot, A., Picandet, V.: Assessment of asymmetrical rheological behavior of cementitious material for 3D printing application. Cem. Concr. Res. **140**, 106305 (2021). https://doi.org/10.1016/j.cemconres.2020.106305

On the Bonding Characteristics of Clays and Biopolymers for Sustainable Earthen Construction

Rebecca A. Mikofsky(✉), Samuel J. Armistead, and Wil V. Srubar III

University of Colorado at Boulder, Boulder, CO 80309, USA
remi8173@colorado.edu

Abstract. Sustainable earthen building materials provide a pathway to mitigating the environmental impacts of the modern construction sector. While the application of these materials has been limited due to the inherent heterogeneity, erosivity, and weak mechanical properties of soil, the physical and thermal properties can be improved through the addition of ubiquitous, non-toxic, sustainable biopolymers. Yet, the fundamental understanding of the physiochemical bonding mechanisms between clays and biopolymers in this system is limited. In this work, a 'micro to macro' methodological approach was applied to investigate the bonding characteristics of common clays and clay-stabilizing biopolymers. At the micro-scale, fundamental interactions of clays (*i.e.,* kaolinite, bentonite) with biopolymer additives (*i.e.,* xanthan gum, guar gum, sodium alginate, microcrystalline cellulose) were assessed through mineral binding characterization techniques, including Fourier-transform infrared spectroscopy (FTIR) and thermogravimetric analysis (TGA). The findings were used to interpret unconfined compressive strength (UCS) tests results for macro-scale soil-biopolymer composites samples (1% biopolymer by soil mass). The results from this study illustrate the utility of understanding the mechanisms of clay-biopolymer interactions for improving the design of strong and durable earthen materials and structures.

Keywords: Earthen materials · Soil stabilization · Biopolymer · Clay · Binding affinity

1 Background

Earthen materials present an opportunity to reduce the high environmental impact of building construction by replacing conventional structural building materials such as concrete. Earthen materials and assemblies (*e.g.,* cob, rammed earth, adobe) are locally sourced and require minimal processing, therefore mitigating the high temperatures and transportation energy costs typically associated with traditional building materials. In addition to their low environmental impacts, these materials have been shown to have valuable material properties (*e.g.,* thermal, hygroscopic) and practical characteristics (*e.g.,* low cost, global availability). Despite these promising qualities, earthen materials often exhibit heterogeneity, erosivity, and weak mechanical properties, preventing their widespread use [1, 2].

S. Amziane et al. (Eds.): ICBBM 2023, RILEM Bookseries 45, pp. 280–292, 2023.
https://doi.org/10.1007/978-3-031-33465-8_22

Current research in geotechnical, civil engineering, and architectural design has shown that biopolymer additives can be used to improve the physical and chemical properties of soil and earthen materials. Biopolymer additives are non-toxic, low cost, and derived from naturally occurring sources. While many biopolymers have been investigated *via* trial and error ('top down' exploration), there is a fundamental lack of understanding of the molecular interactions responsible for the binding (or lack thereof) between soils and biopolymers. This study investigates these molecular interactions through a 'micro to macro' approach, in order to establish selection criteria for optimal soil-biopolymer composites.

Earthen materials are typically composed of subsoil containing a majority of sand 'aggregates' (50% or less) with a minority of cohesive clay 'binders'. According to the literature, soil for earth construction should include 15–30% clay content [3]. Clay minerals are classified as aluminum phyllosilicates with layers of octahedral alumina oxide and tetrahedral silica oxide sheets. The main categories of clays are illite, smectite, and kaolinite clays based on the arrangement of these sheets (2 tetrahedral : 1 octahedral, 2:1 and 1:1, respectively). Within the smectite category, bentonite (also known as montmorillonite) is the most commonly found variety and is highly expansive due to its 2:1 structure, which allows water to bind the groups of Si-Al-Si layers. Within the kaolin category, kaolinite is the most commonly available kaolin clay mineral and has a more tightly bound 1:1 structure. Bentonite and kaolinite were chosen in this work to serve as model clay systems. Clay (15% by weight) and sand (85% by weight) mixtures were designed to function as model earthen material systems for structural material property testing.

Table 1. Biopolymer Properties.

Biopolymer	Source	Stability	Water Solubility	Charge
Cellulose	plants	excellent thermal	mostly insoluble	non-ionic
Guar Gum	seeds	excellent	cold	non-ionic
Sodium Alginate	algae	limited	cold	anionic
Xanthan Gum	microbes	excellent	cold	very anionic

Biopolymers represent a broad category of biologically derived organic materials. Biopolymers are extracted from a variety of sources (*e.g.*, plant, bacteria, algae, fungi) and exhibit a wide range of chemical and physical properties. Polysaccharides, for example, are highly abundant in nature and are currently employed in large-scale commercial use (*e.g.*, food, packing, tissue engineering). They contain a plethora of chemically distinct monosaccharide monomer units linked through glycosidic bonds to create linear, branched, and network biopolymers. Four polysaccharides (*i.e.*, cellulose, sodium alginate, guar gum, and xanthan gum) were selected for this study based on their differing properties summarized in Table 1. Cellulose (C) is a linear polymer of β-(1 $\rightarrow$ 4)-linked D-glucose. The linear structure of cellulose causes it to become crystalline and insoluble in water. Guar gum (GG) has a β-(1 $\rightarrow$ 4)-linked D-mannose backbone with an α-(1 → 6)-linked D-galactose side chain on alternating mannose units [4–6]. Guar gum is nonionic and stable across a large pH range [4–6]. Sodium alginate (SA) is a sodium salt of alginic acid, which is (1 → 4)-linked β-D-mannuronic acid and α-L-guluronic acid [7]. Sodium alginate is anionic and sensitive to pH, affecting viscosity and solubility [7]. Xanthan gum has a backbone of β-(1 → 4)-linked D-glucose backbone with trisaccharide (glucuronic acid, mannose, mannose) side-chains on alternating glucose units [8]. Xanthan gum is negatively charged and highly soluble due to its branched structure and side chain units [8]. These biopolymers represent a range of properties including different charges, different sources, different impacts due to change in pH, and different monosaccharide subunits.

The aim of this study was to explore the molecular interactions between two clay minerals, bentonite and kaolinite, and four biopolymers, namely cellulose, guar gum, sodium alginate, and xanthan gum. Doing so uncovered the chemical interactions of these biopolymers (or lack thereof) with common clay minerals. Mineral binding characterization methods developed in Armistead *et al.* (2020) were employed using thermogravimetric analysis (TGA) and attenuated total reflection Fourier-transform infrared (ATR-FTIR) spectroscopy [9]. Following micro-scale investigations, the mechanical properties of biopolymer-stabilized model earth systems were characterized using unconfined compressive strength (UCS) testing. By performing a 'micro to macro' approach, this study aimed to identify which of these biopolymers bind to specific clay minerals and which would be effective as property-improving additives for earthen materials.

2 Materials and Methods

2.1 Materials

Ultra-pure water was obtained from a Milli-Q purification system. Kaolinite (K) and bentonite (B) clay minerals were obtained from Sigma Aldrich. Ottawa sand (80% passing 30–40 mesh, resulting in a 0.42–0.6 mm particle size) was obtained from VWR Chemicals BDH. Guar gum from *Cyamopsis tetragonoloba* seeds, xanthan gum from *Xanthomonas campestris* bacteria, sodium alginate salt from brown algae, and microcrystalline cellulose were purchased from Sigma Aldrich. Molar mass was determined using average molar mass for the polymeric sub-units giving 162.14, 180.16, 176.12, 204.56 g/mol for cellulose (C), guar gum (GG), sodium alginate (SA), and xanthan gum (XG) [9, 10]. Commercial 37% Hydrogen Chloride and LabChem 25% Ammonium Hydroxide were used for pH adjustments.

2.2 Clay Characterization

Clays were dried in a furnace at 110 ± 10 °C for 5 h, pestled in small batches for 1 min, then sieved to obtain particles < 45 microns using a Gilson SS-3 sieve shaker. The median particle size (d50) was measured to be 4.75 μm for B and 8.66 μm for K using a Malvern Panalytical Mastersizer 3000 with a HydroMV dispersion accessory (Fig. 1). Stirring (2500 rpm) and sonication (50% power) were used to disperse clay in water and achieve a laser obscuration (5–15%) based on the particle size range.

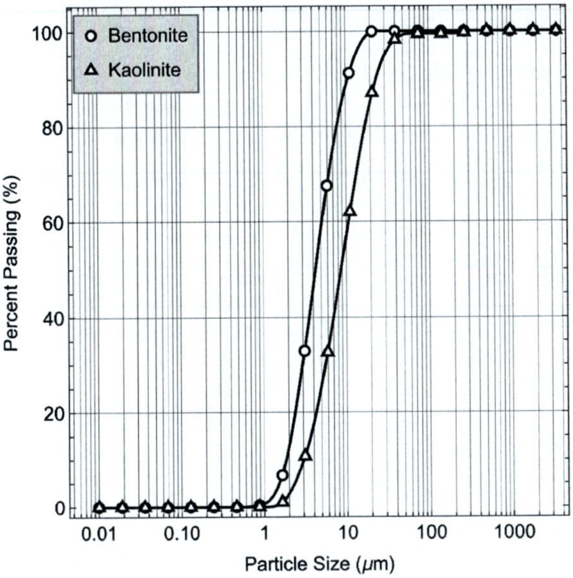

Fig. 1. Particle size distribution for bentonite and kaolinite.

Clay compositions were confirmed using X-ray diffraction spectroscopy (Bruker D8 single-crystal X-ray diffractometer) with a step size of 0.019, 2 s/step, and a 0.02 mm divergence slit. Clay samples for XRD were prepared by suspending clay in water and drying in air on a quartz slide. Diffractograms showed that the bentonite clay sample included montmorillonite, quartz, and plagioclase feldspar (Fig. 2a), and the kaolinite clay sample included kaolinite and quartz (Fig. 2b) [11–13]. ATR-FTIR confirmed these results [14, 15].

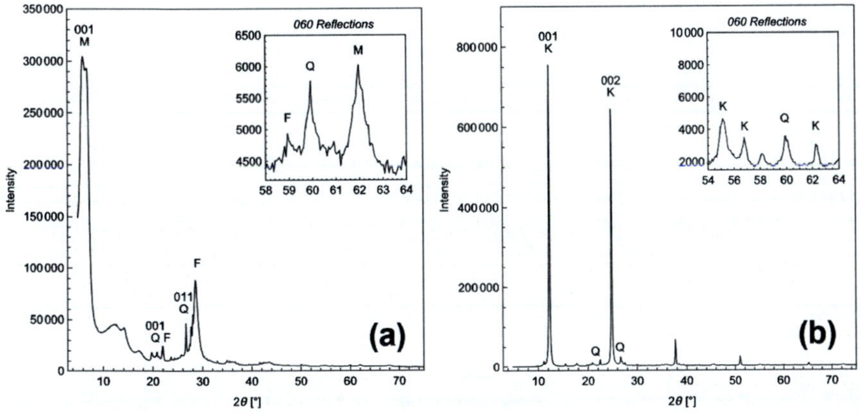

Fig. 2. XRD analysis for (a) bentonite and (b) kaolinite. M = montmorillonite, Q = quartz, F = feldspar, K = kaolinite.

2.3 Mineral Binding Characterization

Mineral binding characterization (MBC) was performed in accordance with Armistead *et al.* [9]. Compositions for each mineral binding characterization sample is listed in Table 2. Biopolymer solutions (0.01 M) were first prepared by slowly adding biopolymer powder to heated (40 °C) ultra-pure water whilst under simultaneous agitation (stir bar, 600 rpm). Then, the solution was held at 40 °C for 10 min with continuous stirring. The solution (20 mL) was poured into centrifuge tubes (50 mL) and ultrasonicated (Branson 2510) for 10 min. Solutions were used within 24 h of preparation and stored away from light. Clay (50 mg) was added to described biopolymers solutions, followed by ultrasonication for 10 min. Then, pH was adjusted to ±1.0 of the desired pH value

Table 2. Sample Composition for MBC Testing

Sample	Clay	Biopolymer	pH
B-C-5	Bentonite (50 mg)	Cellulose (20 mL 0.01M)	5
B-C-7			7
B-C-9			9
B-GG-5		Guar Gum (20 mL 0.01M)	5
B-GG-7			7
B-GG-9			9
B-SA-5		Sodium Alginate (20 mL 0.01M)	5
B-SA-7			7
B-SA-9			9
B-XG-5		Xanthan Gum (20 mL 0.01M)	5
B-XG-7			7
B-XG-9			9
K-C-5	Kaolinite (50 mg)	Cellulose (20 mL 0.01M)	5
K-C-7			7
K-C-9			9
K-GG-5		Guar Gum (20 mL 0.01M)	5
K-GG-7			7
K-GG-9			9
K-SA-5		Sodium Alginate (20 mL 0.01M)	5
K-SA-7			7
K-SA-9			9
K-XG-5		Xanthan Gum (20 mL 0.01M)	5
K-XG-7			7
K-XG-9			9

(*i.e.*, 5, 7 or 9) using a few drops of 0.5 M HCl or 0.1 M NH_4OH. These pHs were chosen to investigate sample stability or changes in binding affinity, through a range of pH values. Samples were mixed with a Labnet Mini LabRoller Rotator for 30 min. Samples were washed four times to remove unbound biopolymer. For the washing procedure, the sample was first centrifuged at 4000 rpm for 10 min. Then, the supernatant was removed, and 20 mL of ultra-pure water was added. The process was repeated four times for a total of five centrifuge cycles. After final centrifuging and supernatant removal, samples were left to dry in air at room temperature. After drying, samples were characterized using TGA and ATR-FTIR.

TGA (TA Discovery TGA 5500) was performed in N_2 gas from 50 to 800 °C with a ramp rate of 5 °C/min. Percent mass loss was calculated from 200 to 400 °C in the range in which the combustion of organic materials is most likely to occur. ATR-FTIR (Thermo Scientific Nicolet iS20 with a Ge Crystal ATR accessory) was performed using 64 scans, 4 cm^{-1} resolution. For comparison, pure clay was measured using TGA and FTIR, and air-dried biopolymer samples (0.1 M) were measure using FTIR.

2.4 Compressive Strength Testing

Cubic samples were created for compressive strength testing (Table 3). Model earthen materials were created by mixing clay (15 wt_{soil} %) with Ottawa sand (85 wt_{soil} %). First, dry biopolymer powder (1 wt_{soil} %) was mixed with the clay-sand mixture. Then, ultra-pure water was added achieve a moldable workability for the K-(22.5 wt_{soil} %) and B-containing mixtures (45 wt_{soil} %) and thoroughly mixed. Each clay-sand-biopolymer mixture was placed into 1.25 in (31.75 mm) cubic silicone molds. All samples were prepared at the same time and left to cure in an ambient environment. The surface was

Table 3. Sample Composition for Mechanical Testing

Sample	Clay (wt_{clay} /wt_{soil})	Sand (wt_{sand} /wt_{soil})	Biopolymer (wt_{bp} /wt_{soil})	Water (wt_{water} /wt_{soil})
B-NONE	15% Bentonite	85% Ottawa sand	-	22.5% water
B-C			1% Cellulose	
B-GG			1% Guar Gum	
B-SA			1% Sodium Alginate	
B-XG			1% Xanthan Gum	
K-NONE	15% Kaolinite	85% Ottawa sand	-	45% water
K-C			1% Cellulose	
K-GG			1% Guar Gum	
K-SA			1% Sodium Alginate	
K-XG			1% Xanthan Gum	

smoothed using a spatula, and the samples were left to dry in air until firm enough to be pressed out of the mold (at least 24 h). Mechanical testing on triplicate samples was performed 96 h after casting, when the samples achieved constant mass (fully dried). Unconfined compressive strength was measured using an Instron Universal Testing machine with a 5 kN load cell and 1.5 mm/min displacement-controlled rate. Maximum compressive strength was obtained for each sample.

3 Results and Discussion

3.1 Mineral Binding Characterization

TGA results, shown in Fig. 3, illustrate that all samples exhibited a characteristic loss in mass with increasing temperature. Mass loss between 200–400 °C indicates the presence of biopolymer and higher biopolymer-clay affinity. For both bentonite (B) and kaolinite (K), the addition of cellulose resulted in the highest mass loss, with B-C pH 7 having the highest mass loss overall for bentonite (26.98%) and K-C pH 9 having the greatest mass loss for kaolinite (13.79%). Guar gum resulted in the second highest mass loss for both bentonite (11.67% for pH 5) and kaolinite (9.58% for pH 5). Values for sodium alginate and xanthan gum (< 1.1%) suggest that negligible quantities of those biopolymers bound to the clays.

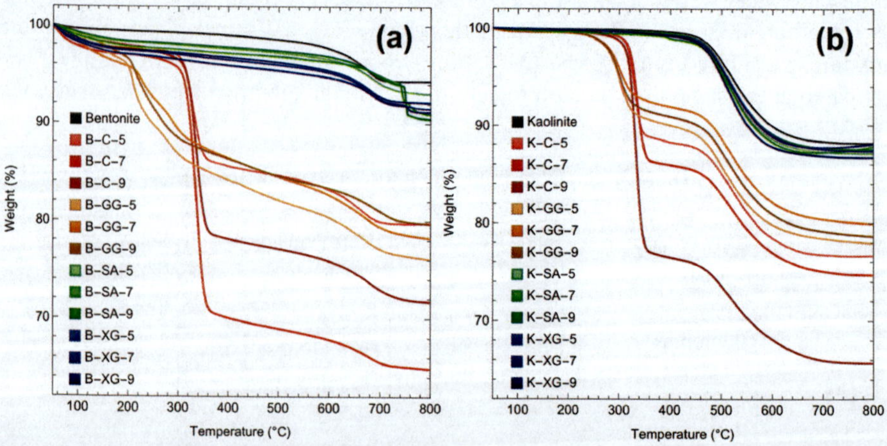

Fig. 3. TGA results for (a) bentonite- and (b) kaolinite-biopolymer mixtures. B = bentonite, K = kaolinite, C = cellulose, GG = guar gum, SA = sodium alginate, XG = xanthan gum.

All FTIR spectra are shown in Fig. 4. The comparison of FTIR spectra of pure clay, clay-biopolymer, and pure biopolymer samples could indicate the presence of biopolymer bound to the clay after washing. When analyzing bentonite and kaolinite control samples, a set of peaks was observed between 600–1200 cm^{-1}, attributable primarily to the Si-O stretching in kaolinite and bentonite [14, 15]. Within this region, peaks due to the octahedral oxide sheets (*e.g.*, AlMg-OH, AlFe-OH, AlAl-OH) or OH of inner hydroxyl groups in kaolinite can occur depending on the specific clay sample [14, 15]. Therefore changes to the 1200–1800 cm^{-1} and 2800–3800 cm^{-1} regions were investigated for evidence of biopolymer binding. These regions contain peaks characteristic of polysaccharides: C-H bending (1330–1470 cm^{-1}), COO- asymmetric stretching (weak at 1400 cm^{-1}, strong at 1600 cm^{-1}), and O-H stretching (3000–3500 cm^{-1}) [16, 17].

FTIR results for the bentonite-guar gum samples (Fig. 4b) indicate biopolymer presence within the 1200–1800 cm^{-1} and 2800–3800 cm^{-1} regions. However, these changes from the pure bentonite FTIR spectra can be seen in all bentonite-biopolymer samples. Notably, a sharp peak can be observed at 1630 cm^{-1} and a broad peak at 3300 cm^{-1}, which are associated with moisture in the bentonite structure [18]. This result is an indication that physiosorbed water, and perhaps biopolymer, is still present within the samples. This concept will be explored in further studies.

The FTIR results corroborate the TGA results that also indicated the presence of bound guar gum in the bentonite and kaolinite samples. While the TGA results suggested that more cellulose was bound to the clays than guar gum, the FTIR results in Fig. 4a suggest that cellulose does not chemically bind to kaolinite. These findings suggest (a) that the cellulose may physisorb onto clay surfaces through van der Waal's forces rather than chemically bond and/or (b) the cellulose was not removed during the washing cycles due to its insolubility in water.

For kaolinite samples (Fig. 4e-h), in line with TGA results, the introduction of new peaks can be observed within guar gum spectra, an indication of chemisorption between the two components. This result is consistent with the TGA results, which also indicated high biopolymer affinity. In contrast, considering sodium alginate, xanthan gum and cellulose additives imparted little change in the FTIR spectra throughout the pH range indicating little bonding to kaolinite.

One key factor to inform whether physisorption or chemisorption is occurring is to interpret these results through surface charge. Bentonite has an isoelectric point of 2.5 pH, meaning it will be negatively charged across the pH range investigated (5–9) [19, 20]. Similarly, kaolinite has a negative zeta potential in the range, but a higher isoelectric point of 3.5–4.6 pH [19, 21]. As noted in Table 1, guar gum is the only biopolymer that is neutral, highly soluble and branched. These properties, as shown in the results, allow it to bind more readily with the negatively charged surface of the clays than the negatively charged sodium alginate or xanthan gum. While cellulose is similarly nonionic, it does not bind due to its linear, insoluble nature, preventing binding interactions with the clay surface.

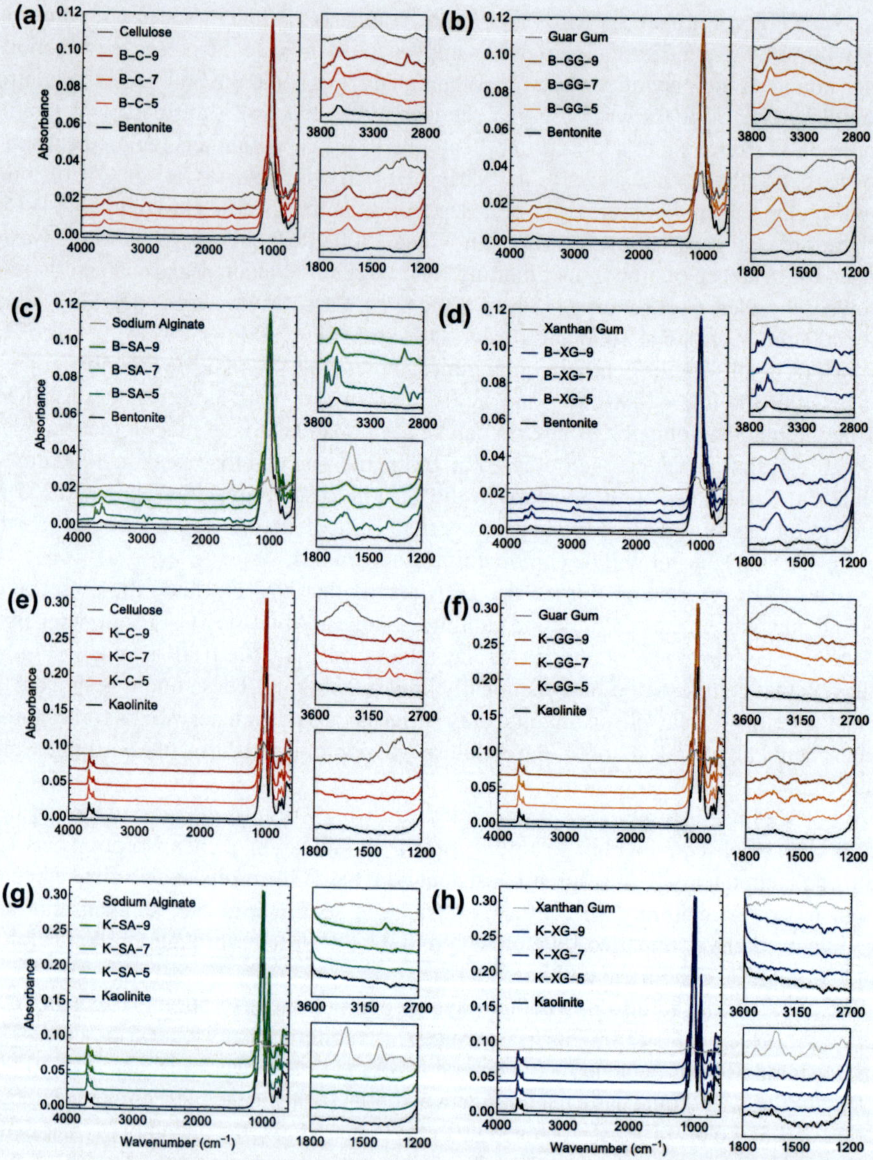

Fig. 4. ATR-FTIR of (a-d) bentonite and (e-h) kaolinite-biopolymer samples. B = bentonite, K = kaolinite, C = cellulose, GG = guar gum, SA = sodium alginate, XG = xanthan gum.

3.2 Compressive Strength Testing

Unconfined compressive strength testing was performed on sand-clay-biopolymer samples 96 hr after casting. Figure 5 shows stress strain curves (with slack removed) for all clay-sand-biopolymer samples. The curves indicate that the choice of biopolymer influences both stress and strain properties.

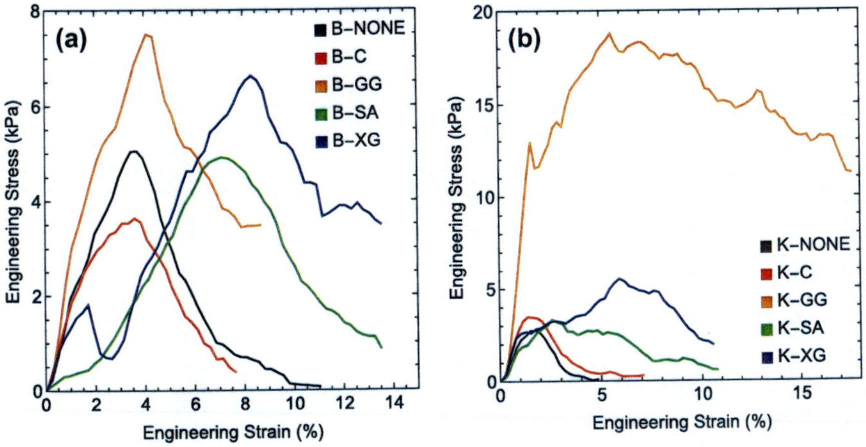

Fig. 5. Representative engineering stress-strain curves for Bentonite (a) and Kaolinite (b) samples with slack removed from loading curve.

Figure 6 shows maximum compressive strength. These results indicate that cellulose (C) and sodium alginate (SA) do not show a significant improvement in strength. Data suggests that these biopolymers neither bind to clays or sand nor do they undergo sufficient cross-linking to impart additional property changes to the soil. The addition of guar gum, in contrast, showed a 512% improvement for kaolinite-sand samples and 136% for bentonite-sand samples. These results are consistent with increased binding affinity to bentonite and kaolinite as measured *via* TGA for both bentonite and kaolinite and FTIR for kaolinite. Xanthan gum exhibited a 182% improvement for K-containing samples and the greatest increase in strength for bentonite-containing samples (152%).

It is hypothesized that xanthan gum has the greatest strength increase for bentonite-sand samples due to its high viscosity at low concentrations. The FTIR results indicate that there is some binding with clay, so while the binding affinity is lower than guar gum, it could have a greater than or similar due to the increased viscosity and the pseudoplastic properties of xanthan gum. For kaolinite-sand samples, xanthan gum resulted in a similar increase showing that it is not forming significant interactions with either kaolinite or bentonite but rather it is the basic properties of xanthan gum as a binder. Further experiments need to be performed to differentiate these effects and investigate the impacts of water and time on strength.

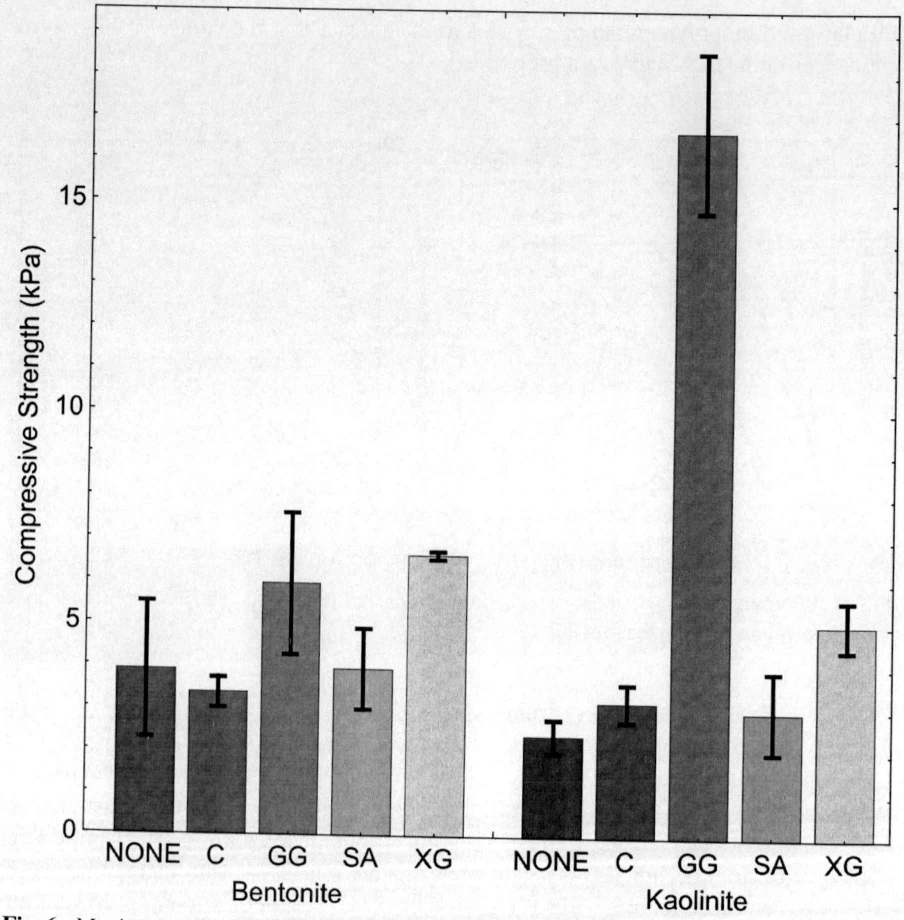

Fig. 6. Maximum compressive strength values sample for each clay. B = bentonite, K = kaolinite, NONE = no biopolymer, C = cellulose, GG = guar gum, SA = sodium alginate, XG = xanthan gum.

4 Conclusion

This paper presents an ongoing study on clay-biopolymer binding characteristics with the aim to enhance the physical properties of soils for construction. Earthen materials, a critical sustainable alternative for conventional materials such as concrete, exhibit superior environmental features but are not fully integrated in construction due to their inherent variability and weak mechanical properties. Using a 'micro to macro' approach, this research investigates the bonding characteristics of bentonite and kaolinite using four biopolymer additives: xanthan gum, guar gum, sodium alginate, and microcrystalline cellulose. The characterization includes Fourier-transform infrared spectroscopy (FTIR), thermogravimetric analysis (TGA), and unconfined compressive strength testing (UCS). The results show that kaolin-guar gum outperforms other clay-biopolymer mix designs in terms of its mineral binding characterization analysis and mechanical testing. This finding is in line with theoretical knowledge based on surface charge of guar gum and kaolinite.

These preliminary results need to be verified using additional testing. Further experiments would include investigating the effect of different pH values and biopolymers, and testing the synergy of more than one biopolymer in combination with others. Methodological changes would include increasing FTIR signal for control biopolymers to improve comparability and integrating steps into the preparation procedure to ensure any physisorbed or insoluble biopolymer is removed.

Acknowledgements. This research was made possible by the Department of Civil, Environmental, and Architectural Engineering, the Materials Science and Engineering Program, the College of Engineering and Applied Sciences, and the Living Materials Lab at the University of Colorado Boulder. The editorial comments and suggestions received from Dr. Lola Ben-Alon at Columbia University are gratefully acknowledged. This work was supported by the National Science Foundation (Grant No. CMMI-2134488). Any opinions, findings, and conclusions or recommendations expressed in this material are those of the authors and do not necessarily reflect the views of the National Science Foundation.

References

1. Ben-Alon, L., Loftness, V., Harries, K.A., DiPietro, G., Hameen, E.C.: Cradle to site Life Cycle Assessment (LCA) of natural vs conventional building materials: a case study on cob earthen material. Build. Environ. **160**, 106150 (2019). https://doi.org/10.1016/j.buildenv.2019.05.028
2. Ben-Alon, L., Loftness, V., Harries, K.A., Hameen, E.C., Bridges, M.: Integrating earthen building materials and methods into mainstream construction. J. Green Build. **15**, 87–106 (2020). https://doi.org/10.3992/1943-4618.15.1.87
3. Jiménez Delgado, M.C., Guerrero, I.C.: The selection of soils for unstabilised earth building: a normative review. Constr. Build. Mater. **21**, 237–251 (2007). https://doi.org/10.1016/J.CONBUILDMAT.2005.08.006
4. Mudgil, D., Barak, S., Khatkar, B.S.: Guar gum: processing, properties and food applications—a review. J. Food Sci. Technol. **51**(3), 409–418 (2011). https://doi.org/10.1007/s13197-011-0522-x
5. Thombare, N., Jha, U., Mishra, S., Siddiqui, M.Z.: Guar gum as a promising starting material for diverse applications: a review. Int. J. Biol. Macromol. **88**, 361–372 (2016). https://doi.org/10.1016/j.ijbiomac.2016.04.001
6. Sharma, G., et al.: Guar gum and its composites as potential materials for diverse applications: a review. Carbohyd. Polym. **199**, 534–545 (2018). https://doi.org/10.1016/j.carbpol.2018.07.053
7. Nordgård, C.T., Draget, K.I.: Chapter 26 - Alginates. In: Phillips, G.O., Williams, P.A. (eds.) Handbook of Hydrocolloids, 3rd edn. pp. 805–829. Woodhead Publishing (2021). https://doi.org/10.1016/B978-0-12-820104-6.00007-3
8. Sworn, G.: Chapter 27 - Xanthan gum. In: Phillips, G.O., Williams, P.A. (eds.) Handbook of Hydrocolloids, 3rd edn. pp. 833–853. Woodhead Publishing (2021). https://doi.org/10.1016/B978-0-12-820104-6.00004-8
9. Armistead, S.J., Rawlings, A.E., Smith, C.C., Staniland, S.S.: Biopolymer stabilization/solidification of soils: a rapid, micro-macro. Cross-Discip. Appr. Environ. Sci. Technol. **54**, 13963–13972 (2020). https://doi.org/10.1021/acs.est.0c02001

10. Armistead, S.J., Smith, C.C., Staniland, S.S.: Sustainable biopolymer soil stabilization in saline rich, arid conditions: a 'micro to macro' approach. Sci. Rep. **12**, 2880 (2022). https://doi.org/10.1038/s41598-022-06374-6
11. Brindley, G.W., Brown, G.: Crystal Structures of Clay Minerals and their X-Ray Identification. Mineralogical Society (1982)
12. Moore, D.M., Reynolds, R.C., Reynolds, R.C.: X-ray Diffraction and the Identification and Analysis of Clay Minerals. Oxford University Press (1989)
13. Chipera, S.J.: Baseline studies of the clay minerals society source clays: powder X-ray diffraction analyses. Clays Clay Miner. **49**, 398–409 (2001). https://doi.org/10.1346/CCMN.2001.0490507
14. Madejová, J.: Baseline studies of the clay minerals society source clays: infrared methods. Clays Clay Miner. **49**, 410–432 (2001). https://doi.org/10.1346/CCMN.2001.0490508
15. Russell, J.D., Fraser, A.R.: Infrared methods. In: Wilson, M.J. (ed.) Clay Mineralogy: Spectroscopic and Chemical Determinative Methods, pp. 11–67. Springer, Dordrecht (1994). https://doi.org/10.1007/978-94-011-0727-3_2
16. Popescu, C.-M., Popescu, M.-C., Singurel, G., Vasile, C., Argyropoulos, D.S., Willfor, S.: Spectral characterization of eucalyptus wood. Appl. Spectrosc. **61**, 1168–1177 (2007). https://doi.org/10.1366/000370207782597076
17. Guo, Q., Ai, L., Cui, S.W.: Fourier Transform Infrared Spectroscopy (FTIR) for carbohydrate analysis. In: Methodology for Structural Analysis of Polysaccharides. SpringerBriefs in Molecular Science. Springer, Cham (2018). https://doi.org/10.1007/978-3-319-96370-9_9
18. Nazdracheva, T., Morozov, A., Yavna, V., Kochur, A.: Data on FTIR spectra of the clays KGa-1b and STx-1b and their mixtures at different moistures. Data Brief. **42**, 108282 (2022). https://doi.org/10.1016/j.dib.2022.108282
19. Schroeder, P.A. (ed.): Critical zone clay geochemistry. In: Clays in the Critical Zone. pp. 133–179. Cambridge University Press, Cambridge (2018)
20. Mekhamer, W.K.: Stability changes of Saudi bentonite suspension due to mechanical grinding. J. Saudi Chem. Soc. **15**, 361–366 (2011). https://doi.org/10.1016/j.jscs.2011.03.014
21. Yukselen, Y., Kaya, A.: Zeta potential of kaolinite in the presence of Alkali, Alkaline earth and Hydrolyzable metal ions. Water Air Soil Pollut. **145**, 155–168 (2003). https://doi.org/10.1023/A:1023684213383

Enhancing the Water and Biodegradation Resistance of Biopolymer Stabilised Soils – Design Concepts

Colin Smith[1(✉)], James McGregor[1], Natalia Martsinovich[1], Samuel Armistead[2], Xinyuan Yu[1], and Nitchamon Siripanich[3]

[1] University of Sheffield, Sheffield, UK
c.c.smith@sheffield.ac.uk
[2] Chalmers University, Gothenburg, Sweden
[3] King Mongkut's University of Technology Thonburi, Bangkok, Thailand

Abstract. The global soil stabilisation market is forecast to grow to $35 billion by 2027 driven primarily by infrastructure and construction activities and exacerbated by the increasingly urgent need to adapt to climate change, flood risk and sea-level rise. Additionally there is anticipated to be increasing demand for stabilised-earth based building products due to their low carbon impact. Cement and lime are widely used to stabilise soil but suffer from significant carbon and energy costs. Naturally sourced biopolymers are a promising low carbon 'green' substitute, achieving higher strength in stabilised soils than cement and at similar cost. However, widespread uptake of biopolymers is impeded by the fact that they typically suffer from (a) poor water resistance and (b) poor resistance to biodegradation over time.

This paper presents an overview of design concepts and pilot results from a research study into novel biopolymer treatment processes which are applied at the soil/biopolymer mixing stage with the aim of enhancing resistance to water and biodegradation. The approaches include investigation of (i) acetylation, whereby vulnerable hydrophilic functional groups in the biopolymers are replaced by natural hydrophobic acetyl groups, (ii) binding vulnerable hydrophilic groups to specific inorganic minerals, thereby blocking interactions with water. The study design includes a comprehensive series of laboratory tests guided by detailed characterisation of the underlying chemical binding processes with the aim of involving the addition of only small volumes of natural materials as part of the treatment process.

Keywords: Biopolymer · Soil stabilisation · Water resistance

1 Introduction

The global soil stabilisation market is forecast to grow from $25 billion in 2019 to $35 billion by 2027 (Emergen Research, 2020) driven primarily by infrastructure and construction activities and exacerbated by the increasingly urgent need to adapt to climate

S. Amziane et al. (Eds.): ICBBM 2023, RILEM Bookseries 45, pp. 293–302, 2023.
https://doi.org/10.1007/978-3-031-33465-8_23

change, flood risk and sea-level rise. Soil chemical stabilisation (where soil is mixed with an additive to enhance its strength) is forecast to be the fastest growing sector of the market, taking just under half the share. In parallel there is anticipated to be a significantly increased demand for stabilised-earth based building products due to their low carbon impact. However, the most widely used additives, cement and lime, can suffer from significant carbon and energy costs and substantial environmental impacts e.g. highly alkaline conditions, toxic cement dust, polluting run-off. Geopolymers offer reduced environmental impacts but still require significant processing of raw materials and the use of highly concentrated alkali activator solutions.

Such concerns have stimulated interest in biologically derived alternatives as a low-carbon sustainable solution. There is existing evidence that naturally sourced biopolymers have significant potential, achieving higher strength in stabilised soils with lower additive mass than cement and at similar cost. However, widespread uptake of biopolymers is impeded by the fact that they often suffer from (a) poor water resistance and (b) poor resistance to biodegradation over time. If these challenges can be addressed, biopolymers could achieve a transformational impact on soil stabilisation. This paper presents design concepts and pilot results from a research study into novel biopolymer treatment processes which are applied at the soil/biopolymer mixing stage.

2 Biopolymer Binding, Water Resistance and Biodegradation

A significant body of literature exists investigating soil stabilisation using biopolymers. Production of soil/biopolymer composites normally involves dissolution of the biopolymer in water to allow thorough mixing with soil and to expose functional groups for binding to soil. When dry (or cured), the biopolymer functional groups within a stabilised soil will bind strongly to other biopolymers or to soil particles and it is typically found that on drying, treated soils display significant gains in strength and stiffness. Chang et al. (2015) used 1% xanthan gum resulting in significant strength gains for soils ranging from sands to clays including an 86% increase in Unconfirmed Compressive Strength (UCS) compared to 10% cement addition for a residual soil. Muguda et al. (2017) identified a 30% higher UCS using 3% guar gum compared to 8% cement addition for a gravel/sand/clay soil. This level of strength gain is attributed to the biopolymer binding to soil particles and the high tensile strength and stiffness of dried biopolymer. There is thus a significant potential for using such materials in environments where design detailing can maintain low humidity and exposure to water.

However, for applications where exposure to high humidity or liquid water is anticipated there are potential challenges since many of the beneficial binding/strengthening mechanisms can be easily reversed on exposure to water. Biopolymer tensile strength and stiffness can diminish significantly with increased water content (Yakimets et al. 2007).

Soil/biopolymer binding mechanisms depend significantly on the soil mineralogy and biopolymer chemistry. Mechanisms proposed in the literature include for example: hydrogen and ionic binding to electrically charged clay particles (Chang 2016); electrostatic adsorption of biopolymer additives to negatively charged SiO_2 surfaces facilitated by aqueous molecules; covalent interactions through acid/base reactions, forming C–O–Fe bonds with Fe_2O_3 particles (Armistead et al. 2020). Some of these binding mechanisms can be weakened or outcompeted by water molecules. Experimental measurements on xanthan gum stabilised soils have previously demonstrated a loss of compressive strength upon exposure to water (Chang et al. 2015; Chang et al. 2020). Elsewhere the addition of chitin - with subsequent reacetylation - has been shown to improve the mechanical properties of either clay-based soils, however this effect is significantly reduced in the presence of water due hydrodegradation (Hataf et al. 2018).

3 Design Concepts

3.1 Background

It is hypothesised that the biopolymer functional groups within a stabilised soil which bind strongly to other biopolymers or to soil particles can be treated post mixing to negate their vulnerability to water and biodegradation without affecting the soil-biopolymer binding. Two different approaches as illustrated in Fig. 1 are being investigated:

- conversion of 'vulnerable' functional groups utilising acetylation,
- binding 'vulnerable' functional groups to inorganic minerals.

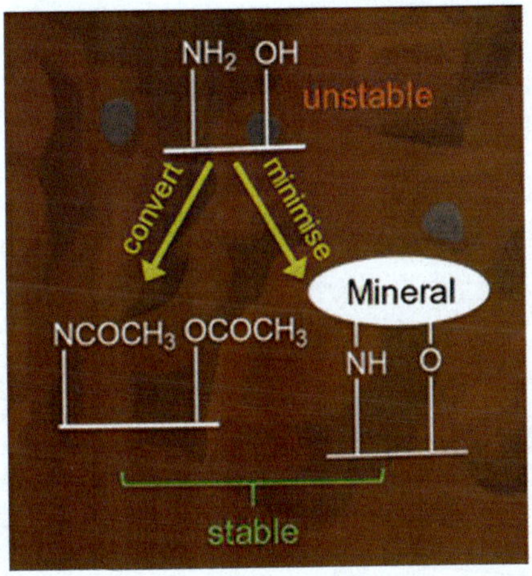

Fig. 1. Enhancing biopolymer resistance to water and biodegradation.

3.2 Acetylation

Acetylation is a chemical process that has been employed as a non-toxic 'green' timber treatment technology, enabling replacement of vulnerable hydrophilic functional groups in the wood biopolymer with natural hydrophobic acetyl groups (Mantanis 2017). Elimination of the hydrophilic groups in this way inhibits the enzymatic digestion of wood and has been demonstrated to achieve exceptional timber lifetimes of >50 years in exposed outdoor conditions. The potential for such a treatment process to be applied to biopolymers employed in soil-strengthening will be investigated.

A strong candidate biopolymer for initial investigation for the potential of this process is chitosan, which has shown significant strengthening potential in soil stabilisation (Hataf et al. 2018; Shariatmadari et al. 2020; Adamczuk and Jozefaciuk 2022). Chitosan is already the product of the deacetylation of chitin, a natural material commonly used by insects and crustaceans for exoskeletons and which has inherent water resistant properties. It is hypothesised that re-acetylation of chitosan, post soil treatment, will revert chemical bonds unused for soil binding to the chitin form and benefit from its associated properties such as high stability.

Re-acetylation can be carried out through the addition of acetic anhydride, acetic acid and mixtures of the two. Acetylation of wood typically avoids use of catalysts, but operates at relatively high temperatures (100 to 120 °C). However previous studies have shown effective (partial) acetylation of chitosan can occur at ambient conditions e.g. Lavertu et al. (2011), Qun et al. (2007). The quantity of acetic acid/acetic anhydride employed will be minimised in order to limit any residue. The degree of deacetylation and subsequent re-acetylation will be a key metric by which the success of the chemical process is evaluated. This will be quantified through Fourier transform infrared (FTIR) spectroscopy; a well-established technique for this purpose. X-ray diffraction (XRD) will be employed to identify changes to the crystal structure of chitin/chitosan effected by the de/re-acetylation procedures.

3.3 Inorganic Binding

The alternative, 'binding' approach will evaluate the potential to bulk bind the vulnerable hydrophilic groups to specific inorganic minerals, thereby blocking interactions with water. This would provide a potentially transformative approach to design bespoke biopolymer mixes for specific soil mineralogies, enhanced where required by addition of specific minerals. The challenge and associated risk is to identify whether the most advantageous binding is achievable in a multi mineral/biopolymer functional group system. Methods developed in previous work (e.g. Armistead et al. 2020) will allow the systematic exploration of the nature of the binding. Exploration of the interactions between naturally occurring silica, alumna and iron compounds and biopolymers containing 'hard' cationic/anionic and 'soft' neutral ligand functional groups will establish the understanding required for this approach.

For example, a recent study by Lalonde et al. (2012) found that 21% ± 8.6% of organic carbon within sediments are associated with reactive iron phases. Further investigation by Barber et al. (2017) found that this phenomena was driven by the formation of irreversible covalent C–O–Fe complexes, which reduce both the biopolymer enzymatic hydrolysis and biodegradation. The implications of this phenomena are promising areas for further study.

4 Study Design

The core aim of the current research is to demonstrate that the hypothesized approaches work under controlled laboratory conditions. Once achieved, the aim is then to refine the process to minimise process complexity, volume of additional additives and residual chemicals, and to investigate the viability for field application. A brief overview of the planned testing programme is given below.

4.1 Materials

Five standardised soil types will be investigated covering a range of mineralogies and specific surface areas: silica sand, calcareous sand, silt (silica based), kaolinite and montmorillonite.

Initial work will target polysaccharide biopolymers, due to their high-volume commercial availability (e.g. a co-product of shellfish processing) and ease of extraction/co-production, focusing on chitosan and galactomannans in the first instance.

4.2 Sample Preparation

Biopolymer solutions will be mixed mechanically with the relevant soil and the soil compacted using a cylindrical drop hammer within a 202 mm × 42 mm hollow cylindrical sample mould. These will then be extruded and allowed to cure under controlled temperature and humidity and at a range of time durations. The treatment agents may be combined into the original mix prior to soil mixing, after soil mixing over a range of time intervals or (if liquid) allowed to permeate a sample once cured in order to assess the feasibility of implementing this process in the field.

4.3 Characterisation Tests

Following curing, samples will undergo a range of characterisation tests, as follows:

- **Water resistance.** Samples will undergo either immersion tests followed by weight and volume change measurement or erodibility tests using the jet index method (ASTM 2017).
- **Biodegradation.** Biodegradation tests will be carried using an approach based on ISO (2019), originally developed for determining the aerobic biodegradability of plastics in soil. Specifically, mass spectrometry will be used to qualitatively monitor the evolution of CO_2 from soil/biopolymer samples over time.
- **Compression testing**. Full stress strain/response (pre- and post-water immersion and biodegradation) will be carried out on unconfined specimens. Selected samples will undergo confined triaxial testing.
- **Chemical/physical characterisation** Both initial reagents/soils and final samples will undergo selected chemical/physical characterisation as appropriate using particle sizing, Atterberg limit determination, optimum moisture content for geotechnical

interpretation and XRD, GPC, zeta potential, FTIR spectroscopy, TGA, SEM to identify key chemical binding characteristics.

In addition to the final performance of the treated soils, additional process related issues to be investigated include:

- Required and residual volumes of treatment agents:
- Effect of treatment reagents on curing rates.

4.4 Initial Results

An initial pilot study (Siripanich 2023) investigated the strength of re-acetylated chitosan in the presence of coarse grained SiO_2 sand particles using two different approaches:

1. Re-acetylation with sand present in the solution
2. Re-acetylation before mixing with sands, i.e. conducting re-acetylation process first to obtain a chitin gel, followed by mixing of the gel with sands

Figure 2 shows the uniaxial stress-strain response of the two samples. It can be seen that Approach 1 where the re-acetylation agents were mixed with the chitosan and sand gave improved strength. However further testing is required to confirm the observed trends. Figure 3 presents a more direct comparison of strength (UCS) and stiffness (E_{50}). The incorporation of the biopolymer into the sand was confirmed by FTIR spectroscopy, Fig. 4, where characteristic peaks for chitin (e.g. C-H stretching peaks, 2900–3000 cm-1) are observed. Figure 5 shows an example of one of the generated samples.

For these tests, the re-acetylation was conducted using different proportions of methanol-water mixture and molar ratios of acetic anhydride to (chitosan) amine. While the amine group reacts with acetic anhydride, this reaction is not 100% selective as the hydroxyl group may also undergo conversion. The addition of methanol however improves yields of the reacetylation process through modulating the pKa of acetic acid and reducing the degree of ionisation of the biopolymer. Therefore, methanol was used as a co-solvent in this process.

Preliminary observations indicate that the chitin-sand samples are resistant to immersion in water however this requires confirmation by a more comprehensive testing programme.

5 Discussion

The proposed investigation is high risk. While the concept is attractive, the proposed solutions may simply not function in the soil chemical environment, may inhibit the desired soil-biopolymer binding, or leave significant residues. One or more of the techniques investigated may work, but either only partially or only in carefully controlled laboratory conditions that are not scalable to the field. This may limit application to factory produced bricks and panels or to specific applications under specific environmental conditions. However it is anticipated that any demonstration of the viability of the core concepts would catalyse a new field for further research and innovation.

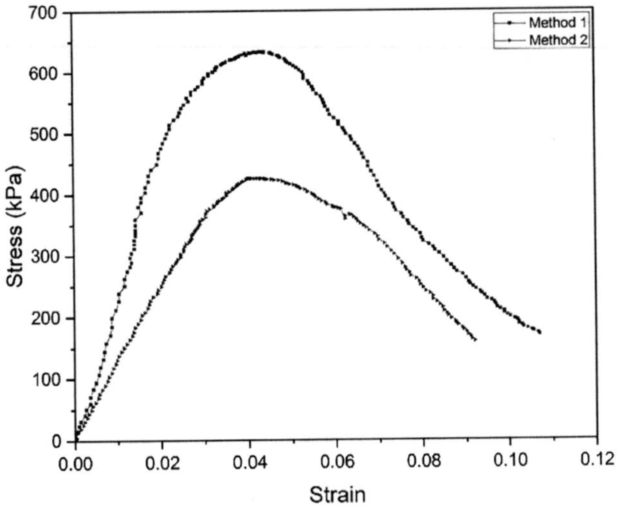

Fig. 2. Uniaxial stress-strain response of sand-re-acetylated chitosan mixes.

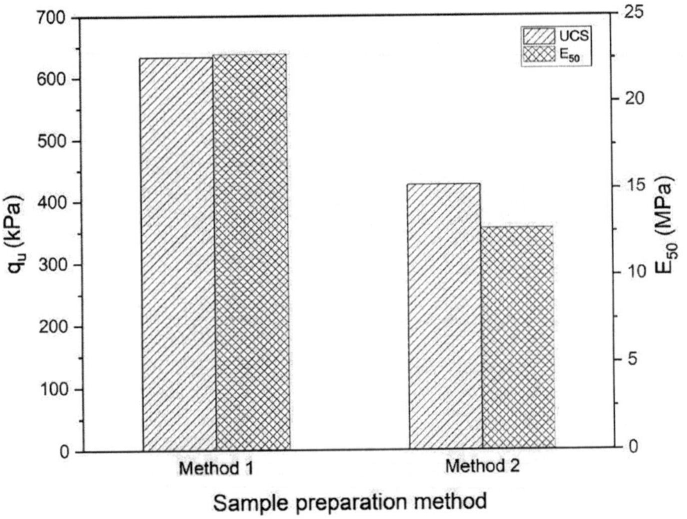

Fig. 3. Strength (UCS or q_u) and stiffness (E_{50}) comparison for sand-re-acetylated chitosan mixes.

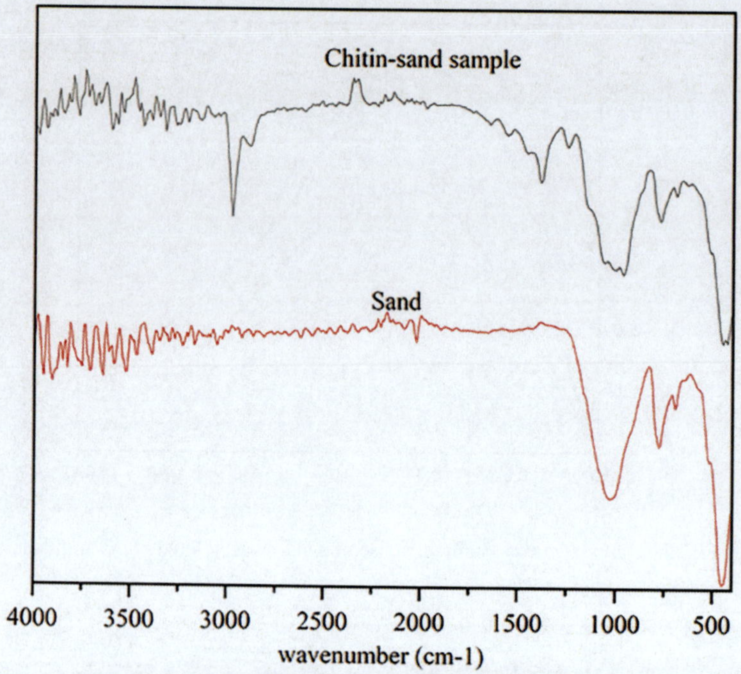

Fig. 4. FTIR spectrum of sand and chitin-sand sample.

Fig. 5. Example of chitin-sand sample employed for UCS testing.

6 Conclusions

1. Biopolymer treated soils typically display high gains in strength but conversely low resistance to biodegradation and water.

2. Conversion of 'vulnerable' functional groups utilising acetylation, and binding 'vulnerable' functional groups to inorganic minerals, have significant potential to enhance biopolymer resistance to biodegradation and water. Selected potential mechanisms and candidate materials have been discussed.
3. An outline test methodology for the investigation of techniques for the enhancement of water and biodegradation resistance of biopolymer-soil mixes is described.
4. Pilot tests involving the re-acetylation of chitosan have shown promising results. However further studies are required.

References

Adamczuk, A., Jozefaciuk, G.: Impact of Chitosan on the mechanical stability of soils. Molecules **27**, 2273 (2022). https://doi.org/10.3390/molecules27072273

Armistead, S.J., Rawlings, A.E., Smith, C.C., Staniland, S.S.: Biopolymer stabilization/solidification of soils: a rapid, micro-macro, cross-disciplinary approach. Environ. Sci. Technol. **54**(21), 13963–13972 (2020)

ASTM: Standard Test Method for Erodibility Determination of Soil in the Field or in the Laboratory by the Jet Index Method 2017. D5852-95, https://www.astm.org/d5852-95.html

Barber, A., et al.: Preservation of organic matter in marine sediments by inner-sphere interactions with reactive iron. Sci. Rep. **7**(1) (2017). https://doi.org/10.1038/s41598-017-00494-0

Chang, I., Im, J., Prasidhi, A.K., Cho, G.-C.: Effects of Xanthan gum biopolymer on soil strengthening. Constr. Build. Mater. **74**, 65–72 (2015). https://doi.org/10.1016/j.conbuildmat.2014.10.026

Chang, I., Im, J., Cho, G.C.: Introduction of microbial biopolymers in soil treatment for future environmentally-friendly and sustainable geotechnical engineering. Sustainability **8**(3), 251 (2016). https://doi.org/10.3390/su8030251

Chang I., et al.: Review on biopolymer-based soil treatment (BPST) technology in geotechnical engineering practices, Transportation Geotechnics **24**, 100385 (2020)

Emergen Research, Soil Stabilization Market By Application (Agriculture, Industrial, Non-Agriculture), By Method (Chemical, Mechanical), By Additive (Mineral & Stabilizing agents, Polymers), Forecasts to 2027 (2023). https://www.emergenresearch.com/industry-report/soil-stabilization-market

Hataf, N., Ghadir, P., Ranjbar, N.: Investigation of soil stabilization using chitosan biopolymer. J. Clean. Prod. **170**, 1493 (2018). https://doi.org/10.1016/j.jclepro.2017.09.256

ISO 17556, Plastics — Determination of the ultimate aerobic biodegradability of plastic materials in soil by measuring the oxygen demand in a respirometer or the amount of carbon dioxide evolved (2019). https://www.iso.org/standard/74993.html

Lalonde, K., Mucci, A., Ouellet, A., Gelinas, Y.: Preservation of organic matter in sediments promoted by iron. Nature **483**(7388), 198–200 (2012). https://doi.org/10.1038/nature10855

Lavertu, M., Darras, V., Buschmann, M.D.: Kinetics and efficiency of chitosan reacetylation. Carbohydrate Polymers **87**(2), 1192–1198 (2011)

Mantanis, G.I.: Chemical modification of wood by acetylation or furfurylation: a review of the present scaled-up technologies. BioRes **12**(2), 4478–4489 (2017)

Muguda, S., et al.: Mechanical properties of biopolymer-stabilised soil-based construction materials. Géotech. Lett. **7**(4), 309–314 (2017). https://doi.org/10.1680/jgele.17.00081

Qun, G., Ajun, W., Yong, Z.: Effect of reacetylation and degradation on the chemical and crystal structures of chitosan 2007. J. Appl. Polym. Sci. **104**, 2720–2728 (2007)

Shariatmadari, N., Reza, M., Tasuji, A., Ghadir, P., Javadi, A.: Experimental study on the effect of chitosan biopolymer on sandy soil stabilization. In: E3S Web of Conferences, vol. 195, pp. 06007 (2020)

Siripanich, N.: Application of chitin as a sustainable material for soil stabilization process. M.Sc thesis. King Mongkut's University of Technology Thonburi, Thailand (2023)

Yakimets, I., Paes, S.S., Wellner, N., Smith, A.C., Wilson, R.H., Mitchell, J.R.: Effect of water content on the structural reorganization and elastic properties of biopolymer films: a comparative study. Biomacromol **8**, 1710–1722 (2007)

Improvement of the Mechanical Properties of P300 Kaolinite Using MICP in the Low Water Content Range

O. Maston[1(✉)], T. Ouahbi[1], A. Dadda[1], A. El Hajjar[1], S. Taibi[1], L. Sapin[2], A. Esnault Filet[2], H. Souli[3], and J.-M. Fleureau[4]

[1] Normandie University, UNIHAVRE, Laboratory of Waves and Complex Media, UMR 6294, CNRS, Le Havre, France
olivier.maston@univ-lehavre.fr

[2] Soletanche Bachy, Rueil-Malmaison, France

[3] Laboratoire de Tribologie et Dynamique des Systèmes - CNRS UMR 5513, École Nationale d'Ingénieurs de Saint-Étienne, Saint Étienne, France

[4] Université Paris-Saclay, CentraleSupélec, Laboratoire de Mécanique des Sols, Structures et Matériaux, UMR 8579, CNRS, Gif-sur-Yvette, France

Abstract. Reducing greenhouse gas emissions is one of the major concerns of our time. That is why eco-building materials such as raw earth are gaining interest. It requires low grey energy due to its manufacture and presence on site. The problem is that the dehydration of its clay fraction causes its shrinkage, generating cracks. Many researchers have tried to solve this problem. Microbially induced calcite precipitation (MICP) appears as a potential solution. Its use has been extensively studied to improve inert soils such as sands, but only some studies have focused on its use in loaded and low-permeability clays.

In this study, we aim to demonstrate the proper functioning of MICP and its positive impact on clays. To do so, we prepare low water content clay samples with the "Soil-Mixing" technique, mixing our dry powder of yellow kaolinite P300 with a bacterial and calcifying solution to form a paste that will be compacted to a fixed dry density. These cylindrical samples are kept at a fixed water content for 3 days and are then subjected to a simple compression test. The comparison of the significant stresses obtained attests to the increase in strength of the material treated by MICP, and the use of the Dietrich Frühling calcimeter confirms the presence of calcite in the MICP treated kaolinite.

The use of the MICP process in clays is very promising. This study has highlighted the improvement of mechanical strength through MICP treatment by "Soil-Mixing" method in clay materials with low water content. We intend to apply this treatment in raw earth construction materials such as compressed earth blocks.

Keywords: Clay · MICP · Biocalcification · Stress · Raw earth construction

S. Amziane et al. (Eds.): ICBBM 2023, RILEM Bookseries 45, pp. 303–309, 2023.
https://doi.org/10.1007/978-3-031-33465-8_24

1 Introduction

There is a great interest in the use of eco-materials in the building sector in order to meet the challenges of the ecological transition. Raw earth is an eco geo material with many advantages, such as its low grey energy, its remarkable properties or its local presence in any part of the world. The limitation of its use comes from its cracking during its desiccation. This is why researchers have been interested for decades in understanding and remedying these cracking mechanisms. The use of vegetable fibers has allowed to reduce the cracking but is not sufficient. The process of Microbially Induced Calcite Precipitation (MICP) is of growing interest. It consists of inducing calcium carbonate precipitation through bacterial enzymatic activity to hydrolyze urea and synthesize carbonate, which will then react with calcium. It has been the subject of numerous studies for its use in inert sandy soils and concrete like the study of [2] or the work of [1]. It is even already applied industrially. On the other hand, very few studies have been carried out on clays. The composition of clays coupled with MICP makes biocalcification more complex and the chemical interactions are still unknown. Some works have been interested in the mechanical improvement that MICP provides in clays, such as the work of [3] who treated a silty clay by injection and observed an increase in its strength. In this study, we use the MICP treatment in kaolinite, with the aim of confirming its good functioning by proceeding by "soil-mixing" method. The evaluation of the increase of resistance and the quantity of precipitated $CaCO_3$ will be the indicative parameters.

2 Materials and Methods

2.1 Materials

To make the samples, we use a yellow clay commercialized by Dousselin, the kaolinite P300. Table 1 summarizes some characteristics of our clays. [4, 5 and 6].

2.2 Methods

Sample Preparation Protocol. The principle of MICP is as follows bacterial enzymatic activity (urease) hydrolyzes urea to synthesize carbonate and ammonium ions. The free calcium ions will react with the previous product of carbonate ions to precipitate calcite.

$$CH_4N_2O + 2H_2O \rightarrow 2NH_4^+ + CO_3^{2-} \quad \text{(Hydrolyzes of urea)}$$

$$Ca^{2+} + CO_3^{2-} \rightarrow CaCO_3 \downarrow \quad \text{(Calcite precipitation)}$$

We use "Soil-mixing" to prepare samples with low water content. We prepare a nutrients solution in which urea and calcium acetate are mixed in demineralized water and a bacterial solution composed of demineralized water and bacteria (Sporosarcina Pasteurii). We mix these two solutions, then pour it little by little on a quantity of clay powder P300 by mixing it manually. The result is a mixture of wet lumps and dry powder is shown. This observation is explained by an uneven distribution of water content in

Table 1. Characteristics of P300 kaolinite

Characteristics			
Designation	**Unit**	**Value Kaolinite**	**Reference**
Specific Area	Sp (m^2/g)	32	Beddiar, 2005
Cation Exchange Capacity	CEC (meq/hg)	15	Beddiar, 2005
Granulometry	(%) <2 μm	83	Hammad, 2010
LPC classification	-	Fine soil (Low plastic clays)	El Hajjar, 2018
Granular Form	-	Rigid hexagonal platelets	Hammad, 2010
Mineral composition	-	Kaolinite, illite and quartz	El Hajjar, 2018
Clay family	-	Phyllosilicates (TO)	El Hajjar, 2018
Liquid Limit	L_L (%)	40	
Plastic Limit	P_L (%)	20	
Plasticity Index	P_I (%)	20	

our clay, due to its low water content during preparation. To remedy this, we sieve our mixture in order to reduce the size of the lumps to 1mm in diameter at most. After sieving, we compact, with a static press, our clay previously that was placed in a mold. When we reach 1.5 dry density, we stop the compaction. The dimensions of our cylindrical sample are 100 mm height and 50 mm diameter. They are kept at a constant water content for 72 h in order to homogenize the water and to let the MICP biocalcify the structure.

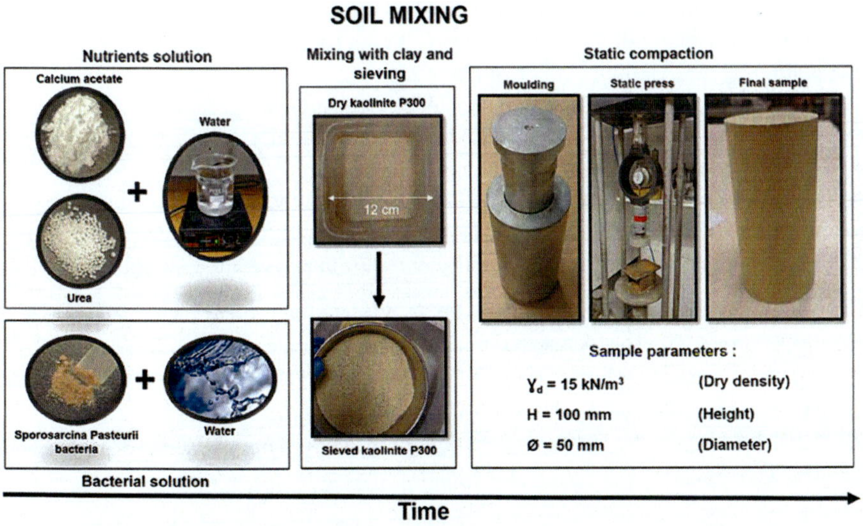

Fig. 1. Protocol for the preparation of samples for Uniaxial compression test

We prepare 2 similar samples to confirm the repeatability of the test and determine the uncertainty of the results. The recap of sample preparation is presented in Fig. 1.

Uniaxial Compression Test (ISO 17892-7:2017). The uniaxial compression test allows us to obtain the major stress of a laterally free and axially stressed sample until it breaks or until its failure. We applied a compression speed of 0.1 mm/min, a value adapted to our material. Figure 2 shows the behavior of a clay sample during the test, with the orientation of the fracture plane following Mohr Coulomb's law.

Fig. 2. Simple compression of a clay sample over time of kaolinite P300 with MICP for w = 16%

Tests to Verify the Functioning of the Bacteria Before the Sample Preparation. As the proper functioning of the MICP depends on the bacterial activity, before the sample preparation, we check their proper functioning.

We first measure the optical density (OD) by emitting a wavelength of 600 nm with a UV spectrophotometer. According to the literature, if the OD is between 0.8 and 1.2 then our bacterial concentration is suitable.

Second, a conductivity meter allows us to measure the electrical conductivity of a solution of urea and bacteria. The monitoring of the electrical conductivity over time, testifies or not the good enzymatic activity and thus ureolysis in our solution.

Tests to Verify the Functioning of MICP After Mechanical Test. We check the quantity of precipitated $CaCO_3$ in our material using the calcimeter of Dietrich-Fruhling. We realize this test following the NF P94-048 standard [7]. We mix a piece of our dried sample, with chloridric acid. The reaction between $CaCO_3$ and chloridric acid in excess will release CO_2. As the volume of CO_2 is known, we can determine the quantity of precipitated calcite.

3 Results and Discussions

3.1 Uniaxial Compression Test

The data provided by the uniaxial compression test allows us to draw the Fig. 3 The maximum stress of the curve corresponds to the major stress of our material that can be associated with a deformation. In addition, we can determine the deformation modulus corresponding to the linear part of the curve.

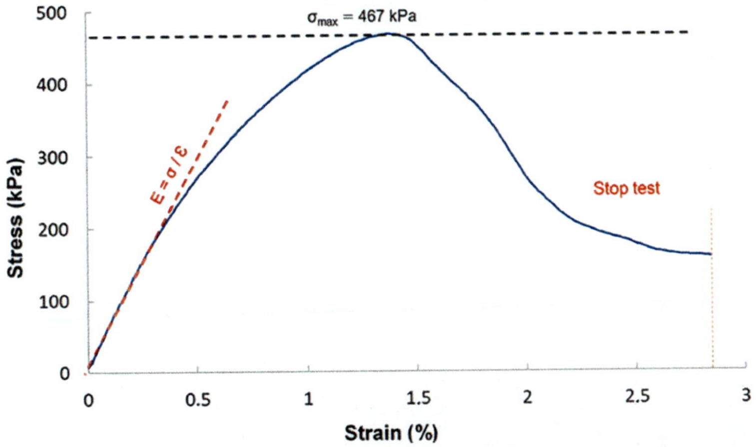

Fig. 3. Evolution of the major stress as a function of the strain of kaolinite P300 with MICP for $w = 16\%$

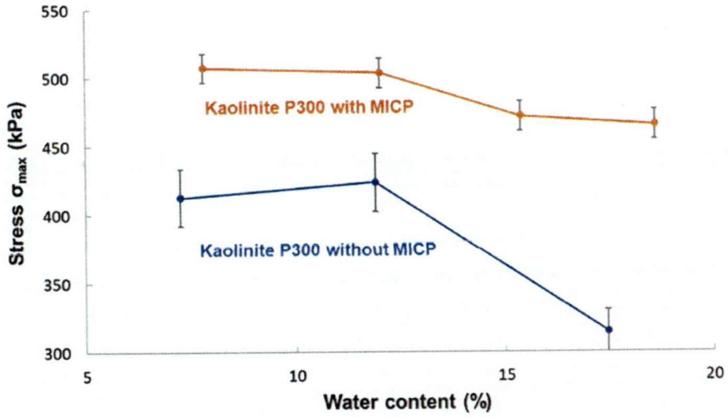

Fig. 4. Evolution of the mechanical strength of kaolinite P300 as a function of the water content

The Fig. 4 presents the major stress of our material as a function of its water content. We notice an increase in strength with decreasing water content, which is physically explained by the increase in friction and cohesion. This graph clearly shows the increase in strength brought by the MICP treatment on our kaolinite.

We can suppose that this mechanical benefit comes from the presence of calcite bridges allowing to rigidify the whole structure.

3.2 Calcite Precipitation Rate

The rate of calcite precipitation obtained by the Dietrich-Fruhling calcimeter is presented by the Fig. 5 highlighting the presence of $CaCO_3$ crystals in our clay treated by MICP. It also shows that the amount of precipitated calcite in the 16% and 20% water content samples is similar if the uncertainty is taken into account. On the other hand, we find a lower precipitation rate in our samples at 10% of water content.

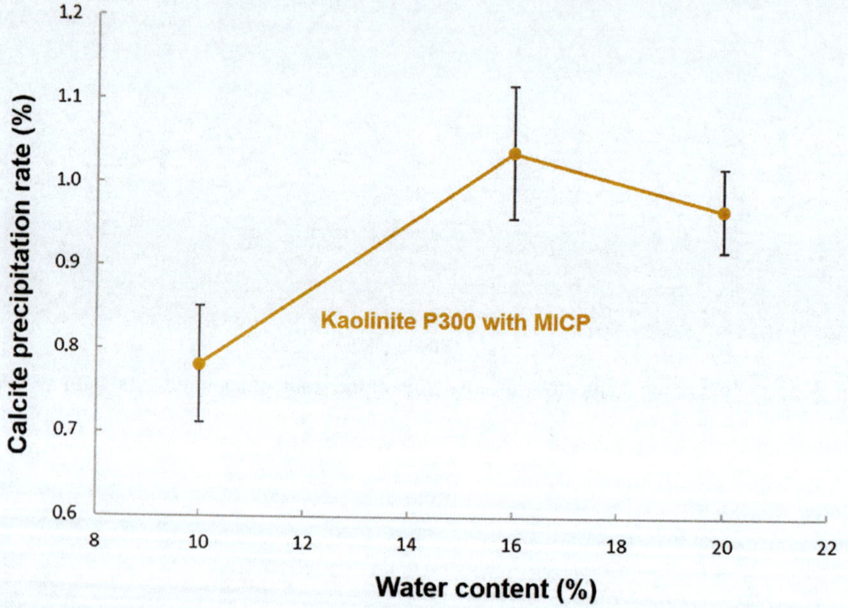

Fig. 5. Evolution of the calcite precipitation rate in kaolinite P300 as a function of the water content

We can assume that these results can explain the strength gained by our treated samples over the untreated ones, with a more significant strength improvement for our clay samples at 16 and 20% water content. The more consistent presence of calcite could probably cement our clay material and thus increase its strength and stiffness.

4 Conclusion

Following our tests on Kaolinite P300, we can conclude that:

1) The uniaxial compression tests showed an increase in the maximum strength of clay sample, thanks to the MICP treatment.

2) The Dietrich-Fruhling calcimeter tests confirmed the presence of calcite in our material and thus the functioning of the calcite precipitation

Biocalcification thus allows the mechanical improvement of P300 kaolinite with low water content.

Many additional studies are to be considered. It would be interesting to follow the same sample preparation protocol as the one presented in this paper, adding an air-drying step, in order to reproduce conditions close to the implementation of raw earth construction such as the preparation of compressed earth bricks. One could also approach the raw earth construction material, by using the local earth of Le Havre. The follow-up of the influence of the MICP treatment on the cracking by desiccation of local clay or silt via various techniques such as the correlation of digital images is also possible.

References

1. Liu, S., Du, K., Wen, K., Armwood-Gordon, C., Li, L.: Influence of rainfall-induced erosion on the stability of sandy slopes treated by MICP. Adv. Civil Eng. **2022**, 1–10 (2022)
2. Nasser, A.-A., Esmail, R.-F., Abbas, R.-Y., Sorour, N.-M.: Effect of Bacillus megaterium bacteria and different calcium source on strength and permeation properties of concrete. ERJ. Eng. Res. J. **45**(3), 401–412 (2022)
3. Teng, F., Sie, Y.-C., Ouedraogo, C.: Strength improvement in silty clay by microbial-induced calcite precipitation. Bull. Eng. Geol. Env. **80**(8), 6359–6371 (2021). https://doi.org/10.1007/s10064-021-02308-0
4. Beddiar, K., Fen-Chong, T., Dupas, A., Berthaud, Y., Dangla, P.: Role of pH in electro-osmosis: experimental study on NaCl–water saturated kaolinite. Transp. Porous Media **61**(1), 93–107 (2005)
5. Hammad, T.: Comportement des sédiments marins de grande profondeur: approche multi-échelle. Thèse de doctorat, Ecole Centrale Paris (2010)
6. El Hajjar, A., Eid, J., Bouchemella, S., Ouahbi, T., Taibi, S.: Retrait et fissuration des sols argileux par dessiccation. Acad. J. Civil Eng. **36**(1) (2018)
7. Sols: reconnaissance et essais - Détermination de la teneur en carbonate - Méthode du calcimètre - NF P 94–048 (1996)

Toughness and Ultimate Compressive Strength of Bio-Based Raw Earth Concrete

Youssef Shamas[1,3(✉)], H. C. Nithin[2], Vivek Sharma[2], S. D. Jeevan[2], Sachin Patil[2], Saber Imanzadeh[1,3], Armelle Jarno[3], and Said Taibi[3]

[1] Normandie Univ., INSA Rouen Normandie, Laboratoire de Mécanique de Normandie, 76801 Saint-Etienne du Rouvray, France
youssef.shamas@insa-rouen.fr
[2] New Horizon College, Marathalli, Ring Road, Near Marathalli, Bangalore 560 103, India
[3] Normandie Université, UNIHAVRE, Laboratoire Ondes et Milieux Complexes, CNRS UMR 6294, Le Havre, France

Abstract. Pollution is widespread around the world, and all the countries started looking for new and green alternative energies, or even started limiting the use of grey energy. Intensive grey energy is used nowadays to produce the modern construction materials made of concrete and steel. Therefore, new environmentally friendly and cost-effective building materials, so-called eco-building materials, are needed, which are mainly made of raw earth materials and have low energy consumption with respect to the modern materials.

However, the mechanical properties of silt-based clays need to be carefully studied and improved to achieve the best possible results with the lowest grey energy consumption, in order to become a better and competitive alternative to typical cement-based concrete. Some of the mechanical properties of this silt-based material have already been studied, such as its compressive strength [1] and ductility [2]. However, several other mechanical properties can also be used to investigate the mechanical properties of raw earth materials, such as deformation modulus and toughness. Generally, toughness is considered as the ability of a material to resist impacts and dynamic loads. It is considered in the literature, as the energy absorbed without cracking; or in other definitions, it is the energy required to slow the propagation of a crack before the material fails. With the exception of metals, studies on the toughness of materials have rarely been carried out, despite the important role they play in determining the energy a material absorbs under load before it breaks. This paper examines the toughness of this bio-based raw material and its relationship with its common known mechanical property: the ultimate compressive strength.

Keywords: Bio-based construction materials · Silt-based raw earth concrete · Stress-Strain curve · Strain Energy · Ultimate Compressive Strength · Toughness

1 Introduction

Raw earth has been used for thousands of years as one of the main building materials and remains usable until now. Raw earth is a material available all the time and everywhere. Its use as a building material exploits one of its properties, cohesion, which acts as

S. Amziane et al. (Eds.): ICBBM 2023, RILEM Bookseries 45, pp. 310–323, 2023.
https://doi.org/10.1007/978-3-031-33465-8_25

a natural binder. There are various construction methods with an infinite number of variants that reflect the identity of places and cultures. Twelve methods of using earth in construction have been identified, six of which are very commonly used and constitute the major technical genres.

Recalling from the 17th century, in south China, the Hakka earth buildings (Kejia tulou) or the roundhouses were built from raw earth materials. These buildings were known by their enormous size and robust architecture with the ability to shelter hundreds of people within one stronghold. They were made of a mixture of earth, sand, lime, glutinous rice, bamboo and wood chips, with walls up to two meters thick [3].

Today, there are more and more modern raw earth buildings around the world. For example, the Chapel of Reconciliation in Berlin was built between 1990 and 2000 [4].

Recently, in 2014, a medical complex called Bayalpata Hospital was constructed in Achham, Nepal, built using rammed earth. It was built as replacement of an old and small hospital in the location, and now with this new building, the hospital will provide low-cost and high-quality care to 100,000 patients a year from Achham which is around more than its original capacity in the old building [5, 6].

Nowadays, the construction materials from concrete and steel need high and intensive grey energy causing high emission of greenhouse gases. These greenhouse gases are manipulating the climate to change on the world level, which gained the attention of all the countries around the world. Hence, a new eco-friendly and more economical construction material is essential to overcome these challenges. The best alternative for the present is the raw earth-based construction material which is considered to be eco-geomaterial with low energy consumption compared to the cement-based construction materials [7, 8].

Raw earth material has the upper hand, since it only requires 1% of energy to be produced with respect to that of the normal concrete [9]. Raw earth material are natural materials which makes them recyclable and reusable. Moreover, they are available everywhere and in large quantities having low and affordable cost.

Various studies were done to improve different properties of the raw earth material related to its mechanical strength, shrinkage and swelling properties, and hygro-thermal properties [10]. Raw earth alone is still not very compatible to be used alone, so, stabilizers such as lime, cement, and/or gypsum and reinforcement using fibers are added to the raw earth mixture enhance its performance [11].

Based on the used raw earth construction technique, adding a small quantity of binders; such as lime and/or cement can improve the material's compressive strength to a certain level in order to be accepted as a construction and building material [10, 12, 13].

The influence of increasing the percentage of flax fibers on the elastic modulus on earth concrete was studied by Kouta et al. 2021 [14]. Where they studied the influence of different proportions content using 0%, 3% and 4.5% and various lengths as12 mm, 24 mm and 50 mm of flax fiber added to earth mixture with constant lime and cement proportions of 3% and 8% respectively.

Imanzadeh et al. 2018 and 2020 studied the unconfined compressive strength of the raw earth material [1] and its ductility [2] and stated the influence of different substituents added in the mixture on these properties. With the help of the design of experiments, they were able to predict and find an optimal range of substituents used in the mixtures to have acceptable mechanical properties as a construction material. Then, to better enhance the mechanical properties of the raw earth and a better understanding with an optimal formulation, additional properties should be studied as toughness which is considered as interesting property to be shown in this paper.

Toughness of a material plays an important in terms of the energy a material can handle, few studies were done on fiber reinforced hydraulic lime mortar [15, 16] and fly ash concrete [17] but none is done on the raw earth concrete. Moreover, toughness is studied for reinforced concrete [18–22] highlighting the fact that adding fibers to a mixture can increase its energy absorption and improve the toughness index of the samples. Although reinforced concrete is a bit different from the raw earth material and can't be compared in terms of constituents, yet, the same definition of toughness referred to the reinforced concrete can be used for earth material too.

In this paper, the toughness of the material from the literature was taken into consideration with some modifications. This work concentrated on studying "toughness" for raw earth which is not done before for this material, to better understand the properties of this material and what ingredients can influence this property to reach to an optimal formulation. Moreover, to relate this toughness to a known mechanical property of the material, a relationship between the toughness of the raw earth concrete and its ultimate compressive strength was shown.

2 Materials

Raw earth concrete is a combination of several materials made of soil, flax fibers, water and binders: cement and lime. Natural silt was chosen as a building material for being plentifully available locally. According to the grading size distribution curve of the soil, its Atterberg limits, and based on the LPC-USCS (ASTM D2487-17e1) standards [23], this soil was categorized as silty sand (SM). Imanzadeh et al. 2018 described in details the properties of this silty sand [1].

The binders from cement and lime were used in the mixture of the raw earth concrete. The quantities of binders in the mixture was limited by 16% maximum for economical and ecological reasons. Where some detailed properties of these binders were well described by Eid 2017 [24].

As the Normandy region is responsible for 55% of the total production of flax in France [25], these flax fibers were locally extracted and chosen to be added to the mixture. The fiber content in the mixture varied between 0.3% and 0.45% in mass.

Limited water content was considered in the mixture to minimize the shrinkage of the material. Additives as superplasticizer were added as an alternative to preserve the consistency during the manufacturing process of the concrete for a better workability of the material. These superplasticizers were used as lubrification of the solid surfaces and it decreases the friction stresses between particles [26]. 5 ml/m^3 of additives has been added for each sample preparation. A laboratory mixer of 4L capacity is used for the mixing procedure in order to obtain homogeneous samples with a random distribution of the materials. Additional details are mentioned by Imanzadeh et al.2020 [2].

2.1 Ranges for Mixing

Different constraints from economical, ecological, and environmental ones were respected constraints. As for the fundamental constraint, the summation of the mass of all constituents used must be 100% in the mixture. As explained before, binders were limited by maximum of 16% for economical and ecological reasons. Finally, the workability constraint which was assured for a better mechanical property [27] by including the fluidity conditions with the help of the concrete slump test fulfilling the S3 consistency condition following the standard NF EN 12350-8 [28]. Considering these constraints, the mixing range (lower and upper limit) of the constituents in the mixture is defined as follows in Table 1.

Table 1. Mixing range of each constituent

x_i	Photos	Lower Limit (%)	Upper Limit (%)
x_5: Silt		47	75
x_1: Fiber	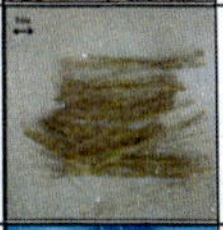	0.3	0.45
x_4: Water		20	25
x_3: Cement		4	16
x_2: Lime		0	12

2.2 Formulations

Considering the mixing limits of the constituents used listed above, 25 different formulations were established varying all constituents at the same time respecting all the constraints highlighted before. Table 2 shows the four formulations that were considered in this study, where each one has different percentages by mass of the constituents.

Only four formulations were taken into consideration with different behaviors an various changes in their constituents. These various behaviors were considered as th most representatives to the main four behaviors that were realised through all the fo mulations that had been experimented. This collection of these four formulations considered to be a good representative of all the other formulations.

Table 2. List of the considered formulations and their constituents

Formulation	Silt (%)	Fiber (%)	Water (%)	Cement (%)	Lime (%)
I	61.31	0.30	22.39	16.00	0.00
II	72.80	0.30	22.90	4.00	0.00
III	60.63	0.45	22.92	16.00	0.00
IV	59.43	0.45	24.12	11.21	4.79

3 Experimental Method

To study the ultimate compressive strength and toughness of the material, unconfined compressive strength test was performed as following:

3.1 Sample Preparation

The majority of the chemical reactions due to binders was ensured by curing the samples for 90 days by storing the produced specimens with a controlled relative humidity RH≈50% and constant temperature T≈22 °C. 75 specimens are produced in total, consisting of 3 trials for each of the 25 formulations considered with different mixing proportions.

3.2 Unconfined Compressive Strength Test

Following the NF P94–420 [29], NF P94–425[30] French standards, an axial Unconfined Compressive Strength test was performed on a 90 days-cured samples.

Imanzadeh et al. 2018 [1] stated in details the experimental device and the procedure. ...e force exerted on each specimen was recorded with its corresponding displacement ...ng corresponding sensors. Where the axial strain is calculated by dividing the axial ...lacement recorded by the initial height of the specimen (Fig. 1) and its corresponding stress is calculated by dividing the force with the specimen's cross-sectional (Fig. 2).

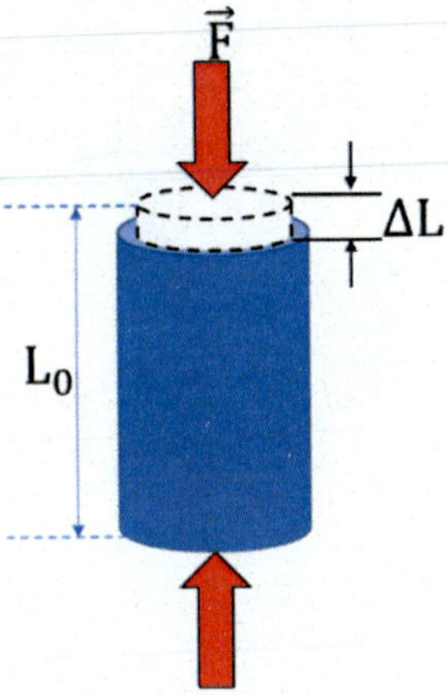

Fig. 1. Deformation of specimen due to applied axial load

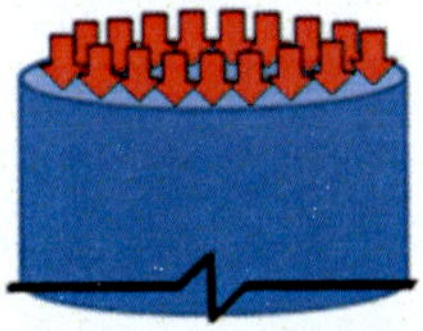

Fig. 2. The applied stress on the top of the specimen

4 Ultimate Compressive Strength (UCS)

Using this raw earth concrete, only nonreinforced structural elements subjected to axial compression [31–34] were intended to be constructed. Where the UCS is the maximum axial compressive stress that a specimen can withstand without applying any confining stress [35–37].

Hence, different parameters could be deduced from the applied axial stress versus axial strain curve obtained from the unconfined compressive strength test. An example of the stress-strain curve relationship is shown in Fig. 3.

In some trials the UCS (the maximum in the stress-strain graph) is not reached so this index couldn't be calculated and replaced by '/'.

5 Toughness

From the obtained stress-strain curves, a threshold of 15% of strength loss after reaching the peak (Fig. 4) is considered as enough to analyze the plastic behavior of the material Imanzadeh et al. 2020 [2]. Hence the study of toughness is done according to this threshold.

In general, toughness is the ability of the material to endure impacts and dynamic loads. It is defined as the energy needed to crack a material before fracturing, or in other

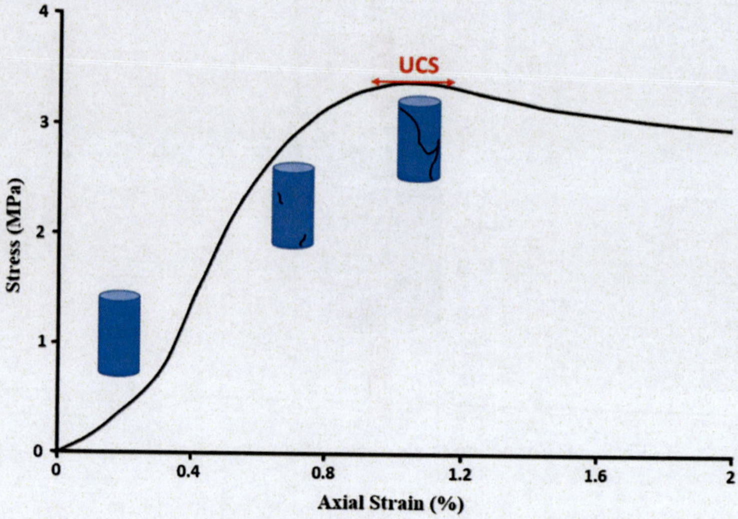

Fig. 3. An example of the stress-strain curve of raw earth concrete

words, it indicates the total amount of strain energy per unit volume a material can handle just before it fractures [38].

Toughness is the combination of strength and plasticity [39], where a tough material can take hard blows without rupturing. So, a material should withstand high loads and be strong and should have high deformations to be considered as plastic and ductile.

The toughness of the material was calculated as the area under the stress-strain curve until reaching the threshold of 15% of strength loss after reaching the maximum. This quantity represents the entire area under the stress-strain diagram presented in Fig. 4 considering the explained threshold.

As said before, in some trials the UCS is not reached and hence the threshold cannot be defined and hence, the toughness of these special trials couldn't be calculated and replaced by '/'.

6 Results and Discussion

Various behaviors of the raw earth material were studied (Fig. 5). Different behaviors could be seen between Formulations I and III even though they show a compatible toughness. Formulation I is considered as strong material with high ultimate compressive strength compared to the other formulations but more fragile. This causes the loss in toughness of FI compared to FIII. Hence, two specimens can have same toughness but one is stronger (FI – T1 = 93.7 kJ/m^3) while the other is more ductile (FIII – TIII = 98,3 kJ/m^3).

The importance of higher strength on the toughness is highlighted when Formulations I and IV are compared. Where, Formulation I has almost the double ultimate compressive strength compared to that of Formulation IV leading for FI to have a very much higher toughness.

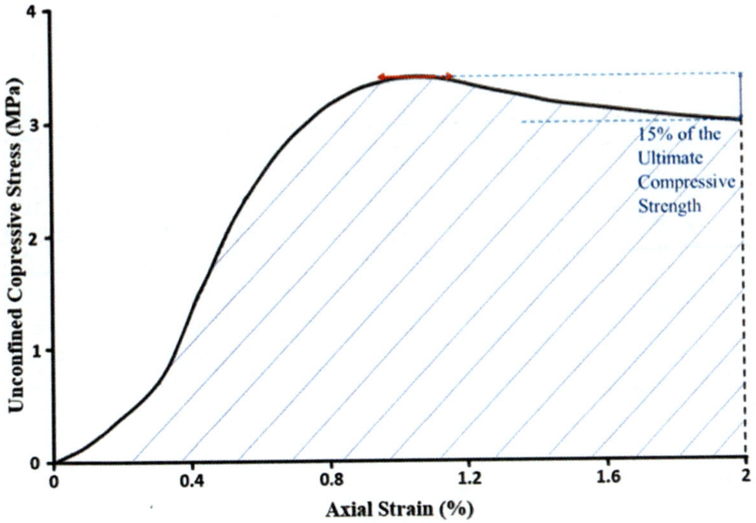

Fig. 4. An example of the raw earth material's stress-strain curve with threshold

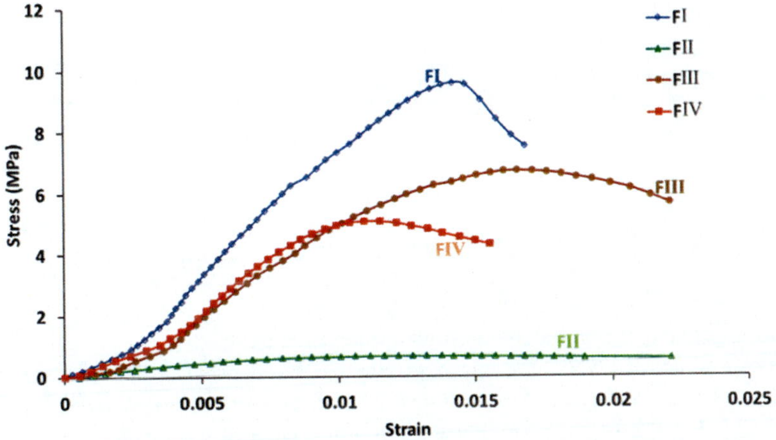

Fig. 5. Stress- strain curves of the Formulations 1, II, III, and IV

The UCS values of the three trials of the four considered formulations obtained from the stress-strain curves are presented in Table 3 with the mean value of the UCS for each specimen. In the same manner, the toughness obtained from the areas under the stress-strain curves for each of the three trials of each formulation is presented in Table 4.

Toughness of raw earth material ranges from 11.8 kJ/m^3 (for formulation with very low lime and cement contents and considered to be highly deformable) to 98.3 kJ/m^3 for 90 curing days period (Table 4).

Table 4 UCS of raw earth material ranges from 0.7 MPa (for formulation with very low binders' content and considered to be weak) to 9.2 MPa for 90 curing days period (Table 3).

Table 3. UCS Values for the 3 trials of the considered formulations

Formulation	UCS_1 [MPa]	UCS_2 [MPa]	UCS_3 [MPa]	MeanvalueofUCS [MPa]
I	8.9	9.2	9.6	9.2
II	0.7	/	/	0.7
III	6.7	/	6.7	6.7
IV	5.0	5.1	5.1	5.1

Toughness of raw earth material ranges from 11.8 kJ/m^3 (for formulation with very low lime and cement contents and considered to be highly deformable) to 98.3 kJ/m^3 for 90 curing days period (Table 4).

Table 4. Toughness values for the 3 trials of the considered formulations

Formulation	T_1 [kJ/m^3]	T_2 [kJ/m^3]	T_3 [kJ/m^3]	Meanvalueof$T$ [kJ/m^3]
I	90.5	103.6	86.8	93.7
II	11.8	/	/	11.8
III	101.5	/	95.1	98.3
IV	41.2	49.7	51.7	47.5

As expected, cement content has high influence on the UCS of the material. Figure 6 shows that as the cement content increases, the UCS of the raw earth concrete increases too. As a maximum amount of cement was used of 16% by total mass, two formulations are found to have different UCS (Formulation I has UCS of 9 MPa and Formulation III with UCS of 6.7 MPa). This shows that other constituents can affect the UCS of the raw earth at the same time, while increasing the cement content causes in high increase in the UCS.

As the water to cement ratio increases and by that the quantity of water increases in the sample, the UCS of the material decreases (Table 5). This is due to the increase in the fluidity of the material and hence decreases its strength to withstand the load applied. As this ratio increase to reach around 6, the UCS decreases to reach 0.7 MPa for a very weak material.

Material with very low cement content causes the material to lose its ability to withstand high loads and hence its toughness is very low as for Formulation II with T around 10 kJ/m^3 (Table 5). The higher the cement content, the higher the toughness of the material is. Until reaching the maximum cement content used of 16% due to the

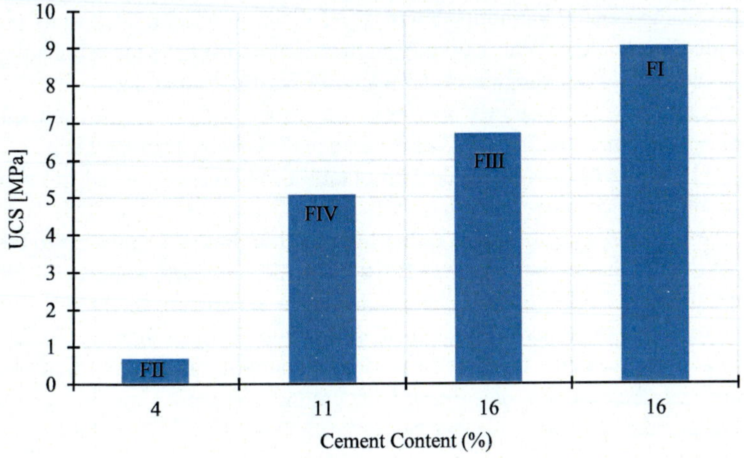

Fig. 6. UCS vs cement content

Table 5. Formulations and corresponding content and mechanical properties

Formulation	Fiber	Cement	Water/Cement	Density [kg/m^3]	UCS [MPa]	T [kJ/m^3]
I	0.0030	0.1600	1.3995	1390	9,05	88,67
II	0.0030	0.0400	5.7241	1397	0,67	11,81
III	0.0045	0.1600	1.4323	1387	6,71	98,29
IV	0.0045	0.1121	1.6626	1342	5,06	50,68

constraints considered, T reaches 98 kJ/m^3 for Formulation III showing that cement has high influence on this property but it is not the only influencing constituent.

As the water to cement ratio increases in the specimen, increasing the material's fluidity and hence losing its combined structure, causes the toughness to decrease. As this ratio is below 2, it gives a well combined structure having high toughness T above 50 kJ/m^3, whereas, the increase in this ratio where water is more than the cement, the material fluidity increases and decreases its toughness to reach a minimum of 10 kJ/m^3 for Formulation II having the value of water/cement ratio near to 6 (Table 5).

The UCS of the material can be related to the toughness of the material (Table 5). It is almost a linear relationship while the ability of the material to withstand higher deformations can increase the area under the stress-strain curve of the material leading to a higher toughness even though its UCS is not very high. As in the case of the Formulation III that has higher toughness (98 kJ/m^3) compared to formulation I (90 kJ/m^3) but it has lower UCS as explained before.

This means that withstanding higher deformations but with weaker material can give a higher toughness. A strong material means that the material is tough, but also a less fragile material but that can withstand higher deformations can mean the same thing. Hence a tough material can be strong and/or withstand high deformations.

Moreover, the density variation shows no clear information related to its influence on the studied parameters. Where both formulations I and II have close densities while the former has the highest toughness and the latter has the lowest one.

Additional study is considered which is the carbon intensity index related to the compressive strength of the material. This index is defined by Damineli et al. 2010 and 2013 [40, 41] as the total emitted CO2 to obtain a material compared to its strength or performance. To compare our material with the literature, the UCS is considered to define the material's performance rather than the toughness. Compared to the curve given by the Van Damme and Houben 2018 [42], the raw earth concrete studied lies down throughout different domains. This is due to the fact that in the literature, 40% of the carbon footprint of cement is not considered which changed their results compared to this material. As it can be seen that the raw earth material mixed with small amount of plasticizers, can decrease the footprint of CO2 emitted to manufacture the material and increases its performance which is its ultimate compressive strength in this case (Fig. 7).

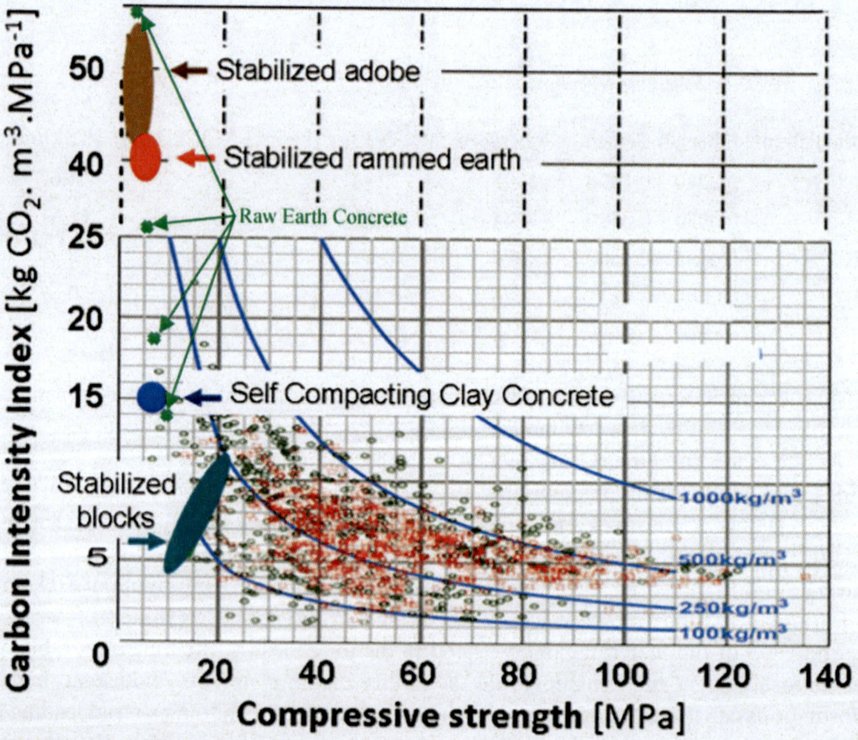

Fig. 7. Adjusted Carbon intensity index vs compressive strength [40, 42]

7 Conclusion

Toughness is an importantly significant characteristic that is needed to be studied for a construction material like raw earth concrete. The aim is to obtain a tough material that can withstand high loads and maintain its behavior at the same time at large strains. The durability of the material is ensured by its toughness. Varying the cement and water contents in the material, showed high effect on its toughness but should be verified with design of experiment that has been proved by other papers as suitable method to obtain optimal formulation having relatively good mechanical properties. All the data analyzed should be validated by the design of experiment to optimize the mechanical properties required.

Even though only four formulations were considered in our study, but these four formulations having different behaviors and difference in their constituents; it gave us an idea about the effect of some constituents on the UCS and toughness of raw earth concrete. These various behaviors considered, correspond to the main four behaviors we obtained through all the formulations had been experimented. Where we can say that this sample of the four formulations is quite representative for all other formulations.

It has been shown the positive effect of binders and specially cement on increasing the toughness of the material. On the other hand, water has a negative effect on this property specially when this content increases with respect to the cement content. Additional study is required with the use of Design of Experiments to have a better analysis of the effect of multiple constituents at the same time on one characteristic, and studying their influence on each other is interesting using a valid model for the toughness characteristic.

References

1. Imanzadeh, S., Hibouche, A., Jarno, A., Taibi, S.: Formulating and optimizing the compressive strength of a raw earth concrete by mixture design. Constr. Build. Mater. **163**, 149–159 (2018). https://doi.org/10.1016/j.conbuildmat.2017.12.088
2. Imanzadeh, S., Jarno, A., Hibouche, A., Bouarar, A., Taibi, S.: Ductility analysis of vegetal-fiber reinforced raw earth concrete by mixture design. Constr. Build. Mater. **239**, 117829 (2020). https://doi.org/10.1016/j.conbuildmat.2019.117829
3. The Hakka earth buildings of Fujian (2010). https://www.lonelyplanet.com/articles/the-hakka-earth-buildings-of-fujian
4. Kapelle der Versöhnung 'La chapelle de la réconciliation. https://www.berlin.de/mauer/fr/lieux/memoriaux/le-memorial-du-mur-de-berlin/kapelle-der-versoehnung-la-chapelle-de-la-reconciliation-527774.fr.php
5. sharon davis Design: Rammed earth bayalpata hospital opens in nepal, by sharon davis design (2020). https://www.designboom.com/architecture/bayalpata-hospital-sharon-davis-design-achham-nepal-04-10-2020/
6. Block, I.: Sharon Davis Design builds rammed-earth hospital in Nepal (2020). https://www.dezeen.com/2020/06/10/bayalpata-hospital-sharon-davis-design-rammed-earth/
7. Přikryl, R., Török, M., Theodoridou, M., Gomez-Heras, Miskovsky, K.: Geomaterials in construction and their sustainability: understanding their role in modern society. Geol. Soc. Spec. Publ. **416**(1), 1–22 (2016). https://doi.org/10.1144/SP416.21
8. Hibouche, A.: Présentée à Sols traités aux liants Performances hydro-mécaniques hydro et hygro-thermique Applications en BTP (2013)

9. Zak, P., Ashour, T., Korjenic, A., Korjenic, S., Wu, W.: The influence of natural reinforcement fibers, gypsum and cement on compressive strength of earth bricks materials. Constr. Build. Mater. **106**, 179–188 (2016). https://doi.org/10.1016/j.conbuildmat.2015.12.031
10. Ashour, T., Korjenic, A., Korjenic, S., Wu, W.: Thermal conductivity of unfired earth bricks reinforced by agricultural wastes with cement and gypsum. Energy Build. **104**(July), 139–146 (2015). https://doi.org/10.1016/j.enbuild.2015.07.016
11. Delgado, M.C.J., Guerrero, I.C.: Earth building in Spain. Constr. Build. Mater. **20**(9), 679–690 (2006). https://doi.org/10.1016/j.conbuildmat.2005.02.006
12. Kanema, J.M., Eid, J., Taibi, S.: Shrinkage of earth concrete amended with recycled aggregates and superplasticizer: Impact on mechanical properties and cracks. Mater. Des. **109**, 378–389 (2016). https://doi.org/10.1016/j.matdes.2016.07.025
13. Al-Mukhtar, M., Lasledj, A., Alcover, J.F.: Lime consumption of different clayey soils. Appl. Clay Sci. **95**, 133–145 (2014). https://doi.org/10.1016/j.clay.2014.03.024
14. Kouta, N., Saliba, J., Saiyouri, N.: Fracture behavior of flax fibers reinforced earth concrete. Eng. Fract. Mech. **241**(August 2020), 107378 (2021). https://doi.org/10.1016/j.engfracmech.2020.107378
15. Chan, R., Bindiganavile, V.: Toughness of fibre reinforced hydraulic lime mortar. Part-1: quasi-static response. Mater. Struct. Constr. **43**(10), 1435–1444 (2010). https://doi.org/10.1617/s11527-010-9598-4
16. Chan, R., Bindiganavile, V.: Toughness of fibre reinforced hydraulic lime mortar. Part-2: dynamic response. Mater. Struct. Constr. **43**(10), 1445–1455 (2010). https://doi.org/10.1617/s11527-010-9599-3
17. Golewski, G.L.: Estimation of the optimum content of fly ash in concrete composite based on the analysis of fracture toughness tests using various measuring systems. Constr. Build. Mater. **213**, 142–155 (2019). https://doi.org/10.1016/j.conbuildmat.2019.04.071
18. Laloui, L., Nuth, M., Vulliet, L.: Experimental and numerical investigations of the behaviour of a heat exchanger pile. Int. J. Numer. Anal. Methods Geomech. **30**(8), 763–781 (2006). https://doi.org/10.1002/nag.499
19. Zeyad, A.M.: Effect of fibers types on fresh properties and flexural toughness of self-compacting concrete. J. Mater. Res. Technol. **9**(3), 4147–4158 (2020). https://doi.org/10.1016/j.jmrt.2020.02.042
20. Barr, B.I.G., Liu, K., Dowers, R.C.: A toughness index to measure the energy absorption of fibre reinforced concrete. Int. J. Cem. Compos. Light. Concr. **4**(4), 221–227 (1982). https://doi.org/10.1016/0262-5075(82)90025-2
21. Jihad, A., Rami, H., Aghajan, J., Lecturer, S., Candidate, P.: Flexural and flexural toughness of fiber reinforced concrete-american standard specifications review. GRD Journals-Global Res. Dev. J. Eng. **4**(3) (2019)
22. Ye, Y., Liu, J., Zhang, Z., Wang, Z., Peng, Q.: Experimental study of high-strength steel fiber lightweight aggregate concrete on mechanical properties and toughness index 2020 (2020)
23. Standard Practice for Classification of Soils for Engineering Purposes (Unified Soil Classification System) - ASTM D2487-17e1 (2017)
24. Eid, J.: New construction material based on raw earth: cracking mechanisms, corrosion phenomena and physico-chemical interactions. Eur. J. Environ. Civ. Eng. **22**(12), 1522–1537 (2018). https://doi.org/10.1080/19648189.2017.1373707
25. Abbar, B., et al.: Experimental investigation on removal of heavy metals (Cu2+, Pb2+, and Zn2+) from aqueous solution by flax fibres. Process Saf. Environ. Prot. **109**(June), 639–647 (2017). https://doi.org/10.1016/j.psep.2017.05.012
26. Pierre, A., Mercier, R., Foissy, A., Lamarche, J.M.: The adsorption of cement superplasticizers on to mineral dispersions. Adsorpt. Sci. Technol. **6**(4), 219–231 (1989). https://doi.org/10.1177/026361748900600405

27. Bui, Q.B., Morel, J.C., Hans, S., Walker, P.: Effect of moisture content on the mechanical characteristics of rammed earth. Constr. Build. Mater. **54**, 163–169 (2014). https://doi.org/10.1016/j.conbuildmat.2013.12.067
28. AFNOR: Testing fresh concrete - Part 8 : self-compacting concrete - Slump-flow test - Nf en 196-10 (2017)
29. "Determination de la resistance a la compression uniaxiale - NF P94-420.pdf (2000)
30. Roches - Détermination du module de Young et du coefficient de Poisson - Nf p94–425 (2002)
31. Park, S.S.: Unconfined compressive strength and ductility of fiber-reinforced cemented sand. Constr. Build. Mater. **25**(2), 1134–1138 (2011). https://doi.org/10.1016/j.conbuildmat.2010.07.017
32. Maher, M.H., Ho, Y.C., Associate Members, ASCE: Mechanical properties of Kaolinite/fiber soil composite **120**(8), 1381–1393 (1995)
33. Groeneveld, A.B., Ahlborn, T.M., Crane, C.K., Burchfield, C.A., Landis, E.N.: Dynamic strength and ductility of ultra-high performance concrete with flow-induced fiber alignment. Int. J. Impact Eng. **111**, 37–45 (2018). https://doi.org/10.1016/j.ijimpeng.2017.08.009
34. Consoli, N.C., Montardo, J.P., Donato, M., Prietto, P.D.M.: Effect of material properties on the behaviour of sand-cement-fibre composites. Gr. Improv. **8**(2), 77–90 (2004). https://doi.org/10.1680/grim.8.2.77.36370
35. Claramunt, J., Ventura, H., Fernández-Carrasco, L.J., Ardanuy, M.: Tensile and flexural properties of cement composites reinforced with flax nonwoven fabrics. Materials (Basel) **10**(2) (2017). https://doi.org/10.3390/ma10020215
36. Rakhshanimehr, M., Esfahani, M.R., Kianoush, M.R., Mohammadzadeh, B.A., Mousavi, S.R.: Flexural ductility of reinforced concrete beams with lap-spliced bars. Can. J. Civ. Eng. **41**(7), 594–604 (2014). https://doi.org/10.1139/cjce-2013-0074
37. Fantilli, A.P., Chiaia, B., Gorino, A.: Fiber volume fraction and ductility index of concrete beams. Cem. Concr. Compos. **65**, 139–149 (2016). https://doi.org/10.1016/j.cemconcomp.2015.10.019
38. Hibbeler, R.: Mechanics of Materials Tenth Edition in SI Units (2017)
39. S.P.S. Incorporated: Metal Properties: Hardness, Toughness, & Strength. https://www.polymersolutions.com/blog/defining-metal-properties/
40. Damineli, B.L., Kemeid, F.M., Aguiar, P.S., John, V.M.: Measuring the eco-efficiency of cement use. Cem. Concr. Compos. **32**(8), 555–562 (2010). https://doi.org/10.1016/j.cemconcomp.2010.07.009
41. Damineli, B.L., Pileggi, R.G., John, V.M.: Lower binder intensity eco-efficient concretes. Eco-Effic. Concr. 26–44 (2013). https://doi.org/10.1533/9780857098993.1.26
42. Van Damme, H., Houben, H.: Earth concrete. Stabilization revisited. Cem. Concr. Res. **114**, 90–102 (2018). https://doi.org/10.1016/j.cemconres.2017.02.035

Mechanical and Thermal Analysis of Performance of Compressed Earth Blocks with Sawdust Material Stabilized with Cement

Ryad Bouzouidja(✉), Tingting Vogt Wu, and Mehdi Sbartai

UMR CNRS 5295, Université de Bordeaux, 2M Bordeaux, Bordeaux, France
{ryad.bouzouidja,tingting.vogt-wu,mehdi.sbartai}@u-bordeaux.fr

Abstract. The aim of this study is to use local resources, namely raw earth and sawdust wood to be used in building materials, in order to contribute to solving the housing problem and to reduce the energy consumption, through the use of raw earth constructions. To achieve this objective, the mechanical, physical properties (Young's modulus, maximum compression stress, and flexural strength) and thermal properties (thermal conductivity and heat capacity) of raw earth bricks physically stabilized with cement (5%) and incorporating sawdust wood (3%; 5% and 10%) were studied. The results show that the sawdust-raw earth has interesting behavior with a maximum compressive strength of about 4 MPa. In addition, the results show that sawdust reduces the mechanical properties of earthen blocks but increases their ductility. The incorporation of 10% of sawdust decreases the Young's modulus of raw earth of 30% comparing to 3% of sawdust. In addition, the results showed that the thermal conductivity decreased from 1.1 W m^{-1} K^{-1} for the 3% sawdust sample, to 0.8 W m^{-1} K^{-1} for the 5% sawdust sample, and 0.7 W m^{-1} K^{-1} for the 10% sawdust sample, indicating an improvement in thermal insulation. Therefore, the results of this study give additional critical feedback to assist in evaluating the potential applications of sawdust as an insulating material for unfired clay blocks.

Keywords: Raw Earth Material · Thermal Conductivity · Compressive Strength

1 Introduction

The building sector is currently responsible for the world's greenhouse gas emissions and also almost 36% of global energy consumption (International Energy Agency, 2020). The most common material used in construction is concrete (mostly made by Portland cement). It represents more than 25 billion tons [1–3]. Concrete is widely used in various industries (e.g. construction of bridges, dams, buildings, tunnels and many physical infrastructures) [4, 5]. Large-scale consumption of concrete has strained natural sources of raw materials in recent decades [6]. It is therefore essential to identify solutions in order to reduce the energy demands of buildings without impacting the performance of the building and without altering its structure.

S. Amziane et al. (Eds.): ICBBM 2023, RILEM Bookseries 45, pp. 324–332, 2023.
https://doi.org/10.1007/978-3-031-33465-8_26

The use of concrete is today confronted with strong contradictions. The need in terms of cost and reduced carbon impact. However, identifying the carbon footprint of concrete remains complex. Purnel [7] reported that attributing a value for the quantity of carbon dioxide emitted per unit of production to the major structural materials is not straightforward. Considering that the majority of concrete is reinforced concrete. The carbon impact calculation includes contributions from cement, reinforcing steel, aggregates, water and admixtures. In general, the carbon impact is between 0.07 and 0.45 $kgCO_2\ kg^{-1}$ reinforced concrete [7]. Such increase in cement use will significantly increase the amount of CO_2 emissions in the atmosphere and cause global warming impacts [8]. The production of each ton of cement (e.g. Portland, clinker) releases an equivalent mass of CO_2 into the atmosphere. In reality, this amount released varies considerably depending on the energy efficiency of the plant and its age [9]. Different ways are suggested to minimize the use and the environmental impacts of concrete, namely, reduce, reuse, and recycle. Among these strategies, biosourced concrete is the most widely adopted strategy worldwide, partly due to its ease of implementation and the availability for the biosourced products, namely hemp, wood sawdust or raw earth. Considering only the manufacturing phase, Maskell et al. [10] founded that elying on the Inventory of Carbon and Energy (ICE) [11] and Ecoinvent [12] databases, that the choice of unfired extruded clay brick induces an embodied energy saving of 86% and of 25% as a reduction of 86% and 72% for CO_2 emissions in comparison, respectively, to fired clay and concrete blocks.

Raw earth materials such as brick have several environmental, social and economic benefits, their use contributes in minimizing the negative effects of the building on the environment while increasing energy savings [13]. The processing and handling of raw earth materials requires only 1% of the energy needed to manufacture and process the same volume of cement concrete [14]. Similarly, the manufacture of compressed earth blocks requires almost one third of the energy needed to manufacture conventional fired bricks, i.e. 440 kWh m^{-3} compared to 1300 kWh m^{-3} [15]. It is true that concrete allows the design of self-supporting walls, to have a low thermal inertia. However, one of the disadvantages of concrete is heat transfer. Concrete, being one of the most commonly used construction materials, its thermal conductivity draws importance [16]. For example, the thermal conductivities of concrete components (e.g. rocks) can vary from 1.163 to 8.6 W $m^{-1}K^{-1}$ [17].

The use of raw earth as a construction material can significantly reduce the environmental impact of current building practice. Raw earth can be locally sourced. In order to improve thermal and mechanical behavior of the raw earth material, two stabilization methods, i.e. chemical and physical, are actually applied [10]. The physical method consists of performing a granular correction and then incorporating fibers [18].

The aim of this study is to valorize local resources, namely earth and hemp to be used in building materials, in order to contribute to solving the housing problem and to reduce the energy consumption, through the use of raw earth constructions. To achieve this objective, it was decided to study the mechanical, physical and thermal properties of raw earth bricks physically stabilized with lime and incorporating hemp.

2 Materials and Methods

2.1 Materials

The local materials used are sand, clay and sawdust. The sand used was crushed sand with a particle size distribution of 0–4 mm from the Massif Central region in France according to the standard XP P13–901. The clay used in this study is an *illite* type provided by the CM Quartz Company. It is supplied from the same location as the sand. The sawdust material is a local product provided by the sawmill close to Bordeaux. This sawdust is produced by a chainsaw ref. TV3–15. The cement used in this study is a grey cement (grade 32.5) manufactured by Lafarge (65–79% of clinker).

The properties of the different martials are presented in Table 1.

Table 1. Physical characteristics of the used material. $^{\Delta}$, *, $^{+}$, $^{\blacklozenge}$ are defined as the different formulation (BTC0, BTC3, BTC5, and BTC10 respectively).

Material	Ratio (%)	Initial water content (wt%)	Density (kg m^{-3})
Sand	75^{Δ};72^{*}; 70^{+}; $65^{\blacklozenge}$	0.6	620
Clay	20	3.1	670
Cement	5	29.9	370
Sawdust	0^{Δ};3^{*}; 5^{+}; $10^{\blacklozenge}$	1.6	252

2.2 Sampling

The formulations were mixed using concrete mixer. The mortar preparation was made as follow. First, 30 wt% of sand and 70 wt% of clay were mixed together. Then, the water-saturated sawdust have been added gradually. After homogenization of the mixture using a vibrating table, the material was placed in the mold and compacted immediately. The parallelepiped (10 × 10 × 10 cm) and (23 × 5.6 × 5 cm) specimens were made by pressing using the hydraulic press of 3.7 MPa. The specimens were stored in plastic bags and after 28 days of curing, physical, mechanical, and thermal conductivity tests were performed.

To determine the particle size distribution, the sieving method was used. A sand specimen of about 200g was taken from the sand sample, it was then put in an oven to dry for 24 h. After the drying time, the sand was used to determine the particle size distribution using 14 sieves with different openings [19]. Also, a test for geometrical properties of aggregates – assessment of fines was done based on standard NF EN 933–8. The objective of this test is to rapidly assess the relative portion of clay in the sand that is to be used in construction. A low value of sand equivalent describes the fine aggregate as “dirty” and indicates that possibly the clay materials are harmful.

2.3 Thermal and Mechanical Experiments

In this part of the work, the thermal characterization method was used for the measurement of thermal conductivity λ [W m^{-1} K^{-1}] and volumetric heat ρc [J K^{-1} m^{-3}]. These properties were assessed in accordance with Standard ISO8301 [20] and NF 12667 [21]. These methods are based on a steady state technique. The measurements were made using a heat flux probe (accuracy $\pm$ 20W m^{-2}) and type T thermocouple (accuracy $\pm$ 1.0 °C). The sample is placed between two exchanger plates (see Fig. 1).

To determine thermal conductivity the method consists in in subjecting a sample of thickness l to a temperature gradient, of so as to impose a flow transfer from the hot side to the cold side. The exchanger plates are maintained at different temperatures (θ_1 and θ_2, expressed in °C). It is important to note that when studying wet porous materials, the stresses thermal conditions imposed on the system can cause mass transfers to appear during the handling. In this type of material, it is necessary to impose a limited temperature difference to avoid coupled transfers.

Volumteric heat capacity is determined using the same measurement as the thermal conductivity, only imposed thermal stresses change. The sample-flowmeter system is first subjected to the stabilized temperatures of the exchanger plates until an equilibrium state is reached. It is then made to evolve towards another state of equilibrium by changing one or more instructions of the thermostatic baths.

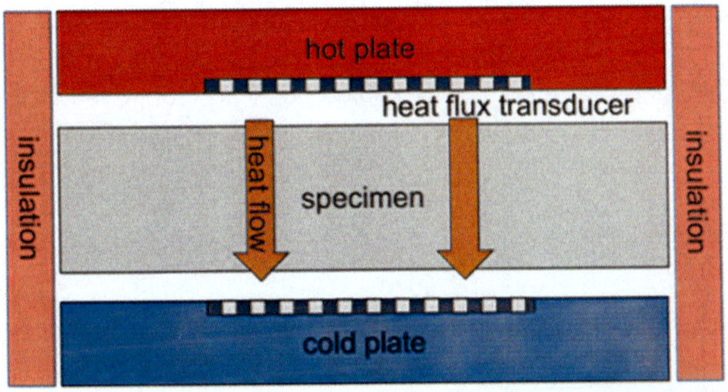

Fig. 1. Schematic diagram of the principle of the heat flow meter based on Kim et al. study [22].

Compressive strength was carried out on cubic blocks (10 × 10 × 10 cm) using hydraulic press. The sample compression test was carried out using a Picsx with a capacity of 100 kN with a loading speed of 0.5 mm min^{-1} and Testwork software. The specimens were equipped with three Linear Variable Differential Transformer (LVDT) sensors to measure the longitudinal deformation [23] (Fig. 2).

Three-point bending tests were also performed on beams of dimension. The tests were performed according to RILEM-TMC 50 recommendations on beams at 28 days of age and kept in a climatic room at 20 °C and 60% RH. Bending tests were performed using an electromechanical machine with a capacity of 100 kN. Unnotched beams were used to avoid stress concentrations at early age that can cause early beam failure. The test was driven with a CMOD (Crack Mouth Opening Displacement) notch opening

Fig. 2. Compressive test device used for (a) compressive strength and (b) three-point bending test.

system with two LVDT sensors (0.01 μm accuracy) placed on both faces of the beam at the bottom center.

These two sensors were placed on the beam by two metal supports glued to the specimen. The aluminum plates, supports for the LVDT sensors, were glued at a distance equivalent to half the height from the middle of the beam to ensure that the crack appears between these two plates. Although this measurement is not exactly the CMOD in the case of unbraced beams, for simplicity it has been called as such. The crack opening velocity was varied during the test with a velocity of 0.05 mm min^{-1} before the peak and a velocity of 0.4 mm min^{-1} towards the end to accelerate the failure in the stable softening phase. The deflection was also measured in the middle with an LVDT sensor and a measurement system to avoid problems related to concrete crushing at the supports.

3 Results and Discussion

3.1 Physical and Mechanical Properties

Figure 3 presents the grain size distribution of the sand. From this figure, the curve shows a regular distribution of particle size and favoring a specific range of diameter greater than or equal to 0.8 mm, which represents 60% in mass proportion. The fine sieved fraction which corresponds to the clayey texture represents a proportion of 4.8%. The clay fraction represents 1.2%. Sand equivalent (SE) test is around 65%. According to the French specifications (60 < SE < 70) is sand is suitable for use in building construction, cleanliness for quality concrete not particularly afraid of shrinkage.

Table 2 shows the results of the compressive strength and Young's Modulus. According to the results, we note that the stress evolves proportionally linearly with the deformation. This evolution verifies Hook's law. In this part the sample has an elastic behavior.

The slope of the linear regression line is called Young's modulus and its value is 10.9 GPa. After 0.0005 deformation, the stress changes with the deformation until it reaches a maximum value of 17.3 MPa before collapsing completely: This is the compressive strength. It turns out that the characterized material has a ductile behavior. It will thus

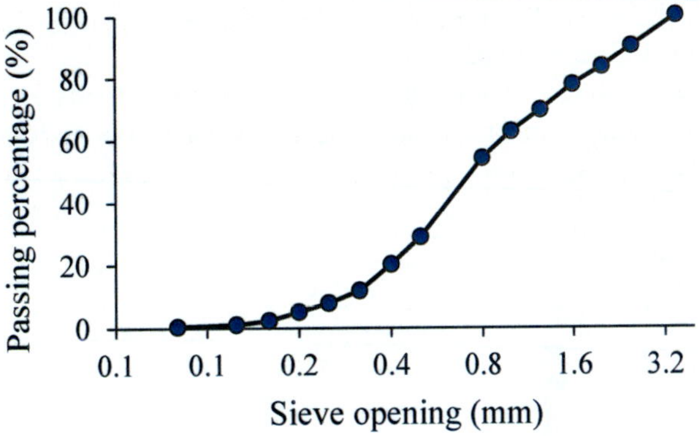

Fig. 3. Granulometery of the sand sample.

Table 2. Compressive stress and young modulus of wood sawdust reinforced concrete.

Ref	Density (kg m^{-3})	Compressive strength (N mm^{-2})	Flexural strength (MPa)	Young's Modulus (GPa)
BTC3	1697.2 ± 55.0	2.4 ± 0.4	0.7 ± 0.1	3.0 ± 0.3
BTC5	1697.3 ± 6.3	2.7 ± 0.3	0.6 ± 0.1	2.2 ± 0.4
BTC10	1624.9 ± 2.4	1.9 ± 0.4	0.4 ± 0.1	1.0 ± 0.2
BTC0	1936.3 ± 1.2	17.3 ± 0.5	2.1 ± 0.1	10.9 ± 0.7

be able to work in compression and withstand very large loads. It could, for example, be used in the construction of load-bearing walls, partitions, carry frames. It noted also a decrease of density of the reference (BTC0) 1936 kg m^{-3} to around 1600–1700 kg m^{-3} (BTC3, BTC5 and BTC10) by raising the treated fibers percentage. The same observation is made for the bending behavior. The reduction in compressive strength could be caused by non-uniform dispersion or segregation of fibers in the earth's matrix [23].

3.2 Thermal Properties

Figure 4 shows the results of a thermal conductivity and density for the different prepared samples. The results show that the thermal conductivity decreases with adding sawdust and reducing sand to the clayey soil.

Sample BTC0 exhibits a conductivity of about 1.201 ± 0.031 W m^{-1} K^{-1}, and after adding 3% of sawdust (BTC3) this value slightly decreases to 1.167 ± 0.046 W m^{-1} K^{-1}. Then by adding 5% of sawdust the value of the thermal conductivity reaches 0.853 ± 0.004 W m^{-1} K^{-1}. This decrease is due to the modification of clay granularity by the reduction of sand in the mixture. Otherwise, the samples with 10% of sawdust (BTC10) exhibit the lowest thermal conductivity (0.766 ± 0.041 W m^{-1} K^{-1}).

This is due to both the incorporation of fibers into the clay matrix and the improvement of their adhesion to the earth matrix. This behavior is in consistency with several authors' results [24] who had observed the decrease in thermal conductivity with the increase in the percentage of fibers.

Figure 5 shows the results of the volumetric specific heat (VSH) of raw earth bricks stabilized with sawdust contents. It was noted that when the sawdust content went from 0% to 10%, the VSH decreases from 2060.6 kJ m $^{-3}$ K^{-1} to 1821.3 kJ m $^{-3}$ K^{-1}.

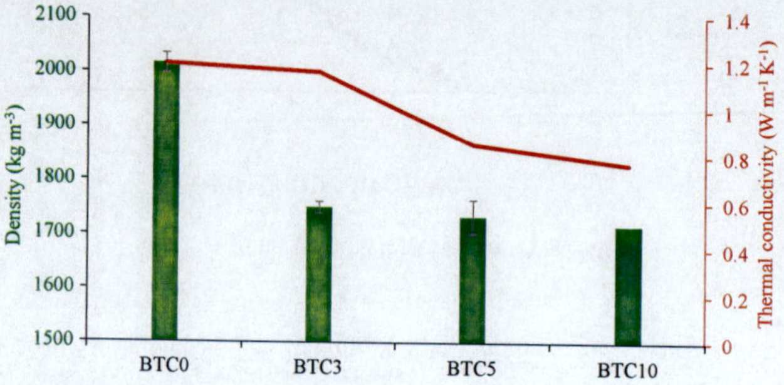

Fig. 4. Density (green bars) and thermal conductivity (red bars) of raw earth materials reinforced with sawdust. BTC0 means that no sawdust is added. BTC3 means that 3% of sawdust is added. BTC5 means that 5% of sawdust is used and BTC10 means that 10% of sawdust is used.

This can be explained by the fact that the incorporation of this fiber whose specific heat (wood family) is higher than that of the control clay block containing sand only.

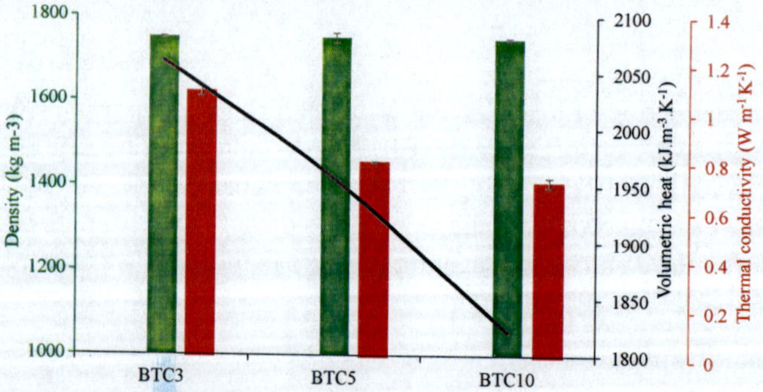

Fig. 5. Density (green bars), thermal conductivity (red bars) and volumetric heat (black continuous line) of raw earth materials reinforced with sawdust. BTC3 means that 3% of sawdust is added. BTC5 means that 5% of sawdust is used and BTC10 means that 10% of sawdust is used.

4 Conclusion

In this paper, the mechanical and thermal properties of cement-stabilized and sawdust-stabilized compressed earth bricks are investigated by experiments, and the relationships between the properties and factors are revealed. The experimental results show that the compressive properties of sawdust cement-stabilized compressed earth bricks are negatively correlated with the percentage of fiber addition. The thermal conductivity and specific heat of the compressed mud brick are positively correlated with the ratio of added sawdust. In this study, we test only the compressive strength, thermal conductivity and specific heat of the main structure. In the future, other indices of mechanical, durability or thermal properties may be added. This information may allow us to improve the understanding of the behavior of this type of material. Also, wall scale tests will be carried out.

References

1. Afroughsabet, V., Biolzi, L., Monteiro, P.J.M.: The effect of steel and polypropylene fibers on the chloride diffusivity and drying shrinkage of high-strength concrete. Compos. Part B Eng. **139**, 84–96 (2018). https://doi.org/10.1016/j.compositesb.2017.11.047
2. Celik, K., Meral, C., Mancio, M., et al.: A comparative study of self-consolidating concretes incorporating high-volume natural pozzolan or high-volume fly ash. Constr Build. Mater. **67**, 14–19 (2014). https://doi.org/10.1016/j.conbuildmat.2013.11.065
3. Cristino, T.M., Lotufo, F.A., Delinchant, B., et al.: A comprehensive review of obstacles and drivers to building energy-saving technologies and their association with research themes, types of buildings, and geographic regions. Renew. Sustain. Energy Rev. **135**, 110191 (2021). https://doi.org/10.1016/j.rser.2020.110191
4. McGinnis, M.J., Davis, M., de la Rosa, A., et al.: Strength and stiffness of concrete with recycled concrete aggregates. Constr. Build. Mater. **154**, 258–269 (2017). https://doi.org/10.1016/j.conbuildmat.2017.07.015
5. Yang, S.T., Li, K.F., Li, C.Q.: Numerical determination of concrete crack width for corrosion-affected concrete structures. Comput. Struct. **207**, 75–82 (2018). https://doi.org/10.1016/j.compstruc.2017.07.016
6. Tiwari, A., Singh, S., Nagar, R.: Feasibility assessment for partial replacement of fine aggregate to attain cleaner production perspective in concrete: a review. J. Clean. Prod. **135**, 490–507 (2016). https://doi.org/10.1016/j.jclepro.2016.06.130
7. Purnell, P.: The carbon footprint of reinforced concrete. Adv. Cem. Res. **25**, 362–368 (2013). https://doi.org/10.1680/adcr.13.00013
8. Nath, P., Sarker, P.K., Biswas, W.K.: Effect of fly ash on the service life, carbon footprint and embodied energy of high strength concrete in the marine environment. Energy Build. **158**, 1694–1702 (2018). https://doi.org/10.1016/j.enbuild.2017.12.011
9. Thomas, M., Barcelo, L., Blair, B., et al.: Lowering the carbon footprint of concrete by reducing clinker content of cement. Transp. Res. Rec. **2290**, 99–104 (2012). https://doi.org/10.3141/2290-13
10. Maskell, D., Heath, A., Walker, P.: Comparing the environmental impact of stabilisers for unfired earth construction. Key Eng. Mater. **600**, 132–143 (2014). https://doi.org/10.4028/www.scientific.net/KEM.600.132
11. Hammond, G., Jones, C.: Inventory of carbon & energy: ICE. Sustainable Energy Research Team, Department of Mechanical Engineering (2008)
12. Hischier, R., Weidema, B., Althaus, H., et al.: Implementation of life cycle impact assessment methods. Ecoinvent report no. 3, v2. 2 (2010)
13. Pittet, D., Jagadish, K.S., Kotak, T., et al.: Environmental impacts of building technologies: a comparative study in Kutch District, Gujarat State, India. In: Bolay, J.-C., Schmid, M., Tejada, G., Hazboun, E. (eds.) Technologies and Innovations for Development: Scientific Cooperation for a Sustainable Future, pp. 113–128. Springer, Paris (2012). https://doi.org/10.1007/978-2-8178-0268-8_9

14. Deboucha, S., Hashim, R.: A review on bricks and stabilized compressed earth blocks. Sci. Res. Essays **6**, 499–506 (2011). https://doi.org/10.5897/SRE09.356
15. Little, B., Morton, T.: Building with earth in Scotland: Innovative design and sustainability. Scottish Executive Central Research Unit Edinburgh (2001)
16. Khan, M.I.: Factors affecting the thermal properties of concrete and applicability of its prediction models. Build. Environ. **37**, 607–614 (2002). https://doi.org/10.1016/S0360-1323(01)00061-0
17. Asadi, I., Shafigh, P., Abu Hassan, Z.F.B., Mahyuddin, N.B.: Thermal conductivity of concrete – a review. J. Build. Eng. **20**, 81–93 (2018). https://doi.org/10.1016/j.jobe.2018.07.002
18. Losini, A.E., Grillet, A.C., Bellotto, M., et al.: Natural additives and biopolymers for raw earth construction stabilization – a review. Constr. Build. Mater. **304**, 124507 (2021). https://doi.org/10.1016/j.conbuildmat.2021.124507
19. Dishman, K.L.: Sieving in particle size analysis. In: Encyclopedia of Analytical Chemistry. John Wiley & Sons, Ltd. (2006)
20. ISO 8301 (1991). https://viewerbdc.afnor.org/Pdf/Viewer/AXlEBc-V7yA1. Accessed 1 June 2021
21. NF EN 12667 (2001). https://viewerbdc.afnor.org/Pdf/Viewer/C4G_g4acsMA1. Accessed 1 June 2021
22. Kim, D., Lee, S., Yang, I.: Verification of thermal conductivity measurements using guarded hot plate and heat flow meter methods. J. Korean Phys. Soc. **78**(12), 1196–1202 (2021). https://doi.org/10.1007/s40042-021-00177-0
23. Shiming, S., Yupu, S.: Dynamic biaxial tensile–compressive strength and failure criterion of plain concrete. Constr. Build. Mater. **40**, 322–329 (2013). https://doi.org/10.1016/j.conbuildmat.2012.11.012
24. Laborel-Préneron, A., Magniont, C., Aubert, J.-E.: Hygrothermal properties of unfired earth bricks: effect of barley straw, hemp shiv and corn cob addition. Energy Build. **178**, 265–278 (2018). https://doi.org/10.1016/j.enbuild.2018.08.021

3D Printable Earth-Based Alkali-Activated Materials: Role of Mix Design and Clay-Rich Soil

Pitabash Sahoo and Souradeep Gupta(✉)

Centre for Sustainable Technologies, Indian Institute of Science, Bangalore, Bengaluru 560012, India
souradeep@iisc.ac.in

Abstract. Earthen constructions are regaining popularity in the building and construction industry due to its low embodied carbon and recyclability. Often, inorganic binders, for instance, Portland cement, are used to stabilize earth-based materials to reduce dimensional instability and impart durability. However, with growing emphasis on decarbonization technologies in the construction sector, there is a need to develop low-carbon binders that are compatible with earth-based materials. Furthermore, additive manufacturing (commonly called 3D printing) method in construction industry is gaining popularity due to its potential to reduce material wastage, increase production rate and improve quality. While 3D printing of clay has been demonstrated, there is a dearth of investigation into printing of earth-based alkali-activated materials.

The research herein aims to develop two classes of 3D printable earth-based alkali-activated slag-fly ash concrete, where (i) 8 molar (M) NaOH solution, and (ii) 4M NaOH solution are used as activators at varying solution to binder ratio to achieve flowability for 3D printing depending on the clay content in the mix. The used earth contained 48–50% clay, primarily Kaolinite with some traces of Montmorillonite. In each of these mixes, the ratio of (slag + fly ash) to aggregate were 1:1, 1:2 and 1:3 by mass where, the clay content varied from 12% to 36% by mass of total fine aggregate. Through several trials, it was found that flowability of 40–50%, determined according to ASTM C1437, offered the desired consistency and rheological properties for extrusion and buildability. Experimental results show that 8M 1:2 mix offered highest buildability of about 460 mm with first layer settlement of only 0.21%. Buildability was found to reduce with reduction of molarity, evident from highest buildability of 377 mm for 4M 1:1 mix. This was attributed to the reduced viscosity of the solution and lower activator concentration, resulting in prolonged setting by 40–60% than 8M 1:1 and 1:2 mixes. On-site ultrasonic pulse velocity (UPV) measurement on the printable material showed 7.50–12% lower velocity through earth-based 4M mixes than the counterparts prepared with 8M, suggesting lower rate of densification and stiffening due to reduced molarity. Introduction of clay plays an important part in 3D printing by improving the thixotropic behaviour and reducing inter-particle friction otherwise existing between manufactured sand (M-sand) particles. While mixes containing combinations of clay-rich soil and M-sand could be extruded at flow values of 40–50%, the mixes with only M-sand (no soil) either could not be unextruded except for 8M 1:2 or offered poor printing quality mix even at a much higher flow

S. Amziane et al. (Eds.): ICBBM 2023, RILEM Bookseries 45, pp. 333–352, 2023.
https://doi.org/10.1007/978-3-031-33465-8_27

value of 88–90%. The designed mixes offered compressive strength of 19–34 MPa and 17–28 MPa at 28-day age for 8M and 4M of activator solutions respectively. Although a reduction in strength compared to control (only sand and no soil) was observed with higher reduction with increase in clay content, the designed materials satisfy the strength requirement as per Indian standard, IS 1725 for soil-based block constructions.

Keywords: Earth-based 3D printing · Alkali-activated materials · Strength · Setting time · Clay

1 Introduction

3D printing of cementitious materials enables formwork-free layer-wise construction, which accelerates the construction process and offers flexibility with lower lead-time than traditional process. While 3D printing in construction is an emerging technology, multiple investigations have reported that 3D printing of building components can be a "greener" and economical way of constructing future buildings compared to conventional construction methods. For example, a comprehensive life cycle impact assessment carried out by Mohammad et al. [1] showed that global warming potential, eutrophication potential, acidification potential and fossil fuel depletion of a 3D printed cement-based wall component are reduced by 21.60%, 61%, 56% and 41.50% respectively compared to conventionally cast wall system with hollow-core mould cast masonry blocks. Similarly, Weng et al. [2] also reported reduction of 25.40% in overall cost, 85.90% in CO_2 emission and 87.10% in energy consumption for a 3D printed prefabricated bathroom unit compared to traditional precast method currently practiced in the construction industry. Such benefits are mainly attributed to saving in material, reduction in construction time and optimized forms and geometries [2].

One of the important aspects of 3D printing in construction is the development of low-carbon materials, which are extrudable, buildable and satisfy the mechanical and durability requirements as per relevant standards. India has set an ambitious target to decarbonize the economy by 2070. Considering the increasing infrastructural demand and use of Ordinary Portland cement in Indian building sector, the greenhouse gas (GHG) emission and associated global warming potential is expected to increase in the coming decades. Production of 1 kg of Ordinary Portland cement (OPC) is associated with emission of 0.80–0.90 kg of CO_2, attributed to calcination of limestone. In this context, Reserve bank of India has recommended to explore green technologies to reduce CO_2 emission from the construction sector [3]. Although OPC-based materials are economical due to scaled-up production and high subsidies to boost construction projects, chemical admixtures (for instance, superplasticizers, accelerators and viscosity modifiers) and energy intensive supplementary mineral admixtures (for instance, limestone and silica fume) are necessary to develop printable materials. These lead to an increase in different life cycle impacts and the cost of printed material. Secondly, 3D printed cementitious composites are rich in binder with sand to cement ratio varying from 1:1 to 1.50:1 [4–6]. Increasing the aggregate content with respect to the binder without increasing environmental impacts can improve the economics associated with low-carbon 3D printing.

Combination of earth-based materials and alternative binders can be a potential option to develop low-carbon 3D printable material as an alternative to traditional OPC-based materials. Earth-based materials is regaining attention due to its low environmental impact, easy accessibility and recyclability. Construction of earth-based buildings rely on the access to suitable raw materials, largely comprised of natural subsoils. Although non-renewable, the raw materials are available in large quantities necessary to scale up earth-based constructions. However, there are several challenges associated with utilization for earth-based materials, including moisture sensitivity and dimensional instability due to clay. Therefore, stabilization of clay-rich soil is often practiced to address these challenges. For instance, Preethi and Reddy [7] and Marsh [8] have conducted extensive investigations into stabilization of clay-rich soil using alkali-activated binders, cement and combination of cement and lime. Although the findings show that alkali-activated binders can stabilize illite-rich soil, it is reported that activation of illite results in structural alteration of the clay mineral and produces blocks of poor physical properties than activated montmorillonite clays. However, it is to be noted that the physical properties of alkali-activated materials largely depend on the mix compositions including precursor compounds and the type of activators used. Furthermore, the flowability required for extrusion and buildability are different from that of earth-based building blocks, and therefore, necessitates further investigations. The flowability and buildability will change with activator content, aggregate-to-binder ratio in earth-based alkali-activated materials due to change in fluid demand with introduction of non-calcined clay into the mixture. There is a dearth of investigation into these areas, which are essential for developing printable mixture formulations.

In the context of Indian construction sector, it is reported that the global warming potential (GWP) of Portland cement concrete varies in the range of 597–660 kg CO_2-eq/ton, while geopolymer concrete entails significantly lower GWP, about 134–208 kg CO_2-eq/ton [9]. While this is encouraging, life environmental evaluation of geopolymer concrete by Habert et al. [10] shows that sodium silicate is the most energy intensive component of the activator system which leads to higher environmental impact of geopolymer than cement-based materials in several categories including human toxicity, fresh water toxicity and acidification. Including only sodium hydroxide (NaOH) at relatively low molar concentration as activator can mitigate the environmental impacts, but it is yet to be investigated how this will influence the mechanical strength, setting time, hydration and shrinkage of 3D printable earth-based materials. Therefore, this research aims to explore the flow, buildability, setting time, micro-structural build-up, hydration and compressive strength of earth-based alkali-activated materials, prepared at two different molar concentrations of NaOH – 4M and 8M and aggregate to solid (slag and fly ash) of 1:1, 1:2 and 1:3 by mass respectively.

2 Materials and Methods

2.1 Properties of Raw Materials

For alkali-activated mixes, the binder consisted of ground granulated blast furnace slag (GGBS from Jindal Southwest) and fly ash (FA) collected from a local power plant. The oxide composition of GGBS and FA, determined by X-ray fluorescence spectroscopy, are presented in Table 1. Slag with CaO to SiO_2 ratio (indicator of basicity) of 0.50–2 and Al_2O_3 to SiO_2 ratio of 0.10 to 0.60 is considered to be most suitable as precursor for AAM [11]. GGBS used in this research has CaO to SiO_2 ratio of 1.11 and Al_2O_3 to SiO_2 ratio of 0.43, which lie within the recommended limits. The total content of Al_2O_3, SiO_2 and Fe_2O_3 in FA is 90.87% (Table 1). SO_3 content is 1.36% (<5%). Therefore, the fly ash used in this research can be qualified as class F according to ASTM C618.

Sodium hydroxide (NaOH) solutions of 8M and 4M were used as alkali activators. NaOH pellets (97% purity) were dissolved in tap water, and then the solution was covered with parafilm for 18–20 h to let it cool down before use. M-sand conforming to ASTM C33 was used as fine aggregate. Locally available loamy red soil was used to replace 25% of M-sand in earth-based alkali activated mixes. The clay content, determined from hydrometer analysis, was 46–48%. The silt and the sand fraction are 13% and 39% respectively. The soil was wet-sieved through a 75 μm sieve, and the collected residue was subjected to XRF. The results are presented in Table 1. The pH of the soil diluted in DI water was 8.30.

X-ray diffractions of the used GGBS, FA and soil are presented in Fig. 1. The silica present in FA is partly crystalline, evident from the peaks of mullite and quartz. The soil is rich in kaolinite (K) with traces of montmorillonite. The specific surface area and micro-pore volume of GGBS, FA and soil were determined by nitrogen adsorption and computed using Brunauer-Emmett-Teller (BET) and Barrett-Joynar-Holanda (BJH) theories respectively. The results presented in Table 2 shows that FA has higher specific surface area and pore volume than GGBS. The specific surface area of soil is about 5 times higher than that of FA, while the pore volume is about 27% higher. This is attributed to the abundance of clay mineral (Kaolinite) which increases the meso-porosity and internal surface area of the soil. The results imply that introduction of soil will increase the fluid demand for maintaining the required flowability for the printed mixtures as the aggregate to (GGBS + FA) ratio is increased.

Based on Fig. 2, the particle gradation of M-sand is within the upper bound (UB) and lower bound (LB) for fine aggregates recommended in ASTM C33. The soil used in this study is finer than sand with median particle size of 182 μm.

Table 3 shows the mix composition and flowability values for 3D printable mixtures. The composition is normalized with respect to 100 g of the total mass of GGBS and fly ash. GGBS and FA were added at 3:1 by mass. The control mixes designed with 4M and 8M NaOH solution consist of only M-sand as aggregate, where sand to (GGBS+FA) ratio were 1:1, 1:2 and 1:3 by mass. In earth-alkali activated mixes (EAAM), soil was used to replace 25% of M-sand while the aggregate to (GGBS+FA) ratios were similar as that of control mixes. The control mixes were not extrudable at the similar flowability range (35–50%) as that of EAAM; the flow had to be increased to 87–94% for extrudability by adjusting the solution to (GGBS+FA) ratio by mass.

Printing was done using a gantry-based 3D printer fitted with a hopper and an extrusion screw. Extrusion was done through a circular nozzle of diameter 20 mm. The feed extrusion rate was 5 mm/s and printing was carried out at a speed of 10 mm/s. The extruded layer height was fixed at 20 mm for all the mixes. The printed shape was a hollow cuboid of footprint of 350 mm × 350 mm. The buildability of the extrudable mixes was measured from the total height that could be printed before collapse. The

Table 1. Oxide composition (%) of GGBS and FA and soil (<75 μm)

	GGBS	FA	Soil (wet sieved <75 μm and dried)
Calcium oxide (CaO)	38.60	3.23	0.38
Aluminium oxide (Al_2O_3)	16.60	25.40	33.90
Silica (SiO_2)	34.80	60.30	53.10
Ferric oxide (Fe_2O_3)	0.39	5.17	9.39
Sulfide (SO_3)	1.61	1.36	0.23
Sodium oxide (Na_2O)	–		–
Potassium oxide (K_2O)	0.46	1.42	1.34
Magnesium oxide (MgO)	5.40	0.89	–
Chlorine (Cl)	0.14	0.16	–
Phosphorus pentoxide (P_2O_5)	0.48	0.46	0.31
Titanium oxide	0.73	1.40	1.20
Other trace metal oxides	<0.80%	<0.20%	<0.15%

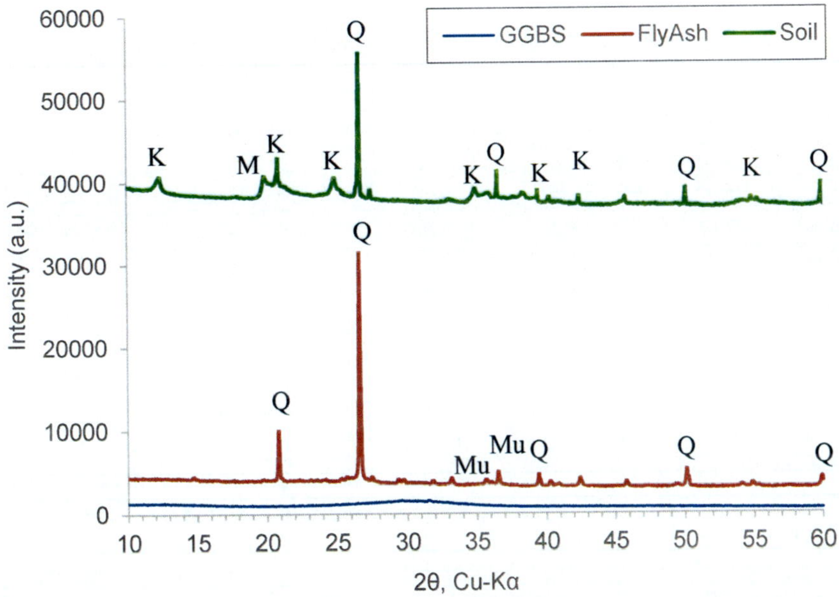

Fig. 1. X-ray diffractogram of GGBS, FA and soil (K: Kaolinite; M: Montmorillonite; Q: Quartz: Mu: Mullite)

stability of the printed filaments was investigated by measuring the change in layer height using a set of calipers. Excessive deformation of a bottom layer due to dead load of the layers above will lead to premature failure; therefore, the lower the change in

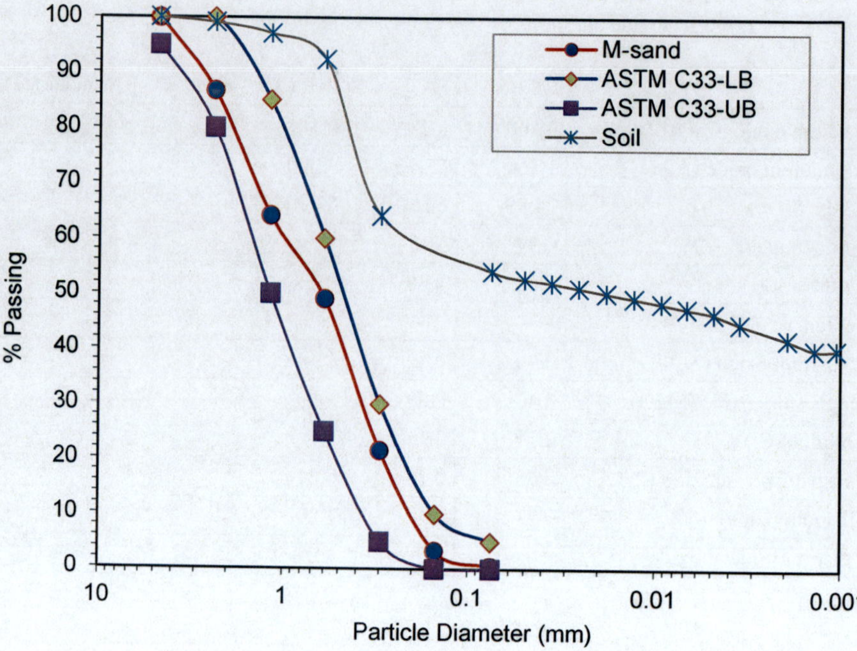

Fig. 2. Particle size distribution of M-sand and soil

Table 2. BET surface area and pore volume

	Specific surface area (m^2/g)	Pore volume (cm^3/g)	Average pore diameter (nm)
FA	6.79	0.0466	26.33
GGBS	0.53	0.0024	14.06
Soil	35.06	0.0596	4.90

height the better is the stability. Yuan et al. [12] suggested that the critical value for change in height of each printed layer shall be restricted to 0.20%. Although this may be applicable for Portland cement-based materials, further investigations are necessary for earth-based alkali-activated materials which have different dimensional stability than Portland-cement based composites.

Static yield stresses were determined at constant shear rate of 0.1 s^{-1}, while dynamic yield stress and plastic viscosities were determined in the shear rate range of 0.10–100 s^{-1}. The thixotropy of the materials were determined by three-interval thixotropy (3iTT) tests, consisting of shearing at 0.10 s^{-1}, 100 s^{-1} and back on to 0.10 s^{-1}. Shearing at each interval was done for 90 s. Initial and final setting times were monitored using an automatic Vicat instrument. Buildability, stability and development of strength depend on the densification of the matrix by hydration products, which is influenced by the rate of micro-structural build-up. Micro-structural build-up can be assessed by using

Table 3. Mix designs of the control and EAAM mixtures

Molarity	Mix code	GGBS	FA	M-sand	Soil	Clay content (% of GGBS+FA)	Solution to (GGBS+FA)	Flow (%) as per ASTM C1437
4M	Control-4M-1:1	75	25	100	0	0	0.47	87
	Control-4M-1:2	75	25	200	0	0	0.56	94
	Control-4M-1:3	75	25	300	0	0	0.66	91
8M	Control-8M-1:1	75	25	100	0	0	0.51	87
	Control-8M-1:2	75	25	200	0	0	0.66	93
	Control-8M-1:3	75	25	300	0	0	0.78	90
4M	25S-4M-1:1	75	25	75	25	12	0.46	47
	25S-4M-1:2	75	25	150	50	24	0.65	48
	25S-4M-1:3	75	25	225	75	36	0.90	53
8M	25S-8M-1:1	75	25	75	25	12	0.46	46
	25S-8M-1:2	75	25	150	50	24	0.74	34
	25S-8M-1:3	75	25	225	75	36	1.00	53

ultrasonic pulse velocity through the 3D printable fresh mixes during the initial 24 h from the addition of solution. The method is described elsewhere [13].

Compressive strength and shrinkage were determined according to ASTM C109 and ASTM C596 respectively. The samples were moist cured using wet burlap for the initial 7 days, considering prefabricated 3D printed building components are generally cured in the factory before deployment at site. After the initial moist curing, samples were exposed to constant temperature and relative humidity (30 °C and 65% RH) until further testing. Compressive strength under dry and saturated wet conditions were measured considering the sensitivity of earth-based building materials to moisture. For each mix, 4–5 samples were investigated.

3 Results and Discussions

3.1 Extrudability and Buildability

Table 4 shows the total number of layers and height of a hollow cuboid shape that could be printed with the developed materials before collapse. Although the flow spread (%) is increased in case of control mixes for extrudability, only Control-8M-1:1 is extrudable but not buildable. The extruded material shows discontinuities in the filaments (Fig. 3a), and noticeable deformation of the first three layers can be observed after printing of only seven layers (Fig. 3b). In comparison, Control-8M-1:2 could be printed up to 15 layers with a total height of 239 mm. This is attributed to two factors – (i) higher solution content which provided a lubrication effect, and (ii) reduction in the amount of fines due to dilution of the mixture by addition of more fine aggregates. M-sand used as fine aggregates has low water absorption capacity (0.60–0.70%), and therefore, the solution can wet and lubricate the surface of the aggregates and other solids, leading to improved extrudability and buildability than Control-8M-1:1. None of the 4M Control mixes is extrudable and buildable at similar flow spread as 8M control mixes.

In comparison, addition of soil as 25% replacement of sand positive contributes to extrudability and buildability. All the mixes containing soil are extrudable and buildable to varying extents as evident from Table 4 and Fig. 3d to 3j. The improvement in buildability of EAAM compared to control is attributed to the difference in flow retention (Fig. 4). For example, Fig. 4a and 4b show that control mixes, prepared with 8M and 4M NaOH solutions, show steep loss in flow retention within the first 60 min of addition of the solution. The flow retention varies between 70–76% for 8M control mixes and 57–60% for 4M control mixes. Addition of soil leads to higher flow retention, ranging from 94–99% for 25S-8M mixes and 86–95% for 25S-4M mixes. Better retention in flowability ensure longer open workable time, that contributes to higher quality of extrusion and buildability compared to control.

Table 4. Number of layers and total printable height with the developed mixes ('N' means not printable; 'Y' means printable; "–" means not applicable)

Molarity	Mix code	Extrudable	Buildable	Number of layers	Maximum height (mm)
4M	Control-4M-1:1	N	N	–	–
	Control-4M-1:2	N	N	–	–
	Control-4M-1:3	N	N	–	–
8M	Control-8M-1:1	Y*	N	–	–
	Control-8M-1:2	Y	Y	15	239
	Control-8M-1:3	N	N	–	–
4M	25S-4M-1:1	Y	Y	23	375
	25S-4M-1:2	Y	Y	15	240
	25S-4M-1:3	Y	Y	11	165
8M	25S-8M-1:1	Y	Y	15	243
	25S-8M-1:2	Y	Y	30	457
	25S-8M-1:3	Y	Y	10	162

*Control-8M-1:1 was extrudable but not buildable. It failed at 6th layer due to poor quality of the printed filaments.

Mix ratios play an important role in the buildability. For example, as the mix ratio of binder (GGBS+FA) to aggregates is changed from 1:1 to 1:3 with 4M NaOH solutions, the buildability is reduced from 23 layers (375 mm) (Fig. 3g) to 11 layers (165 mm) (Fig. 3j). It can be observed from Fig. 3h (25S-4M-1:2), that during printing of the 16th

Fig. 3. 3D printed hollow shapes of (a) Discontinuities in extruded filaments in Control-8M-1:1; (b) Deformation in printed layers due to poor quality of extruded material; (c) Control-8M-1:2; (d) 25S-8M-1:1; (e) 25S-8M-1:2; (f) 25S-8M-1:3; (g) 25S-4M-1:1; (h) 25S-4M-1:2 at 16th layer just before collapse (i) Reprinted 25S-4M-1:2 to 13 layers*; (j) 25S-4M-1:3. * 25S-4M-1:2 could be printed up to 15 layers and collapsed during printing of 16th layer as shown in Fig. 3h

layer, deformations on the sides and bottom layer have taken place, suggesting buckling failure. The shape remains steady until 15th layer (240 mm) and collapsed as the 16th layer is being printed. The same material formulation has then been used to print up to 13 layers (Fig. 3i), which shows formation of shear bands on the printed face. This is attributed to the faster structural build-up in 25S-4M-1:1 compared to 25S-4M-1:2 and 25S-4M-1:3 respectively. (Fig. 5). It can be seen from Fig. 5 (inset) that during the first 10–12 min, UPV through 25S-4M-1:1 mix starts increasing, while it takes about 45 min and 90 min for the first change in UPV through 25S-4M-1:2 and 25S-4M-1:3 to take place. The initial 30 min coincide with the printing window, meaning during this time 25S-4M-1:1 will show a higher gain in stiffness. Buildability of 3D printed structures is influenced by stability of the first printed layer – higher the stiffness, more stable will be the stacking up of the consecutive layers. Therefore, faster densification, as evident from higher evolution of UPV, in 25S-4M-1:1 ensure better buildability the other two 4M EAAM mixes.

For EAAM made with 8M NaOH solutions, 25S-8M-1:2 shows superior buildability with a buildable height of 457 mm (30 layers, Fig. 3e) compared to 243 mm (15 layers, Fig. 3d) and 162 mm (10 layers, Fig. 3f) for 25S-8M-1:1 (Fig. 3d) and 25S-8M-1:3 (Fig. 3f) respectively. From Fig. 5b, it can be observed that the UPV through 25S-8M-1:2 is initially higher than 25S-8M-1:1 mix, which can be attributed to the higher content of fine clay particles. As the clay fraction increases due to increase in soil content, it forms a solid flocculated network [14] for transmission of the ultrasonic pulses. However, due to high proportion of reactive solids (GGBS+FA), 25S-8M-1:1 rapidly gains static yield stress (Fig. 6a) besides achieving high dynamic yield stress and plastic viscosity (Fig. 5b and 5c). Rapid gain in yield stress results in suppression of the layer below, leading to instability during stacking up of the layers [15]. In comparison, 25S-8M-1:3 has 62–64% and 68–75% reduction in static yield stress (Fig. 6a) and plastic viscosity (Fig. 6c) respectively than 25S-8M-1:1although it possesses superior thixotropic behaviour (Fig. 6d and 6e) evident from highest viscosity recovery based on 3iTT. Clay is known to impart thixotropy to alkali-activated and cement-based 3D printable mixes [14, 16]; however, stability of the printed layer is an interplay between thixotropy, dynamic yield stress and static yield stress. Besides lower static and dynamic yield stress of 25S-8M-1:3, it is worth noting that the dynamic yield stress of this mix reduces with time – 603 Pa at 20 min, 559 Pa at 35 min and 430 Pa at 60 min (Fig. 6b) with slight increase of 18% in plastic viscosity from 20 min to 60 min (Fig. 6c). These suggest that continuous shearing leads to deformations in the mix, which can be attributed to the weakness imparted by the flocculation of clay particles. In comparison, 25S-8M-1:2 shows higher static yield stress and maintains dynamic yield stresses of 503 Pa, 563 Pa and 500 Pa at 20 min, 35 min and 60 min from the addition of solution. The plastic viscosities at 20 min, 35 min and 60 min are also fairly similar ranging from 1.21 to 1.26. The results suggest that 25S-8M-1:2 has more rheological stability than 25S-8M-1:3, which contributes to better shape retention and improvement in buildability. The recovery in viscosity of 25S-8M-1:2 ranges from 95% to 106%, which is slightly higher and similar to 25S-8M-1:1, suggesting that reducing the reactive solids (GGBS+FA) by adding in more aggregates (Sand + soil) does not affect the thixotropic behaviour of the printable mix. This approach will reduce the material cost while offering better buildability.

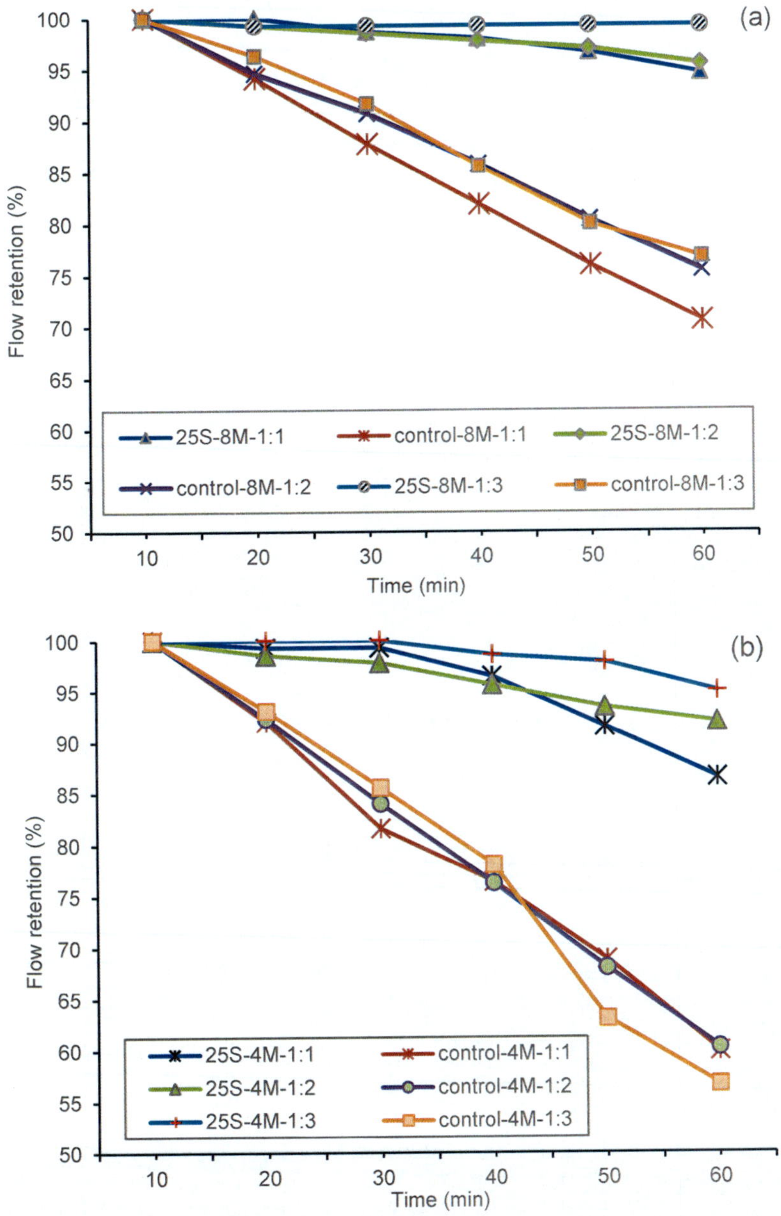

Fig. 4. Flow retention of earth-based alkali-activated mixes during initial 60 min of hydration with different NaOH concentrations of (a) 8 molar (M); b. 4 molar (M)

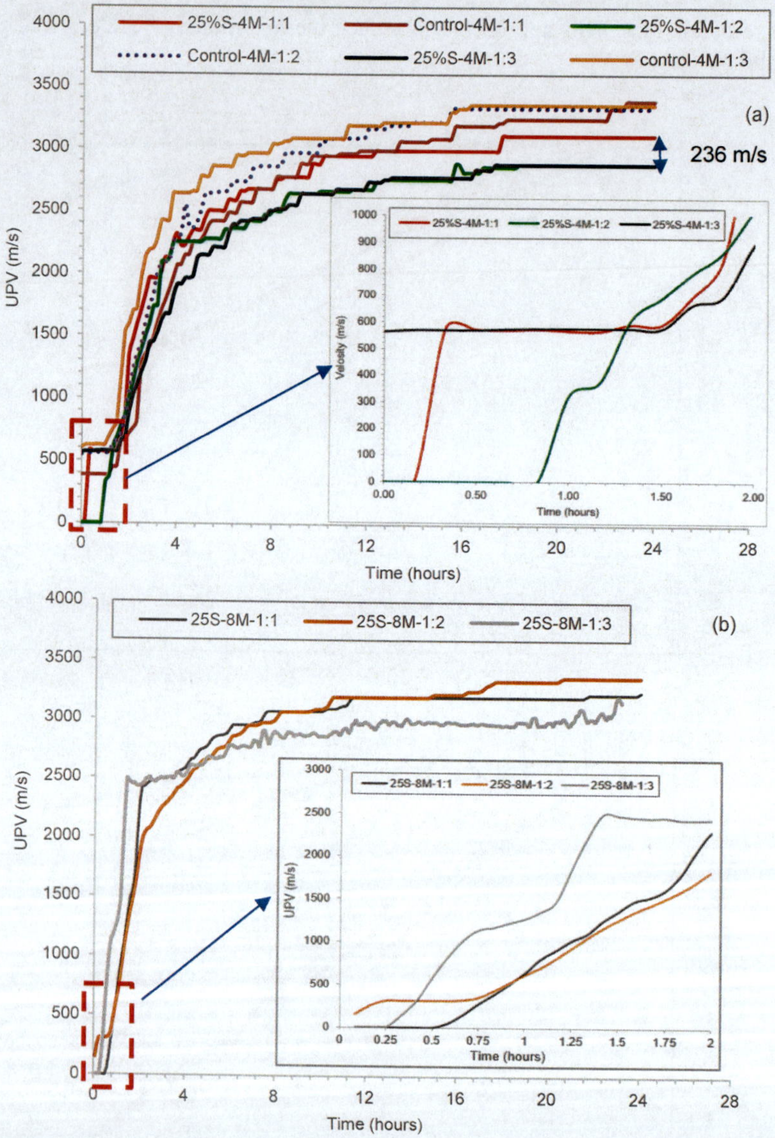

Fig. 5. Evolution of UPV through a. different 4M mixes, and b. 8M mixes during the initial 24 h from the addition of NaOH solution

3.2 Setting Time and Compressive Strength Development

Based on Fig. 7a, the initial setting times of 25S-8M-1:1 and 25S-8M-1:3 are higher than the respective controls. The final setting time of 25S-8M-1:3 is also 27 min longer than control, which is statistically significant. The delay in initial and final setting are more prominent in 4M mixes compared to 8M mixes irrespective of (GGBS+FA) to aggregate ratio. Similar effect can be seen in case of final setting: as the clay content

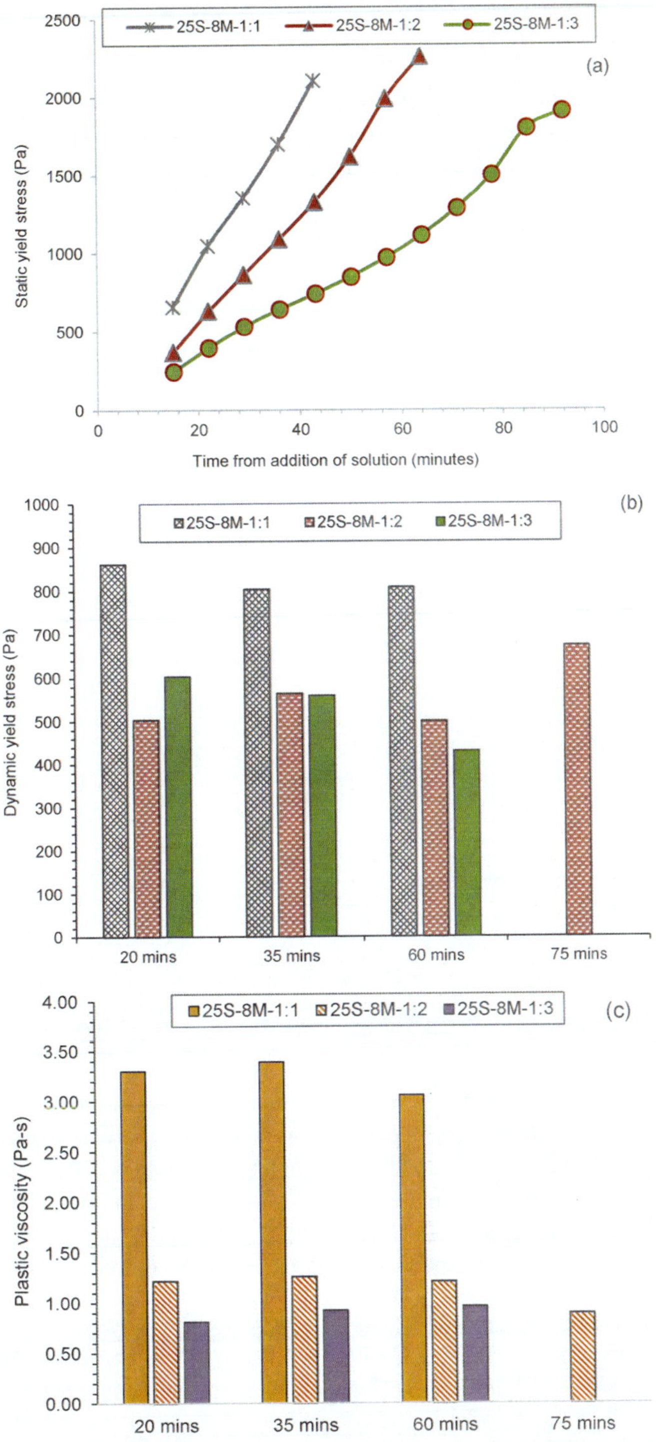

Fig. 6. a. Time-dependent static yield stress of 8M EAAM; b. Dynamic yield stress of 8M EAAM; c. Plastic viscosities of 8M EAAM; d. 3iTT of 8M EAAM; e. Recovery of viscosity at 60 s in the third leg of 3iTT for 8M EAAM up to 60 min from addition of NaOH solution.

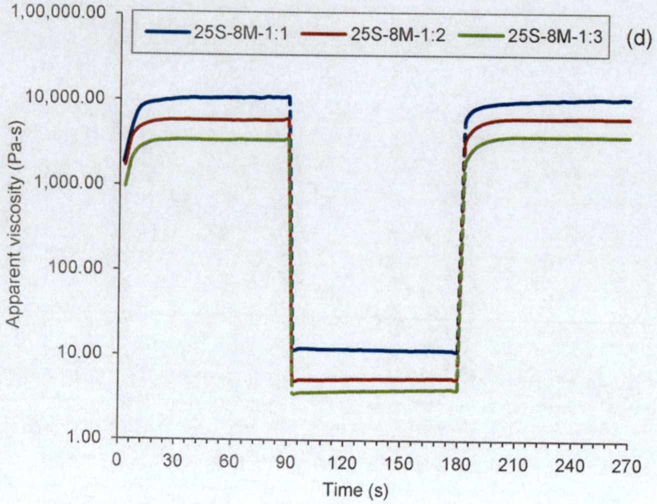

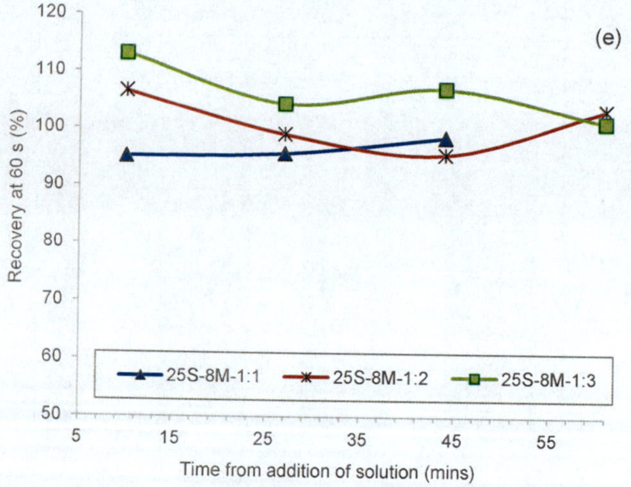

Fig. 6. (*continued*)

increases from 12% (25S-4M-1:1) to 36% (25S-4M-1:3) delay in final setting by 46 min and 52 min compared to respective controls is observed. This is attributed to lower concentration of NaOH, which reduces the precipitation rate of hydration products. Prolonged setting due to addition of soil is attributed to the retardation action of clay on alkali activation. Clay particles can bind with OH^- through hydrogen bonding, which reduces the availability of OH groups for alkali activation. At lower molarity of NaOH, there would be lesser fraction clay particles bound or wrapped in hydration products, meaning the unbound clay particles will form more hydrogen bonds with OH groups. For sufficient open mixing time, delay in initial setting is desired. 25S-4M-1:1, which has highest buildability among the 4M mixes, demonstrates initial setting at 127 min compared to 92 min in case of control, which is beneficial to attain a longer printing window.

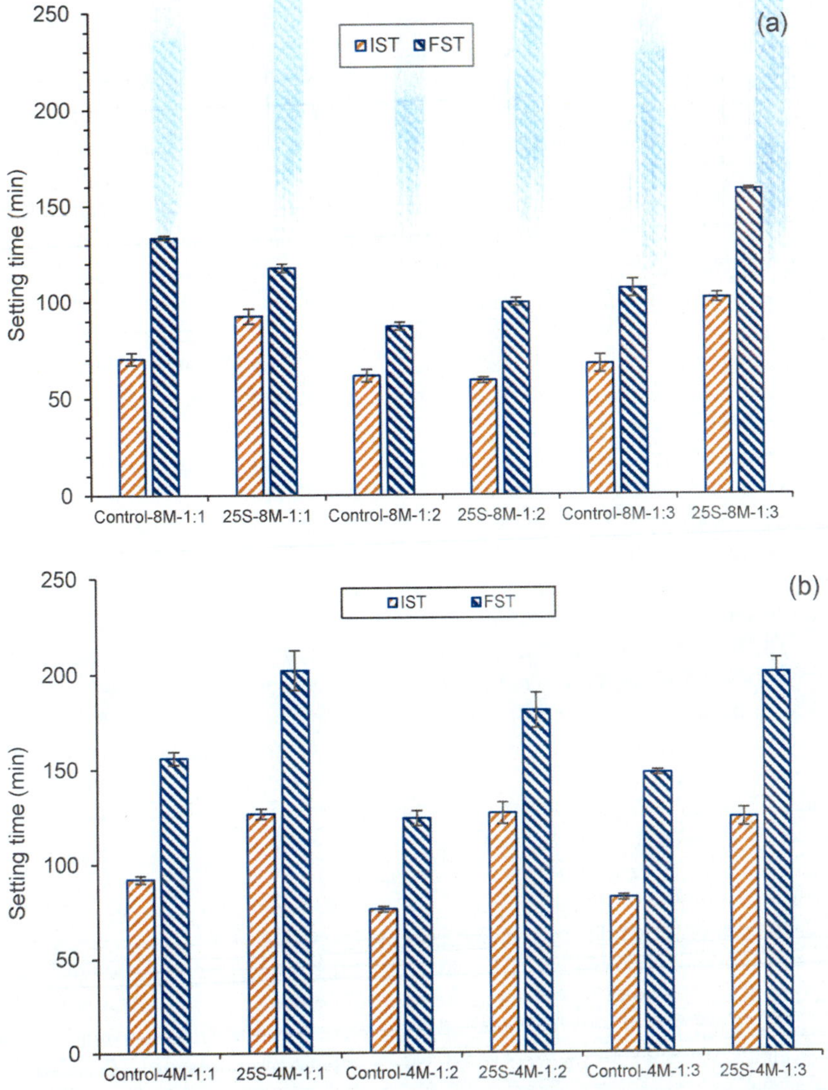

Fig. 7. Initial setting time (IST) and final setting time (FST) of a. 8M EAAM mixes; b. 4M EAAM mixes

Figure 8a and 8b show the 28-day compressive strength development of EAAM with 4M and 8M solutions. Increase in (GGBS+FA) to aggregate in EAAM content leads to reduction in 28-day compressive strength, which is strongly correlated to UPV recorded for the 28-day samples just before testing for strength. Increase in aggregate to (GGBS+FA) ratio from 1:1 to 3:1 triples the clay content of the mixture, resulting in increase in solution demand and volumetric instability. This leads to increase in porosity, thereby reducing the pulse velocity through the samples. Thus, reduced densification due to increase in clay content results in reduction in 28-day strength.

Control-4M-1:1 and 25S-4M-1:1 show similar 28-day compressive strength. Both of these mixes are made at similar solution to binder ratio (0.46 and 0.47, Table 3), meaning addition of 12% clay by mass of reactive solids (GGBS+FA) do not reduce compressive strength. However, the presence of clay has a noticeable influence on the ratio between dry and wet compressive strength (R_{dw}), evident from $R_{dw} = 1.18$ in case of 25S-4M-1:1 compared to 0.90 for Control-4M-1:1. Clay particles can swell upon exposure to moisture which imparts internal tensile stresses in the matrix, leading to reduction in wet strength relative to dry strength. To the contrary, M-sand has no measurable clay content and therefore, less sensitive to the moisture state of Control AAM. As the clay content is increased to 36% (25S-4M-1:3), a marked reduction in 28-day compressive strength by 22% is observed, which is statistically significant at 95% confidence interval. This is attributed to higher solution-binder ratio for the EAAM mix (0.90, Table 3) compared to control, due to higher moisture retention by the clay fraction. Furthermore, R_{dw} is also higher (1.26) than 25S-4M-1:1 (1.18), attributed to increased volumetric instability due to increase in clay content. It is interesting to note that 25S-4M-1:2 shows 24% higher dry compressive strength than Control-4M-1:2 (Fig. 8a). Similar, trend can be seen for EAAM made with 8M solutions – while 25S-8M-1:1 and 25S-8M-1:3 show reductions in strength by 30% and 27% respectively than the respective controls, the compressive strength of 25S-8M-1:2 is similar to that of the Control-8M-1:2. It is reiterated that 25S-8M-1:2 offers the highest buildability among all the investigated mixes. The findings show that there is apparently an optimized clay content ranging from 12–24% for EAAM, which offers higher or similar compressive strength than control. While clay imparts volumetric instability, they can also be activated by NaOH solutions resulting in precipitation of hydrosodalite and sodium-alumino-silicate (N-A-S-H) based geopolymer phases [17]. Therefore, within an optimal range of clay content, activation of clay can generate additional binder gel which can contribute to strength development. Content of clay higher than the optimal will increase the fraction of unreacted and unstabilized clay, leading to reduction in strength.

It is interesting to note that R_{dw} varies between 0.90 and 1.09 for EAAM made with 8M NaOH solutions, while the variation is in the range of 1.18 to 1.26 for EAAM made with 4M solutions. The ratio is closer to 1 for 8M EAAM, meaning that the relative reduction in wet strength due to moisture exposure is lower compared to 4M EAAM. This is attributed to a more densified hydrated matrix with 8M NaOH, which is evident from higher UPV at 1-day age (3151–3363 for 8M EAAM compared to 2879–3115 for 4M EAAM, Fig. 5a and 5b) and 28-day age (Fig. 8). More precipitation of hydration products lead to more effective stabilization of the clay particles, reducing the sensitivity of wet strength to moisture.

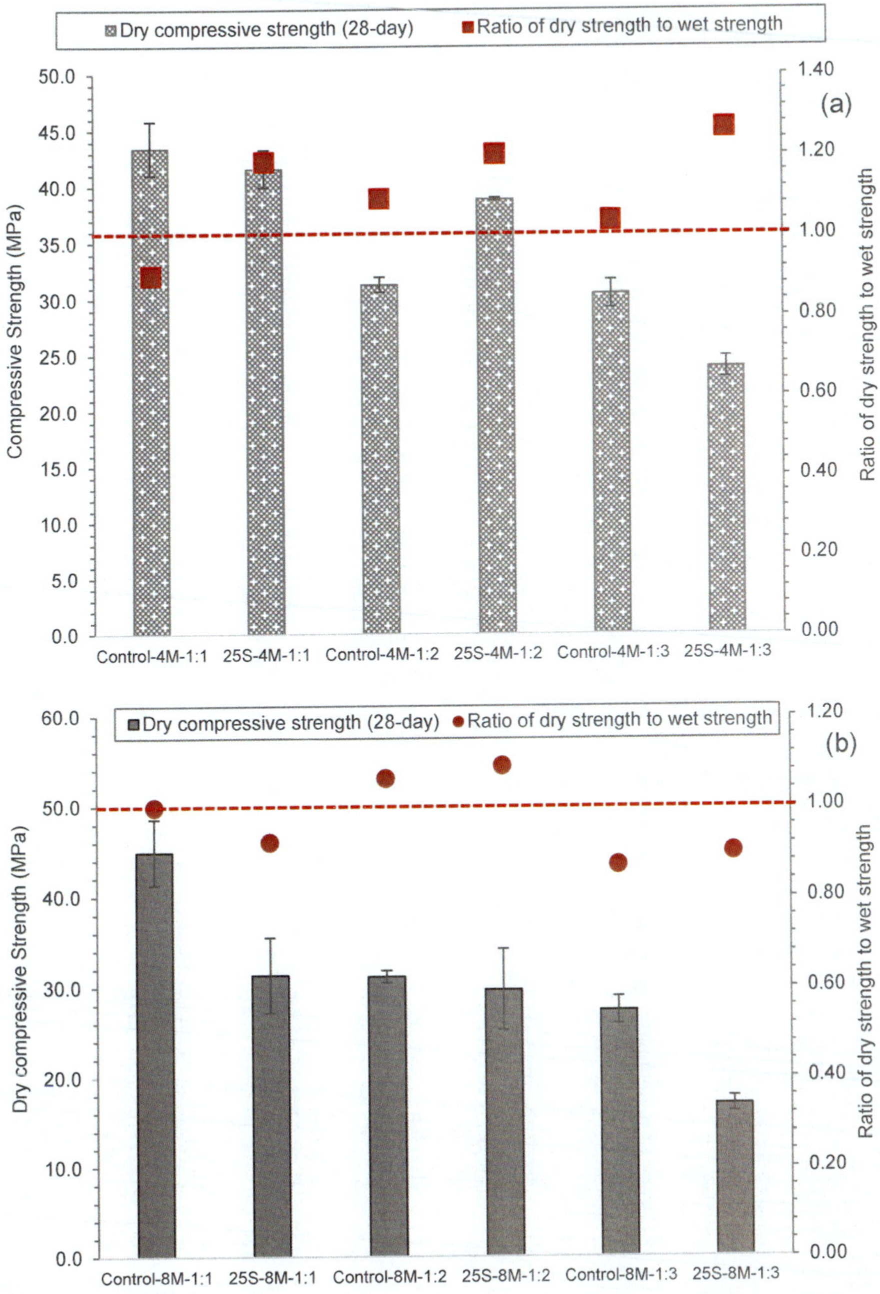

Fig. 8. a. Dry compressive strength and dry strength-to-wet strength of 4M mixes at 28-day; b. Dry compressive strength and dry strength-to-wet strength of 8M mixes at 28-day; c. Relation between UPV and 28-day dry compressive strength

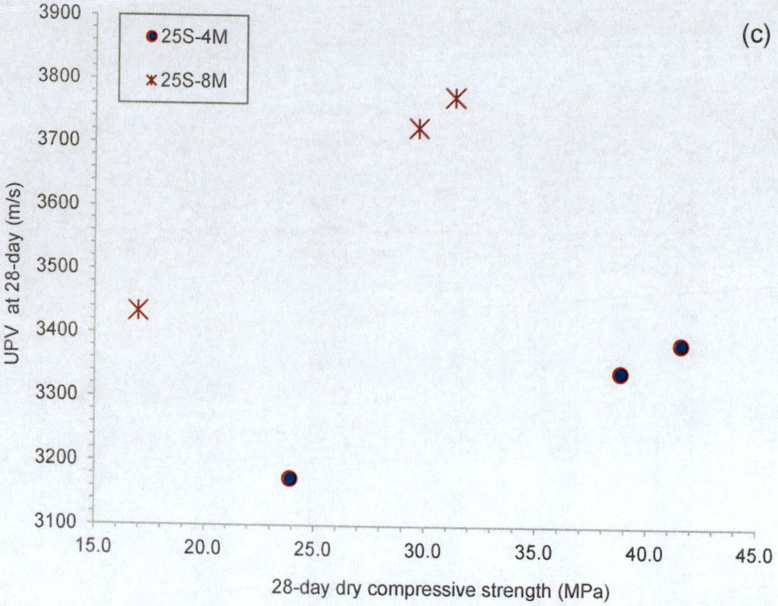

Fig. 8. (*continued*)

4 Conclusion

The following conclusions can be drawn based on the findings:

a. Addition of soil to replace 25% of sand contributes to thixotropic properties and desired yield stress and viscosity for extrudability and buildability during 3D printing compared to control with only sand.
b. The buildability of earth-based alkali-activated mixes is dependent on the concentration of the activator and the ratio of binder (GGBS and fly ash) to aggregate (B/A). Increasing the B/A from 1:1 to 1:3 increases the clay content and solution demand, leading to lower static yield stress. While B/A = 1 (1:1 mix) has highest buildability for 4M NaOH solutions, B/A = 0.50 (1:2 mix) can be reached with higher molarity (8M) of NaOH. This is attributed to higher viscosity, hydration and densification with increase in molarity of NaOH, evident from higher UPV compared to EAAM with 4M solutions.
c. Addition of clay prolongs initial and final setting of EAAM, which will provide longer printing window than control.
d. Dry compressive strength development of EAAM compared to that of control is influenced by B/A and molarity of NaOH. Considering buildability and strength development, 25S-4M-1:1 and 25S-8M-1:2 are most optimal, where dry compressive strength is similar to that of control at 28-day. The ratio between dry strength and wet strength is slower to 1 as the molarity of NaOH is increased from 4M to 8M, attributed to more effective stabilization of the unreacted clay particles.

References

1. Mohammad, M., Masad, E., Al-Ghamdi, S.G.: 3D concrete printing sustainability: a comparative life cycle assessment of four construction method scenarios. Buildings **10**, 245 (2020). https://doi.org/10.3390/buildings10120245
2. Weng, Y., et al.: Comparative economic, environmental and productivity assessment of a concrete bathroom unit fabricated through 3D printing and a precast approach. J. Clean. Prod. **261**, 121245 (2020). https://doi.org/10.1016/j.jclepro.2020.121245
3. Aggarwal, M.: RBI recommends technology for India's cement industry to reduce carbon emissions. Mongabay (2022)
4. Ding, T., Xiao, J., Zou, S., Wang, Y.: Hardened properties of layered 3D printed concrete with recycled sand. Cem. Concr. Compos. **113**, 103724 (2020). https://doi.org/10.1016/j.cemconcomp.2020.103724
5. Panda, B., Unluer, C., Tan, M.J.: Investigation of the rheology and strength of geopolymer mixtures for extrusion-based 3D printing. Cem. Concr. Compos. **94**, 307–314 (2018). https://doi.org/10.1016/j.cemconcomp.2018.10.002
6. Le, T.T., Austin, S.A., Lim, S., Buswell, R.A., Gibb, A.G.F., Thorpe, T.: Mix design and fresh properties for high-performance printing concrete. Mater. Struct. **45**, 1221–1232 (2012). https://doi.org/10.1617/s11527-012-9828-z
7. Preethi, R.K., Venkatarama Reddy, B.V.: Experimental investigations on geopolymer stabilised compressed earth products. Constr. Build. Mater. **257**, 119563 (2020). https://doi.org/10.1016/j.conbuildmat.2020.119563
8. Marsh, A., Heath, A., Patureau, P., Evernden, M., Walker, P.: Alkali activation behaviour of un-calcined montmorillonite and illite clay minerals. Appl. Clay Sci. **166**, 250–261 (2018). https://doi.org/10.1016/j.clay.2018.09.011
9. Bajpai, R., Choudhary, K., Srivastava, A., Sangwan, K.S., Singh, M.: Environmental impact assessment of fly ash and silica fume based geopolymer concrete. J. Clean. Prod. **254**, 120147 (2020). https://doi.org/10.1016/j.jclepro.2020.120147
10. Habert, G., d'Espinose de Lacaillerie, J.B., Roussel, N.: An environmental evaluation of geopolymer based concrete production: reviewing current research trends. J. Clean. Prod. **19**, 1229–1238 (2011). https://doi.org/10.1016/j.jclepro.2011.03.012
11. Winnefeld, F., ben Haha, M., le Saout, G., Costoya, M., Ko, S.-C., Lothenbach, B.: Influence of slag composition on the hydration of alkali-activated slags. J. Sustain. Cem. Based Mater. **4**, 85–100 (2015). https://doi.org/10.1080/21650373.2014.955550
12. Yuan, Q., et al.: A feasible method for measuring the buildability of fresh 3D printing mortar. Constr. Build. Mater. **227**, 116600 (2019). https://doi.org/10.1016/j.conbuildmat.2019.07.326
13. Gupta, S., Mahmood, A.H.: A multi-method investigation into rheological properties, hydration, and early-age strength of cement composites with admixtures recovered from inorganic and bio-based waste streams. Constr. Build. Mater. **347**, 128529 (2022). https://doi.org/10.1016/j.conbuildmat.2022.128529
14. Kawashima, S., Chaouche, M., Corr, D.J., Shah, S.P.: Influence of purified attapulgite clays on the adhesive properties of cement pastes as measured by the tack test. Cem. Concr. Compos. **48**, 35–41 (2014). https://doi.org/10.1016/j.cemconcomp.2014.01.005
15. Manikandan, K., Wi, K., Zhang, X., Wang, K., Qin, H.: Characterizing cement mixtures for concrete 3D printing. Manuf. Lett. **24**, 33–37 (2020). https://doi.org/10.1016/j.mfglet.2020.03.002

16. Chougan, M., et al.: Effect of natural and calcined halloysite clay minerals as low-cost additives on the performance of 3D-printed alkali-activated materials. Mater. Des. **223**, 111183 (2022). https://doi.org/10.1016/j.matdes.2022.111183
17. Marsh, A., Heath, A., Patureau, P., Evernden, M., Walker, P.: Phase formation behaviour in alkali activation of clay mixtures. Appl. Clay Sci. **175**, 10–21 (2019). https://doi.org/10.1016/j.clay.2019.03.037

Bio-Based Insulation Materials and Constructions

Hygroscopic and Thermal Inertia Impact of Biobased Insulation in a Wood Frame Wall

Vincent Claude(✉), Evelyne Nguyen, André Delhaye, Antonin Mayeux, and Stéphane Charron

Buildwise (ex-BBRI), 1342 Limelette, Belgium
vincent.claude.bbri@gmail.com

Abstract. In this study, large scale experiments are performed with walls implemented in a climatic chamber in order to assess either the hygroscopic impact or the thermal phase shift of flexible bio-based insulating materials and clay panels. The conditions are set to be as close as possible to the field. Before the large scale tests, the materials are characterized individually with moisture-buffering, sorption-desorption and water vapour permeability tests.

The aim of the first part is to assess the moisture buffering ability of a flexible bio-based insulating material set behind a OSB board. Conventional mineral wool (rock wool) is set inside wall n°1 and flax is set inside wall n°2. The outer side of the chamber is set at T = 5 °C/RH = 80% in order to simulate winter climate. The inner side of the chamber is set at T = 23 °C and different variations of water relative humidity between 40–95% are applied. The results show that the hygroscopic impact of the insulating material set behind a OSB is very limited. The inner relative humidity is mostly absorbed and released by the OSB board that acted both as water vapor barrier and moisture buffer. Characterizations and real tests show that the moisture buffering capacity of clay is slightly higher than gypsum.

In the second part, summer thermal phase shifts tests are performed with either conventional mineral wool or wood wool, and either gypsum or clay inner panels. In the outer side of the chamber temperature cycles of 24 h are applied (T = 15–40–15 °C). The temperature in the inner side is not monitored. The results show that wall made with wood wool and clay have higher thermal phase shift than wall made with mineral wool and gypsum. The highest impact is due to the use of a panel of clay instead of gypsum. The type of insulation has limited influence. The maximum difference between the two walls is 0.9 °C, which corresponds to an increase of performance of about ~5%. This value of ~5% is similar to the conclusion obtained by previous modellings performed by the BBRI and the EMPA. In order to prevent summer overheating, one should at first guarantee overnight ventilation and sunscreen during the daytime, and thereafter care about the type of insulating material.

Keywords: Biobased insulation · moisture buffering · thermal phase shift

S. Amziane et al. (Eds.): ICBBM 2023, RILEM Bookseries 45, pp. 355–372, 2023.
https://doi.org/10.1007/978-3-031-33465-8_28

1 Introduction

The use of bio-based insulating materials is getting more and more attention from the construction field. In this way, official studies showed that it currently represents 8% of the insulation business turnover in France and is expected to grow at a rate of 10%/year until 2030. From initially small and medium size enterprises, the market is now being invested by large companies who also produces conventional materials [1, 2].

The development of these materials in modern buildings was initially driven by their ecological assets. However the recent commercial spread is mostly due to several studies performed by academics, manufacturers and laboratories delivering certifications who showed that bio-based insulating material performances can compete to conventional materials such as mineral wools or synthetic panels [3–7].

It is usually claimed that bio-based insulating materials are famous for their ability to maintain a comfortable indoor relative humidity (RH between 30 and 60%), thanks to their moisture buffering capacity; and to maintain an good temperature, thanks to their high thermal phase shifts.

The high hygroscopic properties of bio-based materials is known for decades. Sorption-desorption tests and moisture buffering value (MBV) tests have shown at lab scale that, taken individually, the bio-based insulating materials can indeed efficiently absorb and release the water vapour from the air [3, 5, 8, 9]. This has also been proved for hempcrete used as blocks or renders for inside insulation [10–12]. However, few experimental studies investigated the behaviour of flexible bio-based materials once integrated in a whole structure.

Among them, the study of Mortensen et al. [13] investigated the moisture buffering ability of cellulose compared to mineral wool in a large climatic chamber. When set behind a plaster board, the cellulose was able to effectively regulate the indoor relative humidity. According to the author the moisture buffer capacity of hygroscopic materials can be used to moderate peaks in the relative humidity (RH) of indoor air as well as moisture content variations in building materials and furnishing which can help to ensure healthier indoor environments by preventing many processes that are harmful such as growth of house dust mites, surface condensation and mould growth. However, the author also showed that the presence of non-permeable to water vapor paint on the gypsum board reduces this effect and concluded that more study had to be performed to assess the real influence on the indoor humidity depending on the construction layers.

One other significant study on the subject of hygroscopic regulation is the study made by Palumbo et al. [14]. The authors studied the performances of calcium alginate, wood fibres and EPS insulating materials at lab scale and then compared the results with a study performed at large scale on an ETICS wall. It was concluded that though the hygrothermal properties were significant at lab scale, no clear difference was visible at large scale.

Even fewer studies are available concerning the influence of flexible bio-based insulation materials on the indoor thermal inertia during hot summer conditions. With their high density and high thermal capacity values, the bio-based flexible insulation materials theoretically have better thermal phase shift and thermal effusivity values than mineral wools and especially compared to synthetic insulation. In this way for 200 mm thickness the thermal phase shift value of cellulose is theoretically 7–8 h whereas it only reaches

2–3 h for EPS and 3–4 h for mineral wools [9]. However, more and more questions are being raised about the real impact once they are set in the field mostly for two reasons: 1) several modellings performed by either the Belgian Building Research Institute (now named Buildwise) [15, 16] or the Swiss Federal Laboratories for Materials Science and Technology (EMPA) [17] have shown that the type of insulating material set in a roof with windows have a very limited effect. The type of insulating material (either rigid PIR, loose cellulose, flexible wood wool or flexible mineral wool) has only a very limited impact in the overheating in summer in rooms located behind pitched roof with skylight windows (5 to 10% of the impact). The most impacting factors were the overnight ventilation and the presence of sunscreen on the windows during the daytime (up to 66% of the impact); 2) It is generally accepted that the Cp values of bio-based materials are around 1800–2100 J/K.kg, whereas those of mineral wools are ~800–1000 J/K.kg, those of PUR/PIR are ~1400 J/K.kg and those of XPS are ~1300 J/K.kg. However, despite a large screening performed in 2022 on the technical data sheet of commercially available materials (hemp, grass, flax…), only one manufacturer provided an officially certified value of Cp for only one of his product (wood wool). More important, this value was lower than expected: a Cp of only 1220 J/K.kg. Discussions are ongoing between stakeholders about the necessity to adopt a new standard for the measurement of the Cp of bio-based materials. Due to their high hygroscopicity and heterogeneity, the interpretation of the results can greatly vary depending on the laboratory. However, since almost no certified Cp value are available, doubts about the high impact on thermal inertia of flexible bio-based materials seem to be justified. It is to notice that no doubts are emitted for the thermal inertia potential of rigid bio-based materials, such as rigid wood fiber board used for sarking applications, which have much higher densities.

In order to get a better understanding of the real impact of the bio-based materials, experiments were performed at large scale with a climatic chamber. In the first series of tests, the hygroscopic behaviours were investigated. In the second series of tests, the thermal inertias were investigated.

2 Methods

2.1 Materials

The materials used for the tests in the climatic chamber are listed in Table 1. Some of them where used for both moisture migration or thermal transition tests.

2.2 Characterization of Materials

Sorption-desorption tests were performed following the EN 12571 standard. These tests are essential to determine the hygroscopic behaviour of bio-based materials. Three samples of each material of 100 * 100 * 100 mm were dried until they reached a constant mass. Thereafter the humidity was set at either 20, 40, 60, 80 or 95%. At each step, the samples were weighted. When the mass was constant for each samples (no variation of more than 0,1% in 3 days), a different level of relative humidity was set.

Moisture buffer value tests were performed following the Rilem TC 275 HDB report. Three samples of each material of 150 * 150 * 150 mm were dried until they reached a

Table 1. Materials used in the study.

Material	Thickness (mm)	Density (ρ, kg/m^3)	Thermal conductivity (λ, W/mK)	Thermal capacity (Cp, J/kg.K)	Water vapour permeability	Individual thermal phase shift (φ, h)
Insulation						
Mineral wool	200	30	0,035	1030	$\mu = 1$–2	4,3 h (200 mm)
Flax	200	23	0,038	1550	$\mu = 1$–2	4,5 h (200 mm)
Wood wool	200	47	0,038	2100	$\mu = 1$–2	7,4 h (200 mm)
Membranes						
Closed membrane	0,31	–	–	–	$S_d = 2400$ m	–
Permeable membrane	0,32	–	–	–	$S_d = 5$ m	–
Panels						
OSB board	18	600–700	0,100–0,130	1350	$\mu = 114$*	1,1 h (18 mm)
Gypsum board	13	600	0,220	900	$\mu = 10$*	0,5 h (13 mm)
Clay board	22	1450	0,353	1100	$\mu = 16$*	1,1 h (22 mm)
External panel (wood fiber bitumen cladding)	18	250	0,048	2100	$\mu = 5$*	1,4 h (18 mm)

*: *values measured at Buildwise.*

constant mass. Once the dry point was reached, the specimen were stabilized at 23 °C/ 50%RH until they reached a constant mass. The specimen were sealed with aluminum tape on all faces, excepted the exchange surface. The target air velocity was 0.10 ± 0.05 m/s. It was measured with an anemometer in the three directions in representative conditions of the test (door close, specimen in place…). The samples were set in a basket to enable a correct air velocity. The samples were set in a circle with their exposed faces toward its centre. The test consisted in measuring the weight variation along daily cyclic variation of relative humidity (8 h at RH = 75%, then 16 h at RH = 33%). The weightings were performed manually 3 times a day. Analytical balance, capable of weighing the test specimens with an accuracy of 1% was used. The mass variation between the highest and the lowest value was calculated in absorption and in desorption for each cycle. The temperature and the relative humidity were recorded continuously with suitable sensors (each 5 five minutes) placed near the specimens. The average relative humidity was calculated for each cycle during absorption and desorption steps. The test ended when the stabilization criterion was met. The stabilization criterion required was a variation of mass variation lower than 5% between the last three cycles. The MBV of a specimen is the average value of the three stabilized cycles. The MBV of the material is the average value of the three specimen. The MBV is calculated as follows for each cycle and each specimen:

$$MBV = \frac{m_{high} - m_{low}}{\left(RH_{high,av} - RH_{low,av}\right)A} \quad (1)$$

$m_{high/low}$: higher/lower value of mass during the cycle (g),
$RH_{high,av/low,av}$: average value of relative humidity at high/low relative humidity level (% RH),
A: exchange surface area (m^2).

2.3 Measurements in Climatic Chambers

Relative humidity sensors were "Honeywell" brand and type "HIH-4021". These sensors provide values in volt that are thereafter converted to percentage of relative humidity.

The thermocouples were made of one wire of copper and one wire of aluminum. They provide a signal in volt. The wood humidity measurement were measured with two nails inserted in the wood rafter. An electrical current flow between these nails and the value of the electrical resistance (MΩ) can then be converted to wood humidity value (%) (Fig. 1).

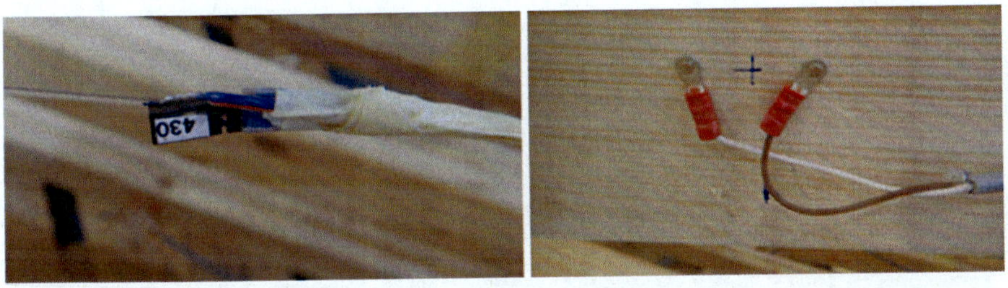

Fig. 1. Left: relative humidity sensor and thermocouple/Right: wood humidity sensors.

The values measured by the sensors were sent to an Agilent monitoring apparatus, converted into physical units and then sent to Graphana software where the data could be read and collected (Fig. 2).

The aim of the tests was to simulate different outdoor and indoor conditions and to study the hygric and thermal behaviours of the two walls. In order to do so, two walls of 1,8 × 1,2 × 0,2 m were built between two climatic chambers.

Figure 3 shows the general scheme of the implementation. Figure 4 shows pictures of the walls at different building steps.

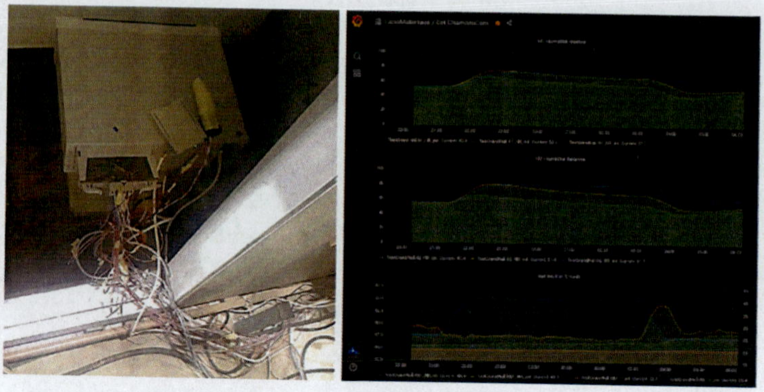

Fig. 2. Left: Agilent monitoring apparatus/Right: Graphana interface.

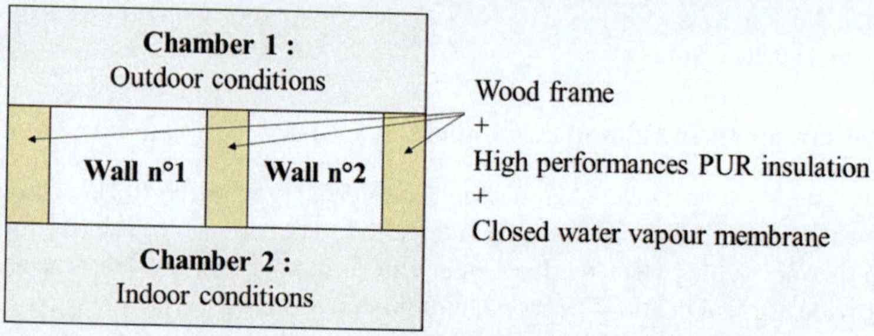

Fig. 3. Overview of the test implementation.

Fig. 4. Walls at different building steps.

For both panels, each type of sensors was set either in the indoor chamber, inside the panels or in the outdoor chamber. Additional temperature sensors were also set next to the render of each panel in the indoor side. Figure 5 shows some pictures of the implementation of the sensors.

The temperature and relative humidity were controlled by the internal apparatus of the climatic chambers. For the first series of tests, the moisture migration was studied. In

Fig. 5. Pictures of sensors.

order to do so, only a defined amount of water vapour was injected in the indoor climatic chamber. In order to do so, a HU-45 ultrasonic humidifier was used.

2.4 Tests Conditions

Two types of tests were performed: moisture migration or thermal transition tests. Two walls were set up for both tests. The indoor chamber had a volume of 21 m^3.

Moisture Migration Tests

The aim of the moisture migration tests were to study how the water vapour is absorbed, diffused and released by the bio-based insulating materials. These tests were performed to investigate the moisture buffering of the bio-based materials. For this test, flax was used as bio-based insulation. Indeed, this materials proved to be the most hygroscopic and the one with the highest sorption/desorption capacity. Mineral wool was used as comparison since it has no hygroscopicity.

Two types of tests were performed:

1) *"Temporary humidity"*: in that case the chamber n°1 was kept at 5 °C/RH = 80% during the whole procedure in order to simulate external cold weather. In the chamber n°2 the temperature was maintained at T = 23 °C in order to simulate indoor conditions. At first the chamber was conditioned for 48 h at RH = 40%. Then the conditioning was stopped and the relative humidity was sharply increased from 40% to 95% thanks to the injection of a define amount of water vapour (see Table 2) injected in 20 min with a humidifier apparatus. The aim of the test was to create a temporary high increase of the water vapour and to observe how this amount of water vapour migrated and was absorbed through the wall. These conditions could for example be closed to the ones encountered when having a shower in a bathroom or when cooking without ventilation;
2) *"Continuous humidity"*: in that case the chamber n°1 was kept at 5 °C/RH = 80% during the whole procedure. In the chamber n°2 the temperature was maintained at T = 23 °C. At first the chamber was conditioned for 48 h at RH = 40%. Then the

conditioning was set at RH = 95% for 60 h. These tests were performed to getter a better understanding of the dynamic of the water vapour migration through the walls.

Table 3 lists the different configurations of walls tested. It is to notice that for the *"Temporary humidity"* tests, the OSB boards were set in the indoor side directly next to the insulating material, whereas it is recommended for a proper implementation to always add a water vapour membrane between the OSB and the insulation. The aim of the water vapour membrane is to prevent a migration of a too high amount of water vapour throuh the insulating materials and thus prevent the possibility of condensation. However, in the present study, for the temporary tests, the water vapour membrane was not install in order to get a scientific understanding of the direct influence of the materials.

Table 2. Volume of liquid water for one injection during *"Temporary humidity"* tests.

Conditions	Vapour pressure (Pa)	Absolute humidity (g_{water}/kg_{dryair})	Absolute humidity (g_{water}/m^3))	g_{water} for a volume of 21 m^3
23 °C/40%	1123	7,0	8,4	176,3
23 °C/95%	2667	16,8	20,3	425,3
→ Volume of liquid water to be vaporized during 1 injection: 249 mL				

Table 3. Configurations of the walls for all humidity tests.

Test n°	Configurations of the walls (from indoor to outdoor)			
"Temporary humidity" tests				
1	/	OSB	Mineral wool	External panel
2	Gypsum board	OSB	Mineral wool	External panel
3	/	OSB	Flax	External panel
4	Clay	OSB	Flax	External panel
5	/	Permeable membrane	Flax	External panel
"Continuous humidity" tests				
6	OSB	Permeable membrane	Mineral wool	External panel
7	OSB	Permeable membrane	Flax	External panel

Thermal Transition Tests

The thermal transition tests were performed in order to assess the influence between either flexible bio-based insulating material vs. conventional mineral wool; and also between either clay board vs. conventional gypsum. For these tests, flexible wood wool was used since it is the material with the highest thermal phase shift (see Table 1). The *"Thermal Transition"* tests were performed as follow: Firstly the chambers were conditioned for 48 h. The chamber n°1 was conditioned at T = 16 °C/RH = 50% and the chamber n°2 was conditioned at T = 19 °C/RH = 50%. The test was thereafter launched and no further regulation of temperature and humidity was applied to the chamber n°2. The RH in chamber n°1 was maintained at 50% and the temperature underwent 4 stages of 6 h at T = 25 °C, 40 °C, 25 °C and finally 16 °C. The aim of this test was to simulate the indoor and outdoor conditions during a hot summer day. The Table 4 shows the different configurations of the walls for thermal transition tests.

Table 4. Configurations of the walls for thermal transition tests.

Test n°	Configurations of the walls (from indoor to outdoor)			
9	Clay board	OSB	Mineral wool	External panel
10	Gypsum board	OSB	Mineral wool	External panel
11	Clay board	OSB	Flax	External panel
12	Gypsum board	OSB	Flax	External panel

3 Results

3.1 Characterizations of Materials

Sorption-Desorption

Figure 6 shows the sorption-desorption curves for the materials studied during moisture migration tests. The mineral wool having absolutely no hygroscopic properties, its curves is not shown.

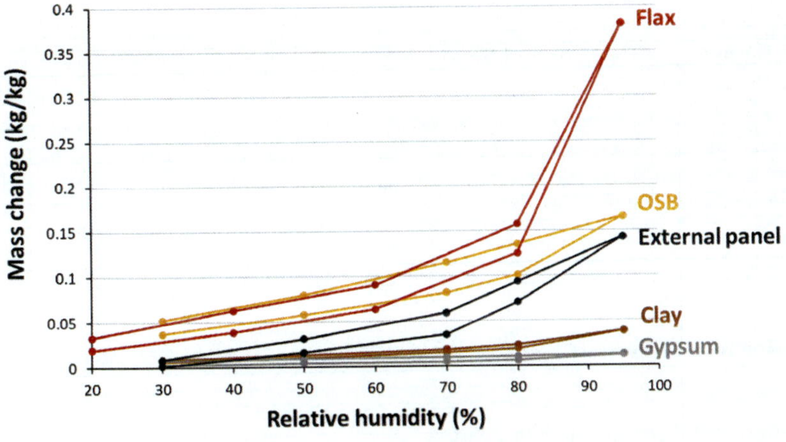

Fig. 6. Sorption-desorption curves.

It can be observed in Fig. 6 that flax is the material with the highest hygroscopicity, then comes OSB, external panel, clay and finally gypsum. Noticeably the OSB has similar properties as flax when the RH is below 80%. For this test, the total mass of flax in a wall is 13 kg, and the total mass of OSB is 25 kg. Below RH = 80%, the OSB is therefore expected to have a higher impact than flax on the regulation of the humidity inside the room.

Table 5 shows the moisture buffering values obtained for the different samples. For a better interpretation, Fig. 7 shows the MBV ranks. It can be observed that in line with sorption-desorption measurement, mineral wool has the lowest hygroscopicity. The gypsum board show a MBV value very close to what Rode et al. [18] listed in their review (between 0.55–0.65 g/(m^2.%RH)). The MBV value of gypsum can be classified as "moderate" according to the same authors. This is in line with the poor values obtained during sorption-desorption. Surprisingly the OSB board showed a "moderate" MBV value. The value can be considered as correct since it is very close to what was obtained in another study performed by Collet et al. [19] who obtained 0.53 g/(m^2.%RH). This confirms that for a neat understanding of the hygric properties of materials, both sorption-desorption and MBV measurement are needed. The external panel, which is made from wood fibers and bitumen, and which had relatively similar sorption-desorption curves as OSB, shows in contrary a "good" MBV value. This could be explained by the fact that OSB panel are much more dense compared to external panel (ρ_{OSB} = 600–700 kg/m^3 instead of $\rho_{External.panel}$ = 250 kg/m^3). Flax insulation is another material that shows disparities between the static and dynamic measurements: despite it showed the highest sorption-desorption performances, its MBV value could only be considered as "moderate-good". In another way, whereas the clay showed relatively low sorption-desorption values, its MBV values were considered as "good".

Table 5. Moisture buffering values obtained from 3 samples.

Material	Mineral wool	Flax	OSB board	Gypsum board	Clay board	External panel
MBV average value (g/(m^2.%RH))	0.29	1.05	0.52	0.66	1.27	1.61
MBV standard deviation g/(m^2.%RH))	0.03	0.05	0.09	0.02	0.04	0.02

3.2 Moisture Migration Tests

Influence of the Insulating Material

Tests n°1 and n°3 enabled to compare the influence of the flexible insulating material. In test n°1 the insulating material was mineral wool and in test n°3 it was flax.

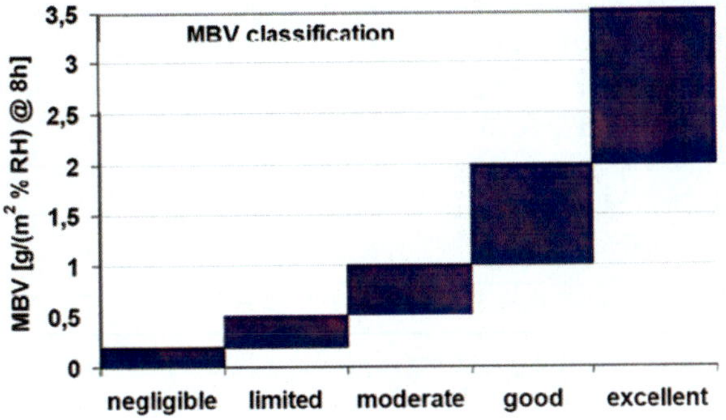

Fig. 7. Moisture buffering values classification according to Rode et al. [18].

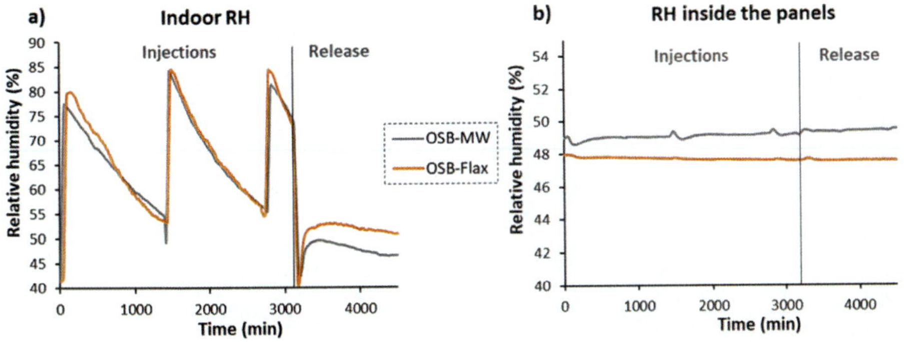

Fig. 8. Influence of insulating material: comparison of *"Temporary humidity"* tests n°1 and n°3. a) indoor relative humidity; b) relative humidity inside the insulation.

It can be observed in Fig. 8a that the behaviors during the injection of water vapour were very similar whatever the type of insulating material. Though the amount of water vaporised was set to reach a relative humidity of 95%, the indoor RH did not exceeded 85%. This could be explained by the fact that the relative humidity was very quickly absorbed by the OSB board in the range between RH = 80 and 100%. Indeed, the sensors were able to perform only one acquisition every 15 min. The relative humidity values then slowly decrease at the same rate for both tests: it decreases almost linearly from RH ~80% to ~55% in 24 h. The slow and almost identical decrease of RH for both tests could be explained by the fact that the OSB boards act as water vapour barriers (μ = 114/Sd = 2 m) and prevent the insulating material behind them to have a real impact on the moisture absorption. Furthermore, despite the OSB had high sorption-desorption performances (Fig. 6), its MBV value was considered as "moderate" (Table 5), which explain the very slow decrease.

Figure 8b shows tiny jolts in the relative humidity of the mineral wool wall at the times where the injection of moisture was high (t = 10, 1600 and 2900 min). In contrary, this was not observed for the flax wall. Despite this, the relative humidity remained globally very stable during all the test steps for walls with mineral wool or flax. It is supposed that the major amount of moisture was slowly absorbed by the OSB board and did not go through the insulation.

During the step of release, the RH values of the indoor chamber increased back for both tests. However, the release of humidity was more important and last longer for the wall with flax than for one with mineral wool. In this way, after 24 h of release, the relative humidity decreased to RH = 46,9% with mineral wool wall whereas it was still at RH = 50,9% with flax.

Influence of the Render

Tests n°1–2 and tests n°3–4 were compared to study the influence of the render.

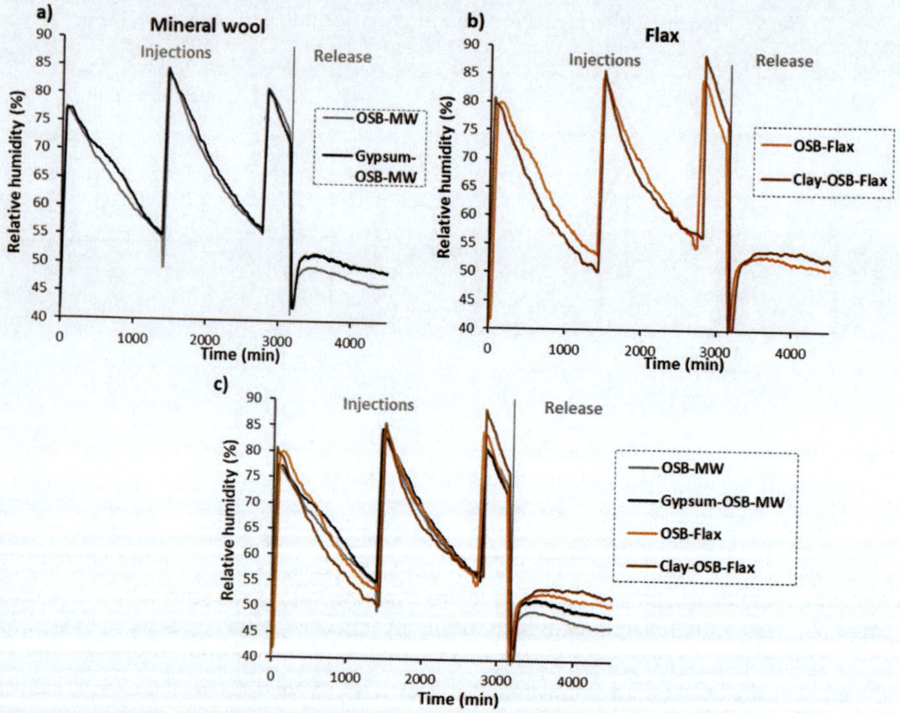

Fig. 9. Influence of render: comparison of *"Temporary humidity"* tests n°1–2 and n°3–4. a) influence of gypsum; b) influence of clay; c) comparison for all compositions.

It can be observed in Fig. 9 that the presence of gypsum and clay increases the amount of moisture released during the last steps of the tests. Interestingly the relative humidity in the room decreased more quickly in the presence of clay during the injection steps.

Influence of the Membrane

Figure 10 shows the moisture migration tests for walls with and without membranes. For a better understanding, comparisons were performed with the results obtained for the other panels.

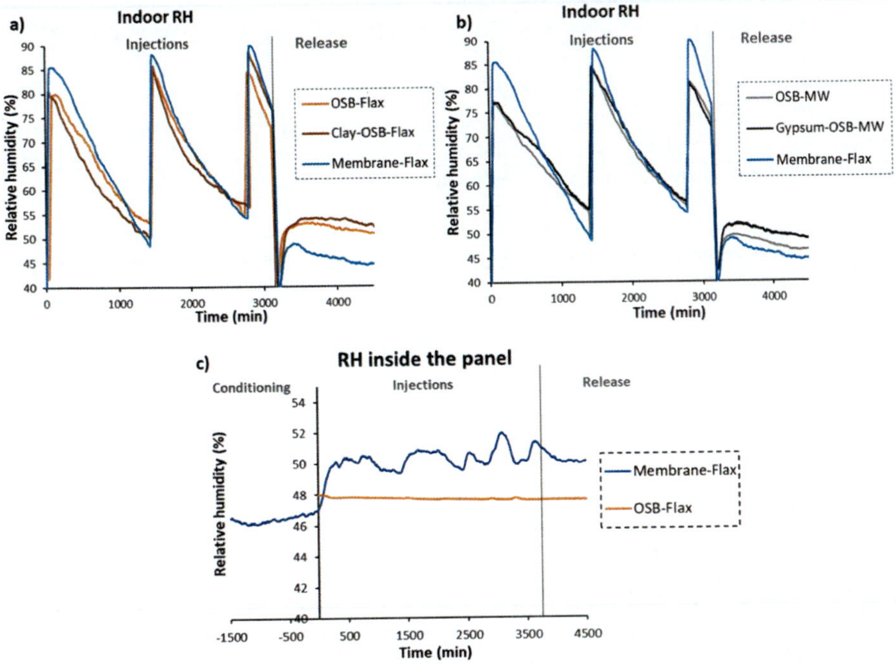

Fig. 10. Influence of membrane: comparison of "*Temporary humidity*" tests. a) comparison between panels using flax; b) comparison between panels using mineral wool; c) relative humidity inside the insulating panels using either a permeable membrane or a OSB.

It can be observed in Fig. 10a that the relative humidity values goes to higher values in the indoor chamber during the injection when a membrane is present. Indeed, in the other cases the moisture was directly in contact with the OSB board or the render (gypsum/clay) and was absorbed very quickly in ranges comprised between RH = 80 and 100%. With the presence of a membrane, the moisture had more difficulty to penetrate through the membrane which explains the longer time for the relative humidity to decrease.

It can be observed that with a membrane the release of humidity value during the step of release is very low. In fact, it is the lowest release of moisture observed for all tests (Fig. 10a & b). This is logical since the membrane prevent the moisture to go through the other side.

Figure 10c shows the relative humidity value inside the insulation. It can be observed that the relative humidity value increases when the membrane is present. This is not the case if the OSB board is present. This confirm the theory that the moisture is at first absorbed by the OSB board and then partially diffuse through the insulating material.

Noticeably, the relative humidity value remained stable inside the insulation during the step of release. It is supposed that in that step, the release of relative humidity only comes from the superficial layer of the insulating material.

In conclusion, in case of temporary rise of humidity inside a room, the regulation of the indoor humidity is therefore impacted by order of influence by:

OSB board >> clay render > gypsum render > bio-based insulating material.

Continuous Injection of Water Vapour

The tests n°6 and 7 were performed to observe the evolution of the relative humidity values inside the insulation when the indoor relative humidity is very high and the panel has an OSB board and a permeable membrane (Sd = 5 m).

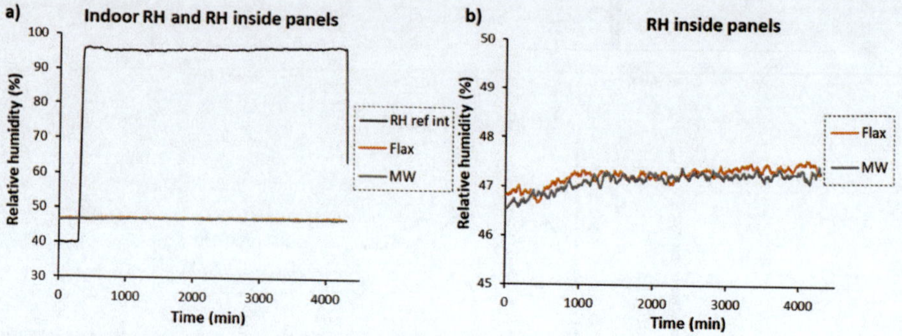

Fig. 11. *"Continuous humidity"* tests. a) Overview; b) Zoom on relative humidity values inside the insulation.

It can be observed in Fig. 11b that the relative humidity inside the insulation of walls using both flax and mineral wool evolved very lightly (+2% after 60 h). Despite the fact that the OSB and the membrane were permeable to water vapour (Sd_{OSB} = 2 m and $Sd_{Membrane}$ = 5 m), the migration of relative humidity from the indoor chamber at RH = 95% through the wall was therefore very limited. These result are in accordance with the sorption-desorption curves (Fig. 6) and with the *"Temporary humidity"* test showed in Fig. 10. The OSB boards act as both water vapour barriers and moisture absorbent and prevent its diffusion through the walls. Furthermore, the MBV value of OSB being "moderate" and its sorption-desorption ability being high, the OSB board can intake water vapour humidity for a long time without letting it pass.

It is expected that eventually with relative humidity kept at 95%, the moisture would penetrate through the walls and increase the relative humidity inside the insulating material. However, the relative humidity would have to remain high for more than 60 h, which would be far from domestic and realistic use. Behind both an OSB board and a permeable membrane (Sd = 5 m), the insulating material seems therefore almost not impacted by the quick and short indoor relative humidity variations.

3.3 Thermal Transitions Tests

Tests Summary

Figure 12 shows an overview of the temperatures that have been observed for the outdoor conditions and for the indoor conditions during the thermal transition tests. It is to remind that before the tests, the outdoor chamber was conditioned 24 h at 16 °C/RH = 50%, and the indoor chamber was conditioned at 24 h at 19 °C/RH = 50%.

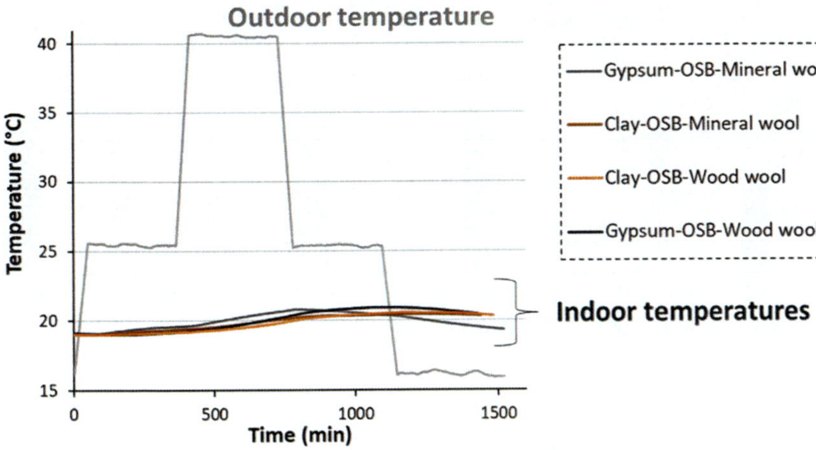

Fig. 12. Overview of the "*Thermal transition*" tests.

Figure 13 shows the graphs of the temperatures obtained at the indoor surface of the renders for the different wall compositions. Figure 14a shows the maximal temperatures. Figure 14b shows the time needed for the different walls to reach T = 20,81 °C, which is the maximum temperature that has been reached by the wall with clay and wood wool. The time shift values are defined by the additional time needed by the walls to reach T = 20,81 °C compared to the wall with gypsum and mineral wool.

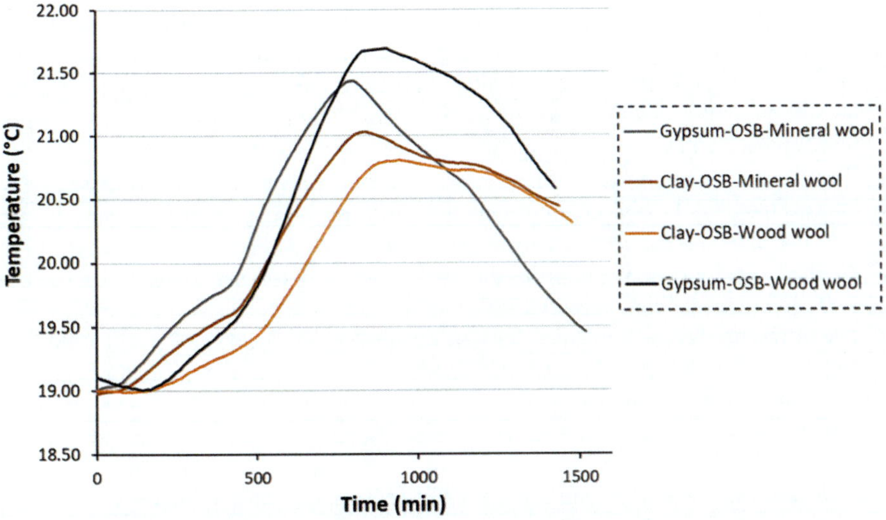

Fig. 13. Temperature on the indoor side of renders for all compositions.

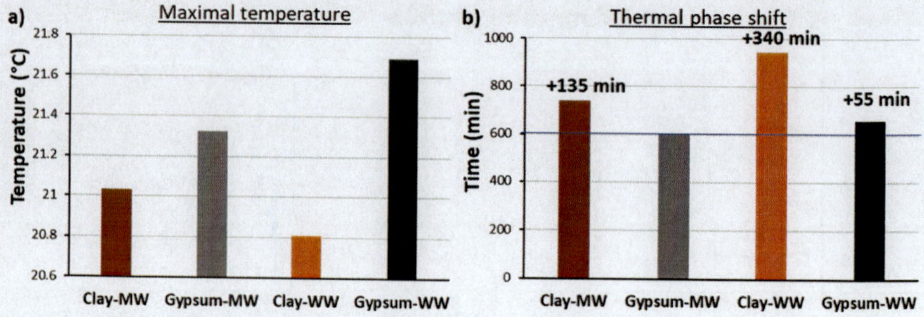

Fig. 14. Results for the temperature on the renders. a) maximal temperature; b) time needed to reach T = 20,81 °C compared to gypsum-MW wall.

Several observations can be made:

- The wall with gypsum+mineral wool showed the lowest thermal phase shift. This result is logical since both gypsum and mineral wool have low φ values (see Table 1).
- The maximal temperature is reached for the wall with gypsum+wood wool (T = 21,69 °C). This can be explained by the combination of two facts: 1) the wood wool have a higher lambda value than the mineral wool ($\lambda_{\text{Wood wool}}$ = 0,038 W/m.K, whereas $\lambda_{\text{Mineral Wool}}$ = 0,035 W/m.K); 2) the gypsum has the lowest thermal phase shift (φ_{Gypsum} = 0,5 h, whereas φ_{Clay} = 1,1 h). Furthermore, these are the commercial lambda values that have been determined at RH = 0%. It is suspected that in the present case at RH ~ 50% and higher temperatures, the $\lambda_{\text{Wood Fiber}}$ might be even higher [3, 9].
- The comparisons between walls with clay+mineral wool and gypsum+mineral wool, and also between clay+wood wool and gypsum+wood wool in Fig. 14 highlight that the decrease of the maximal temperature value and the longer thermal phase shift are mostly brought by the presence of a clay board. This is strengthened by the graphs trends presented in Fig. 13.
- Results show that the wall with clay+wood wool is the one with the lowest maximal temperature and the highest thermal phase shift. In that case, there is a synergy of the high thermal phase shifts of both clay and wood wool which enables to bring a better thermal inertia of the wall during the cycle. This better thermal inertia compensate the higher thermal conductivity of the wood wool.
- The difference between the outdoor temperature and the wall clay+wood wool is the highest: ΔT = 40–20,8 = 19,2 °C. The difference between the outdoor temperature and the wall gypsum+wood wool is: ΔT = 40–21,7 = 18,3 °C. The difference between these two walls is 0,9 °C, which corresponds to an increase of performance of 4,6%. This value is similar to the conclusion obtained by the modellings performed by Buildwise (ex-BBRI) and the EMPA that compared different types of material (mineral wool, cellulose, wood fibres) for roof insulation [15–17]. In these studies the authors concluded that effective cooling was mostly impacted by the combination of overnight ventilation and the presence of sunscreen on the windows during the daytime rather than by the type of insulating material. The modelling showed that

the impact on the summer comfort of the type of insulating material was comprised between 5 and 10%, which is close to the 4,6% obtained in this study.

- It can be observed in Fig. 13 that during the last step, the temperatures decrease much quicker for the walls with gypsum than for the wall with clay. It could be judicious in a future work to perform similar tests with several continuous cycles. This would enable to see if the indoor temperature stabilizes itself at an acceptable value or continue to increase cycle after cycle.

4 Conclusions

This study enabled to collect interesting information about the hygroscopic and thermal behaviours of bio-based materials implemented in a wood frame wall.

It was observed that the moisture-buffering effect of a wall without permeable membrane was impacted by order of influence mostly by the OSB board, then by the clay render (or with lower impact by the gypsum render), and finally by the bio-based insulating material. However, despite its high sorption-desorption capacity, the OSB showed a low MBV value. OSB could therefore be interesting for a long-term regulation of the hygric properties of a building, but seems not very interesting for deep and intense day-to-day regulation.

It was shown that behind an OSB board and a permeable membrane (Sd = 5 m), the hygroscopic impact of the insulating material on quick and short indoor relative humidity variations was almost insignificant.

Experiments revealed that the presence of thick clay board (22 mm) instead of conventional gypsum board (13 mm) is the factor that has the highest impact on the thermal inertia of the wall. It also showed that, in order to optimize the configuration and manage to obtain the lowest temperature and the longer thermal phase shift, wood wool should be associated with clay board rather than gypsum board.

In correlation with previous modellings, the experiments showed that the impact of flexible bio-based insulating materials on the thermal phase shift was relatively low (~ 5%) compared to flexible mineral wool. If one want to improve summer comfort, one should at first act on the sunscreen and nightcooling, and only after on the type of flexible insulation.

If, in spite of everything, the optimization of the thermal inertia remains an objective, one will then turn to bio-based insulating materials with densities higher than 100 kg/m^3 (type of rigid wood fibers, hemp-lime, hempcrete) or one will adapt the type/thickness of the interior finishing (plaster, clay…).

References

1. CODEM-BATLAB. Guide de la rénovation de parois à l'aide de matériaux biosourcés (2021)
2. ADEME; DGE. Recensement des produits biosourcés disponibles sur le marché et identification des marchés publics cibles (2015)
3. Ormondroyd, G.A.: An industry prioritised survey of thermal, mechanical, hydro and decay properties of natural fibre insulation materials (2017)

4. Koh, C.H., Gauvin, F., Schollbach, K., Brouwers, H.J.H.: Investigation of material characteristics and hygrothermal performances of different bio-based insulation composites. Constr. Build. Mater. **346**, 128440 (2022). https://doi.org/10.1016/j.conbuildmat.2022.128440
5. Lafond, C., Blanchet, P.: Technical performance overview of bio-based insulation materials compared to expanded polystyrene. Buildings **10** (2020). https://doi.org/10.3390/BUILDINGS10050081
6. Fachagentur Nachwachsende Rohstoffe. Nachwachsende Rohstoffe. Zeitbild (2016)
7. HWK. Untersuchung des Langzeitverhaltens von Dämmstoffen aus nachwachsenden Rohstoffen in Bauteilen (2016)
8. Vėjelis, S., Gnipas, I., Keršulis, V.: Performance of loose-fill cellulose insulation. Mater. Sci. **12**, 338–340 (2006)
9. Jones, D., Brischke, C.: Performance of nio-based building materials (2017)
10. Gourlay, E., Borderon, J.: Modélisation des besoins de chauffage et de refroidissement d'un bâtiment en béton de chanvre (2021)
11. Collet, F., Pretot, S.: Experimental investigation of moisture buffering capacity of sprayed hemp concrete. Constr. Build. Mater. **36**, 58–65 (2012). https://doi.org/10.1016/j.conbuildmat.2012.04.139
12. Asli, M., Brachelet, F., Sassine, E., Antczak, E.: Thermal and hygroscopic study of hemp concrete in real ambient conditions. J. Build. Eng. **44** (2021). https://doi.org/10.1016/j.jobe.2021.102612
13. Mortensen, L.H., Peuhkuri, R., Rode, C.: Full scale tests of moisture buffer capacity of wall materials. In: 7th Nordic Symposium on Building Physics, pp. 662–669 (2005)
14. Palumbo, M., Lacasta, A.M., Giraldo, M.P., et al.: Bio-based insulation materials and their hygrothermal performance in a building envelope system (ETICS). Energy Build. **174**, 147–155 (2018). https://doi.org/10.1016/j.enbuild.2018.06.042
15. Flamant, G., Heijmans, N.: Capacité thermique des isolants et risque de surchauffe (2010)
16. Bock, D., Heijmans, N.: Surchauffe estivale: impact mineur de la nature de l'isolant des toitures à versants (2021)
17. EMPA. Protection thermique estivale des pièces sous les combles Analyse des facteurs d'influence sur le climat intérieur (2008)
18. Rode, C., Peuhkuri, R., Time, B., et al.: Moisture buffer value of building materials. J. ASTM Int. **4**, 100369 (2007). https://doi.org/10.1520/jai100369
19. Collet, F., Prétot, S., Colson, V., et al.: Hygric properties of materials used for isobio wall solution for new buildings. Acad. J. Civ. Eng. **37**, 349–355 (2019)

Development of Various Building Materials Based on Paludiculture Cattail

M. Krus[1](✉), W. Theuerkorn[2], Th. Großkinsky[1], and R. Kraft[3]

[1] Fraunhofer Institute for Building Physics IBP, Fraunhoferstraße 10, 83626 Valley, Germany
martin.krus@ibp.fraunhofer.de
[2] Typha Technik, Baron Riederer Straße 51, 84337 Schönau, Germany
[3] Georg-August-University Göttingen, Büsgenweg 2, 37077 Göttingen, Germany

Abstract. Besides the reduction of gases through the rewetting of peatlands that have been laid dry, cattails are optimally suited as raw material for industrial use due to their enormous growth performance and yield. Typha stands form resilient, natural monocultures with an annual production output of 15 to 20 tons of dry matter per hectare. Wet use of peatland ensures natural peatland functions such as nutrient and water retention and represents an appropriate, sustainable alternative land use for the future. Cultivation brings ecological benefits in terms of stabilisation and renaturation of rewetted fen areas. At the same time, water protection is also achieved through the cultivation of Typha. With the help of cultivated areas planted with Typha, large amounts of nutrients and pollutants can be removed from the surface water and the area can be protected from further progressive soil erosion.

Due to the special structural properties of cattails, a building material was developed, the magnesite-bonded Typhaboard, which offers a unique combination of load-bearing capacity and insulation. This innovative building material has many other positive properties: Building material made from renewable raw material with a very high resistance to mold growth; easy workability with all common tools; low energy consumption during production and its recyclability. This product contains only vegetable ingredients and a purely mineral adhesive without any other additives. As a result, it is easily recyclable and even fully compostable and thus almost complies with the Cradle to Cradle concept.

Starting from the Typhaboard already used in several buildings, new building materials based on cattail have been developed in the meanwhile. In addition to magnesite as a binder, other adhesives have been successfully tested which are also sustainable and do not stand in the way of composting. Even using clay, stable and at the same time, insulating boards can be produced for applications that are not exposed to any particular moisture load. Another example is the development of a board for on-roof insulation, which has a lower bulk density and thermal conductivity due to a special particle geometry and arrangement. Furthermore, a low-dust blown-in insulation made of cattail and a kind of OSB substitute with comparable strengths are developed as well as a special very slim ceiling structure made of Typha. The development of a high load-bearing wood substitute is also in progress.

Keywords: Paludi-culture · Typha angustifolia · Building material

S. Amziane et al. (Eds.): ICBBM 2023, RILEM Bookseries 45, pp. 373–387, 2023.
https://doi.org/10.1007/978-3-031-33465-8_29

1 Background/Origin

1.1 Typha as Raw Material for the Development of the Magnesite-Bonded Typhaboard

Due to its special structural properties, typha angustifolia enables the production of building materials that offer a combination of insulation and load-bearing effect that is unique on the market. The special suitability of the leaf mass of Typha is determined by the structure of the plant. The leaves have a fibre-reinforced and stable support fabric, which is filled with a soft, open-cell sponge fabric (see Fig. 1). This results in both amazing statics and an excellent insulating effect.

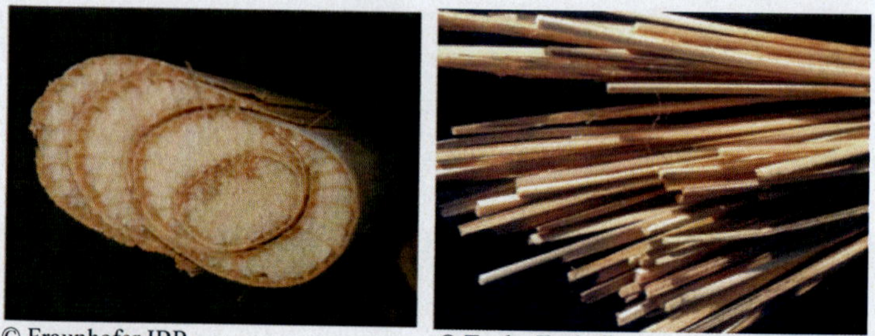

© Fraunhofer IBP © Typha Technik

Fig. 1. Left: Cross section of Typha leaf bundle with well visible sponge tissue and supporting tissue. Right: Typha rod particles produced by longitudinal cutting.

For the production of the panels, uniform, relatively large particles are produced with a specially designed cutting device in a first step, whereby no defibration takes place, but the leaf structure is retained. In this way, both positive properties of the plant, the strength and the insulating effect, are transferred into the product. These particles are then glued together with mineral adhesive (e.g. magnesite) under low pressure and energy input to form panels. In Fig. 2 the simple production process is illustrated by means of photographic images.

By means of the described procedure, the so-called magnesite-bonded typhaboard is produced as a building material with the following advantages:

- Excellent fire protection, no smouldering [1].
- Good sound insulation and thermal insulation in summer
- Natural mould resistance, diffusion openness and capillary activity
- Easy to process with common tools [2].
- Excellent suitability as interior insulation in existing buildings [3].
- Relatively insensitive to production fluctuations [4].
- Practical use proven in pilot project (DBU funding) (see Fig. 3) [4]

© Typha Technik © Typha Technik © Typha Technik

Fig. 2. Production process in pictures: Typha rods after longitudinal cutting, gluing drum and plate press.

© Alexandra Fritsch © Typha Technik © Architect Bruno Franchi

Fig. 3. Application examples of the Typhaboard: Repair of half-timbered building in Nuremberg with energetic and static upgrade (left) [4]; sauna house in Radolfzell as timber frame construction with 4.3 m grid! (in the middle, [5]); Typha pavilion EXPO Milan [6], elemental solid construction as self-supporting construction with shaped parts (right).

1.2 Environmental Aspects

The establishment of cattail fields and the rewetting of lowland moor soils, river floodplains, etc. could achieve the following positive ecological effects [7]:

- Denutriphication
- Reduction of CO_2 emissions
- Long-term binding of CO_2 in the building material
- Protection from erosion
- Creation of water retention areas and floodplains
- Creation of valuable biotopes for typical lowland moor fauna

- Recyclability and thus closed material cycle (cradle to cradle, C2C)

The cultivation potential and environmental benefits of cattails have already been demonstrated in a pilot project funded by the German Federal Foundation for the Environment (DBU) [8].

1.3 Agricultural, Economic and Social Benefits

Cattail is predestined as a raw material for industrial utilization or small-scale local plants due to its worldwide availability, enormous productivity and economic efficiency. Cattail stands are insensitive, long-term and natural monocultures, which produce 15 to 20 tons of dry matter per hectare annually [8]. This can generate an annual yield of about €5000/ha; more than most other agricultural products. Harvesting takes place in the winter months and full yield is possible from the second to third year with harvesting machinery already available. The nutrients required by the plant are preferably provided by nutrient-loaded water. Storage is easy to perform, as only rain protection is required (no pests) [9].

The cultivation of cattail also brings other advantages:

- Maintaining and creating jobs in structurally weak regions
- Securing employment opportunities for small and medium-sized entrepreneurs and agriculture
- Paludiculture as a second pillar to conventional agriculture
- Third pillar with small plant for regional product possible, as relatively tolerant of production fluctuations
- Product supply from the region possible
- Lucrative subsidy-free agriculture achievable in the medium term

2 Development of New Products from Typha Angustifolia

Even though the magnesite-bonded Typhaboard, developed in cooperation with Typha Technik and IBP, already has many positive unique selling points and will have an extremely high market potential in this form, further developments based on Typha angustifolia could be realized for other purposes or applications. The particularly interesting products that have emerged are described below.

2.1 Variation of the Binder

Due to the relatively low bonding strength of magnesite, relatively large proportions of binder are required to produce a load-bearing material. Although these are small in volume, they represent a mass proportion of approx. 50%. This leads to the relatively high bulk density of approx. 270 kg/m^3 of the standard magnesite board (which is advantageous in terms of sound insulation), but thus also to a higher thermal conductivity (0,055 W/mK). For this reason, tests were also carried out with other binders, as described below. Figure 4 shows a test specimen in which blood albumin, with its significantly higher adhesive strength, was used as a binder. The blood albumin can be recognized by

Fig. 4. Test specimen with blood albumin as binding agent.

the dark discoloration. The variant with blood albumin has a lower thermal conductivity (0,044 W/mK) than the variant with magnesite, because of its lower bulk density of 162 kg/m^3, caused by the lower binder content.

Additional tests with a mixture of magnesite and albumin as a binder had the purpose of combining the advantages of both binders as far as possible. Due to its significantly higher binding strength, albumin requires a significantly lower amount of binder. The addition of magnesite is aimed at improving the fire properties. It was shown that both binders can be combined without any problems.

Experiments were also carried out with geopolymer. Geopolymers are two-component systems consisting of a reactive solid containing silicon and aluminium oxides and a basic activating solution of alkali hydroxides or silicates in water. The solid is a natural rock or mineral, hence the prefix "geo". Figure 5 shows two variants of plates manufactured with geopolymer, white kaolin being used for geopolymer 1 and cheaper red kaolin for geopolymer 2.

An important result of the investigations was that the low binder requirement due to its higher bonding strength with geopolymer enables sufficiently stable Typha boards with very low bulk density and thus low thermal conductivity (Fig. 5, right picture). Thereby, a thermal conductivity of 0.039 W/mK is achieved with an apparent density of 110 kg/m3 (The material on the left shows a higher density of 217 kg/m^3 and thermal conductivity of 0.052 W/mK.) In addition, it is expected that the energy requirement for the production of the binder geopolymer is significantly lower due to the lower firing temperatures of kaolin. However, this results in a loss of the excellent fire protection properties, which are mainly due to the use of magnesite and the associated evaporating of water of crystallization.

Binding with clay was also tested as a particularly sustainable binder variant, since no mineral raw material has to be fired. Figure 6 shows test specimens bound with clay, with one example with a high density of 430 kg/m^3 and a thermal conductivity of 0.062 w/mK (right side). The sample on the left in the picture, marked "Leipheim", was

© Fraunhofer IBP © Fraunhofer IBP

Fig. 5. View of two different geopolymer-bonded typha plates. The binder of the plate shown on the left is made with white kaolin and that of the right plate with cheaper red kaolin.

produced at the Paludiculture Day, in Leipheim, Germany, as an on-site demonstration. Astonishingly high strengths are obtained with this clay, so that even slabs with low bulk density are possible (thermal conductivity of 0.042 W/mK at a density of 155 kg/m^3). The advantage of this type of bonding is the worldwide availability of clay material and the extremely low energy requirement for the bonding agent. However, since no chemical setting reaction takes place, it is absolutely necessary to dry the material immediately afterwards. It should also noted that the material must be protected from direct exposure to rain.

© Fraunhofer IBP © Fraunhofer IBP

Fig. 6. Specimens bound with clay. The sample on the right of the picture, labelled "Leipheim" was produced at the Paludiculture Day as an on-site demonstration.

2.2 Further Development of the Magnesite-Bonded Board to Lower Bulk Density or Thermal Conductivity

The starting point was the consideration of reducing the specific weight of the board material and thus also the thermal conductivity lambda with a three-layer structure of the board compared to the classic Typha board, with the least possible reduction in mechanical and static properties:

1. The core layer of the three-layer cladding structure is provided with considerably larger particles than the cover layers, which means that the particle surfaces and thus the adhesive requirement are smaller overall. This in turn has the effect of reducing the specific weight and thus the lambda value.
2. In order to displace the free air spaces from larger particles, higher pressures are required. Since these pressures also affect the surfacc layer to the same extent, they result in a more rigid and overall stronger structure. Thus, despite the lower mass, the bending stiffness and the compressive strength are achieved.
3. The special growth form of the raw material Typha, especially of Typha angustifolia with its strong leaf bundles with an oval cross-section, offers the possibility of cutting off the areas of the leaf bundle with less aerenchyma on the two flat sides of the oval using a special cutting technique. Since the aerenchyma in Typha is the primary carrier of the thermal insulation property (it has values of less than 0.035 W/mK measured on its own), the sole use of the central area produced in this way for the core layer of the sandwich significantly promotes the thermal insulation capacity of the material. In addition, due to the crescent-shaped blade cross-section, the blades in the central area of the blade bundle are thicker and thus produce larger particles when cut. This also helps to minimize the total surface area of the particles and thus also the adhesive consumption.

These three aspects were achieved in the following way:

- The core layer particles were prepared with a cutting grid of 6 mm cutting width, the surface layer with 3 mm cutting width. In the board, the surface layers each account for 20% of the sheet mass used, the core layer for 60%.
- Due to the larger particles in the core layer, the grouting pressures could be increased by approx. 40% without significantly squeezing the aerenchyma, with the effect that the surface layers became more compacted and thus more resistant to tension and compression.
- The machine specially designed for the project to separate the aerenchyma- and sclerenchyma-containing regions of Typha leaf bundles was oriented so that about 60% of the total leaf mass was separated from the aerenchyma-rich middle region by about 2 times 20% sclerenchyma-rich side regions. These separated side regions can then be used to produce Typha materials, where strength is more important than insulating properties

In Fig. 7 is an example of such a three-layer panel and in Table 1 the essential material characteristics of this panel are shown.

2.3 Bulk Insulation from Typha Angustifolia

The leaf structure of Typha angustifolia with its sponge and supporting tissue (aerenchyma and sclerenchyma) can also be used to advantage in the production of an insulating fill. One problem of loose-fill insulation that requires special attention, especially for the craftsmen working with it, is the formation of dust. However, if the

© Fraunhofer IBP

Fig. 7. Photographic view of the three-layer Typha insulation board.

Table 1. Tabular summary of the essential characteristic values of the Typha insulation board used as a three-layer board.

Parameter	Unit	Value
Binder	–	Magnesite
Bulk density	kg/m3	140–160
Thermal conductivity	W/mK	<0,045
Compressive strength	kPa	>100
Flexural strength	kPa	>300

typha sheet bundle is not crushed in a mill, but cut in a special cutting device, the dust formation is reduced quite considerably. In addition, there is also the possibility of improving the insulation properties by removing the lateral sclerenchyma-rich portions with little sponge tissue. These segregated sclerenchyma-rich portions can then be used preferentially for the production of building panels where strength is the primary concern. Typha angustifolia with its open-pored sponge tissue also offers the possibility of relatively simple storage of fire protection agents. The simple fire test carried out (see Fig. 8 right) shows a significantly lower flammability. A special possibility to improve the settlement behaviour of such a fill is the admixture of Typha seed parachutes (Fig. 8 left). After very positive tests for the production of a largely dust-free blow-in insulation made of Typha angustifolia, attempts were made to further optimize the bulk insulation by varying the parameters of particle cutting method, particle size and distribution as well as the addition of fibers (e.g. Typha seed umbrellas) (see Fig. 9). Thermal conductivities of less than 43 mW/mK were achieved in some cases.

Fig. 8. Left: Typha insulation fill with seed flight shield Right: Simple fire test to demonstrate the lower flammability of the typha particles after shaking the fire protection agent into the open-pored sponge fabric.

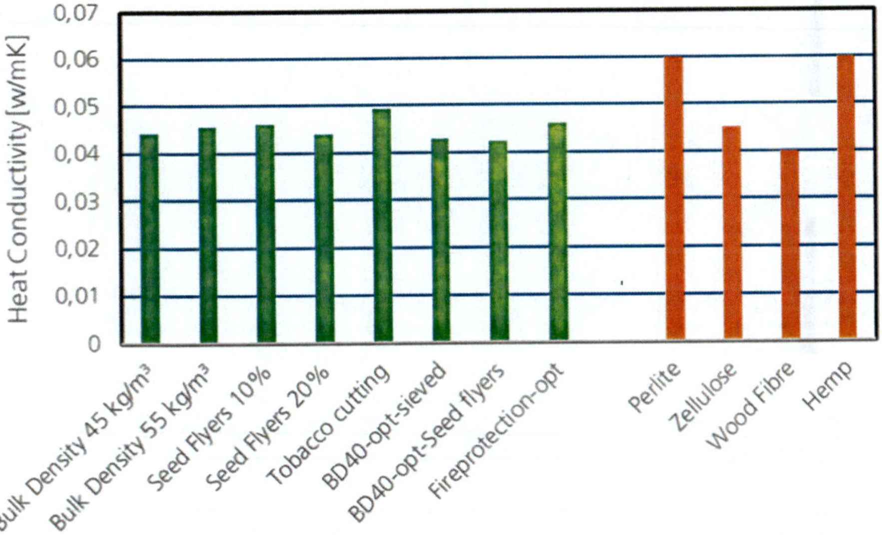

Fig. 9. Thermal conductivities obtained for bulk insulation made from Typha angustifolia under variation of density, cutting method and addition of Typha seed umbrella as well as fire retardant. In comparison, the four bars on the right show the thermal conductivities of a selection of conventional bulk insulation materials.

2.4 Wood Substitute from Typha

In addition, developments have been made in the direction of wood substitutes with the following approach. In a parallel arrangement, the bonding can take place particularly well, since the orderly adjacent rod-shaped Typha leaf particles have a large contact area even at comparatively low pressure. There is thus also little air in the area between the rod-shaped Typha leaf particles. This leads to a low thermal conductivity across the rod-shaped Typha leaf particles. In the longitudinal direction, the many rod-shaped Typha leaf particles lying next to each other lead to a high load-bearing capacity. For good thermal insulation perpendicular to the leaf axes and with high load bearing capacity in the longitudinal direction, this embodiment with a low weight, low typha leaf mass and adhesive requirement can provide a very suitable structural element capable of supporting very high loads.

© Fraunhofer IBP

© Fraunhofer IBP

Fig. 10. Circular test specimen with parallel alignment of typha particles to increase compressive strength with a density of ca. 435 kg/m^3; left before and right after conducting the compressive strength test, indicated by the cracks along the axis.

The very first test for this already resulted in more than a tenfold increase in compressive strength compared to the standard board with 6.28 N/mm^2 and makes further development up to comparable strengths of wood appear realistic. Figure 10 shows the round test specimen with parallel alignment of the typha particles to increase the compressive strength before and after carrying out the compressive strength test. Compared to wood, this would have two significant advantages. On the one hand, this wood substitute will have only about half the thermal conductivity and thus significantly alleviate the thermal bridge problem. An even more significant positive property will be the associated improvements in fire protection, which is probably the biggest problem of conventional lightweight constructions made of wood.

2.5 Typha Ceiling Element

As a further development, a panel element curved on the underside was produced, which can be used as a ceiling element similar to the so-called Prussian caps. Since, in addition to its load-bearing capacity and its thermal insulation properties, it also provides impact sound protection, it is possible to use it for ceiling structures with a significantly lower construction height, which are also much simpler to construct and thus likely to be cheaper than standard ceiling constructions. Figure 11 shows the view of the underside of the ceiling structure made of cob with the characteristic curvature. With this construction, there is no need for a structure based on the principle of floating screed.

© Fraunhofer IBP

Fig. 11. Bottom view of the ceiling structure with the characteristic camber.

In addition to the possibility of simpler slab construction, the significantly higher sustainability of this approach should be mentioned here, but above all also the associated high level of fire protection. However, there is still potential for improvement here in the area of the ceiling beam supports.

2.6 OSB Replacement from Typha

A substitute material made of Typha was also developed for the so-called Oriented Strand Boards (OSB). This is a sandwich construction made up of top layers of typha strands and a middle layer of poplar chips. These panels were made in one step. With 90% Sorel cement and 10% albumin as binder, the results of both flexural and transverse tensile strength as well as thickness swelling after 24 h of water storage reached a good level nearly comparative to that of OSB Boards. The proportion of the middle layer made of poplar chips is 60%, while the top layers made of typha strands have a weight proportion of 40%. The chips were glued in a weight ratio of 1:1 with a binder liquor of sorel cement (90%) and 10% albumin. Obviously, both binder proportions complement each other here. While pure albumen leads to good strength properties, the thickness swelling of boards produced with it after 24 h of water storage is rather negative; while pure sorel cement leads to good swelling values of boards produced with it and is also to be evaluated positively with regard to fire protection, but is unfavorable with regard to the strength properties. In the above-mentioned dosage, both binders complement each other.

Table 2 gives a tabular summary of the essential characteristic values of the "T-board" newly developed as a replacement for OSB. Figure 12 shows T-boards with a groove milled into the side for connecting the T-boards to each other using spring elements.

With comparable strengths, the T-board has the following advantages over OSB:

- Greater sustainability through the use of more environmentally friendly binders
- Better fire protection
- Lower thermal conductivity
- Reduced input of grey energy

Table 2. Tabular summary of the main characteristic values of the "T-board" newly developed as a replacement for the OSB board.

Property	Unit	Values
Core layer	Mass fraction	Poplar 60%
Top layer	Mass fraction	Typha 40%
Binder	Mass fraction	Magnesite 90% Albumin 10%
Bulk density	kg/m^3	≈850
Flexural strength	N/mm^2	≈9
Transverse tensile strength	N/mm^2	≈0,12
Thermal conductivity	W/mK	0,083

3 Summary

With the rewetting of drained fens and the cultivation of cattails (Typha angustifolia), a significant reduction in greenhouse gas emissions from conventional cultivation and the use of the plant as a building material with long-term fixation of the stored carbon makes it possible to relieve the environment like no other agricultural product. In addition, the cleaning of nutrient-polluted surface waters and the possibility of using them as retention areas and for improving the water balance are environmental benefits. A large-scale switch from dry to wet farming will only succeed in Germany and other European countries if this is also associated with financial benefits for farmers, and preferably not in the form of permanent public subsidies. However, this will only be possible if a high-quality and thus comparatively high-priced product is produced from the raw material, which also allows relatively high raw material prices to be paid to the farmer.

© University Göttingen © University Göttingen

Fig. 12. T-boards with groove milled into the side for connecting the T-boards to each other (left) using spring elements (right picture).

With the magnesite-bonded Typhaboard already developed and used in demonstration buildings the development of a building material with numerous positive unique selling points has already been achieved in an impressive manner. It offers a unique combination of load-bearing capacity and insulation, contains only plant-based ingredients and a purely mineral adhesive, and is thus easily recyclable and even fully compostable, In order to open up additional areas of application, further product developments have been driven forward. For example, other binding agents that are also sustainable and do not stand in the way of composting, such as blood albumin or geopolymers with higher binding strength, have been successfully used to reduce the amount of binding agent and thus the bulk density and thermal conductivity. Even using clay, stable and at the same time insulating boards can be produced for applications that are not exposed to any particular moisture load. With these binders, however, the excellent fire protection properties, which are largely due to the magnesite, are partially lost.

For use in areas where a high bulk density is a hindrance, e.g. as on-roof insulation, and where the insulating effect plays a greater role than the strength properties, this objective was achieved with a three-layer board structure and this while retaining the binding agent magnesite and the associated good fire protection properties. In addition, a bulk insulation made of Typha was developed which, due to the special manufacturing technique, is far less dusty than comparable bulk insulations. Due to the open-pored sponge fabric, it is also possible, as will be shown, to store ecologically uncritical fire protection agents relatively easily.

Developments have also been made in the direction of wood substitutes by arranging the cut, rod-shaped typha particles in parallel for particularly effective bonding. In the longitudinal direction, this leads to a high load-bearing capacity with good thermal insulation at the same time. With this design a very sustainable building element can be created that can take very high loads.

As a further development, a panel element curved on the underside was produced which, in addition to its load-bearing capacity and its thermal insulation properties, also provides impact sound protection. This construction eliminates the need for a structure based on the principle of floating screed. Ceiling structures with a significantly lower construction height are thus possible, which are also likely to be much simpler to construct and thus less expensive than standard ceiling constructions. Added to this is the higher sustainability of this approach, but above all the associated high level of fire protection.

Last but not least, a substitute material made of Typha was also developed for the so-called OSB. This is a sandwich construction made up of top layers of typha strands and a middle layer of poplar chips. With comparable strengths, the T-board is more sustainable than the OSB due to the use of more environmentally friendly binders, better fire protection and lower thermal conductivity.

All the developments mentioned here also serve to move back in the direction of "simple construction" instead of the increasingly complex construction methods. Instead of creating wall structures with a separate material layer for each structural-physical task, the magnesite-bonded Typhaboard, for example, can combine many construction-relevant properties in itself (thermal protection, statics/stiffening, fire protection, moisture protection, plaster base…). All these products are suitable for cost-effective, element-based construction and for the creation of lightweight construction for inner-city redensification by adding storeys, where fire protection plays a major role.

However, they are also excellently suited for building in existing structures, e.g. for the renovation of half-timbered houses; interior insulation; on-roof insulation and other areas of application. Another major advantage is that they are usually easy to dismantle and can be reused directly, as in many areas of application these products can simply be fastened with screw connections instead of glueing.

References

1. Research report EG2050: RokoDaMi - Demonstration of an innovative and sustainable rooftop insulation made of bulrush. Funding reference: 03EGB0008A-B
2. Fritsch, A., Theuerkorn, W.: Fachwerksanierung und Energieeffizienz. In: Denkmalpflege Informationen, no. 154, p. 15 (2013)
3. Krus, M., Bichlmair, St., Kilian, R.: Comparative investigation of different interior insulation systems applied in the "Alte Schäfflerei" and their evaluation. In: Proceedings EEHB2022; The 4th International Conference on Energy Efficiency in Historic Buildings, Benediktbeuern, Germany, 4–5 May 2022 (2022)
4. Theuerkorn, W., et al.: New building material for environmentally friendly and structural renovation in monument conservation. DBU report (funding code AZ 27918) (2022)
5. Krus, M., Theuerkorn, W., Großkinsky, Th., Georgiev, G.: A new load-bearing insulation material - Typha tubular piston panels; A New Load-Bearing Insulation Material, Typha Panels; in Detail Jg. **54**(1/2), 74–80 (2014). ISSN 0011-9571
6. Franchi, B.: Ecological and aesthetic building in the garden/pavilion made of cattail. In: Drexel, Th. (ed.) Gartenhäuser - Lauben Lounges Pavillons Baumhäuser Hütten Carports. Deutsche Verlags-Anstalt, pp. 68–71. ISBN 978-3-421-04014-5

7. Ing, M.: Faulstich: SRU Environmental Report 2012, Responsibility for a Limited World
8. Pfadenhauer, J., Heinz, S.: Multitalent Rohrkolben -Ökologie, Forschung, Verwertung Brochure zum Abschlussbericht des DBU-Projektes "Rohrkolbenanbau in Niedermooren - Integration von Rohstoffgewinnung, Wasserreinigung zu einem nachthaltigen Nutzungskonzept" im Donaumoos 1998–2001, TU München, Lehrstuhl für Vegetationsökologie
9. Krus, M.: Cattail cultivation in agriculture. A novel, load-bearing and insulating building material. PlanerIn H, pp. 33–35 (2014)

Experimental Study of the Long-Term Effect of Natural Additives, After Recycling Biocomposites: Mechanical and Thermal Properties

Z. Alshndah[1,2(✉)], F. Becquart[3,4], and N. Belayachi[1]

[1] Université d'Orléans, INSA Centre Val de Loire, Université de Tours, Laboratoire de Mécanique Gabriel Lamé (LaMé) Polytech Orléans, 8 rue Léonard de Vinci, 45072 Orléans, France
zeyad.alshndah@etu.univ-orleans.fr
[2] Department of Civil Engineering, Engineering College, Tikrit University, Tikrit 34001, Iraq
[3] IMT Nord Europe, Institut Mines Télécom, Centre for Materials and Processes, 59000 Lille, France
[4] Univ. Lille, Institut Mines-Télécom, ULR 4515 – LGCgE, Laboratoire de Génie Civil et géo-Environnement, 59000 Lille, France

Abstract. Thermal insulation biocomposite materials are based on vegetal aggregates and mineral binders. Reducing the quantity of binder used contributes to reducing the carbon emissions related to its manufacture, and, at the same time, decreases the thermal conductivity of these biocomposites. Previous research has enhanced thermal behavior by including additives in the biocomposites mixture (Ismail et al. 2020) to reduce the binder quantity in the mixes developed initially by (Belayachi et al. 2013). The objective of this present study is to evaluate the possibility to recycle vegetal aggregates with additives used in the first straw-lime blocks made three years ago. The studied biocomposites are composed of recycled aggregates obtained by crushing naturally aged biocomposites blocks, without the addition of any additives. In the previous blocks, three additives have used the replacement of a lime part to improve the binder porosity with hemoglobin or the binding of the aggregate-binder interface with casein and gelatin. Recycled wheat straw (RWS) with additives used in this study was produced by crushing previous biocomposite blocks in the laboratory using a manual process for aggregate manufacturing. Then, biobased samples were produced by mixing RWS and normal wheat straw in mass proportions of (30, 50, 70, and 100%) with a water-binder ratio of 1.1, 0.8, and a straw-binder ratio of 0.2. After 28 and 90 days of curing under constant temperature and humidity conditions (20 °C and 65%, respectively), mechanical and thermal tests were performed. The results show that increasing the amount of RWS increases the thermal conductivity and compressive strength of the biocomposites. This increase was confirmed by water porosity tests and microscopic examination. Additives have no negative effect on recycling. Microscopic observations confirm that casein and hemoglobin have no impact due to the presence of needle-like calcite crystals by comparing the results with the previous works (Ismail et al. 2020).

S. Amziane et al. (Eds.): ICBBM 2023, RILEM Bookseries 45, pp. 388–398, 2023.
https://doi.org/10.1007/978-3-031-33465-8_30

Keywords: Biocomposite materials · mineral binder · additives · lime · recycled straw

1 Introduction

Carbon emissions cause serious environmental impacts. The construction industry emits more than 120 million tons of CO2 in the world each year, making it one of the most important sectors in the battle against global warming and a significant actor in the energy transition (Tlaiji et al. 2022). Buildings in France consume about 44% of the total energy generated and contribute about 25% of the greenhouse gas emissions that have a significant impact on the environment (Omeme Ada et al. 2021). The use of bio-based materials is one of the most important strategies for reducing both carbon emissions and energy consumption in the construction sector; in addition, it permits safeguarding and reducing the depletion of non-renewable natural granular materials. The three most important factors in sustainable construction are comfort, environmental impact, and health (Kashtanjeva et al. 2015). These sustainable concepts may be achieved by using some bio-based materials in buildings (Tlaiji et al. 2022). The use of plant aggregates in the production of thermal insulation materials, such as wheat straw, hemp, flax, and other agricultural residues helps to improve the hygrothermal behavior of buildings. Indeed, these materials can regulate the moisture content of the surrounding interior and exterior environment (Douzane et al. 2016).

Binders such as cement and lime are often used in the production of biocomposites to improve their mechanical properties and their durability. The use of these two binders has environmental consequences, as evidenced by the carbon emissions associated with their production. The use of eco-friendly natural binders is one of the best solutions for reducing the environmental impact of mineral binders. As a result, researchers are interested in the use and development of environmentally acceptable binders and in improving the thermal properties of biocomposites. (Brouard et al. 2018) compared the thermal and mechanical characteristics of several materials based on vegetable aggregates and clay to determine the performance of four distinct biocomposites to highlight the agricultural waste potential. (Rim et al. 1999) discusses the effect of the wood aggregate fraction on the thermal and mechanical performance of a clay-cement-wood composite. According to the results, adding wood to clayey concrete enhances its insulating properties, decreases its mechanical strength, and increases its deformability. (Vitola et al. 2022) investigated three bio-aggregates for their potential use in the creation of biocomposites using potato starch as a binder. The obtained results have been like those of other biocomposites, with thermal conductivity ranging from 0.05 to 0.062 W/m.k and compressive strength between 0.2 to 0.8 MPa. (Yaping et al. 2022) developed a novel insulation material from rice straw. A sodium alginate binder, produced from brown algae, is used. Thermal conductivity and water vapor permeability are investigated. For an average density of 100 to 200 kg/m^3, thermal conductivity values for composite materials range from 0.039 to 0.045 W/m.k. All of the composite samples may be used as building insulators. They also found that the binder ratio has a minimal influence on thermal conductivity. (Mati-Baouche et al. 2014) developed biocomposites-based sunflower stalks using chitosan as a bio-based eco-friendly binder. The research examined how compaction stress, chitosan/sunflower ratio, and particle size grading influenced thermal and mechanical properties. The results show the composites containing 4.3% chitosan (w/w) had 0.056W/m.k thermal conductivity, and 2 MPa maximum stress. No chitosan binder penetrated sunflower cells, maintaining sunflower stalk particle characteristics. These mechanical and thermal characteristics are equivalent to those of existing insulating bio-based materials.

Additives are used to improve the properties of the binder used in the production of the biocomposites as well as reduce the amount of the mineral binder to reduce the carbon emissions associated with its production. Specific previous research on this topic in the laboratory (Ismail et al. 2020) used three additives. First, a hemoglobin powder was added as an air entrainer to the material components to produce additional porosity in the matrix and, therefore, decrease the thermal conductivity of the biocomposite. A second time, by substituting a proportion of the binder, casein powder was used to act on the porosity matrix and the adhesion of the matrix aggregates interface. A third additive, gelatin, aimed to replace a part of the mineral resulting in stronger bonding between the plant-based aggregates and binder and allowing the introduction of a larger fraction of aggregates.

The objective of this present research is to investigate the possibility of using RWS aggregates with additives used in the original straw-lime blocks (produced three years ago) (Ismail et al. 2020) to make new materials after recycling and analyze their thermal and mechanical properties. Biocomposites are made from RWS aggregates with the different additives obtained by crushing naturally aged biocomposite blocks without the addition of any additives in the new mixture. Two mixtures were used from previous work (Belayachi et al. 2013). The samples were performed, with two different water-to-binder ratios of 1.1 and 0.8, and the same straw-to-binder ratio of 0.2. Four mass ratios of RWS (30, 50, 70, and 100%) were mixed with normal wheat straw. After 28 to 90 days of curing, mechanical and thermal properties were investigated.

2 Experimental Methodologies

2.1 Materials

The classic lime binder PF70 is used for the biocomposite formulations. This pre-formulated binder includes about 75% aerial lime, 15% hydraulic binders, and 15% pozzolan. Normal straw aggregates (Fig. 1a) were produced in 2020 in France (Centre Val de Loire); recycled wheat straw (RWS) aggregates (Fig. 1b) with additives are obtained from the crushing of biocomposite blocks produced in the previous study in 2018 (Ismail et al. 2020). By comparison with natural wheat straw, the presence of residual lime on recycled aggregates implies a color aspect with more grey tint.

Particle Size Distribution

This technique included scanning the wheat straw particles, and using the image in software Images Analysis to determine various metrics (length, width, area, etc.). As recommended by RILEM (Amziane et al. 2017), a sample of about 2000 particles was distributed and scanned on a black surface. The collected images were processed using Matlab code. After segmenting the pictures, each detected particle was ellipse-adjusted.

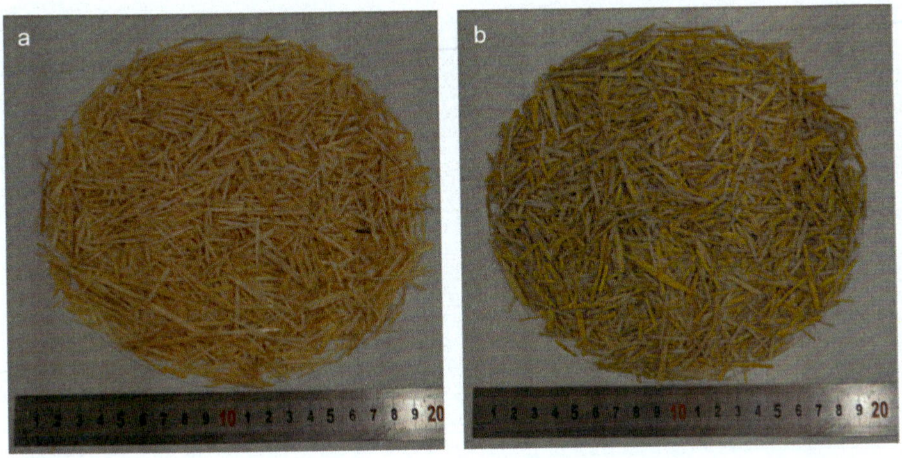

Fig. 1. Types of wheat straw (a) Natural straw (b) RWS with residual lime and additives.

Based on the data, the length and width of each particle were calculated. The large and small axes of the adjusted ellipse. From Fig. 2, the majority of RWS particles vary between 0.05–3 cm (length) and 0.05–0.7 mm (width). So, it is important to notice that the length of RWS particles is between 1 and 4.3 times greater than their width. From Fig. 2, the length of 80% of the RWS is 0.5–1.5 cm, whereas the length of 90% of the natural straw from the first straw-lime blocks (made three years ago) is 3.5 cm (Ismail et al. 2020). This physical evolution is due to straw particles being broken during the manufacture and demolition of the blocks to obtain recycled straw. This size of recycled straw particles enhances the compaction of biocomposites. This will have an impact on the thermal and mechanical behavior of the materials based on recycled aggregates.

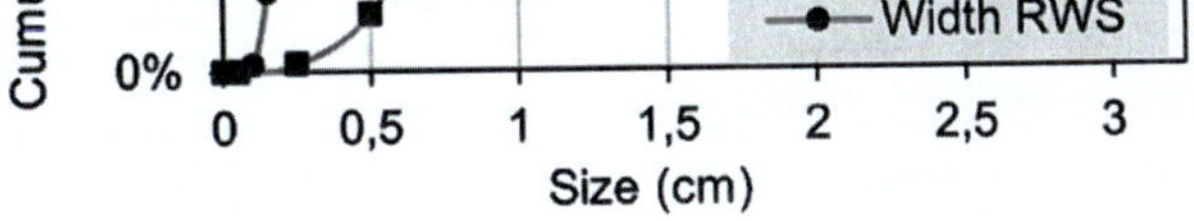

Fig. 2. Particle size distribution of RWS.

Water Absorption

The measurement of the absorbed water is an important parameter for the calculation of the effective water content of the mixture. For this purpose, an amount of RWS is put in an oven until the mass stabilizes at 60 °C. The sample is then weighed at 50 g. This sample is immersed in water for several minutes (1, 5, 10, 15, 20, 30, 45, 60), with the weight recorded at each period. The test is repeated three times with different samples of shives. Equation 1 is used to calculate the absorption coefficient:

$$A(t) = \frac{m(t) - m(t_o)}{m(t)} \tag{1}$$

With A(t) (%) water absorption rate.
m(t) (kg) sample's weight at t time.
$m(t_o)$ (kg) sample's initial weight.

Figure 3 shows the water absorption of RWS. According to the curves, the water absorption of RWS after 10 min is around 212% and 232% after one hour. The presence of lime around the RWS particles caused some of the existing pores to close and prevented water from accessing the inner pores, resulting in a reduction in the water absorption rate for RWS in comparison with normal straw (Bouasker et al. 2014).

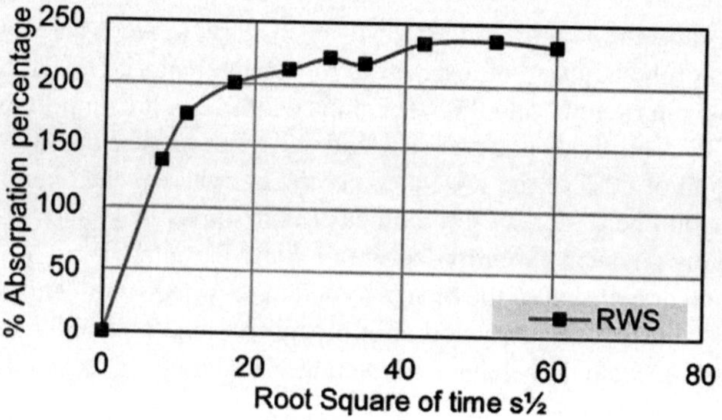

Fig. 3. Rate of absorption with time for RWS.

2.2 Mixing Design and Sample Preparation

Two previously optimized mixtures were used (Belayachi et al. 2013) for the formulation parameters: for the first case, water/lime = 1.1 and straw/lime = 0.2; for the second case, water/lime = 0.8 and straw/lime = 0.2. A reference mixture with 100% normal straw was systematically mixed for each group; for all other formulations, RWS aggregates were added in proportion by mass (30, 50, 70, 100%). Ten formulations of biocomposites were thus produced as shown in Table 1.

The mixtures are made in two steps. The first step is to prepare the binder, with the appropriate quantity of water and lime depending on the approved mixing ratios. As wheat straws absorb water, aggregates are moistened with 20% of the water mixture before mixing to maintain the workability of the mixture. The residual water is mixed with the lime. The mixing process starts with an electric vertical mixer set at 140 rpm for 5 min. The straw is prepared in the second step. Straw and hydration water are combined

Table 1. Formulation parameters for mixtures and biocomposites designation (by mass).

Samples	S/B	W/B	NS%	RWS%
S1R	0.2	1.1	100	0
S2	0.2	1.1	70	30
S3	0.2	1.1	50	50
S4	0.2	1.1	30	70
S5	0.2	1.1	0	100
S6R	0.2	0.8	100	0
S7	0.2	0.8	70	30
S8	0.2	0.8	50	50
S9	0.2	0.8	30	70
S10	0.2	0.8	0	100

S/B: Straw/Binder. W/B: Water/Binder. N.S: Normal straw. RWS: Recycled wheat straw with additives.

for one minute in a mixer. The lime is then put into the mixer, and 5 min of mixing starts. The mixture is placed in 40 × 10 × 10 cm metal prismatic molds without compaction or vibration. Two groups of samples are produced for 28-day and 90-day curing. After 24-h mold removal, the samples are stored in an air-conditioned room at 20 °C and 65% relative humidity. After curing, samples of 10 × 10 × 20 cm^3 are prepared for mechanical compressive; samples of 10 × 10 × 10 cm^3 are cut for thermal tests. The samples are dried in an oven at 60 °C until mass stability before testing.

2.3 Testing Methods for Biocomposites

Porosity

The water-accessible porosity was evaluated by hydrostatic weighing according to the standard NF P18–459. Three samples of 5 × 5 × 5 cm of each bio-composite were dried at 60 °C until mass stability (*m* solid). Samples are then put under a vacuum for 24 h to remove air from the pores. After 24 h of saturation in water, samples were weighed in water to obtain the hydrostatic mass (*m* hydrostatic) and then in the air to obtain the saturated mass (*m* saturated. The total porosity of a sample is the volume of voids to the total sample volume Eq. 2. This amount is represented as a percentage:

$$N = \frac{m_{saturated} - m_{solid}}{m_{saturated} - m_{hydrostatic}} \tag{2}$$

Thermal Conductivity

Thermal conductivity was measured using a NeoTIM conductivity meter using the hot wire method. The tests were performed on 10 × 10 × 10 cm samples. The probe was extended to 10 positions. The test was carried out in a dry state.

Compressive Strength

The compressive strength of the biocomposites was evaluated using prismatic samples with 20 × 10 × 10 cm in dimension. An Instron mechanical press with a 250 kN load cell was used to perform tests at 0.5 mm/min. Figure 4 shows a biocomposite sample before/after rupture. The test is carried out perpendicular to the direction of sample molding. A total of three samples were prepared for each combination and age, and the result presented for each age is the average of the three samples tested. At 28 and 90 days, the compressive strength was evaluated.

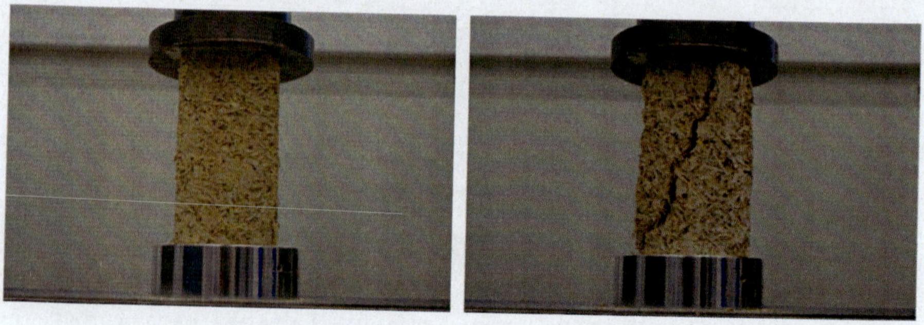

Fig. 4. Compressive strength test on a prismatic biocomposite (20 × 10 × 10 cm).

Scanning Electron Microscopy (SEM)

Electron Scanner Microscopic observations were performed on a sample of biocomposite materials (1 × 1 cm). The samples used to observe mineralogical changes are selected for 90 days of curing at 20 °C. Samples are dried at 60 °C until mass stability then placed under a vacuum in a Hitachi SEM chamber.

3 Results and Discussion

3.1 Porosity

Figure 5 shows that increasing the percentage of RWS in the biocomposites causes a decrease in porosity for both W/B ratios (1.1 and 0.8). This is due to the residual lime presence in the RWS, which improved the density of the biocomposites, resulting in a reduction in porosity. Additives did not appear to influence significantly the porosity of the biocomposites. Figure 6 shows the cross-section of biocomposites. The section of the sample PF1 (Ismail et al. 2020) containing 5% hemoglobin showed a porous matrix and cellular structure. The explanation is that hemoglobin surfactant properties create a stable foam when mixed with lime and water (Ismail et al. 2020). While observing the sections of samples S5-S10 of the present study (containing 100% RWS), there is very little pore with a high lime content. This is due to the residual lime content of the RWS contributing to enhancing density. This solid structure with limited pores shows no effect of hemoglobin.

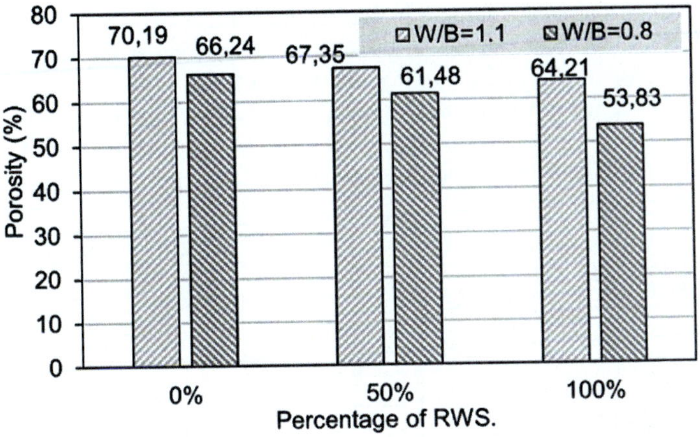

Fig. 5. Biocomposite porosity for W/B = 1.1 and 0.8 after 90 days, with three percentages of RWS.

Fig. 6. Cross section in bio-composite samples PF1 (Ismail et al. 2020) and S5-S10 for present work.

3.2 Thermal Conductivity and Compressive Strength

The effect of RWS on the thermal conductivity and compressive strength of biocomposites for different ages and W/B ratio is presented in Fig. 7. In general, results show that the thermal conductivity and compressive strength increase up to an RWS percentage increase. The presence of residual lime in the RWS increases the binder ratio in the biocomposites and therefore, thermal conductivity and strength increase. This indicates that previous additives of casein and hemoglobin have no significant impact. Figure 8 compares previously formed biocomposites (with natural wheat straw) PF0-PF5 and PF10 from (Ismail et al. 2020) with the formulation S5 of the present work. PF0 was the reference formed with S/B = 0.2 and without additives; PF5 considered S/B = 0.3 and 5% hemoglobin - 2.5% of casein in substitution of lime; PF10 was formulated with S/B = 0.4 in which 10% of the lime was substituted by gelatin.

The thermal conductivity of biocomposites PF5 decreased to 0.065 W/m.k when casein and hemoglobin were added, and the porosity increased resulting in a reduction in thermal conductivity. For F10, thermal conductivity decreased from the reference to 0.069 W/m.k, due to the percentage of straw being increased with the presence of gelatin. S5 has a high thermal conductivity of 0.166 W/k.m. This is due to the residual lime in the RWS, which causes an increase in density and a decrease in porosity. These results indicate that the previously used additives have no significant impact when recycling the old biocomposite. Figure 8b shows that the additives reduced the strength of PF5 and PF10 formulations, because of the increase in porosity and reduction in density. By comparison, the strength of S5 increased to 1.041 MPa. Previously used additives did not influence the properties of new compounds with recycled wheat straw aggregates.

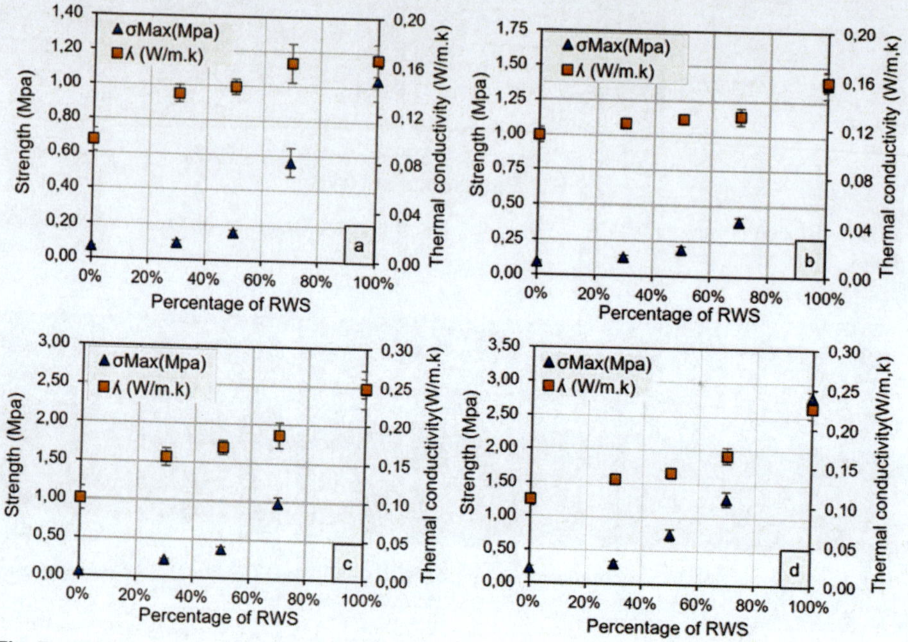

Fig. 7. Thermal conductivity and compressive strength with the percentage of RWS for biocomposite (a) W/B = 1.1 for 28 days (b) W/B = 1.1 for 90 days (c) W/B = 0.8 for 28 days (d) W/B = 0.8 for 90 days.

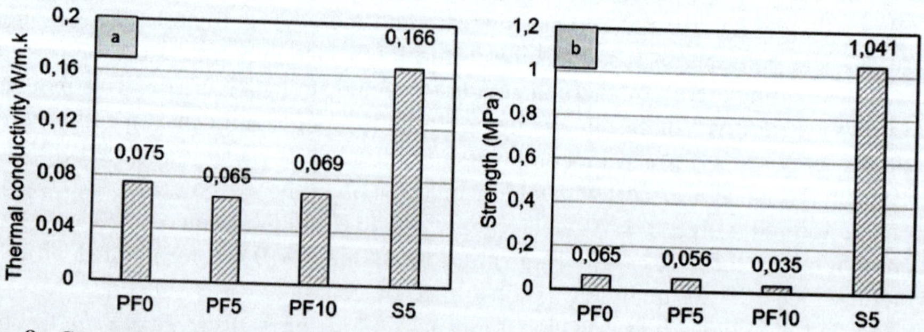

Fig. 8. Comparison of biocomposites PF0, PF5, PF10 (Ismail et al. 2020) and S5 for (a) thermal conductivity and (b) strength.

3.3 Scanning Electron Microscope (SEM)

A comparison of microscopic observations of biocomposites PF5 (Fig. 9a) (Ismail et al. 2020) and S5 (Fig. 9b) is investigated. PF5 contains additives like casein and hemoglobin. Casein additives changed the morphology of the calcite crystals, causing them to become smaller and more rounded (Ismail et al. 2020). For the biocomposite S5 with 100% RWS, the presence of long needles is observed, like calcite, that did not form a round shape and were smaller because of the casein effect. This fact confirms that the additives used previously do not significantly affect the microstructure of the new biocomposites with RWS.

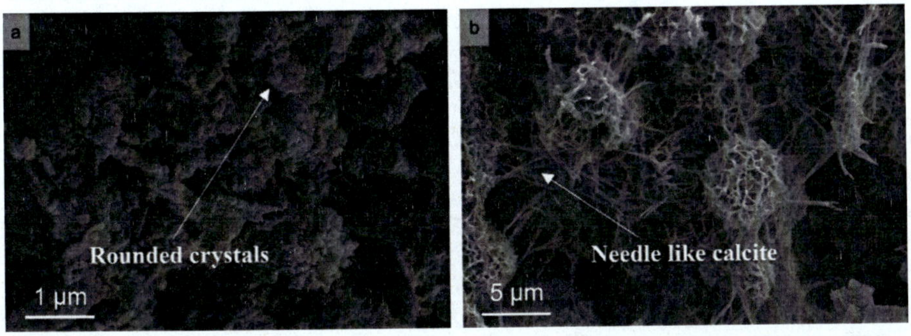

Fig. 9. Scanning electron microscope images of the bio-composite samples (a) PF5 (Ismail et al. 2020) (b) S5 (present work).

4 Conclusion

Recycling construction materials has become common and important due to environmental and economic benefits. Ecological additives are a good solution for improving the properties of construction materials. As a result, it may be included in the development and production of building materials to improve some properties. Understanding its future influence on recycling processes requires study and analysis of its long-term effects, after the building demolition. In this present study, the effects of additives previously used in the preparation of wheat straw blocks with lime as thermal insulation materials were investigated by using recycled aggregates (RWS) from these blocks formulated three years ago.

No significant impact of additives contained in recycled wheat straw (RWS) aggregates was observed on the thermal and mechanical properties of new biocomposites: thermal conductivity and compressive strength are significantly impacted by the residual lime existing with the RWS.

Acknowledgments. The authors gratefully acknowledge the "REGION CENTRE VAL DE LOIRE" for the financial support of the research program MATBIO.

References

Amziane, S., Collet, F., Lawrence, M., Magniont, C., Picandet, V., Sonebi, M.: Recommendation of the RILEM TC 236-BBM: characterisation testing of hemp shiv to determine the initial water content, water absorption, dry density, particle size distribution and thermal conductivity. Mater. Struct. **50**(3), 1–11 (2017). https://doi.org/10.1617/s11527-017-1029-3

Belayachi, N., Bouasker, M., Hoxha, D., Al-Mukhtar, M.: Thermo-mechanical behaviour of an innovant straw lime composite for thermal insulation applications. AMM **390**, 542–546 (2013). https://doi.org/10.4028/www.scientific.net/AMM.390.542

Bouasker, M., Belayachi, N., Hoxha, D., Al-Mukhtar, M.: Physical characterization of natural straw fibers as aggregates for construction materials applications. Mater. **7**, 3034–3048 (2014). https://doi.org/10.3390/ma7043034

Brouard, Y., Belayachi, N., Hoxha, D., Ranganathan, N., Méo, S.: Mechanical and hygrothermal behavior of clay – Sunflower (Helianthus annuus) and rape straw (Brassica napus) plaster bio-composites for building insulation. Constr. Build. Mater. **161**, 196–207 (2018). https://doi.org/10.1016/j.conbuildmat.2017.11.140

Douzane, O., Promis, G., Roucoult, J.-M., Le Tran, A.-D., Langlet, T.: Hygrothermal performance of a straw bale building: in situ and laboratory investigations. J. Build. Eng. **8**, 91–98 (2016). https://doi.org/10.1016/j.jobe.2016.10.002

Ismail, B., Belayachi, N., Hoxha, D.: Optimizing the performance of insulation materials based on wheat straw, lime, and gypsum plaster composites using natural additives. Constr. Build. Mater. **254**, 118959 (2020). https://doi.org/10.1016/j.conbuildmat.2020.118959

Kashtanjeva, A., Sonebi, M., Amziane, S., Pascal, U.B.: Investigation of the mechanical performance and drying shrinkage of hemp concrete. **33**, 7 (2015)

Mati-Baouche, N., et al.: Mechanical, thermal, and acoustical characterizations of an insulating bio-based composite made from sunflower stalk particles and chitosan. Ind. Crops Prod. **58**, 244–250 (2014). https://doi.org/10.1016/j.indcrop.2014.04.022

Omeme Ada, A.M., et al.: Hygrothermal performance of hemp lime concrete embedded with phase change materials for buildings. J. Phys.: Conf. Ser. **2069**, 012005 (2021). https://doi.org/10.1088/1742-6596/2069/1/012005

Rim, K.A., Ledhem, A., Douzane, O., Dheilly, R.M., Queneudec, M.: Influence of the proportion of wood on the thermal and mechanical performances of clay-cement-wood composites, p. 8 (1999)

Tlaiji, G., Ouldboukhitine, S., Pennec, F., Biwole, P.: Thermal and mechanical behavior of straw-based construction: a review. Constr. Build. Mater. **316**, 125915 (2022). https://doi.org/10.1016/j.conbuildmat.2021.125915

Vitola, L., Vilnitis, M., Pundiene, I., Bajare, D.: Sustainable building materials based on hemp shives and geopolymer paste. In: Journal of Physics: Conference Series, vol. 2162, p. 012015 (2022). https://doi.org/10.1088/1742-6596/2162/1/012015

Yaping, Z., Abdelkrim, T., Mohamed, E.M. (eds.): Development and characterization of thermal insulation materials based on rice straw and natural binder (2022)

Bio-based Solutions for the Retrofit of the Existing Building Stock: A Systematic Review

Giorgio Castellano(✉), Ingrid Maria Paoletti, Laura Elisabetta Malighetti, Olga Beatrice Carcassi, Federica Pradella, and Francesco Pittau

Department of Architecture Built Environment and Construction Engineering (ABC), Politecnico di Milano, Via Ponzio 31, 20133 Milan, Italy
giorgio.castellano@polimi.it
https://www.dabc.polimi.it/

Abstract. The European Green Deal has posed an urgent challenge by requiring a radical reduction of carbon emissions in the following decades. In 2050 most of the existing European building stock will still stand responsible for high global environmental, economic, and social impacts if the deep renovation rate will not double. There are multiple challenges, from increasing renovation rates and bringing in deep renovation projects the emerging technologies and skills. The need for a more holistic process, resulting in innovative integrated (seismic-energy-architectural) renovation solutions, pushes construction production and assembly to be part of the design process from the early stage of the design.

This paper aims to analyze the state-of-the-art of retrofit solutions to reclad the external envelope of non-historic buildings that act as energy-efficient skin, improve the architectural image, and may work under some circumstances as structural restraint. The case studies focus on bio-based components for building envelope applications that combine nature-based solutions as novel regenerative materials. Using bio-based materials is beneficial for climate mitigation due to the reduced energy intensity and attitude to store carbon during service life, providing an innovative alternative fabrication paradigm based on circular regeneration rather than an open emission process. Finally, the review has been classified into different retrofit approaches and methodology strategies from the already on-the-market industry solutions, research, and on-development stage projects as base knowledge for building designers.

Keywords: building retrofit · energy retrofit · building envelope · carbon footprint · bio-based materials · off-site construction · prefabricated components

1 Introduction

The European Green Deal has posed an urgent challenge by requiring a shift in CO_2 emissions. Buildings currently account for 40% of total energy use and 36% of total GHG emissions [1].

S. Amziane et al. (Eds.): ICBBM 2023, RILEM Bookseries 45, pp. 399–419, 2023.
https://doi.org/10.1007/978-3-031-33465-8_31

The research industry must be fully mobilized to achieve a climate-neutral circular economy. It will take 25 years to change an industrial sector and all value chains [2]; therefore, straightforward strategies must be adopted in the following years to ensure readiness in 2050; as per the European agenda, carbon emissions should be cut by 60% by 2030 [3].

Since conventional methods have proven, inadequate, experimentation and collaboration across sectors and disciplines are essential.

Moreover, all the building's manufacturing procedures are linear rather than circular. In 2016 in terms of volume, architecture, engineering, and construction (AEC) waste was the most significant stream, with 30% of all generated, with only 12% of the materials coming from recycling [4]. The PNRR plan [5] has also explicitly targeted these issues to develop innovative and low-impact technologies to support the energy supply and distribution network's stability and resilience for a smooth transition to a green circular economy.

By following the circular economy paradigms, the reuse of waste from the agricultural sector can be employed as a high-insulated building component producing a beneficial effect for both the buildings and the environment, such as in virtuous cases where agricultural waste was used as insulation material [6, 7].

Given the status mentioned above, bio-based materials are beneficial thanks to their reduced energy consumption and usage as a carbon sink [8]. They provide an innovative alternative fabrication paradigm based on material regeneration rather than extraction. Bio-fabrication is a cross-disciplinary field focusing on digital manufacturing processes and a range of emerging technologies [9].

Aimed to investigate the currently available scientific research and application in the intersection between two main fields, bio-based building materials (3BM) and the retrofit of the existing multi-housing building stock, this review of retrofit projects pursues a holistic approach to building renovation. The purpose is to search for retrofit technology to clad the external envelope of existing non-historic buildings in reinforced concrete with a prefabricated shell that acts as energy-efficient skin, improves the architectural image and may act under some circumstances as structural restraint.

2 The Building Retrofit and the Use of Bio-based Materials Issue

2.1 The Retrofit Issue

There are multiple surveys on the existing building stock that need attention in terms of retrofit requirements. They are mainly the result of European-funded projects, such as Tabula (https://episcope.eu/iee-project/tabula/), which aims to create a harmonized structure for European building typologies. Most of the existing European residential buildings are reinforced concrete from the post-war years. This stock needs protection from earthquakes, weathering, and degradation [10]; this data is also essential for the structural connection the retrofit system must have with the existing buildings.

Retrofit means recovering and improving some of the features of a building [11]. There are multiple types of possible measures that we can take on the buildings. It is possible to divide them into the following categories [12]:

- Energetic retrofit – to improve the thermal capacity of the building and therefore reduce the carbon emissions and improve the quality of life for the inhabitants.
- Seismic retrofit – to improve the structural behavior of the building under seismic actions.
- Architectural retrofit – to improve the general aspects of the building. Protecting the building from agents of deterioration and usury.

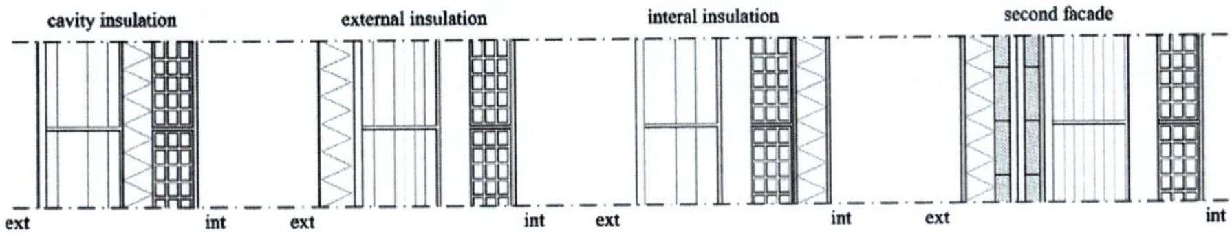

Fig. 1. Different approaches on the refurbishment

Today the retrofit approach is more of a general holistic method that can improve different aspects. In this paper, we will focus on second façade retrofit systems (Fig. 1) or, in other words, when there is an external layer on the façade. This method allows the installation of prefabricated solutions, enables the use of the building during the renovation, and reduces the impact on the inhabitants.

2.2 Reasons for and Applications of Bio-based Materials

Bio-based products are wholly or partly derived from materials of biological origin, excluding materials embedded in geological formations or fossilized (https://single-market-economy.ec.europa.eu/sectors/biotechnology/bio-based-products_en). In the European specifications, to define a bio-based product, only 25% of bio-based content [13] is necessary for its composition. The use of bio-based materials in the retrofit is essential due to the impact that the retrofit has on the building environment since bio-based materials are beneficial thanks to their reduced energy consumption and usage as a carbon sink [8]. They provide an innovative alternative fabrication paradigm based on material regeneration rather than extraction. The result of the dynamic LCA showed that fast-growing bio-based materials have a higher potential to act as a carbon sink for the *GWP*-global warming potential [14]. The choice of building materials to reduce energy consumption and emissions is the focus of contemporary research to reduce building environmental impacts. In this context, the solution of bio-based building materials is becoming increasingly relevant as they have a very low or even positive impact on the environment (renewable resource, CO_2 storage, and reduction of greenhouse gas emissions, etc.) [15]. Some are also characterized by excellent multi-physical properties (heat insulation, sound absorption, transmission loss) and require very low resource production techniques [16].

Moreover, by lowering embodied carbon, they can also create healthier buildings since their environmental performance contains both the production process and their capability to preserve a healthy indoor environment during their lifetime [17].

From an architectural retrofit perspective is possible to imagine the use of bio-based material in two main applications: structure and insulation material [18].

Concerning the structural materials, the bio-based solutions are mostly in wood. They are either straight from the plants, such as timber slats, boards or poles, or have undergone an industrial transformation, turning them into engineered products such as fiberboards or glue-laminated timber (GLT) [9]. Wood is the most promising in terms of space efficiency (relatively highest carbon storage in buildings while occupying the smallest land), has a vast application, and can perform as a structural material. According to [19], the amount of land currently available for growing wood is sufficient to fulfil the whole material demand for construction in the EU. Moreover, an advanced infrastructure network is already developed for timber supply.

As an external retrofit methodology, preassembled timber components have multiple advantages for building constructions. The seriality and modularity of the production can decrease manufacturing costs by optimizing processes and reducing fossil-carbon emissions during the assembly phase [20]. The rapidity of assembly ensures a short on-site construction duration without reducing the structural stiffness and the technical performance. The modular timber construction, primarily based on wooden elements (Plywood, Laminated Veneer Lumber LVL, Panels, Cross Laminated Timber CLT), allows for storing a significant amount of carbon in the structure (50% of the mass) [21]. It also allows for minimizing the construction site-related discomfort for the occupants and integrating building components such as windows, building services systems, and ready-made surfaces of the façade elements.

The main non-structural application is for the insulation material. There is a wide variety of bio-based materials used for insulation and façade component purposes (e.g., hemp and cotton fibers, wood) [6, 22–24].

Among the different options of biomasses, the interesting ones for insulation are those with a porous structure (e.g., corn cob has a honeycomb structure, corn stalks). At the same time, those with long fibre show good mechanical properties and can improve the tensile strength of many materials [17].

Thermal insulations made of biomasses come mainly from [22]:

- agricultural waste (corn/wheat/rice straw, hemp, rice husk, linseed, kenaf, flax, jute, olive fiber, corn stalk, reed, sunflower)
- forest residues (wood, bark, coniferous residues, bamboo, coconut fibers, cork
- farming (sheep wool)
- living organism (mycelium).

Straw can be strongly recommended for large-scale construction and renovation since it has the highest cumulative carbon storage potential until 2050 [7]. Using it for construction should not change the amount of land used or the intensity of wheat production. RiceHouse s.r.l. Società Benefit (https://www.ricehouse.it/en-home) is an Italian company developing new building materials based on rice straw from this loss of rice cultivation and enhancing the organization of its whole supply chain.

Hemp is not sufficiently available across the geo clusters. The land needed for hemp shives production covers only 2% of the demand, and the legislation limits its future development in many countries. Cork is only available in dry southern regions; only 4% of cork can be supplied for building insulation, which leads to slight potential carbon storage in the building stock [25].

Materials derived from animals or living organisms (e.g. sheep wool or mycelium) have much more durability and degradation problems, including ageing, weathering, mould and algae growth, decay, waterlogging, and insect infestation [26].

2.3 Applications

According to multiple sources [27], on-site renovation looks problematic and old as an approach, with too many technicians and craftsmen involved, complex coordination on-site, low-quality production level, and inefficient construction process. Contrary to the prefabricated elements concept that sees fewer companies involved, easier to coordinate modules, better quality control, and rapid construction processes. Prefabricated construction technologies optimize the manufacturing process and decrease carbon emissions during material processing and assembly, mainly with the second façade approach, a system for external application on the building.

There are two primary prefabricated systems: one is based on a solid cross-laminated panel, and one is on a timber frame panel structure. These two different structural approaches are used individually or combined according to the specific need of the case building (Fig. 2).

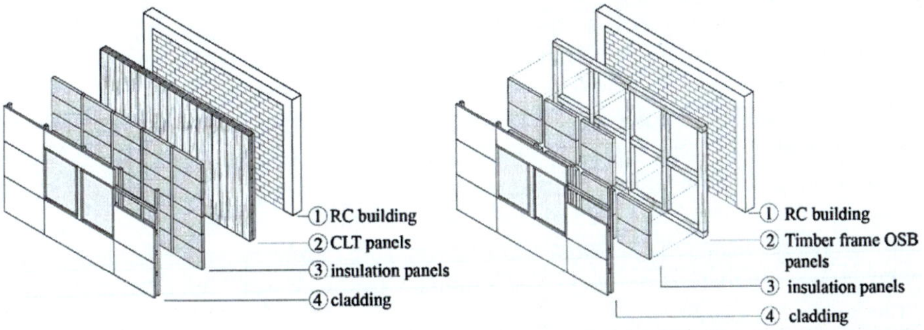

Fig. 2. Prefabricated system: with solid cross laminated panel and a timber frame panel structure

As an example of these methods, AdESA is a holistic retrofit technique for the integrated renovation of existing buildings [1]. Funded by the European project 4RinEU it considers buildings at the end of their lifetime with poor energy performance and low living quality. The AdESA solution is designed to be easily mountable by adopting macro-prefabricated dry components and standardized connections. It consists of a dry, modular, and flexible shell exoskeleton in CLT, behaving as a structural layer, thermal insulation panels for improving energy performance and a cladding system for the architectural appearance. It can be arranged both vertically and horizontally. The use of

CLT as a structural layer and the presence of alternatives to EPS insulation layers (cork, cellulose, glass wool and rock wool) allows for the reuse-recycling of its components at the end of the life. These materials showed fire resistance, durability, availability, cost/Performance and Eco-Friendly approach as pros.

The e-SAFE [12] system comprises two prefabricated elements: e-CLTs and e-Panels applied to the outer envelope of a potential building. The structural *e-CLTs* composed of Cross laminated timber boards are applied to the exterior blind walls and are connected to the existing reinforced concrete beams through the friction dampers. The *e-CLTs* aim to increase the seismic and dissipative capacity of the existing structure by exploiting the high mechanical properties of CLT and the additional energy dissipation source of the dampers. The extra stiffness can be controlled by modulating the thickness of the CLT panel. The system is also externally coupled with bio-based insulation and the desired finishing material to meet thermal and acoustic requirements.

The non-structural *e-PANELS* are applied to all the windowed walls and to the blind walls having a smaller width. The *e-PANELS* integrate new high-isolated windows frame which replaces the existing ones.

The panels will be prefabricated off-site and installed through mobile lifting equipment from the building outside to avoid demolition interventions, minimize waste production and reduce implementation costs and time and the occupants' disruption.

The idea of preassembling and prefabrication was also in the Habitat panel, a timber panel case study for the retrofit solutions [28]. Characterized by eco-friendly materials, with a high content of recycled material and wood-based products, the panel is suitable when a certification of a high level of environmental quality is needed and it is easy to manage for assembly due to its lightness and modularity.

From an energetic and architectural point of view, Naturwall [29] is a multifunction timber system for a sustainable refurbishment process, aiming to improve existing buildings' energy, environmental and aesthetic qualities. It is a prefabricated, modular, lightweight, cost-effective solution positioned in the existing envelope between structural beams fixed on the floors. The panel concept is an adaptive system to reduce the heat flow that crosses the façade. While during the winter it is used as a solar collector, the solar heat is attenuated in the summer by naturally driven ventilation in the cavity, avoiding overheating. The tested module comprises bio-based materials, such as wood and cardboard. It includes external stratified glazing, a natural ventilated cavity with honeycomb cardboard material, and an insulated sandwich panel mounted on a timber structure.

In the following Table 1 an extended review summary of the retrofit technologies is provided.

3 Findings and Open Issues

The external retrofit technologies are made of elements fixed to the existing structure (Fig. 3). These modular elements are composed of a structural layer, an insulation or energy layer and a plaster or a cladding system layer. In Table 2 all the references are divided according to the layer where the retrofit is applied. The holistic approaches are the ones that consider the retrofit from all the three aspects at same time.

Table 1. Summary review of bio-based retrofit technologies.

Existing building technology	Retrofit technology	Description	Bio-based solution	Reference
URM (unreinforced masonry walls)	CLT panel external and internal	New seismic retrofit technique for masonry walls using CLT panels fastened on the existing wall. CLT also provide thermal and acoustic insulation It's a dry and removable technology that can be used externally and internally	The internal shell of CLT walls is used for performing experimental tests on masonry examples of seismic retrofit	[30]
RC (reinforced concrete) framed building	CLT panels	AdESA project is a holistic retrofit technique that consists of a dry, modular, and flexible shell exoskeleton. CLT panels for the structural retrofit, thermal insulation panels for energy efficiency, and claddings for the architectural restyling. The solution is designed to be easily mountable by adopting macro-prefabricated dry components and standardized connections	The use of CLT as a structural layer and the presence of an alternative to EPS insulation layers: cork. Cellulose They allow for the reuse-recycling of components at the end of their life	[1]

(*continued*)

Table 1. (*continued*)

Existing building technology	Retrofit technology	Description	Bio-based solution	Reference
RC framed buildings	CLT panels	Seismic and energy retrofit for RC frame using CLT panels. The strengthening of the RC frame system allows lower drift values and higher peak load	CLT shear walls as a replacement for existing masonry infill walls	[31]
RC framed buildings	CLT elements	Example of seismic retrofit using CLT panels connected externally to RC frame structure by friction dampers. The system is named e-CLT. Friction dampers partially dissipate earthquake's energy and limit the internal force transmitted by CLT panel, avoiding possible failure	CLT panels prove the strength and lateral stiffness	[32]
RC framed buildings	Timber cladding	NATURWALL is a façade system for both energy and architectural retrofit. It's a prefabricated, modular, lightweight, cost-effective solution. It's positioned in the existing envelope between structural beams fixed to the floors	It's made of external stratified glazing, ventilated cavity, honeycomb cardboard, and an insulated sandwich panel	[29] [33]

(*continued*)

Table 1. (*continued*)

Existing building technology	Retrofit technology	Description	Bio-based solution	Reference
Concrete masonry-clad buildings	Insulated timber cladding	Energy and architectural retrofit with TRADA system. Timber cladding is fixed to timber battens and nailed to the existing structure. On-site construction, not prefabricated	TRADA Technology: Timber cladding systems of three types Hanging on the existing wall or self-supporting	[34]
RC framed buildings	Timber cladding	Evaluation of the energy savings of wooden prefabricated façade elements with a simple assembly process. The advantages are the dry production process, on-site assembly, and reduced number of thermal bridge's external applications	Timber-based cladding system Use of recyclable materials	[35]
Timber, concrete, or brick structure	CLT wood panel, exterior	Energy and architectural retrofit are studied with prefabricated timber elements to provide environmentally friendly, energy- and cost-efficient solutions. The system is flexible. It can be applied to pre-existing timber, concrete, or brick structures	"Modular wall system": consists of two outer multilayer solid wood panels with insulation in between connected to each other by wooden rods	[36]

(*continued*)

Table 1. (*continued*)

Existing building technology	Retrofit technology	Description	Bio-based solution	Reference
RC framed buildings	Timber Frame	The Ri.Fa.Ree project consists of an innovative timber-based panel for energy and seismic retrofit. The study aims to provide a new process based on parametric design and lean construction principles. The process is composed of design, development, and testing	Timber-based solution: wall panels are placed on timber beams (called console beams) directly anchored on the main concrete structure	[37]
RC framed building	Textile reinforced concrete with EPS core Timber panel for pitched roofs	Example of energy and architectural retrofit of a multi-story residential building For the façade a preassembled insulated panel based on two textile-reinforced concrete thin precast layers rigidly connected to an EPS core has been installed on the outer side. Instead, the existing pitched roof has been refurbished with preassembled timber panels	Preassembled timber panel	[21]

(*continued*)

Table 1. (*continued*)

Existing building technology	Retrofit technology	Description	Bio-based solution	Reference
RC framed building	Precast multilayer panel	EASEE project: envelope approach to improve sustainability and energy efficiency in existing multi-owner residential buildings. The system proposes both energy and architectural retrofit. Fundamental characteristics are prefabrication and quick installation. The system is a multi-layer panel made of two textile-reinforced concrete thin precast layers rigidly connected to an EPS core	x	[38] [39] [21]
RC framed building – roof system	Preassembled timber Panel	The work aims to develop a prefabricated retrofit solution. The base version for the roof system is the non-load bearing panel "HABITAT" "Evo-HABITAT" is the optimized panel. The roof's primary timber structure was integrated into the preassembled panel, with significant benefits in terms of structural performance and optimization of the construction process	HABITAT panel: preassembled timber panel with eco-friendly materials and wood-based product: multi-cell box by OSB panels Evo-Habitat is a structural version whit a GLT structure that allows covering bigger spans	[28] [21]

(*continued*)

Table 1. (*continued*)

Existing building technology	Retrofit technology	Description	Bio-based solution	Reference
RC framed building	Bio-based panels Sustainable adhesive	An energy retrofit solution is proposed by comparing different bio-based insulating materials. The insulating panels are directly attached to the existing façade through a sustainable adhesive Experimental energy performance and solutions comparison between various insulating materials and changing thickness	Bio-based panel for insulation Case studies: 5cm hemp insulation, 5cm straw insulation, 10cm hemp insulation, 10cm straw insulation, 12cm hemp insulation	[40]
RC framed concrete	Timber frame for façade system	The aim is to test a new integrated façade system to improve the overall energy performance: It consists of both energy and architectural retrofit. The system is prefabricated, well-insulated, cost-effective, flexible, and equipped with HVAC	BERTIM: Building Energy Renovation Through Timber Prefabricated Module. Hephaestus system: off-site manufactured curtain wall module	[41]

(*continued*)

Table 1. (*continued*)

Existing building technology	Retrofit technology	Description	Bio-based solution	Reference
RC frame building	Prefabricated timber-based panels CLT Timber frame	Seismic, energy and architectural retrofit is proposed as a new innovative holistic approach It is based on the addition of CLT panels to the outer walls able to strengthen the existing building for seismic action through friction dampers (e-CLT system), in combination with wooden-framed panels for insulation and architectural renovation (e-PANEL design), hosting high-performing windows	CLT panels Wooden-frame panels Wood screws Wood fibre thermal insulation	[12] [42]
RC building	Bio-based insulation materials Timber frame	Different energy retrofits for existing façade are proposed to test the best solution acting as carbon storage. Five alternative construction solutions are compared. Fast-growing bio-based materials have a higher potential for carbon storage than timber	Materials inventory: I-joist frame with pressed straw, preassembled frame with injected hempcrete, timber frame, Hempcrete blocks, EPS	[7]

(*continued*)

Table 1. (*continued*)

Existing building technology	Retrofit technology	Description	Bio-based solution	Reference
RC building	Prefabricated panel systems: Timber frame panels RC panels Steel panels for seismic retrofit	Energy and architectural retrofit systems are proposed The three renovation models are made of façade panels. Structural framing, insulation, windows, doors, and air sealing are all included in a unitized panel nailed with hooks to the existing building	Timber-based panel with wood, cellulose, recycled paper, and mineral wool Insulation is then blown into gaps between the existing and new facades	[43]
Light-framed wood building	Zero-carbon Bio-based wall panel	An energy retrofit solution is proposed to reduce carbon emissions. A bio-based retrofit panel system is optimally sized, efficient for installation with net-zero embodied carbon and a low-cost installation process, fabricated from a low-density wood-based substrate of three-ply cross-laminated timber (CLT) technology supporting the bio-based insulation in wood fibers	CLT Wood fibre insulation systems are compared with different bio-based insulating materials with different thicknesses, according to climatic zones	[44]

(*continued*)

Table 1. (*continued*)

Existing building technology	Retrofit technology	Description	Bio-based solution	Reference
URM buildings RC buildings	Panel with a timber frame with bio-based thermal insulation	A seismic and energy retrofit solution is proposed combining timber mechanical properties and bio-based material's thermal insulation properties. Timber frame and insulation material are attached to the existing façade, internally or externally. Three retrofitting systems are investigated	Timber frame for seismic retrofit Two bio-based thermal insulation layers are used: cork boards and pressed grass panels	[45]
Concrete wall	Laminated wood panels	The main contribution of this paper is to propose a new business model with standardized products, logistics and project management, discussing different aspects of efficient business models considering prefabricated wooden elements It can be achieved only with façade standardized elements	Termowood system solution: Laminated wood panels connected with wooden dowels and insulation. It can come in two versions: structural and no structural	[46]

(*continued*)

Table 1. (*continued*)

Existing building technology	Retrofit technology	Description	Bio-based solution	Reference
URM walls	Internal and external CLT/LVL panels and timber strongbacks	The system proposes structural and energy retrofit. The structural retrofit consists of timber-based components (panels and strong-backs) connected to the masonry walls. Instead, the energetic performance is guaranteed by the installation of insulation layers	Seismic retrofit with CLT/LVL panels and timber strong backs: the best solution is obtained with a CLT panel of 60 mm thickness, insulation layer and coating	[47]
URM cavity-wall, precast RC panels	Internal timber frame connected to the URM through steel profiles	The aim is to provide a large-scale application of timber retrofit for the URM walls. The system discussed is lightweight, reversible, cost-effective, and bio-based The seismic performance of an empty building and a retrofitted building have been compared, demonstrating the benefit of the timber frame retrofit	Seismic retrofit with timber frames (80 × 60 mm): vertical timber posts, horizontal nogging (spacing 60 mm) connected to the masonry and floor systems and OSB panels nailed to them	[48]

The retrofit is necessary to reduce the building's carbon emissions. The actual retrofit approaches are not considering the use of bio-based materials enough and not taking full advantage of the digital production processes for optimizing the prefabrication systems.

Low-emission technologies may facilitate the logic of the circular economy, and some requirements in achieving these objectives could be:

- The modularity of the components – items (modular coordination, identifiable part of a device that performs a specific function) has an essential role in the reversible connections of the components (disassembly and reassembly).

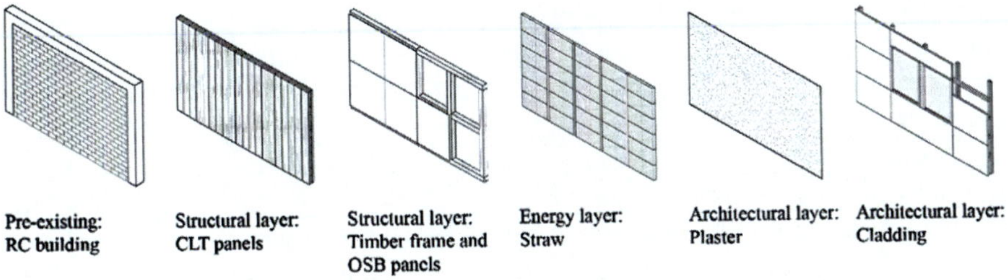

Fig. 3. Timber retrofit system classification

Table 2. Bio-based retrofit solutions.

Holistic approach	Structural layer	Energy layer	Architectural layer
- AdESA [1]	- CLT panels [1, 12, 30, 31, 32, 42, 44, 47]	- cork [1, 45]	- curtain wall modules [29, 33, 41, 46]
- Naturwall [29, 33]	- LVL panels [47]	- cellulose [1, 43]	- timber cladding system with timber battens [34]
- e-SAFE [12, 32, 42]	- Timber frame [21, 28, 34, 35, 36, 43, 45, 47, 48]	- cardboard [29, 33]	
- BERTIM [41]		- wood fiber [12, 42, 44]	
		- hemp [40]	
		- hempcrete [7]	
		- recycled paper [43]	
		- pressed grass panels [45]	
		- straw [7, 40]	

- Modular timber construction, primarily based on wooden elements (Plywood, LVL, Panels, CLT), allowing storing a significant amount of carbon in the structure (50% of the mass) [21]; the modular prefabricated system also allows for minimizing the construction site-related discomfort for the occupants.
- The internal joints of the wooden frame may be studied for dry assembly. The connection with the existing building must face the irregular geometry of the existing building's uneven surfaces, which meant that the steel connectors of the façade module needed to be calibrated and adjusted to the façade geometry and its variations. Producing and integrating connectors for an off-site prefabricated façade module is challenging and time-consuming.
- Lightweight transportable material systems.
- System interchangeability and reversibility of components (replacement with a similar element), flexibility (ability to adapt to different situations or needs), also through dry assembled connections.
- Integrating building components such as windows, building services systems, and ready-made surfaces of the façade elements in the controlled ergonomic working environment allows customization based on the specific needs of individual buildings.
- Optimization of construction materials by providing multiple functions to the same element (e.g., reducing the thickness of the energy layer by exploiting the thermal properties of the structural panels)

One of the main drawbacks of bio-based solutions is their durability; by interacting with biotic and abiotic factors, biomaterial surface and bulk undergo several changes affecting the original material properties [26]. The mechanism of such changes is somewhat complex and may result from diverse processes co-occurring or in succession. The most common processes of biomaterial degradation include ageing, weathering, mould and algae growth, decay, waterlogging, and insect infestation [26].

Fire regulation is generally problematic for a bio-based material. In the case of panels, it is always needed a non-bio-based break fire element (e.g. Paraflam® is a perimeter fire stop designed for curtain walling and cavity barrier external rain-screen facades. Paraflam® is a non-combustible stone wool-based product with aluminium foil. https://fsiltd.com/product/paraflam%C2%A8-seb/).

4 Conclusion

The scientific community is now responsible for studying strategies and methodologies to meet the EU targets on energy efficiency in buildings, as requested by the Directive (EU) 2018/844, with particular attention to the national building stock. This study offers a general overview of the most common approaches tested in the field of building retrofits. Researchers are encouraged to develop faster, more reliable, flexible, and straightforward decision support systems to help asset holders to enhance existing stocks.

Some successful technological solutions are being developed (TES [49, 50], BERTIM, MORE-CONNECT https://www.more-connect.eu/more-connect, CLT panel façade system). However, there has not been a general leading technique for the broader use of such solutions.

The above-mentioned prefabricated solutions have significant potential, yet there are still gaps in the process of optimizing them for the use of renewable materials. Moreover, industry-optimized processes increase cost-effectiveness reducing time on the construction sites in the building refurbishment. This could minimize mistakes, improve modules' quality, and reduce construction waste.

Differences in costs between retrofitting based on ETICS (external thermal insulation composite system) and prefabricated panels are generally small; prefabricated panels can provide a higher standard of living since they may combine other characteristics beyond the thermal insulation function (mechanical ventilation, new heating piping, electrical wiring and other systems). However, such solutions have special requirements regarding worker skills and techniques needed for assembling the system that are somewhat more complex than standard practices.

Designing, building, and constructing new possibilities can often lead to the idea of a complete break from the present conditions. At the same time, the review has shown that supply chain, industrial practices and legislative frameworks are pre-existing conditions that need to bond with new materials use.

This work is intended as the starting point to imagine a decision-making procedure for managing bio-based components in architecture that is not always straightforward. Working with less standardized elements means making designers and builders more accountable for their decisions.

References

1. Zanni, J., et al.: Integrated deep renovation of existing buildings with prefabricated shell exoskeleton. Sustainability **13**(20), 1–27 (2021). https://doi.org/10.3390/su132011287
2. European Commission: "The European Green Deal," Report (2019)
3. Roscini, A.V., Rapf, O., Kockat, J.: On the way to a Climate Neutral Europe 2030 - Contributions from the building sector to a strengthened 2030 climate target, p. 24 (2020). https://www.bpie.eu/wp-content/uploads/2020/12/On-the-way-to-a-climate-neutral-Europe-_Final.pdf
4. EU construction sector: in transition towards a circular economy European Construction Sector Observatory Trend Paper Series (2019). http://appsso.eurostat.ec.europa.eu/nui/show.do?lang=en&dataset=env_wasgen
5. Governo Italiano. Presidenza del Consiglio dei Ministri, "Piano nazionale di ripresa e resilienza," p. 269 (2021). https://www.governo.it/sites/governo.it/files/PNRR.pdf
6. Platt, S.L., et al.: Sustainable bio & waste resources for thermal insulation of buildings. Constr. Build. Mater. **366**, 130030 (2023). https://doi.org/10.1016/j.conbuildmat.2022.130030
7. Pittau, F., Lumia, G., Heeren, N., Iannaccone, G., Habert, G.: Retrofit as a carbon sink: the carbon storage potentials of the EU housing stock. J. Clean. Prod. **214**, 365–376 (2019). https://doi.org/10.1016/j.jclepro.2018.12.304
8. Kuittinen, M., Zernicke, C., Slabik, S., Hafner, A.: How can carbon be stored in the built environment? A review of potential options. Archit. Sci. Rev. (2021). https://doi.org/10.1080/00038628.2021.1896471
9. Jones, D.: Introduction to the performance of bio-based building materials. In: Performance of Bio-based Building Materials, pp. 1–19. Elsevier Inc. (2017). https://doi.org/10.1016/B978-0-08-100982-6.00001-X
10. Corrado, V., Ballarini, I., Corgnati, S.P., Talà, N.: Building Typology Brochure–Italy (2011)
11. Eames, M., et al.: Retrofit 2050: Critical Challenges for Urban Transitions Retrofit 2050: Critical Challenges for Urban Transitions Authors, pp. 1–14 (2014)
12. Margani, G., Evola, G., Tardo, C., Marino, E.M.: Energy, seismic, and architectural renovation of RC framed buildings with prefabricated timber panels. Sustainability **12**(12), 4845 (2020). https://doi.org/10.3390/SU12124845
13. Sherwood, J., Clark, J.H., Farmer, T.J., Herrero-Davila, L., Moity, L.: Recirculation: a new concept to drive innovation in sustainable product design for bio-based products. Molecules **22**(1) (2017). https://doi.org/10.3390/molecules22010048
14. Caldas, L.R., Da Gloria, M.Y.R., Pittau, F., Andreola, V.M., Habert, G., Toledo Filho, R.D.: Environmental impact assessment of wood bio-concretes: evaluation of the influence of different supplementary cementitious materials. Constr. Build. Mater. (2020). https://doi.org/10.1016/j.conbuildmat.2020.121146
15. Gourlay, E., Glé, P., Marceau, S., Foy, C., Moscardelli, S.: Effect of water content on the acoustical and thermal properties of hemp concretes. Constr. Build. Mater. **139**, 513–523 (2017). https://doi.org/10.1016/j.conbuildmat.2016.11.018
16. Aditya, L., et al.: A review on insulation materials for energy conservation in buildings. Renew. Sustain. Energy Rev. **73**, 1352–1365 (2017). https://doi.org/10.1016/J.RSER.2017.02.034
17. Liu, L., et al.: The development history and prospects of biomass-based insulation materials for buildings. Renew. Sustain. Energy Rev. **69**, 912–932 (2017). https://doi.org/10.1016/J.RSER.2016.11.140
18. Carcassi, O.B., Paoletti, I., Malighetti, L.E.: Reasoned catalogue of biogenic products in Europe. An anticipatory vision between technical potentials and availability. Techne **22**, 63–70 (2021). https://doi.org/10.36253/techne-10578

19. Pittau, F., Göswein, V., Habert, G.: Healing the European building stock with bio-based materials: do we have enough available land? (2021)
20. Carcassi, O.B., De Angelis, E., Iannaccone, G., Malighetti, L.E., Masera, G., Pittau, F.: Bio-based materials for the italian construction industry: buildings as carbon sponges. In: Della Torre, S., Cattaneo, S., Lenzi, C., Zanelli, A. (eds.) Regeneration of the Built Environment from a Circular Economy Perspective. RD, pp. 237–247. Springer, Cham (2020). https://doi.org/10.1007/978-3-030-33256-3_23
21. Pittau, F., Malighetti, L.E., Iannaccone, G., Masera, G.: Prefabrication as large-scale efficient strategy for the energy retrofit of the housing stock: an Italian case study. Procedia Eng. **180**, 1160–1169 (2017). https://doi.org/10.1016/j.proeng.2017.04.276
22. Asdrubali, F., D'Alessandro, F., Schiavoni, S.: A review of unconventional sustainable building insulation materials. Sustain. Mater. Technol. **4**(2015), 1–17 (2015). https://doi.org/10.1016/j.susmat.2015.05.002
23. Marques, B., Tadeu, A., Almeida, J., António, J., de Brito, J.: Characterization of sustainable building walls made from rice straw bales. J. Build. Eng. **28** (2020). https://doi.org/10.1016/j.jobe.2019.101041
24. Palumbo, M., Lacasta, A.M., Giraldo, M.P., Haurie, L., Correal, E.: Bio-based insulation materials and their hygrothermal performance in a building envelope system (ETICS). Energy Build. **174**, 147–155 (2018). https://doi.org/10.1016/j.enbuild.2018.06.042
25. Göswein, V., Reichmann, J., Habert, G., Pittau, F.: Land availability in Europe for a radical shift toward bio-based construction. Sustain. Cities Soc. **70** (2021). https://doi.org/10.1016/j.scs.2021.102929
26. Sandak, A., Sandak, J., Brzezicki, M., Kutnar, A.: Portfolio of bio-based façade materials. In: Sandak, A., Sandak, J., Brzezicki, M., Kutnar, A. (eds.) Bio-based Building Skin. EFEPP, pp. 155–177. Springer, Singapore (2019). https://doi.org/10.1007/978-981-13-3747-5_6
27. Zweifel, G.: Prefabricated Systems for Low Energy Renovation of Residential Buildings: Retrofit Simulation Report IEA ECBCS Annex 50, p. 95 (2011). http://www.empa-ren.ch/A50.htm
28. Pittau, F., Malighetti, L.E., Masera, G., Iannaccone, G.: A new modular preassembled timber panel for the energy retrofit of the housing stock (2016). https://www.researchgate.net/publication/318902538
29. Callegari, G., Spinelli, A., Bianco, L., Serra, V., Fantucci, S.: NATURWALL© - a solar timber façade system for building refurbishment: Optimization process through in field measurements. Energy Procedia **78**, 291–296 (2015). https://doi.org/10.1016/j.egypro.2015.11.641
30. Pozza, L., Evangelista, F., Scotta, R.: CLT used as seismic strengthener for existing masonry walls CLT used as seismic strengthener for existing masonry walls. XVII Xonvegno ANIDIS, pp. 210–220 (2017)
31. Stazi, F., Serpilli, M., Maracchini, G., Pavone, A.: An experimental and numerical study on CLT panels used as infill shear walls for RC buildings retrofit. Constr. Build. Mater. **211**, 605–616 (2019). https://doi.org/10.1016/j.conbuildmat.2019.03.196
32. Barbagallo, F., Margani, G., Marino, E.M., Moretti, A., Tardo, C.: Impact of retrofit of RC frames by CLT panels and friction dampers. In: COMPDYN Proceedings, vol. 2021-June, no. September (2021). https://doi.org/10.7712/120121.8636.18851
33. Spinelli, A., Callegari, G.: Naturwall - timber multifunctional systems in refurbishment sustainable process. In: Proceedings of the 2nd International Conference on Structures and Architecture, Concepts, Applications and Challenges, ICSA 2013, pp. 1121–1128 (2013). https://doi.org/10.1201/b15267-156
34. Sheet, W.I.: Wood information sheet cladding for timber frame buildings, no. 41, pp. 1–9 (2018)

35. Ruud, S., Östman, L., Orädd, P.: Energy savings for a wood based modular prefabricated façade refurbishment system compared to other measures. Energy Procedia **96**, 768–778 (2016). https://doi.org/10.1016/j.egypro.2016.09.139
36. Sandberg, K., Orskaug, T., Andersson, A.: Prefabricated wood elements for sustainable renovation of residential building façades. Energy Procedia **96**, 756–767 (2016). https://doi.org/10.1016/j.egypro.2016.09.138
37. Malacarne, G., Monizza, G.P., Ratajczak, J., Krause, D., Benedetti, C., Matt, D.T.: Prefabricated timber façade for the energy refurbishment of the italian building stock: the Ri.Fa.Re. project. Energy Procedia **96**, 788–799 (2016). https://doi.org/10.1016/j.egypro.2016.09.141
38. Salvalai, G., Sesana, M.M., Iannaccone, G.: Deep renovation of multi-storey multi-owner existing residential buildings: a pilot case study in Italy. Energy Build. **148**, 23–36 (2017). https://doi.org/10.1016/j.enbuild.2017.05.011
39. Salvalai, G., Iannaccone, G., Sesana, M.M., Pizzi, E., Sesana, M.M., Pizzi, E.: Outer facade retrofitting trough precast insulation panels: method and planning tool applied to an Italian residential building (2017)
40. Rosso, F., Peduzzi, A., Diana, L., Cascone, S., Cecere, C.: A sustainable approach towards the retrofit of the public housing building stock: energy-architectural experimental and numerical analysis. Sustainability **13**(5), 1–19 (2021). https://doi.org/10.3390/su13052881
41. Pan, W., Iturralde, K., Bock, T., Martinez, R.G., Juez, O.M., Finocchiaro, P.: A conceptual design of an integrated Façade system to reduce embodied energy in residential buildings. Sustainability **12**(14), 1–23 (2020). https://doi.org/10.3390/su12145730
42. Tardo, C., Fiore, G., Margani, G.: Technological analysis of prefabricated timber-based panels for the integrated renovation of RC framed buildings (2020)
43. Egerter, A., Martha, C.: Prefabricated Zero Energy Retrofit Technologies: A Market Assessment (2020). https://rmi.org/wp-content/uploads/2020/04/prefabricated-zero-energy-retrofit-technologies.pdf
44. Kurzinski, S., Crovella, P., Razkenari, M., Smith, W.B.: A zero-carbon bio-based wall panel as an energy retrofit solution for buildings (2022)
45. Zuluaga, S.S., Kallioras, S., Tsiavos, A.: Optimization of synergetic seismic and energy retrofitting based on timber beams and bio-based infill panels: application to an existing masonry building in Switzerland. Buildings **12**(8) (2022). https://doi.org/10.3390/buildings12081126
46. Bystedt, A., Östman, L., Knuts, M., Johansson, J., Westerlund, K., Thorsen, H.: Fast and simple - cost efficient façade refurbishment. Energy Procedia **96**, 779–787 (2016). https://doi.org/10.1016/j.egypro.2016.09.140
47. Busselli, M., Cassol, D., Prada, A., Giongo, I.: Timber based integrated techniques to improve energy efficiency and seismic behaviour of existing masonry buildings. Sustainability **13**(18) (2021). https://doi.org/10.3390/su131810379
48. Miglietta, M., Damiani, N., Guerrini, G., Graziotti, F.: Full-scale shake-table tests on two unreinforced masonry cavity-wall buildings: effect of an innovative timber retrofit. Bull. Earthq. Eng. **19**(6), 2561–2596 (2021). https://doi.org/10.1007/s10518-021-01057-5
49. Lattke, F., Larsen, K., Ott, S., Cronhjort, Y.: TES Energy Facade – prefabricated timber based building system for improving the energy efficiency of the building envelope (2011). funded by: Woodwisdom Net, Research project from 2008-2009
50. Cronhjort,Y., Lattke, F.: Energy Retrofits with TES Energy Façade as an Opportunity for Architectural Regeneration, Athens: ATINER'S Conference Paper Series, no. ARC2014-1150 (2014)

Impact of Solar Radiation on Hygrothermal Behavior of a Washing Fine Hemp Wall

Naima Boumediene[1,2](✉), Sylvie Prétot[1], Florence Collet[1], and Geoffrey Promis[2]

[1] Laboratoire Génie Civil et Génie Mécanique, Université de Rennes, 35704 Rennes, France
[2] Laboratoire de Technologie Innovantes, Université de Picardie Jules Verne, 80025 Amiens, France
naima.boumediene@u-picardie.fr

Abstract. The hygrothermal behavior of Washing Fines Hemp Composites was investigated experimentally and numerically at the wall scale for several French and Tunisian climates without considering solar gains nor rain. The stabilization phase leads to experimental and numerical thermal conductive resistance consistent with the theoretical one. The hygric resistance is consistent for simulation performed up to steady state. The profiles at different time during daily climate cycles show that two thirds of the thickness of the wall on the exterior side are active and sorption-desorption phenomena occur in the wall.

This paper investigates the impact of solar radiation on the hygrothermal behavior of a WFH wall under typical French summer climate. The experimental study is performed with a dual-climatic rooms that simulates the indoor and outdoor climates where the solar radiation is taken into account with equivalent outdoor temperature. The test wall is instrumented with temperature and Relative Humidity sensors and with heat flux sensors.

The responses of the wall are analyzed from the temperature and vapor pressure kinetics and profiles. Thermal and hygric penetration depth are similar as the ones obtained without solar gains. Heat storage and sorption-desorption phenomena are enhanced by solar gains.

Keywords: Bio-based material · bi-climatic set up · summer climate

1 Introduction

Sustainable development is one of the main challenges of recent years. It encourages the use of building materials with low environmental impact and high hygrothermal performance in order to reduce energy consumption. Bio-based materials, made of particles or plant fibres, are one of these materials, which offer competitive thermal and environmental performance (Amziane et al. 2017). In this context, Washing Fines Hemp composites have been developed and characterised at the material scale at LGCGM (Mazhoud et al. 2017). They show good thermal and hygric performances such as low thermal conductivity about 0.11 $W.m^{-1}.K^{-1}$ and high moisture buffer ability with MBV equal to 2.24 $g.m^{-2}.\%RH^{-1}$ for a density of 448 $kg.m^{-3}$ (Mazhoud et al. 2021). Their

S. Amziane et al. (Eds.): ICBBM 2023, RILEM Bookseries 45, pp. 420–429, 2023.
https://doi.org/10.1007/978-3-031-33465-8_32

characterisation at the wall scale is a necessary intermediate step before moving onto building-scale construction (Medjelekh et al. 2017; Seng et al. 2021). In a previous step, the hygrothermal behavior of this material was investigated experimentally and numerically at the wall scale for several French and Tunisian climates without considering solar gains nor rain (Boumediene et al. 2022).

Thus, this study focuses on the impact of solar radiation on the hygrothermal characterisation of a Washing Fines Hemp wall under typical French summer climate. The experimental study is performed with a bi-climatic rooms that simulates the indoor and outdoor climates where the solar radiation is taken into account with equivalent outdoor temperature.

The first part of this paper presents the experimental device. Then, the kinetic are presented all along the test. The results are analyzed in the dynamic phase from temperature and vapor pressure kinetics at several depth in the wall and temperature and vapor pressure profiles at serval times in both cases with and without solar radiation. Finally, this study underline that heat storage and sorption-desorption phenomena are enhanced by solar gains.

2 Methods

2.1 Experimental Device and Metrology

The experimental set-up consists in dual-climatic rooms (Fig. 1a): each room is 2.35 m deep, 2.78 m wide and 2.4 m high. The floor is built of concrete; the outside walls and the ceiling are insulated from heat and moisture with air proof polyurethane panels. The two rooms are controlled in temperature and relative humidity to simulate the indoor and the outdoor climates. Their respective temperature and humidity ranges are 18 to 27 °C / −5 to 35 °C and 30 to 60%RH / 30 to 90%RH. Each parameter (T and RH) is regulated by a DR4020 universal controller that acts on convectors for heating, cooling unit for dehumidification and refrigeration and ultrasonic humidifiers for humidification.

These rooms are separated by wall that can be as a whole (Collet and Prétot 2014) or divided into several parts to compare various wall compositions simultaneously. The studied wall is 1.1 cm long and 1 m high. It consists in a 28 cm thick layer of washing fine – hemp composite (WFH), coated with a 0.3 cm thick skincoat clay plaster (SCP) on the indoor side and with 3 cm thick hemp-lime render (HLR) on the outdoor side (Fig. 1b).

The hygrothermal behavior of the wall is supervised with ten SHT35 Sensirion sensors used as ambience sensors and distributed along thickness of the wall (Fig. 1c, d). The accuracy of these sensors is ±0.1 °C for a range from 20 °C to 60 °C of temperature and ±1.5% of relative humidity up to 80%RH at 25 °C. The time step for data acquisition is five minutes. The heat flux is measured on the indoor and on the outdoor surfaces of the wall with HFP01 heat flux sensors from Hukseflux (80 mm in diameter and 5 mm thick). The sensors are oriented so that the heat flux is positive from the indoor side to the outdoor side of the wall.

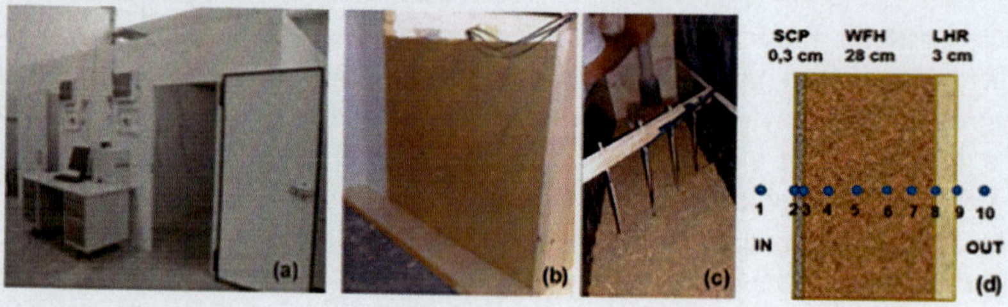

Fig. 1. Experimental device (a), WFH wall (b), implementation of hygrothermal sensors (c), sensors positions (d)

2.2 Solicitations

This study aims to investigate the effect of solar radiation on the hygrothermal behavior of the WFH wall for a sub-oceanic climate corresponding to the city of Rennes, in France, during the summer. The indoor conditions are constants, with temperature of 23 °C and relative humidity of 50% RH. The outdoor conditions correspond to a typical day that is chosen to be representative of the minimum and maximum 30-year average temperatures in August (1981–2010, Méteo France. (2022). The effect of solar radiation is taken into account with an equivalent temperature calculated from the solar radiation of the cities (Fig. 2). The relative humidity is calculated from vapor pressure and equivalent temperature (Eq. 1).

$$T_{eq} = T_a + \alpha\, \phi_r R_{se} \tag{1}$$

T_{eq}: equivalent temperature (°C).
T_a: outdoor ambiant temperature (°C).
α absorption coefficient .
ϕ_r: solar radiation (W/m^2).
R_{se}: exchange surface resistance (m^2.K.W^{-1}).

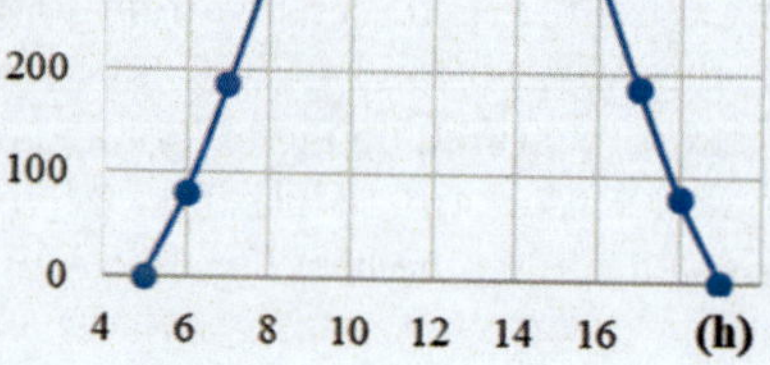

Fig. 2. Solar radiation for one typical day in Rennes in August ('Diagramme Solaire Intéractif' 2022)

Firstly, the wall is stabilized at night conditions until the temperature/vapor pressure variation is lower than 1% / 2% over the last 24 h at each position. This needs up to 15 days. The wall is then exposed to daily cyclic variations. The cycles are performed several times until a repeatable response of the hygrothermal behavior of the wall is obtained. The repeatability is reached when the temperature/vapor pressure variation is lower than 1% / 2% between the last two cycles at each position. This needs four to five cycles. Due to the experimental device regulation, the climate cycles are defined with four temperature and relative humidity steps as shown on Fig. 3. The cycles given on the left are considered representative of cloudy days, north exposition and sunscreen effect. The cycles given on the right include the effect of solar radiation.

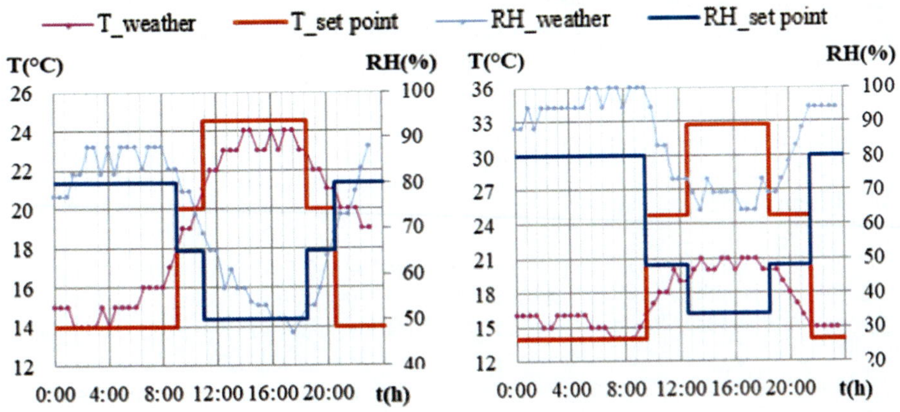

Fig. 3. Left: Ambient outdoor temperature and relative humidity without solar gains, Right: Equivalent temperature and relative humidity with solar gains for one typical day in Rennes in August

2.3 Results Analysis

For representative cycles, the kinetics of temperature, relative humidity and vapor pressure are given at each position and allow to plot profiles at several times along a day.

The results under dynamic solicitation are analyzed regarding dynamic parameters as the time lag and the relative amplitude variation, also called damping factor for the temperature (Fig. 4). The time lag is the time between the excitation at the outdoor surface and the response to the excitation. The relative amplitude variation is the ratio of the amplitude at the given point to the amplitude at the outdoor surface over the wall thickness.

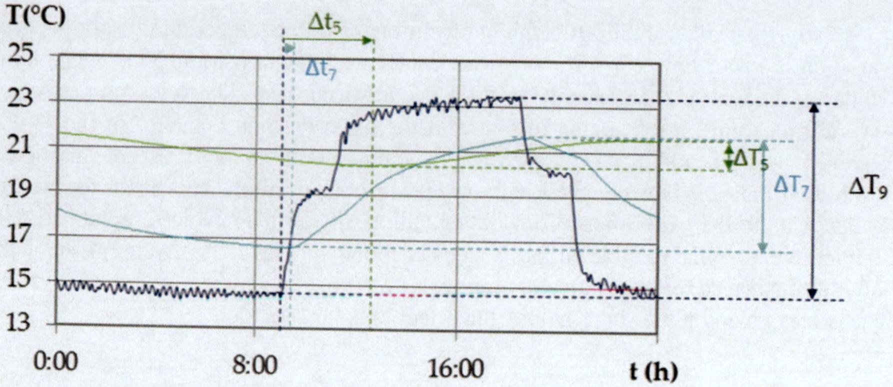

Fig. 4. Identification of the time lag Δt_i between the surface and a given point, and of the amplitude at the outdoor surface (ΔT_9) and at a given point (ΔT_i).

3 Results

3.1 Impact on Temperature

Figure 4 displays the last cycle, it shows the indoor and outdoor ambient conditions are in good agreement with the set points. Over the repetition of cycles, the temperature beams increase, the wall heats up over its entire thickness in connection with the set points.

At night conditions, the outdoor temperatures are the same with or without solar radiation during the day. The heat flux induced by the temperature gradient goes from the inside to the outside. At daytime, without solar radiation, the indoor and outdoor temperatures are close and the heat flux is almost zero. Taking into account the solar radiation leads to equivalent outdoor temperature about 10 °C higher than those without solar radiation, and the heat flux goes from the outside to the inside (Fig. 5).

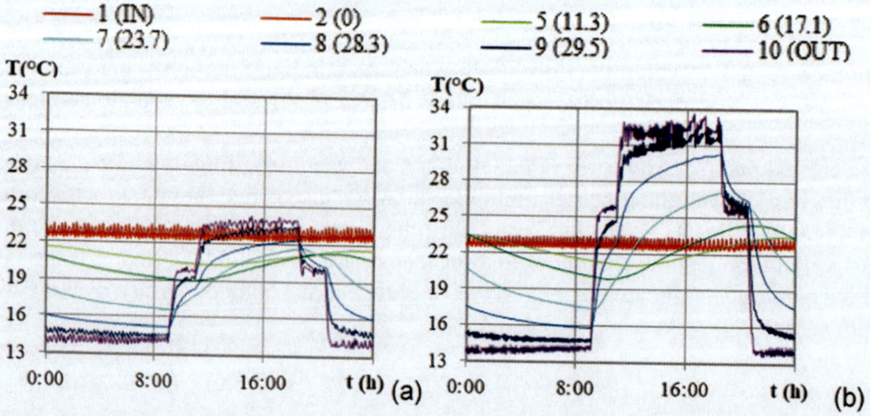

Fig. 5. Kinetics of temperature at several depth in the wall – a: without solar gains, b: with solar gains

The time lag and the temperature damping are similar for both cases, with or without taking into account the solar radiation. In fact, they depend only on the heat diffusion in the wall. The increase of the amplitude of the signal on the surface is translated by an increase of the amplitude of the signal in the thickness of the wall (Fig. 6).

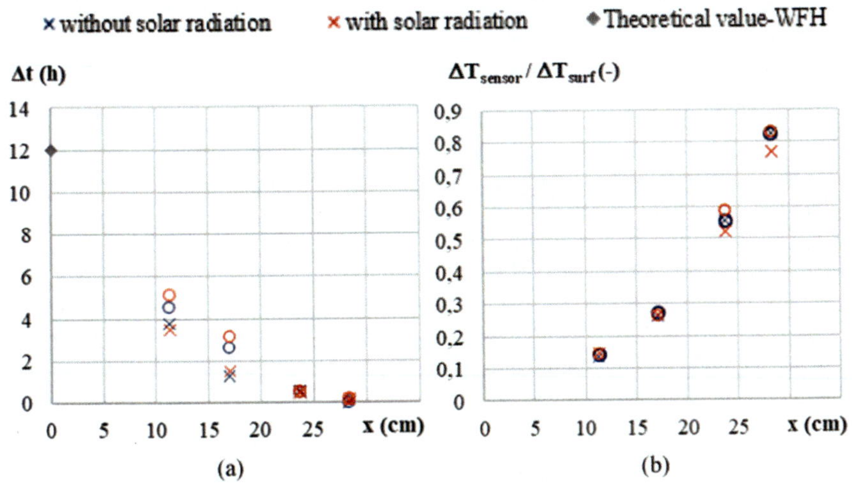

Fig. 6. Dynamic thermal parameters versus location (x/o: calculated during increasing/decreasing temperature step): (a) Shift of temperature (b) damping factor.

The temperature profiles are presented each hour during the last day of the dynamic study in Fig. 7. During the day, the temperature is almost constant on the inside surface of the wall, whereas on the outside surface, it evolves according to the set points.

Without solar radiation, the temperature profile shows a daily variation amplitude that evolves over the thickness of the wall. It is limited to the third of the thickness on the inner side and then increases towards the exterior. On the interior side, the temperature gradient is small to negligible but always negative. At the change of set point, the temperature increases on the outdoor surface. Then, in the core of the wall, the gradient of the temperature and the heat flow are reversed. The absorption of the heat flow on the outside leads to storage with a rapid increase in temperature in the outer layers.

With solar radiation, the increase of the temperature in the outside ambience during the day leads to an increase of the average temperature over the entire thickness of the wall. The increase of temperature on the outdoor surface leads to a propagation of heat that reaches the indoor side of the wall and induces a slight temperature gradient. This gradient of temperature reverses during the day and leads to slight delayed gains at night. This effect was not observed in the case where the sunlight was not taken into account. The two thirds of the wall thickness on the outdoor side are active and contribute to limit the heat gains on the indoor side.

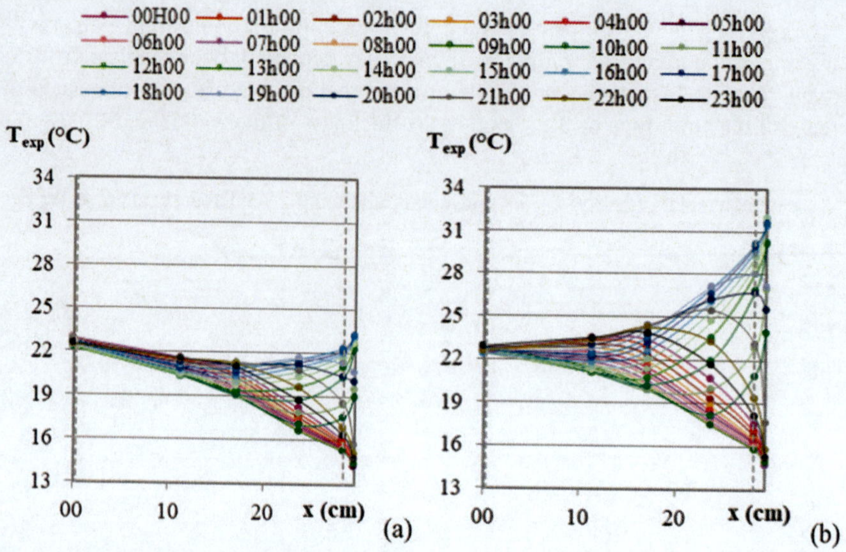

Fig. 7. Temperature profiles during the last day of the dynamic solicitation phase for the climate of Rennes; (a) without solar radiation, (b) with solar radiation

3.2 Impact on Vapor Pressure

Figure 8 shows the vapor pressure kinetics at the different positions in the wall. For the vapor pressures, when considering solar radiation, the levels are higher than those observed during the solicitations without solar radiation, with higher amplitudes of variation. In addition, the vapor pressures within the wall are higher than those in the ambient conditions.

Figure 9 shows that the time lags are similar without and with solar radiation. The relative variations of the vapor pressure amplitude are higher when the effect of the solar radiation is considered. These values were higher than 1 on the external third in the case of solicitations without solar radiation and become higher than 1 on the whole thickness with solar radiation. This underlines the amplification of the desorption phenomena induced by the rise in temperature.

Figure 10 shows the vapor pressure profiles. During the daily cycle, the vapor pressure gradient reverses on the outdoor side of the wall, but not on the indoor side of the wall. When the effect of solar radiation is taken into account, the amplitude increases over the outdoor two-thirds of the wall. The maximum vapor pressure reached increases significantly due to the desorption phenomena related to the solar radiation. This leads to an increase of the vapor pressure all over the wall thickness, and a vapor pressure gradient on the indoor side about 2 times higher.

3.3 Impact on Heat Storage

Figure 11 shows the evolution of the heat fluxes on the indoor and outdoor surfaces of the wall under daily cyclic variations, without and with solar radiation. The experimental

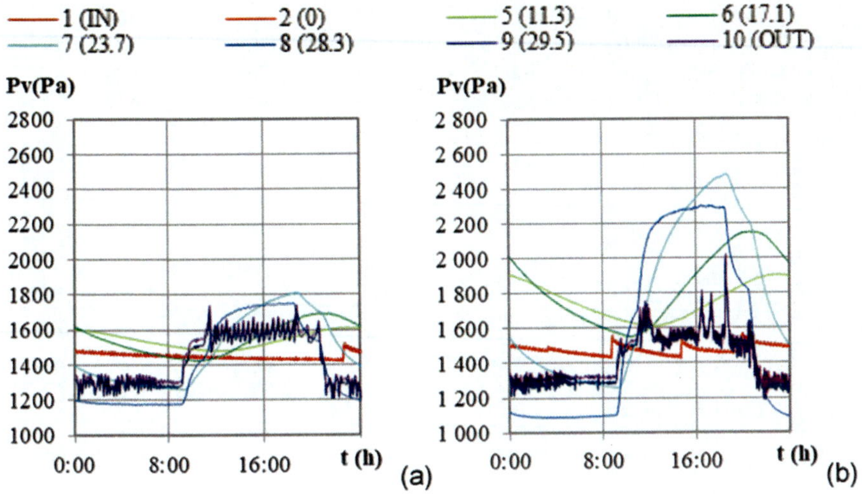

Fig. 8. Kinetics of vapor pressure at several depth in the wall – a: without solar radiation, b: with solar radiation

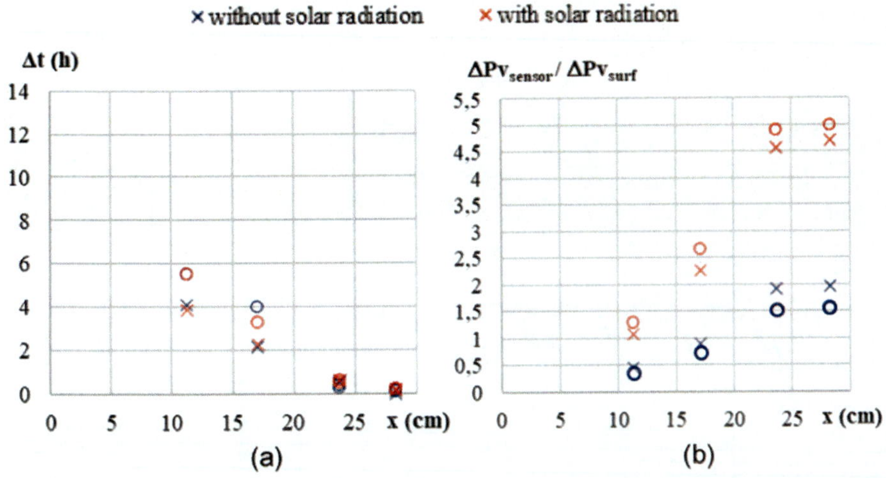

Fig. 9. Dynamic hygric parameters versus location (x/o: calculated during increasing/ decreasing vapor pressure step): (a) Shift of vapor pressure (b) damping factor

results show that the heat flux is repeated during the cycles in both cases. The change of the flux sign corresponds to the heat storage and release. The solar radiation enhances the heat exchanges on the outdoor surface. On the indoor surface, the heat flow remains very low because the heat flow does not reach this side of the wall, as seen on the profiles.

The integration of the fluxes on each surface gives the amount of heat stored and released by the wall. Figure 12 gives the total amount of heat stored and released over the last cycle. The effect of solar radiation leads to heat storage 3 times higher and heat release 1.2 times higher than without sun. The heat balance is nearly zero without solar

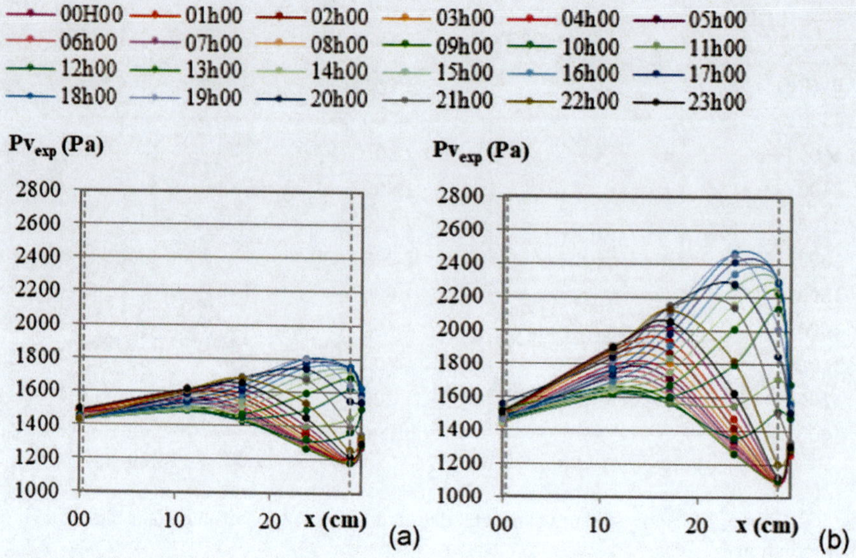

Fig. 10. Vapor pressure profiles during the last day of the dynamic solicitation phase for the climate of Rennes; (a) without solar radiation, (b) with solar radiation

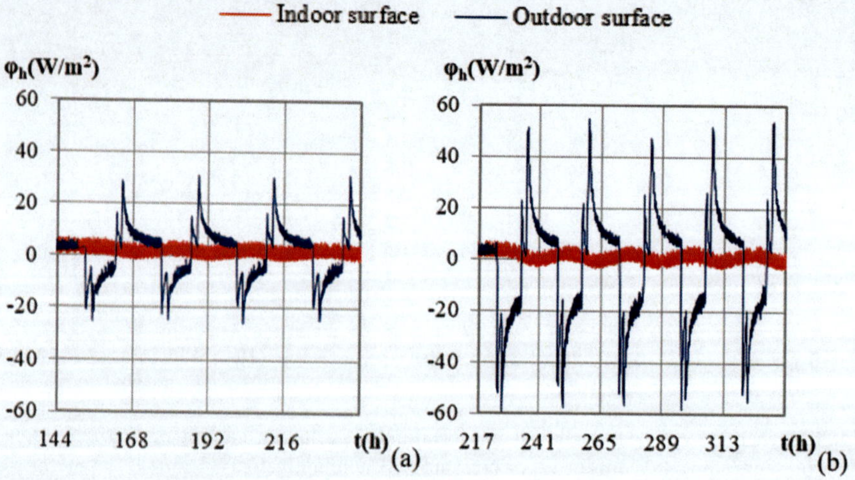

Fig. 11. Heat flux for the climate of Rennes; (a) without solar radiation, (b) with solar radiation

radiation and widely positive in the case with solar radiation, that induces an increase in the temperature of the wall.

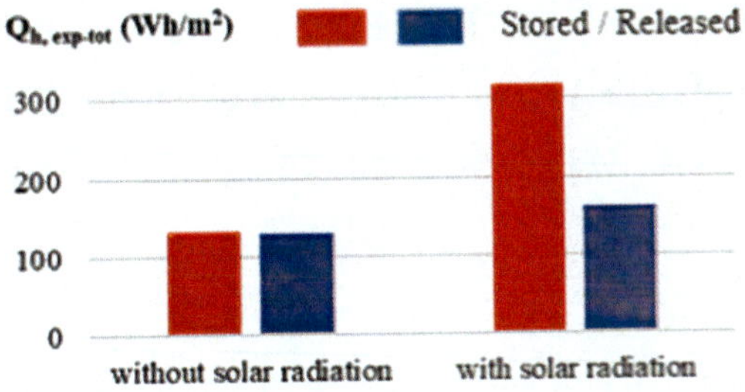

Fig. 12. Heat storage for the climate of Rennes without and with solar radiation

4 Conclusion

The investigation of the effect of the solar radiation on the hygrothermal behavior of a washing fines hemp wall highlights several key points.

With and without solar radiation effect, the main variations of temperature and vapor pressure occur over the two outdoor third of the thickness of the wall. Sorption-desorption phenomena are highlighted. Both temperature and vapor pressure variation and sorption-desorption phenomena are magnified with solar radiation.

References

Amziane, S., et al.: Bioaggregates based building materials, Springer Netherlands, Amziane Sofiane and Collet Florence (2017). http://www.springer.com/la/book/9789402410303. Accessed 24 February 24

Boumediene, N., Collet, F., Pretot, S., Elouad, S.: Hygrothermal behavior of a washing fines—hemp wall under French and Tunisian summer climates: experimental and numerical approach. Materials **15**, 1103 (2022). https://doi.org/10.3390/ma15031103

Collet, F., Prétot, S.: Experimental highlight of hygrothermal phenomena in hemp concrete wall. Build. Environ. **V82**, 459–466 (2014)

Diagramme Solaire Intéractif (2022). http://diagsol.herokuapp.com/

Mazhoud, B., Collet, F., Pretot, S., Lanos, C.: Mechanical properties of hemp-clay and hemp stabilized clay composites. Constr. Build. Mater. **155** 1126–1137 (2017). https://doi.org/10.1016/j.conbuildmat.2017.08.121

Mazhoud, B., Collet, F., Prétot, S., Lanos, C.: Effect of hemp content and clay stabilization on hygric and thermal properties of hemp-clay composites. Construction and Building Materials **300**(September), 123878 (2021). https://doi.org/10.1016/j.conbuildmat.2021.123878

Medjelekh, D., Ulmet, L., Dubois, F.: Characterization of hygrothermal transfers in the unfired earth. Energy Procedia. **139,** 487–492 (2017). https://doi.org/10.1016/j.egypro.2017.11.242

Méteo France (2022). https://meteofrance.com/

Seng, B., Magniont, C., Gallego, S., Lorente, S.: Behavior of a hemp-based concrete wall under dynamic thermal and hygric solicitations. Energy Build. **232** 110669 (2021). https://doi.org/10.1016/j.enbuild.2020.110669

Preliminary Results of Thermal Conductivity Test on Earth and Fibers Building Elements, from Waste Materials to Thermal Constructive Solution

Amanda Rivera Vidal(✉) and Maddalena Achenza

LABTERRA, Dipartimento di Ingegneria civile, ambientale e architettura, Università degli Studi di Cagliari, Cagliari, Italy

amandariverav@gmail.com, maddalena.achenza@unica.it

Abstract. Earth and natural fibers are local materials that can be find worldwide, both obtained from other production processes, commonly considered as waste. The present work describes the development and assessment of building materials made of earth and vegetable fibbers with diverse thermal characteristics, based on thermal conductivity test. The reuse soil and natural fibers as widely available materials was proposed because both do not need transportation and they require a minimum expenditure of energy for their transformation. Furthermore, they maintain natural and healthy characteristics in their whole life cycle: in the construction process, during their use and also at the moment of dismission. This paper presents the initial stage of research and test of bio and geo based materials, a comparison of the thermal capacity in various fiber-earth composite building elements developed in Sardinia (Italy), that could represent an interesting contribution both for new sustainable construction and to improve the thermal performance of existing buildings.

Keywords: natural fibers · raw earth · soil for building · thermal performance · waste material

1 Introduction

The use of natural available materials, such as natural fibers and earth, have been widely used by traditional cultures to build their habitat, and we can find countless examples of these elements coupled together in practically the whole world.

Different building elements based on the mixture of earth and fibers are found in different vernacular cultures. Its use has been observed mainly in buildings where the bearing structure is made of soil, such as adobe, cob or rammed earth buildings; but also in buildings with a bearing wooden structure, such as wattle and dub, and other filled partition walls or walls with earthen finishes. On the other hand, mixes of earth and fibers, such as renders and mortars have been observed in stone and brick buildings. The distribution and massive use of earth and fibers as construction materials extends

S. Amziane et al. (Eds.): ICBBM 2023, RILEM Bookseries 45, pp. 430–441, 2023.
https://doi.org/10.1007/978-3-031-33465-8_33

geographically and temporally, but also in the diversity of their uses: as a structure, as a filling, as a finish, and, also, as a thermal regulatory component. The present work will focus on this last use of the material, both in vertical elements (walls), and in horizontal ones (mezzanines and roofs).

The use of the mixture of earth and natural fibers for the creation of construction materials is mainly due to the wide availability and diversity of both materials. On the other hand, the integration of fibers to earth in the development of constructive elements optimizes their performance, since it improves its fragile and tensile behavior, reduces the cracking, at in addition it significantly increases their thermal insulation capacity [1].

Buildings that mixed both elements can be seen in archaeological sites such as Caral (5000 BC), Babylon (600 BC) and the great wall of China [2].

More recently, the particular development of the so-called "lightened earth" technique, according to Volhard, took place after the First World War in Germany (since 1920), as it represented a cheaper alternative for filling wooden bearing structures, with the wattle and daub construction technique, known in German as "Leichtlehmbau" [3].

In recent years, as a result of the emergency arised by climate change and the consequent promotion of a more ecological architecture, construction with fibers and earth represent a good opportunity for contemporary architecture (Fig. 1). Vegetable fibers offer great ecological advantages, since they not only do not release CO2, but also store carbon during their growth. Furthermore, thin fibers are fast-growing plant parts, so their renewal is constant [4].

Fig. 1. Utilization of earth-fiber mix on diverse uses and architectures. Left:detail of a funerary structure called Chullpa (Bolivia 1000 AD) with earth and local fibbers [5]. Center: thermic regulator system in a roof in Valparaiso (Chile s.XX) with earth and wood chip © Claudio Vega. Right: Contemporary construction with lightened earth inside of a wooden and steel structure in Coliumo (Chile, 2018) © Amanda Rivera Vidal.

2 Raw and Waste Matters

The raw matters used in this research are components considered waste from other production processes. The one is the excavated soil resulting both from infrastructure and building works, and the second are various natural fibers residual from agricultural processes. This matters have been defined by Stulz and Mukerij as by-products, as not all wastes are useless garbage and also not freely available. It is mostly a question of definition: from one point of view, a material can be of no use, while from another it could be a valuable resource [6].

This research focused on the development and assessment of different composites elements produced using always the same soil mixed with different natural fibers, with the aim to generate building elements with significant thermal performance.

Three different fibers have been used: hemp, wheat straw and loofah.

2.1 Earth

Soil or earth is a material that results of the disaggregation of rocks due to the interaction of climatic factors (sun, wind, rain, frost) and the physicochemical evolution [7]. It's a worldwide available material with endless diversity of characteristics and one of the oldest building material used by humankind, transform into infinite architectural forms and building techniques. It's composed by a diversity of grains being the biggest stones and gravels, follow by coarse and fine sand and the smallest silts and clays. Each grain fulfills a function in the matter, being the clays responsible for cohesion due to its laminar geometry with electrostatic charges on their faces, which generate attraction between them. When extracted from excavation, earth as a waste material starts a virtuous cycle of use as a by-product; reducing even more its energy consumption when a short supply chain is apply with short distance between production supply, production place and construction site. On the other hand, its non-stabilized used assure its infinitely reuse [8].

For this research a locally available earth has been chosen, taken from the excavation of a road construction in Serrenti (Sardinia, Italy). Through the granulometric analysis of a 178,88gr of raw material 92,45% are smaller than 0,063 mm. From this finest part a X-ray diffraction (XRD) analysis showed the presence of Montmorillonite (MMT), clay that belongs to the smectite family with a structure of two tetrahedral silica layers separated by an octahedral alumina layer [9]. It's a with high percentage of silt and a medium cohesive capacity.

2.2 Fibers

All fibers used are vegetal fibers available in Sardinia (Italy), and represent a left-over of other production process. These fibers consist mainly of cellulose, representing 50% of the biomass [10]. All three of them are fast growing plants with annual cycles (Table 1).

Hemp. A annual herbaceous plant native from the Himalayas, hemp is one of the first plants domesticated by humanity and a member of the Cannabaceae plant family.

It has been cultivated since the Neolithic period and adapted to temperate climates; it grows very quickly since it matures four months after seeding. The hemp plant has a single main stalk, grows to a height of about 1.8 to 4.5 m and will flower and seed if pollination occurs between male and female plants [11]. Historically, hemp is cultivated to manufacture textiles, ropes and paper, for which the peripheral fibers detached from the heart of the stem are used. The center of the stem is then transformed into hemp shive or hurd. This is a very porous element with a high water absorption capacity. The fibers used, have a very diverse dimension up to 30 mm length and 5mm diameter (Fig. 2).

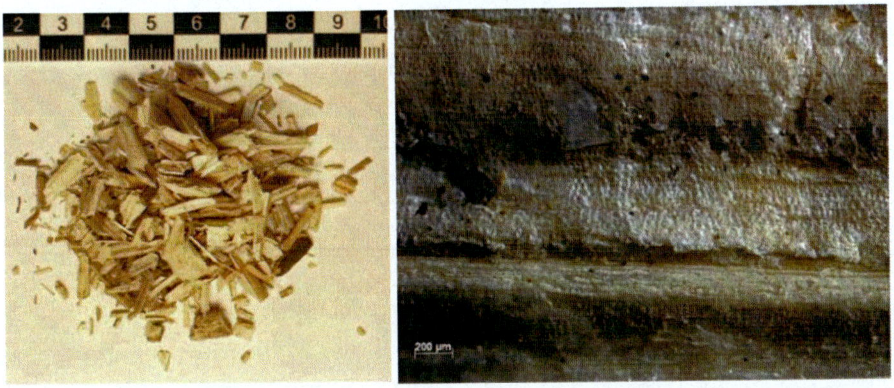

Fig. 2. Images of the hemp used for the panels construction. Left: picture of the bundle of fibers; Right: microscope photography

Wheat Straw. Annual herbaceous plant with alternate leaves, formed of a culm bearing an ear made up of two rows of spikelet containing the grains. The wheat was domesticated in the Near East from a wild grass and cultivated as food supply. Once harvested, the stem of the grain is cut, becoming a waste for agriculture. The stems are culms, cylindrical, often hollow by resorption of the central marrow [10].

They look like fluted tubes with long and numerous sub-conducting bundles. The culms are interrupted by nodes. There is a concentration of multiple microscopic but very hard clusters of silica in the epidermis, making the culms sharp. The fibers used are up to 40 mm long with dimeter up to 3 mm (Fig. 3).

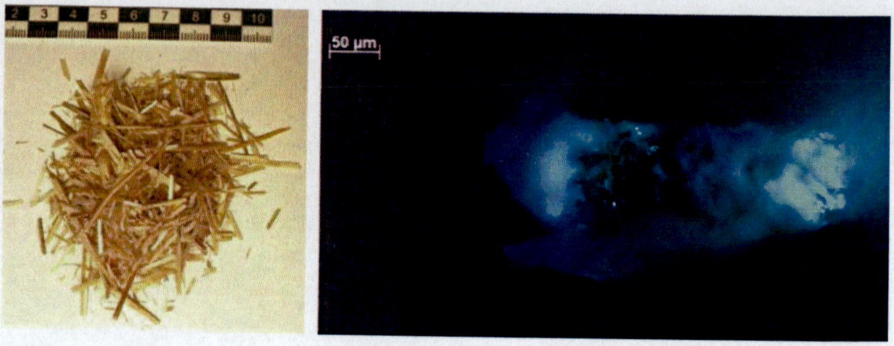

Fig. 3. Images of the wheat straw used for the panels construction. Left: picture of the bundle of fibers; Right: microscope photography

Loofah. An annual tropical and subtropical Cucurbitaceae plant, cultivated for centuries in the Middle East and Asia. The dry fruit has a hard shell that surrounds a stiff, dense network of cellulose fibers [12]. Luffa or Loofah its used as sponge and its cultivation was widely disseminated until the beginning of the XX century. In recent years, its cultivation has expanded again, due to the promotion of the ecological qualities of its sponge. Loofah products are made just from the outer part of the fruit, leaving the heart of the fruit as a residue. This central part has been shredded for this experimentation into 20 mm long fibers with a diameter of about 0,5 mm (Fig. 4).

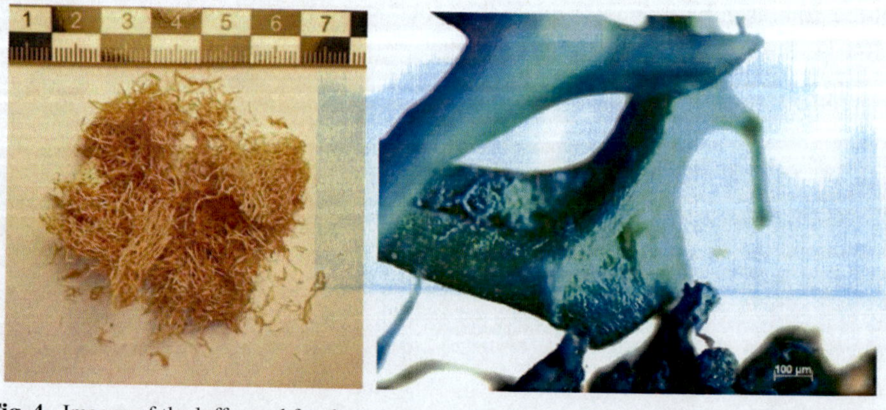

Fig. 4. Images of the luffa used for the panels construction. Left: picture of the bundle of fibers; Right: microscope photography

3 Methodologies and Composition of the Fiber-Earth Materials

Fibers, due to their porous characteristics made up of cavities of various sizes, manage to encapsulate air in a tight manner, thus acting as thermal insulators in fiber-earth composites. For this same quality, fiber-earth composites produce quite light building elements. As a limitation, vegetable fibers are highly vulnerable to fire, but earth being a fireproof element, bears the task of encapsulate the fibers protecting them. Finally, earth provides cohesion to the fibers, allowing the mix to become a compact and coherent system.

3.1 Earth-Fiber Elements Construction

As a part of the initials tests earth-fibers elements were tested for their thermal performance, to carry out these tests 200x200x50mm parallelepiped panels have been produced. The development of the panels was carried on in three phases: mix-design development, manufacture of panels, thermal testing (Table 1).

Mix-design Development. For each of the fibers, a series of mix-tests were carried out. Initially, the necessary volume of fibers to fill the mold was determined, then the fiber was mixed with a determined amount of water and then mixed with diverse percentages of earth sieved to 1 mm. For each fiber the volume to fill the mold was different, as well as the amount of added water. The amount of soil needed to achieve a cohesive panel also varied. The initial search aimed to arrive to the possible lighter panel, to reach a low density thus a better thermal insulation. In this search, both wheat straw and the hemp shives tended to lose cohesion. For this reason, two samples where made with these materials: a lighter with less cohesion and a heavier with more cohesion (Table 2). These panels have been produced between August and September in Cagliari, Italy (summer), thus the drying process was quite fast.

Table 1. Fibers geometric characteristics.

	MAX LENGTH (mm)	DIAMETER (μm)	PLANT PART	SKELETAL DENSITY (g/cm3)
Hemp	30	17 – 5000	Stem	1,4161
Wheat Straw	40	50 – 3000	Stem	1,2656
Loofah	20	20 – 500	Fruit	1,6553

Manufacture of Selected Samples. Once the composition of the five samples was determined (HE24, HE14, LUF7, WST14, WST4, named after the percentage of earth contained) we proceeded to manufacture 5 samples for each composition. For this procedure, the initial volumes were multiplied by five, then the 18 mm thick wooden mold was filled with the mixture, pressing once with a 12mm wooden plate, and then removing

Table 2. Table of composition of the selected panels, where E (earth), F (fiber), E% (earth percentage) W (water), HE (hemp), LUF (loofah), WST (wheat straw).

EATH-FIBER PANEL COMPOSITION							
	COMPOSITION VOLUME (L)				RESULTS		
	E	F	E%	W	DRY WEIGHT (gr)	DENSITY (Kg/m3)	OBS
HE24	0,96	4	**24,0%**	2000	861	430,50	Heaviest
HE14	0,32	2,3	**13,9%**	750	679	339,50	Lighter
LUF7	0,36	5	**7,2%**	1000	620	310,00	Lighter
WST11	0,5	3,5	**11,4%**	850	799	399,50	Heaviest
WST4	0,144	3,2	**4,5%**	850	405	202,50	Lighter

the mold. The manufacturing of the panels was carried out during the months of October and November in Cagliari, Italy (autumn), being the drying process much slower than in the preliminary stage, which caused the germination of sprouts in all the samples, which dried together with the panels without damage.

Preparation for Thermal Testing. The dry samples had relatively defined borders, but with very heterogeneous faces, with an important quantity of cavities and therefore not completely parallel between. In order to place these samples in the thermal conductivity measuring device (TCA 300 TAURUS), it has been necessary to prepare every sample individually. The samples haven been then placed in a thermally insulated 300x300x60 mm closed cell expanded polystyrene ΛD 0,035 W/MK frame. The latter had a hole in the center of 200x200 mm size where the fiber-earth panels were inserted. Each exposed face has been then coated with an earth and sand render made with the same earth used for the samples (Serrenti) in volume proportions of 1:2, all sieved to 0.7 mm. The render has been applied by layers with a drying process in-between, to avoid fissuration. The first layer was applied before the sample was introduced in the polystyrene frame and the last one once inside. With this procedure, we obtained samples with two parallel faces with flat and homogeneous surfaces. Because of the irregularity of the fiber-earth samples, the coating thickness varied from 6 to 12 mm (the addition of both sides of coating) (Fig. 5).

Also a sample of render had to be prepared to be tested, in order to have information of its thermal conductivity thus the information of the fiber-earth panel alone also. In this case also a panel of 300x300x60 mm of expanded polystyrene was prepared but with a central perforation of 100x100 mm, inside of with a layer by layer render was inserted (R_ES).

Fig. 5. Process of preparation of the samples. Left: panel result; center: panel with the first layer of coating; right: panel inside the polystyrene frame with the final coating.

4 Tests and Results

The tests carried out on the fiber-earth panels concerned thermal conductivity and were done with the TCA 300 TAURUS INSTRUMENTS. This is a measuring device that helps to determine the thermal conductivity of materials by means of a heat flow meter in accordance with ISO 8301, DIN 52616, DIN EN 1946- 3, EN 12664, EN 12667 and EN 12939. This device is equipped with an insulated testing chamber in order to avoid the interference with the external temperature [13]. This instrument has two 300x300 mm horizontal, flat and parallel surfaces inside, between which the material to be tested is enclosed. Each of these surfaces will change temperature to measure the heat flow on a given surface. For this test, a measurement area of 100x100 mm has been established in the center of the panel.

For the correct execution of the test, the material to be tested must have faces that are completely flat and parallel to each other, without irregularities that could allow air to flow inside the chamber and not through the tested material. Due to this specificity, the fiber-earth panels have been coated with the aforementioned render which has been also tested in order to estimate data of the complete panels. Additionally, the samples must have 0% relative humidity (RH), since the amount of water retained inside the material would vary the results, having a high thermal conductivity capacity. To ensure this, each of the samples, after drying in a natural environment, was oven-dried at 60ºC for 24 h and covered with a cellophane film to avoid re-wetting.

The testing cycle lasted 15 h. On the basis of the research, the coefficients λ of individual samples were determined for the temperatures of 10 °C, 20 °C and 30 °C, with a 5 h cycle for each one of them. By analyzing the results, it can be observed that the stable mean of temperature was reached after one hour for each temperature cycle (Fig. 6).

In reference to the mean of temperature of 10ºC, the measured thermal conductivity coefficient λ of the sample of hemp, 13,9% of earth (HE14) and 8,3 mm of render was equal to λ = 0,080 W/mK. In comparison, the sample made with the same material but a higher percentage of earth 24% (HE24) and with a 6,3 mm render resulted λ =

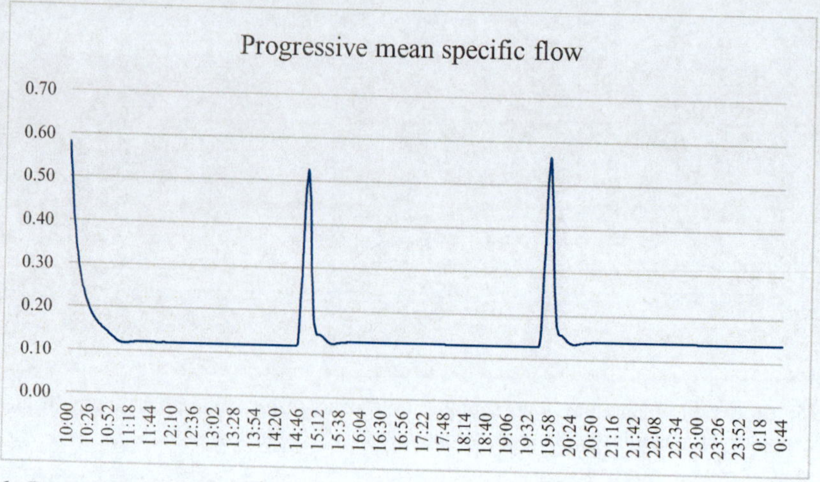

Fig. 6. Progressive mean specific flow showing the stability of 4 h of each temperature mean cycle on the sample HE24.

0,097 W/mK. The loofah sample with a 7,2% of earth (LUF7) and with 11,8 mm of render was equal to λ = 0,101 W/mK. The sample made with wheat straw, a 4,5% of earth (WST4) and a 8 mm render was equal to λ = 0,061 W/mK while the sample with 11,4% of earth (WST11), a 9,3 mm render resulted λ = 0,090 W/mK. As each of the samples has a different render thickness and the render has a very high thermal conductivity (λ = 0,285 W/mK), this influences results considerably increasing thermal conductivity of tested panels (Fig. 7).

In order to achieve the thermal conductivity coefficient λ of each sample without considering the thickness (L) of the render, it was decided to apply the formula $R = L/\lambda$.

Comparing the results of the thermal conductivity (Table 3) it is evident that neither the density of the material nor the percentage of earth contained are directly linked with its thermal conductivity, even if the sample WST4 with less percentage of soil (4%) had the less density (231 kg/m3) and the less thermal conductivity of λ = 0,0541 W/mK. An example for this statement are the results of the loofah panel LUF7 performance, being the second sample with less earth (7%), the third sample by density (317,5 kg/m3) and the fourth by lower thermal conductivity (0,087 W/mK).

5 Perspectives on Further Works

The temperature conductivity tests indicated in this work are the initial of a series, where each material will be tested 5 times to obtain average results that verify trends.

On the other hand, it is of great importance to carry out other tests to determine their performance integrally and evaluate diverse possible use of this geo- and bio- source material. The tests to be carried on will be focus on three axes:

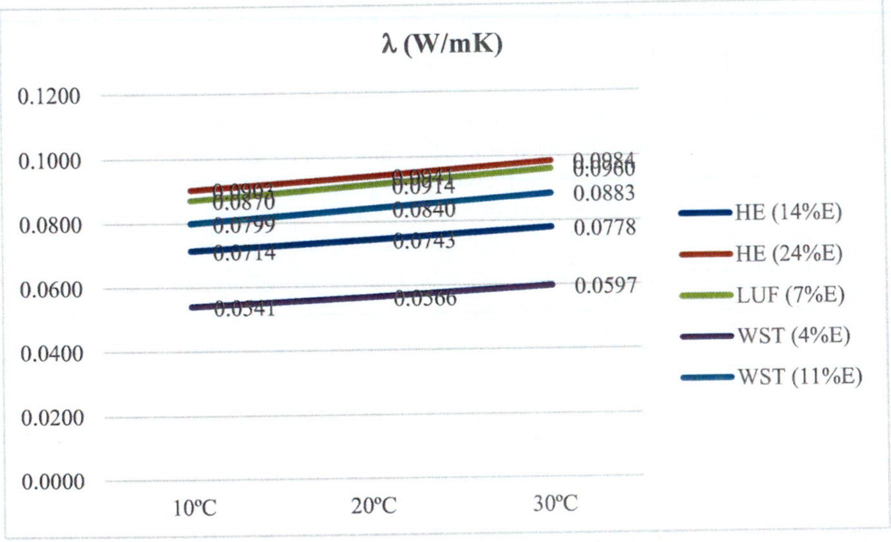

Fig. 7. Comparison of the thermal conductivity coefficient λ of each sample in the three means of temperature (10°C, 20°C and 30°C)

Table 3. Results of the thermal test with the sample information of the fibber-earth panels and the earth-sand render for a mean of temperature of 10°C. Where E (earth), E% (earth percentage), R_ES (render of earth and sand), HE (hemp), LUF (loofah), WST (wheat straw).

	SAMPLE INFO		COATING INFO		THERMAL PERFORMANCE [mean of temperature of 10°C]		
SAMPLE	E%	DENSITY (Kg/m3)	% COATING in volume	L Coating THICKNESS (m)	λ Thermal conductivity (W/mK)	R Heat Flow resistance (m2K/W)	U Thermal transmit-tance (W/m2K)
R_ES	100%	1552,50	100%	0,0401	0,285	0,14089	7,0975
HE14	13,9%	301,50	14%	0,0083	0,0714	0,7202	1,3885
HE24	24,0%	437,50	11%	0,0063	0,0903	0,5901	1,6946
LUF7	7,2%	317,50	20%	0,0118	0,0870	0,5500	1,8182
WST4	4,5%	231,00	13%	0,0080	0,0541	0,9579	1,0439
WST11	11,4%	402,00	16%	0,0093	0,0799	0,6262	1,5970

- Hygrothermal performance, permeability to water vapor, thermal phase shift
- Acoustic performance, acoustic absorption and impedance.
- Behavior against fire, reaction and resistance to fire.

It is presumed that the permeability to water vaper will be high, as bio- and geo- bases materials are highly porous. This can be a great input to the hydrometric regulation of buildings as well as a solution for the thermal improvement of traditional building with a compatible and not waterproof layer, since many problems have been observed in the incorporation of polymers waterproof insulating elements that generate condensation inside the walls of stones, wood, earth, and handmade bricks, causing the appearance of fungus and degradation, including structural materials degradation of its original fabric.

Also, the development of the life cycle of building materials analysis [14], from the sourcing and production stages to use, consumption and disposal, would be an important complement to the comparison of the use of natural materials.

6 Conclusions

The tests carried out on samples made with various fibers and same soil aimed to compare the thermal performance of composite elements, as well as their manipulation. It has been noted that handling with lower percentages of soil is highly complex when using longer fibers (wheat straw and hemp), as panels have little cohesion and are highly fragile. Reflecting on the applicability of these elements in buildings, the possibility of elaborating materials to fill in frame structures has arisen, as the samples that achieve greater cohesion could be further developed as prefabricated panels.

Regarding the diversity in thermal conductivity, it has been observed that the fibers with the greatest diversity of voids (wheat straw and hemp) tend to decrease the value, despite containing a higher percentage of earth and higher density.

As by the preliminary results obtained, thermal conductivity values can be comparable with those of synthetic materials, which have a huge carbon footprint, little chance of reuse or recycling and decomposition times of hundreds of years.

Considering that:

- the use of materials exceeding from other production processes has practically no impact on the environment;
- the use of fast-growing vegetable fibers, which in their annual cycle store significant amounts of CO2, contributes to offsetting the carbon footprint in the field of construction, which currently contributes 23% of total greenhouse gas emissions [15];

We can assume that the use of earth and natural fibers, and their consequent testing, is an important contribution to the future of construction with a conscious and respectful sense of belonging, where local materials represent an opportunity for the development of technologies for building. These materials can offer important changes in new constructions, and they also represent great opportunities for the thermal adaptation of existing buildings that today do not meet the necessary standards but are not compatible with the use of synthetic materials, due to their characteristics of impermeability. Fibers and earth are local, thus inclusive; and represents a future of sustainable and regenerative architecture and construction materials.

Acknowledgements. The authors acknowledge LABTERRA intern engineering student Lisa Asté who was actively involved in the early development of both conceptual and practical development of the research. We also acknowledge the collaboration of the Laboratorio di Didattica e

Ricerca per la Salvaguardia dei Beni Culturali Bonaria, Professor Paola Meloni, and the Laboratorio di Fisica Tecnica ed Energetica, Professor Roberto Ricciu, who have collaborated respectively in carrying out the characterization of the raw matters and in the tests of the thermal performance of the materials.

References

1. Minke, G.: Lehmbau Handbuch. Der Baustoff Lehm und seine Anwendung, Rastede (2017)
2. Achenza, M., Rivera Vidal, A.: Terra cruda e fibre vegetali: un sodalizio antico per la contemporaneità. In: Bonetto, J., Previato, C. (eds.) Construire nel mondo antico, pp. 1249–1256. Università degli Studi di Padova (2022)
3. Volhard, F.: Construire en terre allegée. Editions actes-sud, Paris (2016)
4. Gauzin-Müller, D.: Architecture en fibres végétales d'aujourd'hui. Museo Editions, Paris (2019)
5. Delaveris, I., Mamani, G., Yampara P.: Chullpawawa, una nueva unidad constructiva de tierra: redescubriendo la historia prehispánica. In: Neves, C., Salcedo Gutierrez, Z., Borges Faria, O. (eds.) Seminario Iberoamericano de Arquitectura y Construcción conTierra, 19. Memorias, San Salvador, pp. 303–320. Fundasal (2019)
6. Stulz, R., Mukerij, K.: Appropriate Building Materials, A Catalogue of Potential Solutions, 3rd edn. SKAT Publications, Gallen (1998)
7. Houben, H., Guillaud, H.: Traité de construction en terre. Editions Parentheses. Marseille (2006)
8. Gauzin-Müller, D., Vissac, A.: TerraFibra Architectures. Éditions du Pavillon de l'Arsenal, Paris (2021)
9. Król-Morkisz, K., Pielichowska, K.: Thermal decomposition of polymer nanocomposites with functionalized nanoparticles. In: Pielichowski, K., Majka, T. (eds.) Polymer Composites with Functionalized Nanoparticles. Elsevier (2019)
10. Vissac, A.: Matière en fibres. Amàco, Villefontaine (2014)
11. Lewis, P., Tsurumaki, M., Lewis, D.: Manual of Biogenic House Sections. ORO Editions, San Francisco (2022)
12. Davis, J.M: Luffa Sponge Gourd Production Practices for Temperate Climates. HORTSCIENCE 29(4), 263–266 (1994)
13. Nakielska, M., Chalamoński, M., Pawłowski, K.: Investigation of effect additive phase change materials on the thermal conductivity. Mater. Sci. Eng. **245**, 022048 (2017)
14. Ben-Alona, L., Loftnessa, V., Harriesb, K., DiPietroc, G., Cochran Hameena, G.: Cradle to site Life Cycle Assessment (LCA) of natural vs conventional T building materials: a case study on cob earthen material. Build. Environ. **160**, 106150 (2019)
15. Keena, N., Duwyn J., Dyson, A.: Biomaterials supporting the transition to a circular built environment in the global south. Yale Center for Ecosystems + Architecture and UN Environment Programme. Connecticut (2020)

Binder Formulation and Properties of Hemp Concrete

Mathieu Bendouma, Philippe Fortin, Daniel Perraton, and Claudiane Ouellet-Plamondon(✉)

École de Technologie Supérieure of Montreal, Montreal, QC H3C 1K3, Canada
Claudiane.ouellet-plamondon@etsmtl.ca

Abstract. The building sector is one of the largest consumers of energy and emitters of greenhouse gases (GHG). The problems linked to climate change push us to design buildings that are more respectful of the environment, whether during the design stage, the life span, or the end of life of the works. The question then arises as to the use of insulating materials of petrochemical and mineral origin, non-renewable. To fight against environmental problems, it is thus necessary to develop new materials that are more eco-responsible and have less impact on the environment. In this sense, hemp concrete is of great interest in the scientific literature. The objective of this research is to determine the feasibility of designing hemp concrete by limiting the addition of hydraulic binder, not available in Canada which promotes GHG emissions (because of transportation) and adding materials available in North America. By combining it with other structural materials, such as OSB, it could replace gypsum board, which is widely used in Canada but often buried instead of recycled. This work will focus on the thermal and hydric properties of hemp concrete developed with metakaolin, to substitute hydraulic lime.

Keywords: Hemp · concrete · thermal conductivity · moisture buffering value

1 Problematic

Hemp concrete has a good thermal conductivity (λ) compared to other building materials such as oriented strand board or gypsum plasterboard. The order of magnitude of the thermal conductivity of this material is around 0.10 W/m/K in dry state, but can be affected by parameters such as its density, temperature or water content [1–4]. Literature on hemp concrete underlined that the mass proportions between the binder and the plant particles also affect the porosity, and thus density, thermal, and hydric properties. Different density for this material makes it adaptable: a low-density hemp concrete can be used for attic or wall insulation, inversely a high-density hemp concrete has better mechanical properties and makes it suitable for floor slabs or insulating wall coatings. This article focus on hemp concrete for thermal insulation. The goal is to develop a mixture without hydraulic binders to lower the embodied carbon of the material.

Most developed hemp concretes are based on hydraulic binders, which are largely available in Europe, for example, but are not locally accessible in Canada. Importing

S. Amziane et al. (Eds.): ICBBM 2023, RILEM Bookseries 45, pp. 442–449, 2023.
https://doi.org/10.1007/978-3-031-33465-8_34

hydraulic binders for the manufacture of hemp concrete would penalize the carbon footprint of this material, while it is one of its major benefits. In Canada hydrated lime is available to develop binders for hemp concrete. This type of lime sets by carbonation, and, at a young age, a proportion of the hydrated lime is not carbonated which does not allow to obtain the mix's cohesion. It is possible to address this challenge with the introduction of pozzolans or hydraulic materials (i.e., cements, natural cement and hydraulic limes with hydrated lime). In order to develop a viable solution for the manufacture of low carbon hemp concrete, it is then necessary to develop a formulation of local binder, adapted to Canada.

Metakaolin, a competitive pozzolan, is a potential candidate to ensure concrete cohesion and is available in North America. In order to optimize the curing of hemp concrete, it is necessary to understand the behaviour of the lime-metakaolin type mixture. This knowledge allows a better comprehension of the impact of introducing metakaolin on the thermal and hydric properties of hemp plant-based concretes. In this work, a thermal and hydric characterization of hemp concrete, manufactured in the laboratory with metakaolin, is presented and compared to the literature and conventional formulation. Other properties, such as mechanical performances of the samples can be found in the work of Fortin and are outside the scope of this article [5].

2 Materials and Method

2.1 Material

Hemp shive particles are harvested in Quebec in the Saguenay-Lac-Saint-Jean region. The hemp cultivar used is USO31 was provided by Agrofibre in Lavaltrie, Quebec, with a length between 180 and 220 cm. The plant was defibred before final bagging of 15 kg packed hemp shives. Table 1 presents the composition of the USO31 hemp fiber variety, and was taken from the literature [6]. The hemp has a bulk density of 125.2 kg/m^3 $\pm$ 2.9 kg/m^3 and an absolute density of 1336 kg/m^3 ($\pm$0.206%).

Table 1. Chemical composition of USO31 hemp fiber [6].

	Cellulose (m%)	Hemicelluloses (m%)	Pectin (m%)	Lignin (m%)
USO31	64–71	4–6	6–10	3–8

In North America, hydraulic lime is mainly accessible by imports from Europe. In this study, and for comparison purposes, a natural hydraulic lime from the company Secil, in Portugal (NHL 3.5) is used. The nomenclature "3.5" indicates that a mortar made with this lime will have a minimum strength of 3.5 MPa, according to the manufacturer's prescribed procedures. The hydraulic and carbonatation reactions by this type of lime allow plant concrete in a young age to attain that minimum mechanical resistance.

In Canada, the company Graymont can provide hydrated lime (CHX, *Graymont's High Calcium Lime)* which makes it a local available product, limiting the transportation and associated environmental impacts of this resource, necessary for the manufacture of hemp-based concrete.

Some distributors can supply metakaolin in eastern North America, among them the company *Poraver* is located in Ontario, Canada. The company's commercial product is called *Metapor* and is generally used nowadays as a supplementary cementing material in the manufacture of reinforced and unreinforced cement concrete elements. *Metapor* is mainly composed of an impure metakaolin and an addition of expanded recycled glass. The chemical composition of *Metapor* can be found in recent literature [7]. Theoretical calculation of the proportion of metakaolin (MK) and hydrated lime (CHX) for a total consumption of hydrated lime to occur during the process is based on Eq. (1), based on the work of Murat, 1983 [8].

$$\mathrm{AS_2 + 6CH + 9H_2O \rightarrow C_4AH_{13} + 2CSH} \tag{1}$$

This calculation of the proportion of meta kaolinite ($\mathrm{AS_2}$) constituents based on their molar mass are $\mathrm{Al_2O_3/AS_2} = 0.45899$ and $\mathrm{2SiO_2/AS_2} = 0.54101$, a detail can be found in Table 2.

Table 2. Calculation of stoichiometric ratios to determine the proportion of metakaolinite.

Molar masses of chemical elements:		
$H = 1.008g$	$O = 16.0g$	$Al = 26.98g$
$Si = 28.09g$	$Ca = 40.08g$	
Molar masses of constituents:		
$Al_2O_3 = 2 \cdot Al + 3 \cdot O = 101.96g$	$SiO_2 = Si + 2 \cdot O = 60.09g$	
$AS_2 = Al_2O_3 + 2 \cdot SiO_2 = 222.14g$		
Stoichiometric ratios:		
$\frac{Al_2O_3}{AS_2} = 0.45899$	$\frac{2 \cdot SiO_2}{AS_2} = 0.54101$	

We can note that $CH = Ca(OH)_2 = Ca + 2 \cdot (O + H) = 74.096g$, whose proportion in CaO is noted as $\%CaO = \frac{Ca+O}{Ca(OH)_2} = 0.757$.

According to the equation of Murat, 1983, the optimized proportion to totality consume the free hydrated lime is close to a 50%/50% ratio. Precisely, 54.8% calcium hydrated lime (CHX) and 45.2% metakaolin (MK). A summary of the binders used in this study is presented in Table 3.

Table 3. Binder used for the formulation of hemp concrete samples.

	MK-45	NHL
Hydrated lime (CHX)	54,8%wt	–
Metakaolin (MK)	45,2%wt	–
Natural hydraulic lime (NHL)	–	100%wt

For the hydric characterization, both formulations are studied, but for the measurement of thermal conductivity, only the formulation based on hydrated lime and metakaolin is considered. The formulation of the studied hemp concrete samples (MK-45) is presented in Table 4.

Table 4. Constituents for the formulation of hemp concrete samples submitted to thermal conductivity tests.

	MK-45
Dry hemp shives (kg)	1
Dry binder (kg)	1,5
Water (kg)	2,5
Dry density (kg/m^3)	326 ± 4

2.2 Method

2.2.1 Thermal Characterization

The hemp concrete is a heterogeneous material which does not allow a measurement with transient or modified transient plane source (TPS/MTPS) available in the laboratory. The thermal conductivity was thus determined with a needle probe and the Transient Line Source method (TLS, Fig. 1), with C-Therm trident, in accordance with ASTM D5334 and ASTM D5930 [9–11]. The sample was drilled with small holes, in different orientations, corresponding to the diameter of the probe and a series of 6 measurements is done. The thermal conductivity is determined using the temperature rise, which varies linearly with the logarithm of time, according to the Eq. (2). The procedure has an associated uncertainty of ± 3% + 0.02 W.m^{-1}.K^{-1} at 20 °C.

$$\lambda = \frac{C.q}{4.\pi.L.\Delta T}.\ln\left(\frac{t2}{t1}\right) \tag{2}$$

With:

- C a calibration factor,
- q the heat flux,

- L the length of the probe,
- ΔT the temperature, and
- t the time.

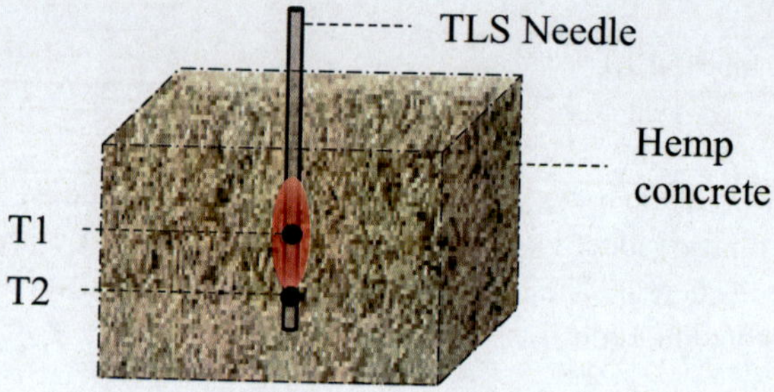

Fig. 1. C-Therm (left) and TLS needle principle (right).

2.2.2 Hydric Behaviour

In order to evaluate the moisture buffering value (MBV) of the samples, the NORDTEST method is used [12]. This method consists of evaluating the moisture absorption and desorption of samples through humidity cycles. The samples are sealed on every surface, except one (with a surface area of at least 0.01 m^2) which is exposed to the controlled environment. Before the tests, the sample are conditioned, and their mass are measured with a balance (with a precision of ±0.01 g). The samples are then submitted to successive high and low humidity cycles, at ambient temperature (23 ± 2 °C). Between each cycle, the mass of the samples is measured and moisture buffering value can be calculated with Eq. (3).

$$MBV = \frac{\Delta m}{A \cdot (RH_{high} - RH_{low})} \quad (3)$$

With:

- m the weight of the sample,
- A the exposed surface,
- RH_{high} and RH_{low} are respectively the high and low relative humidity of the cycle.

The velocity and the air film above the samples are neglected and may have an impact on the actual moisture content of the material. The humidity cycles are performed in a commercial *Thermotron* environmental chamber (Fig. 2).

Fig. 2. Moisture Buffering Value Test in Thermotron environmental chamber.

3 Results

3.1 Thermal Conductivity

Thermal conductivity measurements led to values between 0.097 and 0.105 W/m/K for hemp concrete manufactured with metakaolin and a mean of 0.102 ± 0.004 W/m/K. It corresponds to the measured values of hemp concrete in literature for this material [13], which underline that the thermal performance is not impacted by the introduction of metakaolin. With respect to commonly used mixes, this mix is adapted and confirms the observations of the literature [1, 2]. Furthermore, the impact of introducing metakaolin in the mix, leads to similar observations of Walker and Pavía, as no major impact is observed [14]. As already mentioned, in the case of hemp concrete, thermal conductivity can be dependent on several parameters, other tests would allow determining fiber orientation and compaction impacts. As an example, Zerrouki et al. measured thermal conductivity between 0.062 and 0.125 W/m/K with the consideration of these parameters [15]. The measurement probe was introduced in different orientations, however low standard deviations were calculated (around 4%), which remains relatively low as it would have been expected.

3.2 Moisture Buffering Value

Figure 3 shows the results obtained from the MBV measurements, for the two binders under study (MK-45 and NHL). The calculated averages for the last three moisture cycles (8 to 10), are in the order of 2.64 and 2.90 $g.m^{-2}.\%RH^{-1}$ for the hemp concretes formulated with the metakaolin (MK) and the hydraulic binder respectively (NHL). The values obtained for the metakaolin-based sample are slightly lower. However both formulations can be qualified as excellent hydric regulators. These results obtained for this material corroborate with Latif et al. (2015) who attempted the test also on a hemp plant concrete rich in hydrated lime. Thus, the binder formulated with metakaolin seems to be a very good alternative for the manufacture of hemp concrete with a low carbon footprint in Canada.

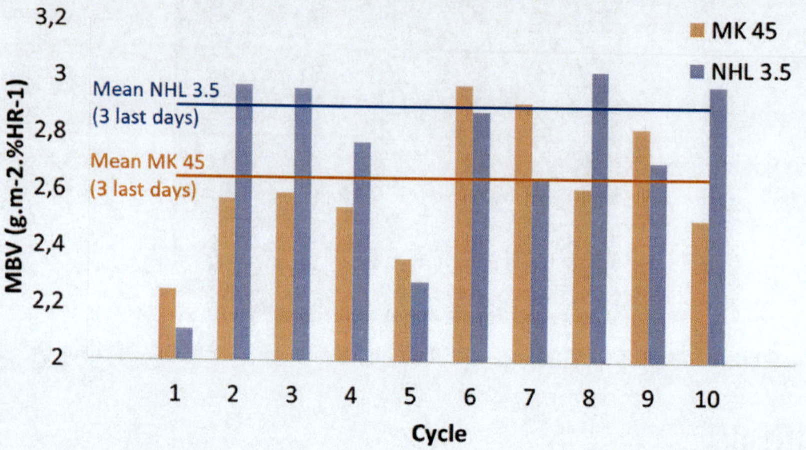

Fig. 3. Moisture Buffering Value measured through each cycle for hemp concrete with NHL and MW

4 Conclusion

The important challenge of this work was to develop a formulation of local binder to replace hydraulic lime. In this study, the thermal conductivity and the moisture buffering values of different samples of hemp concrete were studied. The thermal conductivity of the manufactured hemp concrete samples with metakaolin was in the same range as other hemp concrete presented in the literature. Considering the hydric performance of the formulation, the samples manufactured with metakaolin were a bit less performant as hemp concrete produced with conventional hydraulic binder. The manufactured hemp concrete formulated with a local binder is adapted for the building sector.

References

1. Barbhuiya, S., Das, B.B.: A comprehensive review on the use of hemp in concrete. Constr. Build. Mater. **341**, 127857 (2022). https://doi.org/10.1016/j.conbuildmat.2022.127857
2. Nguyen, T.T., Picandet, V., Carre, P., Lecompte, T., Amziane, S., Baley, C.: Effect of compaction on mechanical and thermal properties of hemp concrete. Eur. J. Environ. Civil Eng. **14**(5), 545–560 (2011). https://doi.org/10.1080/19648189.2010.9693246

3. Arnaud, L., Cérézo, V.: Qualification physique des matériaux de construction à base de chanvre. CNRS, Lyon, Rapport final de recherche 0711462 (2001)
4. Evrard, A., De Herde, A.: Bioclimatic envelopes made of lime and hemp concrete. In: Présenté à CISBAT 2005 - Renewables in a Changing Climate - Innovation in Building Envelopes and Environmental Systems (2005). https://dial.uclouvain.be/pr/boreal/object/boreal:73818
5. Fortin, P.: Formulation d'un liant pouzzolanique pour la confection d'un béton végétal hygroscopique. École de Technologie Supérieure, Montréal, Mémoire en Génie de la Construction (M. Sc. A.) (2022)
6. Mazian, B.: Approche intégrée du procédé de rouissage des fibres de chanvre : Vers une amélioration de la qualité des intrants pour la fabrication des matériaux biocomposites. phdthesis, IMT - MINES ALES - IMT - Mines Alès Ecole Mines – Télécom (2018). Consulté le 15 novembre 2022. https://tel.archives-ouvertes.fr/tel-03019451
7. Brylewska, K., Rożek, P., Król, M., Mozgawa, W.: The influence of dealumination/desilication on structural properties of metakaolin-based geopolymers. Ceram. Int. **44**(11), 12853–12861 (2018). https://doi.org/10.1016/j.ceramint.2018.04.095
8. Murat, M.: Hydration reaction and hardening of calcined clays and related minerals. I. Preliminary investigation on metakaolinite. Cem. Concr. Res. **13**(2), 259–266 (1983). https://doi.org/10.1016/0008-8846(83)90109-6
9. ASTM D5334: Standard Test Method for Determination of Thermal Conductivity of Soil and Rock by Thermal Needle Probe Procedure (2022). https://doi.org/10.1520/D5334-22
10. ASTM D5930: Standard Test Method for Thermal Conductivity of Plastics by Means of a Transient Line-Source Technique (2017). https://doi.org/10.1520/D5930-17
11. IEEE 442: Guide for Soil Thermal Resistivity Measurements. IEEE Stand. Assoc. (1981). https://doi.org/10.1109/IEEESTD.1981.81018
12. Rode, C., et al.: Nordtest project on moisture buffer value of materials, p. 7 (2005)
13. Seng, B., Magniont, C., Lorente, S.: Characterization of a precast hemp concrete. Part I: physical and thermal properties. J. Build. Eng. **24**, 100540 (2019). https://doi.org/10.1016/j.jobe.2018.07.016
14. Walker, R., Pavía, S.: Moisture transfer and thermal properties of hemp–lime concretes. Constr. Build. Mater. **64**, 270–276 (2014). https://doi.org/10.1016/j.conbuildmat.2014.04.081
15. Zerrouki, R., Benazzouk, A., Courty, M., Hamed, H.B.: Potential use of matakaolin as a partial replacement of preformulated lime binder to improve durability of hemp concrete under cyclic wetting/drying aging. Constr. Build. Mater. **333**, 127389 (2022). https://doi.org/10.1016/j.conbuildmat.2022.127389

Assessing Hygrothermal Parameters of Plant-Based Building Materials for Simulation: A Mini Review

Amer Bakkour(✉), Salah-Eddine Ouldboukhitine, Pascal Biwole, and Sofiane Amziane

Université Clermont Auvergne, Clermont Auvergne INP, CNRS, Institut Pascal, 63000 Clermont–Ferrand, France
amer.bakkour.95@gmail.com

Abstract. Recently, plant-based materials have gained significant attention in construction and renovation projects due to their eco-friendly nature, superior thermal insulation properties, and humidity regulation capabilities. However, accurately predicting the hygrothermal behavior of these materials in various environmental conditions can be challenging due to their hygroscopic and anisotropic nature. The difficulty in implementing hygrothermal transfer models lies in the lack of knowledge of the intrinsic characteristics of the plant materials. The objective of this paper is to provide an up-to-date review of the hygrothermal characteristics of plant materials to efficiently implement them in simulation tools. The review begins by discussing the characteristics of raw plant fibers and aggregates and then moves on to their characteristics at the composite material scale.

Keywords: Plant-based materials · Hygrothermal parameters · Simulation

1 Introduction

The high energy consumption in buildings has become a significant environmental concern. Buildings are one of the highest energy-consuming sectors, accounting for almost 36% of the world's energy consumption and responsible for 37% of CO_2 emissions. The majority of this energy is utilized to maintain indoor hygrothermal comfort through space heating [1]. To overcome this energy problem, the focus has been paid recently on using plant-based materials in buildings to enhance energy performance and reduce energy demand.

Depending on the formulation used, plant-based materials can be used as thermal insulators and in concretes. New composite materials based on plant fibers and aggregates have been developed for the construction sector, such as hemp [2, 3], straw [4, 5], corn stalk [6], rice husk [7], etc.

The literature showed that plant-based materials have interesting hygrothermal properties. They possess a significantly porous microstructure, which enables them to retain a substantial quantity of moisture and exchange water vapor with the surroundings.

S. Amziane et al. (Eds.): ICBBM 2023, RILEM Bookseries 45, pp. 450–464, 2023.
https://doi.org/10.1007/978-3-031-33465-8_35

Additionally, they contribute to the energy efficiency of constructions by providing high thermal resistivity. Besides, plant-based materials offer several benefits compared to traditional materials, including lower environmental and carbon footprint, and the availability of renewable raw resources. For instance, straw bale walls have a negative "global warming potential (GWP)" as the straw absorbs CO_2 during its lifespan, whereas brick walls with Styrofoam insulation have an exceedingly high global warming potential of 435.02 kg $CO_2.m^{-2}$ [8]. Moreover, hemp concrete has a grey energy value of 90 $kWh.m^{-3}$, which includes the energy required for extraction, production, packaging, and transportation, whereas concrete and cellular concrete have grey energy values of 500 $kWh.m^{-3}$ and 200 $kWh.m^{-3}$, respectively [9]. As a result, the use of plant-based materials becomes an inevitable solution to address the present environmental issues.

As energy efficiency becomes a more critical aspect of building design, numerical simulation is increasingly used to predict the hygrothermal transfer within plant-based walls and the energy efficiency of plant-based buildings. However, the lack of knowledge of the intrinsic characteristics of porous plant materials makes it highly challenging to implement the input parameters in transfer models. Therefore, it is essential to understand the hygrothermal properties of plant materials and their dependency on temperature and relative humidity variations to better predict their hydrothermal behaviors. This can provide useful input parameters to be implemented in numerical modeling tools and simulation software for thermal and hydric studies.

The objective of this paper is to provide a comprehensive overview of the current state of knowledge on the hygrothermal properties of plant-based materials. First, an analysis of the hygrothermal properties of raw plant fibers and aggregates is summarized. Then, the intrinsic properties of various plant-based concretes and plant-based insulation boards are presented.

2 Raw Plant Fibers and Aggregates

To distinguish between fibers, aggregates, and straw particles, Fig. 1 illustrates two geometrical dimensions: minor axis and the inverse aspect ratio (1/AR), which is the ratio of major to minor axis. Fibers exhibit the smallest 1/AR (lower than 0.017), whereas aggregates have the highest 1/AR values (greater than 0.10). The 1/AR values of straw are between those of fibers and aggregates (between 0.01 and 0.3).

2.1 Thermal Properties of Raw Plant Materials

The relationship between thermal conductivity and density varies between plant fibers and aggregates. For plant fibers, the thermal conductivity initially decreases with increasing density until it reaches a critical point, which depends on the fiber manufacturing process [28]. Once this critical point is reached, the thermal conductivity of fibers increases with density. Figure 2 displays how the thermal conductivity of various plant fibers changes with density.

The reason for this phenomenon is that when the density is low, heat transfer is mainly facilitated by convection and radiation while conduction heat transfer is restricted due to the low number of fibers. However, as the density increases, the impact of radiation

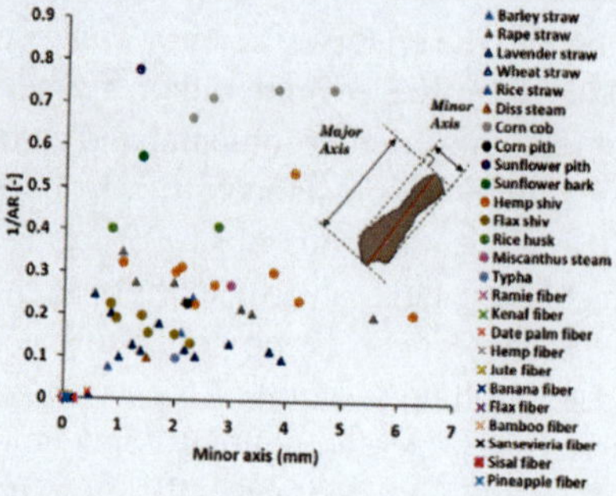

Fig. 1. Inverse aspect ratio (1/AR) of various plant raw materials versus their minor axis [6, 7, 10–27].

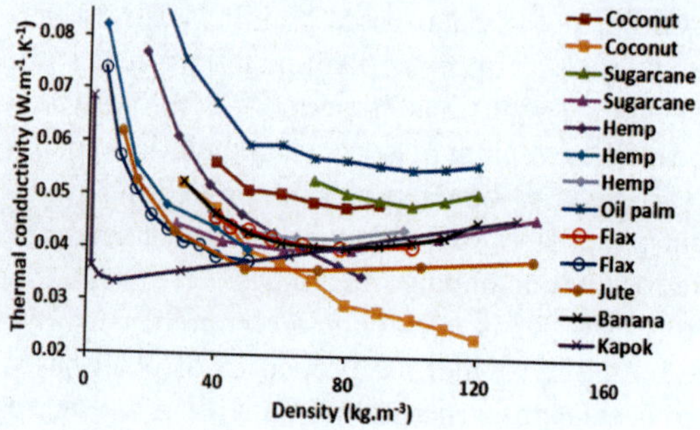

Fig. 2. Thermal conductivity of various raw plant fibers as a function of density [29–35].

and convection becomes less significant, and radiation heat transfer becomes negligible at the critical point [29, 36]. At this point, conductive heat transfer becomes dominant, leading to an increase in thermal conductivity.

Further, Fiber length, temperature, and mechanical processing can also affect the thermal conductivity of fibers. For instance, long flax and hemp fibers have lower thermal conductivity compared to chopped ones due to an increase in empty spaces between particles [30]. Additionally, the study carried out by [28] discovered that the mechanical processing of hemp and flax fibers not only alters their density but also changes their microstructure. Lastly, the experiment performed by [31] showed that the thermal conductivity of plant fibers increases significantly with temperature.

In contrast, the thermal conductivity of plant aggregates increases with the increase of density, as presented in Fig. 3. The thermal conductivity is greatly influenced by the type of raw aggregates, which varies from 0.036 to 0.096 $W.m^{-1}.K^{-1}$ for densities between 15.9 to 497 $kg.m^{-3}$.

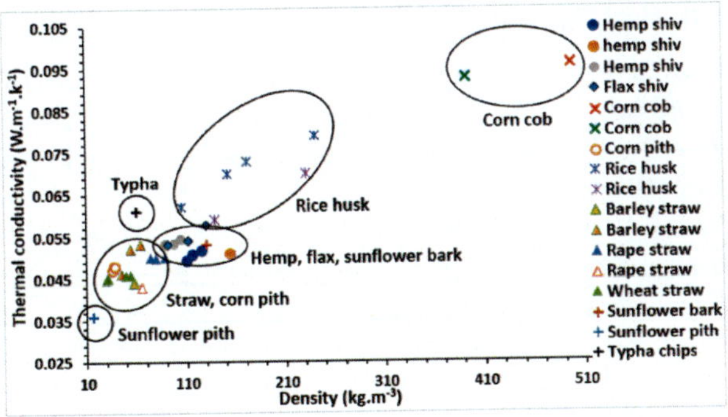

Fig. 3. Thermal conductivity of various plant aggregates as a function of density [10, 13, 37–41].

2.2 Hydric Properties of Raw Plant Materials

2.2.1 Water Absorption Kinetics

RILEM TC 236-BBM [42] recommends the immersed water method for assessing the water absorption capacity of plant fibers and aggregates. This method offers valuable information about their hygroscopic behavior during the first few minutes of exposure to water or a binder mixture, as well as when they reach saturation.

Figure 4 shows the water absorption rates of various raw plant fibers and aggregates. Aggregates show a two-stage water absorption process. The initial stage involves rapid water absorption through capillary action due to their hydrophilic properties and takes from seconds to minutes depending on the material's porosity. In contrast, the second stage is slower and may take hours or days to reach saturation. Plant fibers, on the other hand, also exhibit a two-stage water absorption behavior similar to that of plant aggregates, but due to their lower porosity, the absorption rate and capacity are comparatively slower.

Fig. 4. Water absorption of plant aggregates (on left) [7, 10, 41, 43–46] and fibers (on rights) [47] as a function of time.

2.2.2 Vapor Sorption Capacity

Figure 5 shows that all plant-based materials, fibers and aggregates, demonstrate a significant increase in moisture content at high relative humidity levels (above 80%) compared to low and medium levels. The sorption isotherms also exhibit a hysteresis throughout the entire range of relative humidity.

The mechanisms of moisture absorption in plant materials are not fully understood, and further research is needed for the development of plant-based composite insulation materials. The Literature shows that plant aggregates have higher surface area and porosity than fibers, but the moisture content evolution is similar for both, except for sunflower and corn bark and pith. This suggests that surface area and pores in aggregates do not significantly affect moisture adsorption.

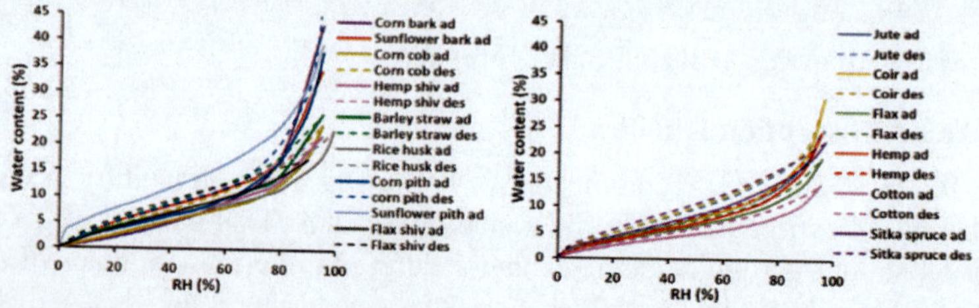

Fig. 5. Vapor sorption capacity of plant aggregates (on left) [10, 39, 45, 48, 49] and fibers (on right) [49, 50].

3 Plant-Based Composite Materials

3.1 Plant-Based Insulation Boards

Plant-based boards have been developed as an alternative to inorganic and synthetic composites for thermal insulation. These materials can be combined with natural or synthetic binders or made without any chemical binders through a hot-pressing process. They possess good hygrothermal behavior, allowing them to regulate temperature and humidity.

3.1.1 Thermal Characterization

The thermal performance of insulation materials is mainly determined by thermal conductivity. The relationship between thermal conductivity and density is clear, and the thermal conductivity of boards is affected by pressure and binder content. Heavier boards have lower insulation efficiency due to fewer air-filled voids[51, 52]. The temperature has a significant impact on thermal conductivity, with an increase in temperature leading to an increase in thermal conductivity [53]. Humidity also affects thermal insulation properties, leading to an increase in thermal conductivity. Studies conducted on different plant-based insulation boards showed that the thermal insulation properties were negatively impacted by humidity [54, 55].

In Table 1, the thermal properties of various plant-based insulation boards are presented. The majority of these boards fall within the optimal range for thermal insulation, with conductivity values ranging from 0.02 to 0.065 $W.m^{-1}.K^{-1}$. These values meet the requirements set by Turkish and German institutions [56], as well as Portuguese thermal regulation [57].

Table 1. Hygrothermal properties of various plant-based insulation boards [54].

Plant materials	Composition	Thickness of board (mm)	Density ($kg.m^{-3}$)	Thermal conductivity ($W.m^{-1}.K^{-1}$)	Thermal capacity ($kJ.kg^{-1}.K^{-1}$)	Water vapor resistance [-]
Hemp fiber	Hemp fiber, adhesive, additives	40	41.1	0.0353	1.5	2.3–3
Corn pith	Corn (89%), sodium alginate (11%)	40	48.1	0.0316	0.858	3.2–5.7
Barley straw	Straw (81%), corn starch (19%)	40	107.5	0.0380	0.633	3.4–3.9
Wood fiber	Wood fiber, additives	40	212.2	0.0536	0.836	4.2–6.5
Hemp shiv	Hemp shiv (50%), lime (50%)	40	286	0.0557	0.42	2.5–3

3.1.2 Hydric Characterization

When evaluating the performance and durability of plant-based insulation materials, it's essential to consider both water vapor diffusion resistance and water vapor absorption. These factors have an inverse relationship, with lower diffusion resistance leading to higher absorption. Table 1 also shows the water vapor diffusion resistance values for various plant-based insulation materials, which are generally low, indicating their hydrophilic nature and high-water vapor absorption capacity. This is supported by the study of [55] that found significant water absorption capacity for different plant boards. Additionally, the type of binder used can also impact water absorption, with hydrophobic binders limiting moisture absorption and hydrophilic binders enhancing it [58].

3.2 Plant-Based Concretes

Plant-based concretes have high porosity, which provides them with thermal insulation and sound-absorbing properties, as well as the ability to exchange water vapor with surroundings. Additionally, they contribute to indoor comfort and enhance building energy performance. However, density, water content, temperature, and relative humidity are factors that can affect the hygrothermal properties of plant-based concretes.

3.2.1 Thermal Properties

Thermal conductivity

Typically, the thermal conductivity of plant-based materials increases as their density increases (Fig. 6). Their low thermal conductivity is due to their porous structure. The thermal conductivity of this type of concretes varies from 0.058 $W.m^{-1}.K^{-1}$ to 0.21 $W.m^{-1}.K^{-1}$ for density ranges between 212 $kg.m^{-3}$ and 1048 $kg.m^{-3}$. For straw bale, the conductivity ranges between 0.0615–0.079 $W.m^{-1}.K^{-1}$ for a density range of 68.1–122.7 $kg.m^{-3}$.

The linear relationship between the dry density and the thermal conductivity of plant-based concretes can be written as the following equation:

$$\lambda = 1.836.10^{-4}.\rho + 0.01647\left(R^2 = 0.8501\right) \tag{1}$$

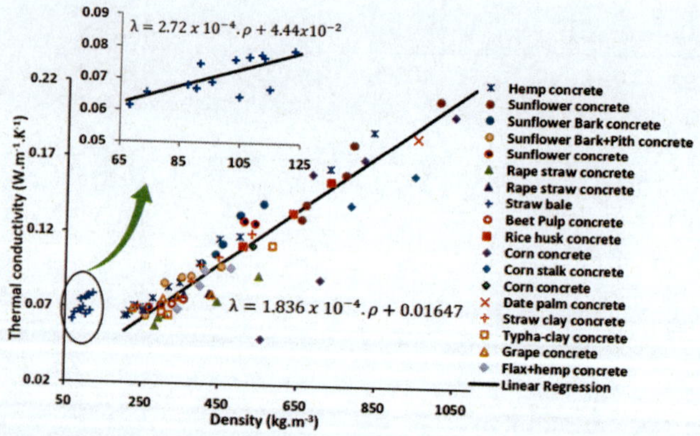

Fig. 6. Thermal conductivity of various plant-based concretes vs dry density [7, 45, 59–71].

In addition, the thermal conductivity of plant-based concretes is negatively affected by RH and T(°C). Figure 7 illustrates the changes in thermal conductivity concerning water content and temperature for hemp-lime, straw rape-lime, and flax-lime concretes. The relationship between thermal conductivity, water content, and temperature can thus be expressed as follows:

$$\lambda = a + b.T(^{\circ}C) + c.w(kg.kg^{-1}) \tag{2}$$

where, the coefficients a, b, and c must be determined experimentally.

Specific heat capacity, thermal effusivity, and thermal diffusivity

In fact, the thermal properties of plant-based concretes are very similar. Table 2 compares the thermal properties of different vegetable concretes in the dry state with those of commonly used materials in construction

.

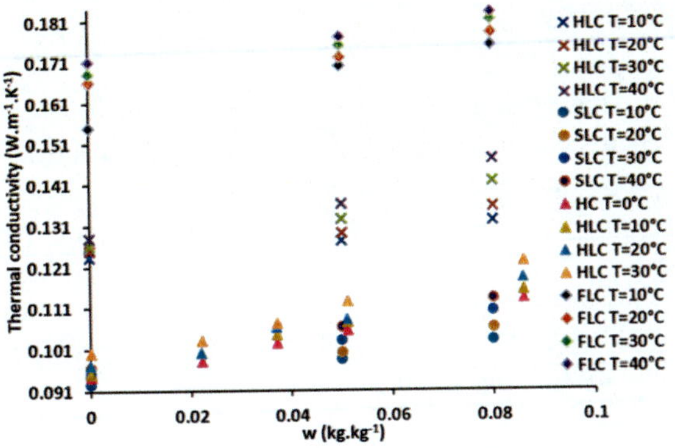

Fig. 7. The evoution of thermal conductivity versus temperature and water content for hemp [72, 73], straw [72], and flax concretes [72].

Table 2. Overview of thermal properties of plant concretes vs. traditional concretes.

Material	Density kg.m^{-3}	Specific heat J.kg^{-1}.K^{-1}	Thermal conductivity W.m^{-1}.K^{-1}	Thermal diffusivity 10^{-7} m^2.s^{-1}	Thermal effusivity J.m^{-2}.K^{-1}.s$^{-1/2}$
Typha concrete [69]	323	958	0.065	2.1	142.2
Beet pulp concrete [64]	360	1430	0.0757	1.47	196.76
Straw rape concrete [72]	462	1168	0.094	1.74	225
Hemp concrete [72]	478	1109	0.123	2.33	256
Flax concrete [72]	598	1144	0.168	2.46	339
Cellular concrete	600	850	0.14	2.7	267.2
C12/C25 concrete	2200	850	1.60	8.6	1729.7

3.2.2 Hydric Properties

Sorption isotherm

Vegetable concretes have a higher capacity to store moisture compared to other building materials, which helps dampen fluctuations in RH. Figure 8 shows sorption curves for various plant-based concretes, which have almost identical curves up to 80% relative humidity. The phenomenon of hysteresis can persist throughout the entire range of relative humidity.

Different models have been proposed to describe the sorption isotherms of plant-based materials, including the GAB, Merakeb, and Van Genuchten models. The GAB model is commonly used in literature.

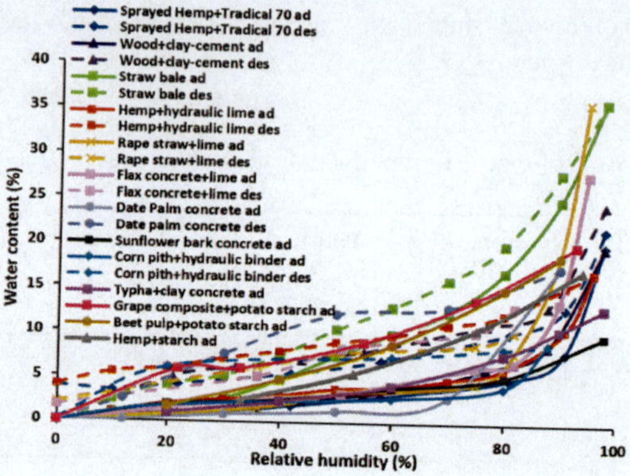

Fig. 8. Sorption isotherm of different vegetable concretes [45, 64, 69, 70, 74–82].

Vapor permeability

Since plant-based materials have a high capacity to absorb moisture, it is important to assess the evolution of water vapor permeability as a function of water content in order to accurately predict hygrothermal transfer in these materials. The vapor permeability of hygroscopic materials is significantly influenced by ambient humidity and increases exponentially as humidity levels increase. However, only few studies have addressed this problem. [83] showed that the exponential relationship between vapor permeability and water content of hempcrete can be written as follows:

$$\delta_p(w) = 3.67.10^{-10}\left(w_{adsorption}\right)^{0.488} \tag{3}$$

Additionally, the power relationship for wood concrete proposed by [84] can be expressed as follows:

$$\delta_p(w) = \delta_0 \exp\left(k.w_{ads}\right) \tag{4}$$

where, $\delta_0 = 2.19.\ 10^{-12}$ and $k = 36$.

Table 3 presents the water vapor permeability of various plant-based concretes.

Table 3. Vapor permeability of vegetable concretes.

Material	Density kg.m^{-3}	Vapor permeability (x 10^{-11}) kg.m^{-1}.s^{-1}.Pa^{-1}
Hemp-lime concrete [81]	478	2.23
Straw-lime concrete [81]	487	2.56
Flax-lime concrete [80]	598	2.34
Typha-clay concrete [69]	304–586	2.83–6.15
Corn stalk-cement concrete [6]	546.13	1.263
Grape-starch concrete [70]	227–433	1.4–1.79
Date palm-cement concrete [79]	–	3.16–3.59
Sunflower-lime concrete [45]	539.64	7.67
Beet pulp-starch concrete [64]	271.4–360	0.686–0.89

4 Conclusion

The analysis of 79 articles revealed that the hygrothermal properties of plant-based materials are not constant, but rather depending on several parameters such as density, temperature, and water content, making the evaluation of hygrothermal transfer more complex. However, there are still several gaps in the literature that need to be addressed in future research. For the majority of plant-based composites, the thermal conductivity, thermal capacity, effusivity, and diffusivity have not been studied under varying temperature and RH conditions, except for hemp. Therefore, further investigations should be conducted on various materials to expand our knowledge in this field. Further, the water vapor permeability of plant-based materials can vary depending on ambient humidity and therefore the water content of the material, so it is crucial to take the evolution of water content into account. In addition, the choice between using fibers or aggregates for thermal insulation is not fully understood. Both materials provide similar thermal performance in terms of thermal resistivity. Therefore, further research is necessary to examine the thermal diffusivity, effusivity, and moisture buffer value (MBV) of plant insulation boards.

References

1. U.N.E. Programme. 2021 global status report for buildings and construction. https://globalabc.org/resources/publications/2021-global-status-report-buildings-and-construction
2. Benkhaled, M., Ouldboukhitine, S.-E., Bakkour, A., Amziane, S.: Sensitivity analysis of the parameters for assessing a hygrothermal transfer model HAM in bio-based hemp concrete material. Int. Commun. Heat Mass Transfer **132**, 105884 (2022)
3. Benkhaled, M., Ouldboukhitine, S.-E., Bakkour, A., Amziane, S.: A 1D model for predicting heat and moisture transfer through a hemp-concrete wall using the finite-element method. Materials **14**(22), 6903 (2021)

4. Tlaiji, G., Pennec, F., Ouldboukhitine, S., Ibrahim, M., Biwole, P.: Hygrothermal performance of multilayer straw walls in different climates. Constr. Build. Mater. **326**, 126873 (2022)
5. Tlaiji, G., Ouldboukhitine, S., Pennec, F., Biwole, P.: Thermal and mechanical behavior of straw-based construction: a review. Constr. Build. Mater. **316**, 125915 (2022)
6. Ahmad, M.R., Chen, B.: Influence of type of binder and size of plant aggregate on the hygrothermal properties of bio-concrete. Constr. Build. Mater. **251**, 118981 (2020)
7. Chabannes, M., Bénézet, J.-C., Clerc, L., Garcia-Diaz, E.: Use of raw rice husk as natural aggregate in a lightweight insulating concrete: an innovative application. Constr. Build. Mater. **70**, 428–438 (2014)
8. Krasny, E., Klarić, S., Korjenić, A.: Analysis and comparison of environmental impacts and cost of bio-based house versus concrete house. J. Clean. Prod. **161**, 968–976 (2017)
9. Mon habitat positif. https://monhabitatpositif.com/blog/constructeur-maison-bois/ecologique/energie-grise-batiment/
10. Laborel-Préneron, A., Magniont, C., Aubert, J.-E.: Characterization of barley straw, hemp shiv and corn cob as resources for bioaggregate based building materials. Waste Biomass Valor 9(7), 1095–1112 (2018)
11. Abbas, M.S., McGregor, F., Fabbri, A., Ferroukhi, M.Y.: The use of pith in the formulation of lightweight bio-based composites: Impact on mechanical and hygrothermal properties. Constr. Build. Mater. **259**, 120573 (2020)
12. Giroudon, M., Laborel-Préneron, A., Aubert, J.-E., Magniont, C.: Comparison of barley and lavender straws as bioaggregates in earth bricks. Constr. Build. Mater. **202**, 254–265 (2019)
13. Viel, M., Collet, F., Lanos, C.: Chemical and multi-physical characterization of agro-resources' by-product as a possible raw building material. Ind. Crops Prod. **120**, 214–237 (2018)
14. Sellami, A., Merzoud, M., Amziane, S.: Improvement of mechanical properties of green concrete by treatment of the vegetals fibers. Constr. Build. Mater. **47**, 1117–1124 (2013)
15. Maddison, M., Mauring, T., Kirsimäe, K., Mander, Ü.: The humidity buffer capacity of clay–sand plaster filled with phytomass from treatment wetlands. Build. Environ. **44**(9), 1864–1868 (2009)
16. Arufe, S., Hellouin de Menibus, A., Leblanc, N., Lenormand, H.: Physico-chemical characterisation of plant particles with potential to produce biobased building materials. Ind. Crops Prod. **171**, 113901 (2021)
17. Ming-Zhu, P., Chang-Tong, M., Xu-Bing, Z., Yun-Lei, P.: Effects of rice straw fiber morphology and content on the mechanical and thermal properties of rice straw fiber-high density polyethylene composites. J. Appl. Polym. Sci. **121**(5), 2900–2907 (2011)
18. Chin, C.S., Nepal, B.: Material properties of agriculture straw fibre-reinforced concrete. In: Ecological Wisdom Inspired Restoration Engineering, pp. 109–120 (2019)
19. Kalita, D.B.B., Gogoi, D.N., Kalita, S.: Properties of ramie and its blends. Int. J. Eng. Res. Gener. Sci. **1**(2), 1–6 (2013)
20. Millogo, Y., Morel, J.-C., Aubert, J.-E., Ghavami, K.: Experimental analysis of Pressed Adobe Blocks reinforced with Hibiscus cannabinus fibers. Constr. Build. Mater. **52**, 71–78 (2014)
21. Wang, B., Sain, M., Oksman, K.: Study of structural morphology of hemp fiber from the micro to the nanoscale. Appl Compos Mater **14**(2), 89–103 (2007)
22. Islam, M.S., Ahmed, S.J.: Influence of jute fiber on concrete properties. Constr. Build. Mater. **189**, 768–776 (2018)
23. Kulkarni, A.G., Satyanarayana, K.G., Rohatgi, P.K., Vijayan, K.: Mechanical properties of banana fibres (Musa sepientum). J. Mater. Sci. **18**(8), 2290–2296 (1983)
24. Joffe, R., Andersons, J., Wallström, L.: Strength and adhesion characteristics of elementary flax fibres with different surface treatments. Compos. A Appl. Sci. Manuf. **34**(7), 603–612 (2003)

25. Yu, Y., Wang, H., Lu, F., Tian, G., Lin, J.: Bamboo fibers for composite applications: a mechanical and morphological investigation. J. Mater. Sci. **49**(6), 2559–2566 (2013). https://doi.org/10.1007/s10853-013-7951-z
26. Widyasanti, A., Napitupulu, L.O.B., Thoriq, A.: Physical and mechanical properties of natural fiber from Sansevieria trifasciata and Agave sisalana. IOP Conf. Ser.: Earth Environ. Sci. **462**(1), 012032 (2020)
27. Johny, V., Mani, A.K., Palanisamy, S., Rajan, V.K., Palaniappan, M., Santulli, C.: Extraction and physico-chemical characterization of pineapple crown leaf fibers (PCLF). Fibers **11**(1), 1 (2023)
28. Stapulionienė, R., Vaitkus, S., Vėjelis, S., Sankauskaitė, A.: Investigation of thermal conductivity of natural fibres processed by different mechanical methods. Int. J. Precis. Eng. Manuf. **17**(10), 1371–1381 (2016). https://doi.org/10.1007/s12541-016-0163-0
29. Finck, J.L.: Mechanism of heat flow in fibrous materials. Bur. Stand. J. Res. **5**(5), 973–984 (1930)
30. Vaitkus, S., Karpavičiūtė, R., Vėjelis, S., Lekūnaitė, L.: Development and research of thermal insulation materials from natural fibres. In: Key Engineering Materials, 285–288. Trans Tech Publications Ltd. (2014)
31. Manohar, K.: Experimental investigation of building thermal insulation from agricultural by-products. Brit. J. Appl. Sci. Technol. **2**(3), 227 (2012)
32. Kosiński, P., Brzyski, P., Szewczyk, A., Motacki, W.: Thermal properties of raw hemp fiber as a loose-fill insulation material. J. Nat. Fibers **15**(5), 717–730 (2018)
33. Wieland, H., Murphy, D.P.L., Behring, H., Jäger, C., Hinrichs, P., Bockisch, F.-J.: Perspektiven für Dämmstoffe aus heimischen nachwachsenden Rohstoffen. LANDTECHNIK **55**(1), 1 (2000)
34. Manohar, K., Adeyanju, A.A.: A comparison of banana fiber thermal insulation with conventional building thermal insulation. Brit. J. Appl. Sci. Technol. **17**, 1–9 (2016)
35. Bui, H., Sebaibi, N., Boutouil, M., Levacher, D.: Determination and review of physical and mechanical properties of raw and treated coconut fibers for their recycling in construction materials. Fibers **8**(6), 6 (2020)
36. Manohar, K., Kochhar, G.S.: Experimental investigation of the influence of air conduction on heat transfer across fibrous materials. J. Mech. Eng. Res. **3**(9), 319–324 (2011)
37. Kosiński, P., Brzyski, P., Tunkiewicz, M., Suchorab, Z., Wiśniewski, D., Palczyński, P.: Thermal properties of hemp shives used as insulation material in construction industry. Energies **15**(7), 7 (2022)
38. Chabi, E. : Étude de la formulation et des propriétés mécaniques et thermiques du béton de balles de riz. Université de Lorraine and Université d'Abomey-Calavi (Bénin) (2017)
39. Fernández, M.P.: Contribution to the development of new bio-based thermal insulation materials made from vegetal pith and natural binders: hygrothermal performance, fire reaction and mould growth resistance. University Politècnica de Catalunya (2015)
40. Brouard, Y., Belayachi, N., Hoxha, D., Ranganathan, N., Méo, S.: Mechanical and hygrothermal behavior of clay – Sunflower (Helianthus annuus) and rape straw (Brassica napus) plaster bio-composites for building insulation. Constr. Build. Mater. **161**, 196–207 (2018)
41. Ba, L., et al.: Experimental investigation of thermal and mechanical properties of clay reinforced with typha australis: influence of length and percentage of fibers. Waste Biomass Valor **12**(5), 2723–2737 (2021)
42. Amziane, S., Collet, F., Lawrence, M., Magniont, C., Picandet, V., Sonebi, M.: Recommendation of the RILEM TC 236-BBM: characterisation testing of hemp shiv to determine the initial water content, water absorption, dry density, particle size distribution and thermal conductivity. Mater. Struct. **50**(3), 1–11 (2017). https://doi.org/10.1617/s11527-017-1029-3

43. Nozahic, V., Amziane, S., Torrent, G., Saïdi, K., De Baynast, H.: Design of green concrete made of plant-derived aggregates and a pumice–lime binder. Cement Concr. Compos. **34**(2), 231–241 (2012)
44. Bouasker, M., Belayachi, N., Hoxha, D., Al-Mukhtar, M.: Physical characterization of natural straw fibers as aggregates for construction materials applications. Materials **7**(4), 4 (2014)
45. Lagouin, M., Magniont, C., Sénéchal, P., Moonen, P., Aubert, J.-E., Laborel-préneron, A.: Influence of types of binder and plant aggregates on hygrothermal and mechanical properties of vegetal concretes. Constr. Build. Mater. **222**, 852–871 (2019)
46. Magniont, C., Escadeillas, G., Coutand, M., Oms-Multon, C.: Use of plant aggregates in building ecomaterials. Eur. J. Environ. Civ. Eng. **16**(1), s17–s33 (2012)
47. Célino, A., Fréour, S., Jacquemin, F., Casari, P.: Characterization and modeling of the moisture diffusion behavior of natural fibers. J. Appl. Polym. Sci. **130**(1), 297–306 (2013)
48. Follain, N., et al.: Hydrophobic surface treatments of sunflower pith using eco-friendly processes. Cellulose **22**(1), 245–259 (2014). https://doi.org/10.1007/s10570-014-0490-1
49. Jiang, Y., Lawrence, M., Hussain, A., Ansell, M., Walker, P.: Comparative moisture and heat sorption properties of fibre and shiv derived from hemp and flax. Cellulose **26**(2), 823–843 (2018). https://doi.org/10.1007/s10570-018-2145-0
50. Hill, C.A.S., Norton, A., Newman, G.: The water vapor sorption behavior of natural fibers. J. Appl. Polym. Sci. **112**(3), 1524–1537 (2009)
51. Panyakaew, S., Fotios, S.: New thermal insulation boards made from coconut husk and bagasse. Energy Build. **43**(7), 1732–1739 (2011)
52. Xu, J., Sugawara, R., Widyorini, R., Han, G., Kawai, S.: Manufacture and properties of low-density binderless particleboard from kenaf core. J. Wood Sci. **50**(1), 62–67 (2004). https://doi.org/10.1007/s10086-003-0522-1
53. Wei, K., Lv, C., Chen, M., Zhou, X., Dai, Z., Shen, D.: Development and performance evaluation of a new thermal insulation material from rice straw using high frequency hot-pressing. Energy Build. **87**, 116–122 (2015)
54. Palumbo, M., Lacasta, A.M., Holcroft, N., Shea, A., Walker, P.: Determination of hygrothermal parameters of experimental and commercial bio-based insulation materials. Constr. Build. Mater. **124**, 269–275 (2016)
55. Korjenic, A., Petránek, V., Zach, J., Hroudová, J.: Development and performance evaluation of natural thermal-insulation materials composed of renewable resources. Energy Build. **43**(9), 2518–2523 (2011)
56. Dilmac, S., Kesen, N.: A comparision of new Turkish thermal insulation standard (TS 825), ISO 9164, EN 832 and German regulation. Energy Build. **35**(2), 161–174 (2003)
57. DIÁRIO DA REPÚBLICA - I SÉRIE-A N.o 67 - 4 de Abril de 2006, Portugal. https://files.dre.pt/1s/2006/04/067a00/24682513.pdf
58. Nguyen, D.M., Grillet, A.-C., Diep, T.M.H., Thuc, C.N.H., Woloszyn, M.: Hygrothermal properties of bio-insulation building materials based on bamboo fibers and bio-glues. Constr. Build. Mater. **155**, 852–866 (2017). https://doi.org/10.1016/j.conbuildmat.2017.08.075
59. Cerezo, V. : Propriétés mécaniques, thermiques et acoustiques d'un matériau à base de particules végétales : approche expérimentale et modélisation théorique. Doctoral Thesis, INSA de Lyon (2005)
60. Costes, J.-P., et al.: Thermal conductivity of straw bales: full size measurements considering the direction of the heat flow. Buildings **7**(1), 1 (2017)
61. Magniont, C. : Contribution à la formulation et à la caractérisation d'un écomatériau de construction à base d'agroressources. Doctoral Thesis, Toulouse (2010)
62. Brouard, Y., Belayachi, N., Hoxha, D., Méo, S., Abdallah, W.: Hygrothermal behavior of clay - sunflower (helianthus annuus) and rape straw (brassica napus) plaster bio-composites for building insulation. Adv. Eng. Forum **21**, 242–248 (2017)

63. Ismail, B.: Contribution au développement et optimisation d'un système composite biosourcé-enduit de protection pour l'isolation thermique de bâtiment. PhD Thesis, Orléans University (2020)
64. Karaky, H., et al.: Hygrothermal and acoustical performance of starch-beet pulp composites for building thermal insulation. Materials **11**(9), 9 (2018)
65. Ahmad, M.R., Chen, B., Oderji, S.Y., Mohsan, M.: Development of a new bio-composite for building insulation and structural purpose using corn stalk and magnesium phosphate cement. Energy Build. **173**, 719–733 (2018). https://doi.org/10.1016/j.enbuild.2018.06.007
66. Wei, Y., Song, C., Chen, B., Ahmad, M.R.: Experimental investigation on two new corn stalk biocomposites based on magnesium phosphate cement and ordinary Portland cement. Constr. Build. Mater. **224**, 700–710 (2019)
67. Haba, B., Agoudjil, B., Boudenne, A., Benzarti, K.: Hygric properties and thermal conductivity of a new insulation material for building based on date palm concrete. Constr. Build. Mater. **154**, 963–971 (2017)
68. Labat, M., Magniont, C., Oudhof, N., Aubert, J.-E.: From the experimental characterization of the hygrothermal properties of straw-clay mixtures to the numerical assessment of their buffering potential. Build. Environ. **97**, 69–81 (2016)
69. Niang, I., et al.: Hygrothermal performance of various Typha–clay composite. J. Build. Phys. **42**(3), 316–335 (2018)
70. Badouard, C., Bogard, F., Bliard, C., Lachi, M., Abbes, B., Polidori, G.: Development and characterization of viticulture by-products for building applications. Constr. Build. Mater. **302**, 124142 (2021)
71. Page, J., Sonebi, M., Amziane, S.: Design and multi-physical properties of a new hybrid hemp-flax composite material. Constr. Build. Mater. **139**, 502–512 (2017)
72. Rahim, M., Douzane, O., Le Tran, A., D., Langlet, T.: Effect of moisture and temperature on thermal properties of three bio-based materials. Constr. Build. Mater. **111**, 119–127 (2016)
73. Pierre, T., Colinart, T., Glouannec, P.: Measurement of thermal properties of biosourced building materials. Int. J. Thermophys. **35**(9–10), 1832–1852 (2013). https://doi.org/10.1007/s10765-013-1477-0
74. Yin, X., Lawrence, M., Maskell, D., Ansell, M.: Comparative micro-structure and sorption isotherms of rice straw and wheat straw. Energy Build. **173**, 11–18 (2018)
75. Collet, F., Chamoin, J., Pretot, S., Lanos, C.: Comparison of the hygric behaviour of three hemp concretes. Energy Build. **62**, 294–303 (2013)
76. Bouguerra, A., Sallée, H., de Barquin, F., Dheilly, R.M., Quéneudec, M.: Isothermal moisture properties of wood-cementitious composites. Cem. Concr. Res. **29**(3), 339–347 (1999)
77. Abbas, M.S., McGregor, F., Fabbri, A., Ferroukhi, Y.M.: Influence of origin and year of harvest on the performance of pith mortars. Acad. J. Civil Eng. **37**(2), 2 (2019)
78. Latif, E., Lawrence, M., Shea, A., Walker, P.: Moisture buffer potential of experimental wall assemblies incorporating formulated hemp-lime. Build. Environ. **93**, 199–209 (2015)
79. Chennouf, N., Agoudjil, B., Boudenne, A., Benzarti, K., Bouras, F.: Hygrothermal characterization of a new bio-based construction material: Concrete reinforced with date palm fibers. Constr. Build. Mater. **192**, 348–356 (2018)
80. Rahim, M., et al.: Characterization of flax lime and hemp lime concretes: hygric properties and moisture buffer capacity. Energy Build. **88**, 91–99 (2015)
81. Rahim, M., Douzane, O., Le Tran, A.D., Promis, G., Langlet, T.: Characterization and comparison of hygric properties of rape straw concrete and hemp concrete. Constr. Build. Mater. **102**, 679–687 (2016). https://doi.org/10.1016/j.conbuildmat.2015.11.021
82. Maalouf, C., Umurigirwa, B.S., Viens, N., Lachi, M., Mai, T.H.: Study of the hygric behaviour and moisture buffering performance of a hemp-starch composite panel for buildings. BioResources **10**(1), 336–347 (2014)

83. Ouméziane, Y.A., Moissette, S., Bart, M., Lanos, C.: Effect of coating on the hygric performance of a hemp concrete wall. In: Presented at the Proceedings of the 5h International Building Physics Conference IBPC (2012)
84. Medjelekh, D. : Caractérisation multi-échelle du comportement thermo hybride des enveloppes hygroscopiques. Doctoral Thesis, Université de Limoges (2015)

Natural Fibre Reinforcement

Concrete with Natural Fibres (Bio Concrete)

Marianna Coelho(✉), Aidin Gibbons, Boril Nestorov, Wilson Zangue, Isabelle Oosthoek, and Tren Fijnaut

HZ University of Applied Sciences, Het Groene Woud 1-3, 4331NB Middelburg, The Netherlands
marianna.coelho@hz.nl

Abstract. The carbon footprint of cement is well known in the civil construction area as one of the most impactful. Concrete is already consolidated as a strong and versatile material for construction. One of the disadvantages of concrete is its high self-weight and carbon footprint. Adding a natural fibre to the concrete mixture reduces the weight due to the lightweight natural fibres and the amount of cement, sand and gravel. Previous research in our research group on the pretreatment of the Miscanthus fibres showed that cement pretreatment improved the strength of concrete, probably due to the increase of bond between the fibres and concrete. The increase in the number of fibres in volume decreases the compressive strength but still has expressive values that can be used for structural elements and increases the tensile strength. This research shows a mixture design for concrete with natural fibres, with a percentage of natural fibres around 20%. The results of the fresh concrete test showed that there was a decrease in the self-weigh for the addition of natural fibres in concrete. Also, the compressive strength decreased with the addition of natural fibres, but on the other hand, the tensile strength increased. The thermal insulation test showed an improvement in the thermal insulation properties with the natural fibres. The fire resistance test proved that the concrete with natural fibres had a good performance.

Keywords: Concrete · Miscanthus fibres · Fire resistance · Thermal insulation

1 Introduction

Carbon dioxide emissions policies and laws, such as [1, 2], and [3], establish specific targets for the reduction in the European Union. Additionally, in 2011, the European Commission published the "Roadmap for moving to a competitive "low-carbon economy in 2050" [4] and, for the industrial sectors, the objective is to reduce CO2 emissions from 83% to 87% by 2050 compared to 1990 CO_2 emissions.

Concrete production is responsible for 6% of the annual global CO_2 emissions. Adding biomass into concrete would significantly reduce CO_2 generation. Further, with the ongoing political development in the European Union, the thermal conductivity of concrete must be improved. This will lead to a reduction in energy used by households. Introducing biomass in the concrete mixture makes the material lighter and increases its air content, affecting the thermal conductivity properties [5].

S. Amziane et al. (Eds.): ICBBM 2023, RILEM Bookseries 45, pp. 467–477, 2023.
https://doi.org/10.1007/978-3-031-33465-8_36

One of the first steps in reducing carbon dioxide levels is replacing some common aggregates, sand and gravel, with more biobased products, like natural fibres. The change is because the more common aggregates are becoming harder to find, resulting in them needing to be transported in from further away, again adding to the CO_2 levels. One means of decreasing the use of the common aggregates and, in turn, reducing the carbon dioxide levels is by replacing the sand or gravel aggregates with sustainable materials, such as Elephant grass (Miscanthus x giganteus) which is dried and used as a fine aggregate replacement [6].

Adding natural fibres to concrete has some advantages as increasing the tensile strength, reducing the self-weight, and reducing plastic shrinkage and fatigue strength. On the other hand, it can reduce compressive strength.

An experiment was conducted by [7] to reinforce concrete with different cement dosages, a water/cement ratio of 0.5 and different biomass input ratios with varying styles of comminution. The Miscanthus fibres were ground, cut, embedded as reinforcement, and introduced as a mixture. The results are that the Miscanthus fibre loading (2% and 4%) improved the strength of the sample. The compressive strength increased by 9 to 25% by increasing the cement dosage, where the limit was 46 N/mm^2. The splitting and bending strength increased by 4% to 8%.

The effect of adding industrial hemp fibres on plain concrete was studied by Awwad et al. [8]. They concluded that the compressive strength tends to decrease in the presence of hemp fibres. The flexure results showed that the presence of hemp fibres allows coarse aggregate reduction and results in ductile behaviour. The splitting tensile results were not affected in the presence of sufficient fibres. The modulus of elasticity results assured the ductility of the hemp concrete mixes. The thermal conductivity results showed the effect of hemp fibres in reducing heat flow. The hemp fibre's water absorption reduced the slump results. The density results were slightly decreased with the coarse aggregate reduction.

Abdalla et al. [9] analyzed concrete containing natural fibres, including hemp, kenaf, and bamboo. It is shown that using hemp fibres improves the compressive strength, flexural tensile strength, and toughness of concrete. Bamboo fibres can be used with different concentrations of concrete to achieve targeting ductility. Overall, bamboo fibres improve concrete samples' tensile strength and torsional toughness. It was also found that using kenaf fibres of different lengths and diameter improves the mechanical properties of concrete. But the durability studies of concrete incorporating these natural fibres remain a research gap, and more studies are needed for its efficient use.

A work based on an acid attack test in concrete with jute fibres performed by Nambiar and Haridharan [10] observed that there was a decrease in the compressive strength by 40.625% from 30 to 90 days. When fibre was added at 1% and 2%, a reduction of compressive strength was seen by 35.77% and 39.39%.

Ren et al. [11] evaluated ultra-high performance concrete (UHPC) with sisal fibre at elevated temperatures. As a renewable, biodegradable, and low-cost plant fibre, natural sisal fibres and steel fibres hybridization to produce eco-friendly UHPC with high-temperature resistance has great potential.

Grubeša et al. [12] evaluated another type of natural fibre (hemp fibres) for improving the fire resistance of concrete. According to visual observations (SEM images and micro images), it was concluded that hemp fibres become partially disintegrated within the concrete after exposure to a temperature of 400 °C. As this is the temperature at which crack formation starts, the unmelted hemp fibres could be useful in preventing crack propagation at higher temperatures, thus improving the fire resistance of concrete in general.

Although the effort in researching concrete with natural fibres, there is still a research gap in understanding the durability of concrete with natural fibres. Also, there are different types of natural fibres with other properties. Therefore one of the objectives of the present work is to evaluate the fire resistance of concrete with Miscanthus fibres.

Also, other objectives of the present research are to optimize the concrete mixture with natural fibres and to evaluate the compressive and tensile strength and thermal insulation.

2 Experimental Program

2.1 Materials

The materials used for the bio concrete were cement CEM I 52.5 R grey Portland. This is the most commonly used cement worldwide, and it has a high mechanical strength, which is reached rapidly. The coarse aggregate is gravel with granulometry in the range of 4 mm to 20 mm. The fine aggregate is natural sand from 0 mm to 4 mm.

The natural fibre used is Miscanthus X Giganteus, ranging from 2 mm to 10 mm. The Miscanthus fibres have a great potential to be a great material for reinforcement if a good bond between the fibre and the matrix is established. The Miscanthus fibres have a tensile strength of 373 N/mm^2 and a compressive strength of 56.9 N/mm^2. Additionally, the Miscanthus biomass is silicate rich, where 74% of the ash is composed of silica and porous parenchyma. These composites are considered to have potential benefits in a matrix out of cement. The parenchyma content and the concordant pore structure reduce the concrete's thermal conductivity [5].

Concrete Mixture Design

The concrete design has a constant amount of cement, sand, gravel, and water for the initial mix design. The number of pretreated fibres was increased respectively in 1 kg. Because the fibres were pretreated with cement, in the end, the cement coated to the fibres was added to the total amount of cement, and the real amount of fibres was calculated. In other words, we kept the number of materials of Table 1, and we added the fibres. As the amount of fibres increases, the final volume of concrete also increases. The volume of each material of the mixture was calculated considering the cement coated in the fibres in the cement part. The result section shows those values.

Table 1 shows the density and material amount used for the concrete design.

Four mixtures were produced, the control mixtures without Miscanthus fibres and three mixtures with Miscanthus fibres (B1P, B2P, and B3P) with the number of pretreated fibres of 5.6 kg, 6.6 kg and 7.6 kg, respectively. The mixtures were cast in cube samples with 15 cm.

Table 1. Density and material amount in mass for the concrete design

Materials	Density: (kg/m^3)	Amount: (kg)
CEM I 52.5 R	3100	20.755
Sand 0–4 mm	2640	32.076
Gravel 4–20 mm	2640	44.930
Pretreated Fibers	160	0.000–7.600
Water	1000	10.616

2.2 Pretreatment of Natural Fibres

In the experiment conducted by Boix et al. [13], the particles were pretreated by alkaline leaching (NaOH) and salinization at different pH levels (pH4, pH6 and pH10). The results were that the compressive strength of the concrete increased (from 2.2 to 11 N/mm^2) for particle treatment, as mentioned above. Furthermore, a negative relationship was observed between the compressive strength and the number of sugars released into the concrete water.

Different pretreatments to reduce water absorption were also researched, and according to Ezechiels [14], the most efficient ones were water glass pretreatment (reduction by 140%) and cement slurry pretreatment (reduction by 210%).

In the present research, the natural fibres are pretreated using cement-water slurry. The mixture consists of 5000 g of cement (CEM I 52.5 R), 5000 g of water and 500 g of Miscanthus fibres (*M.* × *giganteus*). After the fibres are mixed with the cement-water slurry, they are left to dry for seven days. After this period, the batches are weighed individually to determine the average natural fibre content.

Table 2 shows the cement amount in the pretreated Miscanthus fibres for the different design mixtures.

Table 2. Cement amount in the pretreated Miscanthus fibres

Mixture code	Fibres (g)	Cement (g)	Natural fibre/cement ratio
B1P	500	1439.50	0.347
B2P	500	1703.75	0.293
B3P	500	1249.40	0.400

2.3 Fire Resistance – Fire Furnace Test

The fire resistance test was performed according to the Eurocodes EN 1992-1-2 [15] and EN 1363-1 [16] for the concrete cubes. The furnace's temperature increases relatively at the same rate as the standard fire curve, where the temperature would be over 800 ºC within 30 min and continues to rise to just under 1000 °C within one hour. Other curves can be programmed and tested with the furnace, such as the external fire or hydrocarbon curves. During the test, the surface temperature, radiation and flame density are assessed [17].

The test was done on each mixture of two cubes (15 cm × 15 cm × 15 cm). The cubes were inserted in holes in an insulated concrete slab. The concrete slab with the cubes was on top of the fire furnace. Thermocouples were inserted in the external face of the cubes to measure the heat transfer during the fire furnace test.

2.4 Thermal Insulation Test – Electrical Conductivity Test

The sample is then placed into a chamber of EPS on top of an aluminium plate, and another aluminium plate is placed on top of the sample. Two thermometer devices are connected to the system to monitor the temperature during the test. The first device is to measure the temperature of the aluminium plates under (T1) and above the sample (T2). The power unit generates heat at 1,6 W. When first conducting a test with a new material, the temperature values shown on the screen of both thermometers must be recorded every 2–5 min throughout the whole test duration. When the temperature stops increasing, or there is only no variation between a lower and a higher value for over 10 min, the steady state is considered reached, and the generator may be run off. The size of the samples is 12 cm × 12 cm × 2 cm.

The thermal conductivity (λ) is calculated with Eq. 1:

$$\lambda = \frac{Q}{DT} \cdot \frac{L}{A} \tag{1}$$

in which Q is the power at which the heat is generated, A is the area of the sample, L is the thickness of the sample, and DT is the difference between T2 and T1 measured at the end of the test (at steady state).

3 Results

3.1 Material Percentage in Volume

After the mixtures were created, the natural fibre content in mass [kg] was calculated using the information from the pretreatment analysis (Table 2). The results of the volume percentage for each mixture are presented in Fig. 1.

As seen on the graph (Fig. 1), Mixture B1P and Mixture B2P are not different, although the amount in mass of pretreated fibres was different. In Mixture B2P, the pretreated fibres mostly originated from the batches with a high amount of cement. Hence the low amount of natural fibres got into the mixture (see Table 2).

The amount of aggregates used in the mixture is also getting lower. As could be seen, there was a decrease of 10% in the gravel and 7.13% in the sand. The natural fibres replace the aggregates in the concrete mixture. Also, there was a small change in the amount of cement, although the use of cement pretreatment.

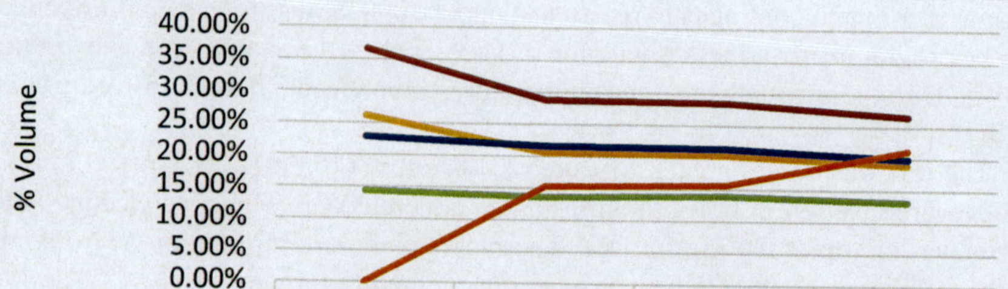

	Control Mix	Mixture B1P	Mixture B2P	Mixture B3P
Sand	26.14%	20.61%	20.38%	19.01%
Gravel	36.62%	28.86%	28.55%	26.62%
Cement	14.40%	13.63%	13.99%	13.21%
Water	22.84%	21.60%	21.37%	19.92%
Natural Fibres	0.00%	15.30%	15.70%	21.24%

Fig. 1. Material content in percentage of the total volume

3.2 Fresh Concrete Tests

Table 3 present the results of the fresh concrete tests. For the control mixture, the workability was very high that the slump test collapsed. And during the slump test, the slump collapses as the number of fibres increases in the mixture the value of the slump test decreases.

There was a small change in the water/cement ratio as the number of fibres increased. The fresh density decreases, and the air content increases with natural fibres. These results show that more voids are immediately present when fibres are added. In turn, this correlates to the results of the fresh density test. The more fibres are added, the lighter the mixture becomes as more voids are present.

Table 3. Results of the fresh concrete test

Mixture	Slump test	Water/cement ratio	Fresh density (kg/m^3)	Air content (%)
Control	Collapsed	0.511	2243.25	2.5%
B1P	160	0.511	1999.31	8.0%
B2P	160	0.493	2022.75	7.7%
B3P	115	0.486	1889.75	10.0%

3.3 Hard Concrete Tests

Table 4 shows the results of compression and tensile strength of the concrete mixtures. The first column is the results of the standard compression strength at 28 days, as was expected. With the increase of fibres in the mixture, the strength decreases. For the tensile strength, the behaviour is the opposite. With the increase of fibres, the tensile strength increases because of the strength of natural fibres in tension.

The third column shows the result of the cubes samples after the fire furnace test, and the fourth column shows the cubes that were not submitted to the fire test. After the fire furnace test, the control mixture performance was better than the mixture with natural fibres. The compression strength after the fire furnace test was around 64% of the samples that were not submitted to the test. The values of the mixtures with natural fibres were around 50% to 55%.

Table 4. Compression and tensile strength of concrete mixture for different ages and situations

Mixture	Compression strength 28 days (MPa)	Tensile strength 28 days (MPa)	Compression strength after fire furnace test (MPa)	Compression strength without fire furnace (MPa)
Control	44.90	5.14	30.28	47.56
B1P	26.62	4.16	15.63	31.43
B2P	25.44	5.35	14.56	26.82
B3P	16.66	10.36	9.37	16.90

3.4 Fire Furnace Test

Figure 2 shows the difference between the temperature in the furnace's interior and the temperature on the unexposed side of the concrete cubes (15 cm). The temperature difference is noticeable around 850 °C.

Figure 3 presents only the temperature of the concrete cubes on the unexposed side. At the start of the test, all mixtures behaved similarly as the rate of temperature increase was stable till about 20 min into the test. At 20 min into the test, the temperature difference is 5 °C, from 50 °C–55 °C for all four mixtures, where the furnace temperature is at 783 °C.

From 20 to 45 min, each mixture takes an individual line, which correlates with the number of fibres in each mixture. As more fibres are added to the mixture, more heat is transferred through the cube. However, they are all lower than the control mix. At 32 min, the furnace temperature is 859 °C, and the pattern between the mixtures can be seen. The control mix with the highest temperature at 78 °C shows that conventional concrete transfers heat faster and in a shorter time than that concrete with fibres. Second is B3P at a temperature of 74 °C and has the highest percentage of fibres within the mixture. 20% after that is B2P at a temperature of 68 °C, which has 18% of fibres present within the mixture, and lastly is B1P (orange) with a temperature of 62 °C, as well as having the lowest fibre content of 16%, which shows that bio concrete does not allow heat to travel as fast as conventional concrete from one side of the cube to the other.

For the last 15 min of the test, the cubes with elephant grass fibres continue to increase steadily. When looking at the end of the test at 60 min, the furnace temperature

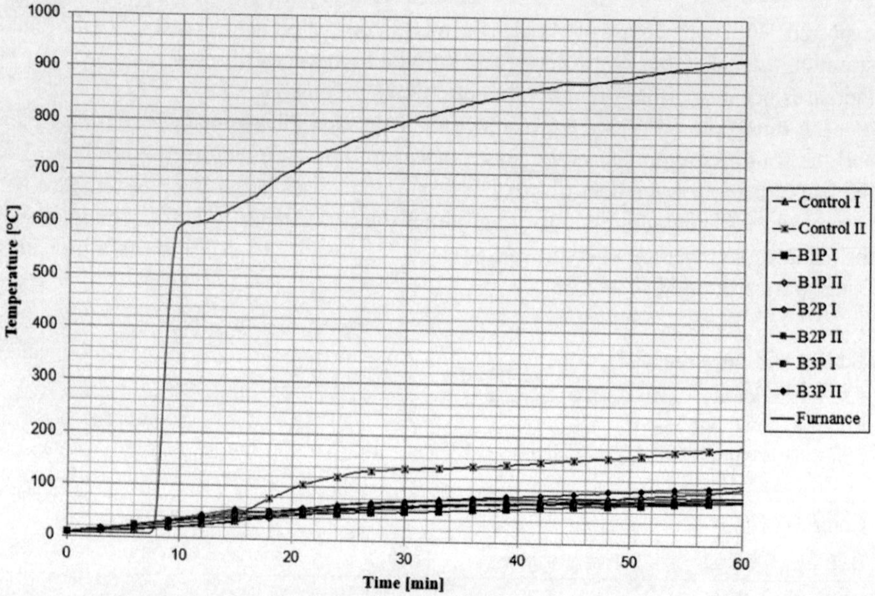

Fig. 2. Furnace temperature and temperature of the concrete cubes on the unexposed side

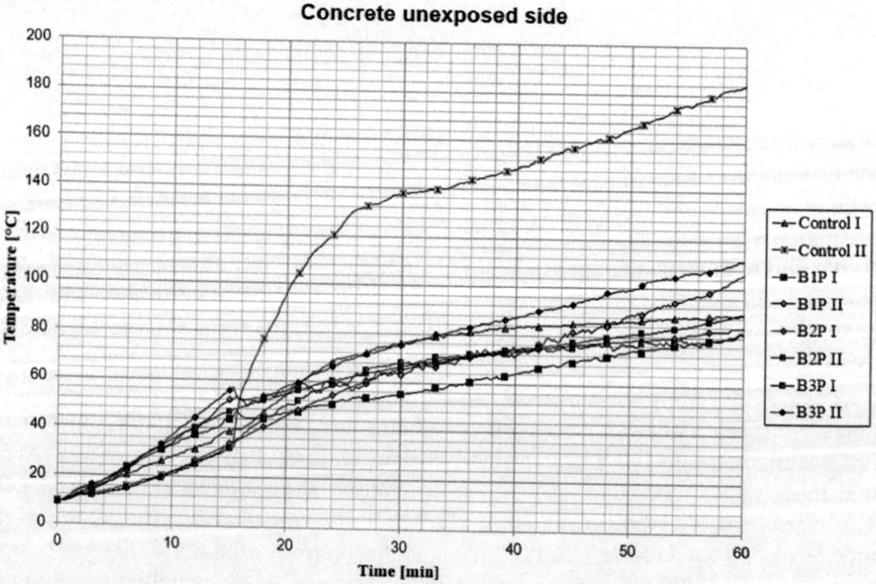

Fig. 3. Temperature of the concrete cubes on the unexposed side

reaches a maximum of 945 °C, whereas, on the unexposed side of the cubes, none passed 100 °C. Starting with B3P at the top at 96 °C, also having the most fibre content within the mixture at 20%. Next is B1P at 93 °C, the mixture with the least amount of fibres being 16%, both of these mixtures show a steady temperature increase at 60 min. Where both mixtures ended above the control mixture, which is at 89 °C, the control mixture decreased its rate of temperature increase. This can be seen in Fig. 3. At approximately 40 min into the test, the increase in temperature was not as fast as it was before. In the last 20 min of the test, the temperature increased by 6 °C compared to the 20 min before, where the temperature difference was 29 °C. Lastly is mixture B2P, which had the lowest temperature at the end of the test at 60 min with a temperature of 87 °C with a fibre content of 18%. However, similar to that of the other two mixtures with fibres, the rate of temperature increase stays constant. This means that conventional concrete will take a longer time to allow high temperatures to transfer through it compared to bio concrete with Miscanthus fibres.

3.5 Thermal Insulation – Thermal Conductivity Test

Table 5 presents the results of the thermal conductivity coefficient and compressive strength for the control mixture and concrete with Miscanthus fibres. The percentage of fibres in the bio concrete was around 20%. And it tested two samples for the control mixture and two for bio concrete. The average values for the compressive strength are similar to Table 4 for control and B3P for the concrete with Miscanthus fibres. The improvement of the values with the addition of Miscanthus fibres is noticeable for the thermal conductivity. As the value of thermal conductivity decreases, it is more difficult to conduct temperature resulting in a material that has better thermal insulation.

Table 5. Thermal conductivity coefficient and compressive strength for control mixture and mixture with Miscanthus fibres

Mixture	Compressive strength (MPa)	Thermal conductivity (λ)
Control I	46.80	0.908
Control II	47.47	0.867
Control (average)	**47.14**	**0.888**
Miscanthus I	12.59	0.771
Miscanthus II	14.18	0.722
Miscanthus (average)	**13.39**	**0.747**

3.6 Production Cost and Environmental Cost Indicator (ECI)

For the calculation of the cost of production for concrete with natural fibres, the values presented in Table 6 were used.

The cost for 1 m^3 of concrete of the control mixture was 114,35 euros and for the concrete with natural fibres (mixture B1P) was 129,48 euros.

The environmental cost indicator (ECI) is a value that expresses the material's impact on the environment. The ECI value is depicted here in euros as the value gets high, which

Table 6. Price per ton in Euros for the concrete materials

Material	Price per ton in Euros	ECI per ton in Euros
Sand 0–4 mm	20	0,30
CEM I 52.5 R	110	60
Water	0,10	0
Gravel 4–20 mm	25	0,25
Miscanthus fibres	1094	9,47

indicates a higher impact on the environment. The ECI values used for each component present in concrete are shown in Table 6.

The ECI calculation follows the steps of the Life Cycle Assessment (LCA). The first step considers the origin of the material, the energy spent in the production and the transportation. The second step considers the emissions and pollutants. The third step is the impact category, for example, climate change. Then the values are weighted, and the ECI is calculated per euro. For example, for global warming, the unit is kg CO_2-eq, and the value is 0,05 €/ unit.

The ECI for the concrete with natural fibres equals 25,52 euros/m^3, and for conventional concrete (control mixture) equals 27,24 euros/m^3.

4 Conclusion

A mixture design for concrete with Miscanthus fibres was developed. The cement pre-treatment used in the natural fibres proved efficient and increased the bond between the fibres and the cementitious material.

Incorporating natural fibres into a concrete mixture design proved beneficial and sustainable. First, the percentage of sand and gravel in volume had a big reduction of around 27%; for cement, it was around 8% for the mixture with 21.24% of Miscanthus fibres. Second, the fresh density decreased, and the air content increased, resulting in lighter concrete. Third, based on the results obtained with the thermal conductivity test, the concrete with natural fibres results in smaller values which means that it is a better insulator than conventional concrete.

On the other hand, with the increase of Miscanthus fibres in concrete, the compressive strength decreased. Also, for the fire furnace test, the conventional concrete showed to take a longer time to allow high temperatures to transfer through it compared to bio concrete with Miscanthus fibres. But overall, the concrete with natural fibres showed a good performance transferring 96 °C for the sample with the worst performance.

The production cost for concrete with natural fibres is more expensive due to the higher fibre price, but the environmental cost indicator was lower than conventional concrete.

Lastly is important to highlight that natural fibres are also renewable and biodegradable, the main difference with synthetic fibres commonly used in concrete.

References

1. Directive 2003/87/EC, of the European Parliament and of the Council of 13 October 2003. Establishing a scheme for greenhouse gas emission allowance trading within the Community and amending Council Directive 96/61/EC (2003)
2. Directive 2009/29/EC, of the European Parliament and of the Council of 23 April 2009. Amending Directive 2003/87/EC so as to improve and extend the greenhouse gas emission allowance trading scheme of the Community (2009)
3. Directive 2012/27/EU, of the European Parliament and of the Council. on ergy efficiency, amending Directives 2009/125/EC and 2010/30/EU and repealing Directives 2004/8/EC and 2006/32/EC (2012)
4. COM 2011, .f.: A roadmap for moving to a competitive low carbon economy in 2050. Technical report (2011)
5. Moll, L., Wever, C., Völkering, G., Pude, R.: Increase of miscanthus cultivation with new roles in materials production - a review. Agronomy **10**(2), 308 (2020)
6. Prakash, R.,Sudharshan, N., Raman, C., Subramanian, N.D.: 6 - Eco-friendly fiber-reinforced concretes. In: Handbook of Sustainable Concrete and Industrial Waste Management. Woodhead Publishing. (2022)
7. Acikel, H.: The Use of Miscanthus (Giganteus) as A Plant Fiber In Concrete Production. Selcuk University, Faculty of Engineering and Architecture, Engineering, Konya (2011)
8. Awwad, E., Mabsout, M., Hamad, B., Farran, M.T., Khatib, H.: Studies on fiber-reinforced concrete using industrial hemp fibers. Constr. Build. Mater. **35**, 710–717 (2012)
9. Abdalla, J.A., Thomas, B.S., Hawileh, R.A.: Use of hemp, kenaf and bamboo natural fiber in cement-based concrete. Mater. Today: Proc. **65**, 2070–2072 (2022)
10. Nambiar, R.A., Haridharan, M.K.: Mechanical and durability study of high performance concrete with addition of natural fiber (jute). Mater. Today: Proc. **46**, 4941–4947 (2021)
11. Ren, G., Gao, X., Zhang, H.: Utilization of hybrid sisal and steel fibers to improve elevated temperature resistance of ultra-high performance concrete. Cem. Concr. Compos. **130**, 104555 (2022)
12. Grubeša, I.N., Marković, B., Gojević, A., Brdarić, J.: Effect of hemp fibers on fire resistance of concrete. Constr. Build. Mater. **184**, 473–484 (2018)
13. Boix, E., Gineau, E., Narciso, J.O., Höfte, H., Mouille, G., Navard, P.: Influence of chemical treatments of miscanthus stem fragments on polysaccharide release in the presence of cement and on the mechanical properties of bio-based concrete materials. Cem. Concr. Compos. **105**, 103429 (2020)
14. Ezechiels, J.: Design of An Innovative Bio-Concrete using Miscanthus Fibres. Technical University Eindhoven, Eindhoven (2017)
15. EN 1992–1–2, Eurocode 2: Design of concrete structures – Part 1.2: General rules – Structural fire design
16. EN 1363–1, Fire resistance tests – Part 1: General Requirements
17. PEUTZ.: Laboratory for fire safety. peutzgroup.com (2022). https://www.peutzgroup.com/node/57

Mechanical Properties of Geopolymer Composites Reinforced with Natural Fibres

Masakazu Terai(✉)

Kindai University, 1Ban, Takaya-Umenobe, Higashi-Hiroshima, Japan
terai@hiro.kindai.ac.jp

Abstract. Concrete is a material that is prone to cracking due to external forces, such as heat, dryness, and earthquakes, due to its low tensile strength. In recent years, the price of steel materials has soared, and in developing countries, it is difficult to obtain expensive steel materials, and their mass use is restricted. As a result, there are many areas in the world where concrete structures, which originally required reinforcement, are not sufficiently reinforced. In addition, since large amounts of fossil fuels are used in the production of steel materials, it is conceivable that production will be restricted in the future from the perspective of carbon dioxide emissions and consideration of environmental issues.

In this study, by mixing natural plant and animal fibres as reinforcing materials into construction materials using geopolymers as solidifying materials, we developed low-cost, low-technology, and environmentally friendly building materials. Fundamental experiments were conducted on natural fibre-reinforced cement composites with improved mechanical properties. In this experiment, three types of plant fibres (sisal, coir, and bamboo) and three types of animal fibres (horse hair, pig hair, and sheep wool) were used as reinforcing materials. Several experiments were conducted on the strength properties (compression, tension, and bending) of hardened fibre-reinforced cement composites with different fibre types, lengths, and mixed amounts.

Keywords: Fibre-Reinforced Cement Composite · Natural Fibre · Geopolymer · Strength · Plant · Animal

1 Introduction

1.1 Fibre Reinforcement Material

Concrete is a material that is prone to cracking due to external forces, such as heat, dryness, and earthquakes, due to its low tensile strength. Therefore, structures made of concrete are usually made of steel or other reinforcement materials to reduce cracks (reinforced concrete structures, fibre reinforced concrete, etc.). Normally, synthetic fibres and steel fibres are mixed to suppress cracks, but synthetic fibres are expensive, expensive steel materials are difficult to obtain in developing countries, and their use in large quantities is restricted. Therefore, there are many areas in the world where concrete structures that originally required reinforcement have not been sufficiently reinforced.

S. Amziane et al. (Eds.): ICBBM 2023, RILEM Bookseries 45, pp. 478–489, 2023.
https://doi.org/10.1007/978-3-031-33465-8_37

In addition, since a large amount of fossil fuels are used in the production of steel materials, it is possible that production will be restricted in the future from the viewpoint of carbon dioxide emissions due to consideration of environmental issues. Therefore, the development of structures that do not use steel materials is meaningful for the future of humankind.

In this study, the effective use of natural plant and animal fibres is examined for the reinforcement of mortar cracks. The current problems are (1) cost and durability, (2) workability, and (3) the fibre reinforcement effect. Natural fibres are inexpensive and available locally in many countries. Moreover, they are easy to process and require very little use as construction materials to improve the properties of composites (compared to the total cost of composites, they cost almost nothing). On the other hand, in terms of durability, when cellulose is mixed with fibres mainly composed of cellulose in alkaline cement, fats and sugars are decomposed, and this decomposition component inhibits the hydration reaction, does not increase strength, and long-term cracking suppression is insufficient [1].

1.2 Fibres

Familiar and inexpensive materials such as stone, soil, wood, plants, and adobe bricks have been used as construction materials, but since the middle of the 19th Century, steel, glass, cement, aluminium, etc., have been produced rapidly, and these have become indispensable materials for manufacturing modern structures. Fibre materials have also been used as building materials for a long time, but they are rarely used alone and have been mixed with soil and used as composite materials. By adding fibre material, the shrinkage rate of the material can be reduced. This is due to the decrease in relative clay content and the absorption of some moisture into the pores of the fibrous material. In addition, the fibre material increases the binding force of the mixture, thereby reducing the occurrence of cracks. Fibre is often used as a material to strengthen matrix materials such as soil and plastic, but in recent years, it has also been used as a reinforcement material for mortar and concrete.

The most commonly used fibre materials for concrete reinforcement are steel, organic polymers, glass, carbon, asbestos, and cellulose. Fibres can be divided into "artificial fibres" made from various chemically treated materials and "natural fibres" taken from parts of minerals, plants, and animals. Furthermore, natural fibres can be classified into organic fibres and inorganic fibres, but plant fibres are representative of organic fibres. Plant fibre is inexpensive and readily available locally in many countries. Their use as construction materials to improve the properties of composites is negligible, and when compared to the total cost of composites, they cost very little. However, previous studies [2–4] have noted that the mixing of cellulose-based plant fibres in alkaline cement decomposes fats and sugars, that these decomposition components inhibit the hydration reaction and do not increase strength, or that long-term cracking suppression is insufficient [1]. Therefore, there are high expectations for animal fibres [5, 6], which are mainly composed of protein. In addition, since geopolymers have lower alkalinity than cement, it is expected that the above reaction will be weakened. For this reason, the use of natural fibres as reinforcement materials has been widely used in recent years. It is detailed in the literature published in 2009 [7], and recent reviews are also introduced in [8–12].

1.3 Geopolymer

Geopolymers are solidified materials produced by the condensation polymerization reaction between an industrial byproduct containing aluminium silicate, which is called an active filler, and an alkaline solution and are one of the relatively new materials developed in the 1970s. Since it is a new material that is still under research in various countries around the world, various raw materials for geopolymers are used, and as a result, the reaction products constituting the cured body are also very different, so universal formulation design and strength management methods have not been established. However, since geopolymers do not use cement, they are expected to be an alternative material to conventional cement concrete. It is known that cement used in concrete emits a large amount of CO_2 in its manufacturing process. Although the demand for cement in developed countries has plateaued, the increase in cement production and usage in emerging countries in recent years shows that if CO2 during cement production cannot be suppressed or the use of cement itself cannot be reduced, it is necessary to develop alternative materials. It has been reported that the construction of the same structure using geopolymers can reduce CO_2 emissions by approximately 80% compared to cement, and if this figure is reliable, geopolymers may be an effective construction material in the future.

In this study, we carried out several fundamental experiments on a natural fibre-reinforced cement and geopolymer composite material that improves tensile strength and mechanical properties as a low-cost, low-technology, and environmentally friendly building material by mixing natural plant fibres and animal fibres as reinforcement materials into construction materials using geopolymers as solidified materials. In this experiment, three types of plant fibres, sisal, bamboo, and coir, and three types of animal fibres, horse hair, pig hair, and sheep wool, were used as reinforcements. Several experiments were conducted on the strength characteristics (compression, tensile, bending) of hardened fibre-reinforced cement composite materials depending on the fibre type, fibre shape, and amount of contamination. In addition, to quantitatively evaluate the reinforcement effect of fibres, we also attempted mechanical evaluation by applying the mixed law of fibres and mortar matrices.

2 Experimental Program

Table 1 and Table 2 show the physical properties of the materials and the mortar mix composition of the test specimens, respectively. Testing variables are mortar type (ordinary Portland cement mortar and geopolymer mortar), fibre type (plant; sisal, coir, and bamboo, animals; horse, pig, and sheep), fibre content (1, 1, and 2% by volume), and fibre length (10 mm and 40 mm). Fresh properties are tested for consistency by a flow table. The experimentally observed effects of mortar and fibre type and fibre content on the strength properties for compressive, tensile, and flexural tests are discussed.

The mix proportion of the experiment is shown in Table 2. The water-cement ratio (W/C) and air content (4% by volume) are common. Chemical admixture is not being used because the chemical effects on bamboo fibre by admixture are of concern. The specimen's description is shown in Table 3.

Table 1. Physical Properties Materials

Specimen	Materials		Description	Density (g/cm^3)	Specific surface area (cm^2/g)
Geopolymer Mortar GP	Binder B	Glass Powder	GLP	2.41	2,500
		Granulated Blast Furnace Slag	BFS	2.90	6,800
	Alkline Activator Solutions AAS	Sodium Silicate	SS	1.53	–
		Sodium Hydroxide	NaOH	1.33	–
		Water	W	1.00	–
	Fine aggregate		S	2.64	–
	Fibre		F		–
Ordinary Portland Cement Mortar CE	Ordinary Portland Cement		C	3.15	9,100
	Water		W	1.00	–
	Fine aggregate		S	2.64	–
	Fibre		F		–

2.1 Materials

Ordinary Portland cement was used as cement mortar. In the geopolymer mortar, the solution/powder ratio was set at 0.6, and 20% of the entire powder was replaced with ground granulated blast furnace slag (BFS). Many studies have confirmed that replacement with BFS is effective for hardening at room temperature. BFS has the advantage that it can be hardened at room temperature and has a high strength, but it has the disadvantage that the pot life is short due to its high viscosity. The alkaline solution was prepared by mixing sodium silicate and 10 mol-sodium hydroxides at a ratio of 1:3 and then adding water to adjust the viscosity. The formulation table is shown in Table 2.

Table 2. Mortar Mix Composition

<table>
<tr><td rowspan="2">Type of Mortar (Series)</td><td rowspan="2">Fibre Content (volume %)</td><td rowspan="2">Fibre</td><td rowspan="2">Air content (volume %)</td><td rowspan="2" colspan="2">Water Cement Ratio W/C (%)</td><td colspan="4">Unit Weight (kg/m^3)</td></tr>
<tr><td>W</td><td>C</td><td>S</td><td>F</td></tr>
<tr><td rowspan="3">OPCM</td><td>0.0</td><td>–</td><td rowspan="3">4</td><td rowspan="3" colspan="2">50</td><td rowspan="3">266</td><td rowspan="3">532</td><td rowspan="3">1596</td><td>–</td></tr>
<tr><td>1.0</td><td rowspan="2">3Plant, 3Animal Fibres</td><td>*</td></tr>
<tr><td>2.0</td><td>**</td></tr>
<tr><td rowspan="3">Type of Mortar (Series)</td><td rowspan="3">Fibre Content (volume %)</td><td rowspan="3">Fibre</td><td colspan="7">Unit Weight (kg/m^3)</td></tr>
<tr><td colspan="3">AAS</td><td colspan="2">AF</td><td rowspan="2">S</td><td rowspan="2">F</td></tr>
<tr><td>NaOH</td><td>SS</td><td>W</td><td>BS</td><td>GLP</td></tr>
<tr><td rowspan="3">GEPM</td><td>0.0</td><td>–</td><td rowspan="3">106</td><td rowspan="3">35</td><td rowspan="3">211</td><td rowspan="3">242</td><td rowspan="3">442</td><td rowspan="3">1431</td><td>–</td></tr>
<tr><td>1.0</td><td rowspan="2">3Plant, 3Animal Fibres</td><td>*</td></tr>
<tr><td>2.0</td><td>**</td></tr>
</table>

* Bamboo 13.2, Sisal 15.0, Coir 15.0, Horse 13.5, Pig 13.5, Wool 13.2
** Bamboo 26.4, Sisal 30.0, Coir 30.0, Horse 27.0, Pig 27.0, Wool 26.4

Table 3. Specimen Description

Specimens	Fibre		Fibre Percentage Content - Length
CEP: Cement Mortar GPP: Geopolymer Mortar	Plant	-0: No fibre	110 1%-10 mm
		-B: Bamboo	140 1%-40 mm
		-C: Coir	210 2%-10 mm
		-S: Sisal	240 2%-40 mm
CEA: Cement Mortar GPA: Geopolymer Martar	Animal	-0: No fibre	110 1%-10 mm
		-H: Horse	140 1%-40 mm
		-P: Pig	210 2%-10 mm
		-W:Wool (Sheep)	240 2%-40 mm

Table 4 shows the properties of the fibres used in this study. The fibre diameter was measured with a Vernier calliper. The specific gravity of the fibres was referred to the value of previous research [7]. All fibres were dried and then cut to length.

Table 4. Fibre Properties

Fibre		size(mm) Diameter	size(mm) length	specific gravity ρ (g/cm³)	Aspect ratio	Photo
Plant	Sisal	φ0.14	10	15.0	71	
			40		2857	
	Coir	φ0.15	10	15.0	67	
			40		2667	
	Bamboo	φ0.50	10	13.2	20	
			40		800	
Animal	Horse	φ0.78	10	13.5	13	
			40		513	
	Pig	φ0.13	10	13.5	77	
			40		3077	
	Wool (Sheep)	φ0.02	10	13.2	500	
			40		20000	

2.2 Specimen Preparation and Curing

Figure 1 shows the process of test specimen preparation. Fibre-reinforced mortar was manufactured using a mortar mixer. The mixing procedure adopted is as follows: (1) add cement, sand, and mix for approximately 1 min; (2) gradually add the fibres as the mixer is running over a period of 2 min, taking care that no fibre balls are formed; (3) add water and mix for 3 min; and (4) stop the mixer. After the mixing process, the fresh fibre-reinforced mortar was tested for flow measurement. For the geopolymer mortar, in step (1), two powders and sand are mixed for 1 min, and in step (3), alkaline solution is added and mixed for 3 min. All the specimens are compacted through the table vibrator for 2 min. After curing in a laboratory for a few days, they are then demoulded and cured in a laboratory with a room temperature of 10 °C and a humidity of approximately 55% until the testing age of 28 days.

From each mix, moulded specimens are manufactured for compressive, tensile and flexure tests in the hardened state. The tensile specimens are cylinders with φ100 mm cross sections and heights of 200 mm. Flexural specimens are prisms with 40 mm square cross sections and a total length of 160 mm tested by 3-point loading on a span of 100 mm. Three specimens for testing in each condition are thus obtained from each batch.

Fig. 1. Specimen preparation

3 Results and Discussion

3.1 Fresh State

Figures 2, 3 and 4 present the test data on the flow test. The flow value of the fibre-reinforced mortar decreases as the total fibre amount increases (see Fig. 3). This is thought to be because the fibres entrained air, and the air could not be fully removed even after compaction. From Fig. 4, B240 and C240 show that the flow values are small. Here, the total fibre amount was calculated by multiplying the fibre content (% by volume) and the fibre length (mm).

In addition, there is no change in this trend between plant fibres and animal fibres. However, each type of fibre has its own characteristics. Flow tends to be small for coir and sisal fibres, and even for animal fibres, regardless of the total amount, the flow decreases when fibre is added. It is thought that the mortar became dry since the fibre lost water immediately after mixing.

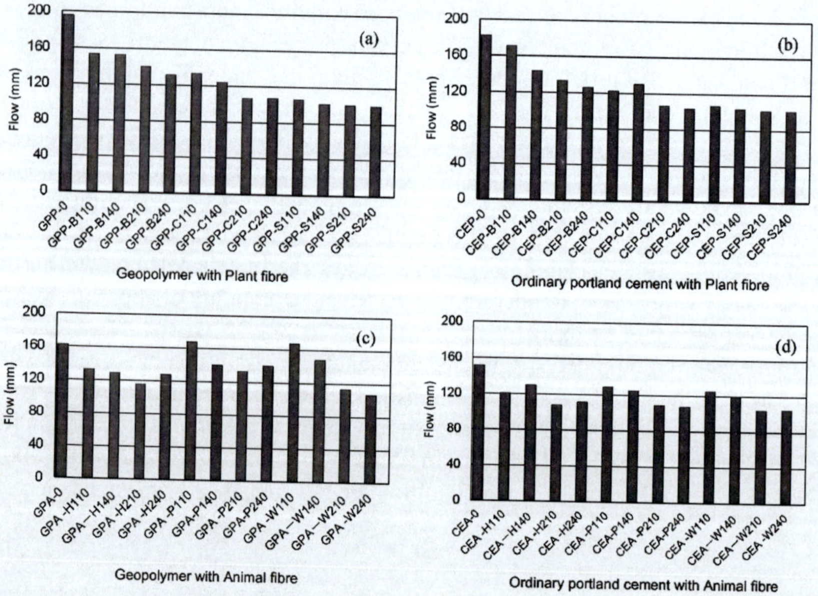

Fig. 2. Flow (a) GPP (b) CEP (c) GPA (d) CEA

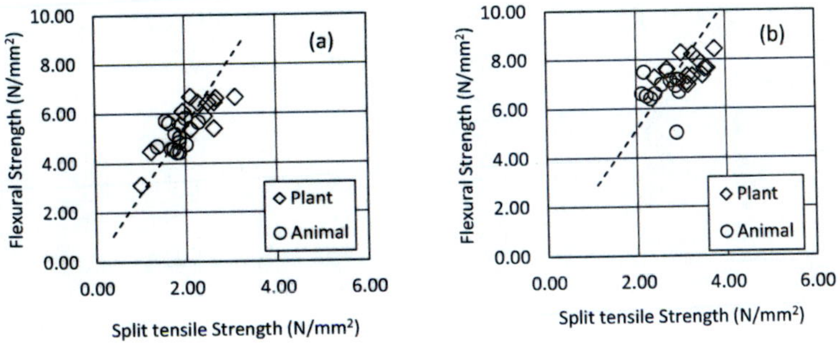

Fig. 7. Comparison of the flexural and split tensile strength of (a) cement mortar and (b) geopolymer mortar for plant and animal fibre

elastic modulus due to fibre reinforcement. In addition, the elastic modulus of animal fibre-reinforced mortar is slightly higher than that of plant fibre-reinforced mortar.

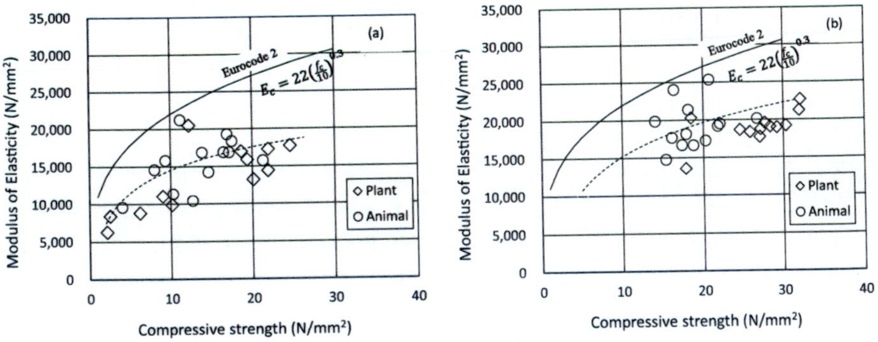

Fig. 8. Relations between the Compressive Strength and Elastic Modulus of Mortar: (a) Cement; (b) Geopolymer with the equation of Eurocode 2 [13]

Flexural Strength

Figure 9 presents the value of flexural strength for plain and fibre-reinforced mortar. The minimum and maximum measured values for each case are also shown in each figure. In Fig. 9, the dashed line presents the value of plain mortar (no fibre). In terms of flexural strength, the effect of adding fibres is small in the case of geopolymer mortar, but the value of decrease is large in the case of cement mortar. Moreover, it turns out that this is large with the cement mortar with plant fibre. It can be said that this is because plant cellulose is decomposed by the alkali of cement and loses its durability.

Figure 10 shows the relationship between the compressive strength and flexural strength. The flexural strength has a strong relationship with the compressive strength, and the experimental results of this study show a similar tendency.

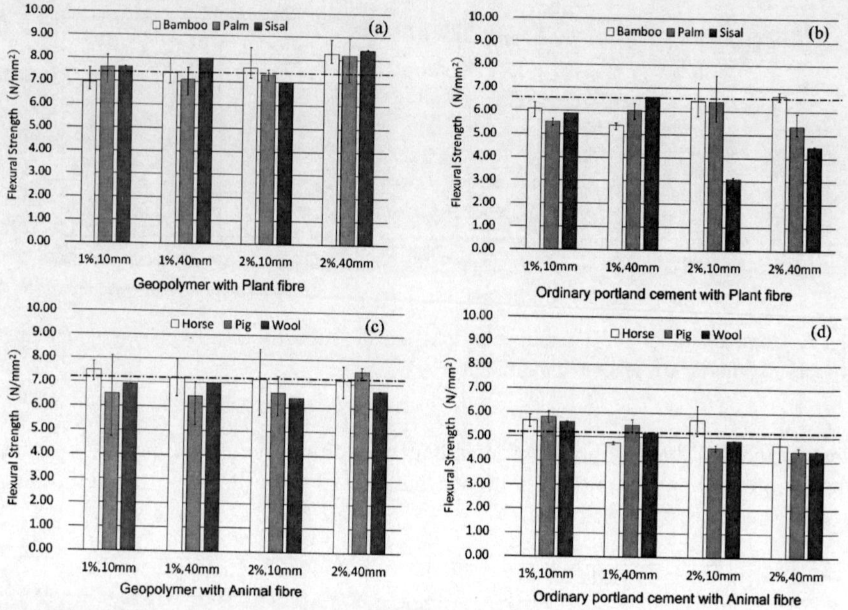

Fig. 9. Variation in flexural strength with fibre: (a) GPP, (b) CEP, (c) CEA, and (d) GPA

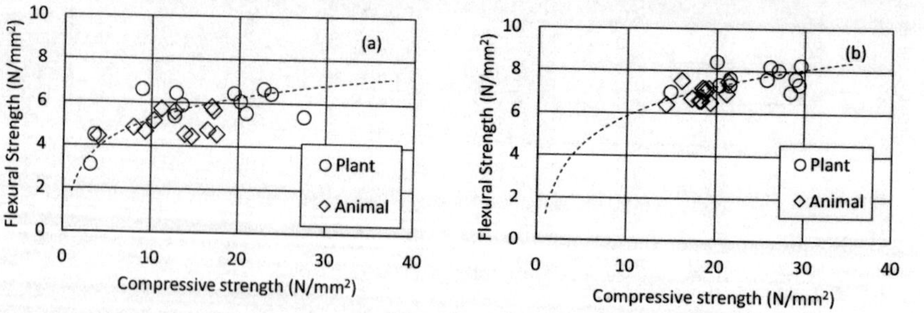

Fig. 10. Relations between Compressive Strength and Flexural Strength: (a) Cement; (b) Geopolymer for Plant and Animal Fibre

4 Conclusions

In this study, by mixing natural plant and animal fibres as reinforcing materials into construction materials using geopolymers as solidifying materials, we developed low-cost, low-technology, and environmentally friendly building materials. Fundamental experiments were conducted on natural fibre-reinforced cement composites with improved mechanical properties. In this experiment, three types of plant fibres (sisal, coir, and bamboo) and three types of animal fibres (horse hair, pig hair, and sheep wool) were used as reinforcing materials. Several experiments were conducted on the strength properties (compression, tension, and bending) of hardened fibre-reinforced cement composites with different fibre types, lengths, and mixed amounts. Based on the evaluation of the experimental data, the following was observed:

1) The compressive strength shows a reduction of less than 20–40% for cement mortar and 10% for geopolymer mortar. This reduction could be related to the increase in porosity of the fibre-reinforced mortar. In some specimens of geopolymer mortar, the compressive strength is slightly higher.
2) The flow value of the fibre-reinforced mortar decreases as the total fibre amount increases. This is thought to be because the fibres entrained air, and the air could not be fully removed even after compaction.
3) In flexural strength, the effect of adding fibres is large in the case of cement mortar with plant fibre. It can be said that this is because plant cellulose is decomposed by the alkali of cement and loses its durability.

References

1. Ghavami, K.: Bamboo as reinforcement in structural concrete elements. Cem. Concr. Compos. **27**, 637–649 (2005)
2. Adeniyi, A.G., Onifade, D.V., Ighalo, J.O., Adeoye, A.S.: A review of coir fiber reinforced polymer composites. Compos. Part B: Eng. **176** (2019)
3. Terai, M., Minami, K.: Fracture behavior and mechanical properties of bamboo fiber reinforced concrete members. Procedia Eng. **10**, 2967–2972 (2011)
4. Filho, R.D.T., Joseph, K., Ghavami, K., England, G.L.: The use of sisal fibre as reinforcement in cement based composites. Rev. Bras. De. Eng. Agrícola E Ambient **3**, 245–256 (1999)
5. Araya-Letelier, G., Antico, F.C., Parra, P., et al.: Fiber-reinforced mortar incorporating pig hair. Adv. Eng. Forum **21**, 219–225 (2017)
6. Maia Pederneiras, C., Veiga, R., de Brito, J.: Rendering mortars reinforced with natural sheep's wool fibers. Mater. **12**(22) (2019)
7. Thomas, S., Pothan, L.A.: Natural Fibre Reinforced Polymer Composites: From Macro to Nanoscale, 1st edn. Old City Publishing (2009)
8. Ranjbar, N., Zhang, M.: Fiber-reinforced geopolymer composites: a review. Cem. Concr. Compos. **107** (2020)
9. de Azevedo, A.R.G., et al.: Natural fibers as an alternative to synthetic fibers in reinforcement of geopolymer matrices: a comparative review. Polymers **13** (2021)
10. Karimah, A., et al.: A review on natural fibers for development of eco-friendly bio-composite: characteristics, and utilizations. J. Market. Res. **13**, 2442–2458 (2021)
11. Kavitha, S., Felix Kala, F.T.: A review on natural fibres in the concrete. Int. J. Adv. Eng. Technol. 32–35 (2017)
12. Shadheer Ahamed, M., Ravichandran, P., Krishnaraja, A.R.: Natural fibers in concrete – a review. In: IOP Conference Series: Materials Science and Engineering, vol.1055 (2020)
13. European Committee for Standardization. EN 1992-1-1 Eurocode 2: Design of Concrete Structures Part 1-NA, 1. General Rules and Rules for Buildings; European Committee for Standardization: Brussels, Belgium (2004)

Fire Resistance of Wood Fiber Reinforced Concrete

Selina Vaculik(✉) and Thomas Matschei

Institute of Building Materials Research, RWTH Aachen University, 52062 Aachen, Germany
vaculik@ibac.rwth-aachen.de

Abstract. One of many possible approaches to reduce the carbon footprint and improve the resource efficiency of building materials is to use wood fibers as alternative fibers and aggregate substitutes from wood processing in concrete. In addition to the interaction between wood fibers, concrete matrix, and the resulting structures, the fire resistance of a hybrid system is of particular interest. Two test rigs were developed to test the material behavior under high-temperature conditions. In one, the 4-point bending tensile strength was tested at 105 °C, 600 °C, and 800 °C with a heating rate of 10 K/min to simulate the residual load-bearing behavior due to the temperature strain in the case of fire. A second test was set up to simulate flame exposure in case of fire. The specimens were exposed to flames at predetermined intervals using a high-propane burner. At a maximum test duration of 70 s, temperatures up to 1450 °C were reached at the specimen surface. Surface tensile strength measurements and structural analyses then investigated the influence of the added wood fibers.

To assess how the novel concrete formulation affects fire resistance, flexural strength tests and flame exposure trials were carried out on different test specimens of a reference concrete and several wood fiber-reinforced concretes with varying fiber contents. Particularly in the surface tensile tests exposed to flame, no decrease in strength was observed for the wood fiber-reinforced specimens, whereas for the reference test, the strength decreased by about 10%. Based on the test results, the influence of the wood fibers on the fire resistance could be determined more precisely and enable a better understanding of this type of new concrete construction.

Keywords: Fire resistance · wood fiber · concrete

1 Introduction

Sustainability is one of the megatrends of current research [1]. With the social movement "Fridays for future", the UN Climate Change Conference series with the latest COP 27 in Egypt, and the restructuring of many companies, it can be substantiated that sustainability is becoming an increasingly important topic for society, politics, and the economy. Sustainability is also of particular interest to the construction industry. As a result of the vast material consumption, this branch is considered one of the main contributors to global warming [1]. In addition, the world population continues to rise [2]. Accordingly,

S. Amziane et al. (Eds.): ICBBM 2023, RILEM Bookseries 45, pp. 490–500, 2023.
https://doi.org/10.1007/978-3-031-33465-8_38

the demand for sustainable building materials is also increasing, and their potential is thus of utmost importance for industry and society. With more sustainable building materials, in which recycled or renewable raw materials are utilized, the urgently needed decarbonization of the construction sector can be advanced. Concrete is currently one of the most important and widely used structural building materials. However, sustainable and economical alternatives must be found due to the high CO_2 emissions associated with the production, transport, and use of concrete and its raw materials [3].

One possibility would be to supplement and partially replace aggregates in concrete with recycled or renewable raw materials. Therefore, the use of wood fibers in concrete represents a promising method for producing sustainable concrete. Wood is considered a natural carbon store and, in the form of fibers or chips, can replace the aggregate in concrete to a certain extent. However, since wood particles have hardly any use in concrete [4], the effects of using wood fibers are not sufficiently understood. In particular, because of the unknown influence of wood particles on various physical properties of concrete and the influence of hemicellulose on the chemical reactions [5]. In the past wood, fibers were used to produce wood-wool lightweight panels, wood particle concrete, or cement-bonded particle boards.

The influence of different wood fiber contents in reinforced concrete specimens were investigated during an ongoing research project at the Institute of Building Materials Research, RWTH Aachen University (ibac). This project aims the development of sustainable concrete formulation in combination with a new hybrid construction method (wood + concrete) to reduce carbon footprint by up to 40%. A critical component of these new wood-concrete composites to be investigated is their fire and temperature resistance. Within this work, the fire resistance of wood fiber reinforced concrete was investigated.

The high-temperature behavior of concrete is composed of the various reactions of the individual constituents, cement paste, and aggregate. Figure 1 [6–12] summarizes the temperature-relevant chemical and physical reactions available in literature. In addition, there are interactions between the aggregate and the hardened cement paste matrix. These factors lead to different mechanical behaviors of the building material after temperature stress [7, 13, 14].

When describing strengths of fiber and high temperature exposed concrete or mortar specimens, a distinction must first be made between hot strength and residual strength. The specimen is tested in the heated state for heat resistance, while the residual strength is determined after the cooling process. In general, the hot strength is higher than the residual strength, which is caused by further destruction mechanisms during the cooling phase. Since this work focuses on determining the heat resistance, this will be mainly considered in the further course [15].

In the experimental investigation of materials under temperature exposure, a strict distinction must be made between fire exposure and high-temperature exposure. In the case of fire exposure, the load-bearing capacity of an entire system component is investigated in a fire test, i.e., it is subjected to a standardized temperature curve. On the other hand, high-temperature tests characterize the material under a defined temperature exposure.

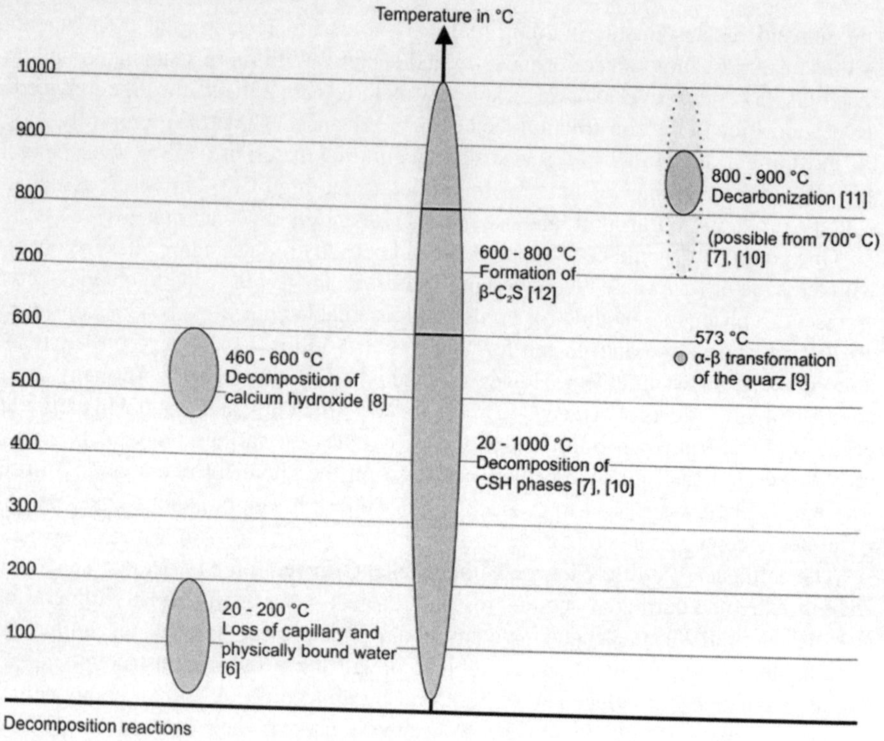

Fig. 1. Essential chemical and physical reactions of concrete during temperature treatments.

A constant heating rate is usually applied to heat to the desired temperature. According to DIN EN 1992-1-2 [16], heating rates between 2 K/min and 50 K/min are used. A distinction is made in high-temperature tests between two types of tests: steady-state and transient tests. In steady-state tests, the temperature is first applied, and then the specimen is loaded [17].

During heating, the specimen should be supported in a force-free manner [18]. After reaching the target temperature, the specimen is loaded until failure. Any creep deformations that may occur cannot be reproduced with the test; instead, an isothermal stress-strain line is determined [19]. In transient or transient creep tests, a constant load is applied, and then the temperature is increased.

The tests carried out here were performed with a steady-state test setup.

2 Materials and Methods

For the batch production of the three investigated concretes, a common Portland fly ash cement with silica-rich fly ash was used in all experiments. The mixture designs are given in Table 1.

A particle size distribution similar to A/B16 according to DIN 1045-2 [20] was selected as the grading curve. The wood fiber content is 7.5% by volume of the mixture

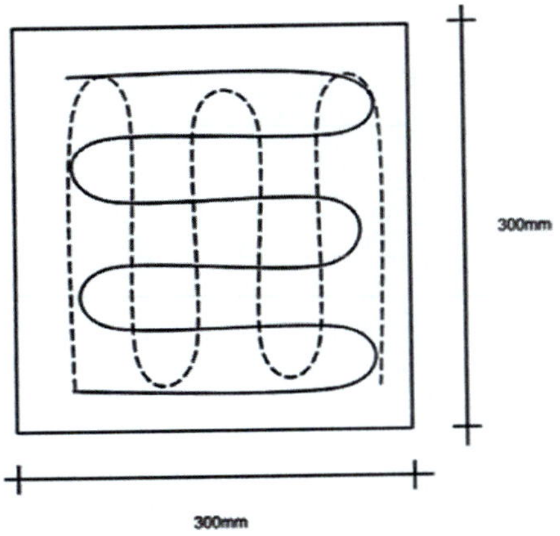

Fig. 3. Movement pattern of the flame exposure of the plates

2.2 High Temperature 4-Point Bending Tensile Test

The test setup developed in [23] was used to investigate the residual strength under the relevant temperature load. In deviation from the standard, this is a small-scale 4-point bending tensile test setup that works with adapted dimensions. The deviating dimensions were chosen to test in the constricted conditions of the high-temperature furnace (Fig. 4).

For the temperature load, a blast furnace, also used in [24] and [25], was installed in a universal testing machine with a maximum load of 150 kN. The furnace has a volume of 30.2 dm^3 and a maximum heating rate of 10 K/min. All concrete mixtures samples were tested at either a temperature of 600 °C or 800 °C. A constant heating rate of 10K/min was used to increase the temperature furnace to the appropriate temperature. Then the temperature was kept until all thermocouples indicated an exact temperature and no temperature gradients could be measured within the oven. Subsequently, the specimens were loaded to failure by 4-point bending. After that, the furnace was cooled down to 20 °C until the next test cycle.

3 Results and Discussion

3.1 Flame Exposure

The slices of the reference compound showed very few visible optical changes. In some cases, individual fragments of the matrix had splintered off. Under the influence of the direct flame, isolated spots glowed on the surfaces of the compounds with wood fiber addition. However, these were extinguished immediately after the removal of the gas burner. Thereafter, a slight odor of burnt wood was detected in some areas, but visible smoke development did not occur at any time.

At a maximum test duration of 70 s, temperatures up to 1450 °C were reached at the specimen surface. After cooling, the surfaces showed burned wood fibers as shown in Fig. 5. These observations occurred regardless of the loading duration but more clearly

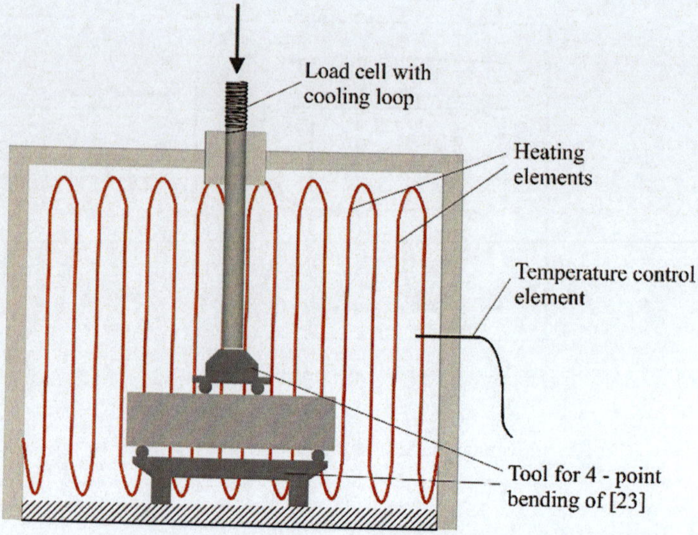

Fig. 4. Setup of the high-temperature furnace with 4- point bending tensile test;

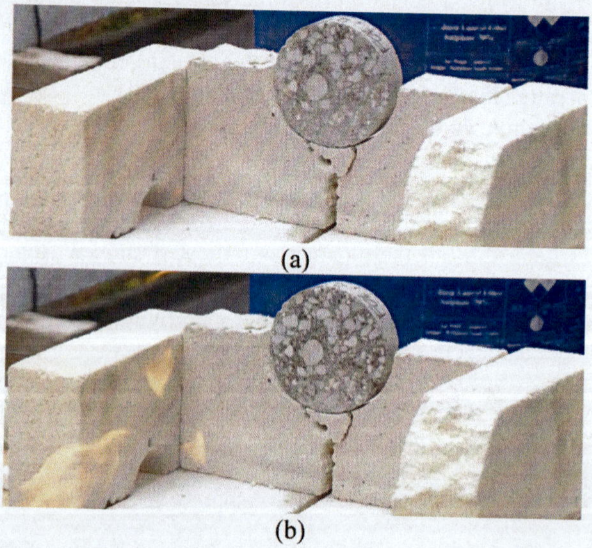

Fig. 5. Optical surface change at the beginning (a) of flame exposure and after 70 s (b)

at 70 s. For all specimens, the mass loss was 0.25 wt % (±0.01%) at 35 s flame exposure and 50 cm spacing and 1 wt % (±0.01%) at 70 s and both 50 and 40 cm spacing.

The results of the determined surface tensile strength are shown in Fig. 6. It is specifically noticeable that the fracture surfaces of the wood-fiber-reinforced specimens (M1 M and M2 M) are located outside the stressed area at a depth of 10–15 mm inside the specimen. A near-surface failure would be expected for all flame-loaded sheets since the most severe microstructural damage is expected in the flame load's immediate area. However, only the reference mixture without wood fibers and the mix with the water glass-impregnated fibers (M1 W) had their fracture surfaces in the surface region. All others failed in the area of the maximum drilling depth.

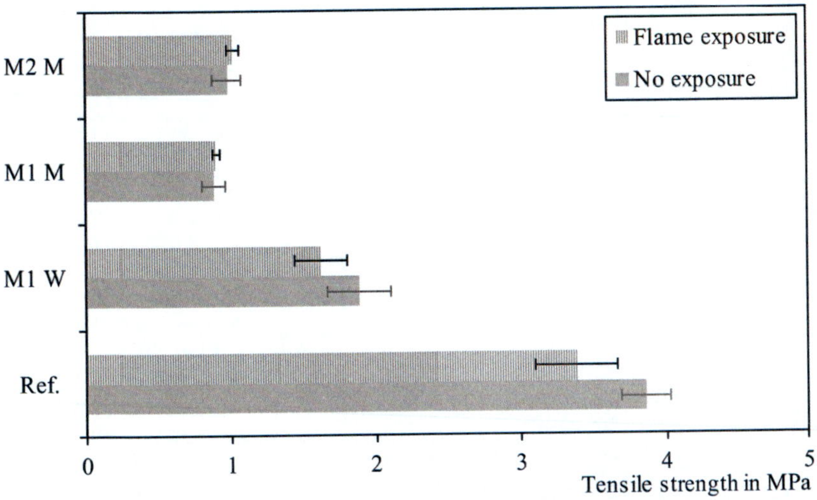

Fig. 6. Results from surface tensile strength tests in MPa

In general, it can be observed that the wood reinforced concrete specimens, whether flamed or unflamed, have lower strengths compared to the reference. Further improvements must be made here to achieve higher strengths However, a decrease in strength between the fire-stressed and non-fire-stressed specimens is still visible for the reference specimen and the M1 W mixtures (Fig. 6). In contrast, no significant change in tensile strength between the fire-stressed and the non-fire-stressed slabs was observed for the mineral-impregnated specimens (M1 M and M2 M).

3.2 High Temperature 4-point Bending Tensile Test

Table 2 shows the residual load capacities after exposure to 600 °C and 800 °C. In each case, the residual load-bearing capacity refers to the load-bearing capacity of the respective mixture of pre-tempered specimens at 105 °C. The sample matrix studied was used to compare the M1 and M2 mixtures and to relate the strength reduction to a sample without wood fibers.

A similar decrease in load-bearing capacity can be observed for all concrete mixes. Mixture M1 shows the highest strength reduction. Since the only difference between mix design M1 and M2 is the inclusion of wood fibers in the particle size distribution of M2, the improved results can probably be explained by this. The proportional load drop of M2 is comparable to that of the reference compound, whereas the flexural strength at 800 °C was even determined to be higher at 0.97 N/mm^2. Presumably, additional pore space was released by the burned wood fibers, which restricted the expansion of the

Table 2. Parameters and results of the 4-point bending tests

Concrete mixture	Testing temperature	Residual strength in 4-point bending test
	°C	%
M2 M	105	100
	600	66
	800	27
M1 M	105	100
	600	46
	800	20
Ref.	105	100
	600	67
	800	14

specimen to a reduced extent under the temperature load and thus led to lower restraint stresses.

4 Conclusion

Based on the test results, the influence of wood fibers on fire resistance was investigated, enabling a better understanding of this new type of concrete construction. No drastic loss of strength was identified by determining the residual bearing strength at the corresponding high temperatures compared to the unreinforced concrete. The fact that no afterburning behavior occurred during the flame tests has a positive effect in favor of using the wood fiber reinforced concrete hybrid system in the construction sector. Whether an assured improvement of the fire resistance, similar to the addition of polypropylene fibers, is given by the addition of wood fibers could be assumed from the results obtained, but this must still be sufficiently pursued and validated with a larger sample size and further investigations.

References

1. Horx-Strather, O.: Home report 2022 - the future of housing and construction. Zukunftsinstitut (2021)
2. Unnerstall, T.: Sustainability Fact Check - Ecological Crises and Resource Consumption Under the Microscope. Springer, Heidelberg (2021)
3. Lübbert D.: CO_2 footprints of different energy sources in comparison - about the climate friendliness of fossil energies and renewable energies. Wissenschaftliche Dienste des deutschen Bundestags (2007)
4. Garrecht, H., Klatt, A.: Concrete from renewable resources. In: 2. Darmstädter Ingenieurkongress Bau und Umwelt (2013)

2.2 Hemp Fibres

Hemp fibres are classified as bast fibres [5]. The cultivation of hemp is one of the oldest non-food producing plants, which have been cultivated for over 8500 years [9]. Today, hemp is cultivated especially in Canada, China, and Europe [10]. Similar to other bast fibres, the natural fibres in hemp are bundled. A distinction is made between primary and secondary fibres. The primary fibres are used for textile applications, but the secondary fibres are not used because of their short length. During fibre extraction, the epidermis is dissolved and the fibres are detached from the inner core. The resulting by-products of fibre extraction are used for animal feed [6].

The fibre properties of hemp vary depending on the region, the cultivation of the plant and due to variably formed fibres in a plant. The hemp plant is stable to fluctuating temperatures and drought. Due to a distinctive root system, the roots can absorb water up to a depth of two metres, which only interrupts growth in periods of drought. Furthermore, hemp thrives at temperatures between 5–27 °C and with a water demand of 500–700 mm per year [6].

The cultivation of hemp limits the spread of fungi and weeds and provides an optimal soil for other plants or for hemp itself. Hemp, unlike flax, is self-tolerant and does not require crop rotation. However, crop rotation is beneficial as, for example, the yield of wheat could be increased by 10–20% after cultivating hemp. Furthermore, hemp can be used as a plant for land reclamation and for the absorption of heavy metals in the soil during the growth phase. No pesticides are used in the cultivation of hemp in the UK, Germany, and the Netherlands [11].

For cultivation, 150–450 seeds per m^2 are sown, which are sown from the end of March to mid-April in order to obtain a high yield of fibre. For this purpose, the harvested hemp is cut with the help of harvesting machines and laid down in rows on the field for dry roasting, similar to flax harvesting. Afterwards, the roasted plants are broken and swung to separate the shives from the fibres [6].

2.3 Coconut Fibres

Coconut fibres are assigned to the category of fruit fibres [5, 6]. Coconut fibres are obtained from the mesocarp of the coconut. This part of the coconut is located between the outer layer (exocarp) and the inner, edible and drinkable, part (endocarp) of the nut. There are two types of coconut fibres: The white and the brown fibres. The white fibres are softer, finer, and have lower mechanical properties compared to brown fibres. Brown fibres are thicker, firmer, and more abrasion-resistant compared to white fibres. White fibres are obtained from unripe coconuts, which are between 10 and 11 months old [6]. Brown fibres are obtained from ripe coconuts that are 12 months old [6]. Due to the above-mentioned properties and the better suitability of the fibres as a composite material, the evaluation only considers brown coconut fibres [9].

90% of the world's coconut production is produced in India and Sri Lanka as permacultures. The main growing regions of coconut palms are located along coastal areas in humid tropical regions. Coconut palms require temperatures between 20–32 °C, with an optimum temperature of 27 °C. The pH value of the soil should be neutral, at seven [7]. For high yields, 1500–2500 mm of precipitation are needed annually. Because the cultivation regions are located in monsoon areas, high annual precipitation is guaranteed. Nevertheless, coconut palms suffer severely from drought after a period of three to six months [9].

Coconut palms can live up to 120 years and produce coconuts from the age of 12 to 80. However, due to the high trunk height and the associated difficulty in harvesting, palms are often felled from the 30th year of standing. Coconuts are harvested either by palm climbers or from the ground. Brown fibres are used in particular for brushes, doormats, bags, car poles or as geotextiles [6].

2.4 Sisal Fibres

The fibres of the sisal plant are classified as leaf fibres [5]. The largest growing countries of sisal are Brazil, Venezuela, Kenya, Indonesia, Mozambique, and Madagascar. Sisal grows all year round and has a height of 0.6 m two years after sowing. During the growth phase, which lasts between seven and 12 years, a sisal plant produces 120 to 180 leaves, which are used for fibre production. The number of leaves depends on the type of plant, the growing region, the growing altitude, and the amount of rainfall. The leaves of the sisal plant can grow up to 1.20 m long, 20 cm wide and 10 cm thick when fully grown. A subtropical climate is needed to cultivate the sisal plant. Temperatures of 10–32 °C and an annual rainfall of 600–1500 mm are prerequisites for the plant's growth. Sisal is not frost-resistant and cultivation in rainy regions can be problematic due to weed development. The soil should have a pH of four to six [12].

During cultivation, no fertiliser is needed for a newly cultivated field, whereas fertiliser in small amounts is beneficial for existing fields. The use of pesticides is not needed as sisal is resistant to diseases. However, care must be taken to control weeds during the first two years of cultivation. Weed control can be done manually, mechanically or chemically. Artificial irrigation is not required for the cultivation of sisal, as the plant is drought-resistant and cultivated as a rain fed crop [12].

The sisal plant is ready for harvest for the first time about three to four years after sowing. After harvesting, the process of dehulling or a water-roasting is carried out [12].

2.5 Properties of Natural Fibres

The following Table 1 lists the mechanical and physical properties of the before mentioned natural fibres. These are compared with the properties of glass and steel fibres.

The listed natural fibres have a lower density and modulus of elasticity compared to glass and steel fibres. The tensile strength of natural fibres is comparable to steel fibres. The elongation at break and water absorption behaviour is higher than that of glass and steel fibres. Especially the water absorption behaviour is relevant for the processing with concrete. Hemp and flax fibres were selected for the following study, because of the higher tensile strength and the lower moisture absorption compared to the other bio-based fibres.

Table 1. Properties of natural fibres in comparison to glass and steel fibres [3, 9, 13–16].

Fibre	Density (g/cm^3)	Tensile Strength (N/mm^2)	E-Modulus (GPa)	Elongation (%)	Moisture absorption (%)
AR-Glass (Roving)	2,7	1800–2000	73	4,3	<0,1
Steel	7,8	200–1600	210	1–2	0
Flax	1,4–1,52	345–1500	8–100	2,7–3,2	7
Hemp	1,4–1,6	310–1110	3–90	1,6	8
Sisal	1,45	468–700	9–38	3–7	11
Coconut	1,15–1,46	131–220	2,8–6	15–51,4	10

3 Challenges and Opportunities of Natural Fibres in the Construction Industry

The use of natural fibres offers the possibility of obtaining a sustainable, environmentally friendly, and energy-saving reinforcement in concrete. Comparable properties of conventional reinforcement can be achieved in certain areas. However, the cultivation and processing of natural fibres, and the use of natural fibres as reinforcement in concrete is fraught with challenges. In this chapter, the challenges and opportunities of natural fibres are discussed. The special challenges in the concrete environment and the reinforcement in concrete are discussed below.

3.1 Challenges

Typical challenges of natural fibres or biogenic materials in general lie in the non-uniform properties of the materials (Table 1). In general, the influencing factors of the fibre properties can be divided into two categories: The cultivation, and the natural origin, such as the type of plant. Some factors that have an influence on the properties of natural fibres are listed below as examples: Cultivation area, cultivation region, time of sowing, type/variety, temperature, precipitation, moisture, fertiliser use, nutrient situation, harvest time, harvesting method, fibre extraction method, processing of the fibres, storage, and other influences. These factors result in different mechanical properties and chemical compositions. For example, the yields of coconuts per year and per palm of the Kalpa Sreshta species are twice as high as those of the Kalpa Sankara [17]. The timing of sowing has similar effect. If hemp is sown in late March, it produces twice as much yield as hemp sown in mid-May [6]. In addition, the measured values depend on the condition of the plants or the fibre as well as on the measuring method itself. The moisture content, the contamination of the fibre or the infestation of mould can have an impact on the measured values of the fibres. Synthetic fibres, on the other hand, can be produced in large quantities with the same properties. For example, the production speed of polyamide fibres is about 4000 m/min [18]. In contrast to synthetically produced fibres, it is difficult to obtain fibres of consistent quality and properties [16].

Further challenges arise due to poor fire resistance, susceptibility to UV light, bacterial and fungal infections. In addition, natural fibres have a low durability due to natural decomposition and a low thermal resistance [19]. The land mass used for agriculture worldwide is 37%, of which 12% is cultivated as arable land and 25% as permanent meadows and pastures [20]. However, climate change is shifting cropping regions, increasing water scarcity and droughts, which in turn increases soil degradation. The agricultural sector in particular requires large amounts of water for the cultivation of crops in certain regions of the world and especially in regions already suffering from water scarcity [21]. Inconsistent weather patterns and future climatic shifts will make it increasingly difficult to create a crop or fibre with uniform properties.

Another problem or challenge for the cultivation of natural fibres is the clearing of forests to create arable land for crops. Such clearing is used, for example, for the production of large pineapple plantations [22]. In addition, the roasting of plants for fibre production leads to health and environmental damage. The level of impact depends on the length and type of roasting [6]. In addition, it must be mentioned that the fibres used for reinforcement are currently extracted using machines and processes that are only used for home and clothing textiles [23]. One concern that arises in the context of natural fibre production is the issue of displacement of farmland for food production. In the following, various reasons are given to dispel this concern. Natural fibres are sometimes grown in regions where it is difficult to cultivate food crops. Grain plants are susceptible to drought and heat, in contrast to plants such as abacá, sisal, cotton or hemp. Thus, despite the difficulty or impossibility of growing food crops, plants can be grown in these regions, which in turn can be used economically. In addition, plants for fibre production and plants for food production are grown as crop rotation. In this way, crop yields can be increased and the food supply can be ensured. In particular, the combined cultivation of different plants on one field leads to a positive development of yields and to a reduction of the pesticides and fertilisers used [8]. In particular, the by-products of the plants from fibre production can be used as food. This is possible, for example, with coconut palms or pineapple plants. In addition, the seeds of plants such as flax or hemp can be used in the food industry [6].

Another challenge arises from the strong alkaline environment to which the natural fibres in the concrete are exposed. In the following, the material clay and its environment will be discussed. In addition, procedures and methods are presented on how natural fibres can be protected from the environment. Concrete has a pH value of 12.5 to 13.5 and is therefore alkaline [24]. The high pH value arises during the setting process of the calcium hydroxide ($Ca(OH)_2$) [25]. The setting process creates a bond between the aggregate and the matrix, which consists of cement crystals. The high alkaline environment attacks the natural fibres and decomposes them over time. The main cause of the decomposition of the fibres is the alkaline solution, which dissolves the lignin and hemicellulose that bind the individual cells of the fibres [4]. The pH of the environment can be adjusted by pozzolanic or latent hydraulic additives. In addition, the selection of certain types of cement, such as blast-furnace or alumina cement, which have a low alkali content, can lower the pH value of the matrix as described in DIN 1164–10 [26]. Furthermore, the pH value can be lowered by carbonation. Such carbonation can occur naturally or be artificially induced. Carbonation has no disadvantages with regard to the cement [27].

3.2 Opportunities

Natural fibres are sustainable, biodegradable, and renewable. The cultivation and fibre extraction of natural fibres is environmentally and energy friendly [6]. Natural fibres absorb more CO_2 from the atmosphere during their cultivation than they produce during their extraction, transport and storage within the value chain [10]. The impact measured by the CO_2/eq of glass fibre production compared to the extraction process of natural fibres is four and a half times higher [10]. In terms of composites, natural fibres have a 10–50% lower greenhouse gas emission. If the CO_2 uptake during cultivation is also taken into account, the reduction in greenhouse gas emissions is increased to 30–70% [10]. The use of natural fibres could limit the use of fossil raw materials and thus protect the environment [7]. The by-products of fibre production do not have environmentally harmful effects and do not have to be specially disposed of [7]. In contrast, the by-products can be used as natural fertiliser, in the food industry, in the pharmaceutical industry or in other areas [6]. Often the fibres are a by-product of a cultivated plant, such as the coconut palm or the pineapple plant, which can reduce waste [6]. Furthermore, the cultivation of plants for the production of natural fibres creates jobs, especially in economically weak countries. For example, more than eight million families in India and Bangladesh are employed in the production of jute [6]. As the cultivation of jute plays a particularly important role for Bangladesh, the largest jute-growing country with an export volume of US$ 106 million [28].

Furthermore, the cultivation of natural fibres can clean up wasteland or chemically polluted soil. Natural fibres such as hemp or kenaf can filter heavy metals and chemicals from the soil during their growth phase and fertilise a nutrient-poor soil [11]. This can prepare the soil for other crops, which has been shown to have positive effects on crop yields [11]. Nowadays, several natural fibres have been established in areas such as the automotive industry as interior linings or in the construction industry as geotextiles or as thermal insulation [6, 7, 10]. The use of natural fibres as thermal insulation in particular has become established due to their thermal conductivity and acoustic insulation properties [19]. Here, the fibres themselves or as a composite material have similar or better properties than conventionally used materials. In particular, the advantages of natural fibres compared to conventional materials lie in a more environmentally friendly production and the reduction of the use of fossil, non-biodegradable materials [10]. Due to the push for sustainable products in all economic sectors, the use of natural fibres as reinforcement in concrete could be further advanced in the future. However, this use also has challenges and problems that need to be overcome, as discussed before.

4 Materials and Methods

Based on the previous results, flax and hemp fibres were selected for the tests. These were introduced into the concrete in various ways as short fibre reinforcement. Important parameters for fibre concretes, which significantly influence the properties of the material, are the fibre content, length, and orientation. For the fibre alignment, three different types of dry fibre insertion and alignment were investigated (Fig. 1).

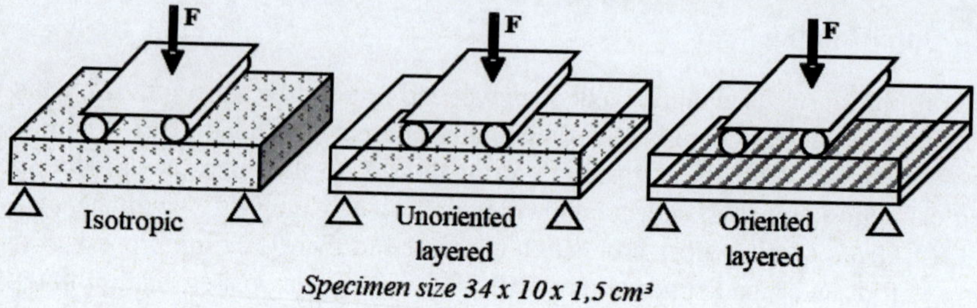

Fig. 1. Method of Contribution and texting method

It can generally be stated that with increasing fibre content (up to a certain proportion) of the total mass, the tensile strength increases and the compressive strength decreases [29]. For the following experiments, the concrete specimens with a fibre content of 0.5% of the total mass are examined. The fibre lengths varied depending on the type of fibre. The hemp fibres varied considerably in length with 8–14 cm, sometimes even up to 24 cm. The flax fibres, on the other hand, were cut in advance to a uniform length of 5.5 cm. Furthermore, parts of the woody plant stem are present in the flax fibres, which makes the fibre mixture coarser and woodier. This could lead to a good adhesion between fibre and matrix. The following table lists the different test series.

Table 2. Specimen.

Fibre Material	Method of contribution	Fibre content [M%]
Flax fibre	Isotropic	0.5
	Unoriented layered	0.5
	Oriented layered	0.5
Hemp fibre	Isotropic	0.5
		1.0
	Unoriented layered	0.5
	Oriented layered	0.5
none	–	–

A total of eight test specimens are produced (cf. Table 2). On the one hand, a control specimen without fibres is produced as a comparative value. Of the seven other concrete specimens, three are made with flax fibre and four with hemp fibre. For each fibre type, a specimen with 0.5% fibre content is produced with an isotropic distribution, with an oriented and with an unoriented layered orientation. The layered oriented fibres are placed in the tension zone, at 1/3 of the specimen height. In addition, a test specimen with hemp fibres with a fibre content of 1.0% and an isotropic distribution is examined.

In this work, flexural test was carried out on flax und hemp fibre reinforced concrete. The Four-point bending test is carried out in accordance with DIN EN 1170-5 for measuring the bending tensile strength [30]. Four each test series, samples with the dimension 34 × 10 × 1,5 cm^3 were produced. The used fine concrete mix was developed in sub-project C1 of the special research area (SFB) 532 "Textile reinforced concrete" at RWTH Aachen University. The concrete mix consists of solid and liquid components. The size of the aggregate is 0.2 mm–0.6 mm. The quantity and volume proportions are listed in the following Table 3.

Table 3. Fine Concrete Mixture.

Component	Quantity [g/dm^3]	Vol. %
Cement CEM I 42.5 R	490	22
Fly ash	175	8
Silica fume (ElkemMicrosilica® 940U)	35	3
Quartz powder	500	23
Sand 0.2–0.6	713	32
Water	280	11
Superplasticizer	7	0.3

The produced samples are stored for 24 h in the mould. Afterwards the samples were stored for 6 days in water and for a further 21 days in room climate before the flexural test were carried out.

5 Results

First, the fibre distribution is evaluated optically at the fracture line before the mechanical test results of the four-point bending test are analysed. The following Fig. 2 shows the cross-section of three different test specimens. The first image is from the flax oriented layered test series, the middle image is from the flax isotropic series and the bottom image is from the hemp isotropic test series.

The layered flax fibres are evenly distributed in a layer at 1/3 of the specimen height between the lower and upper concrete layer. The flax fibres do not break at the edge of the concrete failure but are mainly pulled out of the concrete. In this method of contribution, the fibres sometimes form a separating layer, as the concrete of the upper and lower layer do not bond completely. This is especially visible on the right side, where the fracture edge in both concrete layers is stepwise.

In contrast, the isotopically distributed fibres do not lie in one layer. The flax fibres have partially accumulated at the encircled area and are present in agglomerations. But, the fibres are also pulled out of the concrete and did not break at the edge of the concrete failure. The agglomeration of the hemp fibres is even more distinctive than the flax fibres as shown in the Fig. 2. In the following, the results of the four-point bending test are

Fig. 2. Cross-section of three test specimen

summarized. A representative curve for each set of specimen was selected and can be seen in Fig. 3.

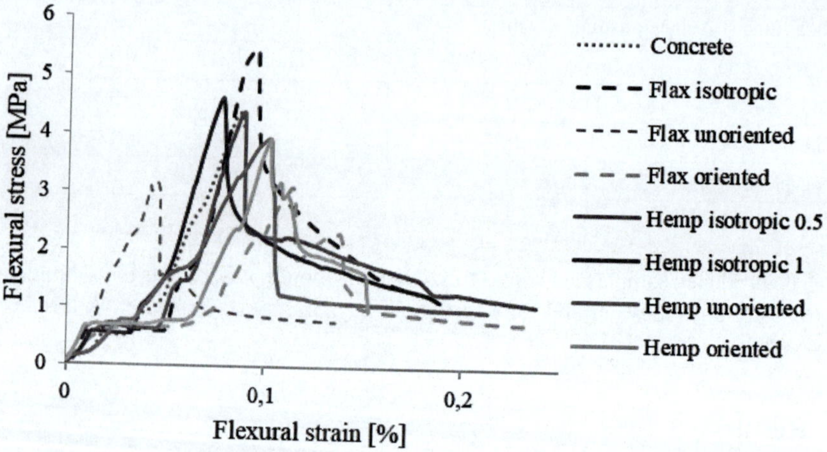

Fig. 3. Flexural strength and strain curves of hemp and flex reinforced concrete

In comparison to unreinforced concrete the flax fibre reinforced concrete specimen are not breaking directly and are showing an elongation. Especially the layered specimen having a gradual curve with two peaks. The isotropic flax fibre reinforced specimen has the highest flexural stress. In comparison the hemp fibre reinforced specimen are having a similar curve with one peak and a mostly quick degradation afterwards. The maximum flexural stress is similar or slightly higher than in the unreinforced concrete specimen.

For the test specimens with hemp fibres, the following evaluation results compared to unreinforced concrete are obtained. The average values for the maximum load (FMOR) and the flexural tensile strength (σMOR) of hemp oriented specimens are on average 2–3% higher. The values for the deflection (ΔMOR) and the elongation at the maximum load (εMOR) are on average 3% lower than those of hemp unoriented. The maximum values of hemp oriented layered specimen show on average between 11–23% better values than for hemp unoriented layered specimen. The discrepancy between the average and maximum values could indicate poor or difficult orientation and/or distribution of the hemp fibres.

For the test specimens with flax fibres, the following evaluation results are obtained. For flax oriented layered specimen, the average and maximum values for FMOR, σMOR, ΔMOR, and εMOR are higher than the values for flax unoriented layered specimen. On average, the values for σMOR show an increase of 9% and for εMOR an increase of 40%. For the maximum deflection and elongation, the specimens with oriented fibres show an improvement of 100% compared to the specimens with unoriented fibres. These results could be due to the provision of the fibres. The flax fibres were of similar length and less entangled, which allowed the fibres to be uniformly oriented.

According to Jonathan Page et al. [29], an optimum ratio of compressive and tensile strength was found to be a flax fibre content of 0.3% and a fibre length of 12 mm [29]. In a comprehensive study by Jawad Ahmad and Zhiguang Zhou [19] on the mechanical properties of natural and synthetic fibres as reinforcement in concrete, showed that a fibre content of up to 1% can improve the mechanical properties of concrete [19]. In other studies, such as by A. C. Da Costa Santos and P. Archbold [31], experiments on hemp, and flax fibres with a proportion of 0.5% and 1.0% of the total mass of the concrete are examined. It was found that a fibre content of 0.5% gave better results than 1.0% [31].

6 Conclusion and Outlook

In conclusion, plant fibres can influence the properties of composite materials, but in the conducted test the influence is still small. This was shown by the use of flax and hemp fibres as reinforcement in concrete. Nevertheless, no long-term tests were carried out on the composites. In particular, the decomposition of the fibres by the alkaline environment and which methods can be used to protect the fibres in the long term need to be researched.

Concluding the results flax fibre reinforced concrete with layered fibre insertion generates a gradual and no direct failure. The Curves are showing two peaks and the specimens are breaking gradually. All hemp fibre reinforced concrete specimens have a similar curve with one peak and an elongation afterwards. Isotropic flax fibre reinforcement shows the highest flexural stress in our investigation. Probably, because of the better fibre-matrix bond and no occurring separating layers compared to the specimens with the layered fibre insertion. Comparing our results with the previous mentioned studies a layered insertion of the fibres was tested. For these specimens, the fibres with a fibre content of 0.5% generate a separating layer. In further investigations, a lower fibre content than 0.5% could be more advantageous for incorporation in the layer. This could prevent the formation of a separation layer and achieve a better fibre-matrix bond. Additionally, further investigations on the fibre-matrix bond with single-fibre pull-out test would be interesting. Furthermore, with regard to the long-term behaviour of the fibres combined tests in a concrete with a lower alkalinity would be important.

Acknowledgments. The work was carried out under the aegis of the PhD-programme "Verbund.NRW", supported by the North-Rhine-Westphalian funding scheme "Forschungskolleg" of the ministry of culture and science.

References

1. Gärtner, C.: C^3 – carbon concrete composite (2019). https://www.bauen-neu-denken.de/c%C2%B3-projekt/. Accessed 28 July 2022
2. Hensen, F., Kolbmüller, H., Stahr, M., Weber, J., Wild, U.: Beton- und Stahlbeton. In: Stahr, M.; Hensen, F.: Bausanierung. 4., vollst. überarb. und aktualisierte Aufl. Vieweg + Teubner, Wiesbaden (2009)
3. Friese, D., Scheurer, M., Hahn, L., Gries, T., Cherif, C.: Textile reinforcement structrures for concrete construction applications – a review. J. Compos. Mater. **56**(26), 4041–4064 (2022). https://doi.org/10.1177/00219983221127181 1
4. Raghoo, Y., Ramasawmy, H., Gooroochurn, M., Chummun, J., See-boo, A., Brown, N.: A further milestone to the use of natural fibres in concrete – past findings, barriers and novel research avenues. In: IOP Conference Series: Materials Science and Engineering, vol. 1203, p. 22038 (2021). https://doi.org/10.1088/1757-899X/1203/2/022038
5. DIN EN ISO 6938: Textiles - Natural fibres - Generic names and definitions (ISO 6938:2012); German version EN ISO 6938:2014, DIN German Institute for Standardization , Germany (2014)
6. Kozlowski, R.M., Mackiewicz-Talarczyk, M. (eds.): Handbook of natural fibres. Volume 1: Types, Properties and Factors Affecting Breeding and Cultivation. Woodhead Publishing, Duxford, Cambridge, Kidlington (2020)
7. Holzmann, G.: Natürliche und pflanzliche Baustoffe. Rohstoff - Bau-physik - Konstruktion. Vieweg+Teubner Verlag, Wiesbaden (2012)
8. Cui, Z., et al.: Agronomic culti-vation measures on productivity of oilseed flax: a review. Oil Crop Sci. **7**, 53–62 (2022). https://doi.org/10.1016/j.ocsci.2022.02.006
9. Muthu, S.S., Gardetti, M.A. (eds.): Sustainability in the Textile and Apparel Industries. Sourcing Natural Raw Materials. Springer, Cham (2020). https://doi.org/10.1007/978-3-030-38541-5
10. De Beus, N., Carus, M., Barth, M.: Carbon Footprint and sustainability of different natural fibres for biocomposites and insulation material. Study providing data for the automotive and insulation industry (2019)
11. Piotrowski, S., Carus, M.: Ecological benefits of hemp and flax cultivation and products (2022)
12. Department of Agriculture, Forestry and Fisheries South Africa: Sisal. Production Guideline, Pretoria (2015)
13. Mechanical and physical testing of biocomposites, fibre-reinforced composites and hybrid composites.Woodhead Publishing, Duxford (2019)
14. Jawaid, M., Mavinkere Rangappa, S., Siengchin, S. (eds.): Bamboo Fiber Composites. Processing, Properties and Applications. Springer, Singapore (2021). https://doi.org/10.1007/978-981-15-8489-3
15. Peças, P., Carvalho, H., Salman, H., Leite, M.: Natural fibre composites and their applications: a review. J. Compos. Sci. **2**, 66 (2018). https://doi.org/10.3390/jcs2040066
16. Türk, O.: Stoffliche Nutzung nachwachsender Rohstoffe. Grundlagen - Werkstoffe - Anwendungen ; mit 128 Tabellen. Springer Vieweg, Wiesbaden (2014)
17. Subramanian, P., et al.: Coconut (2020)
18. Gries, T., Veit, D., Wulfhorst, B.: Textile Fertigungsverfahren. Eine Einführung. Hanser, München (2019)
19. Ahmad, J., Zhou, Z.: Mechanical properties of natural as well as synthetic fiber reinforced concrete: a review. Constr. Build. Mater. **333**, 127353 (2022). https://doi.org/10.1016/j.conbuildmat.2022.127353

20. FAO: Land use statistics and indicators. Global, regional and country trends 1990–2019, Rome (2021)
21. The state of the world's land and water resources for food and agriculture - systems at breaking point. Synthesis report 2021. Food and Ag-riculture Organization of the United Nations, Rome (2021)
22. Jawaid, M., et al. (eds.): Pineapple Leaf Fibers Processing. Properties and Applications. Springer, Singapore (2020). https://doi.org/10.1007/978-981-15-1416-6
23. Kozłowski, R.M. (eds.): Handbook of natural fibres. Volume 2: Processing and Applications. Woodhead Publishing, Duxford (2020)
24. Benedix, R.: Bauchemie. Einführung in die Chemie für Bauingenieure und Architekten. Springer, Heidelberg, (2020)
25. Wietek, B.: Faserbeton. Im Bauwesen. Springer Fachmedien Wiesbaden GmbH, Wiesbaden (2015)
26. DIN 1164-10: Special cement - Part 10: Cement with low effective alkali content, composition and requirements, DIN German Institute for Standardization
27. Bergmeister, K., Fingerloos, F., Wörner, J.-D. (eds.): Beton-Kalender. Taschenbuch für Beton-, Stahlbeton und Spannbeton…/Beton-Kalender 2017. Schwerpunkte: Spannbeton, Spezialbetone. Ernst Wilhelm & Sohn, Berlin (2016)
28. Export Value of Jute in the Asia-pacific region in 2020, by country (2022)
29. Page, J., Khadraoui, F., Boutouil, M., Gomina, M.: Multi-physical properties of a structural concrete incorporating short flax fibers. Constr. Build. Mater. **140**, 344–353 (2017). https://doi.org/10.1016/j.conbuildmat.2017.02.124
30. DIN EN 1170-5:1998-01: Precast concrete products - Test method for glass-fibre reinforced cement - Part 5: Measuring bending strength; "complete bending test" method; German version EN 1170-5:1997, DIN German Institute for Standardization, Germany (1998)
31. Da Costa Santos, A.C., Archbold, P.: Experimental investigation on the fracture energy and mechanical behaviour of hemp and flax fibre FRC compared to polypropylene FRC. In: Bio-Based Building Materials, pp. 326–331. Trans Tech Publications Ltd., Switzerland (2022)

Biobased Fiber Reinforced Composite Materials for Construction 3D-Printing

Material Selection Based on Strength, Embodied Environmental Impact, and Cost

Katy Bradford(✉)

Georgia Institute of Technology, Atlanta, GA 30332, USA
kbradford8@gatech.edu

Abstract. Construction 3D-printing (C3DP) provides new opportunities to improve the mechanical properties of biobased fiber-reinforced construction materials. The extrusion process used in many C3DP systems has been shown to induce fiber alignment along the printing path, which can improve the strength of composites in certain loading directions. Several studies have demonstrated this with synthetic short fiber-reinforced C3DP materials; however, very few studies have considered the addition of biobased fibers in extrusion-based fabrication processes. Biobased fibers are known to have lower environmental impacts and cost when compared to synthetic fibers, but their tensile strength tends to be lower. Thus, the use of C3DP with biobased fibers is a prospective solution to improve the mechanical properties of these sustainable construction materials. This review evaluates the performance of 3D-printed fiber-reinforced construction materials in terms of strength, cost, and embodied environmental impact. Using a material selection process to compare between the selected properties, the most proficient biobased fiber mix designs for C3DP are then identified based on fiber type, fiber content, and binder type. By establishing a set of criteria for the performance of biobased fiber reinforced C3DP materials, future research can be compared to these results to work towards more sustainable, high-performance biomaterials for construction applications.

Keywords: Biobased fibers · plant-based fibers · natural fibers · construction 3D-printing · biobased construction materials

1 Introduction

The building sector is responsible for the largest contributions of greenhouse gas (GHG) emissions and primary energy consumption in the world [1]. Consequently, an increasing demand has been placed on the construction industry to improve the sustainability of the built environment. Biobased materials, such as natural fibers and wood products, have the potential to significantly reduce the embodied carbon and energy of buildings due to carbon sequestration during their growth periods [2, 3]. In fact, construction with biobased fibers has been shown to absorb more CO_2 than timber due to faster growth

S. Amziane et al. (Eds.): ICBBM 2023, RILEM Bookseries 45, pp. 514–544, 2023.
https://doi.org/10.1007/978-3-031-33465-8_40

Table 3. (*continued*)

Material	Density (g/cm^3)	GWP ($kgCO_2eq/kg$)	CED (MJ/kg)	Cost ($/kg)	Tensile Strength (MPa)
Sand [19, 37, 45, 67, 68, 81, 83, 87, 100–102]	2.62	0.008	0.084	$0.02	
silica fume [66, 68, 73, 92, 103]	0.52	0.010	0.100	$0.84	
sisal fiber (SF) [6, 34, 36, 46, 53, 57]	1.39	0.473	10.312	$0.66	544.60
slag [45, 66–68, 81, 87, 104–106]	2.91	0.070	11.040	$0.10	
sodium hydroxide [68, 93, 107–111]	2.13	0.635	11.100	$90.37	
sodium silicate [68, 93, 107, 112, 113]	2.60	0.831	5.371	$391.39	
soil (S) [30, 36]	2.38	0	0	$0	
steel fiber (SS) [40, 45, 46, 87, 114–116]	7.85	1.465	13.093	$1.70	2132.00
straw (ST) [117, 118]	0.11	−1.250	0.850	$0.57	
type 1 Portland cement (PC) [119–121]	3.16	0.923	3.553	$0.13	

* properties representative of material most closely related
** values with biogenic carbon considered

absorb moisture, which can be harmful to the fiber-matrix interaction, depending on the constituents of the matrix [124]. All of these aspects of the mix design must be considered if a higher strength is to be achieved with the additions of short fiber reinforcements.

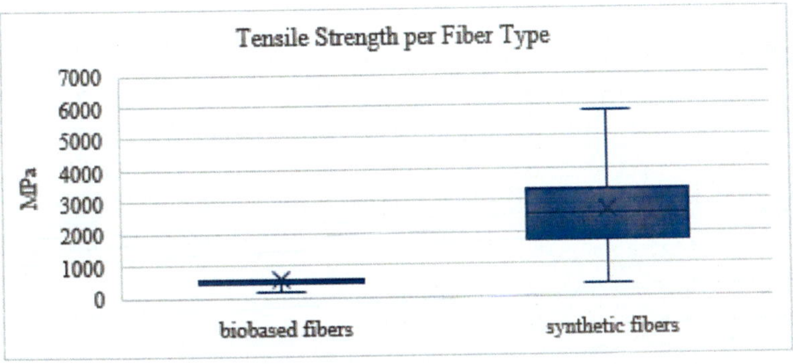

Fig. 2. Reported tensile strength (MPa) of fibers per fiber type

Short fibers may also be added to a matrix material to improve the environmental impact of a composite if the impacts of the fiber are lesser than that of the matrix [6]. Biobased fibers exude less GWP than synthetic fibers, with emissions as low as -1.73 kg CO_2eq/kg (hemp fibers, considering biogenic carbon [65]). Synthetic fibers emit anywhere from 0.40 kg CO2eq/kg (basalt fiber, biogenic carbon considered [48]) to 2.69 kg CO_2eq/kg (PP fiber, no biogenic carbon considered [19]) from the data reviewed in this study.

Likewise, biobased fibers typically consume less energy in manufacturing. From the data evaluated in the study, biobased fiber CED is anywhere from 0.85 MJ/kg of fiber produced (straw [117]) to 57.93 MJ/kg of fiber produced (cotton [76]), with a mean value of 18.67 MJ/kg. In contrast, synthetic fibers can consume between 6.63 MJ/kg of fiber produced (basalt [48]) and 339 MJ/kg (carbon fiber [51]) on average, with a mean value of 94.46 MJ/kg. Consequently, the addition of biobased fibers is more likely to reduce the environmental impact of a composite material than the addition of synthetic fibers, depending on the specific fibers used and the manufacturing processes used to create them (Fig. 3).

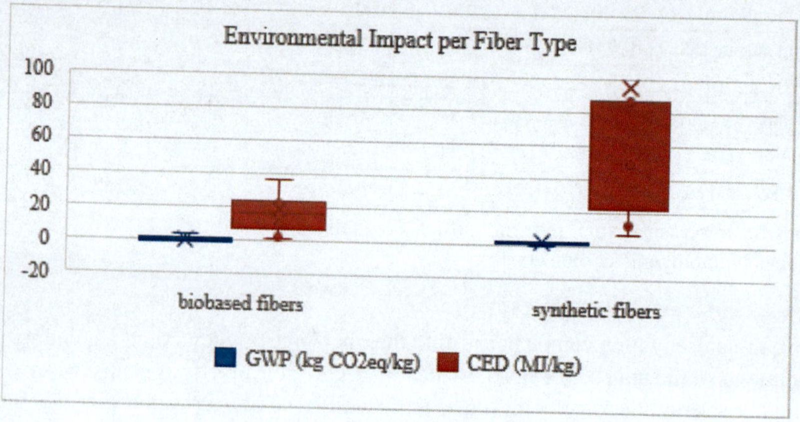

Fig. 3. Cradle-to-gate GWP and CED of fibers per fiber type

Lastly, the addition of fibers to a composite material can also affect its cost. Biobased fibers tend to be less expensive than synthetic fibers, with cost per kg in the range of \$0.02-\$4.95. From the data evaluated in this study, synthetic fibers cost between \$1.70 per kg and \$14.88 per kg, in comparison. Thus, depending on the specific fiber used, the manufacturer purchased from, and the quantity purchased, either type of fiber could

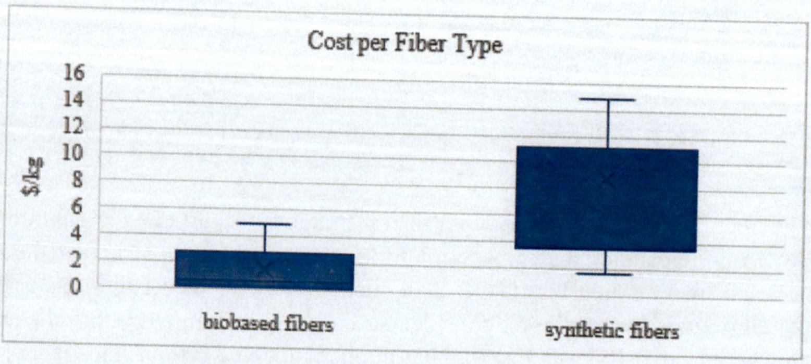

Fig. 4. Upfront cost of fibers per fiber type

be the more economical option; however, biobased fibers are more likely to reduce the overall cost of the composite (Fig. 4).

3.2 Mixes

Several types of binders were used in the studies reviewed, including Portland cement (PC), geopolymer (GP), and soil (S). Composites were identified using the binder abbreviation, followed by the fiber abbreviation, as seen in some of the following figures. The strength results are displayed per fiber type (biobased, synthetic, or none); however, the strength of the binder plays a large role in the resulting properties of the composites, both reinforced and unreinforced with fibers. Although statistical analyses of the entire dataset are beneficial to understanding the performance of the various types of materials, comparisons are also made between composites of the same matrix to better compare the impacts of fiber addition and extrusion on the resulting strength.

As shown in Fig. 5, composites reinforced with biobased fibers tend to have a lower perpendicular bending strength than those reinforced with synthetic fibers. Interestingly, while 3D-printing tends to decrease the strength of composites without fiber reinforcements, it tends to increase the strength of those with fibers. This could be due to the increased fiber packing and alignment found in 3D-printed composites. It should be noted that some outliers were removed in this figure, especially regarding synthetic fiber reinforced materials. Without these outliers, it appears that the addition of fiber reinforcements to cast samples tends to decrease the strength of the evaluated samples; however, this is most likely due to the evaluation of various fiber contents per mix design, many of which were not ideal and actually decreased the strength of the composites.

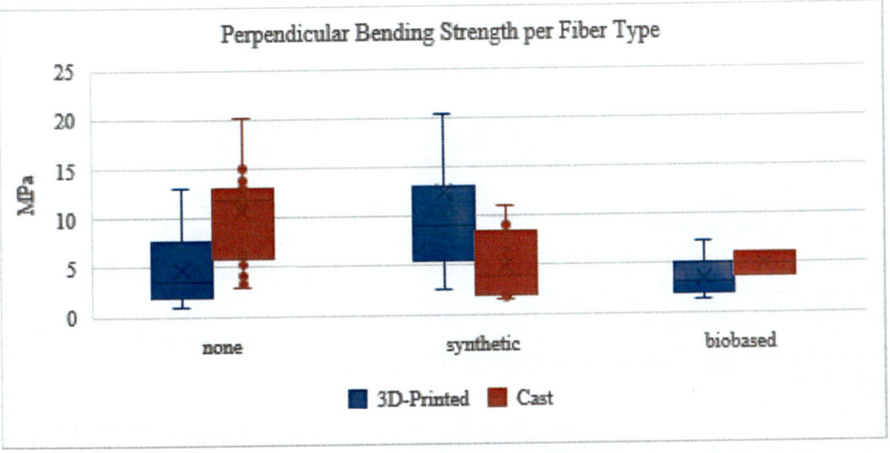

Fig. 5. Results from all studies that reported perpendicular bending strength per fiber type and fabrication process.

When specific mix designs are evaluated with and without fibers, a more obvious trend of perpendicular bending strength improvement due to the addition of fibers can be observed. In Fig. 7, studies that compared the perpendicular bending strength of composites with identical matrix mix designs are presented. The strength of the fiber-reinforced composites shown are the maximum value per study, which is assumed to be the optimal fiber content that each study found. On average, these results show that the addition of fibers, when added in the optimal % volume of matrix material, can improve the bending strength by 4.38 MPa on average, which represents a 151% average improvement. These results disregard one outlier result, in which the perpendicular bending strength of a type 1 Portland cement mixture reinforced with carbon fiber increased from 8.3 MPa to 119.6 MPa [24] (Fig. 6).

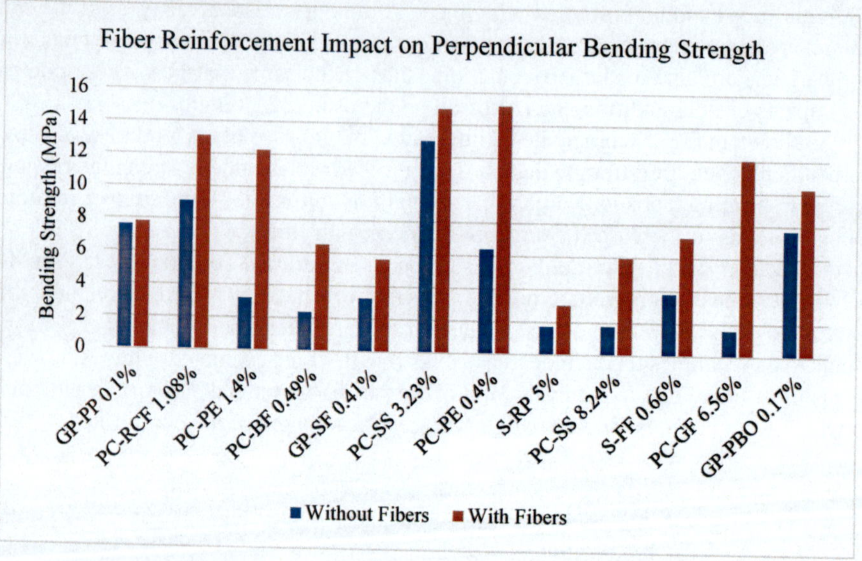

Fig. 6. Bending strength results from studies considering unreinforced mixes and mixes reinforced at the optimal fiber content with identical matrix mix designs.

It can be observed that the composites made with Portland cement are the strongest, and these materials tend to be reinforced with synthetic fibers due to the incompatibility of biobased fibers with the alkaline environment of cement [125]. Biobased fiber reinforced-composite binders, on the other hand, tend to be composed of geopolymer, clay, and soil, resulting in weaker overall bending strength. These composites do exhibit an increase in bending strength due to fiber addition though, with an average improvement of 2.39 MPa, or 79.85%.

The results from studies that compared the perpendicular bending strength of cast and 3D-printed samples with identical mix designs are displayed in Fig. 7. These mixes also represent the optimal fiber content per study, displayed as % volume of the matrix. In most cases, at least a slight improvement in perpendicular bending strength arose from 3D-printing in comparison to the cast samples. On average, these results represent a 17.6% improvement in perpendicular bending strength, or 0.476 MPa. In some cases, 3D-printing decreased the perpendicular bending strength. This could be attributed to many factors, including the fiber-binder interaction of the mix and the parameters of the 3D-printing process employed, such as nozzle size, printing speed, and material flow rate.

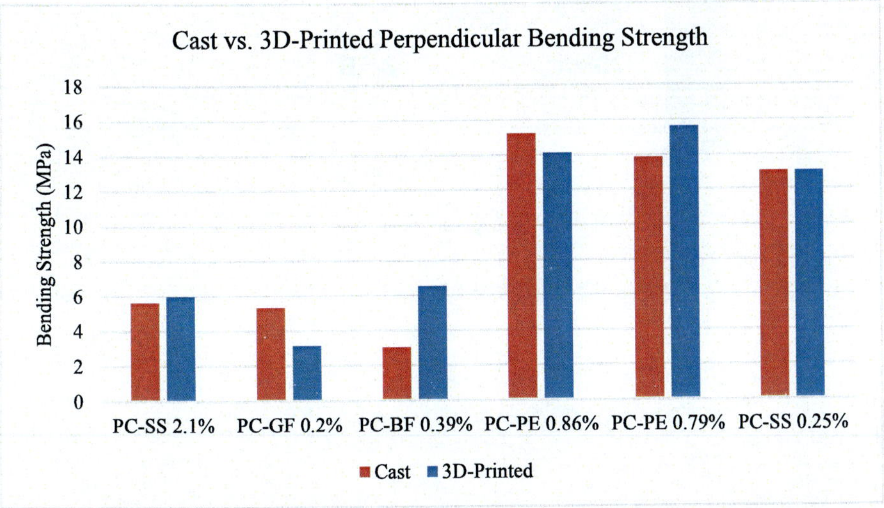

Fig. 7. Perpendicular bending strength results from studies considering cast samples with identical mix design and optimal fiber content.

When only considering the biobased fiber mixes that were tested for perpendicular bending strength, mixes reinforced with longer fibers (flax fiber and sisal fiber) exhibit a higher perpendicular bending strength, up to 7.25 MPa, than those reinforced with pulp (bamboo pulp and recycled paper pulp), which only reach 4.56 MPa. The binders used in the three studies represented here vary; however, the reinforcement of soil with both long fibers (flax fiber) and pulp fibers displays a much higher strength improvement when flax fibers are used [18]. This corresponds to the strength of these fibers themselves, as displayed in Table 3. None of these studies made comparisons between cast and 3D-printed samples.

As observed in Fig. 9, the addition of synthetic fibers does tend to increase the perpendicular compressive strength of the mix designs tested, but not always. The addition of biobased fibers, on the other hand, whether cast or 3D-printed, tends to decrease the perpendicular compressive strength. Again, the comparison between cast and 3D-printed samples must account for the difference in the binders used in the studies being compared. With that in mind, it does appear that 3D-printing with both synthetic and biobased fiber-reinforced materials tends to decrease the perpendicular compressive strength in comparison to cast materials of the same category. This can similarly be attributed the range of fiber content present in the studies considered. More often than not, a suboptimal fiber content will decrease the strength, while an optimal fiber content will increase it (Fig. 8).

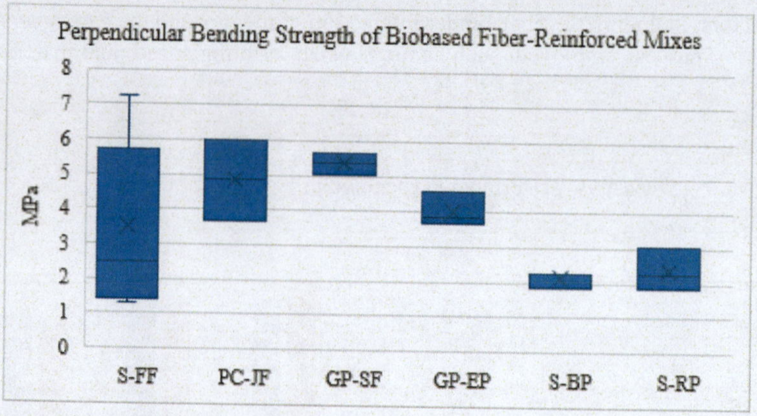

Fig. 8. Perpendicular bending strength results from biobased fiber-reinforced materials, both 3D-printed and cast.

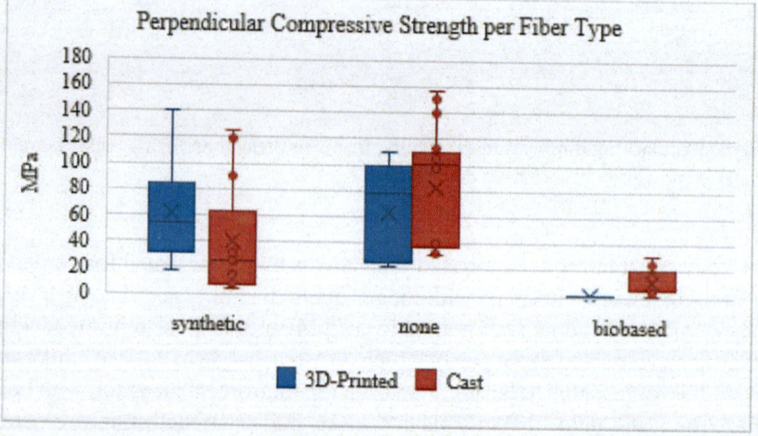

Fig. 9. Results from all studies reporting perpendicular compressive strength per fiber type and sample fabrication method.

In Fig. 10, studies that compared the perpendicular compressive strength between fiber-reinforced and plain matrix materials are represented. The results from the fiber-reinforced mix design that resulted in the maximum strength are shown, with the optimum fiber content displayed as % volume of the matrix. It can be observed that the addition of fibers tends to increase the perpendicular compressive strength when the appropriate volume percentage of fiber are added. On average, the mixes displayed experienced a 18.1% increase in strength due to the addition of fibers at the optimal fiber content.

The effect of 3D-printing on the perpendicular compressive strength is displayed in Fig. 11. In many cases, 3D-printing decreases the perpendicular compressive strength of the mixes. On average, however, 3D-printing caused a 10.77% increase in perpendicular compressive strength at the optimal fiber content of these studies. Counterintuitively, there is also an average decrease of strength by −4.11 MPa. This is due to the magnitude with which some perpendicular compressive strengths were improved, which tended to be higher than that of which was damaging.

The longitudinal compressive strength of several studies that compared identical mixes with and without fiber reinforcements at the optimal fiber content is shown in Fig. 12. Again, in some cases, the longitudinal compressive strength is decreased by the addition of fibers; however, an average increase of 13.3% is experienced amongst these mix designs. This equates to an average increase of 6.04 MPa of the mix designs displayed in Fig. 12. Only one study investigated the effects of 3D-printing on the longitudinal compressive strength of a single mix design [21]. Using glass fiber to reinforce a Portland cement mix, the authors found an 8 MPa improvement in longitudinal compressive strength from 3D-printing the samples, which is a 19.75% improvement.

The longitudinal compressive strength also brings some insight into the fiber-binder interactions due to the loading direction, which is parallel to the direction of fiber alignment in 3D-printed samples. Compressing the samples in tis direction causes the matrix material to be pulled away from the surface of the fiber rather than pushed against it. Therefore, a stronger fiber-matrix bond will cause a higher longitudinal compressive strength. Taking this into consideration, the results in Fig. 12 suggest that the bond between the Portland cement mixes used with steel fibers was more sufficient than those of the Portland cement mixes used with basalt fibers and carbon fibers, as well as the bond between the geopolymer mix used with sisal fibers. There are many factors that could influence these results beyond the material interactions however, such as the printing parameters used and the specific constituents of each matrix material.

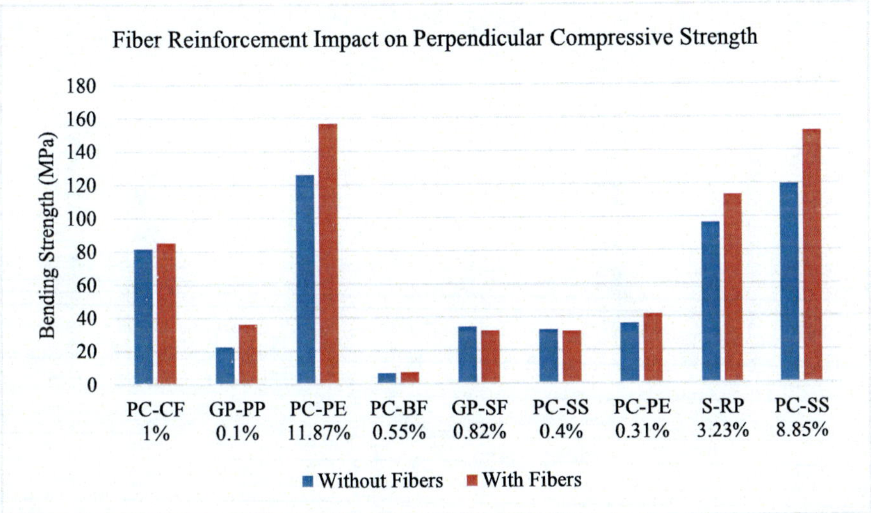

Fig. 10. Perpendicular compressive strength results from studies considering optimal fiber-reinforced and plain samples with identical matrix.

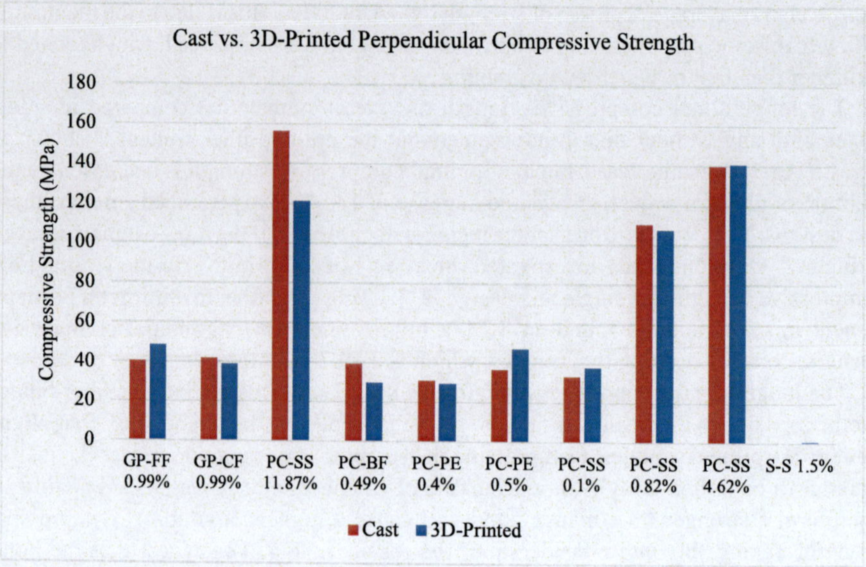

Fig. 11. Perpendicular compressive strength results from studies considering cast samples with identical mix designs with optimal fiber content.

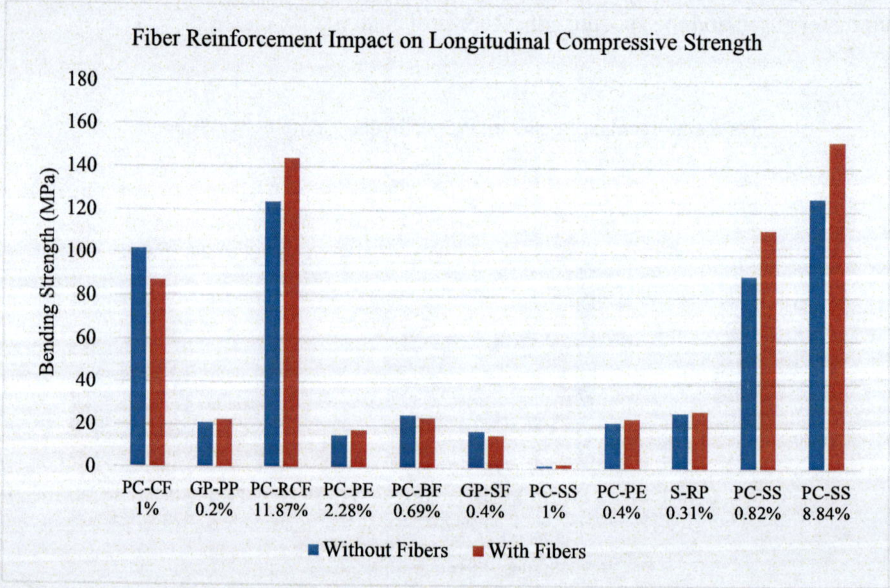

Fig. 12. Longitudinal compressive strength results from studies considering optimal fiber-reinforced (%weight) and plain samples with identical matrix.

It should be noted that the optimal fiber content for perpendicular bending strength is not always the same as the optimal fiber content for perpendicular or longitudinal compressive strength. For example, in one study evaluating a geopolymer mix with PP fiber, the optimal % volume of fiber for perpendicular bending strength was 0.1%, but 0.25% for perpendicular compressive strength and 0.5% for longitudinal compressive strength [25]. Likewise, in another study evaluating soil reinforced with flax fiber, the optimal fiber content for perpendicular bending strength was 0.67%, while it was 3% for longitudinal compressive strength [18]. This emphasizes the need for consideration of the application when choosing the ideal mix design based on strength in various directions of an anisotropic material.

4 Material Selection

When choosing a material for a given function, it is important to determine a material selection strategy based on the constraints and objectives of the application at hand [6]. Structural engineering constraints include a minimum compressive strength, bending strength, and tensile strength of the material used [6]. For extrusion-based C3DP, the material must be pumpable to be moved from the mixing pump to the extruder where it is deposited. It must also be extrudable when it reaches the nozzle, such that it can flow properly onto the printing bed and subsequent printed layers. Then, it must be buildable, or present a high enough yield strength to support the weight of the printed layer and the layers deposited on top of it [126]. These properties are highly dependent on the water content of the mix, as well as the additives used to alter the plasticity and set time. Additionally, in short-fiber reinforced C3DP materials, a material will not be pumpable or extrudable if the volume ratio of fibers is too high, or if the fibers are too long [52].

The main objectives of this study are to minimize the embodied energy and carbon of the mix design, while also minimizing cost and maximizing strength. Minimizing the embodied environmental impact and cost of the mix depends on the materials used and their individual impacts, first and foremost. However, it also depends on the strength of the resulting hardened composite. A stronger material will require less volume to meet the same loading condition than a weaker material will, thus resulting in a lower overall embodied energy and carbon, as well as cost, per application. Ultimately, the results of these comparisons amongst cost, cradle-to-gate global warming potential (GWP) and cumulative energy demand (CED), and strength are used to determine the best possible mix designs for biobased fiber-reinforced C3DP applications.

4.1 Cost vs. Strength

From the data reviewed in this study, it is apparent that synthetic fiber-reinforced materials are both the strongest and the most expensive. Some of those mixes, however, appear to be economical for the high strength they exhibit. For example, the strongest reported mix in terms of perpendicular bending strength (119.6 MPa [24]), which is considered an outlier and therefore not displayed in Fig. 13, is composed of type 1 Portland cement and carbon fiber and costs \$0.60 per kg. In comparison, the strongest biobased fiber mix, which is composed of soil reinforced with flax fiber [18], exhibits a perpendicular bending strength of 7.25 MPa and costs $0.09. This equates to a strength-specific cost of $0.005/MPa and $0.0124/MPa, respectively. Overall, the biobased fiber mixes reviewed cost the least, with the cheapest being $0.02 per kg of mixture. The highest perpendicular bending strength of the mixes at this cost (soil reinforced with recycled paper pulp) is 3.05 MPa [22], which is 116.55 MPa less than the strongest synthetic fiber-reinforced mix while $0.58 per kg less expensive.

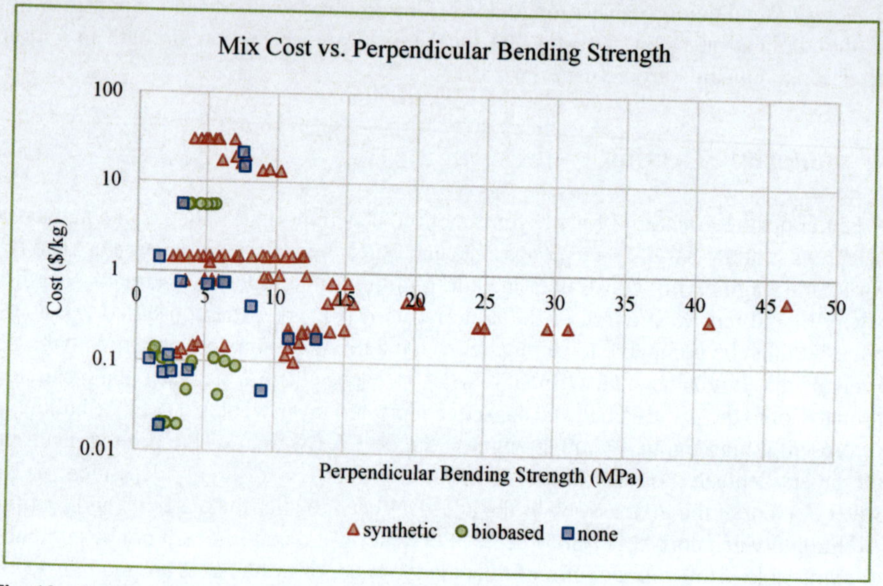

Fig. 13. Comparison of the cost (log scale) of mixes with reported perpendicular bending strength per fiber type.

The comparison of perpendicular compressive strength shows slightly different results. The synthetic-fiber reinforced materials are still the strongest; however, the biobased-fiber reinforced are just as expensive in some cases. The mix with the highest strength is Portland cement reinforced with steel fiber, which resulted in 156.4 MPa in this direction [27]. This mix costs $0.94 per kg. For comparison, the most expensive estimation for a mix is $29.98 per kg, which is a geopolymer binder reinforced with glass fiber.

The biobased-fiber reinforced mix with the highest perpendicular compressive strength is a slag cement binder reinforced with eucalyptus pulp, resulting in 31.3 MPa. This mix costs an estimated value of $0.04. In general, Fig. 14 shows that biobased-fiber reinforced mixes display the lowest estimated cost, but also the lowest estimated strength.

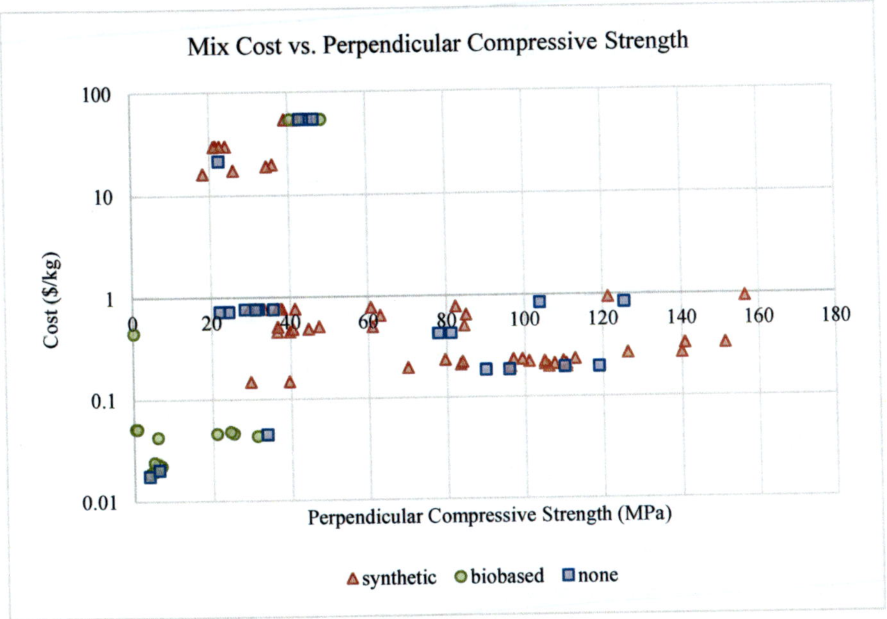

Fig. 14. Comparison of the cost (log scale) of mixes with reported perpendicular compressive strength per fiber type.

4.2 Environmental Impact vs. Strength

The biobased-fiber reinforced mixes reviewed in this study have the lowest perpendicular bending and compressive strength, but they exhibit lower estimated GWP and CED per kg when compared to the synthetic fiber-reinforced mixes. The best performing mix design in terms of GWP and CED with results for perpendicular bending strength is a geopolymer mix, which emits 0.03 kg CO_2eq/kg and consumes 0.37 MJ/kg of energy in the cradle-to-gate stages. This mix resulted in a perpendicular bending strength of 3.25 MPa [20]. Reinforcing it with sisal fiber increased the perpendicular bending strength to 5.67 MPa [20] without increasing the GWP, but slightly increasing the CED to 0.41 MJ/kg. The strongest biobased-fiber reinforced mix is soil reinforced with flax fiber, which resulted in 7.25 MPa in perpendicular bending strength [18]. This mix has a GWP of 0.19 kg CO_2eq/kg and a CED of 1.17 MJ/kg.

In contrast, the synthetic fiber-reinforced mix with the highest perpendicular bending strength (119.6 MPa [24]), Portland cement reinforced with carbon fiber, emits and estimated 0.62 kg CO_2eq/kg and consumes 7.05 MJ/kg. The highest strength mix displayed in Fig. 16 is also composed of Portland cement and carbon fiber, with a perpendicular bending strength of 46.5 MPa [24], a GWP of 0.62 kg CO_2eq/kg, and a CED 5.64 MJ/kg. The synthetic fiber-reinforced mix with the lowest GWP (0.03 kg CO_2eq/kg) and CED (0.6 MJ/kg) is geopolymer reinforced with PP fiber (3.31 MPa) (Fig. 15).

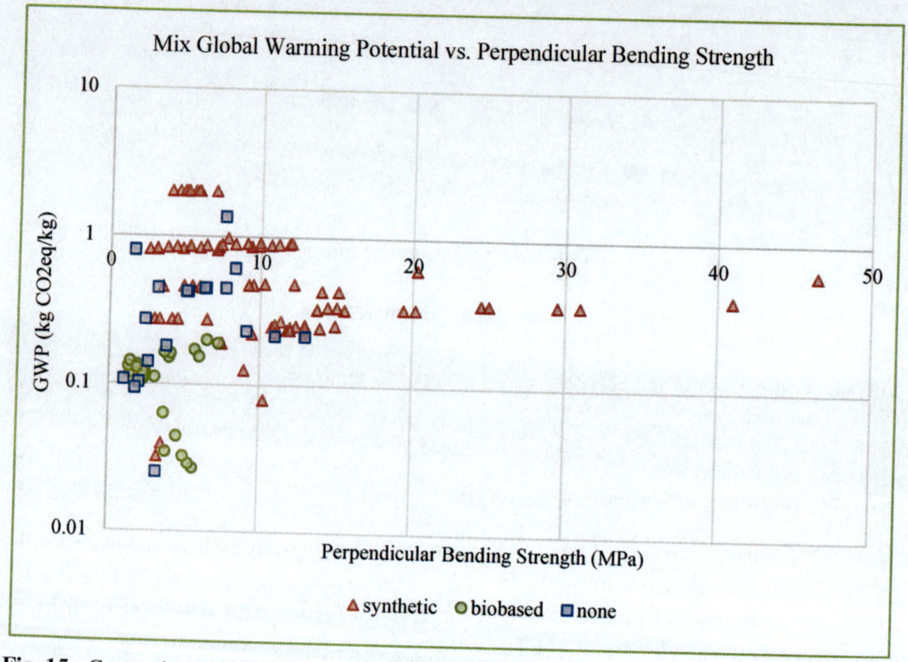

Fig. 15. Comparison of the GWP (log scale) of mixes with reported perpendicular bending strength.

The strongest mix in terms of perpendicular compressive strength (Fig. 17) is a Portland cement mix reinforced with steel fiber [27]. The strength of this mix reach 156.4 MPa, while the GWP was 0.57 kg CO_2eq/kg and the CED was 4.93 MJ/kg. In comparison, the strongest biobased fiber-reinforced mix (slag cement and eucalyptus pulp [34]), which reached 31.3 MPa, has a GWP of 0.07 kg CO_2eq per kg and a CED of 3.41 MJ/kg.

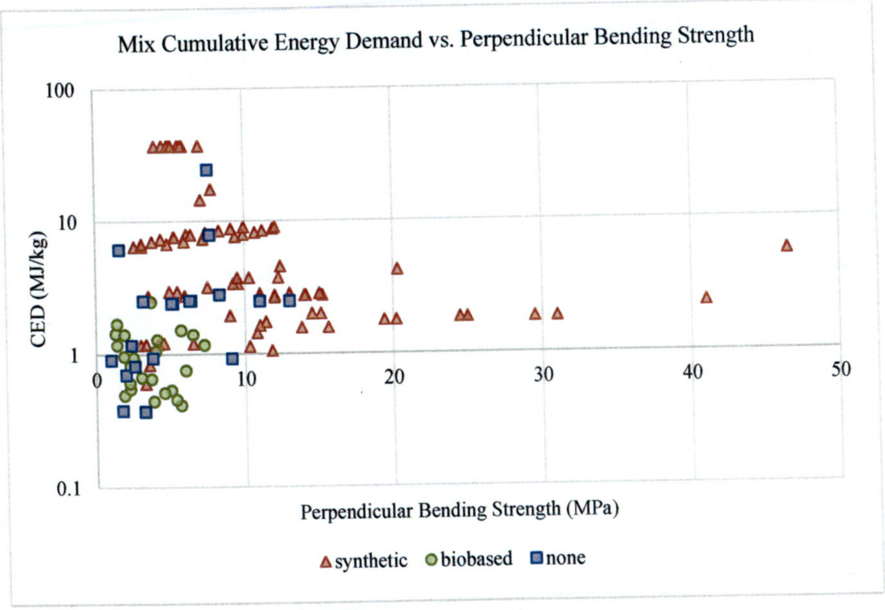

Fig. 16. Comparison of the CED (log scale) of mixes with reported perpendicular bending strength per fiber type.

The biobased fiber-reinforced mix tested for perpendicular compressive strength with the lowest GWP is slag cement reinforced with flax fiber [33], which resulted in a GWP of 0.04 kg CO_2eq/kg and a CED of 2.89 MJ/kg. The biobased fiber-reinforced mix tested for perpendicular compressive strength with the lowest CED is soil reinforced with coir fiber, which resulted in a perependicular compressive strength of 3.89 MPa, a GWP 0.06 kg CO_2eq/kg, and a CED of 0.28 MJ/kg.

The mix tested for perpendicular compressive strength with the overall lowest GWP was plain slag cement, which resulted in a maximum strength of 15.1 MPa [33]. The GWP and CED of this mix were estimated to be 0.03 kg CO_2eq/kg and 2.6 MJ/kg, respectively. For synthetic fiber-reinforced mixes, the mix with the lowest GWP is a geopolymer mix reinforced with PP fiber with a strength of 18.2 MPa [25]. This mix has an estimated GWP of 0.19 kg CO_2eq/kg and CED of 3.18 MJ/kg. The synthetic fiber-reinforced mixes tested for perpendicular compressive strength with the lowest CED, on the other hand, was type 1 Portland cement reinforced with basalt fiber, with strength of 39.5 MPa [16], an estimated GWP of 0.27 kg CO_2eq/kg and CED of 1.2 MJ/kg (Figs. 17 and 18).

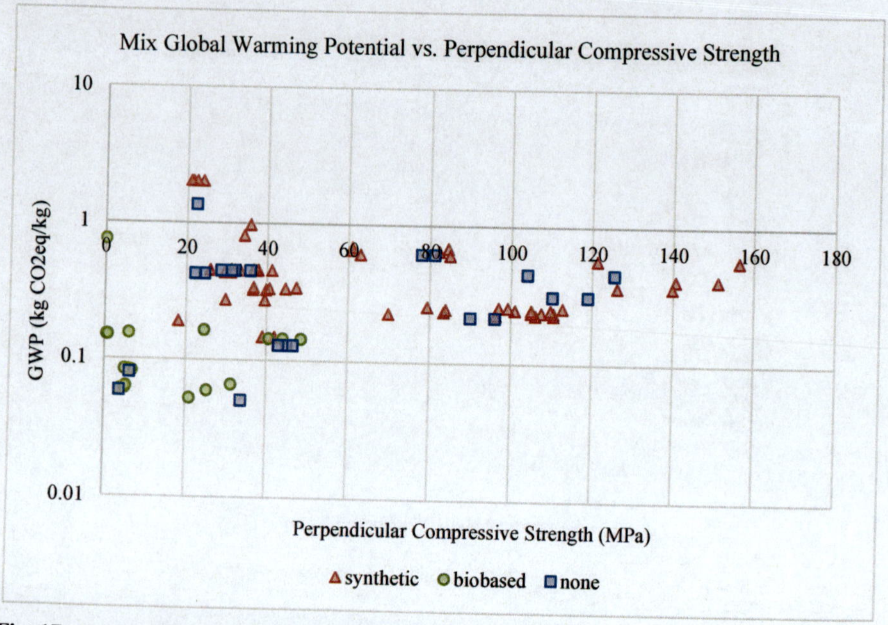

Fig. 17. Comparison of the GWP (log scale) of mixes with reported perpendicular compressive strength per fiber type.

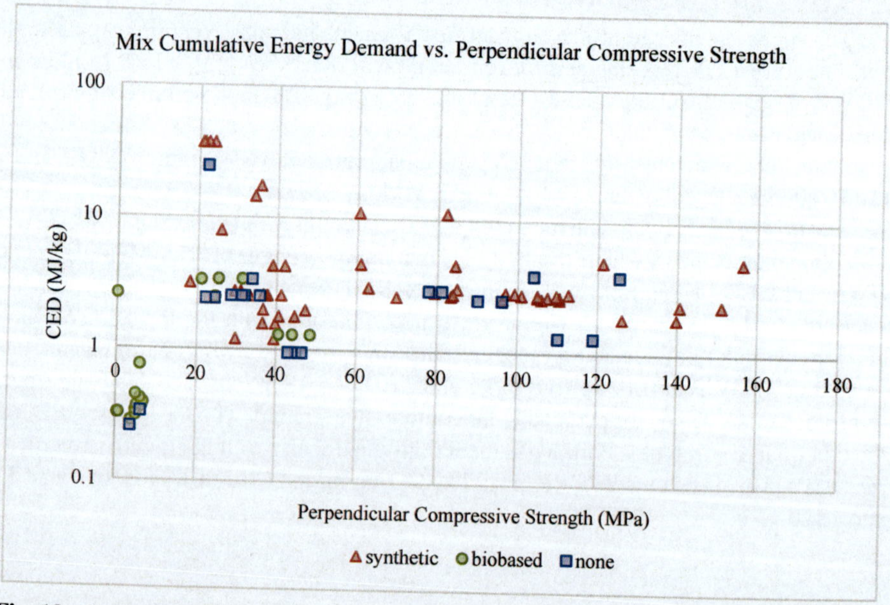

Fig. 18. Comparison of the CED (log scale) of mixed with reported perpendicular compressive strength per fiber type.

4.3 Environmental Impact vs. Cost

The lowest cost biobased fiber-reinforced mix is soil reinforced with coir fiber. This mix was estimated to cost \$0.02 per kg and have a GWP and CED as low as 0.06 kg CO_2eq/kg and 0.28 MJ/kg, respectively. The biobased fiber-reinforced mix with the lowest GWP, on the other hand, is soil reinforced with sisal fiber, which has an estimated cost of \$0.03 per kg, GWP of 0.02 kg CO_2eq/kg, and CED of 45.1 MJ/kg. The lowest CED, 0.28 MJ/kg, is found in a soil mix reinforced with coir fiber, which costs approximately \$0.02 per kg and emits an estimated 0.06 kg CO_2eq/kg.

For synthetic fiber-reinforced mixes, the lowest cost of \$0.10 is attributed to both a soil mix reinforced with glass fiber and a type 1 Portland cement mix reinforced with glass fiber. These mixes have an estimated GWP of 0.13 kg CO_2eq/kg and 0.27 kg CO_2eq/kg, and a CED of 1.27 MJ/kg and 1.74 MJ/kg, respectively. The mix with both the lowest GWP and CED, 0.03 kg CO_2eq/kg and 0.6 MJ/kg, is a geopolymer reinforced with PP fiber that costs \$5.66 per kg. Overall, aside from plain soil, the most cost-effective mixes (\$0.02/kg) were soil reinforced with coir fiber, bamboo pulp, and recycled waste paper pulp, which have estimated GWP between 0.06–0.13 kg CO_2eq/kg and CED between 0.26–0.96 MJ/kg (Figs. 19 and 20).

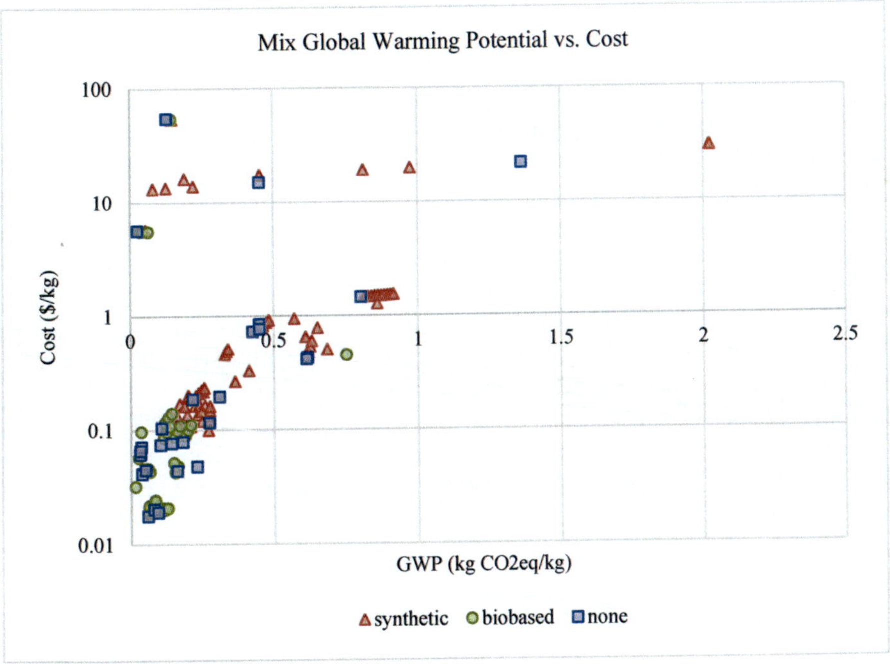

Fig. 19. Comparison of the cost (log scale) with the GWP of mixes per fiber type.

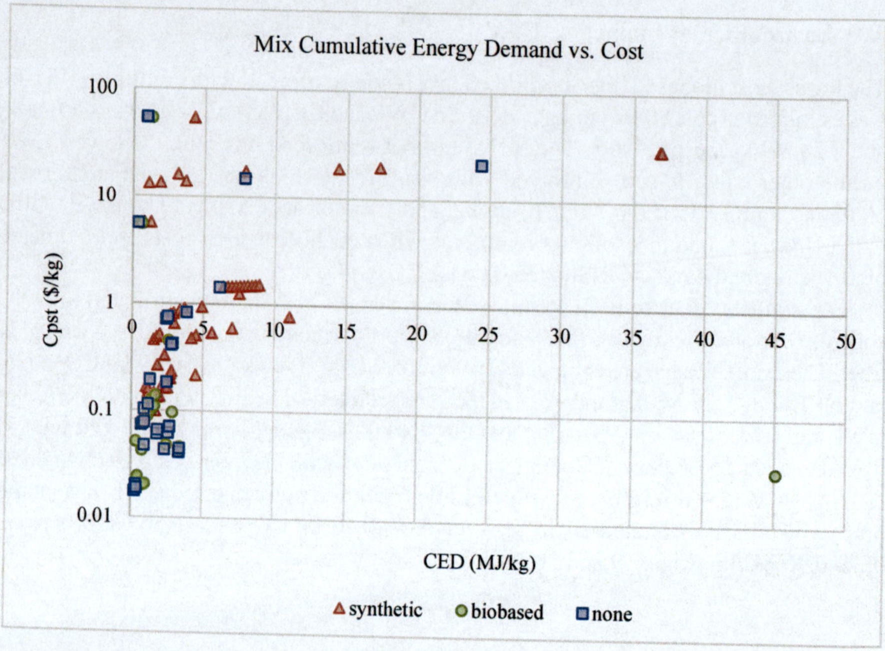

Fig. 20. Comparison of the cost (log scale) with the CED of mixes per fiber type.

Figure 21 displays the properties (log scale) of the biobased fiber-reinforced mixes with optimal fiber content for perpendicular bending strength. No one mix performs the best in each category, thus a level of importance must be associated with each property to choose the best mix design for a given application. The strongest mix (7.25 MPa [18]), which is soil reinforced with 1% flax fiber, costs about $0.09 per kg and has a GWP and CED of 0.19 kg CO_2eq/kg and 1.17 MJ/kg, respectively. The most cost-effective mixes are soil reinforced with recycled waste paper pulp and bamboo pulp, which have an estimated price of $0.02/kg. The GWP of these mixes is 0.11 kg CO_2eq/kg, the CED is 0.67 MJ/kg and 0.55 MJ/kg, respectively, and the perpendicular bending strength is 3.05 MPa and 2.275 MPa, respectively. The lowest GWP of 0.03 kg CO_2eq/kg, on the other hand, is associaed with the soil mix reinforced with sisal fiber, which costs approximately $5.63/kg, has a CED of 0.41 MJ/kg, and a perpendicular bending strength of 5.67 MPa.

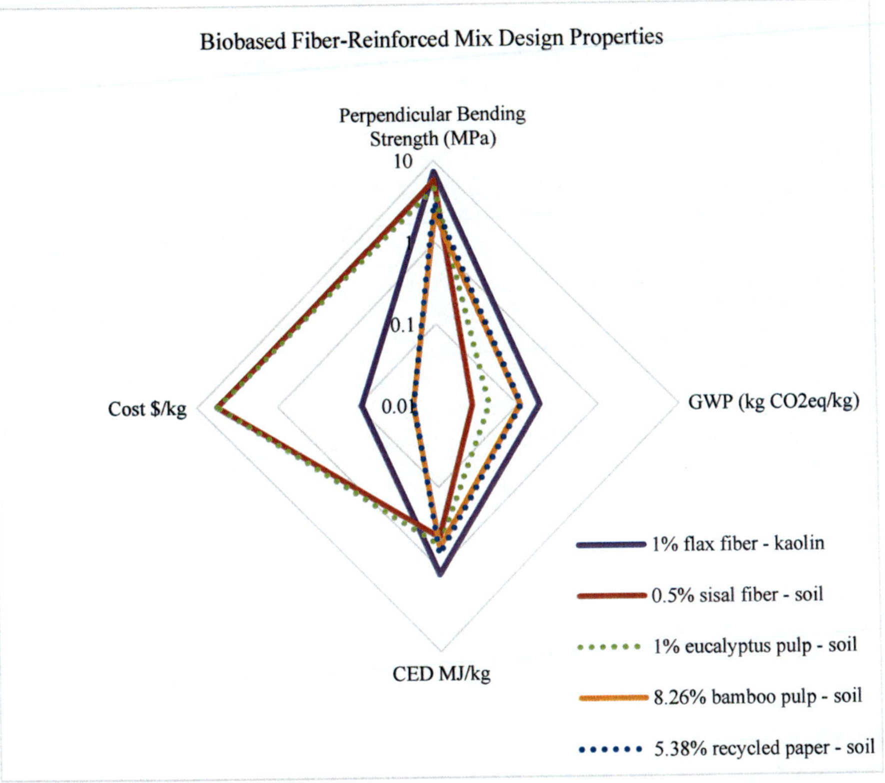

Fig. 21. Logarithmic scale comparison of 3D-printed biobased fiber-reinforced composites at optimal fiber content.

5 Discussion

The results of this study show that the addition of biobased fibers can improve the strength of construction materials by up to 90.8%, depending heavily on the fiber content present in the mix. At the optimal fiber content, the addition of fibers, whether synthetic or biobased, increased the perpendicular bending and compressive strength of the mixes reviewed in this study by 151% and 18.1%, respectively, at the optimal fiber content. Additionally, of the studies reviewed that compared cast samples and 3D-printed samples of the same mix design, 3D-printing increased the perpendicular bending and compressive strength by 17.6% and 10.77%, respectively. However, none of the biobased fiber-reinforced mixes in this study made this comparison. Therefore, the results would suggest that the use of extrusion-based C3DP processes with biobased fiber-reinforced materials could improve their strength even further. Since these materials tend to be weaker than their synthetic counterparts, the use of C3DP fabrication processes could make their mechanical properties more competitive in their consideration as structural materials, especially when long fibers are used.

Aside from strength, the other objectives of this study (reducing environmental impact and cost) are favorable to biobased fiber-reinforced materials over synthetic fiber-reinforced materials. On average, the GWP, CED, and cost of biobased fiber-reinforced materials reviewed in this study were 0.12 kg CO_2eq/kg, 4.01 MJ/kg, and $4.30/kg. Alternatively, that of the synthetic fiber-reinforced mixes was 0.56 kg CO_2eq/kg, 6.7 MJ/kg, and $5.29/kg, which is higher in every category. When observing the entire range of data collected, however, it becomes apparent that the specific choice of materials is paramount in the reduction of these properties. The estimated GWP, CED, and cost of biobased fiber-reinforced materials ranged from 0.02–0.75 kg CO_2eq/kg, 0.28–90.1 MJ/kg, and $0.02–$53.86/kg, respectively. The estimated GWP, CED, and cost of synthetic fiber-reinforced materials ranged from 0.03–2.02 kg CO_2eq/kg, 0.6–36.92 MJ/kg, and $0.10–$53.99. Therefore, the mixes with the lowest cost and environmental impacts should be chosen of the biobased-fiber reinforced mixes.

Based on these results, the best biobased fiber-reinforced mix designs reviewed in this study for further research in extrusion based C3DP are composed of clay and soil, reinforced with flax fiber, jute fiber, and sisal fiber. These mixes resulted in the highest strength while maintaining cost and environmental impact in the lowest 10% of all mix designs reviewed. There were no studies found that examined the use of hemp fibers as reinforcements, which were found in the literature review to have the lowest GWP. Thus, more research is needed to make a broader comparison amongst biobased fibers and their respective composites for C3DP applications.

6 Conclusion

This study revealed a great potential for C3DP to improve the strength of biobased fiber-reinforced composite materials in structural material applications. Due to the lack of existing data regarding these materials, however, more research will be needed in the future to validate these results and draw stronger conclusions about the applicability of biobased fibers as reinforcements for C3DP materials. The variability of LCA and EPD data in this study should also be considered in the interpretation of these results. The evaluation of LCAs and EPDs did not distinguish between attributional and consequential LCA, nor did it consider the allocation of byproducts amongst individual studies [127].

Additionally, not all studies considered biogenic carbon in their GWP evaluation, which could impact the GWP results of the mix designs. As more LCA information becomes available for the materials used in the mix designs reviewed in this study, more research must be conducted to refine the results presented here. In particular, the biobased fibers used should all have the carbon sequestration associated with their growth captured in their GWP value. Lastly, a cradle-to-grave analysis considering the performance of these materials in the use phase in terms of insulative properties and durability would provide a more detailed understanding of the environmental performance of these mix designs in the long term.

Overall, much more research is needed to understand the effects of C3DP on the strength, environmental impact, and cost biobased fiber-reinforced materials. With a larger dataset in mind, building-specific studies could provide more insight into the amount of material required for structural applications based on the resulting properties of these composites, which can contribute greatly to the embodied environmental impacts of construction and the upfront costs.

References

1. United Nations Environment Programme. 2021 Global Status Report for Buildings and Construction: Towards a Zero-emission, Efficient and Resilient Buildings and Construction Sector (2021)
2. Arehart, J.H., Hart, J., Pomponi, F., D'Amico, B.: Carbon sequestration and storage in the built environment. Sustain. Prod. Consum. **27** (2021). https://doi.org/10.1016/j.spc.2021.02.028
3. Churkina, G., et al.: Buildings as a global carbon sink. Nat. Sustain. **3** (2020). https://doi.org/10.1038/s41893-019-0462-4
4. Pittau, F., Krause, F., Lumia, G., Habert, G.: Fast-growing bio-based materials as an opportunity for storing carbon in exterior walls. Build. Environ. **129** (2018)
5. Goswein, V., Arehart, J., Phan-huy, C., Pomponi, F., Habert, G.: Barriers and opportunities of fast-growing biobased material use in buildings. Build. Cities **3**, 745–755 (2022)
6. Ashby, M.F.: Materials and the environment: eco-informed material choice. Materials and the Environment: Eco-informed Material Choice(2021). https://doi.org/10.1016/B978-0-12-821521-0.01001-X
7. Fan, M., Fu, F.: Introduction: a perspective - natural fibre composites in construction. In:Advanced High Strength Natural Fibre Composites in Construction (2016). https://doi.org/10.1016/B978-0-08-100411-1.00001-7
8. Pandey, J.K., Ahn, S.H., Lee, C.S., Mohanty, A.K., Misra, M.: Recent advances in the application of natural fiber based composites. Macromol. Mater. Eng. **295** (2010). https://doi.org/10.1002/mame.201000095
9. Fu, F., Lin, L., Xu, E.: Functional pretreatments of natural raw materials. In:Advanced High Strength Natural Fibre Composites in Construction (2017). https://doi.org/10.1016/B978-0-08-100411-1.00004-2
10. Kabir, M.M., Wang, H., Lau, K.T., Cardona, F.: Effects of chemical treatments on hemp fibre structure. Appl. Surf. Sci. **276** (2013)
11. Fan, M., Weclawski, B.: Long natural fibre composites. In: Advanced High Strength Natural Fibre Composites in Construction (2017). https://doi.org/10.1016/B978-0-08-100411-1.00006-6
12. Mechtcherine, V., et al.: Extrusion-based additive manufacturing with cement-based materials – Production steps, processes, and their underlying physics: a review. Cem. Concr. Res. **132** (2020). https://doi.org/10.1016/j.cemconres.2020.106037
13. Ge, Z., et al.: Influence of an extrusion approach on the fiber orientation and mechanical properties of engineering cementitious composite. Constr. Build. Mater. **306**, (2021)
14. Arunothayan, A.R., et al.: Fiber orientation effects on ultra-high performance concrete formed by 3D printing. Cem. Concr. Res. **143** (2021)
15. Zhang, H., Zhu, L., Zhang, F., Yang, M.: Effect of fiber content and alignment on the mechanical properties of 3D printing cementitious composites. Materials **14** (2021)
16. Ma, G., Li, Z., Wang, L., Wang, F., Sanjayan, J.: Mechanical anisotropy of aligned fiber reinforced composite for extrusion-based 3D printing. Constr. Build. Mater. **202** (2019)
17. Bradford, K.: Fiber-reinforced 3D-printed construction material properties. SMARTech - Georgia Tech Library [database] (2023). http://hdl.handle.net/1853/70329
18. Khelifi, H., Lecompte, T., Perrot, A., Ausias, G.: Mechanical enhancement of cement-stabilized soil by flax fibre reinforcement and extrusion processing. Mater. Struct. **49**(4), 1143–1156 (2015). https://doi.org/10.1617/s11527-015-0564-z

19. Akbar, A., Liew, K.M.: Multicriteria performance evaluation of fiber-reinforced cement composites: an environmental perspective. Compos. B Eng. **218** (2021)
20. Ojo, E.B., et al.: Effects of fibre reinforcements on properties of extruded alkali activated earthen building materials. Constr. Build. Mater. **227** (2019)
21. Korniejenko, K., et al.: Mechanical properties of short fiber-reinforced geopolymers made by casted and 3D printing methods: a comparative study. Materials **13** (2020)
22. Stanislas, T.T., et al.: Effect of cellulose pulp fibres on the physical, mechanical, and thermal performance of extruded earth-based materials. J. Build. Eng. **39**, 102259 (2021)
23. Hambach, M., Volkmer, D.: Properties of 3D-printed fiber-reinforced Portland cement paste. Cem. Concr. Compos. **79** (2017)
24. Hambach, M., Möller, H., Neumann, T., Volkmer, D.: Portland cement paste with aligned carbon fibers exhibiting exceptionally high flexural strength (> 100 MPa). Cem. Concr. Res. **89** (2016)
25. Nematollahi, B., et al.: Effect of polypropylene fibre addition on properties of geopolymers made by 3D printing for digital construction. Materials **11** (2018)
26. Ding, T., Xiao, J., Zou, S., Zhou, X.: Anisotropic behavior in bending of 3D printed concrete reinforced with fibers. Compos. Struct. **254** (2020)
27. Arunothayan, A.R., Nematollahi, B., Ranade, R., Bong, S.H., Sanjayan, J.: Development of 3D-printable ultra-high performance fiber-reinforced concrete for digital construction. Constr. Build. Mater. **257** (2020)
28. Panda, B., Chandra Paul, S., Jen Tan, M.: Anisotropic mechanical performance of 3D printed fiber reinforced sustainable construction material. Mater. Lett. **209** (2017)
29. Bos, F.P., Bosco, E., Salet, T.A.M.: Ductility of 3D printed concrete reinforced with short straight steel fibers. Virtual Phys. Prototyp. **14** (2019)
30. Sri Bhanupratap Rathod, R., Venkatarama Reddy, B.V.: Strength and stress–strain characteristics of fibre reinforced cement stabilised rammed earth. Mater. Struct./Materiaux Constr. **54** (2021)
31. Sinka, M., Spurina, E., Korjankins, A., Bajare, D.: Hempcrete – CO_2 neutral wall solutions for 3D printing. Environmenal and Climate Technologies **26**, 742–753 (2022)
32. Nematollahi, B., Xia, M., Sanjayan, J., Vijay, P.: Effect of type of fiber on inter-layer bond and flexural strengths of extrusion-based 3D printed geopolymer. In: Materials Science Forum, vol. 939 (2018)
33. D'haese, R., et al.: Composite for additive manufacturing including flax by-products and quarry fines. In: 2nd EuroMaghrébine Conference of BioComposites (2018)
34. Savastano, H., Agopyan, V., Nolasco, A.M., Pimentel, L.: Plant fibre reinforced cement components for roofing. Constr. Build. Mater. **13** (1999)
35. Zhou, X., Ghaffar, S.H., Dong, W., Oladiran, O., Fan, M.: Fracture and impact properties of short discrete jute fibre-reinforced cementitious composites. Mater. Des. **49** (2013)
36. Silva, G., et al.: Eco-friendly additive construction: analysis of the printability of earthen-based matrices stabilized with potato starch gel and sisal fibers. Constr. Build. Mater. **347**, 128556 (2022)
37. Xiao, J., Zou, S., Ding, T., Duan, Z., Liu, Q.: Fiber-reinforced mortar with 100% recycled fine aggregates: a cleaner perspective on 3D printing. J. Clean. Prod. **319**, 128720 (2021)
38. Ye, J., Cui, C., Yu, J., Yu, K., Dong, F.: Effect of polyethylene fiber content on workability and mechanical-anisotropic properties of 3D printed ultra-high ductile concrete. Constr. Build. Mater. **281**, 122586 (2021)
39. Singh, A., Liu, Q., Xiao, J., Lyu, Q.: Mechanical and macrostructural properties of 3D printed concrete dosed with steel fibers under different loading direction. Constr. Build. Mater. **323** (2022)
40. Pham, L., Tran, P., Sanjayan, J.: Steel fibres reinforced 3D printed concrete: Influence of fibre sizes on mechanical performance. Constr. Build. Mater. **250** (2020)

41. Alqenaee, A., Memari, A.: Experimental study of 3D printable cob mixtures. Constr. Build. Mater. **324** (2022)
42. Baumann, H., Tillman, A.-M.: The hitch hiker's guide to LCA. Studentlitteratur Lund (2004)
43. United States Environmental Protection Agency, O. of A. and R. & Office of Atmospheric Programs, C. C. D. Framework for Assessing Biogenic CO2 Emissions from Stationary Sources (2014)
44. Institut Bauen und Umwelt e.V. (IBU). Environmental Product Declaration: Concrete admixtures – Hardening Accelerators European Federation of Concrete Admixtures Associations Ltd. (EFCA) (2015)
45. RSMeans Data Online. Preprint at (2011)
46. Luhar, S., et al.: Sustainable and renewable bio-based natural fibres and its application for 3d printed concrete: a review. Sustain. (Switz.) **12** (2020) https://doi.org/10.3390/su12241 0485
47. Technical Data Sheet - Basalt Chopped Strand
48. Fořt, J., Kočí, J., Černý, R.: Environmental efficiency aspects of basalt fibers reinforcement in concrete mixtures. Energies (Basel) **14** (2021)
49. Basalt Engineering. Chopped Fiber Bags 30 lbs. https://www.basalt-usa.com/product/chopped-fiber-bags-30-lbs/
50. Man, Y., Han, Y., Li, J., Hong, M.: Review of energy consumption research for papermaking industry based on life cycle analysis. Chin. J. Chem. Eng. **27**, 1543–1553 (2019)
51. Wang, J., Shi, S.Q., Liang, K.: Comparative life-cycle assessment of sheet molding compound reinforced by natural fiber vs. glass fiber. J. Agric. Sci. Technol. B J. Agric. Sci. Technol. **3** (2013)
52. Lotfi, A., Li, H., Dao, D.V., Prusty, G.: Natural fiber–reinforced composites: a review on material, manufacturing, and machinability. J. Thermoplast. Compos. Mater. **34** (2021). https://doi.org/10.1177/0892705719844546
53. Dittenber, D.B., Gangarao, H.V.S.: Critical review of recent publications on use of natural composites in infrastructure. Compos. Part A: Appl. Sci. Manuf. **43** (2012). https://doi.org/10.1016/j.compositesa.2011.11.019
54. Arvidsson, R., Nguyen, D., Svanström, M.: Life cycle assessment of cellulose nanofibrils production by mechanical treatment and two different pretreatment processes. Environ. Sci. Technol. **49** (2015)
55. Safety Data Sheet - Sodium Carboxymethyl Cellulose 419273 (2021)
56. Carboxymethylcellulose Sodium USP MSDS (2015)
57. Broeren, M.L.M., et al.: Life cycle assessment of sisal fibre – exploring how local practices can influence environmental performance. J. Clean. Prod. **149** (2017)
58. Liu, S., Cao, W., Fang, J., Shang, S.: Variance analysis and performance evaluation of different crumb rubber modified (CRM) asphalt. Constr. Build. Mater. **23** (2009)
59. Farina, A., Zanetti, M.C., Santagata, E., Blengini, G.A.: Life cycle assessment applied to bituminous mixtures containing recycled materials: crumb rubber and reclaimed asphalt pavement. Resour. Conserv. Recycl. **117** (2017)
60. Xu, W., Becker, G.: Environmental impact assessment of wood pulp from a eucalyptus plantation in south china by using life-cycle analysis. For. Prod. J. **62** (2012). https://doi.org/10.13073/0015-7473-62.5.365
61. Summerscales, J., Dissanayake, N.P.J.: Allocation in the life cycle assessment (LCA) of flax fibres for the reinforcement of composites. In: ICCM International Conferences on Composite Materials, vol. 2017-August (2017)
62. Dissanayake, N.P.J., Summerscales, J., Grove, S.M., Singh, M.M.: Energy use in the production of flax fiber for the reinforcement of composites. J. Natural Fibers **6** (2009)
63. González-García, S., Hospido, A., Feijoo, G., Moreira, M.T.: Life cycle assessment of raw materials for non-wood pulp mills: hemp and flax. Resour. Conserv. Recycl. **54** (2010)

64. Stevulova, N., Kidalova, L., Junak, J., Cigasova, J., Terpakova, E.: Effect of hemp shive sizes on mechanical properties of lightweight fibrous composites. In: Procedia Engineering, vol. 42 (2012)
65. Zampori, L., Dotelli, G., Vernelli, V.: Life cycle assessment of hemp cultivation and use of hemp-based thermal insulator materials in buildings. Environ. Sci. Technol. **47** (2013)
66. Kuruşcu, A.O., Girgin, Z.C.: Efficiency of structural materials in sustainable design. J. Civ. Eng. Archit. **8** (2014)
67. Thilakarathna, P.S.M., et al.: Embodied carbon analysis and benchmarking emissions of high and ultra-high strength concrete using machine learning algorithms. J. Clean. Prod. **262,** 121281 (2020)
68. Ouellet-Plamondon, C., Habert, G.: Life cycle assessment (LCA) of alkali-activated cements and concretes. In: Handbook of Alkali-Activated Cements, Mortars and Concretes (2015). https://doi.org/10.1533/9781782422884.5.663
69. Du, J., et al.: Utilization of off-specification fly ash in preparing ultra-high-performance concrete (UHPC): mixture design, characterization, and life-cycle assessment. Resour. Conserv. Recycl. **180**, 106136 (2022)
70. Safety Data Sheet - Fly Ash Product (Class F Fly Ash) (2018)
71. ChemSrc. Hydroxypropyl Methyl Cellulose - CAS Number 9004-65-3
72. Laveglia, A., Sambataro, L., Ukrainczyk, N., de Belie, N., Koenders, E.: Hydrated lime life-cycle assessment: current and future scenarios in four EU countries. J Clean Prod **369**, 133224 (2022)
73. Maddalena, R., Roberts, J.J., Hamilton, A.: Can Portland cement be replaced by low-carbon alternative materials? A study on the thermal properties and carbon emissions of innovative cements. J. Clean. Prod. **186**, 933–942 (2018)
74. Safety Data Sheet - Calcium Hydroxide 31219 (2021)
75. Safety Data Sheet - Calcium Hydroxide 239232 (2021)
76. la Rosa1, A.D., Grammatikos, S.A.: Comparative life cycle assessment of cotton and other natural fibers for textile applications. Fibers **7** (2019)
77. National Center for Biotechnology Information. Compound Summary: Kaolin 56841936. National Library of Medicine (2023)
78. Lecompte, T., Perrot, A., Subrianto, A., le Duigou, A., Ausias, G.: A novel pull-out device used to study the influence of pressure during processing of cement-based material reinforced with coir. Constr. Build. Mater. **78**, 224–233 (2015)
79. Kaolinite Products – Sector EPD (2021)
80. U.S. average price of kaolin 2010–2021. https://www.statista.com/statistics/248194/average-price-of-kaolin/
81. Mohan, M.K., et al.: Performance criteria, environmental impact and cost assessment for 3D printable concrete mixtures. Resour. Conserv. Recycl. **181**, 106255 (2022)
82. Bulk Material Density Table
83. Galan-Marin, C., Rivera-Gomez, C., Garcia-Martinez, A.: Use of natural-fiber bio-composites in construction versus traditional solutions: operational and embodied energy assessment. Materials **9** (2016)
84. Hammilll & Gillespie Specialty Clays and Minerals. HC - 11 Magnesium Aluminosilicate (2021)
85. GCP Technologies. EPD 402 - High-Range Water Reducer (2022)
86. EFCA. EPD - Concrete admixtures – Plasticizers and Superplasticizers (2021)
87. EMI Supply. Sikament 686 High Range Water Reducer and Superplasticizer Normal Slump 55 Gallon Drum. https://www.emisupply.com/
88. OrganoMetallics. Polycarboxylate Ether. https://www.concreteplasticizer.com/
89. Safety Data Sheet - MasterGlenium 3030 (2020)

Fig. 3. Mecopress machine and different steps in the manufacture of a block

After compaction, the blocks are manually taken carefully after demolding; they are then covered by a plastic film for 24 h so as to ensure its water content homogeneity, its sealing and avoid rapid drying. Then, the blocks are dried by exposure to air (T = 20 ± 2 °C, Relative Humidity (RH) = 50 ± 10%) until reaching constant mass (average 4 weeks). The different formulations are noted as follows: CEB_X stands for compacted earth brick reinforced with algae fibers, where X = addition ratio (%). In order to study the effect of fiber addition on the plasticity of the earth mix and to determine the water content suitable for compaction process, the Atterberg limits (EN ISO 17892-12) and the plunger penetration test (EN 1015-4) with a modification of the surcharge were carried out in the fresh state. The mechanical strengths of the CEB (compressive and flexural) were measured in the hardened state.

4 Results and Discussions

4.1 Fresh State Characterization: Atterberg Limits and Plunger Penetration with Surcharge

Table 1. Plasticity of the earth mix as a function of fiber content

Fibers (%)	LL (%)	PL (%)	PI (%)
0	42.13 ± 0.41	20.65 ± 0.50	21.48 ± 0.09
0.5	43.39 ± 0.34	20.87 ± 0.37	22.52 ± 0.03
1	44.39 ± 0.38	21.52 ± 0.25	22.87 ± 0.13
1.5	45.76 ± 0.50	22.44 ± 0.45	23.32 ± 0.05
2	47.10 ± 0.45	23.02 ± 0.30	24.08 ± 0.15
3	49.03 ± 0.23	24.89 ± 0.46	24.14 ± 0.23

The effect of adding algae fibers with different content ratio on the plasticity of the earth mix are presented in Table 1. The increase in the addition rate of algae fibers does not affect the plasticity of the earth mix. The fiber mixtures represent a low plasticity that does not exceed a PI value of 24%. Figure 4 shows the influence of adding the algae fibers on the plunger penetration depth of the earth mix. The manufacturing water content of

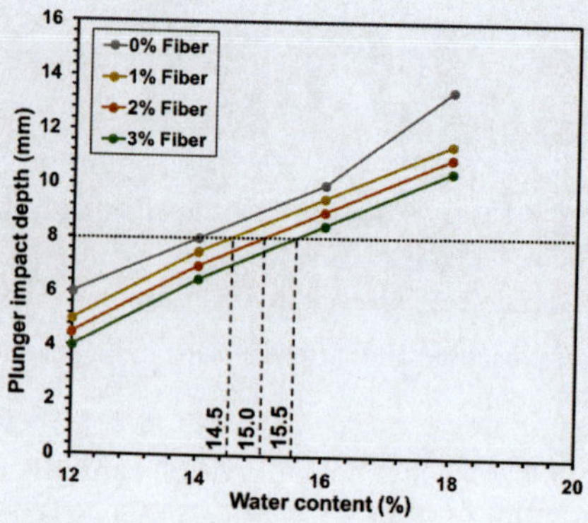

Fig. 4. Plunger impact depth as a function of water content

the CEB control is 14%, which corresponds to an overloaded plunger depth of 8 mm. The introduction of the algae fibers does not significantly affect the plunger penetration depth, which is in agreement with the plasticity results. Based on these results, a water correction to be made during fabrication was determined according to Eq. (1):

$$W_a(\%) = 0.5 \times \%F + W_i \tag{1}$$

W_a: Added water content; % F: Fibers content; W_i: Initial water content, which is 14%.

4.2 Characterization of CEB in the Hardened State: Compressive and Flexural Strength

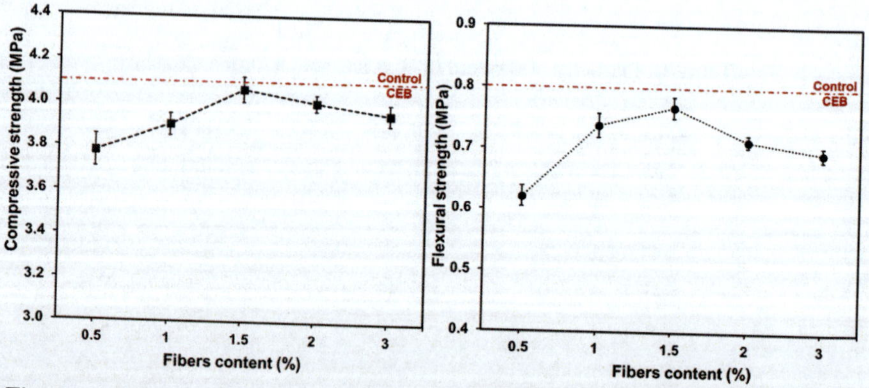

Fig. 5. Compressive and flexural strengths of unreinforced and algae fiber-reinforced CEB

Figure 5 shows the compressive and flexural strength as a function of algae fibers addition rate. The control CEB (CEB_0) has a compressive strength of 4.10 MPa and a flexural strength of 0.8 MPa. Compared to CEB_0, the incorporation of algae fibers reduces the mechanical strengths. Indeed, $CEB_{0.5}$ has a compressive strength of 3.78 MPa, a decrease of 7% compared to CEB_0. A decrease in flexural strength of 22% was also observed for this formulation. However, a slight decrease in mechanical strength was observed for $CEB_{1.5}$. Therefore, the addition ratio of 1.5% seems to be the best compared to the other addition content. This reduction in mechanical strengths could be related to the random distribution of fibers in the earth mix, which leads to the creation of voids inside the clay matrix. Similar results were observed by [6] with the addition of barley straw fibers. This behavior was attributed to the hydrophilic character of the fibers which enhances the porosity produced during the drying of the CEB. Based on the results obtained, it appears necessary to modify the surface of the algae fiber in order to decrease its water absorption capacity (hydrophilic properties) and improve its adhesive properties.

4.3 Effect of Fiber Treatments on the Mechanical Properties of CEB

According to the literature, the purpose of the treatments is to improve the adhesion properties of the fibers [16], which can strongly influence the fiber/matrix interface, and improve the mechanical properties of the composite. Alkaline treatment by sodium bicarbonate, double coating treatment by linseed oil and metakaolin, and pre-saturation of the fibers are carried out to improve the performance of the fiber within the clay matrix. CEB containing treated and untreated fibers are performed with a 1.5% mass fraction of algae fibers.

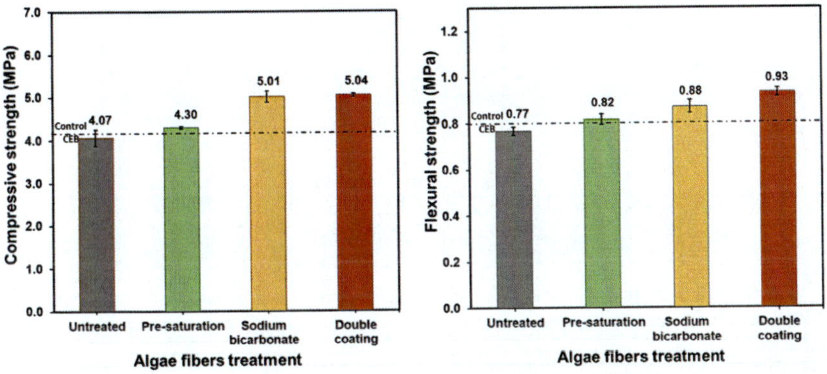

Fig. 6. Compressive and flexural strengths of CEB reinforced with treated algae fibers

Figure 6 shows the compressive and flexural strength of the CEB reinforced with the treated and untreated algae fibers. The incorporation of the raw algae fibers into the CEB leads to decrease the mechanical strengths compared to the control CEB. However, the treated algae fibers showed better behavior inside the matrix. An improvement in mechanical strengths was observed for all three treatments. A slight increase of 5.6% in compressive strength and 6.5% in flexural strength of the CEB containing pre-saturated fibers compared to the CEB with untreated fibers was observed. The CEB reinforced with the sodium bicarbonate treated fibers have a compressive strength of 5.01 MPa and a flexural strength of 0.88 MPa, an increase of 23% and 14% over the CEB reinforced with the untreated fibers respectively. Similarly, the CEB containing the double coated algae fibers showed a 24% improvement in compressive strength and 20% enhancement in flexural strength. This proves the effectiveness of both treatments. Sodium bicarbonate treatment leads to the partial removal of some non-cellulosic compounds present in the fibers structure [17]. Therefore, the mechanical properties were improved due to the enhancement of fiber adhesion properties after treatment. The treatment by double coating allows to isolate the fibers from the matrix preventing its water absorption, in order to decrease their hydrophilic characters [9].

Fig. 7. Video microscope observation of the algae fiber and clay matrix interface: (a) untreated fiber;(b) pre-saturated fiber; (c) fiber treated by sodium bicarbonate, and (d) fiber treated by double coating

Figure 7 presents the fractured surface of CEB with untreated and treated algae fibers. The addition of algae fibers creates voids inside the matrix which leads to poor adhesion between the untreated algae fiber and the matrix (Fig. 7-a). The CEB reinforced with the treated algae fibers (Fig. 7-(b-d)) show a less porous aspect compared to the CEB containing the untreated fibers. In addition, the treated fibers show better adhesion with the clay matrix, which confirms the effectiveness of the treatment, and confirms the mechanical strength results.

5 Conclusion

The aim of this work is to study the influence of the addition of untreated and treated algae fibers on the mechanical properties of compacted earth bricks (CEB). The following main conclusions could be:

- The addition of the algal fibers did not affect the plasticity of the soil mixture, and did not significantly influence the manufacturing water content of the CEB.
- The incorporation of the algae fibers decreased the mechanical strengths. However, the addition of 1.5% seems to be the best compared to the other rates.
- The treated fibers showed a better adhesion with the clay matrix, this was confirmed by video microscope images. In addition, an increase in the mechanical strengths of the CEB reinforced with the treated algae fibers compared to the CEB reinforced with untreated fibers was observed.

From these results, it is possible to assert that treated algae fibers with sodium bicarbonate solution and double coating (linseed oil/metakaolin) leads to achieve highest interfacial adhesion with clay matrices thus leading to better mechanical properties of the CEB. The behavior of CEB containing treated fibers subjected to ageing tests will be studied to assess the durability performance of CEB. This study will present the objective of the following paper.

References

1. Pacheco-Torgal, F., Jalali, S.: Earth construction: lessons from the past for future eco-efficient construction. Constr. Build. Mater. **29**, 512–519 (2012)
2. Waziri, B.S., Lawan, Z.A., Mustapha, M.M.: Properties of compressed stabilized earth blocks (CSEB) for low-cost housing construction: a preliminary investigation. Int. J. Sustain. Constr. Eng. Technol. **4**(2), 39–46 (2013)
3. Meukam, P., Jannot, Y., Noumowe, A., Kofane, T.C.: Thermo physical characteristics of economical building materials. Constr. Build. Mater. **18**(6), 437–443 (2004)
4. Millogo, Y., Morel, J.C.: Microstructural characterization and mechanical properties of cement stabilised adobes. Mater Struct. **45**(9), 1311–1318 (2012)
5. Millogo, Y., Morel, J.C., Aubert, J.E., Ghavami, K.: Experimental analysis of pressed adobe blocks reinforced with hibiscus cannabinus fibers. Constr. Build. Mater. **52**, 71–78 (2014)
6. Laborel-Préneron, A., Aubert, J.E., Magniont, C., Maillard, P., Poirier, C.: Effect of plant aggregates on mechanical properties of earth bricks. J. Mater. Civ. Eng. **29**(12), 04017244 (2017)
7. Ajouguim, S., Talibi, S., Djelal-Dantec, C., Hajjou, H., Waqif, M., Stefanidou, M., et al.: Effect of Alfa fibers on the mechanical and thermal properties of compacted earth bricks. Mater. Today: Proc. **37**, 4049–4057 (2021)
8. Page, J., Khadraoui, F., Gomina, M., Boutouil, M.: Influence of different surface treatments on the water absorption capacity of flax fibres: rheology of fresh reinforced-mortars and mechanical properties in the hardened state. Constr. Build. Mater. **199**, 424–434 (2019)
9. Page, J., Amziane, S., Gomina, M., Djelal, C., Audonnet, F.: Using linseed oil as flax fibre coating for fibre-reinforced cementitious composite. Ind. Crops Prod. **161**, 113168 (2021)
10. Silva, E.J., Marques, M.L., Velasco, F.G., Junior, C.F., Luzardo, F.M., Tashima, M.M.: A new treatment for coconut fibers to improve the properties of cement-based composites–combined effect of natural latex/pozzolanic materials. Sustain. Mater. Technol. **12**, 44–51 (2017)

11. Seo, Y.B., Lee, Y.W., Lee, C.H., You, H.C.: Red algae and their use in papermaking. Biores. Technol. **101**(7), 2549–2553 (2010)
12. Ouhssine, K., Ouhssine, M., Elyachioui, M. : L'application des déchets traités de l'algue Gelidium sesquipédale dans la culture du Maïs. Afrique Sci.: Rev. Int. Sci. Technol. **3**(2) (2007)
13. El Achaby, M., Kassab, Z., Aboulkas, A., Gaillard, C., Barakat, A.: Reuse of red algae waste for the production of cellulose nanocrystals and its application in polymer nanocomposites. Int. J. Biol. Macromol. **106**, 681–691 (2018)
14. Tayibi, S., et al.: One-pot activation and pyrolysis of Moroccan Gelidium sesquipedale red macroalgae residue: production of an efficient adsorbent biochar. Biochar. **1**(4), 401-412 (2019)
15. Talibi, S., Page, J., Djelal, C., Waqif, M., Saâdi, L.: Study of earth-based materials for manufacturing compaction process. J. Build. Eng. **64**, 105546 (2023)
16. Ajouguim, S., Page, J., Djelal, C., Waqif, M., Saadi, L.: Performance of Alfa fibres in cementitious materials exposed to diverse surface treatments. Constr. Technol. Archit. **1**, 660–667 (2022)
17. Fiore, V., Scalici, T., Nicoletti, F., Vitale, G., Prestipino, M., Valenza, A.: A new eco-friendly chemical treatment of natural fibres: effect of sodium bicarbonate on properties of sisal fibre and its epoxy composites. Compos. B Eng. **85**, 150–160 (2016)

Developing 3D-Printed Natural Fiber-Based Mixtures

Tashania Akemah and Lola Ben-Alon(✉)

School of Architecture, Planning and Preservation, Columbia University, New York, NY 10027, USA
rlb2211@columbia.edu

Abstract. Natural earth-fiber building assemblies such as light straw clay, hempcrete, and clay-plastered straw bales incorporate vegetable by-products that are mixed with geological binders, traditionally used as an insulative infill in building construction. As a geo- and bio-based insulative infill method composed mostly of fiber, heat transfer coefficients are lower than mass materials, making it a compatible assembly that meets energy code requirements. Furthermore, due to their permeability, these materials exhibit high hygric capacity, providing regulated indoor temperatures and relative humidity levels, thus showing a promising future for socially just and healthier built environments. Despite these advantages, the use of earth-fiber building materials in digital construction is still underdeveloped. In the past few years, 3D-printed earth has gained an increasing interest, however, high contents of fibers in earth mixtures have yet to be fully tested and characterized. This paper presents an experimental workflow to characterize fiber-earth composites for 3D printed assemblies, using natural soils infused with natural fibers. The paper begins with a literature review of a range of fibers: straw, hemp, kenaf, sisal, and banana leaves, as well as naturally occurring biopolymer additives. The experimental setup includes manual extrudability and buildability tests, to identify optimal mix designs that are then tested for their printability and buckling using clay 3D printers. As a final deliverable, first pass geometric studies showcase the lightweight and structural possibilities of each material. The significance of this research lies in the development of a methodology for identifying novel mix design for digital fabrication, by increasing carbon storing vegetable fiber content within digital earth, and by creating a range of natural 3D printed assembly types: from mass-insulation walls to paper-thin lightweight partition assemblage.

Keywords: Natural building materials · Earth- and Bio-based building materials · Light straw clay · 3D printing · Additive manufacturing

1 Introduction

Earth-based construction materials have been gaining interest in recent years due to their environmental impacts and affordability (Van Damme and Houben 2018). Earth-based construction (interchangeably termed as earthen materials, or generally addressed

S. Amziane et al. (Eds.): ICBBM 2023, RILEM Bookseries 45, pp. 555–572, 2023.
https://doi.org/10.1007/978-3-031-33465-8_42

as "natural building") is a field of building materials and construction techniques that incorporate clay-rich soils as a base matter, infused with other geological products (such as sand, larger stone aggregate, and lime), and fibers (such as wheat straw, rice straw, and hemp). Earth-based construction has been used for thousands of years and is still prevalent in many parts of the world. Earth-based materials have been shown to exhibit significantly lower embodied energy and emissions as opposed to conventional materials such as concrete due to their minimal thermal and chemical processing (Ben-Alon et al. 2021). Further ways to increase the environmental opportunities of earthen construction is to increase fiber content within the mix design due to its carbon storage and lighter weight for transportation.

The use of fiber proposes advantages throughout production to disposal (cradle to grave). During the growth cycle of fibers, carbon dioxide is consumed through photosynthesis, removing it from the atmosphere (Walker et al. 2019). Furthermore, the cultivation of bast fiber crops requires fewer pesticides and fertilizers, thus reducing risks for acidification and eutrophication (Fernando et al. 2015). In building construction, the use of locally available fibers is ideal to reduce carbon emissions from transportation due to their low weight and packaging possibilities. Natural fibers also improve ductility (Miccoli et al. 2014) and thermal properties (Holzhueter and Itonaga 2017). During the operational phase, earth assemblies with high contents of natural fibers help reduce operational emissions by increasing thermal resistivity, coupled with the mass capacity (Rempel and Rempel 2016; Shang and Tariku 2021). Some examples of natural fibers that are commonly used in building construction are straw, hemp, flax, fique, banana, jute, sisal, animal hair and wool.

Despite these benefits, the use of earth-fiber assemblies remains a marginal field in sustainable construction due to a range of constraints, including the labor intensity of construction using these materials and their techno-mechanical-thermal properties that requires further characterization. The use of 3D printing can therefore introduce accuracy (due to the machine manufacturing) while reducing labor needs (Correa et al. 2015).

This paper presents a workflow for identifying earth-fiber mixtures suitable for 3D printed architectural applications. Illustrated in Fig. 1, this research proposes a methodology for developing earth-fiber mixtures for 3D printing of architectural artifacts, starting with a performance synthesis, to material selection, the tests include manual extrusion and buildability, microstructural electron scanning diagnosis, and digital printability using simple and complex geometries.

The proposed methodology is initiated with a literature review of a range of fibers, followed by a series of manual extrudability and buildability tests, and machine printability tests and processing parameters development. The final demonstration exhibits a successful implementation of earth-fiber materialities in 3D printed applications in small scale walls and paper-thin lightweight partition elements.

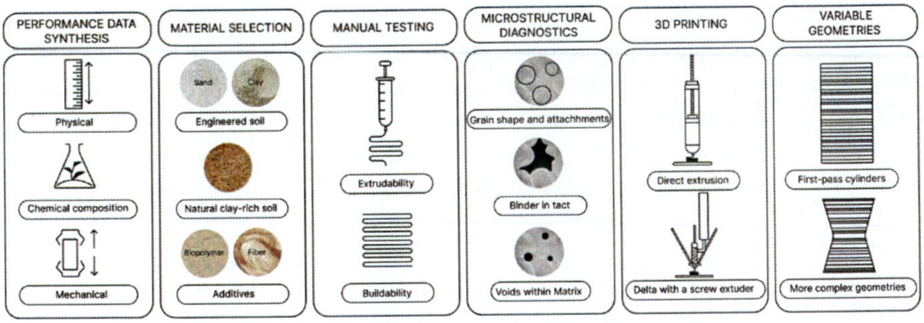

Fig. 1. The methodology developed as part of this research.

2 Background on Natural Fibers: From Traditional to Digital Construction

2.1 Traditional Earth-Fiber Assemblies

Earth-fiber materials use locally-adjusted mix designs in which clay-rich soils act as a natural binder, fibers act as the reinforcement (and, often, aggregate), with or without additives that may provide enhanced water resistivity (such as cow dung, cactus juice, flaxseed oil). Hempcrete, for instance, is an infill material made from hemp hurds (hemp shives) as a plant aggregate, and lime as a geological binder. As a construction material, hempcrete is a fire resistant, robust, and non-toxic building product, with a sequestering power of 135 $kgCO_{2e}/m^3$, primarily from the growth phase of the hemp plant and carbonation phase of the lime binder.

Fig. 2. Light straw clay workshop by The Year of Mud; Light straw clay monolithic wall construction, and; Manually constructed earth-fiber composites. Image sources: (Goodhew et al. 2021; Baker-Laporte, Laporte 2015).

Most relevant to this work, light straw clay (also known as light clay, straw clay, slip straw, rammed straw and leichtlehmbau), is an infill material that can be applied as a modular block or tamped within a structural frame. Light straw clay serves as an insulating material which can be mixed and packed into a variety of densities but is not load bearing. Shown in Fig. 2, the production process of light straw clay includes tamping the long stalks of fiber, mixed with clay slurry (very wet clay) which helps stiffens the mixture after compaction before drying. Besides acting as an insulation material, light straw clay is excellent for retrofit insulation due to its compatibility with structural wood frame (Ben-Alon 2020). Furthermore, light straw clay can easily be applied around windows, doors, and other openings. Light straw clay is also a healthy building material that serves as an alternative for individuals with sensitivity to mold and chemicals. Over their embodied phase, light straw clay assemblies were shown to reduce up to 70% in global climate change impacts, 55% in energy demand, 57% in air acidification, and 27% in air particle pollution as opposed to fiberglass insulated wood frame assembly (Ben-Alon et al. 2021). Over their use phase, light straw clay assemblies were shown to reduce energy consumption by 32–59% in hot desert climates, 29–55% in semi-arid climates, 46–73% in Mediterranean climates, 34–55% in temperate climates, and 27–50% in cold continental climates, as opposed to conventional building assemblies.

2.2 Reviewing Natural Fibers and Their Properties for Digital Construction

In order to identify natural fibers for 3D printed earth, a review study of the plant physical structure and composition was obtained, as shown in Fig. 3. Factors were chosen based on their influence on the final structure and 3D printing process.

Physical and Mechanical Properties. Fiber length was assumed to affect mixture production more than the dry state mechanical performance of the earthen composite. Previous studies have shown that fiber length has a minimal influence on the flexural strengths of earth materials such as cob (Pullen and Scholz 2011). However, the length of the fiber is important to the extrusion of mixtures through the nozzle during 3D printing. Similarly, smaller fiber diameter was assumed to reduce the risk of clogging within the extrusion flow mechanisms of the printers. Fiber with shorter natural lengths reduce the need for fiber processing prior to printing.

The flexural strengths of earth-fiber composites were shown to increase as a function of the fiber tensile strength (Pullen and Scholz 2011). Additionally, high absorption rates were shown to correlate with shrinkage and dimensional variation of the material caused by water evaporation (Laborel-Préneron et al. 2016).

Fiber Composition. There are three naturally occurring biopolymer components found in plants that affect the final material: cellulose, hemicellulose, and lignin. Cellulose is the primary element found in the cell wall of plants and is a polymer of carbohydrates ($C_6H_{10}O_5$). Cellulose is the most abundant organic polymer on Earth and can be found in all agricultural products and waste. It is a linear polymer of $\beta - (1 \rightarrow 4)$–linked D-glucose. Secondary to cellulose, hemicellulose (also known as polyose) is found in almost all plant cell walls, however, compared to cellulose structural make-up, hemicellulose molecules are branched shorter in length and show a tendency to crystallize (Smith et al. 2022). Lignin is the third most abundant reproducible natural resource and the only renewable aromatic polymer in nature (Suhas et al. 2007). Lignin is a cementing material found mostly in wood plant fibers, and is a polymerized hydrophobic amorphous substance, which is three-dimensional in structure and highly branched. Lignin is considered a waste product in certain industries (such as paper making), where it is removed by acid or alkali treatment, for its dissolution and fiber separation.

The complex structure and composition of lignin provide unique properties versus those of cellulose and hemicellulose. Lignin fills the cell walls between the cellulose, hemicellulose, and pectin molecules. As a structural component, lignin act as a support tissue of most plants (Saake and Lehnen 2007) thus greatly impacting the structural strength of plants. Lignin resists degradation, as opposed to cellulose and hemicellulose that are prone to biodegradation (Vane et al. 2003). Shrinking and swelling behavior of fibers is dictated by the movement mechanism of bound water, which escape from between cellulose and hemicellulose molecules (Reeb 2009). Lignin, however, reduces absorption rates, functioning as a water barrier, a critical characteristic for construction elements that are required to withstand weather forces and moisture. Despite this advantage, high lignin contents that decrease absorptive ability might also affect the fiber's ability to integrate within a mix design matrix, presumably because of fibers that absorb water may soften within the mix design.

From the fiber analysis, six fibers were deemed to be suitable for printability testing due to their chemical composition and source efficiency: straw, hemp, banana leaf, sisal, and kenaf, given their wide representation of physical and chemical composition properties.

3 Materials and Methods

The overall research procedure presented in this paper included three main steps, as shown in Fig. 1: (1) material selection following the literature review; (2) manual extrudability and buildability tests using a manual 2 mm syringe, and; (3) digital printability tests using two types of 3D clay printers.

The soil used in the research was of a dark grey-brown composition, sourced from a local quarry (Goshen, New York USA). This soil, characterized to its mineralogical content and particle size distribution, is presented in Bryson et al. (2021). Additionally, "engineered" soils were created by mixing bentonite and kaolinite clays with Ottawa sand. Throughout this study, naturally occurring biopolymers were used in quantities of 2% or less, due to their ability to modify the rheology of the material (Dourado et al. 2016). The main purpose of using biopolymers is to assist with the flowability (Malaeb et al. 2019). Additionally, biopolymers were shown to increase green strengths and the structural buildup of a 3D printed earth materials (Perrot et al. 2018). The biopolymers used included methyl cellulose, sodium alginate, chitosan, locust bean gum, guar gum, and xanthan gum.

3.1 Manual Extrudability and Buildability Tests

The extrudability and buildability were tested manually, using a 2 mm syringe. The hypothesis that underlies this test is that the 3D printed mixture is dependent on the cohesion between the binder (the clay-additive paste) and the plant fiber reinforcement. An optimal mix design would therefore have a high static yield stress which would result in shape retention and minimal buckling, thus supporting the weight of subsequent layers (Tay et al. 2019).

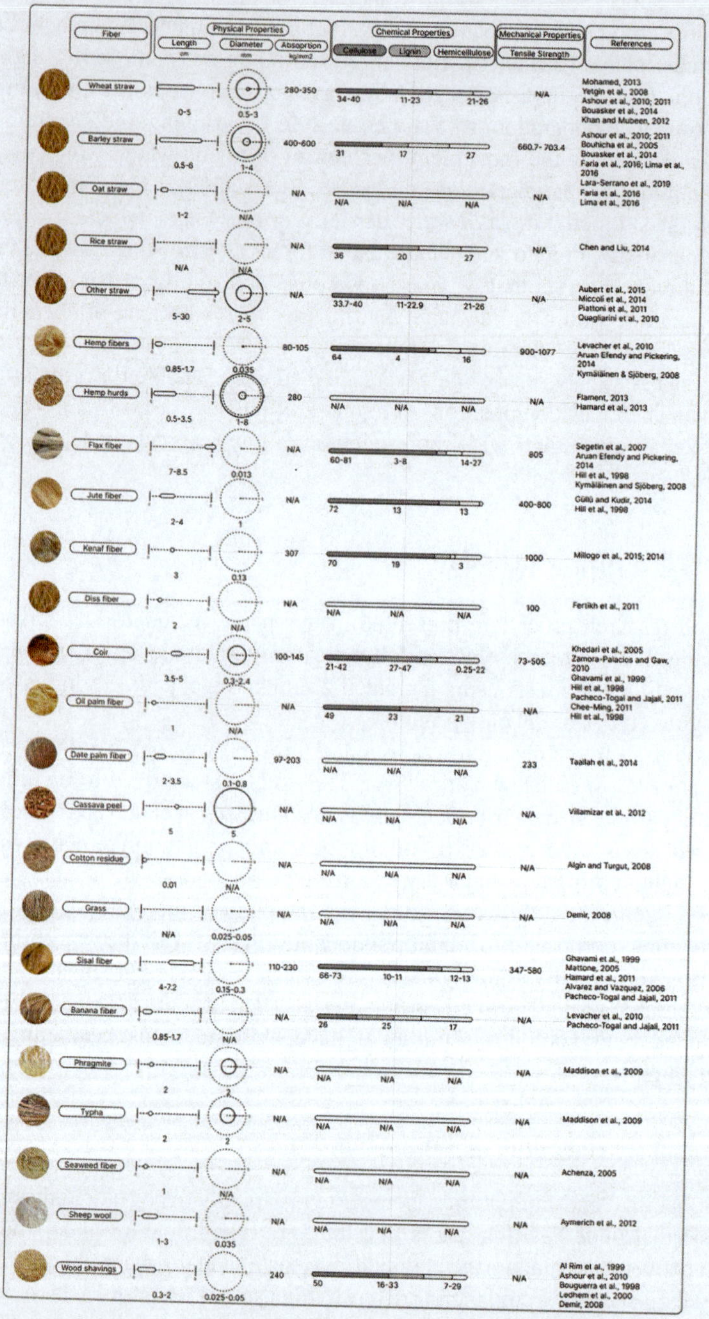

Fig. 3. Synthesized physical, compositional, and mechanical performance for natural fibers that have been previously used in earth-based construction. * N/A means not applicable.

The extrudability through the nozzle was evaluated qualitatively, assessing the flow and clogging or blockage of the mixture during the extrusion process. The buildability of the samples was assessed quantitatively by counting the number of vertically depositing 50 mm layers on top of one another. The buildability was also evaluated qualitatively by analyzing the adhesion between layers, the shape retention of the freshly extruded specimen, and the tendency of the built layers to collapse.

3.2 Microstructural Imaging

The microstructural imaging was conducted using a Zeiss Sigma VP scanning electron microscope (SEM), which combines electrostatic and magnetic fields in its optical performance. The aim of this step was to observe the extruded samples and correlate the extruded results with microstructural observations. The samples were observed to obtain insights about the arrangement of the clay and sand particles, and fibers, in relation to the extrusion directionality, the paste-aggregate interactions, and the quality and cohesion of the matrix at large.

3.3 Machine Printability Tests

Using the successful mixtures from the manual test, the printability was assessed using two types of 3D printers: a direct extruder clay printer (3D PotterBot 4 Pro), and a ceramic delta extruder with a screw system (Delta WASP 40100 Clay). Compared to the direct extruder printer, the delta extruder a printing system of three arms on a rail that operate vertically to move the print head, thus allowing higher geometrical precision. The nozzle radius used for the machine tests included 4, 5, 6, and 8 mm.

File Preparation. To conduct the machine 3D printing tests, a series of vase forms were created in Rhino3D modeling software, imported into a slicer software (3D version 5.0 by All3DP, and Cura version 4.3 by Ultimaker) to generate the G-code for printing. Nozzle sizes Simplify were also indicated within the slicer software to correlate to the material flow and speed. Additional parameters dictated within the slicer software included the primary layer height (the vertical distance between one layer to the next), which was set to half of the radius of the nozzle (2 mm to 4 mm, respectively), a value that was found to be optimal to ensure proper bonding between layers. Increased layer heights were also tested, showing a possible deposition of subsequent layers, though resulted in reduced grip between one layer to another. The preliminary geometry included basic cylindrical extrusions, followed by more complex forms that aimed to examine the geometrical variability and structural stability of the material.

4 Results

4.1 Manual Extrudability and Buildability Tests

The manual extrudability and buildability tests results showed, on an immediate observation, that only the straw and hemp fibers deemed to be soft enough to be extruded. Other fibers with high lignin content were too stiff to orient correctly in the direction of extrusion flow. Shredding the fibers using a Vitamix Blender (Vitamix 7500 machine, Vitamix Corp., Cleveland, OH, USA) was required in order to obtain the desired length, however, processing the fique, banana, and sisal fibers did not yield in shorter lengths but rather in lumps of aggregated fiber.

The printability tests shown in Table 1 and Table 2 indicate that the best extrudability and buildability are achieved for the alginate, cellulose, and locust bean gum additives. Mixtures with natural soils outperformed the engineered soils, assumingly due to the

Table 1. Manual extrudability and buildability tests for the soil-additive pastes

	Code	Composition	Soil (vol %)	Clay (vol %)	Sand (vol %)	Additive (vol%)	Water (vol%)	Extrudability*	Buildability **	Qualitative Assessment Notes
Engineered soil	NbS	Nanoclay Bentonite, Sand	-	29	29	-	42	3	4	Hardened sample is brittle
	BS	Bentonite, Sand	-	27	27	-	46	3	4	Hardened sample is brittle
	NkS	Nanoclay Kaolinite, Sand	-	38	38	-	24	3	1	Hardened sample is brittle. Extrudability is limited, fluids pushed down first.
	KS	Kaolinite, Sand	-	40	40	-	20	1	1	Hardened sample is brittle. Extrudability is limited, fluids pushed down first.
	BA	Bentonite, Sand, Sodium Alginate	-	13	50	1	35	3	2	Mixture is too dry to extrude but additional water would render the mixture unbuildable
	KA	Kaolinite, Sand, Sodium Alginate	-	15	60	2	23	1	0	Low water retention does not allow the mixture to be extrudable or buildable
	BGg	Bentonite, Sand, Guar Gum	-	12	46	1	41	4.5	3	Too wet for enhanced buildability results
	KGg	Kaolinite, Sand, Guar Gum	-	13	49	1	38	4.5	4	Too wet for enhanced buildability results – material quickly collapses
	BLbg	Bentonite, Sand, Locust Bean Gum	-	12	46	1	41	3.5	5	Result is buildable, however clogs in extrusion
	KLbg	Kaolinite, Sand, Locust Bean Gum	-	13	52	1	33	3.5	4	Result is buildable, but extrudability requires more water.
	BXg	Bentonite,Sand, Xanthan Gum	-	13	52	1	33	3.5	5	Result is buildable, material quickly stiffens
	KXg	Kaolinite,Sand, Xanthan Gum	-	14	56	1	29	4	4	Doesn't retain shape as well as the bentonite mix
	BCe	Bentonite, Sand, Methyl Cellulose	-	13	52	1	33	4	4	Extrudes and builds well
	KCe	Kaolinite, Sand, Methyl Cellulose	-	13	50	1	35	5	5	Extrudes and builds well
Natural soil	SCe	Soil, Cellulose	61			9	30	3	3	Hard to extrude
	SXg	Soil, Xanthan gum	64			9	27	2	5	Sticky while extruded, which allows cantilevering
	SGg	Soil, Guar gum	56			8	36	4.5	5	Extrudes and builds well
	SA	Soil, Sodium Alginate	64			9	27	5	5	Easily extrudable, successfully buildable, sample hardens quickly
	SLbg	Soil, Locust bean gum	64			9	27	5	5	Easily extrudable, successfully buildable, sample hardens quickly

*Based on qualitative assessment results, rated from 1(poor extrusion) to 5 (successful extrusion)

** Based on quantitative assessment, counting the number of vertically deposited layers wider particle size distribution. The maximum quantity of fibers for a mixture to be deemed extrudable and printable was 13% (total wt.%) for both the hemp (55% by total volume) and wheat straw (63% by total volume). The marked 8 mixtures were selected for further evaluation through the 3D additive manufacturing process using the printers due to its higher fiber content.

Table 2. Manual extrudability and buildability tests for the earth-fiber mixtures

	Mix design	Composition	Soil (wt%)	Fiber (wt%)	Additive (wt%)	Water (wt%)	Extrudability*	Buildability **	Qualitative Assessment Notes
Soil-fiber	SF	Soil, Hemp	43	2	-	55	5	5	Extrudes and builds well
	SSt	Soil, Wheat straw	43	2	-	55	5	5	Extrudes and builds well
	SSi	Soil, Sisal	43	2	-	55	-	-	Non extrudable
	SK	Soil, Kenaf	43	2	-	55	-	-	Non extrudable
	SB	Soil, Banana leaf	43	2	-	55	-	-	Non extrudable
Soil hemp	SLbgH1	Soil, Locust Bean Gum, Hemp Fiber	39	1	7	52	5	5	Extrudes and builds well
	SLbgH2	Soil, Locust Bean Gum, Hemp Fiber	38	2	7	53	5	5	Extrudes and builds well
	SLbgH3	Soil, Locust Bean Gum, Hemp Fiber	36	3	7	54	5	5	Extrudes and builds well
	SLbgH5	Soil, Locust Bean Gum, Hemp Fiber	30	5	6	58	5	5	Extrudes and builds well
	SLbgH7	Soil, Locust Bean Gum, Hemp Fiber	27	7	7	60	5	5	Extrudes and builds well
	SLbgH8	Soil, Locust Bean Gum, Hemp Fiber	27	8	5	60	5	5	Extrudes and builds well
	SLbgH10	Soil, Locust Bean Gum, Hemp Fiber	27	10	6	57	5	5	Extrudes and builds well
	SCAH13	Soil, Cellulose Sodium Alginate, Hemp Fiber	3	13	4	79	5	5	Extrudes and builds well, fiber clogging is initiated
Soil straw	SS1	Soil, Wheat straw, Cellulose, Alginate	49	3	2	24	5	5	Extrudes and builds well
	SS2	Soil, Wheat straw, Cellulose, Alginate	44	8	2	46	5	5	Extrudes and builds well
	SS3	Soil, Wheat straw, Cellulose, Alginate	3	13	4	79	5	5	Extrudes and builds well

Shown in Fig. 4, the mix designs generated exhibit a range of colors and textures, depending on the clay type in the engineered soil, or the natural soil. The resulting textures depended on fiber ratios, and the presence of sand in the soil.

4.2 Material Microstructural Imaging

Selected extrudable and buildable mixture from Sect. 4.1 were tested for their microstructural composition and macrostructural printability aspects.

Microstructural. The electron scanning conducted on four selected samples exhibit the improved matrix obtained by the soil-fiber-biopolymer composite. Shown in Fig. 5, the soil scans illustrate a clay plate structure, as anticipated from the clay-rich soils. For the engineered soil, Fig. 6 and Fig. 7 illustrate the uniformity of the composition, with the paste sand grain shape embedded within the matrix. The grains of sand are shown to be held by the clay-biopolymer paste, which forms a binding matrix. Lastly, the electron scanning of the soil-fiber-biopolymer composite, shown in Fig. 8, exhibits an

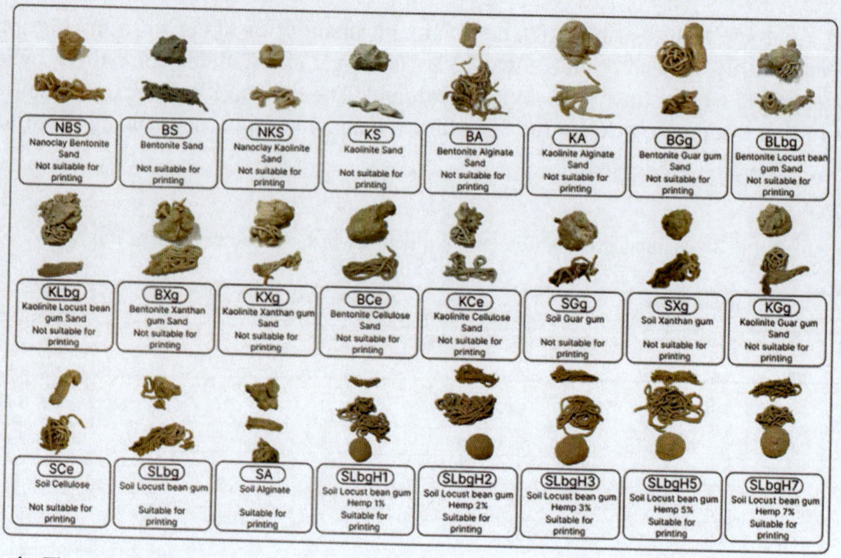

Fig. 4. The extrudability and buildability samples generated as part of this research, exhibiting the range of material colors and textures.

extremely heterogeneous composite, with fibers integrated throughout the matrix. This result, presumably explain the enhanced performance of the latter mixture in regards to strength (variable grain size) and printability (grain size smaller than the angular sand).

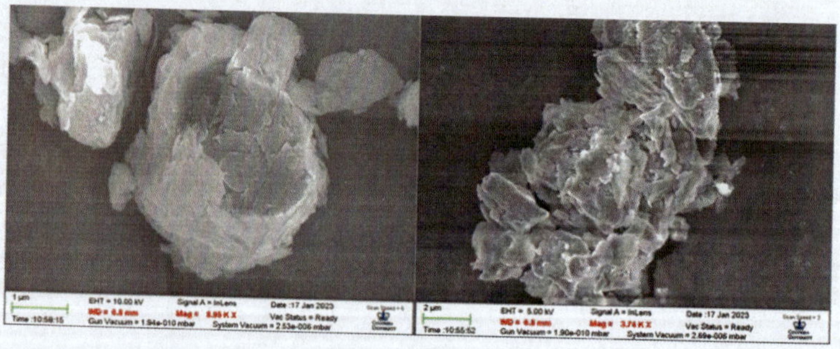

Fig. 5. Electron scanning of the soil used in this study, showing its clay mineral plates.

4.3 Machine Printability Tests

The selected mixtures from the manual extrudability and buildability tests were further evaluated through a 3D additive manufacturing process using the printers. While the soil-straw mixtures exhibited significant mold growth on the 24 h after printing, the soil-hemp mixtures experienced minimal molding within the drying process. This result might be explained by the fungal decay that is mostly attracted mainly to the cellulose components

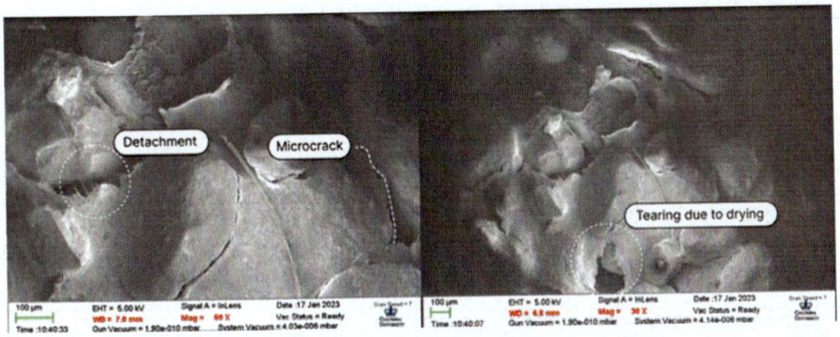

Fig. 6. Electron scanning of the Electron scanning of the sample BA (Bentonite, Sand, Alginate), showing significant detachments of the sand grains from the clay-additive paste.

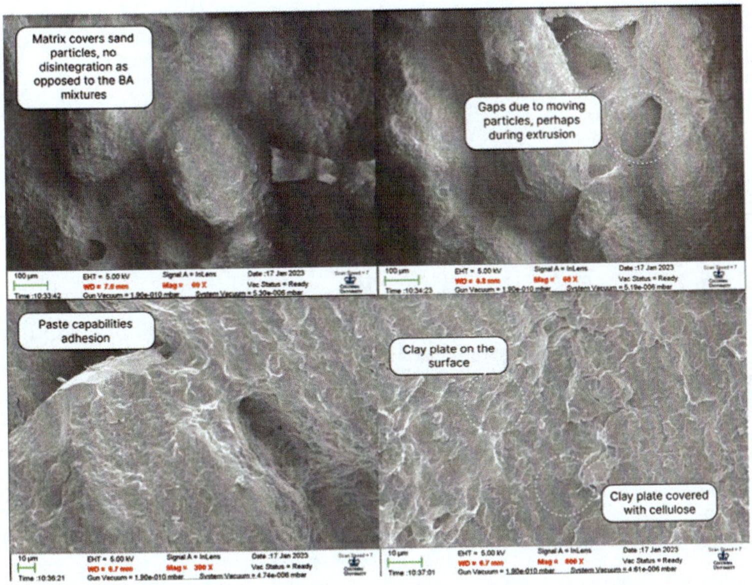

Fig. 7. Electron scanning of the KCe (Kaolinite, Sand, Cellulose), showing the matrix consistency and coating of the sand grains.

in the plant (Baldrian and Valášková 2008). The glucose units found in cellulose are consumed by the fungi agents to produce energy, and biomass in an anabolic process, while releasing CO_2 to the environment through catabolic processes (Lynd et al. 2002). Furthermore, the hemp clay exhibited higher ductility during the wet stage and decreased fragility in its dry form. The printed results, illustrated and interpreted in Fig. 9, exhibit the soil-straw and soil-hemp material as produced by the 3D printing machines.

Fiber Clog within Extrusion Flow Screw. As part of the experimental iterations, it was observed that fibers longer than 3 mm led to printing blockages due to clogging along the flow screw (observed in the WASP Delta 40100). Similarly, sand grains exhibited a similar phenomenon. The WASP Delta 40100 is engineered as a clay 3D Printer. Compared to clay particles, the sand grains were shown clog the edges of the screw due

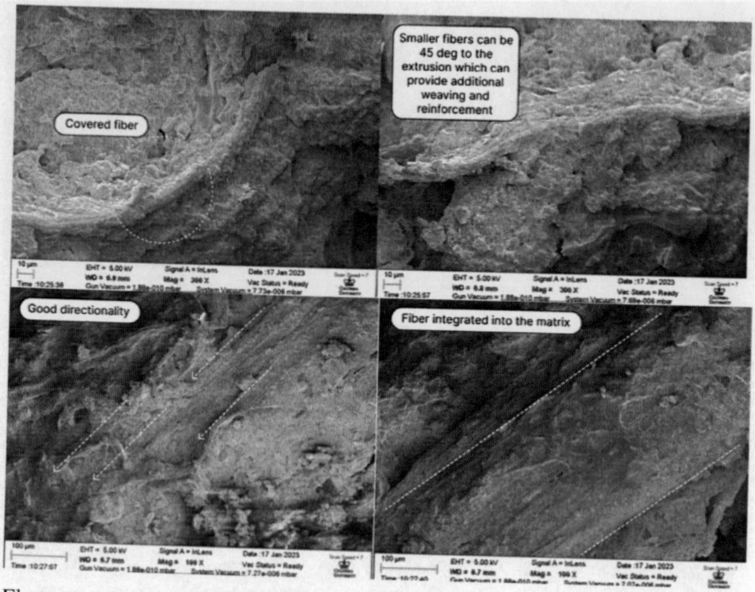

Fig. 8. Electron scanning of the SLbgH (Soil, Locust Bean Gum, Hemp), showing the comprehensive coating of the fibers by the paste, and the directionality of fibers according to the extrusion.

to their larger sizes as opposed to clay particles. Similarly, the 3mm hemp fibers were observed to clog in the screw system due to their length. Further observation also showed that the hemp fibers between the screw and the extruder wall caused friction that in turn clogged the machine.

Printing Parameters. The printing parameters were obtained for both mixtures, showing that different air pressure and layer heights were required, depending on the fiber type. It was observed that the first 3–4 layers should be printed in lower speed, to allow proper adhesion of the print "base". Once the first layers are successfully deposited, the speed is increased successfully to the machine's 100% rate. The printing process showed that the air pressure needed for the hemp clay mixture (5–6 psi) was much higher than the required pressure for light straw clay (3–4 psi). As a finer fiber, the hemp showed tendency to clog in small lumps within the mixture, which caused blockage of the printer if not mixed thoroughly.

5 Conclusions

This paper presents a series of experiments aimed to develop novel earth-fiber mix designs for 3D printed building materials. Earth materials are an emerging, sustainable alternative to cementitious materials because of their low embodied carbon, affordability, and indoor air quality characteristics. To date, however, most 3D printed earth research has been applied in nature, lacking consistent, reliable technical information related to material properties or design methodologies. For example, 3D printed earth mixture design research has been limited to mix designs that contain relatively low fiber content.

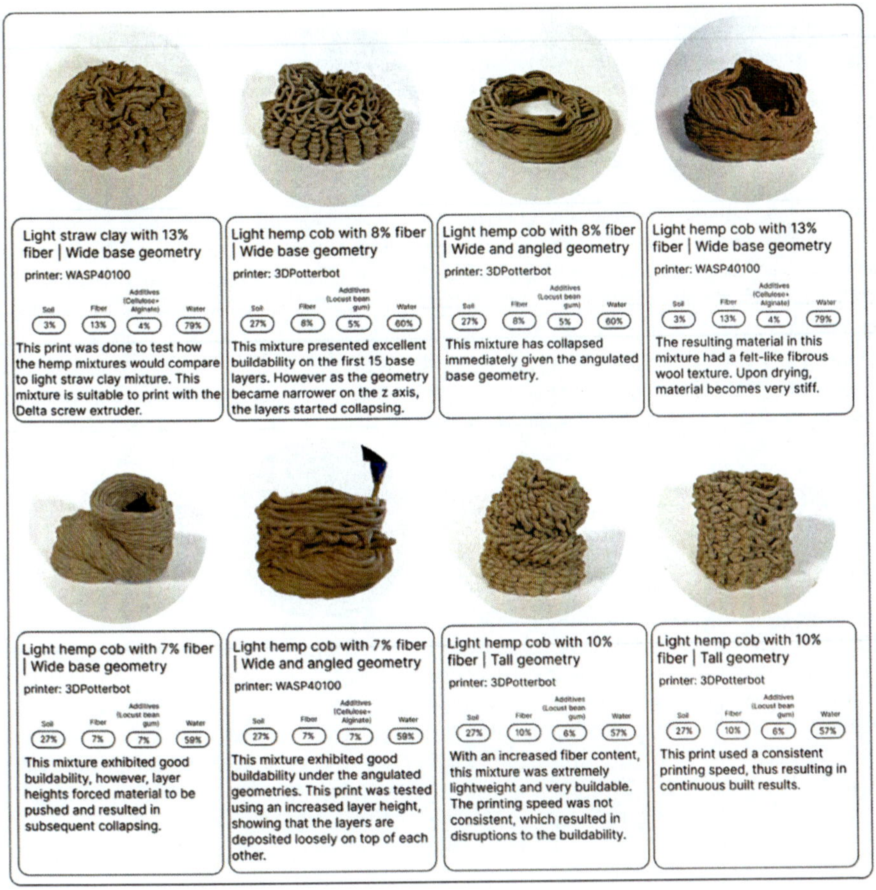

Fig. 9. Printed samples from the selected mix designs.

Using biopolymer binding agents and natural fibers, this project develops optimized mix designs for 3D printed earth while maximizing fiber content, and thus, carbon storage and thermal resistivity. Following a synthesis of the literature performance on fibers used for earth construction, the workflow approach developed as part of this work includes a manual extrudability and buildability assessment, a microstructural diagnosis using a scanning electron microscope, and a geometrical evaluation using machine printability tests.

The conclusions of the study point to enhanced mixtures of natural soils, infused with locust bean gum, alginate, and cellulose biopolymers, with up to 13% wheat straw or hemp fibers. The extrudability and buildability tests prove successful in projecting printability outcomes using two 3D printer types: a direct extrusion and delta screw extrusion machines. The final demonstration exhibits a successful implementation of earth-fiber materialities in 3D-printed applications in small scale samples.

This research critically catalyzes earth- and bio-based materials formulations and digital fabrication techniques. As a long-term deliverable, this work contributes to developing future fabrication techniques using carbon-storing earth composite wall assemblies, while also gearing towards a broader nonconventional geo-bio-based mixtures for digital construction.

Acknowledgement. This research is based upon work supported by the National Science Foundation under Grant No. # 2134488. The authors would like to thank the Natural Materials Lab students who have contributed to this work. In particular, Zackary Bryson who assisted with the early-stage development of the straw-clay mixtures and the 3D printing apparatus developed as part of this ongoing line of research. We thank Rebecca Mikofsky for her review and contribution to this manuscript. Lastly, the authors acknowledge the use of facilities and instrumentation (i.e. Zeiss Sigma VP) supported by the Columbia Nano Initiative at Columbia University, as well as the technical assistance from Dr. AmirAli ZangiAbadi in acquiring and analyzing the SEM images.

References

Achenza, M.: On earth stabilization with natural polymers for earth masonry construction. Mater. Struct. **39**, 21–27 (2005). https://doi.org/10.1617/14098

al Rim, K., Ledhem, A., Douzane, O., Dheilly, R.M., Queneudec, M.: Influence of the proportion of wood on the thermal and mechanical performances of clay-cement-wood composites. Cem. Concr. Compos. **21**(4), 269–276 (1999). https://doi.org/10.1016/S0958-9465(99)00008-6

Algin, H.M., Turgut, P.: Cotton and limestone powder wastes as brick material. Constr. Build. Mater. **22**(6), 1074–1080 (2008). https://doi.org/10.1016/J.CONBUILDMAT.2007.03.006

Alvarez, V.A., Vázquez, A.: Influence of fiber chemical modification procedure on the mechanical properties and water absorption of MaterBi-Y/sisal fiber composites. Compos. Part A: Appl. Sci. Manuf. **37**(10), 1672–1680 (2006). https://doi.org/10.1016/J.COMPOSITESA.2005.10.005

Arehart, J.H., Nelson, W.S., Srubar, W.V.: On the theoretical carbon storage and carbon sequestration potential of hempcrete. J. Clean. Prod. **266**, 121846 (2020). https://doi.org/10.1016/j.jclepro.2020.121846

Aruan Efendy, M.G., Pickering, K.L.: Comparison of harakeke with hemp fibre as a potential reinforcement in composites. Compos. Part A: Appl. Sci. Manuf. **67**, 259–267 (2014). https://doi.org/10.1016/J.COMPOSITESA.2014.08.023

Ashour, T., Georg, H., Wu, W.: An experimental investigation on equilibrium moisture content of earth plaster with natural reinforcement fibres for straw bale buildings. Appl. Therm. Eng. **31**(2–3), 293–303 (2011). https://doi.org/10.1016/J.APPLTHERMALENG.2010.09.009

Ashour, T., Wieland, H., Georg, H., Bockisch, F.J., Wu, W.: The influence of natural reinforcement fibres on insulation values of earth plaster for straw bale buildings. Mater. Des. **31**(10), 4676–4685 (2010). https://doi.org/10.1016/J.MATDES.2010.05.026

Aubert, J.E., Marcom, A., Oliva, P., Segui, P.: Chequered earth construction in south-western France. J. Cult. Herit. **16**(3), 293–298 (2015). https://doi.org/10.1016/J.CULHER.2014.07.002

Baker-Laporte, P., Laporte, R.: The EcoNest Home: Designing and Building a Light Straw Clay House. New Society Publishers (2015)

Baldrian, P., Valášková, V.: Degradation of cellulose by basidiomycetous fungi. FEMS Microbiol. Rev. **32**(3), 501–521 (2008). https://doi.org/10.1111/J.1574-6976.2008.00106.X

BeMiller, J.N.: Hydrocolloids. Gluten-Free Cereal Products and Beverages, pp. 203–215 (2008). https://doi.org/10.1016/B978-012373739-7.50011-3

Ben-Alon, L., Loftness, V., Harries, K.A., Cochran Hameen, E.: Life cycle assessment (LCA) of natural vs conventional building assemblies. Renew. Sustain. Energy Rev. **144**, 110951 (2021). https://doi.org/10.1016/J.RSER.2021.110951

Bhojvaid, P.P., Rai, A.K.: PAPERMAKING | paper raw materials and technology. Encyclopedia Forest Sci. 701–707 (2004). https://doi.org/10.1016/B0-12-145160-7/00262-3

Bouguerra, A., Ledhem, A., de Barquin, F., Dheilly, R.M., Quéneudec, M.: Effect of microstructure on the mechanical and thermal properties of lightweight concrete prepared from clay, cement, and wood aggregates. Cem. Concr. Res. **28**(8), 1179–1190 (1998). https://doi.org/10.1016/S0008-8846(98)00075-1

Bouhicha, M., Aouissi, F., Kenai, S.: Performance of composite soil reinforced with barley straw. Cement Concr. Compos. **27**(5), 617–621 (2005). https://doi.org/10.1016/J.CEMCONCOMP.2004.09.013

Bryson, Z.E., Srubar, W.V., Kawashima, S., Kawashima, S.: Towards 3D printed earth- and bio-based insulation materials: a case study on light straw clay (2021). https://doi.org/10.5281/ZENODO.6611395

Chand, N., Fahim, M.: Natural fibers and their composites. Tribol. Nat. Fiber Polymer Compos. 1–59 (2021). https://doi.org/10.1016/B978-0-12-818983-2.00001-3

Chen, H.Z., Liu, Z.H.: Multilevel composition fractionation process for high-value utilization of wheat straw cellulose. Biotechnol. Biofuels **7**(1) (2014). https://doi.org/10.1186/S13068-014-0137-3

Choi, E., Ostadrahimi, A., Kim, W.J.: Enhancement of compressive strength and strain ductility of SMA fiber reinforced concrete considering fiber's aspect ratios. Constr. Build. Mater. **345**, 128346 (2022). https://doi.org/10.1016/J.CONBUILDMAT.2022.128346

Correa, D., et al.: 3D-printed wood: programming hygroscopic material transformations. 3D Printing Additive Manuf. **2**(3) (2015). https://doi.org/10.1089/3dp.2015.0022

Demir, I.: Effect of organic residues addition on the technological properties of clay bricks. Waste Manag. **28**(3), 622–627 (2008). https://doi.org/10.1016/J.WASMAN.2007.03.019

Dourado, F., Leal, M., Martins, D., Fontão, A., Cristina Rodrigues, A., Gama, M.: Celluloses as food ingredients/additives: is there a room for BNC? Bacterial Nanocellulose: Biotechnol. Bio-Econ. 123–133 (2016). https://doi.org/10.1016/B978-0-444-63458-0.00007-X

Faria, P., Santos, T., Aubert, J.-E.: Experimental characterization of an earth eco-efficient plastering mortar. J. Mater. Civil Eng. **28**(1) (2016). https://doi.org/10.1061/(ASCE)MT.1943-5533.0001363

Fernando, A.L., Duarte, M.P., Vatsanidou, A., Alexopoulou, E.: Environmental aspects of fiber crops cultivation and use. Ind. Crops Prod. **68**, 105–115 (2015). https://doi.org/10.1016/J.INDCROP.2014.10.003

Ghavami, K., Toledo Filho, R.D., Barbosa, N.P.: Behaviour of composite soil reinforced with natural fibres. Cement Concr. Compos. **21**(1), 39–48 (1999). https://doi.org/10.1016/S0958-9465(98)00033-X

Goodhew, S., Boutouil, M., Streiff, F., Le Guern, M., Carfrae, J., Fox, M.: Improving the thermal performance of earthen walls to satisfy current building regulations. Energy and Build. **240**, 110873 (2021)

Güllü, H., Khudir, A.: Effect of freeze-thaw cycles on unconfined compressive strength of fine-grained soil treated with jute fiber, steel fiber and lime. Cold Reg. Sci. Technol. **106–107**, 55–65 (2014). https://doi.org/10.1016/J.COLDREGIONS.2014.06.008

Hamard, E., Morel, J.C., Salgado, F., Marcom, A., Meunier, N.: A procedure to assess suitability of plaster to protect vernacular earthen architecture. WIT Trans. Built Environ. **118**, 507–516 (2011). https://doi.org/10.2495/STR110421

Hill, C.A.S., Khalil, H.P.S.A., Hale, M.D.: A study of the potential of acetylation to improve the properties of plant fibres. Ind. Crops Prod. **8**(1), 53–63 (1998). https://doi.org/10.1016/S0926-6690(97)10012-7

Holzhueter, K., Itonaga, K.: An evaluation of WUFI-Bio to predict mold growth in straw bale walls in Japan. J. Asian Archit. Build. Eng. **16**(2), 357–362 (2017). https://doi.org/10.3130/jaabe.16.357

Khan, T.S., Mubeen, U.: Wheat straw: a pragmatic overview. Curr. Res. J. Biol. Sci. **4**(6), 673–675 (2012)

Khedari, J., Watsanasathaporn, P., Hirunlabh, J.: Development of fibre-based soil-cement block with low thermal conductivity. Cem. Concr. Compos. **27**(1), 111–116 (2005). https://doi.org/10.1016/J.CEMCONCOMP.2004.02.042

Kumar, V.G., Ramadoss, R., Rampradheep, G.S.: A study report on carbon sequestration by using Hempcrete. Mater. Today: Proc. **45**, 6369–6371 (2020). https://doi.org/10.1016/J.MATPR.2020.11.012

Kymäläinen, H.R., Sjöberg, A.M.: Flax and hemp fibres as raw materials for thermal insulations. Build. Environ. **43**(7), 1261–1269 (2008). https://doi.org/10.1016/J.BUILDENV.2007.03.006

Laborel-Préneron, A., Aubert, J.E., Magniont, C., Tribout, C., Bertron, A.: Plant aggregates and fibers in earth construction materials: a review. Const. Build. Mater. **111**, 719–734 (2016). https://doi.org/10.1016/j.conbuildmat.2016.02.119

Lara-Serrano, M., Morales-delaRosa, S., Campos-Martín, J.M., Fierro, J.L.G.: Fractionation of lignocellulosic biomass by selective precipitation from ionic liquid dissolution. Appl. Sci. (Switzerland) **9**(9) (2019). https://doi.org/10.3390/APP9091862

le Duigou, A., Barbé, A., Guillou, E., Castro, M.: 3D printing of continuous flax fibre reinforced biocomposites for structural applications. Mater. Des. **180**, 107884 (2019). https://doi.org/10.1016/J.MATDES.2019.107884

le Duigou, A., Correa, D., Ueda, M., Matsuzaki, R., Castro, M.: A review of 3D and 4D printing of natural fibre biocomposites. Mater. Des. **194**, 108911 (2020). https://doi.org/10.1016/J.MATDES.2020.108911

Ledhem, A., Dheilly, R.M., Benmalek, M.L., Quéneudec, M.: Properties of wood-based composites formulated with aggregate industry waste. Constr. Build. Mater. **14**(6–7), 341–350 (2000). https://doi.org/10.1016/S0950-0618(00)00037-4

Levacher, D., Wang, F., Liang, Y.: Co-valorisation de matériaux fins et sédiments. 869–876 (2010). https://doi.org/10.5150/JNGCGC.2010.094-L

Li, Y., Mai, Y.W., Ye, L.: Sisal fibre and its composites: a review of recent developments. Compos. Sci. Technol. **60**(11), 2037–2055 (2000). https://doi.org/10.1016/S0266-3538(00)00101-9

Lima, J., Faria, P.: Eco-efficient earthen plasters: the influence of the addition of natural fibers. In: Fangueiro, R., Rana, S. (eds.) Natural Fibres: Advances in Science and Technology Towards Industrial Applications. RB, vol. 12, pp. 315–327. Springer, Dordrecht (2016). https://doi.org/10.1007/978-94-017-7515-1_24

Loureiro dos Santos, L.A.: Natural polymeric biomaterials: processing and properties ☆. Reference Module in Materials Science and Materials Engineering (2017). https://doi.org/10.1016/B978-0-12-803581-8.02253-0

Lynd, L.R., Weimer, P.J., van Zyl, W.H., Pretorius, I.S.: Microbial cellulose utilization: fundamentals and biotechnology. Microbiol. Mol. Biol. Rev. **66**(4), 739 (2002). https://doi.org/10.1128/MMBR.66.4.739.2002

Maddison, M., Mauring, T., Kirsimäe, K., Mander, Ü.: The humidity buffer capacity of clay-sand plaster filled with phytomass from treatment wetlands. Build. Environ. **44**(9), 1864–1868 (2009). https://doi.org/10.1016/J.BUILDENV.2008.12.008

Malaeb, Z., AlSakka, F., Hamzeh, F.: 3D concrete printing. Concr. Printing Technol. **3D**, 115–136 (2019). https://doi.org/10.1016/B978-0-12-815481-6.00006-3

Maraldi, M., Molari, L., Molari, G., Regazzi, N.: Time-dependent mechanical properties of straw bales for use in construction. Biosys. Eng. **172**, 75–83 (2018). https://doi.org/10.1016/J.BIOSYSTEMSENG.2018.05.014

Mattone, R.: Sisal fibre reinforced soil with cement or cactus pulp in bahareque technique. Cem. Concr. Compos. **27**(5), 611–616 (2005). https://doi.org/10.1016/J.CEMCONCOMP.2004.09.016

Miccoli, L., Müller, U., Fontana, P.: Mechanical behaviour of earthen materials: a comparison between earth block masonry, rammed earth and cob. Constr. Build. Mater. **61**, 327–339 (2014). https://doi.org/10.1016/J.CONBUILDMAT.2014.03.009

Millogo, Y., Aubert, J.E., Hamard, E., Morel, J.C.: How properties of kenaf fibers from Burkina Faso contribute to the reinforcement of earth blocks. Materials **8**(5), 2332–2345 (2015). https://doi.org/10.3390/MA8052332

Millogo, Y., Morel, J.C., Aubert, J.E., Ghavami, K.: Experimental analysis of Pressed Adobe Blocks reinforced with Hibiscus cannabinus fibers. Constr. Build. Mater. **52**, 71–78 (2014). https://doi.org/10.1016/J.CONBUILDMAT.2013.10.094

Mohamed, A.E.M.K.: Improvement of swelling clay properties using hay fibers. Constr. Build. Mater. **38**, 242–247 (2013). https://doi.org/10.1016/J.CONBUILDMAT.2012.08.031

Pegg, A.M.: The application of natural hydrocolloids to foods and beverages. Nat. Food Addit. Ingredients Flavourings 175–196 (2012). https://doi.org/10.1533/9780857095725.1.175

Perrot, A., Rangeard, D., Courteille, E.: 3D printing of earth-based materials: processing aspects. Constr. Build. Mater. **172**, 670–676 (2018). https://doi.org/10.1016/J.CONBUILDMAT.2018.04.017

Piattoni, Q., Quagliarini, E., Lenci, S.: Experimental analysis and modelling of the mechanical behaviour of earthen bricks. Constr. Build. Mater. **25**(4), 2067–2075 (2011). https://doi.org/10.1016/J.CONBUILDMAT.2010.11.039

Pullen, Q.M., Scholz, T.V.: Index and engineering properties of oregon cob. J. Green Build. **6**(2), 88–106 (2011). https://doi.org/10.3992/jgb.6.2.88

Quagliarini, E., Lenci, S.: The influence of natural stabilizers and natural fibres on the mechanical properties of ancient Roman adobe bricks. J. Cult. Herit. **11**(3), 309–314 (2010). https://doi.org/10.1016/J.CULHER.2009.11.012

Readle, D., Coghlan, S., Smith, J.C., Corbin, A., Augarde, C.E.: Fibre reinforcement in earthen construction materials. Proc. Inst. Civil Eng.: Constr. Mater. **169**(5), 252–260 (2016). https://doi.org/10.1680/JCOMA.15.00039

Rempel, A., Rempel, A.: Intrinsic evaporative cooling by hygroscopic earth materials. Geosciences **6**(3), 38 (2016). https://doi.org/10.3390/geosciences6030038

Reeb, J. E.: Wood and Moisture Relationships. Oregon State University Extension Catalog (2009). https://ir.library.oregonstate.edu/downloads/td96k297v

Saake, B., Lehnen, R.: Lignin. Ullmann's Encyclopedia of Industrial Chemistry (2007). https://doi.org/10.1002/14356007.A15_305.PUB3

Sabbà, M.F., Tesoro, M., Falcicchio, C., Foti, D.: Rammed earth with straw fibers and earth mortar: mix design and mechanical characteristics determination. Fibers **9**(5) (2021). https://doi.org/10.3390/FIB9050030

Sanjay, M.R., Siengchin, S., Parameswaranpillai, J., Jawaid, M., Pruncu, C.I., Khan, A.: A comprehensive review of techniques for natural fibers as reinforcement in composites: preparation, processing and characterization. Carbohyd. Polym. **207**, 108–121 (2019). https://doi.org/10.1016/J.CARBPOL.2018.11.083

Segetin, M., Jayaraman, K., Xu, X.: Harakeke reinforcement of soil-cement building materials: manufacturability and properties. Build. Environ. **42**(8), 3066–3079 (2007). https://doi.org/10.1016/J.BUILDENV.2006.07.033

Shang, Y., Tariku, F.: Hempcrete building performance in mild and cold climates: Integrated analysis of carbon footprint, energy, and indoor thermal and moisture buffering. Build. Environ. **206**, 108377 (2021). https://doi.org/10.1016/J.BUILDENV.2021.108377

Smith, P.J., et al.: Enzymatic synthesis of xylan microparticles with tunable morphologies. ACS Mater. Au **2**(4), 440–452 (2022). https://doi.org/10.1021/ACSMATERIALSAU.2C00006

Suhas, Carrott, P.J.M., Ribeiro Carrott, M.M.L.: Lignin - from natural adsorbent to activated carbon: a review. Bioresource Technol. **98**(12), 2301–2312 (2007). https://doi.org/10.1016/J.BIORTECH.2006.08.008

Taallah, B., Guettala, A., Guettala, S., Kriker, A.: Mechanical properties and hygroscopicity behavior of compressed earth block filled by date palm fibers. Constr. Build. Mater. **59**, 161–168 (2014). https://doi.org/10.1016/J.CONBUILDMAT.2014.02.058

Tay, Y.W.D., Qian, Y., Tan, M.J.: Printability region for 3D concrete printing using slump and slump flow test. Compos. B Eng. **174**, 106968 (2019). https://doi.org/10.1016/J.COMPOSITESB.2019.106968

Van Damme, H., Houben, H.: Earth concrete. stabilization revisited. Cem. Concr. Res. **114**, 90–102 (2018). https://doi.org/10.1016/j.cemconres.2017.02.035

Vane, C.H., Drage, T.C., Snape, C.E.: Biodegradation of oak (Quercus alba) wood during growth of the shiitake mushroom (Lentinula edodes): a molecular approach. J. Agric. Food Chem. **51**(4), 947–956 (2003). https://doi.org/10.1021/JF020932H

Viera, R.G.P., Filho, G.R., de Assunção, R.M.N., da Carla, C., Vieira, J.G., de Oliveira, G.S.: Synthesis and characterization of methylcellulose from sugar cane bagasse cellulose. Carbohyd. Polym. **67**(2), 182–189 (2007). https://doi.org/10.1016/J.CARBPOL.2006.05.007

Vigneshwaran, S., Uthayakumar, M., Arumugaprabu, V., Sundarakannan, R.: Numerical and experimental assessment of water absorption of red mud-an industrial waste reinforced sisal/polyester hybrid polymer composite. In: Jawaid, M., Hamdan, A., Hameed Sultan, M.T. (eds.) Structural Health Monitoring System for Synthetic, Hybrid and Natural Fiber Composites. Composites Science and Technology, pp. 217–229. Springer, Singapore (2021). https://doi.org/10.1007/978-981-15-8840-2_15

Villamizar, M.C.N., Araque, V.S., Reyes, C.A.R., Silva, R.S.: Effect of the addition of coal-ash and cassava peels on the engineering properties of compressed earth blocks. Constr. Build. Mater. **36**, 276–286 (2012). https://doi.org/10.1016/J.CONBUILDMAT.2012.04.056

Walker, P., Thomson, A., Maskell, D.: Straw bale construction. Nonconventional Vernacular Constr. Mater.: Characterisation Prop. Appl. 189–216 (2019). https://doi.org/10.1016/B978-0-08-102704-2.00009-3

Yadav, M., Saini, A.: Opportunities & challenges of hempcrete as a building material for construction: an overview. Mater. Today: Proc. **65**, 2021–2028 (2022). https://doi.org/10.1016/J.MATPR.2022.05.576

Yetgin, Ş, Özlem, Ç., Çavdar, A.: The effects of the fiber contents on the mechanic properties of the adobes. Constr. Build. Mater. **22**(3), 222–227 (2008). https://doi.org/10.1016/j.conbuildmat.2006.08.022

Zhao, X., et al.: Poplar as biofiber reinforcement in composites for large-scale 3D printing. ACS Appl. Bio Mater. **2**(10), 4557–4570 (2019). https://doi.org/10.1021/ACSABM.9B00675/ASSET/IMAGES/MEDIUM/MT9B00675_0008.GIF

Effect of Adding Phragmites-Australis Plant on the Chemical Shrinkage and Mechanical Properties of Mortar

Rawan Ramadan[1(✉)], Jamal Khatib[1,2], Elhem Ghorbel[3], and Adel Elkordi[1,4]

[1] Department of Civil Engineering, Beirut Arab University, P.O. Box 11 - 50 - 20 Riad El Solh, Beirut 11072809, Lebanon
r.ramadan@bau.edu.lb

[2] Faculty of Science and Engineering, University of Wolverhampton, Wolverhampton WV1 1LY, UK

[3] Université de Cergy-Pontoise/Paris Seine (CY Cergy Paris Université), Cergy, France

[4] Department of Civil Engineering, Alexandria University, Alexandria, Egypt

Abstract. Building structures would be impossible without concrete, which is the world's most commonly used construction material after water. Resources, recycling, durability, and environmental quality are just a few of the many areas of concrete production that have seen advancements. The use of natural fibers in concrete is one of these methods. Plant-based materials derived from natural fibers are readily available, low-cost renewable resources. In this study, the use of natural fiber from the Phragmites-Australis plant (PA) has been investigated. The effect of incorporating PA in different volume percentages (0 to 2%) on the chemical shrinkage and mechanical properties of mortar has been examined. In this experiment, the PA stems were cut into pieces 1 cm long and 2 cm wide and treated with 4% sodium hydroxide solution (NaOH). The w/c was kept constant for all mixes. Mortar samples were tested for 28 days. Testing involved chemical shrinkage, density, ultra-pulse velocity (UPV), and compressive and flexural strengths. Results showed that the addition of PA reduced the chemical shrinkage of mortars. The addition of 1% PA increased the UPV, compressive strength, and flexural strength of mortar samples. Furthermore, a strong correlation was found between chemical shrinkage and compressive strength, indicating that as the pozzolanic reaction progressed, the chemical shrinkage increased. Therefore, the inclusion of PA plant as an eco-friendly material in mortar mixtures contributes to a sustainable building.

Keywords: Phragmites-Australis · density · UPV · compressive strength · flexural strength · mortar

1 Introduction

In order to protect and save our environment, natural fiber reinforcement have been introduced to modern research to ensure the sustainability of the construction industry and protection of the environment. Such natural reinforcement fibers to be tested upon as a

S. Amziane et al. (Eds.): ICBBM 2023, RILEM Bookseries 45, pp. 573–584, 2023.
https://doi.org/10.1007/978-3-031-33465-8_43

reinforcement agent and a substitute material for cement in concrete. The incorporation of natural fibers in concrete could provide a sustainable concrete at low cost. Several researches have been investigated. As stated by some authors in previous studies, the use of roselle, sisal, coconut, sugarcane bagasse, hemp, and jute fiber have been found to increase the compressive and tensile strength of concrete composites [1–5]. Ferreira et al. [6] studied the Effect of fiber treatments on the Sisal fiber properties and fiber-matrix bond in cement-based systems, the pullout test demonstrated that the various treatments reduced water absorption capacity, increased tensile strength and stiffness, and significantly improved fiber-matrix interference. Furthermore, the chemical treatment improved the elastic and fictitious bond, while the polymer and hybrid treatments caused fiber slip hardening. Another study conducted by Toledo et al. [7] showed that sisal fibers reduce the plastic shrinkage and restrained shrinkage cracking in concrete. Additionally, the incorporation of hemp fiber in concrete has been investigated. Comak et al. [8] studied the effects of hemp fibers on characteristics of cement-based mortar. Results showed that the largest water absorption ratio was reached by a mix of 18 mm hemp fibers and a 1% hemp fiber (19% absorption), while the least water absorption ratio was obtained by samples with 1% hemp fiber with 12 mm hemp length (<2%). The greatest compressive strength was found for 6 mm and 2% ratio of hemp fibers (39.48 MPa). Another study showed that the addition of 10% of flax fibers (by weight of hemp shives) enhanced the ductility of concrete and reduced the capillary water absorption in hemp concrete mixes [9]. Ataie, [10] stated that samples containing washed rice straw fiber (RSF) had lower drying shrinkage compared to control mix. Machaka et al. [11] found that the capillary water absorption coefficient and total water absorption increase with the increase of palm fibers. Last but not least is the Phragmites-Australis fiber (PA). This plant is a tall, perennial grass that can grow to heights of 4.6 m or more. Leaves are 15–60 cm long, 1–6 cm wide, flat and glabrous. The flower heads are dense, fluffy, gray or purple in color, 15–40 cm long. PA is found on every continent except Antarctica. It is available in the eastern part of Lebanon (Bekaa). This plant grows close to waterways and rivers. Machaka et al. [12] studied the effect of adding 0.5%, 1% and 1.5% PA fiber by volume of mix to experiment upon the samples' compressive strength, density, ultrasonic pulse velocity and sorptivity index. The results showed that adding up to 1.5% PA fibers improved the stability and durability of mixes while reducing the overall water absorption of concrete. According to many authors [13–15], PA fibers help to regulate and minimize the shrinkage of concrete. However, there haven't been enough studies done on this plant's potential uses. This study aims to investigate how PA fibers can enhance the mechanical properties of mortar and control cracking.

2 Experimental Program

2.1 Materials and Mix Proportions

In this experiment, four mortar mixes were performed. The PA plant was collected from the eastern part of Lebanon (Bekaa). The plant must be sun-dried. After removing the flowers, a cut machine was used to cut the stems into fibers that were 1 cm long and 2 mm wide. After that, the fibers were treated with a 4% sodium hydroxide solution (NaOH) for 24 h. The steps how to obtain these fibers are listed in Fig. 1. These fibers have a density of almost 665 kg/m^3. The percentages of PA fibers in the mix were 0, 0.5, 1, and 2%. The water-to-cement (w/c) and sand-to-cement (s/c) ratios were 0.55 and 2, respectively. Furthermore, Portland cement type I (PA-L-42.5 N) with a density of 1440 kg/m^3 and specific gravity of 3.15 was used in all mortar mixes. Siliceous Sand was used with a fineness modulus of 2.8. Table 1 presents the mortar mix ratio.

(a) (b) (c) (d)

Fig. 1. PA fibers procedure: a): green PA plant; (b): Sun-dried stems;(c): Cut-fiber machine;(d):PA fibers in 1 cm length.

Table 1. Mortar mix proportions.

Mortar code	Cement	Sand	PA	w/c ratio
M-0%PA	1	2	0	0.55
M-0.5%PA	1	2	0.005	0.55
M-1%PA	1	2	0.01	0.55
M-2%PA	1	2	0.02	0.55

2.2 Testing Methods

2.2.1 Shrinkage Measurement

Chemical shrinkage was carried out in accordance with ASTM C1608 [16]. Previous research [17–24] mentioned the test protocol. Mortar mixture was poured in 250 ml glass containers and filled to a height of 2 cm. After that, this container was filled with water all the way to the top and soaked with a rubber stopper. The rubber was then injected with a 2 ml pipette through a hole. The rubber was then squeezed to allow water to enter the pipette. A drop of oil was added as a final step to avoid water evaporation. Two duplicate samples were taken for each batch. The volume change is represented by the drop in water level (chemical shrinkage). The drop was recorded for 28 days.

2.2.2 Mechanical Strength Measurement

Flexural and compressive strength test were performed according to I.S. EN 998-2. Mortars were cast in 40*40*160 mm steel molds and kept for 24 H. After demolding, mortars prisms were placed in water and cured for 28 days. At the day of testing (at 1, 3, 7 and 28 days), the prisms were firstly weighted to record the density, then were subjected to the ultra-pulse velocity test (UPV) to identify the quality of the mortar mix. For testing, the bending strength was firstly applied on the mortar beams. After that, compressive strength test was performed on the two broken parts from flexural testing as shown in Fig. 2(a) and Fig. 2(b).

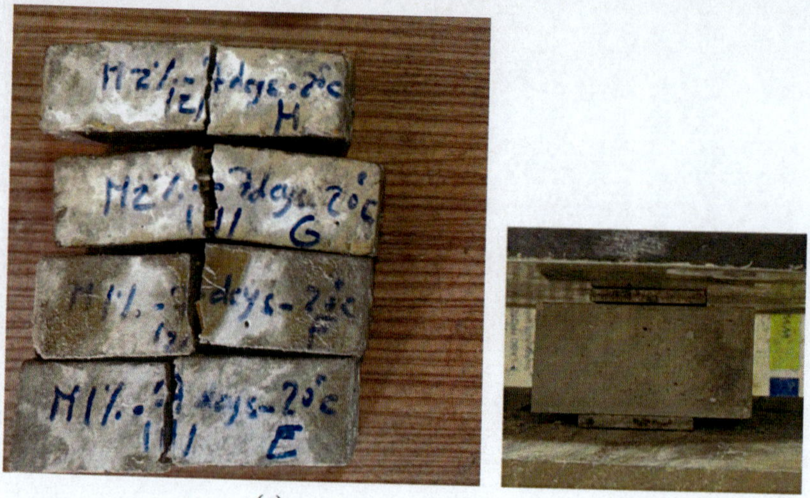

(a) (b)

Fig. 2. Flexural and compressive strength test: a): Beams after applying the flexural strength; b): the broken part from flexure subjected to compressive strength.

3 Results and Discussion

3.1 Density

The density of mortars at 1, 3, 7 and 28 days were recorded and listed in Table 2. As shown, the density increases with the increase in the curing time. The density increases with the increase of PA fibers up to 1% then drops for the addition of 2%PA. For example, at 28 days, the density is 2.17 g/cm^3 at 0% PA. This value goes up to the peak value of 2.36 g/cm^3 for an addition of 1% PA succeeded by a decrease at 2% PA addition. This is directly related to the hydration of cement. As cement reacts with water, the chemical reaction increases as well. Therefore, the skeleton condenses and became more rigid [22].

Table 2. Density for mortars with different percentages of PA fibers.

	Density g/cm^3			
Days	M-0%PA	M-0.5%PA	M-1%PA	M-2%PA
1	2.04	2.18	2.29	1.94
3	2.07	2.32	2.32	1.98
7	2.09	2.34	2.34	2.03
28	2.17	2.35	2.36	2.05

3.2 Ultra-Pulse Velocity (UPV)

The UPV test is used to check the concrete quality. Figure 3 show the UPV for mortars versus curing ages for different percentages of PA fibers. As shown, mix 3 with an addition of 1%PA displays a higher UPV value with a good quality of mortar (3.62 km/s, at 28 days) among other mixes. Adding more than 1% PA deteriorates the quality of mortars. This can be elucidated by the fact that when 1% PA fibers are added, these random fibers are distributed in an isotropic shape and permit the passage of the pulse in an easy way. Additionally, adding a high amount of PA fibers (2%) could create voids inside the skeleton, which would delay the passage of the pulse [25].

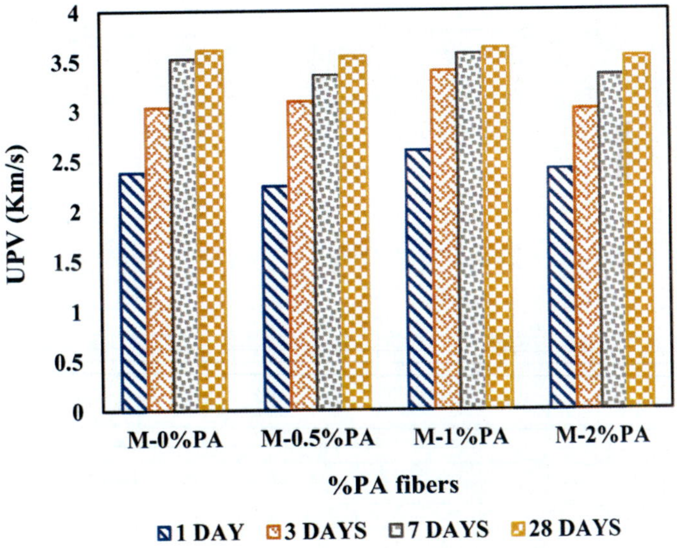

Fig. 3. UPV for mortars at different percentages of PA fibers

3.3 Mechanical Strength

Flexural strength and compressive strength results for mortars with different percentages of PA fibers for 1, 3, 7, and 28 days are shown in Fig. 4 and Fig. 5 respectively. As shown, in Fig. 4, the addition of 0.5% PA fibers slightly decreases the flexural strength. This is mainly due to the lack of compaction during casting. After this decrease, the addition of 1% PA fibers increases again the flexural strength where it achieves the highest value (9.6 MPa, at 28 days) then drops. As shown in Fig. 5, the addition of 0.5% PA fibers decreases the compressive strength of mortars. After this decrease, the value increases again and reaches the highest value with the inclusion of 1% PA fibers (12.28 MPa, at 28 days) then drops. These results are consistent with those of density and UPV where the inclusion of 1%PA fibers achieves the highest value. A correlation between flexural strength and compressive strength is illustrated in Fig. 6. As shown, there is high correlations exist between them with high coefficient of determination ($0.76<R^2>0.94$). The addition of 1% PA fibers attains the highest flexural and compressive strength values. This is well observed in Fig. 7 which shows the slope of linear regression between compressive and flexural strength. As shown, adding 1%PA fibers gives the highest slope which means that it has the highest compressive and flexural strength value. Two hypotheses can be elucidated: 1) the microfibrils presents in PA fibers are enclosed by the matrix elements lignin and hemicelluloses. It becomes aligned with fiber axis during the load application [25]. 2) the washing of PA fibers after chemical treatment can remove some of organic compounds and inorganic impurities [26].

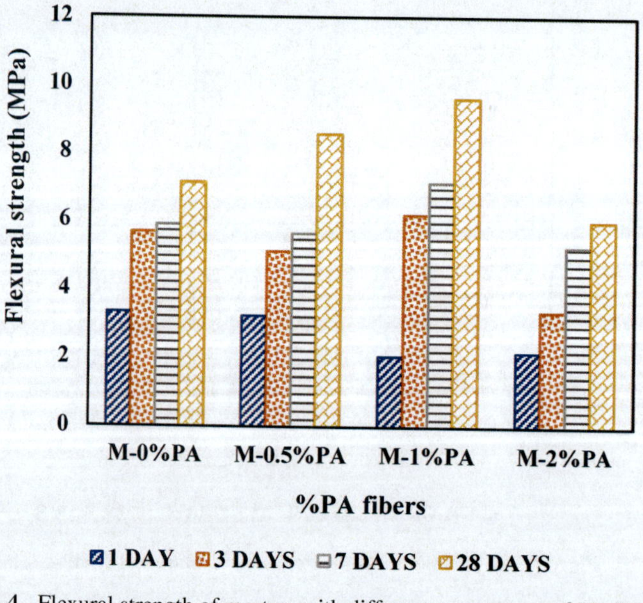

Fig. 4. Flexural strength of mortars with different percentages of PA fibers.

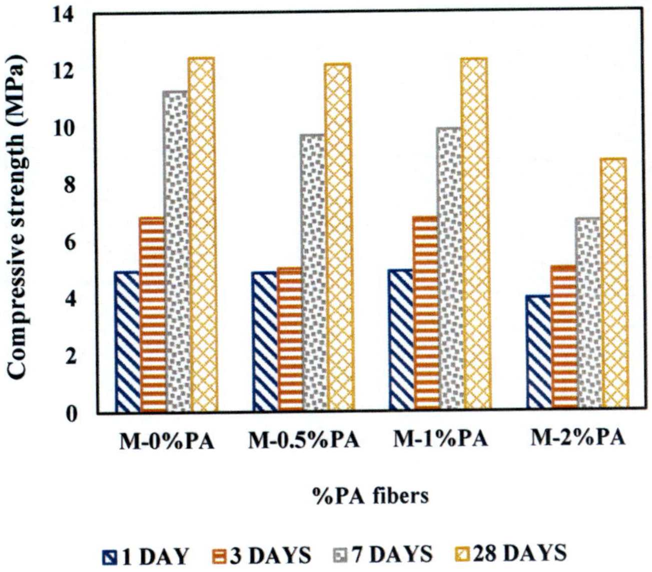

Fig. 5. Compressive strength of mortars with different percentages of PA fibers.

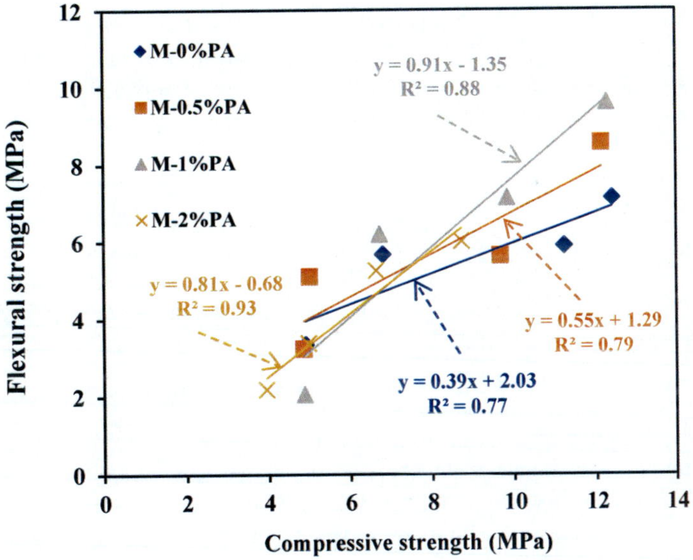

Fig. 6. Correlation between flexural strength and compressive strength for mortars at different percentages of PA fibers.

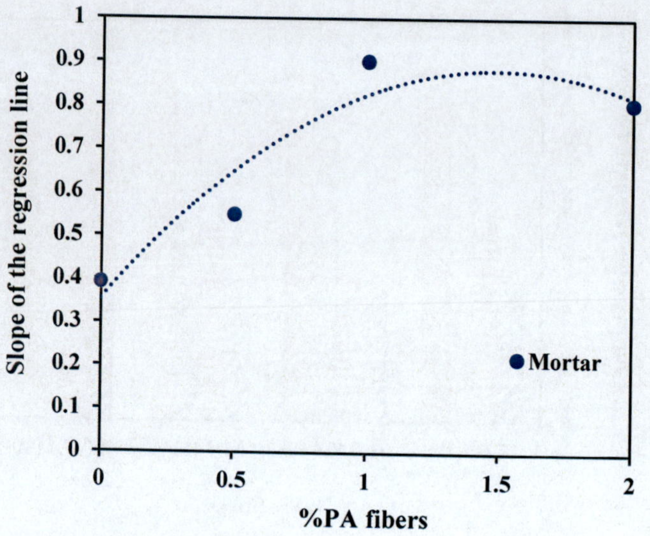

Fig. 7. Slope of the regression line versus PA fibers.

3.4 Chemical Shrinkage

Figure 8 depicts the chemical shrinkage data of mortar samples for different percentage of PA fibers for a total period of 28 days. As shown, chemical shrinkage reduces with the increase in PA percentage. For example, at 28 days, the chemical shrinkage for control mix is equal to 0.023 ml/g of binder). This value slightly decreases with the addition of 2% PA (0.015 ml/g of binder). The percentage decreases are 13.04%, 26.09% and 34.78% for 0.5%, 1% and 2% of PA respectively. This is explained by the fact that PA fibers increase the matrix's stiffness through bonding, which reduces cement's chemical reaction. Furthermore, the chemical elements in PA fibers postpone the cement's first hydration. These chemical processes result in the production of acids, which are then neutralized by hydroxyl ions derived from cement hydration reactions [27–29].

3.5 Correlation Between Compressive Strength and Chemical Shrinkage

Figure 9 displays the correlation between compressive strength and chemical shrinkage for different percentages of PA fibers (0, 0.5, 1, and 2%). A high correlation exists between them for all mixes with high coefficient of determination R^2 of 0.85, 0.8, 0.97 and 0.98 for 0, 0.5,1 and 2%PA fibers respectively. This indicates that as chemical shrinkage increases, compressive strength increases as well. This is due to the chemical reaction exists between cement and water. As pozzolanic reactions increases, the chemical shrinkage increases as well. Moreover, the reaction between calcium silicate (C_3S and C_2S), which produces the silicate hydrate CSH, causes a heat increase (heat increase also caused by other minor reactions). During this phase, the formation of CSH has a significant impact on the mortar strength [30].

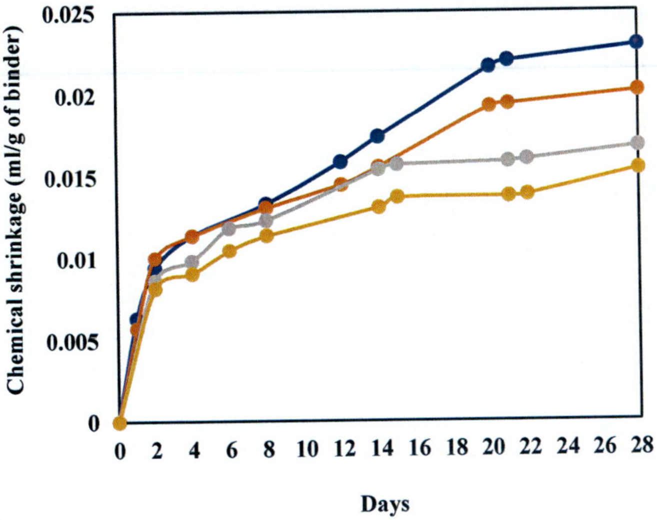

Fig. 8. Chemical shrinkage of mortars with different percentages of PA fibers.

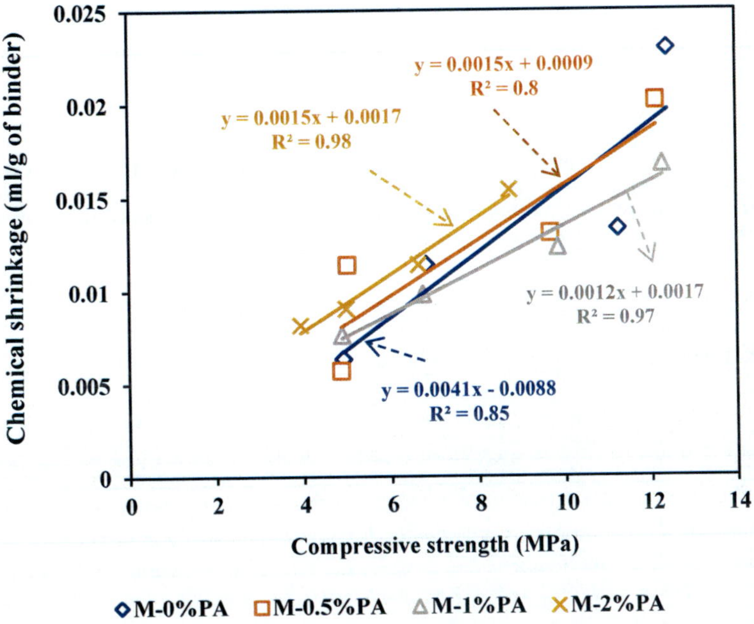

Fig. 9. Correlation between compressive strength and chemical shrinkage for mortars at different percentages of PA fibers

3.6 Correlation Between Compressive Strength and UPV

Figure 10 shows the correlation between compressive strength and UPV. As shown a high correlation exists between both parameters with high coefficient of determination R^2 of 0.7. This correlation means that as UPV increases, the compressive strength increases as well. This is clarifying by the fact that as mortar samples are confined and have less voids, it gives a high compressive strength value.

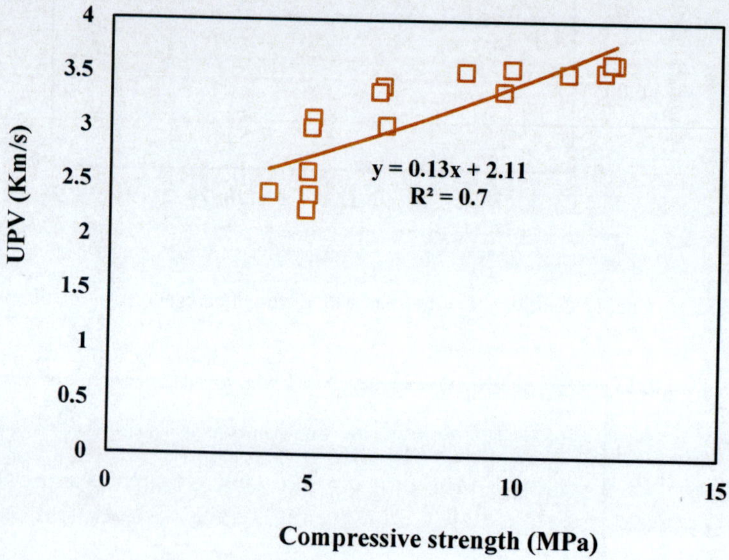

Fig. 10. Correlation between compressive strength and UPV for mortars at different percentages of PA fibers.

4 Conclusions

This paper illustrated the effect of adding Phragmites-Australis fibers on the chemical shrinkage and mechanical properties of mortar. Several conclusions could be derived:

- PA fiber has a crucial role in enhancing the mechanical properties of mortar. The addition of 1% PA increases the density, UPV, compressive and flexural strength. At 28 days, With the inclusion of 1% PA, the density attains a value of 2.36 g/cm^3, the UPV reaches a value of 3.62 km/s, the compressive strength and flexural strength attain a value of 12.28 and 9.6 MPa respectively.
- Chemical shrinkage of mortars decreases with the increase in PA fibers percentages. It decreases by 34.78% with the addition of 2%PA fibers.
- A high positive correlation exists between compressive strength and chemical shrinkage. Compressive strength is directly related to chemical shrinkage. As pozzolanic reactions increases, the chemical shrinkage increases. Therefore, compressive strength increases as well.

- A high correlation exists between compressive strength and UPV. Compressive strength and UPV are proportional to each other. As UPV increases, the compressive strength increases as well. This is directly related to the quality of concrete. Mortars with less voids gives the highest compressive strength.

References

1. Ramakrishna, G., Sundararajan, T.: Impact strength of a few natural fibre reinforced cement mortar slabs: a comparative study. Cem. Concr. Compos. **27**(5), 547–553 (2005). https://doi.org/10.1016/j.cemconcomp.2004.09.006
2. Li, Z., Wang, X., Wang, L.: Properties of hemp fibre reinforced concrete composites. Compos. A Appl. Sci. Manuf. **37**(3), 497–505 (2006). https://doi.org/10.1016/j.compositesa.2005.01.032
3. Popov, M., Zakrevskaya, L., Vaganov, V., Hempel, S., Mechtcherine, V.: Performance of lightweight concrete based on granulated foamglass. In: IOP Conference Series: Materials Science and Engineering, vol. 96, no. 1, p. 012017. IOP Publishing (2015). https://doi.org/10.1088/1757-899X/96/1/012017
4. Branca, T.A., et al.: Reuse and recycling of by-products in the steel sector: recent achievements paving the way to circular economy and industrial symbiosis in Europe. Metals **10**(3), 345 (2020). https://doi.org/10.3390/met10030345
5. Ahmad, J., Aslam, F., Martinez-Garcia, R., Ouni, M.H.E., Khedher, K.M.: Performance of sustainable self-compacting fiber reinforced concrete with substitution of marble waste (MW) and coconut fibers (CFs). Sci. Rep. **11**(1), 1–22 (2021). https://doi.org/10.1038/s41598-021-01931-x
6. Ferreira, S.R., Lima, P.R.L., Silva, F.A., Toledo Filho, R.D.: Effect of sisal fiber hornification on the fiber-matrix bonding characteristics and bending behavior of cement-based composites. In: Key Engineering Materials, vol. 600, pp. 421–432. Trans Tech Publications Ltd. (2014). https://doi.org/10.4028/www.scientific.net/KEM.600.421
7. Toledo Filho, R.D., Ghavami, K., Sanjuán, M.A., England, G.L.: Free, restrained and drying shrinkage of cement mortar composites reinforced with vegetable fibres. Cem. Concr. Compos. **27**(5), 537–546 (2005). https://doi.org/10.1016/j.cemconcomp.2004.09.005
8. Çomak, B., Bideci, A., Bideci, Ö.S.: Effects of hemp fibers on characteristics of cement-based mortar. Constr. Build. Mater. **169**, 794–799 (2018). https://doi.org/10.1016/j.conbuildmat.2018.03.029
9. Page, J., Sonebi, M., Amziane, S.: Design and multi-physical properties of a new hybrid hemp-flax composite material. Constr. Build. Mater. **139**, 502–512 (2017). https://doi.org/10.1016/j.conbuildmat.2016.12.037
10. Ataie, F.: Influence of rice straw fibers on concrete strength and drying shrinkage. Sustainability **10**(7), 2445 (2018). https://doi.org/10.3390/su10072445
11. Machaka, M., Elkordi, A., Ghanem, H., Khatib, J., Baalbaki, O.: Selected properties of concrete containing palm fibers. Acad. J. Civ. Eng. **37**(2), 279–286 (2019)
12. Machaka, M., Khatib, J., Baydoun, S., Elkordi, A., Assaad, J.J.: The effect of adding phragmites australis fibers on the properties of concrete. Buildings **12**(3), 278 (2022). https://doi.org/10.3390/buildings12030278
13. Amacha, N., et al.: Assessment of the efficiency of a pilot constructed wetland on the remediation of water quality; case study of Litani River, Lebanon. Environ. Pollut. Clim. Chang. **1**, 119 (2017)

14. Milke, J., Gałczyńska, M., Wróbel, J.: The importance of biological and ecological properties of *Phragmites australis* (Cav.) Trin. Ex Steud., in phytoremendiation of aquatic ecosystems—the review. Water **12**(6), 1770 (2020). https://doi.org/10.3390/w12061770
15. Rezania, S., Park, J., Rupani, P.F., Darajeh, N., Xu, X., Shahrokhishahraki, R.: Phytoremediation potential and control of *Phragmites australis* as a green phytomass: an overview. Environ. Sci. Pollut. Res. **26**(8), 7428–7441 (2019). https://doi.org/10.1007/s11356-019-04300-4
16. ASTM C 1608: Standard test method for chemical shrinkage of hydraulic cement paste. American Society for Testing and Materials, West Conshohocken (2007)
17. Wild, S., Khatib, J.M., Roose, L.J.: Chemical shrinkage and autogenous shrinkage of Portland cement—metakaolin pastes. Adv. Cem. Res. **10**(3), 109–119 (1998)
18. Ghanem, H., Zollinger, D., Lytton, R.: Predicting ASR aggregate reactivity in terms of its activation energy. Constr. Build. Mater. **24**(7), 1101–1108 (2010). https://doi.org/10.1016/j.conbuildmat.2009.12.033
19. Ghanem, H., Zollinger, D., Lytton, R., Ghanem, N.: Determining ASR characteristics using dilatometer method. Constr. Build. Mater. **36**, 1008–1015 (2012). https://doi.org/10.1016/j.conbuildmat.2012.06.027
20. Ghanem, H., Zollinger, D., Lytton, R., Ghanem, N.: Determining aggregate reactivity in various alkaline solutions. Proc. Inst. Civ. Eng. Constr. Mater. **167**(3), 151–161 (2014). https://doi.org/10.1680/coma.12.00005
21. Khatib, J.M., Ramadan, R., Ghanem, H., Elkordi, A., Sonebi, M.: Effect of limestone fines as a partial replacement of cement on the chemical, autogenous, drying shrinkage and expansion of mortars. Mater. Today Proc. **58**, 1199–1204 (2022). https://doi.org/10.1016/j.matpr.2022.01.336
22. Khatib, J., Ramadan, R., Ghanem, H., ElKordi, A.: Effect of using limestone fines on the chemical shrinkage of pastes and mortars. Environ. Sci. Pollut. Res. **30**(10), 25287–25298 (2022). https://doi.org/10.1007/s11356-022-18496-5
23. Khatib, J.M., Ramadan, R., Ghanem, H., Elkordi, A., Baalbaki, O., Kırgız, M.: Chemical shrinkage of paste and mortar containing limestone fines. Mater. Today Proc. **61**, 530–536 (2022). https://doi.org/10.1016/j.matpr.2022.01.288
24. Khatib, J., Ramadan, R., Ghanem, H., Elkordi, A.: Volume stability of cement paste containing limestone fines. Buildings **11**(8), 366 (2021). https://doi.org/10.3390/buildings11080366
25. Komuraiah, A., Kumar, N.S., Prasad, B.D.: Chemical Composition of natural fibers and its influence on their mechanical properties. Mech. Compos. Mater. **50**(3), 359–376 (2014). https://doi.org/10.1007/s11029-014-9422-2
26. Thyavihalli Girijappa, Y.G., Mavinkere Rangappa, S., Parameswaranpillai, J., Siengchin, S.: Natural fibers as sustainable and renewable resource for development of eco-friendly composites: a comprehensive review. Front. Mater. **6**, 226 (2019). https://doi.org/10.3389/fmats.2019.00226
27. Yang, B.Y., Montgomery, R.: Alkaline degradation of glucose: effect of initial concentration of reactants. Carbohydr. Res. **280**(1), 27–45 (1996)
28. Guo, Y., Wu, P.: Investigation of the hydrogen-bond structure of cellulose diacetate by two-dimensional infrared correlation spectroscopy. Carbohydr. Polym. **74**(3), 509–513 (2008). https://doi.org/10.1016/j.carbpol.2008.04.005
29. De La Grée, G.D., Yu, Q.L., Brouwers, H.J.H.: Assessing the effect of CaSO4 content on the hydration kinetics, microstructure and mechanical properties of cements containing sugars. Constr. Build. Mater. **143**, 48–60 (2017). https://doi.org/10.1016/j.conbuildmat.2017.03.067
30. Thongsanitgarn, P., Wongkeo, W., Sinthupinyo, S., Chaipanich, A.: Effect of limestone powders on compressive strength and setting time of Portland-limestone cement pastes. In: Advanced Materials Research, vol. 343, pp. 322–326. Trans Tech Publications Ltd. (2012). https://doi.org/10.4028/www.scientific.net/AMR.343-344.322

The Role of Animal Fibers on the Pore Structure of One-Year-Age Concrete

S. Dosque[1], J. M. Tulliani[2], B. Chiaia[3], and F. C. Antico[1](✉)

[1] Facultad de Ingeniería y Ciencias, Universidad Adolfo Ibáñez, Avenida Padre Hurtado, 750 Viña del Mar, Chile
sdosque@alumnos.uai.cl, federico.antico@uai.cl

[2] Department of Applied Science and Technology, Politecnico di Torino, 10129 Turin, Italy
jeanmarc.tulliani@polito.it

[3] Department of Structural, Geotechnical and Building Engineering (DISEG) – Politecnico di Torino, 10129 Turin, Italy
bernardino.chiaia@polito.it

Abstract. The construction industry is considered one of the industries that pollute the most worldwide, since it contributes to pollution directly and indirectly, each of its steps can be classified as harmful to the environment. At the same time, 900,000 million metric tons of pig waste are produced per year in Europe with a management cost of 20 million euros per year. This work aims to quantify the fiber-size pore structure of a concrete matrix when randomly distributed valorized animal fibers are added to the mixture. One-year-age concrete specimens plain and reinforced with 8 kg/m^3 of animal fibers were tested following an experimental multiscale approach. The pore structure of concrete between 60 and 180 microns was addressed using mercury intrusion porosimetry. The electrical resistivity and flexural strength of all concrete samples were also measured to address the tortuosity and mechanical strength of one-year-age reinforced concrete subject to the different curing conditions. Two curing conditions were considered for concrete samples: 1) 22 °C and 50% relative humidity, and 2) water saturated at 22 °C. Our results show that the external curing conditions influence the fiber-size pore structure of one-year-age reinforced concrete. While reviewing the reinforced samples at an age of one year, traces of degraded hair were found.

Keywords: Animal fiber · Reinforced concrete · Mechanical properties · Pore structure

1 Introduction

Reinforcing cementitious materials with natural fibers has shown to be effective at crack control at early ages, but there are still questions about their durability [1]. For some countries like Chile, only a decade ago, one of the first challenges to tackling concrete durability was that the regulations for concrete mixtures were based on their 28-day compressive strength [2]. The latter disregards the durability variable making concrete structures fail prematurely [3]. The transition to a circular economy of the construction

S. Amziane et al. (Eds.): ICBBM 2023, RILEM Bookseries 45, pp. 585–593, 2023.
https://doi.org/10.1007/978-3-031-33465-8_44

industry requires a full understanding of how adding valorized materials modified the geometrical scales of concrete after 28 days of age. The durability of concrete depends on the mixture design and the curing conditions. Regarding mixture design, using low water/cement ratio (w/c) mixtures is a way of increasing the durability of concrete to reduce porosity [3]. Reducing w/c of concrete also contributes to the rational use of freshwater in industry construction that, in the past decades, has been annually growing up to 1 trillion liters worldwide [3]. Most of the production of concrete in the twentieth century used w/c ratios higher than 0.5, and a change in this point with new materials for concrete means a significant saving in fresh water and results in stronger concrete [3]. Regarding curing conditions, the variable environmental temperatures produce cycles of volumetric changes that develop excessive plastic shrinkage to the concrete that eventually cracks during the first hours if it is not placed and cured correctly [3]. An effective way to control crack growth is embedding fibers into cement-based material [6]. Synthetic fibers are the principal fibers production and used in the industry. They are manufactured from polypropylene (PP), glass, and steel and have been constantly studied for their capacity to improve properties of cement-based materials such as tensile and impact strength, fire and abrasion resistance, crack control, shrinkage, and freeze-thaw resistance [7–12]. Commercial PP fibers based on olefins are made from non-renewable resources. The environmental impact of the production of PP was addressed by many studies through different environmental variables. Commercial PP fibers are difficult to dispose of where recycling and re-use are not feasible, representing an economic and environmental cost at the end of the life of the structures. Adding PP fibers to extend durability and control cracks at early ages represents a major challenge to recycling concrete, making it difficult to meet the standards imposed by the European Union that state that by 2020 a minimum of 70% (by weight) of non-hazardous construction and demolition waste (CDW), e.g. concrete and aggregates, shall be prepared for re-use, recycled or undergo another material recovery [13]. For the aforementioned reasons, the use of natural fibers has become an attractive option. They are commonly made from materials such as wood, hemp, or sisal from vegetals and pig hair from the animal industry [1, 4, 14–18]. Natural fibers used to strengthen mortars boost their toughness [4]. According to previous studies, a pig hair content of 8 kg/m^3 is effective in the prevention of cracks at early ages and plastic shrinkage [4].

The novelty of this work is the quantification of the pore structure and mechanical properties of one-year-age reinforced concrete matrix with randomly distributed valorized animal fibers. Considering different initial curing conditions in this work we aim to provide insights into the importance of the external curing conditions of these composites and address the remaining pore structure of concrete after the natural fibers degrade.

2 Materials and Methods

2.1 Mixture Design and Curing Conditions

Concrete was mixed as follows: 1) aggregates were oven-dried for 24 h at 105 °C first and then cooled down to laboratory temperature inside the oven. Then mixed with 50% of the water of the mixture for one minute. Fine aggregates and approximately 25% of the remaining water of the mixture were added to the mixer and mixed for one minute. The rest of the water was poured and finally, the cement of the mixture was added and mixed for one minute. 2) the plasticizer was diluted into a small fraction of the water of the mixture, added to the mixture, and homogenized, 3) surface-dried saturated fibers were added to the corresponding mixtures, 4) fresh concrete was poured into previously oiled 15 × 15 × 15 cm^3-molds and vibrated externally. Later, all the samples of this work were stored in sealed conditions for 24 h. At the age of one day, samples were de-molded and stored according to the design of each test as will be described later. Table 1, presents the specifications of the materials used to cast the concrete for this work. Two types of fine aggregates were used to obtain particle size distribution according to ASTM C33/C33M-18. The mixture proportions of mortar are presented in Table 2 for convenience.

The samples were split into three groups according to three different curing conditions to explore the long-term properties of concrete reinforced with pig hair: 1) monitored room temperature and humidity (i.e. 22 °C and 50% relative humidity, average), 2) saturated condition in water at 22 °C until an age of 77 days, then monitored room temperature and humidity conditions, and 3) saturated condition in water at 22 °C until the age of 156 days, then monitored room temperature and humidity conditions. All the samples at environmental conditions were preserved with one face of the cube exposed to diffusion and the rest sealed to address differential changes in the microstructure due to the humidity content inside the samples. At the age of one year, a slice 5 cm wide was cut next to the diffusion surface of each cube. Then the slice was cut into three prisms of 5 × 5 × 15 cm^3 to perform testing.

2.2 Flexural Strength

Flexural strength of each concrete prism was assessed in accordance with the standard ASTM C348 [20] at an age of one year. Flexural strength was determined as the average of three concrete prisms corresponding to those at the same distance of the diffusion face of the original cube.

2.3 Surface Resistivity

Surface resistivity measurements, typically reported in kOhm.cm, were implemented using Proceq Resipod equipment that induces an electrical current and measures voltage using four probes at a relative distance of 38 mm from each other. As described by Castro et al. [21], a more porous solid causes lower resistivity as more and larger pores are present in the mixture with a higher degree of connectivity. The latter can be used as an indication of concrete permeability. Therefore, as the current is transferred by ions in the mortar pore structure, resistivity is an indirect measure of durability. An adaptation of the procedure recommended by AASHTO T 38-15 [22] was adopted for this work. Three prisms of concrete samples for each mortar mixture were measured. In this work, resistivity measurements were taken at an age of 365 one year.

Table 1. Constituents of concrete. Material properties.

Material	Properties		
Portland-pozzolana cement	Type IP		
Fine aggregates	Type	(i)	(ii)
	Water absorption (%):	4.24	3.28
	Fineness modulus:	2.78	0.99
	Density (g/cm^3):	2.26	2.46
Coarse aggregates	Water absorption (%):	1.15	
	Maximum aggregate size (mm)	20	
	Density (g/cm^3):	2.36	
Water	Tap water		
Plasticizer	Density: 1.15 (g/cm^3). Lignosulfonate based		
Fibers: Pig hair	Water absorption (%)	95	
	Mid-section average diameter (mm) [SD[1] (mm), COV[2] (%)]	0.16 [0.05, 31]	
	End-section average diameter (mm) [SD[1] (mm), COV[2] (%)]	0.13 [0.05, 38]	

[1]Standard deviation; [2]Coefficient of variation.

Table 2. Concrete mixture proportions.

Fine sand	135.4 kg/m^3	Coarse sand	541.6 kg/m^3
Coarse aggregate	748 kg/m^3	Cement	405.0 kg/m^3
Plasticizer	0.81 kg/m^3	Tap water	170.2 kg/m^3

2.4 Mercury Intrusion Porosimetry

Mercury intrusion porosimetry (MIP) is a method used to measure the pore size distribution in a porous material. The basic principle behind MIP is that mercury is forced into the pores of a sample by applying increasing pressure, and the amount of mercury that adheres to the sample is proportional to the size of the pores. A porosimetry chamber is used to measure the sample after it has been cleaned of any undesirable particles. By adding more mercury to the chamber, the sample is put under increasing pressure. The mercury starts to enter the tiniest holes in the sample as the pressure rises. A constant measurement of the mercury sticking to the sample is made. The amount of mercury that is still affixed to the sample is measured once the maximum pressure has been reached. A mercury intrusion curve is recorded using the data gathered that shows the correlation between the size of the pores and the amount of mercury that has adhered to the sample. MIP is a reliable technique for determining the distribution of pore sizes in a wide range of porous materials, such as rocks, soils, ceramics, and composites. It is crucial to remember that the MIP only reports on accessible pores or those that are sizable enough to accommodate mercury. In this work, small pieces mainly composed of concrete paste (see Fig. 1) were recovered from the prisms after being broken at the flexural test and performed MIP.

Fig. 1. Sample preparation for the mercury intrusion porosimetry test.

3 Results and Analysis

3.1 Mercury Intrusion Porosimetry

Figure 2 depicts the incremental mercury intrusion versus pore size of concrete samples, separated by saturated cured condition, in the range of 180 μm to 60 μm. Within this range, pores are classified as entrapped air voids by IUPAC [23], which correspond to entrained and entrapped air voids and pre-existing microcracks. Specifically, voids left by organic fiber after degradation are expected in the range of air-entrained voids as the average diameter of pig hair was 160 μm [1, 4]. Figure 2 presents in gray colors the results of MIP corresponding to plain concrete under the three different external curing conditions. The MIP results of reinforced concrete with 8 kg/m^3 are also presented in Fig. 2 with different dark lines according to the external curing condition.

The results from Fig. 2 show that, reinforced concrete not saturated (i.e., "Reinforced 0 Days) presents the highest pore volume between 60 to 150 μm. This is expected as after visual inspection, pig hair inside concrete samples was completely degraded after one year. As the time in saturated conditions increases, the pore volume of reinforced

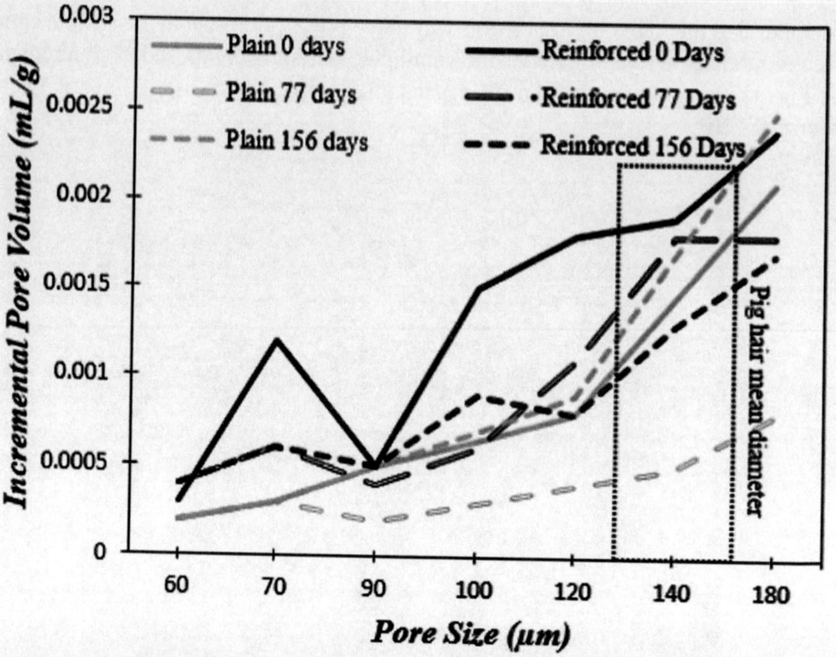

Fig. 2. Mercury intrusion porosimetry test (MIP) in the range of entrained air voids.

concrete decreases (see "Reinforced 77 Days" and "Reinforced 156 Days"). The result from the plain sample not saturated (i.e. "Plain 0 Days") shows a similar pore volume distribution as the reinforced samples cured in saturated conditions up to 77 and 156 days. More repetitions need to be done to study the difference between the pore volume distribution of plain concrete cured in saturated conditions for 77 and 156 days.

3.2 Flexural Strength

Figure 3a represents the flexural strength as a function of the saturated curing time; each result is the average of three identical samples. The error bars represent the standard deviation of the results. It can be observed that the external curing condition is a significant variable that affects flexural strength, but a larger sample size is necessary for definitive conclusions. As expected, as the time of the saturating condition increases, the average flexural strength increases. Based on the standard deviation of the results, there are no statistical differences between the flexural strength achieved by samples saturated up to an age of 77 days versus those saturated up to an age of 156 days. The previous could be related to a deceleration with the age of the hydration process. If the flexural strength of plain and reinforced concrete is compared, there is no statistical difference between these two. This is an interesting finding, as it is not expected that the natural fibers increase the flexural strength, as previously stated by Araya-Letelier et al. [4], and is consistent with the MIP results that show that the void structure of reinforced concrete paste at an age of one year, does not change compared to plain concrete paste.

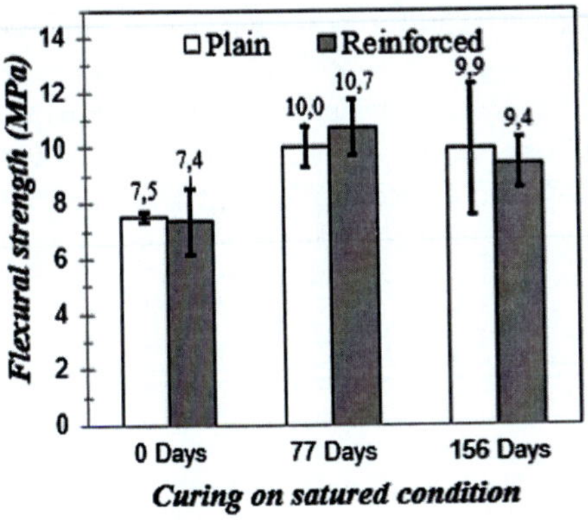

(a)

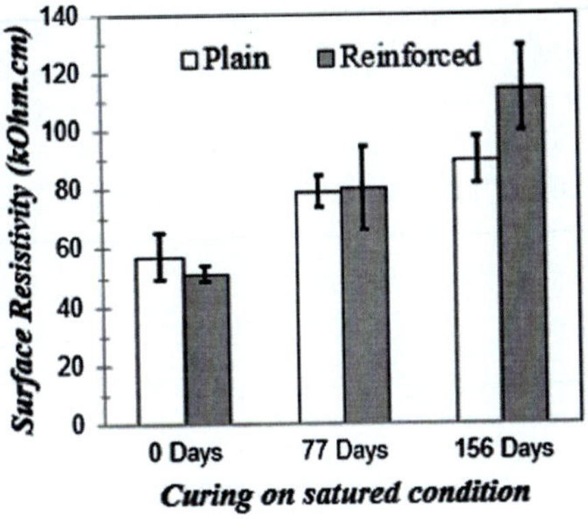

(b)

Fig. 3. (a) Bending on mortar samples. (b) Surface resistivity on mortar samples.

3.3 Surface Resistivity

Figure 3b represents the surface resistivity when the saturating curing time is considered. To obtain these results, each concrete cube was cut into nine prisms, identical to the prisms used to measure its flexural strength. Each result is the nine-sample average. The error bars indicate the standard deviation of the collected data. If compared to plain concrete, there is a statistical increase in the surface resistivity of concrete paste reinforced with pig hair of the samples cured for 156 days under saturated conditions. Concrete externally cured at laboratory conditions and saturated for the first 77 days of age do not show statistical differences between reinforced and plain samples. Yet, as the number of days in saturated curing condition increased, it can be observed that the surface resistivity increased. More work needs to be done to understand the reasons why reinforced concrete saturated up to 156 days of age presents higher surface resistivity than plain concrete. One possible explanation could be the internal curing caused by the saturated fibers [1].

4 Final Comments and Conclusions

This work provided an initial characterization of the void structure of one-year-age reinforced concrete and correlated those results with flexural strength and electrical resistivity as a measure of mechanical properties and durability respectively. We have confirmed that pig hair degrades inside one-year-age concrete leaving voids that sometimes could be observed with the naked eye. The results presented in this work show that one-year-age reinforced concrete can have different air-entrained pore-size volume to plain concrete depending on the curing condition. Only plain concrete saturated for 77 days presents lower pore volume within the studied pore-size. As the saturated curing conditions of reinforced concrete extends, the pore-volume between 60 and 180 microns reduces. We claim that this could be due to the continuous hydration process of the cementitious materials in presence of water. Flexural strength and electrical resistivity did not show statistical differences between plain and reinforced concrete, regardless of the external curing conditions proposed in this work. One possible explanation of the similarities between the results of plain and reinforced concrete in this work could be that natural fibers could hold the internal humidity of concrete for long periods of time, improving the hydration process, and counteracting the effect of the voids left due to its degradation.

References

1. Antico, F.C., Rojas, P., Briones, F., Araya-Letelier, G.: Animal fibers as water reservoirs for internal curing of mortars and their limits caused by fiber clustering. Constr. Build. Mater. **267**, 120918 (2021). https://doi.org/10.1016/j.conbuildmat.2020.120918
2. Instituto Nacional De Normalización (INN). NCh 170 "Hormigón Requisitos Generales" (2016)
3. Aïtcin, P.-C., Mindess, S.: Sustainability of Concrete. CRC Press (2011)
4. Araya-Letelier, G., Antico, F.C., Carrasco, M., Rojas, P., García-Herrera, C.M.: Effectiveness of new natural fibers on damage-mechanical performance of mortar. Constr. Build. Mater. **152**, 672–682 (2017)
5. Mindess, S., Young, J.F., Darwin, D.: Concrete, 2nd edn. Prentice Hall (2002)
6. Grzybowski, M., Shah, S.P.: Shrinkage cracking of fiber reinforced concrete. ACI Mater. J. **87**(2), 138–148 (1990)
7. Gopalaratnam, V.S., Shah, S.P.: Tensile failure of steel fiber-reinforced mortar. J. Eng. Mech. **113**(5), 635–652 (1987)
8. Firmo, J.P., Correia, J.R., Bisby, L.A.: Fire behaviour of FRP-strengthened reinforced concrete structural elements: a state-of-the-art review. Compos. Part B Eng. **80**, 198–216 (2015)
9. Corinaldesi, V., Nardinocchi, A.: Mechanical characterization of engineered cement-based composites prepared with hybrid fibres and expansive agent. Compos. Part B Eng. **98**, 389–396 (2016)
10. Olivito, R.S., Zuccarello, F.A.: An experimental study on the tensile strength of steel fiber reinforced concrete. Compos. Part B Eng. **41**(3), 246–255 (2010)
11. Çavdar, A.: Investigation of freeze-thaw effects on mechanical properties of fiber reinforced cement mortars. Compos. Part B Eng. **58**, 463–472 (2014)
12. Horszczaruk, E.K.: Hydro-abrasive erosion of high performance fiber-reinforced concrete. Wear **267**(1–4), 110–115 (2009)
13. European Parliament and Council: Directive 2008/98/EC of the European Parliament and of the Council of 19 November 2008 on waste and repealing certain directives (Waste framework). LexUriServ. do, 3–30 (2008)
14. Araya-Letelier, G., Concha-Riedel, J., Antico, F.C., Valdés, C., Cáceres, G.: Influence of natural fiber dosage and length on adobe mixes damage-mechanical behavior. Constr. Build. Mater. **174**, 645–655 (2018)
15. Ramakrishna, G., Sundararajan, T.: Impact strength of a few natural fibre reinforced cement mortar slabs: a comparative study. Cem. Concr. Compos. **27**(5), 547–553 (2005)
16. Pacheco-Torgal, F., Jalali, S.: Cementitious building materials reinforced with vegetable fibres: a review. Constr. Build. Mater. **25**(2), 575–581 (2011)
17. Vantadori, S., Carpinteri, A., Zanichelli, A.: Lightweight construction materials: mortar reinforced with date-palm mesh fibres. Theor. Appl. Fract. Mech. **100**, 39–45 (2019)

18. Araya-Letelier, G., Concha-Riedel, J., Antico, F.C., Sandoval, C.: Experimental mechanical-damage assessment of earthen mixes reinforced with micro polypropylene fibers. Constr. Build. Mater. **198**, 762–776 (2019)
19. Standard A: C595. Standard specification for blended hydraulic cements (2003)
20. ASTM International: Standard test method for flexural strength of hydraulic-cement mortars; ASTM C348-14: West Conshohocken, PA, USA (2014)
21. Castro, J., Spragg, R., Kompare, P., Weiss, W.J.: Portland cement concrete pavement permeability performance (2010)
22. American Association of State Highway and Transportation Official: AASHTO T 358-15 standard method of test for surface resistivity indication of concrete's ability to resist chloride ion penetration (2015)
23. Sing, K.S.W.: Reporting physisorption data for gas/solid systems with special reference to the determination of surface area and porosity (Recommendations 1984). Pure Appl. Chem. **57**(4), 603–619 (1985). https://doi.org/10.1351/pac198557040603

Check for updates

Experimental Investigation on the Strengthening of Gypsum Board by Flax Fibres

Oladikpo Gatien Agossou(✉) and Sofiane Amziane

Université Clermont Auvergne, Clermont Auvergne INP, CNRS, Institut Pascal, 63000 Clermont-Ferrand, France
{Oladikpo_gatien.agossou,sofiane.amziane}@uca.fr

Abstract. Gypsum board is a material used in construction due to its numerous qualities and functional properties, such as: low density, fire resistance, thermal insulation, moisture control, decorative appearance, etc. From an environmental point of view, the production of one tonne of gypsum board causes about 355 kg of CO_2 emissions into the atmosphere. The aim of this study is to reduce this CO_2 footprint while improving the mechanical and thermal properties of gypsum board. Flax fibres with a length of 12 mm, 24 mm and 36 mm were used to replace gypsum in the volume fraction. The addition of 10% flax fibres neutralises the carbon footprint of the gypsum board. However, due to the possible formation of fibre strands during mixing, which may affect the mechanical strength, the addition of flax fibres is limited to 1, 2 and 3%. An improvement in flexural strength was observed in the fibre-reinforced composites. The fibres changed the mechanical behaviour, e.g., toughness, compared to unreinforced gypsum. In addition to the improvement in mechanical properties, the addition of the fibres lowered the thermal conductivity value of the reinforced composites, resulting in improved thermal insulation.

Keywords: Gypsum board · strengthening · flax fibres · carbon footprint · mechanical and thermal properties

1 Introduction

Gypsum boards have long been used in the construction industry due to their numerous qualities and functional properties, such as low density, fire resistance, thermal and acoustic insulation, moisture control, low production costs, etc. However, they are very fragile and have low mechanical strength, which may limit their use in certain applications. This has led to the exploration of reinforcement with synthetic fibres [9] (glass, carbon, basalt fibres etc.) to improve mechanical properties, especially toughness, which is of interest for bolted joints. However, the high cost and environmental impact of synthetic fibres are an obstacle to their use in gypsum.

On the other hand, the production of gypsum board releases about 355 kg of carbon dioxide into the atmosphere. In the context of sustainable development and reducing the environmental impact of building materials, the use of plant fibres instead of synthetic

S. Amziane et al. (Eds.): ICBBM 2023, RILEM Bookseries 45, pp. 594–600, 2023.
https://doi.org/10.1007/978-3-031-33465-8_45

fibres in gypsum board seems to be a promising solution for the well-being of the atmosphere. It should also be noted that plant fibres are a renewable resource. In the context of the fight against global warming and for the reasons mentioned above, this work aims, on the one hand, to reduce the carbon footprint associated with the production of gypsum boards by replacing part of the gypsum with plant fibres. On the other hand, this work focuses on the mechanical and thermal characterization of gypsum boards reinforced with flax fibres.

2 Environmental Impact: Assessment of the Carbon Footprint of Gypsum Board

In a first approach to this study, understanding the reduction of the carbon footprint of gypsum board by plant fibres was of great importance for the production of fibre composites. Some data were collected from Life Cycle Assessment (LCA) BA 13 and used to quantify the mass of carbon dioxide that could be released into the atmosphere by fibre composites. This calculation is based on the establishment of mathematical equations derived from the balance of the phenomenon of photosynthesis and the mass of carbon sequestered in a plant species. It should also be remembered that the mass of carbon dioxide in a plant fibre depends on the carbon content of the fibre.

It should be noted that the effect of the plates on the carbon content is almost zero when 10% fibre is added (see Fig. 1). It is therefore necessary to verify the experimental feasibility of producing a gypsum board with 10% flax fibres. Some preliminary tests have allowed us to determine the limits of production under better processing conditions. As a result, a limit of 3% fibre admixture allowed a blend without fibre strands with a good processability of the composite, which could ensure a good mechanical behaviour and a reduction of the carbon footprint by 29%.

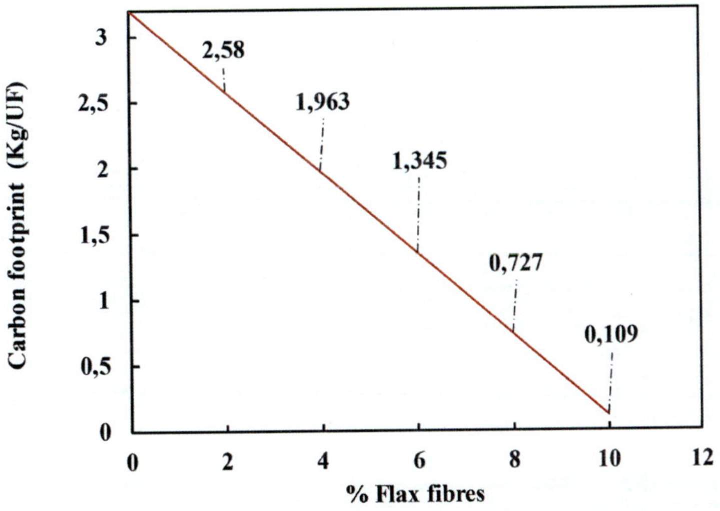

Fig. 1. Evaluation of the carbon footprint of flax fibre-reinforced gypsum boards

3 Materials and Methods

3.1 Materials

The gypsum used in this study consists of semi-hydrated calcium sulphate type β with chemical formula $CaSO_4{\cdot}2H_2O$, obtained by grinding gypsum at high temperature without additives, and was supplied by Siniat (France) of the Etex Building Performance group.

Flax fibres of different lengths (12, 24 and 36 mm) were supplied by Teillage Vandecandelaere of Depestele Group (France). In this study, the fibres were not modified (chemical or biological treatment) to better reflect environmental conditions (Table 1).

Fig. 2. Flax fibres

Table 1. Mechanical properties of the flax fibres (J. Page 2017) [11]

	Average	Typical deviation
Diameter (μm)	14,66	±2,95
Rupture stress (MPa)	1254	±456
Elongation at rupture (%)	1,86	±0,60
Young Module (GPa)	65,5	±14,8

3.2 Methods: Manufacture and Testing Procedures of Flax Fibre Reinforced Gypsum

The mixture is essentially based on the water/gypsum ratio. The density of the flax fibres used is 1302 kg/m^3, considering an intergranular porosity of 7%. The flax fibres were used as a volume fraction of the gypsum matrix. The fibres have a great influence on the homogenization and workability of our composites due to their high absorption capacity. In some preliminary tests, phenomena such as the formation of fibre strands occurred. Therefore, the fibre incorporation was fixed at 1 to 3%. In the production of the test specimens, mixing is carried out manually.

The mixtures are poured into prismatic test specimens $4 \times 4 \times 16$ cm^3 according to the standard NF EN 13279-2. The latter are stored in a cold room at 20 °C and demoulded after 24 h and dried at 50 °C for 24 h before performing the various tests (Table 2).

Table 2. Composition of mixtures

Samples	Volume fraction of fibres (%)	Density (Kg/m^3)	Gypsum (g)	Fibres (g)	Water (g)	W/C (C = composite)
GBR	0	900	230,4	0	184,32	0,8
GBRF	1	1302	228,10	3,33	185,14	0,8
	2	1302	225,79	6,66	185,97	0,8
	3	1302	223,49	10	186,79	0,8

- GBR: Reference gypsum board (without flax fibres)
- GBRF: Gypsum board reinforced with flax fibres

4 Results and Discussion

4.1 Effect of Flax Fibres on the Mechanical Properties of Composites

Figure 2 shows the mechanical behaviour (in 4-point bending) of the various composites. Unreinforced gypsum had the lowest strength with a value of 2.15 MPa compared to fibre reinforced gypsum. It is important to note that the flax fibres provided a real increase in the flexural strength of the reinforced specimens. The addition of 2% 12 mm flax fibres resulted in an increase in strength of about 98%. The same finding of increasing strength applies to the different fibre lengths. The reference specimen shows a linear behaviour followed by a sudden fracture due to e- to instantaneous crack propagation. This behaviour shows the typical fracture of a non-ductile, brittle material. The post-fracture behaviour of the 12-mm specimen made of 1% flax fibres is significantly affected by the presence of the fibres. There is a marked difference between the mechanical behaviour of the reference specimen and the fibre composite. The reinforced composites show ductile behaviour compared to the brittle behaviour of the reference specimen. It is observed that after the fracture of the gypsum matrix, the fibres resumed the forces, which prevented a sudden fracture of the material. Thus, the cracks that develop are inhibited by the presence of flax fibres, limiting their propagation (see Fig. 4). This result demonstrates the effectiveness of flax fibres in bridging microcracks. The fibres lead to a controlled and progressive fracture of the reinforced specimen [10]. This can be explained by the ability of the fibres to stitch the edges of the crack to prevent the material from collapsing abruptly (see Fig. 3). On the other hand, it is also observed that the reference gypsum has a higher stiffness than the reinforced composites. According to Dalmay et al. [12], the presence of fibres reduces the stiffness of the reinforced specimens [3, 9, 10].

4.2 Effect of Flax Fibres on Thermal Conductivity

The values obtained from the thermal conductivity tests of the flat fibre reinforced renders are shown in Fig. 5. It can be seen that the reference sample has the highest thermal

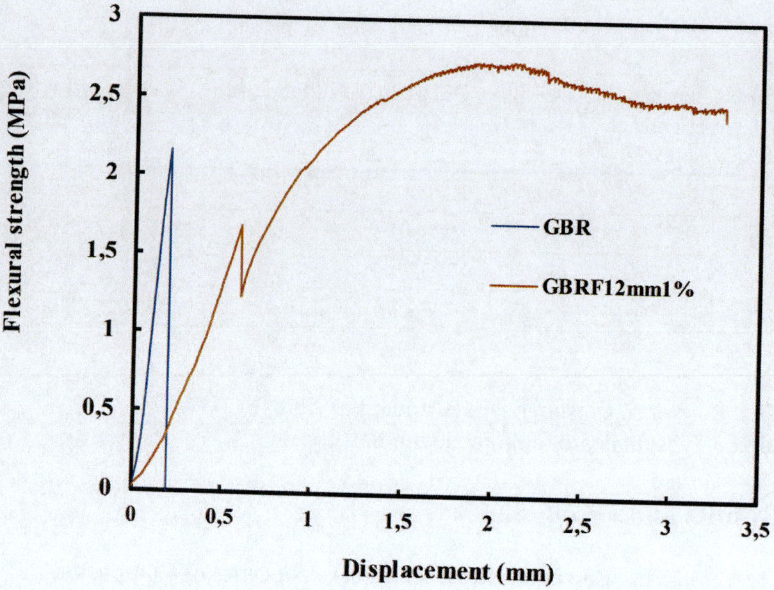

Fig. 3. Comparative curve of the flexural behaviour of reference plaster and flax fibre reinforced Gypsum

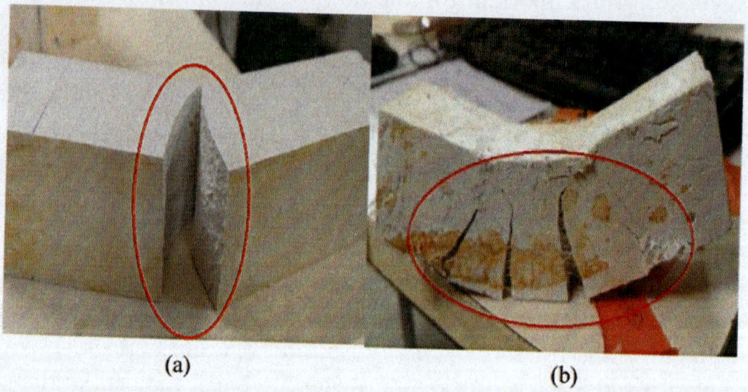

Fig. 4. Failure mode of composites: a) Reference gypsum b) Fibre gypsum

conductivity value. The flax fibres decrease the thermal conductivity. It is important to note that the thermal conductivity decreases as the volume fraction of the fibres increases. This decrease can be related to the high porosity (microporous structure) of the plants. The results show that the gypsum board with 3% flax fibres has the lowest value of thermal conductivity with an improvement of 47% (from 0.256 to 0.174 W/m.K) of thermal insulation compared to the reference gypsum. Other authors have also demonstrated the positive effect of plant fibres in thermal insulation, ensuring the comfort of the occupants [6].

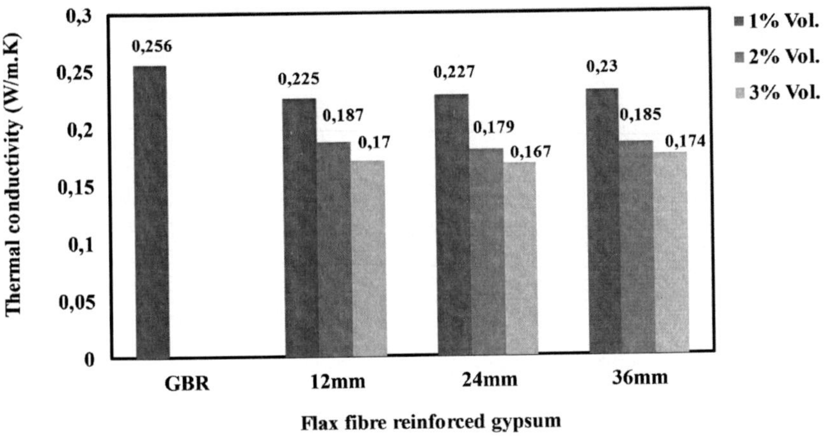

Fig. 5. Influence of flax fibres on the thermal conductivity of gypsum composites

5 Conclusion

In this study, the effect of flax fibres on the environmental impact and thermal and mechanical properties of gypsum board was investigated. The incorporation of flax fibres in gypsum board is a way to promote the production of more environmentally friendly gypsum board and expand its application in construction.

Considering the obtained results, the following conclusions can be drawn:

- The incorporation of 10% flax fibres in gypsum reduced the carbon footprint of gypsum board production to almost zero. However, the constraint of forming fibre strands, which limited the amount added to 3%, resulted in a 29% reduction in the carbon impact of gypsum.
- Flax fibres improved the mechanical behaviour of gypsum board. Four-point bending tests showed that the addition of the fibres increased the flexural strength and toughness of the gypsum boards but decreased the stiffness of the composites. The addition of 3% flax fibres resulted in the highest flexural strength (5 MPa for 12 mm fibres).
- In addition to the observed increase in strength, the increase in fibre content led to a gradual decrease in the thermal conductivity value of the composites, resulting in better thermal insulation.

References

1. Hamza, S., Saad, H., Charrier, B., Ayed, N., Charrier-El Bouhtoury, F.: Physico-chemical characterization of Tunisian plant fibers and its use as a reinforcement for plaster-based composites. Ind. Crops Prod. **49**, 357–365 (2013)
2. Iucolano, F., Caputo, D., Leboffe, F., Liguori, B.: Mechanical behavior of plaster reinforced with abaca fibers. Constr. Build. Mater. **99**, 184–191 (2015)
3. Iucolano, F., Liguori, B., Aprea, P., Caputo, D.: Evaluation of bio-degummed hemp fibers as reinforcement in gypsum plaster. Compos. B Eng. **138**, 149–156 (2018)

4. Hernández-Olivares, F., Oteiza, I., de Villanueva, L.: Experimental analysis of toughness and modulus of rupture increase of sisal short fiber reinforced hemihydrated gypsum. Compos. Struct. **22**(3), 123–137 (1992)
5. Iucolano, F., Liguori, B., Aprea, P., Caputo, D.: Thermomechanical behavior of gypsum coatings reinforced with hemp fibers. Constr. Build. Mater. **185**, 256–263 (2018)
6. Charai, M., Mezrhab, A., Karkri, M., Moga, L.: Study of the thermal performance of plaster reinforced with Alfa fibers. In: Dans AIP Conference Proceedings, vol. 2429, no. 1, p. 020005. AIP Publishing LLC (2021)
7. Charai, M., Sghiouri, H., Mezrhab, A., Karkri, M.: Thermal insulation potential of non-industrial hemp fibers (*Moroccan cannabis sativa* L.) for green plaster-based building materials. J. Clean. Prod. **292**, 126064 (2021)
8. Pinto, C.C., Carvalho, R.F.: Evaluation of the thermal performance of a low-cost housing ceiling prototype composed of gypsum board and sisal fibers. In: IOP Conference Series: Earth and Environmental Science, vol. 296, no. 1, p. 012015. IOP Edition (2019)
9. Williams, D.W., Mould, A.E.: Felxural strength of glassfibre reinforced gypsum and plasterboard. Build. Sci. **10**(1), 79–83 (1975)
10. Morlier, P.M.M., Khenfer, M.M.: Effect of fiber length on the mechanical properties of cements reinforced with cellulosic fibers. Suppress. Struct. **24**, 185–190 (1991)
11. Page, J.: Formulation and characterization of a biofibrated cementitious composite for prefabricated construction processes. Doctoral dissertation, Normandy University (2017)
12. Dalmay, P., Smith, A., Chotard, T., Sahay-Turner, P., Gloaguen, V., Krausz, P.: Properties of plaster reinforced with cellulosic fibers: influence of hemp or flax fibers on the properties of hardened gypsum. J. Mater. Sci. **45**(3), 793–803 (2010)

Reinforced Bio-Based Concrete by Natural FRCM

Rafik Bardouh(✉), Omayma Homoro, and Sofiane Amziane

Clermont Auvergne INP, Institut Pascal UMR 6602 – UCA/CNRS, 63178 Aubière, France
Rafik.bardouh@doctorant.uca.fr

Abstract. Biobased building materials are increasingly used in Europe thanks to their interesting hygrothermal properties, and sustainable development. The main issue results in the low mechanical performance of green concrete that limits their use in the construction industry due to safety standards. An innovative approach to solve this problem is to reinforce bio-based concrete with FRCM (Fabric Reinforced Cementitious Matrix) using two distinct modes of reinforcement: Flexural and compressive reinforcement. The flexural reinforcement consists of the fabrication of a composite sandwich made of hemp concrete core and FRCM skins. The Compressive reinforcement consists of confining hemp concrete specimens by FRCM. The influence of textile pre-impregnation and reinforcement ratio was critically investigated in the case of flexural reinforcement. Flexural reinforced specimens were tested by a 4 point bending tests while compressive reinforced specimens were tested by an axial compression test. The results shown a significant improvement in the mechanical properties of the reinforced specimens compared to reference hemp concrete specimens. For example, flexural reinforcement was found to increase the maximum load capacity by up to 17530%. Similarly, an increase in mechanical strength of 257% was achieved with compressive reinforcement. The results are promising and summarize the effectiveness of FRCM in mechanically reinforcing plant-material concrete to open new opportunities for the increasing use of these materials in the construction industry.

Keywords: FRCM · Biobased materials · Impregnation · Reinforcing · Mechanical properties

1 Introduction

The building industry is the responsible for 40% of primarily energy use and 36% energy related to CO_2 emissions [1]. Therefore, it was found important to develop an environmentally sustainable, low embodied energy materials to substitute conventional cement-based materials in the construction sector has increased [2]. Nowadays, plant-based materials especially the hemp-lime concrete or so called hempcrete was the most developed type of bio-aggregate based building materials. [3]. In Europe, hemp shiv sector was the most studied to be incorporated in bio-based building materials [4]. One of the most remarkable environmental qualities of hemp lime concrete is its carbon

S. Amziane et al. (Eds.): ICBBM 2023, RILEM Bookseries 45, pp. 601–612, 2023.
https://doi.org/10.1007/978-3-031-33465-8_46

negativity. In this process, lime carbonation equation allows CO_2 to be absorbed from the atmosphere according to the following equation [5].

$$Ca(OH)_2 + CO_2 \rightarrow CaCO_3 + H_2O \quad (1)$$

Subsequently, as the hemp grows, it traps CO_2 molecules to the structure of the plant due to photosynthesis cycle. However, the production of lime results in the emission of CO_2, yet this release is countered by the carbon sequestration of hemp. Boutin et al. (2006) [6] determined that 1 m^2 of hemp lime wall (260 mm thick) requires 370–394 MJ of energy for its production and sequesters 14–35 kg of CO_2 over its 100-year life span.

Biobased building materials has demonstrated an interesting physical, hygrothermal and acoustic properties. For example, these materials can be used as filling materials in walls that can perform as insulation layers due to their low thermal conductivity range from 0.06 to 0.19 W/m.K [7]. Hemp concrete was known by its lightness as its density was varied in the range of 256 kg/m^3 to 782 kg/m^3 [7]. This lower density was mainly caused by the porous structure of hemp shives, their size, the compaction energy, and the formulation process of hemp concrete [8].

However, the mechanical strength of bio-based concrete is very low compared to conventional building materials [9]. For instance, the compressive strength of hemp-lime concrete is less than 2 MPa [3]. Many factors have been attributed to explain the weak mechanical performance of bio-based concrete. These factors were related to the chemical incompatibility between lignocellulosic aggregates and cement, the high flexibility of the plant aggregates, lightness of plant aggregates, the imperfect arrangement of aggregates, and the casting process including the compaction energy [10, 11].

Many treatments were addressed on the modification of the physical, chemical, and thermal properties of bio-aggregates to bring enhancement on the mechanical response of bio-based concrete. Among these treatments, Asasutjarit et al. (2007) [12] has proven that the best pretreatment to enhance the mechanical properties of coir fibers was to boil and wash them. In the same context, Nozahic and Amziane [13] have shown that surface treatment on sunflower aggregates can promote aggregate/binder adhesion bond. Treatments were also addressed to the manufacturing process, S. Elfordy et al. [14] have proven that a new manufacturing technique consists on projection process of a mixture of lime, shives, and water can bring an increase in both thermal and mechanical properties.

Although, these treatments have contributed to slightly increase in the mechanical properties of plant-based concrete, the amount of mechanical gain was still insufficient for characterizing bio-based concrete as load bearing materials.

On the other side, FRCM have shown an effectiveness in reinforcing and retrofitting of concrete and masonry structures [15]. In the literature, FRCM was used to strengthen masonry structures by 3 different modes of reinforcement, compression [16, 17], shear [18], and flexion [19]. In compression reinforcement, studies showed that wrapping columns with FRCM composites has a dual effect: (1) an increase in the axial load capacity and (2) an improvement in the axial deformation capacity [16, 17]. Ferrera et al. (2020) have tested the shear capacity of masonry walls externally strengthened using flax-TRM composite systems [20]. The results have shown that the shear capacity was increased by 118% in the case of the system with one layer of flax fabric (SW-1L) and by 136% in the case of the system with two superimposed layers of flax fabric (SW-2L). Moreover, the flexural reinforcement consists of applying the FRCM composites to lightweight sandwich panels that are used as facade elements, and interior or exterior walls [19]. During the last decade, FRCM was applied to strengthen an arch bridge in Rome, Italy [21]. In 2016, Pino et Vanessa A, have studied the use of FRCM as a repair system for transportation infrastructure on the application of a RC (Reinforced Concrete) tunnel [22].

The design principle of FRCM reinforcement into building structures generally favors high strength and stiffness over high deformability [15]. Natural fibers such as flax were found to be a good alternative to synthetic and organic fibers in FRCM composite structures due to their high mechanical properties, cost efficiency, eco-environmental appeal, and renewability [23]. Hence, the present study implies an innovative mode of reinforcement by natural FRCM in order to enhance the mechanical properties of green concrete.

2 Materials and Methods

2.1 Materials

Hemp Shives

Hemp shives were the bio aggregates used in the fabrication of the bio-based concrete. The physical and hygrothermal characteristics of the porous by-products were determined by a series of tests performed at laboratory and were listed in the Table 1 below. The thermal conductivity was measured at the bulk density at dry point.

Table 1. Physical and hygrothermal properties of Hemp shives

Bulk density (kg/m^3)	112
Water retention capacity (%)	225
Dust content (%)	1.35
Thermal conductivity (W/(m.K))	0.048

Calcium Sulfo-Aluminate Cement

The calcium Sulfo-aluminate cement (CSA) was the mineral binder employed in the manufacturing of the flexural reinforced specimens. CSA cement mixture consists of combining 82% of CSA clinker and 18% of anhydrite. The hydration equation of CSA cement displayed below allows the production of ettringite rather which is the main reason that lower the alkalinity level of this mineral binder to make it more compatible with the lignocellulosic products.

$$C_4A_3\ \check{S} + 2C\check{S} + 38H \rightarrow C_6A\check{S}\ H_{32} + 2AH_3 \quad (2)$$

Three-point bending tests were performed on three prismatic specimens of 4 × 4 × 16 cm^3 of CSA matrix according to the French standard NF EN 196-1 [24] in order to determine the tensile properties. The average flexural strength determined after 28 days of curing time was 5.6 MPa (CoV 12%).

Prompt Natural Cement

The prompt natural cement (PNC) was the mineral binder employed in the manufacturing of the compressive reinforced specimens. An interesting feature of using PNC is its ability to reduce CO_2 emission by 20% compared to Portland cement [25]. For a W/C ratio of 0.6, the compressive strength and Young's modulus after a setting time of 28 days was 10 MPa and 20 GPa respectively [26].

Flax Textile

Flax textile was used in the FRCM composite structure to reinforce the bio-based concrete. The commercial name of the flax textile used was ampliTex5025 issued by Bcomp company. The fabric geometry of the flax textile used was unidirectional. Direct tensile tests were conducted on the flax textile in accordance to the ISO 13934-1 standard [27] to determine the mechanical properties of the fabric. The values of average tensile stress and elastic modulus were 123.1 MPa and 7 GPa respectively. The ranges of stress/strain that were used to calculate the elastic modulus were between 1.5 to 2% for strain and 70 to 100 MPa for stress.

2.2 Specimens Manufacturing

Reference Hemp Concrete

The mixing proportion included in the design of hemp concrete consists of binder/aggregates/water ratio of 1:0.5:1, respectively. The same proportions were applied for the two different types of molds, parallelepiped molds and the cylindrical molds. The dimensions of parallelepiped molds were 40 cm length, 10 cm width, and 10 cm height. On the other hand, the cylindrical molds were 12 cm diameter, and 22 cm height. The only difference between the mixing process of the two type of molds was the choice of the mineral binder incorporated in the hemp concrete design. The fabrication procedure was detailed in Fig. 1. Firstly, the hemp shives were placed in a plastic container (Fig. 1.a) followed by dropping 1/3 of the amount of water weighted to humidify the aggregates (Fig. 1.b) by continuous hand mixing. Then, the cement powder was added (Fig. 1.c) and slightly mixed with the hemp shives to improve the intramolecular porosity between the constituents. Final step in the mixture results in adding the remaining water quantity and continuously mixing the constituents all together (Fig. 1.d) until the desired adherence between hemp shives and cement paste was obtained (Fig. 1.e). The resulting mixture was then filled in molds whether parallelepiped or cylindrical. The casting process consists of energy compaction loads using a wooden stick under three successive layers of the resulting mixture (Fig. 1.f). After 48 h, the reference hemp concrete specimens were demolded in the case of the parallelepiped molds (Fig. 1.g) and the cylindrical molds (Fig. 1.h) and stored in pre-conditioned chamber (20 °C and 50% RH).

Flexural Reinforced Specimens

Flexural reinforcement was applied for the parallelepiped specimens of 10 × 10 ×

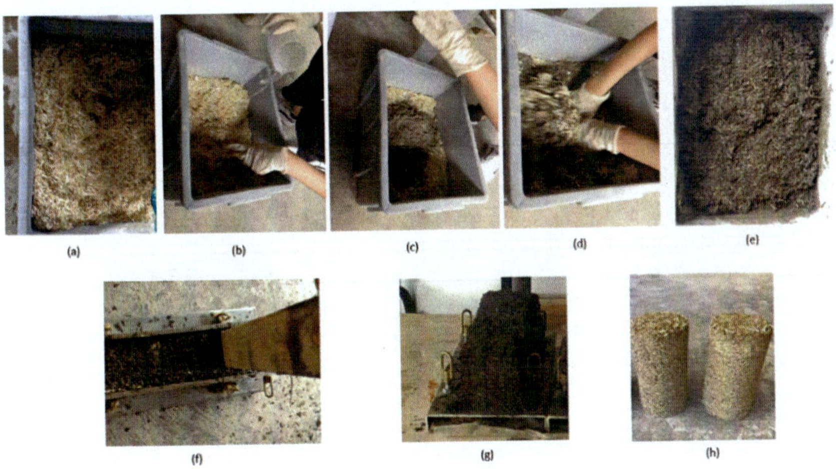

Fig. 1. Process of hemp concrete design for bending specimens; (a): dropping of hemp shives, (b): wetting of hemp shives, (c): put on CSA cement, (d): hand mixing of hemp shives with water and cement, (e): status of the resulting mixture, (f): compaction loads on hemp concrete inside the mold, (g): hemp concrete after demolding for the parallelepiped specimens, (h): hemp concrete after demolding for the cylindrical specimens.

40 cm^3. Flexural reinforcement consists of a composite sandwich fabrication using natural FRCM as upper and lower skins. The manufacturing procedure was detailed in Fig. 2. First of all, a demolding compound was placed at the bottom and the surfaces of the metallic mold for the ease of demolding at the appropriate time. The steps of lower FRCM fabrication were described as follow, a first CSA matrix that was prepared according to a W/C ratio of 0.35 was laid at the bottom of the mold through a spatula (Fig. 2.a), then a precut flax textile having the appropriate dimensions of the molds was placed above the matrix layer (Fig. 2.b). Hand pressure was applied on the flax textile to assure better penetrability of CSA matrix to the flax yarns. Another CSA matrix layer was placed above the flax textile to form the 1 cm thick lower FRCM (Fig. 2.c). Reference hemp concrete designed in the previous section was placed as the core of the composite sandwich with 3 successive layers followed by compaction loads over each layer (Fig. 2.d). The upper FRCM skin then consists of depositing a fresh CSA matrix layer over the core by ensuring a well bounded interphase between hemp core and matrix layer (Fig. 2.e). Another pre-cut flax textile was placed above the CSA matrix layer (Fig. 2.f). Likewise, the manual pressure was applied on the flax textile to promote the bond interphase between cementitious matrix and flax fabric. A second matrix layer was then laid to cover the flax textile and to form the 1 cm thick upper FRCM skin (Fig. 2.g). 48 h later, the flexural reinforced specimens were demolded (Fig. 2.h) and placed in a pre-conditioned chamber (20 °C and 50% RH) for 28 days before testing.

Compressive Reinforced Specimens

Compressive reinforcement was applied for the cylindrical specimens. The confining procedure of the cylindrical specimens was detailed in Fig. 3. First of all, the PNC matrix was prepared according to a W/C ratio of 0.5. A first fresh layer of the prepared

Fig. 2. Composite sandwich fabrication process; (a): first layer of cementitious matrix at the bottom of the mold, (b):flax textile placement, (c): second matrix layer to form the lower skin of the sandwich, (d): core of the sandwich by placing hemp concrete, (e): first matrix layer of the upper FRCM skin, (f): flax textile placement, (g): second matrix layer to form the upper skin of the sandwich, (h): demolded flexural reinforced specimens.

PNC matrix was applied on the reference hemp concrete surface (Fig. 3.a), then a precut flax textile of (44 cm long & 22 cm wide) was wrapped all around the matrix layer with an overlap length of 10 cm (Fig. 3.b). Flax textile was then overlapped with a second PNC matrix layer to finalize the confinement process (Fig. 3.c). The surface finishing of the confined cylinder was accomplished by a spatula in order to reach a proper distribution of PNC matrix and to obtain a smooth outer surface (Fig. 3.d). After an hour, the compressive reinforced specimens were placed in a pre-conditioned chamber at 20 °C and 50% relative humidity for 28 days before testing.

Fig. 3. Confined cylinder fabrication process; (a): first layer of cementitious matrix, (b): flax textile wrapping with overlap distance, (c): second layer of cementitious matrix over the placed textile, (d): final state of the confined cylinder

2.3 Experimental Configurations

The experimental configurations of the flexural-reinforced specimens were tailored according to the condition of the outer FRCM skins based on two main parameters: the reinforcement rate and textile pre-impregnation.

On the first side, F.R. 1L and F.R.2L denoted flexural reinforced specimens with FRCM comprising of 1 and 2 textile layers, respectively. The effect of impregnation was represented by the status of the flax textile. Two different status of flax textile were used in this study. F.R.NP and F.R.PM denoted flexural reinforced specimens with FRCM consisting of flax reference non-impregnated textile, and mineral matrix impregnated textile, respectively.

On the other side, the compressive reinforced specimens were studied based on only one experimental configuration denoted by C.R that represents the compressive specimens with one reference flax textile.

Aside from the reinforced specimens, F.REF and C.REF represent the reference specimens without the use of FRCM in flexion and compression, respectively.

Hence, 7 different configurations were employed in this study as follow, F.REF, F.R.1L.NP, F.R.2L.NP, F.R.1L.PM, F.R.2L.PM, C.REF, and C.R.

For statistical validation, 3 specimens were fabricated for each configuration resulting in a total of 21 specimens to be tested.

3 Results and Discussion

3.1 Flexural Reinforced Specimens

The mechanical properties for the flexural reinforced specimens and the reference hemp concrete specimens are listed in the Table 2. F_1 and F_{max} represents the force at the initiation of the first crack on FRCM skin and the maximum load carried by the composite sandwich respectively. K is the stiffness coefficient defined as the slope of the curve Force-deflection at the linear section.

Table 2. Mechanical properties for the different flexural reinforced configurations and reference hemp concrete.

		F.R.1L.NP	B.R.1L.PM	F.R.2L.NP	F.R.2L.PM	F.REF
F_1 (KN)	Mean	1.34	1.95	2.05	2.82	–
	CV (%)	15	10	7	3	–
F_{max} (KN)	Mean	2.36	4.33	6.33	8.11	0.046
	CV (%)	17	11	11	5	6
K (KN/mm)	Mean	0.53	1.14	1.3	1.37	0.025
	CV (%)	16	9	14	19	13

In general, a considerable increase in F_{max} can be noticed between the reference and flexural reinforced specimens. For instance, flexural reinforced specimens brought an increase of 5030% and 9313% compared to the reference specimens in terms of F_{max} for the configurations of F.R.1L. NP and F.R.1L.PM respectively. The tremendous increase in F_{max} reached a maximum of 17530% when the attributed configuration of the composite sandwich is F.R.2L.PM. The two key reasons for this meaningful improvement were firstly, the very poor strength of the reference hemp concrete and secondly, the performance of the FRCM composite structures to strengthen lightweight structures. The second textile layer added to the reinforced configurations led to an overall gain in F_l and F_{max}. For example, a gain in F_l of 53% and 45% was obtained when the flexural reinforced configurations were shifted from 1L to 2L for the F.R.NP and F.R.PM respectively. More gain was noticed in terms of F_{max} when the reinforced configurations were supported by another flax textile due to the heavy influence of flax fabric on the third zone of the mechanical behavior of the flexural reinforced specimens. Due to this reason, the increase in F_{max} was about 168% and 87% for the F.R.NP and F.R.PM configurations respectively.

Mineral impregnation has shown an efficiency in promoting mechanical resistance, as matrix impregnated reinforced specimens exhibited a higher strength than reference non-impregnated reinforced specimens. For example, it can be noticed in Fig. 4, an increase of 83% and 28% was found between the non-impregnated specimens and the matrix impregnated specimens for the Reinforced configurations of 1L and 2L respectively. This evidence was due to the influence of the mineral impregnation on the bond interphase at the level of the composite FRCM structure which led to general enhancement in the mechanical response of this latter.

According to Fig. 4, a tremendous improvement can be noticed in terms of stiffness coefficient between the reference specimens and the flexural reinforced specimens. This huge increase has reached up a maximum of 2020% compared to the ideal configuration of F.R.2L.PM. This was due to the very weak resistance to flexion of the reference specimens and to the effective influence of FRCM in stiffening the composite sandwiches. The stiffness of the reinforced configurations increase as the textile layer involved was doubled. However, the enhancement obtained was more remarkable in the case of non-impregnated specimens compared to the matrix impregnated specimens. For instance, K increases by 113% in the case of non-impregnated specimens whereas this increase was limited only to 20% in the case of the matrix-impregnated specimens. This difference can be explained by the poor contribution of the 1L flax textile at the dry state to the mechanical response of the reinforced specimens. This poor contribution was related to the very low number of cracks propagated at the level of the lower FRCM skin in the case of F.R.1L.NP.

Similarly, the mineral impregnation induce an increase in the stiffening coefficient as this latter went up by 115% compared to flexural non-impregnated specimens in the case of 1L. This enhancement was related to the better adherence the mineral impregnation provides between the inner filaments and the cementitious matrix.

3.2 Compressive Reinforced Specimens

The values of the mechanical strength σ_{max}, and the strain to rupture ε_{max} for the compressive reinforced specimens and the reference hemp concrete are listed in Table 3.

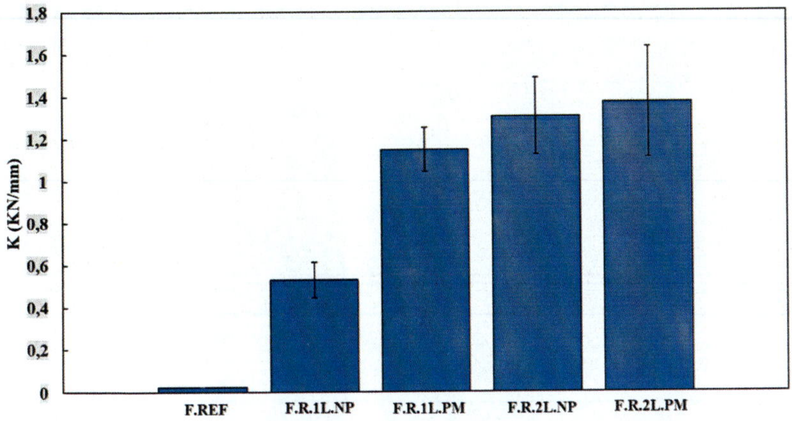

Fig. 4. Evolution of stiffness coefficient according to the different flexural reinforced configurations.

Table 3. Mechanical properties for the compressive reinforced specimens and reference hemp concrete.

		C.REF	C.R
σ_{max} (MPa)	Mean	0.26	0.93
	CV (%)	12	12
E (MPa)	Mean	4.73	29.83
	CV (%)	7	10

Figure 5 illustrates the evolution of the mechanical strength for the compressive reinforced specimens compared to the reference specimens. A gain of 257% was reached for the C.R specimens compared to C.REF specimens. This significant gain was attributed to the effectiveness of FRCM to support high uniaxial load owing to the additional mechanical properties of the flax textile to the reinforced specimens. Another reason was due to the very low compressive strength of reference hemp concrete described in literature.

Figure 6 illustrates the evolution of the strain to rupture for the compressive reinforced specimens compared to reference specimens. An increase of 34% was obtained between the reference specimens and the compressive reinforced specimens. The reference specimens exhibit a ductile behavior due to the flexibility of the hemp particles all along the cylinder. However, this ductility zone was further enlarged in the case of the compressive specimens due to the flexible reaction that flax textile generated under the compressive loads. This flexibility prevents the crushing of hemp particles during the tests for a longer period by absorbing the loads as form of cracks which were generated all along the FRCM later.

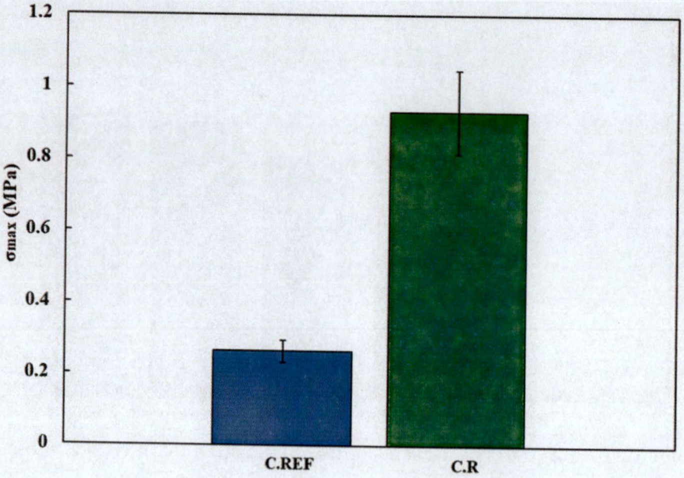

Fig. 5. The mechanical strength for the compressive reinforced specimens compared to reference specimens.

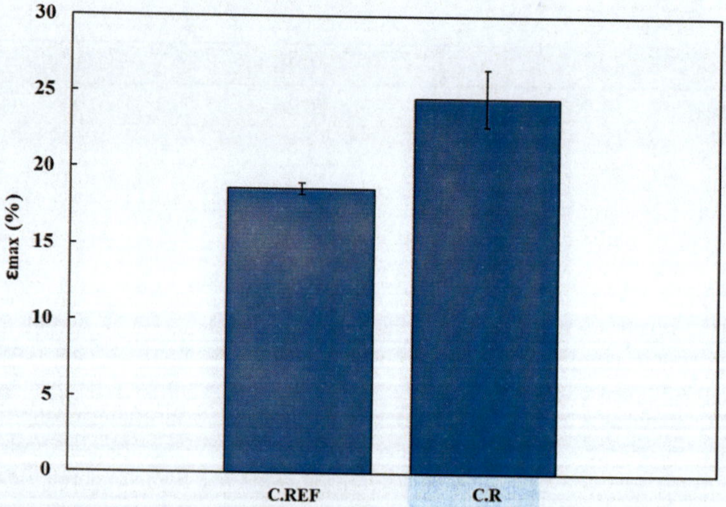

Fig. 6. The strain to rupture for the compressive reinforced specimens compared to reference specimens.

4 Conclusion

As a result of the experimental evidence obtained, the following conclusion points can be reached:

- In both flexural and compressive behavior, a ductile trend was obtained for the reinforced specimens compared to brittle behavior for the reference specimens owing to the flexibility of flax textile.

- FRCM reinforcement has proven a very significant effectiveness in improving the mechanical properties of hemp concrete by flexural and compressive reinforcement.
- The mineral impregnation technique led to better mechanical properties for the flexural reinforced specimens owing to the promoted bond adhesion between the inner flax yarns and the cementitious matrix that allows flax textile to contribute more efficiently to the mechanical response.

References

1. Intergovernmental Panel on Climate Change (IPCC): Climate change 2001: mitigation, contribution of working group III to the third assessment report of the Intergovernmental Panel on Climate Change. United States of America: Cambridge University Press, Cambridge (2001)
2. Walker, R., Pavia, S., Mitchell, R.: Mechanical properties and durability of hemp-lime concretes. Constr. Build. Mater. **61**, 340–348 (2014)
3. Tronet, P., Lecompte, T., Picandet, V., Baley, C.: Study of lime hemp concrete (LHC)–mix design, casting process and mechanical behaviour. Cem. Concr. Compos. **67**, 60–72 (2016)
4. Ratsimbazafy, H.H., Laborel-Préneron, A., Magniont, C., Evon, P.: A review of the multi-physical characteristics of plant aggregates and their effects on the properties of plant-based concrete. Recent Prog. Mater. **3**(2), 69 (2021)
5. de Bruijn, P.B., Jeppsson, K.H., Sandin, K., Nilsson, C.: Mechanical properties of lime–hemp concrete containing shives and fibres. Biosyst. Eng. **103**(4), 474–479 (2009)
6. Boutin, M.-P., Flamin, C., Quinton, S., Gosse, G., Inra, L.: Étude des caracteristiques environnementales du chanvre par l'analyse de son cycle de vie [Study of environmental characteristics of hemp by means of an analysis of its lifecycle]. Ministère de l'Agriculture et de la Pêche [Ministry of Agriculture and Fishery] (2005)
7. Véronique, C.: Propriétés mécaniques, thermiques et acoustiques d'un matériau à base de particules végétales: approche expérimentale et modélisation théorique. Institut National des Sciences Appliquées, Lyon (2005)
8. Niyigena, C., et al.: Variability of the mechanical properties of hemp concrete. Mater. Today Commun. **7**, 122–133 (2016)
9. Sellami, A., Merzoud, M., Amziane, S.: Improvement of mechanical properties of green concrete by treatment of the vegetals fibers. Constr. Build. Mater. **47**, 1117–1124 (2013)
10. Nozahic, V., Amziane, S., Torrent, G., Saïdi, K., De Baynast, H.: Design of green concrete made of plant-derived aggregates and a pumice–lime binder. Cem. Concr. Compos. **34**(2), 231–241 (2012)
11. Tai Thu, N.: Contribution à l'étude de la formulation et du procédé de fabrication d'éléments de construction en béton de chanvre. Université de Bretagne Sud, Diss (2010)
12. Asasutjarit, C., Hirunlabh, J., Khedari, J., Charoenvai, S., Zeghmati, B., Shin, U.C.: Development of coconut coir-based lightweight cement board. Constr. Build. Mater. **21**(2), 277–288 (2007)
13. Nozahic, V., Amziane, S.: Influence of sunflower aggregates surface treatments on physical properties and adhesion with a mineral binder. Compos. A Appl. Sci. Manuf. **43**(11), 1837–1849 (2012)
14. Elfordy, S., et al.: Mechanical and thermal properties of lime and hemp concrete ("hempcrete") manufactured by a projection process. Constr. Build. Mater. **22**(10), 2116–2123 (2008)
15. Hallonet, A., Michel, L., Ferrier, E.: Investigation of the bond behavior of flax FRP strengthened RC structures through double lap shear testing. Compos. B Eng. **100**, 247–256 (2016)

16. Carloni, C., Mazzotti, C., Savoia, M., Subramaniam, K.V.: Confinement of masonry columns with PBO FRCM composites. Key Eng. Mater. **624**, 644–651 (2014)
17. Fossetti, M., Minafò, G.: Strengthening of masonry columns with BFRCM or with steel wires: an experimental study. Fibers **4**(2), 15 (2016)
18. Askouni, P.D., Papanicolaou, C.G.: Experimental investigation of bond between glass textile reinforced mortar overlays and masonry: the effect of bond length. Mater. Struct. **50**(2), 1–15 (2017). https://doi.org/10.1617/s11527-017-1033-7
19. Mercedes, L., Bernat-Maso, E., Martínez, B.: Bending behaviour of sandwich panels of vegetal fabric reinforced cementitious matrix: experimental test and numerical simulation. Constr. Build. Mater. **340**, 127820 (2022)
20. Ferrara, G., Caggegi, C., Martinelli, E., Gabor, A.: Shear capacity of masonry walls externally strengthened using Flax-TRM composite systems: experimental tests and comparative assessment. Constr. Build. Mater. **261**, 120490 (2020)
21. Nanni, A.: A new tool for concrete and masonry repair. Concr. Int. **34**(4), 43–49 (2012)
22. Pino, V.A.: Fabric reinforced cementitious matrix (FRCM) composites as a repair system for transportation infrastructure. University of Miami (2016)
23. Homoro, O., Michel, M., Baranger, T.N.: Improvement of the mechanical properties of a glass multifilament yarn reinforced ettringitic matrix using an innovative pre-impregnation process. Eur. J. Environ. Civ. Eng. **26**(5), 1977–1992 (2022)
24. EN, N. F. 196-1-Septembre 2016. Méthodes d'essais des ciments—Partie, 2020, vol. 1
25. The Louis Vicat Technical Center: Technical specifications - Prompt natural cement - The roman cement of Grenoble (2007)
26. http://www.romanportland.net/files/doc/cahier_technique_cr_cnp_eng.pdf
27. ISO, CENEN: 13934-1: Textiles-Tensile Properties of Fabrics–Part 1: Determination of Maximum Force and Elongation at Maximum Force Using the Strip Method. British Standard: Brussels, Belgium (1999)

Application of Natural FRCM Composites in Structural Strengthening/Rehabilitation: State-of-the-Art Analysis

Oladikpo Gatien Agossou(✉), Omayma Homoro, and Sofiane Amziane

Université Clermont Auvergne, Clermont Auvergne INP, CNRS, Institut Pascal, 63000 Clermont-Ferrand, France
{Oladikpo_gatien.agossou,Omayma.homoro,sofiane.amziane}@uca.fr

Abstract. In recent decades, the vulnerability of structures in the building sector has led to the need to improve the structural performance of existing and new structures. Composite materials referred to as textile-reinforced mortars (TRM) and known in the literature as textile-reinforced concrete (TRC) or fabric-reinforced cementitious matrix (FRCM), have been extensively investigated as a promising alternative to fibre-reinforced polymers (FRP) for strengthening concrete or masonry elements.

These cementitious matrix composites offer substrate compatibility, lower cost, and better performance at high temperatures compared to FRP. In addition, the use of natural fibres in FRCM has increased significantly due to their good mechanical properties and lower the environmental impact of construction materials.

This paper presents a brief overview of the work on strengthening structural elements with natural FRCM. It gives an overview of the different parameters affecting the effectiveness of natural FRCM, the different strengthening techniques and some applications to improve mechanical performance.

Keywords: Fabric-reinforced cementitious matrix (FRCM) · Strengthening · Natural fibres · Masonry · Mechanical properties

1 Introduction

The ageing and vulnerability of civil engineering structures have been of particular concern to engineers and those involved in construction for several decades. In earthquake-prone areas, masonry buildings have suffered significant damage in previous earthquakes [1–3]. The issue of rehabilitation and strengthening of existing structures is of great importance due to their deterioration, lack of maintenance, or the need to meet current modernization requirements. Developing composite materials or designing effective strengthening systems [4–6] to solve current problems and improve the structural behaviour of structures has attracted increasing interest and triggered important research programs. Fibre-reinforced polymers (FRP) have long been used as the primary external strengthening application for existing structural elements. Studies [84–87] have shown

S. Amziane et al. (Eds.): ICBBM 2023, RILEM Bookseries 45, pp. 613–626, 2023.
https://doi.org/10.1007/978-3-031-33465-8_47

that FRP have advantageous properties such as good tensile behaviour (high strength), corrosion resistance, ease and speed of application. However, some disadvantages of strengthening technique using FRP have also been found: use of epoxy resins (high cost, carbon impact, hearth risk), poor compatibility with the substrate (concrete or masonry), poor adhesion to wet surfaces, lack of vapour permeability, and poor performance at high temperatures [83].

To overcome the above-mentioned shortcomings of using FRP, some authors [7–9] suggested using mortar (inorganic matrix), which promotes effective bonding between the reinforcing fibres and the matrix, as well as between the composite and the substrate, instead of epoxy resin (organic matrix). However, the problem of adhesion between the fibres and the cementitious matrix has existed for a long time, regardless of the grain size of the mortar. Thus, the solution to the problem of impregnation of the mortar into the fibres has been to replace the fibres with textiles. As a result, new textile-based materials are being developed around the world. These new textile-based composites have been given various names, such as textile-reinforced concrete (TRC), textile-reinforced mortar (TRM), or fabric-reinforced cementitious matrix (FRCM), etc. The increasing use of FRCM composites is due to their good fire resistance compared to FRP, good mechanical properties, permeability, reversibility of the intervention, etc. Other studies have also shown that FRCM composites are more compatible with masonry and concrete substrates than FRP composites [7], 8, 9]. It should also be noted that the mechanical behaviour of FRCM composites is significantly different from that of FRP composites due to the nature of the matrix. Several authors have studied the mechanical behaviour of FRCM composites, including the tensile properties of FRCM and the adhesion of textiles to the cementitious matrix, identifying several failure modes [10, 11, 16–23].

As for the application of FRCM in strengthening systems, the longest used high strength textiles are polybenzoxazole (PBO), carbon, steel, glass, basalt, aramid, etc. [10–15]. However, in recent decades, sustainable development goals have led to an interest in using natural fibres to develop innovative materials and technological solutions in the construction sector. The choice of natural fibres is justified by their good mechanical properties, their answers to the questions of durability and limiting the environmental impact of building materials. Moreover, the use of natural fibres (such as hemp, flax, jute, sisal, miscanthus, etc.) reduces or eliminates some of the problems associated with the low recyclability and energy consumption associated with the production of glass and carbon fibres in conventional composites [25–27]. The use of natural fibres as strengthening in FRCM composites has shown that there is a good bond between the fibres and the mortar matrix [19, 24]. However, variability in the results was observed, which was explained by the type of fibres and the low stiffness values.

In order to verify the structural effectiveness of FRCM composites based on natural textiles, several studies have addressed their application in the reinforcement of structural elements such as masonry (brick, stone, etc.) and concrete (beams, columns, concrete walls, etc.). It should also be noted that the vulnerability of masonry structures is increased when they are subjected to eccentric, seismic and out-of-centre loads. Experimental campaigns have also been carried out to study the structural behaviour of masonry elements [28–33, 33–35]. Regarding the shear behaviour, several studies have been carried out on masonry structures subjected to diagonal compression tests [30, 31,

34, 35]. For example, Olivito et al. [35] investigated the strengthening of brick slabs under eccentric loading using natural textile FRCM composites. The results of these different experimental campaigns showed the high efficiency of FRCM composite systems in improving the shear behaviour of masonry units. Moreover, an increase in the ductility of reinforced structures under in-plane loading was also demonstrated [29]. In addition to the numerous improvements observed, the durability of these natural textile-based reinforced composites remains a concern in recent studies. These strengthening materials may be exposed to different environmental conditions, which may lead to the deterioration of their mechanical properties.

The durability of these strengthening materials is currently under investigation, as they may tend to deteriorate their mechanical properties under certain environmental conditions [36, 37]. The procedure for the strengthening of structural elements by FRCM composites made of natural textiles can be summarised as follows: 1°) preparation of the substrate surface; 2°) application of a first layer of the matrix on the wetted substrate surface; 3°) application of the textile (this process is repeated until all textile layers are well impregnated with the matrix); and 4°) application of a final layer of the matrix on the last textile layer.

This paper reviews the state of the art of various works on the strengthening of structural elements by FRCM composites made of natural textiles based on concrete applications. It also provides a critical review of previous studies to identify deficiencies and gaps in the literature that could be the subject of future studies. This introductory section is followed by a description of the mechanical behaviour (tensile and matrix-textile adhesion) of natural textile FRCM composites. An overview of the different applications (shear, flexural or confinement) is given, followed by the different failure modes.

2 Natural FRCM Composites

2.1 Materials

Today, the growing interest in the use of natural textile FRCM composites as reinforcement and restoration materials for masonry structures is mainly due to their high strength-to-weight ratio, minimal invasiveness, good resistance to high temperatures, the possibility of application on irregular surfaces or in high humidity, vapour permeability, easy and safe application for workers, chemical and mechanical compatibility with various masonry substrates, and reversibility of the intervention, etc. [43, 47, 48].

Natural FRCM composites consist of textiles of plant fibres (flax fibres, hemp fibres, sisal fibres, jute fibres, etc.) embedded in an inorganic matrix, e.g., a cement-based matrix or hydraulic lime [41], etc. Studies [17, 37] have shown that the cementitious matrix does not have the same adhesive properties as an organic matrix (epoxy resin), resulting in a different bonding mechanism between the textile and the matrix than is the case with FRP [7]. The mortar does not allow good penetration and impregnation of the yarn filaments due to the large size of the granules in the matrix. Bonding at the interface between textile and matrix determines the mechanical behaviour of the composite, as it is responsible for load transfer between filaments within a yarn and between the yarn and the matrix. The impregnation of fabrics with polymeric matrices (resin) or mineral fillers has been proposed to improve the bonding properties and thus enhance the mechanical performance of natural FRCM composites [10, 44].

2.2 Mechanical Behaviour of Natural FRCM

The tensile behaviour of FRCM composites has been studied in detail by several authors. FRCM composites are usually used to reinforce masonry components with low tensile strength, so the tensile behaviour of FRCM systems has been studied by several authors [17, 37–40, 45, 46]. Tensile tests on FRCM specimens allowed the determination of the mechanical properties, in particular the ultimate stress and strain and the modulus of elasticity. The typical response describing the mechanical tensile behaviour of FRCM composites is represented by the idealised trilinear shape in Fig. 1, where E1, E2, and E3 are the elastic moduli of phases I, IIa, and IIb, respectively [17, 18, 69]. The uncracked phase I, characterised by a linear behaviour, determines the mechanical properties of the matrix. The appearance of the first crack in the matrix from the first stress drop, followed by several cracks, represents phase II. The opening, number and spacing of the cracks depend mainly on parameters such as the geometry of the textile (mesh size, weave properties), the matrix and the curing conditions. The adhesion between the fabric and the matrix and the number of fibres also determine the extent of phase IIa [70]. Other studies on FRCM made of flax have shown that increasing the number of layers in the fabric reduces the maximum crack widths [28, 46]. The stress-strain curve's third phase (phase IIb) is nearly linear and has a lower slope (modulus of elasticity) than phase I. The composite is completely cracked, and its behaviour is determined by the mechanical properties of the textile. The stiffness of this phase is like that of the textile. The critical cracks propagate to failure, followed by abrupt failure due to the rupture of the textile fibres. Several failure modes can be observed: (i) fibre breakage near the clamps, (ii) fibre breakage at the free part (the unclamped part at the corners), and (iii) fibre slip without breakage [70, 71]. The mechanical performance of the natural FRCM composite is strongly influenced by the reinforcement ratio, textile properties, and adhesion performance at the fibre/mortar interface [45, 46, 71].

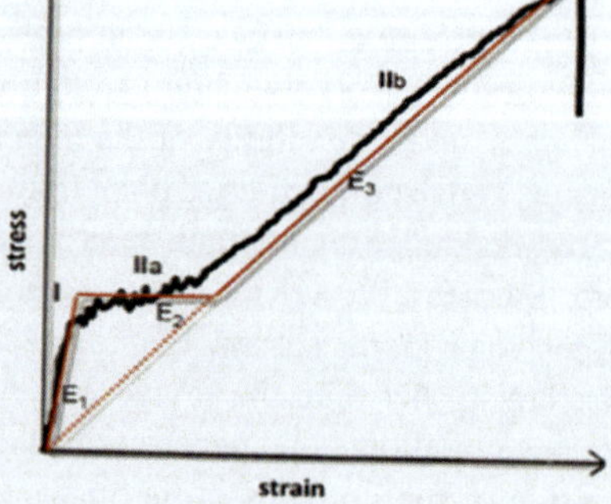

Fig. 1. Tensile behaviour of an FRCM composite [82]

2.3 Adhesion Between Natural FRCM Composite and Substrate

In a reinforcement system, the bond between the different elements plays an indispensable role to achieve the set objectives. In fact, the adhesion between the matrix and the textile, on one hand, and the natural FRCM composite and the substrate, on the other hand, is of great importance, allowing to realise of the effectiveness of the reinforcement. The adhesion between the textile matrix and the FRCM composite and the substrate has been studied in detail by the authors using shear tests (pull-out test, pull-in test…) [21, 46]. It is necessary that poor adhesion leads to detachment, which in turn leads to failure. These failures limit the loading of the natural FRCM composites, especially the activation of the natural textile [46]. Therefore, when there is good adhesion between the natural FRCM composite and the substrate, the following failure modes can be observed: Slippage or breakage of the textile through the matrix and detachment of part of the substrate or matrix from the natural FRCM composite [28, 29]. In addition, six failure modes can be identified (see Fig. 2) [21, 28, 46]. These failure modes are observed in most studies by examining parameters such as the effect of bond length, the effect of bond width, and the effect of substrate surface preparation [11, 14, 15, 19, 21, 22, 24, 28–32, 35].

Trichoutsou et al. (2021) [46] studied the effect of the number of plies for two different geometries of linen fabric. The results showed that increasing the number of plies may or may not change the failure mode. Insufficient impregnation of the yarns may prevent the effective transfer of forces from the yarns to the matrix. As a result, significant deformation/slippage of the textile occurs, possibly due to pronounced telescoping behaviour, followed by failure of individual filaments or yarns (failure modes D, E1, and E2). However, the increase in the number of plies resulted in a non-proportional increase in the maximum load. This non-proportional increase in load capacity is due to the better stress distribution within the composite due to the more uniform distribution of the wires in the matrix. In addition, a change in the failure mode of fibre slip through the matrix was observed upon the detachment of the natural FRCM layers with a portion of the substrate. The same observations were made in the studies of [72], and [73] where synthetic fibres were used.

Bond length was a parameter investigated in most studies. It should be noted that it is important to determine the sufficient bond length of the sheathing for better transfer of tensile loads from the masonry substrate or concrete to the natural FRCM to avoid the different types of premature failure [21, 28, 46, 72, 73, 75]. The capacity of the bond is related to the length of the bond between the substrate and the natural FRCM composite. The results of these studies have shown that the bonding capacity increases nonlinearly with increasing bond length and tends to stabilise at a certain length, called the "effective bond length" (see Fig. 3). In a recent study by [46], the results showed that the effective bond length of the tested flax/substrate FRCM systems ranged from 150 to 210 mm and the observed failure was dominated by fibre breakage.

The effects of bond width have also been studied in the literature [28, 46, 74–76]. This work has shown that an increase in bond width results in a linear increase in axial load capacity but has no effect on the change in failure mode.

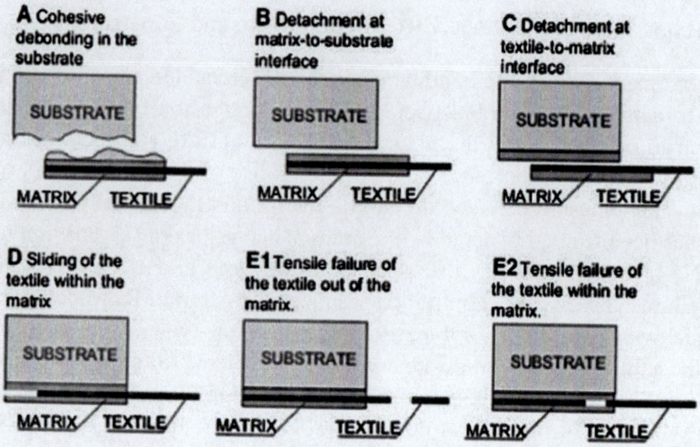

Fig. 2. Different failure modes of FRCM-substrate [21]

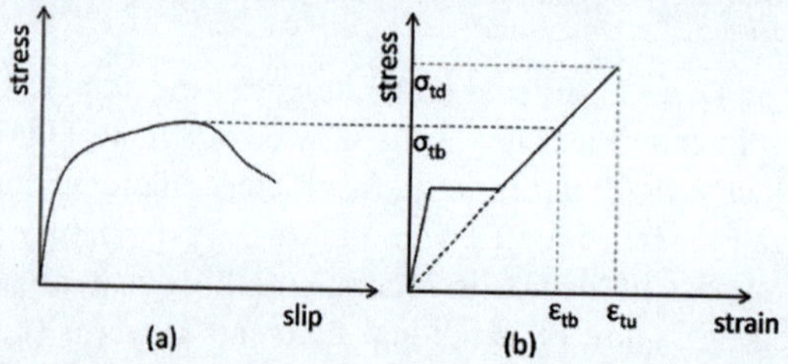

Fig. 3. Definition of the maximum bond a) stress and b) strain of the FRCM [82]

3 Strengthening of Structural/Masonry Elements by Natural FRCM Composites

The use of FRCM composites has been shown to be effective in strengthening masonry structures such as.: vertically loaded columns [49–52], in-plane loaded walls [53–57], out-of-plane loaded walls [58–61], vaults and arches [62–68], sandwich panels [77].

3.1 Compression Strengthening of Masonry Structures

The load-bearing walls of masonry are designed to support both vertical and lateral loads. It should be noted that masonry has complex behaviour and is influenced by various parameters such as the mechanical and geometric properties of the components and the connection between the components and the mortar. Nowadays, the strengthening of these building elements (masonry/concrete) with natural FRCM composites is an innovative topic. Studies have shown that the enclosure of columns with FRCM composites has a double effect: (1) an increase in axial load-bearing capacity and (2) an improvement in axial deformation capacity (ultimate strength) [49–52]. KHALEEL et al. (2021) [78] used a jute textile FRCM composite to reinforce a brick masonry prism in the following manner (see Fig. 4): A 2 mm mortar layer was applied to the four sides of the prism, then the jute textile was applied so that the mortar passes through the meshes of the textile, once the textile is properly wound, another mortar layer of about 2 mm is applied to the surface of the fibre to cover it completely. The results show that the compressive strength increases. Different failure modes were observed such as longitudinal cracking of the masonry or detachment of the outer layer of the FRCM matrix (see Fig. 5).

Fig. 4. Application of FRCM in confinement [78]

Fig. 5. Modes of rupture [78]

3.2 Shear Strengthening

Masonry units generally have a low shear capacity caused by seismic actions parallel to the walls or by sliding. These actions make the masonry vulnerable. The importance of strengthening these walls with FRCM composites based on plant fibres fits well with the goals of sustainable development and the promotion of eco-materials. Strengthening can be accomplished by placing FRCM layers on one or both sides of the wall, depending on the load on the wall or the possibility of intervention. Diagonal compression tests (see Fig. 7) have long been used to assess the shear strength of walls [75, 78, 79]. Ferrara et al. (2020) [28] proposed an experimental study on the shear strength of masonry walls externally reinforced with flax-FRCM composite systems. The results show that the application of the flax-FRCM system significantly improves the mechanical response of brick masonry (see Fig. 6) tested under diagonal compression, increasing the shear capacity by 118% in the case of the system with one layer of flax fabric (SW -1L) and by 136% in the case of the system with two superimposed layers of flax fabric (SW -2L). The increase in shearing capacity is not linear with the increase in the number of layers of the textile, as other authors have found [46, 72, 73].

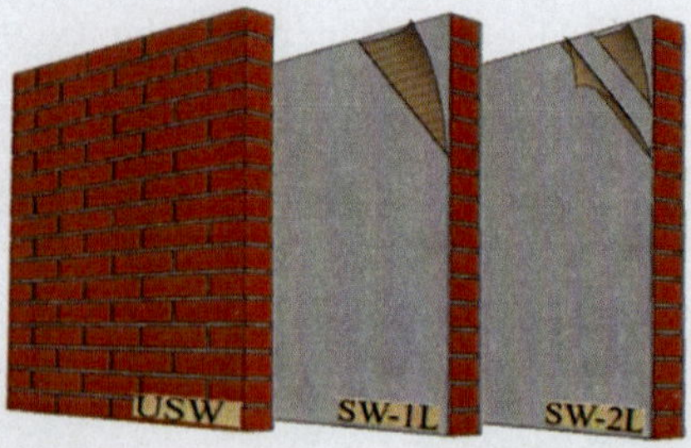

Fig. 6. Walls reinforced with flax-FRCM [28]

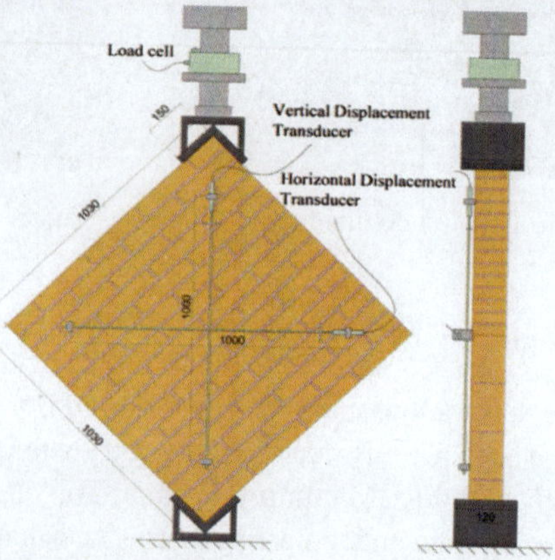

Fig. 7. Diagonal compression test [28]

3.3 Flexural Strengthening of Sandwich Panels

Sandwich panels are a lightweight solution with additional insulating properties compared to conventional solutions used for building envelopes. These panels are used as facade elements, and interior or exterior walls. They can also be used as structural or non-structural elements. In recent decades, the use of FRCM composites as reinforcement systems has increased to achieve better effectiveness against tensile and shear stresses and to expand the use of these panels in the construction sector [77, 80, 81]. Mercedes et al. (2022) [77] recently conducted a study on the reinforcement of sandwich panels (see Fig. 10) using FRCM composites made of different plant fabrics (hemp and sisal) as sandwich skin and an extruded polystyrene core. The two composite skins were joined with different types of connectors, including plant-based ones (see Fig. 8). The results of their work showed that the sandwich panel with hemp-FRCM composite had higher flexural strength without causing fabric cracking, while the panels with sisal and glass fibre FRCM had fabric cracking. A better distribution of cracks was observed with hemp-FRCM compared to sisal-FRCM. This failure is related to the properties of the textile as shown by other authors [28, 46]. Shear failure was also observed despite the use of connectors (Fig. 9).

Fig. 8. Connectors [77]

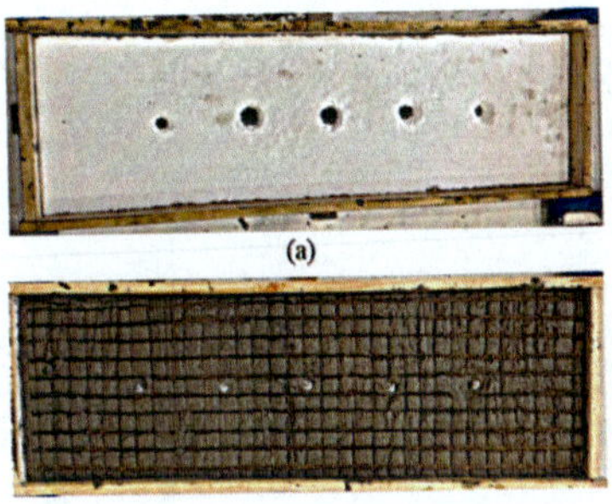

Fig. 9. Sandwich panels with connector system [77]

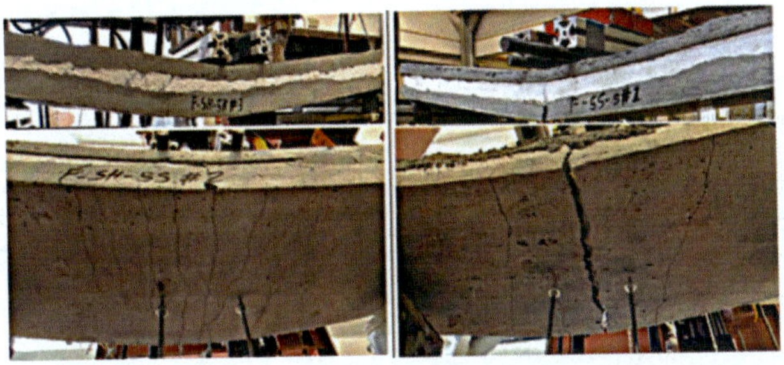

Fig. 10. Failure modes of sandwich panels [77]

4 Conclusion

The application of natural FRCM composites in the field of reinforcement and rehabilitation of existing and new structures is briefly presented in this paper. The focus is on the mechanical behaviour of natural FRCM composites, the adhesion phenomenon of FRCM and its bond to the substrate, and some of the different reinforcement modes (bending, shear, column confinement). The key parameters affecting the effectiveness of the reinforcement systems have been described for better optimization in future works.

In all the studies reviewed in this manuscript, the strengthening of structural elements, masonry or sandwich elements by natural FRCM composites was found to be successful. This strengthening technique was found to be effective in increasing the mechanical capacity (flexural, shear or compressive strength) and stiffness of existing or new structures (concrete or masonry, etc.) and consequently their performance under service loads. On the other hand, several rather complex failure modes have been identified, mainly related to the adhesion of the natural FRCM and to the properties of the matrix and the textile (plant fibres). Although several works have already addressed this issue, the topics of durability and the effect of high temperatures certainly deserve further research.

It is important to note that the studies conducted to date on the practical applications of natural FRCM (plant fibres) provide only sufficient data to allow the development of a design guideline and recommendations for standardising their use in construction.

References

1. Ceci, A.M., et al.: Structural performance of the historic and modern buildings of the University of L'Aquila during the seismic events of April 2009. Eng. Struct. **32**(7), 1899–1924 (2010)
2. Indirli, M., Kouris, S.L.A., Formisano, A., Borg, R.P., Mazzolani, F.M.: Seismic damage assessment of unreinforced masonry structures after the Abruzzo 2009 earthquake: the case study of the historical centers of L'Aquila and Castelvecchio Subequo. Int. J. Archit. Herit. **7**(5), 536–578 (2013)
3. Penna, A., Morandi, P., Rota, M., Manzini, C.F., da Porto, F., Magenes, G.: Performance of masonry buildings during the Emilia 2012 earthquake. Bull. Earthq. Eng. **12**(5), 2255–2273 (2013). https://doi.org/10.1007/s10518-013-9496-6
4. Lagomarsino, S.: Vulnerability assessment of historical buildings. Assessing and Managing Earthquake Risk: Geo-scientific and Engineering Knowledge for Earthquake Risk Mitigation: developments, tools, techniques, 135–158 (2006)
5. Rota, M., Penna, A., Strobbia, C.L.: Processing Italian damage data to derive typological fragility curves. Soil Dyn. Earthq. Eng. **28**(10–11), 933–947 (2008)
6. Ferretti, F., Ferracuti, B., Mazzotti, C., Savoia, M.: Destructive and minor destructive tests on masonry buildings: experimental results and comparison between shear failure criteria. Constr. Build. Mater. **199**, 12–29 (2019)
7. Papanicolaou, C.G., Triantafillou, T.C., Papathanasiou, M., Karlos, K.: Textile reinforced mortar (TRM) versus FRP as strengthening material of URM walls: out-of-plane cyclic loading. Mater. Struct. **41**(1), 143–157 (2008)
8. Papanicolau, C.G., Triantafilou, T.C., Kyriakos, K., Papathanasiou, M.: Textile-reinforced mortar versus FRP as strengthening material of URM walls: in-plane cyclic loading. Mater. Struct. **40**, 1081–1097 (2007)
9. Papanicolaou, C., Triantafillou, T., Lekka, M.: Externally bonded grids as strengthening and seismic retrofitting materials of masonry panels. Constr. Build. Mater. **25**(2), 504–514 (2011)
10. Caggegi, C., et al.: Experimental analysis on tensile and bond properties of PBO and aramid fabric reinforced cementitious matrix for strengthening masonry structures. Compos. Part B Eng. **127**, 175–195 (2017)
11. Carozzi, F.G., et al.: Experimental investigation of tensile and bond properties of Carbon-FRCM composites for strengthening masonry elements. Compos. Part B Eng. **128**, 100–119 (2017)
12. De Santis, S., et al.: Round robin test on tensile and bond behaviour of steel reinforced grout systems. Compos. Part B Eng. **127**, 100–120 (2017)
13. Saidi, M., Gabor, A.: Experimental analysis of the tensile behaviour of textile reinforced cementitious matrix composites using distributed fibre optic sensing (DFOS) technology. Constr. Build. Mater. **230**, 117027 (2020)
14. Lignola, G.P., et al.: Performance assessment of basalt FRCM for retrofit applications on masonry. Compos. Part B Eng. **128**, 1–18 (2017)
15. Ferrara, G., Caggegi, C., Gabor, A., Martinelli, E.: Experimental study on the adhesion of basalt textile reinforced mortars (TRM) to clay brick masonry: the influence of textile density. Fibers **7**(12), 103 (2019)
16. Ferrara, G., Pepe, M., Martinelli, E., Tolêdo Filho, R.D.: Tensile behavior of flax textile reinforced lime-mortar: Influence of reinforcement amount and textile impregnation. Cement Concr. Compos. **119**, 103984 (2021)
17. Trochoutsou, N., Di Benedetti, M., Pilakoutas, K., Guadagnini, M.: Mechanical characterisation of flax and jute textile-reinforced mortars. Constr. Build. Mater. **271**, 121564 (2021)

18. De Santis, S., Carozzi, F.G., de Felice, G., Poggi, C.: Test methods for textile reinforced mortar systems. Compos. Part B Eng. **127**, 121–132 (2017)
19. Cevallos, O.A., Olivito, R.S.: Effects of fabric parameters on the tensile behaviour of sustainable cementitious composites. Compos. Part B Eng. **69**, 256–266 (2015)
20. de Felice, G., et al.: Experimental characterization of composite-to-brick masonry shear bond. Mater. Struct. **49**(7), 2581–2596 (2015). https://doi.org/10.1617/s11527-015-0669-4
21. Leone, M., et al.: Glass fabric reinforced cementitious matrix: tensile properties and bond performance on masonry substrate. Compos. Part B Eng. **127**, 196–214 (2017)
22. Akhoundi, F., Vasconcelos, G., Lourenço, P., Silva, L.M., Cunha, F., Fangueiro, R.: In-plane behavior of cavity masonry infills and strengthening with textile reinforced mortar. Eng. Struct. **156**, 145–160 (2018)
23. de Felice, G., et al.: Recommendation of RILEM technical committee 250-CSM: test method for textile reinforced mortar to substrate bond characterization. Mater. Struct. **51**(4), 1–9 (2018). https://doi.org/10.1617/s11527-018-1216-x
24. Codispoti, R., Oliveira, D.V., Olivito, R.S., Lourenço, P.B., Fangueiro, R.: Mechanical performance of natural fiber-reinforced composites for the strengthening of masonry. Compos. B Eng. **77**, 74–83 (2015)
25. Lebrun, G., Couture, A., Laperrière, L.: Tensile and impregnation behavior of unidirectional hemp/paper/epoxy and flax/paper/epoxy composites. Compos. Struct. **103**, 151–160 (2013)
26. Placet, V., Trivaudey, F., Cisse, O., Gucheret-Retel, V., Boubakar, M.L.: Diameter dependence of the apparent tensile modulus of hemp fibres: a morphological, structural or ultrastructural effect. Compos. Part A Appl. Sci. Manuf. **43**(2), 275–287 (2012)
27. Coroller, G., et al.: Effect of flax fibres individualisation on tensile failure of flax/epoxy unidirectional composite. Compos. Part A Appl. Sci. Manuf. **51**, 62–70 (2013)
28. Ferrara, G., Caggegi, C., Martinelli, E., Gabor, A.: Shear capacity of masonry walls externally strengthened using Flax-TRM composite systems: experimental tests and comparative assessment. Constr. Build. Mater. **261**, 120490 (2020)
29. Menna, C., Asprone, D., Durante, M., Zinno, A., Balsamo, A., Prota, A.: Structural behaviour of masonry panels strengthened with an innovative hemp fibre composite grid. Constr. Build. Mater. **100**, 111–121 (2015)
30. Carvalho Bello, C.B., Cecchi, A., Meroi, E., Oliveira, D.V.: Experimental and numerical investigations on the behaviour of masonry walls reinforced with an innovative sisal FRCM system. In: Key Engineering Materials, vol. 747, pp. 190–195. Trans Tech Publications Ltd. (2017)
31. Cevallos, O.A., Olivito, R.S., Codispoti, R., Ombres, L.: Flax and polyparaphenylene benzobisoxazole cementitious composites for the strengthening of masonry elements subjected to eccentric loading. Compos. Part B Eng. **71**, 82–95 (2015)
32. Cevallos, O.A., Olivito, R., Codispoti, R.: Experimental analysis of repaired masonry elements with flax-FRCM and PBO-FRCM composites subjected to axial bending loads. Fibers **3**(4), 491–503 (2015)
33. Loccarini, F.: Comportement des ouvrages en pisé : matériaux durables et techniques de renforcement. Thèse, p. 151, 2017 (2017)
34. Emami, A., Fehling, E., Schlimmer, M.: Renforcement externe des structures de maçonnerie avec des fibres naturelles, vol. 35, non. 2, pp. 125–135 (2011)
35. Olivito, R.S., Dubois, F., Venneri, A., Zuccarello, F.: Analyse expérimentale et numérique des macroéléments de maçonnerie renforcés par des matériaux composites en fibres naturelles. In: 6e International Conference Compositions PRF Civil Engineering, pp. 1–8 (2012)
36. Faruk, O., Bledzki, A.K., Fink, H.P., Sain, M.: Rapport d'étape sur les composites renforcés de fibres naturelles. Macromol. Mater. Ing. **299**(1), 9–26 (2014)
37. Wei, J., Meyer, C.: Degradation rate of natural fiber in cement composites exposed to various accelerated aging environment conditions. Corros. Sci. **88**, 118–132 (2014)

38. Mercedes, L., Gil, L., Bernat-Maso, E.: Mechanical performance of vegetal fabric reinforced cementitious matrix (FRCM) composites. Constr. Build. Mater. **175**, 161–173 (2018)
39. Olivito, R.S., Cevallos, O.A., Carrozzini, A.: Development of durable cementitious composites using sisal and flax fabrics for reinforcement of masonry structures. Mater. Des. **57**, 258–268 (2014)
40. Dalalbashi, A., Ghiassi, B., Oliveira, D.V., Freitas, A.: Fiber-to-mortar bond behavior in TRM composites: effect of embedded length and fiber configuration. Compos. Part B: Eng. **152**, 43–57 (2018). ISSN 1359-8368. https://doi.org/10.1016/j.compositesb.2018.06.014
41. Mechtcherine, V., Schneider, K., Brameshuber, W.: Mineral-based matrices for textile-reinforced concrete. In: Triantafillou, T. (ed.) Textile Fibre Composites in Civil Engineering, Woodhead Publishing, 2016, pp. 25–43 (2016). ISBN 9781782424468, https://doi.org/10.1016/B978-1-78242-446-8.00003-3
42. Donnini, J., Corinaldesi, V., Nanni, A.: Mechanical properties of FRCM using carbon fabrics with different coating treatments. Compos. Part B Eng. **88**, 220–228 (2016). https://doi.org/10.1016/j.compositesb.2015.11.012
43. Valluzzi, M.R., Modena, C., de Felice, G.: Current practice and open issues in strengthening historical buildings with composites. Mater. Struct. **47**(12), 1971–1985 (2014). https://doi.org/10.1617/s11527-014-0359-7
44. Mechanical properties of FRCM using carbon fabrics with different coating treatments. Compos. Part B Eng. **88**, 220–228 (2016). https://doi.org/ https://doi.org/10.1016/j.compositesb.2015.11.012
45. BRAMESHUBER, Wolfgang, HINZEN, Marcus, DUBEY, A., *et al.* RILEM Recommendation TC 232-TDT: Test methods and sizing of textile reinforced concrete: Uniaxial tensile test: test method for determining the bearing behaviour of tensile test pieces made of textile reinforced concrete. Mater. Struct. Materiaux et Constructions **49**(12), 4923–4927 (2016)
46. Trochoutsou, N., Di Benedetti, M., Pilakoutas, K., et al.: Bonding of linen textile reinforced mortars to the masonry. Construct. Build. Mater. **284**, 122849 (2021)
47. Valluzzi, M., Modena, C., de Felice, G.: Current practice and open problems in the reinforcement of historic buildings with composites. Suppress. Struct. **47**, 1971–1985 (2014)
48. De Felice, G., et al.: Papanicolaou, CG Mortar-based systems for external reinforcement of masonry. Suppress. Struct. **47**, 2021–2037 (2014)
49. Carloni, C., Mazzotti, C., Savoia, M., Subramaniam, K.V.: Containment of masonry columns with PBO FRCM composites. In: Mechanics of Masonry Structures Strengthened with Composite Materials, Proceedings of the 4th International Seminar on the Mechanics of Masonry Structures Reinforced with Composite Materials (MuRiCo4 2014), Ravenna, Italy, 9–11 September 2014; Trans Tech Publications Ltd., Zurich, Switzerland, 2014; Volume 624, pp. 644–651 (2014)
50. Fossetti, M., Minafò, G.: Reinforcement of masonry columns with BFRCM or with steel wires: an experimental study. Fiber **4**, 15 (2016)
51. Maddaloni, G., Cascardi, A. Balsamo, A., di Ludovico, M., Micelli, F., Aiello, M.A., Prota, A.: Containment of large-scale masonry columns with FRCM systems. In: Mechanics of Masonry Structures Strengthened with Composite Materials II, Proceedings of the International Conference on the Mechanics of Masonry Structures Reinforced with Composite Materials (MuRiCo5), Bologna, Italy, 28–30 June 2017; Trans Tech Publications Ltd.: Zurich, Switzerland, 2017, Volume 747, pp. 374–381 (2017)
52. Aiello, M.A., Cascardi, A., Shadows, L., Glass, S.: Containment of masonry columns with the FRCM system: theoretical and experimental study. Infrastructure **5**, 101 (2020)
53. Prota, A., Marcari, G., Fabbrocino, G., Manfredi, G., Aldea, C.: Experimental behavior in the plan of tuff masonry reinforced with matrix-grid cementitious composites. J. Compos. Constr. **10**, 223–233 (2006)

54. Papanicolaou, C.G., Triantafillou, T.C., Karlos, K., Papathanasiou, M.: Textile reinforced mortar (MRT) versus FRP as wall reinforcement material URM: cyclic loading in the plane. Suppress. Struct. **40**, 1081–1097 (2007)
55. Augenti, N., Parisi, F., Prota, A., Manfredi, G.: Lateral response in the plane of a large-scale masonry subassembly with and without an inorganic matrix-grid reinforcement system. J. Compos. Constr. **15**, 578–590 (2011)
56. Parisi, F., Iovinella, I., Balsamo, A., Augenti, N., Prota, A. : Behavior in the plane of tuff masonry reinforced with inorganic matrix-grid composites. Compos. Part B Eng. **45**, 1657–1666 (2013)
57. Kałuża, M., Galman, I., Kubica, J., Agneloni, C.: Diagonal tensile strength of AAC block masonry with thin joints superficially reinforced by reinforcement with GFRP mesh coating. In: Mechanics of Masonry Structures Strengthened with Composite Materials, Proceedings of the 4th International Seminar on the Mechanics of Masonry Structures Reinforced with Composite Materials (MuRiCo4 2014), Ravenna, Italy, 9–11 September 2014; Trans Tech Publications Ltd.: Zurich, Switzerland, 2014; Volume 624, pp. 363–370 (2014)
58. Papanicolaou, C.G., Triantafillou, T.C., Papathanasiou, M., Karlos, K.: Textile Reinforced Mortar (TRM) versus FRP as wall reinforcement material URM: cyclic loading out of plane. Suppress. Struct. **41**, 143–157 (2008)
59. Papanicolaou, C., Triantafillou, T., Lekka, M.: Externally related grids as materials for reinforcement and seismic modernization of masonry panels. Constr. Build. Suppress. **25**, 504–514 (2011)
60. Valluzzi, M., da Porto, F., Garbin, E., Panizza, M.: Out-of-plane behaviour of infill masonry panels reinforced with composite materials. Suppress. Struct. **47**, 2131–2145 (2014)
61. Bellini, A., Incerti, A., Bovo, M., Mazzotti, C.: Effectiveness of FRCM reinforcement applied to masonry walls subjected to axial force and out-of-plane loads evaluated by experimental and numerical studies. Int. J. Archit. Inherit. **12**, 376–394 (2018)
62. Bednarz, Ł, Gorski, A., Jasieńko, J., Rusiński, E.: Simulations and analyses of arched brick structures. Autom. Constr. **20**, 741–754 (2011)
63. Garmendia, L., Larrinaga, P., García, D., Marcos, I.: Textile reinforced mortar as reinforcement material for masonry arches. Int. J. Archit. Inherit. **8**, 627–648 (2014)
64. Hojdys, Ł, Krajewski, P.: Laboratory tests on masonry vaults with extrados-reinforced backfill. Eng. Mater Key **624**, 510–517 (2014)
65. Giamundo, V., Maddaloni, G., Balsamo, A., Prota, A., Manfredi, G.: Experimental investigation of the seismic performance of IMG reinforcements on curved masonry elements. Compos. Part B Eng. **70**, 53–63 (2015). Lignola, general practitioner;
66. Hojdys, Ł., Krajewski, P.: Fiberglass grids incorporated into a cement-based matrix as reinforcement of masonry structures. In: Balen, K.V., (ed.) Structural Analysis of Historical Constructions: Anamnesis, Diagnosis, Therapy, Controls, Proceedings of the 10th International Conference on Structural Analysis of Historical Constructions (SAHC) Leuven, Belgium, 13–15 September 2016; CRC Press/Balkema: Leiden, Netherlands, 2016, pp. 372–376 (2016)
67. Alecci, V., et al.: Experimental survey of masonry vaults reinforced with PBO-FRCM composite. Compos. Part B Eng. **100**, 228–239 (2016)
68. De Santis, S., Roscini, F., de Felice, G.: Large-scale tests on masonry vaults reinforced with steel-reinforced grout. Compos. Part B Eng. **141**, 20–36 (2018)
69. Triantafillou, T.: Reinforcement of existing concrete structures: concepts and structural behaviour. In: Textile Fibre Composites in Civil Engineering. Woodhead Publishing, 2016, pp. 303–322 (2016)
70. Trochoutsou, N., Di Benedetti, M., Pilakoutas, K., Guadagnini, M., Bond of flax textile-reinforced mortars to masonry. Constr. Build. Mater. **284**, 122849 (2021). ISSN 0950-0618.https://doi.org/10.1016/j.conbuildmat.2021.122849

71. Caggegi, C., Lanoye, E., Djama, K., Bassil, A., Gabor, A.: Tensile behaviour of a basaltic MRT reinforcement system: influence of mortar and textile reinforcement ratios. Compos. Part B Eng. **130**, 90–102 (2017). https://doi.org/10.1016/j.compositesb.2017.07.027
72. D'Antino, T., Carloni, C., Sneed, L.H., Pellegrino, C.: Matrix-fibre bond behaviour in PBO FRCM composites: a fracture mechanics approach. Eng. Fract. Mech. **117**, 94–111 (2014). https://doi.org/10.1016/j.engfracmech.2014.01.011
73. Ombres, L., Mazzuca, S.: Confined concrete elements with cement-based composites: confinement effectiveness and prediction models. J. Compos. Constr. **21**(3), 04016103 (2016). https://doi.org/10.1061/(ASCE)CC.1943-5614.0000755
74. D'Antino, T., Sneed, L.H., Carloni, C., Pellegrino, C.: Influence of the substrate characteristics on the bond behaviour of PBO FRCM-concrete joints. Constr. Build. Mater. **101**, 838–850 (2015). https://doi.org/10.1016/j.conbuildmat.2015.10.045
75. Askouni, P.D., Papanicolaou, C.G.: Experimental investigation of bond between glass textile reinforced mortar overlays and masonry: the effect of bond length. Mater. Struct. **50**(2), 1–15 (2017). https://doi.org/10.1617/s11527-017-1033-7
76. Askouni, P.D., Papanicolaou, C.G.: Textile-reinforced mortar-masonry bonding: experimental study of critical bonding parameters. Constr. Construction. Mater. **207**, 535–547 (2019). https://doi.org/10.1016/j.conbuildmat.2019.02.102
77. Mercedes, L., Bernat-Maso, E., Martínez, B.: Bending behaviour of sandwich panels of vegetal fabric reinforced cementitious matrix: experimental test and numerical simulation. Constr. Build. Mater. **340**, 127820 (2022)
78. Khaleel, S., Madhavi, K., Basutkar, S.M.: Mechanical characteristics of brick masonry using natural fiber composites. Mater. Today Proc. **46**, 4817–4824 (2021)
79. Ismail, N., Ingham, J.M.: In-plane and out-of-plane testing of unreinforced masonry walls reinforced using polymer textile reinforced mortar. Eng. Struct. **118**, 167–177 (2016)
80. Shams, A., Horstmann, M., Hegger, J. : Experimental investigations on Textile-Reinforced Concrete (TRC) sandwich sections. Compos. Struct. **118**, 643–653 (2014)
81. Portal, N.W., Flansbjer, M., Zandi, K., Wlasak, L., Malaga, K.: Bending behaviour of novel Textile Reinforced Concrete-foamed concrete (TRC-FC) sandwich elements. Compos. Struct. **177**, 104–118 (2017)
82. Kouris, L.A.S., Triantafillou, T.C.: État de l'art sur le renforcement des structures de maçonnerie avec du mortier armé textile (TRM). Construction et matériaux de construction **188**, 1221–1233 (2018)
83. Hollaway, L.C., Cadei, J.: Progress in the technique of upgrading metallic structures with advanced polymer composites. Progress Struct. Eng. Mater. **4**(2), 131 à 148 (2002)
84. Valluzzi, M.R., Tinazzi, D., Modena, C.: Shear behavior of masonry panels strengthened by FRP laminates. Constr. Build. Mater. **16**(7), 409–416 (2002)
85. Marcari, G., Manfredi, G., Prota, A., Pecce, M., et al.: In-plane shear performance of masonry panels strengthened with FRP. Compos. Part B Eng. **38**(7–8), 887–901 (2007)
86. Prota, A., Manfredi, G., Nardone, F. : Assessment of design formulas for in-plane FRP strengthening of masonry walls. J. Compos. Constr. **12**(6), 643–649 (2008)
87. Hollaway, L.C.: A review of the present and future utilisation of FRP composites in the civil infrastructure with reference to their important in-service properties. Constr. Build. Mater. **24**(12), 2419–2445 (2010)

Bio-susceptibility of Building Materials

Self-healing Concrete with Fungi: An Exploration on Nutritional Sources to Sustain Fungal Growth in a Cementitious Environment

Aurélie Van Wylick[1,2,3(✉)], Eveline Peeters[1], Hubert Rahier[2], and Lars De Laet[3]

[1] Research Group of Microbiology, Department of Bioengineering Sciences, Vrije Universiteit Brussel, Pleinlaan 2, B-1050 Brussels, Belgium
aurelie.van.wylick@vub.be

[2] Research Group of Physical Chemistry and Polymer Science, Department of Materials and Chemistry, Vrije Universiteit Brussel, Pleinlaan 2, B-1050 Brussels, Belgium

[3] Research Group of Architectural Engineering, Department of Architectural Engineering, Vrije Universiteit Brussel, Pleinlaan 2, B-1050 Brussels, Belgium

Abstract. Concrete plays a central role as building material in the construction sector, however it is plagued by crack formation which leads to reinforcement corrosion and material degradation. To address these durability issues in a sustainable way, fungi have been introduced as self-healing agents to seal concrete cracks by promoting the precipitation of $CaCO_3$ on their cell walls. In order to enable fungi to perform self-healing in concrete, nutritional sources need to be present that sustain the growth of the organism. This research therefore investigates potential nutritional sources to sustain the fungi *Trichoderma reesei* and *Neurospora crassa* in a cementitious environment. Samples with cotton, hay, hemp (lignocellulosic natural fibres), corn and oat (starch-based cereal grains) were prepared. Different grain sizes were considered by mixing the fibres and cereals. The influence of gelatinization of the starch-based cereals on the fungal growth was assessed. Additionally, the inoculation method was investigated by comparing a solid and liquid inoculation. Finally, a selection of fibres and cereals was combined with cement paste in three different ways: laid on cured cement paste, laid on fresh cement paste and mixed with cement paste. Results show that *Trichoderma reesei* thrives on cotton and (mixed) hay. *Neurospora crassa* thrives on starch but grows equally well on cotton and to a lesser extent on the other natural fibres. When combined with cement paste, fungal growth is mostly observed on corn, oat and cotton, however more experiments need to be conducted.

Keywords: Filamentous fungi · Lignocellulose and starch · Self-healing concrete

1 Introduction

Although concrete has a central role in the built environment and construction sector, used in nearly all construction cases, it is plagued by the formation of cracks. These cracks are the result of the material's shrinkage during hardening, limited tensile strength and

S. Amziane et al. (Eds.): ICBBM 2023, RILEM Bookseries 45, pp. 629–639, 2023.
https://doi.org/10.1007/978-3-031-33465-8_48

brittle behaviour. Over time, the durability of the material is severely threatened. Water, oxygen and CO_2 find their way through the cracks, causing reinforcement corrosion and further material degradation [1]. Self-healing concrete is therefore of great interest. Though concrete is able to heal itself autogenously, the crack width that can be healed is limited to 0.2 mm [2, 3]. Research on autonomous self-healing techniques have therefore emerged, such as microbial induced calcite precipitation (MICP) [4–8]. Calcite, or calcium carbonate ($CaCO_3$) is known to be a highly compatible self-healing agent as it is autogenously present in concrete [9]. Microbial precipitation of $CaCO_3$ can be promoted by certain bacterial and fungal species on their cell walls [3, 6, 9–15]. With bacteria, cracks with a width of up to 1 mm can already be healed [16], while research on fungi has only recently emerged [3, 10, 17]. With their extensive three-dimensional mycelial growth, larger sized cracks could potentially be targeted. Concrete however presents a very harsh environment for microorganisms to grow due to its alkaline conditions and lack of nutrients. This research therefore investigates nutritional sources to sustain fungal growth in a cementitious environment. Based on the research on mycelium composites [18, 19], the concept of using natural fibres as carriers for the fungal spores emerged. Mycelium composites are grown by inoculating a lignocellulosic substrate with a fungal species. The fungal hyphae grow in a mycelial network by degrading the substrate and bind the material together. The incorporation of natural fibres in the concrete mix is already a familiar approach to assist the formation of multiple small cracks instead of large cracks, thus promoting the autogenous self-healing efficiency [20, 21]. Both the fungi *Trichoderma reesei* and *Neurospora crassa* are considered in this research, both fungi have been described in literature as potential self-healing agents [3, 22]. *T. reesei* and *N. crassa* belong to the phylum Ascomycota and are soft rot fungi [23, 24]. In mycelium composites, the lignocellulosic substrate is effectively degraded by white rot fungi [18], which belong to the phylum Basidiomycota [25]. The phyla differ both in morphological and phylogenetic features [26, 27]. While white rots possess higher lignin abilities, and are thus highly suitable for lignin degradation, soft rots are better equipped for the degradation of cellulose, hemicellulose and pectin [23]. The fungus *T. reesei* is well-known for its high production levels of cellulases used in biotechnological processes such as in the textile and food industry [28, 29]. Cellulase is an enzyme that is capable of degrading cellulose into glucose [30], which is a carbon source for the fungi. Also *N. crassa* is known to have the ability to degrade lignocellulosic materials [31]. Next to lignocellulosic feedstock, this research also considers starch-based nutrients as a second carbon source. The aim is to define suitable carbon sources as potential nutrients to sustain fungal growth in a cementitious environment. This study therefore explores several natural fibres and cereal grains, two inoculation techniques, two preparation methods and different grain sizes. Additionally, the substrates are also combined with cement to assess the fungal growth in an environment relevant to concrete. This study is the first fundamental step in the research, therefore all experiments are conducted in sterile conditions and in a controlled environment to avoid contaminations. The translation of the results to more natural and realistic conditions will take place in a later phase of the research. The aim of this study is to make a selection of suitable nutritional sources for the implementation of fungi in the concrete mix. To protect the fungal spores from the mixing and hardening of concrete, as well as to avoid adverse effects on concrete's mechanical properties, encapsulation of the spores is essential. The end goal is thus to encapsulate the fungal spores together with nutrients and possible additives to ensure the spores' survivability, as research has demonstrated for bacteria [6]. Results are evaluated in a qualitative way.

2 Materials and Methods

2.1 Fungal Strains

The fungi *Trichoderma reesei* (DSM 768, Leibniz Institute DSMZ-German Collection of Microorganisms and Cell Cultures, ATCC13631) and *Neurospora crassa* (WT FGSC # 2489, Fungal Genetics Stock Centre (FGSC), Kansas, U.S.A.) were maintained on malt extract agar (MEA) in 90 mm diameter Petri dishes at 30 °C with 12 h light/dark cycles.

2.2 Substrate Preparation

For the standard preparation of the substrate, fibres, cereals and distilled water were autoclaved at 121 °C for 15 min. Once cooled down, the nutrients were humidified with the water in sterile conditions and placed in 54 mm Petri dishes. The amount of water added was dependent on the fibre or cereal as the water absorption was different for each nutritional source (Table 1). Too wet or too dry conditions would adversely affect the fungal growth, therefore water was added gradually to reach a humid state.

For the nutritional sources rich in starch, an additional preparation method was considered to assess the influence of gelatinization on the fungal growth. Starch is a polysaccharide composed of glucose units joined by glycosidic bonds and its properties vary with heat [32, 33]. When water is added to starch and the mixture is heated, the intermolecular bonds between the starch molecules breakdown, and the hydrogen bonding sites have the ability to engage more water molecules [32, 33]. This process, in which the starch becomes gelatinous, is called starch gelatinization. The starch granules become dissolved in water irreversibly, the alignment of the molecules becomes amorphous, and the starch acts as a plasticizer [32, 33]. The cooldown period after autoclaving was therefore omitted, sterile water was added to the starch-based nutrients directly after autoclaving. The preparation of the samples was performed with heat-resistant gloves. The gelatinized substrate was placed as well in 54 mm Petri dishes in sterile conditions.

The nutrients were considered in their original grain size and smaller grain size(s). The mixed nutrients were first sieved through a 4.5 mm sieve (type M1) followed by a 0.85 mm sieve (type M2). Mixing was done prior to autoclaving.

An overview of the used nutrients and ratio of water added is given in Table 1.

2.3 Inoculation Method

All media were inoculated with 7 days-old plates. Both solid and liquid inoculation techniques were explored. For the solid inoculation, cylindrical disks were made with a 200 μL pipette tip in the 7 days-old cultures to inoculate the media. A liquid inoculum was prepared by adding sterile water to the cultures and scraping of the spores with a 1 mL pipette. Samples were then inoculated with a 0.5 mL spore solution. All samples were grown in an incubator at 30 °C with 12 h light/dark cycles for one week.

2.4 Cement

Nutrients were also combined with cement in three different ways: laid on cured cement paste, laid on fresh cement paste and mixed with cement paste. Cement (CEM III/B 42,5 N-LH/SR LA) was obtained from Holcim [34], the cement paste was prepared with w/c factor 0.4. The samples were made in 90 mm Petri dishes. For the cured and fresh cement samples, the nutrient/water ratios from Table 1 were used. When the nutrients were mixed with cement, more water was added depending on the fibre/cereal type. Cured cement samples were 7 days old, fresh cement samples were made at the moment of inoculation. Only the standard preparation method was considered.

Table 1. Overview of the nutrient types and the ratio of sterile water added. The abbreviations M1 and M2 refer to a grain size between 4.5 mm and 0.85 mm (M1), and smaller than 0.85 mm (M2). *Oat refers to sprouted oat flakes.

Type	Preparation method	Nutrient source	Nutrient/water ratio
Cellulose	Standard	Cotton	2/3
		Hay	3/4
		Hay M1	3/4
		Hemp	1/2
		Hemp M1	1/2
Starch	Standard	Oat*	2
		Oat M1	2
		Oat M2	2
		Corn M1	3
		Corn M2	3
	Gelatinization	G - Oat	1
		G - Oat M1	2
		G - Oat M2	1
		G - Corn M1	3/2
		G - Corn M2	3/2

3 Results

Preliminary experiments already excluded beechwood, sisal and potato starch as both fungi grew significantly less on these nutrients compared to the others.

Only qualitative assessments were made. The use of cellophane was considered, yet both *T. reesei* and *N. crassa* grew through the foil after respectively 5 and 3 days, making it difficult to weigh the grown biomass correctly. The development of a method to quantify the results is currently in progress.

Trichoderma reesei. The samples with *T. reesei* are given in Fig. 1. The abbreviations 'S', 'L' and 'G' refer respectively to solid inoculation, liquid inoculation and gelatinization. *T. reesei* grew notably better on a lignocellulosic substrate compared to the starch-based nutrients, with an abundant growth on cotton and hay. This can be explained by the fungus' cellulase activity [28, 29]. Cotton and hay contain respectively 82–96% and 31.5% of cellulose [35–37]. Little growth was noticed on hemp, which has however a cellulose content of 70–92% [35]. An overview of the fibres' chemical composition is given in Table 2. The lack of growth on hemp could be related to the fibre type of the nutrient, hemp fibres are extracted from the plant stem and are known as 'bast' fibres bundles. Cotton and hay are respectively 'seed' fibres and 'grass' fibres. [35] Being a soft rot fungus, *T. reesei* generally prefers low lignin contents. The hemp fibres are therefore too hard to degrade.

Table 2. Chemical composition of the natural fibres cotton, hay and hemp.

Nutrient	Fibre type	Cellulose [%]	Hemicellulose [%]	Lignin [%]
Cotton [35]	Seed	82–96	2–6.4	0–5
Hay [36, 37]	Grass	30–35	10–30	4–5
Hemp [35, 38]	Bast	70–92	18–22	3–21

When comparing the solid and liquid inoculation, no significant difference can be seen for the lignocellulosic fibres. For the starch-based nutrients, dark spots can be observed on the samples with liquid inoculation. These could indicate that the samples were contaminated. Liquid inoculation was therefore discarded.

Gelatinized substrates were all inoculated with a mycelial disk. Compared to the standard preparation, better growth was obtained for the samples containing corn.

To draw conclusions on the fibre/grain size, each nutrient is considered separately. For corn, M1 proved to be a better grain size than M2 in all three conditions (S, L and G). For oat, it is difficult to make conclusions as almost no growth was observed. Cotton was not mixed due to its texture. The growth on hay M1 seemed to be denser compared to normal hay. For hemp, only very limited growth was observed on both types. Mixing the nutrients could thus improve the fungal growth, however a limit should be placed on the minimal fibre/grain size as type M1 led to better results than type M2.

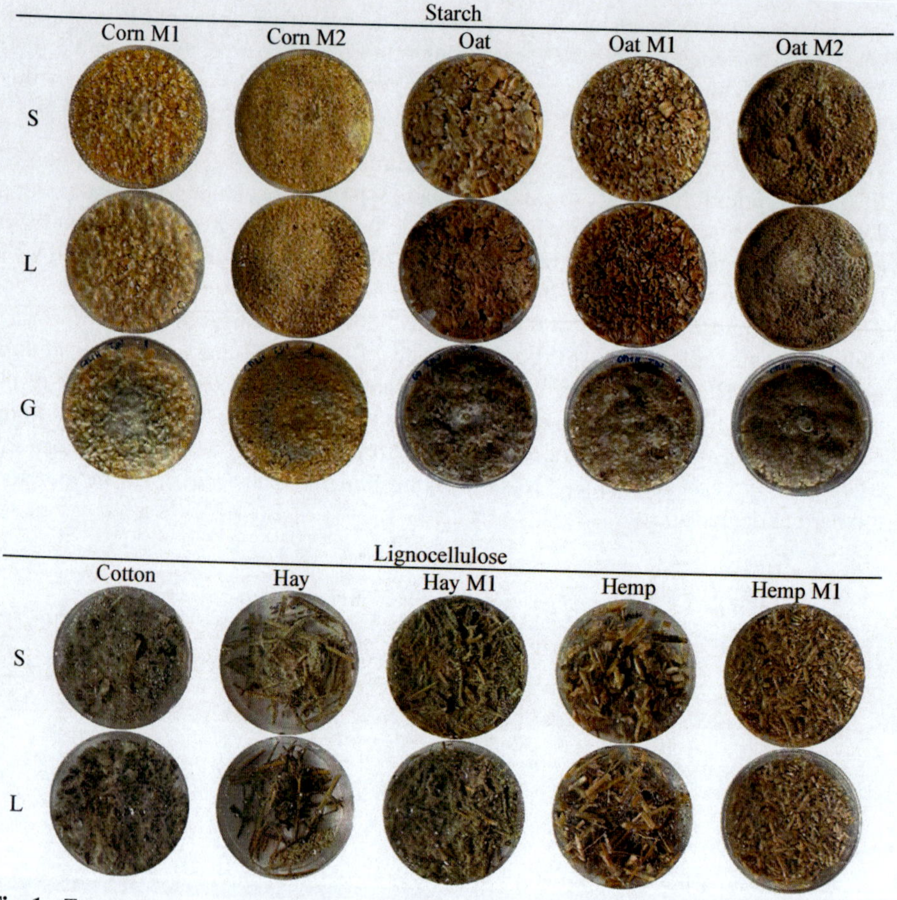

Fig. 1. *T. reesei* grown on starch-based and lignocellulosic nutrients. Pictures were taken after 7 days of growth.

Neurospora crassa. The samples with *N. crassa* can be seen in Fig. 2. Liquid inoculation resulted in a darkening of the starch-based samples, as noticed for *T. reesei*. This caused a notable reduced growth when comparing to the solid inoculation. For the lignocellulosic substrates however, no negative influence on the growth was witnessed. Growth on hay and hemp fibres was even improved with the liquid inoculation. Nevertheless, due to the darkening of the samples with starch, the liquid inoculation technique was discarded as well.

Growth on gelatinized samples was comparable to the standard preparation with a solid inoculum, although corn M2 showed a possible contamination.

A smaller grain size for the starch-based nutrients seemed to be detrimental for oat (oat M2) and to a lesser extend for corn (corn M2 – L and G).

Fig. 2. *N. crassa* grown on starch-based and lignocellulosic nutrients. Pictures were taken after 7 days of growth.

To assess the fungal growth on fibres and grains in a cementitious environment, nutrients were combined with cement paste in three different ways using the standard preparation method: laid on cured cement paste, laid on fresh cement paste and mixed with cement paste. It must be noted that the results were very scattered: biological replicates often showed different growth patterns (varying from no growth to abundant growth). All fibres and grains were explored. Samples with cotton, mixed corn (M1) and oat are shown in Fig. 3. As seen for the samples without cement, *T. reesei* thrived on cotton. Growth was witnessed in every sample when using cured cement (Fig. 3a). However, when cotton was laid on fresh cement and mixed with cement, results were conflicting and scattered. Although corn M1 showed little to no growth for *T. reesei* in the first experiments, growth was witnessed when combined with cement (Fig. 3 d-e). *N. crassa* seemed to be more tolerant to the cementitious conditions. It was able to grow on the cement layer itself, as can be seen for corn M1 (Fig. 3 f-g), and even abundantly when mixed with cement (Fig. 3 h). Inconsistencies however occurred as well for *N. crassa,* as for example in Fig. 3 (i-j), these samples were biological replicates.

Additional experiments should thus be conducted to draw conclusions. Moreover, gelatinized corn and oat should be tested as well.

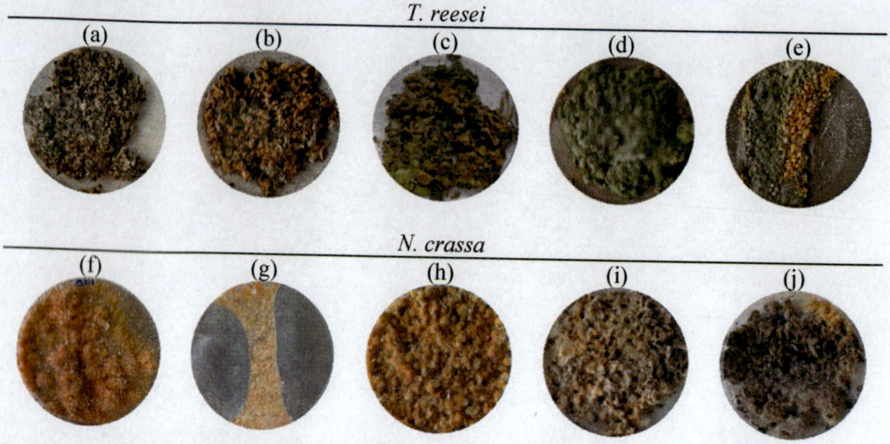

Fig. 3. Top row: *T. reesei* grown on (a) cotton laid on cured cement, (b) cotton laid on fresh cement – no growth, (c) cotton mixed with cement paste, (d-e) corn M1 laid on and placed in between fresh cement paste. Bottom row: *N. crassa* grown on (f-g) corn M1 in between fresh cement paste, image (g) shows the back of the sample, (e) oat mixed with cement, (i-j) cotton mixed with cement.

4 Conclusion

The fungi *Trichoderma reesei* and *Neurospora crassa* were able to grow on a ligno-cellulosic and starch-based substrate. Results show that *T. reese*i thrives on cotton and (mixed) hay. To a lesser extent, the fungus also grows on mixed corn, gelatinization also improves the growth. *N. crassa* thrives on starch but grows equally well on cotton. Mixing the fibres positively influences the growth, while liquid inoculation tends to darken the substrate causing little to no growth. When combined with cement paste, fungal growth is mostly observed on mixed corn, oat and cotton, however more experiments need to be conducted due to scattered and inconsistent results. Additionally, gelatinized substrates should be considered as well. Next steps should also include the promotion of $CaCO_3$ precipitation on the fungal hyphae. Nevertheless, this research opens the field towards the use of natural fibres and cereal grains as a nutritional source and carrier for fungal spores in a cementitious environment.

cknowledgements. This work was supported by the Vrije Universiteit Brussel and funded by ie Research Foundation Flanders (FWO-Vlaanderen). [FWO-SB PhD fellowships to Aurélie an Wylick (Grant Number 1SA9721N)]. The authors would like to thank master thesis students nber Dehaen and Simon Vandersanden for their dedicated contributions to this study, Professor

David Cannella for providing a laboratory work environment at the Université Libre de Bruxelles and the company Holcim for the supply of cement.

References

1. Ahmad, S.: Reinforcement corrosion in concrete structures, its monitoring and service life prediction—a review. Cement Concr. Compos. **25**(4), 459–471 (2003)
2. Tittelboom, K., Snoeck, D., Wang, J., Belie, N.: Most recent advances in the field of self-healing cementitious materials (2013)
3. Luo, J., Chen, X., Crump, J., Zhou, H., Davies, D., Zhou, G., et al.: Interactions of fungi with concrete: significant importance for bio-based self-healing concrete. Constr. Build. Mater. **164**, 275–285 (2018)
4. Castro-Alonso, M.J., et al.: Microbially induced calcium carbonate precipitation (MICP) and its potential in bioconcrete: microbiological and molecular concepts. Front. Mater. **6**, 126 (2019)
5. De Muynck, W., De Belie, N., Verstraete, W.: Microbial carbonate precipitation in construction materials: a review. Ecol. Eng. **36**(2), 118–136 (2010)
6. Fahimizadeh, M., Diane Abeyratne, A., Mae, L.S., Singh, R.K.R., Pasbakhsh, P.: Biological self-healing of cement paste and mortar by non-ureolytic bacteria encapsulated in alginate hydrogel capsules. Materials. **13**(17), 3711 (2020)
7. Wiktor, V., Jonkers, H.M.: Bacteria-based concrete: from concept to market. Smart Mater. Struct. **25**(8), 084006 (2016)
8. Wu, M., Johannesson, B., Geiker, M.: A review: self-healing in cementitious materials and engineered cementitious composite as a self-healing material. Constr. Build. Mater. **28**(1), 571–583 (2012)
9. Seifan, M., Samani, A.K., Berenjian, A.: Bioconcrete: next generation of self-healing concrete. Appl. Microbiol. Biotechnol. **100**(6), 2591–2602 (2016). https://doi.org/10.1007/s00253-016-7316-z
10. Menon, R., Luo, J., Chen, X., Zhou, H., Liu, Z., Zhou, G., et al.: Screening of fungi for potential application of self-healing concrete. Sci. Rep. **9**, 2075 (2019)
11. Sharma, T.K., Alazhari, M., Heath, A., Paine, K., Cooper, R.M.: Alkaliphilic Bacillus species show potential application in concrete crack repair by virtue of rapid spore production and germination then extracellular calcite formation. J. Appl. Microbiol. **122**(5), 1233–1244 (2017)
12. Tziviloglou, E., Wiktor, V., Jonkers, H.M., Schlangen, E.: Bacteria-based self-healing concrete to increase liquid tightness of cracks. Constr. Build. Mater. **30**(122), 118–125 (2016)
13. Zhang, W., Zheng, Q., Ashour, A., Han, B.: Self-healing cement concrete composites for resilient infrastructures: a review. Compos. B Eng. **15**, 189 (2020)
14. Jonkers, H.M., Thijssen, A., Muyzer, G., Copuroglu, O., Schlangen, E.: Application of bacteria as self-healing agent for the development of sustainable concrete. Ecol. Eng. **36**(2), 230–235 (2010)
15. Van Wylick, A., Monclaro, A.V., Elsacker, E., Vandelook, S., Rahier, H., De Laet, L., et al.: A review on the potential of filamentous fungi for microbial self-healing of concrete. Fungal Biol. Biotechnol. **8**(1), 16 (2021)
16. Basilisk. Basilisk Self-Healing Concrete [Internet]. Basilisk. [cited 2020 Nov 18]. https://www.basiliskconcrete.com/en/
17. Zhang, X., Fan, X., Li, M., Samia, A., Yu, X.B.: Study on the behaviors of fungi-concrete surface interactions and theoretical assessment of its potentials for durable concrete with fungal-mediated self-healing. J. Clean. Prod. **292**, 125870 (2021)

18. Elsacker, E., Vandelook, S., Van Wylick, A., Ruytinx, J., De Laet, L., Peeters, E.: A comprehensive framework for the production of mycelium-based lignocellulosic composites. Sci Total Environ. **10**(725), 138431 (2020)
19. Elsacker, E., Vandelook, S., Brancart, J., Peeters, E., Laet, L.D.: Mechanical, physical and chemical characterisation of mycelium-based composites with different types of lignocellulosic substrates. PLoS ONE [Internet] **14**(7) (2019)
20. Snoeck, D., De Belie, N.: From straw in bricks to modern use of microfibers in cementitious composites for improved autogenous healing – a review. Constr. Build. Mater. **1**(95), 774–787 (2015)
21. Snoeck, D., Tittelboom, K., Steuperaert, S., Dubruel, P., Belie, N.: Self-healing cementitious materials by the combination of microfibres and superabsorbent polymers. J. Intell. Mater. Syst. Struct. **1**(25), 13–24 (2014)
22. Zhao, J., Dyer, T., Csetenyi, L., Jones, R., Gadd, G.M.: Fungal colonization and biomineralization for bioprotection of concrete. J. Clean. Prod. **1**(330), 129793 (2022)
23. Sista Kameshwar, A.K., Qin, W.: Comparative study of genome-wide plant biomass-degrading CAZymes in white rot, brown rot and soft rot fungi. Mycology **9**(2), 93–105 (2018)
24. Zabel, R., Morrell, J.: Chemical changes in wood caused by decay fungi, pp. 215–44 (2020)
25. Tuomela, M., Hatakka, A.: Oxidative fungal enzymes for bioremediation. In: Moo-Young, M. (ed.) Comprehensive Biotechnology (Second Edition) [Internet]. Burlington: Academic Press; 2011 [cited 2022 Dec 20], pp. 183–96 (2011). https://www.sciencedirect.com/science/article/pii/B9780080885049003706
26. Liu, X., Zhang, Q., Xia, X., Liu, X., Ge, L., Yang, L.: Phylogenetic relationships of ascomycetes and basidiomycetes based on comparative genomics analysis. Genes Genom. **39**(12), 1307–1316 (2017). https://doi.org/10.1007/s13258-017-0595-7
27. Hibbett, D.S., Blackwell, M., James, T.Y., Spatafora, J.W., Taylor, J.W., Vilgalys, R.: Phylogenetic taxon definitions for Fungi, Dikarya, Ascomycota and Basidiomycota. IMA Fungus. **9**(2), 291–298 (2018)
28. Zhao, C.H., Liu, X., Zhan, T., He, J.: Production of cellulase by Trichoderma reesei from pretreated straw and furfural residues. RSC Adv. **8**(63), 36233–36238 (2018)
29. Znameroski, E.A., Glass, N.L.: Using a model filamentous fungus to unravel mechanisms of lignocellulose deconstruction. Biotechnol. Biofuels **6**(1), 6 (2013)
30. Najjarzadeh, N., Matsakas, L., Rova, U., Christakopoulos, P.: How Carbon Source and Degree of Oligosaccharide Polymerization Affect Production of Cellulase-Degrading Enzymes by Fusarium oxysporum f. sp. lycopersici. Frontiers in Microbiology [Internet] (2021). [cited 2022 Dec 20];12. https://www.frontiersin.org/articles/10.3389/fmicb.2021.652655
31. Wang, B., Cai, P., Sun, W., Li, J., Tian, C., Ma, Y.: A transcriptomic analysis of Neurospora crassa using five major crop residues and the novel role of the sporulation regulator rca-1 in lignocellulase production. Biotechnol. Biofuels **8**(1), 21 (2015)
32. Madhu. Difference between gelatinization and retrogradation [Internet]. Difference between gelatinization and retrogradation. 2020 [cited 2022 Dec 8]. https://www.differencebetween.com/difference-between-gelatinization-and-retrogradation/
33. Eliasson, A.C.: Gelatinization and retrogradation of starch in foods and its implications for food quality. In: Skibsted, L.H., Risbo, J., Andersen, M.L., (eds.) Chemical Deterioration and Physical Instability of Food and Beverages [Internet]. Woodhead Publishing; 2010 [cited 2022 Dec 8], pp. 296–323. (Woodhead Publishing Series in Food Science, Technology and Nutrition) (2010). https://www.sciencedirect.com/science/article/pii/B9781845694951500101
34. Holcim Belgium [Internet]. Holcim Belgium - Ciment, béton prêt à l'emploi, granulats. [cited 2022 Nov 10]. https://www.holcim.be/nl/home

35. Ansell, M.P., Mwaikambo, L.: The structure of cotton and other plant fibers. Handb. Text. Fibre Struct. **31**(2), 62–94 (2009)
36. Kraszkiewicz, A., Kachel, M., Lorencowicz, E., Przywara, A.: Influence of cellulose content in plant biomass on selected qualitative traits of pellets. Agric. Agric. Sci. Procedia **31**(7), 125–130 (2015)
37. Wedig, C.L., Jaster, E.H., Moore, K.J.: Composition and digestibility of alfalfa and orchard-grass hemicellulose monosaccharides by holstein steers. J. Dairy Sci. **69**(5), 1309–1316 (1986)
38. Liu, M., Thygesen, A., Summerscales, J., Meyer, A.S.: Targeted pre-treatment of hemp bast fibres for optimal performance in biocomposite materials: a review. Ind. Crops Prod. **1**(108), 660–683 (2017)

Evaluating the Humidity Responsiveness of Bacterial Cellulose for Application in Responsive, Breathable Building Skins

Natalia Pynirtzi[2](✉), Kumar Biswajit Debnath[1], Giannis Lantzanakis[3], Karolina Bloch[1], Jane Scott[1], Colin Davie[2], and Ben Bridgens[1]

[1] Hub for Biotechnology in the Built Environment, School of Architecture Planning and Landscape, Newcastle University, Newcastle Upon Tyne NE1 7RU, UK
[2] School of Engineering, Newcastle University, Newcastle Upon Tyne NE1 7RU, UK
n.pynirtzi2@ncl.ac.uk
[3] Foundation for Research and Technology, 70013 Heraklion, Greece

Abstract. Breathable building skins have potential to reduce the operational energy requirements of buildings, which currently accounts for 28% of global greenhouse gas emissions. Moisture-responsive bio-based materials, such as wood and nanocellulosic fibres, can provide mechanical actuation due to their reversible shape change, through continuous wetting and drying process enabling the building envelope to passively respond to the environment while avoiding the need for mechanical ventilation. This study was conducted to evaluate the moisture responsiveness of bacterial cellulose materials that would lead to their potential use in the built environment. Thin sheets of bacterial cellulose were grown via the 'kombucha' process and six square samples were exposed to high and low relative humidity in a climate chamber at a constant temperature of 25 °C. . The results from eight, 4-day long, consecutive wetting-drying cycles (10% to 90% to 10% relative humidity), showed that bacterial cellulose presents significant, rapid and reversible dimensional changes. Changing from 10% to 90% relative humidity resulted in a thickness change around 15% and a weight change of 10% within one hour. The maximum changes detected were 97% in weight and 116% in thickness after 48 h at 90% relative humidity. After 24 days of cyclic testing, the cyclic dimensional changes were maintained indicating no reduction in moisture responsiveness. The outputs of this study will inform the development and integration of bacterial cellulose within breathable wall constructions, in combination with insulation, thermal mass and cladding, to provide a passive, responsive, variable porosity building skin.

Keywords: Bacterial Cellulose · Moisture-Responsiveness · Bio-based Materials · Cyclic Behavior

S. Amziane et al. (Eds.): ICBBM 2023, RILEM Bookseries 45, pp. 640–652, 2023.
https://doi.org/10.1007/978-3-031-33465-8_49

1 Introduction

The building and construction industries use a vast amount of fossil fuels, which contributes to atmospheric carbon emissions and account for about 40% of the annual global CO_2 emissions [1]. 11% are embodied emissions and 28% are operational carbon emissions, which are related to heating, cooling and ventilation. For climate change mitigation the focus over the past few years has been set on the exploration of pathways to achieve 'net-zero' CO_2 emissions in every sector and in the case of the built environment this involves both reduction of embodied emissions, which can be achieved by the use of bio-based materials [2], and reduction of operational energy by improving insulation and air-tightness, controlling solar gain and use of heat recovery ventilation.

Adaptive/responsive facades show great potential for reducing building energy use [3]. Current adaptive/responsive facades use materials with high embodied energy (e.g., fiberglass, bio-adaptive concrete [4]), and have complex control and actuation systems, which are expensive, difficult to retrofit, require energy to operate, need maintenance and can lack reliability. Use of the inherent responsiveness of many biobased materials to moisture, addresses many of these concerns [5]. Recent research on adaptive building skins has shown that the inclusion of moisture-responsive porous materials with increased moisture buffering capacity and air-permeability properties allow vapour mobility and enhance wall performance, improving the energy efficiency of a building [6]. Response driven bio-based materials that have the ability to shrink and expand in response to changes in moisture, such as hydrophilic hydrogels and nano-cellulosic water sensors reported in literature, are well-known for their use in other industries (biomedical [7], mechanical engineering [8], electronic applications [9]). The performance profile of such moisture-responsive materials is promising and they could possibly be scaled up to suit building applications.

The thermal and moisture responsive properties of hydrophilic nano-cellulosic foams [10], as well as nano-cellulose aerogels [11] have been studied in the past and have shown that the incorporation of moisture-responsive materials in construction, provides thermal insulation and/or can enable building skins to 'breathe' through a nano-porous network structure that controls the porosity with ambient humidity changes. Moreover, the transition towards an adaptive/active building system is not new, as the concept of living skins in architecture [3, 12] is well-known as a sustainable solution for energy performance optimization within a building. This means that the living skins concept can be explored further by incorporating moisture responsive biobased materials in breathable building skins, which will probably provide promising results.

Natural polysaccharides, such as bacterial cellulose (BC) [13], cellulose hydrogels [14], chitin-based hydrogels [15], organic hemp [16], natural textiles and wood [17], present the potential to be incorporated in a breathing wall, which will be defined by their moisture responsiveness and air permeability properties. Cellulose is the most abundant biomaterial in the world [18], as it is mainly sourced from plant cell walls [19] and can be extracted easily from wood pulp. Cellulosic films are known to be highly hygroscopic and they can withstand long cyclic periods of wetting and drying [20]. Bacterial cellulose (BC) belongs to the environmentally benign materials and is a cellulose derived material synthesized by bacteria [21] (e.g., *Acetobacter xylinum, Gluconobacter)* whose advantages for biological compatibility and hygroscopic mechanisms have been thoroughly investigated [22]. BC exhibits a hygroexpansion mechanism, as it shares the same functional attributes with most cellulosic materials [5], although, it presents higher crystallinity and purity [23] than plant cellulose. While its functionality as a biomedical, biosensor and packaging component has already been established [24], there has been little exploration of BC features for a wider range of applications, mainly due to its biodegradability [22]. So far, the few attempts to embody BC in the built environment are related to cementitious applications [25], where BC and other nanocellulose substances were infused to either create more sustainable cementitious mixtures or achieve good rheological properties while maintaining high strength and improving sustainability to the cement pastes [26].

This study aims to establish a background for the potential benefits of using BC produced by kombucha as a moisture-responsive natural material for the construction of breathable building skins. As mentioned earlier, cellulose can be produced by certain bacteria using different methods. The easiest and most affordable method to produce BC at large scale is via kombucha [27]. Kombucha is a fermentation tea beverage made of sugar and SCOBY (Symbiotic Culture of Bacteria and Yeast). BC pellicles are formed on the surface of kombucha as it ferments. Sugar is used as the food substance for the bacteria and yeast, whereas tea is a valuable ingredient that in combination with sugar help enhance the BC pellicle growth. The BC pellicles have a thick gelatinous membrane form, as they grow on the air-exposed surface of the liquid medium [24]. The choice of BC over other natural materials was due to its sustainability [28, 29], biodegradability [30], and high-water holding capacity. Furthermore, it has only been explored to a limited content in the built environment [31, 32]. BC from kombucha is relatively cheap and presents good response to moist habitats in terms of durability [33], which can be beneficial for adaptive environments. However, it is noted that, for large scale applications, the use of high quantities of sugar may present a challenge in terms of sustainability, as wastewater and emissions from sugar production have a negative impact on the environment [34]. Thus, a detailed LCA analysis is required in order to show the environmental impact of kombucha-derived BC compared to other cellulose based materials. Here, BC is being investigated because of its unique structure and properties (i.e., very large swelling and shrinkage with changes in humidity).

For this study, the water responsivity of BC will be determined by measuring the dimensional changes, considering swelling, shrinking and recovery rates under different climatic conditions to determine the specific environmental behaviour of this material. The results from this study will show whether the obtained moisture responsive properties of BC are useful for serving the purpose of incorporating it in a breathable building skin.

2 Materials and Methods

2.1 Materials

Organic green tea (*Clipper Teas, UK*) and organic apple cider vinegar (*Pure Source Nutrition, UK*) were used as the substrate for the Bacterial Cellulose fermentation process. Granulated sugar from sugar beet (*Silver Spoon, UK*) was used as the carbon-food source for the bacterial cellulose and SCOBY (Symbiotic Colony of Bacteria and Yeast) pellicles of 2 cm thickness and 15 cm in diameter (*Funny shaped scobies, Happy Kombucha, UK*) were used as the starter culture of the kombucha fermentation process.

2.2 Preparation of Media – Fermentation Process

Bacterial cellulose growth medium was made by the kombucha process. Several batches were made during a period of 2 months, where each batch contained 2 L of boiled distilled water, 160 gr of sugar, 100 ml of apple cider vinegar, 6 green tea bags and 1 SCOBY. A 5-L cylindrical plastic beaker was used to brew the tea. The 6 green tea bags and 2 L of boiled distilled water were placed into the beaker. The brewing process lasted approximately 15 min before removing the tea bags from the water. Sugar was added while the tea was still warm enough for the sugar to fully dissolve. The media were then left to cool at room conditions for around 3 h and afterwards the apple cider vinegar was added.

2.3 Cultivation of BC Pellicles

Once the media had cooled down to room temperature, it was poured into a 50 cm × 35 cm × 10 cm rectangular plastic box. The box and the lid of the box had been sterilized using ethanol and distilled water to prevent any contamination. A piece of SCOBY was added to the media in the box. The box was then covered with the lid and was placed on a shelf in the OME building at Newcastle University, at 28° and 70% relative humidity (Fig. 1). The box remained closed for about 3 weeks to achieve growth of about 3 cm thickness of the BC pellicles. All the batches were made using the same recipe and process, to produce replicable samples for testing. The growth of the BC pellicles during the fermentation process and the purification steps (harvesting, washing, sterilizing and drying) were monitored and controlled, which resulted in uniform bacterial cellulose samples. The reproducibility of the bacterial cellulose properties was verified by repeating the tests on samples from 3 different batches.

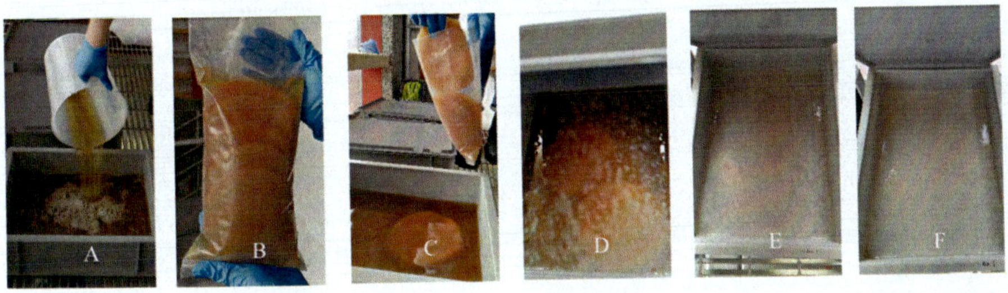

Fig. 1. Steps of the BC pellicle growth. A) Day 1: Kombucha media is poured into the box where BC will grow, B) Day 1: Each purchased bag of SCOBY contains 6 pieces of SCOBY, C) Day 1: 1 SCOBY is added to the media of each box, D) Day 4: Bacterial growth is visible on the surface of the media, E) Day 7: A thin sheet of BC is being formed, F) Day 9: The BC pellicle has already been formed and is growing.

2.4 Preparation of BC Thin Films

At a constant temperature of 28 °C the BC pellicles grow at a rate of approximately 1 cm per week. Thus, after around 3 weeks, sheets with a thickness of around 3 cm were recovered. The BC pellicles were harvested and then washed for 1 h in a bath of distilled water to remove any residual alkali impurities, such as sodium hydroxide or potassium hydroxide, which can result from the use of alkaline water or other ingredients in the fermentation process. The distilled water washing will also remove any other bacterial cells that may have been attached to the pellicles. The washing step was repeated several times until neutral pH of the pellicles was achieved. At this point, the pellicles were rinsed with hot tap water for 5 min and sprayed with ethanol. After harvested and washed, the mass of the 3 BC pellicles was determined by weighing the pellicles and it was found to be 97 g, 103 g and 112 g. To dry the surface of BC sheets, paper towels were used for 1–2 days (depending on their thickness) and subsequently the pellicles were placed on thin acrylic sheets. The BC pellicles stick to the acrylic sheet surface, which prevents them from bending while drying. The BC pellicles need to remain around two weeks at room temperature and relative humidity of 50% to dry. Afterwards, they can be removed from the acrylic sheet, resulting in a flat homogenous surface – thin film – (Fig. 2). After drying, the pellicles reduced in thickness to around 0.4 mm. The resulted BC films were cut in six square specimens of 7 cm × 7 cm and were placed in a climate chamber for cyclic humidity tests.

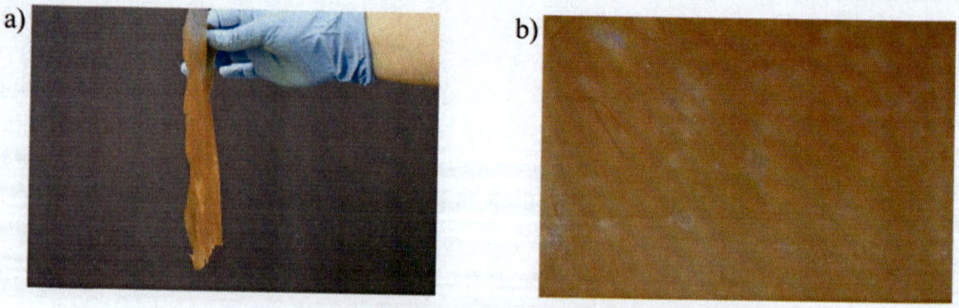

Fig. 2. a) Dry BC thin sheet, b) moist BC thin sheet.

2.5 Moisture Absorption Tests

At room conditions the BC samples had reached equilibrium but in order to make sure that the samples would maintain the same weight before placing them into the climate chamber for tests, all the specimens were left at room conditions for a month prior to testing and were weighed every day using a Muaket TL-series model, digital milligram balance scales with high precision (0.001g). Once equilibrium was achieved, the specimens were placed into a Memmert HPP IPP plus climate chamber for moisture absorption tests. This climate chamber has a relative humidity range of 10% to 90% and a temperature range between 5 °C and 70 °C. The time required to achieve the maximum temperature is half an hour whereas in order to reach the maximum relative humidity

10 min are required. Also, the RH is slightly changing when opening the door for some seconds to one minute. However, the scale of this change is negligible when the door is only open for 1 min. The moisture absorption measurements were performed under 10% RH and 90% RH (relative humidity) at 25 °C temperature. The specimens were taken out from the chamber at fixed intervals shown in Table 1. Weight and dimensional measurement timeline of each cycle and the weight (g) of the specimens was measured before and after changing the relative humidity conditions inside the climate chamber. The moisture uptake was calculated using Eq. (1) below:

$$\text{Moisture absorption degree}(\%) = \frac{W_{wet} - W_{dry}}{W_{dry}} \times 100\% \tag{1}$$

where W_{wet} and W_{dry} represent the weight of humid specimens and the initial weight of materials prior to exposure to moisture absorption (dry value), respectively. Four different mass measurements of each one of the 6 specimens were taken. Hence, the results were reported as the mean of these four different measurements.

2.6 Dimensional Tests

The dimensional changes of the BC specimens were measured manually, using an RS PRO Digital Calliper to measure the length and width, and a micrometer from the same measuring set to measure the thickness. The dimensions of the specimens were measured in mm and were collected at the same time as the weight measurements. The length, width and thickness changes were calculated using Eq. (2) below:

$$\text{(Dimensional) change } (\%) = \frac{D_{wet} - D_{dry}}{D_{dry}} \times 100\% \tag{2}$$

where D_{wet} and D_{dry} represent the dimension (width, length or thickness) of a humid specimen and its' initial dimension prior to exposure to moisture absorption (dry value), respectively. Four different measurements of thickness at different locations on the surface of each one of the 6 specimens were taken. The same locations were measured each time; therefore, the results were reported as the mean of these four different measurements.

2.7 Cyclic Behaviour Tests

The cyclic tests lasted for 32 consecutive days and each cycle lasted 4 days in total. In the beginning of the tests, the first cycle started at RH = 90% and the materials stayed at these conditions for two days. When changing the RH settings in the climate chamber, a period of about 10 min is needed until the desired RH level reaches equilibrium. Measurements of the weight, width, length and thickness of the specimens were taken every hour during the first 6 h and then the specimens would be measured again at 24 and 48 h as seen in Table 1. When the RH was changed to 10%, the weight and dimensional measurements were taken every 2 h for the first 6 h and then again at 24 and 48 h (Table 1). The next cycle would start when the samples remained for 48 h at RH = 10%, so each cycle lasted 4 days in total and the results from 8 consecutive cycles were analysed.

Table 1. Weight and dimensional measurement timeline of each cycle

Time (h)	1 2 3 4 5 6 24 48	50 52 54 72 96
RH (%)	90	10
T (°C)	25	25

3 Results

3.1 Moisture Absorption

For the moisture absorption tests, the climate chamber was used for a period of 4 months at constant temperature (T) of 25 °C. The moisture responsivity of the BC samples was determined by taking accurate weight and dimensional measurements of 6 specimens throughout that period. Equation (2) was used for the swelling calculations and the results are presented in Fig. 3. All dimensional changes are plotted on the same scale to show that the change in length and width is very small compared to the change in thickness. The response was fast and the thickness change of the specimens at humid state was found to be 85% within 24 h. At dry state (10% RH) the thickness change dropped to 1.6% on average, also within 24 h. In contrast to the significant change in thickness, there was a very small change in length and width, with the maximum increase at 90% RH being 2.6% and an average fluctuation of 1.5% throughout the experiment.

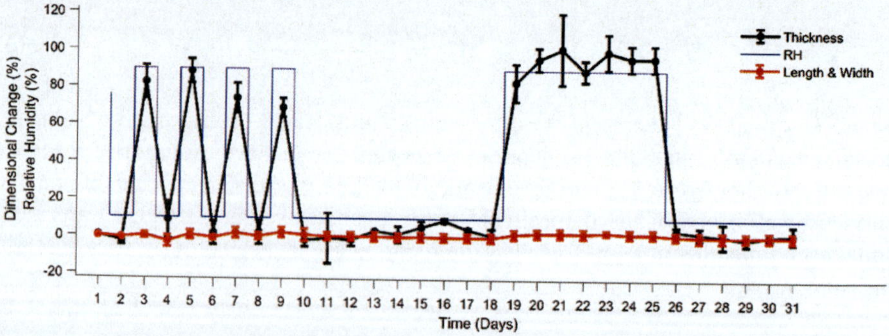

Fig. 3. Thickness change (%), length and width average change (%) of BC over time at a constant T = 25 °C and two different RH levels (10%–90%). The error bars show the standard deviation.

The specimens needed to reach equilibrium at each RH state, however, it wasn't possible to achieve equilibrium within 24 h. Thus, the experimental pattern was changed to the following settings: 8 days at RH = 10%, 8 days at RH = 90% and then again 6 days at RH = 10%. After 5 days at RH = 90% the BC films seemed to have reached equilibrium, however, likely because of errors in the manual measurements, some apparent fluctuation in the results is seen at constant RH levels. Compared to wool fibres and wood, which swell in response to changing moisture content by 35% [35] and 12% [33] respectively, here, the moisture absorption tests revealed that BC has a significantly higher swelling capacity than other natural fibrous materials. More precisely, the results in Fig. 3a reveal an average swelling in thickness of about 96% and a maximum swelling of 101%. Previous research on water absorption BC derived from pure culture media process has shown equivalent results [23, 36], however, comparisons between these two methods are invalid due to the differences in the methodology and the manufacturing process. The total swelling in length and width dimensions did not exceed 3% in humid conditions.

It is noticeable that BC can reach about 85% of its full moisture capacity within only 24 h in a humid environment, which is validated from previous research work [37]. When BC is wet, the film becomes flexible, sticky and soft, whereas in its dry state it is stiff and rough, as can be seen in Fig. 2.

3.2 Repeatable Moisture Tests

Here, the repeatability of the moisture responsive behaviour of BC was examined at different RH levels (10%–90%–10%). Eight repeating cycles were conducted, which followed the pattern explained in 0. During wetting, the specimens presented an average 13.4% swelling in thickness and an average 13.6% moisture weight gain within the first hour. At Day 1 (24 h) the swelling and weight change had almost reached the total maximum point, as the specimens changed only by a further 10% on Day 2 (48 h). Figure 4 shows that both weight and thickness follow a pattern where the rate of moisture absorption decreases over time upon wetting, meaning that BC initially absorbs moisture at a higher rate, which decreases after the 1st day. Equally, BC releases moisture in a non-linear drying process, shrinking by 67% and loosing 60% of its gained weight within the first 2 h of drying. The maximum thickness swelling observed was 116.3% on Day 2 of the 3rd cycle, whereas the minimum thickness swelling upon drying was found to be 1.5% at the end of Day 4 of the 4th cycle (Fig. 4a). The total maximum weight gain was 97% on Day 2 of the 5th cycle (Fig. 4b). It is noteworthy to mention that the thickness calculations present big fluctuations because of the materials softness. The surface of BC gets slightly indent when measuring, thus it was difficult to know when the micrometer first touches the surface. The weight and thickness changes of BC upon drying are more consistent than in wetting. The response times of BC wetting and drying is on the order of minutes [38], whereas the response time of wood [33] varies between minutes to hours depending on the wood type and thickness. Wood veneer, and other natural fibres such as wool [35] and flax [39] present similar fast moisture response to BC. The dry conditions in the climate chamber forced BC to dry further as the RH reached a minimum point of 10% (less than the RH of room conditions) resulting in a negative weight change.

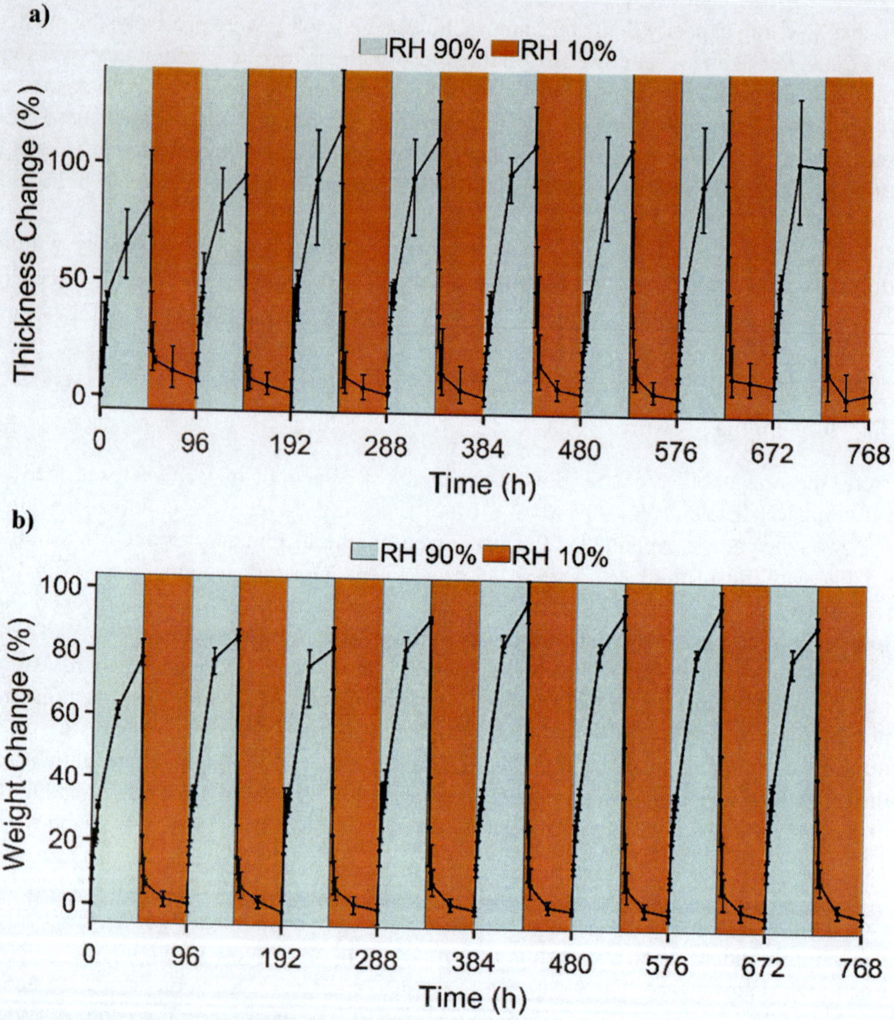

Fig. 4. Cyclic tests on BC moisture responsiveness measuring the a) thickness change (%) over time (h), b) weight change (%) over time (h).

4 Discussion

The in-plane dimensional changes (length and width) of the BC sheet are two orders of magnitude smaller than the out-of-plane (thickness) dimensional changes because of the unique structure of BC. The fibres in the BC pellicle are randomly distributed across the plane of the film [40] and the molecular orientation is parallel to the direction of the length of the fibrils [41]. This structure makes the cellulose fibres stiff along their length, which stops the material expanding and contracting in plane (i.e., in the length and width), but much less stiff across the width. Therefore, when the fibres absorb water and expand, they expand across their width, increasing the thickness of the sheet [22].

In comparison to wood, which also presents an anisotropic structure [42], BC has higher swelling properties because it presents intra and intermolecular hydrogen bonds in its structure that are 100 times smaller than those of plant cellulose (wood).

The results from BC weight change show an average water absorption capacity of 90% and a maximum weight gain of 97%. These results show a higher moisture absorption capacity from previous research [18, 40] where it has been reported that the moisture content of BC at 90% RH is 14%. In this work, BC absorbs a moisture capacity equal to 1x its dry weight [5]. The response of BC thin sheets is fast (achieving a thickness swelling of 13% within one hour), reversible and repeatable over eight cycles without presenting mechanical reduction nor physical degradation. With respect to the literature [23], the results from the swelling in thickness tests revealed that BC produced by kombucha has a higher swelling capacity (96%) than BC produced by pure culture media process (87.2%). Compared to other moisture-responsive natural fibrous materials BC presents big anisotropic dimensional changes and fast rate of response during wetting and drying as discussed in Sect. 3.1.

5 Conclusions

Moisture-responsive materials change their volume when exposed to different levels of relative humidity or direct wetting and drying. In order to understand the level of BC moisture-responsiveness, dimensional and weight changes were studied by conducting moisture absorption and cyclic tests. The purpose of this work was to investigate the extent to which BC expands and shrinks at a constant temperature (T25 °C) and between two distinct relative humidity levels: 10% and 90%. The response time of BC specimens under these circumstances was also studied. The moisture absorption tests showed that bacterial cellulose can absorb up to 97% of its own dry weight in water and increases about 116% from its original thickness. The cyclic tests showed that BC presents the same cyclic behaviour in both weight and thickness change upon wetting and drying. The high level of dimensional change in one direction, as well as the rapid, reversible response to varying levels of humidity gives scope for using this material in responsive, variable porosity facade systems. One of the main challenges of the future research will be the combination of BC sheets with other responsive and non-responsive bio-based materials for the creation of a responsive, variable porosity façade that incorporates both insulation and thermal mass.

References

1. United Nations Environment Programme: Global Alliance for Buildings and Construction 2020 GLOBAL STATUS REPORT FOR BUILDINGS AND CONSTRUCTION Towards a zero-emissions, efficient and resilient buildings and construction sector EXECUTIVE SUMMARY 2020 GLOBAL STATUS REPORT FOR BUILDINGS AND CONSTRUCTION Towards a zero-emissions, efficient and resilient buildings and construction sector (2020)
2. Yadav, M., Agarwal, M.: Biobased building materials for sustainable future: an overview. Mater. Today Proc. **43**, 2895–2902 (2021). https://doi.org/10.1016/j.matpr.2021.01.165
3. Dewidar, K.M., Mohamed, N.M., Ashour, Y.S.: Living skins: a new concept of self active building envelope regulating systems. In: SB 13 Dubahi (2013)

4. Student, T.N.: Smart Materials Build Smart Cities: ExploringMaterials for Smart wall facades; Smart Materials Build Smart Cities: ExploringMaterials for Smart wall facades (2020)
5. Holstov, A., Farmer, G., Bridgens, B.: Sustainable materialisation of responsive architecture. Sustainability (Switzerland). **9** (2017). https://doi.org/10.3390/su9030435
6. Reichert, S., Menges, A., Correa, D.: Meteorosensitive architecture: biomimetic building skins based on materially embedded and hygroscopically enabled responsiveness. CAD Comput. Aided Des. **60**, 50–69 (2015). https://doi.org/10.1016/j.cad.2014.02.010
7. Guo, Y., et al.: A biodegradable functional water-responsive shape memory polymer for biomedical applications. J. Mater. Chem. B. **7**, 123–132 (2019). https://doi.org/10.1039/c8tb02462f
8. Zhu, X., Yang, C., Jian, Y., Deng, H., Du, Y., Shi, X.: Ion-responsive chitosan hydrogel actuator inspired by carrotwood seed pod. Carbohydr. Polym. **276**, 118759 (2022). https://doi.org/10.1016/j.carbpol.2021.118759
9. Zhang, C., Cha, R., Zhang, P., Luo, H., Jiang, X.: Cellulosic substrate materials with multi-scale building blocks: fabrications, properties and applications in bioelectronic devices. Chem. Eng. J. **430**, 132562 (2022). https://doi.org/10.1016/j.cej.2021.132562
10. Apostolopoulou-Kalkavoura, V., Gordeyeva, K., Lavoine, N., Bergström, L.: Thermal conductivity of hygroscopic foams based on cellulose nanofibrils and a nonionic polyoxamer. Cellulose **25**(2), 1117–1126 (2017). https://doi.org/10.1007/s10570-017-1633-y
11. Ahankari, S., Paliwal, P., Subhedar, A., Kargarzadeh, H.: Recent developments in nanocellulose-based aerogels in thermal applications: a review. ACS Nano **15**, 3849–3874 (2021). https://doi.org/10.1021/acsnano.0c09678
12. Lv, Q., Dqg, O., Fruehwirth, S., Development, L.I.: LIVING envelope (2008)
13. Chiaoprakobkij, N., Seetabhawang, S., Sanchavanakit, N., Phisalaphong, M.: Fabrication and characterization of novel bacterial cellulose/alginate/gelatin biocomposite film. J. Biomater. Sci. Polym. Ed. **30**, 961–982 (2019). https://doi.org/10.1080/09205063.2019.1613292
14. Guan, Q.F., et al.: Sustainable cellulose-nanofiber-based hydrogels. ACS Nano. **15**, 7889–7898 (2021). https://doi.org/10.1021/acsnano.1c01247
15. Tang, H., Chen, H., Duan, B., Lu, A., Zhang, L.: Swelling behaviors of superabsorbent chitin/carboxymethylcellulose hydrogels. J. Mater. Sci. **49**(5), 2235–2242 (2013). https://doi.org/10.1007/s10853-013-7918-0
16. Christian, S.J., Billington, S.L.: Mechanical response of PHB- and cellulose acetate natural fiber-reinforced composites for construction applications. Compos. B Eng. **42**, 1920–1928 (2011). https://doi.org/10.1016/j.compositesb.2011.05.039
17. Xiao, S., et al.: Lightweight, strong, moldable wood via cell wall engineering as a sustainable structural material. Science **1979**(374), 465–471 (2021). https://doi.org/10.1126/science.abg9556
18. Izakura, S., Koga, H., Uetani, K.: Humidity-responsive thermal conduction properties of bacterial cellulose films. Cellulose **28**(9), 5363–5372 (2021). https://doi.org/10.1007/s10570-021-03888-6
19. Tammelin, T., Abburi, R., Gestranius, M., Laine, C., Setälä, H., Österberg, M.: Correlation between cellulose thin film supramolecular structures and interactions with water. Soft Matter **11**, 4273–4282 (2015). https://doi.org/10.1039/c5sm00374a
20. Reyssat, E., Mahadevan, L.: Hygromorphs: from pine cones to biomimetic bilayers. J. R. Soc. Interface **6**, 951–957 (2009). https://doi.org/10.1098/rsif.2009.0184
21. Gao, X., Shi, Z., Liu, C., Yang, G., Sevostianov, I., Silberschmidt, V.V.: Inelastic behaviour of bacterial cellulose hydrogel: in aqua cyclic tests. Polym. Test. **44**, 82–92 (2015). https://doi.org/10.1016/j.polymertesting.2015.03.021
22. Ullah, M.W., Manan, S., Kiprono, S.J., Ul-Islam, M., Yang, G.: Synthesis, structure, and properties of bacterial cellulose. In: Nanocellulose: From Fundamentals to Advanced Materials, pp. 81–114. Wiley (2019). https://doi.org/10.1002/9783527807437.ch4

23. Seifert, M., Hesse, S., Kabrelian, V., Klemm, D.: Controlling the water content of never dried and reswollen bacterial cellulose by the addition of water-soluble polymers to the culture medium. J. Polym. Sci. Part A Polym. Chem. 463–470 (2004). https://doi.org/10.1002/pola.10862
24. Lu, H., Jiang, X.: Structure and properties of bacterial cellulose produced using a trickling bed reactor. Appl. Biochem. Biotechnol. **172**(8), 3844–3861 (2014). https://doi.org/10.1007/s12010-014-0795-4
25. Guo, A., Sun, Z., Sathitsuksanoh, N., Feng, H.: Review on the application of nanocellulose in cementitious materials (2020). https://doi.org/10.3390/nano10122476
26. Hisseine, O.A., Basic, N., Omran, A.F., Tagnit-Hamou, A.: Feasibility of using cellulose filaments as a viscosity modifying agent in self-consolidating concrete. Cem. Concr. Compos. **94**, 327–340 (2018). https://doi.org/10.1016/j.cemconcomp.2018.09.009
27. Aditiawati, P., et al.: The Nanocellulose Fibers from Symbiotic Culture of Bacteria and Yeast (SCOBY) Kombucha: Preparation and Characterization. Nanofibers - Synthesis, Properties and Applications (2021). https://doi.org/10.5772/intechopen.92484
28. Bungay, H.R., Serafica, G., Mormino, R., Isermann, H.P.: Environmental implications of microbial cellulose. Global Environ. Biotechnol. (1997). Elsevier Science
29. Margarita, A., Gallegos, A., Carrera, S.H., Parra, R., Keshavarz, T., Iqbal, H.M.N.: Bacterial cellulose (2016)
30. Kamaruddin, I., Dirpan, A., Bastian, F.: The novel trend of bacterial cellulose as biodegradable and oxygen scavenging films for food packaging application: an integrative review. In: IOP Conference Series: Earth and Environmental Science. IOP Publishing Ltd (2021). https://doi.org/10.1088/1755-1315/807/2/022066
31. Birch, E., Bridgens, B., Zhang, M., Dade-Robertson, M.: Bacterial spore-based hygromorphs: a novel active material with potential for architectural applications. Sustainability (Switzerland). **13** (2021). https://doi.org/10.3390/su13074030
32. Yang, L., Cui, J., Zhang, L., Xu, X., Chen, X., Sun, D.: A moisture-driven actuator based on polydopamine-modified mxene/bacterial cellulose nanofiber composite film. Adv. Funct. Mater. **31** (2021). https://doi.org/10.1002/adfm.202101378
33. Holstov, A., Bridgens, B., Farmer, G.: Hygromorphic materials for sustainable responsive architecture. Constr. Build. Mater. **98**, 570–582 (2015). https://doi.org/10.1016/j.conbuildmat.2015.08.136
34. Environmental Impacts of Sugar Production The Cultivation and Processing of Sugarcane and Sugar Beet
35. Dénes, O., Florea, I., Manea, D.L.: Utilization of sheep wool as a building material. In: Procedia Manufacturing, pp. 236–241. Elsevier B.V. (2019). https://doi.org/10.1016/j.promfg.2019.02.208
36. Schrecker, S.T., Gostomski, P.A.: Determining the water holding capacity of microbial cellulose. Biotechnol. Lett. **27**, 1435–1438 (2005). https://doi.org/10.1007/s10529-005-1465-y
37. Lin, W.C., Lien, C.C., Yeh, H.J., Yu, C.M., Hsu, S.H.: Bacterial cellulose and bacterial cellulose-chitosan membranes for wound dressing applications. Carbohydr. Polym. **94**, 603–611 (2013). https://doi.org/10.1016/j.carbpol.2013.01.076
38. Ciecholewska-Juśko, D., et al.: Superabsorbent crosslinked bacterial cellulose biomaterials for chronic wound dressings. Carbohydr. Polym. **253** (2021). https://doi.org/10.1016/j.carbpol.2020.117247
39. Hearle, J.W.S., Morton, W. E.: Physical Properties of Textile Fibres, 4th edn. Woodhead Publishing Series in Textiles, USA (2008)
40. Goh, W.N., Rosma, A., Kaur, B., Fazilah, A., Karim, A.A., Bhat, R.: Microstructure and physical properties of microbial cellulose produced during fermentation of black tea broth (Kombucha). II. Int. Food Res. J. **19**, 153–158 (2012)

41. Yamanaka, S., et al.: The structure and mechanical properties of sheets prepared from bacterial cellulose. (1989)
42. Li, J., Chen, C., Zhu, J.Y., Ragauskas, A.J., Hu, L.: In situ wood delignification toward sustainable applications. Acc Mater Res. **2**, 606–620 (2021). https://doi.org/10.1021/accountsmr.1c00075

Exploring the Potential of Mycelium Composites as Natural Board Materials

Ilaria La Bianca, Joost F. Vette(✉), and Nisalyna Bontiff

Avans University of Applied Sciences, Hogeschoollaan 1, 4818 CR Breda, The Netherlands
jf.vette@avans.nl

Abstract. Mycelium composites are an attractive alternative to turn to more sustainable construction materials. Mycelium, the vegetative growth of fungi, is formed by a network of hyphae (elongated fungal cells). This mycelium network has the capacity to partly degrade carbon-rich fibres while binding them together. In this way, a biological process can serve to create biobased composite materials, in which the fungal biomass functions as inherent adhesive while the fibres provide the structural support. Finished mycelium composites can be heat-pressed into sheet materials. In this research, mycelium composites are heat-pressed to explore their potential as board material. We studied the influence of the substrate, heat-press process and increased thickness on the flexural strength of the materials. The used mycelium composites consisted of straw fibres, or a combination of rapeseed straw (RS) and cellulose, bound together with the fungus *Ganoderma resinaceum*. Through systematic study of the process parameters pressure, time and temperature, an optimized heat-press process was determined by flexural strength performance. As opposed to literature this project found flexural strengths within the range of glue-bound board materials such as particleboard and MDF. Mycelium composites made 100% of RS reached a flexural strength of 19 MPa, while mixed cellulose and RS showed a flexural strength of 12 MPa. Thicker materials were also produced and showed a flexural strength similar to thinner boards. This study showed that substrate composition, and heat-press process can influence material properties of mycelium boards.

Keywords: heat-press · biobased building · mycelium-material research

Graphical Abstract

S. Amziane et al. (Eds.): ICBBM 2023, RILEM Bookseries 45, pp. 653–664, 2023.
https://doi.org/10.1007/978-3-031-33465-8_50

1 Introduction

Increased environmental awareness has led to the rising use and production of sustainable materials to substitute conventional, often (environmentally) toxic, products. Among these, mycelium composites are a novel and attractive product. They are formed by the growth of filamentous fungi on humidified fibres [1]. Mycelium, the vegetative growth of fungi, is formed by a network of hyphae responsible for the microorganism development. These hyphae utilize the fibres as carbon source and simultaneously bind them together forming a three-dimensional matrix [2]. Filamentous fungi from the basidiomycetes group are mainly used for production of mycelium composites. These fungi naturally grow on trees and present the capacity of degrading complex molecules such as lignin, hemicellulose and cellulose [3]. Agricultural residues such as hemp shives, rapeseed straw and flax shives can be used as substrate to make 100% bio-based mycelium composites [4].

Mycelium composites can be post-processed with a heat-pressing process. This process results in an increased density and increased flexural strength [5]. This could be an interesting process to create heat-pressed mycelium composites as a sustainable alternative to particleboards, a class of wood products manufactured through chemically binding and heat-pressing of wood particles. The resins used in particleboards are non-biodegradable, petroleum-based products, which can increase waste production at the end of the life of products [6]. These resins often contain urea-formaldehyde, which emissions can be detrimental to human health [6].

The main parameters of the heat-press process are temperature, time and pressure. A variety of used substrates and species, process parameters and ranges of density and flexural strength is found in literature (Table 1). Flexural strength is a frequently used material property to explore the potential heat-pressed mycelium-composites as alternative for (particle) board materials (Table 1).

The aim of this research is to further explore the potential of heat-pressed mycelium composites as an alternative to conventional wood particle and glue-based board materials focussing on the flexural strength of the composites, with studies on the following aspects: 1. Cellulose as supplementary substrate; 2. Effect of pressing time and temperature on the flexural strength; 3. Effect of pressing thicker boards on the flexural strength of mycelium composites; 4. Potential of residual substrates from oyster mushroom production as heat-pressed mycelium composites.

Cellulose as supplementary substrate. A high cellulose content in the substate mix can speed up the mycelium growth process and can reduce the risk of infections with other organisms [16]. Cellulose can be obtained from waste-waster treatment, as solid residue [8]. Added cellulose has no statistical impact on the flexural strength of a mycelium composite pressed at 70 °C [7]. However, higher pressing temperatures can result in higher flexural strength [7]. This research includes a study into the effect of cellulose as substrate for mycelium composites pressed at higher temperatures.

Effect of pressing time and temperature on the flexural strength. The input parameters of the heat-press process are temperature, pressure and time. Based on available literature research in Table 1, a knowledge gap was identified considering the effects of varying the heat-press process parameters. This research includes a study into the effect of

Table 1. Overview of heat-pressed mycelium composite literature

Resources	Fungal species	Tempera-ture (°C)	Pressure (MPa)	Time (minutes)	Obtained density (g/cm^3)	Obtained flexural strength (MPa)	Source
Beech sawdust, rapeseed straw, cotton fibre	*Trametes ochracea, Pleurotus ostreatus*	150	0,26	20	0,35	0,62–0,87	[3]
Cotton stalk	*Ganoderma lucidum*	160–200	3,5–4	6	0,6	1,4–4,6	[5]
Palm fruit fibre, sawdust	*Ganoderma lucidum*	120	20	50		2,68	[6]
Bacterial cellulose, hemp fibres	*Trametes versicolor*	200			0,45–0,47	2,9	[7]
Hempfibre and nanoclay	*Trametes versicolor*	200	30	60	0,45–0,53	0,98–1,94	[15]

variations in both time and temperature on the flexural strength of heat-pressed mycelium composites.

Effect of pressing thicker boards on the flexural strength of mycelium composites. Based on existing literature (Table 1), no conclusions can be drawn on the effect of heat-pressing various samples thicknesses, and the impact of increased thickness on the flexural strength of the mycelium composite. The effect of the thickness of the heat-pressed samples on the flexural strength of the mycelium composite is also studied, taking into consideration the most optimal heat-press parameters and substrates, resulting from our investigation.

Potential of residual substrates from oyster mushroom production as heat-pressed mycelium composites. *Pleurotus ostreatus*, the oyster mushroom, is cultivated on lignocellulosic substrates. The colonised substrate is a waste stream after the mushroom harvest. These materials can be dried and heat-pressed into mycelium composites, with the benefit that these are an existing waste stream [14]. In this research, the flexural strength of heat-pressed mycelium composites from residual substrates from oyster mushroom cultivation are compared to lab-grown heat-pressed mycelium composites.

2 Material and Methods

2.1 Spawn and Composite Preparation

Ganoderma resinaceum (M9730) was purchased from Mycelia [9]. The millet grains were purchased at Versele-laga [10]. The rapeseed straw (RS) was purchased from Gedizo Trading Inc [11]. The cellulose (C) was provided by Recell® [8]. The tested spent mycelium substrate (SMS) (residual blocks from mushroom cultivation) were provided by Verbruggen Paddestoelen BV [12].

For spawn preparation, *G. resinaceum* M9370 was grown on potato dextrose agar (PDA) medium for 7 days. Ten pieces of 5 × 5 mm were cut from the agar and used to inoculate millet grains. The grains were previously soaked in water, let to rest at 60 °C for 24 h, rinsed and sterilized for 25min at 121 °C and 1 bar pressure. The millet grains were then cooled down at room temperature and inoculated with the fungal species. The inoculated cultures were allowed to grow at 25 °C and kept in the dark. After two and four days from inoculation the spawn was shaken to allow thorough distribution of hyphae and let to grow for 3 more days. After the total growth period of 10 days the spawn was ready to be used. To slow down fungal growth the spawn was kept at 4 °C in the dark for a maximum time of one month.

For the composite preparation, the fibres were humidified with demineralized water with final water content of 62% (w/w). Fibres moisturization was followed by their sterilization for 20 min at 121 °C and 1 bar pressure. Each substrate was then inoculated with a ratio equal to 3% of the dry weight (amount of dry fibres) for the substrate and heat-press process study and of 5% total weight (fibres and water) for the post-process study. Fungus-substrate combinations were grown in polypropylene (PP) boxes [13] with dimensions of 185 × 185 × 50 mm (L × W × H). For the material dimensions study, the samples were grown in the dark at 30 °C for a total period of 14 days. At the end of the growth period, the specimens were oven dried at 65 °C for 24 h. For the substrate and heat-press study, a total of 3 growth boxes per condition were used, from which 3 test samples were obtained, for a final total of 9 testable replicates per condition. For the post-process study, a total of 2 growth boxes per condition were used, from which 3 test samples were obtained, for a total of 6 testable replicates per condition. The samples dimensions and the bending test setup were determined following the NEN-EN 310:1993 standard. The samples were tested with the SHIMADZU - Autograph AGS-X 50 kN (max. Capacity) with operating program TRAPEZIUM X. Samples dimension, weight and density are summarized in Supplementary data file 1.

2.2 Substrate Study

Rapeseed straw and cellulose were combined in different rates (85% RS: 15% C and 70% RS: 30% C). The composites' water content was adjusted to the optimal water contents of rapeseed straw and cellulose. The final amount of water content in the composite was: 58.8% (w/w). The oven dried samples were heat-pressed at 150 °C, 2.8 MPa and 12 min, laser cut and tested in their bending test as described above.

2.3 Thermogravimetric Analyses (TGA) and Heat Press Process Study

TGA was performed with Q500 TA Instruments. 10 mg of substrate (RS and cellulose) were heated up from 20 °C to 600 °C, 20 °C/min. The heat press PHI (max. 170 kN) was used to study the heat press process parameters. The constant pressure of 2.8 MPa was used. The temperatures 150 and 180 °C, and the time of 6 and 12 min were considered. In brief, the heat press was heated at the desired temperature, the dried samples were positioned between the heated plates and pressed for the desired amount of time. The pressed samples were then laser cut in samples with dimensions of 50 x 150 mm (WxL) with a thickness of 2–3 mm (determined by the heat-press process).

2.4 Material Dimensions Study

The heat press Fontijn Holland (max. Temperature: 300 °C and max. Pressing force 200 kN) and heat press PHI (max. 170 kN) were used for this study. Increased thickness of the heat-pressed samples was achieved in the post-growth phase. The mycelium blocks were grown in the regular dimensions (185 × 185 × 40 mm L × W × H), oven dried and stacked together when pressed (2 mycelium blocks, with final dimensions 185 × 185 × 80 mm L × W × H). The samples were pressed at 180 °C, 2.8 MPa and 12 min. The pressed samples were then sawn to the desired dimensions. The residual blocks from mushroom cultivation had initial dimensions of 515 × 340 × 155 mm, the blocks were sawn to ~150 × 150 × 50 mm (L × W × H) dimensions varied ± 5 mm for length and width, and treated following the methods for the post processed mycelium blocks.

2.5 Statistical Analyses

The statistical analysis was performed with the software R studio (RStudio Version 2022.07.1-554 RStudio, Inc.) using a p-value ≤ 0.05. Anova followed by Tukey post-hoc test was performed for all the experiments.

3 Results

3.1 100% Rapeseed Straw Substrate Results in Stronger Composite

For the substrate study, *G. resinaceum* M9730 was grown on pure RS, or cellulose combined to RS for 14 days at 30 °C in the dark. Two ratios were investigated: 30% cellulose plus 70% RS and 15% cellulose plus 85% RS. The combination with higher cellulose content (30%) was also checked for speed of growth. For this purpose, additional samples were grown for 7 days only and compared to the specimens grown for 14 days, in their flexural strength. The comparison was made only between the samples showing the same substrate composition. The composites were oven dried, heat pressed at 150 °C for 12 min and laser cut in the desired dimensions. The obtained samples were tested for flexural strength (Fig. 1).

The composites made of pure RS showed the highest flexural strength (average 16.74 MPa) compared to the mixed substrates which showed average flexural strengths of 8.52 and 10.55 MPa for 15% and 30% cellulose content respectively. Growing the composites for only 1 week resulted in slightly higher flexural strength (mean = 13.24 MPa), however, the difference was not significant (p. value = 0.38). Comparison of mycelium-bound composites with composites made solely of fibres, without mycelium, (C_30%Cel_70%St) showed that the presence of the fungus attributes strength to the material, resulting in higher flexural strength.

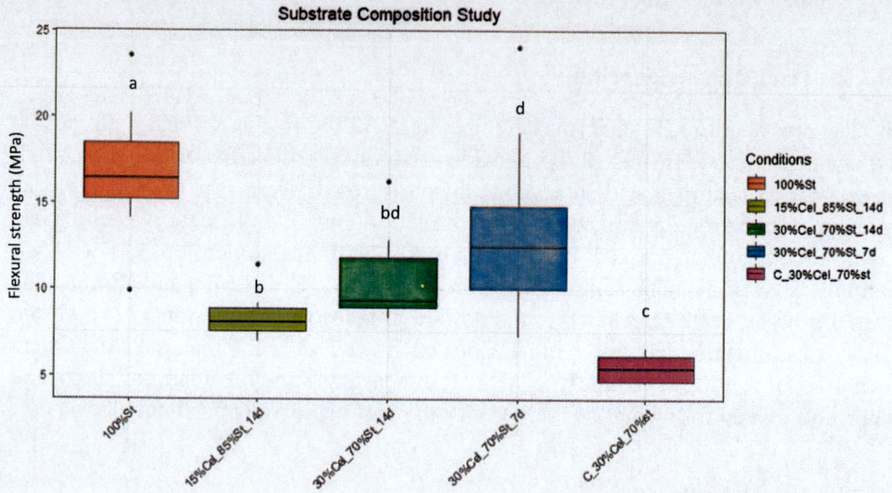

Fig. 1. Comparison of the effect on flexural strength from substrate composition and from time of growth.

3.2 Time of Pressing Rather Than Temperature Impacts Heat-Press Process

G. resinaceum M9730 was grown on humidified RS for 14 days in the dark at 30 °C and oven dried at 65 °C. The dried samples were then heat pressed in different combinations of heat and time at a constant pressure of 2.8 MPa. The used temperatures were 150 and 180 °C, while the durations of pressing were 6 and 12 min. The hot temperature limit of pressing was determined by TGA, which indicated the maximum temperature endured by the fibers before degradation (±200 °C) (Suppl. Fig. 1). The samples were then laser cut into the desired dimensions and tested for bending strength (Fig. 2).

Longer time of pressing resulted in higher flexural strength, with average strength of 19.26 and 16.74 MPa for samples pressed at 180 and 150 °C respectively. Temperature variance did not result in significant difference, however, temperature 180 °C showed higher flexural strength. Not heated mycelium-bound boards (board pressed at room temperature) showed a drastically lower flexural strength equal to 0.32 MPa. As demonstrated in the substrate study, the presence of mycelium resulted in significant difference with the mycelium-free specimens.

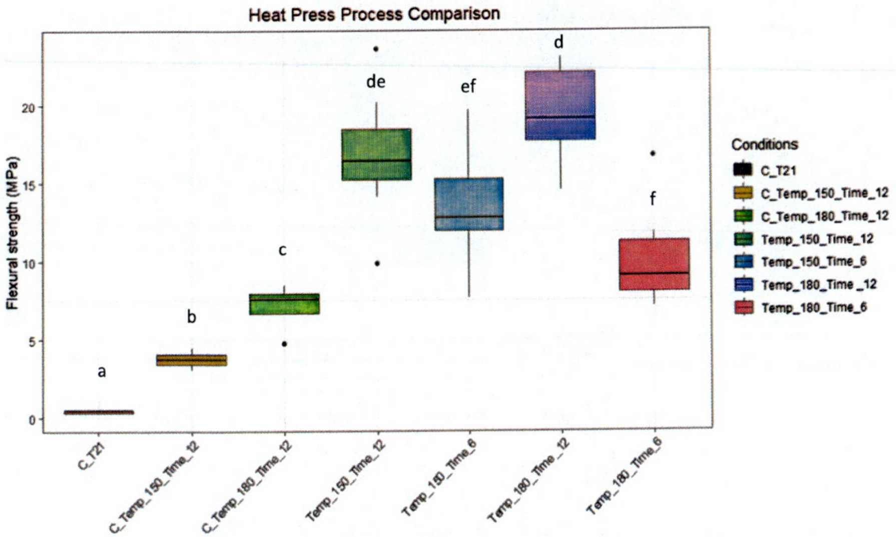

Fig. 2. Heat-press process study. The tested temperatures are 180 and 150 °C, the times are 6 and 9 min. Two controls are present: mycelium-bound composites pressed without heat (12 min) and mycelium-free fibres pressed at 180 °C for 12 min. The controls are solely compared to the sample with process: temperature 180 °C and time 12 min.

3.3 The Effect of Sample Thickness (for Lab Grown and SMS Composites) and Spawn Percentage (for Lab Grown Composites) on Flexural Strength

G. resinaceum M9730 was grown on humidified RPS for 14 days in the dark at 30 °C and oven dried at 65 °C. The samples were heat-pressed at 180 °C for 12 min, laser cut and tested for flexural strength. Increase in thickness was obtained by stacking 2 dry samples at the moment of pressing, almost doubling the original thickness. Two spawn percentages were also analyzed in their resulting flexural strength (Fig. 3).

Increase in material thickness did not have significant impact on the flexural strength of the material. The sets of composites made from SMS have a thickness of 4–5 and 7 mm, and an average flexural strength of 4.14 MPa and 3.01 MPa respectively. The average flexural strength of both sets is significantly lower than the average flexural strength of the sets of lab grown samples. Spawn content seemed to significantly impact the bending strength of the material, resulting in a difference of 6.21 MPa between the tested rates.

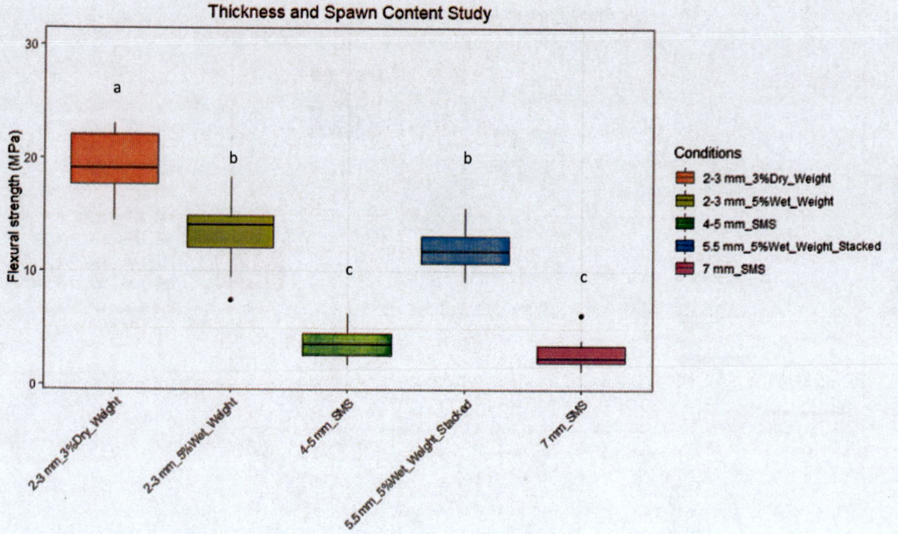

Fig. 3. Thickness study. Different thicknesses comparison (2–3 mm and 5–7 mm). Spent mycelium substrate (SMS) resulted in thicker board (7 mm). Spawn content (3 and 5%) were also tested.

4 Discussion

Mycelium board composites were studied in the substrate composition, heat-press process and dimension alteration. *G. resinaceum* M9730 was selected as sole fungal species and tested in the production of mycelium boards on rapeseed straw and cellulose. Cellulose is among the natural energy sources of basidiomycetes and was selected as a source of easily degradable carbohydrates, compared to RS [3]. The chemical composition of the used substrates can impact mycelium composite properties. The rates of cellulose, hemicellulose and cellulose of the selected feedstock can affect fungal development in both morphology and cellular chemical composition [19] and influence the material final properties [20] This study shows that pure RS substrate results in higher flexural strength than mixed RS and cellulose, which can be possibly due to the intrinsic fibres properties and chemical composition [21].

The performed systemic study of heat-press parameters showed that the production process can also make a difference in the final material properties. The combination of heat, pressure and time resulted in varying mechanical strengths, which also highly differ from studies found in literature (Table 1). Direct comparison is challenging because of the variance in used fungal species, production process, heat-press parameters and test method/standard to determine the flexural strength. However, the obtained results in this research place mycelium composites closer to conventional board materials (considering flexural strength), for example particle board which show bending strength 9–20 MPa [17, 18].

This research also verified whether increasing the thickness of mycelium boards would decrease their mechanical properties. The study aimed at producing boards with larger thickness and density equal to the thinner boards (0.08–0.11 g/cm^3), to accurately compare the outcoming mechanical properties. Gain in sample thickness did not seem to decrease the material's bending strength, rendering the produced boards interesting for further study. SMS composites showed lower flexural strength than lab grown composites, possibly due to several factors, including: 1. High degradation of fibres (due to the mushroom cultivation), 2. Differences in substrate composition; and 3. Variance in production process. The materials' mechanical property seemed influenced by the amount of used spawn to inoculate. The assessed percentages showed a great difference in amount of millet grains which were introduced in the substrate (11 g and 2.6 g). The presence of larger quantities of grains could have impacted the materials' mechanical properties.

5 Conclusions

Various parameters were tested to produce mycelium board composites, focusing on substrate composition, heat-press process parameters, spawn content and material dimensions. It was found that 100% RS resulted in higher flexural strength than mixing RS and cellulose. Process parameters appeared to make a difference in the final flexural strength, with time of pressing more determining than temperature. Lower spawn content seemed to result in higher strength, and increase in thickness did not decrease the composites' flexural strength. These results need to be corroborated with additional research, aimed at understanding why lower spawn content results in higher strength, how to achieve thicker boards and including more material properties tests. SMS composites showed lower flexural strength in both thicknesses compared to lab grown composites.

Appendix

Supplementary Figure 1

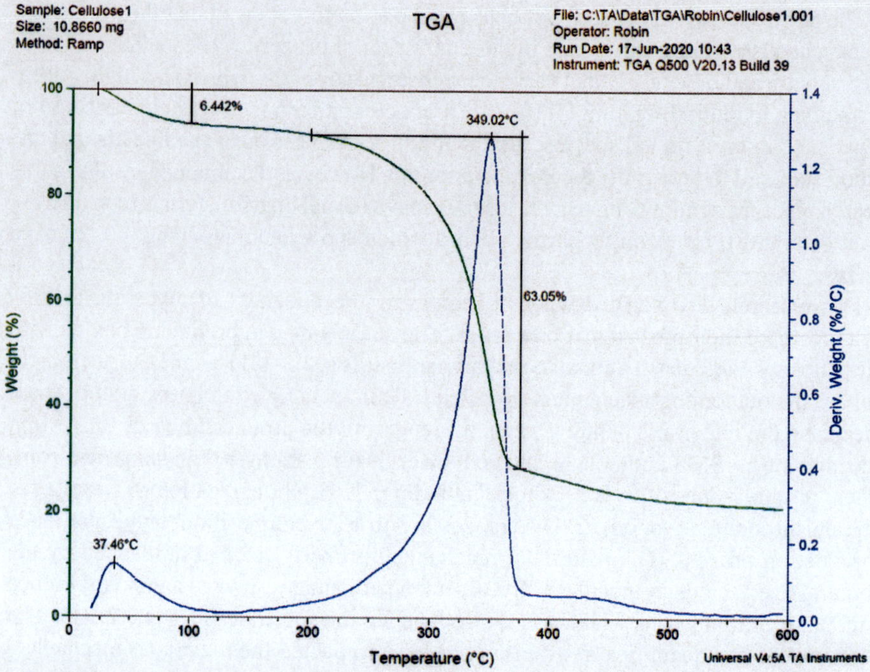
Sample: Cellulose1
Size: 10.8660 mg
Method: Ramp
TGA
File: C:\TA\Data\TGA\Robin\Cellulose1.001
Operator: Robin
Run Date: 17-Jun-2020 10:43
Instrument: TGA Q500 V20.13 Build 39
6.442%
349.02°C
63.05%
37.46°C
Weight (%)
Deriv. Weight (%/°C)
Temperature (°C)
Universal V4.5A TA Instruments

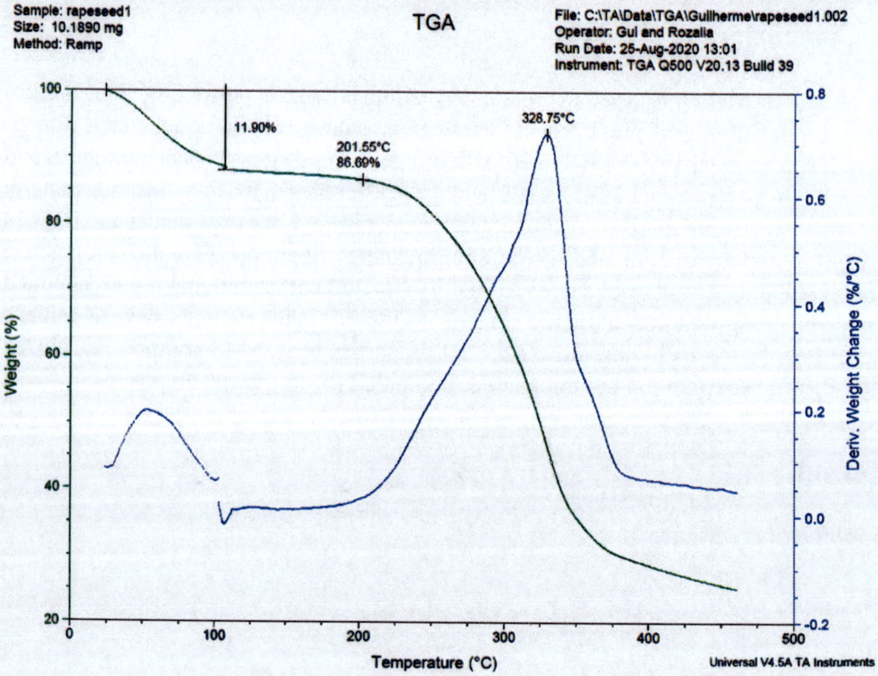
Sample: rapeseed1
Size: 10.1890 mg
Method: Ramp
TGA
File: C:\TA\Data\TGA\Guilherme\rapeseed1.002
Operator: Gui and Rozalia
Run Date: 25-Aug-2020 13:01
Instrument: TGA Q500 V20.13 Build 39
11.90%
201.55°C
86.69%
328.75°C
Weight (%)
Deriv. Weight Change (%/°C)
Temperature (°C)
Universal V4.5A TA Instruments

Supplementary Data File 1

Supplementary data
file 1.xlsx

References

1. Pelletier, M.G., Holt, G.A., Wanjura, J.D., Bayer, E., McIntyre, G.: An evaluation study of mycelium based acoustic absorbers grown on agricultural by-product substrates. Industr. Crops Prod. (2013). https://doi.org/10.1016/j.indcrop.2013.09.008
2. Jones, M., Tien, H., Sabu J.: Inherent species characteristic influence and growth performance assessment for mycelium composite applications. Adv. Mater. Lett. (2018). https://doi.org/10.5185/amlett.2018.1977
3. Appels, F.V.W., et al.: Fabrication factors influencing mechanical, moisture- and water-related properties of mycelium-based composites. Mater. Des. **161**, 64–71 (2019)
4. Jones, M.P., et al.: Agricultural by-product suitability for the production of chitinous composites and nanofibers utilising trametes versicolor and polyporus brumalis mycelial growth. Process Biochem. (2019). https://doi.org/10.1016/j.procbio.2019.01.018
5. Liu, R., et al.: Preparation of a kind of novel sustainable mycelium/cotton stalk composites and effects of pressing temperature on the properties. Ind. Crops Prod. **141**, 111732 (2019)
6. Chan, X.Y., Saeidi, N., Javadian, A., Hebel, D.E., Gupta, M.: Mechanical properties of dense mycelium-bound composites under accelerated tropical weathering conditions. Sci Rep **11**, 22112 (2021)
7. Elsacker, E., Vandelook, S., Damsin, B., Van Wylick, A., Peeters, E., De Laet, L.: Mechanical characteristics of bacterial cellulose-reinforced mycelium composite materials. Fungal Biol. Biotechnol. **8**, 18 (2021)
8. https://recell.eu/
9. https://mycelia.be/
10. https://www.versele-laga.com/nl
11. https://gedizo.nl/en/
12. https://verbruggen-paddestoelen.nl/
13. https://saco2.com/
14. Khoo, S.H., et al.: Development of formaldehyde-free bio-board produced from mushroom mycelium and substrate waste. J. Hazard. Mater. **400**, 123296 (2020)
15. Elsacker, E., De Laet, L., Peeters, E.: Functional grading of mycelium materials with inorganic particles: the effect of nanoclay on the biological chemical and mechanical properties. Biomimetics **7**(2), 57 (2022)
16. Lelivelt, R.J.J., Lindner, G., Teuffel, P., Lamers, H.: The production process and compressive strength of Mycelium-based materials. In: First International Conference on Bio-based Building Materials 2015, Clermont-Ferrand, France, pp. 1–6 (2016)
17. Bekhta, P., Marutzky, R.: Bending strength and modulus of elasticity of particleboards at various temperatures. Eur. J. Wood Wood Prod. **65**(2), 163–165 (2007)
18. https://www.rowood.net/Portals/0/Documenten/P6-Spaanplaat.pdf. Accessed 23 Dec 2022

19. Haneef, M., Ceseracciu, L., Canale, C., Bayer, I.S., Heredia-Guerrero, J.A., Atnanassiou, A.: Advanced materials from fungal mycelium: fabrication and tuning of physical properties. Sci. Rep. **7**, 41292 (2017). https://doi.org/10.1038/srep41292
20. Pickering, K.L., Aruan Efendy, M.G., Le, T.M.: A review of recent developments in natural fibre composites and their mechanical performance. Compos. A Appl. Sci. Manuf. **83**, 98–112 (2016). https://doi.org/10.1016/j.compositesa.2015.08.038
21. Elsacker, E., Vandelook, S., Van Wylick, A., Ruytinx, J., De Laet, L., Peeters, E.: A comprehensive framework for the production of mycelium-based lignocellulosic composites. Sci. Total Environ. **725**(4), 138431 (2020). https://doi.org/10.1016/j.scitotenv.2020.138431

Biobased Alternative Binders from Agar for Civil Engineering Applications: Thermal, Biodeterioration, and Moisture Sorption Properties

Melissa R. Frey(✉), Sarah L. Williams, Cristina Torres-Machi, and Wil V. Srubar

University of Colorado Boulder, Boulder, CO 80309, USA
{melissa.frey,wsrubar}@colorado.edu

Abstract. This work evaluated the properties of agar, a biopolymer extracted from red seaweeds, as a direct material alternative to petroleum-based construction binders and materials. The thermal and moisture sorption properties, along with their ability to resist biodeterioration by fungi, were evaluated and compared to petroleum-based materials. Thermogravimetric analysis indicated that the onset of thermal decomposition occurs at a slightly lower temperature for agar-based binders as compared to petroleum-based materials and binders. The addition of limestone filler was shown to significantly improve the thermal stability. When subjected to fungal deterioration testing, only traces of growth were observed on samples 28 days after inoculation, indicating that agar did not serve as a primary carbon source for the growth of heterotrophic microorganisms, which is known to lead to biodeterioration. Isothermal moisture sorption measurements were used to characterize sorption behavior of agar-based binders. In general, agar-based binders absorbed a large percentage of moisture relative to the dehydrated mass of samples. Chemically cross-linking agar with maleic anhydride was shown to reduce moisture absorption.

Keywords: Agar · Non-petroleum · Alternative Binder · Sustainable Construction

1 Motivation

Petroleum is an integral component in the production of numerous industrial and domestic commodities. Within the construction industry, products derived from petroleum include fuel, plastics (e.g., polyvinyl chloride, high density polyethylene) [1], adhesives (e.g., polyurethane, epoxies, poly(vinyl acetate), silicones) [2], and binders (e.g., asphalt for roadways, coatings, and waterproofing) [3]. Despite the widespread use and acceptance of these materials, petroleum-derived products are associated with significant price fluctuations, due to the volatility of the petroleum market, as well as negative environmental impacts associated with petroleum sourcing, refining, and manufacturing [4]. Petroleum is a nonrenewable resource, and oil reserves are dwindling, requiring a transformative shift to direct alternatives (*i.e.,* 100% replacement) for petroleum-based construction products.

S. Amziane et al. (Eds.): ICBBM 2023, RILEM Bookseries 45, pp. 665–675, 2023.
https://doi.org/10.1007/978-3-031-33465-8_51

To meet the global environmental and economic need for sustainable and resilient construction materials, recent work has focused on the use of algae as an environmentally friendly, biorenewable feedstock for energy [5, 6] and materials production [7, 8]. One algae-derived biopolymer that may be well suited as an alternative material for construction applications is agar. Agar is a linear polysaccharide extracted from red seaweed belonging to the Rhodophyceae class [2] where it fulfills a function analogous to cellulose in terrestrial plants [9]. These macroalgae have been recognized as promising candidates for carbon capture as they produce more biomass per acre than terrestrial farmed crops [3]. Agar gels form *via* hydrogen bonding between agarose molecules [10], resulting in a transparent, thermoreversible material that exhibits hysteresis. Gelling occurs at temperatures between 38 and 45 °C and melting occurs between 85 to 90 °C [11]. Due to its very high hysteresis (*i.e.,* 45 °C) and gel strength at minimal concentrations, low concentrations of agar (*i.e.,* 0.5–2.0%) are commonly used as additives to food, microbiology culture media, or pharmaceutical applications [12].

Recently, agar and agar blends have been investigated as a transparent film material for the replacement of plastic food packaging [10, 13] and as soil-strengthening materials [14]. This work explores the suitability of high concentration agar gels (*i.e.,* 5% and 7.5% w/w) as a direct replacement material for traditional petroleum-based binders in composites for construction. Specifically, the thermal stability (*i.e.,* thermogravimetric analysis), fungal biodeterioration, and isothermal moisture sorption properties of agar-based binders were evaluated and compared to traditional petroleum-based binders and materials.

2 Materials and Methods

2.1 Materials

Agar, maleic anhydride, and sodium hydroxide powder were supplied by Sigma Aldrich (St. Louis, MO, USA). Reagent-grade limestone (calcium carbonate, CaCO3) with a particle size $< 250\ \mu m$ and polyurethane coatings were supplied by Research Products International (Mt Prospect, IL, USA) and Republic Powdered Metals, Inc (Medina, OH, USA), respectively. Fungal cultures were supplied by the American Type Culture Collection (ATCC, Manassas, VA, USA).

2.2 Methods

Sample Preparation. Agar gels were prepared at 5% and 7.5% w/w concentration (i.e., relative to deionized water) with and without limestone additives. In order to increase the dimensional stability of samples, ground limestone was investigated as a filler material at a 15% w/w concentration. Sample naming convention is the letter A (for agar) followed by the % w/w agar concentration. Samples are denoted with a C for control formulations without limestone and an L for samples modified with limestone filler.

To prepare samples for thermogravimetric analysis and fungal biodeterioration testing, 200-mL of deionized water was added to 500-mL media bottles and the appropriate amount of agar powder (*i.e.,* 10 or 15-g) and limestone powder (*i.e.,* 0 or 30-g) was dissolved at ambient laboratory conditions (22 ± 2 °C) under continuous agitation using a magnetic stir bar. The media bottles were placed in a Yamato autoclave (Santa Clara, CA, USA) and heated to 100 °C for 20-min. After autoclaving, samples were poured into 55-mm diameter molds to create disc-shaped samples with a depth of approximately 10-mm.

Agar gels have been shown to be highly hydrophilic, absorbing as much as 1,600% relative to their dehydrated mass [15]. In order to investigate functional improvements in the hydrophilicity of agar-based binders, 5% agar-based binder samples with and without limestone and two additional sample subsets were investigated in moisture sorption testing: Physically modified (*i.e.*, polyurethane coated) and chemically modified (*i.e.*, chemically crosslinked with maleic anhydride). Physically modified samples were coated in three layers of oil-based (denoted OB) or water-based (denoted WB) polyurethane after sample conditioning. Chemically modified samples were cross-linked with 10%, 30%, and 50% maleic anhydride (relative to the agar concentration) using methods outlined in work conducted by Ye et al. [16]. The appropriate amount of maleic anhydride powder (*i.e.*, 1, 3 or 5-g) was dissolved in 200-mL sample solutions at ambient laboratory conditions under constant agitation with a magnetic stir bar. The pH of solutions was adjusted to 7.0 using 2.0 M NaOH solution. Solutions were then heated to 70 °C for 30-min on a Corning hot plate before sample molding.

Thermogravimetric Analysis. Thermogravimetric Analysis (TGA) was performed with a TA Discovery 5500 (TA instruments, New Castle, DE, USA). After gelling for 30-min at ambient laboratory conditions, samples were place in incubation at 30 °C until samples reached constant mass defined as <1% change in mass in 24-h. A 10–20 mg sample was collected from the center of dehydrated agar-based binder samples and placed in a platinum pan. Testing was conducted using continuous nitrogen gas flow at a rate of 10 mL/min. Samples were acclimated at 30 °C and then heated to 900 °C using a heating rate of 10 °C /min. The temperature and sample weight loss were recorded for analysis and each sample was tested in duplicate.

Fungal Biodeterioration. Resistance to fungal biodeterioration was determined according to ASTM G21 [17]. The following fungal strains, shown in Fig. 1, were used to create the testing spore solution: Aspergillus brasiliensis (ATCC 9642), Penicillium funiculosum (ATCC 11797), Chaetomium globosum (ATCC 6205), Trichoderma virens (ATCC 9645), and Aureobasidium pullulans (ATCC 15233). Samples were placed on the top surface of the nutrient-salts agar plates and inoculated by spreading a spore solution over the entire exposed surface with a sterile cell spreader.

A negative control (*i.e.*, blank nutrient-salts agar plates without a sample) and a positive control (*i.e.*, nutrient-salts agar plates with sterilized cellulose paper) were also prepared to validate the experiment. Samples were sealed with parafilm and incubated at 30 °C for 28 days. Samples were then visually examined and evaluated based on the percentage of sample area covered in fungal growth as shown in Table 1. All samples were tested in triplicate.

Moisture Sorption. The water sorption properties of agar-based binders were determined according to a modified ASTM D570 [17]. The procedure was modified by lowering the conditioning temperature from 50 °C to 40 °C. After gelling for 30-min at

Fig. 1. Fungal strains cultivated for fungal biodeterioration testing of agar-based binders. Scale bar = 100 mm.

Table 1. Rating system used for the evaluation of agar-based binder fungal resistance.

Observed Growth on Specimens	Rating
None	0
Traces of growth (< than 10% of sample area)	1
Light growth (10–30% of sample area)	2
Medium growth (30–60% of sample area)	3
Heavy growth (60% to complete coverage)	4

ambient laboratory conditions, samples were weighed and then conditioned in incubation at 40 °C until reaching constant mass, defined as <1% decrease in mass in a 24-h period, and then immersed in 20 ± 2 °C distilled water. At set testing intervals, samples were removed from the distilled water, surface moisture was removed with a dry cloth, and the mass of each sample was recorded. Sample masses were recorded every 24-h until samples again reached constant mass. The intermittent moisture content, M, was calculated for each increment of immersion time using Eq. 1:

$$M(\%) = (m_t - m_c)/m_c * 100\% \tag{1}$$

where m_t and m_c are the sample mass at a given time of immersion and the conditioned mass of a sample, respectively. The moisture content was plotted against the square root of time. Each agar-based binder was tested in triplicate.

3 Results and Discussion

3.1 Thermogravimetric Analysis (TGA)

Results from TGA are displayed in Fig. 2. All agar-based binder formulations were thermally stable up to 210 °C, followed by a steep mass loss due to decomposition of agar. Similar thermal behavior for agar composite materials has been reported previously [9, 18, 19]. For the agar-based formulations containing limestone, a second steep mass loss was observed beginning at 650 °C due to the decarbonation of limestone. Although the thermal stability of agar-based binders appeared to be minimally affected by agar concentration, the addition of limestone increased thermal stability. Control 5% and 7.5% agar-based binder formulations retained 29.60% and 29.54% of their original mass, while limestone formulations retained 43.55% and 44.33% of their original mass, respectively.

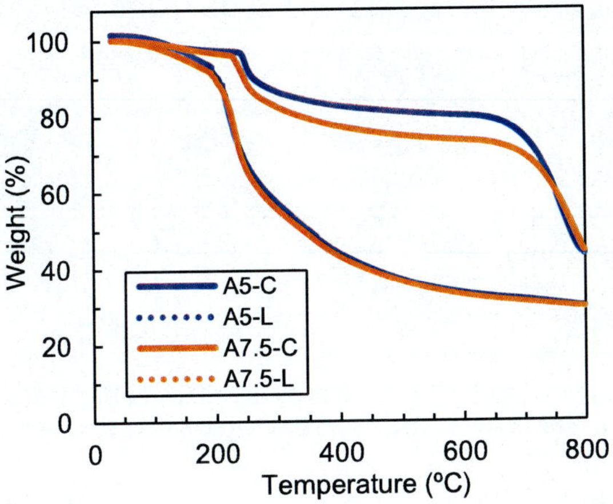

Fig. 2. Representative results from thermogravimetric analysis for 5% and 7.5% agar specimens. Each sample was tested in duplicate.

Petroleum-based plastics including polyethylene, polypropylene, polystyrene, and polyethylene terephthalate begin to thermally degrade between 250 and 450 °C, with the maximum thermal degradation taking place between 420 and 490 °C [20]. Petroleum-based asphaltic binders for roadway or roofing applications begin to degrade between 300 to 350 °C [21]. Comparatively, agar-based binders exhibit partial thermal degradation at lower temperatures than petroleum-based plastics and binders commonly utilized in the construction industry (*i.e.,* beginning at 210 °C versus 250 to 300 °C). However, these data suggest that the addition of limestone filler can significantly improve the thermal stability of agar-based binders allowing for a tailorability of thermal durability. Previous work on asphalt mastics with additives including aluminum hydroxide, magnesium hydroxide, and limestone filler have shown similar improvements in thermal stability with the addition of a material that is inert until higher temperatures [21].

3.2 Fungal Biodeterioration

Representative biodeterioration results for each agar-based binder are displayed in Fig. 3. The control agar-based binder samples exhibited good fungal resistance and only traces of fungal growth (*i.e.,* <than 10% of sample area) were observed after 28 days. Although the fungal resistance of agar-based binders decreased slightly with an increase in agar concentration (*i.e.,* 5% vs 7.5%), the growth rating of both the A5-C and A7.5-C samples were rated as a 1. The fungal resistance of samples increased with the addition of limestone and less sporulating growth was observed on A5-L and A7.5-L samples as compared to control counterparts. After 28 days, A5-L and A7.5-L samples exhibited

trace fungal growth and were rated as a 1. The experiment was validated using both a negative control (*i.e.*, no sample) and positive control (*i.e.*, cellulose). At 28 days, no growth was observed on the negative control specimens (rating of 0), and full growth coverage was observed on the positive control specimens (rating of 4).

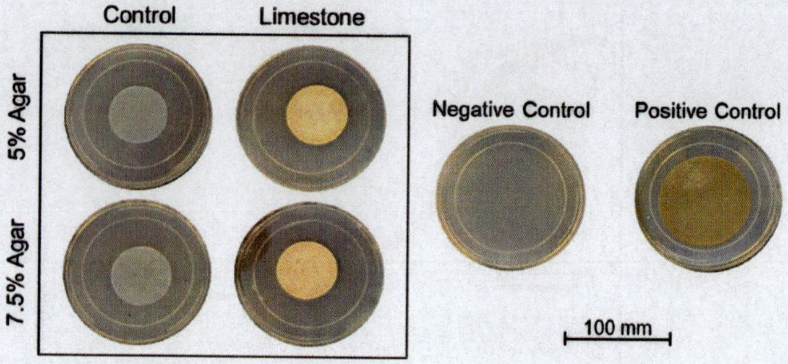

Fig. 3. Representative photographs of agar-based binder samples subjected to ASTM G21 biodeterioration resistance testing. All photographs were taken at the end of a 28-day incubation period at 30 °C. Scale bar = 100 mm.

A major concern in relation to developing biobased building or construction materials is biological deterioration due to microorganisms. Organic materials including wood, cellulose, and other biopolymers are vulnerable to biodeterioration due to undesirable hydrophilic behavior that can provide pathways of moisture for the introduction of harmful microorganisms [1]. Biodeterioration risks pose challenges related to the appearance and mechanical properties of biobased materials as well as impact human health and the service life of built infrastructure these materials are integrated into [22].

Biodegradation is a biochemical process where microorganisms metabolically degrade complex materials into natural compounds such as water, CO_2, and biomass [23]. The ASTM G21 [17] procedure is designed to provide an environment where heterotrophic microorganisms are provided with all the necessary components for growth (e.g., moisture, salts, minerals) except for excess organic carbon. In the current study, because only traces of growth were observed on the samples 28 days after inoculation, it was concluded that agar and limestone did not serve as a primary carbon source for the growth of heterotrophic microorganisms, which is known to lead to biodeterioration. Agar is used as a microbiology culture media because it is non-nutritive for the vast majority of microorganisms, making it a promising material candidate for construction applications where fungal durability is of utmost importance.

3.3 Moisture Sorption

The experimental sorption curves of agar-based binders are shown below in Fig. 4. The moisture content was calculated using Eq. 1 and plotted against the square root of time ($\sqrt{}$hr). In general, pure agar-based binders absorbed a large percentage of moisture relative to the conditioned mass of samples (*i.e.*, samples incubated at 40 °C until constant mass). This behavior was expected due to the chemical structure of agar and the hydrophilic behavior of hydrogels [24]. Upon immersion, the initial rate of hydrogel sorption is very high due to the filling of capillary spaces withing the hydrogel structure that initiates hydrogen bonding of water with the -OH groups present in agar. As free space within the hydrogel structure is filled with solvent, water molecules displace the polymer chains and the hydrogel structure is able to swell [25].

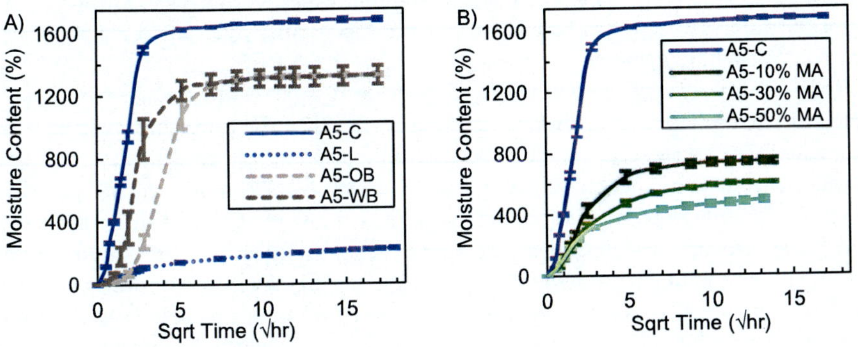

Fig. 4. Characteristic sorption curves of 5% agar-based binders. A) Agar-based binders without modification (C) and with limestone filler addition (L), and with oil-based (OB) and water-based (WB) polyurethane coating at 20 ± 2 °C. B) Agar-based binders without physical modification (C) and with 10%, 30%, and 50% maleic anhydride chemical modification at 20 ± 2 °C. Error bars represent one standard deviation of triplicate measurements.

As shown in Fig. 4A, the final equilibrated moisture content of A5-C and A5-L samples were 1,682% and 215%, respectively, while the final equilibrated moisture content of A5-OB and A5-WB samples were 1,325% and 1,327%, respectively. The inclusion of limestone significantly reduced the moisture sorption of agar-based binders and these findings align with literature on hydrogel-based composites. Rigid fillers can increase composite density, mechanical stability, and alter the swelling capacity of hydrogels [26, 27].

Coatings and chemical crosslinking with maleic anhydride were investigated as potential solutions to further mitigate the propensity of water sorption. As shown in Fig. 4, oil-based (OB) and water-based (WB) polyurethane coated samples initially performed well upon immersion, absorbing virtually no moisture. Polyurethane is susceptible to crack formation on high stress application areas which can lead to hydrolytic degradation [28]. After approximately 3–5 h, imperfections in the polyurethane coating formed around sample edges, which initiated the ingress of moisture and eventual swelling of the sample beneath the coating. Notably, the WB polyurethane samples failed before the OB polyurethane samples, indicating higher hydrolytic durability of OB polyurethane, comparatively. The difference in final equalized moisture content between A5-C and both WB and OB polyurethane coated samples is likely due to the mass of the polyurethane coatings, which absorb trivial amounts of moisture.

Figure 4B presents results for 5% agar-based binder formulations chemically modified by cross-linking with maleic anhydride. The addition of 10%, 30%, and 50% maleic anhydride decreased moisture sorption from 1,682% to 744%, 605%, and 512%, respectively. There are four -OH groups present on the chemical structure of agar. When introduced into the agar solution, maleic anhydride forms two carboxylic acid groups. One of these groups is neutralized by sodium hydroxide while the remaining carboxylic acid group reacts with a hydroxyl group of agar to form an ester. Cross-linking with maleic anhydride decreases the number of available hydroxyl groups and increases the rigidity of the agar structure, which can reduce the water attraction and hydrophilicity of agar-based binders [16]. Although agar-based binders are not moisture resistant, these results indicate that steps can be taken to chemically modify agar-based materials and improve hydrophilic behavior.

The use of polyurethane and maleic anhydride in agar-based materials is accompanied by trade-offs in the environmental impact and material properties, respectively. Although polyurethane materials can be derived from natural sources like lignin and vegetable oils including caster, soybean, and sunflower oil [29], polyurethane materials are more commonly produced using petroleum-based raw materials [30]. Additionally, although maleic anhydride can decrease agar hydrophilicity, it can come at the cost of reducing other functional properties like gel strength [16]. Notably, the addition of 15% w/w limestone filler led to the largest decrease in moisture sorption as compared to physical modification with polyurethane coating and chemical modification by cross-linking with maleic anhydride and is accompanied by less significant deleterious environmental or function impacts on agar-based binder composites.

4 Conclusions

This work evaluated agar-based binders for use as direct alternatives for traditional petroleum-based binders in structural composites. The thermal, biodeterioration, and moisture sorption properties of 5 and 7.5% w/w agar-based materials were characterized and compared to traditional petroleum-based materials. The experimental data indicated the following conclusions:

- Thermogravimetric analysis illustrated that agar-based materials were thermally stable up to 210 °C. The onset of thermal decomposition occurred at a lower temperature for agar-based materials as compared to petroleum based plastic and asphaltic binders. However, the addition of limestone filler was shown to significantly improve the thermal stability of agar-based binders allowing for thermal tailorability.
- When subjected to fungal biodeterioration testing, agar-based materials exhibited improved resistance to biodeterioration as compared to a positive control (*i.e.,* cellulose). Samples with and without limestone had only trace fungal growth after 28 days, illustrating that agar and limestone did not serve as effective carbon sources for the growth of heterotrophic microorganisms.
- In moisture sorption testing, agar-based materials were found to absorb over 1,600% in moisture relative to their dehydrated mass. The hydrophilicity of agar-based binders could be modified through the use of limestone filler, polyurethane coating, and chemical cross-linking with maleic anhydride. Of the following methods, the addition

of 15% w/w limestone filler was found to be the most effective and reduced moisture sorption from 1,682% to 215%.

While further development is needed, the results from this study substantiate that agar could be a viable material candidate for use as a direct alternative material for traditional, petroleum-based construction products in select applications. Although results from this work are promising, future work is needed to evaluate aging mechanisms, durability, scalability, environmental benefits, and the life cycle cost of agar-based materials before they can be implemented at large scale.

Acknowledgements. This research was made possible by the Department of Civil, Environmental, and Architectural Engineering, the College of Engineering and Applied Sciences, and the Living Materials Laboratory at the University of Colorado Boulder. The financial support of the first author through United States Department of Education Graduate Assistance in Areas of National Need Grant P200A180024 and collaboration with the Colorado Department of Transportation Materials Laboratory is gratefully acknowledged. This work represents the views of the authors and not necessarily those of the sponsors.

References

1. Billington, S.L., Srubar, W.V., Michel, A.T., Miller, S.A.: Renewable biobased composites for civil engineering applications. Sustain. Compos. Fibers Resins Appl. **2**, 313–348. DEStech Publications, Inc. (2014)
2. Magalhães, S., et al.: Brief overview on bio-based adhesives and sealants. Polymers **11**(10), Art. no. 10 (2019). https://doi.org/10.3390/polym11101685
3. Speight, J.G.: Chapter 8 - uses of asphalt. In: Speight, J.G. (ed.) Asphalt Materials Science and Technology, Butterworth-Heinemann, Boston, pp. 341–359 (2016). https://doi.org/10.1016/B978-0-12-800273-5.00008-8
4. Eneh, O.C.: A review on petroleum: source, uses, processing, products, and the environment. J. Appl. Sci. **11**(12), 2084–2091 (2011). https://doi.org/10.3923/jas.2011.2084.2091
5. Tabassum, M.R., Xia, A., Murphy, J.D.: Potential of seaweed as a feedstock for renewable gaseous fuel production in Ireland. Renew. Sustain. Energy Rev. **68**, 136–146 (2017). https://doi.org/10.1016/j.rser.2016.09.111
6. Milledge, J.J., Smith, B., Dyer, P.W., Harvey, P.: Macroalgae-derived biofuel: a review of methods of energy extraction from seaweed biomass. Energies **7**(11), Art. no. 11 (2014). https://doi.org/10.3390/en7117194
7. Pahlavan, F., Rajib, A., Deng, S., Lammers, P., Fini, E.H.: Investigation of balanced feedstocks of lipids and proteins to synthesize highly effective rejuvenators for oxidized asphalt. ACS Sustain. Chem. Eng. **8**(20), 7656–7667 (2020). https://doi.org/10.1021/acssuschemeng.0c01100
8. Audo, M., et al.: Subcritical hydrothermal liquefaction of microalgae residues as a green route to alternative road binders. ACS Sustain. Chem. Eng. **3**, 583–590 (2015). https://doi.org/10.1021/acssuschemeng.5b00088
9. Madera-Santana, T.J., Robledo, D., Azamar, A., Rios-Soberanis, C.R.: Preparation and characterization of low density polyethylene-agar biocomposites: torque-rheological, mechanical, thermal and morphological properties. Polym. Eng. Sci. **50**(3), 585–591 (2010). https://doi.org/10.1002/pen.21574

10. Mostafavi, F.S., Zaeim, D.: Agar-based edible films for food packaging applications - a review. Int. J. Biol. Macromol. **159**, 1165–1176 (2020). https://doi.org/10.1016/j.ijbiomac.2020.05.123
11. Armisen, R., Galatas, F., Spain, H.S.A.: Agar. In: Handbook of Hydrocolloids, 2nd ed., Woodhead Publishing, 2009, pp. 82–105 (2009)
12. Lee, W.-K., Lim, Y.-Y., Leow, A.T.-C., Namasivayam, P., Abdullah, J.O., Ho, C.L.: Biosynthesis of agar in red seaweeds: a review. Carbohyd. Polym. **164**, 23–30 (2017). https://doi.org/10.1016/j.carbpol.2017.01.07
13. Guo, Y., Zhang, B., Zhao, S., Qiao, D., Xie, F.: Plasticized starch/agar composite films: processing, morphology, structure, mechanical properties and surface hydrophilicity. Coatings **11**(3), 311 (2021). https://doi.org/10.3390/coatings11030311
14. Chang, I., Prasidhi, A.K., Im, J., Cho, G.-C.: Soil strengthening using thermo-gelation biopolymers. Constr. Build. Mater. **77**, 430–438 (2015). https://doi.org/10.1016/j.conbuildmat.2014.12.116
15. Nasef, M., Elhefian, E., Saalah, S., Yahaya, A.: Preparation and properties of non-crosslinked and ionically crosslinked chitosan/agar blended hydrogel films. J. Chem. **8** (2011). https://doi.org/10.1155/2011/513204
16. Ye, S., et al.: Synthesis and properties of maleic anhydride-modified agar with reversibly controlled gel strength. Int. J. Biol. Macromol. **201**, 364–377 (2022). https://doi.org/10.1016/j.ijbiomac.2021.12.096
17. ASTM International, "ASTM G21: Standard Practice for Determining Resistance of Synthetic Polymeric Materials to Fungi." American Society for Testing and Materials (2021). https://www.astm.org/Standards/G21.htm
18. Nishinari, K., Fang, Y.: Relation between structure and rheological/thermal properties of agar. A mini review on the effect of alkali treatment and the role of agaropectin. Food Struct. **13**, 24–34 (2017). https://doi.org/10.1016/j.foostr.2016.10.003
19. Taharuddin, N.H., et al.: Effect of agar on the mechanical, thermal, and moisture absorption properties of thermoplastic sago starch composites. Materials **15**(24), 8954 (2022). https://doi.org/10.3390/ma15248954
20. Miandad, R., et al.: Catalytic pyrolysis of plastic waste: moving toward pyrolysis based biorefineries. Front. Energy Res. **7**, 27 (2019). https://doi.org/10.3389/fenrg.2019.00027
21. Bonati, A., Merusi, F., Polacco, G., Filippi, S., Giuliani, F.: Ignitability and thermal stability of asphalt binders and mastics for flexible pavements in highway tunnels. Constr. Build. Mater. **37**, 660–668 (2012). https://doi.org/10.1016/j.conbuildmat.2012.07.096
22. Stefanowski, B.K., Curling, S.F., Ormondroyd, G.A.: A rapid screening method to determine the susceptibility of bio-based construction and insulation products to mould growth. Int. Biodeterior. Biodegrad. **116**, 124–132 (2017). https://doi.org/10.1016/j.ibiod.2016.10.025
23. Folino, A., Karageorgiou, A., Calabrò, P.S., Komilis, D.: Biodegradation of wasted bioplastics in natural and industrial environments: a review. Sustainability **12**(15), 6030 (2020). https://doi.org/10.3390/su12156030
24. Liu, Y., et al.: Polymer–water interaction enabled intelligent moisture regulation in hydrogels. J. Phys. Chem. Lett. **12**(10), 2587–2592 (2021). https://doi.org/10.1021/acs.jpclett.1c00034
25. Yang, L., Paulson, A.T.: Mechanical and water vapour barrier properties of edible gellan films. Food Res. Int. **33**(7), 563–570 (2000). https://doi.org/10.1016/S0963-9969(00)00092-2
26. Utech, S., Boccaccini, A.R.: A review of hydrogel-based composites for biomedical applications: enhancement of hydrogel properties by addition of rigid inorganic fillers. J. Mater. Sci. **51**(1), 271–310 (2015). https://doi.org/10.1007/s10853-015-9382-5
27. Dave, P.N., Gor, A.: Chapter 3. Natural polysaccharide-based hydrogels and nanomaterials: recent trends and their applications. In: Handbook of Nanomaterials for Industrial Applications, C. M. Hussain, Elsevier Publishing, 2018, pp. 36–66 (2018). https://doi.org/10.1016/B978-0-12-813351-4.00003-1

28. Sastri, V.R.: Chapter 7 - engineering thermoplastics: acrylics, polycarbonates, polyurethanes, polyacetals, polyesters, and polyamides. In: Sastri, V.R. (ed.) Plastics in Medical Devices, William Andrew Publishing, Boston, 2010, pp. 121–173 (2010). https://doi.org/10.1016/B978-0-8155-2027-6.10007-8
29. Reghunadhan, A., Thomas, S.: Chapter 1 - polyurethanes: structure, properties, synthesis, characterization, and applications. In: Thomas, S., Datta, J., Haponiuk, J.T., Reghunadhan, A. (eds.) Polyurethane Polymers, Elsevier, Amsterdam, pp. 1–16 (2017). https://doi.org/10.1016/B978-0-12-804039-3.00001-4
30. Enderus, N.F., Tahir, S.M.: Green waste cooking oil-based rigid polyurethane foam. IOP Conf. Ser. Mater. Sci. Eng. **271**, 012062 (2017). https://doi.org/10.1088/1757-899X/271/1/012062

Development of a Bio-Hybrid Insulation Material – Connection by Growth and Interlinking

Sabine Giglmeier(✉), Wolfgang Karl Hofbauer, Christian Kaiser, Nicole Krueger, Martin Krus, and Regina Schwerd

Fraunhofer Institute for Building Physics IBP, Fraunhoferstraße 10, 83626 Valley, Germany
sabine.giglmeier@ibp.fraunhofer.de

Abstract. The project presentedhere investigates the extent to which fungal mycelium can be used to replace conventional fossil or mineral binders in building materials. Both biological raw materials and residual materials were used as structural building materials and bonded together by mycelial growth. The material to be developed should be applicable as building insulating material. In tentative preliminary tests, different material combinations of nutrient, fungus and building structure material were examined for their potential. Based on these findings, larger prototypes measuring 25 × 25 × 3 cm^3 were then manufactured. Their material properties were subsequently determined in the laboratory. It was found that the manufactured prototypes consistently exhibited very low diffusion resistance. The resulting thermal conductivity lies within ranges from 0.049 to 0.068 W/(mK), depending on the type of fungus, the substrate and the aggregates. This is rather high compared to conventional insulation materials. However, the experiments show the potential of myco-composites as substitute for mineral oil based products.

Keywords: myco-composite · fungus · insulation

1 Introduction

In order to achieve a biological transformation, material flows have to be considered as a whole and bio-intelligent solutions have to be developed. Closed material cycles are essential in this context. Nowadays, the use of biological principles for the production of building materials still is a niche compared to the use of fossil fuel based raw materials. At the same time, construction waste (more than 220 million tons in Germany in 2016) represents a huge part of the overall waste. Hence, there is an urgent need to reuse residues and waste materials and to integrate them into new construction materials.

Within this project, it was investigated if a bio-hybrid construction product can be developed by replacing chemical binding processes by biological principles. Both mineral waste from the construction industry and biological raw materials are used as structural building materials. The crosslinking between the particles takes place by

S. Amziane et al. (Eds.): ICBBM 2023, RILEM Bookseries 45, pp. 676–689, 2023.
https://doi.org/10.1007/978-3-031-33465-8_52

mycelial growth. After achieving sufficient stability and strength, the mycelial growth has to be stopped. It is crucial that this process has no negative influence on the material properties and at the same time guarantees a hygienically harmless product since, in the long term, the newly developed material is to be used as an insulation material.

Recently a lot of different approaches on fungal based materials were realized. Examples are fungal based packaging materials e.g. [1], myco-leather e.g. [2], high-performance foams e.g. [2], and building materials e.g. [3–6]. If products derived from fungal growth shall be realized, a suitable organic substrate as source for growth is necessary. In exceptional cases the substrate could be a laboratory growth medium. For sustainability reasons the use of organic waste substrates and by-products, e.g. from agriculture, forestry or food production, is strongly preferred. Because of their material properties, especially organic materials with fiber content are suitable. Typical substrates which are also used in the production of edible mushrooms are cereal straw, *Miscanthus* straw, saw dust [5, 7], bamboo fibers [6], hemp shives or wood chips [4] as well as other wastes, by-products or recycling materials [8, 9].

Various fungal species have been used in attempts to produce mycelium derived products. With regard to myco-leather *Ganoderma* spp, *Trametes versicolor*, *Fomes* spp., *Pycnoporus* spp. And *Perenniporia* spp. Are or have been in the focus [2]. Fungi used for the production of biocomposites are typically white rot fungi (e.g. *Ganoderma lucidum, Ganoderma resinaceum. Pleurotus ostreatus, Trametes ochracea (multicolor), Trametes versicolor*) [4–7, 9, 10], because they leave the fiber structure of the organic substrate relatively intact. Only seldom brown rot fungi like *Gloeophyllum sepiarium* have been used [8]. But even saprotrophic fungi originating from soil habitats like *Agaricus bisporus* or mold fungi like *Trichoderma asperellum* were applied in trials to produce myco-composites [11].

2 Materials and Methods

As the objective of this project consists in developing a bio-hybrid insulation material, different mixtures of aggregates, nutrients and fungi as well as different methods to stop the mycelial growth process were investigated. Basic requirements to the resulting product are sufficient strength and low thermal conductivity, combined with a plate-like shape.

2.1 Materials

Some attempts on fungal based insulation materials have been made or are currently under way [7, 8]. The basis for the project presented within this contribution is a bio-intelligent approach due to the raw materials which were used for the mycelium bound bio hybrid insulation products. Firstly, the extent to which fungal mycelium can be used to replace conventional fossil or mineral binders in building materials like insulation products was investigated. Secondly, both biological raw materials and residual materials from recycling processes were used as substrates. Biological raw materials and by-products, consisting mainly of ligno-cellulose, demand some effort until they can serve as substrate for fungal growth. In order to enhance mycelium production it can be helpful to use an additional nutrient source. Therefore, additional use of a yeast granulate was investigated. The different raw materials are listed in Table 1.

Table 1. Materials used for trials and prototypes.

Material	Particle size	Function	Characteristic
Typha (Common cattail) particles coarse	Up to 100 mm	substrate	Leafs from common cattail possess a unique porous structure and can be used as an insulation material by themselves already
Typha (Common cattail) particles fine	2–3 mm	substrate	Smaller particle size than the coarse version for more homogenous structure in the finished product
Typha (Common cattail) seed hairs	2–3 mm	additional substrate	Seed hairs from common cattail represent a very fine material with a large surface and could improve the nutrient quality and potentially provide better fungal growth
Spelt hulls	Up to 15 mm	substrate	Agricultural byproduct, also used as poultry litter
Wheat straw coarse	Up to 200 mm	substrate	„Classical" substrate for mushroom cultivation
Wheat straw fine	Up to 40 mm	reference substrate	Provides a more homogenous structure and allows good growth for many fungi, therefor used as reference substrate
Torrefied material	2–10 mm	filler	Similar to charcoal. Only weak substrate, therefor used as filler
Aerated concrete	3–5 mm	filler	Recycle product with no nutrient content, used as filler
Yeast granulate	Very fine (<1 mm)	additional nutrients („booster")	Provision of additional nutrients – might enhance mycelial growth

Typha particles present an innovative substrate with respect to insulation products because of their high porosity and relatively low density but have never been used in bio-composites so far. Other substrates like wheat straw and spelt hulls are typical byproducts from agriculture. In addition, wheat straw also served as a reference substrate, since experience with culture on wheat straw is available for the test fungi. Particles from aerated concrete as well as from torrefied wood are derived from waste streams. As a preliminary test showed, torrefied material and aerated concrete do not possess any remarkable nutrient content, hence they were used as a filler. Dried yeast filtrate (yeast granulate) is a byproduct from breweries and forms an ideal nutrient supplement also containing vitamins and micro nutrients.

2.2 Organisms

The fungi used for the experiments are given in Table 2. Apart from *Schizophyllum commune*, from which a proprietary strain (after separation from a damaged hemp fiber insulation) was used, the test fungi were obtained from the Mushroom Research Center Austria GmbH MRCA, Innsbruck. Fungi were propagated by mushroom spawn. Spawn was also used to inoculate the substrate mixtures and was prepared according to [12] with the exception that the mineral vermiculite was added in order to facilitate better intermixture with the substrates.

Table 2. Selected test fungi

Fungus common name	Fungus scientific name	Characteristic
Oyster mushroom	*Pleurotus ostreatus* (Jacq.) P. Kumm	White rot fungus. Widely cultured edible mushroom. Easy to grow on a very great variety of substrates
Split gill mushroom	*Schizophyllum commune* Fr	White rot fungus. Very widespread, may dry out, but recovers. Affinity to plant fiber substrates. Proprietary isolate used
Reishi mushroom	*Ganoderma lucidum* (Curtis) P. Karst	White rot fungus. Widely used and cultivated as medicinal mushroom. Mycelium development often dense and therefore often used in building products and myco-leather
Turkey Tail fungus	*Trametes versicolor* (L.) Lloyd	White rot fungus. Very widespread. Also cultivated as medicinal mushroom. Mycelial characteristics comparable to *Ganoderma lucidum*, also used for building products and myco leather production
Shiitake mushroom	*Lentinula edodes* (Berk.) Pegler	White rot fungus. Widely cultured edible mushroom. Affinity to hardwood substrates. Used strain also can grow on straw

2.3 Preparation of Test Specimens

Different kinds of test samples containing different combinations of the above described materials were produced. For all specimen (orienting preliminary tests and plate-shaped prototypes), in a first step, all raw materials were soaked in water containing 1g gypsum powder for 16 to 20 h according to [12]. Soaking is necessary to provide the sufficient amount of water for fungal growth and gypsum has revealed to be a good Calcium source for good hyphal development. Then the specimen were manufactured by packing the mixtures into specific shapes by use of different containers and partly additionally packing the material between two sets of membranes. After this, the filled containers were autoclaved (121 °C, 1 h) and were inoculated with the test fungi after cooling down by use of fungus spawn. Inoculation with the test fungi was provided by application of approx. 10% of fungus spawn ("Pilzbrut") which was mixed in thoroughly to achieve a homogenous mixture. As test containers, autoclavable bags with air filters or plastic molds were used for orienting trials. In Table 3 the material combinations are given which were used for orienting trials. Figure 1 shows one specimen (T.1) of these orienting trials.

For further measurements, panel shaped prototypes were developed using specially designed containers in combination with fiber mats. As Fraunhofer IBP has facilities and research infrastructure that allow panels with a base area of 25 cm by 25 cm to be used to determine both the thermal conductivity and the hygric material parameters, the size of the future prototypes was chosen in such a way that four test specimens with an area of 10 × 10 cm^2 can be cut out of each prototype for the hygrothermal characteristic value determinations without edge areas. However, special press molds had to be designed, that withstand autoclaving or steam application of the samples. The fungi are inoculated into the substrate as well as into further aggregate from recycled material in the corresponding laboratory. The inoculated substrate mixture is placed in biological incubators where mycelium growth ties the different particles together. During this process special measures preclude mold growth and bacteria infestation. To avoid an unnecessary additional work step, it should also be possible to kill the fungus in the mold using the selected procedures. These requirements excluded the use of wooden components for the press mold as wood is not suitable to be used for autoclaving.

Particularly because of the potential exposure to steam, a temperature-resistant filament with fiber reinforcement was used for the molds. They were produced using 3D printing and can tolerate temperatures above 100 °C. Since the available press can only accommodate molds up to a width of just over 30 cm, particularly slim but still sufficiently strong wall thicknesses of only 15 mm were used. To allow the gas exchange necessary for fungal growth and subsequent drying, perforated plates were made for the bottom and the top plate. The entire construction consists of an outer frame with the permanently installed base plate and a retractable inner frame, which is placed in the outer frame. It is filled with the "growth mixture" and then fitted with the lid that can be locked at the selected height, thus determining the sample thickness. For illustration purposes, Fig. 2 shows photographic images of the molds as a top view in the closed state (left) and after the test specimen has been stripped (right).

In Table 4 the material combinations are given which were used for the prototypes (production in duplicates). The prototypes were prepared in the especially developed containers with height lock (as described above).

Table 3. Conducted trials.

ID	Substrate	Filler	Additional nutrients	Fungus	Container	Inactivation of growth	Result
T.1	Typha coarse	-	-	*Schizophyllum commune*	Autoclavable bag with filter	autoclaving + drying	almost no stability
T.2	Typha coarse	torrefied material	-	*Pleurotus ostreatus*	autoclavable bag with filter	autoclaving + drying	almost no stability
T.3	Typha coarse	aerated concrete	-	*Pleurotus ostreatus*	autoclavable bag with filter	autoclaving + drying	no stability
T.4	Typha coarse	aerated concrete	-	*Schizophyllum commune*	autoclavable bag with filter	autoclaving + drying	no stability
T.5	Typha particles	-	-	*Pleurotus ostreatus*	autoclavable bag with filter	autoclaving + drying	almost no stability
T.6	Typha particles	-	-	*Schizophyllum commune*	autoclavable bag with filter	autoclaving + drying	no stability
T.7	Typha particles + seed hairs	-	-	*Pleurotus ostreatus*	autoclavable bag with filter	autoclaving + drying	almost no stability
T.8	Typha particles + seed hairs	-	-	*Schizophyllum commune*	autoclavable bag with filter	autoclaving + drying	no stability
T.9	Wheat straw coarse	-	-	*Pleurotus ostreatus*	autoclavable bag with filter	autoclaving + drying	mediocre stability
T.10	Wheat straw coarse	-	Yeast granulate	*Pleurotus ostreatus*	autoclavable bag with filter	autoclaving + drying	mediocre stability
T.11	Wheat straw fine	-	-	*Pleurotus ostreatus*	autoclavable bag with filter	autoclaving + drying	mediocre stability
T.12	Wheat straw fine	-	-	*Schizophyllum commune*	autoclavable bag with filter	autoclaving + drying	no stability

(*continued*)

Table 3. (*continued*)

ID	Substrate	Filler	Additional nutrients	Fungus	Container	Inactivation of growth	Result
T.13	Wheat straw fine	-	-	*Ganoderma lucidum*	autoclavable bag with filter	autoclaving + drying	promising
T.14	Wheat straw fine	-	-	*Trametes versicolor*	autoclavable bag with filter	autoclaving + drying	good stability
T.15	Wheat straw fine	-	-	*Lentinula edodes*	autoclavable bag with filter	autoclaving + drying	no stability
T.16	Wheat straw fine	-	Yeast granulate	*Pleurotus ostreatus*	autoclavable bag with filter	autoclaving + drying	mediocre stability
T.17	Wheat straw fine	-	-	*Pleurotus ostreatus*	plastic container with membrane	autoclaving + drying	crumbling
T.18	Wheat straw fine	-	-	*Pleurotus ostreatus*	plastic container with mesh	autoclaving + drying	crumbling
T.19	Wheat straw fine	-	-	*Ganoderma lucidum*	plastic container with mesh	autoclaving + drying	good stability
T.20	Wheat straw fine	-	-	*Ganoderma lucidum*	plastic container with mesh	Only drying	good stability
T.21	Wheat straw fine	-	-	*Trametes versicolor*	plastic container with mesh	autoclaving + drying	good stability
T.22	Wheat straw fine	-	-	*Trametes versicolor*	plastic container with mesh	Only drying	good stability
T.23	Wheat straw fine	Aerated concrete	-	*Pleurotus ostreatus*	autoclavable bag with filter	autoclaving + drying	almost no stability
T.24	Wheat straw fine	Aerated concrete	-	*Schizophyllum commune*	autoclavable bag with filter	autoclaving + drying	no stability

(*continued*)

Table 3. (*continued*)

ID	Substrate	Filler	Additional nutrients	Fungus	Container	Inactivation of growth	Result
T.25	Spelt hulls	-	-	*Pleurotus ostreatus*	autoclavable bag with filter	autoclaving + drying	mediocre stability
T.26	Spelt hulls	-	Yeast granulate	*Pleurotus ostreatus*	autoclavable bag with filter	autoclaving + drying	good stability
T.27	Wheat straw fine + Spelt hulls	-	-	*Pleurotus ostreatus*	autoclavable bag with filter	autoclaving + drying	mediocre stability

Fig. 1. Specimen with coarse Typha particles and *Schizophyllum commune* (T.1). On the left: preparation of the mixture in an autoclave bag with filter, on the right: specimen after finishing including the autoclavation process.

After inoculation the mycelium was allowed to grow for six to eight weeks into the substrate mixture under environmental conditions favorable for fungal growth (25 °C).

Growth of fungi usually results into a layered structure. While in the core of the substrate sample, the branched hyphal structures are effectively binding the particles by penetration and encapsulation, the fungus usually forms a dense mat at the surface. Additional stability can be achieved by applying a fiber membrane at the lower and upper surface. An extremely diffusion-open film was placed on the base plate and under the top plate of the molds to enhance stability and to prevent the fungal mycelium from growing through it. This film may be left as part of the composite structure or stripped at the end of the production process.

Once the substrate particles are sufficiently bound by the mycelial network the fungal growth must be stopped. For this purpose, different methods are at hand. In this project

Fig. 2. View of the molds made from high-temperature filament in 3D printing to produce the prototypes. The left image shows a top view in the closed state and the right one after stripping the test specimen.

Table 4. Material combinations of the prototypes

ID	Substrate	Filler	Fungus	Inactivation of growth	Result
PT.1	Typha particles	-	*Ganoderma lucidum*	autoclaving + drying	Good stability
PT.2	Typha particles	Aerated concrete	*Ganoderma lucidum*	autoclaving + drying	Good stability
PT.3	Wheat straw fine	Aerated concrete	*Ganoderma lucidum*	autoclaving + drying	Good stability

autoclaving and the application of dry heat (60 h at 80 °C) to inactivate the fungi were employed.

2.4 Determination of Hygrothermal Characteristics

The test specimens removed from the molds (prototypes) were primarily used to determine the hygrothermal material parameters. Four test specimens with dimensions of 10 cm x 10 cm were prepared out of each panel, whereby no edge areas are deliberately used. The water vapor diffusion resistance number was determined at first on these test specimens using the dry cup method and then using the wet cup method (according to DIN EN ISO 12572). Subsequently, these samples were used to determine the water absorption coefficient (according to DIN EN 772–4). Both the determination of the porosity (DIN 66137) and the determination of the sorption properties (DIN EN ISO 12571) were carried out using smaller residual pieces.

3 Results and Discussion

3.1 Preparation of Test Specimens and Prototypes

The extensive trials (Table 3) were the basis for the optimization of the growth process and the choice of the prototypes for further measurements.

As expected wheat straw usually supported the growth of the test fungi very well, so it proved to be a good reference material. Growth with fine Typha particles was sufficient, but the use of fillers did not improve the quality of the final product, nor did the addition of torrefied material. Additional nutrients (yeast granulate) did not consistently improve mycelial growth. The test fungi *Pleurotus ostreatus* showed to be a good reference organism, because it grew on all substrates very well. But often the structure of the bound material with *Pleurotus* was not very firm. Also *Schizophyllum commune* and *Lentinula edodes* resulted in unstable composites. Best structures and firmness of the myco-composites were reached when *Ganoderma lucida* and *Trametes versicoloir* were used.

A fiber membrane was considered to be conducive for surface stabilization and therefore the prototypes were equipped with it on both sides (Table 4). Although this may enhance the stability of the final product, it did complicate the process of removing the prototype from the mold. Before stopping the growth the fiber membrane was removed. For future improvement of the prototypes it should be considered to use natural fiber membranes or other techniques to obtain a dense and smooth surface.

All the trials were conducted in a microbiological laboratory and the specimens were autoclaved to stop the mycelial growth for hygienic purposes. However, mold growth could not be ruled out completely. In some trials (Table 3), dry heat was applied with success. Since there was no obvious difference in the properties of the materials with respect to their stability depending on the inactivation treatment (autoclaving or heating), only autoclaving was used to inactivate the test fungi in the further course. As the myco-composites were moist after this treatment, the specimens required drying after autoclaving to prevent mold development.

Concerning the thermoplastic molds for the prototypes, degradation could be observed during the autoclaving process. Obviously 121 °C were too high a temperature for the selected material. The tests showed that also dry heat (80 °C) applied for several hours (lasting at least until the material is dry) could be used to inactivate the

Fig. 3. View of the prototypes PT1.1 (left, Typha - *Ganoderma*), PT2.1 (middle, Typha - aerated concrete - *Ganoderma*) and PT3.2 (right, wheat straw fine - aerated concrete - *Ganoderma*).

fungi production. Thus, it should be possible to avoid destruction of the molds in the future.

The resulting panels (Table 4) showed sufficient stability for further hygrothermal measurements. Figure 3 shows the specimen PT1.1 (left) and PT2.1 (middle) and PT3.2 (right).

3.2 Hygrothermal Characteristics of the Biohybrid Materials (Prototypes)

Table 5 lists the results regarding diffusion resistance coefficient (dry and wet cup), sorption characteristics, water absorption coefficient and porosity. This table also indicates which aggregates and substrates and which types of fungi were used.

Table 5. Tabular compilation of the hygric material parameters determined

Sample	ID	Bulk density [kg/m^3]	Dry cup (s_d value) [m]	Wet cup (s_d value) [m]	Sorption [Vol.-%]	A-Value [kg/m$^2\sqrt{h}$]	Porosity [%]
PT.1.1 Typha Ganoderma	PT.1.1a	57	0.094	0.092	0.64–65% RH	1.16	95.68
	PT.1.1b	63	0.097	0.088	0.98–80% RH	0.89	
	PT.1.1c	79	0.096	0.096	1.60–93% RH	1.29	
	PT.1.1d	84	0.102	0.086	No meas. at 97% RH	1.17	
PT.1.2 Typha Ganoderma	PT.1.2a	69	0.079	0.074	0.64–65% RH	1.44	95.85
	PT.1.2b	66	0.096	0.093	1.01–80% RH	1.57	
	PT.1.2c	84	0.09	0.089	1.55–93% RH	1.06	
	PT.1.2d	74	0.09	0.09	No meas. at 97% RH	1.39	
PT.2.1Typha Aerated concrete Ganoderma	PT.2.1a	107	0.108	0.094	0.69–65% RH	1.35	94.69
	PT.2.1b	114	0.108	0.097	1.03–80% RH	1.04	
	PT.2.1c	111	0.1	0.088	1.60–93% RH	1.27	
	PT.2.1d	97	0.104	0.091	No meas. at 97% RH	0.91	
PT.2.2 Typha Aerated concrete Ganoderma	PT.2.2a	92	0.122	0.109	0.7–65% RH	0.95	95.41

(continued)

Table 5. *(continued)*

Sample	ID	Bulk density [kg/m³]	Dry cup (s_d value) [m]	Wet cup (s_d value) [m]	Sorption [Vol.-%]	A-Value [kg/m² √h]	Porosity [%]
	PT.2.2b	98	0.125	0.114	1.06–80% RH	0.88	
	PT.2.2c	110	0.126	0.108	1.81–93% RH	1.05	
	PT.2.2d	93	0.123	0.112	No meas. at 97% RH	0.75	
PT.3.1 Wheat straw Aerated concreteGanoderma	PT.3.1a	105	0.157	0.13	0.87–65% RH	1.91	94,33
	PT.3.1b	123	0.163	0.137	1.25–80% RH	1.47	
	PT.3.1c	110	0.159	0.136	1.9–93% RH	1.29	
	*PT.3.1d**						
PT.3.2 Wheat straw Aerated concreteGanoderma	PT.3.2a	102	0.144	0.132	0.85–65% RH	1.5	94.71
	PT.3.2b	94	0.125	0.115	1.21–80% RH	1.19	
	PT.3.2c	114	0.142	0.128	1.79–93% RH	1.61	
	PT.3.2d	115	0.145	0.133	No meas. at 97% RH	1.82	

* *The PT3.1d specimen broke during cutting and could therefore not be measured.*

The determination of the thermal conductivity could only be carried out on two prototypes (PT.1 and PT.3), as all other specimens had a surface that was too uneven. The following results were obtained:

PT.1 with Typha and Ganoderma: 0.048 W/(mK).

PT.3 with straw, aerated concrete and Ganoderma: 0.068 W/(mK)

The measured thermal conductivities are surprisingly high, despite low bulk densities. The lowest measured thermal conductivity of sample PT.1 is slightly above 0.045 W/(mK). As expected, the value for sample PT.3 with aerated concrete aggregate is significantly higher than for variant PT.1.

The results of the water vapor diffusion resistance measurement are particularly striking. On the one hand, there are only minor differences between dry and wet cup measurements, which means that there is only a minor influence from surface diffusion. This also correlates with the relatively low sorption moisture contents. Most importantly, all the samples had extremely low water vapor diffusion resistance values, below 4.5 down to 3. This means that the material cannot be used as interior insulation without application of a vapor control layer to avoid condensation problems in the construction.

4 Conclusion

The extensive trials within the study greatly helped to optimize the processes needed for the production of mycelium-bound construction materials at a laboratory scale. Based on the broad variety of material combinations, it was possible to identify promising mixtures of biological raw materials, mineral aggregates and fungi for the production of prototypes, which can serve as a basis for further examinations.

It could be shown that autoclaving as well as the application of dry heat stopped the binding process reliably. In future, it should be kept in mind that the autoclaving process is a very energy consuming process, nevertheless it is essential for hygiene and safety purposes. If proper industrial hygiene can be achieved in a fungal-composite producing factory also less energy consuming techniques could be employed.

The thermal conductivities which were measured for the prototypes were still rather high compared to conventional insulation materials. In some approaches compressing the myco-composite led to good results. If it is better to compress the composites during [9] or after [10] growth remains unanswered so far.

Even if the suitability of the newly prepared bio-intelligent materials as building insulation could not be proven by these first examinations described within this work, the experiment nevertheless shows that myco-composites using biological raw materials and mineral waste materials have some potential as replacement of mineral oil based products. For the application as building insulation, the material characteristics of the myco-composites have to be further enhanced. However, alternative fields of application such as acoustically active panels for noise reduction are possible.

Acknowledgment. This project was part of "Ideenwettbewerb Biologisierung der Technik" and funded by VDI Technologiezentrum on behalf of the Bundesministerium für Bildung und Forschung BMBF (funding code 13XP5157).

References

1. Ziegler, A.R., Bajwa, S.G., Holt, G.A., McIntyre, G., Bajwa, D.S.: Evaluation of physico-mechanical properties of mycelium reinforced green biocomposites made from cellulosic fibers. Appl. Eng. Agric. **32**(6), 931–938 (2016)
2. Vandelook, S., Elsacker, E., Van Wylick, A., De Laet, L., Peeters, E.: Current state and future prospects of pure mycelium materials. Fungal Biol. Biotechnol. **8**(1), 1–11 (2021). https://doi.org/10.1186/s40694-021-00128-1
3. Meyer, V., Basenko, E.Y., Benz, J.P., Braus, G.H., Caddick, M.X., Csukai, M., et al.: Growing a circular economy with fungal biotechnology: a white paper. Fungal Biol. Biotechnol. **7**(1), 5 (2020)
4. Zimele, Z., Irbe, I., Grinins, J., Bikovens, O., Verovkins, A., Bajaree, D.: Novel mycelium-based biocomposites (MBB) as building materials. J. Renew. Mater. **8**(9), 1067–1076 (2020)
5. Appels, F.V.W., et al.: Fabrication factors influencing mechanical, moisture- and water-related properties of mycelium-based composites. Mater. Des. **161**, 64–71 (2019)
6. Soh, E., Chew, Z.Y., Saeidi, N., Javadian, A., Hebel, D., Le Ferrand, H.: Development of an extrudable paste to build mycelium-bound composites. Mater. Des. **195**, 109058 (2020)

7. Dias, P.P., Jayasinghe, L.B., Waldmann, D.: Investigation of Mycelium-Miscanthus composites as building insulation material. Results Mater. **10**, 12 (2021). https://doi.org/10.1016/j.rinma.2021.100189
8. Schritt, H., Pleissner, D.: Recycling of organic residues to produce insulation composites: a review. Cleaner Waste Syst. **3**, 100023 (2022)
9. Elsacker, E., Vandelook, S., Brancart, J., Peeters, E., De Laet, L.: Mechanical, physical and chemical characterisation of mycelium-based composites with different types of lignocellulosic substrates. PLoS ONE **14**(7), e0213954. https://doi.org/10.1371/journal.pone.0213954
10. Shakir, M.A., Azahari, B., Yusup, Y., Yhaya, M.F., Salehabadi, A., Ahmad, M.I.: Preparation and characterization of mycelium as a bio-matrix in fabrication of bio-composite. J. Adv. Res. Fluid Mech. Thermal Sci. **65**(2), 253–263 (2020)
11. Tacer-Caba, Z., Varis, J.J., Lankinen, P., Mikkonen, K.S.: Comparison of novel fungal mycelia strains and sustainable growth substrates to produce humidity-resistant biocomposites. Mater. Des. **192**, 108728 (2020)
12. Stamets, P.: Growing Gourmet and Medicinal Mushrooms, 3rd edn. Ten Speed Press, Berkeley (2000)

Numerical Investigation of Mold Growth Risk in Vapor-Permeable Building Envelopes with Bio-Based Insulation in Cold Climates

Leonardo Delgadillo Buenrostro[1], Louis Gosselin[1](✉), and Pierre Blanchet[2]

[1] Department of Mechanical Engineering, Université Laval, Quebec City, Qubec, Canada
louis.gosselin@gmc.ulaval.ca

[2] Department of Wood and Forestry Sciences, Université Laval, Quebec City, Qubec, Canada

Abstract. It is generally accepted that a vapor barrier is necessary to ensure the durability of building envelopes, especially in cold climates where the risk of winter condensation is high. However, the water vapor resistance of the typical vapor barrier —polyethylene foil— is considered by many to be overly high because it also reduces the drying capacity of the envelope. However, installing an excessively permeable vapor barrier could trigger condensation and mold growth. Organic insulation materials could allow reducing the vapor resistance of the vapor barrier in a safer way, as these are more hygroscopic than petroleum-based insulations. This study numerically investigated the risk of condensation and mold growth of three wall assemblies, from which two were insulated with bio-based materials and the other one was a reference standard wall. Hygrothermal simulations considered different levels of permeability of the vapor barrier and ventilation rates in the air gap behind the exterior cladding. Results indicated that walls with bio-insulation materials have the potential to function in a cold climate without triggering mold growth, especially when using a vapor retarder. Ventilation of the exterior air layer proved to be effective when in contact with organic materials, but practically useless otherwise. It was found that the polyethylene foil performed similarly to a smart vapor barrier in all assemblies, except for the highly hygroscopic wall.

Keywords: Vapor permeability · building envelope · mold growth · hygrothermal simulation

1 Introduction

It is known that a vapor-impermeable layer on the warm side of a building envelope in a cold climate is necessary to hinder water vapor diffusion and avoid excessive moisture accumulation, which can lead to problems such as the earlier deterioration of building materials, damage from expansion of materials, odors and mold growth on surfaces [1]. Mold proliferation does not only render building materials visibly unesthetic, but it also makes wood-based materials unreliable as it induces rot [2]. Moreover, the mold by-products can be diffused into the interior air, where they can threat the occupants'

S. Amziane et al. (Eds.): ICBBM 2023, RILEM Bookseries 45, pp. 690–704, 2023.
https://doi.org/10.1007/978-3-031-33465-8_53

health with allergic reactions, pronounced asthma, and respiratory distress [3]. Hence, the interest of having an adequate moisture protection for the building envelope is evident. The polyethylene foil is one of the most popular materials used as a vapor barrier; however, its low water vapor permeability can backfire by reducing the drying potential of the envelope towards the interior side of the building [4]. In this way, the moisture that found its way into the enclosure has only one way out regardless of the transport mechanism that got it in (vapor diffusion, liquid flow, capillary suction, or air convection). A more permeable vapor barrier, hereinafter called vapor retarder, could allow this moisture to escape towards the interior along with the outward drying, thus decreasing the drying time of the assembly [5]. Furthermore, building materials with a greater hygroscopicity could absorb and redistribute larger amounts of water, therefore reducing the risk of having moisture related issues [6]. For instance, Geving and Holme [7] numerically investigated the maximum vapor resistance of the vapor retarder of two timber-framed assemblies, given that at least a quarter of the total drying should be directed towards the inside to increase the drying speed. It was concluded that vapor retarders with constant vapor resistance did not substantially quicken the drying, and that a smart vapor barrier could be more useful. However, a vapor retarder without enough vapor resistance could have a negative impact over the durability of the building materials because the risk of condensation increases. Condensation greatly increases the risk of mold proliferation, and it should be avoided at all costs. Birjukovs et al. [8] assessed the mold growth risk in test buildings situated in Latvia and found that biodegradable materials were at a greater risk than materials with little organic content. Vinha [9] studied the minimum acceptable vapor resistance of the vapor retarder needed to avoid mold growth in timber-framed wall assemblies. This study covering numerical, laboratory, and field tests found that the best way to improve the hygrothermal performance of the envelopes is to have a thermally insulating and highly permeable sheathing. Likewise, Rode [10] reported that if a hygroscopic insulation material is not directly in contact with the interior air, its moisture buffering properties become negligeable. Moreover, Bastien and Winther-Gaasvig [11] compared the impact of rain loads, capillarity, and vapor diffusion on a hygroscopic and vapor permeable building envelope. What they found is that the capillarity uptake of the exterior cladding and the driving rain have a significantly greater impact on the moisture content of the building envelope than vapor diffusion; and that a generous overhang is essential to ensure the materials durability.

Bio-based insulation materials often have a higher hygroscopicity than petroleum-based ones, and thus can absorb and release humidity more easily. Using organic insulation materials along with a permeable vapor barrier could buffer the interior air humidity in an energy-effective way, particularly when the interior relative humidity (RH) is low [12]. Simonson et al. [13, 14] studied this topic in a house with a hygroscopic vapor-permeable building envelope with wood fiber insulation. It was found that moisture storage in the building envelope substantially improved the level of indoor humidity while keeping an acceptable moisture performance in a cold environment, as long as the vapor diffusion resistance on the warm side was at least three times greater than on the cold one. Their results also showed that the vapor resistance of the vapor barrier can be lower than that of polyethylene foil if the building airtightness is good (i.e., 3 air changes per hour (ACH) at 50 Pa). Similarly, Geving et al. [15] studied the moisture conditions of 15 different wall configurations, with either glass wool or wood fiber insulation. Their results showed that configurations with both kinds of insulation performed rather similarly. The factors that had a positive influence on the hygrothermal conditions were having a 50 mm wood fiber board as wind barrier and installing a smart vapor barrier. In addition, a factor that has proven to enhance the drying process of a building envelope is the air convection in the air layer behind the exterior cladding, as reported by Karagiozis and Künzel [16]. Their simulation results showed the improved moisture performance of a ventilated brick cavity over an unvented wall system. The latter numerical model was validated by Künzel et al. [17], who modelled the effect of air movements in a flat roof. The present work numerically compared the hygrothermal behavior of three wall assemblies with various types of insulation materials and vapor barriers under the influence of air movements in the air cavity behind the exterior cladding. The purpose was to identify ventilation rates and vapor resistance values that yield acceptable moisture conditions and avoid mold growth in building envelopes with organic insulation materials in cold climates, as well as determining how the bio-based wall assemblies with higher hygroscopicity compared to the reference standard wall.

2 Materials and Methods

2.1 Studied Assemblies

Three different building envelopes were considered for the present study. The first one was a standard lightweight wall assembly recommended by the energy transition program Novoclimat of the Canadian province of Quebec, which played the role of a reference wall. It was developed to be equipped with a vapor barrier of 0.1 perms or less (e.g., polyethylene foil) along with petroleum-based insulating materials, that hinder mold growth thanks to their claimed hypoallergenic properties. It was proven that the wooden materials of this wall assembly could be at risk of mold growth if the wall was exposed to a cold climate with unfavorable moisture conditions (i.e., considering a 1% rain infiltration behind the cladding, a 0.02 $L/s \cdot m^2$ air infiltration, and a high initial moisture content of the sheathing) [18]. The second studied assembly was a highly hygroscopic (HH) wall insulated with organic materials, more precisely, with wood fiber batt and wood fiber board. Field experiments of this wall assembly were conducted and presented by Simonson et al. [13, 14]. The third one was a cross-laminated timber (CLT) wall from an existing and occupied residential building located in Quebec City. Figure 1 displays the studied assemblies with their various layers listed from the exterior to the interior (top to bottom of the wall detail).

The thermal bridges created by the wood framing and the metal supports of the exterior siding were neglected. All the studied assemblies considered a ventilated air layer adjacent to the exterior cladding/siding, whereas no air movements were considered in the interior air layer of the Novoclimat and the CLT walls.

2.2 Numerical Calculations

Simulations were carried out with the program COMSOL Multiphysics, which has a module specifically designed for heat and moisture transport in building materials. The

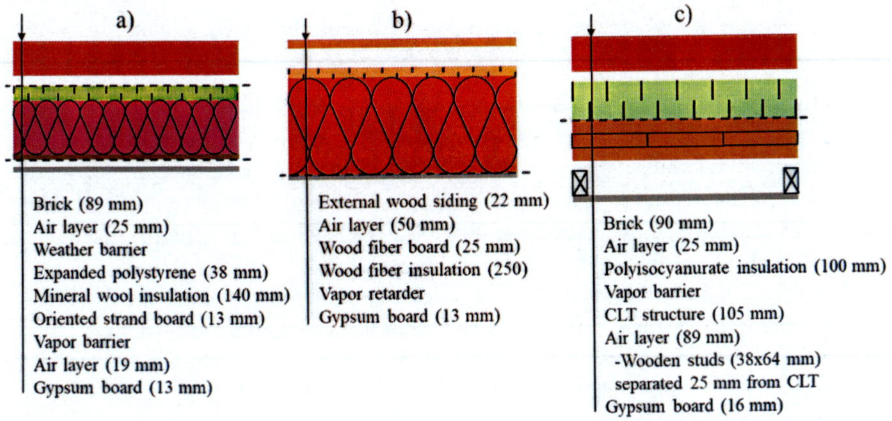

Fig. 1. Studied building envelopes: a) Wall assembly suggested by the energy transition program Novoclimat, b) highly hygroscopic (HH) wall with organic insulation studied by [13], and c) CLT wall of an existing residential building.

use of COMSOL as a hygrothermal tool has been validated by [19, 20]. The coupled partial differential equations (PDEs) that describe the transient heat and moisture transport in building materials are:

$$(\rho c_p)_{eff}\frac{\partial T}{\partial t} - \nabla\cdot\left(k_{eff}\nabla T + L_v\delta_p\nabla(\phi_w p_{sat})\right) = Q \quad (1)$$

$$\frac{\partial w}{\partial \phi_w}\frac{\partial \phi_w}{\partial t} - \nabla\cdot\left(D_w\frac{\partial w}{\partial \phi_w}\nabla\phi_w + \delta_p\nabla(\phi_w p_{sat})\right) = G \quad (2)$$

where T is temperature, ϕ_w is relative humidity, ρ is the density, c_p is the specific heat capacity at constant pressure, k_{eff} is the effective thermal conductivity, L_v is the latent heat of evaporation, δ_p is the vapor permeability, p_{sat} is the vapor saturation pressure, D_w is the liquid transport coefficient, w is the moisture content, and lastly, Q and G respectively represent the HAM sources. COMSOL uses the finite-element method to solve the coupled PDEs, more precisely, it employs a damped version of Newton's method. The convergence was determined by a tolerance factor, if the estimated relative error was smaller than 0.01, the iteration was considered as terminated. The maximum number of iterations was set to one thousand, and the size of the mesh elements was set to the finest possible that could be automatically generated by the program, to consider the temperature and RH variations that could occur within a small distance.

Material properties were obtained from the extensive material library of WUFI 2D [21]. The studied assemblies were considered as one-dimensional domains with HAM sources in the exterior air layers. These HAM sources represented the gains or losses of thermal energy and humidity of the building envelope when it got in contact with the moving exterior air. This approach was respectively developed and employed by references [16, 17] to study the hygrothermal behavior of ventilated air cavities. The dependence of these sources on the air change rate per hour (ACH) is shown by:

$$Q = (\rho c_p)_{air,out}\frac{\text{ACH}}{3600}(T_{\infty,out} - T) \quad (3)$$

$$G = \frac{\text{ACH}}{3600}\left(c_{\infty,out} - c\right) \quad (4)$$

where c refers to the water vapor mass concentration in the air. In these equations, ρ, c_p, T_∞ and c_∞ were evaluated at the outside ambient conditions, while T and c represent the temperature and water vapor mass concentration at each mesh node in the source domain. Please notice that the Novoclimat and the CLT wall assemblies also had an air layer on the interior side. Nevertheless, these did not have an associated HAM source and therefore the air in this cavity was considered stagnant.

The city of Quebec was chosen for exterior climatic conditions (zone 7A according to [22]). The data used for the exterior boundary conditions came from the ASHRAE 2017 meteorological data incorporated in the COMSOL library. Solar radiation effects were neglected for the calculations, and therefore the wall orientation had no impact over the hygrothermal performance. As for the interior boundary conditions, these came from in-situ measurements of a test house located in the selected city. Please notice that this test house was built by a partner company of our research team, and it was not the object of study of this paper. The building materials had an initial temperature and RH of 20 °C and 72% RH, respectively. Simulations considered a one-year period with time steps of one hour, starting on late September as the available experimental data coming from the test house began on this date. Various combinations of the vapor resistance factor of the vapor barrier and ACH ventilation rates in the external air cavities were studied for all wall assemblies. More precisely, three different vapor barriers were investigated, the first one was a vapor retarder with a high vapor permeability, the second one was a smart vapor barrier that shifted its vapor permeability as a function of RH, and the last one was a polyethylene foil with a low vapor permeability. Table 1 displays these barriers along with their dimensionless vapor resistance factor (μ-value), considering a 1 mm thick layer. The mentioned smart vapor barrier had a high water vapor resistance from zero to 70% RH, value after which it rapidly reduced its vapor resistance to become permeable and allow inward drying, as stated by its manufacturer [23].

Table 1. Water vapor resistance factor (μ-value) of the considered vapor barriers.

Vapor barrier	μ-value [-]
Vapor retarder	2000
Smart vapor barrier	250–25 000
Polyethylene foil	40 000

As to the ACH, there is a wide range of values reported for external air cavities adjacent to the cladding. However, Rahiminejad and Khovalyg [24] found that most of the ACH values obtained from theoretical, numerical, and experimental tests are lower than 400 h^{-1}. Therefore, this study considered ACH values ranging between zero (the unvented case) and 400 h^{-1}, with steps of 50 h^{-1}. A parametric sweep was applied to all vapor barriers considering all ACH values, generating a total of 27 different simulations for each wall assembly. The ACH values remained constant in time since Karagiozis and Künzel [16] determined that this simplification yielded good agreement between numerical results and field measurements.

2.3 Mathematical Mold Growth Model

Mean temperature and RH values were calculated at the insulation layers to evaluate the risk of mold growth on the wooden materials of the various assemblies. The mathematical model developed by Hukka and Viitanen [25] was used for this task. That method presents a threshold curve relating the critical relative humidity RH_{crit} to the temperature T. Any point above this curve in the T-RH plane generates conditions that favor mold proliferation. It is described by the following piecewise function defined in the domain [0, 50] °C:

$$RH_{crit} = \begin{cases} -0.00267T^3 + 0.16T^2 - 3.13T + 100 \text{ if } T \in [0, 20] \\ 80\% \qquad \text{if } T \in (20, 50] \end{cases} \tag{5}$$

It should be mentioned that crossing this curve does not necessarily yield to an immediate apparition of mold. Depending on the time and intensity of the exposure there is a spectrum between mold growth on a microscopic scale and the full coverage of the surface of a wooden material [26].

3 Results and Discussion

3.1 Novoclimat Wall

The hourly average temperature and RH values of the insulation layers of the Novoclimat assembly (see Fig. 1a) were determined with the different vapor barriers and ventilation rates. Figure 2 shows these results at the expanded polystyrene (EPS) layer for three different vapor barriers and considering two ACH values (the maximum and the unvented cases). Each point represents the average temperature and RH of the entire layer at a given time. The starting point (i.e., the first time step of the simulation) was situated at 20 °C and 72% RH. The points then followed a counterclockwise path over time. The red dashed line in Fig. 2 represents the critical RH boundary for mold growth apparition described by Eq. (5). Figure 2 allows a direct assessment of the mold growth risk at the EPS layer, which presented some critical conditions during a certain period, corresponding to autumn. The initial conditions affected the results of a part of this season, as the hygrothermal behavior at the beginning was almost identical regardless of the utilized vapor barrier.

It can be observed that in all cases the critical temperature and RH conditions were reached at the EPS regardless of the vapor resistance of the vapor barrier of the Novoclimat wall. However, the vapor retarder generated the conditions with the lowest RH. In this case the maximum was around 90% RH, whereas it was around 95% when the polyethylene foil and the smart vapor barrier were used. Interestingly, the temperature-RH profiles yielded by the polyethylene foil and the smart vapor barrier were almost identical, meaning that the magnitude of the vapor resistance of the vapor barrier can

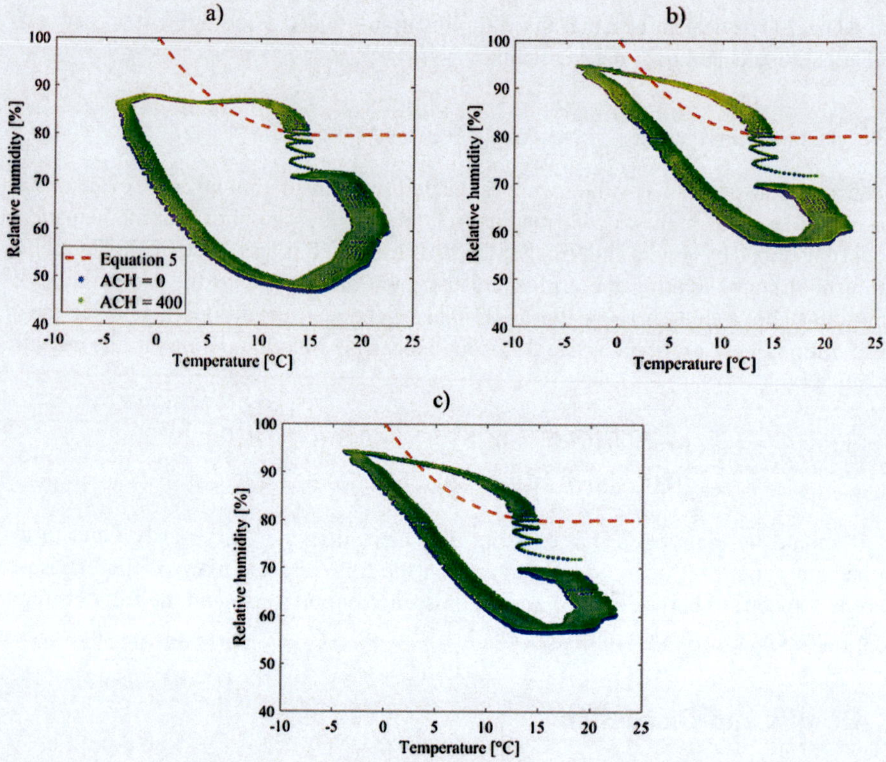

Fig. 2. Hourly average temperature and relative humidity of the EPS layer of the Novoclimat assembly for two ACH values in the air gap behind the cladding and with the a) vapor retarder, b) polyethylene foil, and c) smart vapor barrier.

be below that of the polyethylene foil without making much difference. Even though all cases crossed the critical RH boundary, the assembly might not be at a great risk as the mold growth mathematical model was developed for wooden materials, and the Novoclimat wall was insulated with petroleum-based insulations that are less susceptible to mold issues. Nonetheless, there could be some microscopic mold proliferation on the wooden studs adjacent to the EPS. Increasing the permeability and hygroscopicity of the EPS could probably help reduce the RH levels. Moreover, the magnitude of the ACH in the exterior air layer seemed to have little effect over the drying capacity of the Novoclimat wall. This is further evidenced by Fig. 3, which shows the average RH of the EPS during January and July with the tested vapor barriers. The three curves in the upper half of the figure represent the mean RH yielded during January, whereas the three below are for July. A particular attention was paid to the moisture conditions in the wall throughout these months to represent a cold and a hot period. The monthly standard deviation of the RH is also given for each ACH value.

The above figure also shows that the RH levels are practically the same for the cases with the polyethylene foil and the smart vapor barrier during either season. The vapor retarder presented considerably lower RH values in January. Nonetheless, its

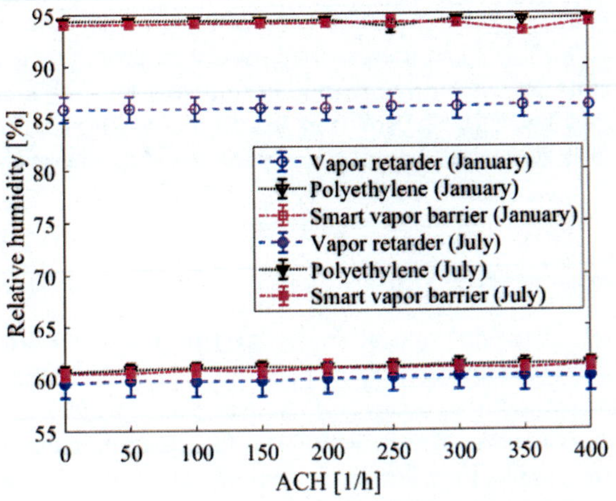

Fig. 3. Mean relative humidity at the EPS layer of the Novoclimat wall with different vapor barriers as a function of the ACH during the months of January and July.

performance was very similar to the other two barriers during July. An almost identical behavior was observed at the mineral wool (MW) insulation layer. This can be observed in Fig. 4, which shows the mean RH of the MW and the EPS layers throughout the whole year considering all ACH values. The RH curves yielded by the various ACH magnitudes are overlapped for each vapor barrier and are represented by a shaded area. However, these shaded zones are imperceptible for all vapor barriers and all insulation layers, once again confirming that the ventilation behind the cladding of the Novoclimat wall had practically no effect over the drying capacity of the assembly.

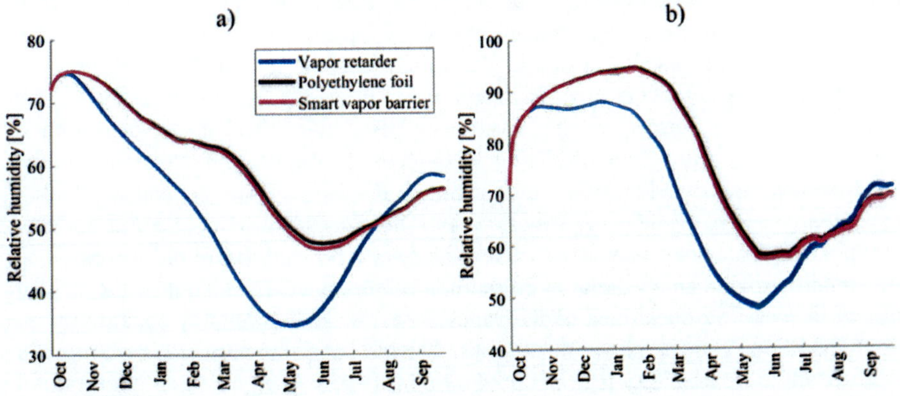

Fig. 4. Mean relative humidity at the a) MW and b) EPS layers of the Novoclimat wall with different vapor barriers as a function of time and considering all ACH values. The shaded areas are the maximum and minimum limits of the RH obtained with different ACH magnitudes.

On the other hand, no critical moisture conditions were met in the MW insulation, as it never exceeded 80% RH and it even dried out after a year. It would be interesting to study the assembly for a longer time period to assess the wall's drying capacity, especially at the EPS layer. In the Novoclimat wall, the vapor retarder generated the lowest RH levels during most of the year except for summer, where it yielded slightly larger values.

3.2 Highly Hygroscopic (HH) Wall

In this wall assembly the ventilation of the air layer behind the external wood siding did make a difference in the drying process. For instance, Fig. 5 shows the mold growth risk assessment at the wood fiber board (WFB) layer of the HH wall (see Fig. 1b) with different vapor barriers and considering two ACH values (the maximum and the unvented cases). Each point represents the average temperature and RH of the entire WFB layer at a given time. It was found that all the tested vapor barriers in the HH wall presented adequate temperature and RH conditions to avoid mold growth at the WFB layer. In contrast to the Novoclimat wall, the HH assembly presented a less pronounced moisture hysteresis and thus the RH in the exterior insulation layer fluctuated less. The moisture conditions in the wood fiber insulation (WFI) layer were also out of the risk zone, and the effect of the magnitude of the ACH was less notorious, as illustrated by Fig. 6. It shows the mold growth risk assessment at the WFI when using a vapor retarder, however, the other considered vapor barriers yielded similar results.

Figure 7 displays the mean RH as a function of time in the WFI and the WFB when using different vapor barriers and considering all the ventilation rates of the exterior air layer. The mean values are represented by the solid lines, while the shaded areas represent the maximum and minimum RH levels yielded by the studied ACH values. The upper limit of the shaded areas corresponds to the unventilated case, whereas the lower one is composed of various ACH values; meaning that the ventilation of the exterior air cavity effectively improved the drying capacity of the HH wall.

Likewise, it was found that during the whole year the lowest RH levels in both the WFI and the WFB were attained when using the vapor retarder. In addition, the smart vapor barrier performed similarly to the polyethylene foil up to the summer. From this season on the yielded RH showed a greater variation, and depending on the insulation layer (i.e., WFI or WFB) the maximum RH values were attained with either the polyethylene foil or the smart vapor barrier. These results once more confirmed that the vapor resistance of the vapor barrier can be further down than that of the polyethylene foil and still provide an adequate hygrothermal performance. Interestingly, the shaded areas grow over time regardless of the vapor barrier or layer, probably because of the initial conditions and the changing seasons. For instance, the warmer air of summer could absorb more humidity than the cold air of winter, so a higher ventilation rate of the exterior air layer could enhance the drying process. Even though the safest choice in terms of the moisture conditions in the HH wall was obtained with the vapor retarder, the other two choices also presented a secure alternative in terms of moisture content.

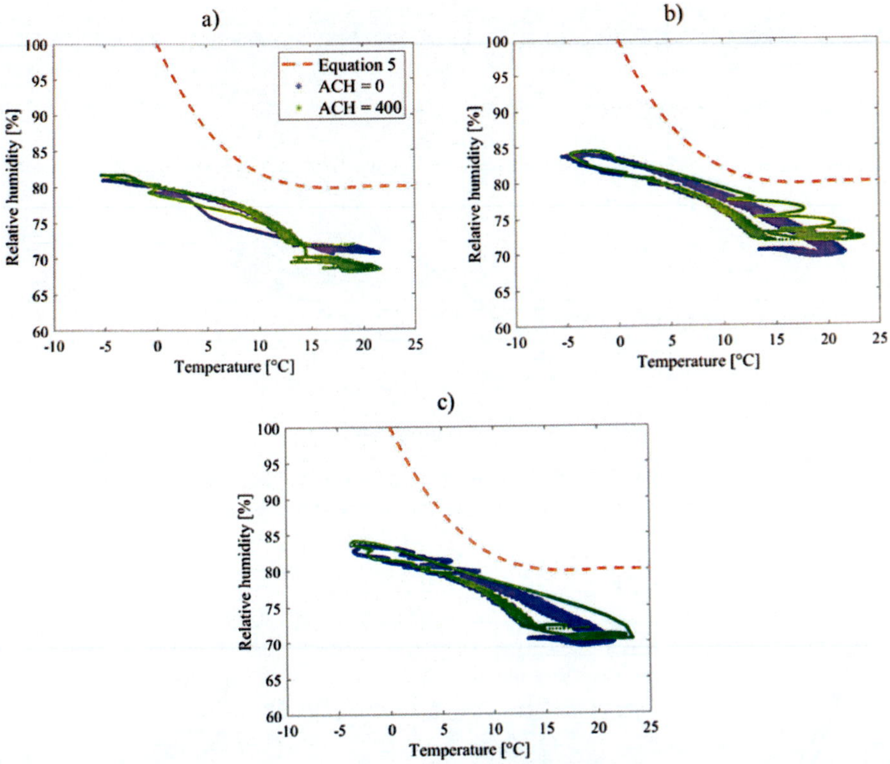

Fig. 5. Hourly average temperature and relative humidity of the WFB layer of the HH wall for two ACH values in the air gap behind the cladding and with the a) vapor retarder, b) polyethylene foil, and c) smart vapor barrier.

3.3 Cross-Laminated Timber (CLT) Wall

The layer of the CLT wall (see Fig. 1c) insulated with polyisocyanurate (ISO) presented safe moisture conditions with all the tested vapor barriers, with and without convection in the air layer adjacent to the external cladding. Figure 8 shows the results of the mold growth risk assessment at the ISO layer of this wall assembly considering all vapor barriers and two ventilation rates (the maximum and the unvented cases). Each point represents the average temperature and RH of the entire ISO layer at a given time. Again, the piecewise function describing the critical conditions for mold growth is denoted by the red dashed line. The RH yielded by the vapor retarder fluctuated less than the cases with the polyethylene foil and the smart vapor barrier, which also generated very similar results. In the ISO layer the magnitude of the ventilation rate seemed to have little effect over the hygrothermal performance of the material.

Figure 9 displays the mean RH as a function of time in the CLT and ISO layers with distinct vapor barriers and considering all the ventilation rates of the exterior air layer. The solid lines represent the mean RH, whereas the shaded areas are the maximum and minimum RH levels yielded by the several ACH magnitudes. It was found that the ACH values had little to no effect over the relative humidity of the CLT panel, probably due

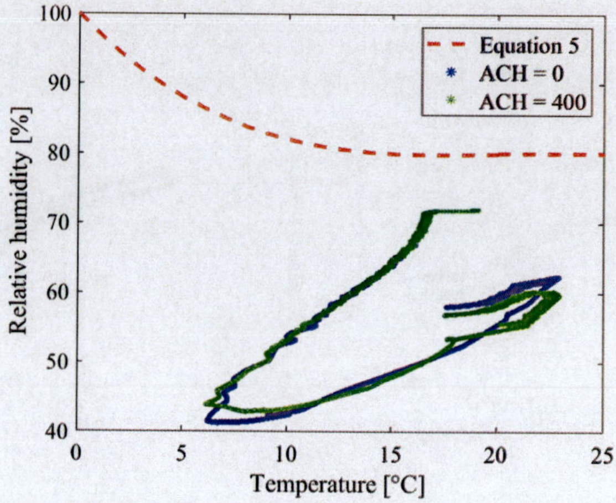

Fig. 6. Hourly average temperature and relative humidity of the WFI layer of the HH wall for two ACH values in the air gap behind the cladding and with a vapor retarder.

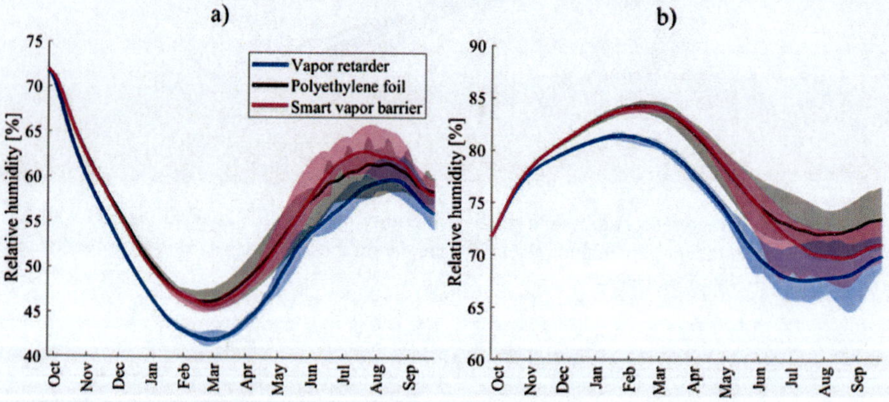

Fig. 7. Mean relative humidity at the a) WFI and b) WFB layers of the HH wall with different vapor barriers as a function of time and considering all ACH values. The shaded areas are the maximum and minimum limits of the RH obtained with different ACH magnitudes.

to its considerable distance from the exterior air layer. Also, regardless of the vapor resistance of the vapor barrier, the CLT panel dried out over time. Likewise, in both the CLT and ISO layers the polyethylene foil and the smart vapor barrier presented a very similar behavior, being nearly indistinguishable in the CLT panel. Interestingly, in the ISO layer the polyethylene foil yielded the lowest RH values during most of the year.

In addition, when the vapor retarder was used the RH in the ISO layer evolved significantly slower than with the other barriers, keeping some more stable moisture conditions. This was most probably due to the hygroscopic inertia of the CLT panel, whose high hygroscopicity permitted the absorption and storage of moisture that could then be slowly diffused towards the ISO layer, and as the permeability of the vapor

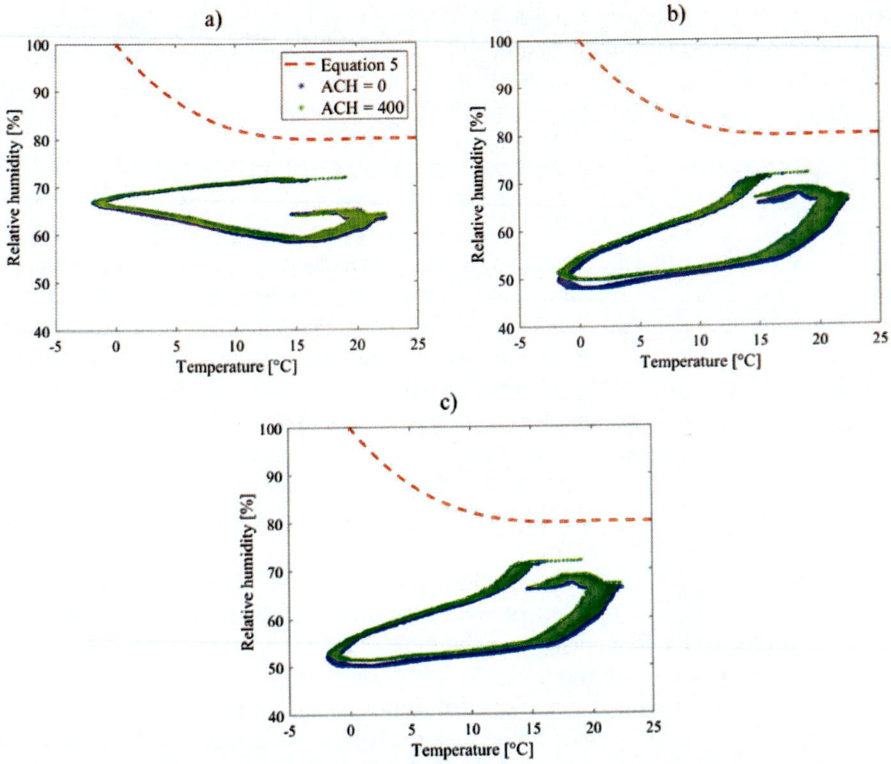

Fig. 8. Hourly average temperature and relative humidity of the ISO layer of the CLT wall for two ACH values in the air gap behind the cladding and with the a) vapor retarder, b) polyethylene foil, and c) smart vapor barrier.

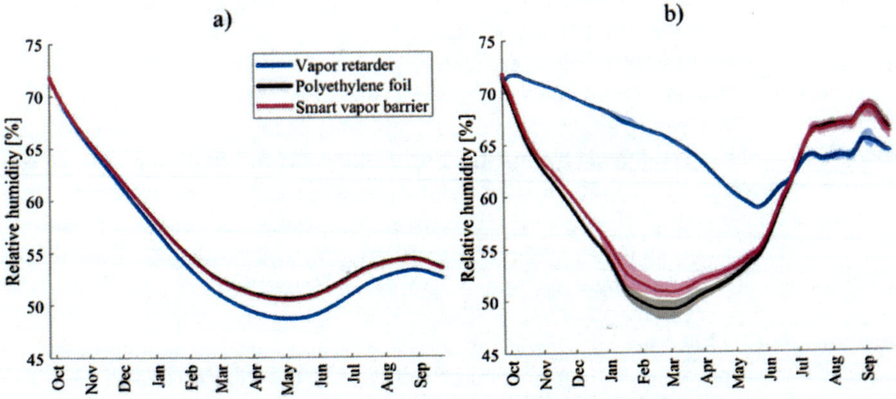

Fig. 9. Mean relative humidity at the a) CLT and b) ISO layers of the CLT wall with different vapor barriers as a function of time and considering all ACH values. The shaded areas are the maximum and minimum limits of the RH obtained with different ACH magnitudes.

barrier was reduced. In contrast to the HH wall (see Fig. 7), the greatest shaded area in Fig. 9b was found during winter, and the effect of ventilation during summer over the RH value was negligeable; most probably due to the moisture buffering capacity of the CLT layer.

4 Conclusion

In general, it was found that the use of a vapor retarder yielded safe moisture levels in most of the cases, except for the Novoclimat wall. However, it is important to notice that critical conditions were attained in this assembly with all the studied vapor barriers, and that the case with the vapor retarder exited the danger zone the quickest and with the lowest RH maximum value, compared to the other two choices. It is also important to notice that the mold growth assessment criterion was initially developed for wooden materials, which is not the case of the EPS nor the MW. However, the structural lightweight wood frame of the Novoclimat wall could be at risk of undergoing some level of mold proliferation. It would be interesting to study the assembly during a longer period of time to determine if it could eventually dry out.

In all the considered layers of the studied assemblies the polyethylene foil and the smart vapor barrier yielded results that were very much alike. The greater deviations were found in the HH wall after the beginning of summer. This proved that the vapor resistance of the polyethylene foil is excessively higher than required to appropriately protect the building envelope from vapor diffusion in cold climates. Likewise, it was found that the ventilation of the exterior air layer behind the cladding did not have a great impact over the drying capacity of the assemblies, except for the HH wall. It was probably due to the higher hygroscopicity of its materials that the moisture transfer was enhanced, when exposed to the exterior air convection.

Both walls insulated with bio-based materials (i.e., HH and CLT) proved to be safe choices for a cold climate, even when using permeable vapor barriers. In fact, not only did the vapor retarder was the safest choice for moisture control, but its impact over the ISO layer of the CLT wall assembly also generated considerably stable moisture conditions during the whole year. This was almost certainly due to the elevated hygroscopicity of the CLT panels, which showed a sluggish response to moisture changes. This research could benefit from further development by investigating the time-dependent drying capacity of the tested assemblies, for instance, by considering a water leak, a higher initial moisture content or a longer period of simulation. Likewise, the consideration of solar radiation could have an impact over the hygrothermal profiles of the wall assemblies, depending on their orientation and tilt angle.

Acknowledgements. The authors are grateful to Natural Sciences and Engineering Research Council of Canada for the financial support through its IRC and CRD programs (IRCPJ 461745-18 and RDCPJ 533629–18) as well as the industrial partner Oïkos Construction of the NSERC industrial chair on eco responsible wood construction (CIRCERB).

References

1. American Society of Heating, Refrigerating and Air-Conditioning Engineers, 2017 ASHRAE Handbook-Fundamentals, Inch-Pound edition. Atlanta, GA: American Society of Heating, Refrigerating and Air-Conditioning Engineers (2017)
2. Irbe, I., Andersone, I.: Wood decay fungi in latvian buildings including cultural monuments. In: International Conference held by COST Action IE0601, Braga, pp. 1000–1007 (2010)
3. Afshari, A., et al.: WHO guidelines for indoor air quality: dampness and mould. Copenhagen, Denmark: World Health Organization (2009)
4. May, N.: Breathability: The Key to Building Performance, pp. 1–30 (2005)
5. Lstiburek, J.: Moisture control for buildings, ASHRAE J. **44**, 8 (2002)
6. Simonson, C.J., Salonvaara, M., Ojanen, T.: Heat and mass transfer between indoor air and a permeable and hygroscopic building envelope: part I - field measurements. J. Therm. Envel. Build. Sci. **28**(2), 161–185 (2004)
7. Geving, S., Holme, J.: Vapour retarders in wood frame walls and their effect on the drying capability. Front. Archit. Res. **2**(1), 42–49 (2013). https://doi.org/10.1016/j.foar.2012.12.003
8. Birjukovs, M., Apine, I., Jakovics, A.: Numerical models for long-term performance assessment of lightweight insulating assemblies. In: IOP Conference Series Material Science Engineering, vol. 660, p. 012024 (2019). https://doi.org/10.1088/1757-899X/660/1/012024
9. Vinha, J.: Analysis method to determine sufficient water vapour retarder for timber-framed walls. In: Proceedings of 8th Symposium Building Physics Nordic Ctries, pp. 1245–1252 (2008)
10. Rode, C.: Organic insulation materials: effect on indoor humidity and necessity of a vapor barrier. Proc. Therm. Perform. Exter. Envel. Build. **7**, 109–121 (1998)
11. Bastien, D., Winther-Gaasvig, M.: Influence of driving rain and vapour diffusion on the hygrothermal performance of a hygroscopic and permeable building envelope. Energy **164**, 288–297 (2018). https://doi.org/10.1016/j.energy.2018.07.195
12. Padfield, T.: The role of absorbent building materials in moderating changes of relative humidity. The Technical University of Denmark (1998)
13. Simonson, C.J., Salonvaara, M.H., Ojanen, T.: Moisture content of indoor air and structures in buildings with vapor-permeable envelopes. Perform. Exter. Envel. Whole Build. VIII (2001)
14. Simonson, C.J., Ojanen, T., Salonvaara, M.: Moisture performance of an airtight, vapor-permeable building envelope in a cold climate. J. Therm. Envel. Build. Sci. **28**(3), 205–226 (2005). https://doi.org/10.1177/1097196305048628
15. Geving, S., Lunde, E., Holme, J.: Laboratory investigations of moisture conditions in wood frame walls with wood fiber insulation. Energy Procedia **78**, 1455–1460 (2015). https://doi.org/10.1016/j.egypro.2015.11.170
16. Karagiozis, A.N., Künzel, H.M.: The effect of air cavity convection on the wetting and drying behavior of wood-frame walls using a multi-physics approach. J. ASTM Int. **6**(10), 15 (2009). https://doi.org/10.1520/JAI101455
17. Künzel, H. M., Zirkelbach, D., Schafaczek, B.: Modelling the effect of air leakage in hygrothermal envelope simulation. In: Proceedings of the Building Enclosure Science & Technology (BEST3) Conference (2012)
18. Caron-Rousseau, A., Blanchet, P., Gosselin, L.: Parametric study of lightweight wooden wall assemblies for cold and subarctic climates using external insulation. Buildings **12**(7), 1031 (2022). https://doi.org/10.3390/buildings12071031
19. Knarud, J.I., Geving, S.: Comparative study of hygrothermal simulations of a masonry wall. Energy Procedia **132**, 771–776 (2017). https://doi.org/10.1016/j.egypro.2017.10.027
20. Allué Hoyos, C., Vinha, J.: Applicability of COMSOL Multiphysics to combined heat, air and moisture transfer modelling in building envelopes. Master of Science Thesis, Tampere University of Technology, Tampere, Finland (2014)

21. Fraunhofer Institute for building physics, "WUFI 2D," WUFI. https://wufi.de/en/software/wufi-2d/. Accessed 02 Dec 2022
22. Canadian Commission on Building and Fire Codes, National Energy Code of Canada for Buildings: 2020, Fifth Edition. Canada: National Research Council of Canada (2022)
23. Proclima, Product information about Intello. www.proclima.com Accessed 28 Nov 2022
24. Rahiminejad, M., Khovalyg, D.: Review on ventilation rates in the ventilated air-spaces behind common wall assemblies with external cladding. Build. Environ. **190**, 107538 (2021). https://doi.org/10.1016/j.buildenv.2020.107538
25. Hukka, A., Viitanen, H.A.: A mathematical model of mould growth on wooden material. Wood Sci. Technol. **33**(6), 475–485 (1999). https://doi.org/10.1007/s002260050131
26. Viitanen, H.: Factors affecting the development of biodeterioration in wooden constructions. Mater. Struct. **27**(8), 483–493 (1994). https://doi.org/10.1007/BF02473453

Potential Use of Waste as a Source of Calcium for Microbial Induced Carbonate Precipitation (MICP)

Olivia Totis, Giuseppe Moita(✉), Vitor Liduino, Eliana Flávia Servulo, João Paulo Bassin, and Romildo Dias Toledo Filho

Federal University of Rio de Janeiro, Rio de Janeiro, Brazil
giuseppe.moita@coc.ufrj.br

Abstract. Microbial induced Calcium carbonate precipitation (MICP) is a promising new technology in the pursuit of sustainable concrete improvement. For MICP to occur, a medium with alkaline pH, dissolved inorganic carbon and a source of calcium must be used, associated with bacteria that serve as nucleation sites through their cell walls. The use of this technology on a large scale is linked to a high financial and environmental cost, so replacing traditional sources of carbon and calcium with waste is a way to mitigate these expenses and possible environmental liabilities using circular economy. The objective of this work is to evaluate the potential use of alternative sources of calcium in the MICP process from eggshells and mussel shells. For this, these materials were collected, washed and ground to carry out a chemical characterization using X-ray Fluorescence spectrometry (XRF), thermogravimetry (TGA) and X-ray diffraction (XRD). The source of calcium was produced from the mixture of waste processed with acids (hydrochloric, lactic and nitric) and vinegar. The solubilized mass was confirmed from the filtration of the remaining solid material. The chemical characteristics of the materials and its solubilization in various acids show that residues have the potential to be used for Microbial induced carbonate precipitation.

Keywords: MICP; waste · calcium carbonate

1 Introduction

1.1 Waste Amended Concrete

Concrete is the most used construction material in the world due to its characteristics as high compression resistance and ease of finding the raw materials and proper work [1, 2]. The housing and infrastructure deficit around the world should increase the demand for building materials in the coming years, especially in underdeveloped countries [3]. This is a serious environmental problem, considering that cement is responsible for 8–9% of anthropogenic carbon dioxide emission [4]. In this context, many studies are focused on the application of waste in replacement of concrete elements [5]. The environmental benefit of this use is broad: it gives an adequate destination to something that would be discarded and reduces the use of raw materials.

S. Amziane et al. (Eds.): ICBBM 2023, RILEM Bookseries 45, pp. 705–715, 2023.
https://doi.org/10.1007/978-3-031-33465-8_54

The use of concrete waste by -products has already become common in some cases and with very advanced research on them. Fly ash, steel slag, silica fumes and rice husk ash are examples of industrial or agricultural wastes that can be used as a substitute for cement with potential to reduce concrete carbon footprint [6, 7]. Recycled aggregates can in even some percentages replace natural aggregates so mitigating the environmental impact of the removal of large volumes of quarries [8].

1.2 Microbial Induced Carbonate Precipitation (MICP)

Aiming at the development of more sustainable alternatives, the application of biotechnology in cementitious materials, such as the use of concrete waste, emerges as a possible solution. Some microorganisms, under adequate conditions, are capable of producing calcium carbonate, a chemically compatible mineral with considerable mechanical resistance. This feature is the focus of several research in areas such as soil stabilization, contaminated water and soil remediation, and especially concrete self-healing [9].

Besides Self-healing many other studies have been carried out with MICP in cementitious materials as ancient buildings surface protection [10], concrete strength enhancement [11], recycled aggregates improvement [12] and development of Living Building Materials (LMB) [13]. This technology, in addition to improving the durability and strength of materials, has great development potential to reduce carbon emissions and energy consumption [14].

Calcium carbonate precipitation technology by microorganisms (MICP) is a promising option in search of sustainable concrete improvement. For MICP to occur, a medium with alkaline pH, dissolved inorganic carbon and a source of calcium, associated with bacteria that serve as nucleation sites through their cell walls, are needed [15]. The use of this large-scale technology is linked to a high financial and environmental cost, so replacing traditional carbon and calcium sources with waste is a way to mitigate these expenses and possible environmental liabilities using the circular economy.

There are already researches that show a possible applicability of some residues in substitution to the expensive industrial nutrients used in the laboratory. Lentil seed powder [16]; corn-steep liquor [17, 18]; wheat bran [19]; lactose mother liquor [20]; chicken manure [21] and tofu wastewater [22] were already studied as substitutes of regular nutrient media. A broader work analyzed diverse alternative sources for carbon and pointed out barley malt powder and mixed grain powder the best ones from a range of 21 agricultural products and by-products [23].

In order to provide an alternative source of nutrients for the bacteria and at the same time to ensure a correct destination to the sewage, mortar was produced with industrial wastewater [24] and concrete with domestic wastewater [25], both of which obtained satisfactory results. A bio-brick with human urine as a urea source, offering a friendlier use for that human waste [26].

All works cited so far seek to replace only the MICP carbon source. Few have studied ways to replace the source of calcium in the process, focusing mainly on eggshell, a material rich in calcium carbonate [27–30]. Thus, this work aims to evaluate the potential use of two wastes (egg and mussel shells) as a source of calcium in the MICP based on a chemical characterization of these materials and on the solubilization capacity of this calcium source in water through its reaction with different types of acids.

2 Materials and Methods

2.1 Waste Collection and Processing

Mussel shells from the species *Mytilus perna* were collected directly at Mariculturists Free Association located at Jurujuba, on the shores of Guanabara Bay, Niterói - Brazil. Figure 1 shows the visual appearance of the material collected. The eggshells were collected in restaurants at the Technology Center of the Federal University of Rio de Janeiro (UFRJ). The eggshells were collected daily at UFRJ restaurants and stored at 4 °C for two weeks to preservation and avoid deterioration of organic matter.

(a) (b)

Fig. 1. (a) mussel shells and (b) eggshells in natura

The collected material needed to be cleaned as it contained a large amount of organic residues that could decompose and influence the solubilization of calcium in acids. In addition, the presence of this large load of organic matter could be detrimental to bacterial growth, as there would be an imbalance in the proportion of carbon and nitrogen present in the medium, compromising the MICP technique.

The total mass of the eggshells was 10.5 kg. The cleaning was carried out with tap water in the concrete mixer at a ratio of 6:1 in mass for 4 washing cycles of 10 min each. After the end of each washing cycle in the mixer, a water sample was collected to observe the color. Thus, it was found that after 4 cycles of changing water and washing in the concrete mixer, the material was considered clean, as there was no significant change in color in the water, as shown in Fig. 2. Pre-drying was performed at 40 °C on sieves for 72 h, then the material was transferred to an oven at 100 °C for 12 days until the weight stabilizes. The final weight after processing was 8.18 kg, 22% of the original mass was organic waste.

The mussel shells total mass was 18.60 kg. Cleaning was carried out with water in the concrete mixer at a ratio of 2:1 for 4 washing cycles of 10 min each. Pre-drying was performed at 40 °C on sieves for 24 h, then the material was transferred to an oven at 100 °C for 7 days until the weight stabilizes. The final weight after processing was 17.95 kg, 3,5% of the original mass was organic waste.

The mussel shell was crushed first by a cone crusher model DENVER n° 12 then by a roller crusher model DENVER 10"x6. The eggshell was crushed only by a roller crusher

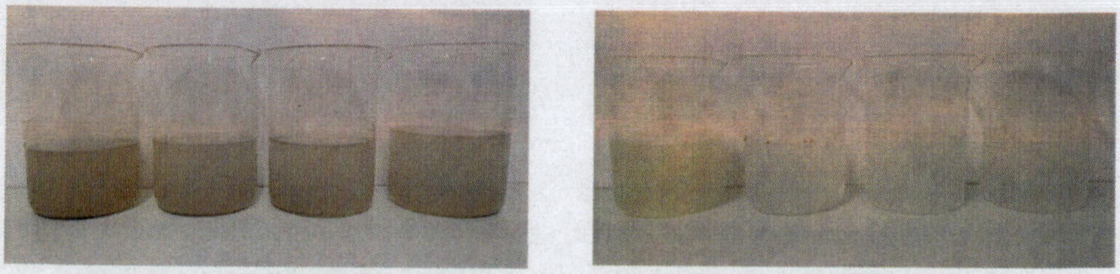

(a) (b)

Fig. 2. Water from each cleaning cycle (a) mussel shells and (b) eggshells.

model DENVER 10"x6". Both were separated by sieves until reaching a size smaller than 0.3 mm. Figure 3 shows the visual appearance of the powders. For the chemical characterization of the materials, they were comminuted in a mortar and ceramic pestle until the particles reached a size capable of passing through a 75 μm mesh sieve.

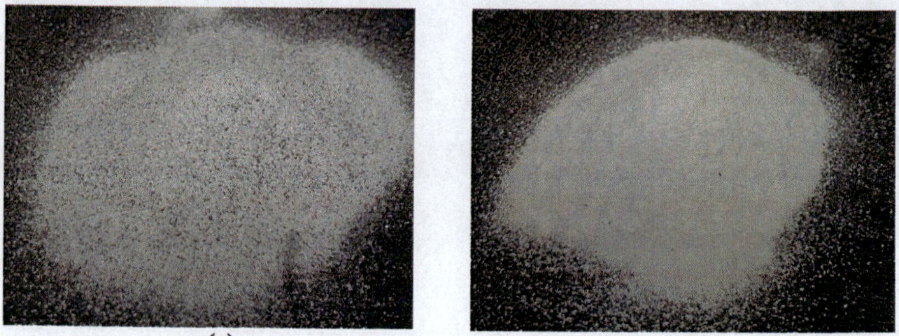

(a) (b)

Fig. 3. (a) mussel shells and (b) eggshells powder

2.2 Chemical Characterization

The chemical composition in oxide content was determined using X-Ray Fluorescence (XRF), model Rigaku Primini with palladium X-ray generating source. The crystalline phases present in the samples were determined by X-Ray Diffraction Analysis (XRD) in a Rigaku Miniflex II with graphite monochromator and using CuKα radiation (30 kV and 15 mA). The scans were performed in the range of 10–60° (2Θ), with an angular step of 0.02° and increment of the 1s step. Both assays were performed in Hydrogen Technology Laboratory of Federal university of Rio de Janeiro.

Thermogravimetric analysis was performed to calculate the amount of calcium carbonate in the studied materials. The test was carried out at the Professor Ivo Giolito Thermal Analysis Laboratory using the NETZSCH STA 409PG equipment. The heating rate used was 10 °C/min between temperatures of 30 and 1000 °C in an atmosphere of N2. 7.720 and 7.272 mg of sample were used for mussel shell and eggshell respectively.

2.3 Calcium Solubilization

The calcium carbonate present in the samples can be solubilized using liquid acids. Forming new compounds that contain calcium molecules capable of ionizing in aqueous solution and providing a usable source of calcium for all types of MICP (Eq. 1–4). In this study, acetic acid -AA (vinegar), nitric acid -NA, lactic acid -LA and hydrochloric acid -HA were used with 5% acidity. The fine powders of eggshell and mussel shell obtained earlier were mixed individually with each of the four acids in a ratio of 1:8 in 500mL Erlenmeyer (Fig. 4) and placed in a shaker for 7 days at 30 °C.

$$CaCO_3 + 2HCl \rightarrow CaCl_2 + CO_2 + H_2O \quad (1)$$

$$CaCO_3 + 2C_2H_4O_2 \rightarrow Ca(C_2H_3O_2)_2 + CO_2 + H_2O \quad (2)$$

$$CaCO_3 + 2HNO_3 \rightarrow Ca(NO_3)_2 + CO_2 + H_2O \quad (3)$$

$$CaCO_3 + 2C_3H_6O_3 \rightarrow Ca(C_3H_5O_3)_2 + CO_2 + H_2O \quad (4)$$

(a) (b)

Fig. 4. (a) mussel shells and (b) eggshells during solubilization process

After this time, the materials were filtered with a pump to quantify the remaining precipitate in the solution and, consequently, calculate what was solubilized. For this, the Whatman® quantitative filter paper, ashless, Grade 42 was used. After filtering, the material was dried for 24 h for weighing.

3 Results and Discussion

3.1 Chemical Characterization

The chemical composition obtained by X-ray fluorescence spectrometry is presented in the Table 1 and shows that in both cases, there is a significant composition of CaO, present in calcium carbonate, indicating the great presence of the mineral of interest in this work.

This result corroborates those obtained in the analysis of XRD and TGA. X-ray diffraction allows knowing the predominant mineralogical composition in the sample

Table 1. Chemical composition of wastes by XRF

Compounds (% w/w)/Material	Eggshell	Mussel shell
CaO	90,69	83,32
Sc2O3	-	11,2
SiO2	2,77	3,29
Al2O3	2,44	-
Fe2O3	1,44	1,37
MgO	0,83	-
SO3	0,73	-
SrO	0,39	0,41
P2O5	0,39	-
K2O	0,25	0,27
ZnO	-	0,08
CuO	-	0,07
Cl	0,06	-
Total	100,00	100,00

and its crystalline arrangements. Both materials presented calcium carbonate, but in different crystal arrangements. The eggshell showed only the polymorph calcite while the mussel shell showed only aragonite (Fig. 5 and Fig. 6). This difference can be visible in the color of the material, calcite is white, and aragonite is yellow. The existence of a smaller amount of calcium carbonate in the mussel shell also influences the color difference.

The use of TGA allows a much more accurate calculation of the amount of calcium carbonate present in the materials. Because at high temperatures calcium carbonate tends to decompose releasing carbon dioxide into the environment and leaving calcium oxide.

This can be seen in the Fig. 7 and Fig. 8 between temperatures 500 and 800 °C there is a significant loss of mass. In the case of mussel shells, there is a constant mass loss of 7.40% before 500 °C, indicating the existence of organic matter in the material still adhered. This does not occur in the eggshell, indicating that its washing was more efficient.

Between 500 and 800 °C, the mussel shell lost 41.40% of mass and the eggshell 41.70%. Indicating that both have a similar amount of calcium carbonate, but the eggshell has more non-volatile solids and less organic matter incorporated into the material. Equation (5) allows to calculate the total amount of calcium carbonate in each of these materials. Thus, mussel shell and eggshell have 94.15% and 94.83% calcium carbonate, respectively.

$$\%ofCaCO3 = \frac{Molar\ mass\ of\ CaCO3}{Molar\ mass\ of\ CO2} \times \%\ mass\ loss \tag{5}$$

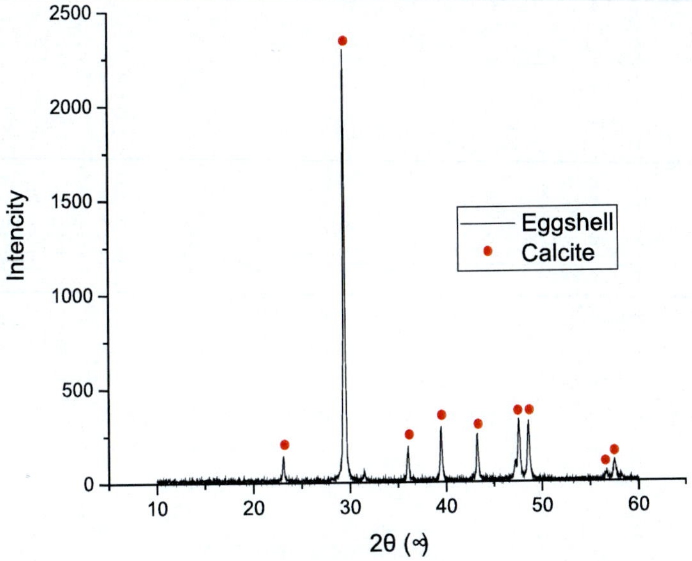

Fig. 5. Eggshell X-ray diffractogram.

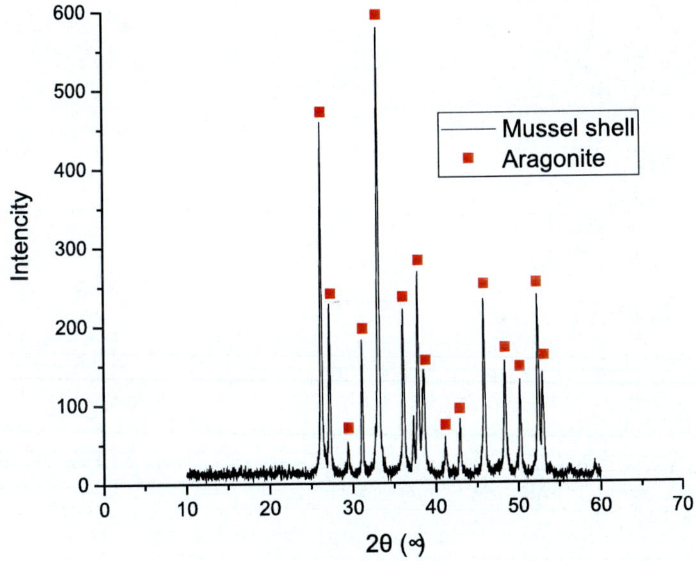

Fig. 6. Mussel shell X-ray diffractogram.

These results corroborate what was expected, showing that both eggshell and mussel shell are residues whose constitution allows them to be used as a possible source of calcium for the production of bacterial calcium carbonate.

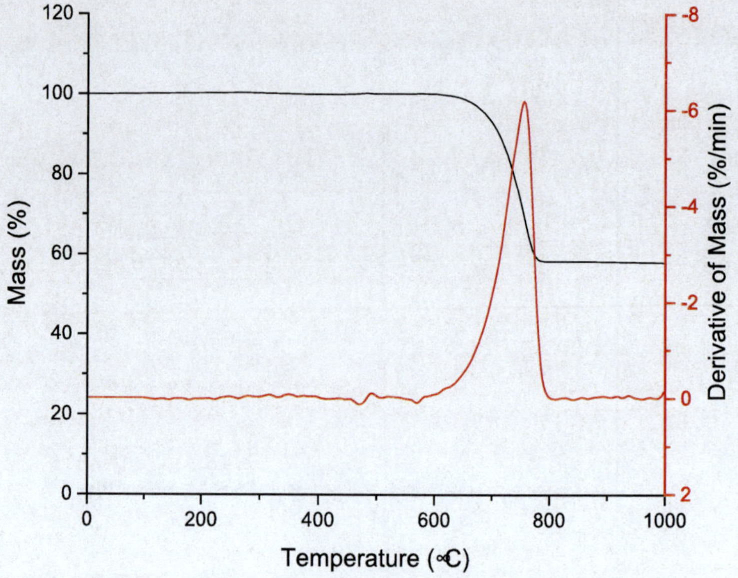

Fig. 7. Eggshell thermogravimetry

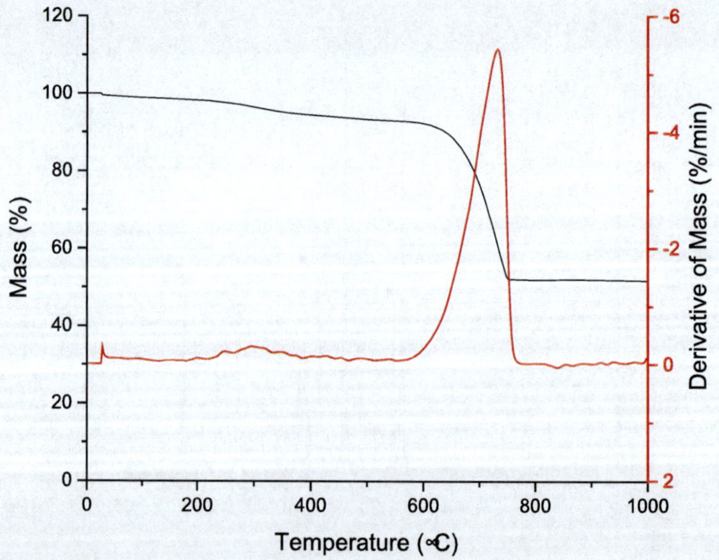

Fig. 8. Mussel shell thermogravimetry

3.2 Calcium Solubilization

Comparing the results obtained from the sample solubilization experiment with the calculations considering the complete reaction (Table 2), one may conclude that nitric acid was more reactive, since with eggshell and mussel shell, this acid presented the greatest mass loss. The high reactivity of nitric acid to the materials was visually observed from the first moments of mixing, when the material almost overflowed from the flask. Lactic and acetic acids presented results close to the theoretical value considering the complete reaction, indicating that they may be good alternatives for more controlled reactions. Hydrochloric acid had the worst performance among the four, considering that even with a mass loss similar to that of the other acids the result presented is far from representing a possible complete reaction of the acid with the calcium sources.

Table 2. Waste solubilization

Material	Acid	Remaining waste (g)	Solubilization (%)
Eggshell	HA	6,40	36,04
	AA	6,84	31,57
	LA	7,22	27,76
	NA	3,35	66,48
Mussel shell	HA	6,65	33,50
	AA	6,28	37,21
	LA	7,60	24,05
	NA	4,75	52,48

The experiments did not differ much in relation to the source of calcium used, eggshell and mussel shell showed similar results. The factor that had the greatest impact on the weight loss of the precipitate was the type of acid used. The use of a more equal proportion between acids and calcium sources, as well as using stronger acids, can be alternatives to achieve better results in material solubilization and should be studied to evaluate its effect on the reaction.

4 Conclusion

The chemical characterization of the study materials shows that there is great potential for using eggshell and mussel shell as an alternative source of calcium for MICP due to its large amount of calcium carbonate that can be solubilized through various acids, including vinegar. The solubilization process via acids still needs to be optimized so that it consumes all or most of the carbon-rich waste. Changes in the ratio between acid and solid matter and reduction of acid dilution are some viable alternatives.

In the next steps, in addition to optimizing the process, we also seek to better understand the process. The results showed that nitric acid, followed by hydrochloric acid had better performance in solubilizing the residues. The use of analytical techniques such as the use of ethylenediaminetetraacetic acid (EDTA) can show the amount of calcium ions solubilized in the liquid. Finally, the application of solubilized calcium for MICP will prove the potential of the residues pointed out in this work, then, to be applied for the improvement of cementitious materials.

References

1. Mehta, P.K., Monteiro, P.J.M.: Concrete: Microstructure, Properties and Materials, 3rd edn. McGraw-Hill (2006)
2. Scrivener, K.L., Nonat, A.: Hydration of cementitious materials, present and future. Cem. Concr. Res. **41**, 651–665 (2011). https://doi.org/10.1016/j.cemconres.2011.03.026
3. (2019) Global Material Resources Outlook to 2060. OECD
4. Monteiro, P.J.M., Miller, S.A., Horvath, A.: Towards sustainable concrete. Nat. Mater. **16**, 698–699 (2017). https://doi.org/10.1038/nmat4930
5. Joshi, S., Goyal, S., Sudhakara Reddy, M.: Influence of biogenic treatment in improving the durability properties of waste amended concrete: a review. Constr. Build. Mater. **263**, 120170 (2020). https://doi.org/10.1016/j.conbuildmat.2020.120170
6. Miller, S.A., John, V.M., Pacca, S.A., Horvath, A.: Carbon dioxide reduction potential in the global cement industry by 2050. Cem. Concr. Res. **114**, 115–124 (2018). https://doi.org/10.1016/j.cemconres.2017.08.026
7. Maddalena, R., Roberts, J.J., Hamilton, A.: Can Portland cement be replaced by low-carbon alternative materials? A study on the thermal properties and carbon emissions of innovative cements. J. Clean. Prod. **186**, 933–942 (2018). https://doi.org/10.1016/j.jclepro.2018.02.138
8. Behera, M., Bhattacharyya, S.K., Minocha, A.K., et al.: Recycled aggregate from C&D waste & its use in concrete - A breakthrough towards sustainability in construction sector: a review. Constr. Build. Mater. **68**, 501–516 (2014). https://doi.org/10.1016/j.conbuildmat.2014.07.003
9. Anbu, P., Kang, C.-H., Shin, Y.-J., So, J.-S.: Formations of calcium carbonate minerals by bacteria and its multiple applications. Springerplus **5**(1), 1–26 (2016). https://doi.org/10.1186/s40064-016-1869-2
10. De Muynck, W., Leuridan, S., Van Loo, D., et al.: Influence of pore structure on the effectiveness of a biogenic carbonate surface treatment for limestone conservation. Appl. Environ. Microbiol. **77**, 6808–6820 (2011). https://doi.org/10.1128/AEM.00219-11
11. Basaran Bundur, Z., Kirisits, M.J., Ferron, R.D.: Biomineralized cement-based materials: impact of inoculating vegetative bacterial cells on hydration and strength. Cem. Concr. Res. **67**, 237–245 (2015). https://doi.org/10.1016/j.cemconres.2014.10.002
12. Qiu, J., Tng, D.Q.S., Yang, E.H.: Surface treatment of recycled concrete aggregates through microbial carbonate precipitation. Constr. Build. Mater. **57**, 144–150 (2014). https://doi.org/10.1016/j.conbuildmat.2014.01.085
13. Heveran, C.M., Williams, S.L., Qiu, J., et al.: Biomineralization and successive regeneration of engineered living building materials. Matter **2**, 481–494 (2020). https://doi.org/10.1016/j.matt.2019.11.016
14. Deng, X., Li, Y., Liu, H., et al.: Examining energy consumption and carbon emissions of microbial induced carbonate precipitation using the life cycle assessment method. Sustainability **13**, 4856 (2021). https://doi.org/10.3390/su13094856
15. Seifan, M., Berenjian, A.: Microbially induced calcium carbonate precipitation: a widespread phenomenon in the biological world. Appl. Microbiol. Biotechnol. **103**(12), 4693–4708 (2019). https://doi.org/10.1007/s00253-019-09861-5
16. Bhutange, S.P., Latkar, M.V., Chakrabarti, T.: Role of biocementation to improve mechanical properties of mortar. Sādhanā **44**(2), 1–8 (2019). https://doi.org/10.1007/s12046-018-1023-7
17. Amiri, A., Bundur, Z.B.: Use of corn-steep liquor as an alternative carbon source for biomineralization in cement-based materials and its impact on performance. Constr. Build. Mater. **165**, 655–662 (2018). https://doi.org/10.1016/j.conbuildmat.2018.01.070
18. Joshi, S., Goyal, S., Reddy, M.S.: Corn steep liquor as a nutritional source for biocementation and its impact on concrete structural properties. J. Ind. Microbiol. Biotechnol. **45**(8), 657–667 (2018). https://doi.org/10.1007/s10295-018-2050-4

19. Krishnapriya, S., Venkatesh Babu, D.L., Prince Arulraj, G.: Isolation and identification of bacteria to improve the strength of concrete. Microbiol. Res. **174**, 48–55 (2015). https://doi.org/10.1016/j.micres.2015.03.009
20. Achal, V., Mukherjee, A., Basu, P.C., Reddy, M.S.: Lactose mother liquor as an alternative nutrient source for microbial concrete production by Sporosarcina pasteurii. J. Ind. Microbiol. Biotechnol. **36**, 433–438 (2009). https://doi.org/10.1007/s10295-008-0514-7
21. Yoosathaporn, S., Tiangburanatham, P., Bovonsombut, S., et al.: A cost effective cultivation medium for biocalcification of Bacillus pasteurii KCTC 3558 and its effect on cement cubes properties. Microbiol. Res. **186–187**, 132–138 (2016). https://doi.org/10.1016/j.micres.2016.03.010
22. Fang, C., He, J., Achal, V., Plaza, G.: Tofu Wastewater as efficient nutritional source in biocementation for improved mechanical strength of cement mortars. Geomicrobiol. J. **36**, 515–521 (2019). https://doi.org/10.1080/01490451.2019.1576804
23. Hong, M., Kim, W., Park, W.: Low-cost cultivation and sporulation of alkaliphilic Bacillus sp. strain AK13 for self-healing concrete. J. Microbiol. Biotechnol. **29**, 1982–1992 (2019). https://doi.org/10.4014/jmb.1908.08034
24. Huang, Y.H., Chen, H.J., Maity, J.P., et al.: Efficient option of industrial wastewater resources in cement mortar application with river-sand by microbial induced calcium carbonate precipitation. Sci. Rep. **10**, 1–11 (2020). https://doi.org/10.1038/s41598-020-62666-9
25. Saxena, S., Tembhurkar, A.R.: Developing biotechnological technique for reuse of wastewater and steel slag in bio-concrete. J. Clean. Prod. **229**, 193–202 (2019). https://doi.org/10.1016/j.jclepro.2019.04.363
26. Lambert, S.E., Randall, D.G.: Manufacturing bio-bricks using microbial induced calcium carbonate precipitation and human urine. Water Res. **160**, 158–166 (2019). https://doi.org/10.1016/j.watres.2019.05.069
27. Liang, S., Chen, J., Niu, J., et al.: Using recycled calcium sources to solidify sandy soil through microbial induced carbonate precipitation. Mar. Georesour. Geotechnol. **38**, 393–399 (2020). https://doi.org/10.1080/1064119X.2019.1575939
28. Choi, S-G., Wu, S., Chu, J.: Biocementation for Sand Using an Eggshell as Calcium Source. Journal of Geotechnical and Geoenvironmental Engineering 142, 06016010 (2016). https://doi.org/10.1061/(ASCE)GT.1943-5606.0001534
29. Hong, M., Jang, I., Son, Y., Yi, C., Park, W.: Agricultural by-products and oyster shell as alternative nutrient sources for microbial sealing of early age cracks in mortar. AMB Express **11**(1), 1–15 (2021). https://doi.org/10.1186/s13568-020-01166-5
30. Kulanthaivel, P., Soundara, B., Selvakumar, S., Das, A.: Application of waste eggshell as a source of calcium in bacterial bio-cementation to enhance the engineering characteristics of sand. Environ. Sci. Pollut. Res. **29**, 66450–66461 (2022). https://doi.org/10.1007/s11356-022-20484-8

Biological Durability of Bamboo Bio-Concretes

Vanessa Maria Andreola[1](✉), Rayane de Lima Moura Paiva[1], Beatriz Palermo Lepine[1], Daniele Oliveira Justo dos Santos[1], Keyna Proença[2], Bruno Menezes da Cunha Gomes[1], Aurea Moraes[2], Simone Quinelato[2], Nicole Pagan Hasparyk[3], and Romildo Dias Toledo Filho[1]

[1] Federal University of Rio de Janeiro, Rio de Janeiro, RJ 21941-972, Brazil
{vanessa.andreola,daniele.justo}@numats.coc.ufrj.br, {rayane, toledo}@coc.ufrj.br, bpalermolepine@poli.ufrj.br

[2] Oswaldo Cruz Institute, Rio de Janeiro, RJ 21040-900, Brazil
{aurea,simone.quinelato}@ioc.fiocruz.br

[3] Eletrobras FURNAS, Goiânia, GO 74993-600, Brazil
nicole@furnas.com.br

Abstract. Studies on bio-based building materials are promising due to their insulating characteristics, low density and good hygrothermal performance. However, range of factors can influence its durability, including biological processes. For that reason, this study investigates the susceptibility to the growth of microorganisms in bamboo bio-concretes (BBC) with different finishes: (i) natural samples and (ii) samples waterproofed. The bio-concretes, after curing until 28 days (23 °C; 55% RH), were sterilized for 24 h in an oven (100 °C) and the hydrogenion potential (pH) determined. Subsequently, the samples were sprayed with suspensions of conidia of the fungus *Aspergillus niger* (IOC 0207) and a control suspension without fungi. After, the bio-concretes were conditioned in two controlled conditions of temperature (T) and relative humidity (RH), simulating environments of possible uses: (1) without water blade (23 ± 2 °C; 74 ± 2% RH) and (2) with water blade (23 ± 2 °C; 98 ± 2% RH). Biological durability was evaluated according to visual analyses by photographs and optical microscopy at 0, 7, 14, and 28 days. For conditioning 1, it was not possible to observe fungal proliferation on the surfaces of any sample during the test. For conditioning 2, the high RH environment and the presence of the water layer caused a color change, saturation and growth of microorganisms. The results show that the higher the RH, the greater the susceptibility to the growth of microorganisms and changes in appearance. Despite this type of study being introductory, bio-concrete is a promising material, with potential to be used in civil construction, showing durability to the growth of the fungus *A. niger.*

Keywords: Durability · Bio-degradation · Bio-concrete · *Aspergillus niger*

S. Amziane et al. (Eds.): ICBBM 2023, RILEM Bookseries 45, pp. 716–728, 2023.
https://doi.org/10.1007/978-3-031-33465-8_55

1 Introduction

Currently, the quest for sustainability is a universal challenge, both in industrial technology to automate processes and equipment and in different sectors of research [1, 2]. Civil construction, especially the concrete industry, aims to create strategies to increase sustainability and reduce impacts on the built environment [3–5].

The use of plants residues, generated in the industry, which can be used in developing new product, is a strategy that has been adopted [6, 7]. From this, researches haves been developed to explore the potential of these innovative materials [8–14] obtained from the use of unconventional constituents in mortars and concrete [15, 16, 18].

Bio-concrete stands out for its versatility, enabling production with different plant sources, such as bamboo waste, wood shavings, rice husks [2, 10, 19, 20], corn stalk, peach peel and damask [8, 14, 21]. Normally, the residues are chosen according to the availability of vegetal residues in each region/country. In the case of bamboo, residues can be obtained during the primary processing of the culms to laminates, which, depending on the species, can generate 65% to 85% of leftovers and pieces [20, 23].

In this context, bio-concrete and bio-mortar produced with a cement-based matrix and a lime-based matrix stand out, mainly because those composites use vegetal residues in the form of bio-aggregates. In the literature, studies are found with bio-concrete between 15% [22] and 80% [12], but most of the researches test volumes of bio-aggregates over than 50% [10, 15, 19, 20]. These materials improve thermal characteristics and reduce energy consumption when applied in buildings [4, 5, 14]. They are promising materials due to their insulating characteristics [9, 10, 16], low density [11, 14], good mechanical properties [2, 8, 10] and good hygrothermal performance [17, 22].

The high volume of plant residues used in bio-concrete production can make them sensitive to the growth of microorganisms, as they are lignocellulosic materials, made up of chemical substances [13, 16, 24]. According to the researches available in the literature [14, 25, 26], many factors can influence durability and performance, and one of the main issues related to the use of bio-concrete is their biological durability, including processes that can directly affect human health [26, 27]. Therefore, before being made available on the market, it was particularly important to evaluate biological durability.

In this context, the present study analyzed the susceptibility to the growth of microorganisms in bamboo bio-concretes (BBC) in natural samples and waterproofed with a water-repellent agent. The samples were conditioned in two conditions of temperature (T) and relative humidity (RH), simulating environments of possible uses of bio-concrete: (1) without water depth (23 $\pm$ 2 °C; 74 $\pm$ 2% RH) and a critical environment: (2) with water depth (23 $\pm$ 2 °C; 98 $\pm$ 2% RH). The biological evaluation was analyzed according to visual analysis by photographs (0, 7, 14 and 28 days) and optical microscopy at 28 days.

2 Materials and Methods

2.1 Bio-Aggregates: Raw Material

The bamboo bio-aggregates used in this study were acquired from the municipality of Petrópolis (Rio de Janeiro, Brazil). First, the culms were cut with a table saw to reduce their length from 15 cm. After the cut, the segments were processed in an industrial crusher (twice) to obtain the particles with lengths between 1.18 to 5.60 mm. Then, the residues were submitted to mechanical sieving using sieves with openings of 5.60 mm and 1.18 mm. All material retained between the sieves was used (Fig. 1). After sieving, the bio-aggregates were treated by immersion in an alkaline calcium hydroxide solution ($Ca(OH)_2$) at a concentration of 1.85 g/L. The proportion, by mass, of water/bamboo used was 1:7. The bio-aggregates were immersed in the solution for 1 h, washed with water and air-dried until reaching to constant mass. The apparent density and moisture content were 580 kg/m^3 and 12.3%, respectively, according to the standards [28, 29]. Water absorption was 65%, considering the mixing conditions during bio-concrete production [2].

Fig. 1. Bamboo bio-aggregates used in this study.

2.2 Fabrication and Waterproofing of Bamboo Bio-Concrete (BBC)

The bio-concretes were produced with a matrix composed of 45% (CP II-F-40 Brazilian cement), 25% rice husk ash (RHA) and 30% fly ash (FA). A volumetric fraction of bio-aggregates of 45% was used, and the water-to-blinder ratio was set at 0.30. In addition, was used 2% of calcium chloride ($CaCl_2$) in relation to the mass of the binders. $CaCl_2$ was used to accelerate cement setting and minimize the inhibitory effect of bio-aggregates, prior to the action of extractives in the mixture [30]. The proportion of the mixtures produced in this work is summarized in Table 1.

The rational dosage of bio-concrete was used for the production of mixtures [2]. Initially, the cementitious materials were homogenized (using a 20 L mixer - 136 rpm), followed by the gradual addition of 90% of the total water with the calcium chloride solution. After 1 min, the bio-aggregates were gradually inserted into the cementitious paste and mixed for 1:30 min. After which, the process was completed by adding the remaining water until reaching a total of 7 min of mixing.

Table 1. Mix proportion of BBC, with materials by mass (kg/m^3).

Bamboo	CP II	RHA	FA	*H_w	**W_c	$CaCl_2$
232.00	341.43	151.75	265.56	227.62	150.80	19.12

* H_w = hydration water; ** W_c = Water of compensation.

The material was cast in prismatic molds with 400 × 100 × 30 mm (length × width × thickness) and mechanically vibrated on a vibrating table (68 Hz) for 15s. After casting, the samples were protected from moisture loss and demoulded after 24 h. Subsequently, with the aid of a circular saw, the samples were cut into dimensions of 100 mm × 100 mm × 25 mm (length × width × thickness), leaving the bio-aggregates scraped and visible on all sides. After cutting, the BBC was dry cure (21 ± 2 °C; 55 ± 5% RH), until reaching 28 days of age.

After complete curing, a surface application (using a brush) of a waterproofing agent with a concentration of 5% silane/siloxane was performed on the bio-concrete samples. The surface protector was applied in two application coats, totaling a consumption of 13.7mL/m^2. The waterproofing agent has a density of 1.01kg/L, is non-flammable and has low emission of volatiles.

2.3 Biological Durability and Characterization Tests

Sample Preparation. After curing for 28 days (21 °C; 55% RH), the BBC was sterilized for 24 h in an oven (100 ± 2 °C), in order to sterilize the surface of the samples in terms of the presence of fungi [13, 17, 30].

Determination of the Hydrogenation Potential (pH). After 28 days of curing, the bio-concretes were milled, macerated and diluted in distilled water (in triplicates). From manual homogenization, the samples solution were sealed and conditioned in a controlled environment (21 °C; 55% RH) for 24 h. The solutions were analyzed using a benchtop pH meter (Nova Instruments, model NI PHM).

The Aqueous Suspensions Used in the Tests. For the biological analysis, the objective was to analyze the susceptibility of bioconcretes to colonization by a fungus species. The samples were sprayed with a high concentration of conidia. After that, they were kept under controlled conditions of temperature and humidity. Thus, the fungus *Aspergillus niger* was selected because it is anemophilous and widely distributed in indoor and outdoor environments, with excellent growth at 30 °C [11, 13, 27]. The fungus strain *A. niger* (IOC 0207) was acquired from the Filamentous Fungi Culture Collection (CCFF) IOC/FIOCRUZ. Two types of aqueous suspensions were prepared for the tests. A control suspension containing only water and 0.1% Tween 80 and a conidial suspension containing water, 0.1% Tween 80 and conidia of the fungus *A. niger* at a concentration of 1.5 × 10^7 conidia.mL^{-1}.

Biological Durability by Spraying Method. For the analysis, the methodology of Andreola [10] and Paiva [17] was followed. The procedure consisted of measuring the suspensions using a beaker and transferring the material to a sterile spray bottle. For samples, equal volumes of conidial suspension and control suspension were used.

Sample Incubation Conditions. After spraying the aqueous suspensions, the samples were kept in glass containers, inside the laboratory environment and the temperature and relative humidity inside the container were controlled by an automatic sensor, which continuously acquired the data (sensor model Impac), as follows:

1) *Without water blade (23 ± 2 °C; 74 ± 2% RH).* Simulated high humidity. The method consisted of keeping the samples with the conidial suspension/control suspension in a container without water. The samples remained in the same condition shown in Fig. 2, with the only difference being that the container was kept dry throughout the test.
2) *With water blade (23 ± 2 °C; 98 ± 2% RH).* Simulated extreme humidity. The method consisted of keeping the samples with the conidial suspension/control suspension in a container that simulated an aquarium. For this, a glass container was used (Fig. 2) which had a plastic mesh at the base. The samples were placed on the mesh, allowing free access to water to their bases. The water level remained constant (5 mm) above the base of the sample. After placing the water slide, the top of the container was sealed with plastic film and small openings in the plastic were made.

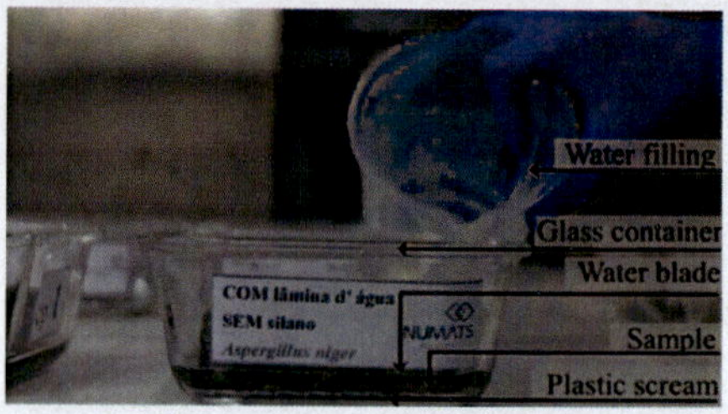

Fig. 2. Water blade testes of the BBC samples.

Visual Analysis by Photographs. The procedure was performed using pictures taken at the ages of 0, 7, 14 and 28 days after suspension spray. A camera attached to a fixed experimental bench was used, with three points of light [10, 15, 17]. The photographs were always taken with the same camera (EOS 80D, Canon) and the base of the bench was configured to receive the glass container, avoiding handling the sample (Fig. 3). The images obtained are always compared with the control samples, which are the photographs taken after spraying the suspensions (day 0).

Analysis by Optical Microscopy (OM). The analysis was carried out to observe the proliferation of microorganisms on the surface of the BBC, at 28 days of the start of the tests. Magnifications of 10x and 100x were adopted using a stereoscopic microscope, with reflected light (Model SMZ-800, Nikon) in order to obtain the images, with a parallel optical zoom system. In the microscope, a wooden base with a glass container fitting was used (Fig. 4). This base served as a support to move the set (container and

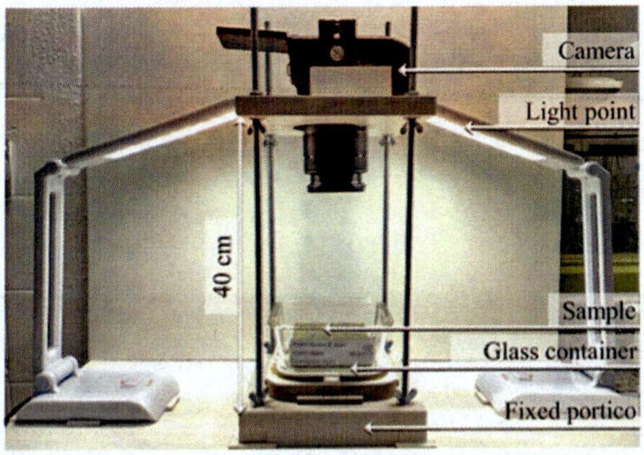

Fig. 3. Photo recording on experimental bench.

sample), without handling the BBC. In the sites where microorganisms were observed (with the naked eye), two points of light were directed for microscopic visualization.

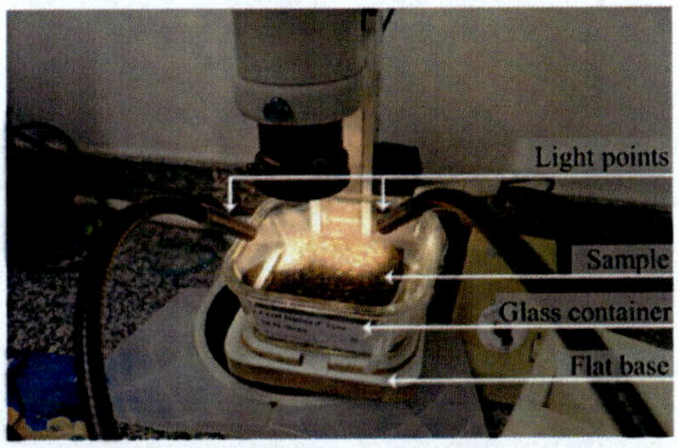

Fig. 4. Microscopic analysis setup: Sample on flat base, using two light points.

3 Results and Discussion

The following abbreviations were used to facilities the results presentation:

- For identify the type of aqueous suspension: (CS) control aqueous suspension and (FS) fungal aqueous suspension.
- For the incubation conditions: (1) Without water blade or (2) With water blade.
- For the finish used in bio-concretes: (N) Natural or (W) Waterproofed.

3.1 Determination of pH

The pH values obtained were 11.5 ± 0.04 and 11.3 ± 0.01 for natural and waterproofed bio-concrete, respectively, i.e.; both samples have an alkaline pH. The high pH values found are directly related to the properties of the binders used, mainly the presence of cement, whose pH varies between 12 and 13.5. The presence of the waterproofing agent used did not significantly alter the pH value of the evaluated BBC.

3.2 Visual Analysis by Photographs

The images obtained for the samples with control aqueous suspension (CS) and fungal aqueous suspension (FS) are presented in the Table 2 and Table 3, respectively.

Table 2. Visual analysis by photographs of the control aqueous suspension (CS).

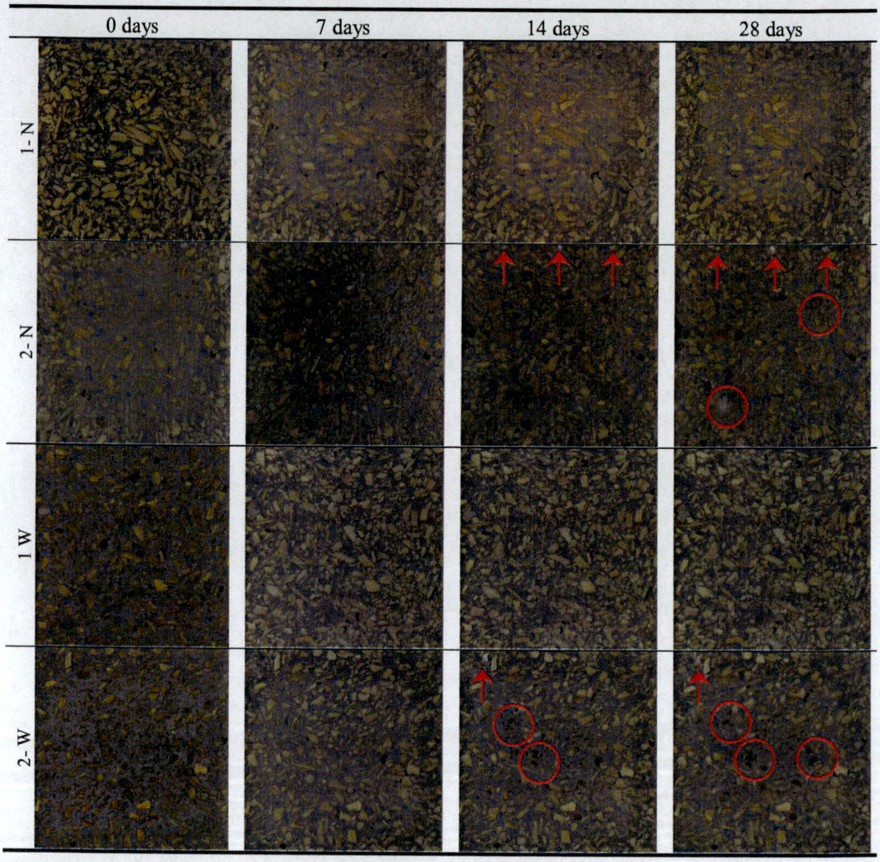

As shown in Table 2 and Table 3, the appearance of the bio-concretes was not changed for the mixtures without a water blade (1), both for the control suspension and for the fungal suspension. The difference in color and saturation is only observed on the day of spraying (0 day), because the suspensions are aqueous, which humidify the sample. The saturated sample showed greater vividness of color, especially the bio-aggregate. This tendency was not observed for samples kept in a water blade, which was expected due to the higher humidity in which the bio-concrete was kept.

The 2-N (CS) samples showed surface saturation on day 0, a more saturated (soaked) appearance on day 7 and remained saturated on days 14 and 28. By day 7 these samples showed fungal growth on the surface, expanding growth points as the assay time

Table 3. Visual analysis by photographs of the fungal aqueous suspension (FS).

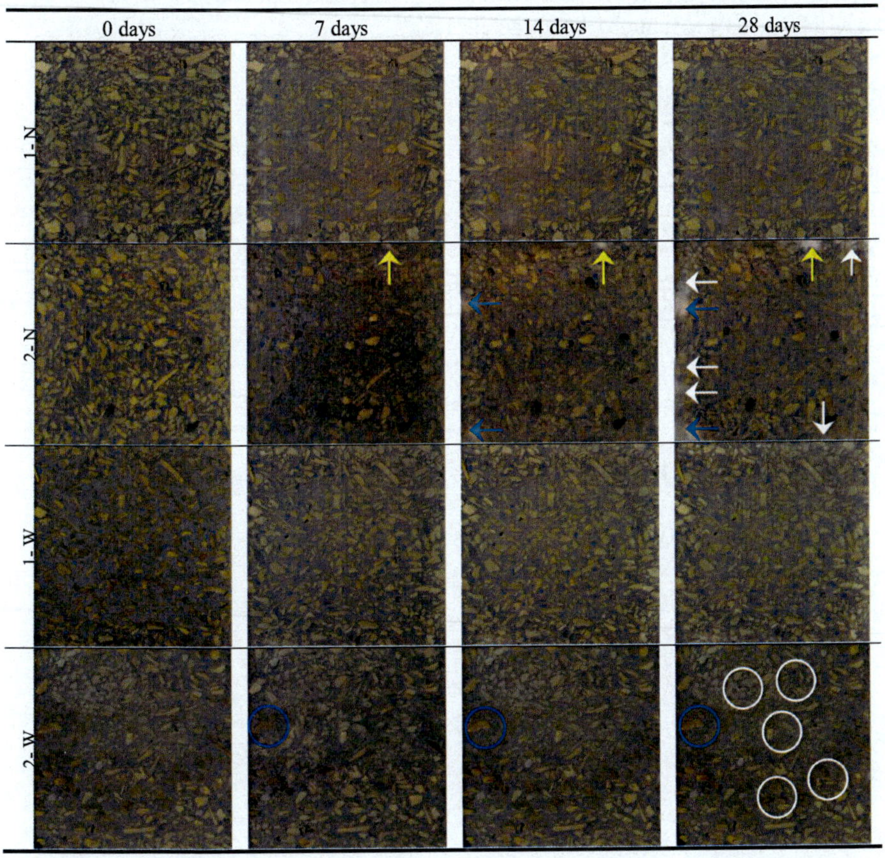

increased (see Table 2, as indications in red). However, the greatest growth of microorganisms was observed in the 2-N (FS) sample and, even on day 7, fungal growth was already observed (see Table 3, the identification in yellow) surrounding a particle of bamboo. From 7 to 14 days it was observed, in all the ends of the sample (see Table 3, the indications in blue), growth of filamentous fungi. From the 14th to the 28th day of testing, it was noted that the growth region increased, spreading to the central region of the specimen (see Table 3, the white arrows).

About the bio-concrete 2-W (CS), the growth of microorganisms was also observed after the first week of the test. On the 14th day, it was observed that the bio-aggregates were saturated, with punctual water droplets around their geometry (see Table 2, circles in red). In addition, spots began to appear in the upper left corner of these samples (Table 2, the red arrows). From 14 to 28 days, there was a greater concentration of these spots, while the water droplets expanded to other bio-aggregates. The 2-W (FS) showed a similar trend (droplet concentration), however, more intensely than the reference. Even on the 7th test day, it was possible to observe the presence of fungal structures on the bio-aggregate (Table 3, the blue circles). From 7 to 14 days after the contamination, the saturated particles became stained. On the 28th day, dark spots appeared at different points in the sample (Table 3, the white circles).

3.3 Microscopy Analysis

Table 4and Table 5, show the MO images of the bio-concrete samples after 28 days of conditioning under exposure of incubation with water depth (2) (23 ± 2 °C; 98 ± 2% RH) with aqueous control suspension (CS) and the aqueous suspension of the fungus *Aspergillus niger* (FS), respectively. This analysis was carried out only for this age because the surfaces of the samples showed greater susceptibility to fungal proliferation, as previously evidenced in the visual analysis by photographs.

Table 4. Visual analysis by optical microscopy (OM) of the aqueous control suspension (CS).

	10x magnification	100x magnification
2-N		
2-W		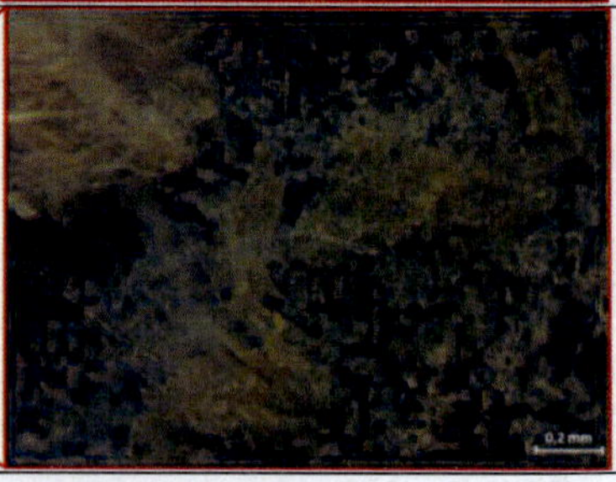

From the simplified MO images, it was possible to verify the presence of the fungus in all tested samples, and observation of its morphological structures, which was evidenced through hyphae, structures in the form of elongated filaments in white color and conidial heads, spherical structures in dark color. This pattern was observed mainly on the bamboo bio-aggregates that are exposed on the surface of the samples and at the interface between the bio-aggregates and the cementitious material. The high susceptibility to fungal patterns observed at these points is directly correlated to: (i) the chemical composition of the bio-aggregates, which is rich in Carbon (C), thus favoring a greater supply of nutrients, (ii) the high hygroscopic characteristic of the BBC due to the multiscale porosity which favors the a moisture concentration and also (iii) the roughness at

Table 5. Visual analysis by optical microscopy (OM) of the fungal aqueous suspension (FS).

	10x magnification	100x magnification
2-N	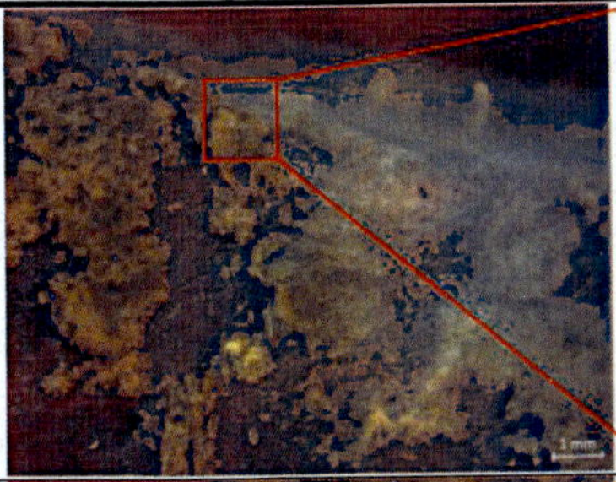	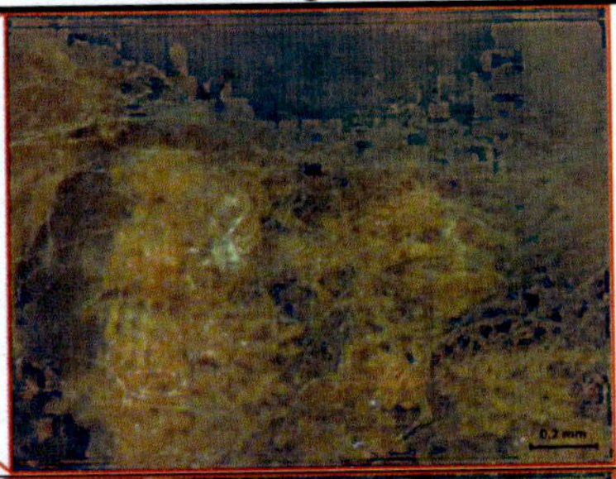
2-W		

the ends of the specimens caused during the process of stripping the samples prepared for the test.

In this case, as already highlighted in subtitle 3.2, the bio-concrete waterproofed showed less biological incidence when compared to natural samples, that is, the waterproofing agent delayed the proliferation of fungi on the surface of the samples exposed to incubation conditions 2 (with water blade), however, its use was not enough to completely inhibit its proliferation.

The results found in this study show the same pattern of behavior found by Laborel *et al.* [13].The research evaluated the biological growth under different environmental conditions in earth-based materials, with and without the addition of bio-aggregates. The inoculum used was *Aspergillus brasiliensis* and the samples were incubated for 12 weeks at 76, 84 and 93% RH and 30 °C or 20 °C. Results showed that earth-based materials were more sensitive to fungi when enriched with 3% plant aggregates. Fungal growth was observed in soil material containing bioaggregates after 4 weeks of exposure (93% RH and 30 °C). While for the raw earth material, under the same conditions, growth was observed after 8 weeks. It should be noted that, despite the samples having a pH of 7.8, in the present study a fraction of biomass equal to 45% was used, and an extreme incubation condition (with water depth) justifying the susceptibility to biological growth, even for waterproofed samples.

Santos *et al.* [32] evaluated the biological susceptibility of mortars and the pH values obtained were between 8.5 and 9.6. The authors found that the partial replacement of fine sand by PCM reduced the pH of the mortars, the addition of fibers did not provide relevant changes, while the addition of lime increased the pH of the mortars. In fact, regardless of the presence of fibers or PCM, the addition of lime and the consequent increase in pH prevented, although not completely, the colonization and development of fungi. In this research it can be seen that the presence of the waterproofing agent used did not significantly change the pH value of the evaluated BBC, this fact is related to the chemical characteristics of the waterproofing agent used and also to the volume used for the surface coating of the samples.

In addition to the analyzes presented here, colonies that grew on the surfaces of the samples were collected. The collection with a swab will allow identifying if there was growth of other fungi, in addition to the *A. niger.* Previously sprayed on the bio-concrete.

4 Conclusion

This work evaluated the resistance to the tolerance of bio-concretes to the susceptibility to colonization by a fungus species. The BBC was stored in two conditions: (1) High moisture (23 ± 2 °C; 74 ± 2% RH) and (2) Extreme moisture (23 ± 2 °C; 98 ± 2% RH). Based on the results, the following conclusions may accordingly be drawn:

(1) In the condition of high moisture, it was not possible to observe the proliferation of fungi on the sample surfaces, as well was no change in their appearance or color. This can be explained by the moisture and temperature levels used, which did not provide a favorable environment. Furthermore, the fact that the BBC has an alkaline pH (11.5 and 11.3), the proliferation of fungi on the surfaces is hindered.
(2) For the extreme moisture condition, the growth of microorganisms was favored, being intensified by the high volume of bio-aggregates used in the bio-concrete, which consists of a greater supply of nutrients for the proliferation of fungi. In addition, there is the hygroscopic capacity of the material as well as the greater roughness of samples due to the procedure used during preparation and cutting. Regarding appearance and color, the photographs showed that the natural BBC (1-N) indicated greater saturation, color change, loss of brightness, and greater growth of microorganisms when compared to samples with waterproofing (2-N). The extreme moisture condition (98 ± 2%) provoked the greatest proliferation of microorganisms due to growth at different locations on the surface of the sample (from 7 to 14 days as well as from 14 to 28 days).

Thus, the results of this research show the importance of evaluating the biological durability of bio-concretes in different exposure conditions and treatments. Despite the high volume of biomass in the composition of bio-concretes, those studies can prove that BBC is a versatile material that can be used safely and without harm to human health. The method proposal can be used to evaluate the biological resistance of other materials, since the choice of a standard and reliable method is a problem encountered by other authors, as reported in this article. With these advancements, it is possible to indicate in which uses and applications the bio-based materials are more appropriate. Tests are easily reproducible as they do not require advanced equipment.

It is important to highlight that the present study is in progress to evaluate the biological durability of bio-concrete over time, at advanced ages.

References

1. Beltrami, M., Orzes, G., Sarkis, J., Sartor, M.: Industry 4.0 and sustainability: towards conceptualization and theory. J. Clean. Prod. **312**, 127733 (2021)
2. Da Gloria, M.H.Y.R., Andreola, V.M., dos Santos, D.O.J., Pepe, M., Toledo Filho, R.D.: A comprehensive approach for designing workable bio-based cementitious composites. J. Build. Eng. **34**, 101696 (2021)
3. Barrelas, J., Ren, Q., Pereira, C.: Implications of climate change in the implementation of maintenance planning and use of building inspection systems. J. Build. Eng. **40**, 102777 (2021)
4. Caldas, L.R., Saraiva, A.B., Andreola, V.M., Toledo Filho, R.D.: Bamboo bio-concrete as an alternative for buildings' climate change mitigation and adaptation. Constr. Build. Mater. **263**, 120652 (2020)
5. Mahecha, R.E.G., Caldas, L.R., Garaffa, R., Lucena, A.F., Szklo, A., Toledo Filho, R.D.: Constructive systems for social housing deployment in developing countries: a case study using dynamic life cycle carbon assessment and cost analysis in Brazil. Energy Build. **227**, 110395 (2020)
6. Pervez, H., Ali, Y., Petrillo, A.: A quantitative assessment of greenhouse gas (GHG) emissions from conventional and modular construction: a case of developing country. J. Clean. Prod. **294**, 126210 (2021)
7. Silva, L.C., Caldas, L.R., Paiva, R.L.M., Toledo Filho, R.D.: Role of Bio-based Building Materials in Climate Change Mitigation: Special Report of the Brazilian Panel on Climate Change. Rio de Janeiro: Brazilian Panel on Climate Change (2018). Special Report of the Brazilian Panel on Climate Change
8. Ahmad, M.R., Pan, Y., Chen, B.: Physical and mechanical properties of sustainable vegetal concrete exposed to extreme weather conditions. Constr. Build. Mater. **287**, 123024 (2021)
9. Amziane, S., Collet, F. (eds.): Bio-aggregates Based Building Materials. RSR, vol. 23. Springer, Dordrecht (2017). https://doi.org/10.1007/978-94-024-1031-0
10. Andreola, V.M.: Durabilidade ao intemperismo natural e ao ataque biológico de bio-concretos de bambu. Universidade Federal do Rio de Janeiro, RJ, Brasil, Tese de Doutorado (2021)
11. Arizzi, A., Viles, H., Martín-Sanchez, I., Cultrone, G.: Predicting the long-term durability of hemp–lime renders in inland and coastal areas using mediterranean, tropical and semi-arid climatic simulations. Sci. Total Environ. **542**, 757–770 (2016)
12. Da Gloria, M.H.Y.R., Toledo Filho, R.D.: Innovative sandwich panels made of wood bio-concrete and sisal fiber reinforced cement composites. Constr. Build. Mater. **272**, 121636 (2021)
13. Laborel-Préneron, A., et al.: Laboratory test to assess sensitivity of bio-based earth materials to fungal growth. Build. Environ. **142**, 11–21 (2018)
14. Wu, F., Yu, Q., Liu, C.: Durability of thermal insulating bio-based lightweight concrete: understanding of heat treatment on bio-aggregates. Constr. Build. Mater. **269**, 121800 (2021)
15. Andreola, V.M., Da Gloria, M.Y.R., Toledo Filho, R.D.: Durability of Bamboo Bio-Concretes Exposed to Natural Aging. In: Construction Technologies and Architecture. Trans Tech Publications Ltd, pp. 834–841 (2021)
16. Claude, S.: Étude expérimentale et numérique de solutions basées sur les éco-matériaux pour la rénovation thermique du patrimoine bâti urbain. Tese de Doutorado. Toulouse, INSA (2018)
17. Paiva, R.L.M.: Avaliação do desempenho físico, mecânico, higrotérmico, ambiental e durabilidade biológica de argamassas com terra e partículas de bambu. Universidade Federal do Rio de Janeiro, Rio de Janeiro, RJ, Brasil, Tese de Doutorado (2022)
18. Paiva, R.D.L.M., Caldas, L.R., de Souza Martins, A.P., Toledo Filho, R.D.: Potentiality of earth-based mortar containing bamboo particles for GHG emissions reduction. Constr. Build. Mater. **317**, 125971 (2022)

19. Santos, D., da Gloria, M., Andreola, V., Pepe, M., Toledo Filho, R.: Compressive stress strain behavior of workable bio-concretes produced using bamboo, rice husk and wood shavings particles. Acad. J.Civil Eng. **35**(2), 211–217 (2017)
20. Silva, N.A.: Estudo da aderência bambu – bioconcrete.: Dissertação de Mestrado, Universidade Federal do Rio de Janeiro, Rio de Janeiro, RJ, Brasil (2019)
21. Belayachi, N., Hoxha, D., Slaimia, M.: Durability studies on lightweight plaster and straw fiber based material for building thermal insulation. Acad. J. Civil Eng. **33**(2), 203–207 (2015)
22. Haba, B., Agoudjil, B., Boudenne, A., Benzarti, K.: Hygric properties and thermal conductivity of a new insulation material for building based on date palm concrete. Constr. Build. Mater. **154**, 963–971 (2017)
23. Kai, Z., Xuhe, C.: Potential of bamboo-based panels serving as prefabricated construction materials (2014)
24. Kallavus, U., Järv, H., Kalamees, T., Kurik, L.: Assessment of durability of environmentally friendly wood-based panels. Energy Procedia **132**, 207–212 (2017)
25. Tobon, A.M., Andres, Y., Locoge, N.: Impacts of test methods on the assessment of insulation materials' resistance against moulds. Build. Environ. **179**, 106963 (2020)
26. Viel, M., et al.: Resistance to mold development assessment of bio-based building materials. Compos. B Eng. **158**, 406–418 (2019)
27. Verdier, T., Coutand, M., Bertron, A., Roques, C.: A review of indoor microbial growth across building materials and sampling and analysis methods. Build. Environ. **80**, 136–149 (2014)
28. Associação Brasileira de Normas Técnicas. Agregado Miúdo–Determinação da Massa Específica e Massa Específica Aparente, NBR NM 52. Rio de Janeiro, RJ (2009)
29. ISO 22157: Bamboo Structures: Determination of Physical and Mechanical Properties of Bamboo Culms: Test Methods
30. Wolfe, R.W., Gjinolli, A.: Durability and strength of cement-bonded wood particle composites made from construction waste. For. Prod. J. **49**, 24–31 (1999)
31. Giroudon, M., Laborel-Préneron, A., Aubert, J.E., Magniont, C.: Comparison of barley and lavender straws as bio-aggregates in earth bricks. Constr. Build. Mater. **202**, 254–265 (2019)
32. Santos, T., Nunes, L., Faria, P.: Production of eco-efficient earth-based plasters: influence of composition on physical performance and bio-susceptibility. J. Clean. Prod. **167**, 55–67 (2017)

Isolation of Bacterial Strains from Concrete Aggregates and Their Potential Application in Microbially Induced Calcite Precipitation

Giuseppe Moita(✉), Vitor Liduino, Eliana Flávia Servulo, João Paulo Bassin, and Romildo Dias Toledo Filho

Federal University of Rio de Janeiro, Rio de Janeiro, Brazil
giuseppe.moita@coc.ufrj.br

Abstract. The search for more sustainable technologies in the civil construction sector has led to development of alternative practices with less environmental impact. The use of microbial activity, such as in the biocalcification process, also known as microbiologically induced calcite precipitation (MICP), other than being considered a low-cost process, is a promising technology capable of improving the characteristics of natural and recycled aggregates. This work aimed to isolate bacteria strains from a mixture of concrete aggregates and analyze their potential for calcium carbonate (CaCO3) mineralization. Eleven different bacterial strains were isolated in an appropriate medium culture. They were characterized based on their colony and cell morphologies. Although colony characteristics varied, most of them were Gram-positive rod-shaped bacteria. Then, the isolates were individually inoculated in Erlenmeyer flasks containing 100 mL of a production medium suitable for the CaCO3 precipitation. All flasks were under the agitation of 150 rpm at 30 °C for 7 days. After incubation, the CaCO3 precipitate was visible to the naked eye in all bacteria-inoculated flasks. This study shows that concrete aggregates are a great source of microorganisms that can be used in biotechnologically produced materials.

Keywords: MICP · bacterial isolation · biomineralization · calcium carbonate

1 Introduction

Concrete is the most used construction material in the world due to its durability, fire resistance, mechanical strength, and other important characteristics [1, 2]. This raises several environmental concerns, as cement alone is responsible for 7 to 9% of carbon gas emissions in the world [3]. In addition, the prospects are for a significant increase in concrete consumption to remedy the infrastructure and housing deficit [4]. Many studies have been carried out to replace cement in whole or in part, in order to mitigate this environmental impact [5, 6].

Although a little less targeted, the volume of aggregate needed for the next few years is also a problem, due to its environmental impact of removing the raw material, which consists of about 75% of the total volume of concrete [7, 8]. There is also the increasing

S. Amziane et al. (Eds.): ICBBM 2023, RILEM Bookseries 45, pp. 729–738, 2023.
https://doi.org/10.1007/978-3-031-33465-8_56

difficulty of extracting it near large urban centers [9]. The use of Recycled Concrete Aggregates (RCA) can be doubly advantageous: it avoids waste production and raw material extraction.

Using RCA can be very advantageous in this regard, but its characteristics impose some limitations. This material has greater porosity due to the attached mortar in it, interfering with the mechanical strength and durability of Recycled Aggregate Concrete (RAC) [10–12]. Studies have been made to improve the characteristics of this material with the removal of the attached mortar, coating of the aggregate or with new mixtures. These solutions have been shown to be environmentally controversial due to the energy expenditure or environmental impact of the materials used for this [8].

In this context, microbial induced calcite precipitation (MICP) appears as a biobased alternative with the potential to reduce environmental damage for aggregate improvement [13]. Some microorganisms under suitable conditions have been shown to be capable of precipitating calcium carbonate ($CaCO_3$), a material that is chemically compatible with concrete and has considerable resistance. In the aggregate, recent studies have shown that microorganisms can produce a layer of CaCO3 to reduce their porosity, improving their quality [14, 15].

To microbiologically induced calcite precipitation (MICP) occurs, bacteria need an inorganic source of carbon, a source of calcium, alkaline pH and nucleation points. The temperature and nutritional media of the bacteria can also affect this process. The carbon source depends directly on the metabolism of the bacteria, which will process it from an organic source, transforming it into an inorganic one. It, as well as the source of calcium, can be incorporated into the nutrient medium [16, 17]. The alkaline pH of the medium must be controlled to allow the formation of $CaCO_3$ whose nucleation point is the bacteria itself [16, 17].

Researchers are looking for more local and more specific solutions for their desired use and for different bacteria with greater efficiency in biomineralization. Thus, several works in addition to studying the MICP itself, previously sought to find a bacterium that was better adapted to its object of study.

The places of collection of material for the identification of biomineralizing bacteria are diverse and depend on the author's intention. To be surer that there will be $CaCO_3$-producing bacteria, collection can be done in a limestone cave [18] and immature calcareous soils [19]. There are also records of bacteria collected even in deep-sea sediment [20]. The soil also seems to be a favourable place to find bacteria producing $CaCO_3$ mainly due to the nitrogen present in it, which tries to develop several species of ureolytic bacteria. Thus, there are records of collections made in alluvial soils [21], mangrove [22], gardening soil [23] and wetland [24].

In other situations, such as the objective of reducing the lag phase of bacterial growth, the collection of material was done in materials similar to those used as walls [25], concrete [26] and frescoes [27]. The isolation of microorganisms from the material to which it will be applied later tends to improve the quality of the future process by being in an already adapted environment.

In this sense, this work aims to isolate bacteria from natural and recycled concrete aggregates to verify their potential application in microbially induced calcite precipitation for application in the improvement of these aggregates.

2 Materials and Methods

2.1 Sampling and Bacterial Isolation

Samples of natural and recycled sand and gravel were collected from the Teaching and Research Center for Low Environmental Impact Materials and Technologies in Sustainable Construction at the Federal University of Rio de Janeiro. For the isolation of bacteria, a poll was performed with samples of the same weight as each of the samples. The coarse aggregates were previously macerated with a mortar and pestle until reaching a granulometry similar to sand.

10 g of this mixture was placed in a 500 mL Erlenmeyer containing 100 mL of sterile nutrient broth (1.0 g/L meat extract, 2.0 g/L yeast extract, 5.0 g/L peptone and 5.0 g/L sodium chloride) supplemented with 2.5 g/L calcium acetate. The pH was adjusted to 9.0 with sodium hydroxide. Parallel, 10 g of aggregates were placed in an Erlenmeyer with the same volume capacity but containing two-fold diluted nutrient broth in order to simulate an environment with a shortage of nutrients. Both Erlenmeyer flasks were sealed and incubated under agitation of 180 rpm for 2 days at 30 °C.

After incubation, samples from both Erlenmeyer flasks were diluted until 10^{-5} and all aliquots were inoculated in Petri plates containing nutrient agar (1.0 g/L meat extract, 2.0 g/L Yeast extract, 5.0 g/L Peptone, 5.0 g/L sodium chloride and 15.0 Agar) added with 2.5 g/L calcium acetate. The plates were incubated for 1 day at 30 ºC.

The microbial colonies that showed visible morphological differences among themselves were isolated in new Petri plates with the same medium mentioned above. In total, 11 different isolates were obtained, which in this work are coded from I to XI. The cell shape and purity of each isolate were examined by Gram staining using an optical microscope. All isolates were individually replicates in agar slant tubes and preserved in a refrigerator at 4ºC.

2.2 CaCO3 Precipitation

Before CaCO3 precipitation test, all isolates were reactivated by growing in agar slant tubes containing nutrient agar added with 2,5 g/L of calcium acetate at pH 9 for 2 days at 30 °C. The grown biomass was scraped using a sterile loop, suspended in 10 mL saline solution (0.85 g/L of NaCl), and homogenized with a vortex.

3 mL of the suspended cell solution were inoculated into a 500 mL Erlenmeyer containing 100 mL of nutrient broth supplemented with 10.0 g/L calcium acetate at pH 9. The inoculated flasks were then incubated at 30 °C on a shaker operating at 150 rpm for 7 days. Each isolate was tested singly and in triplicate. The same medium without bacteria was also incubated under the same conditions as the experiment control.

After the experiment time, the $CaCO_3$ produced was harvested by centrifugation at 4000 rpm for 5 min. The material deposited at the bottom was washed with distilled water, placed in an open petri dish and dried in an oven at 100 °C for 24 h.

The dried material was then transferred to a porcelain crucible to be burned at a temperature of 550 °C for one hour in order to eliminate all organic matter. This temperature is not enough to decompose calcium carbonate. The precipitate was weighed before and after the burning process. Finally, the remaining material was analyzed by the Fourier Transform Infrared Spectroscopy (FTIR) technique using the Cary 630 FTIR Spectrometer to qualitatively assess the presence of post-firing calcium carbonate.

3 Results and Discussion

3.1 Isolates Characterization

The macro and micro-observations of the isolates allow to classify all belong to the bacteria domain. Selected images of each isolate can be seen in the Table 1, as well as a short description. Isolate I was identified as Gram-Negative and all other isolates were identified as Gram positive. IV, VI, VIII and XI colonies are similar to each other, but some visual differences in cell morphology were observed. Future genetic analysis will be required to differentiate these isolates.

3.2 CaCO3 Precipitation

After removing the Erlenmeyer flasks from the shaker, it was possible to observe the existence of a white precipitate, the characteristic color of $CaCO_3$, in all those in which the bacteria were inoculated (Fig. 1a). There was no precipitate in the control tests (Fig. 1b).

In centrifuged material, it was observed a white layer at the bottom of the Falcon tube. After transferring to a Petri dish and drying the material, it became more visible. A yellowish coloration was also observed in the precipitate, which indicates the presence of aragonite or remains of dead bacterial (organic matter).

Analyzing precipitates with different shades of yellow before and after burning the material in a kiln at 550 °C to eliminate organic matter (Fig. 2), it was observed that the greater amount of yellowish material before burning, the greater amount of carbonized material afterward.

Table 2 shows the value of the dry weight of the average of the precipitates, as well as the loss on ignition, and the matter remaining after this process. Considering that all the remaining material is calcium carbonate, the conversion percentage of each isolate was calculated from the maximum possible conversion of calcium acetate in the medium (calculated from Eq. 1).

$$\text{g of CaCo}_3 = \frac{\text{Molar mass of C}_4\text{H}_6\text{CaO}_4}{\text{Molar mass of CaCo}_3} \times \text{g C}_4\text{H}_6\text{O}_4 \tag{1}$$

The results showed that the ignition loss had a significant value, so even performing the centrifugation of the material to separate the carbonate from the organic matter, a part of it remained in the studied material. This fact makes it difficult to measure and analyze the material in depth, as seen in the Fig. 2. Even burning at 550 °C, the black and grayish color indicates that there were still residues of this material with the sediment.

Thus, centrifugation of the material more than once and using filters that allow the passage of organic matter are a solution to avoid the accumulation of organic matter

Table 1. Characterization of isolates

Isolate	Colony observation	Cell morphology	Description
I	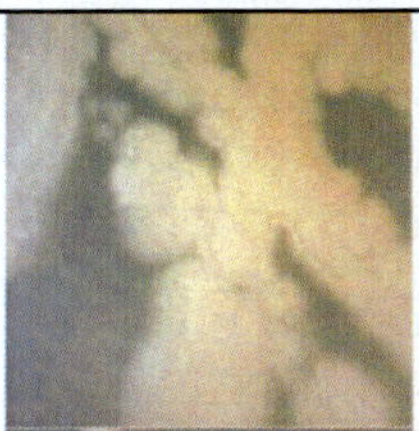	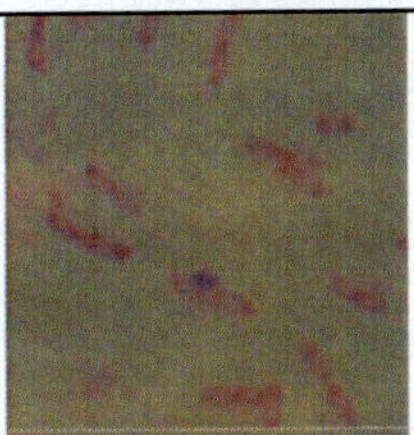	Large colonies withirregular edge, flat, opaque and white. Small rods with endospore.
II	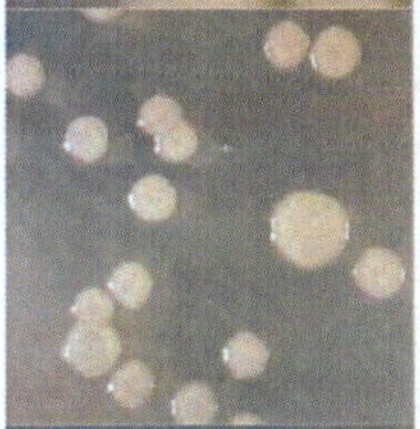	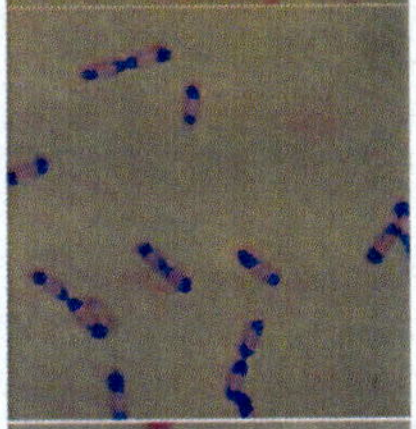	Colonies are round, raised, glistening and white color. Rod shaped cells with endospore.
III	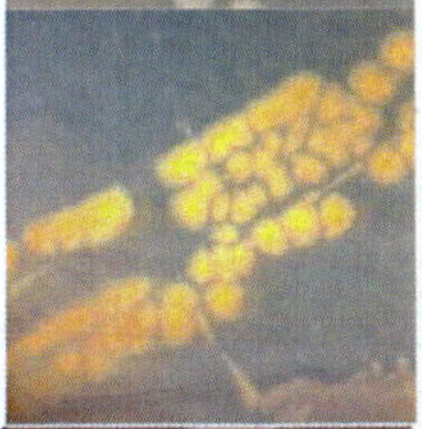	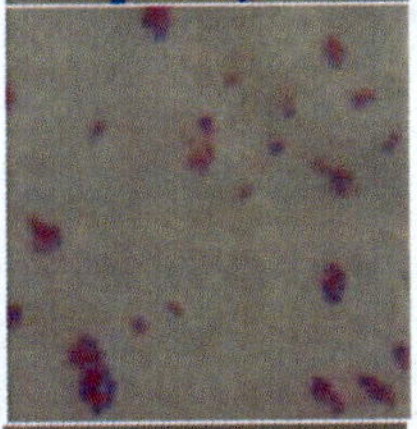	Small colonies, round, flat, wet and yellow/orange pigmented. Small cells shaped like coccobacilli
IV	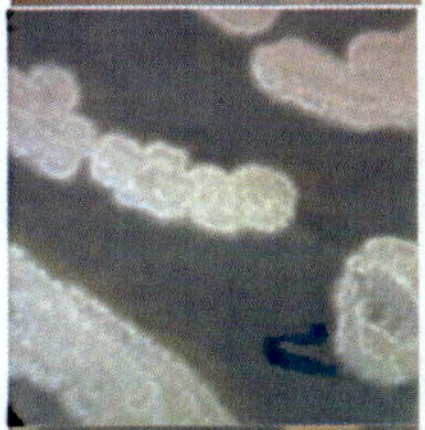		Colonies with irregular edge, flat, dull and white. Alongated rods occurring in long chains.
V	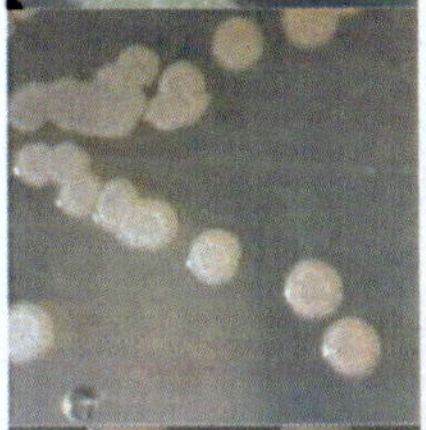	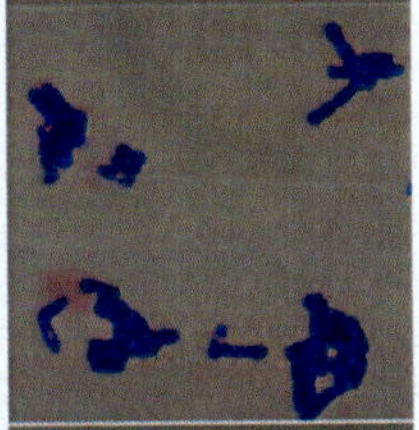	Round colonies, raised, mucoid and white. Short rods with some endospore.
VI	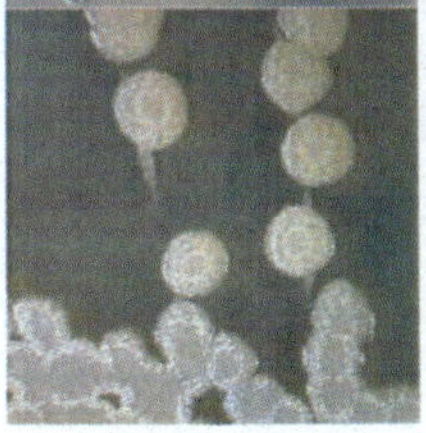	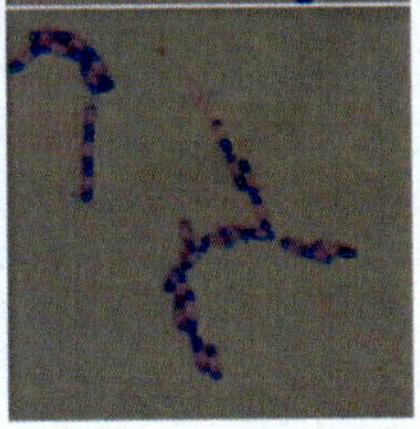	Round colonies, rough, flat, dry and white. Thin rods in chains with ocorrence of endospore.

(*continued*)

Table 1. (*continued*)

VII	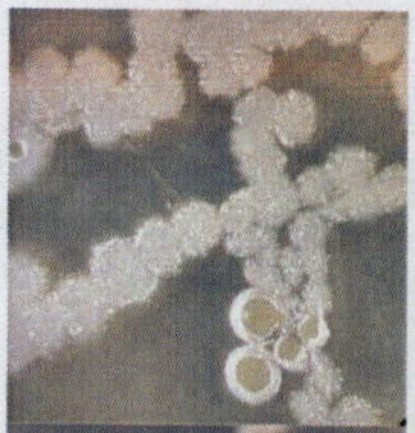	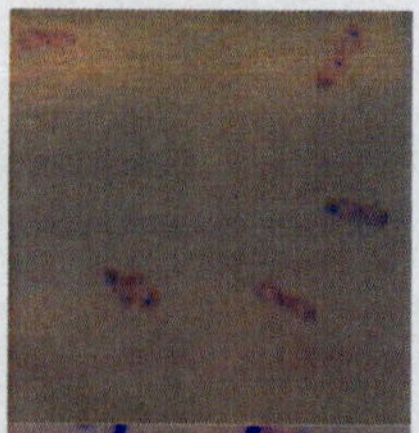	Colonies withirregular edge, flat, rough and white Endospore-forming rods.
VIII	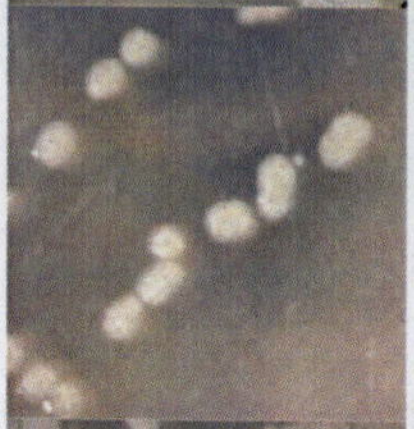	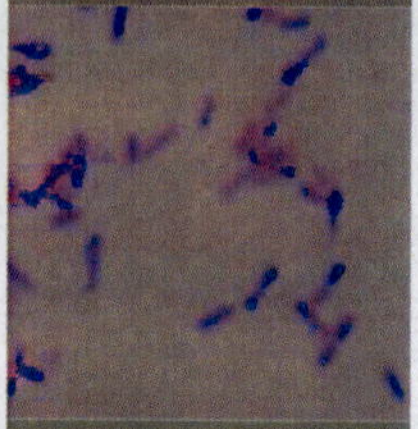	Small colonies, round, raised, wet and white Endospore-forming coccobacilli
IX	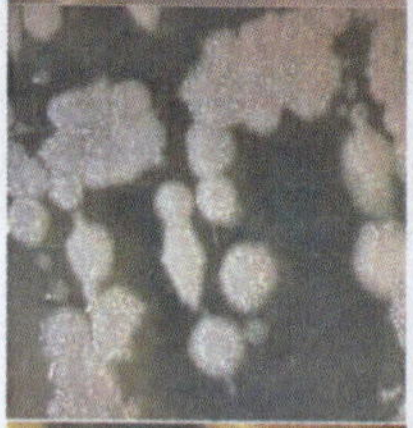	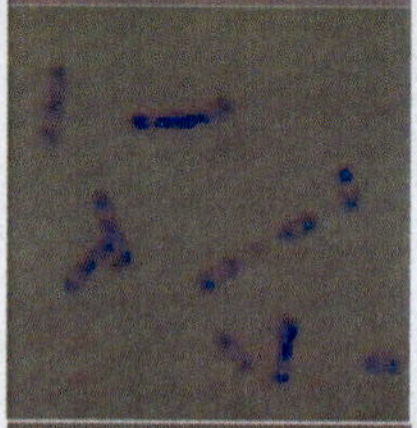	Colonies with irregular edge, flat, dry and white. Cells appear as small rods with endospore.
X	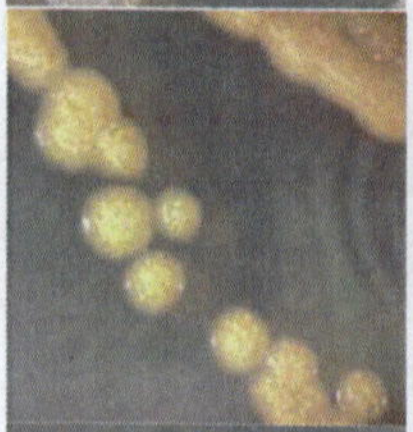	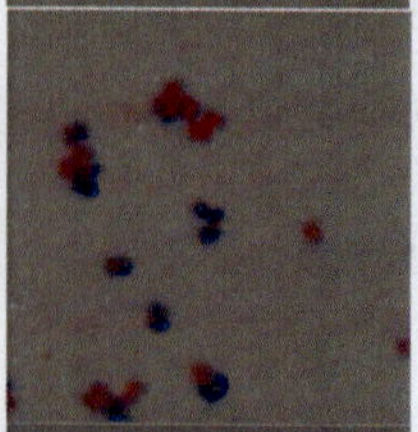	Colonies with convex elevation, round, mucoid and beige color. Spherical-shaped cells.
XI	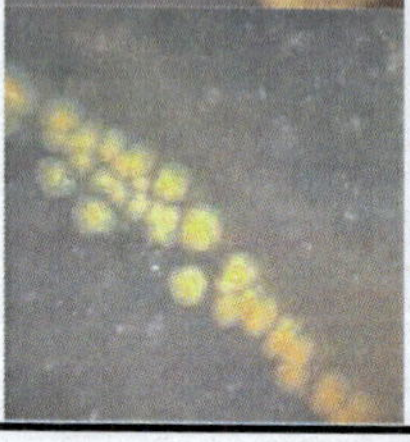	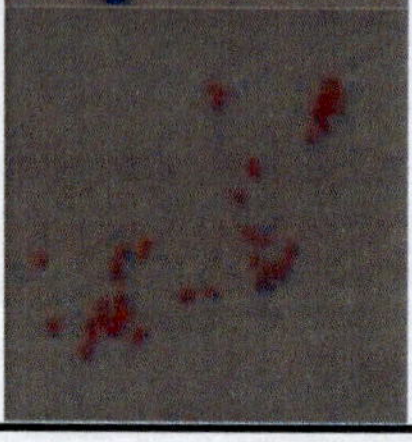	Small colonies, round, flat, smooth and light yellow color. Spherical-shaped cells.

in the sediment to be studied. Anyway, considering the amount of material precipitated and remaining after burning, all isolates were able to produce, at some level, a white precipitate.

FTIR results in the Fig. 3 show that all precipitates from bacteria activity are composed of calcium carbonate. The program software identified the precipitate of isolates I, IV, V, VII and IX as $CaCO_3$. The other precipitates had bands in similar places to those that were identified, and may not have been identified readily due to the low transmittance.

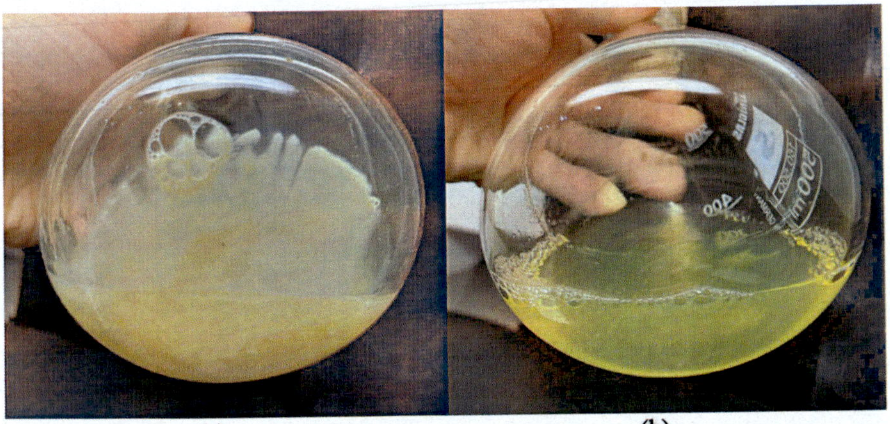

(a) (b)

Fig. 1. Erlenmeyer bottom (a) with isolate precipitate VIII and (b) without control precipitate

Fig. 2. Precipitate before and after burning

All samples presented the absorption bands at 1409–1387, 1088–1050, 873–871 and 745–742 cm^{-1}. Most of them also presented absorption bands at 710 cm^{-1}. These values are significantly similar to other results found in the literature for $CaCO_3$ bioprecipitation. [27–30].

The results show that the isolates I, II and VIII are capable of generating greater conversion. Being of great interest for further studies, as using bioprecipitate to enhance

Table 2. Summary of carbon precipitation tests

Isolate	Precipitate (g)	Loss on ignition (g)	Loss on ignition (%)	Remaining Matter (g)	Conversion (%)
I	0,5541 ± 0,1293	0,0635	11,45	0,4906	77,54
II	0,6506 ± 0,0109	0,1107	17,02	0,5399	85,32
III	0,1829 ± 0,0965	0,0483	26,43	0,1345	21,26
IV	0,4876 ± 0,0509	0,0930	19,06	0,3946	62,37
V	0,5130 ± 0,2535	0,1381	26,93	0,3749	59,24
VI	0,4029 ± 0,1899	0,1214	30,12	0,2816	44,50
VII	0,5256 ± 0,0685	0,0997	18,97	0,4259	67,31
VIII	0,6361 ± 0,0349	0,1271	19,98	0,5090	80,43
IX	0,4840 ± 0,0230	0,0492	10,16	0,4348	68,72
X	0,6383 ± 0,0649	0,2202	34,50	0,4181	66,07
XI	0,2319 ± 0,0181	0,0626	27,00	0,1693	26,75

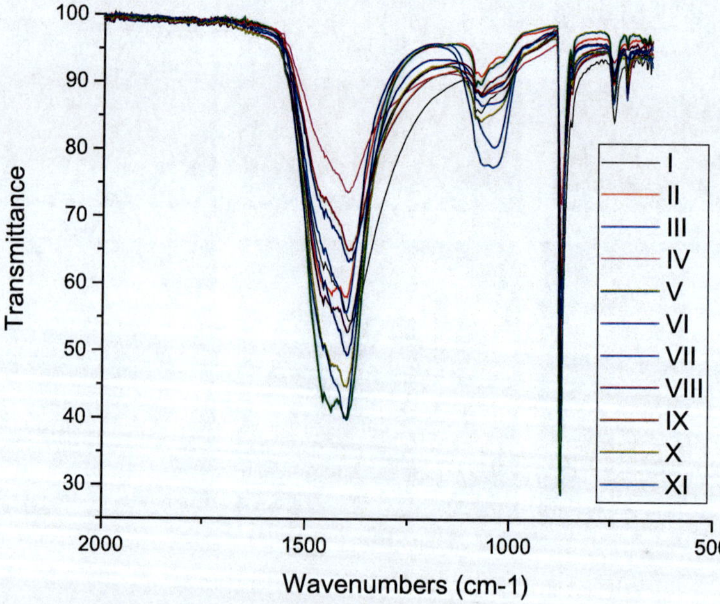

Fig. 3. FTIR analysis of precipitates after firing at 550 °C.

the properties of natural e recycled aggregates. For this, it will be necessary to identify the selected bacteria, study their growth and verify the ability of this precipitate to adhere to the aggregates.

4 Conclusion

This study showed that it was possible to isolate calcium carbonate-producing bacteria from a pool of different aggregates (natural, recycled, coarse and fine). The isolates have different rates of conversion of the calcium source (calcium acetate) into the product of interest. Thus, isolates I, II and VIII show more promising results for the study sequence. The continuation of this study involves the identification of these isolates, the study of their kinetics and the ability of calcium carbonate to adhere to the aggregates.

Some problems must also be overcome, such as the excess of organic matter in the precipitate, which can interfere with a more in-depth study of its physical and chemical characteristics. For this, it is suggested to perform more than one centrifugation with washings between them and to use filters capable of retaining only calcium carbonate .

References

1. Mehta, P.K., Monteiro, P.J.M.: Concrete: Microstructure, Properties and Materials, 3rd edn. McGraw-Hill (2006)
2. Scrivener, K.L., Nonat, A.: Hydration of cementitious materials, present and future. Cem. Concr. Res **41**, 651–665 (2011). https://doi.org/10.1016/j.cemconres.2011.03.026
3. Monteiro, P.J.M., Miller, S.A., Horvath, A.: Towards sustainable concrete. Nat. Mater. **16**, 698–699 (2017). https://doi.org/10.1038/nmat4930
4. OECD Global Material Resources Outlook to 2060: Economic Drivers and Environmental Consequences, OECD Publishing, Paris (2019). https://doi.org/10.1787/9789264307452-en
5. Scrivener, K.L., John, V.M., Gartner, E.M.: Eco-efficient cements: potential economically viable solutions for a low-CO2 cement-based materials industry. Cem. Concr. Res. **114**, 2–26 (2018). https://doi.org/10.1016/j.cemconres.2018.03.015
6. Nidheesh, P.V., Kumar, M.S.: An overview of environmental sustainability in cement and steel production. J. Clean. Prod. **231**, 856–871 (2019). https://doi.org/10.1016/j.jclepro.2019.05.251
7. Brito, J., Saikia, N.: Recycled Aggregate in Concrete, 1st ed Springer, London (2013). https://doi.org/10.1007/978-1-4471-4540-0
8. Wang, R., Yu, N., Li, Y.: Methods for improving the microstructure of recycled concrete aggregate: a review. Constr. Build. Mater. **242**, 118164 (2020). https://doi.org/10.1016/j.conbuildmat.2020.118164
9. Magnusson, S., Lundberg, K., Svedberg, B., Knutsson, S.: Sustainable management of excavated soil and rock in urban areas - A literature review. J. Clean. Prod. **93**, 18–25 (2015). https://doi.org/10.1016/j.jclepro.2015.01.010
10. Thomas, C., Setién, J., Polanco, J.A., et al.: Durability of recycled aggregate concrete. Constr. Build. Mater. **40**, 1054–1065 (2013). https://doi.org/10.1016/j.conbuildmat.2012.11.106
11. Tam, V.W.Y., Soomro, M., Evangelista, A.C.J.: A review of recycled aggregate in concrete applications (2000–2017). Constr. Build. Mater. **172**, 272–292 (2018). https://doi.org/10.1016/j.conbuildmat.2018.03.240
12. Bai, G., Zhu, C., Liu, C., Liu, B.: An evaluation of the recycled aggregate characteristics and the recycled aggregate concrete mechanical properties. Constr. Build. Mater. **240**, 117978 (2020). https://doi.org/10.1016/j.conbuildmat.2019.117978
13. Deng, X., Li, Y., Liu, H., et al.: Examining energy consumption and carbon emissions of microbial induced carbonate precipitation using the life cycle assessment method. Sustainability (Switzerland) **13**, 4856 (2021). https://doi.org/10.3390/su13094856

14. Sahoo, K.K., Arakha, M., Sarkar, P., et al.: Enhancement of properties of recycled coarse aggregate concrete using bacteria. Int. J. Smart Nano Mater. **7**, 22–38 (2016). https://doi.org/10.1080/19475411.2016.1152322
15. Zeng, W., Zhao, Y., Poon, C.S., et al.: Using microbial carbonate precipitation to improve the properties of recycled aggregate. Constr. Build. Mater. **228**, 116743 (2019). https://doi.org/10.1016/j.conbuildmat.2019.116743
16. Iqbal, D.M., Wong, L.S., Kong, S.Y.: Bio-cementation in construction materials: a review. Materials **14**, 2175 (2021). https://doi.org/10.3390/ma14092175
17. Mondal S, (Dey) Ghosh A.: Review on microbial induced calcite precipitation mechanisms leading to bacterial selection for microbial concrete. Constr. Build. Mater. 225, 67–75 (2019). https://doi.org/10.1016/j.conbuildmat.2019.07.122
18. Li, M., Fang, C., Kawasaki, S., et al.: Bio-consolidation of cracks in masonry cement mortars by acinetobacter sp. SC4 isolated from a karst cave. Int. Biodeterior Biodegradation **141**, 94–100 (2019). https://doi.org/10.1016/j.ibiod.2018.03.008
19. Reeksting, B.J., Hoffmann, T.D., Tan, L., et al.: EMIn-depth profiling of calcite precipitation by environmental bacteria reveals fundamental mechanistic differences with relevance to application. bioRxiv 86:1–16 (2019). https://doi.org/10.1101/850883
20. Rangamaran, V.R., Shanmugam, V.K.: Biocalcification by piezotolerant bacillus sp. NIOTVJ5 isolated from deep sea sediment and its influence on the strength of concrete specimens. Mar. Biotechnol. **21**(2), 161–170 (2018). https://doi.org/10.1007/s10126-018-9867-8
21. Vashisht, R., Attri, S., Sharma, D., et al.: Monitoring biocalcification potential of Lysinibacillus sp. isolated from alluvial soils for improved compressive strength of concrete. Microbiol. Res. **207**, 226–231 (2018). https://doi.org/10.1016/j.micres.2017.12.010
22. Zhang, J., et al.: Enhanced calcite precipitation for crack healing by bacteria isolated under low-nitrogen conditions. Appl. Microbiol. Biotechnol. **103**(19), 7971–7982 (2019). https://doi.org/10.1007/s00253-019-10066-z
23. Manzur, T., Rahman, F., Afroz, S., et al.: Potential of a microbiologically induced calcite precipitation process for durability enhancement of masonry aggregate concrete. J. Mater. Civ. Eng. **29**, 1–12 (2017). https://doi.org/10.1061/(asce)mt.1943-5533.0001799
24. Saxena, S., Tembhurkar, A.R.: Developing biotechnological technique for reuse of wastewater and steel slag in bio-concrete. J. Clean. Prod. **229**, 193–202 (2019). https://doi.org/10.1016/j.jclepro.2019.04.363
25. Kim, H.J., Eom, H.J., Park, C., et al.: Calcium carbonate precipitation by bacillus and sporosarcina strains isolated from concrete and analysis of the bacterial community of concrete. J. Microbiol. Biotechnol. **26**, 540–548 (2015). https://doi.org/10.4014/jmb.1511.11008
26. Montaño-Salazar, S.M., Lizarazo-Marriaga, J., Brandão, P.F.B.: Isolation and Potential biocementation of calcite precipitation inducing bacteria from Colombian buildings. Curr. Microbiol. **75**(3), 256–265 (2017). https://doi.org/10.1007/s00284-017-1373-0
27. Tomassetti, M.C., Cirigliano, A., Arrighi, C., et al.: A role for microbial selection in frescoes' deterioration in Tomba degli Scudi in Tarquinia, Italy. Sci. Rep. **7**, 1–8 (2017). https://doi.org/10.1038/s41598-017-06169-0
28. Li, Y., Wen, K., Li, L., et al.: Experimental investigation on compression resistance of biobricks. Constr. Build. Mater. **265**, 120751 (2020). https://doi.org/10.1016/j.conbuildmat.2020.120751
29. Feng, J., Su, Y., Qian, C.: Coupled effect of PP fiber, PVA fiber and bacteria on self-healing efficiency of early-age cracks in concrete. Constr. Build. Mater. **228**, 116810 (2019). https://doi.org/10.1016/j.conbuildmat.2019.116810
30. Wu, M., Hu, X., Zhang, Q., et al.: Growth environment optimization for inducing bacterial mineralization and its application in concrete healing. Constr. Build. Mater. **209**, 631–643 (2019). https://doi.org/10.1016/j.conbuildmat.2019.03.181

Building on Mycelium

Ilse (C.A.) Rovers[1](✉), Francisco Ortega Exposito[1], Ilaria la Bianca[1], Joost Vette[1,3], Jordi Pelkmans[2], Wasabii Ng[4], and Willem Böttger[5]

[1] Avans University of Applied Sciences, Centre of Expertise Material and Energy Transition MNEXT, Lovendijkstraat 63, 4818 AJ Breda, The Netherlands
ca.rovers@avans.nl
[2] Utrecht University, Padualaan 8, 3584 CH Utrecht, The Netherlands
[3] Fairm, Schieweg 15A, 2627 AN Delft, The Netherlands
[4] Caradt, Post Office Box 90116 4800 RA, Lovendijkstraat 63, Breda, The Netherlands
[5] NPSP, Moezelhavenweg 9, 1043 AM Amsterdam, The Netherlands

Abstract. Conventional building materials contribute to climate change due to their CO_2-footprint and often decrease the health of a building. Mycelium bio-composites (MBCs) solve this as they are a renewable, circular and a healthy alternative. MBCs are produced by colonizing a fungal species through a fibre-like substrate. In this research growth conditions and characteristics of MBCs were determined for combinations of 5 different fungi and 10 different agricultural residue streams.

The material characteristics of the best developed samples were tested: compressive strength, thermal conductivity, moisture buffering value, sound absorption, water absorption, moisture absorption and desorption, as well as material density, according to common standards and procedures. The best combination was *Ganoderma resinaceum* on rapeseed straw and results indicated that MBCs have similar characteristics to conventional mainstream insulation materials.

Furthermore, growing parts of living material together without external attachment agents for repair or construction purposes revealed new opportunities. Apart from showing a list of well-performing MBC-combinations, it can be concluded that MBCs are a promising material for insulation purposes.

The research continues, both on improving the material characteristics experimentation, as well as on new material research, to show the application opportunities in interior and construction purposes as an alternative to current mainstream materials: see www.mnext.nl.

Keywords: mycelium bio-composites · mycelium-based materials · mycelium · material properties · compressive strength · thermal conductivity · moisture buffering value · sound absorption · water absorption/desorption

Abbreviations

G. resinaceum	Ganoderma resinaceum
T. hirsuta	Trametes hirsuta
L. betulina	Lenzites betulina

S. Amziane et al. (Eds.): ICBBM 2023, RILEM Bookseries 45, pp. 739–757, 2023.
https://doi.org/10.1007/978-3-031-33465-8_57

S. commune	Schizophyllum commune
P. ostreatus	Pleurotus ostreatus
S. esculentes	Strobilurus esculentes
B. adusta	Bjerkandera adusta
MBC	Mycelium-based Bio-Composites
EPS	Expanded Polystyrene
XPS	Extruded Polystyrene
MBV	Moisture Buffer Value
MNEXT	Materials and Energy Transition
EAD	European Assessment Document
EOTA	European Organization for Technical Assessment

1 Introduction to Building on Mycelium

With the building industry being one of the largest CO_2-emitters in the world, the need for CO_2-neutral materials grows, such as biodegradable materials which are part of the circular economy [1]. One of the promising fully biodegradable materials are mycelium bio-composites (MBCs) [2, 3]. In this paper, MBCs are composed of an agricultural residue, a non-living material, colonized by a fungus. The agricultural residue is rich in lignin and/or cellulose and thereby functions as a nutrient source for the fungus. The fungus partially degrades and colonizes this substrate from a network of mycelium, thereby binding the fibers of the substrate. The fungus providing the mycelium is alive when producing the MBC but dormant after drying due to low humidity. Most fungi are dead after being exposed to a high temperature, like with baking or heat-pressing.

Figure 1 below details the production process of the MBCs. The MBC is grown solid state in a mold, following the mycelium foam pathway, which keeps the product spongy, relatively light and with a low density. The basis of the research is the locally available agricultural residue stream, combined with a fungal species endemic or at least native to Northwestern Europe.

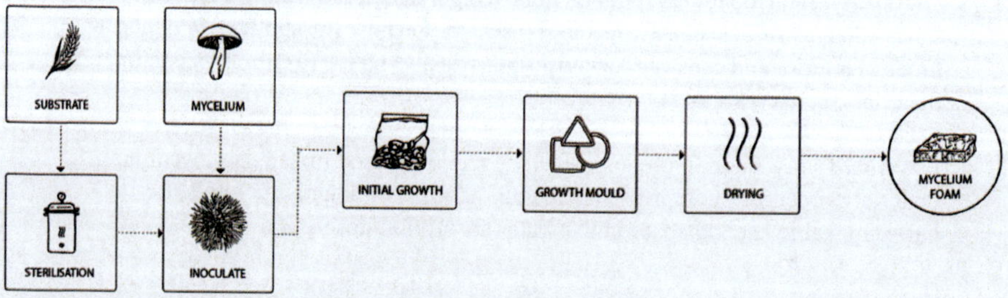

Fig. 1. MBC production process

Most research is done on a few fungal species only [4]. This research project aims to identify well growing combinations of fungal species and local agricultural residual streams for producing MBCs. The composites have promising applications in the building industry as insulation materials. The focus of the research was on developing lightweight and low-density bio-composite combinations. The 5 best combinations were tested and ranked on diverse mechanical properties in relation to their application. These results were then compared to mainstream insulation materials and their diverse applications.

1.1 Substrate – Fungus Combinations

The combinations of agricultural residue and fungal species to produce MBCs are shown in Table 1. Each of the ten substrates were combined with each of the five fungal species, producing fifty different potential MBC combinations.

Table 1. Substrate-fungal species MBC combinations

Substrate	Fungal species
Rapeseed straw fibres	*G. resinaceum*
Hemp fibres	*T. hirsuta*
Hemp shives	*L. betulina*
Reed fibres (from maintenance in nature areas)[a]	*P. ostreatus*
Raw cellulose (from wastewater treatment)[b]	*S. esculentes*
Processed cellulose (from wastewater treatments)[c]	
Flax fibres	
SMS (spent mushroom substrate)[d]	
Elephant grass	
Wood chips	

[a]Waterschap de Dommel.
[b]RECELL.
[c]Waterschap de Dommel.
[d]Verbruggen Paddestoelen.

1.2 Partners

The research was conducted by the MNEXT, the research institute at Avans University of Applied Sciences, together with different research partners (Utrecht University, Caradt, Zeeland University of Applied Sciences) and commercial partners. The SME partners represented different phases along the MBC production chain. Table 2 shows the SME collaboration partners for this project.

1.3 Reading Guide

This paper has been divided into four content sections. The first and second section show the methodology for the research, and the results, which are divided into preliminary research, material characteristics, and the design & application of MBCs. The next part is the discussion, examining the methodology, results and a comparison of the material characteristics of MBCs to the reference materials. The last sections of the paper are the conclusion to the research, together with the follow-up research.

Table 2. SME collaboration partners

Partners	
V8 architects	Gispen
Verbruggen-paddenstoelen	Wagro
Mogu	Waterschap De Dommel
Impershield	Kok Houtagenturen
Isolco B.V.	Agrodome B.V.
Recell®	CBM

2 Methods

The methods section establishes the methodology of the research, separated firstly into preliminary research, namely growth conditions and characteristics, secondly material properties and finally design and application.

2.1 Preliminary Research

In the MBCs study, growth conditions and characteristics were examined. Various parameters were analyzed to find the optimal conditions for producing desirable MBCs. The growth characteristics of different MBC combinations were evaluated and rated.

Growth Conditions

All possible combinations of the five species and ten substrates selected (see Table 1) were analyzed. The research focused on the optimal water content of the substrate and temperature for growing the different MBCs. Combinations that would not grow quickly or grew non-homogeneously were excluded.

Temperature

Each fungal species was grown on potato dextrose agar (PDA) on a polycarbonate membrane at 20, 25 and 30 degrees Celsius in the dark. The mycelium was harvested, dried and weighed to determine biomass.

Maximum Water Retention

Moisture levels of the substrate play a key role on the colonization rate of fungi. Therefore, two levels of water saturation were chosen to monitor growth at: 80% and 90% of total water retention of the substrate. To this point, first maximum water retention (the amount of water the substrate can absorb) was determined. Substrate was weighed at 0% (dried for 3 days at 60 °C) (Wd) and 100% water content (submerged for 24 h and drained from excess water for 1 h) (Ww). The maximum water retention percentage (Wm) was then calculated according to the following formula:

$$\mathrm{Wm} = 100\frac{\mathbf{Ww} - \mathbf{Wd}}{\mathbf{Ww}} \quad (1)$$

Growth Characteristics

To produce a high quality MBC, it is important to achieve increased fungal growth throughout the substrate. This study evaluated growth characteristics by assessing the colonization rate and mycelial coverage on the outside of the substrate. The colonization was determined by the grey value, which is an indicator of colonization, and allows for determination of the time required for full coverage (A). These data, together with the mycelium coverage of the substrate (mycelium skin), provide insight into the homogeneity of substrate colonization. In addition, shrinkage after drying and material density are also important factors that affect the material properties of the MBCs.

Grey Value

Mycelium growth through the substrate was measured by high-resolution flatbed scanning and subsequently determining the grey value of the samples. This was done by inoculating samples and growing them in micro-boxes for 28 days, during which the bottom of the box was scanned daily for 14 days and then every other day until the 28th day. Next, the images were analyzed using ImageJ and R studio. The grey value was normalized to plot typical growth curves which were used to determine the lag- (λ), exponential- (μ) and stationary (A) phase in the growth curve according to the Gompertz model (Fig. 2) [5].

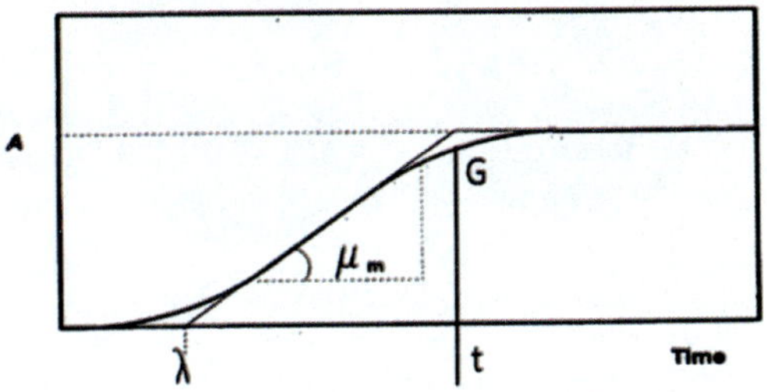

Fig. 2. Gompertz model

The data was used in the following formula;

$$y = A\exp\left\{-\exp\left[\frac{\mu_m e}{A}(\lambda - t) + 1\right]\right\} \tag{2}$$

where A is the maximum growth value, μ_m is the point at which growth is fastest (where the slope is steepest) and λ is the point at which the lag phase (the first and more horizontal part of the curve) changes into the exponential growth phase.

Days to Reach A

The days (T) to reach A was determined by using the grey value. The following formula calculated the number of days to reach the maximum growth value. A low value of T indicates faster colonization.

$$T = \frac{A + (\mu_m * \lambda)}{\mu_m} \tag{3}$$

Skin Coverage Development

The mycelium growing on the outside of the MBC forms a layer of aerial hyphae known as the mycelium skin. Its development is evaluated based on factors like thickness, color, texture, and coverage. Figure 3 shows two hemp shives MBC samples with similar grey values after 28 days. The mycelium skin development on the right sample was more homogeneous with a smoother surface structure and fewer indentations.

Fig. 3. Skin coverage examples

Shrinkage and Density

The shrinkage was calculated by percentual decrease of diameter of the dried samples compared to the initial diameter upon filling the micro-box. Density was calculated after drying by weight and (shrunken) volume.

Ranking the MBCs on Growability

After the preliminary testing was finalized, a multi-criteria analysis (MCA) was performed to rank the different substrate-fungus combinations. Included in the MCA were days to reach total colonization (A), skin coverage, shrinkage after drying and density. Table 3 shows the scoring per criterium and the weight appointed to each criterium. The MCA was used to choose the best growing combinations, which were then examined for material properties.

Table 3. Parameter scoring and weight for MCA

	Criteria 1: days to reach A	Criteria 2: skin coverage	Criteria 3: shrinkage (%)	Criteria 4: density (g/cm^3)	Score
	5	Completely covered, no substrate visible	3	0.05	10
	7	Completely covered, substrate visible	4	0.078	9
	9	Partly covered, top, sides not covered	5	0.106	8
	11		7	0.134	7
	13		9	0.162	6
	15	Partially covered, only top	11	0.19	5
	17		13	0.218	4
	19	No fungal skin visible	15	0.246	3
	21		20	0.274	2
	23		35	0.3	1
Weight:	4	3	1.5	0.5	

2.2 Material Properties

The mechanical properties of MBCs are highly dependent on interactions of the specific substrate-fungus combination.

Compressive Strength
Compressive strength was tested using EN 826 [6]. Samples were subjected to the test before water absorption, after a moisture absorption test (MAT) and after a long-term water absorption test (WAT).

Thermal Conductivity
Thermal conductivity of the samples was measured using the transient method as per ASTM D5334-08, with an ISOMET model 2104 thermal needle probe [7].

Moisture Buffering Value
The NORDTEST method was used to evaluate the practical moisture buffering value

(MBV) [8]. Gypsum and EPS were also tested as reference materials to compare results with MBCs directly.

Sound Absorption
The MBCs were tested according to protocol ASTM E1050-98 for acoustic testing, using an impedance tube [9].

Water Absorption
The long-term water absorption was tested for the MBCs using standard EN 12087 [10].

Moisture Absorption/desorption
The moisture absorption and desorption of the MBCs was tested based on standard EN 12571 [11].

Dimension Stability
Dimensional stability testing was conducted utilizing the method EN 2796 [12].

pH Resistance
Due to the lack of established pH resistance assessment protocols for material testing, a new method was employed by submerging MBCs in pH 4 or pH 10 solutions for 48 h and measuring their dimensional changes. The EN 2796 method was used to evaluate the dimensional change when submerged in a media with different pH's for a certain time, quantifying pH resistance. In the meantime, two control samples were immersed in water to distinguish water absorption from pH exposure [12].

Vapor Diffusion
Vapor diffusion was tested according to the standard ISO 12572 - Cup method [13].

2.3 MBC Design and Application

MBCs are unique, consisting only of biological materials, with the fungus developing throughout the substrate, giving them the unique quality of being alive [14]. Although MBCs may not seem alive after the last production stage, they can be restored for repairs or used to build 3D structures that diverge from their original form [14, 15].

This study aimed to explore the healing potential of MBCs for both unique aesthetic expressions and everyday applications [14]. Two forms of inoculation, grains and liquid cultures, were investigated for joining MBC parts [14]. The growth characteristics of mycelium were analyzed to determine the time and distance required for closing gaps and cracks [14] (Fig. 4).

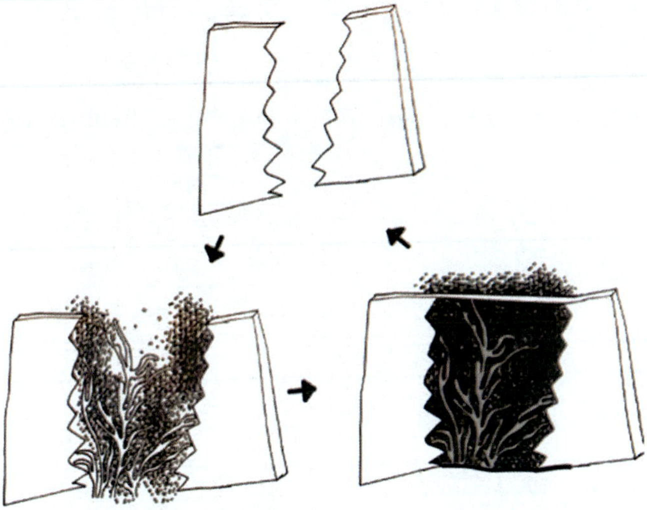

Fig. 4. Three stages of healing with fungi

3 Results

This section shows the results, divided into preliminary research, material characteristics and MBC design & application.

3.1 Preliminary Research

Under the growth conditions tested, the majority of species exhibited optimal biomass production at 30 °C, with the exception of L. betulina which displayed a preference for 25 °C. Hence 30 °C was established at the optimal temperature, together with 80% and 90% as optimal moisture content. The growth characteristics were used to score the MBC combinations, which ultimately led to the identification of a MBC top-5. Table 4 shows the scoring and raking for the top-5 MBC combinations.

The fungi which scored best were G. resinaceum, T. hirsuta, and L. betulina. The best scoring substrates were rapeseed straw, hemp shives and hemp fibres. Therefore, these three fungal species and three substrates were used in further material properties experiments.

3.2 Material Characteristics

The MBC material characteristics are displayed in this sub-chapter.

Compressive Strength

The compressive strength results are shown in Table 5, differentiating between the different water exposure tests.

The MBCs (using various combinations) achieved the compressive strength results of 0.064 ± 0.0057 MPa prior to undergoing the MAT and WAT. However, following the

Table 4. Ranking of MBC combinations, with results of the different preliminary tests.

				Score				
Ranking	Species	Substrate	Humidity	Days to reach A	Grey value/Skin coverage	Shrinkage	Density	Total
1	*G. Resinaceum*	Rapeseed straw	80	9	9	8	10	22.0
2	*G. Resinaceum*	Hemp shives	80	9	9	8	9	21.9
3	*T. Hirsuta*	Straw	80	9	9	7	10	21.4
4	*L. betulina*	Hemp fibres	90	6	10	9	10	20.9
5	*T. hirsuta*	Hemp fibres	90	7	9	9	10	20.8
			Weight	3.5	3.5	2.5	0.5	10

Table 5. Compressive strength testing

Material	Compressive strength (MPa)	Compressive strength after MAT (MPa)	Compressive strength after WAT (MPa)
Mycelium bio-composites	0.064 ± 0.0057	0.057 ± 0.0055	0.050 ± 0.0035

MAT test, the compressive strength reduced slightly to 0.057 ± 0.0055 MPa, while the WAT resulted in a decrease to 0.050 ± 0.0035 MPa. A notable difference exists between the initial compressive strength and the strength after the WAT, but there is no significant variance between the initial compressive strength and the MAT or between the MAT and WAT. This indicates that moisture has no impact on the compressive strength of MBCs, but exposure to water could lead to a decrease in strength.

Thermal Conductivity

The results of the thermal conductivity tests are shown in the Table 6 below.

Table 6. Thermal conductivity of MBCs

Sample	Thermal conductivity ($WM^{-1}k^{-1}$)
Mycelium bio-composites	$0.057–0.085 \pm 0.0035$

The thermal conductivity of the MBCs (different combinations) was determined to be between $0{,}057–0{,}085 \pm 0.0035$ $WM^{-1}k^{-1}$.

Moisture Buffering Value

Table 7 shows the results of moisture buffering value for the MBCs, gypsum and EPS (reference materials).

Table 7. Moisture buffering value

Sample Type	Mean $MBV_{practical}$	Standard Deviation
Mycelium bio-composites	1.632	0.024
Gypsum	0.547	0.005
EPS	0.002	0.000

The mean practical moisture buffering value ($MBV_{practical}$) of the MBCs (different combinations) 1.632 ± 0.024, which is significantly the highest MBV when compared to gypsum (0.547 ± 0.005) and EPS (0.002 ± 0.000).

Sound Absorption

Figure 5 shows the sound absorption spectrum across the frequency range of the MBCs.

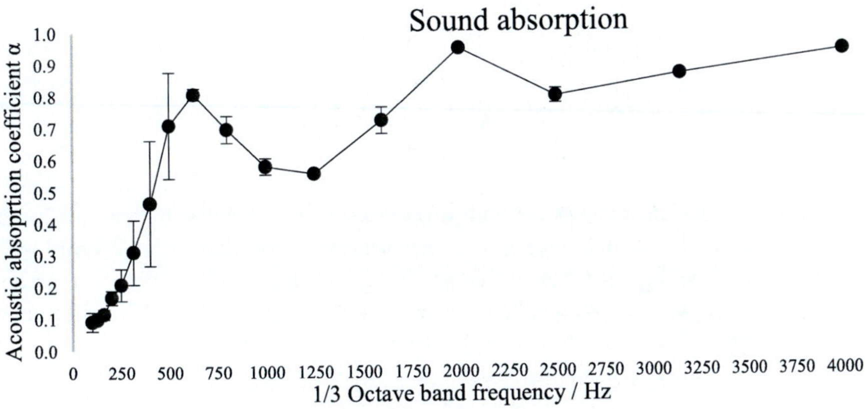

Fig. 5. Sound absorption spectrum of MBCs

Figure 5 visualizes the highest sound absorption spectrum produced by the MBCs (different combinations). The graph displays the sound absorption coefficient spectrum curves for MBCs at 1/3 octave band frequency intervals. The sound absorption coefficient at low frequencies is below 0.5 until the 500 Hz range. The sound absorption coefficient at mid-frequencies 500–2000 tends to fluctuate between the ranges of 0.8 at 750Hz, dipping to 0.6 at 1250 Hz to then increase to 0.95 at 2000 Hz. At high frequencies 2000–4000 Hz, the sound absorption coefficient drops from 0.95 at 2000 Hz to around 0.8 at 2500 Hz, to then rise to around 0.95 again at 4000 Hz. This shows that MBCs are a good sound absorber in low and middle frequencies, allowing for potential usage as vibration insulation and soundproofing.

Water Absorption

Figure 6 shows the water absorption of the MBCs after submersion for a certain amount of hours.

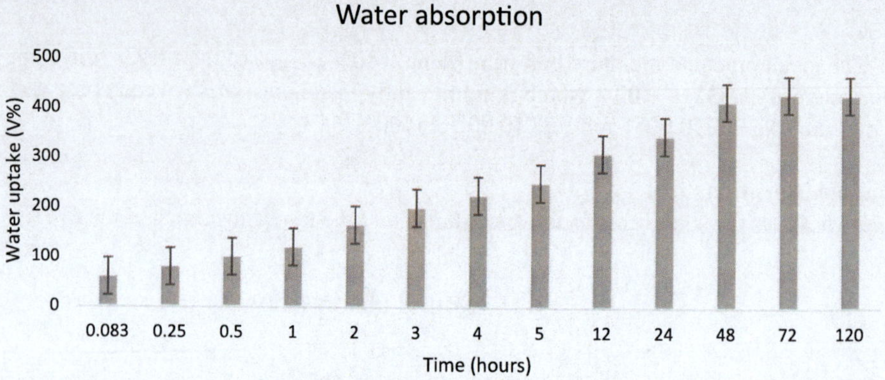

Fig. 6. Water absorption graph

Figure 6 indicates the highest water absorption spectrum produced by the MBCs per varying hour intervals. The increase in water absorption for the MBCs seems to follow a logarithmic rate. The highest measured water absorption was obtained after 72 h, where the water uptake equals around 450% of the initial volume of the MBC. The difference in water absorption between 48 h and 120 h seems to be insignificant, meaning that at 48 h, the MBCs were saturated with water.

Moisture Absorption and Desorption

Figure 7 and Fig. 8 show the moisture absorption and desorption curves of the MBCs respectively.

The moisture absorption increases from around 1.5% uptake to the highest uptake, around 6.5% after 5 h, to later fluctuate between 5–6% from 24 h to 144 h. The moisture desorption starts at around −1.2%, decreasing steadily until the 48 h mark, where it has decreased to the most desorption at −5.3%. Lastly, the moisture desorption increases again to −3%.

Dimensional Stability

Table 8 shows the results for the dimensional stability of MBCs.

According to the results, at 50 °C, MBCs do not thermally expand, contrarily they shrink. On average, the change in length is $-1.74 \pm 1.62\%$, the change in width is –

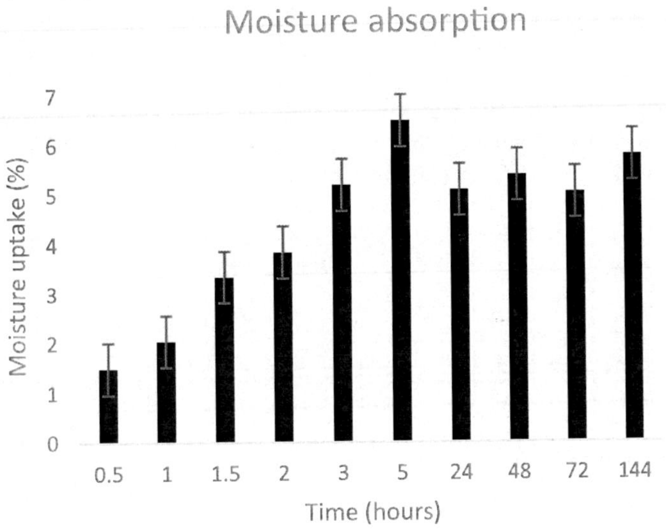

Fig. 7. Moisture absorption graph

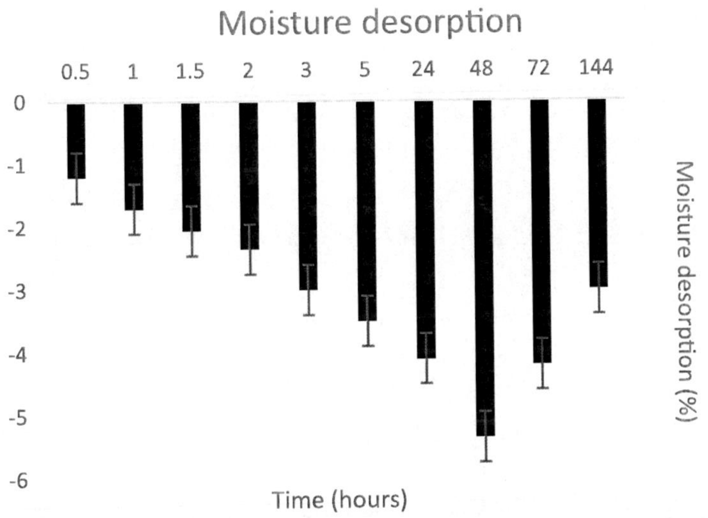

Fig. 8. Moisture desorption graph

Table 8. Dimensional stability results

Sample Type	Δεl (%)	Δεb (%)	Δεd (%)	Δε (%)
Mycelium bio-composites	−1.74 ± 1.62	−1.12 ± 0.60	−5.25 ± 3.59	−2.70 ± 0.82

1.12 ± 0.60% and the change in thickness is −1.44 ± 3.59%. Overall, the dimensional stability related to the thermal expansion of the MBCs is −2.70 ± 0.82%.

pH Resistance

Table 9 presents the results of the pH resistance analysis for MBCs, including measurements taken at pH 4 and pH 10, as well as the control.

Table 9. pH resistance results

Sample Type	Δεl (%)	Δεb (%)	Δεd (%)
Mycelium bio-composites at pH 4	0.76 ± 0.36	0.67 ± 0.35	2.13 ± 1.01
Mycelium bio-composites at pH 10	0.74 ± 0.30	0.57 ± 0.25	2.50 ± 2.05
Control	0.57 ± 0.07	0.62 ± 0.20	1.23 ± 0.91

The change in length of MBCs was found to be similar at pH 4 (0.76 ± 0.36%) and pH 10 (0.74 ± 0.30%), with no significant difference compared to the control (0.57 ± 0.07%). Similarly, there was no significant difference in the change of width for MBCs at pH 4 (0.67 ± 0.35%) and pH 10 (0.57 ± 0.25%) compared to the control (0.62 ± 0.20%). Finally, the change in thickness was also found to be not significantly different for MBCs at pH 4 (2.13 ± 1.01%) and pH 10 (2.50 ± 2.05%) compared to the control (1.23 ± 0.91%). The results indicate that MBCs can withstand submersion in these pHs for 48 h, with no significant difference between the samples and the control.

Vapor Diffusion

Table 10 tabulates the results derived from the vapor diffusion experiments.

Table 10. Vapor diffusion results

Sample Type	W ($ng \cdot s^{-1} \cdot m^{-2} \cdot Pa^{-1}$)	δ (Perm)	μ
Mycelium bio-composites	0.32 ± 0.0050	4.57 ± 0.76	64.00 ± 10.81

The MBCs had a water vapor permeance of 0.32 ± 0.0050 $ng \cdot s^{-1} \cdot m^{-2} \cdot Pa^{-1}$, a water vapor permeability of 4.57 ± 0.76 perm and a water vapor resistance factor of 64.00 ± 10.81. This means that the material has a high resistance to water vapor which can help prevent moisture accumulation and allows for use in high moisture environments.

3.3 MBC Design and Application

Figure 9 displays the mending of MBCs using mycelium.

Figure 9, show that mycelium can bridge gaps and fill empty spaces, enabling the repair of MBCs. Non-living MBCs can be healed by adding substrate to fill cracks and

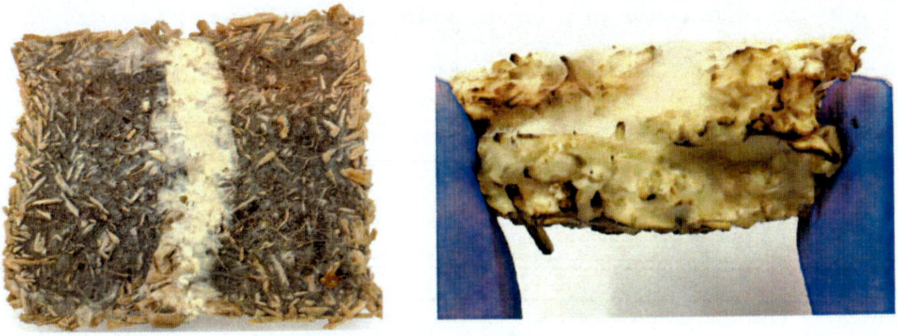

Fig. 9. Mending with mycelium

introducing living inoculum under appropriate conditions in a sterile, moist environment [14]. These findings suggest that MBCs have promising potential as a sustainable and versatile material in various applications [14], either to build shapes that can not be molded that way, or to repair items.

4 Discussion

The discussion section details the analysis of the results, and a comparison to the reference materials.

4.1 Discussion of Results

The reliability of the experiments for growth conditions and for the material characteristics is strong, as all the experiments were carried out with multiple repeats. Notably the MBCs were produced with locally available substrates and abundant fungal species in the Netherlands which have already been used in MBC research previously [16, 17].

The reliability of the experimentation was affected by the fact that MBCs are inherently a biological material and therefore difficult to standardize, also due to the limited size of the samples. Despite following experimental standards and selecting a suitable number of repeats, the MBCs exhibited significant variation in terms of their coverage, density, and shape within a combination. It is worth noting that in the used production method, no two MBCs can be produced exactly the same. To improve the homogeneity of the MBCs, it is recommended to carry out experiments with samples that are more homogenous.

MBCs are still an innovative material, hence often lacking data and directed protocols to test the materials. Protocols and standards are selected related to the resemblance of MBCs to the material for which the protocol is originally directed for. This could lead to a difference to the real material characteristic value being tested. To improve this aspect, the standards should be selected according to the EAD published by EOTA on insulation products made of vegetable fibres [18].

Based on the information presented, repetition of all the experiments with the highest ranked MBCs should be carried out with larger and more homogenous samples, utilizing the most relevant standards and protocols available [18].

4.2 Material Characteristics' Comparison to Reference Materials

The reference materials selected for this research were gypsum and EPS. These are mainstream construction materials regularly used as indoor construction and insulation materials. Table 11 shows the material characteristics of MBCs, gypsum and EPS.

Table 11. Material characteristics comparison between MBCs, gypsum and EPS

Property	Mycelium bio-composite	Gypsum	EPS
Compressive strength	0.035–0.13 MPa	0.24 MPa [19]*	0.47 MPa [20]
Thermal conductivity	0.057–0.085 $WM^{-1}k^{-1}$	0.079 $WM^{-1}k^{-1}$ [19]*	0.035 $WM^{-1}k^{-1}$ [20]
Moisture buffering value	1.63 g/m^2%RH	0.547 g/m^2%RH	0.02 g/m^2%RH
Sound absorption	0.95 at maximum	-	0.55 at maximum [22]
Water absorption	450% (V/V)	37% [21]	43.7% [20]
Moisture absorption	6.5%	10% [19]*	0.5–3% [20]
Dimensional stability	-2.70 ± 0.82	-0.396 ± 0.0084 [23]	<2% [24]
pH resistance	Resistant	Resistant	Resistant [24]
Vapor resistance factor	64.00 ± 10.81	8.4 ± 3.2 [25]^	20–60 [26]

*Using ½ (regular gypsum) measurements
^Using Knauf White

When comparing MBC properties to gypsum, a regular construction (insulating) material, and EPS, a common insulation material, it becomes evident that MBCs have similar properties to these two materials. These materials were selected for comparison because they were investigated during the moisture buffering value, and because they are also conventional construction materials.

In comparison to gypsum and EPS, MBCs exhibit lower compressive strength, higher thermal conductivity, and higher density. However, MBCs have a higher moisture buffering value, higher sound absorption, and comparable thermal insulation potential to gypsum, but potentially lower than EPS. Additionally, MBCs have a higher water absorption/desorption value than EPS, but less than gypsum. The pH resistance of MBCs is similar to the reference materials, and MBCs exhibit a vapor resistance factor that is comparable to that of EPS and higher than gypsum. The dimensional stability results show that MBCs shrink when subjected to heat and have slightly more variation in dimensions than gypsum and EPS.

At first glance, MBCs seem to have a suitable thermal conductivity value, while the rest of the values are similar to the values of gypsum and EPS. In terms of sound absorption and MBV, MBCs are more suitable than EPS and gypsum. This suggests that MBCs have suitable characteristics for the use as insulation materials, particularly in environments where air humidity fluctuates.

Additional experiments are needed to determine if the other following properties are also suitable for the use as insulation material. According to two different sources, important properties to investigate further are: fire resistance, electrical conductivity and thermal conductivity [27, 28]. New reference materials should be tested alongside the MBCs in all the experiments, which should include EPS, straw insulation and cotton fibre insulation. Moreover, the industry is interested in research on pricing, design and emissions to ensure MBCs are competitive in the building materials market, in addition to being environmentally friendly [29].

5 Conclusion

In conclusion, this study analyzed the growth conditions and characteristics of different fungal species combined with different substrates. This resulted in a top-5 of fungal species and substrates as determined by a MBC, which were then tested for their material characteristics. Additionally, the design and application of the MBCs were investigated.

The top-5 best growing MBCs consisted on the following fungal and fibre combinations: *G. resinaceum* and rapeseed straw, *G. resinaceum* and hemp shives, *T. hirsuta* and rapeseed straw, *L. betulina* and hemp fibres, and *T. hirsuta* and hemp fibres.

The material characteristics compressive strength, thermal conductivity, moisture buffering value, sound absorption, water absorption, moisture absorption and desorption, dimensional stability, pH resistance and vapor diffusion properties were successfully tested using their respective standards and procedures. The screening shows that MBCs have similar material characteristics to mainstream insulation materials, namely gypsum and EPS.

For design, the material-driven explorations demonstrate that the living fungus can fill in the material substrate to repair and reappropriate MBCs and while doing so, extend their lifetime and make shapes that can not be molded that way, even if the MBC elements themselves are not considered alive at the moment of application.

6 Follow-Up Research

The initial screening of material characteristics of MBCs suggests that MBCs are potentially suitable as an insulation material, which also is substantiated by the comparison of the material characteristics between MBCs, gypsum and EPS. For application in buildings, further experimentation on fire resistance, electrical conductivity and thermal resistance is needed. Moreover, experiments should be repeated with bigger and homogenous samples, utilizing the standards suggested by EOTA with all the high ranked MBC combinations. This research also should include reference materials to reliably conclude if the MBCs have the required material characteristics compared to the mainstream construction materials.

Therefore, the research on foam-like MBCs continues with the focus on an environmentally friendly production process, while also meeting the legal and constructive requirements for insulation materials in the building industry. In the project, true pricing of biobased materials and design for indoor application are the main emphasis.

References

1. UNEP: Global status report for buildings and construction (2020)
2. Gandia, A., van den Brandhof, J.G., Appels, F.V.W., Jones, M.P.: Flexible fungal materials: shaping the future. Trends Biotechnol. **39**(12), 1321–1331 (2021)
3. Livne, A., Wösten, H.A.B., Pearlmutter, D., Gal, E.: Fungal mycelium bio-composite acts as a CO2-sink building material with low embodied energy. ACS Sustain. Chem. Eng. (2022). https://doi.org/10.1021/acssuschemeng.2c01314
4. van den Brandhof, J.G., Wösten, H.A.B.: Risk assessment of fungal materials. Fungal Biol. Biotechnol. **9**(3), (2022)
5. Tjørve, K.M.C., Tjørve, E.: The use of Gompertz models in growth analyses, and new Gompertz-model approach: an addition to the Unified-Richards family. Commun. Soil Sci. Plant Anal. **48**(17), (June 2017)
6. (CEN), European Committee for Standardization: EN 826 - Thermal insulating products for building applications - Determination of compression behaviour. CEN (2013)
7. ASTM International: ASTM D5334-08 - Standard Test Method for Determination of Thermal Conductivity of Soil and Soft Rock by Thermal Needle Probe Procedure. ATSM International (2008)
8. Rode, C., Peuhkuri, R.H., Hansen, K.K., Time, B., Svennberg, K.: NORDTEST Project on Moisture Buffer Value of Materials. In: AIVC 26th Conference (2005)
9. ASTM International: ASTM E1050-98 - Standard Test Method for Impedance and Absorption of Acoustical Materials Using a Tube, Two Microphones, and a Digital Frequency Analysis System. ASTM (1998)
10. (CEN), European Committee for Standardization: EN 12087:2013 - Thermal insulating products for building applications - Determination of long term water absorption by immersion. CEN (2013)
11. (CEN), European Committee for Standardization: EN 12571:1998 - Thermal and moisture-resistant properties of building materials and products - Determination of hygroscopic absorption properties. CEN (2021)
12. International Organization for Standardization: ISO 2796:1986 - Cellular plastics, rigid—Test for dimensional stability. ISO (1986)
13. ISO (International Organization for Standardization): Hygrothermal performance of building materials and products—Determination of water vapour transmission properties—Cup method (2016)
14. Ng, W., Barati, B., Karana, E.: Healing with fungi: unique aesthetic expressions for mycelium-based materials through patch and mend. In: IASDR 2021 International Design Research Conference: The Ninth Congress of the International Association of Societies of Design Research (2021)
15. Barvir, J.: 3D Printing with Mycelium, 27 February 2022. https://www.fungiakuafo.com/3d-printing-with-mycelium/. Accessed 01 Mar 2023
16. Appels, F.V., et al.: Fabrication Factors Influencing Mechanical, Moisture- and Water-Related Properties of Mycelium-Based Composite. Elsevier, Amsterdam (2018)
17. Sydor, M., Cofta, G., Doczekalska, B., Bonenberg, A.: Fungi in Mycelium-Based Composites: Usage and Recommendations. National Center of Biotechnology Information, Poznań, Poland (2022)
18. European Organisation for Technical Assessment: Factory-made thermal and/or acoustic insulation products made of vegetable or animal fibres (2015)
19. Gypsum Association (2019). https://www.americangypsum.com/sites/default/files/2022-01/ga-235_gypsum_board_typical_mechanical_and_physical_properties.pdf. Accessed 2022

20. Matweb (2022). https://www.matweb.com/search/DataSheet.aspx?MatGUID=5f099f2b5eeb41cba804ca0bc64fa62f
21. Mungyo: Absorption of Gypsum (2016). http://www.mgplaster.com/en/technical/absorb. Accessed 01 Mar 2023
22. Sokol, E.: Better Soundproofing, 22 May 2022. https://bettersoundproofing.com/is-styrofoam-good-for-soundproofing/. Accessed 2022
23. Aljubori, O.M., Aljafery, A.M.A., Al-Mussawi, R.M.: Evaluation of the Linear Dimensional Changes and Hardness of Gypsum Product/Stone Type IV after Adding Silica Nanoparticles, 11 August 2020. http://nanobe.org/Data/View/655?type=100. Accessed 01 Mar 2023
24. Omnexus by SpecialChem: Expanded Polystyrene (EPS) Foam Insulation: Selection Guide. https://omnexus.specialchem.com/selection-guide/expanded-polystyrene-eps-foam-insulation. Accessed 01 Mar 2023
25. Richter, J., Staněk, K.: Measurements of water vapour permeability – tightness of fibreglass cups and different sealants and comparison of μ-value of gypsum plaster boards. Procedia Eng. **151**, 277–283 (2016)
26. Ediltec: Water vapour permeability (2020). https://italy.ediltec.com/en/technical-notes/-3. Accessed 01 Mar 2023
27. Electrical Workbook, 31 July 2021. https://electricalworkbook.com/properties-of-insulating-materials/#:~:text=The%20desirable%20properties%20of%20a%20good%20heat%20insulating,as%20a%20result%20of%20exposure%20to%20surrounding%20conditions. Accessed 2022
28. Electrical Area, 11 December 2021. https://electricalarea.com/properties-of-insulating-materials/#:~:text=Mechanical%20properties%20Strength%3A%20A%20key%20aspect%20of%20the,is%20used%20based%20on%20volume%20and%20net%20weight. Accessed 2022
29. Agrodome (2022). https://www.agrodome.nl/. Accessed 2022
30. Peuling, S.: Moisture resistance on dried mycelium-based composites; determining compressive strength and dimension stability. CoEBBE, Breda (2021)
31. Centre of Expertise Biobased Economy: Centre of Expertise Biobased Economy (2022). www.mnext.nl. Accessed 2022
32. British Gypsum: Product Data Sheet Gyproc WallBoard Duplex 12.5 mm, 27 September 2022. https://www.british-gypsum.com/documents/product-data-sheet-pds/british-gypsum-pds-gyproc-wallboard-duplex-12-5mm.pdf. Accessed 01 Mar 2023

Innovative Bio-Binders and Additives

Physical and Mechanical Properties of Mortar Made with Recycled Fine Aggregate

Margareth Silva da Magalhães[1](✉) and Wallace Alves Pinto[2]

[1] Post-Graduate Programme in Civil Engineering, State University of Rio de Janeiro, Rua São Francisco Xavier, 524, 5032E, Rio de Janeiro, RJ CEP 20550-900, Brazil
margareth.magalhaes@eng.uerj.br

[2] Faculty of Engineering, State University of Rio de Janeiro, Rio de Janeiro, Brazil

Abstract. Construction and demolition waste recycling is generally limited to the use of the coarser fraction as aggregate for new concrete. The recovery of fine aggregates requires a cleaning by removing the hydrated cement waste. These fines, however, are normally discarded, despite being the by-product of the highest costing component of the original concrete, both in economic and in environmental terms. This paper investigates the properties of mortar containing recycled fine aggregate (RFA) as natural aggregate replacement. The RFA was obtained in laboratory by crushing mechanically hydrated cement waste, and then sieved. Five mixtures were made with natural and/or RFA and physical, mechanical and durability (based on drying shrinkage test) behavior was studied. Natural aggregate was replaced with RFA contents between 25% and 100% and a control mixture (with 100% natural aggregate) was also developed. Results indicate an elevated porosity, water absorption and shrinkage of mortars by using RFA, but a strong reduction in density, flexural and compressive strength. However, even with these reductions, the mixtures studied obtained the minimum values established in standards for the use as masonry mortars or other purposes with lower density or strength level. In addition, by using 100% RFA in mortar mixture can be an effective strategy to maximize the hydrated cement waste recycling and enhancing the sustainability index of mortars.

Keywords: Recycled fine aggregate · hydrated cement waste · recycled mortar · mechanical properties · physical properties

1 Introduction

Nowadays, there is a growing trend on the development of cement-based materials with reduced amount of natural resources, which has been substituted with wastes or by-products. This leads to a new generation of cement-based materials, with low carbon footprint, and significant economic advantages. Additionally, peculiar performances can be obtained according to the waste type, allowing potentially to obtain new materials with improved mechanical or durability properties. Among the wastes, some studies dealt with the valorization of construction and demolition waste (CDW), which has been commonly used as a replacement of natural aggregates [1–3] or binder [4] in concrete mixtures.

S. Amziane et al. (Eds.): ICBBM 2023, RILEM Bookseries 45, pp. 761–770, 2023.
https://doi.org/10.1007/978-3-031-33465-8_58

Coarse recycled aggregates from CDW are commonly used in partial replacement of natural coarse aggregates in concrete [1–3]. However, the quality of recycled concrete aggregate is generally lower than that of natural aggregates, due to presence of residual mortar [5]. Thus, techniques have been made to minimize the residual mortar content on the recycled coarse aggregate. On the contrary, recycled fine aggregates (RFA) are less useful in concrete mixtures, as they can be highly detrimental for what concern mechanical properties, workability, and durability [6–10] of concrete.

Recent studies have indicated that the finer portion of recycled aggregates can be used as partial replacement of natural sand in mortar mixtures [11–15]. However, the characteristics of the mortars containing RFA are strongly dependent of the RFA quality and source, and on the specific mortar composition, therefore it is expected that different outcomes emerge from different studies [11].

For example, Corinaldesi and Moriconi [12] replaced 100% natural sand in mortars with RFA from different sources (prefabricated concrete and recycled bricks). They indicated that the recycled mortars have always poorer mechanical behavior than the reference one, but it gained better mortar–brick bond strength. Among the recycled sand used, the sand coming from concrete waste led to mortars with the best performances. Instead, Neno et al. [11] showed that mortar containing 20% RFA, from concrete waste, performed better than the reference mortar, in terms of water absorption and compressive and flexural strength, while adhesive strength and dimensional stability was slightly poorer. Jimènez et al. [13] replaced up to 40% natural sand with RFA obtained from ceramic waste. They also found an improvement of compressive and flexural strength of recycled mortars, and only a small decrease in bulk density and workability of the hardened mortars.

Braga et al. [14] instead used the fine fraction of RFA (size < 0,150 mm), from concrete waste, in the preparation of mortars. They demonstrated that replacement ratios up to 15% could improve mortar performances, in terms of flexural and compressive strength and to reduce water absorption. Vegas et al. [15] used wastes from concrete and ceramic debris, crushed below 2 mm in size, and suggested that a 25% replacement of RFA led to mortars with no loss of compressive strength, workability and shrinkage in respect to the reference mortar.

In contrast to previous studies, this research is intended to provide additional information about using RFA in a sustainable mortar. The RFA used in this work was obtained from hydrated cement waste. Hence, the aim of this study is to replace natural fine aggregate (between 25% and 100%) with RFA in the mortar production. The effects of replacing natural fine aggregate with RFA on porosity, density, water absorption and flexural and compressive strength of RFA mortars are investigated and compared with the reference mortar, with only natural sand. In addition, drying shrinkage of RFA mortars are also investigated.

2 Experimental Program

2.1 Materials and Mixture Details

Portland cement CPII F-32 composed with 6% of calcareous filler [16], two fine aggregates (natural and recycled) and tap water were used to produce the mortars mixtures. Recycled fine aggregate (RFA) was obtained from hydrated cement waste. The waste was obtained by mixing a Brazilian Portland cement (CP III), that is composed with 35–70% of blast furnace slag, and water. At 28 days, the waste was crushing mechanically and then sieved to produce the RFA. Only the fraction below 2.36 mm of RFA was used in the RFA mixtures. It is important to keep the similar particle size distribution of aggregates, since the goal is to identify the influence of the recycled aggregate on the mixture, and the aggregate size distribution can influence the characteristics of the fresh and hardened mortar mixture. The physical characteristics of aggregates are presented in Table 1 and the particle distribution curves are shown in Fig. 1.

Table 1. Physical characteristics of aggregates.

Characteristics	Natural sand	RFA
Maximum aggregate size (mm)	2.36	2.36
Specific gravity (g/cm^3)	2.37	2.29
Fineness modulus	2.20	2.35
Water absorption (%)	10.00	29.00
Particles smaller than 150 μm (%)	0.46	4.56

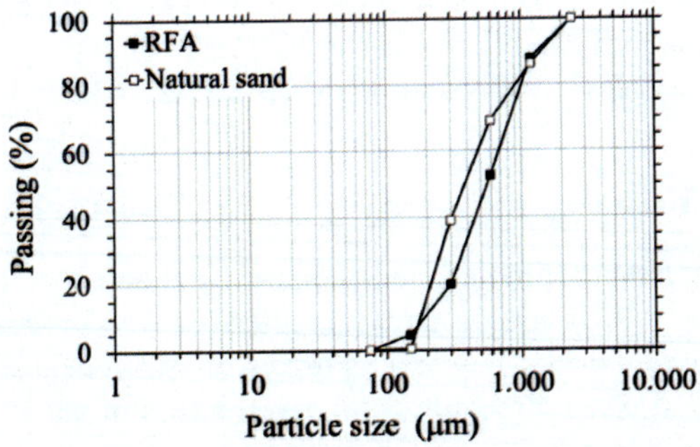

Fig. 1. Particle size distribution of aggregates.

To investigate the effect of using low and high content of RFA, four mixtures were produced with 25%, 50%, 75% and 100% of RFA, substituting the natural fine aggregate, and one mixture was produced with only natural fine aggregate, as a control mixture. Substitution ratios are expressed in weight percentage.

A typical cement to total aggregate ratio equal to 1:3 (by mass) was used in this work, and it was maintained constant in all mixtures. The total aggregate is defined as the sum of all aggregates (Natural aggregate and RFA). The free water/cement rate was 0.74 for all mortar mixtures. As the water absorption of the recycled aggregate is larger than that of the natural aggregate (see Table 1), water equivalent to 15% RFA was added to the M02-M05 mixtures to avoid the initial water absorption by RFA. The proportions and quantities of materials used in the mixtures are listed in Table 2. The ingredients quantities (kg per 1 m^3 of mortar) in Table 2 were calculated using the density (SSD) of each material, without considering the void content.

Table 2. Designation and composition of mortar mixtures.

Mix	% RFA[1]	Cement (kg/m^3)	Natural sand (kg/m^3)	RFA[1] (kg/m^3)	Water (kg/m^3)
Ref	0.00	429.09	1287.28	0.00	317.53
M02	25	427.07	960.90	320.30	316.03
M03	50	425.06	637.59	637.59	314.54
M04	75	423.07	317.30	951.91	313.07
M05	100	421.10	0.00	1263.31	311.62

[1]RFA: Recycled fine aggregate

The preparation of the mixtures was performed in a mechanical mixer with a 20 L capacity. After the mixing, the specimens were cast in steel moulds and demoulded 24 h after casting. The specimens were cured until the time of testing, in a water tank at temperature of 21 ± 1 °C.

A small flow cone for conventional flow table test was used to quantify the deformability of the fresh mixtures, according to NBR 13276 [17] standard, and the density of the fresh mixtures was measured according to NBR 13278 [18] standard. All mortars achieved a plastic consistency with values between 220–270 mm.

2.2 Experimental Methods

Three cylindrical specimens with diameter of 50 mm and height of 100 mm were used on compression tests at 7,14 and 28 days, according to the procedures described in NBR 7215 [19] standard. Further, the flexural tests were carried out at 28 days, on three specimens with dimensions of 40 × 40 × 160 mm for each mixture, according to NBR 13279 [20] standard.

At 28 days, dry bulk density, total porosity and water absorption were also measured. The tests were based on the NBR 9778 [21] standard. Three cylindrical specimens with diameter of 50 mm and height of 100 mm were used for each mixture.

The drying shrinkage tests were carried out on three prismatic specimens (25 mm wide, 25 mm thick and 285 mm long) per mixture according to NBR 15261 [22] standard. Measuring studs were placed on opposite ends of the mould before casting. After 2 days, the initial length and weight readings of the specimens were immediately recorded, and the specimens were placed in an environmental room at 23 ± 2 °C with 60 ± 5% RH.

The lengths and the weights of the mortar bars were measured using the dial indicator and an electronic balance, respectively. Subsequent length and weight variations were conducted at different ages up to 60 days.

3 Results and Discussion

3.1 Fresh and Dry Bulk Density, Water Absorption and Porosity of Mortars

A reduction of the bulk density of the fresh mortars, shown in Fig. 2a, occurred when the RFA content increases. This reduction is caused by the lower density of the recycled aggregate (2290 kg/m^3) with respect to natural sand (2370 kg/m^3). This trend is similar with those obtained by Jimenez et al. [13], Silva et al. [23] and Ledesma et al. [24] in all of which the mortars with recycled aggregates from ceramic waste had lower fresh density than the reference mixture.

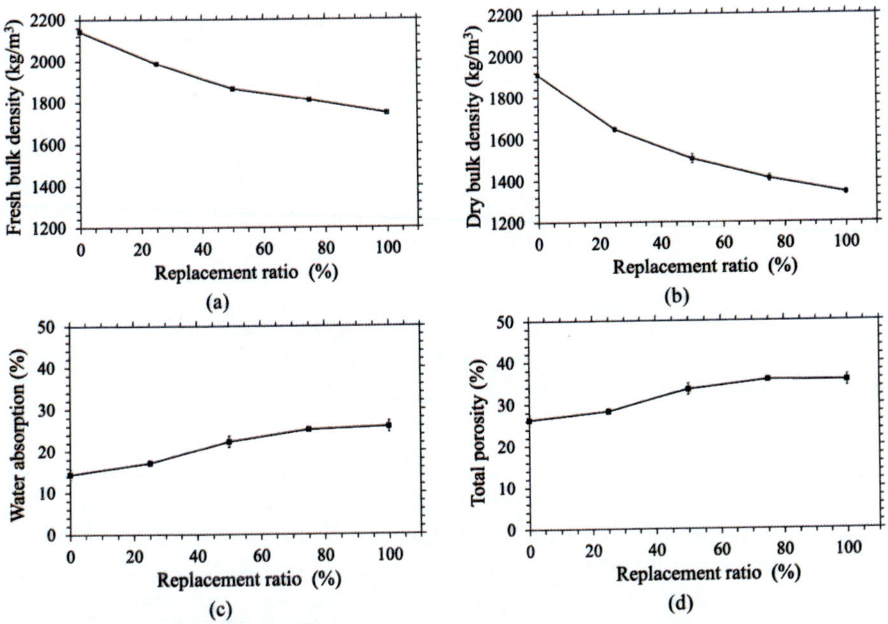

Fig. 2. Physical properties of mortars vs. RFA content.

The Brazilian standard NBR 13281 [25] specifies a D4 class for masonry mortars with fresh density values between 1600–2000 kg/m^3 and the European standards do not set a minimum or maximum value for the masonry mortars' fresh density, so the use of RFA is not a limiting factor for this property. According to Ledesma et al. [24], the main advantage of a lighter mortar is that for the same volume of mortar, the amount of mass to be transported is smaller.

A similar trend in the bulk density of the dry mortars was also obtained (Fig. 2b). Values of dry density decreased from approximately 1910 kg/m^3, for reference mortar

(0% of RFA), to approximately 1400 kg/m^3, for mixtures containing 100% RFA. These values are lower than 2000 kg/m^3 and could be classified in various classes according to Brazilian standard classification [25].

Test results also showed that the reference mortar has a porosity and water absorption of approximately 26% and 15%, respectively (Figs. 2c and 2d) and a strong increase of porosity and water absorption (Figs. 2c and 2d) were observed when natural sand was replaced by up to 75% RFA. Such increase can be estimated as up to 36%. However, for higher content (100%), total porosity and water absorption does not seem to increase.

3.2 Compressive and Flexural Strength

Figure 3 shows the development of compressive strength with curing period (7, 14, and 28 days) for mortar containing 0%, 25%, 50%, 75%, and 100% RFA levels. The rate of compressive strength development is relatively similar for all mortars, showing an increase in mechanical strength as a function of time between 7 and 28 days. All mixtures showed compressive strength values ranging from 8 to 13 MPa after 7 days of curing, which correspond to 54–59% of the strength attained at 28 days.

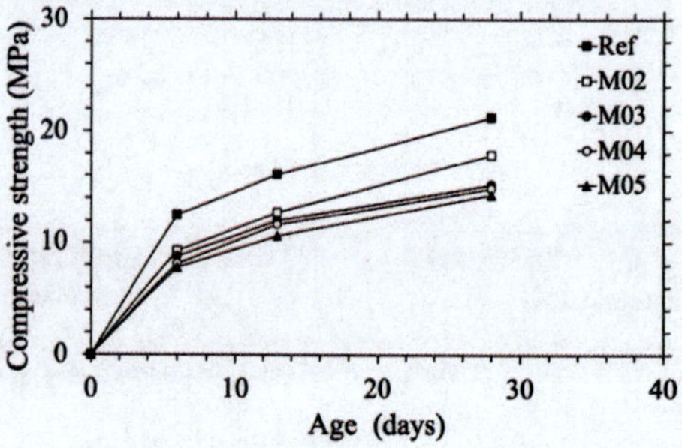

Fig. 3. Compressive strength vs. age curve of mortars.

The compressive strength of the mortar reduced with the replacement of natural sand by 25% (M02), 50% (M03), 75% (M04) and 100% (M05) RFA at all curing times (7, 14 and 28 days). However, the P-value (<0.05) of the F-test (ANOVA) indicated that mixture fabricated using 50%, 75% and 100% RFA was not statistically different on the compressive strength, at 95% confidence level. For instance, at 7 days the average compressive strength of mortars reduced 21.8% (M02), 30% (M03), 35.2% (M04) and 37.8% (M05). At 14 days the reductions were 21% (M02), 24.7% (M03), 25% (M04) and 34.2% (M05) and at 28 days were 15.8% (M02), 28.3% (M03), 29.9% (M04) and 32.6% (M05).

This result is comparable with what normally happens in mortar made with ceramic masonry waste aggregates [13, 23], in which increasing the recycled aggregate content leads to decreasing compressive strength. A possible reason for a poorer mechanical performance of the recycled mortar is the higher porousness of hardened cement recycled particles (RFA), compared that one natural aggregate. Figure 4 shows that the compressive strength is directly related with dry density of mortars, i. e, compressive strength decreases with the reduction in dry density.

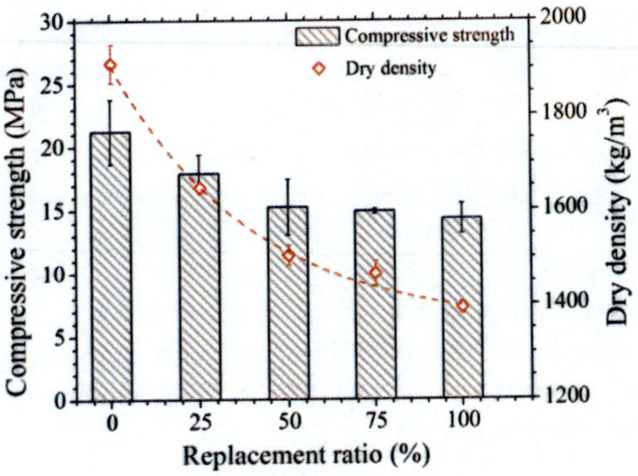

Fig. 4. 28-day compressive strength of mortars and its relationship with dry density.

Figure 5 shows the effect of using RFA on the flexural strength of mortars. The flexural strength also reduced with increasing RFA content. The mixture that presented the lowest flexural strength value is one that was fabricated using 100% RFA (M05). Flexural strength of this mixture is approximately 4 MPa, which means a loss of strength of approximately 32% with respect to the reference mixture (0% RFA). The P-value (<0.05) of the F-test also indicated that mixture fabricated using 50%, 75% and 100% recycled sand are not statistically different on the flexural strength at 95% confidence level.

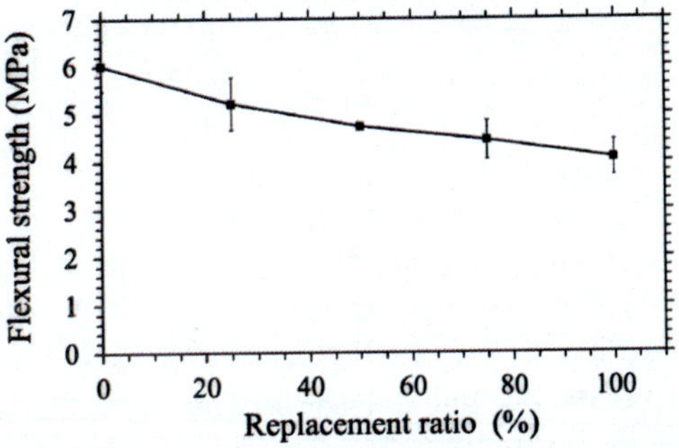

Fig. 5. Flexural strengths with respect to RFA content.

In this work, the brittleness factor of mortars was calculated. Brittleness factor was calculated as compressive strength/flexural strength ratio (f_c/f_f). No influence was observed on the ratio f_c/f_f, when natural aggregate was replaced by RFA. For mortars studied, the brittleness factors are 3.5 (M01), 3.2 (M02), 3.2 (M03), 3.4 (M04) and 3.5 (M05). It means that compressive and flexural strength are influenced similarly.

3.3 Drying Shrinkage

Drying shrinkage was evaluated after 2 days of curing and its increase in time can be seen in Fig. 6a. Each point in the curve represents the average values from three specimens.

The five mortars showed a similar trend, the drying shrinkage increased with time, more rapidly in the initial ages and slower later. This is in agreement with the shrinkage mechanisms more relevant. The drying shrinkage of mixtures with RFA increased with the augment of RFA content. An increment in shrinkage of 13% (M02), 47% (M03), 66% (M04) and 64% (M05) was observed when comparing to the reference mortar. This trend is similar to that observed in studies with ceramic masonry waste aggregate [13, 23] and recycled concrete aggregates [11, 15]. Some of the studies attributed this phenomenon to the lower modulus of elasticity of the recycled mortar compared to natural aggregate mortar [13, 23]. According to Aitcin et al. [26], the porosity plays a significant role in influencing the drying shrinkage of cement mortar. Higher porosity in the mortar lead to increased drying shrinkage due to the presence of more capillary voids that become filled with water during mixing and later evaporate, causing shrinkage.

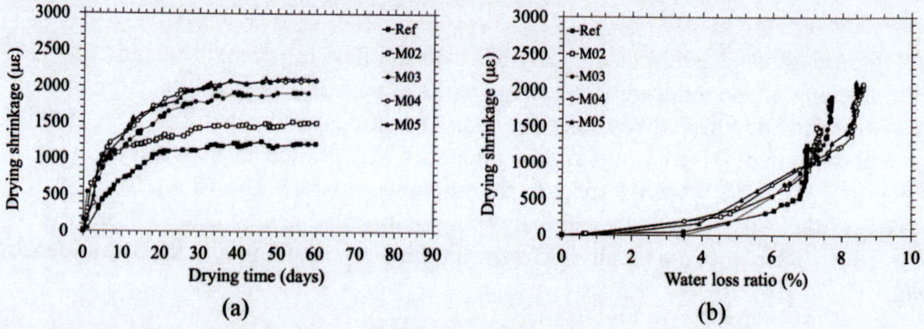

Fig. 6. Drying shrinkage as a function of drying time (a) and water loss ratio (b).

In Fig. 6b, the drying shrinkage is plotted against the corresponding water loss ratio. There are two defined stages in the curves. In the early-age mass loss stage, which takes at about 5 days, the free water in the mortars mainly comes from macropores (greater than 50 nm). Its removal does not cause any volume change in the specimens. After approximately 5 days, the slope of the curve gradually increased, indicating that the water in the small capillaries pore (with diameter of < 50 nm) begin to dissipate, resulting in a gradual increase in capillary pressure and eventually a gradual increase in drying shrinkage [27].

4 Conclusions

The following conclusions can be drawn from the study reported in this paper:

1. The bulk density of fresh and hardened mortars decreases progressively as the use of recycled aggregate increase because of the lower density of recycled aggregate compared to natural aggregate. This does not appear to be a limitation to the use of recycled fine aggregate from hydrated cement waste in mortar production.
2. The water absorption of mortar increases as the use of up to 75% recycled fine aggregate because of the higher porosity of recycled mortars. However, for higher content (100%), water absorption does not seem to increase compared to mortar with 75% recycled aggregate.

3. There is no change on the compressive strength development as the use of recycled fine aggregate.
4. Compressive and flexural strength are influenced similarly, i.e., the strength of mortar specimens reduced as the use of recycled fine aggregate at all curing times (7, 14 and 28 days), however, strength of mortars fabricated with 50%, 75% and 100% recycled aggregate are not statistically different.
5. A severe increment in shrinkage was observed when used recycled aggregate from hydrated cement waste on cement mortar.

Acknowledgments. The authors acknowledge the Brazilian Agency FAPERJ for the financial support.

References

1. Ravindrarajah, R.S., Tam, C.T.: Recycled concrete as fine and coarse aggregate in concrete. Mag. Concr. Res. **12**, 214–220 (1978)
2. Ryu, J.S.: An experimental study on the effect of recycled aggregate on concrete properties. Mag. Concr. Res. **54**, 7–12 (2002)
3. Limbachiya, M.C., Leelawat, T., Dhir, R.K.: Use of recycled concrete aggregates in high-strength concrete. Mater. Struct. **33**, 574–80 (2000)
4. Shui, Z., Xuan, D., Wan, H., Cao, B.: Rehydration reactivity of recycled mortar from concrete waste experienced to thermal treatment. Construct. Build. Mater. **22**, 1723–1729 (2008)
5. de Juan, M.S., Gutiérrez, P.A.: Study on the influence of attached mortar content on the properties of recycled concrete aggregates. Constr. Build. Mater. **23**, 872–877 (2009)
6. Khatib, J.M.: Properties of concrete incorporating fine recycled aggregate. Cem. Concr. Res. **35**, 763–769 (2005)
7. Bravo, M., de Brito, J., Pontes, J., Evangelista, L.: Mechanical performance of concrete made with aggregates from construction and demolition waste recycling plants. J. Clean. Prod. **99**, 59–74 (2015)
8. Bravo, M., de Brito, J., Pontes, J., Evangelista, L.: Durability performance of concrete made with aggregates from construction and demolition waste recycling plants. Construct. Build. Mater. **77**, 357–369 (2015)
9. Rodrigues, F., Carvalho, M.T., Evangelista, L., de Brito, J.: Physical chemical and mineralogical characterization of fine aggregates from construction and demolition waste recycling plants. J. Clean. Prod. **52**, 438–445 (2013)
10. Zhao, Z., Remond, S., Damidot, D., Xu, W.: Influence of recycled concrete aggregates on the properties of mortars. Construct. Build. Mater. **81**, 179–186 (2015)
11. Neno, C., de Brito, J., Veiga, R.: Using fine recycled concrete aggregate for mortar production. Mater. Res. **17**(1), 168–177 (2014)
12. Corinaldesi, V., Moriconi, G.: Behaviour of cementitious mortars containing different kinds of recycled aggregate. Constr. Build. Mater. **23**, 289–294 (2009)
13. Jimenez, J.R., Ayuso, J., Lopez, M., Fernandez, J.M., de Brito, J.: Use of fine recycled aggregates from a ceramic waste in masonry mortar manufacturing. Constr. Build. Mater. **40**, 679–690 (2013)
14. Braga, M., de Brito, J., Veiga, R.: Incorporation of fine concrete aggregates in mortars. Constr. Build. Mater. **36**, 960–968 (2013)

15. Vegas, I., Azkarate, I., Juarrero, A., Frias, M.: Design and performance of masonry mortars made with recycled concrete aggregates. Mater. Constr. **95**(295), 5–18 (2009)
16. Brazilian Standard NBR 16697. Cimento Portland-Requisitos. Associação Brasileira de Normas Técnicas (ABNT) (2018)
17. Brazilian Standard NBR 13276. Argamassa para assentamento e revestimento de paredes e tetos - Determinação do índice de consistência. Associação Brasileira de Normas Técnicas (ABNT) (2016)
18. Brazilian Standard NBR 13278. Argamassa para assentamento e revestimento de paredes e tetos - Determinação da densidade de massa e do teor de ar incorporado. Associação Brasileira de Normas Técnicas (ABNT) (2005)
19. Brazilian Standard NBR 7215. Cimento Portland - Determinação da resistência à compressão de corpos de prova cilíndricos. Associação Brasileira de Normas Técnicas (ABNT) (2019)
20. Brazilian Standard NBR 13279. Argamassa para assentamento e revestimento de paredes e tetos - Determinação da resistência à tração na flexão e à compressão. Associação Brasileira de Normas Técnicas (ABNT) (2005)
21. Brazilian Standard NBR 9778. Argamassa e concreto endurecidos - Determinação da absorção de água, índice de vazios e massa específica. Associação Brasileira de Normas Técnicas (ABNT) (2009)
22. Brazilian Standard NBR 15261. Argamassa para assentamento e revestimento de paredes e tetos - Determinação da variação dimensional (retratação ou expansão linear). Associação Brasileira de Normas Técnicas (ABNT) (2005)
23. Silva, J., de Brito, J., Veiga, R.: Recycled red-clay ceramic construction and demolition waste for mortars production. J. Mater. Civ. Eng. **22**(3), 236–244 (2010)
24. Ledesma, E.F., Jiménez, J.R., Ayuso, J., Fernández, J.M., de Brito, J.: Maximum feasible use of recycled sand from construction and demolition waste for eco-mortar production – Part-I: ceramic masonry waste. J. Clean. Prod. **87**(15), 692–706 (2015)
25. Brazilian Standard NBR 13281. Argamassa para assentamento e revestimento de paredes e tetos – Requisitos. Associação Brasileira de Normas Técnicas (ABNT) (2005)
26. Aitcin, P.C., Nevile, A.M., Acker, P.: Integrate view of shrinkage deformation. Concr. Int. **19**(9), 35–41 (1997)
27. Hansen, W.: Drying shrinkage mechanisms in Portland cement paste. J. Am. Ceram. Soc. **70**(5), 323–328 (1987)

Development of Rice Husk Ash as a Cement Substitute for Environmental Conservation and Its Effective Use in Green Infrastructures

Tetsuya Suzuki[1] (✉) and Yuma Shimamoto[2]

[1] Niigata University, Niigata 980-2181, Japan
suzuki@agr.niigata-u.ac.jp
[2] Tokyo University of Agriculture and Technology, Tokyo 183-8509, Japan

Abstract. In recent years, there has been a growing demand for the development of structural materials capable of introducing vegetation in order to solve environmental problems and achieve sustainable development of society. The authors have been developing a cement-based greening base material using "rice husk ash", which is incinerated ash from rice husks, an agricultural by-product. The reason for focusing on rice husks is that most of them are not utilized despite their abundance in the world. In addition, the advantage of rice husk ash is not only the reuse of resources but also its ability to accelerate hardening reactions as a cement substitute due to its pozzolanic activity. The greening base material is a porous material that must have both stability as a soil material and a pore structure that does not inhibit the development of plant root systems. The problem with the mechanical properties of porous materials is that fracture occurs due to stress concentration at the site of minimum strength in various stress fields. Therefore, in addition to the strength value, which is the maximum stress, it is necessary to clarify the cracking kinetics that appear during the fracture process. Previous studies have quantified the cracking behavior using fracture mechanics parameters and evaluated the material properties based on particle bonding. However, these studies have not been sufficiently conducted for greening base materials. In this study, we report the results of cracking and tensile strength tests of cementitious greening base materials with five different rice husk ash replacement ratios, and the evaluation of material properties using acoustic emission (AE) energy indices. In addition, planting tests were conducted on these materials and the relationship between the materials and the development of plant rhizosphere was discussed. Thus, the results indicated that the particle binding strength decreased as the rice husk ash replacement ratio increased, and that the change in fracture behavior could be evaluated from the AE energy index. The planting tests revealed a close relationship between the development of the rhizosphere and the material represented by the pore distribution by the replacement ratio. Therefore, rice husk ash is useful as a cement substitute, enabling both material strength and plant growth in service green infrastructure.

Keywords: Bio-Material · Rice Husk Ash · Greening Substrate Material · Rhizosphere · Fracture Mechanics Parameter · Material Physical Parameter

S. Amziane et al. (Eds.): ICBBM 2023, RILEM Bookseries 45, pp. 771–781, 2023.
https://doi.org/10.1007/978-3-031-33465-8_59

1 Introduction

In recent years, there has been a growing demand for the development of structural materials capable of introducing vegetation in order to solve environmental problems and achieve sustainable development of society. The authors have been developing a cement-based greening base material using "rice husk ash" which is incinerated ash from rice husks, an agricultural by-product [1]. The reason for focusing on rice husks is that most of them are not utilized despite their abundance in the world. According to a survey by the Food and Agriculture Organization of the United Nations (FAO), about 150 million tons of rice husks were discharged annually worldwide in 2014 [2]. In addition, the advantage of rice husk ash is not only the reuse of resources but also its ability to accelerate hardening reactions as a cement substitute due to its pozzolanic activity [3]. The cementitious materials develop strength through C-S-H. The pozzolanic reaction is an alternative cement chemical reaction. In this study, we will focus on rice husk ash based on decrease of cementitious materials and acceleration of pozzolanic reaction. The greening base material is a porous material that must have both stability as a soil material and a pore structure that does not inhibit the development of plant root systems. The problem with the mechanical properties of porous materials is that fracture occurs due to stress concentration at the site of minimum strength in various stress fields [4]. Therefore, in addition to the strength value, which is the maximum stress, it is necessary to clarify the cracking kinetics that appear during the fracture process. Previous studies have quantified the cracking behavior using fracture mechanics parameters [5] and evaluated the material properties based on particle bonding [6]. However, these studies have not been sufficiently conducted for greening base materials. By the authors, development of evaluation method for mechanical properties and damage level of various cementitious materials using acoustic emission (AE) method and related elastic wave method [7–12]. We focus on energy of detected AE waves in mechanical test, such as compression, tensile and bending tests. The AE energy generated during the loading process is related to the stress-strain behavior of the testing material and is highly correlated with the modulus of elasticity and strain energy [10].

In this study, we focus on the results of cracking and tensile strength tests of cementitious greening base materials with five different rice husk ash replacement ratios, and the evaluation of material properties using AE energy indices. In addition, planting tests were conducted on these materials and the relationship between the materials and the development of plant rhizosphere is discussed.

2 Experiments and Analysis Methods

2.1 Evaluation of Pore Distribution by X-ray CT and Pore Size Distribution Tests

In this study, pore size distribution is evaluated by pore size distribution test below 360 μm. Pores larger than 360 μm are evaluated by X-ray CT method. The measurement conditions of the X-ray CT method were referred to previous studies [10]. The cylindrical specimens, 50 mm in diameter and 100 mm in height, were prepared according to JGS 0811, "Method of preparing specimens by ramming of stabilized soil".

The mixing table is shown in Table 1. The rice husk ash was incinerated ash discharged from a rice husk gasification plant in Niigata Prefecture [13]. The density of the rice husk ash is 1.79 g/cm^3. The specific surface area is 3,670 cm^2/g, and XRD/Rietveld analysis showed that the amorphous mass is 61.73%. For the pore size distribution test, a 10 mm thick disk-shaped piece was cut from the center of the specimen in the height direction, and the pore size distribution was measured using a micromeritics AutoPore IV 9520 mercury injection porosimeter. Pore size distribution tests were conducted at 28 days of age. The X-ray CT measurements were taken using a Toshiba whole-body X-ray computed tomography system (Toshiba Medical Systems Corporation) on specimens at 7 and 28 days of age. The imaging conditions were the same as in previous studies [8, 10, 11]. A longitudinal section of the center of the specimen was acquired, and porosity and void shape were calculated by binarization. Porosities were extracted from those with lightness in the range of 0 to 40.

Table 1. Mix proportion of testing samples.

Case	Sand	Cement	Rice husk ash	Water
	(g)	(g)	(g)	(g)
40%	350.00	14.70	9.80	32.55
50%	350.00	12.25	12.25	35.55
60%	350.00	9.80	14.70	32.55
70%	350.00	7.35	17.15	32.55
80%	350.00	4.90	19.60	32.55

2.2 Tensile Strength Test with AE and DIC Methods

Overview of Experiments. In order to obtain more AE signals in the AE measurement, loading was performed at 5.0 × 10–3 N/mm2 /sec, which is smaller than the JIS standard. AE measurements and image analysis were performed during the loading process. The AE measurements were performed using SAMOS (PAC). A total of six 150 kHz resonant sensors were installed. The threshold and amplification values were set to 40 dB and 60 dB, respectively, and the frequency range was 5 kHz to 400 kHz. The deformation behavior of the specimens was measured using two CCD cameras (CRAS-14S5M-C: Point Grey Research). The resolution was 1384 × 1036 pixel. The CCD cameras were set to capture images at an interval of 2 Hz.

Estimation of Particle Binding Force using AE Energy Parameter. In this section, the relationship between the cumulative AE energy and the particle bonding strength of porous materials is discussed. The relationship between the tensile strength σ of the granulated material and the porosity P is expressed by the following equation, assuming that the particles are equal-sized spheres [14].

$$\sigma = \frac{(1-P)kH}{\pi d^2} \tag{1}$$

where k: number of contacts with other particles, H: particle size condensation force, d: diameter of constituent particles.

In this study, the Eq. (1) is modified to obtain the product of the number of contacts k with other particles and the particle size condensation force H as the particle binding force. The diameter d of the constituent particles was 0.69 mm, which is 50% of the particle diameter.

Evaluation of Fracture Mechanics Parameter: Energy Release Rate. The energy release rate G is expressed by the following equation [5].

$$G = -\frac{d\Pi}{dA} \tag{2}$$

where dA: the amount of increase in crack area, $d\Pi$: the amount of potential energy release. In this study, the amount of potential energy release is calculated from the area O_{ac} as shown in Fig. 1. The crack area is obtained by binarizing the captured image and then differencing it with the binarized image before loading. The threshold value for the binarization process is set to 137. The differencing process is used to distinguish between black spots and cracks.

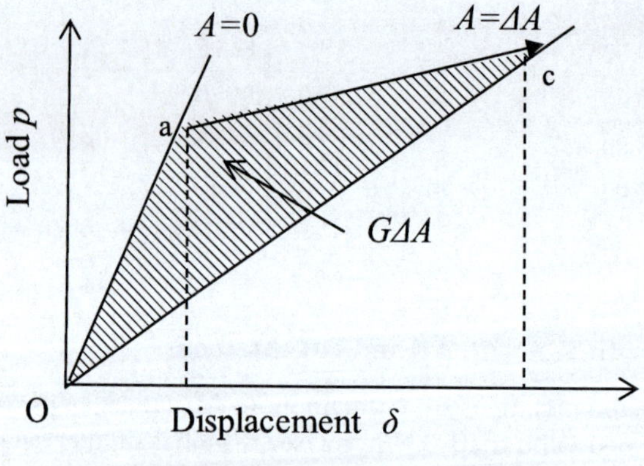

Fig. 1. Method of calculating potential energy.

AE Energy Analysis. In this study, the relationship between AE energy and energy release rate is examined. The AE energy EAE is calculated using the AE signal amplitude as ap and the following equation [15].

$$E_{AE} = (a_p)^2 \tag{3}$$

2.3 Planting Test for Material Evaluation by *Lippia Canescens Kunth*

The test plant was selected by *Lippia canescens Kunth*. The test samples were made of transparent PET resin plates and transplanted with seedlings of the test plants. The sample size was set to 3. The test period duration was 45 days. After 45 days, the above-ground parts of the test plants were cut and harvested. The underground parts (roots) were manually sorted as much as possible. The above- and below-ground parts were each dried at 80 °C for 48 h using a draft dryer, and the dry matter weight was determined. Binarization is used to determine the depth of root reach (maximum root growth) and the direction of root growth. Figure 2 shows a visual image of the side of the specimen and the image after the binarization process. Only the roots are extracted by binarization using the brightness of the image (1400 x 1400 pixel). In this analysis, the threshold is set between 43 and 75.

Fig. 2. Calculation of root elongation direction by binarization process

3 Results and Discussion

3.1 Evaluation of Pore Distribution

Evaluation of Pore Size Distribution. Examination revealed that total pore volume is associated with an increase in rice husk ash replacement rate. Total pore volume is measured at 40% rice husk ash replacement rate: 0.1684 ml/g, 50%: 0.1619 ml/g, 60%: 0.1751 ml/g, 70%: 0.1873 ml/g and 80%: 0.1920 ml/g.

Evaluation of Pore Shape by X ray CT Method. Figure 3 shows comparative images of pore distribution according to the age of the lumber. Figure 4 shows the relationship between the total pore length and average aspect ratio. A positive correlation is observed between the total pore length and the average aspect ratio. The specimens at 7 days of age have longer total pore lengths and larger aspect ratios than those at 28 days of age. The smaller total pore length at 28 days of age is presumably due to the progress of the hydration and pozzolanic reactions. The difference in the pore shape depending on the replacement ratio become more pronounced at 7 days of age. The higher the replacement ratio, the longer the total pore length, the larger the aspect ratio, and the more open pores are detected.

3.2 Estimation of Fracture Mechanics Parameter by AE of Crack Stress Field

Figure 5 shows the results of the comparison of particle binding strength for different rice husk ash replacement rates. It is clear that the particle binding strength decreases with increasing substitution ratio. The particle binding force did not decrease linearly, but decreased to about 2/3 between 40% and 50% substitution ratio. The crack fracture behavior was examined using the energy release rate and AE energy. Image analysis results showed that crack initiation was observed at a distance of 0.25 to 0.30 d (d: specimen diameter) from the specimen center. Binarization results showed that the crack width and crack length were 0.35 mm and 1.02 mm, respectively, at maximum load. The displacement at the initial crack initiation location was used to calculate the energy release rate.

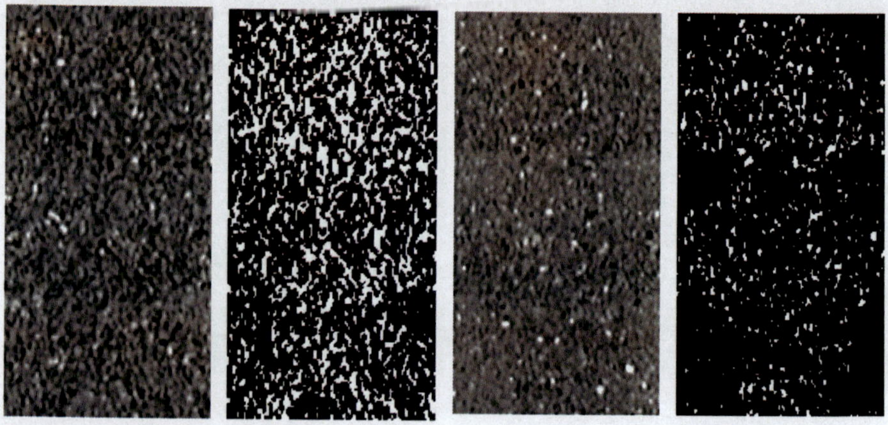
(a) 7 days aging. (b) 28 days aging.
* White color indicates voids.

Fig. 3. Comparison of pore distribution by age of timber (80% rice husk ash replacement rate).

The relationship between cumulative AE energy and energy release rate is shown in Fig. 6. The cumulative AE energy is the value obtained by adding up the AE energy detected from all six sensors. It is clear that the AE wave with a high amplitude value of more than 80 dB (Fig. 7b) was detected just before the maximum load, and that the energy release rate and the cumulative AE energy increased rapidly almost simultaneously. This suggests that AE energy is a useful indicator for estimating the energy release rate. The accumulated AE energy increased even when the displacement was 0.15 mm. Cracks that cannot be detected inside the specimen or by image analysis may be detected by AE measurement.

The relationship between the crack tensile strength and the center of gravity of AE energy is shown in Fig. 8. The center of gravity of AE energy is a relative value with the maximum load set at 1.0. As an overall trend, the higher the displacement ratio, the more it changed in the early stage of loading, and the more pronounced AE energy release was detected under low stress. This may be attributed to the fact that the particle binding force decreases with increasing replacement ratio. In conclusion, it is clear that

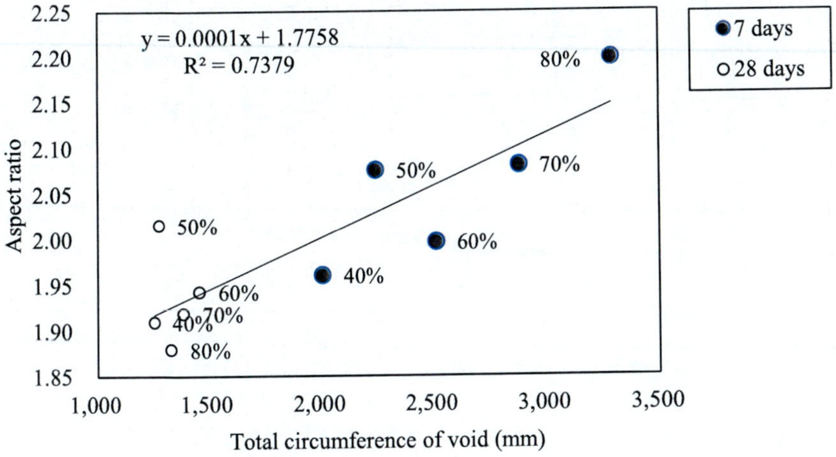

Fig. 4. Relationship between total circumference of void and aspect ratio by X-ray CT.

differences in cementitious greening base materials alter the AE energy release. The possibility of using AE energy characteristics to evaluate materials was demonstrated.

In order to discuss, the long-term durability of the greening substrate after its construction, it would be possible to estimate the crack development process and fracture mechanics index by measuring AE energy indices in various stress fields in addition to the strength properties.

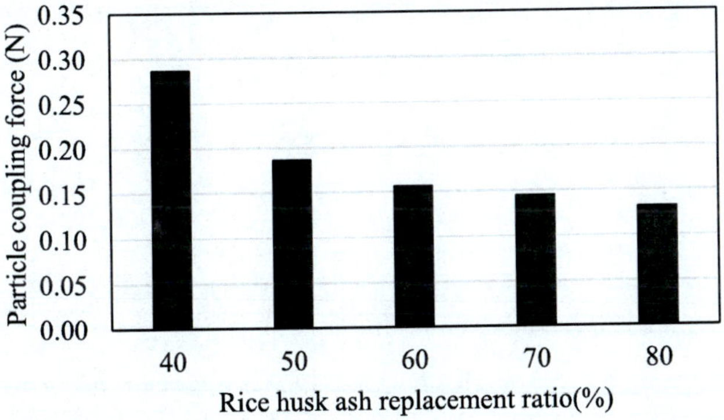

Fig. 5. Comparison of particle binding force for different rice husk ash replacement ratio.

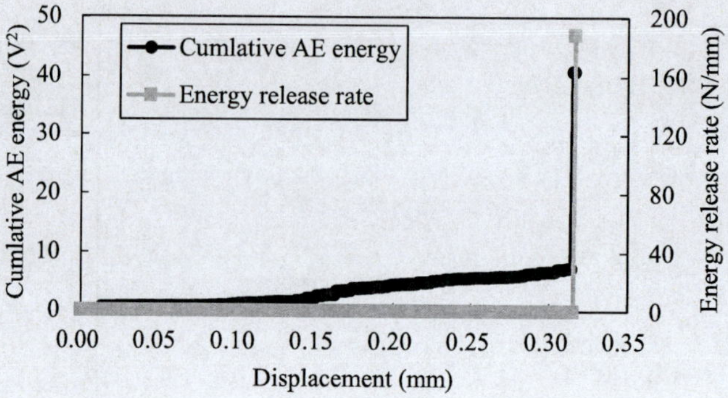

Fig. 6. Relationship between cumulative AE energy and energy release rate.

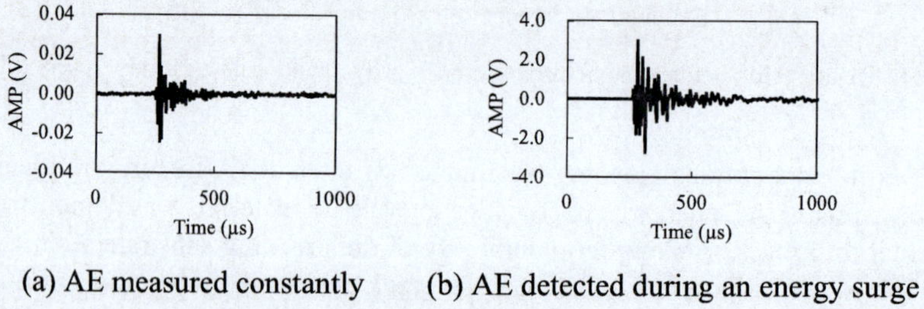

Fig. 7. Change in detected AE waveforms.

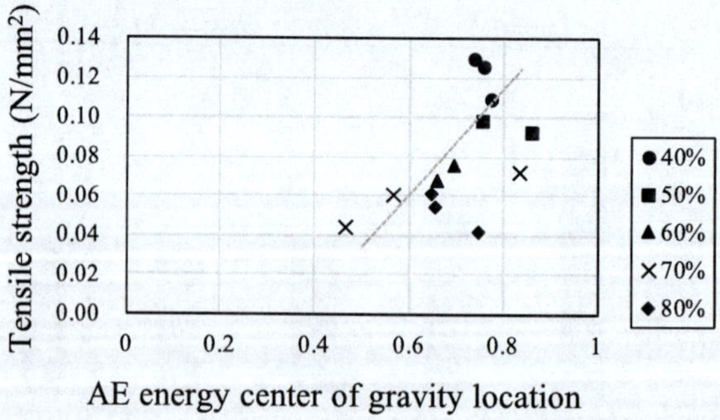

Fig. 8. Relationship between tensile strength and AE energy center of gravity location.

3.3 Vegetation Introduction into Cementitious Green Base Materials and the Effect of Porous Characteristics

Comparative images of plant growth 45 days after sowing are shown in Fig. 9. Comparative results of root dry matter weight and root reach depth at different rice husk ash replacement rates are shown in Fig. 10. It is clear that the root dry matter weight decreases with decreasing replacement rate. Compared to the root dry matter weight at 80% replacement rate, the root dry matter weight at 40% and 50% replacement rate

(a)Rice husk ash replacement rate: 40% (b)Rice husk ash replacement rate: 80%

Fig. 9. Change in root elongation due to differences in rice husk ash replacement rate.

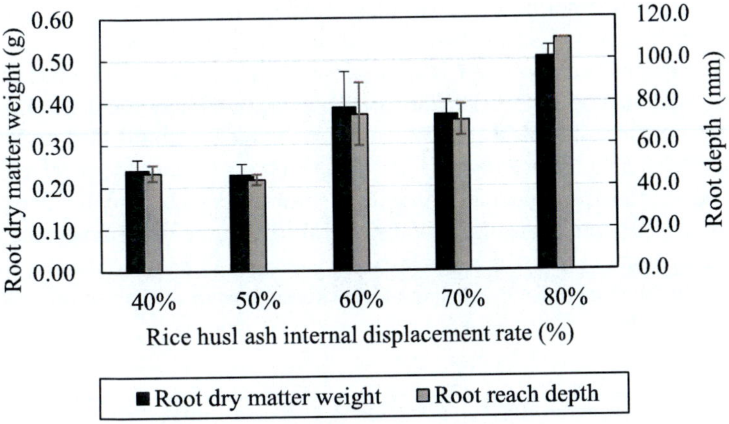

Fig. 10. Comparison of root dry matter weight and root depth at different rice husk ash replacement rates.

decreased by 45–52%. At 60% and 70% replacement rates, the reduction ranged from 73 to 84%. A significant difference in the depth of root penetration was observed between the three replacement rates of 40% and 50%, 60% and 70%, and 80%. The reasons for the differences in root penetration depths with replacement rate could be attributed to the effects of substrate hardness, pH, and water holding capacity. A decrease in replacement rate may have caused an increase in the hardness of the substrate material. It has been shown that roots grown in soils with high mechanical resistance exhibit specific morphological changes, resulting in slower root elongation and shorter root length [16]. In the present experimental results, the 40% and 50% replacement rates resulted in the shallowest depths attained among the experimental cases. Comparing the porosity, the 40% replacement ratio was about 4% less than that of the 70% replacement ratio and above. Compressive strength was detected 2.3 to 6.3 times greater for the 40% replacement rate than for the 50% replacement rate and above. The shallow-rootedness at 40% replacement rate could be attributed to the hardness of the substrate material. Another reason for the differences could be the differences in pH and water retention depending on the replacement rate. It has been shown that in alkaline soils, iron deficiency symptoms are more likely to occur due to excess calcium, and plant growth is suppressed [16]. The pH measured in this study was 12.18 at 40% replacement rate, 12.22 at 50%, 11.04 at 60%, 11.97 at 70%, and 11.88 at 80%. A trend of increasing alkalization was observed as the rice husk ash replacement rate decreased. This suggests that alkalization of the substrate material was one of the factors that inhibited root growth. The pore size distribution test results indicated that the capillary porosity was approximately 2.3 to 3.2% higher at 60% and 70% replacement rate compared to 50%. The increased water holding capacity and larger porosity may have promoted root elongation.

Figure 11 shows a comparison of the direction of root growth at different rice husk ash replacement rates. The experimental results show that in all cases, the largest percentage of roots were located in the 75–90° direction. The root system of dicotyledonous plants is formed around a single main root that develops from a juvenile root, and most of the main roots show strong normal gravitropism and elongate directly below the plant body [16]. Therefore, a large proportion of the 75–90° roots were considered to be 75° to 90°. At 80% replacement rate, roots were shown to spread in all directions, unlike the other cases. This result may be attributed to the low mechanical resistance and the large initial pore space. Comparing the 60% and 70% replacement rates with the 50% replacement rate, the 60% and 70% replacement rates had a greater proportion of roots between 60 and 75°. Lateral roots showed inclined gravitropism and elongated laterally or obliquely [16]. At 60 and 70% replacement rates, the lateral roots were more developed than at 50%. This development of the rhizosphere is considered to be due to the development of pore structure and increased water retention capacity, as described in the discussion of dry matter weight.

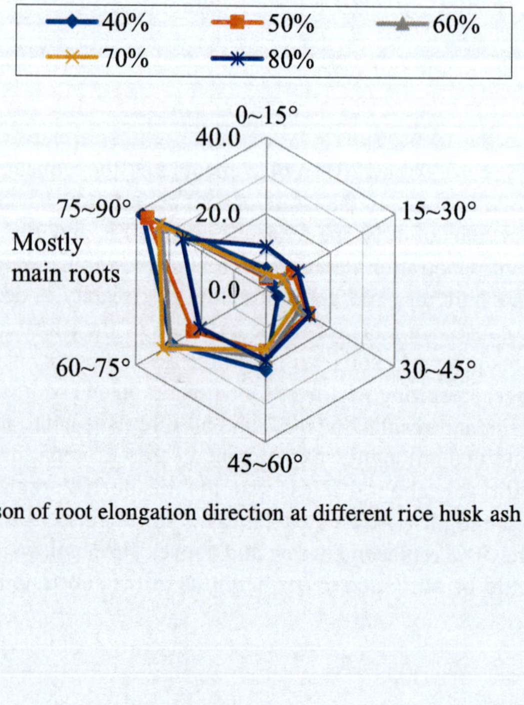

Fig. 11. Comparison of root elongation direction at different rice husk ash replacement rates.

4 Conclusion

In this study, tensile strength tests and planting tests were conducted on cement-based greening base material mixed with rice husk ash. The results indicated that the particle binding strength decreased as the rice husk ash replacement ratio increased, and that the change in fracture behavior could be evaluated from the AE energy index. The planting tests revealed a close relationship between the development of the rhizosphere and the material represented by the pore distribution by the replacement ratio.

References

1. Shimamoto, Y., Suzuki, T., Morii, T.: Ability of cement improved soil mixed with rice husk ash as greening material. J. Re-vegetation Technol. **43**(1), 127–132 (2017)
2. Food and Agriculture Organization of United Nations (2014). http://faostat.fao.org
3. Bagheri, Y., Ahmad, F., Isamail, M.A.: Strength and mechanical behavior of soil-cement-lime-rice husk ash (soil CLR) mixture. Mater. Struct. **47**, 55–66 (2014)
4. Kondo, R. (ed.): Mechanical Properties, Porous Materials, pp. 165–183. Gihoudo Publishing (1973)
5. Togo, K.: Material Strength Analysis, pp. 77–97. Uchida Rozuruho publishing, (2013)
6. Hatanaka, S., Mishima, N., Yuasa, Y.: Compressive strength of porous concrete: Experimental study on the effects of binder strength and coarse aggregate particle size on porosity relationship. J. Architect. Inst. Jpn **70**(594), 17–23 (2005)
7. Suzuki, T., Shigeishi, M., Ohtsu, M.: Relative damage evaluation of concrete in a road bridge by AE rate - process analysis. Mater. Struct. **40**(2), 221–227 (2007)
8. Suzuki, T., Ogata, H., Takada, R., Aoki, M., Ohtsu, M.: Use of acoustic emission and X-ray computed tomography for damage evaluation of freeze-thawed concrete. Constr. Build. Mater. **24**, 2347–2352 (2010)
9. Ishibashi, A., Matsuyama, K., Alver, N., Suzuki, T., Ohtsu, M.: Round-robin tests on damage evaluation of concrete based on the concept of acoustic emission rates. Mater. Struct. **49**, 2627–2635 (2016)
10. Suzuki, T., Shimamoto, Y.: Chapter 1 damage evaluation in concrete materials by acoustic emission. In: Acoustic Emission and Related Non-Destructive Evaluation Techniques in the Fracture Mechanics of Concrete, pp. 1–17. Woodhead Publishing (2019)
11. Suzuki, T., Nishimura S., Shimamoto, Y., Shiotani, T., Ohtsu, M.: Damage estimation of concrete canal due to freeze and thawed effects by acoustic emission and X-ray CT methods. Construct. Build. Mater. **245**(10), 118343 (2020). https://doi.org/10.1016/j.conbuildmat.2020.118343
12. Shimamoto, Y., Tayfur, S., Alver, N., Suzuki, T.: Identifying effective AE parameters for damage evaluation of concrete in headwork: a combined cluster and random forest analysis of acoustic emission data. J. Paddy Water Environ. (2022). https://doi.org/10.1007/s10333-022-00910-w
13. Shimamoto, Y., Suzuki, T.: Recycle of rice husk into agro-infrastructure for decreasing carbon dioxide. Paddy Water Environ. **17**(3), 555–559 (2019)
14. Takeuchi, Y.: Properties of Porous Materials and its Application Technology, pp. 372–392. Fuji Techno System Publishing (1999)
15. Japan Society for Non-Destructive Inspection: Acoustic Emission II, Japan Society for Non-Destructive Inspection, pp. 60–61 (2008)
16. Root Encyclopedia Editorial Committee: Root Encyclopedia, p. 438. Asakura Shoten Publishing (2012)

Revealing Value from Bioderived Polymers: Effects of Locally Sourced Polysaccharides on the Rheology of Limestone Mixtures

Patrick R. Cunningham[1,2], Alexander Mezhov[3](✉), and Wolfram Schmidt[3]

[1] Visiting Fellowship Holder, Federal Institute for Materials Research and Testing (BAM), 12205 Berlin, Germany

[2] Department of Civil and Environmental Engineering, University of California, Davis 95616, USA

[3] Federal Institute for Materials Research and Testing (BAM), 12205 Berlin, Germany

alexander.mezhov@bam.de

Abstract. Admixtures are important constituents to enhance the performance of concrete. They allow for more efficient use of binders which can mitigate negative environmental impacts from producing cement-based materials. Commonly used rheology modifying agents like polycarboxylate ethers or cellulose ethers are synthetic or semi-synthetic, respectively. This requires additional energy consumption for their production and global supply chains particularly for many developing regions, which will be large consumers of concrete in the future. However, many locally available bio-based polysaccharides could be effectively used instead. These polymers are often overlooked by engineers and scientists due to their limited distribution and inherent complexity, yet they represent an under-leveraged source of precursors for admixtures. This study investigates the action mechanisms of some bio-based rheology modifying agents, i.e., acacia gum and miscanthus gum, and provides a comparison to a conventionally modified starch. The results show that the mechanism of action of these polymers is closely related to the composition of the mixture, mixing regime, and the composition of the aqueous phase. Depending on the agent, either plasticizing or stabilizing effects on rheology can be revealed.

Keywords: Polysaccharides · superplasticizers · hydration · cement

1 Introduction

Enhanced durability properties and mechanical behavior of concrete have enabled the construction of high-performance structures and development of a more resilient built environment [1–3]. Over the course of several decades, chemical admixtures have become a key tool to control the fresh properties and the related quality of concrete [4]. These admixtures have enabled workable, higher strength concretes with fewer internal defects [5, 6]. These higher-strength and higher quality concretes can allow for more efficient material use and, thus, increase the sustainability of concrete [2, 4]. However,

S. Amziane et al. (Eds.): ICBBM 2023, RILEM Bookseries 45, pp. 782–792, 2023.
https://doi.org/10.1007/978-3-031-33465-8_60

standard chemical admixtures are expensive and inaccessible to producers in the Global South [7, 8]. In this work, the rheological effects are evaluated for limestone slurries with a variety of gums derived from different regions in the world.

As future regions across the Global South develop, the consumption of building materials, and especially concrete, will increase. Global concrete consumption is inevitably accompanied by a large amount of greenhouse gas emissions [9]. As the availability of supplementary cementitious materials is already restricted [10, 11], access to regionally available and affordable chemical admixtures could be one of the few levers available to drive higher quality concrete production and enable more-sustainable future developments. Furthermore, many established admixtures are based on naphtha and thus depend on global value chains. Hence, alternative admixture feedstocks will be needed to make the products and their supply chains less energy and carbon intensive [12, 13].

Acacia gum and miscanthus gums are two bioderived admixtures evaluated in this work. Acacia gum (which consists to a large extent of gum Arabic) is a complex polysaccharide derived from the sap of acacia trees [14, 15]. Literature describes them as slightly acidic, water-soluble, and containing a mix of mineral salts [14–16]. Predominantly grown in Africa, Senegal and Sudan is the largest exporters of acacia gums [17]. Acacia gums can have both plasticizing and stabilizing effects depending on the dosage, gum composition, and paste system [18–20]. Other works have investigated acacia gum to modify the workability of cement-based systems (e.g., [21–23]) and as a dispersant (e.g., [23, 24]). However, the industrial use of acacia gums is largely limited to the food and cosmetic industries [15, 24].

Miscanthus grass is composed of cellulose, hemicellulose, and lignin [25]. Research has primarily focused on miscanthus and other lignocellulosic biomass as an alternative to petroleum-based fuels (e.g. [26–28]). Miscanthus, specifically, is of great interest because it requires relatively low amounts of input and management. As a C4 plant, meaning it is more efficient at photosynthesis in hot climates with restricted water access, miscanthus has relatively high biomass yields in adverse environments [25, 29, 30]. Valorizing this biomass into fuels or petrochemical resources requires lignin extraction and then bio-refinement through fermentation or hydrolysis. These processes are typical done at elevated temperatures and often with hazardous reagents to maximize extraction and conversion [31, 32]. Currently, most of this processing remains at laboratory scale, but the environmental and economic implications of these systems need to be considered [27, 33]. In rheological studies have investigated carboxymethylcelulloses derived from miscanthus feedstocks [34] and nanofilms made with cellulose nanocrystals derived from miscanthus fibers [35]. However, little attention has been given to miscanthus gums in cement-based systems, though initial work by Schmidt et al., has shown that miscanthus gums could reduce the dynamic and static yield stress in cement pastes [17]. In this study minimally processed acacia gums and miscanthus gums are evaluated in limestone slurries using rheological measurements to better understand the behavior of these compounds.

2 Materials and Methods

2.1 Materials

The behavior of bioderived acacia gum and miscanthus gum were evaluated using a shear rheometer to determine the effect of the biopolymers in a particle-liquid systems. For rheological evaluation, slurries were made using limestone powder, deionized (DI) water, and polycarboxylate ethers (PCE). The water to limestone powder ratio was set to 0.22, which corresponds to the minimum water demand according to the Punkte test [36]. The PCE, Melflux AP 101F (BASF; Ludwigshafen, Germany), was used to disperse the slurry and provide a consistency that can be tested in a cylinder system in a rheometer. A hydroxypropylated starch with a degree of substitution of 0.6, was selected so that the behavior of the unmodified bio-based admixtures could be compared to an industrially modified admixture. Admixture dosages and slurry mixture IDs can be found in Table 1.

Acacia gum, in solid form, was ground with a mortar and pestle into a powder. The water-soluble fraction of miscanthus was extracted from 2 cm long pieces of dried miscanthus by soaking in DI water for 72 h. A non-bleached coffee filter, pre-soaked in DI water, was used to separate the liquid and solid fractions. Then the residual miscanthus was soaked for an additional 24 h and, again, the leachate was collected using the same filter. Both leachates were combined and then dried in a 40 °C oven for 24 h, until only the extracted solid fraction remained.

Table 1. Limestone slurry mixture IDs and admixture dosages used to make slurry mixtures for rheological evaluation.

ID	Bio-based admixture	Admixture dose (% limestone mass)	PCE dose (% limestone mass)
REF	–	–	0.062
AC	Acacia gum	0.042	0.062
Mi	Miscanthus gum	0.042	0.062
St6	Hydroxypropylated starch	0.042	0.062
0.5 × St6	Hydroxypropylated starch	0.021	0.062

2.2 Methods

Limestone slurries were mixed using an STARVISC 200–2.5 control measuring mixer (IKA; Staufen, Germany). To mix the slurry, the PCE and limestone powder were combined, then water was poured in the mixture and mixed at 250 RPMs for 45 s. Then the starch, acacia gum, or miscanthus gum were added to the mixture and mixed for a final 45 s at 250 RPM before being loaded into the testing cell. Three replicates were tested for each slurry mixture.

All testing was performed using a concentric cylinder cell (CC27) with cup dimensions of 28.895 mm (d) x 120.2 mm (height) and a bob measured 26.665 mm (d) x

72.5 mm (height) in a MCR 502 WESP shear rheometer (Anton Paar; Graz, Austria). A three-part testing protocol was employed to evaluate the performance of the admixtures dynamic shearing, to indicate different conditions during mixing and casting, and to measure the effect of pre-shearing. The initial pre-shear ramp was performed by linearly increasing shear rate from 1/s to one of three maximum shear rates over a 60 s interval, maintaining the maximum as a constant for 120 s, and then ramping down linearly over a 60 s interval from the maximum shear rate to 1/s (Fig. 1, region I). The maximum shear rates of 50/s, 100/s, or 150/s were measured for each admixture to understand how pre-shearing affected the structural build-up and static yield stress measured in later regions. Step two involved a resting phase with low amplitude oscillation. This oscillation phase was performed with a constant strain rate of 0.008% and a linear angular frequency ramp from 1 rad/s to 10 rad/s (Fig. 1, region II). The constant strain rate of 0.008% was determined in preliminary tests to be within the viscoelastic region of the St6 slurry. Finally, an amplitude sweep was performed with a constant angular frequency of 2 rad/s and a linear shear strain ramp from 0.008% to 0.1% (Fig. 1, region III).

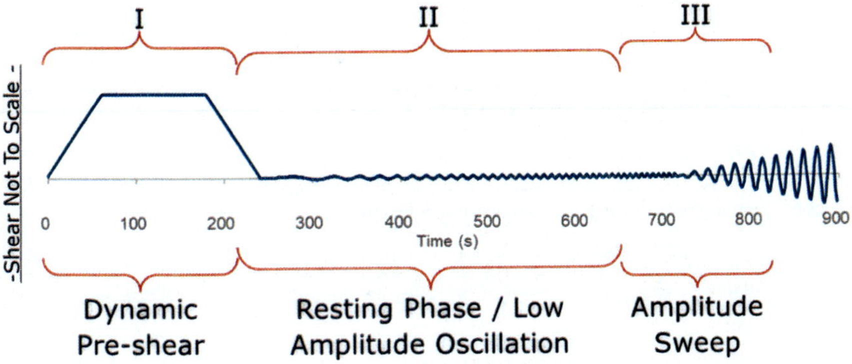

Fig. 1. Illustration of the testing profile for rheological assessment of gums in a limestone slurry showing (I) the dynamic pre-shearing region and (II) the small amplitude oscillatory shear region.

3 Results and Discussion

3.1 Dynamic Pre-shearing

To understand how the studied bio-derived admixture behave in a slurry under dynamic shearing, which can be an indicator of workability properties in a cementitious material, shear stress was measured over a shear ramp (Fig. 2). Increasing the maximum shear rate led to an expected increase in the shear stresses, with St6 showing very large increases across the three shear rates (Fig. 2A-C). During the dynamic pre-shearing region, the shear stresses trends in the Mi and Ac slurries are similar to the lower dosage starch slurry, 0.5 × St6 (Fig. 2A-C). For the REF, Mi, Ac, and 0.5 × St6 slurries, the measured shear stresses for a given strain rate remained similar regardless of the maximum shear rate used. Notably, St6 shows a shear thickening behavior triggered at a shear rate of approximately 50 1/s which increased as shear rate increased (Fig. 3A-B). This led to a large viscosity increase at higher maximum shear rates, which maintained even when the shear rate was reduced below 50/s. On a mass basis, the starch has stronger viscosity modifying affect compared to the other polymers [37]. The strong viscosity increase and shear thickening behavior can be caused by the better dispersed large amylopectin molecules, causing a higher viscosity of the liquid phase and more particle interactions between cement particles, hydrates, and amylopectin, respectively. The hydroxypropylation might improve the viscosity efficiency of starch [38].

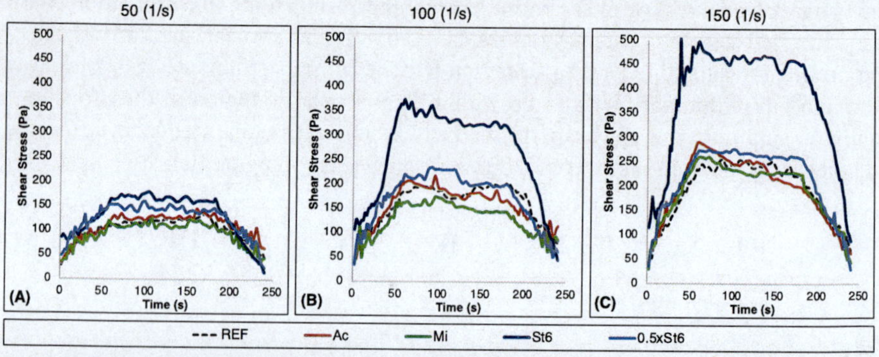

Fig. 2. Shear stress vs. time and shear stress vs. shear rate over the pre-shear ramp phase of testing with A showing results for 50/s maximum shear rate; B showing results for 100/s maximum shear rate; and C showing results for 150/s maximum shear rate.

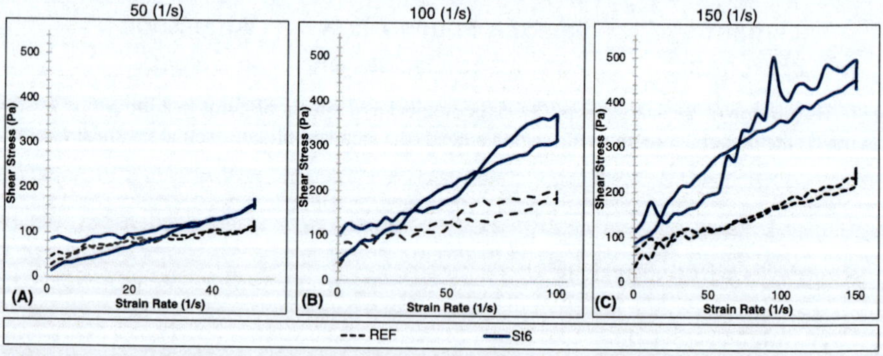

Fig. 3. Shear stress vs strain rate of the full ramp interval in Region I for REF and St6 slurry mixtures.

The relationship between shear stress and strain rate measured during the downward ramp at the end of Region I was observed to follow a linear trend for shear stresses greater than 10 Pa. Thus, a Bingham regression was applied to this region to determine the yield stress and plastic viscosity of the slurry mixtures and these values are presented and discussed below.

3.2 Low Amplitude Oscillation

The slurry mixtures were evaluated with low amplitude oscillation to measure the shear stress and structural build up after pre-shearing at different maximum shear rates (Fig. 4). The control slurry, REF, shows little change with very small decreases as pre-shearing increases. As the pre-shearing increased, St6 showed an increase in shear stress across the shearing conditions. However, 0.5 × St6 shows an initial increase in shear stress values from 50/s to 100/s pre-shearing conditions, but this was followed by lower shear stresses at 150/s. Ac and Mi both show smaller shear stress curves as the pre-shearing increased.

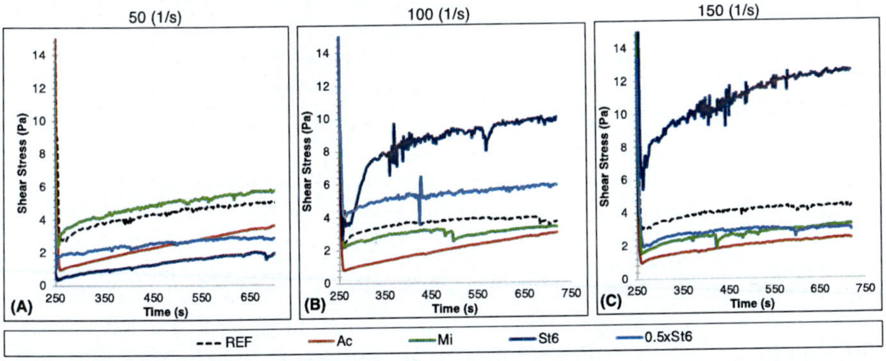

Fig. 4. The shear stress over time of slurry mixtures after (A) 50 /s maximum shear rate, (B) 100 /s maximum shear rate, and (C) 150 /s maximum shear rate.

3.3 Amplitude Sweep

The measurements of the amplitude sweep (Fig. 1, Region III) did not enter the linear viscoelastic regime. Likewise, as the shear stresses in Fig. 4 show a plateauing behavior, with the potential exception of Ac, the shear stresses at the beginning of the Amplitude Sweep (~720 s) can be interpreted as an indicator static behavior. The corresponding storage moduli (G') values were determined and are show in Fig. 6B along with the dynamic yield stress (Fig. 6A) and plastic viscosity (Fig. 6C) determined earlier for the dynamic pre-shearing regime.

For dynamic yield stress, St6 and Mi increased with increased shear rate. REF increases initially approximately 30% from 50/s to 100/s, but then remains the approximately the same at 150/s. Conversely, from 50/s to 100/s Ac shows little change, but then shows a moderate decrease when pre-shearing is increased to 150/s. While the increase in yields stress of Mi appears linear (i.e., increasing approximately 20 Pa for each 50/s increase in shear rate), St6 shows a sharp increase as maximum shear rate increased (i.e., increasing ~ 60 Pa from 50/s to 100/s and then increasing ~ 90 Pa from 100/s to 150/s). However, 0.5 × St6 shows little change across shear rates (ranging from 47–52 Pa) This indicates that the starch dosage influences the observed behavior and that there may be a minimum dose required effectively modify the yields stress. Likewise, higher shear rates appear to improve the effectiveness of the starch. Possibly the starch molecules, when first introduced into the system, are agglomerated and higher shearing aids in breaking up the agglomerants and better dispersing the starch in the system (Fig. 5).

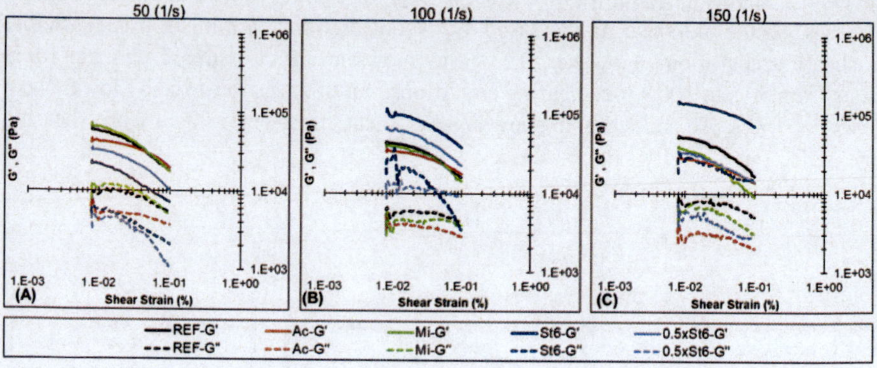

Fig. 5. The storage and loss moduli versus shear strain for the amplitude sweep testing (regime III) for 50/s, 100/s, and 150/s pre-shearing conditions.

As with the results of dynamic yield stress, the Ac and Mi slurries shows a gradual decrease in structural build up at higher pre-shearing rates (Fig. 6B). The REF and 0.5 $\times$ St6 slurries exhibit mix results at different pre-shearing conditions. And St6 shows a sharp increase, indicating a significant increase in structure build for increase pre-shearing conditions, matching the previous observations of improved starch efficiency at higher pre-shearing rates.

The REF, Ac, and Mi slurries, the plastic viscosity showed mixed effects in relation to the shear rate (Fig. 6C). However, both St6 and 0.5 $\times$ St6 slurries showed a decreasing viscosity with increasing maximum shear rate. These differences observed with the strong change of St6 compared to the 0.5 $\times$ St6 (with half the mass of starch) and other admixtures is likely related to the starch having a larger impact in higher solid-volume systems [18]. These results also show that miscanthus gum has plasticizing properties in the system under shear, but at rest leads to structural development like the REF system. Acacia gum, on the other hand, shows an increasing yield stress, but decreases plastic viscosity meaning the structural build up is slower compared to the REF.

4 Outlook

In this work, acacia gum and miscanthus gums are compared to a hydroxypropylated starch in limestone slurries with different maximum pre-shearing rates. Herein, we demonstrate how increasing pre-shearing led to larger storage moduli for the starch admixtures, but also coincided with lower moduli values or mixed effects for the slurries with miscanthus or acacia gums. This shows that higher shearing rates caused the polymers to be more effective in the mixtures. Literature shows that the properties of the polysaccharides and processing (e.g., hydroxypropylation) affect the efficiency of the admixture in the liquid-particle systems. Likewise, higher shearing may help break

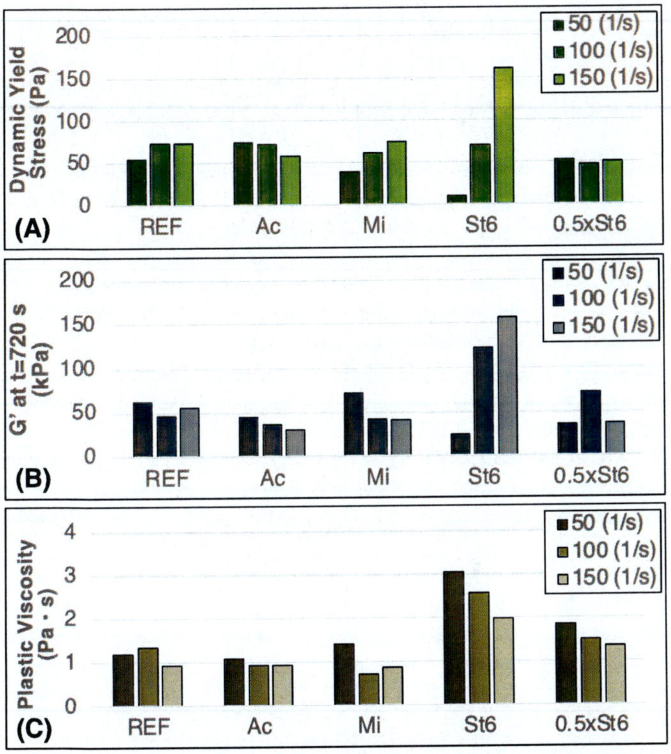

Fig. 6. The (A) yield stress calculated using the Bingham regression for the downward ramp in Region I, (B) the storage modulus at t = 720 s for the tested slurry samples at pre-shearing rates of 50/s, 100/s, and 150/s and (C) plastic viscosity calculated using the Bingham regression for the downward ramp in Region I.

up agglomerated starch and better disperse the starch in a system. At higher shear, miscanthus gums were observed to exhibit plasticizing behaviors and acacia gum increased yield stress. At rest, Mi developed structure similar to the REF slurry, but Ac developed structure slower.

Future work to determine the molecular weight or polymer content of the bioderived admixture could allow for comparison on a polymer weight basis instead of a material mass basis. Additionally, improved processing and purification could increase the effects observed from the bioderived materials. Likewise, observing these admixtures in slurries with ionic solutions or mixtures with cement is imperative future step to identify the behavior of the bioderived admixtures in cement-based systems.

Acknowledgement. P.R.C. gratefully acknowledges financial support by the German-American Fulbright Commission and the Fulbright U.S. Student Program, which is sponsored in part by the U.S. Department of State. The contents of this work are solely the responsibility of the author and do not represent the official views of the Fulbright program, the government of the United States, or the German-American Fulbright Commission.

References

1. Lai, G., et al.: Development of chemical admixtures for green and environmentally friendly concrete: a review. J. Clean. Prod. **389**, 136116 (2023). https://doi.org/10.1016/j.jclepro.2023.136116
2. Kunhi Mohamed, A., Weckwerth, S.A., Mishra, R.K. Heinz, H., Flatt, R.J.: Molecular modeling of chemical admixtures; opportunities and challenges. Cem. Concr. Res. **156**, 106783 (2022). https://doi.org/10.1016/j.cemconres.2022.106783
3. Schmidt, W., Alexander, M., John, V.: Education for sustainable use of cement based materials. Cem. Concr. Res. **114**, 103–114 (2018). https://doi.org/10.1016/j.cemconres.2017.08.009
4. Cheung, J., Roberts, L., Liu, J.: Admixtures and sustainability. Cem. Concr. Res. **114**, 79–89 (2018). https://doi.org/10.1016/j.cemconres.2017.04.011
5. Plank, J., Sakai, E., Miao, C.W., Yu, C., Hong, J.X.: Chemical admixtures - chemistry, applications and their impact on concrete microstructure and durability. Cem. Concr. Res. **78**, 81–99 (2015). https://doi.org/10.1016/j.cemconres.2015.05.016
6. Schmidt, W., Brouwers, H.J.H., Kühne, H.C., Meng, B.: The working mechanism of starch and diutan gum in cementitious and limestone dispersions in presence of polycarboxylate ether superplasticizers. Appl. Rheol. **23**, 1–12 (2013). https://doi.org/10.3933/ApplRheol-23-52903
7. Schmidt, W.: Why Africa can spearhead innovative and sustainable cement and concrete technologies globally. Proc. Second KEYS Symp. Knowl. Exch. Young Sci. (2016)
8. Schmidt, W., Olonade, K.A.: Optimised processes for the production of performance concrete constituents based on agricultural wastes (n.d.)
9. Miller, S.A., Habert, G., Myers, R.J., Harvey, J.T.: Achieving net zero greenhouse gas emissions in the cement industry via value chain mitigation strategies. One Earth. **4**, 1398–1411 (2021). https://doi.org/10.1016/j.oneear.2021.09.011
10. Caltrans, Fly Ash: Current and Future Supply. A Joint Effort Between Concrete Task Group of the Caltrans Rock Products Committee and Industry (2016). https://globalpozz.com/wp-content/uploads/2021/02/Caltrans-Fly-Ash-Future-and-Alternatives.pdf
11. Cunningham,P.R., Wang, L., Thy, P., Jenkins, B.M., Miller, S.A.: Effects of leaching method and ashing temperature of rice residues for energy production and construction materials. ACS Sustain. Chem. Eng. **9**, 3677–3687 (2021). https://doi.org/10.1021/acssuschemeng.0c07919
12. Schmidt, W., Msinjili, N.S., Kühne, H.C.: Materials and technology solutions to tackle the challenges in daily concrete construction for housing and infrastructure in sub-Saharan Africa. African J. Sci. Technol. Innov. Dev. **11**, 401–415 (2019). https://doi.org/10.1080/20421338.2017.1380582
13. Schmidt, W., et al.: Sustainable circular value chains: from rural waste to feasible urban construction materials solutions. Dev. Built Environ. **6**, 100047 (2021). https://doi.org/10.1016/j.dibe.2021.100047
14. Renard, D., Lavenant-Gourgeon, L., Ralet, M.C., Sanchez, C.: Acacia senegal gum: continuum of molecular species differing by their protein to sugar ratio, molecular weight, and charges. Biomacromol **7**, 2637–2649 (2006). https://doi.org/10.1021/bm060145j
15. Sanchez, C., et al.:Acacia gum: history of the future. Food Hydrocoll. **78**, 140–160 (2018). https://doi.org/10.1016/j.foodhyd.2017.04.008
16. Elmanan, M., Al-Assaf, S., Phillips, G.O., Williams, P.A.: Studies on acacia exudate gums: Part VI. Interfacial Rheol. Acacia Senegal Acacia Seyal, Food Hydrocoll. **22**, 682–689 (2008). https://doi.org/10.1016/j.foodhyd.2007.02.008
17. Schmidt, W.: Effect of calcium ions on bio-based polymers for rheology control. Am. Concr. Institute, ACI Spec. Publ. **SP-354**, 75–86 (2022). https://doi.org/10.14359/51736063

18. Schmidt, W., Brouwers, H.J.H., Kühne, H.-C., Meng, B.: Interactions of polysaccharide stabilising agents with early cement hydration without and in the presence of superplasticizers. Constr. Build. Mater. **139**, 584–593 (2017). https://doi.org/10.1016/j.conbuildmat.2016.11.022
19. Schmidt, W., et al.: Innovation potentials for construction materials with specific focus on the challenges in Africa. RILEM Tech. Lett. **5**, 63–74 (2020). https://doi.org/10.21809/rilemtechlett.2020.112
20. Schmidt, W., Ngassam, I.L.T., Olonade, K.A., Mbugua, R., Kühne, H.C.: Plant based chemical admixtures – potentials and effects on the performance of cementitious materials, RILEM Tech. Lett. **3**, 124–128 (2018). https://doi.org/10.21809/rilemtechlett.2018.83
21. Mbugua, R., Salim, R., Ndambuki, J.: Effect of gum arabic karroo as awater-reducing admixture in concrete. Materials (Basel). **9**, 1–16 (2016). https://doi.org/10.3390/ma9020080
22. Elinwa, A.U., Umar, M.: X-ray diffraction and microstructure studies of gum Arabic-cement concrete. Constr. Build. Mater. **156**, 632–638 (2017). https://doi.org/10.1016/j.conbuildmat.2017.08.162
23. Schmidt, W., Olonade, K.A., Mbugua, R.N., Lenz, F.J., Ngassam, I.T.: Bio-based rheology modifiers for high performance concrete – possible modes of actions and case study for cassava starch in West Africa. In: Boshoff, W.P., Combrinck, R., Mechtcherine, V., Wyrzykowski, M. (eds.) SAP 2019. RB, vol. 24, pp. 158–166. Springer, Cham (2020). https://doi.org/10.1007/978-3-030-33342-3_17
24. Zhao, C., Zhao, Q., Zhang, Y., Zhou, M.: The effect of gum Arabic on the dispersion of cement pastes. In: Dong, F. (ed.) Proceedings of the 11th International Congress for Applied Mineralogy (ICAM). SG, pp. 483–494. Springer, Cham (2015). https://doi.org/10.1007/978-3-319-13948-7_47
25. Brosse, N., Dufour, A., Meng, X., Sun, Q., Ragauskas, A.: Miscanthus : a fast-growing crop for biofuels and chemicals production, Biofuels. Bioprod. Biorefining. **6**, 580–598 (2012). https://doi.org/10.1002/bbb.1353
26. Sarker, T.R., Pattnaik, F., Nanda, S., Dalai, A.K., Meda, V., Naik, S.: Hydrothermal pretreatment technologies for lignocellulosic biomass: a review of steam explosion and subcritical water hydrolysis. Chemosphere **284**, 131372 (2021). https://doi.org/10.1016/j.chemosphere.2021.131372
27. Zhang, J., Cai, D., Qin, Y., Liu, D., Zhao, X.: High value-added monomer chemicals and functional bio-based materials derived from polymeric components of lignocellulose by organosolv fractionation, Biofuels. Bioprod. Biorefining. **14**, 371–401 (2020). https://doi.org/10.1002/bbb.2057
28. Baruah, J., et al.: Recent trends in the pretreatment of lignocellulosic biomass for value-added products. Front. Energy Res. **6**, 141 (2018). https://doi.org/10.3389/fenrg.2018.00141
29. Serrano, L., Egües, I., Alriols, M.G., Llano-Ponte, R., Labidi, J.: Miscanthus sinensis fractionation by different reagents. Chem. Eng. J. **156**, 49–55 (2010). https://doi.org/10.1016/j.cej.2009.09.032
30. Irmak, S., Meryemoglu, B., Sandip, A., Subbiah, J., Mitchell, R.B., Sarath, G.: Microwave pretreatment effects on switchgrass and miscanthus solubilization in subcritical water and hydrolysate utilization for hydrogen production. Biomass Bioenerg. **108**, 48–54 (2018). https://doi.org/10.1016/j.biombioe.2017.10.039
31. Yadav, M.P., Hicks, K.B.: Isolation, characterization and functionalities of bio-fiber gums isolated from grain processing by-products, agricultural residues and energy crops. Food Hydrocoll. **78**, 120–127 (2018). https://doi.org/10.1016/j.foodhyd.2017.04.009
32. Rivas, S., Vila, C., Alonso, J.L., Santos, V., Parajó, J.C., Leahy, J.J.: Biorefinery processes for the valorization of Miscanthus polysaccharides: from constituent sugars to platform chemicals. Ind. Crops Prod. **134**, 309–317 (2019). https://doi.org/10.1016/j.indcrop.2019.04.005

33. Periyasamy, S., et al.: Chemical, physical and biological methods to convert lignocellulosic waste into value-added products. a review. Environ. Chem. Lett. **20**(2), 1129–1152 (2022). https://doi.org/10.1007/s10311-021-01374-w
34. Barba, C., Montané, D., Farriol, X., Desbrières, J., Rinaudo, M.: Synthesis and characterization of carboxymethylcelluloses from non-wood pulps II. Rheol. Behav. CMC Aqueous Solution, Cellul. **9**, 327–335 (2002). https://doi.org/10.1023/A:1021136626028
35. El Achaby, M., et al.: Cellulose nanocrystals from Miscanthus fibers: insights into rheological, physico-chemical properties and polymer reinforcing ability. Cellulose **25**(11), 6603–6619 (2018). https://doi.org/10.1007/s10570-018-2047-1
36. Puntke, W.: Wasseranspruch von feinen Kornhaufwerken. Beton-Dusseldorf. **5**, 242–249 (2002)
37. Zhang, K., Schmidt, W., Mezhov, A.: Influence of the Hydroxypropylation of starch on its performance as viscosity modifying agent. In: Superplast. Other Chem. Admixtures Concrete – Conference on Proceedings of the, American Concrete Institute, SP-354, pp. 209–218 (2022). https://doi.org/10.14359/51736074
38. Vieira, M.C., Klemm, D., Einfeldt, L., Albrecht, G.: Dispersing agents for cement based on modified polysaccharides. Cem. Concr. Res. **35**, 883–890 (2005). https://doi.org/10.1016/j.cemconres.2004.09.022
39. Zhang, K., Mezhov, A., Schmidt, W.: Chemical and thixotropic contribution to the structural build-up of cementitious materials. Constr. Build. Mater. **345**, 128307 (2022). https://doi.org/10.1016/j.conbuildmat.2022.128307

Availability and Reactivity of Agricultural Bio-Based Mineral Additives Originating from Vojvodina, Serbia

Slobodan Šupić(✉), Vladan Pantić, Mirjana Malešev, Vlastimir Radonjanin, and Ivan Lukić

Faculty of Technical Sciences, University of Novi Sad, Novi Sad, Serbia
ssupic@uns.ac.rs

Abstract. The present study focuses on evaluating seven different agricultural biomass ashes, generated by combustion of wheat straw, soya straw, cob corn, silo waste and oil rapeseed, as sustainable alternatives to carbon-intensive conventional cement. A detailed characterization was carried out from a chemical, mechanical and physical standpoint, aimed at exploring their potential as supplementary cementitious materials (SCM). The results indicated that all the ashes, where wheat straw and cob corn are the prevailing biomass sources, exhibited greater pozzolanic behavior and higher activity index, depending on the amorphous silica content and level of fineness. Furthermore, the influence of two optimal types of biomass ashes, selected based on the characterization results, on the tested mechanical and durability properties of cement-based mortars is also presented. The results indicate that mortars blended with a high volume of biomass ashes had comparable mechanical properties with reference values, owing to their pozzolanic activity, increased number of nucleation sites for cement hydration and the filler (packing) effect. However, due to the dilution effect, higher water absorption was registered for all blended mortars, pointing to the need for mix design optimization by lowering the effective water-to-binder ratio.

Keywords: Biomass Ash · Characterization · Wheat Straw · Cob Corn · Mortar · Pozzolanic Activity · Sustainable Alternatives

1 Introduction

The idea of sustainability and green production in the construction industry encourages the use of various waste materials to replace depleted natural resources. In an endeavor to minimize the carbon footprint, researchers all around the globe are exploring various alternative materials, generally rich in reactive silica that can be employed as supplementary cementitious materials (SCMs). So far, extensive research has been conducted on effective utilization of granulated blast furnace slag (GBFS), metakaolin (MK), silica fume (SF) and fly ash (FA), among which the application of FA and SF has been standardized.

S. Amziane et al. (Eds.): ICBBM 2023, RILEM Bookseries 45, pp. 793–803, 2023.
https://doi.org/10.1007/978-3-031-33465-8_61

The use of agricultural biomass ashes may be an important alternative in replacing traditional SCMs. Globally, the cob and straw are the main biomass wastes, which can generate a significant amount of ashes with pozzolanic potential when adequately combusted. The pozzolanic activity of biomass ash is associated with the amount of reactive silica, which, in turn, is related to biomass calcination [1]. Another factor that influences pozzolanic activity is the level of fineness of biomass ash. Different grinding procedures have been applied to adjust the particle size of these materials, enhance their reactivity and increase homogeneity [2, 3]. So far, the main focus of numerous researchers has been on exploring the possibility of rice husk ash (RHA) [4], sugar cane straw ash (SCSA) [5] and palm oil fuel ash (POFA) [6] application as SCMs, especially RHA. Few studies have provided the information on less utilized harvest residues ashes such as wheat straw ash (WSA) [7] and corn cob ash (CCA) [8].

Growing quantities of harvest residue ashes are created as waste products during combustion, pyrolysis, and gasification of agricultural biomass in Vojvodina - the northern province of Serbia - currently estimated at 5000 tons per year. Basic data on the ashes, including origin, type of biomass used as an energy source, temperature of combustion and available quantities are provided in Table 1.

These ashes are most commonly disposed of in open landfills or recycled on agricultural farms. Exploring the possibility of the utilization of ashes in engineering practice in Serbia has been recently initiated. Few earlier studies are available on the effect of wheat straw and soya straw ash incorporation on the properties of blended cement mortar [9] and masonry mortar [10]. No studies referring to other biomass ashes application in cement-based composites in Serbia can be found. Hence, the aim of this paper is to: a) assess the performances of available biomass ashes through characterization, which includes testing all relevant chemical, physical, mechanical, and pozzolanic properties, b) investigate the feasibility of the selected ashes employed as partial substitutes of cement up to 50% to prepare blended cement mortars.

2 Materials and Methods

2.1 Materials

Ordinary Portland cement (OPC), originating from Lafarge cement factory in Beočin, Serbia, was used. The cement has a specific gravity of 3.1 g/cm^3 and the Blaine fineness of 4.000 cm^2/g.

For experimental investigation of chemical and physical properties and pozzolanic activity of biomass ashes, the following materials were collected and tested:

- Wheat straw ash (WSA), "Mitrosrem" Sremska Mitrovica,
- Mixed wheat and soya straw ash (WSSA), "Soya-protein" Bečej,
- Cob corn ash (CCA), "IPOK" Zrenjanin,
- Soya straw ash (SSA), "HIPOL" Odžaci,
- Silo waste ash (SWA), "Soya-protein" Bečej,
- Oil rapeseed ash (ORA), "Knot-Autoflex" Bečej,

Table 1. Available quantities of biomass ashes in AP Vojvodina

Company	Biomass type	Temperature of combustion	Types of biomass ashes	Annual production (tons)
Mitrosrem Sremska Mitrovica	wheat straw soya straw	600–650 °C	ash from boiler furnace ash from multicyclone fly biomass ash	15
Soya Protein Bečej	wheat straw soya straw silo waste melasa	700–900 °C	ash from boiler furnace ash from multicyclone fly biomass ash	1100
Hipol Odžaci	soya straw	800–1000 °C	ash from boiler furnace ash from multicyclone fly biomass ash	700
Victoria Starch Zrenjanin	wheat straw soya straw	unknown	ash from boiler furnace	9
Almex-IPOK Zrenjanin	wheat straw soya straw corn cob	700–900 °C	ash from boiler furnace ash from multicyclone fly biomass ash	1100
Knot-Autoflex Bečej	wheat straw soya straw	unknown	ash from boiler furnace	60
DTD Fishery Lovćenac	soya straw	unknown	ash from boiler furnace ash from multicyclone	9
Victoria Oil Šid	sunflower husk	700–1000 °C	ash from boiler furnace ash from multicyclone	720
Sava Kovačević Vrbas	cob corn	500 °C	ash from boiler furnace	30
PTK Panonija Mecker farm	wheat straw soya straw	500 °C	ash from boiler furnace	60
Total				**≈ 5.000**

- Mixed wheat straw and sunflower husk ash (WSSHA), "Soya-protein".

The ashes were roughly sieved, through a 4 mm sieve, to separate un-burnt straw and other large impurities. In order to obtain a material with satisfactory level of fineness, ashes were further ground in a laboratory ball mill for 6 h.

2.2 Methods

The chemical composition of collected ashes was determined using energy dispersive X-ray fluorescence spectrometer (EDXRF 2000 Oxford instruments) according to EN 196-2, 2015 and ISO 29581-2, 2010. The representative samples (100 g) were pulverized in a laboratory vibratory mill prior to the testing. The loss on ignition (LOI) was determined as a weight difference between 20 °C and 950 °C.

Specific surface area of biomass ashes was determined according to Blaine air permeability method given in EN 196-6, 2011, which is widely used for the fineness determination of hydraulic cement. The test is based on the principle of resistance to air flow through a partially compacted sample of powder material.

Soundness of biomass ashes was determined in accordance with EN 196-3, 2010. The method is used for assessing whether this physical property of a SCM material is in conformity with the requirements given in EN 450-1.

The pozzolanic activity was studied on samples prepared according to the procedure given in SRPS B.C1.018, 2015. Mortars were prepared with biomass ash, slaked lime and standard sand, with the following mass proportions: m_{sl}:m_{bash}:m_{qs} = 1:2:9 and water – binder ratio 0.6 (where: m_{sl} – mass of slaked lime; m_{bash} – mass of biomass ash; m_{qs} – mass of CEN standard sand). After compacting, the samples were hermetically sealed and cured for 24 h at 20 °C, then for 5 days at 55 °C. Subsequently, 24 h period was allowed for samples cooling process to reach 20 °C, followed by compressive and flexural strength tests.

The activity index of biomass ashes was examined according to EN 450-1, 2014. Activity index is defined as a ratio (in percent) of the compressive strength of mortar prepared with 75% test cement plus 25% ash by mass, and the compressive strength of standard cement mortar, when tested at the same age. The preparation of mortar test specimens and determination of the compressive strength were carried out in accordance with EN 196-1, 2018.

The consistency of fresh mortar was determined by flow table test and expressed as a measure of horizontal spread that a cement mortar experiences due to successive dynamic impacts, in accordance with EN 1015-3.

Compressive strength of mortar was tested according to EN 196-1:2008. After casting, the samples were cured in tap water until the testing age.

The capillary water absorption of mortars has been measured according to ASTM C1585. The method is used to determine the rate of water absorption by measuring the increase in the mass of a specimen resulting from absorption as a function of time when only one surface of the specimen is exposed to water.

3 Characterization of Biomass Ashes

3.1 Chemical Composition

The results of testing the chemical composition of biomass ashes, after the grinding procedure in a laboratory ball mill, are summarized in Table 2.

Table 2. Chemical composition of OPC and biomass ashes

Material (%)	SiO_2	Al_2O_3	Fe_2O_3	CaO	MgO	Na_2O	K_2O	SO_3	P_2O_5	LOI
Cement	21.25	5.55	25.3	61.6	2.37	0.25	0.72	3.75	0.00	1.77
WSA	69.13	1.12	0.73	5.78	2.50	0.11	13.03	0.20	1.72	5.18
WSSA	56.36	2.03	1.53	7.13	3.54	0.20	20.02	0.18	3.72	4.85
CCA	45.76	5.92	3.38	14.08	8.30	0.00	13.10	1.26	2.81	2.40
SSA	32.62	4.58	1.46	15.78	8.33	0.85	20.96	0.47	3.72	10.73
SWA	43.66	5.41	2.06	15.89	5.05	0.97	14.09	0.76	4.82	6.41
ORA	32.37	6.32	2.45	14.00	2.91	0.89	17.85	3.78	1.44	17.61
WSSHA	44.12	1.99	1.93	10.94	4.91	0.26	16.24	0.45	5.43	13.05

Summarized results of important chemical properties, requirements in relevant standards, as well as criteria fulfillment are given in Table 3. As the European standard for biomass ashes has not been established yet, the assessment of properties was executed as stated in EN450-1.

Table 3. Chemical composition - criteria fulfillment

Material (%)	Total amount of oxides $SiO_2 + Al_2O_3 + Fe_2O_3$ ($\geq$70%)	Free CaO ($\leq$1.5%)	Reactive SiO2 ($\geq$25%)	Total amount of alkalis $Na_2O + 0.658\ K_2O$ ($\leq$5%)
WSA	70.98	0.14	67.13	8.68
WSSA	59.92	0.28	41.23	13.37
CCA	55.05	11.34	35.19	8.62
SSA	38.66	0.39	25.03	14.64
SWA	51.13	1.29	25.14	10.24
ORA	41.14	1.62	15.26	12.63
WSSHA	48.04	3.12	35.18	10.95

The results demonstrate that the chemical composition of different HRA has large variations and this is due to the highly variable contents of bulk inorganic matter and different genetic classes of inorganic matter in biomass varieties.

WSA-based ashes contain considerable levels of important oxides, whereas only WSA satisfies the criterion for total amount of SiO_2, Al_2O_3 and Fe_2O_3. As the share of wheat straw in biomass decreases, amorphous silica content in ashes decreases along. Apart from WSA-based ashes, CCA, SWA and ORA satisfy the criterion for amorphous silica content (>25%), while ORA lacks reactive silica in its chemical compositions, which is expected to influence its pozzolanic activity.

All ashes exceed the limit value for the total amount of alkalis, which could trigger an alkali-aggregate reaction (ASR). ASR refers to a deleterious reaction that occurs over time in cement-based composites between the highly alkaline cement paste and the reactive silica in aggregates. This mechanism should be experimentally verified in cementitious systems incorporating biomass ashes for each type of aggregate.

CCA exceeds the limit for free lime content. Free lime may trigger undesired expansion and volume instability when used in concrete as a cement replacement. A greater share of free CaO is expected to be spent within hydration, and, in addition, a re-carbonation reaction can take place due to the reaction of CaO and CO_2 forming $CaCO_3$. Notwithstanding this expectation, durability issues and time-dependent properties of concrete containing SCMs with elevated percentages of free lime should be taken into account.

3.2 Physical Properties

The physical and pozzolanic properties of raw materials are listed in Tables 4 and 5.

Table 4. Physical and pozzolanic properties of biomass ashes

	Criteria	WSA	WSSA	CCA
Specific gravity (g/cm^3)	/	2.38	2.37	2.44
Specific surface area (cm^2/g)	/	9500	15550	8090
Pozzolanic activity	Class 5: $f_c \geq 5MPa$ $f_{fl} \geq 2MPa$ Class 10: $f_c \geq 10MPa$ $f_{fl} \geq 3MPa$ Class 15: $f_c \geq 15MPa$ $f_{fl} \geq 4MPa$	Class 10	Class 5	Class 5
Activity index	$AI_{28} \geq 75\%$ $AI_{90} \geq 85\%$	$AI_{28} = 104\%$ $AI_{90} = 108\%$	$AI_{28} = 92\%$ $AI_{90} = 92\%$	$AI_{28} = 102\%$ $AI_{90} = 115\%$

All tested biomass ashes are characterized by a greater level of fineness in relation to reference cement, as a consequence of an efficient grinding procedure. However, significant variations in specific surface area can be registered amongst different samples, which may be attributed to different combustion technologies and levels of fineness of

Table 5. Physical and pozzolanic properties of biomass ashes

	SSA	SWA	ORA	WSSHA
Specific gravity (g/cm^3)	2.4	2.39	2.32	2.18
Specific surface area (cm^2/g)	5200	11800	6100	10650
Pozzolanic activity	Insufficient	Class 5	Insufficient	Insufficient
Activity index	$AI_{28} = 70\%$ $AI_{90} = 69\%$	$AI_{28} = 90\%$ $AI_{90} = 100\%$	$AI_{28} = 86\%$ $AI_{90} = 85\%$	$AI_{28} = 76\%$ $AI_{90} = 85\%$

raw materials. In general, the higher level of fineness is expected to induce a higher pozzolanic activity.

The main contributing factor to the pozzolanic activity of material is amorphous silica content. In addition, as previously mentioned, the decrease in particle size can additionally improve the activity. As can be seen from the Table shown above, WSA displays a pozzolanic activity of Class 10, WSSA, CCA and SWA demonstrates a pozzolanic activity of Class 5, while the other tested samples manifest an insufficient activity to achieve this type of class.

According to the criteria given in EN 450-1, the activity index at 28 days and at 90 days shall not be less than 75% and 85%, respectively. Only SSA sample does not meet these requirements. The activity index for all other materials at the age of 28 days exceeds 75%, which classifies these materials as pozzolans. As expected, these SCMs showed a consistent increase in strength as the curing period elapsed, which is a direct consequence of a delayed pozzolanic reaction between portlandite from the hydration of Portland cement with silica from the SCM and the subsequent C-S-H gel formation. Noteworthy, WSA and CCA fulfilled not only the prescribed criteria but they achieved compressive strength higher than that of the reference mortar, which may be ascribed to sufficient reactive silica content and the filler effect of small particles of these biomass ashes.

Based on the obtained results, WSA and CCA were chosen as cement substitutes in the production of cement-based mortars.

4 Properties of Cement Mortars Blended with Biomass Ashes

4.1 Composition of Mortar

The second stage of experimental study was carried out on seven different mortar mixtures. Composition of mixtures was defined in accordance with standard EN 196-1: ratio (by mass) of cement and aggregate was 1:3, while water to binder ratio (w/b) was 0,5. Reference mortar (C) was prepared with cement as binder. In the six remaining mixtures, 10%, 30% and 50% of cement were replaced by WSA or CCA, by mass.

4.2 Consistency of Fresh Mortar

Consistency of fresh mortars, determined by flow table, is shown in Fig. 1.

From the results, it can be clearly noted that the flow value gradually reduced with the increments of the replacement level. This condition may be caused by the high reactivity and higher surface area of WSA and CCA when compared to OPC.

Nevertheless, replacement of cement up to 30%, didn't have a remarkable impact on the workability of mortar. In mixtures with 50% replacement level, superplasticizer has been used for flow correction, in amount of 0,2% (by mass of cement).

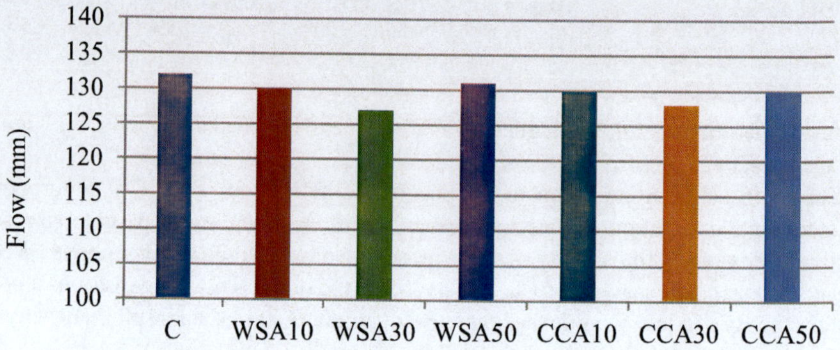

Fig. 1. Consistency of fresh mortars.

4.3 Compressive Strength of Mortar

The compressive strengths of mortars after 28 and 90 days are shown in Figs. 2 and 3.

As displayed in Figures above, the compressive strength of the mortars blended with chosen SCMs was slightly lower than that of the reference mortar at 28 days. This is because their contribution to the compressive strength is not as significant as that of cement early on, due to prolonged pozzolanic reaction of biomass ashes. The only exception is corn cob based mortar with 10% cement substitution, which exceeded the strength of the reference mortar. At this stage, this may be attributed to better pore refinement and higher compactness of the matrix due to the packing (filler) effect of small ash particles.

Over time, a steady strength increase has been registered in all blended mortars. Substantial gains were observed for WSA-blended mortars, as they attained a higher strength value than the reference mortar at 90 days, regardless of the substitution level. This is a direct consequence of the delayed pozzolanic reaction, as reactive silica consumes portlandite and produces more C-S-H gel which increases the strength of mortar samples. CCA-blended mortars generally demonstrated a compressive strength comparable to the reference sample, while CCA10 exceeded that of the reference one.

4.4 Capillary Water Absorption of Mortar

The capillary water absorption of all the mortars prepared with varying contents of the selected SCMs was tested after the curing period of 28 days. The kinetics of capillary

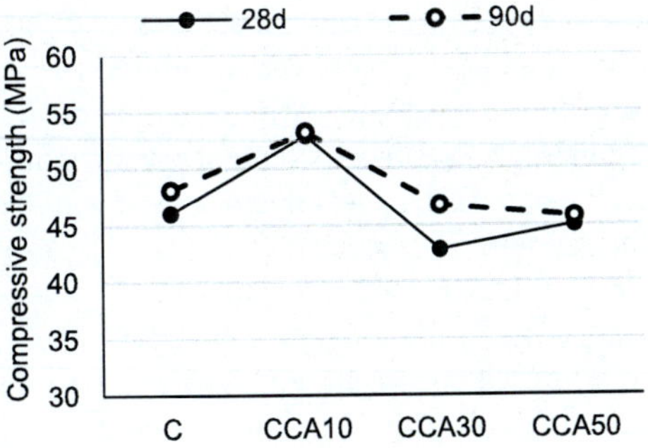

Fig. 2. Compressive strength of mortars blended with CCA.

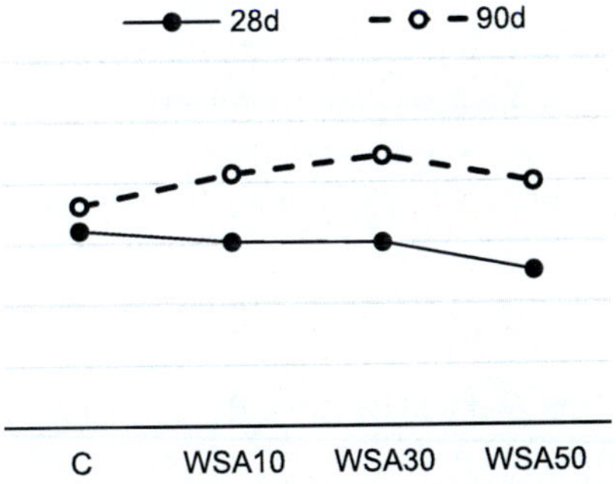

Fig. 3. Compressive strength of mortars blended with WSA.

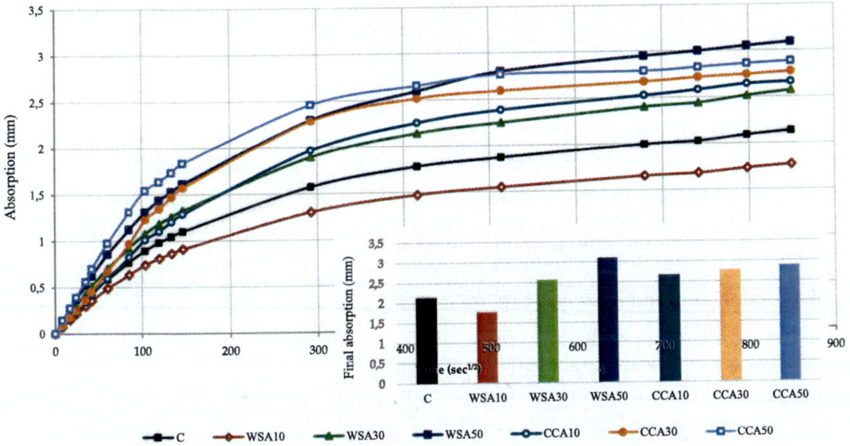

Fig. 4. Capillary water absorption: the kinetics and the final values.

water absorption and the final absorption values for all the mortars prepared with varying contents of the selected SCMs are demonstrated in Fig. 4.

Water to binder ratio (w/b) is one of the key characteristics of cement-based materials, influencing the pore structure and capillary water absorption characteristics. Due to the dilution effect, as the substitution level increased, cement content decreased, while the effective w/b rose correspondingly, leading to the higher absorption – Table 6. As a result, all blended mortars, apart from WSA10, exhibited greater capillary water absorption capacity in relation to the reference sample.

Effective w/b ratios were calculated according to the k-value concept, established for fly ash. The k-value concept, conforming to EN 206-1, permits fly ash to be taken into account by replacing the term "water/cement ratio" with "water/(cement + k × fly ash) ratio". The amount of fly ash that may be taken into account must meet the requirement: fly ash/cement $\leq$ 0.33 by mass. If a greater amount of fly ash is used, the excess shall not be taken into account for the calculation of the water/(cement + k × fly ash) ratio. For concrete containing cement type CEM I, the permitted k-value is 0,4.

Table 6. Effective w/b ratio

	C	WSA10	WSA30	WSA50	CCA10	CCA30	CCA50
Effective w/b	0,5	0,532	0,631	0,844	0,532	0,631	0,844

As shown in Table 6, effective w/b ratios, and consequently, final water absorption values, increased along with the higher cement replacement level. The explanation for this finding is that the total porosity of mortar increased; hence its ability to absorb and accommodate water increased, which increased the water absorption capacity of the cement blends.

5 Conclusions

The aim of this study was to analyze the ashes generated from agricultural biomass combustion, locally available in Vojvodina province in Serbia, as potential SCMs. These waste materials are generated at large and small agricultural farms, derived as by-products within biomass combustion for the purpose of heat energy production. Based on the conducted survey, the following conclusions can be drawn:

- Ashes based on wheat straw and cob corn are good SCM candidates for cement substitution, owing to high amorphous silica content and satisfactory level of fineness (obtained by an appropriate grinding procedure),
- The use of high volume CCA and WSA as a partial substitute for cement decreased the workability. The mortars containing selected SCMs up to 30% maintained satisfactory workability, while the increased water demand due to the addition of higher shares of biomass ashes was compensated by the use of superplasticizer,
- As a result of pozzolanic activity and enhanced compactness (filler effect), all blended mortars displayed satisfactory compressive strength, comparable to the strength of the reference one,

- With increasing proportion of biomass ashes, the capillary water absorption increased, as a consequence of higher effective w//b ratio and greater porosity. Higher replacement level could potentially be achieved by lowering effective water to binder ratio, while maintaining the same consistency of the mortar (by using chemical admixtures), which is a further step of the experimental research.

Acknowledgement. The research has been conducted within the project "Scientific theoretical and experimental research and improvement of educational process in the field of civil engineering", developed at the Department of Civil Engineering and Geodesy, Faculty of Technical Sciences, University of Novi Sad, Serbia.

References

1. Cordeiro, G.C., Sales, C.P.: Influence of calcining temperature on the pozzolanic characteristics of elephant grass ash. Cem. Concr. Compos. **73**, 98–104 (2016)
2. Rithuparna, R., Jittin, V., Bahurudeen, A. Influence of different processing methods on the recycling potential of agro-waste ashes for sustainable cement production: a review. J. Clean. Prod. **31** (2021)
3. Cordeiro, G.C., Toledo Filho, R.D., Tavares, L.M., Fairbairn, E.D.M.R.: Ultrafine grinding of sugar cane bagasse ash for application as pozzolanic admixture in concrete. Cem. Concr. Res. **39**, 110–115 (2009)
4. Mosaberpanah, M.A., Umar, S.A.: Utilizing rice husk ash as supplement to cementitious materials on performance of ultra high performance concrete: – a review. Mater. Today Sustain. 7–8 (2020)
5. Wu, N., Ji, T., Huang, P, Fu, T., Zheng, X., Xu, Q.: Use of sugar cane bagasse ash in ultra-high performance concrete (UHPC) as cement replacement. Const. Build. Mater. **317** (2022)
6. Santosh, K.G., Subhani, Sk.M., Bahurudeen, A.: Sustainable reuse of palm oil fuel ash in concrete, alkali-activated binders, soil stabilisation, bricks and adsorbent: a waste to wealth approach. Industrial Crops and Products **183** (2022)
7. Šupić, S., Malešev, M., Radonjanin, V., Bulatović, V., Milović, T.: Reactivity and pozzolanic properties of biomass ashes generated by wheat and soybean straw combustion. Materials **14** (2021)
8. Shakouri, M., Exstroom, C.L., Ramanathan, S., Suraneni, P.: Hydration, strength, and durability of cementitious materials incorporating untreated corn cob ash. Const. Build. Mater. **243** (2020)
9. Šupić, S., Malešev, M., Radonjanin, V., Radeka, M., Laban, M.: Application of biomass ashes as supplementary cementitious materials in the cement mortar production. World academy of science, engineering and technology. Int. J. Struct. Constr. Eng. **12** (2018)
10. Šupić, S., Bulatović, V., Malešev, M., Radonjanin, V., Lukić, I.: Sustainable masonry mortars with fly ash, blast furnace granulated slag and wheat straw ash. Sustainability **13** (2021)

Mechanical and Hygrothermal Properties of Cement Mortars Including Both Phase Change Materials and Miscanthus Fibers

Franck Komi Gbekou[1](✉), Abderrahim Boudenne[2], Anissa Eddhahak[3], and Karim Benzarti[1](✉)

[1] Univ Gustave Eiffel, Ecole Des Ponts, CNRS, Navier, 77447 Marne La Vallée, France
{komi.gbekou,karim.benzarti}@univ-eiffel.fr
[2] Univ Paris Est Créteil, CERTES, 94010 Créteil, France
[3] Arts Et Métiers ParisTech, PIMM, 75013 Paris, France

Abstract. This work investigates the mechanical and hygrothermal properties of hybrid cement mortar including both microencapsulated phase change materials and micronized miscanthus fibers. An experimental protocol was first developed for preparing the hybrid mortar mix, which allows to maintain the same workability as that of the reference mortar, and to ensure homogeneous distribution of the two types of inclusions within the cement matrix. This hybrid mortar was then characterized at the age of 28 days, in terms of mechanical strength, thermophysical properties (determination of the thermal conductivity/diffusivity by the Hot-Disk method), and hygric properties (evaluation of the Moisture Buffer Value – MBV, according to the NordTest method). Overall, the hybrid mortar showed a significant decrease in thermal conductivity compared to the reference mortar (−68%), suggesting it is a good insulation material. Conversely, the MBV was only slightly improved compared to the reference mortar and remained much below that of a mortar filled with the same amount of miscanthus fibers but without phase change material. Furthermore, a large drop of mechanical properties was observed with respect to the reference (−91%). Hence, the interest of the hybrid solution is questionable, as the presence of phase change materials seems to hinder the beneficial effect of vegetal fibers on the hygroscopic behavior. In the present case, the introduction of single vegetal fibers in the mortar remained the best option for improving both thermal performance and moisture regulation capacity at material scale.

Keywords: Cement Mortar · Vegetal Fibers · Phase Change Materials · Hygrothermal Properties

1 Introduction

The building sector was responsible for ~ 30% of the world's global energy consumption and 27% of total $C0_2$ emissions in 2021 [1], which is partially due to the growing use of energy-consuming devices for temperature regulation in hot and cold regions. Therefore, improving the energy efficiency of buildings has become a major challenge in the current context of climate change and global warming.

S. Amziane et al. (Eds.): ICBBM 2023, RILEM Bookseries 45, pp. 804–816, 2023.
https://doi.org/10.1007/978-3-031-33465-8_62

To address this issue, public policies promote the development of innovative insulation building material which provide significant gain in terms of thermal performances and comfort for building users.

Over the last decade, a lot of research has been focused on building materials including either phase change materials (PCMs) or plant fibers, which are considered as promising solutions.

On one hand, thermal energy storage systems based on PCMs allow to store a large amount of energy in the form latent heat when the temperature increases over the melting point of the PCMs, and to release this energy when the temperature decreases. This capacity to absorb/release energy can be used to regulate the temperature variations of the indoor environment of buildings [2]. An attractive way to incorporate PCMs into conventional building materials like cement mortars or concrete is to use microencapsulated PCMs (noted mPCMs in the following). Numerous studies have shown that the addition of mPCMs can decrease the thermal conductivity of mortars very significantly [3–6], sometimes by more than 77%, and leads to an increase in the thermal mass as well [6]. Nevertheless, such improvements are generally obtained at the expense of mechanical properties, and the dosage of mPCMs must be tuned to obtain an acceptable balance depending on the target application [5].

On the other hand, many authors have reported that the incorporation of plant fibers (hemp, cork, rice husk, date palm fibers, etc.) in ultra-lightweight concretes does not only lead to significant improvement in the thermal performance [7–9], but also provides superior moisture absorption capacity [10–13]. This latter effect is of great interest since the thermal comfort of building occupants is also favored by good moisture regulation of the indoor air.

Conversely, very few studies have investigated the effect of introducing together mPCMs and vegetal fibers into cement mortars, with the objective to combine the benefits of both fillers [14]. Therefore, the present work explores the combined use of phase change materials (in the form of an aqueous dispersion of bio-based mPCM particles) and vegetal fibers (in the form of a micronized powder of miscanthus fibers) in a mortar formulation. In a first stage, an experimental protocol was developed to introduce the two types of fillers in the mortar mix, while ensuring both a homogenous distribution of those inclusions in the cement matrix and a constant workability of the mortar. Then, after hardening, this hybrid mortar was characterized following existing standards to assess its mechanical, thermophysical and hydric properties. The results were compared with those of mortars containing the same amount of each single filler (either mPCMs or miscanthus fibers), in order to discuss the interest of the hybrid solution.

2 Materials and Experimental Methods

2.1 Materials

The PCM used in this study is manufactured by Croda International (Snaith, UK) under the trade name CrodaTherm™ ME29D. It is a liquid dispersion of microencapsulated PCM particles (mPCMs) in water with a solid content of 50 wt.%. These particles consist of an organic PCM core derived from plant-based feedstocks, which is surrounded by an acrylic polymer shell [15]. Table 1 provides characteristics of the melting and crystallization peaks of these mPCMs, as determined by differential scanning calorimetry (DSC) in a previous study [5].This liquid PCM dispersion was chosen because it allows an easy and homogeneous incorporation into cement mortar formulations at an early stage of the mixing process. In addition, the ME29D is certified bio-based and formaldehyde free, which ensures both reduced environmental footprint and low safety/health hazards.

Table 1. Characteristics of the melting and crystallization peaks of the mPCM particles, as determined by DSC at a rate of 0.5 °C min^{-1}

Properties		Mean value	Standard deviation
Characteristics of the melting peak	Latent heat of melting (J g^{-1})	172.5	2.7
	Onset (°C)	26.0	0.2
	Peak maximum (°C)	27.8	0.2
Characteristics of the crystallization peak	Latent heat of crystallization (J g^{-1})	169.1	2.4
	Onset (°C)	25.3	0.1
	Peak maximum (°C)	21.8	0.2

Vegetal fillers used in this work consist of a micronized powder of miscanthus fibers, with particle size distribution in the range 200–500 µm. This powder was supplied by Addiplast Group (Saint-Pal-de-Mons, France) and was used as-received without further chemical treatment. The choice of miscanthus fibers was driven by the plant's potential, and its large local availability and low price. Besides, miscanthus requires low fertilization and pesticide inputs and does not compete with the food sector (human or animals feeding) which makes it highly suitable for the development of bio-based building materials. The use of a micronized powder was mainly intended to promote fine and homogeneous particle distribution within the mortar matrix, opening possibilities for the design of 3D printable (this aspect is not ad-dressed in the present paper).

Due to the 3D printing perspective of our research project, an existing cement mortar formulation developed for additive manufacturing applications was selected from the literature and used as reference mortar (noted M0) [16]. This latter contains two types of cement at a ratio of 93/7 wt.%: an Ordinary Portland Cement (EXTREMAT® CEM I 52,5 N-SR3 SEG), denoted CEM I, and a rapid setting Sulfo-aluminous cement (Alpenat R^2) denoted CSA, both manufactured by VICAT company (L'Isle-d'Abeau, France). CSA acts as a setting accelerator of the fresh mix to ensure its buildability in the case of 3D printing applications [16]. A superplasticizer/high water reducer (VISCOCRETE TEMPO 11) obtained from SIKA Company (Baar, Switzerland), denoted as SP, was also introduced in the mix. Siliceous sand complying with EN 196–1 standard [17] (graded sand with a maximum grain size of 2 mm and moisture content $< 0.2\%$) was supplied by Société Nouvelle du Littoral (Leucate, France). The composition of the reference mortar M0 is summarized in Table 2.

Table 2. Composition of the various mortars considered in this study

Type of mortar	Sand (g)	CEM I (g)	CSA (g)	SP (g)	Mass of solid mPCMs (g)	Dry fibers (g)	Total amount of water (g)	W/C ratio	mPCMs content (wt. %)	Fiber content (wt. %)
M0	1350	1059.33	79.73	2.96	–	–	398.67	0.35	–	–
M10D	1350	1059.33	79.73	2.96	289.07	–	851.36	0.75	7.96	–
M5F	1350	1059.33	79.73	2.96	–	144.53	843.61	0.74	–	4.15
MDF	1350	1059.33	79.73	2.96	289.07	144.53	1068.16	0.94	7.24	3.62

2.2 Mix Design and Sample Preparation

A preliminary study has investigated the incorporations of mPCMs and miscanthus fibers (separately) into the reference mortar M0, as well as their effects on the microstructure, the mechanical properties, and the hygrothermal characteristics of the resulting materials. From this preliminary test program, "optimal" dosages of mPCMs and vegetal fibers were determined, that provide significant improvements in hygrothermal properties of the mortar (thermal resistance and moisture absorption capacity) while ensuring minimum requirement in terms of mechanical strength.

Regarding the mortar containing mPCMs, the optimal dosage of inclusions was ~ 8 wt.% (material denoted M10D in the following). The composition of M10D is also given in Table 2. It is to note that the amount of water was increased compared to the reference composition in order to maintain the workability of the fresh mix constant (as determined by slump tests). After hardening, M10D exhibits a thermal conductivity of 0.85 $W.m^{-1}.K^{-1}$ at 20 °C and a compressive strength of 17 MPa (compared to 2.32 $W.m^{-1}.K^{-1}$ and 72 MPa for the reference mortar M0, respectively) [5]. Hence the decrease in thermal conductivity is about 63% compared to M0, but there is also a loss of mechanical strength of about 76%.

For the mortar including miscanthus fibers, the optimal dosage of inclusions was ~ 4 wt.% (material denoted M5F in the following). The composition of M5F is given in Table 2. Here again, a higher amount of water was used compared to the reference composition, because it was necessary to presoak the miscanthus fiber in water in order to preserve an acceptable workability of the fresh mix. The hardened material exhibits a thermal conductivity of 0.62 $W.m^{-1}.K^{-1}$ at 20 °C and a compressive strength of 7 MPa (values reduced by 73% and 90% compared to the reference mortar M0, respectively). Besides, the moisture buffer value of M5F is 1.93 (much higher than the value of 0.5 obtained for M0), showing that this material is highly hygroscopic due to the presence of vegetal fibers.

Based on the previous results, a hybrid mortar formulation including both mPCMs and miscanthus fibers was designed in the present work, considering the same amounts of inclusions (in g) used in the M10D and M5F materials. The final composition of this hybrid mortar, denoted MDF, is also reported in Table 2. Here again, the miscanthus fibers are presoaked in water prior to their introduction in the fresh mix, and the mixing water is adjusted to maintain the same workability as the reference mortar M0 (based on slump tests). Hence, the "total water" reported in Table 2 includes both the quantity of water absorbed by fibers and the excess mixing water added, which explains its high value. The mass fractions of both types of inclusions in the fresh MDF mix, calculated by considering the amounts of all components (including water), are also given in the table.

In practice, the samples of hybrid MDF mortar were prepared according to the specifications of EN 196–1 standard in terms of rotational speed and duration of the mixing sequences, using a programmable mortar mixer (model E092N from Matest Spa Company, Treviolo, Italy). The mixing process comprised the following steps: **(1)** a mixture of the two cements (OPC and CSA in the proportion of 97/3 parts by weight) was placed in the mixer tank. Water containing the SP and the liquid dispersion of mPCMs (ME29D) (Fig. 1.a) was then added to the cement, and the zero time was counted from the moment the water came into contact with the cement. **(2)** Mixing was started at a low speed and sand was added slowly during the first minute while mixing at low speed (140 rotations per minute). The mortar was then mixed at a high speed (285 rpm) for 30 s. **(3)** The rotation was stopped for 30 s to scrape the mortar from the bottom of the tank. **(4)** Another high-speed mixing was performed for 2.5 min. **(5)** The soaked miscanthus fibers (Fig. 1.b) were then added to the fresh mix and additional low-speed mixing was performed for 30 s followed by high-speed mixing for 1 min to ensure homogenous fiber distribution.

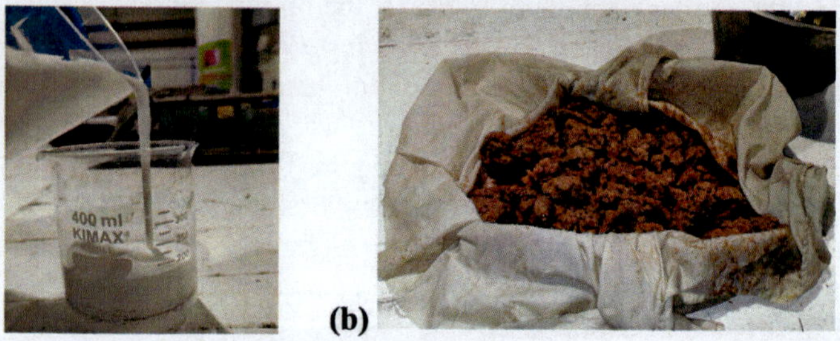

(a) (b)

Fig. 1. (a) Liquid dispersion of mPCMs and (b) presoaked miscanthus fibers after filtration.

After mixing, the mortar was poured into molds of different sizes, depending on the type of characterization intended. $4 \times 4 \times 16$ cm^3 samples were casted for mechanical characterizations, $4 \times 4 \times 8$ cm^3 samples for thermal and microstructural characterization and $10 \times 10 \times 4$ cm^3 samples for Moisture Buffer value determination. The molds were filled in 2 layers, and each layer was subjected to 52 blows (for about 1 min) using a shock table. The samples were demolded 24 h later, hermetically wrapped with a plastic film, and finally conditioned at (20 $\pm$ 1) °C until the characterizations at 28 days.

2.3 Experimental Methods

Scanning Electron Microscopy

SEM observations were first carried out on fractured surfaces of MDF mortar samples, in order to qualitatively assess the distribution of fibers and mPCM inclusions within the cementitious matrix. A QUANTA 400 microscope from FEI Company (Hillsboro, OR, USA) was used for these observations, with an electron voltage of 20 kV and under low vacuum.

Density and Water-Accessible Porosity

The density and water-accessible porosity of the MDF mortar was determined according to NF P18–459 standard [18], on 3 small samples of dimension $2 \times 2 \times 4$ cm^3.

Mechanical Characterizations

At 28 days, hardened MDF mortar samples were subjected to mechanical tests according to EN 196–1 standard [17]. 3-point bending tests were first conducted on 15 samples of dimension $4 \times 4 \times 16$ cm^3 to determine the mean flexural strength. After breakage, the 30 half specimens were retrieved and subjected to compression tests to evaluate the mean compressive strength.

Assessment of Thermophysical Properties

The hot disk (HD) method was used to quantify several thermophysical properties of the MDF mortar. HD allows a rapid, accurate, and nondestructive evaluation of the thermal conductivity (λ in W.m^{-1}.K^{-1}), thermal diffusivity (α in mm^2.s^{-1}), and volumetric heat capacity of a sample (ρ.Cp in J.m^{-3}.K^{-1}) in a single measurement. A TPS 2500 S device from Hot Disk Company (Gothenburg, Sweden) comprising a "Kapton insulated" probe (of radius 6.4 mm) connected to a thermal constant analyzer was used.

Before testing, mortar samples were conditioned in a climatic chamber at 20 °C and 50% RH, in order to reach an equilibrium state. HD measurements were then conducted over a broad range of temperature from 10 °C to 45 °C, to cover the entire phase change domain of the mPCMs. Besides, the RH level was kept constant in the climate chamber (50%) to avoid any influence of moisture variations on the measurements.

Determination of the Moisture Buffer Value

The ability of a building material to regulate humidity can be evaluated through the Moisture Buffer Value (MBV), which express the amount of moisture that can be stored/released by the material when subjected to moisture variations. This parameter was determined following the NORDTEST Protocol [19]. This latter consists in exposing a sample to a cyclic RH variation between high (75% RH) and low (33% RH) levels for periods of 8 and 16 h, respectively, under isothermal condition.

In our study, a Memmert HPP 750 climatic chamber was used to apply the RH cycles. The mortar samples (dimension $10 \times 10 \times 4$ cm^3) were first stabilized at 23 °C and 50% RH until mass stabilization. All faces of the samples were then sealed with plastic film, except the top face to allow a one-directional (1D) moisture transfer. The specimens were weighted twice per cycle, once during absorption phase and once during desorption phase. A minimum of three cycles must be carried out, and the weight amplitude must not vary by more than 5% day to day. The specimens were weighted until the masses are stable over three successive cycles. The MBV value (in kg.(m^2.%RH)$^{-1}$) is then defined by the following expression:

$$MBV = \frac{\Delta m}{A \times (RH_{high} - RH_{low})} \tag{1}$$

where Δm is the mass change during the absorption/desorption phase (kg), A is the exchange area (m^2), RH_{high} and RH_{low} are high and low RH levels (i.e. 75% and 33%)

3 Results

3.1 SEM Observation

Figure 2 presents SEM pictures of a fractured surface of MDF mortar at different magnifications, showing the microstructure of the material. Due to the large difference in size between the mPCM particles (7–10 μm) and the miscanthus fibers (200–500 μm), it is difficult to make a good observation the two types of inclusion on the same image.

(a) (b)

Fig. 2. SEM images (backscattered electron mode) showing the microstructure of the MDF mortar at different magnifications: (a) × 600, (b) × 1200.

Overall, mPCMs can be identified as small black particles and their distribution seems rather homogeneous within the cement matrix. Besides, significant porosity can be observed within this matrix. From the whole set of pictures collected, it was found that the miscanthus fibers are also globally well distributed in the volume of the material. The surface of the miscanthus fibers appears rather smooth on the pictures, suggesting that there are limited interactions between the cement matrix and the fibers.

3.2 Density and Water-Accessible Porosity

The density and water-accessible porosity of the various mortars considered in this study are displayed in Fig. 3. The results of M10D and M5F (including only a single type of inclusion, either mPCMs or miscanthus fibers) are also provided for the sake of comparison with the MDF mortar.

The hybrid MDF mortar exhibits a water-accessible porosity of 41.7%, which is much higher than that of the reference mortar M0 (19,7%) and lies between the values obtained for the M10D mortar (34%) and the M5F mortar (47%). This is due to the high intrinsic porosity of fibers, to the air entrapped in the mix while adding mPCMs and fibers, but also to the excess water used for adjusting the workability and to the water absorbed by the fibers at presoaking stage. This excess water does not contribute to the cement hydration process, evaporates during early shrinkage phase, hence leading to the formation of voids.

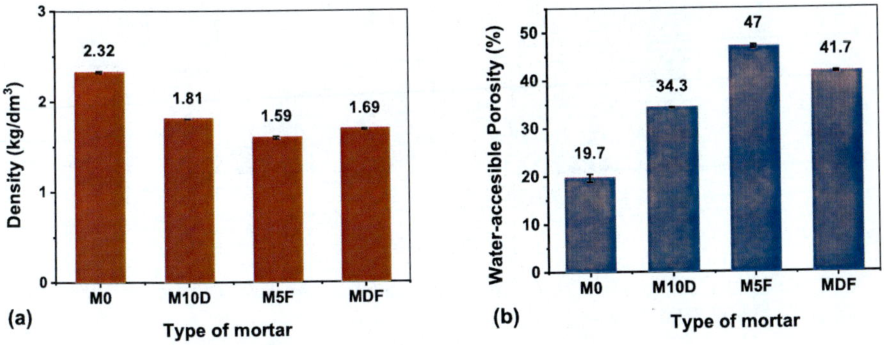

Fig. 3. (a) Density and (b) water-accessible porosity of the various mortars considered in this study (including the hybrid MDF mortar).

Besides, MDF exhibits a density of 1.69 kg.dm^{-3}, which is significantly lower than that of the reference mortar M0 (2.32 kg.dm^{-3}) and close to the values obtained for M10D and M5F mortars. This is due the low density of the mPCMs and miscanthus fibers added compared to the other mineral components of the mortar. This result is also in line with the increased porosity of the modified mortars, as previously highlighted.

Similar effects (increased porosity and reduced density) are generally reported in the literature when adding vegetal fibers or PCMs into cement mortars [2–9].

3.3 Mechanical Properties

The compressive and flexural strength of the various mortars considered in this study are presented in Fig. 4.

The hybrid MDF mortar exhibits a compressive strength of 6.29 MPa and a flexural strength of 2.08 MPa, which are much lower than the values obtained for reference material M0 (reduced by 91% and 75% respectively). This result is consistent with general trends reported in the literature, as most authors observed a decrease in the mechanical properties of cement mortars when adding either mPCM particles or different types of vegetal fibers, like miscanthus, date palm or hemp fibers for in-stance [2–9]. This effect is generally attributed to the low mechanical properties of the inclusions, to the high porosity content of these modified mortars, and to the poor compatibility and low bond between the inclusions and the cement matrix.

Another interesting point is that the compression strength of the hybrid MDF mortar is also lower compared to M10D (-59%) and M5F (-11%) that contain respectively the

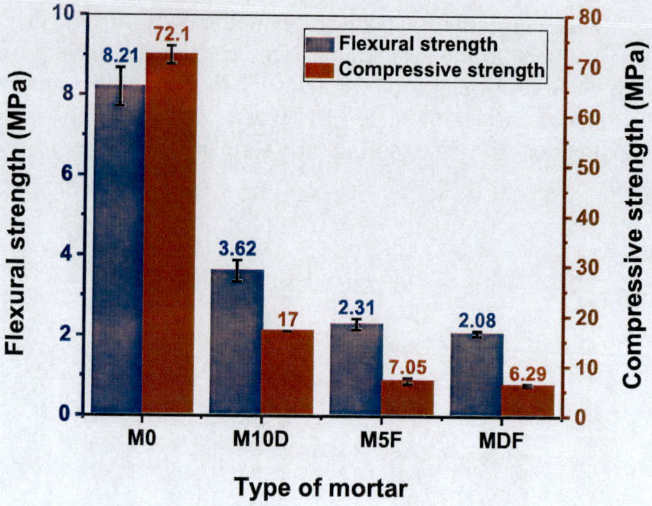

Fig. 4. Flexural and compressive strengths of the various mortars considered in this study (including the hybrid MDF mortar).

same amounts mPCMs or plant fibers. This may be due to the higher global content of inclusions in the MDF mortar compared to M10D and M5F, and to the higher water content in the mix that increases the porosity of the hardened cement matrix.

It is also worth noting that the incorporation of miscanthus fibers in the mortar, even at very low dosage as for M5F, has a more detrimental impact on the mechanical properties than the incorporation of mPCMs (M10D).

3.4 Thermophysical Properties

The evolutions versus temperature of the thermal conductivity (λ) and the thermal diffusivity (α) of the various mortars studied are represented in Fig. 5. Errors bars are displayed on the graphs, based on standard deviations from 3 repeated measurements.

The reference mortar (M0), which has a density of 2300 kg.m^{-3}; exhibits λ and α values of 2.32 W.m.$^{-1}$.K.$^{-1}$ and 1.05 mm^{2}.s^{-1}, respectively, at 20 °C. These values are in accordance with the work of Shafigh et *al.* [20] which reports λ and α values of 2.4 W.m.$^{-1}$.K.$^{-1}$ and 1.03 mm^{2}.s^{-1}, respectively, for cement mortars of similar density (2200 kg.m^{-3}). Besides, λ shows very little variation over the temperature range of 5 °C-45 °C, whereas α shows a slight decreasing trend as the temperature increases.

Overall, the thermal conductivities of M10D and M5F mortars are much lower than that of the reference M0, with values of 0.85 and 0.61 W.m.$^{-1}$.K.$^{-1}$ at 20 °C. This shift is highly correlated to the porosity rate of these mortars, which is much higher compared to M0 as previously seen in Sect. 2.2. It is also related to the low thermal conductivities of the added inclusions themselves (between 0.35 and 0.2 W.m^{-1}.K^{-1} for mPCMs depending on the physical state of the core [5], and about 0.04 W.m^{-1}.K^{-1} for miscanthus fibers [21]). Besides, here again, values of λ are found almost constant over the investigated temperature range.

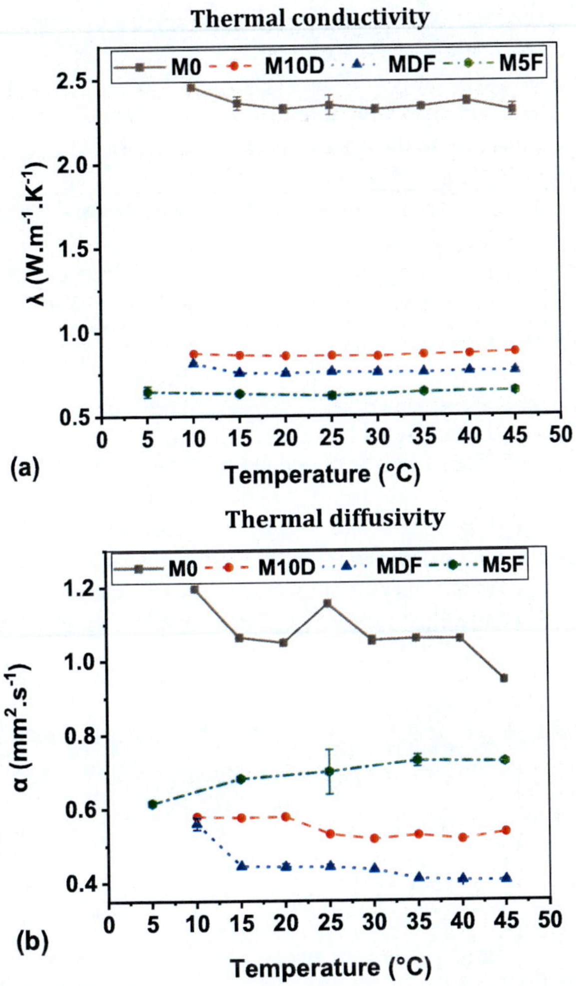

Fig. 5. Evolutions versus temperature of (a) the thermal conductivity and (b) the thermal diffusivity of the various mortars considered in this study (including the hybrid MDF mortar).

The hybrid mortar MDF also exhibits a very low level of thermal conductivity (decreased by 68% compared to the reference M0), with a value of λ close to 0.75 $W.m.^{-1}.K.^{-1}$ over the entire temperature range, and lying between values measured for M10D and M5F. This may be due to the porosity rate of MDF which is intermediate compared to those of M10D and M5F mortars.

In the end, the MDF mortar can be considered a good insulation material.

3.5 Moisture Buffer Value

The moisture buffer values determined for the various mortars are reported in the graph of Fig. 6.The reference mortar M0 has a low MBV value of 0.5 $g.m^{-2}.\%RH^{-1}$, showing that this material exhibits moderate/limited moisture regulation capacity according to the NORDTEST classification.

M10D is characterized by an even lower MBV value (0.5 g.m^{-2}.%RH^{-1}), suggesting that the introduction of mPCMs in the mortar has a detrimental effect on the moisture buffer capacity, probably due to the organic nature of the mPCMs.

Differently, M5F shows a high MBV value of 1.93 g.m^{-2}.%RH^{-1}, and hence this material can be classified as a good moisture regulator. Such a good hygroscopic behavior can be attributed to the high intrinsic porosity and specific surface area of the vegetal fibers added. Similar effect has been reported in the literature for cement mortars including various types of plant fibers (date palm, hemp, etc.) [10–13].

Finally, the hybrid MDF mortar that contains both mPCMs and miscanthus fibers, exhibit an MBV value of 0.66 g.m^{-2}.%RH^{-1}, which highlights very limited improvement in hygric behavior compared to the reference mortar M0. It seems that the presence of mPCMs hinders the beneficial effects of vegetal fibers on the moisture regulation capacity of MDF mortar. Therefore, the combination of mPCM and plant fiber inclusions may not be the best solution to improve the hygric properties of cement mortars.

According to the present study, the introduction of miscanthus fibers alone pro-vides better results, and makes it possible to improve both the thermal resistance and hygroscopic behavior of mortars in a very significant way. Nevertheless, this gain is balanced by a severe impact on mechanical properties, which limits the range of applications.

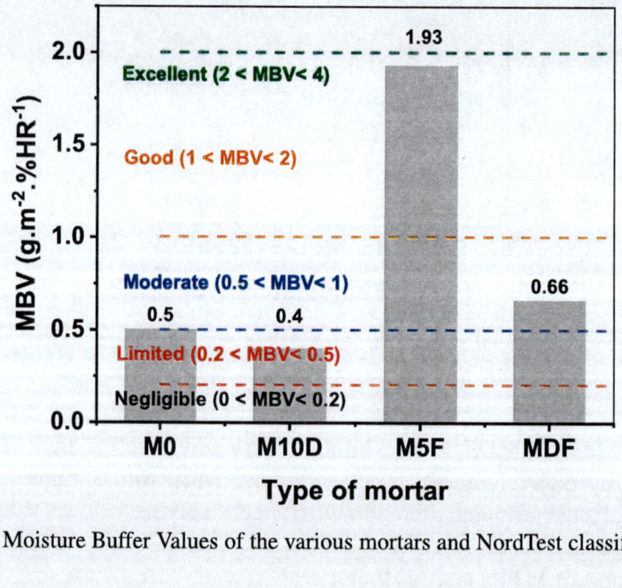

Fig. 6. Moisture Buffer Values of the various mortars and NordTest classification.

4 Conclusion

In this study, we explored the mechanical and hygrothermal properties of a hybrid mortar (noted MDF) containing both mPCMs and micronized miscanthus fibers. In order to evaluate the interest of this hybrid solution, the results obtained were compared to those of both the neat reference mortar, a mortar filled with the same amount of mPCMs only, and another filled with the same amount of miscanthus fibers only.

Overall, the hybrid MDF mortar exhibited a large decrease in thermal conductivity compared to the reference (-68%), and the conductivity level was of the same order of those recorded for the mortars filled with each type of single inclusion. In addition, a severe loss of compressive strength (-91%) was noticed compared to the reference. Both effects are strongly correlated to the high porosity of the hybrid MDF material, due to the air entrapped during the mixing process and to the excess water used in the hybrid formulation (water absorbed by miscanthus fibers at presoaking stage and water added to maintain a constant workability of the fresh mix). This first result suggests that MDF may be used as a non-structural material dedicated to thermal insulation.

On the other hand, the MDF mortar showed very limited improvement in the moisture buffer capacity compared to the reference mortar, and still displayed moderate/limited moisture regulation ability according to the NordTest classification. Differently, the mortar including only the same amount of miscanthus fiber exhibited a high MBV value corresponding to a good moisture regulator, thanks to the intrinsic characteristics of the vegetal fibers. It is likely that the presence of mPCMs particles in the MDF hinders the moisture transport process within the hybrid material, and impedes this beneficial effect of plant fibers on the hygroscopic behavior.

In the present case, the interest of the hybrid solution is therefore not obvious, and the use of plant fibers as single inclusion seems the best option for improving both thermal and hygric properties of mortars, providing the severe impact on mechanical strength remains compatible with the target application. Nevertheless, further investigation should be carried out at the wall scale.

Acknowledgement. The authors would like to acknowledge the Labex MMCD (Multi-Scale Modelling & Experimentation of Materials for Sustainable Construction) for its financial support through ANR Investments for the Future (program ANR-11-LABX-022-01).

References

1. International Energy Agency (IEA), Building sectorial overview, Tracking report (2022). https://www.iea.org/reports/buildings. Accessed 09 Jan 2023
2. Kalnæs, S.E., Jelle, B.P.: Phase change materials and products for building applications: a state-of-the-art review and future research opportunities. Energy Build. **94**, 150–176 (2015)
3. Jayalath, A., et al.: Properties of cementitious mortar and concrete containing micro-encapsulated phase change materials. Constr. Build. Mater. **120**, 408–417 (2016)
4. Dehdezi, P.K., Hall, M.R., Dawson, A.R., Casey, S.P.: Thermal, mechanical and micro-structural analysis of concrete containing microencapsulated phase change materials. Int. J. Pavement Eng. **14**(5), 449–462 (2013)
5. Gbekou, F.K., Benzarti, K., Boudenne, A., Eddhahak, A., Duc, M.: Mechanical and thermo-physical properties of cement mortars including bio-based microencapsulated phase change materials. Constr. Build. Mater. **352**, 129056 (2022)
6. Hunger, M., Entrop, A.G., Mandilaras, I., Brouwers, H.J.H., Founti, M.: The behavior of self-compacting concrete containing micro-encapsulated phase change materials. Cement Concr. Compos. **31**, 731–743 (2009)

7. Amziane, S., Sonebi, M.: Overview on biobased building Material made with plant aggregate. RILEM Tech. Lett. **1**, 31–38 (2016)
8. Benmansour, N., Agoudjil, B., Gherabli, A., Kareche, A., Boudenne, A.: Thermal and mechanical performance of natural mortar reinforced with date palm fibers for use as insulating materials in building. Energy Build. **81**, 98–104 (2014)
9. Panesar, D.K., Shindman, B.: The mechanical, transport and thermal properties of mortar and concrete containing waste cork. Cement Concr. Compos. **34**(9), 982–992 (2012)
10. Benmahiddine, F., Bennai, F., Cherif, R., Belarbi, R., Tahakourt, A., Abahri, K.: Experimental investigation on the influence of immersion/drying cycles on the hygrothermal and mechanical properties of hemp concrete. J. Build. Eng. **32**, 101758 (2020)
11. Haba, B., Agoudjil, B., Boudenne, A., Benzarti, K.: Hygric properties and thermal conductivity of a new insulation material for building based on date palm concrete. Constr. Build. Mater. **154**, 963–971 (2017)
12. Chennouf, N., Agoudjil, B., Boudenne, A., Benzarti, K., Bouras, F.: Hygrothermal characterization of a new bio-based construction material: concrete reinforced with date palm fibers. Constr. Build. Mater. **192**, 348–356 (2018)
13. Chennouf, N., Agoudjil, B., Alioua, T., Boudenne, A., Benzarti, K.: Experimental investigation on hygrothermal performance of a bio-based wall made of cement mortar filled with date palm fibers. Energy Build. **202**, 109413 (2019)
14. Wu, D., Rahim, M., Li, W., Ganaoui, M.E., Bennacer, R.: Simulation of the hygrothermal behavior of a building envelope based on phase change materials and a bio-based concrete. FDMP-Fluid Dyn. Mater. Proc. **18**(5), 1483–1494 (2022)
15. CrodaTherm^{TM} ME29D: Microencapsulated ambient temperature phase change material, technical datasheet (2018). https://www.crodaindustrialspecialties.com/en-gb/product-finder/product/1381-crodatherm_1_me_1_29d. Accessed 09 Jan 2023
16. Khalil, N., Aouad, G., El Cheikh, K., Rémond, S.: Use of calcium sulfoaluminate cements for setting control of 3D-printing mortars. Constr. Build. Mater. **157**, 382–391 (2017)
17. European Committee for Standardization (CEN), NF EN 196–1: Methods of testing cement - Part 1: Determination of strength (2016)
18. AFNOR Editions, NF P18–459, Béton - Essai pour béton durci - Essai de porosité et de masse volumique (2010)
19. Rode, C., et al.: Moisture Buffering of Building Materials. Report Technical University of Denmark (DTU), Denmark (2005)
20. Shafigh, P., Asadi, I., Akhiani, A.R., Mahyuddin, N.B., Hashemi, M.: Thermal properties of cement mortar with different mix proportions. Mater. Constr. **70**, 224 (2020)
21. Schnabel, T., Huber, H., Petutschnigg, A., Jäger, A.: Analysis of plant materials pre-treated by steam explosion technology for their usability as insulating materials. Agron. Res. **17**, 1191–1198 (2019)

Comparative Study of the Thermal Behaviors of a Cement Mortar Wall Including Bio-based Microencapsulated Phase Change Materials and a Reference Wall

Franck Komi Gbekou[1](✉), Abderrahim Boudenne[2], Anissa Eddhahak[3], and Karim Benzarti[1](✉)

[1] Univ Gustave Eiffel, Ecole des Ponts, CNRS, Navier, 77447 Marne la Vallée, France
{komi.gbekou,karim.benzarti}@univ-eiffel.fr
[2] Univ Paris Est Créteil, CERTES, 94010 Créteil, France
[3] Arts et Métiers ParisTech, PIMM, 75013 Paris, France

Abstract. This study aims to evaluate the thermal behavior of a cement mortar wall (denoted M15D) including bio-based microencapsulated phase change materials (mPCMs) in comparison to a reference wall without mPCMs (M0). A bi-climatic test setup was designed and built in order to submit the two sides of the walls to different hygrothermal conditions representing the outdoor and indoor environments. Two scenarios were considered for the outdoor conditions: a cyclic solicitation of 12 h at 40 °C followed by 12 h at 15 °C to simulate a day/night variation, and a step change from 20 °C to 40 °C followed by prolonged exposure at 40 °C until steady-state, both at fixed relative humidity (50% RH). Temperature sensors installed at different depths made it possible to monitor temperature gradients within the walls during the tests. Overall, quite similar responses were collected for the two walls exposed to the same outdoor scenario. However, the maximum temperatures recorded on the external face and at several depth locations were slightly lower for the M15D wall compared to M0, with a gap of about 1 °C. This result suggests that the incorporation of mPCMs in the wall may contribute to dampen the effect of external temperature variations, although this action remains limited. Furthermore, this bio-based mPCM system offers an environmentally friendly alternative to traditional paraffinic mPCMs.

Keywords: bio-based microencapsulated phase change materials (mPCMs) · heat transfer · bi-climatic setup · wall-scale study

1 Introduction

In the current context of global warming, the reduction of both energy consumption and CO_2 emissions has become a major challenge for our modern societies [1]. The building and construction sector is responsible for about 30% of the total energy consumption in France [2], and most part of this energy is used to maintain living comfort (electricity, air conditioning, heating). To reduce this impact, public policies encourage the development

S. Amziane et al. (Eds.): ICBBM 2023, RILEM Bookseries 45, pp. 817–828, 2023.
https://doi.org/10.1007/978-3-031-33465-8_63

of innovative solutions capable of improving the thermal efficiency of buildings [1, 3]. Therefore, significant research is being dedicated to the design of construction materials with improved thermal properties, and preferably including bio-based components to provide more environmentally friendly alternatives [4–6].

Latent heat thermal energy storage systems are considered as a promising approach [7]. These systems allow the storage of certain quantity of energy that can be released later to meet specific needs [6, 8]. They include phase change materials (PCM), which have the ability to changes their physical state (e.g. solid to liquid) upon temperature variations and store/release energy during phases changes [9, 10]. The combination of PCMs (in the form of macro or micro encapsulated materials) with conventional buildings materials like mortar or concrete, has been investigated in many studies and has given promising results, as it resulted in significant improvement in thermophysical properties [11–14]. Studies at larger scales (wall and building scale) have been also carried-out to evaluate the effective gain in thermal performance following the incorporation of PCM into mortars. In general, these studies mainly focused on the use of PCM layers in multi-layered insulation systems. Li et al. [15], showed for instance that the addition of an insulating layer with PCMs to a cementitious mortar wall, allows to attenuate the indoor/outdoor temperature difference in a room and to increase energy saving rates up to 30%.

To the best of our knowledge, most systems are multi-layered and the hygrothermal performance of a thick and homogeneous cement mortar wall including microencapsulated PCMs (mPCMs) has not been investigated so far. Therefore, this study aims to compare the thermal responses of a homogeneous mortar wall incorporating bio-based and formaldehyde-free mPCMs, and a reference wall made of regular mortar, in order to evaluate the gain in performance resulting from mPCM addition. Both walls were tested with a home-made bi-climatic setup, in which the wall sides are exposed to specific hygrothermal conditions representing the outdoor and indoor environments. Temperature variations were recorded at different depth of the wall during testing, in order to determine the global thermal behavior of each wall and discuss the possible effect of mPCM incorporation.

2 Materials and Experimental Methods

2.1 Materials

The study involved two cement mortar walls: one reference wall made of regular mortar, and a second wall including microencapsulated phase change materials (mPCMs). The reference mortar (denoted M0) is an existing formulation from the literature [16] and contains: an ordinary Portland cement EXTREMAT® CEM I 52.5 N (denoted OPC), a fast-setting Sulfo-aluminous cement Alpenat R^2 (denoted CSA), both manufactured by VICAT company (L'Isle-d'Abeau, France), a superplasticizer (SP)/water reducer VISCOCRETE TEMPO 11 from SIKA (Baar, Switzerland), standardized sand with a maximum diameter of 2 mm provided by Société Nouvelle du Littoral (Leucate, France). The composition of the reference mortar M0 is detailed in Table 1.

The mPCM used is a commercial product (CrodaTherm™ ME29D), supplied in the form of an aqueous dispersion of microencapsulated particles in water (solid content of about 50 wt.%). The particles consist of an organic PCM core derived from plant-based feedstocks, surrounded by an acrylic polymer, and display maximum melting and crystallization temperatures of 28.8 °C and 23.5 °C, respectively, and heats of fusion and crystallization of 183 kJ/kg and −179 kJ/kg, according to the product datasheet [17]. Such a liquid dispersion was chosen because it can be easily incorporated into the cement mortars at an early stage of the mixing process (together with the mixing water). In addition, CrodaTherm™ ME29D is certified bio-based and formaldehyde free, which ensures both reduced environmental footprint and low safety/health hazard.

A preliminary material-scale study was conducted regarding the incorporation of these mPCMs into the reference mortar M0 and its influence on the mechanical and thermophysical properties of the resulting materials [11]. This study, carried out on several mortars with mPCM contents in the range 0–13.5 wt.%, made it possible to highlight significant improvement in thermal properties upon mPCM addition, but at the expense of mechanical performance. Introducing 11.03 wt.% of mPCM in the mortar was found to reduce the thermal conductivity by 72% (0.6 $W.m^{-1} \cdot K^{-1}$ compared to 2.3 $W.m^{-1} \cdot K^{-1}$ for M0), while still providing an acceptable compressive strength of 12 MPa. This formulation, denoted M15D, was thus selected to conduct a study at the wall scale, with the objective to compare the thermal responses of a wall made of the reference mortar M0, and another wall based on this M15D formulation. The composition of mortar M15D is also summarized in Table 1.

Table 1. Composition of the two mortar formulations used in this work.

Type of mortar	Sand (g)	OPC (g)	CSA (g)	SP (g)	Mass of mPCM particles (g)	Total water (g)	W/C ratio	Weight fraction of mPCM particles (%)	Volume fraction of mPCM particles (%)
M0	1350	1059.33	79.73	2.96	–	398.67	0.35	0	0
M15D	1350	1059.33	79.73	2.96	433.06	1006.2	0.88	11.03	20.94

2.2 Determination of the Thermophysical Properties of the Mortars

The thermophysical properties of the two mortars M0 and M15D were determined using the Hot Disk method (denoted HD) with a TPS 2500 S device from Hot Disk Company (Gothenburg, Sweden). The sensor used was a "Kapton insulated" model (ref 5501) from Hot Disk® with a radius of 6.4 mm. The HD enables a fast, non-destructive and accurate measurement of the thermal conductivity (λ in $W.m^{-1} \cdot K^{-1}$), the thermal diffusivity (α in $m^2 \cdot s^{-1}$), and the volumetric heat capacity ($\rho.Cp$ in $MJ.m^{-3} \cdot K^{-1}$). The sensor is placed between two samples of the same material, and acts both as a heat source to increase the sample's temperature and as a resistance thermometer to record the time-dependent increase in temperature. To ensure reproducible conditions, measurements were carried out in a controlled climatic chamber at a constant relative humidity of 50%

and over a temperature range from 10 °C to 45 °C to cover the entire phase change domain the mPCMs.

2.3 Experimental Setup for Bi-climatic Wall Test

The Bi-climatic Experimental Setup

A specific setup was designed for the experiment. The objective is to simulate a bi-climatic environment by submitting the wall sides to two different hygrothermal conditions representing the outdoor and indoor environments.

The bi-climatic system combines a climatic chamber (Memmert ICH 260, whose door has been removed) and a metallic frame designed to support the wall and connect this latter to the chamber. The whole device is installed in a laboratory at controlled temperature (20 °C). The climatic chamber allows to simulate the outdoor environment by applying dynamic temperature cycles on the outer face of the wall, while the other face is exposed to the laboratory condition that simulates the indoor environment of a building.

The metallic frame is fixed to the chamber and a 6 cm thick insulation material is placed in-between to insulate the borders of the central open area (40 cm × 45 cm), so that thermal exchanges between the chamber and the wall can only occurs through this open area. The insulation material is then completely covered with an insulating and vapor barrier adhesive plastic film (Fig. 1).

Fig. 1. Picture of the climatic chamber and the designed frame's structure

Preparation and Instrumentation of the Walls

The two mortar formulations M0 and M15D were prepared according to the mixing procedure detailed in a previous article [11]. Walls of dimensions 50 × 50 × 10 cm^3 were then manufactured by casting the fresh mortar mixes in wood formworks. They were demolded after about one week, and were cured for several months in the laboratory at 20 °C and ambient hygrometry, so that the microstructure of the cement matrix is almost stabilized.

The walls were then instrumented with temperature sensors. 3 holes of depth 25 cm were drilled in the thickness of the wall, at distances of 2.5 cm, 5 cm and 7.5 cm respectively from the outer face (Fig. 2), and using a column drilling machine to ensure precise vertical drilling (Fig. 3). The holes were vertically spaced 15 cm from each other to avoid any disturbance on sensor's measurements (Fig. 2).

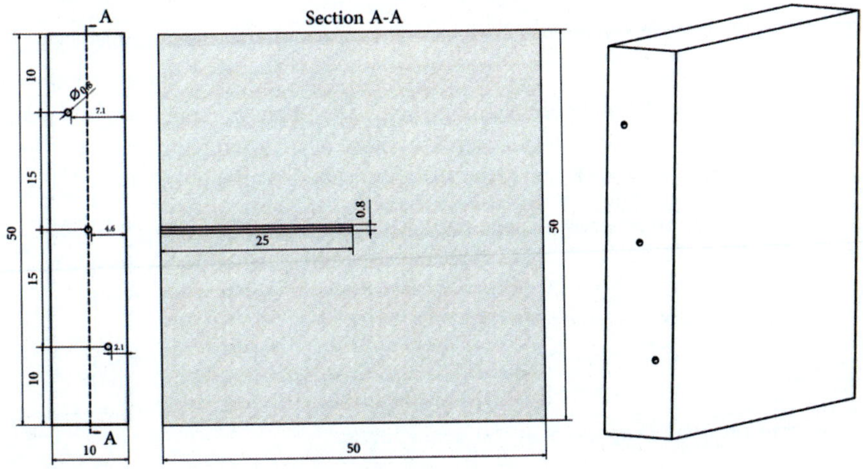

Fig. 2. Locations of holes for the insertion of temperature sensors at different depths of the wall

A total of five sensors were installed on each wall (Fig. 3), three in the thickness and one on each face to record temperature variations at different depths. All lateral faces were insulated with expanded polystyrene, to force unidirectional (1D) thermal transfer to occur through the two exposed surfaces only. In addition, two other sensors are placed in the climatic chamber and in the laboratory to record indoor and outdoor temperature conditions.

Finally, a thermal camera was used to check the quality of the setup insulation, in order to prevent possible heat loss and ensure uniform heat transfer across the observed surface.

Temperature Cycles

After the setup configuration, the wall was pre-conditioned by setting the climatic chamber at 20 °C and 50% RH, until all the sensors provide a stabilized temperature (for

(a) (b) (c)

Fig. 3. Preparation of the wall: (a) drilling stage, (b) sensors' introduction, (c) final setup.

about one week). The dynamic temperature cycles were then programmed on the climatic chamber using Atmo-Control (control software of the Memmert chamber). Two experimental programs were applied to the walls, both at fixed humidity level (50% RH):

- A cyclic temperature program which consists in exposing the wall to a period of 12 h at 40 °C, followed by 12 h at 15 °C, representative of a day/night alternation.
- An initial step change from 20 °C to 40 °C, followed by a prolonged exposure at 40 °C until reaching a steady-state regime.

3 Results

3.1 Thermophysical Properties of the Two Mortars

Thermophysical properties of the two mortars M0 and M15D measured with the HD method at different temperatures are presented in Fig. 4.

The mortar filled with 11.03 wt.% of mPCMs exhibits much lower values of the three measured properties compared to the reference material, whatever the temperature considered. For example, the thermal conductivity at 20 °C is reduced by ~72%, from 2.32 $W.m^{-1} \cdot K^{-1}$ to 0.66 $W.m^{-1} \cdot K^{-1}$. This drop can be explained by the low thermal conductivity of the mPCMs (0.25–0.35 $W.m^{-1} \cdot K^{-1}$ depending on the physical state of the PCM core) [11], and by the large increase in porosity of the mortar which favors thermal transfer by convection in the pores instead of thermal conduction [18–20]. This result is consistent with the observations of Dehdezi *et al.* [18], Hunger et *al.* [19], and Jayalath *et al.* [20]. These authors reported respective decreases in thermal conductivity of ~36%, ~38%, and ~45% with the addition of 5 wt.% mPCMs into cement composites, compared to reference specimens.

3.2 Comparison of the Walls' Behaviors Under Heating/cooling Solicitation

Both walls were conditioned at 20 °C and 50% RH for one week before applying the temperature cycles. The outer face of both walls was then exposed to a periodic variation

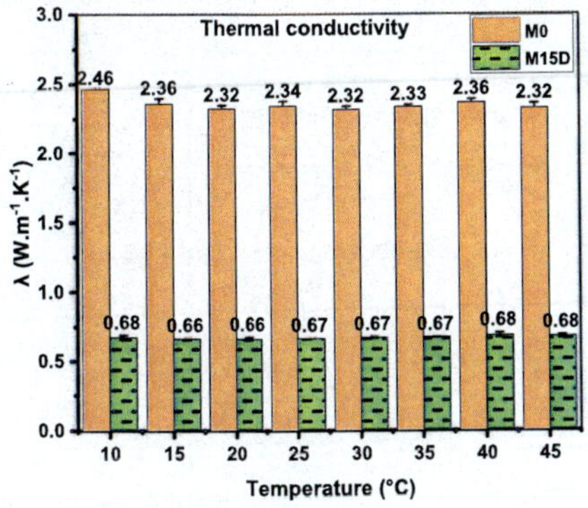

(a)

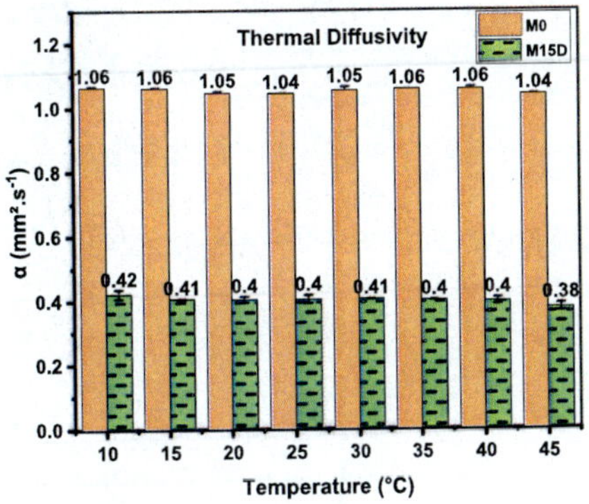

(b)

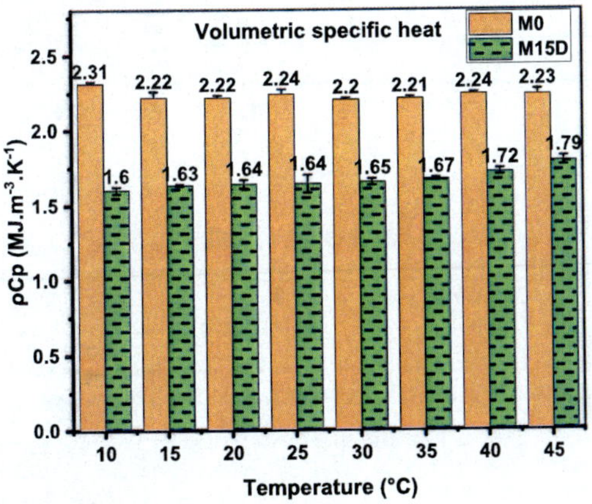

(c)

Fig. 4. Thermophysical properties of M0 and M15D determined by the HD method: (a) thermal conductivity, (b) thermal diffusivity, (c) volumetric heat capacity

of temperature (12 h at 40 °C followed by 12 h at 15 °C). This cycle was repeated 3 times to simulate a 3-day exposure, in order to check the repeatability of the response.

Figures 5a and 5b display the responses of the two walls (M0 and M15D) to the dynamic variation of outdoor temperature at a constant RH level of 50%. The temperature profiles obtained at various depths within the walls are shown in the graphs.

For both walls, a good repeatability is obtained between the successive cycles applied for the same boundaries' conditions, which indicate that only one cycle can represent the behavior of each wall. Therefore, a focus on a 24 h cycle with a comparison between the two walls during the heating/cooling process is presented in Fig. 6.

In Fig. 6, the same evolution trends are observed for the temperature profiles of the two walls over a 24 h cycle. During the heating period at 40 °C, the highest temperature recorded by each sensor is obtained at the end of the heating stage. On the outer face of the wall, this maximum temperature is about 35.9 °C for M0 and 34.7 °C for M15D.

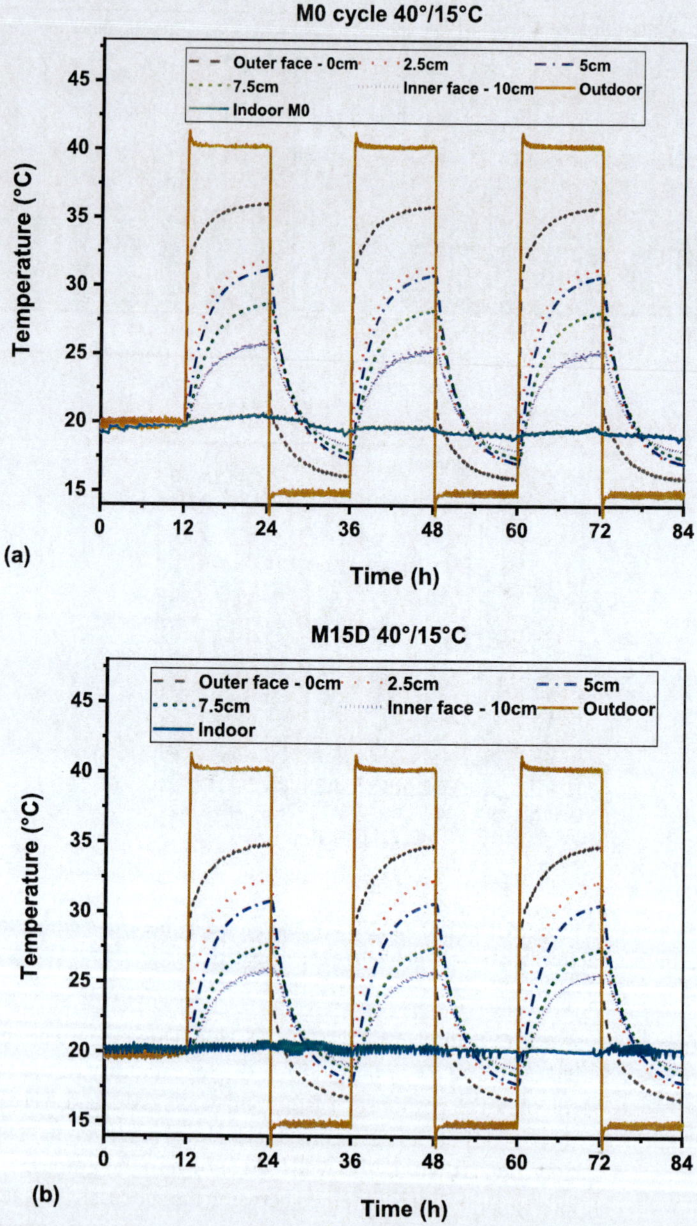

Fig. 5. Responses of the walls to temperature cycles between 40°/15 °C: (a) M0; (b) M15D.

Besides, a gradient of temperature is observed as a function of depth within the two walls, with values decreasing from the outer face to the inner one. Here again, the temperature at a given depth is lower in the M15D wall compared to the reference M0 (with a gap of about 1 °C), except at the depth of 2.5 cm which may result from an error on the position of the corresponding sensors. This gap of ~1 °C between the two walls may reflect the influence of the mPCMs contained in the M15D wall: as temperature increases over the melting point of mPCMs (~28 °C), the core of mPCM particles becomes liquid and some energy is stored in the form latent heat. This phenomenon may slightly reduce the heating kinetics in the M15D wall. Another interesting feature is that this gap of 1 °C is not observed near the inner face of the walls exposed to the indoor laboratory condition at 20 °C. As the local temperature is below the melting transition of mPCMs, the thermal energy storage process is not activated in this case.

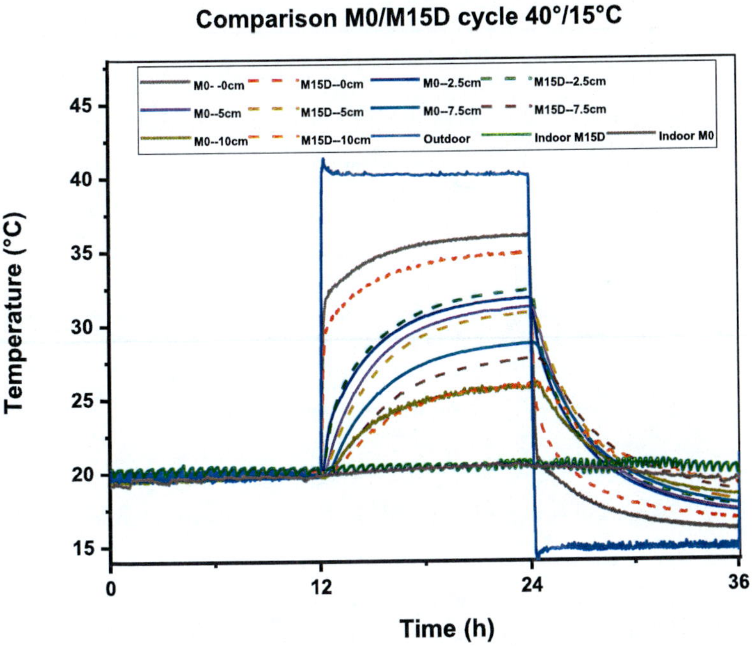

Fig. 6. Compared responses of M0 and M15D walls exposed to a 24 h cycle at 40°/15 °C.

During the cooling stage (12 h at 15 °C), a reverse effect is observed for the temperature gradient evolution. Indeed, the highest temperature is recorded on the inner face (exposed to laboratory condition at 20 °C) and is about 18.3 °C for M0 (19 °C for M15D) and the outer face exhibits the lowest temperature with 15.9 °C for M0 (and 16.8 °C for M15D). The gap of ~1 °C is observed again between the two walls, as M15D exhibits higher temperature than M0 at all depths. Here again, this phenomenon may reflect the influence of the mPCMs: as the temperatures decreases below the crystallization point of mPCMs, a solidification of the PCM core occurs and the stored energy is released, which may dampen the cooling kinetics in M15D compared to M0.

Overall, the damping effect can be evaluated to ~2 °C on the peak-to-peak amplitude during the entire cycle (including both heating and cooling phases).

3.3 Comparison of the Walls' Behaviors Under Steady-State Heating at 40 °C

Figure 7 displays the temperature profiles recorded at various depths within the two walls, when exposed to the second scenario on outdoor conditions (step change from 20 °C to 40 °C followed by a prolonged exposure at 40 °C until steady-state).

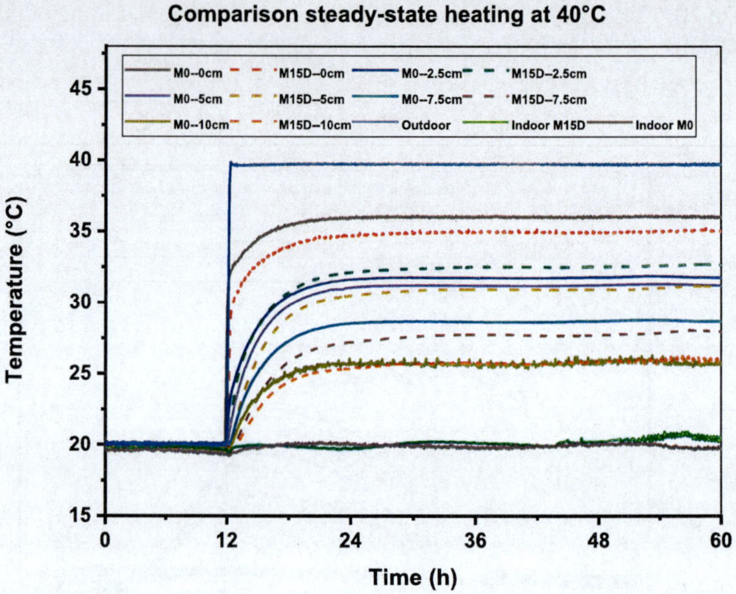

Fig. 7. Comparison between M0 and M15D for a steady-state regime at 40 °C

The temperatures measured at the various depths within the walls increase rapidly after the step change in outdoor temperature from 20 °C to 40 °C; then the evolution kinetics progressively decreases, and a steady-state is reached after about 15 h. As previously observed for scenario 1, the maximum temperature value is recorded on the outer faces of the walls and is about 36 °C for M0 and 35 °C for M15D. A gradient of temperature is also observed as a function of depth in each wall, decreasing from the outer to the inner face. The gap of ~1 °C between the two walls is again observed at depths of 5 cm and 7.5 cm, and can still be attributed to the action of mPCMs contained in M15D. As temperature increases locally over the melting point, the core of mPCM particles becomes liquid, latent heat is stored, and the heating kinetics is slightly decreased.

For the two walls, temperatures measured on the inner faces exposed to the laboratory condition at 20 °C are again almost similar (value of about 25 °C). In this case, the heat storage process of mPCMs is not activated, as local temperature remains below the melting point of the PCM core.

Overall, previous results show that the introduction of mPCMs in the mortar makes it possible to dampen temperature variations in the wall, but only with a limited effect (damping of ~2 °C on peak-to-peak amplitude during cyclic variation of outdoor temperature between 40 °C/15 °C). One could expect a much stronger effect, since thermo-physical properties of the two mortar formulations are very different, especially thermal conductivity (2.32 $W.m^{-1} \cdot K^{-1}$ for M0 and 0.66 $W.m^{-1} \cdot K^{-1}$ for M15D at 20 °C).

Therefore, it may be interesting to conduct complementary investigations by considering alternative scenarios for the variation of the outdoor conditions, or by using specific sensors (to assess the total heat flux across the wall, or the power delivered during heating/cooling phases). Furthermore, finite element (FE) modeling of heat and mass transfer through the two walls could also provide additional information.

4 Conclusion

This study explores the effect of the incorporation of a bio-based and formaldehyde-free mPCMs on the thermal performance of a mortar wall (M15D), compared to a reference wall made of regular mortar (M0). The thermal responses of these walls subjected to two types of dynamic solicitations on a home-made setup were investigated.

During the day/night cyclic loading as well as in a steady-state scenario, the evolutions of temperature gradients exhibit the same trends in both walls. Nevertheless, a gap of about 1 °C is noticed between the walls during the heating phases (and during the cooling phases as well), both on the external face subjected to the outdoor condition and at several depths. Indeed, the incorporation of mPCMs seems to induce a slight damping effect of the temperature variations, due to the energy storage process by latent heat. This damping effect is not observed on the inner face of the M15D wall, as the indoor temperature is below the melting point of the mPCMs and does not allow to activate the heat storage capacity.

It could be interesting to perform additional experiments by applying alternative dynamic scenarios or using flux sensors, and to conduct FE modeling of the heat transfer process, in order to better understand the global thermal behavior of the walls.

Acknowledgement. The authors would like to acknowledge the Labex MMCD (Multi-Scale Modelling & Experimentation of Materials for Sustainable Construction) for its financial support through ANR Investments for the Future (program ANR-11-LABX-022-01).

References

1. Yüksek, I., Karadayi, T.T.: Energy-efficient building design in the context of building life cycle. IntechOpen (2017)
2. Service des données et études statistiques. Bilan énergétique de la France en 2021 - Données provisoires (2022)
3. Chel, A., Kaushik, G.: Renewable energy technologies for sustainable development of energy efficient building. Alex. Eng. J. **57**, 655–669 (2018). https://doi.org/10.1016/j.aej.2017.02.027
4. Amziane, S., Sonebi, M.: Overview on biobased building material made with plant aggregate. RILEM Tech. Lett. **1**, 31–38 (2016). https://doi.org/10.21809/rilemtechlett.2016.9
5. Tyagi, V.V., Buddhi, D.: PCM thermal storage in buildings: a state of art. Renew. Sustain. Energy Rev. **11**, 1146–1166 (2007). https://doi.org/10.1016/j.rser.2005.10.002

6. Kalnæs, S.E., Jelle, B.P.: Phase change materials and products for building applications: a state-of-the-art review and future research opportunities. Energy Build. **94**, 150–176 (2015). https://doi.org/10.1016/j.enbuild.2015.02.023
7. Sarbu, I., Sebarchievici, C.: A comprehensive review of thermal energy storage. Sustainability **10**, 191 (2018). https://doi.org/10.3390/su10010191
8. Sharma, A., Tyagi, V.V., Chen, C.R., Buddhi, D.: Review on thermal energy storage with phase change materials and applications. Renew. Sustain. Energy Rev. **13**, 318–345 (2009). https://doi.org/10.1016/j.rser.2007.10.005
9. Cabeza, L.F., Castell, A., Barreneche, C., de Gracia, A., Fernández, A.I.: Materials used as PCM in thermal energy storage in buildings: a review. Renew. Sustain. Energy Rev. **15**, 1675–1695 (2011). https://doi.org/10.1016/j.rser.2010.11.018
10. Giro-Paloma, J., Martínez, M., Cabeza, L.F., Fernández, A.I.: Types, methods, techniques, and applications for microencapsulated phase change materials (MPCM): a review. Renew. Sustain. Energy Rev. **53**, 1059–1075 (2016). https://doi.org/10.1016/j.rser.2015.09.040
11. Gbekou, F.K., Benzarti, K., Boudenne, A., Eddhahak, A., Duc, M.: Mechanical and thermo-physical properties of cement mortars including bio-based microencapsulated phase change materials. Constr. Build. Mater. **352**, 129056 (2022). https://doi.org/10.1016/j.conbuildmat.2022.129056
12. Das, R., Siva Ranjani Gandhi, I., Muthukumar, P.: Use of agglomerated Micro-encapsulated phase change material in cement mortar as thermal energy storage material for buildings. Mater. Today Proc. **65**, 808–814 (2022). https://doi.org/10.1016/j.matpr.2022.03.316
13. Dehmous, M., Franquet, E., Lamrous, N.: Mechanical and thermal characterizations of various thermal energy storage concretes including low-cost bio-sourced PCM. Energy Build. **241**, 110878 (2021). https://doi.org/10.1016/j.enbuild.2021.110878
14. Lecompte, T., Le Bideau, P., Glouannec, P., Nortershauser, D., Le Masson, S.: Mechanical and thermo-physical behaviour of concretes and mortars containing phase change material. Energy Build. **94**, 52–60 (2015). https://doi.org/10.1016/j.enbuild.2015.02.044
15. Li, Q., et al.: Thermal performance and economy of PCM foamed cement walls for buildings in different climate zones. Energy Build. **277**, 112470 (2022). https://doi.org/10.1016/j.enbuild.2022.112470
16. Khalil, N., Aouad, G., El Cheikh, K., Rémond, S.: Use of calcium sulfoaluminate cements for setting control of 3D-printing mortars. Constr. Build. Mater. **157**, 382–391 (2017). https://doi.org/10.1016/j.conbuildmat.2017.09.109
17. Croda. CrodaTherm^{TM} ME29P: Microencapsulated ambient temperature phase change material, technical datasheet (2018). https://www.crodaindustrialspecialties.com/en-gb/product-finder/product/1381-crodatherm_1_me_1_29d
18. Dehdezi, P.K., Hall, M.R., Dawson, A.R., Casey, S.P.: Thermal, mechanical and microstructural analysis of concrete containing microencapsulated phase change materials. Int. J. Pavement Eng. **14**, 449–462 (2013). https://doi.org/10.1080/10298436.2012.716837
19. Hunger, M., Entrop, A.G., Mandilaras, I., Brouwers, H.J.H., Founti, M.: The behavior of self-compacting concrete containing micro-encapsulated Phase Change Materials. Cement Concr. Compos. **31**, 731–743 (2009). https://doi.org/10.1016/j.cemconcomp.2009.08.002
20. Jayalath, A., et al.: Properties of cementitious mortar and concrete containing micro-encapsulated phase change materials. Constr. Build. Mater. **120**, 408–417 (2016). https://doi.org/10.1016/j.conbuildmat.2016.05.116

The Influence of Biochar on the Flow Properties, Early Hydration, and Strength Evolution of Paste

Wolfram Schmidt[1](✉), Louise Midroit[2], Patrick R. Cunningham[3], Sabbie A. Miller[3], and Sofiane Amziane[4]

[1] Bundesanstalt für Materialforschung und -prüfung (BAM), Berlin, Germany
Wolfram.schmidt@bam.de

[2] Département Génie Civil, Polytech Clermont Ferrand, Aubière, France

[3] Department of Civil and Environmental Engineering, University of California, Davis, Davis, USA

[4] INP Clermont, Institut Pascal, Université Clermont Auvergne, Aubière, France

Abstract. In order to reduce the carbon footprint, biochar can be used as CO_2-negative concrete constituent. The paper shows results of experiments with cement paste and mortar mixtures with 5%, 15% and 30% of cement volume replaced by biochar. The density of the char is measured with a pycnometer I dried and wet conditions, workability is assessed using a spread flow test, and hydration characteristics are measured using isothermal heat flow calorimetry. In addition strength results are presented. The results show that char does not interfere with the cement hydration at early ages, but the water entrapped in the particles can have a slight retarding effect. In general workability and strength were negatively affected by the presence of biochar. However, small amounts of biochar can even have a positive effect on strength without negatively affecting the other properties of the material. The fact that at dosages above 5% the performance specifications of paste and mortar deteriorated has a detrimental effect on the use of biochar as carbon sink in structural concrete.

1 Introduction

The construction industry is responsible for approx. 40% of the global energy demand and carbon emissions [1]. With technology development, the energy demand will decrease in the future, while the issue of embodied carbon gets worse, since the global construction demand will increase dramatically, with pressing challenges of infrastructure and maintenance in the global North, and an urgent need for infrastructure and urbanization in the global South. Therefore, responsible use of resources is key for achieving the Paris Climate goals.

Concrete is the most relevant construction material today, which alone makes out nearly half of everything human beings have ever produced [2]. With 8% carbon emissions, ordinary Portland Cement (OPC), the binder of concrete, has a tremendous climate impact, 60% of which is related to the decomposition of limestone (CaCO3) to reactive calcium oxide (CaO), the reaction of which releases high amounts of CO2. However,

S. Amziane et al. (Eds.): ICBBM 2023, RILEM Bookseries 45, pp. 829–838, 2023.
https://doi.org/10.1007/978-3-031-33465-8_64

the use of any other materials for structural application would even increase the carbon footprint [3]. Therefore, cement-based concrete will remain the major solution for both, the global North and South, but global and regional efforts are urgently required to minimize the materials consumption, and to develop conscious resource supply streams [4].

The most promising way to reduce the carbon footprint of concrete technology today are to:

- Replace the carbon intensive OPC as much as possible by lower carbon footprint materials. Today, depending on cement type between 15% and 25% replacements are common, but research has shown that up to 80% could be replaced by so-called supplementary cementitious materials (SCMs), which are mostly industrial by-products or natural or tempered primary mineral resources [5].
- Reduce the demand for cement without compromising strength and durability by addition of organic superplasticizers (SP), which help to reduce the water in concrete, thus, to increase the material's density [6–8]. SPs are typically based on petrochemical precursors, and their availability is limited and costly, particularly in the global South.

However, these methods are limited, and not capable of creating a carbon neutral concrete industry, which is why future technology developments are urgently required, such as carbon storage and capture and electrification of the entire process chain.

The process technological solutions, however, come along with a variety of challenges, and their effective implementation will take decades. Therefore, the use of biochar as possible CO_2-negative constituent has been discussed more often recently in the area of concrete technology. After pyrolysis up to 80% of the CO_2 that was sequestered by biomass remains in the char. Biomass can roughly sequester CO_2 roughly three times its mass. Due to the decay of biochar of roughly one quarter of its mass when used in agriculture, the immobilization of biochar in construction materials like concrete could potentially become interesting, with the capacity to make a cubic meter of concrete nominally even carbon negative, when large fractions of concrete constituents are replaced by biochar.

The use of biochar is particularly interesting in conjunction with the production of further concrete constituents from biomass such as reactive ashes and polysaccharides. Applying adequate processing technologies [9], biomass from agricultural waste can be converted into a variety of useful high-performance concrete constituents and this way create circular rural to urban synergies, particularly in the global South [10].

Nevertheless, there is hardly a positive contribution of biochar to the fundamental performance values of concrete such as mechanical strength or durability. Therefore, advantages of biochar depend upon the tradeoff between reduced carbon emissions and reduced mechanical performance. In addition, it is important to identify the best use of biochar, since to date its occurrence is limited in most regions of the world and particularly its usefulness in agriculture stands against a use in concrete, as long as the supply is limited.

In order to assess, the potentials and limitations of using biochar in concrete, this paper showcases a variety of experiments that highlight the challenges that come along with the use of biochar in concrete construction technology.

2 Methods and Materials

Biochar for all tests and evaluations was acquired from a variety of resources in different regions in the world. However, the present results are based on woody feedstock for pyrolysis. Ordinary Portland cement type CEM 1 42,5 R (CEMEX; Rüdersdorf, Germany) according to EN 197-1 was used for all mortar and paste mixtures. The choice of CEM I was made to minimize impacts from other additives or SCMs. A powder type polycarboxylic superplasticizer, Melflux AP 101 F (BASF; Ludwigshafen, Germany) was used as a superplasticizer. The mortar mixtures were fully prepared according to EN 196-1. However, for the pastes with biochar the replaced fraction of cement was replaced with char by volume to ensure that the volumetric mixture compositions remained identical. For the char density when stored in room climate 1100 kg/m^3 was assumed as appropriate value. It has to be noted that the determination of the density for char is difficult. The reason for this is the swelling of char in the presence of water, which causes that the volume increase overcompensates the weight increase due to water filled pores to certain degree and with the given char (other chars might have different properties). Deionized (DI) water was used for all sample production. The reference (REF) paste and mortar mixtures were both designed with a w/b ratio of 0.47 following EN 196-1:2016. Mixture IDs and replacement rates are listed in Table 1 where the water and sand content were kept the same as the reference for all mixtures.

Table 1. Mortar / Paste mixture IDs and corresponding replacement rates, superplasticizer dosing, and testing protocols applied.

ID	Cement replacement by volume	Superplasticizer dose	Experimental Protocols used
Ref	0%	–	Isothermal calorimetry, Haegermann cone, rheometry, compressive strength
V5%	5%	–	Isothermal calorimetry, Haegermann cone, rheometry, compressive strength
V15%	15%	–	Isothermal calorimetry, Haegermann cone, rheometry, compressive strength
V30%	30%	–	Isothermal calorimetry, Haegermann cone, rheometry, compressive strength
V30%SP	30%	0.2% of cement by mass	Isothermal calorimetry, Hagerman cone, rheometry, compressive strength

2.1 Pycnometer Density Measurements

The true density of the biochar is an important parameter for mixture designing whether via mass or volume replacement of cement. Thus, two methods were used to experimentally measure the true density of the biochar. The first method was with gas pycnometer and helium gas to determine the true density of the biochar based on the empty volume in the cell not filled by the solid volume of the biochar.

Second, a 400 mL glass pycnometer filled with either DI water or isopropanol alcohol was used to experimentally determine true density using the Archimedes Principle where in a known mass of material is submerged in a liquid-filled pycnometer to determine the volume of the mass based on the volume of liquid displacement. Dry biochar is a very light material which can float and lead to incorrect displacement readings. To overcome this, a fine steel mesh and piece of filter paper was used to keep the particle submerged. The volume of the mesh and filter paper were subtracted from the displacement volume when calculating the experimental true density. To evaluate the potential for the particles to absorb the fluid and swell, and thus change the volume, displacement was measured immediately after submerging in the liquid and 24h after being submerged in the liquid.

2.2 Spread Flow and Rheology Evaluations

Rheological and workability evaluates were performed to understand the effects of biochar, and its reported water absorption characteristics, in Portland cement-based binder systems. Spread flow was determined using a Haegermann cone and a flow test table. First, a Hagerman cone on a flow-test table was filled half-way with mortar and then compacted by rodding 10 times. Then, the remaining half of the cone was filled and, again, compacted by rodding 10 times. The mortar was levelled, and excess material was removed. The cone was then removed, and flow-test table was used to 15 drops. Finally, two spread diameter measurements were taken perpendicular to one another and averaged to determine the spread flow.

From these experiments, the impact of biochar on the static yield stress of pastes was observed.

2.3 Isothermal Calorimetry

To identify how the biochar affect the hydration of cement-based materials, isothermal calorimetry was performed in a TAM Air isothermal calorimeter (TA Instruments) at 20 °C using approximately 14 g of cement paste in plastic 20 mL ampoules.

The powders were mixed dry for 60 s at 200 RPM. Then water was added and the paste mixed for an additional 60 s. The material was placed into the plastic ampoules and the exact weight of cement paste was measured so the heat flow measurements could be normalized on a per gram of paste basis. Heat flow was recorded over 7d.

2.4 Compressive Strength

To measure the effect of biochar on early age compressive strength, mortar prisms of 40 mm × 40 mm × 160 mms were produced using a vibrating table for consolidation. First the mold was filled half-way and then vibrated for 15 s. Then filled the to the top and vibrated for an additional 15 s. After molding, samples were covered for 3 d, then demolded and submerged in water until testing.

Following EN 196-1:2016, a ToniPrax (Toni Technik) was used to load the specimens at 2400 N/s (±200 N/s) until failure occurred. The maximum load applied was used to calculate the ultimate compressive strength. Three replicates of each mixture were tested, and the ultimate strengths averaged. Specimens were tested after 3, 6, 28, 52, and 90 days.

3 Results

3.1 Pycnometry

Figure 1 shows the density results for the char used in the investigations measured in the pycnometer. The char, when stored as received in a closed container contained humidity from its environment. The measured specific gravity was 1.04 and 1.01 after 5 min and after 24 h, respectively. This value is close to the assumed density of 1100 kg/m^3 for the mixture composition. When the char is dried, the density of the char is higher than in humid state, which indicates that the particles' volumes decrease upon drying. After soaking water for 24 h, the specific gravity is in the same order of magnitude as the char as received after the same duration.

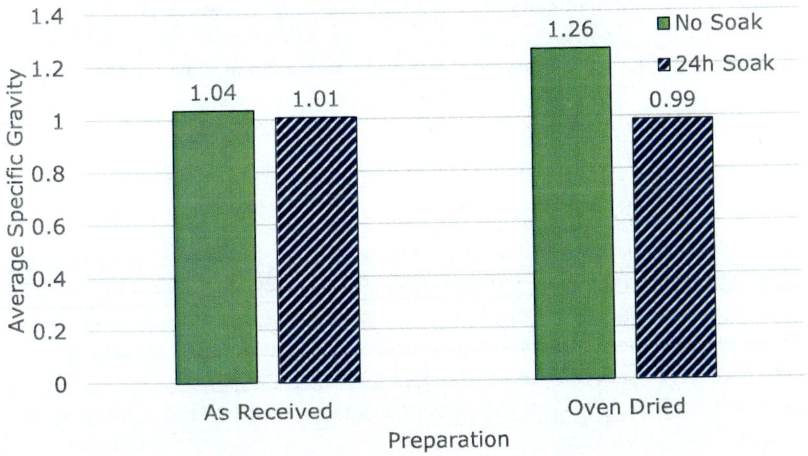

Fig. 1. Measured specific gravities of as received wood char and the same wood char dried in the oven and the specific gravity after of soaking in water for 24 h.

3.2 Rheological and Workability Evaluations

Figure 2 shows the results of the spread flow tests. The spread flow of the reference mortar was 18.8 mm. With increasing char dosage, the spread flow was significantly reduced by 6.0%, 18.2% and 27.4% for 5%, 15% and 39% volume replacement, respectively. The main reason for the flow reduction might be found in the huge specific surface area of the char particles, as well as its porosity, which absorbs water that is no longer availability for flow.

However, by using superplasticizer, the spread diameter could reach the value of the reference again, but the dosage of 0.2% dry polymer content related to cement required to reach this flow property is high, which indicates that the char also captures PCE polymers that will not be involved in dispersion of the overall system.

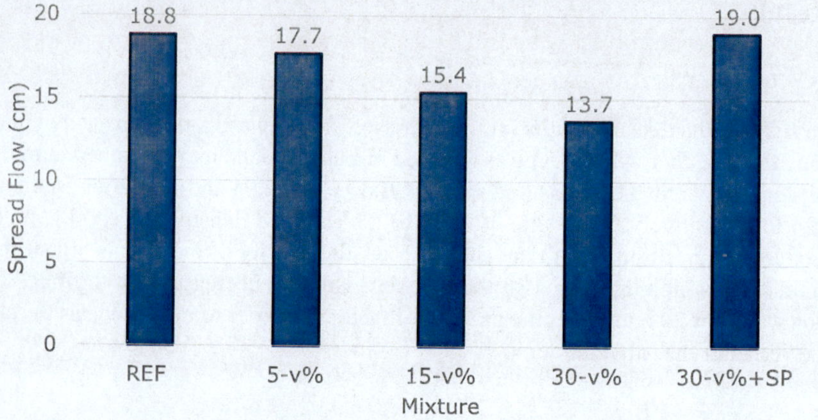

Fig. 2. The spread flow of mortars with no wood char (REF); 5, 15, and 30% volume replacement of cement with wood char; and with superplasticizer and 30% volume replacement of cement with wood char.

3.3 Isothermal Calorimetry

The heat flow curves are shown in Fig. 3 and Fig. 4 related to the overall paste content and the cement inside the system, respectively. The presence of char does not significantly affect the onset of the accelerated hydration, but the maximum of the heat flow curve decreases with increasing char contents (Fig. 3). Furthermore a retardation can be observed.

When the heat flow is related to the actual cement content (Fig. 4), it can be seen that the heat flow curves show similar maximum values, though, for the char samples the maximum is always slightly lower. This indicates that the char does not significantly interfere with the cement hydration actively, but possibly some of the water stored in the char is released during the hydration, causing the slight retardation that can be observed as well. The fact that no significant difference can be observed for systems with 5% and 15% char indicates that the char particles contribute to more stable water distribution during the hydration.

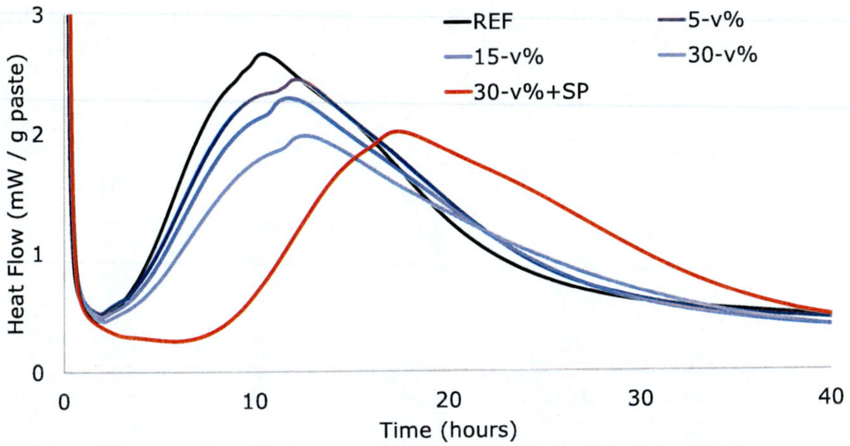

Fig. 3. The heat of hydration of paste samples with the heat flow normalized to the mass of paste for paste with no wood char (REF); 5, 15, and 30% volume replacement of cement with wood char; and with superplasticizer and 30% volume replacement of cement with wood char.

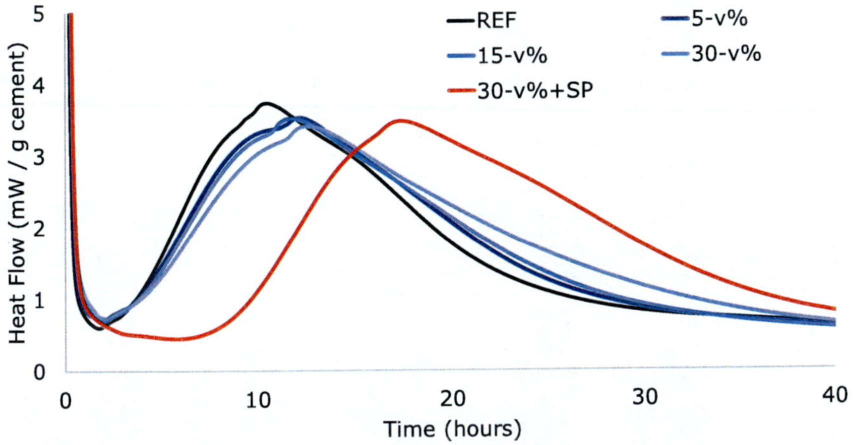

Fig. 4. The heat of hydration of paste samples with the hear flow normalized to the mass of the cement for paste with no wood char (REF); 5, 15, and 30% volume replacement of cement with wood char; and with superplasticizer and 30% volume replacement of cement with wood char.

3.4 Compressive Strength

Figure 5 shows the results of the compressive strength tests after 3, 6, 28, 52, and 90 days for the different mixtures. It can be seen that at early ages the replacement of 5% char does not cause significant performance losses, and that after 28 days the samples with 5% char even show higher compressive strength values. At higher dosages of char it can be clearly seen that the strength significantly goes down. The addition of superplasticizers causes a slight improvement of the strength in comparison to an identical system without, which is most likely related to the better compaction as a result of the enhanced workability.

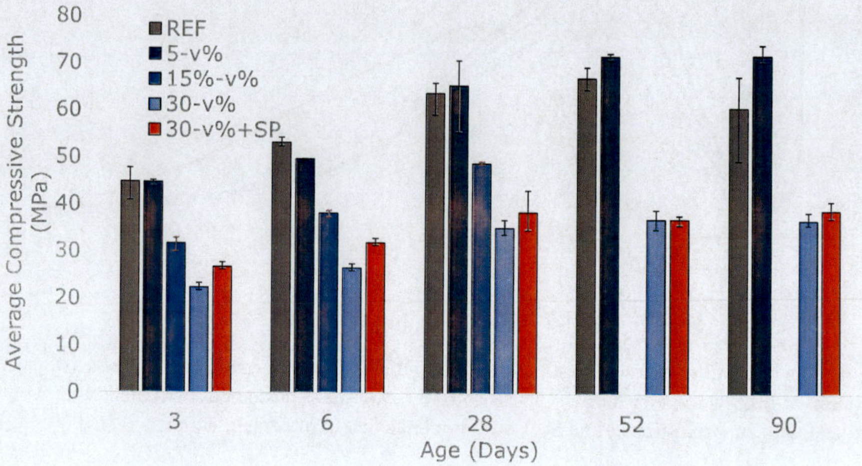

Fig. 5. Compressive strengths of mortar specimens at 3, 6, 28, 52, and 90 days for mortar with no wood char (REF); 5, 15, and 30% volume replacement of cement with wood char; and with superplasticizer and 30% volume replacement of cement with wood char. For 28 days n = 6. All other ages n = 3. Bars show range of values.

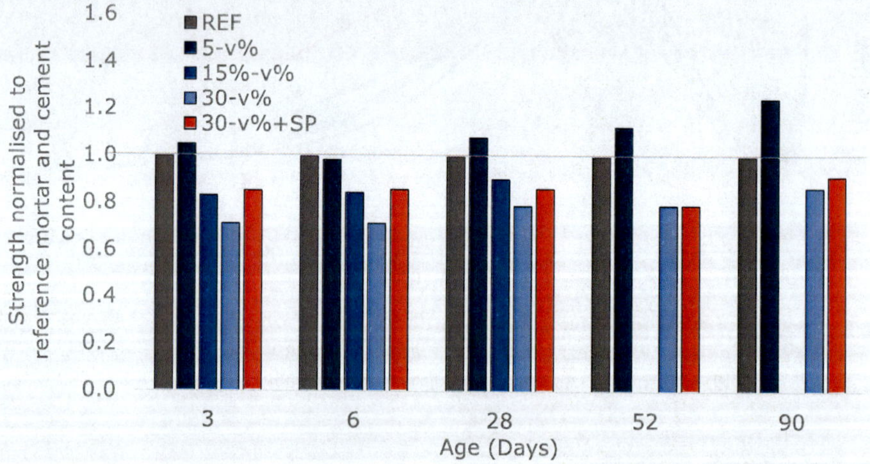

Fig. 6. Compressive strength normalized to REF mortar (without wood char) at 3, 6, 28, 52, and 90 days for mortar with no wood char (REF); 5, 15, and 30% volume replacement of cement with wood char; and with superplasticizer and 30% volume replacement of cement with wood char. For 28 days n = 6. All other ages n = 3.

Figure 6 shows the strength values normalized to the reference sample and the cement content (100% cement content = 1). It shows that indeed the 5% cement replacement by char can enhance the strength properties. The reason for this effect is unclear, but recent research results at BAM with micro-CT under load indicate that at low dosages the char particles enhance the overall pore system of the cement paste, causing a more finely dispersed distribution of air pores, which reduces the number of bigger pores that can be initiator for microcracks.

4 Discussion

Biochar in concrete causes significant technological challenges. First the density determination of the char particles is difficult, and it is questionable, which value must be taken for the volumetric mixture proportioning. The water content of the char before adding it to the mix plays a significant role for density and volume of the particles. The water content inside the char also has implications for the heat of hydration. While the char particles appear to be largely inert, their ability to release and absorb water can have an impact on retardation or acceleration (if a dried char is used). Hence, also the mixing needs further attention, because water trapped in the char particles might get released due to the mixing energy.

In general the char particles seem to reduce workability at early stage, which is assumably due to the water absorption at this point in time. However, compensating for the workability loss requires high amounts of polymers due to the enormous surface of the char particles, which in return causes strong retardation of the hydration. In terms of strength, small amounts of biochar can have a positive effect, but higher amounts of char reduce compressive strength over-proportionally high. This makes it challenging to have a significant reduction of the carbon footprint for structural concrete. However, functionalizing the char could become a solution to overcome this challenge.

5 Summary and Outlook

The present paper shows results on density, workability, heat flow, and strength properties of pastes and mortars where 5%, 15%, and 30% of cement were replaced by biochar. The following conclusions can be made:

- The determination of the density is a major challenge in mixture composition of the char. Since the char particles swell in the presence of water, a wet char has a lower density than a dry char.
- The low density of the char requires that only small weight fractions of char can be incorporated, which limits its capacity to be an efficient carbon sink.
- With increasing replacement of cement by char, the workability is reduced.
- The workability loss can be compensated by addition of superplasticizer, but the required dosage is high.
- The presence of biochar does not significantly affect the cement hydration, but a slight retardation can be observed, which might be due to the release of water entrapped in the char particles with time.
- A 5% replacement of cement with biochar could increase the strength slightly. Higher dosages caused decreased strength properties such way that the presence of char also reduced the effect of the cement.

- Addition of superplasticizer slightly increased the strength properties, assumably due to the enhanced workability, but the effect is marginal.

The paper indicates that biochar in concrete can have a variety of effects on the concrete properties. Except for small dosages of char, where no changes or even improved properties could be observed, at higher dosages of biochar, all materials performance specifications were deteriorated. This indicates that without further processing, char is unlikely to be capable of replacing significant amounts of cement in concrete, thus, the effect on the carbon footprint might be limited. Further investigations are required to identify, how process parameters can enhance the climate effect of biochar in concrete without overly affecting the performance negatively.

References

1. United Nations Environment Programme. 2021 Global Status Report for Buildings and Construction: Towards a Zero-emission, Efficient and Resilient Buildings and Construction Sector. GlobalABC, Nairobi (2021). https://globalabc.org/sites/default/files/inline-files/2020%20Buildings%20GSR_FULL%20REPORT.pdf
2. Venditti, B., Belan, M.: Visualizing the Accumulation of Human-Made Mass on Earth. https://www.visualcapitalist.com/visualizing-the-accumulation-of-human-made-mass-on-earth/. Accessed 05 Jan 2023
3. Barcelo, L., Kline, J., Walenta, G., Gartner, E.: Cement and carbon emissions. Mater. Struct. **47**(6), 1055–1065 (2013). https://doi.org/10.1617/s11527-013-0114-5
4. Schmidt, W., et al.: Specific materials challenges and innovation potentials for minerally bound construction materials in Africa. RILEM Tech. Lett. **4**(50), 63–74 (2020). https://doi.org/10.21809/rilemtechlett.2020.112
5. Scrivener, K., John, V.M., Gartner, E.M.: Eco-efficient cements: potential economically viable solutions for a low-CO_2 cement-based materials industry. United Nations Environment Programme, Paris (2017). (Revised edition)
6. Schmidt, W., Leinitz, S., Mota Gassó, B., Crasselt, C.: Influence of the aqueous phase of the cement paste on the rheology in the presence of PCE. In: Liu, J., Wang, Z., Holland, T., Huang, J., Plank, J. (eds.) Superplasticizers and Other Chemical Admixtures in Concrete, pp. 429–443. American Concrete Institute (2018)
7. Schmidt, W., Brouwers, H.J.H., Kühne, H.-C., Meng, B.: Interactions of polysaccharide stabilising agents with early cement hydration without and in the presence of superplasticizers. Constr. Build. Mater. **139**, 584–593 (2017). https://doi.org/10.1016/j.conbuildmat.2016.11.022
8. Schmidt, W.: Influence of nano scale effects on the macroscopic rheology (2017)
9. Schmidt, W., et al.: Sustainable circular value chains: From rural waste to feasible urban construction materials solutions. Dev. Built Environ. **6**, 100047 (2021). https://doi.org/10.1016/j.dibe.2021.100047
10. Schmidt, W., Olonade, K., Radebe, N., Zando, F., Ssekamatte, V.: Rural employment perspectives from sustainable, green construction materials for urban development. Rural **21**(2/20), 36–38 (2020). https://www.rural21.com/fileadmin/downloads/2020/en-02/rural2020_02-S36-38.pdf

Modeling and Digitalization of Bio-based Building Materials and Constructions

Multiscale Modelling of Bio-composites: Towards Prediction of Their Thermal Conductivity Based on Adequate Knowledge of Their Constituents

S. Rosa Latapie[1](✉), M. Lagouin[2], N. Douk[1], V. Sabathier[1], and A. Abou-Chakra[1]

[1] Université de Toulouse, INSA, UPS, LMDC (Laboratoire Matériaux et Durabilité des Constructions), 135 avenue de Rangueil, 31 077 Toulouse Cedex 04, France
slatapie@insa-toulouse.fr

[2] Laboratoire Génie de Production (LGP), INP-ENIT, Univ. de Toulouse, Tarbes, France

Abstract. To face up to the present energy crisis and the climate challenge, incorporating plant aggregates into reactive or clayey mineral matrices has proved to be an ecological and effective way to improve the thermal performance of buildings. It also helps to reduce both the grey energy involved and the consumption of fossil or non-renewable materials. However, the characterization and optimization of bio-based composites is relatively time-consuming, which may be a hindrance to their development. Our objective is to efficiently predict the thermal conductivity of a hemp particle as this is a fundamental input value for predicting thermal behaviour at the material scale. We found few reliable data on this topic in the literature. This is a major challenge that needs to be addressed as it can lead to erroneous results even when the prediction model is relevant and well developed. An additional difficulty arises from the fact that plant aggregate shows high variability. This paper opens new perspectives by suggesting new procedures for determining the thermal conductivity of hemp shiv using two approaches and considering, in particular, the variability of the raw material. Since existing methods for measuring thermal conductivity do not provide access to the particulate value, an experimental and a theoretical method are proposed to determine it. Both are based on analytical equations obtained by homogenization and both are shown to lead to consistent results. Comparison with the few data available in the literature also highlights a possible overestimation of the values found in previous works.

Keywords: Multiscale modelling · plant aggregates · thermal conductivity

1 Introduction

To address environmental issues, the building sector has to innovate and reduce its energy consumption and greenhouse gas emissions by developing new materials that are environmentally friendly while ensuring the comfort of users. Using plant aggregates is a way to deal with this challenge, as they are identified as agricultural, renewable and locally-available by-products. The use of plant aggregate for construction limits both

S. Amziane et al. (Eds.): ICBBM 2023, RILEM Bookseries 45, pp. 841–858, 2023.
https://doi.org/10.1007/978-3-031-33465-8_65

transport emissions and fossil energy consumption. Plant-based aggregates also allow large quantities of carbon dioxide to be stored in the materials throughout the building's life-cycle thereby reducing its carbon impact [1] and presenting an additional environmental asset. Last but not least, plant particles possess high insulating and hygroscopic capacities. These multiple factors should lead the building industry to massive use of such raw materials. However, the substantial variety of formulations (nature and dosage of aggregates and binder) induces an equally wide range of thermal performance. Therefore, the characterization and optimization of those types of building materials can be costly and time consuming.

In consequence, several models have been proposed to predict thermal properties of bio-based materials. However, the particulate thermal conductivity remains unmeasurable by current techniques while they are needed as model inputs. In addition, while a homogenization model is currently applied by several authors [2–5], considering the particle as an assembly of composite spheres could be questioned, in particular in view of the actual particle geometry. A more faithful representation of microstructure should be investigated. It is also worth noting that hemp shiv is the only plant aggregate with values of particle thermal conductivity available in the literature [3, 6, 7]. The methods by which these values are determined are not always explicitly stated by the authors. Finally, the hygroscopic nature of composites enables ambient hydric variations to be attenuated. However, it makes forecasting their behaviour a more complex task.

Our work is aimed at developing a new tool to guide and optimize bio-based material formulations even before the manufacturing step. This paper presents two different methods for determining the particulate thermal conductivity of dry hemp shiv. In order to validate the suggested procedure and also to limit the raw material variability, only one plant aggregate was chosen for this study: commercial hemp shiv. In addition, to avoid having to consider the effect of water absorption on thermal conductivity, this research was carried out on dry aggregates. Thus, the first method requires measurement of the thermal conductivity of dry particles in bulk, while the second one is more theoretical, being based on the chemical composition and microstructure of the particles. Both methods are founded on analytical equations obtained by homogenization. A protocol to access both thermal conductivity and volume fractions of particles in bulk is therefore established in the following. Indeed, this data is required in the models used. The shape and the microstructure of the particles as well as the anisotropy of the material and its effect on the thermal behaviour are also taken into account. Finally, the values obtained for the thermal conductivity of a hemp shiv particle were compared with those found in the literature. Our models predicted results in reasonable agreement with both isotropic and anisotropic considerations. However, this study may indicate an overestimation of the values determined by the self-consistent scheme [3]. This is discussed in this paper.

2 Materials

2.1 Choice of the Plant Aggregate

Adapted to temperate climates, hemp is a fast growing plant. Peripheral fibres of the plant are used to manufacture textiles and rope. The central part of the stalk, the porous pith, is crushed to obtain hemp shiv, which has been used for several years now as a building material in hemp concrete. This plant aggregate was selected for our study on different criteria. Firstly, the hemp particle is the one that has been most studied in the literature and for which comparative values of particulate thermal conductivity are available, either in scalar form (isotropic considerations) or as a tensor (anisotropic considerations).

Moreover, the hemp particle has high variability [8, 9]. As it is an agricultural by-product, its intrinsic properties depend on the geographical location and method of production, but also on the climate and the harvesting period. This leads to significant irregularity in the raw material extracted. Thus, variability can be observed in the particulate porosity, the density or the chemical composition of the solid skeleton. It is worth noting that the retting process used to separate fibres has an impact on the chemical composition of the aggregates obtained. In particular, the proportion of crystalline phase in the cellulose of hemp fibres depends on the degree of retting. This degree of crystallinity is known to impact the mechanical properties [5]. A possible effect on thermal conductivity will be highlighted in this paper. Finally, in order to limit intrinsic variability (shape factor, size, density, chemical composition) as far as possible, we chose commercial hemp shiv. To assess the consistency of the proposed models, this aggregate is taken as a reference in this study.

2.2 Physical Specificities of Hemp Shiv

Shape of the Hemp Particle. The hemp shiv comes in the form of straw. The commercial product chosen for the study was dusted and calibrated (Fig. 1). Macroscopically, it can be considered as a kind of elongated parallelepiped or even a cylinder. The shape parameters are available in the thesis of Ratsimbazafy [10]. Finally, hemp shiv particles are assumed to be cylindrical in line with the geometrical values of the aggregate studied and according to the literature [11, 12].

Fig. 1. Hemp shiv particle used for this study (size ranging from 2 to 14 mm)

Microstructure of the Hemp Particle. Many capillaries are present in plant aggregates because they allow sap to circulate when the plant is growing. After drying, this structure naturally fills with air, which causes intra-particle porosity. From SEM observations [6, 7], hemp shiv reveals a tubular microstructure, which can be thus assimilated to a cylinder. In consequence, the solid skeleton of the hemp particle is made of cylindrical pores embedded in a continuous matrix.

Porosity. Hemp shiv is employed in construction for its insulating properties. This insulating capacity is a consequence of the high porosity present both inside the particle (intra-particle porosity) and between the particles themselves in the bulk paste (inter-particle porosity). The particle porosity of our samples was 83.5 ± 4.8% [10]. The inter-particle porosity depends on the degree of compaction of the bulk mix.

Density. The density of the particles used in our study was experimentally estimated at $\rho part = 248\ kg \cdot m^{-3}$. This result is consistent regarding the high particle porosity and the density of the solid skeleton constituents [10]. Note that the density of the skeleton of hemp shiv and of cellulose in lignocellulosic biomass are both close to the order of $1540\ kg \cdot m^{-3}$ [10, 13].

2.3 Chemical Characteristics of Hemp Shiv

Chemical composition also influences the properties of plant aggregates and cellulose is well known to be the major component of bio-aggregates. The hemp shiv used in our study was composed of more than 55% cellulose. The chemical components of this plant aggregate are listed in Table 1. These values will be useful to justify the assumptions made at the particle level in the homogenization process.

Table 1. Proportions of the constituents of the hemp shiv used expressed in % of their dry mass according Ratsimbazafy [10].

	Cellulose	Hemicellulose	Lignin	Minerals
Dosage (expressed in % of their dry mass)	55.5 ± 0.2	18.9 ± 0.1	16.4 ± 0.4	1.9 ± 0.0

3 Experimental and Theoretical Methods

3.1 Measurement of the Thermal Conductivity

Thermal conductivity was measured with the hot wire test since it is very suitable for measuring thermal conductivities ranging from 0.02 to $5\ W \cdot m^{-1}K^{-1}$. Values around $0.051–0.070\ W \cdot m^{-1}K^{-1}$ have been reported in the literature for hemp in bulk [10]. Prior testing, the aggregates were dried in an oven at 60 °C until a constant weight was achieved and then cooled in a desiccator to ambient temperature. Measurements were made in a temperature-controlled space at 20 °C. The samples (described in Sect. 2.2) were weighed again after a sequence of measurements to ensure that water adsorption was negligible.

A Neotim-FP2C hot wire apparatus was used for this study. Since this transient method only allows local measurements, thermal conductivity measurements were repeated 5 times. The samples were stirred after each test because the hot wire caused local heating. A first round of measurements was carried out on a bulk mixture. Then, the aggregates were hand-compacted by 30% of the original volume in order to investigate the effect of compaction on the value of particle thermal conductivity.

3.2 Volume Fraction Measurement

One of the difficulties of our protocol was to know the volume fraction of the aggregates in the bulk in which the thermal conductivity measurement was carried out. This was because the granular arrangement was different for each bulk mix. To overcome this difficulty, and based on the recommendations of the RILEM TC 236-BBM [14], particles were weighed and placed in a container for thermal conductivity measurement and, at the same time, the total volume occupied by the mixture was marked. After removing the shiv from the container, it was filled with water to the level of the mark. Finally, the container was weighted and the volume of water, i.e. of bulk aggregates, was calculated.

The volume fraction of aggregates in the bulk arrangement, f_g, was calculated as follows:

$$f_g = \frac{V_g}{V_{tot}} \tag{1}$$

where V_{tot} is the total volume of bulk and V_g that of aggregates.

To gain access to this value experimentally, the previous equation was transformed into:

$$f_g = \frac{M_g}{\rho_{part}} \cdot \frac{1}{V_{tot}} \tag{2}$$

where M_g is the mass of aggregates.

3.3 Choice of Homogenization Schemes

Since it is currently impossible to experimentally measure the thermal conductivity of a single particle, homogenization techniques were used to overcome this difficulty. Note that a self-consistent scheme is generally used in the literature both at the bulk scale [3] and at the composite scale [15]. It is based on an assembly of spherical spheres of various sizes.

Another method was chosen to take the real morphology of the homogenized material into account. Here, we develop the theoretical basis of the techniques chosen for the study. For the heterogeneous material considered, i) the particles in bulk and ii) a single particle, were reduced to an equivalent homogeneous material. A representative volume element Ω (RVE) of the heterogeneous material had to be considered, with δΩ its outer boundary and n the outward unit normal δΩ. The RVE was the minimum volume to be considered as representative of the macroscopic behaviour of the heterogeneous medium. Then homogenization required three steps:

Representation, which consisted of describing the material at the microscopic scale (shape, orientation of inclusions in the matrix).

Localization, which allowed the physical relationships to be expressed at the local scale of the material as a function of the different phases considered. Relations between the quantities at the microscopic scale and those defined at the macroscopic scale were defined.

Homogenization, which finally gave the behaviour of the equivalent homogeneous material thanks to averaging operations on the solutions of the previous step.

Thus, for our study, we considered two RVEs (Fig. 2(a) and Fig. 2(b)) to work on the thermal conductivity tensor of a hemp shiv particle.

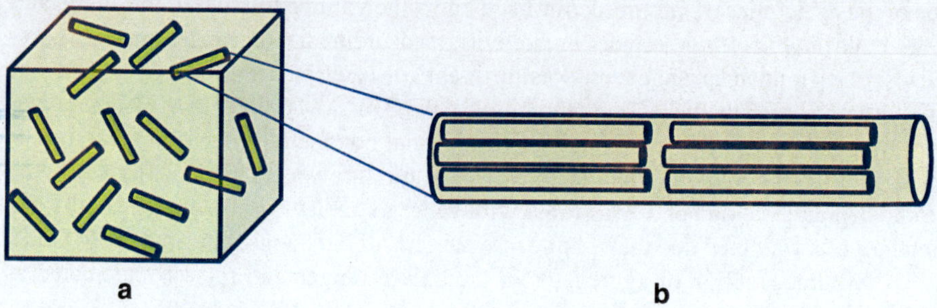

Fig. 2. (a) RVE for the particles in bulk. (b) RVE at the particle scale (5–10 mm)

The RVE considered for particles in bulk comprised two phases: a matrix of air and hemp cylindrical particles as randomly arranged inclusions. The RVE considered at the particle scale was also composed of a two-phase material containing solid matrix and elongated pore-inclusions. In both cases, the matrix was continuous and the interfaces were assumed to be perfect. Thus, we considered two scales: the macroscopic one (bulk or particle) and the microscopic one, at the scale of heterogeneities (inclusions). The fictitious homogeneous medium had the same geometry as the RVE and, subjected to the same stresses, was assumed to have an identical global response (Fig. 3).

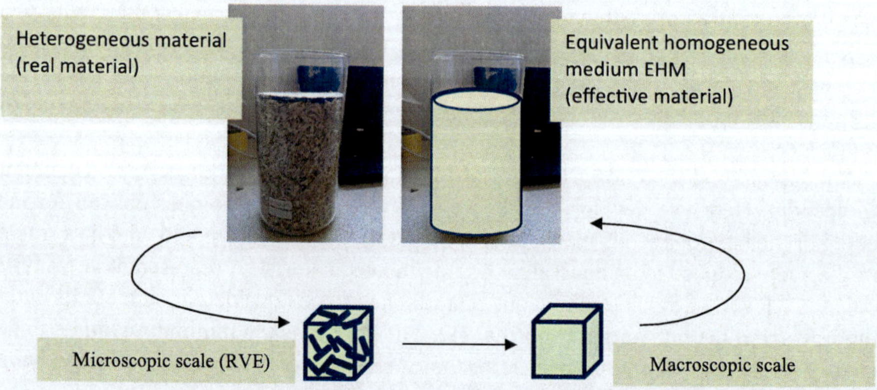

Fig. 3. Homogenization process applied to hemp shiv in bulk

In order to provide uniformity in the notations, the following were adopted for this paper: index 0 to for the matrix, index i for the inclusions.

If the heterogeneous material is a matrix-type composite composed of several inclusions, according to classical theory of Eshelby, this homogenized thermal conductivity tensor is given by:

$$\Lambda^{hom} = \sum_{i=0}^{n} f_i \cdot \Lambda \cdot \mathbb{A}_i = \Lambda_0 + \sum_{i=0}^{n} f_i \cdot (\Lambda_i - \Lambda_0) \cdot \mathbb{A}_i \tag{3}$$

where Λ_0 is the thermal conductivity tensor of the matrix, f_i is the volume fraction of the inclusion i, Λ_i its thermal conductivity tensor and $\mathbb{A}_i$ its second order concentration tensor.

In this study, only two phases were considered: the matrix (the air or the dry particle) and only one type of inclusions (respectively the hemp shiv particles or the air in the pores).

3.4 Estimation of the Effective Thermal Conductivity of a Particle by Inverse Method: Isotropic Consideration

Since the plant aggregates made up a relatively large volume fraction in bulk, the Mori Tanaka approach appeared appropriate to consider the interactions between the inclusions [16]. This homogenization scheme considers heterogeneous medium as an isotropic reinforced matrix (air). Its inclusions behave as isolated inclusions in an infinite matrix. Initially, these inclusions (hemp shiv particles) were considered as spherical. In this case, the thermal conductivity tensor was expressed as:

$$\Lambda_{iso}^{\text{hom}} = \Lambda_0 + f_i \cdot (\Lambda_i - \Lambda_0) \colon \mathbb{A}_i \quad \text{with } \mathbb{A}_i = \left((\mathbb{I} + \mathbb{S}\Lambda_0)^{-1}(\Lambda_i - \Lambda_0)\right)^{-1} \tag{4}$$

where $\mathbb{S}$ is the Eshelby tensor, which depends on the shape of the inclusions and the properties of the matrix.

In the case of a spherical inclusion, the Eshelby tensor is given by:

$$S_{ij} = \begin{cases} 0 \textit{ if } i \neq j \\ 1 \textit{ if } i = j \end{cases} \tag{5}$$

This gives the following analytical formula for the expression of the homogenized thermal conductivity tensor:

$$\Lambda_{iso}^{hom} = \Lambda_0 + \frac{3f_i(\Lambda_i - \Lambda_0)\Lambda_0}{(1-f_i)\Lambda_i + (2+f_i)\Lambda_0} \tag{6}$$

with $\Lambda_{iso}^{hom} = \lambda_{iso}^{hom}\mathbb{I}$

where $\mathbb{I}$ is the second order unit tensor.

The scalar thermal conductivity $\lambda_{iso}^{\text{hom}}$ can be deduced directly from this formula since the distribution is isotropic. The analytical expression obtained is then inverted to lead to the following result:

$$\lambda_i = \frac{\lambda_0(2-3f_i) - 2\lambda_{hom}}{\lambda_{hom} - \lambda_{hom}(1+3f_i)} \tag{7}$$

where λ_0 is the thermal conductivity of the air (0.026 W · m^{-1}K^{-1} at 20 °C), f_i the volume fraction of the hemp particles, and λ_{hom} the measured thermal conductivity of the bulk.

The relation (6) enabled us to determine the thermal conductivity for a single particle of hemp shiv, considering spherical pores and the solid skeleton as the matrix. Expression (7) was calculated to determine the thermal conductivity for a single particle of hemp shiv by an inverse method.

3.5 Estimation of the Effective Thermal Conductivity of a Particle by Inverse Method: Anisotropic Consideration

Because of the oriented thermal conduction of the samples studied, a transversely isotropic thermal conductivity tensor $\Lambda_{ani}^{\text{hom}}$ is considered here. According Nguyen [6], the thermal conductivity tensor at the particle scale, $\Lambda_{ani}^{\text{hom}}$, is given by:

$$\Lambda_{ani}^{\text{hom}} = \begin{pmatrix} \lambda_N^{hp} & 0 & 0 \\ 0 & \lambda_T^{hp} & 0 \\ 0 & 0 & \lambda_T^{hp} \end{pmatrix} \tag{8}$$

The components are detailed in the work of Nguyen [6]. For the specific case of dry particles used here and according to the results of the author, a value of 1.57 was taken for the anisotropy parameter. This coefficient characterizes the ratio between the longitudinal and transverse conductivities of the particle expressed as:

$$\gamma_{dry}^{hp} = \frac{\lambda_N^{hp}}{\lambda_T^{hp}} = 1.57 \tag{9}$$

This method is more advanced than the one presented in Sect. 3.4 as it takes the anisotropy of the particle into account, which is more in agreement with experimental results of the literature [17]. It made it possible to give the values of the components of the tensor $\Lambda_{ani}^{\text{hom}}$ at the particle scale by applying an inverse method.

3.6 A New Method to Estimate the Effective Thermal Conductivity of a Particle Based on Its Chemical Composition

By considering randomly oriented, transversely elongated or isotropic inclusions, the thermal conductivity tensor of hemp shiv particles can be obtained [7]. The conductivity tensor and its components are expressed as:

$$\Lambda_{cyl}^{hom} = \begin{bmatrix} \lambda_N^{hom} & 0 & 0 \\ 0 & \lambda_T^{hom} & 0 \\ 0 & 0 & \lambda_T^{hom} \end{bmatrix} \tag{10}$$

with:

$$\lambda_N^{\text{hom}} = f_i\,\lambda_0 + (1 - f_i)\lambda_i \tag{11}$$

$$\lambda_N^{\text{hom}} = \left(\frac{f_i}{\lambda_i+\lambda_0} + \frac{1-f_i}{2\,\lambda_0}\right)^{-1} - \lambda_0 \tag{12}$$

where λ_0 is the thermal conductivity of the solid skeleton, f_i the volume fraction of air (particular porosity), and λ_{hom} the homogenized thermal conductivity of the hemp particle.

In this part, we are concerned with the thermal conductivity of the particulate solid skeleton since it is not referenced in the literature and is difficult to measure. First, we need to focus on the microscopic description of heat transfer modes in solids. In this context, conductive energy transfer occurs mainly through electronic or vibrational excitations. In the case of a dielectric (insulator), as the electrons are strongly bound to the nucleus, thermal energy is only transmitted by vibration of the crystal lattice. These vibratory motions are generally described by elastic waves and associated with quasi-particles called phonons. Thermal conductivity is therefore governed by the mean free path of phonons [18].The more ordered the atomic lattice (crystalline), the higher the conductivity. According to Sect. 2.3, the solid skeleton of hemp shiv is mainly composed of cellulose, hemicellulose and lignin. However, while the rate of crystallinity observed in cellulose is about 80% [19], hemicellulose and lignin are essentially amorphous polymers [20]. In short, heat energy may be assumed to be transmitted only by cellulose. In the first instance, we considered that the solid skeleton was made of 100% cellulose. We were thus able to apply the formula that relates density and thermal conductivity as a function of temperature [21] for cellulose fibres (CF) from the formula for cellulose nano-fibres (CNF), respectively λ_{CF} and λ_{CNF}:

$$\lambda_{CNF} = \left(0.0787 + 2.73\cdot 10^{-3}T - 7.6749\cdot 10^{-6}T^2 + 8.4637\cdot 10^{-9}T^3\right)\left(\frac{\rho_{sample}}{\rho_0}\right)^2 \tag{13}$$

$$\lambda_{CF} = 1.065\,\lambda_{CNF} \tag{14}$$

where T is the temperature (in Kelvins) of the sample, $\rho_{\text{sample}} = 1543\ \text{kg}\cdot\text{m}^{-3}$ its density (average density of cellulose [10]) and $\rho_0 = 1340\ \text{kg}\cdot\text{m}^{-3}$ (reference value).

Following through on the strong assumption described earlier, the value of λ_0 will be taken to be equal to λ_{CF}. Thus, we theoretically recovered the value of the thermal conductivity of the particles, using the assumptions of, firstly, spherical pores and, secondly, cylindrical pores.

4 Results and Discussion

4.1 Hemp Shiv Considered as a Spherical Inclusion

The results of the thermal conductivity (as a scalar value λ_{iso}^{hp}) for a single particle of hemp shiv and under the assumption of a spherical inclusions are shown in Table 2.

The relative difference between our theoretical and experimental values is 11%, which indicates that they are consistent. Although the limited number of references made comparison difficult, we can see that the relative difference between the average

Table 2. Values of our spherical inclusion models compared to the literature

λ_{iso}^{hp} ($W \cdot m^{-1} \cdot K^{-1}$)	Reference	Working methods
0.085 ± 0.014	Experimental method	Dry particles in bulk, temperature-controlled environment (relation 7)
0.0953	Theoretical method	From the chemical composition and particulate porosity (relation 6)
0.102	Cérézo [3]	Particles in bulk, other precautions not specified
0.1	Pham [22]	Not specified
0.110	Dartois [23]	Iterative micromechanical modelling

literature value of 0.104 $W \cdot m^{-1} \cdot K^{-1}$ and our values is 9% for the theoretical value and 23% for the experimental one. This larger discrepancy may have several origins: if the bulk particle measurement by Cérézo [3] was conducted on non-dry particles, the adsorbed water increased the value of the particulate thermal conductivity and the measurement method could also have a significant effect on the measured values [24]. In addition, the origin of the particle and the variability this implies is likely to have had a significant impact on the value and will be discussed in Sect. 4.5.

Nevertheless, whether they result from the theoretical or the experimental method, our mean values are systematically below those of the literature. It should be noted that Pham [23] concludes that homogenization models systematically overestimate the thermal conductivity of hemp concrete. However, these models use the thermal conductivity value of a single particle, as seen in Sect. 3. It may therefore be possible to suggest that overestimation comes from a non-significant value. Thus, according to our results, the value of 0.1 $W \cdot m^{-1} \cdot K^{-1}$ for a single dry particle of hemp shiv is probably above the real value.

4.2 Hemp Shiv Considered as a Cylindrical Inclusion

The thermal conductivity tensor for a single particle of hemp shiv, under the assumption of cylindrical inclusions, is composed of two types of values: a normal component and a tangential component as detailed in Fig. 4. The transversal conductivity λ_T^{hp} is perpendicular to the plane of the hemp particle and is assumed to be invariant under rotation by angle θ, according to the cylindrical shape. The normal conductivity λ_N^{hp} is oriented parallel to the direction of the tubular pores.

Our results are summarized and compared with the literature in Table 3.

Transversal Component. The relative difference between our theoretical and experimental values is 17%, which indicates they are fairly consistent. Furthermore, the relative difference between the average literature value of 0.915 $W \cdot m^{-1} \cdot K^{-1}$ and our values is 14% for the theoretical value and 37% for the experimental one. Our theoretical value is therefore rather in line with the results of the literature, while our experimental value is clearly different. This discrepancy, which is not significant for the other values, could

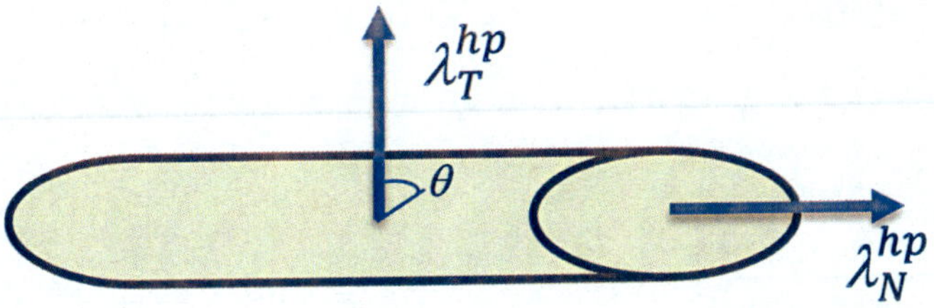

Fig. 4. Components of the conductivity tensor in the case of the cylindrical inclusion hypothesis

Table 3. Values of our cylindrical inclusion models compared to the literature

λ_T^{hp} $(W \cdot m^{-1} \cdot K^{-1})$	λ_N^{hp} $(W \cdot m^{-1} \cdot K^{-1})$	Reference	Working methods
0.067 ± 0.009	**0.105 ± 0.013**	Experimental method	Dry particles in bulk, temperature-controlled environment
0.0805	**0.122**	Theoretical method	From the chemical composition and particulate porosity
0.095	**0.112**	Dartois [23]	Iterative micromechanical modelling
0.088	**0.122**	Tran-Le [7]	Homogenization from a calibrated value for the conductivity of the solid phase

be explained by a combination of several factors: impact of the measurement method, value of the anisotropy coefficient of our sample different from the one considered, overestimation of the literature values. Another bias is discussed in Sect. 4.4.

Normal Component. The relative difference between our theoretical and experimental values is 14%, which indicates again that they are fairly consistent. Furthermore, the difference between the average literature value of $0.117\ W \cdot m^{-1} \cdot K^{-1}$ and our values is 4% for the theoretical value and 19% for the experimental one. Our theoretical value is therefore in very good agreement with the results of the literature, while our experimental value is slightly lower.

4.3 Effects of Compaction

Thermal conductivity measurements on bulk particles were repeated using loose or compacted samples by a maximum of 30% of the initial volume. The particle thermal conductivity values deduced from these measurements by the inverse method are given in Table 4.

Table 4. Comparison of values with and without compaction

Case studied	λ^{hp} $(W \cdot m^{-1} \cdot K^{-1})$ from loose particles	λ^{hp} $(W \cdot m^{-1} \cdot K^{-1})$ from compacted particles	Relative Difference
Isotropic	0.083 ± 0.008	0.086 ± 0.014	3%
Transversal	0.067 ± 0.007	0.066 ± 0.009	2%
Normal	0.105 ± 0.010	0.104 ± 0.015	2%

The difference between the particle thermal conductivity values with and without compaction is very small. Furthermore, no general tendency could be identified for any case. We can therefore conclude that the particulate thermal conductivity may not be impacted by the compaction of the aggregates. It is deduced that the solid skeleton is not altered by compaction forces. This is in agreement with the study by Antlauf [21], which demonstrates that no significant irreversible changes occur for a cellulose structure up to a pressure of about 0.9 GPa. What is true for an exclusively cellulosic skeleton could therefore be extended to a solid skeleton consisting mainly of cellulose, hemicellulose and lignin, under the conditions of our study.

4.4 Back to the Solid Skeleton Hypothesis

A general trend emerges from the comparison of our experimental values with the literature references and with our theoretical results. Our experimental results are consistently lower. A first explanation could be that an offset was induced by the measurement method. However, to date, it is difficult to quantify this discrepancy [24]. Another possible explanation is the fact that the solid skeleton was considered to be composed exclusively of cellulose. This polymer, which is semi-crystalline, participates significantly in heat conduction, much more than an amorphous solid [25]. However, hemicellulose and lignin, which are the two other main components of the plant particle skeleton, are amorphous. In other words, we have overestimated the thermal conductivity of the solid skeleton and, by extension, that of the plant aggregate with this assumption. It would be useful to have access to the thermal conductivity values of the other constituents. This would allow the results of our theoretical method to be refined. Nevertheless, our value for the thermal conductivity of the solid skeleton is consistent with the only estimated value we found in the literature (Table 5).

Table 5. Comparison of thermal conductivity values of the solid skeleton

λ_s^{hp} $(W \cdot m^{-1} \cdot K^{-1})$	Reference	Working methods
0.6109	Theoretical method	From the results of Stelte [20]
0.5755	Nguyen [6]	Inverse analysis

4.5 Variability of Particle Thermal Conductivity

The key strength of the theoretical method is that it quantifies the influence of certain intrinsic parameters of hemp shiv on its particulate thermal conductivity. Thus, the available data [10] on the variability of the cellulose density and the variability of the intra-particle porosity enabled us to set two bounds for theoretical values (Table 6).

Table 6. Extreme values considered for the calculation of the theoretical bounds

Variable parameter	Maximum	Minimum
Density of cellulose ($kg \cdot m^{-3}$)	1450	1630
Particulate porosity (%)	88.30	78.30

These maximum and minimum values obtained were compared with our own values and with those of the literature. The results for the different cases studied in this paper (isotropic and anisotropic) are presented in Figs. 5, 6 and 7.

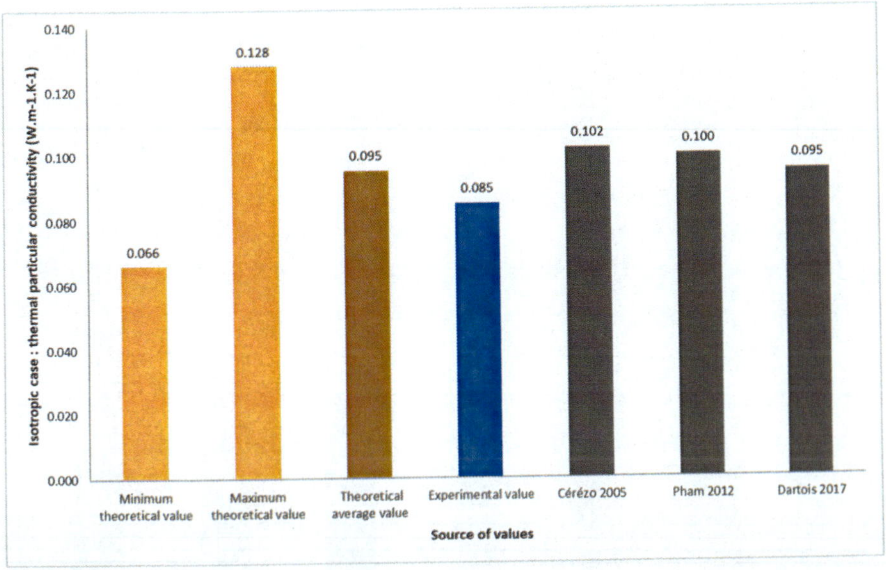

Fig. 5. Isotropic case: extreme theoretical values compared to our results and those of the literature [3, 22, 23].

It is worth noting that the theoretical bounds allow us to include, on the one hand, our experimental value and, on the other hand, those from the literature. Thus, we were able to define a range of particle thermal conductivities for hemp shiv, in both the isotropic and anisotropic cases. This range takes account of the main parameters (particle porosity and density of the solid skeleton), which depend on the very varied conditions of the plant cultivation (growth, place and time of harvesting) and on the nature of the aggregate (cellulosic composition).

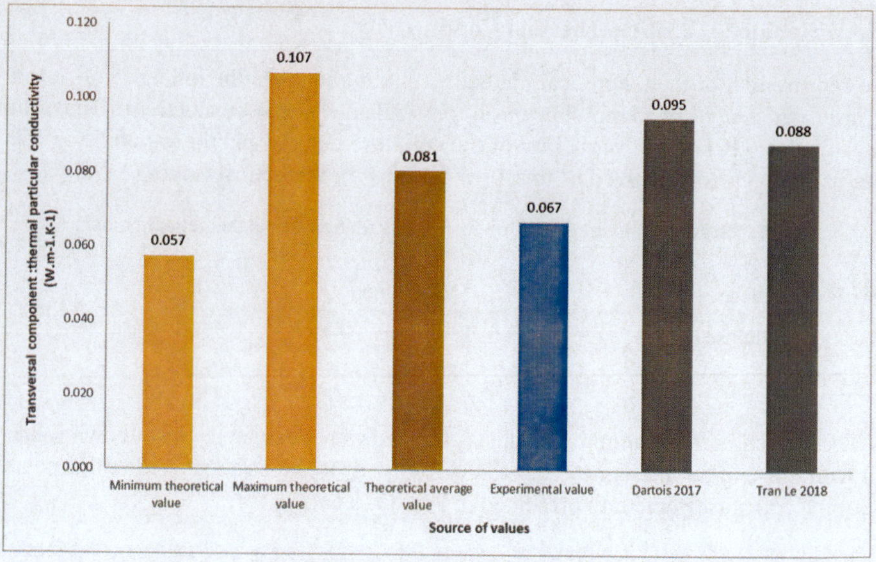

Fig. 6. Transversal component: extreme theoretical values compared to our results and those of the literature [7, 23].

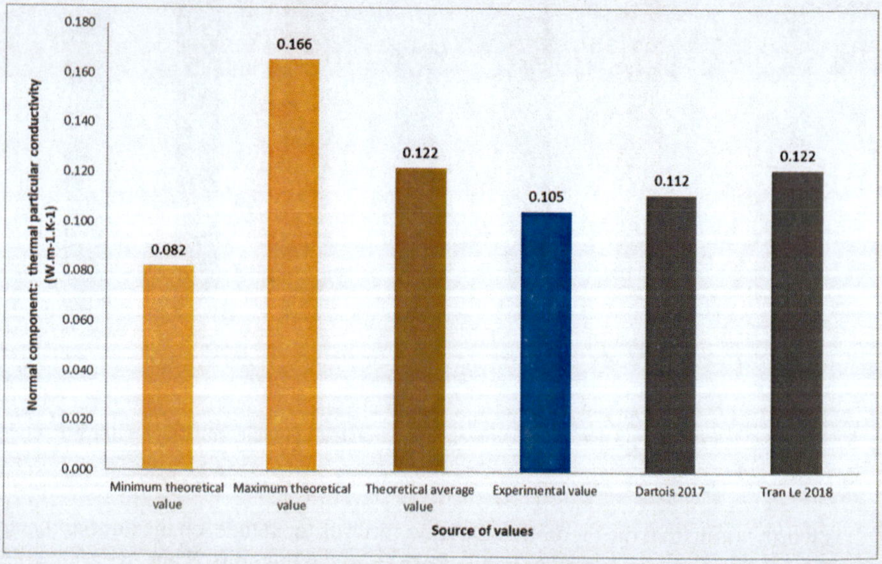

Fig. 7. Normal component: extreme theoretical values compared to our results and those of the literature [7, 23].

Finally, by calculating the ratio of the extreme theoretical thermal conductivities obtained in the anisotropic case, we recover an anisotropy factor consistent with the literature (Table 7).

Table 7. Anisotropy parameter γ^{hp} of hemp shive particle

γ^{hp}	Reference
1.57	Experimental method
1.55	Theoretical method (with maximum values)
1.44	Theoretical method (with minimum values)
1.5	Nguyen [6]
1.18	Dartois [23]
1.53	Tran-Le [7]

Our results are consistent with those of the literature and support a particle anisotropy factor of around 1.5 for hemp shiv particles. This value is important because it determines the anisotropy at the particle scale, but also impacts the anisotropy at the material and wall scale for composites made from hemp shiv. The application process of these materials (sprayed or compacted) leads to a preferential orientation of the particles and thus to anisotropy in the thermal behaviour of the wall.

5 Conclusion

In this study, a better knowledge and a better definition of the thermal properties of hemp-based building materials were being sought. We demonstrated that, in order to reliably predict their behaviour, a deeper understanding of the intrinsic properties of the plant aggregate is required. The development of methods to access local and microstructure-specific values is also essential.

Thus, homogenization techniques have been applied to determine the conductivity tensor at the scale of the dry hemp shiv particle from its microstructural characteristics (morphology of the particle and properties of the constituents). The isotropic (which can be seen as a simplified study) and anisotropic cases have been systematically treated. First, an experimental method was proposed. Unlike some values in the literature for which we did not have the test conditions, the protocol here has been described in detail, and the conditions controlled and explained. The aim of this was to be able to propose a reproducible method. Nevertheless, the values obtained experimentally were below the values of the literature. We therefore suspected an overestimation of the latter, since we were unable to assess the methods of determination.

Furthermore, a new method for estimating the particulate thermal conductivity has been developed. Using the results of a recent study [21] the hypothesis of a solid skeleton consisting exclusively of cellulose has been proposed. This assumption has been discussed and could lead to an overestimation of thermal conductivity values. Nevertheless, the strong point of this method is to be able to quantify the impact of the variability of the resource, coming from the plant world, on the particulate thermal conductivity. Our study also shows that the impact of compaction on the particle thermal conductivity value can be negligible. In addition, it demonstrates that the anisotropy parameter at the particle scale, considering heat transfer, is around 1.5. This determining factor has to be taken into account at the scale of the particle used as a bulk insulator or incorporated in a binding matrix, as the insulating layer at the wall scale will also be thermally anisotropic as a result.

Finally, the proposed methods were found to be consistent with the results of the literature. This highlights their relevance. Given the lack of data on the thermal conductivity of hemp shiv particles, this study provides new ways of assessing this value, which is impossible to measure directly with current measurement techniques. The experimental method used here is relatively easy to implement although it uses complex micromechanical modelling results. This inverse method therefore deserves to be applied to other aggregates from agricultural co-products. The theoretical method has also demonstrated its applicability by providing the range of values within which our theoretical, experimental and literature values fall. The same models should be applied first to other types of hemp shiv to evaluate the impact of the variability of this plant-aggregate. It could then be extended to other raw materials provided that a minimum of data is available on their microstructure. Further work will therefore focus on extending these methods to other aggregates. It will also be useful to consider the impact of adsorbed water on thermal conductivity at the particle level. This study provides thus new perspectives for determining the particulate thermal conductivity value, which is essential for predicting the insulating properties of bio-based materials.

References

1. Pretot, S., Collet, F., Garnier, C.: Life cycle assessment of a hemp concrete wall: impact of thickness and coating. Build. Environ. **72**, 223–231 (2014). https://doi.org/10.1016/j.buildenv.2013.11.010
2. Collet, F.: Caracterisation hydrique et thermique de materiaux de genie civil a faibles impacts environnementaux (2004)
3. Cérézo, V.: Propriétés mécaniques, thermiques et acoustiques d'un matériau à base de particules végétales: approche expérimentale et modélisation théorique. These de doctorat, Lyon, INSA (2005). https://www.theses.fr/2005ISAL0037. Accessed 03 Oct 2022
4. Bederina, M., Marmoret, L., Mezreb, K., Khenfer, M.M., Bali, A., Quéneudec, M.: Effect of the addition of wood shavings on thermal conductivity of sand concretes: experimental study and modelling. Constr. Build. Mater. **21**(3), 662–668 (2007). https://doi.org/10.1016/j.conbuildmat.2005.12.008
5. Mazhoud, B.: Elaboration et caractérisation mécanique, hygrique et thermique de composites bio-sourcés. These de doctorat, Rennes, INSA (2017). https://www.theses.fr/2017ISAR0024. Accessed 03 Oct 2022
6. Nguyen, S.T., Tran-Le, A.D., Vu, M.N., To, Q.D., Douzane, O., Langlet, T.: Modeling thermal conductivity of hemp insulation material: a multi-scale homogenization approach. Build. Environ. **107**, 127–134 (2016). https://doi.org/10.1016/j.buildenv.2016.07.026
7. Tran-Le, A.D., Nguyen, S.-T., Langlet, T.: A novel anisotropic analytical model for effective thermal conductivity tensor of dry lime-hemp concrete with preferred spatial distributions. Energy Build. **182**, 75–87 (2019). https://doi.org/10.1016/j.enbuild.2018.09.043

8. Niyigena, C., Amziane, S., Chateauneuf, A.: Etude de la variabilité des caractéristiques de granulats de chanvre. In: Rencontres Universitaires de Génie Civil, Bayonne, France (2015). https://hal.archives-ouvertes.fr/hal-01167627. Accessed 18 Nov 2022
9. Vinceslas, T.: Caractérisation d'éco-matériaux terre-chanvre en prenant en compte la variabilité des ressources disponibles localement. These de doctorat, Lorient (2019). https://www.theses.fr/2019LORIS538. Accessed 18 Nov 2022
10. Ratsimbazafy, H.H.: Évaluation du potentiel de co-produits agricoles locaux valorisables dans le domaine des matériaux de construction (PALOMAC). These de doctorat, Toulouse 3 (2022). https://www.theses.fr/2022TOU30005. Accessed 01 Oct 2022
11. Couture, A., François, V., Cuillière, J.-C., Pilvin, P.: Modélisation de la microstructure de composites particulaires: application à l'homogénéisation en conduction thermique. In: 13e colloque national en calcul des structures, Giens, Var, France (2017). https://hal.archives-ouvertes.fr/hal-01921977. Accessed 11 Nov 2022
12. Huang, G., Abou-Chakra, A., Absi, J., Geoffroy, S.: Optimization of mechanical properties in anisotropic bio-based building materials by a multiscale homogenization model. J. Build. Eng. **57**, 104890 (2022). https://doi.org/10.1016/j.jobe.2022.104890
13. Nozahic, V., Amziane, S., Torrent, G., Saïdi, K., De Baynast, H.: Design of green concrete made of plant-derived aggregates and a pumice–lime binder. Cem. Concr. Compos. **34**(2), 231–241 (2012). https://doi.org/10.1016/j.cemconcomp.2011.09.002
14. Amziane, S., Collet, F., Lawrence, M., Magniont, C., Picandet, V., Sonebi, M.: Recommendation of the RILEM TC 236-BBM: characterisation testing of hemp shiv to determine the initial water content, water absorption, dry density, particle size distribution and thermal conductivity. Mater. Struct. **50**(3), 1–11 (2017). https://doi.org/10.1617/s11527-017-1029-3
15. Collet, F., Pretot, S.: Thermal conductivity of hemp concretes: variation with formulation, density and water content. Constr. Build. Mater. **65**, 612–619 (2014). https://doi.org/10.1016/j.conbuildmat.2014.05.039
16. Mori, T., Tanaka, K.: Average stress in matrix and average elastic energy of materials with misfitting inclusions. Acta Metall. **21**(5), 571–574 (1973). https://doi.org/10.1016/0001-6160(73)90064-3
17. Nguyen, T.T.: Contribution à l'étude de la formulation et du procédé de fabrication d'éléments de construction en béton de chanvre. Ph.D. thesis, Université de Bretagne Sud (2010)
18. France-Lanord, A.: Transport électronique et thermique dans des nanostructures. Ph.D. thesis, Université Paris Saclay (COmUE) (2016). https://tel.archives-ouvertes.fr/tel-01447655. Accessed 14 Nov 2022
19. Almusawi, A.M.: Mise en œuvre et optimisation des propriétés d'une structure sandwich en matériaux biosourcés (fibres et bois de chanvre) avec une matrice en polystyrène expansé pour le bâtiment. Ph.D. thesis, Université Bourgogne Franche-Comté (2017). https://tel.archives-ouvertes.fr/tel-01870446. Accessed 14 Nov 2022
20. Stelte, W., Clemons, C., Holm, J.K., Ahrenfeldt, J., Henriksen, U.B., Sanadi, A.R.: Thermal transitions of the amorphous polymers in wheat straw. Ind. Crops Prod. **34**(1), 1053–1056 (2011). https://doi.org/10.1016/j.indcrop.2011.03.014
21. Antlauf, M., Boulanger, N., Berglund, L., Oksman, K., Andersson, O.: Thermal conductivity of cellulose fibers in different size scales and densities. Biomacromol **22**(9), 3800–3809 (2021). https://doi.org/10.1021/acs.biomac.1c00643
22. Pham, T., Férec, J., Picandet, V., Tronet, P., Costa, J., Pilvin, P.: Etude expérimentale et numérique de la conductivité thermique d'un composite chaux–chanvre (2012)
23. Dartois, S., Mom, S., Dumontet, H., Hamida, A.B.: An iterative micromechanical modeling to estimate the thermal and mechanical properties of polydisperse composites with platy particles: application to anisotropic hemp and lime concretes. Constr. Build. Mater. **152**, 661–671 (2017). https://doi.org/10.1016/j.conbuildmat.2017.06.181

24. Seng, B., Magniont, C., Lorente, S.: Characterization of a precast hemp concrete. Part I: Physical and thermal properties. J. Build. Eng. **24**, 100540 (2019). https://doi.org/10.1016/j.jobe.2018.07.016
25. Schick, V.: Caractérisation d'une mémoire à changement de phase: mesure de propriétés thermiques de couches minces à haute temperature. These de doctorat, Bordeaux 1 (2011). https://www.theses.fr/2011BOR14280. Accessed 08 Dec 2022

Influences of Geometrical Imperfections on the Buckling Behavior of Slender Bamboo Culms

Henrieke Fritz(✉) and Matthias Kraus

Chair of Steel and Hybrid Structures, Bauhaus University Weimar, Weimar, Germany
henrieke.fritz@uni-weimar.de

Abstract. The European construction sector is increasingly facing the challenge of utilizing ecologically sustainable and renewable resources due to climate change and the increasing scarcity of raw materials. Employing bamboo as a renewable building material can offer significant environmental and economic benefits, supposing it is obtained from local resources. While naturally native to tropical and subtropical areas, bamboo is nowadays also cultivated in temperate regions such as Europe. Although these species generally provide smaller cross sections compared to the bamboo grown in tropical areas, they have excellent mechanical properties for construction purposes. However, due to the increased slenderness, the susceptibility to the failure mode of flexural buckling when applied as compression members is of particular importance, being strongly influenced by natural geometrical imperfections. Thus, in the present study, corresponding influences (e.g. out-of-straightness, stem taper and ovality of the cross section) on the structural behavior of slender bamboo culms are analyzed. For this purpose, measurements of geometrical imperfections taken from Italian Phyllostachys edulis culms are applied to functionally graded numerical models of bamboo members. Subsequently, the influences of the imperfections on the flexural buckling behavior are quantified. Finally, the numerical results are compared with the current recommendations for calculating the buckling load in ISO 22156:2021. In summary, this study contributes to developing safe design rules for slender bamboo culms in load-bearing structures and thus to establishing bamboo construction as a sustainable construction method in Europe.

Keywords: Bamboo structures · Phyllostachys edulis (Moso) bamboo · geometrical imperfections · stability analysis

1 Introduction

Bamboo is one of the oldest construction materials in the world, widely employed for scaffolds, houses, bridges and other structures. The use of bamboo as a renewable building material offers considerable ecological and economic advantages. Amongst others, bamboo significantly contributes to the environmental decarbonization due to its rapid growth, cost-effectiveness, high weight-to-strength ratio, and its possibility to serve as

S. Amziane et al. (Eds.): ICBBM 2023, RILEM Bookseries 45, pp. 859–869, 2023.
https://doi.org/10.1007/978-3-031-33465-8_66

a substitute for timber in case demand increases and supply dwindles [1]. Although bamboo is naturally native to tropical and subtropical zones, it has spread to temperate regions in Europe, such as Italy, where about 2000 ha of giant bamboo grows [2]. European bamboo culms likewise exhibit excellent mechanical properties for construction purposes, as found in [3]. However, due to different growth conditions, they generally dispose of more slender culms.

Bamboo culms are often applied as compression members, whose load-bearing capacities are strongly influenced by their slenderness, i.e., the ratio between column length and gyration radius. When slenderness ratio is high, the stability behavior reduces capacity such that the buckling strength may be significantly lower than its cross sectional compressive strength. In numerous studies, the corresponding member capacity is determined neglecting initial natural geometric imperfections of bamboo culms, such as ovality, culm taper, and pre-deflections. However, the imperfections can strongly affect the stability behavior of bamboo culms, as initial pre-deflections produce additional bending under axial compression for instance.

Thus, this study aims to analyze influences of geometric imperfections on the flexural buckling behavior of slender bamboo culms. For this purpose, geometric imperfections of 90 Italian bamboo culms of the species Phyllostachys edulis (Moso) bamboo are assessed and statistically evaluated. Subsequently, the imperfections measured are applied to a parameterized finite element model with transversal isotropic material behavior. The influences of the geometrical variations on the buckling capacity are investigated by means of numerical stability analyses. Finally, the results are compared with buckling loads calculated according to ISO 22156:2021 [13] and an outlook on future work is provided.

2 Morphology and Geometrical Imperfections of Bamboo Culms

Bamboo is the fastest-growing plant in the world, with a growth rate of up to 100 cm per day [4]. Essential for the plant's rapid growth is its underground rhizome network – the bamboo stem. The above-ground part is called the bamboo culm. The bamboo culm is initially compressed in so-called bamboo shoots that stretch axially during growth. When reaching its maximum height, the lateral branching and leaves unfold, which later produce scars on the bamboo culms. Bamboo culms are periodically divided into segments ("internodes") by transverse nodes and diaphragms (Fig. 1). The hollow culms are permeated by vascular fibers, supplying the culms with nutrients and water. At the nodes, parts of the longitudinally fibers change their orientation integrating orthogonally into the diaphragms, resulting in a change of material properties within the nodal regions.

The number and distribution of nodes are species and growth-dependent. The average internode length is not constant but increases along the culm axis from the base to the top and especially varies at culm ends. Besides varying internodes along the length, bamboo culms exhibit further geometrical variations. Geometrical imperfections of bamboo culms are generally considered to include varying parameters defined as deviations from a straight circular hollow profile. These imperfections include (i) ovality, (ii) culm taper, and (iii) pre-deflections (Fig. 2).

Although bamboo culms are often abstracted as circular hollow sections, they provide rather elliptical cross sections. Therefore, ovality is referred to as the degree of deviation

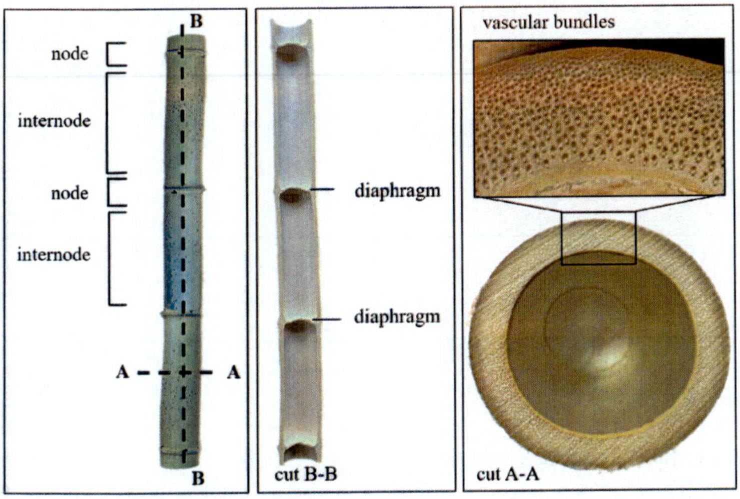

Fig. 1. Morphology of bamboo culms.

of the culm cross section from circularity, given as the deviation between the largest and the smallest diameter $(D_{max} - D_{min})$ related to the mean diameter $(D_{max} + D_{min})/2$, see [12]. Accordingly, the ovality of the outer diameter o_{ext} is defined to be:

$$o_{ext} = 2 \cdot \frac{(D_{max} - D_{min})}{(D_{max} + D_{min})}. \tag{1}$$

Since wall thicknesses may vary, it can be suitable to define the ovality of the inner diameter o_{int} as well, which, however, is not considered in this study.

The diameter of bamboo culms decreases along the culm axis, referred to as culm taper. The culm shape related to taper can be described as a mixture of paraboloid and cone [5]. Its development depends on several characteristics, such as genetics, site and climate [6]. The culm taper is expressed as the degree of change in diameter over member length. The maximum outer taper α_i is determined by

$$\alpha_i = \frac{D_{max,bot.} - D_{max,top}}{L}, \tag{2}$$

where $D_{max,bot.}$ is the maximum diameter at the member base and $D_{max,top}$ the maximum diameter at the top.

Next to ovality and culm taper, pre-deflection or "out-of-straightness" is considered to measure the deviation from a straight culm. Measurements of maximum deflections perpendicular to the ideal (straight) member axis δ_u may be used to subsequently approximate the deflection δ_j at a position x_j. For example, by applying a sine alternation (half-wave) or a 3rd degree polynomial considering the magnitude and position of the maximum bow δ_u as well as the pre-deflections at the culm top and bottom.

It should be noted that both the culm taper and the pre-deflections may be investigated for different axes within the cross sections. However, in the present work, only maximum values are considered without differentiation.

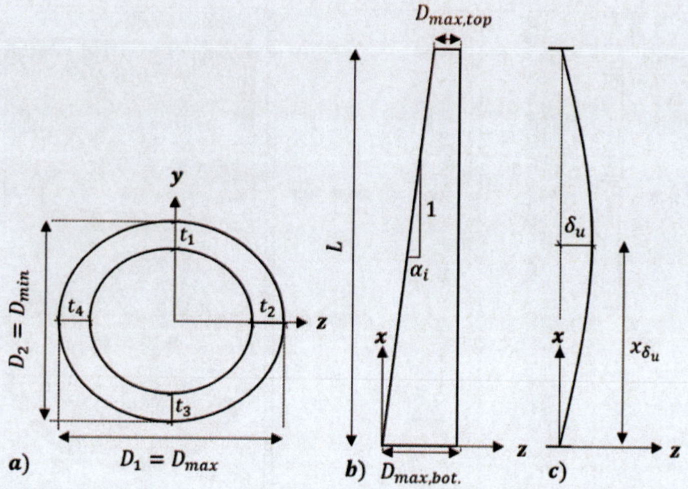

Fig. 2. Geometric imperfections of bamboo culms (a) ovality of cross sections, (b) stem taper, and (c) pre-deflections.

3 Geometry Measurements of Italian Phyllostachys Edulis

To investigate natural geometric imperfections of European bamboo species, geometry measurements of 90 bamboo culms of four meters in length are recorded. The bamboo culms originate from the northern Italian province of Pordenone (Friuli Venezia-Giulia region). On average, the culms are four years old and have subsequently been dried for one to three years after harvesting. The total length of the bamboo culms ranged from 10 to 13 m, with all of the culms examined being lower-end pieces.

Top and bottom cross sections with diameters D_1 and D_2 and wall thicknesses t_1 to t_4 are measured using a digital caliper (cf. Fig. 2). The y axis is defined to be the stronger rotation axis with respect to the bending stiffness, specified by the largest diameter of the cross section at the culm bottom $D_{1,\text{bot.}}$, whereas the z axis is defined to be perpendicular to the y axis, specified by $D_{2,\text{bot.}}$. At the top of the culm, the diameters D_1 and D_2 correspond to the axes defined at the bottom, with approximately $D_{\text{max,top}} = D_1$ and $D_{\text{min,top}} = D_2$ due to overall low rotations of the axes. The diameters D_1 and D_2 are measured with respect to both axes in the middle of each internode. The setup for measuring pre-deflections in terms of member straightness is adopted from [7], where the deflections in y and z direction, respectively, in the middle of each internode is evaluated.

For all measurements x_i, the corresponding mean values $\overline{x}$ and standard deviations s are calculated according to

$$\overline{x} = \frac{\sum_{i=1}^{n} x_i}{\text{n}}, \tag{3}$$

and

$$s = \sqrt{\frac{\sum_{i=1}^{n} (x_i - \overline{x})^2}{\text{n}}}. \tag{4}$$

The results x_i of the $n = 90$ culm measurements including the corresponding mean values, standard deviations as well as the minimum and maximum values of the parameters are summarized in Table 1 (left). In Table 1, the culm thicknesses t_{bottom} and t_{top} are calculated as averages of t_1 to t_4.

Table 1. Standard deviation $\overline{x}$, variance s_{i} and min./max. values of the 90 four-meter bamboo culms and respective values of the bamboo culms considering two-meter length.

	Measured values at the bottom and the top of the culms (culm length 4 m)			Measured values at the bottom of the culms and approximated values at x = 2 m		
parameter	$\overline{x}_i$	s_{i}	min./max	$\overline{x}_i$	s_{i}	min./max
$D_{1,\text{bot.}}$ [cm]	6.889	0.657	5.878–9.501	*6.889*	*0.651*	*5.878–9.501*
$D_{2,\text{bot.}}$ [cm]	6.631	0.634	5.561–9.434	*6.631*	*0.634*	*5.561–9.434*
$D_{1,\text{top}}$ [cm]	4.903	0.591	3.873–6.217	5.992	0.655	4.978–8.600
$D_{2,\text{top}}$ [cm]	4.795	0.563	3.669–6.076	5.729	0.634	4.661–8.534
$t_{\text{bot.}}$ [cm]	1.121	0.158	0.773–1.687	*1.121*	*0.158*	*0.773–1.687*
t_{top} [cm]	0.485	0.085	0.371–0.934	0.681	0.158	0.433–1.347
$o_{\text{ext,bot.}}$ [-]	0.039	0.031	0.001–0.145	*0.039*	*0.031*	*0.001–0.145*
$o_{\text{ext,top}}$ [-]	0.035	0.029	0.001–0.123	0.045	0.036	0.001–0.169
	Measured and calculated parameters (culm length 4 m)			Measured and calculated parameters (culm length 2 m)		
nodes [-]	16.42	1.74	13–21	10.0	1.20	8–13
α_i [%]	0.482	0.148	0.195–0.845	0.450	–	–
δ_u [cm]	6.081	2.640	1.116–14.76	3.124	1.608	0.501–7.964
x_{δ_u} [m]	1.638	0.547	0.030–2.990	0.764	0.304	0.030–1.450

The relationship describing the decrease of cross section diameter from bottom to top of the culm $D(x_i)$ [cm] in y and z direction is based on the diameters measured of each internode of the 90 culms and can be approximated linearly by:

$$D(x_i) = -0.0045 \cdot x_i + D_{bot}. \tag{5}$$

To determine the variation in thicknesses along the culm axis at the culm position x_i [in cm], five of the investigated culms (C1 to C5) are cut into six parts where the thicknesses and diameters are additionally measured. Based on the additional measurements the decrease of the wall thicknesses $t(x_i)$ [cm] along the culm length is assumed to be a 2nd degree polynomial of the following form:

$$t(x_i) = 9 \cdot 10^{-6} \cdot x_i^2 - 0.004 \cdot x_i + t_{bot}. \tag{6}$$

Approximations for culms C1 to C5 based on Eqs. (5, 6) are depicted in Fig. 3 showing good agreement to the measured values.

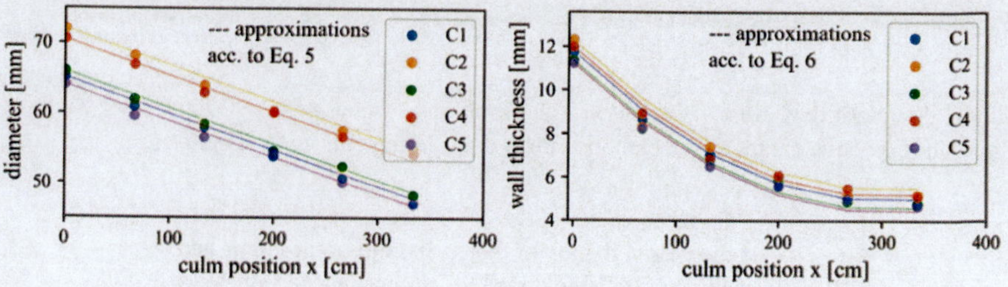

Fig. 3. Variation of wall thicknesses (left) and diameters (right) along the culm axis position *x*.

The spatial pre-deflection of the culms was found to compound of different shapes in different directions, either single or triple waved as shown in Fig. 4. In this paper, only the maximum stich of the corresponding pre-deflections is considered, of which the orientation may vary with respect to the y-z coordinate system. All maximum pre-deflections δ_u measured and presented in Table 1 are connected to single-waved deflection components of the spatial shapes.

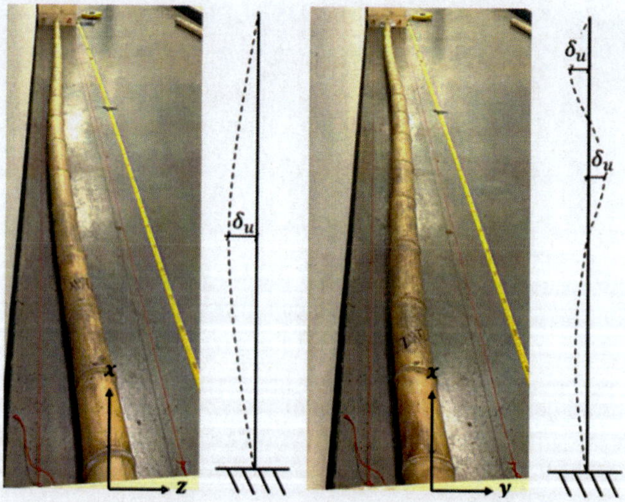

Fig. 4. Example of typical pre-deflections in two different axes of a bamboo culm rotated by 90°.

Next to the measurement results of the four-meter culms, Table 1 (right) also shows the geometrical parameters scaled to culms of two-meter length starting from the bottom and being considered for the stability investigations. The two bottom meters are found to be suitable for culm-construction purposes with European Phyllostachys edulis, as they have larger diameters and wall thicknesses. In addition, there are no leaf scars at the culm nodes, which arise mostly from diameters smaller than approximately 5.0 cm.

Leaf scars considerably influence the geometry of the culm cross section at the nodes and may decrease the buckling capacity of the culms.

The geometrical parameters for the two-meter culms are approximated according to Eq. 5 and Eq. 6 and ovality and stem taper are calculated following Eq. 1 and Eq. 2. To determine the magnitude of the pre-deflections of the two-meter culms, linear functions are set-up for each culm between the base and the measured pre-deflection at two meters. Subsequently, the measured pre-deflections of the culms for the first two meters are customized by subtracting the respective linear functional values (cf. Fig. 5).

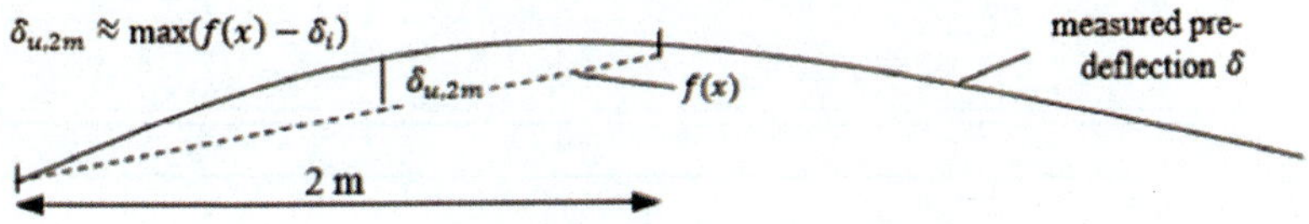

Fig. 5. Calculation of pre-deflections of the culms at x = 2 m.

4 Numerical Modelling of Bamboo Culms with Imperfections

To investigate the influences of the measured geometric variabilities on the stability behavior, a bamboo culm is modeled in the finite element (FE) program Ansys 2021 [8]. Bamboo is an orthotropic material, with high strength along the fibers and significantly lower strength in transverse direction. As a special case of orthotropy, bamboo is modeled as a transversely isotropic material ($E_y = E_z$, $\upsilon_{xy} = \upsilon_{xz}$, $G_{xy} = G_{xz}$). The material parameters for the investigations in the present study are taken from studies on Phyllostachys edulis in [9] and [10]. The Youngs Modules E_x for compression and tension parallel to the fibers are averaged for reasons of simplicity since both compressive and tensile stresses are found to occur in the numerical calculations. To account for the transversally functionally graded behavior of bamboo, three layers are modeled providing different Young's moduli in the transversal directions for inside (index *i*), middle (index *m*) and outside (index *o*), respectively. In the longitudinal direction, the same Youngs modulus is applied in all layers, since no sufficient material studies have yet been performed investigating the variation of E_x. Thus, the material parameters are set to $E_x = 1461$ kN/cm^2, $E_{y,o} = E_{z,o} = 354.6$ kN/cm^2, $E_{y,m} = E_{z,m} = 316.35$ kN/cm^2, $E_{y,i} = E_{z,i} = 278.1$ kN/cm^2, $\upsilon_{xy} = \upsilon_{xz} = 0.3$, $\upsilon_{yz} = 0.44$, $G_{xy} = G_{xz} = 151.45$ kN/cm^2, and $G_{yz} = 138.85$ kN/cm^2. In Ansys, the material is defined to provide unlimited linear elastic behavior. However, in subsequent analyses, it is assumed that the resistance of the material is limited by its maximum strengths, i.e. by a maximum tensile strength of 19.35 kN/cm^2 and maximal compressive strength of 6.87 kN/cm^2, as stated in [9].

The bamboo culm is modeled considering solid elements of type SOLID186 with a mesh size of 5 mm that is chosen following mesh sensitivity studies. To apply the geometrical variations, a parameterized FE model is setup allowing to vary geometrical parameters within the limits given in Table 1. For the stability analysis, the bottom of the culm is fully restraint while the upper end is aerially restricted around the y- and z axis, allowing displacement in culm direction (x axis). To investigate the buckling behavior, centric compression using a surface load is applied to the top cross section of the bamboo culm. The boundary conditions, the load as well as the numerical model including the functionally graded structure of bamboo is shown in Fig. 6.

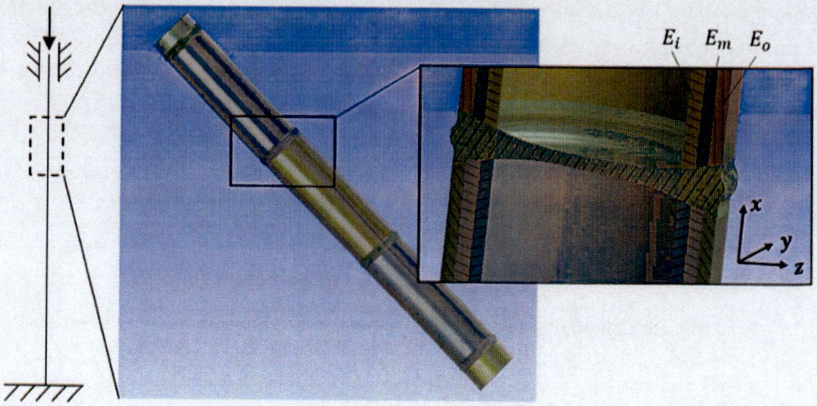

Fig. 6. Boundary conditions and FE model with functionally graded material properties.

5 Stability Analysis of Bamboo Culms with Imperfections

According to ISO 22156:2021 [13], the buckling capacity of one compression member with a centric concentrated load is obtained according to.

$$P_e = \frac{\pi^2 \cdot (EI)_{min} \cdot C_{bow}}{(K \cdot L)^2}, \tag{7}$$

where $(EI)_{\text{min}}$ is the minimum flexural stiffness of the culm, C_{bow} is a reduction factor to account for an initial bow imperfection, and $K \cdot L$ the effective compression member length dependent on the boundary conditions. The boundary conditions chosen are described in Sect. 4 and represent Euler case 4, i.e. laterally restrained and both ends fixed. For fixed-fixed columns, ISO 22156:2021 defines $K = 0.65$, differing from Euler's buckling length coefficient of 0.5 for an ideal, perfect member exhibiting unlimited elastic behavior [11]. The K-factor is recommended greater and more unfavorable than its theoretical comparative value to cover actual conditions in practical applications, which often do not perfectly comply to the ideal situation. However, as these uncertainties do not affect the numerical model, $K = 0.5$ is considered in the subsequent calculations for reasons of comparison. The critical buckling load according to Eq. 7 for a culm with minimum flexural stiffness (i.e., considering the mean values for $D_{2,\text{top}}$ and t_{top} according to Sect. 3, Table 1, right) without any pre-deflection $C_{\text{bow}} = 1$ for a constant circular hollow cross section with $E_{\text{x}} = 1461$ kN/cm^2 gives $P_{\text{e}} = 50.51$ kN. In comparison, considering the maximum diameters for the two-meter culm $D_{1,\text{bot.}}$ And $t_{\text{bot.}}$, $P_{\text{e,max}} =$ 126.24 kN. To account for pre-deflections, ISO 22156:2021 defines the factor C_{bow} to:

$$C_{bow} = 1 - (b_o/0.02), \text{ with } b_o = b_{max}/L \leq 0.02. \tag{8}$$

Considering the mean pre-deflection of $b_{max} = \delta_u = 3.124$ cm, $C_{bow} = 0.219$, whereby the buckling capacities are calculated to be $P_e = 11.06$ kN, and $P_{e,max} = 27.68$ kN. For comparing these buckling loads with numerical simulations, stability analyses are performed. The theoretical buckling loads are determined based on the first eigenmode of the structure, where the mean values of ovality, stem taper and pre-deflections are considered in the modelled geometry as initial culm imperfections according to Table 1.

The critical buckling loads P_{cr} obtained numerically considering the geometrical imperfections are compared to the results of ISO 22156:2921 in Table 2. However, due to the unlimited elastic material model, the numerically determined critical buckling loads P_{cr} only reflect theoretical values and upper limits of the ultimate loads, i.e., capacities of the structure. To obtain more realistic ultimate loads, the material behavior considering nonlinear effects would have to be implemented. However, for approximating lower limits of ultimate capacities for culms with pre-deflections, the load $P_{cr,fc}$ is specified in Table 2, representing the load when reaching the maximum compressive strength in one bamboo fiber according to geometric nonlinear analyses. $P_{cr,fc}$ may be considered as minimum load, as the ultimate capacity of the material and effects of stress redistribution due to the nonlinear material behavior are not considered. In addition to the influences investigated, there are further influences, e.g., pre-damage, cracks or changes in the cross section caused by leaf scars or absent nodal membranes due to treatment methods. In particular, the change in Young's modulus in the longitudinal direction over the culm length is a decisive factor that must be included in future buckling analyses.

Table 2. Buckling capacities obtained by ISO 22156:2021 compared to critical buckling loads obtained numerically with and without considering the max. compressive strength.

Geometry parameters	P_{cr} (numerical)	$P_{cr,fc}$ (numerical)	P_e (Eq. 7/8)
(1) Ideal culm, $D_{2,top}$	**48.07 kN (100%)**	–	50.51 kN (+5%)
(2) Ideal culm, $D_{1,bot}$	116.6 kN (+143%)	–	126.2 kN (+162%)
(3) Ideal culm, $D_{2,top} + \delta_u$	47.71 kN (−1%)	33.36 kN (−31%)	11.06 kN (−77%)
(4) Ovality, $D_{1,top}/D_{2,top}$	50.92 kN (+6%)	–	–
(5) Ovality, $D_{1,top}/D_{2,top} + \delta_u$	50.62 kN (+5%)	36.44 kN (−24%)	–
(6) Stem taper, $D_{2,bot.}/D_{2,top}$	64.57 kN (+34%)	–	–
(7) Stem taper, $D_{2,bot.}/D_{2,top.} + \delta_u$	64.05 kN (+33%)	42.50 kN (−12%)	–
(8) Ovality (4) + stem taper (6)	68.13 kN (+42%)	–	–
(9) Ovality (4) + stem taper (6) + δ_u	67.73 kN (+41%)	45.89 kN (−5%)	–

6 Results and Discussion

The results demonstrate that natural geometric variations have non-negligible effects on the flexural buckling capacities. The average ovality for the European bamboo culms (Table 2, case 4) increased the theoretical buckling load compared to the ideal culm with circular cross section and minimum diameter (Table 2, case 1) by 6%. Considering stem taper and therefore larger diameter $D_{2,bot.}$, the buckling capacities are significantly increased by 34% (Table 2, case 6) and 42%, when additionally considering ovality (Table 2, case 8). Contrary, pre-deflections decrease the critical buckling loads with respect to the maximum capacities $P_{cr,fc}$, as the pre-deflections lead to additional stress related to the culm deformation. However, the magnitude of influence depends on the size, shape, and position of the initial bow imperfection and the maximum stich. For example, in case the shape of the pre-deflection is affine and superimposes to the first eigenmode, the critical buckling load decreases. Therefore, further studies on the influence of size and position of maximum pre-deflections are necessary.

To account for imperfections, ISO 22156:2021 introduced the coefficient C_{bow} for potential pre-deflections. Even neglecting nonlinear material behavior, however, Table 2 shows that the numerical results considering the more accurate geometry of the culms may be higher than the critical buckling loads determined in accordance with the current standard. Amongst others, this may be caused by the buckling length coefficient for tapered structures (fixed on both sides), tending to be smaller than $K = 0.5$. Therefore, large-scale experiments should be conducted to determine to what extent reduction factors for buckling loads due to the boundary conditions and pre-deflections may be too conservative. Furthermore, the influence of non-linear material behavior and varying material properties along the culm length has to be studied.

7 Summary and Conclusions

The difficulty in deriving generally valid design regulations for bamboo as construction material is strongly affected by its natural geometric imperfections. Therefore, this study investigates the influence of geometrical imperfections such as ovality, stem taper and pre-deflections on the stability behavior of bamboo culms. For this purpose, geometrical measurements of 90 European bamboo culms of the species Phyllostachys edulis have been evaluated and examined. Subsequently, the geometric imperfections measured have been captured in a numerical model of a bamboo culm and stability analyses have been performed. The results demonstrate that the average measured values of ovality, stem taper and pre-deflections can significantly affect the stability behavior and buckling loads. However, the comparison with the calculation recommendations for the buckling capacities of axially compressed members currently given in ISO 22156:2021 shows that higher load-carrying capacities may be assumed if geometric variations are kept within acceptable tolerances. In future work, the results should be compared to full-scale tests in the laboratory to derive more realistic design concepts and to advance bamboo construction in Europe.

Acknowledgements. The authors gratefully acknowledge the support by Mr. Sagar Gupta Satrasala in numerically modelling bamboo culms in the scope of his Master Thesis. Furthermore, we would like to thank Prof. Dr. Luisa Molari for her support in obtaining the European bamboo culms. This work is partially funded by the Women's Promotion Fund of the Bauhaus University Weimar.

References

1. Göswein, V., et al.: Wood in buildings: the right answer to the wrong question. In: IOP Conference Series: Earth and Environmental Science, vol. 1078, no. 1, p. 012067 (2022)
2. Donini, G., et al.: Structural design of an Italian bamboo house in an Italian regulatory context: revisiting a small building built in Costa Rica with tropical bamboo. Case Stud. Constr. Mater. **16**, e00891 (2022)
3. Greco, S., Molari, L.: Flexural behavior of six species of Italian bamboo. Constr. Technol. Archit. **1**, 723–729 (2022)
4. Ben-zhi, Z., et al.: Ecological functions of bamboo forest: research and application. J. Forestry Res. **16**(2), 143–147 (2005)
5. Inoue, A., Sato, M., Shima, H.: A new taper index based on form-factor: application to three bamboo species (Phyllostachys spp.). Eur. J. Forest Res. **140**, 1533–1542 (2021)
6. Schneider, R., et al.: Climate-induced changes in the stem form of 5 North American tree species. For. Ecol. Manag. **427**, 446–455 (2018)
7. Bahtiar, E.T., et al.: Experimental study on buckling resistance of Guadua angustifolia bamboo column. Eng. Struct. **228**, 111548 (2021)
8. ANSYS, Release 15.0, Mechanical User's Guide (2013)
9. Molari, L., Mentrasti, L., Fabiani, M.: Mechanical characterization of five species of Italian bamboo. Structures **24**, 59–72 (2020)
10. Al-Rukaibawi, L.S., Omairey, S.L., Károlyi, G.: A numerical anatomy-based modelling of bamboo microstructure. Constr. Build. Mater. **308**, 125036 (2021)
11. Kindmann, R., Kraus, M.: Steel Structures, Design Using FEM. Berlin: Ernst & Sohn. (2011). ISBN 978-3-433-02978-7
12. Kraus, M., et al: Geometric imperfections of additive manufactured members. Eng. Struct. **252**, 113596 (2021)
13. ISO 22156:2021, Bamboo structures – Bamboo culms – Structural design (2021)

Prediction of the Development of Microorganisms on Biocomposites Based on Plant Aggregates

Mohamad El Hajjar[1](✉), Sylvain Bourgerie[2], and Naima Belayachi[1]

[1] Univ. Orléans, Univ. Tours, INSA-CVL, LaMé – EA794, 8 rue Léonard de Vinci, 45072 Orléans, France
mohamad.el-hajjar@univ-orleans.fr

[2] Univ. Orléans, LBLGC – EA1207, INRAe USC1328, rue de Chartres BP6759, 45067 Orléans, France

Abstract. The aim of this paper is to predict the development of microorganisms on biocomposites, proposed for building insulation. These biocomposites are based on different plant aggregates (wheat straw, rapeseed straw, bark, and sunflower's pith) and a lime binder. This prediction was carried out by using two software called WUFI BIO and WUFI VTT. They are based on analytical models developed previously and referred to earlier applications works on other materials as wood. Indeed, they can calculate an index of microorganisms as function of microbial development time. This index is directly related to a percentage of contaminated surface of the material. In order to compare and validate the analytical prediction, an experimental study was performed to determine the contaminated surface of the four studied biocomposites. The samples are exposed to accelerated microbial conditions, to enhance the development of microorganisms on their surfaces. After different time periods for one year of fungal development, an image analysis was carried out in order to calculate an experimental percentage of contaminated surface of the biocomposites, using Leica software. Finally, the goal of this study is to compare the analytical and experimental percentages of the contaminated surface, in order to evaluate the efficiency of the analytical models to predict the real growth of the microorganisms on the biocomposites surface. The WUFI VTT software has made a very efficient prediction of the growth on one material among four, after choosing specific conditions for the software as the material type, the temperature, and the humidity. By setting the values of temperature and humidity at 30 °C and 90% (±3%) respectively, an inverse study was done, consisting of combining the types of materials according to their microbial development sensitivity. These combinations allow the experimental percentages to be adjusted to the analytical ones, so that the WUFI VTT can predict microbial growth on the surface of the four investigated materials in this paper.

Keywords: Bio-based materials · microbial growth · microbial prediction models · image analysis · accelerated aging

S. Amziane et al. (Eds.): ICBBM 2023, RILEM Bookseries 45, pp. 870–884, 2023.
https://doi.org/10.1007/978-3-031-33465-8_67

1 Introduction

The biological degradation, or biodegradation, is a damage due to biological attack by microorganisms. Various research has been carried out to evaluate the risk on building materials to be biodegraded. Many laboratory studies were carried out in order to understand the biodegradation mechanisms, and their factors, to give simpler and more environmentally friendly repair/prevention solutions. (Rabii 2017) studied the microbiological aging of composites reinforced with flax fibers. Different characterizations were realized on the biocomposites to study their durability facing the microorganisms. Furthermore, (Delannoy 2018) worked on the durability of materials based on plant aggregates of hemp's shives. The characterizations of the hemp's shives and the biocomposites were carried out after their exposition to accelerate microbial aging to study their capacity toward the biodegradation. (Nielsen et al. 2004) studied the influence of humidity and temperature on fungal development as well as secondary metabolites acting on building materials. They demonstrated that the wood and the composite need a minimum humidity of 78% at 20–25 °C and increase to 90% at 5 °C to authorize microbial development. In contrast, for the gypsum board a humidity of 86% is necessary. (Urzì et al. 2007) evaluated the effectiveness of biocide and water repellent compounds against microbial colonization of mortars. Due to their high cost and the limit of uses required by the European Community Directive (2004/42/EC), these products are rarely suggested. Their use for the restoration project is preferable where their application is harmless.

Despite the high interest of the biodegradation study in the laboratory to evaluate the risk due to microorganisms, these studies have some drawbacks related to the long time to obtain results and thus their high economic cost. To assess the risk of microorganism's growth on a material, different prediction models have been described in the literature. Hens model (Hens 2006) is based on the isopleth of Aspergillus versicolor given by previous experiments. Such model was developed to predict the critical relative humidity, favorable for the development of microorganisms for long and short periods. Johansson's mold growth index model (Johansson et al. 2014) is a model able to calculate three mold growth indices, based on measurements taken on facades with different conditions, by taking into account the relative humidity, the temperature, necessary conditions for microbial development. These models and others are based on experimental studies from which to derive some equations to assess the risk linked to microorganisms' propagation while proposing simplifying assumptions. Therefore, to achieve a reliable prediction of microbial development, the choice of model has a potential influence.

This study aims to compare the prediction of microbial growth by two analytical prediction models developed in the software WUFI BIO and WUFI VTT. WUFI BIO is based on the biohygrothermal model developed by (Krus and Sedlbauer 2007) to predict by using some isopleth systems the development of microorganisms on a material. These isopleths are created to describe the development of a large range of microorganisms for different temperature and humidity conditions. WUFI VTT, developed by (Hukka and Viitanen 1999), is based on the VTT model to predict the development of microorganisms on wood materials. Different equations are established using experimental studies on the wood. Then, this model was updated to apply it on different types of materials after modifying some parameters depending on the type of the material. This two software, as the other predictions models, need some input data such as the type of the material, the temperature, and the humidity. One major advantage, that specifies these two models, is their ability to calculate a microorganism's index which can describe the development of microorganisms. This index is directly related to a percentage of a contaminated surface of the material. In the present study, several experiments were carried out on four biocomposites based on vegetable aggregates, to assess the development of microorganisms, and in order to validate the analytical models of prediction. This study consists in following microorganisms' development on these materials and in particular, on their surface, under conditions chosen to accelerate microbial growth. Then, an analysis of the images of the materials, taken during the microorganism's development phase, allows to calculate an experimental percentage of contaminated surface of the four studied biocomposites. The second objective of this study is to compare the analytical percentages of contaminated surface, determined by these two analytical models, with the experimental percentages of contamination.

2 Materials and Methods

2.1 Analytical Models for the Prediction of Microorganism's Development

Two different analytical models are used to predict the development of microorganisms on the surface of a material in terms of time. These two models are established by using two software called WUFI BIO and WUFI VTT. This two software are able to calculate an index describing the development of microorganisms named the microorganism's index. This index is correlated to a percentage of the contaminated surface of the material.

2.1.1 WUFI BIO

The software WUFI BIO was firstly used for this study. It is based on the transient biohygrothermal method that can describe the prediction of the germination of spores and refer to the isopleth systems. The isopleths, developed by (Ayerst 1969) and (Smith et al. 1982) are systems determined for a few types of microorganisms, represented by curves (Fig. 1) describing humidity evolution in terms of temperature. They are based on measured biological data and also consider the growth prerequisites of all microorganisms of one hazardous class. (Krus and Sedlbauer 2007) added an expansion of these isopleths, by subdividing in several classes the material types used in building's domain. Four different categories (Cat0 to CatIII) are defined to consider the influence of building substrate, which can be contaminated. Noted that in curves established, all the mold microorganisms are taken into consideration. These categories, described in Table 1, are all derived from experimental examinations. Detailed description of the model can be found in (Sedlbauer 2002) and (Krus and Sedlbauer 2007).

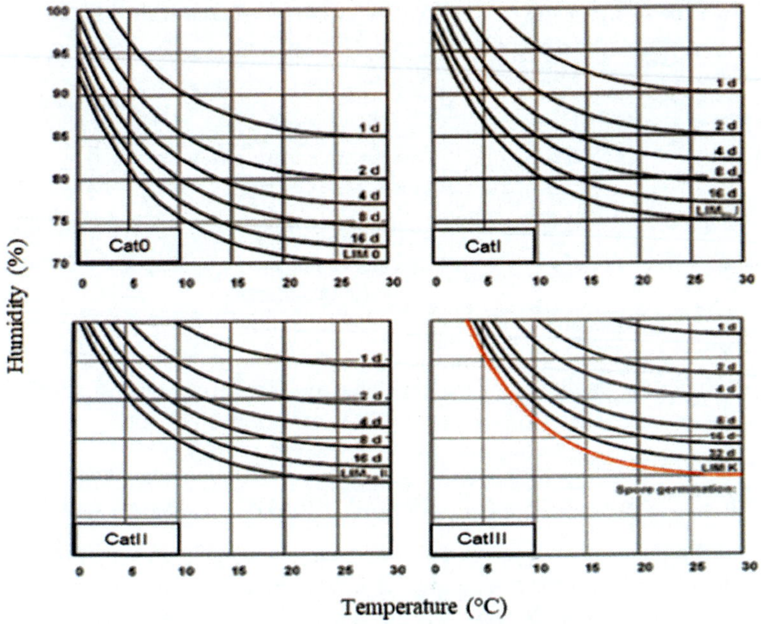

Fig. 1. Isopleth systems for 4 categories of substrates (Krus and Sedlbauer 2007)

Table 1. Categories of WUFI BIO's materials (Krus and Sedlbauer 2007)

Categories	Designation	Descriptions
Substrate category 0	Cat0	Optimal culture medium
Substrate category I	CatI	Biologically recyclable building materials like wallpaper, plaster cardboard, building materials made of biologically degradable raw materials, material for permanent elastic joints
Substrate category II	CatII	Biologically adverse recyclable building materials such as renderings, mineral building material, certain wood as well as insulation material not covered by I
Substrate category III	CatIII	Building materials that are neither degradable nor contains nutrients

2.1.2 WUFI VTT

The WUFI VTT is the second software used for this study. It is based on the VTT model, developed by (Hukka and Viitanen 1999), and used to predict the development of the microorganisms on a material. It was firstly programed to predict microorganism's growth on wood materials. Then, it was later upgraded to handle many types of materials (Viitanen et al. 2010). In their research, they divided the materials into 4 classes to take into consideration their sensibility to microbial growth (Table 2).

Table 2. Categories of WUFI VTT's materials (Viitanen et al. 2010)

	Designation	Material group
1 (Very sensitive)	C1	Untreated wood
2 (Sensitive)	C2	Planed wood; paper-covered products, wood-based panels
3 (Moderately resistant)	C3	Cementitious or plastic materials, mineral fibers
4 (Resistant)	C4	Glass and metal products, materials with effective protective treatments

In this model, the prediction of the development of microorganisms is represented by a microorganism's index (Viitanen et al. 2015). It was given by the Eq. (1).

$$\frac{dI_m}{dt} = \frac{K_1K_2}{7 * \exp(-0.68\,lnT - 13.9lnRH + 0.14W - 0.33SQ + 66.02)} \tag{1}$$

where K_1 is the intensity coefficient depending on microorganism's growth level and K_2 the moderation of the growth intensity when the microorganism's index I_{mana} level approaches the maximum peak value. T is the temperature, RH the relative humidity, W (=0 for pine and 1 for spruce), and SQ the surface quality (=0 for sawn and 1 for kiln dried).

Detailed description of the model can be found in (Hukka and Viitanen 1999), (Viitanen et al. 2010) and (Viitanen et al. 2015).

2.2 Sample Preparation for Experimental Study

Four biocomposites are studied (Fig. 2): (1) wheat straw and lime (WL), (2) wheat straw, additives, and lime (WaL), (3) rapeseed straw and lime (RL), and (4) sunflower and lime (SL). They are composed of a mineral binder (Lime PF70 (L)), and four different aggregates harvested in the region "Centre Val de Loire" in France". The WL material based on wheat straw and lime was developed by (Belayachi et al. 2013). The second material WaL (Ismail et al. 2020) was an upgraded material for WL, it was prepared with the same components as the WL but, by adding two animal-based additives: The hemoglobin (H) (Table 3), creates a significant porosity in the matrix, to improve the thermal behavior. The second additive is the casein (C) (Table 3), it increases the mechanical strength of the material.

The materials RL and SL, are developed by (Brouard 2018) using clay as a binder, but for this study, the binder used is the lime. The first one, RL, is based on rapeseed straw and lime. The second one, SL, is based on a mixture of sunflower bark (119.77(±3) kg/m^3) and pith (28.89 (±0.3) kg/m^3) with percentages of mass 40% and 60% respectively. These percentages are chosen to have the same bulk density as rapeseed straw. To prepare the biobased material, the lime is mixed with water for 5 min to form the binder. The binder is then added to the vegetable aggregates to be subsequently mixed for 5 min. Then, it will be molded in a formwork of 20 * 10 * 10 cm to form the samples which are kept for 28 days at 50% of humidity for the cure.

Table 3. Different biocomposites, A: Aggregates, L: Lime, W: Water, H: Hemoglobin, and C: Casein

Materials	Designation	ρ_{dry} (kg/m^3)	Volumetric percentages (%)				
			A	L	W	H	C
Wheat straw + lime	WL	450 (±8)	59.8	25.9	14.2	0	0
Wheat straw + additives + lime	WaL	350 (±9)	67.8	19.8	10.7	1.0	0.5
Rapeseed straw + lime	RL	500 (±9)	43.4	36.4	20.0	0	0
Sunflower + lime	SL	500 (±8)	43.4	36.4	20.0	0	0

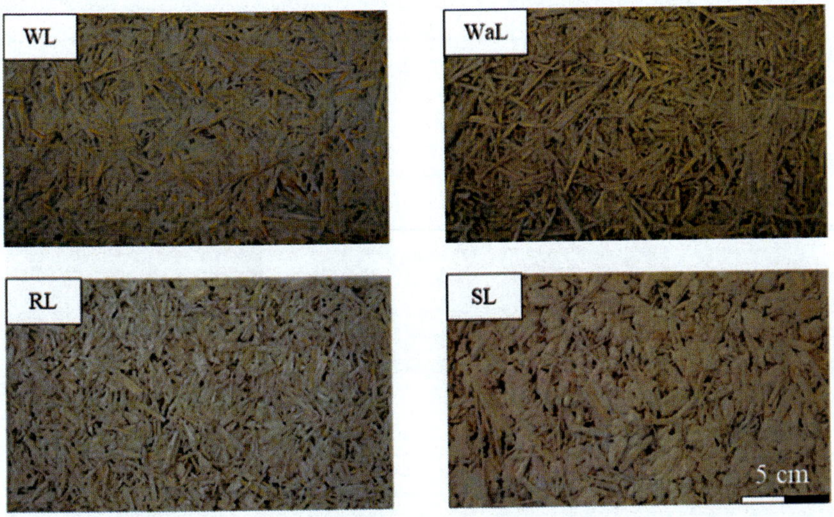

Fig. 2. Photos of biocomposites used: WL, WaL, RL, and SL at t = 0 day

2.3 Conservation Conditions for Accelerated Microbial Aging

The development of microorganisms on a material, depends on some conditions which are favorable for their growth, such as the temperature and the humidity. In building's domain, a temperature between 15 to 45 °C may cover the majority weather of a year, which's correspond to mesophiles category. Moreover, the optimal temperature for many fungal species is between 20 °C and 30 °C (Santos et al. 2017). Furthermore, in steady state conditions, fungal growth begins at around 80% of relative humidity according to (Fog Nielsen 2003). (Krijgsheld et al. 2013) reviewed that a minimum relative humidity for fungal growth is 77%, but the optimal value is 97%. Based on these studies, for the present experiments, the conditions of accelerated microorganism's growth adopted are 30 °C for the temperature and 90% (±3%) for the relative humidity. These two conditions are ensured by stocking the samples in a box with a specific saline solution (potassium nitrate KNO3). Then, this box is conserved in an oven at 30 °C for a one-year follow-up.

2.4 Image Acquisition and Analysis

An analysis of the images of the biocomposites, taken during the phase of development of the microorganisms, allows to calculate an experimental percentage of contaminated surface for the four biocomposites. For each material, photos of 3 different faces, was taken by a professional camera at different times of the study for a one-year follow-up. Then, each photo was treated by a software for image analysis called Leica. It is a software program, derived from "The LEICA Image Overlay Module Application Suite", used to enhance images with a wide range of processing functions. Among these functions, there are contrast, brightness, and gamma, used to adjust the photo to optimize the quality of the photo display. This image analysis leads to a quantitative description of microbial growth. Leica program can calculate on the one hand; the surface contaminated by microorganisms, called Sc and on the other hand, the total surface of the face, called St. Figure 3 shows two examples of the image acquisition of the surface, for WL and WaL materials, at 6 months of microbial development aging.

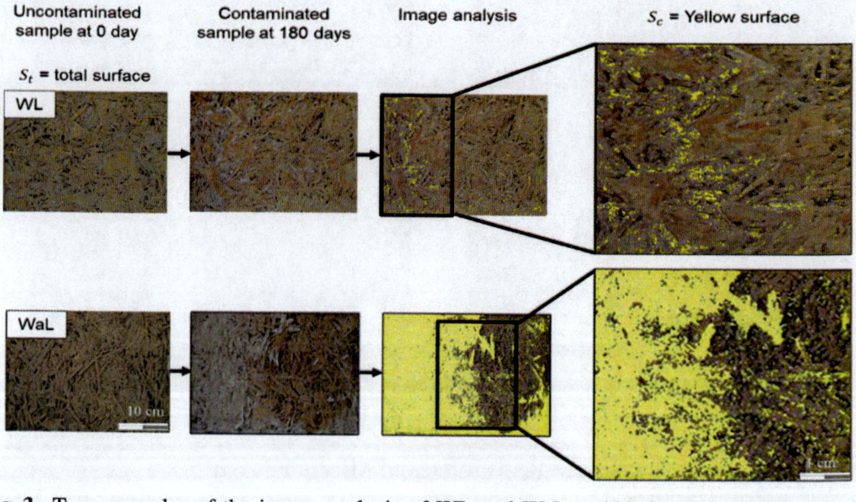

Fig. 3. Two examples of the image analysis of WL and WaL at 180 days of conditioning for microbial development

The experimental percentage P_{exp} of contaminated surface is given by the Eq. (2):

$$P_{exp} = \frac{S_c}{S_t} * 100 \tag{2}$$

With S_c is the contaminated surface and S_t the total face of the material.

3 Results and Discussions

The input data for WUFI BIO and WUFI VTT are the class of the material (CATII and C1 respectively), the temperature (30 °C), the relative humidity (90%), and the duration of the test (one year). In fact, the materials, based on vegetable aggregates such as wheat straw, rape straw, sunflower bark, and pith, can be considered as sensitive and untreated materials, and that justify the choices of the class of the materials. The output of these two software is the microorganism's index (I_{mana}). This index shows the intensity of microorganism's growth, with values between 0 (no growth) and 6 (almost 100% of the surface covered with microorganisms) (Table 4).

Table 4. Description of the levels of the microorganism's index

I_{mana}	Growth
0	No growth
1	Small quantity of microorganisms on the surface
2	<10% of the surface covered with microorganisms
3	10%–30% of the surface covered with microorganisms
4	30%–70% of the surface covered with microorganisms
5	>70% of the surface covered with microorganisms
6	Almost 100% of the surface covered with microorganisms

Referring to this table, the analytical percentage (P_{ana}) of the contaminated surface was deducted from the analytical index of microorganisms (I_{mana}). This table is just limited to 7 values of I_{mana} with their corresponding P_{ana}. A polynomial fitting is applied to the seven pairs of values, deduced from the (Table 4), to assign for any value of I_{mana}, a corresponding value of P_{ana}. These pairs are with following coordinates (I_{mana}; P_{ana}) = (0; 0), (1; 1), (2; 10), (3; 30), (4; 70), (5; 99) and (6; 100). The polynomial fitting is given by the equations below. The fitting and the seven points are shown in Fig. 4.

$$P_{ana} = \begin{cases} \boldsymbol{Imana^1} \text{ if } I_{mana} = [0; 1] \\ \mathbf{-46 + 92,33 * \mathit{Imana}^1 - 61,83 * \mathit{Imana}^2 + 18,16 * \mathit{Imana}^3 - 1,66 *} \\ \boldsymbol{Imana^4} \text{ if } I_{mana} = [1; 5] \\ \mathbf{4 + 1 * \mathit{Imana}} \text{ if } I_{mana} = [5; 6] \end{cases}$$

The analytical percentage of the contaminated surface calculated by WUFI BIO and WUFI VTT are shown in Fig. 5. As an input data, the temperature is 30 °C and the humidity is 90% for both software. The class of the materials chosen for WUFI BIO is the category CATII specific for mineral building materials, meanwhile the class of the material for WUFI VTT is the class C1 specific for the very sensitive untreated wood. The WUFI BIO and WUFI VTT show the same behavior between 0 and 70 days. After 70 days, for WUFI BIO, the percentage continues to increase until attending the

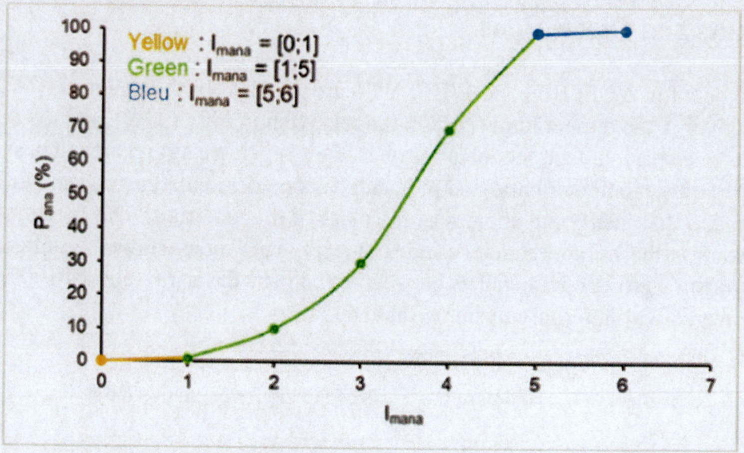

Fig. 4. Fitting of the analytical percentage P_{ana} in terms of microorganism's index I_{mana}

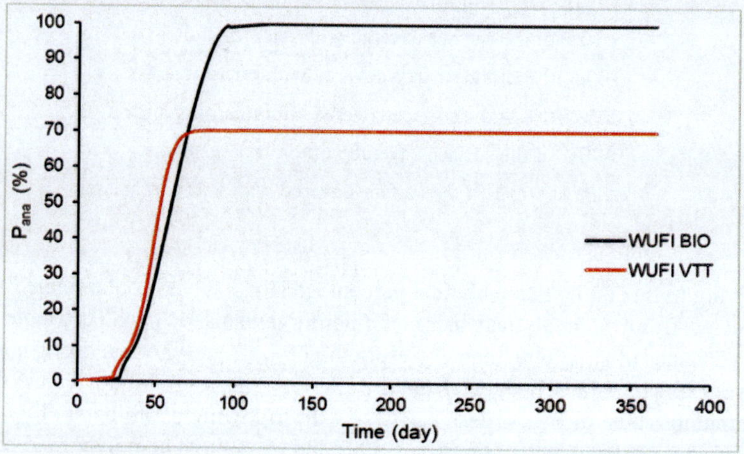

Fig. 5. Evolution of the analytical percentage P_{ana} (WUFI BIO and WUFI VTT) in terms of time, at 30 °C and 90% of humidity

maximum value of $P_{ana} = 100\%$ at 100 days and still constant until 365 days. For WUFI VTT, the maximal $P_{ana} = 70\%$ is achieved at 70 days and still constant until 365 days.

Figure 6 shows the monitoring of the progression of the microorganism's development on the four biocomposites and the corresponding image acquisition performed by Leica program, at 4 different time (days). The WaL, the material composed of wheat straw with lime and the two additives of animal origin (hemoglobin and casein) is the most contaminated compared with the other materials, because of the presence of these two additives, which guarantees a favorable environment for microbial development.

Figure 7 shows the experimental percentage of the contaminated surface for the four biocomposites, calculated after image analysis. We demonstrate that materials formed

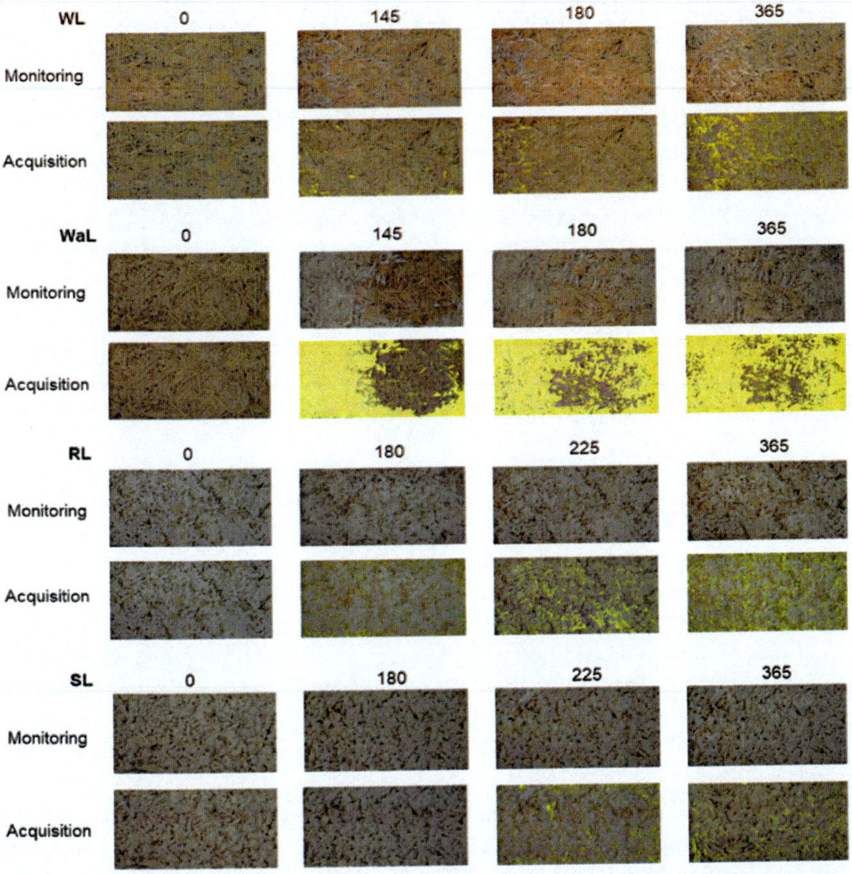

Fig. 6. Photos monitoring and acquisition of the development of microorganisms on the different biocomposites for 4 different periods (t0; 145, 180, 365 days or t0, 180, 225 and 365 days)

by wheat straw, lime and the two additives hemoglobin and casein (WaL) are the most contaminated and consequently achieved a maximal value of $P_{exp} \approx 77.5\%$ at 180 days. Meanwhile, the three other materials show lower values compared to WaL. The first material formed by the wheat straw and lime without additives (WL), the second formed by the rapeseed straw and lime (RL) and the third formed by the mixture of the sunflower bark and pith with lime (SL) attend a maximum value of $P_{exp} = 9\%$ at 150 days, 23.5% and 9.9% at 230 days respectively.

The four materials investigated in this paper contain lime as a binder. Its use increases the pH, which favors a microbial development. (Delannoy 2018), developed materials based on hemp shiv, and lime as a binder and studied the effect of the development of microorganisms on its properties. The effect of the lime on the microbial development was carried out. The alkalinity of the fresh binder reduces the number of viable cells in hemp concrete by a factor of 20, but fungal development is still possible from remaining fragments from the hemp shiv. On one hand, lime can be considered to have favored

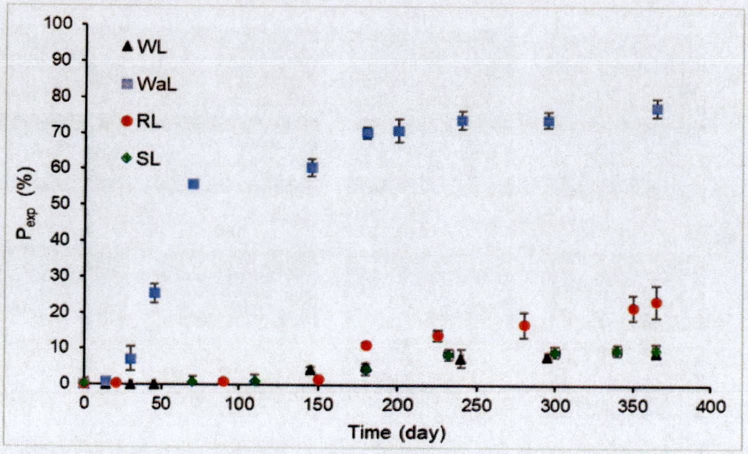

Fig. 7. Evolution of the experimental percentage of contaminated surface P_{exp} as a function of time for 4 biocomposites studied (WL, WaL, RL, and SL)

microbial development on the studied biocomposites, but on the other hand, the microbial development effect was not the same on the four biocomposites. The most contaminated was Wal, which contains, in addition to the aggregates and the lime, two animal-derived additives (hemoglobin, and casein). Their presence in the formulation favored significant microbial development despite the fact that the percentage of lime used in WaL was the smallest (19.8%) compared to those of WL, RL, and SL (25.9%, 36.4%, and 36.4% respectively).

Figure 8a shows the variation of the maximal P_{exp} for each material at 6 months of microbial conditioning. Figure 8b shows the evolution of the bulk density in terms of time for a one-year follow-up at 4 times (0, 3, 6 and 12 months). From these data, we can conclude that the percentage of the contaminated surface is high when the bulk density is low and that is remarkable for the WaL material which shows the upper percentage of contamination while its bulk density is the lower. Note that the density decreased after 6 and 12 months due to material's degradation, which is visible in Fig. 6. On the other hand, when the bulk density is high, the percentage of contamination is low. In fact, when the bulk density is lower, the porosity is higher, and it means that the material is more capable to stock the humidity in its pores and makes it as a favorable environment for microbial development.

Figure 9 shows the comparison between the analytical results of WUFI BIO and WUFI VTT and the experimental ones. The prediction of WUFI BIO is underestimated comparing to the experimental results. The WUFI VTT prediction coincides with the experimental percentage P_{exp} only for the material WaL, but it also underestimates the P_{exp} for WL, RL and SL.

As a result, we can consider that the WUFI VTT is the most efficient tool for the prediction of the development of microorganisms on the WaL material. Otherwise, one of the drawbacks of the VTT model is its limitation to a limited number of materials. They are distributed in 4 classes according to their sensitivity to microbial growth as shown in

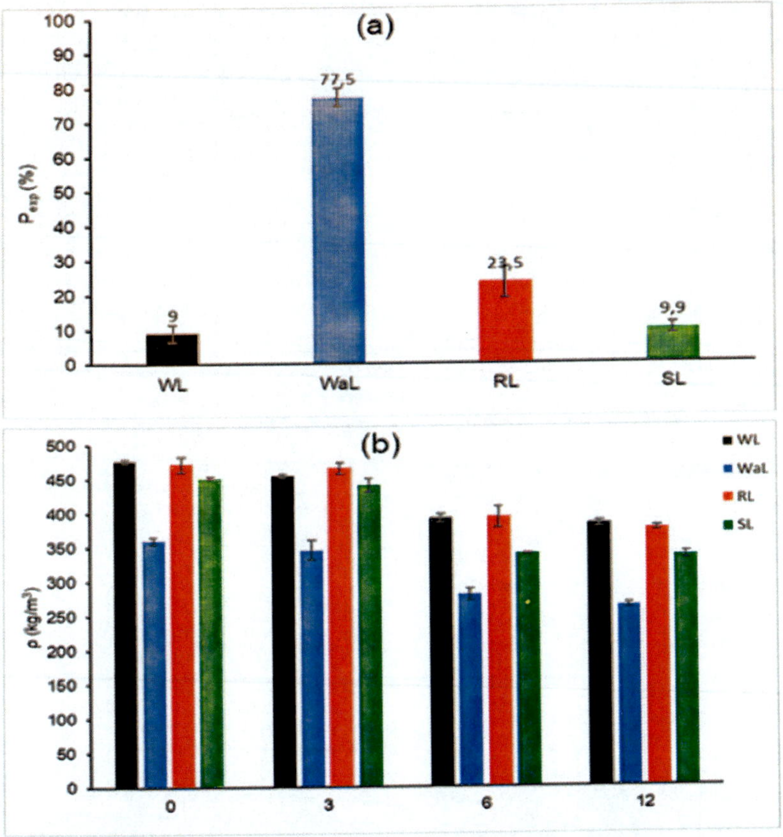

Fig. 8. (a) Experimental percentages P_{exp} at 6 months of microbial conditioning (b) Evolution of the bulk density of the materials WL, WaL, RL and SL as a function of time

Table 2 (C1, C2, C3 and C4). The aim of this study is to propose an intermediate class to classify these three materials that didn't coincide with the experimental results. For that, an inverse analysis, that consists of combining percentages of each class of material, was performed to find a combination, that represents the experimental percentages of WL, RL, and SL, until the error, between experimental and analytical percentages, tends to 0. Six combinations are established between the 4 classes of the materials: (C1, C2), (C1, C3), (C1, C4), (C2, C3), (C2, C4), and (C3, C4). Figure 10 shows an example of the development prediction of microorganisms after choosing the combination between the material classes C1 and C2.

Figure 11 shows the results by using the classes' combinations determined by the inverse analysis. With the software WUFI VTT, with input data 30 °C for the temperature and 90% for the relative humidity, it is able to predict the development of the microorganisms on the surface of the four materials studied, by combining the classes as following: WL with the combination 20%C2; 80%C3, WaL with 100% C1; 0%C2, RL with 50%C2; 50%C3 and SL with 30%C2; 70%C3. However, a disadvantage of the

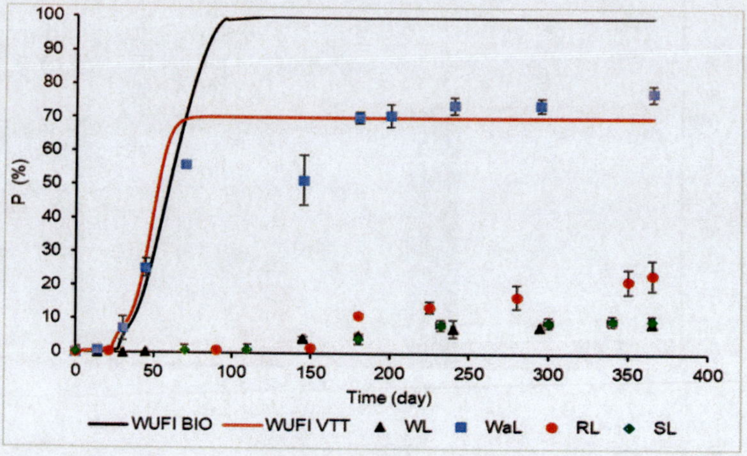

Fig. 9. Comparison between the analytical (WUFI BIO, WUFI VTT) and the experimental percentages

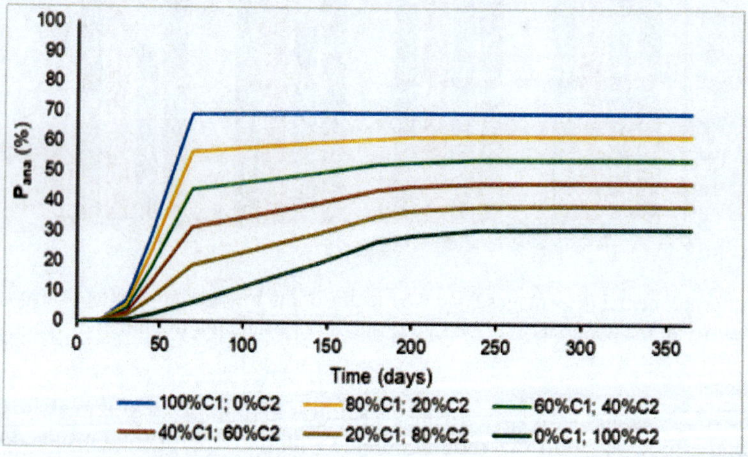

Fig. 10. Example of the inverse analysis of P_{ana} with a combination between C1 and C2 in WUFI VTT

analytical model is that it predicts only a surface development and neglects the core of the material. The choice of a material with a C1 class, described to be very sensitive, is supported by the presence of additives from biomass.

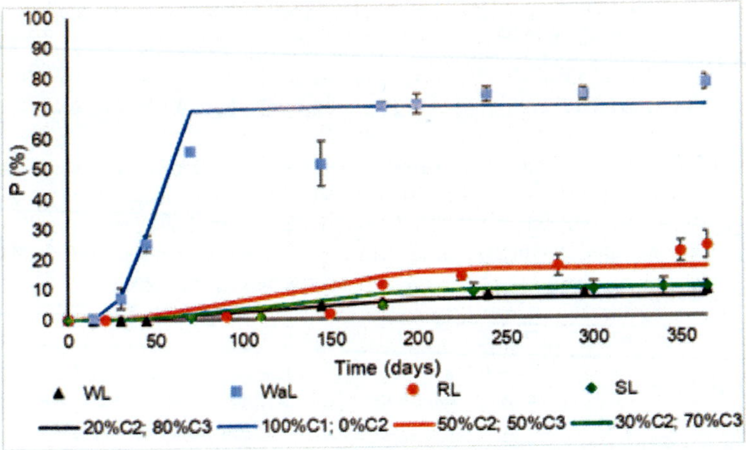

Fig. 11. Predicted results by using the combined material classes determined by inverse study between analytical and experimental percentages of contaminated surface for the four studied biocomposites

4 Conclusion

The objective of the present study was to predict the development of microorganisms on four biocomposites, based on vegetable aggregates and used for insulation for buildings. The prediction was carried out using two models WUFI BIO and WUFI VTT. The prediction consists of calculating a microorganism's index in terms of time. This index represents the evaluation of the development of microorganisms on material, related to a percentage of contamination of its surface. In parallel, an experimental study was carried out to calculate an experimental percentage of contamination of the surface of the biocomposites, exposed to favorable conditions for microbial development. A comparison between the analytical and experimental percentages of contamination was achieved. The WUFI VTT was more efficient with the experimental results of WaL, the material formulated with wheat straw, lime and two animal additives (hemoglobin and casein). To adapt the VTT model with the three other materials WL, RL and SL, an inverse analysis of the analytical model of WUFI VTT was carried out, by varying the types of the material used as an input data. To predict the development of microorganisms on the surface of the four materials, the following combinations can represent the experimental percentage of WL with the combination of material's classes (20%C2; 80%C3), WaL with (100%C1; 0%C2), RL with (50%C2; 50%C3), and SL with (30%C2; 70%C3). These analytical models have some limits of use. They can really predict the development on the surface of material and not on its core. In other word, they can represent a microbial development during time but do not permit to evaluate the materials biodegradation.

Acknowledgments. The authors gratefully acknowledge the "REGION CENTRE VAL DE LOIRE" for the financial support of the research program MATBIO.

References

Ayerst, G.: The effects of moisture and temperature on growth and spore germination in some fungi. J. Stored Prod. Res. **5**(2), 127–141 (1969). https://doi.org/10.1016/0022-474X(69)90055-1

Belayachi, N., Bouasker, M., Hoxha, D., Al-Mukhtar, M.: Thermo-mechanical behaviour of an innovant straw lime composite for thermal insulation applications. Appl. Mech. Mater. **390**, 542–546 (2013). https://doi.org/10.4028/www.scientific.net/AMM.390.542

Brouard, Y.: Caracterisation et optimisation d'un composite biosource pour l'habitat, p. 169 (2018)

Delannoy, G.: Durabilité d'isolants à base de granulats végétaux, p. 269 (2018)

Fog Nielsen, K.: Mycotoxin production by indoor molds. Fungal Genet. Biol. **39**(2), 103–117 (2003). https://doi.org/10.1016/S1087-1845(03)00026-4

Hens, H.: Mold in dwellings: field studies in a moderate climate, p. 8 (2006)

Hukka, A., Viitanen, H.A.: A mathematical model of mould growth on wooden material. Wood Sci. Technol. **33**(6), 475–485 (1999). https://doi.org/10.1007/s002260050131

Ismail, B., Belayachi, N., Hoxha, D.: Optimizing performance of insulation materials based on wheat straw, lime and gypsum plaster composites using natural additives. Constr. Build. Mater. **254**, 118959 (2020). https://doi.org/10.1016/j.conbuildmat.2020.118959

Johansson, P., Ekstrand-Tobin, A., Bok, G.: An innovative test method for evaluating the critical moisture level for mould growth on building materials. Build. Environ. **81**, 404–409 (2014). https://doi.org/10.1016/j.buildenv.2014.07.002

Krijgsheld, P., et al.: Development in aspergillus. Stud. Mycol. **74**, 1–29 (2013). https://doi.org/10.3114/sim0006

Krus, M., Sedlbauer, K.: Mould growth prediction by computational simulation, p. 8 (2007)

Nielsen, K.F., Holm, G., Uttrup, L.P., Nielsen, P.A.: Mould growth on building materials under low water activities. Influence of humidity and temperature on fungal growth and secondary metabolism. Int. Biodeterioration Biodegradation **54**(4), 325–36 (2004). https://doi.org/10.1016/j.ibiod.2004.05.002

Rabii, H.: Étude du vieillissement de composites renforcés par des fibres naturelles: application bâtiment, p. 285 (2017)

Santos, J.L.P., Chaves, R.D., Sant'Ana, A.S.: Estimation of growth parameters of six different fungal species for selection of strains to be used in challenge tests of bakery products. Food Biosci. **20**, 62–66 (2017). https://doi.org/10.1016/j.fbio.2017.08.006

Sedlbauer, K.: Prediction of mould growth by hygrothermal calculation. J. Therm Envelope Build Sci. **25**(4), 321–336 (2002). https://doi.org/10.1177/0075424202025004093

Smith, S.L., Hill, S.T.: Influence of temperature and water activity on germination and growth of Aspergillus restrictus and A. versicolor. Trans. Br. Mycol. Soc. **79**(3), 558–560 (1982). https://doi.org/10.1016/S0007-1536(82)80058-2

Urzì, C., De Leo, F.: Evaluation of the efficiency of water-repellent and biocide compounds against microbial colonization of mortars. Int. Biodeterior. Biodegradation **60**(1), 25–34 (2007). https://doi.org/10.1016/j.ibiod.2006.11.003

Viitanen, H., Krus, M., Ojanen, T., Eitner, V., Zirkelbach, D.: Mold risk classification based on comparative evaluation of two established growth models. Energy Procedia **78**, 1425–1430 (2015). https://doi.org/10.1016/j.egypro.2015.11.165

Viitanen, H., et al.: Moisture and bio-deterioration risk of building materials and structures. J. Build. Phys. **33**(3), 201–224 (2010). https://doi.org/10.1177/1744259109343511

Development of a Fire-Resistant Straw- and Waste Glass-Based Thermal Insulation Board

Dániel Csanády and Balázs Nagy(✉)

Budapest University of Technology and Economics, Műegyetem rkp. 3., Budapest 1111, Hungary
nagy.balazs@emk.bme.hu

Abstract. In the presented research, barley- and wheat straw, waste glass and waterglass were used as the primary raw materials for developing bio-based thermal insulation. An analytic-empirical model was developed to describe the heat transfer process in raw straw bulks based on laboratory experiments dealing with the properties of individual natural straw stems and stem bulks. Based on the key properties of stems and optimal bulk density, which can minimize the thermal conductivity, we developed a treatment method to reduce the thermal conductivity of stems further. The base binder was upgraded with modifiers to meet all the requirements. Mainly the water resistance of waterglass has to be improved. A cyclic test of cured binder samples showed that the dissolution loss could be decreased by 30–40%. Mechanical properties could be improved using different drying methods (conventional oven, microwave). Microwave drying can increase the compressive strength by twice-three times. The thermal conductivity of straw can be reduced by 8–11% and straw-silicate composite by 17–18% using heat treatment and the modified binder. The binder behaves well at high temperatures and protects natural fibres from fire combined with the fibre treatment so that the straw silicate material could be classified A2 based on the EN ISO 1182 standard. As a result of the briefly summarized research, we created a fire-resistant thermal insulation whose thermal conductivity is below 0.039 W/mK, meets the requirements of crotch resistance, and can be applied on floors and facades in the same way as the currently widely used thermal insulation boards.

Keywords: Straw thermal insulation · natural fibers · fire resistant · thermal conductivity · circular economy

1 Introduction

1.1 Natural Fibres and Secondary Raw Materials in the Construction Industry

The construction industry is the largest energy-consumer and waste-generating sector (35.9% of the total waste production in the European Union), and it needs to increase the use of sustainable methods and materials [1–3].

S. Amziane et al. (Eds.): ICBBM 2023, RILEM Bookseries 45, pp. 885–898, 2023.
https://doi.org/10.1007/978-3-031-33465-8_68

The solution offered by material development is to use more secondary raw materials. Using natural-origin by-products is much more efficient, and the products require less energy for further processing. Moreover, the availability of cereals is always provided, as the food supply requires production, and it is available in large quantities and can be used for thermal insulation fabrication [4]. The use of glass waste in the construction industry has high potential. In the EU, on average, 76% of the glass waste is recycled, but in Hungary, for example, only 35% is recycled [5], although could be used as a raw material for manufacturing thermal insulations. A proper thermal insulation of facades can reduce the energy use and CO_2 emission of buildings [3, 6]. Currently, the most common thermal insulations are plastic foams and composites of inorganic fibres and organic binder matrix [7, 8], which are made mostly form non-recycled materials, but the application of natural fibres in the building industry is increasing [9, 10].

These artificial materials are more effective than their natural-based competitors, but their raw material demand and manufacturing energy are high [11], or their recycling technology is not developed [12, 13]. The available environmentally friendly thermal insulations are expensive or uncompetitive due to their material characteristics, and many of them are only environmentally friendly by name.

A solution to the raised problems could be the creation of straw-based thermal insulation material. The porous structure of straw makes it suitable for this purpose. It is used in its natural state in straw bale houses [14, 15] and lightweight panel structures [16, 17]. Measurements on these structures prove that natural straw is suitable for thermal insulation but provides adequate thermal insulation only in relatively great thicknesses. Contrary, it regulates well the vapour balance of interior spaces [18]. With an appropriate density, it is an excellent sound absorber [19] and effectively controls humidity fluctuations between the interior and exterior areas [15, 20, 21]. Questions about their fire risk, higher moisture absorption capability and durability, natural anti-fungal and insect properties, and bacterial activity are often asked. The resistance of straws to these pests can be further improved by using the treatment and appropriate binder described later in the paper. Thomson and Walker [22] and Yin et al. [23] found that mould growth is possible only in fresh straw, and hot and humid climates do not significantly impact the durability of straw bales in the walls and endanger the health with a low probability.

Contrary to popular belief, the straw could behave well against fire. Reaction to fire of straw-based-silicate composite thermal insulation boards was also tested previously, and, with particular binder composition, it is possible to reach excellent performance [24]. Belayachi et al. [25] reported that linseed oil treatment retards the flame spread on the surface of the straw-based composite. Chetehouna et al. [26] mixed wheat and barley straw with lime and plaster to form thermal insulation. They found that the pyrolysis of the two straw (at 300 °C) types generated the same compounds but in different relative percentages affected by the binder. Using current flame-retardant compounds in crop-based building materials dominantly benefits fire retardancy. In contrast, these compounds impair mechanical properties, thermal conductivity, durability, and water resistance [27]. Furthermore, new and more effective flame-retardants must be developed because the current efficient flame-retardant system has a high cost and may cause health issues.

Some investigations stated that their physical properties have higher variability, increasing the variability in the final product [28, 29]. Although, a recent study showed that the variability of composites made of long natural fibre is similar to carbon fibre-based composites [30]. Different sustainable processes can improve the physical properties to meet thermal insulation requirements better. In many cases, there arises the problem that natural-based thermal insulators are not fire-resistant and do not simultaneously have the lowest possible thermal conductivity and high load-bearing capability. Still, in many cases, they have homogeneity problems or contain hazardous additives. However, the biggest problem is that straw-based insulation boards are non-competitive, because of their too high thermal conductivity. Ismail et al. [31] made an insulation material by using straw and composite binder and achieved 0.054 W/(m K) thermal conductivity. Zhou et al. [32] presented a low-density wheat straw board using low-pressure steam-injection pressing, and this product had the result of 0.0481 W/(m K) thermal conductivity. Tůmová et al. [33] mixed straw with casein binder to form insulation boards, the lowest thermal conductivity they reached was 0.045 W/(m K). Babenko et al. [34] developed a flax straw and water glass-based insulation board, which achieved 0.084 W/(m K).

In the case of straw polymer composites providing the proper bonding between stems and binder is also a problem [28]. Common and industrial-scale solutions to adhesion problems usually cause significant damage to the pore structure of the stems [35, 36] (e.g. aggressive alkaline treatment) or reduce the porosity of the stems (e.g. mineralisation), thus also impairing the thermal insulation capacity. Making thermal insulation boards with a binder can solve application and durability problems because the binder gives rigidity to the stem bulks and protects against environmental effects. The binders used in the published research do not meet all the criteria (durability, fire resistance, mechanical behaviour, etc.) of bonded thermal insulation boards. Resins have bad behaviour against fire, and they are not environmentally friendly. Silicate-based binders have potential in this field, but the water resistance and bonding capacity must increase. Sodium silicate combined with some additives (e.g. waste glass) may be suitable for this purpose. A direct reaction of silicon dioxide source and sodium hydroxide can also create liquid sodium silicate. Keawthun et al. [37] prepared a waterglass solution by reacting crystalline NaOH and powdered waste glass at 600 °C and then dissolving the product in water. Seun et al. [38] and Puerta and Torres [39] dissolved powdered waste glass in a concentrated NaOH solution at low temperature ($T < 200$ °C) with continuous stirring, thus obtaining a water glass solution. Vinai and Soutsos [40] treated a sludge prepared from a mixture of crystalline NaOH and pulverized waste glass and water at 300 °C and produced sodium silicate. The research found in this literature confirmed the idea that glass ground to a sufficiently small mesh diameter dissolves in a strongly alkaline medium.

1.2 Research Perspectives, Aims and Goals

The main goal of the research is to create a straw-based thermal insulation board which can be a competitor to the thermal insulations that are widely used today. Proper investigation of natural straw structure (macro, micro) and heat-transport process in stem bulks must be done to achieve that purpose. Based on these data, an analytic-empirical model was developed to find the key properties of straws, which can minimize the thermal conductivity of straw bulks. Knowing these properties, we can develop a treatment method to reduce the thermal conductivity of stems. In a favourable case, the surface adhesion of the stems can also improve in parallel. We must find a binder that can be considered environmentally friendly and able to satisfy some important criteria (e.g. reaction to fire) in the case of thermal insulation. If possible, we wanted to improve the other properties (water resistance) using secondary raw materials. So the effect of several modifiers (e.g. waste glass) on the base binder should be investigated. In the case of sustainable materials, the embodied energy should be minimized in addition to achieving the best possible material properties. For this purpose, we compare the effect of conventional and modern drying technologies on the mechanical properties of thermal insulation. After the development of the components, the thermal properties of the new straw composite are compared with the properties of the initial straw composite (raw straw, pure waterglass). At least the development is evaluated. This paper briefly summarizes the results of this research. The availability of more detailed descriptions was indicated at the end of the related chapter.

2 Materials and Methods

2.1 Materials

In this research, barley and wheat straw were investigated. Both were grown in Hungary but in different regions. The straws were collected from the same harvest period in the Autumn of 2019. These straws are different species, so there were some natural disparities in the structure of stems. Both straw types were stored in the laboratory under the same environmental conditions (23 ± 5 °C, 50 ± 10% relative humidity) until the tests. The base binder was pure, liquid sodium-, potassium-silicate. Modifiers of the binder were powdered waste glass, other silicate-based materials, and some bases and acids.

2.2 Methods

Model

Firstly, we created an analytic-empirical model that describes the thermal performance of the different straw bulks. The thermal conductivity of straw bulks (λ_{meas}) was measured in a completely dry state and in the function of bulk density (between 80–180 kg/m^3) with using EN 12667 [41], EN 1946-2 [42] and ISO 10456 [43] standards. The dry condition of the straw bulks during the measurement was ensured by the foil cover. Through laboratory experiments, the properties of individual natural straw stems and stem bulks were established, for instance, material density, density, porosity, cellulose content, the unit mass of stem by length, and the volume ratio of gaps in the bulk. The model parameters were created based on barley's measurement results. The results of wheat were used for validation of the model. The calculated thermal conductivity (k_{tot})

was separated into conduction in solid *(k_s)*, conduction in gas *(k_g)*, and radiative heat transfer *(k_r)*, see Fig. 1-a.

Stem Treatment

In the pH-dependent wet heat treatments, the thermal conductivity was measured at a bulk density of 80 kg/m^3 in five conditions (natural and four treatment methods). The changes in microstructure were also tracked by mapping the stem cross-sections with a scanning electron microscope (SEM) before and after treatments. The stems were treated in solutions with different pH values, but the mass ratio of stem-to-solution, temperature and treatment time was the same. The straw dried to constant mass and wrapped in a thin layer of gauze (this ensured the constancy of the stem arrangement and eliminated structural changes), then placed in a 150 mm × 150 mm × 20 mm thin-walled paper box and the thermal conductivity of the stem bulk was measured before and after treatment. For both types of straw, the effects of the treatments were tested using Guarded Hot Plate Method according to EN 12667 [41]. The relation between thermal conductivity reduction and microstructural changes was revealed.

Binder Modification

The water resistance of the binder of stems was tested on two silicate binders, sodium *(NaS)* and potassium silicate *(KS)*. These pure binders were modified with waste glass (soda-lime-*G1*, borosilicate-*G2*) and other silicate powders, acids and alkalis. In the present work, the impact of waste glass is presented. Each series of measurements (same composition) was performed with nine specimens. In all of the tested specimens, the mass of the pure wet binder was the same, and the modifier was added as additional material. The binder was dried to constant mass at 100 °C and rested on laboratory air for two days. Each set of specimens was dissolved in the same mass ratio of distilled water in five-hour cycles, and then their mass was then measured after drying to constant mass. This process was repeated until the samples lost their integrity or up to five cycles. The dissolution mass loss of the modifier material was not considered, so the pure and modified specimens can be compared. SEM was used to investigate the temporal changes in the material structure of samples of different compositions. The difference in the surface structure of the pure and modified binders was observed (Development of CSH, CASH crystals), from which the reason for the different behaviour under the water can be deduced. X-ray diffraction analysis (XRD) confirmed the results.

Compressive Strength Tests

Compressive strength is an important mechanical property for thermal insulations, as the step resistance criterion is based on this value. The 50 mm × 50 mm × 50 mm specimens were randomly cut out from a larger board for compressive strength tests, and the tests were performed perpendicular to stem direction according to EN 826 [44]. The specimens were divided into two main groups, conventionally dried *(Cv)* (forced-air tumbler drier) and dried by microwave technology *(Mc)*, to determine the difference in the effect of these two drying methods. Both of the groups contained two significantly different subgroups. In the first, barley straw was used with natural stem length distribution and no treatment, and the binder was pure potassium silicate *(Pure)*. In the second subgroup, wheat straw was used with adjusted stem length and with treatment, the binder was modified by corn starch *(Mod)*. The straw was mixed manually with the binder and then filled and compressed into an HT-PLA formwork, and then a fresh mixture was cured with conventional or microwave methods. The measurements were carried out on dry specimens and at laboratory air temperature (~20 °C). The reasons for the differences in compressive strength were explored using optical microscopy and considering the specificities of the different drying methods.

Reaction to Fire

The performed non-combustibility test and the conditioning of specimens were regulated by ISO 1182 [45], and the insulation classification was based on EN 13501–1 [46]. The specimens were cylinders with a diameter of 45 ± 5 mm and a height of 50 ± 3 mm. In the centre of the specimens, a hole was drilled for the sensor placement with a diameter of 2 mm. The moisture content of the specimens was reduced to near zero in a tumble drier at 60 ± 5 °C, then placed into the non-combustibility device stabilized at 750 °C, manufactured by Fire Testing Technology. One thermocouple measured the temperature of the centre of the specimen, one of the surfaces of the specimen, and one of the interiors of the furnace. The possible presence of flame was observed with the help of a mirror attached to the furnace.

3 Results and Discussion

3.1 Model

Natural stems' fundamental properties are material density, porosity, cellulose content, average diameter, and stem mass by length. These parameters predominantly determine the thermal conductivity of the average individual stem. The most important and changeable properties of stem bulks are gap content (space among stems) void content (space in the tubular stem), flatness factor and reverse flatness factor, which are functions of bulk density. The flatness factor and reverse flatness factor are vector that shows the dominance of the thermal conductivity of the solid part of a stem or the gas in the voids (respectively) at the current density through the length of the heat path in stem wall or amount of enclosed gas in voids (respectively). Adjusting the optimal values of these parameters/properties minimizes thermal conductivity. Between 80–180 kg/m^3, the thermal conductivity values calculated by the model very well approximated the measured values. When the model parameters were defined and adjusted using the measurement values of barley or when the model was validated with wheat measurement values, the difference between the measured and modelled values was less than 2% (Fig. 1-b). Therefore, we considered it acceptable. The detailed description of the analytic-empirical model, equations and the used parameters can be found in a previously published paper of Csanády et al. [47].

3.2 Stem Treatment

Within a given pH value range, the porosity of the stems can be increased, and the average pore diameter of the stem and the thickness of the pore walls (within the stem wall) can be reduced (Fig. 2-a).

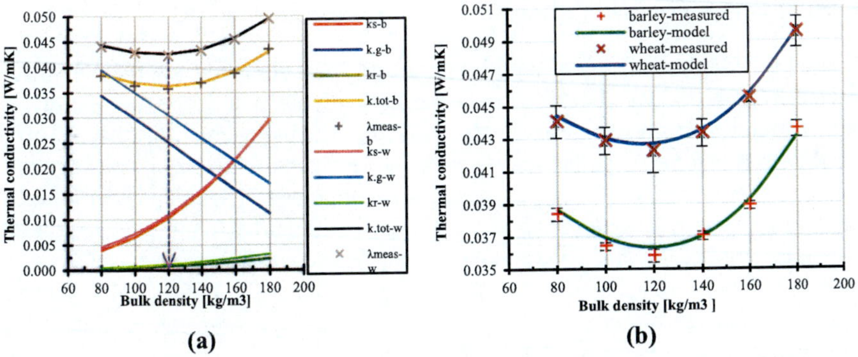

Fig. 1. Thermal conductivity factors from the model in relation to the function of bulk density with added trend lines regression *(b-barley, w-wheat)* (a), Relationship between bulk density and total thermal conductivity in case of the results of laboratory measurements and the analytic-empirical model (b), based on [47]

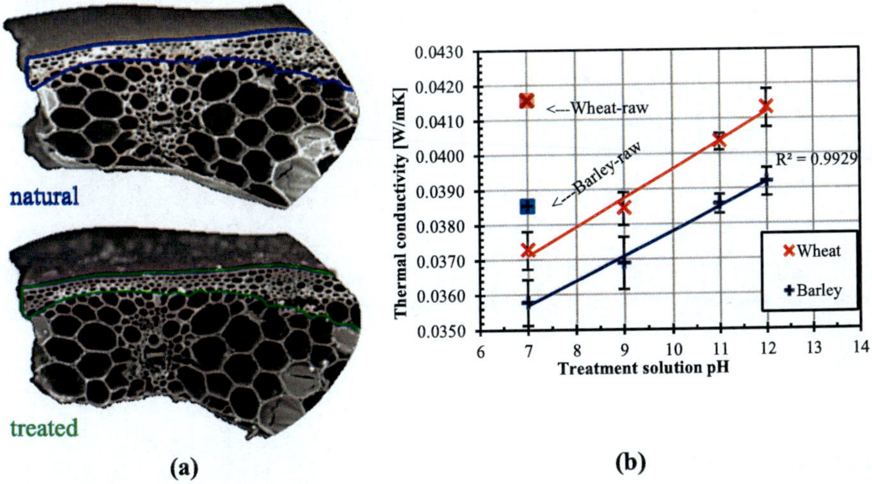

Fig. 2. The adequately performed stem treatment increasing porosity of stem (a), The relation between the pH of the solution and the thermal conductivity of the treated stem bulks (b)

Because of this, under the right circumstances, wet-heat treatment can reduce thermal conductivity. However, in solutions with high pH, the pore structure may be degraded because the strong bases dissolved some pore wall components (e.g. lignin). Thus the segmentation of the pores in the stem wall is damaged, and the porous stem wall (not the pore wall) may be thinned. This increases the possibility of free convection of gases and reduces the thickness of the porous stem wall, where we can assume pores are filled with stationary air in the isolated pores. The thermal conductivity of barley was reduced by 8% and that of wheat by 10% (Fig. 2-b) [48].

3.3 Binder Modification

Because of the high alkalinity of liquid sodium and potassium silicates, the non-alkali-resistant, amorphous, fine silicate powder is partially soluble. It integrates into the silicate matrix, forming a crystal nucleus and acts like a cross-linking agent. Lime that dissolves from the glass can be identified and later forms hexagonal calcium hydrate crystals. Based on experiments, we found that adding a small amount of glass powder to liquid sodium or potassium silicate increases the stability of the dried samples against water and their material quality becomes more uniform (Fig. 3-a). Microstructural analysis has demonstrated that adding powder glass forms calcium silicate hydrate (CSH) crystals in the sodium or potassium silicate matrix. The most spectacular examples are hillebrandite, xonotlite, and other cubic CSH particles (Fig. 3-b) [49].

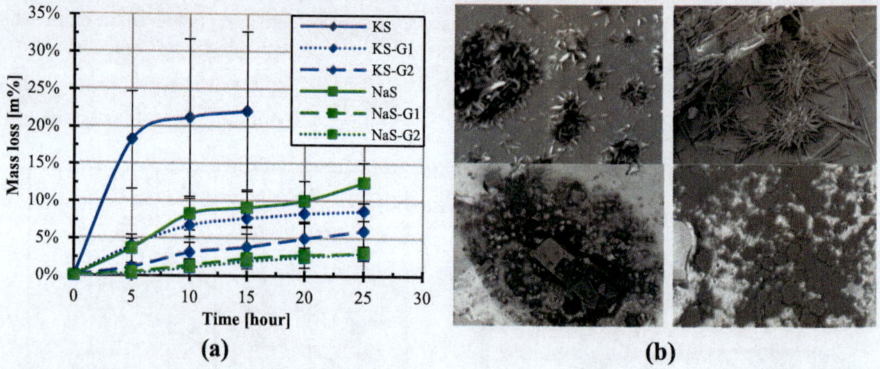

Fig. 3. Mass loss of pure and waste glass powder modified binder specimens during dissolution in water as a function of time (a), CSH crystals growing under the influence of dissolved glass powder (b)

3.4 Compressive Stress

The shrinkage of composites during curing differs along the cross-section (in time and space), depending on the drying method. The nature of the energy input with the rate and direction of heat spread determines it. The drying process also affects the stem-binder adhesion, as it influences the time difference between the final shape of the stems and the final shape of the binder matrix covering the stems. When the conventional method was used, the heat needed more time to reach the inner sections of the specimen, so at the different points of the specimen, the shrinkage was different, and the external coat of the composite was fixed, while the internal part was ductile. In contrast, microwave drying simultaneously generates heat in all points of the specimens. Internal stresses from restrained deformations deteriorate the stem-binder-stem joints. Microwave drying (Mc) significantly reduces shrinkage differences; therefore, it does not form many cracks. The compressive strength of straw-silicate composite increases if the moisture content of the freshly mixed and wet composite, which is placed into formworks, is reduced to a value approaching zero by microwave drying instead of the conventional drying (Cv) method (Fig. 4-b). The composite meets the requirements of crotch resistance [50].

(a) (b)

Fig. 4. Conventionally dried (above) and microwave-dried (below) specimens (a), Compressive strength of conventional and microwave-dried silicate-straw composite specimens (b)

3.5 Utilization of the Experimental Results

In the first stage of the research, the thermal insulations made from natural stems with pure binder were able to achieve 0.045–0.046 W/mK (in laboratory conditioned state, at 23 °C and 50% RH) average thermal conductivity in the case of barley and wheat straws (Fig. 5). With the use of the presented results we can create an insulation that meets the requirements of our time. We showed that the thermal conductivity of raw straw could be reduced with the treatments by 8–11%. The modification of binder and the applied drying method allowed to use less (and more porous) binder furthermore provided better thermal and mehanical properties for the composite. Thus the thermal conductivity of straw-silicate composite is reduced by 17–18% for the current stage of the developement. Therefore, the thermal insulation boards using treated stems and modified binders achieved 0.0372–0.0378 W/mK average thermal conductivity (Fig. 5). This product is competitive, and already applicable in the construction industry.

3.6 Reaction to Fire

In the first stage of the measurements, during the continuous low flame, which is a maximum of 8–12 s during the test, some shorter flash-like flames can be observed. In the remaining part of the test, solid phase combustion was observed, and the material was glowing and charred without any other perceptible flame effect.

The volume reduction of the specimens is obvious, the organic fibres are charred (so they are primarily decomposed into CO_2 and H_2O), and the pores embedded in the stem wall and the tubular stem have contracted. However, the porous structure remained (see

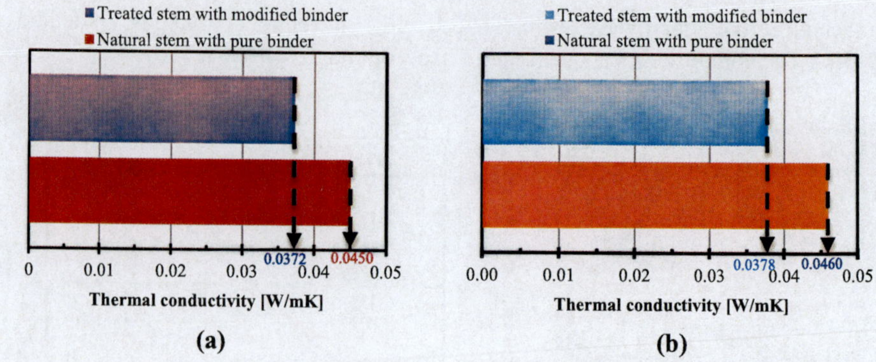

Fig. 5. Reduction of the thermal conductivity achieved during development straw-silicate composite, barley (a), wheat (b)

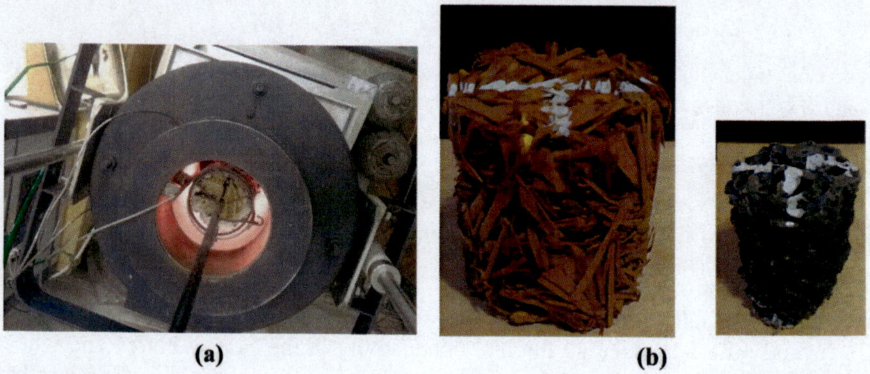

Fig. 6. Specimen in the furnace (a), Specimen before and after the non-combustibility test (b)

Fig. 6). The inorganic silicate embedding matrix melted and coated the fibres, which could be the reason for the lack of flames and solid phase combustion and became vitrified after cooling. The mass loss during the test did not reach 50% for any test specimens. It was only 39%, even in the worst case. The maximum temperature rise of the furnace was 15.5 °C compared to the permissible 50 °C. The thermocouple located in the centre of the specimens did not reach the furnace temperature during the tests, so the charred test specimen shows good insulating ability at high temperatures. Based on EN 13501-1 [46], the insulation may be classified as an A2 material, but further tests, such as a Single Burnin Item test (SBI) needid to confirm it.

3.7 Current Status and Further Perspectives

The previously described research showed that based on the tests mentioned before, a competitive product could be developed suitable for entering the market. To enter, we created the manufacturing process using the developed novel technology. The thermal insulation boards are manufactured now in small batches, with the use of 500 mm ×

1000 mm × 150 mm size and these were built into a family house in 2022. Nowadays, the boards are being tested under real circumstances in Vas county in Hungary (see Fig. 7). In the future, we would like to increase the manufacturing capacity and proportion of buildings covered with the discussed insulation.

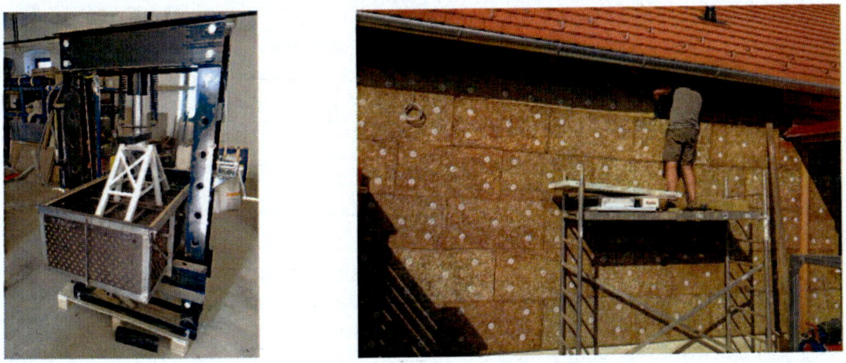

(a) (b)

Fig. 7. Making of the insulation (a) The installation of straw-silicate composite boards on a family house (b)

4 Conclusions

As a result of the briefly summarized research, an analytical-empirical model was developed [47], which can estimate the thermal conductivity of raw straw bulks between 80–180 kg/m^3 bulk density. Both raw straw and straw-silicate composite thermal conductivity can be significantly reduced by applying a treatment method and the modified binder. This happens through porosity increasing of stems and improving the binder quality, therefore reducing the quantity. The waste glass increased the water resistance of the moisture-sensitive silicate adhesive. We created a fire-resistant thermal insulation whose average thermal conductivity is below 0.039 W/mK. It meets the requirements of crotch resistance and can be used on floors and facades in the same way as the currently widely used insulations.

References

1. United Nations Environment Programme. Towards a zero-emissions, efficient and resilient buildings and construction sector. 2019 Global Status report. Glob. Status Rep. (2019)
2. Eurostat. Waste generation [Internet]. Waste Stat. (2018). https://ec.europa.eu/eurostat/statistics-explained/index.php?title=Waste_statistics
3. European Union. Energy, transport and environment statistics—2020 edition. Luxembourg: Publications Office of the European Union (2020)

4. Tuomas, H., Pirkko, V., Heikki, A., Klaus, N.: Deliverable D1.1: availability of lignocellulosic biomass types of interest in the study regions [Internet]. BIOCOmmodity Refin. (2012). http://www.biocore-europe.org/file/D1_1. Availability of lignocellulosic biomass types of interest in the study regions.pdf
5. The European Container Glass Federation. Latest glass packaging recycling rate steady at 76% [Internet]. https://feve.org/glass_recycling_stats_2018/
6. Ürge-Vorsatz, D., Cabeza, L.F., Serrano, S., Barreneche, C., Petrichenko, K.: Heating and cooling energy trends and drivers in buildings. Renew. Sustain. Energy Rev. **41**, 85–98 (2015)
7. Bozsaky, D.: The development of thermal insulation materials from the beginnings to the appearance of plastic foams. In: 37th IAHS World Congress on Housing Science (2010)
8. Grand_view_Research. Building Thermal Insulation Market Size | Industry Report, 2019–2025 [Internet]. (2019). https://www.grandviewresearch.com/industry-analysis/building-thermal-insulation-market
9. Asdrubali, F., Desideri, U.: Handbook of Energy Efficiency in Buildings: A Life Cycle Approach, p. 836. Elsevier Inc. (2018)
10. Ene, A.G., Mihai, C.: Eco-friendly thermal insulation structures based on natural and biodegradable materials for environmentally durable development. Solid State Phenom. **305**, 97–102 (2020)
11. Bull, J.: Embodied Carbon of Insulation [Internet]. Embodied Carbon Insul. (2012). http://www.greenspec.co.uk/building-design/embodied-carbon-of-insulation/
12. Väntsi, O., Kärki, T.: Mineral wool waste in Europe: a review of mineral wool waste quantity, quality, and current recycling methods. J. Mater. Cycles Waste Manag. **16**, 62–72 (2014)
13. Kowatsch, S.: Mineral wool insulation binders. In: Pilato, L. (ed.) Phenolic Resins a Century Progress, pp. 209–242. Springer, Heidelberg (2010). https://doi.org/10.1007/978-3-642-04714-5_10
14. Cornaro, C., Zanella, V., Robazza, P., Belloni, E., Buratti, C.: An innovative straw bale wall package for sustainable buildings: experimental characterization, energy and environmental performance assessment. Energy Build. **208**, 109636 (2020)
15. Douzane, O., Promis, G., Roucoult, J.M., Le Tran, A.D., Langlet, T.: Hygrothermal performance of a straw bale building: In situ and laboratory investigations. J Build Eng. **8**, 91–98 (2016)
16. Yin, X., Dong, Q., Zhou, S., Yu, J., Huang, L., Sun, C.: Energy-saving potential of applying prefabricated straw bale construction (PSBC) in domestic buildings in Northern China. Sustain. **12**(8), 3464 (2020)
17. Beaudry, K., MacDougall, C.: Structural performance of non-plastered modular straw bale wall panels under transverse and gravity loads. Constr. Build. Mater. **2016**, 424–439 (2019)
18. Shi, L., Zhang, H., Li, Z., Man, X., Wu, Y., Zheng, C., et al.: Analysis of moisture buffering effect of straw-based board in civil defence shelters by field measurements and numerical simulations. Build. Environ. **143**, 366–377 (2018)
19. Fan, M., Fu, F.: Advanced High Strength Natural Fibre Composites in Construction. Elsevier Inc. (2016)
20. Ashour, T., Georg, H., Wu, W.: Performance of straw bale wall: a case of study. Energy Build. **43**(8), 1960–1967 (2011)
21. Gallegos-Ortega, R., Magaña-Guzmán, T., Reyes-López, J.A., Romero-Hernández, M.S.: Thermal behavior of a straw bale building from data obtained in situ. A case in Northwestern México. Build. Environ. **124**, 336–341 (2017)
22. Thomson, A., Walker, P.: Durability characteristics of straw bales in building envelopes. Constr. Build. Mater. **68**, 135–141 (2014)
23. Yin, X., Dong, Q., Lawrence, M., Maskell, D., Yu, J., Sun, C.: Research on prediction model for durability of straw bale walls in warm (humid) continental climate-a case study in Northeast China. Materials **13**(13), 3007 (2020)

24. Csanády, D., Fenyesi, O.: Development of new-generation, eco-friendly thermal insulation board. In: Gatuingt, F, Torrenti, J.-M. (eds.) Proceedings of 13th fib International Ph.D. Symposium on Civil Engineering, Paris, pp. 10–17 (2020)
25. Belayachi, N., Hoxha, D., Ismail, B.: Impact of fiber treatment on the fire reaction and thermal degradation of building insulation straw composite. Energy Procedia **139**, 544–549 (2017)
26. Chetehouna, K., Belayachi, N., Lemée, L., Hoxha, D., Rengel, B.: Pyrolysis gases released during the thermal degradation of insulation materials based on straw fibers. J. Therm. Anal. Calorim. **122**(3), 1417–1422 (2015). https://doi.org/10.1007/s10973-015-4584-2
27. Yang, Y., Haurie, L., Wang, D.-Y.: Bio-based materials for fire-retardant application in construction products: a review. J. Therm. Anal. Calorim. **147**, 6563–6582 (2021)
28. Faruk, O., Bledzki, A.K., Fink, H.P., Sain, M.: Biocomposites reinforced with natural fibers: 2000–2010. Prog. Polym. Sci. **37**, 1552–1596 (2012)
29. Virk, A.S., Hall, W., Summerscales, J.: Failure strain as the key design criterion for fracture of natural fibre composites. Compos. Sci. Technol. **70**, 995–999 (2010)
30. Torres, J.P., Vandi, L.J., Veidt, M., Heitzmann, M.T.: The mechanical properties of natural fibre composite laminates: a statistical study. Compos. Part A Appl. Sci. Manuf. **98**, 99–104 (2017)
31. Ismail, B., Belayachi, N., Hoxha, D.: Optimizing performance of insulation materials based on wheat straw, lime and gypsum plaster composites using natural additives. Constr. Build. Mater. **254**, 118959 (2020)
32. Zhou, X.Y., Li, J., Zhou, D.G.: Thermal transfer properties of low density wheat strawboard. J. Nanjing For. Univ. **28**, 1–4 (2004)
33. Tůmová, E., Drochytka, R., Černý, V., Čada, P.: Development of organic and biodegradable insulating material for ETICS. Procedia Eng. **195**, 81–87 (2017)
34. Babenko, M., Estokova, A., Savytskyi, M., Unčík, S.: Study of thermal properties of lightweight insulation made of flax straw. Slovak J. Civ. Eng. **26**, 9–14 (2018)
35. Hýsek, Š, Čermák, J., Lexa, M.: Influence of lignocellulosic waste pre-treatment on the characteristics of bond rupture. Sustain. **11**(17), 4784 (2019)
36. Jiang, D., An, P., Cui, S., Sun, S., Zhang, J., Tuo, T.: Effect of modification methods of wheat straw fibers on water absorbency and mechanical properties of wheat straw fiber cement-based composites. Adv. Mater. Sci. Eng. **2020**, 5031025 (2020)
37. Keawthun, M., Krachodnok, S., Chaisena, A.: Conversion of waste glasses into sodium silicate solutions. Int. J. Chem. Sci. **12**(1), 83–91 (2014)
38. Owoeye, S.S., Abegunde, S.M., Oji, B.: A novel low temperature synthesis of sodium silicate (water glass) from silica-rich wastes using hydrothermal method. Glob. J. Eng. Technol. Adv. **6**, 66–75 (2021)
39. Puertas, F., Torres-Carrasco, M.: Use of glass waste as an activator in the preparation of alkali-activated slag. Mechanical strength and paste characterisation. Cem. Concr. Res. **57**, 95–104 (2014)
40. Vinai, R., Soutsos, M.: Production of sodium silicate powder from waste glass cullet for alkali activation of alternative binders. Cem. Concr. Res. **116**, 45–56 (2019)
41. MSZ EN 12667:2001. Thermal performance of building materials and products - Determination of thermal resistance by means of guarded hot plate and heat flow meter methods—Products of high and medium thermal resistance. Hungarian Standard Institute (2001)
42. MSZ EN 1946-3:1999. Thermal performance of building products and components- Specific criteria for the assessment of laboratories measuring heat transfer properties - Part 3 : Measurements by heat flow meter method. Hungarian Standard Institute (1999)
43. MSZ EN ISO 10456:2008. Building materials and products. Hygrothermal properties. Tabulated design values and procedures for determining declared and design thermal values (ISO 10456:2007). Hungarian Standard Institute (2008)

44. MSZ EN 826 2013. Thermal insulating products for building applications - Determination of compression behaviour. Budapest: Hungarian Standard Institute (2013)
45. MSZ EN ISO 1182:2020. Reaction to fire tests for products—Non-combustibility test (ISO 1182:2020). Budapest: Hungarian Standard Institute (2020)
46. MSZ EN 13501-1:2019. Fire classification of construction products and building elements - Part 1: Classification using data from reaction to fire tests. Budapest: Hungarian Standard Institute (2019)
47. Csanády, D., Fenyvesi, O., Nagy, B.: Heat transfer in straw-based thermal insulating materials. Materials **14**(16), 4408 (2021)
48. Csanády, D., Fenyvesi, O.: Effect of modification of micro- and microstructure of unbound straw fibre clusters on thermal conductivity (in Hungarian). In: 24th International Conference on Civil Engineering Architecture, Csíksomlyó, pp. 32–38. (2020). ISSN 2734-4525
49. Csanády, D., Fenyvesi, O.: Shear behaviour of silicate-based structural adhesives for timber. Epa - J. Silic. Based Compos. Mater. **73**(2), 72–82 (2021)
50. Fenyvesi, O., Csanády, D.: Effect of drying method on compressive strength of straw-based thermal insulations. Mater. Struct. Technol. **3**, 108–116 (2021)

Residual Stresses in Green Wood Based on a Phase Field Model

Jan Bernd Wulf(✉) and Ingo Muench

Institute of Structural Mechanics and Dynamics, TU Dortmund University, August-Schmidt-Straße 8, 44227 Dortmund, Germany
jan.wulf@tu-dortmund.de
https://bmsd.ab.tu-dortmund.de/en/

Abstract. In forested areas, the application of living trees as load-bearing structures is well suited for the construction of residential buildings above ground. Both land consumption and technical building material usage are reduced in this architectural style. However, in that regard, trees become supporting structures and require deterministic modelling to verify their load-bearing capacity. An assessment of residual growth stress provides a basis for improved prediction of the tree's load-bearing capacity. We propose a model that emulates the tree's annual growth cycle for the analysis of residual stresses within an arbitrary trunk geometry. As the properties of wood are determined by fibres and a supporting matrix, our model also captures transverse isotropy.

Keywords: Phase field model · green wood · residual growth-stress · stress-induced growth · transverse isotropy

1 Introduction

With an increasing awareness of the significance of trees in mitigating pressing problems such as climate change, species extinction, flooding and soil erosion, society's appreciation of forests and trees in urban spaces has vastly increased. This poses the question of how the future development of our civilization can be positively influenced by the broad integration of trees into spatial planning and architecture (urban greening).

Therefore, our research group proposes a holistic approach of building with trees to develop sustainable buildings in tree-covered landscapes. Our hypothesis is that such a settlement can be reconciled with the requirements of forestry, environmental protection, recreation and safety. Ecotourism and forest education are already expanding worldwide, using trees to build platforms and simple tree houses. Permanently inhabited housings, however, place higher demands on safety. There are more requirements concerning the technical implementation as well as the risk assessment.

Since real tree houses do not need artificial foundations, levelling is not required and the function of the ground is not affected. Thus, the integrity of the roots and the vitality of the tree can be maintained if it is not overloaded.

S. Amziane et al. (Eds.): ICBBM 2023, RILEM Bookseries 45, pp. 899–910, 2023.
https://doi.org/10.1007/978-3-031-33465-8_69

However, the mechanical behaviour of green wood differs significantly from the behaviour of dried, technically processed wood, so that a statement on the load-bearing capacity cannot be derived from simple engineering formulas. For example, the stem remains in its basic state and retains its growth stresses and moisture content. The associated residual stress state is advantageous for the tree and increases its load-bearing capacity, cf. [1,4].

Furthermore, the local growth of wood layers partially depends on the stress state in the wood matrix, which is influenced by additional loads, e.g. from ropes of technical structures. All these aspects motivate the computational model presented here, which takes the mentioned aspects into account.

The model simulates the growth of wood layers and the associated residual stresses in the wood matrix. The existing wood layers can be mechanically loaded, resulting in a stress state overlapping with the residual stresses. In addition, the direction of fibres is iteratively aligned with the morphology of the tree.

Our model is designed to simulate the stress state in an already existing tree that is additionally loaded, e.g. by a tree house. Thus, the morphology of the tree is known and serves as basic information for the phase field model. The model is capable of simulating growth in the direction of the sprouts (primary growth) as well as in the thickness direction (secondary growth). The growth layers receive initial, natural growth strains that contract the new fibres at the edge of the trunk and compress the existing sapwood core, resulting in a residual stress state discussed in this model and in our example.

2 A Phase Field Model for the Accumulation of Residual Stresses in Trees

The morphology of the trunk is described by the phase field parameter φ. Wood is defined by $\varphi > 1$, and $\varphi < 0$ describes voids. In between lies the phase transition zone, which is particularly important for the model as the formation of residual stresses originates here. In addition, this layer is also responsible for thickness growth. Objective functions of the phase field model generate the required growth pressure to develop a tree topology. In previous papers [6,10] we used stress driven objectives as primary functions for topology optimisation. Here, we apply a direction constrained objective function

$$J = \int_{\mathcal{B}} -\,|\mathbf{d}_\theta|\,\varphi \;\mathrm{d}V = \int_{\mathcal{B}} \gamma(\mathbf{d}_\theta)\,\varphi \;\mathrm{d}V \qquad \rightarrow \text{min w.r.t. } \varphi\,, \tag{1}$$

characterising the growth in the direction of the sprouts. For this purpose, a primary growth direction $\mathbf{d}_\theta$ is defined that also coincides with the direction of fibre within the wood matrix, cf. Fig. 1. The component of the stress tensor in the direction of fibre is given by

$$\sigma_{\theta\theta} = \mathbf{d}_\theta \cdot (\sigma \cdot \mathbf{d}_\theta). \tag{2}$$

2.1 Phytomorphological Description of the Sprout Tip Direction

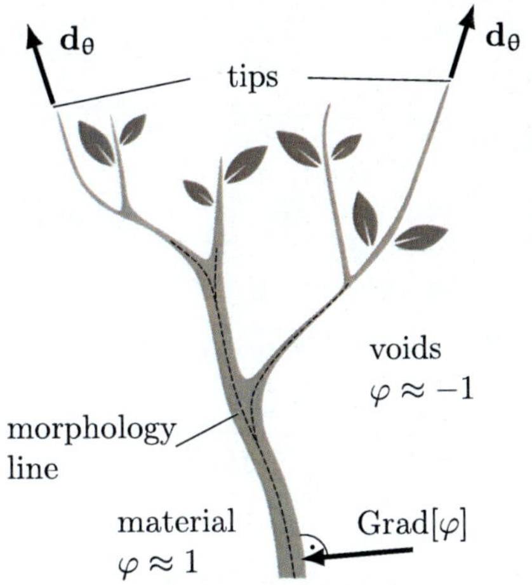

Fig. 1. Illustration of the primary growth direction $\mathbf{d}_\theta$ at sprout tips, the morphology line in the centre of the cross-section, and the gradient of the phase field variable Grad[φ] perpendicular to the surface in the stem area.

Since we intend to simulate the residual stresses in an already grown tree trunk, its morphology is known and can be reduced to the centre line by visual methods. The morphology line is defined as an input via one or more functions $g(x)$, $g_1(x)$, $g_2(x)$, ..., $g_n(x)$ $x \in \mathbb{R}$. It defines the initial primary growth vectors for elements in the regular finite element mesh that is common for phase field models via their slope

$$\mathbf{d}_\theta = \begin{bmatrix} 1/\sqrt{(g')^2+1} \\ g'/\sqrt{(g')^2+1} \end{bmatrix}. \tag{3}$$

The thickness growth creates new material lacking a primary growth direction $\mathbf{d}_\theta$. This is where the phase field gradient Grad[φ] is used to determine a growth direction. The gradient is positioned orthogonally to the surface, see Fig. 1. A rotation by $\beta = \pm\pi$ then identifies a new primary growth direction

$$\mathbf{d}_\theta = \begin{bmatrix} 0 & -\sin(\beta) \\ \sin(\beta) & 0 \end{bmatrix} \frac{\mathrm{Grad}[\varphi]}{|\mathrm{Grad}[\varphi]|}. \tag{4}$$

2.2 The Phase Field Model

The phase field parameter $\varphi \in \mathbb{R}$ defines the density of the wood by

$$f(\varphi) = \frac{\mathrm{e}^{a\,(\varphi-1)}}{\mathrm{e}^{a\,(\varphi-1)}+1} \qquad \rho = f(\varphi)\,\rho_0, \tag{5}$$

interpolating between $\rho = 0$ and $\rho = \rho_0$, the density of the employed green wood material. The slope of the function is specified by the parameter $a = 5$. The stiffness of the material

$$\mathsf{C}_\varphi = f(\varphi)\,\mathsf{C} \tag{6}$$

depends on its density and then amounts to a proportion of the stiffness of the material C based on the phase field parameter. The strain energy for linear elasticity is given by

$$\Psi_{\mathrm{mech}}(\varepsilon, \varphi) = \frac{1}{2}\,\varepsilon(\mathbf{u})\,f(\varphi)\,\mathsf{C}\,\varepsilon(\mathbf{u}). \tag{7}$$

Also typical for phase field models is the double-well potential

$$\Psi_{\mathrm{well}} = \varphi^6 - \varphi^4 - \varphi^2 + 1 \tag{8}$$

that defines equilibrium values of the phase field for void $\varphi = -1$ and solid $\varphi = 1$. The gradient energy

$$\Psi_{\mathrm{Grad}} = \frac{1}{2}\, \mathsf{p}_{\mathrm{d}}\, ||\mathrm{Grad}[\varphi]||^2 \tag{9}$$

sets the extent of the transition zone between void and solid. The here applied interface matrix

$$\mathsf{p}_{\mathrm{d}} = L_{\mathrm{c}}\, \mathbf{1} + p_{\mathrm{s}}\, \mathrm{diag}(\mathbf{d}_{\theta}), \tag{10}$$

additionally ensures an alignment of the transition zone with the primary growth direction $\mathbf{d}_{\theta}$. The interface parameter L_{c}, contributes to a constant part, whereas p_{s} applies to the vector $\mathbf{d}_{\theta}$ that is used for the alignment. The combination of these energy densities forms the elementary representation of the free energy

$$\Psi_{\mathrm{int}}(\varepsilon, \varphi, \mathrm{Grad}[\varphi]) = \Psi_{\mathrm{mech}}(\varepsilon, \varphi) + \Psi_{\mathrm{well}}(\varphi) + \Psi_{\mathrm{grad}}(\mathrm{Grad}[\varphi]). \tag{11}$$

The total energy of the model

$$\begin{aligned} \Pi = & \int_{\mathcal{B}} \Psi_{\mathrm{int}}(\varepsilon, \varphi, \mathrm{Grad}[\varphi])\, \mathrm{d}V + \int_{\mathcal{B}} \gamma\varphi\, \mathrm{d}V - \int_{\mathcal{B}} f(\varphi)\rho_0\, \mathbf{b}\cdot\mathbf{u}\, \mathrm{d}V \\ & - \int_{\partial\mathcal{B}} (\bar{\mathbf{t}}\cdot\mathbf{u} + y\,\varphi)\, \mathrm{d}A, \end{aligned} \tag{12}$$

extends the free energy via objective functions to generate growth. Thus, the equilibrium of momentum and the equilibrium of the phase field is given by

$$\mathrm{Div}[\sigma] + f(\varphi)\rho_0\, \mathbf{b} = 0, \tag{13}$$

$$\eta - \mathrm{Div}[\mathrm{Grad}[\varphi]\cdot \mathsf{p}_{\mathrm{d}}] - f(\varphi)\rho_0\, \mathbf{b}\cdot\mathbf{u} + \gamma = -\omega\dot{\varphi} \qquad \eta := \frac{\partial \Psi_{\mathrm{int}}}{\partial\varphi}. \tag{14}$$

In order to solve each initial distribution of the phase field parameter, a term containing the rate of the phase field parameter $-\omega\dot{\varphi}$ is included to control the velocity of the phase transition, as suggested in [2]. A Newton-Ralphson algorithm is implemented due to the non-linear character of (14). Our approach uses an implicit Euler time integration scheme. Within the procedure, the rate of the phase field parameter

$$\dot{\varphi}(t_{n+1}) = \frac{1}{\Delta t}\, [\varphi(t_{n+1}) - \varphi(t_n)] \tag{15}$$

is computed by using the cumulated time increments Δt within a Newton step.

2.3 Residual Stress Accumulation

Close to the trunk surface, the cambium layer is located underneath the bark and defined by an upper and a lower threshold value $\overline{\varphi}$ and $\underline{\varphi}$ in the model. New cambium layers develop as a result of cell division, cf. [7,9]. As the new cells fill with water, the hydrostatic pressure induces a change in the shape of the cells, causing them to expand in the transverse direction and contract in the vertical direction, as illustrated in Fig. 2. Within the model, this effect is taken into account by a maturation strain $\bar{\varepsilon}_0$ in the direction of $\mathbf{d}_\theta$. The strain state in the newly generated cambial layer is manifested during maturation. Therefore, a strain history field is utilised

$$\underline{\varphi} > \varphi > \overline{\varphi} \qquad \varepsilon_{\mathrm{h}}(t_{n+1}) = \bar{\varepsilon}_0 + \Delta\varepsilon(t_{n+1}), \tag{16}$$

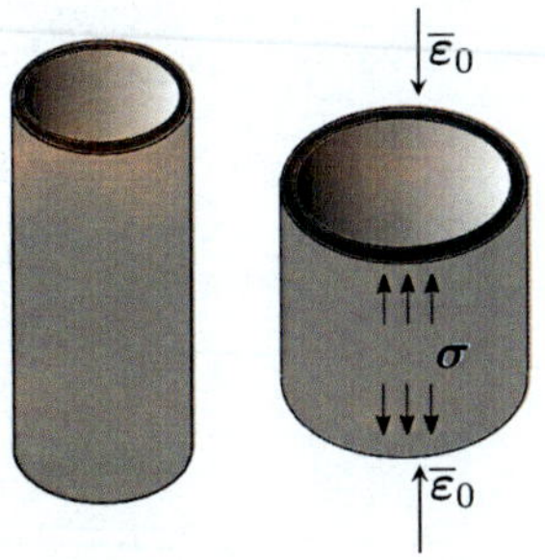

Fig. 2. Simplified illustration of a cell undergoing a change in shape due to the hydrostatic pressure.

with the change of the integral strain state $\Delta\varepsilon(t_{n+1})$ provided at time t_{n+1}. The process is an independent application of a pre-stress condition. However, the growth tip is assumed to be stress free. In each layer of wood, the residual stress state results from the strain history and the change in the integral strain state, yielding

$$\varphi > \overline{\varphi} \qquad \varepsilon_{\mathrm{h}}(t_{n+1}) = \varepsilon_{\mathrm{h}}(t_n) + \Delta\varepsilon(t_{n+1}). \tag{17}$$

The adhesion of the new layer to the already matured sapwood implicates longitudinal tension and transverse compression, known as maturation stress, cf. [9]. Figure 3 shows a flow chart of the residual stress accumulation with consideration of a stress-free growth tip and the assignment of new primary growth directions $\mathbf{d}_\theta$ as described in Sect. 2.1.

2.4 Transverse Isotropic Fibres

The microstructure of the wood material exhibits a preferred orientation due to the orientation of the cell walls of the xylem. At small strains it responds like an isotropic material, at larger strains it behaves anisotropically. In order to map this property, we apply a transversely isotropic constitutive law in the model. The longitudinal direction (L-direction) is the direction of fibre, designated as the preferred direction. The radial (R-direction) and circular direction (C-direction) represent the two transversal directions. The following material parameters describe transverse isotropy involving two isotropic axes

$$E_{\mathrm{L}}, \quad E_{\mathrm{R}} = E_{\mathrm{C}}, \quad \nu_{\mathrm{LR}} = \nu_{\mathrm{LC}}, \quad G_{\mathrm{LR}} = G_{\mathrm{LC}}, \quad \nu_{\mathrm{RC}} = \nu_{\mathrm{CR}} = \frac{E_{\mathrm{R}}}{2\,G_{\mathrm{RC}}} - 1. \tag{18}$$

The stress-strain relationship for linear elasticity in the plane strain state parallel to the preferred direction with the inserted parameters then reduces to

$$\sigma = \begin{bmatrix} (1-\nu_{\mathrm{RC}}^2)E_{\mathrm{L}}/\chi & (1+\nu_{\mathrm{RC}})\nu_{\mathrm{LR}}E_{\mathrm{R}}/\chi & 0 \\ (1+\nu_{\mathrm{RC}})\nu_{\mathrm{LR}}E_{\mathrm{R}}/\chi & (1-\nu_{\mathrm{RL}}\nu_{\mathrm{LR}})E_{\mathrm{R}}/\chi & 0 \\ 0 & 0 & G_{\mathrm{LR}} \end{bmatrix} \varepsilon, \tag{19}$$

with $\chi = (1+\nu_{\mathrm{RC}})(1-\nu_{\mathrm{RC}}-2\nu_{\mathrm{LR}}\nu_{\mathrm{RL}})$ and $\nu_{\mathrm{RL}} = E_{\mathrm{R}}\nu_{\mathrm{LR}}/E_{\mathrm{L}}$.

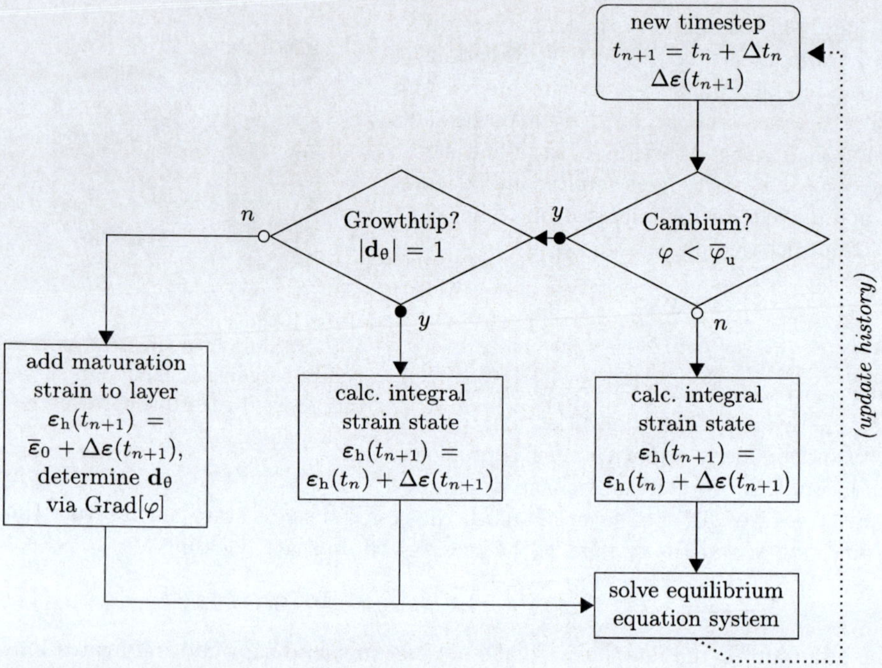

Fig. 3. Flow chart for the accumulation of residual stresses within the model

3 Off-Centric Loading of a Straight Stem

Our example simulates a straight vertical trunk geometry. The morphology line in Fig. 4a specifies the initial growth direction $\mathbf{d}_\theta$. We examine the trunk growth of a common beech over a period of 50 years. The branches of the tree and its crown are not included in the simulation. However, they are exemplarily illustrated in Fig. 5. Table 1 lists the material parameters with transversal isotropic properties typical for the tree species and necessary computational parameters of the model. As initial topology at time $t = 0$ we consider a sapling of 60 cm height and 2 cm diameter free of residual stresses, cf. Figure 4b. At first, we observe the natural growth of the tree without additional technical loads (**scenario 1**). In a second scenario, after 25 years, an off-centred technical load of 3 MN is applied to the tree (residential body), as shown in Fig. 4b (**scenario 2**). After a further 25 years, it is removed (deconstruction).

For both scenarios, Fig. 5 shows successive states of the trunk at significant points in time. The stem diameter increases by about 65 cm within the observed period of 50 years.

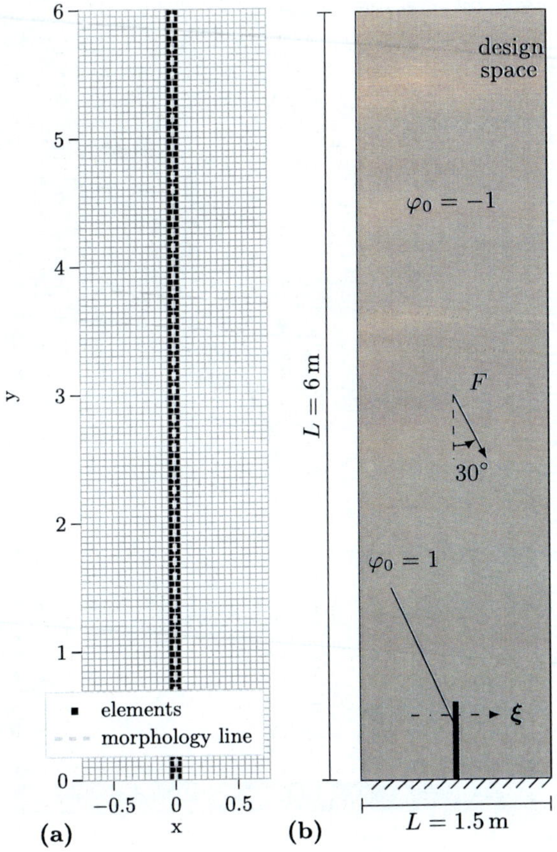

Fig. 4. **a** Discretization of the domain with morphological definition as described in Sect. 2.1. **b** Design space with a phase field parameter of $\varphi = -1$ and a small area with $\varphi = 1$ representing the sapling. The $\boldsymbol{\xi}$ axis is used to map stress curves in the cross-section.

Table 1. Material parameters of *Fagus sylvatica* (common beech) in green condition as rounded mean values from [3,5,7,8]

parameter	symbol	value	unit
density	ρ	833	kg/m^3
Young's modulus longitudinal	E_{L}	1.00×10^4	MPa
shear modulus LR	G_{LR}	1000	MPa
Poisson's ratio	ν_{LR}	0.30	[-]
Young's modulus radial	E_{R}	1000	MPa
shear modulus RC	G_{RC}	300	MPa
maturation strain	$\overline{\varepsilon_0}$	5.00×10^{-3}	[-]
interface parameter 1	L_{c}	3.00×10^{-2}	[-]
interface parameter 2	p_{s}	0.10	[-]

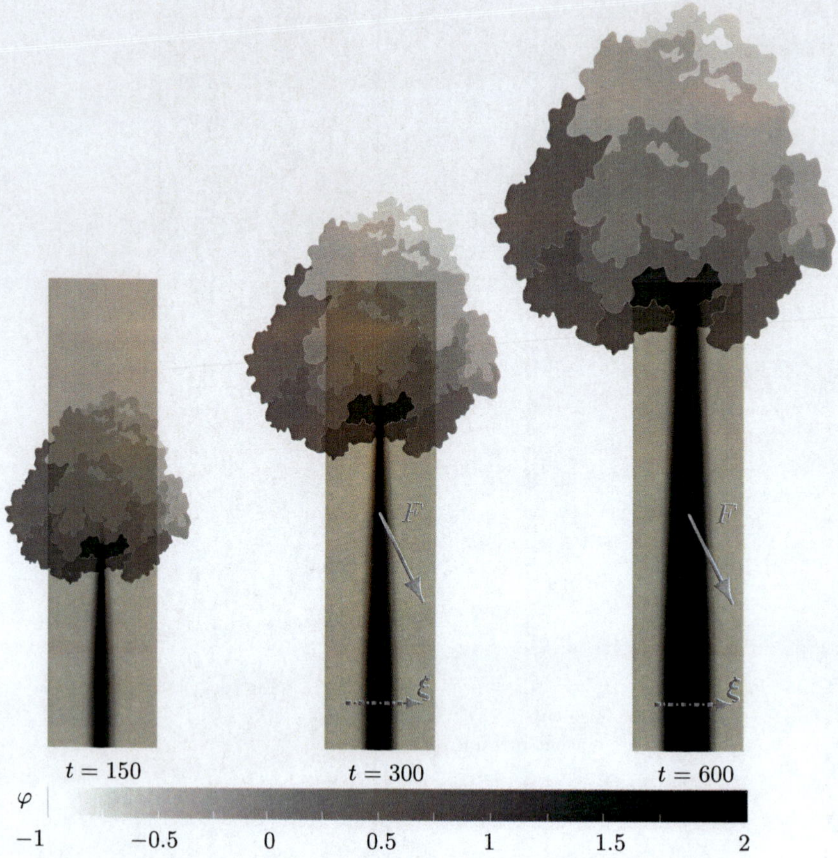

Fig. 5. Growth states of the beech (wood is indicated by $\varphi > 1$) over a period of 50 years (600 months) at exemplary points in time t (indicating months)

3.1 Natural Growth with F = 0 (Scenario 1)

The evolving residual stress state is observed along a log cross-section at height $y = 0.5\,\text{m}$, shown in Fig. 5 by the $\boldsymbol{\xi}$ axis. After 25 and 50 years, residual stress states arise that exhibit a similar shape, cf. Fig. 6. Since the diameter of the trunk increases by 20 cm within 25 years, at the tree age of 50 years a larger cross-sectional area with tensile stresses in the outer layers occurs, while the axial compressive stresses in the heartwood tend to increase.

In the region of the cambium, we model reduced stiffness such that the stress state decreases to zero.

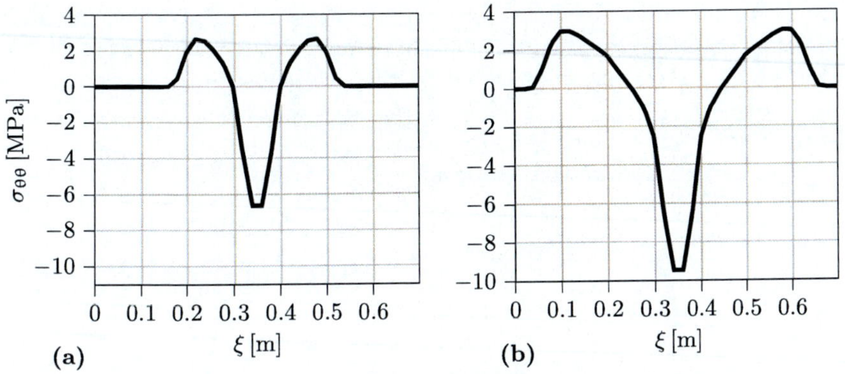

Fig. 6. Stress curves after 25 and 50 years in the trunks cross-section 0.5 m above the ground. **a** $\sigma_{\theta\theta}(t = 300)$, scenario 1. **b** $\sigma_{\theta\theta}(t = 600)$, scenario 1.

3.2 Technical Loading with F (Scenario 2)

In this scenario, the tree grows during the first 25 years as in scenario 1. Afterwards, the tree is loaded by the technical load $F = 3\,\text{MN}$ over a period of 25 years. Figure 7 shows the stress state in the cross-section shortly before and after applying the load. The residual stress state after 25 years and the stresses due to the technical loading superimpose, see Fig. 7. As a result, compressive stresses occur in the right half of the cross section, however, the amount is reduced due to the residual stresses.

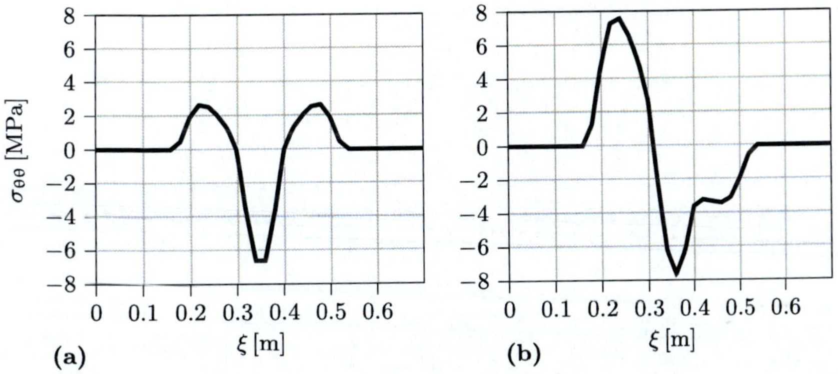

Fig. 7. Stress curves at $t = 300$ (25 years) in the trunks cross-section 0.5 m above the ground before and after technical loading. **a** $\sigma_{\theta\theta}(t = 300)$ without technical load. **b** $\sigma_{\theta\theta}(t = 300)$ with technical load

The compressive strength of wood in the direction of fibres is typically lower than its tensile strength. Thus, the superposition of residual stress and stress resulting from technical load at the edge of the shaft is advantageous as it increases the bending load capacity of the trunk. This scenario demonstrates how the residual stress state preferred by the tree is altered as a result of the technical load.

In the further chronological progression the tree forms new layers of wood, leading to a tendency to restore this preferred state, see Fig. 8.

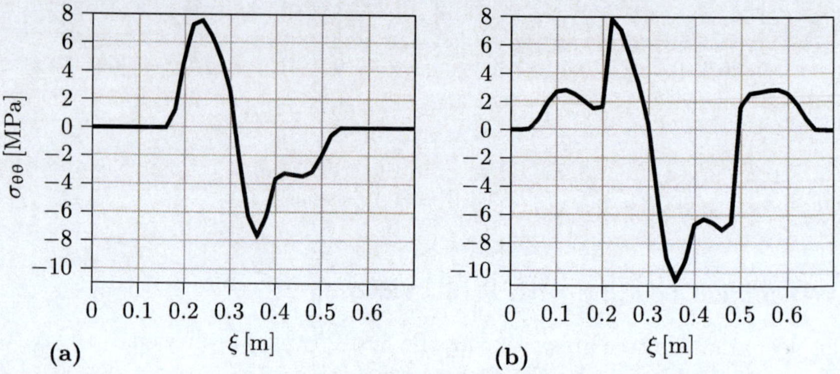

Fig. 8. Stress curves in the trunks cross-section 0.5 m above the ground after 25 and 50 years after technical loading. **a** $\sigma_{\theta\theta}(t = 300)$ with technical load. **b** $\sigma_{\theta\theta}(t = 600)$ with technical load

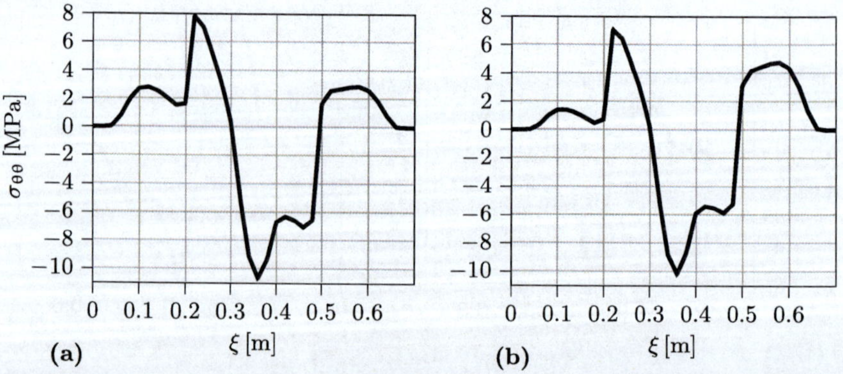

Fig. 9. Stress curves at $t = 600$ (50 years) in the trunks cross-section 0.5 m above the ground before and after technical loading is removed. **a** $\sigma_{\theta\theta}(t = 600)$ with technical load. **b** $\sigma_{\theta\theta}(t = 600)$ without technical load.

After 25 years, when the technical load is removed (unloading at time $t = 600$), the stress state in the cross-section also alters, see Fig. 9. In comparison to the loading at time $t = 300$, however, this alteration is significantly lower, since the cross-section has increased in the meantime due to the growth of wood layers.

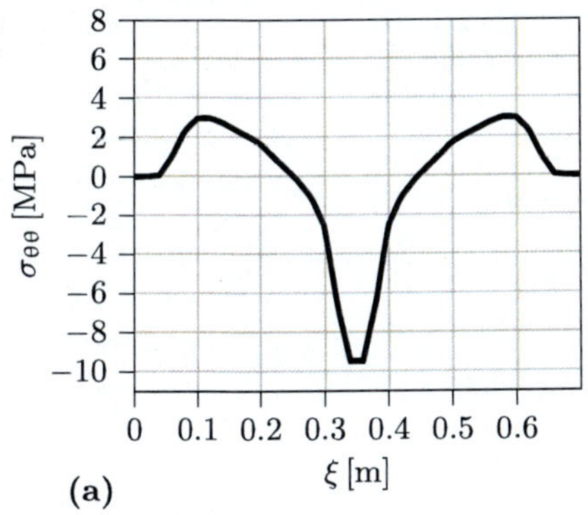

Fig. 10. Stress curves at $t = 600$ (50 years) of the two scenarios in the trunks cross-section 0.5 m above the ground. **a** $\sigma_{\theta\theta}(t = 600)$, scenario 1. **b** $\sigma_{\theta\theta}(t = 600)$, scenario 2.

Comparing both scenarios at time t = 600 (50 years), we observe how the intermediate technical loading in scenario 2 leads to a more distinct residual stress pattern after unloading, see Fig. 10.

4 Conclusion and Outlook

The presented phase field model simulates the growth and stress state of a generating tree trunk. It incorporates growth strains in newly added wood layers and manifests the current strain state at the transition from growth layer to wood layer. An existing tree morphology, also including branches can be simulated, cf. [11].

The example discussed in this paper employs a simple, purely vertically growing shaft geometry that is observed over a 50-year period to study the effect of an engineering load on the stress state in the cross-section. Thereby, the simulation model reveals how a technical load interferes with the natural residual stress state. However, this aberration is mitigated by the growth of new layers of wood. The state of stress due to technical loading is therefore gradually compensated for by newly added layers of wood, meaning that growing trees can evidently be classified as an adaptive load-bearing elements.

This categorisation is highly relevant for the use of supporting trees in the realisation of construction projects. In terms of risk analysis, it implies higher resilience to rare but exceptional loadings such as strong wind events. In strong wind events, optimal load-bearing capacity of the tree shaft is assumed if the natural residual stress state is inherent in the outer layers of the wood. Since this state is also gradually re-attained after technical loading, an increase in technical loading, for instance, after a definable time interval is justifiable. This is intended to support our working hypothesis that building in and with trees can save material resources and allows to protect forested areas. This does not exclude a hybrid construction method as a combination of classical foundation and the utilisation of supporting trees.

References

1. Archer, R.R.: Growth Stresses and Strains in Trees, vol. 3. Springer, Heidelberg (2013)
2. Cahn, J., Allen, S.: A microscopic theory for domain wall motion and its experimental verification in Fe-Al alloy domain growth kinetics. Le J. Phys. Colloques **38**(C7), C7–51 (1977)
3. Green, D.W., Winandy, J.E., Kretschmann, D.E.: Mechanical properties of wood. Wood handbook: wood as an engineering material. Madison, WI: USDA Forest Service, Forest Products Laboratory, 1999. General technical report FPL; GTR-113, vol. 113, pp. 4.1–4.45 (1999)
4. Kübler, H.: Studien über wachstumsspannungen des holzes-zweite mitteilung die spannungen in faserrichtung. Holz als Roh-und Werkstoff **17**(2), 44–54 (1959)
5. Lavers, G.: The strength properties of timbers. Bulletin (Forest Products Research Laboratory (Princes Risborough, England)). H.M. Stationery Office (1969)
6. Muench, I., Gierden, C., Wagner, W.: A phase field model for stress-based evolution of load-bearing structures. Int. J. Numer. Meth. Eng. **115**(13), 1580–1600 (2018)
7. Niklas, K.J., Spatz, H.C.: Plant Physics. University of Chicago Press, Chicago (2012)
8. Ross, R.J., et al.: Wood handbook: wood as an engineering material. USDA Forest Service, Forest Products Laboratory, General Technical Report FPL-GTR-190, 2010: 509, vol. 190, p. 1 (2010)
9. Thibaut, B., Gril, J., Fournier, M.: Mechanics of wood and trees: some new highlights for an old story. Comptes Rendus l'Acad. Sci. - Ser. IIB - Mech. **329**(9), 701–716 (2001)
10. Wulf, J.B., Muench, I.: Topology evolution of composite structures based on a phase field model. PAMM **20**(1), e202000,163 (2021)
11. Wulf, J.B., Muench, I.: Growth of green wood based on a phase field model. PAMM **22**(1), e202200,067 (2022)

On the Acoustical Characterization and Modelling of Typha Reed Aggregates and Stems - TyCCAO Project

Clément Piégay, Philippe Glé(✉), Matteo Buatois, and Hamza Bentounsi

Cerema – Univ. Eifel, UMRAE, Strasbourg, France
{clement.piegay,philippe.gle}@cerema.fr

Abstract. The proliferation of *Typha Australis*, a fast-growing reed considered as an invasive plant, has greatly accelerated in recent years in the Senegal River basin. This is causing numerous negative impacts, such as sanitary, economic and biodiversity impacts, as well as an increase in the risk of flooding. The measures implemented so far have not succeeded in slowing down the progression of Typha and in this context it seems relevant to consider that this reed has a significant potential for valorization as a construction material with insulating properties. It is in this context that the international project Typha Combustible and Construction in West Africa (*TyCCAO*) was developed. It aims to contribute to the ecological transition and the fight against global warming, by developing the use of renewable fuels and energy efficiency in the building industry through the development of biobased materials based on Typha.

The acoustic behavior of Typha stems and aggregates could therefore be studied in this context through characterization and modelling approaches. The main physical characteristics of the materials have first been evaluated, including bulk density, porosity, airflow resistivity, shapes and size distributions. Acoustical performances have then been characterized in normal incidence with a Kundt tube in order to observe the effects of density and characteristic size of the samples. For the stems, the effect of orientation has also been tested.

Results show that these materials present good sound absorption properties and highlight a significant effect of the parameters investigated (density, orientation and size). The modellings with a conventional porous modelling approach (*Johnson-Champoux-Allard-Lafarge* model) yield for both kinds of materials to satisfactory predictions, and enable to compare the behavior of Typha reed to other biobased materials.

Keywords: Typha reed · acoustical performances · characterization · modelling · *TyCCAO* project

1 Introduction

Typha Australis or Typha Domingensis (scientific name), a fast-growing reed, grows along slow-flowing rivers as the Senegal River. In Mauritania and Senegal, the creation of dams on the Senegal River 30 years ago created environmental conditions that were

S. Amziane et al. (Eds.): ICBBM 2023, RILEM Bookseries 45, pp. 911–925, 2023.
https://doi.org/10.1007/978-3-031-33465-8_70

highly conducive to Typha. In this context, the proliferation of Typha Australis has greatly accelerated in recent years in the Senegal River basin, with an estimated average increase in the surface area covered of about 15% per year on more than 130 km of riverbanks. Now this invasive plant already occupies areas estimated at between 60,000 and 80,000 hectares and its proliferation is causing numerous negative impacts, such as health, economic and biodiversity impacts, as well as an increase in flood risks.

Over the last two decades, studies and evaluations of this situation and the methods of regulating these colonizations have been the subject of scientific publications [1–4].

However, the measures implemented so far have not succeeded in curbing the spread of Typha and in this context it seems appropriate to consider that this reed has a significant potential for recovery in terms of vegetal biomass [5]. Thus, two possibilities for using Typha are envisaged. The first concerns the production of energy using its combustion [6] and the second concerns its use as a building material with insulating properties. With these orientations, the Senegalese and Mauritanian governments could contribute both to the energy transition of their countries, but also fight against the proliferation of Typha.

It is in this context that the international Typha Combustible and Construction in West Africa (TyCCAO in French) project was developed. Its objective is to contribute to the ecological transition and the fight against global warming by developing the use of renewable fuels and energy efficiency in the building industry through the development of biobased materials based on both Typha particles and stems.

Indeed, the use of bio-based materials provides innovative, relevant and sustainable solutions in the field of construction. As they are derived from vegetal raw materials, they provide significant storage of atmospheric carbon dioxide [7, 8]. In addition, when they are used for green building insulation, these materials have high levels of performance in terms of sound properties [9–12], contributing significantly to the improvement of acoustic comfort inside buildings in order to fight against noise pollution.

The behaviour of biobased materials has already been studied and several results are presented in the literature. For particles, many works were related to hemp shiv [10–13], but also to sunflower pith and bark, flax shiv, rape straw [14, 15]. Biobased stems have also been investigated acoustically, but only few studies exist in this case: Reeds were examined in [9], and wheat straw in [15]. These papers bring interesting results, from the interesting level of sound absorption or transmission loss performance to the dissipation behavior including particle size and orientation, density, and multiple porosity scale effects.

In this context, the present paper focuses on the acoustical characterization and modelling of both loose particles and stems from Typha, while other properties such as their hygrothermal behavior will be related through other Work Packages of TYCCAO project. It is organized through this introduction and three sections. Materials and methods used in the study are presented in next section. The experimental and modelling results are then detailed and analyzed and a conclusion is finally drawn.

2 Materials and Methods

2.1 Typha Particles and Stems Under Investigation

All the Typha particles and stems characterized in this paper come from the same batch and from Senegal (Elementerre SARL). These resources are intended to be used either loose or in concretes for the particles, and in panel with metal clamping systems for the stems. In the TyCCAO acoustic Work Package, tests were directly carried out on samples of loose particles or loose stems of Typha by varying their thickness and density in the measurement devices.

Particles. The Typha particles studied in this project was declined in 9 variants, represented in Fig. 1:

- The bulk particles were tested as is, and are referred in this paper as 'raw Typha';
- A variant of this first batch was obtained by defibrating (by mechanical sieving combined with manual sorting) the reference batch. This second sample is referred as 'defibrated Typha';
- Finally, 7 additional declinations were obtained by mechanical sieving, as follows:

A sample whose particles were retained by the sieve (square section) of 10 mm, identified as 'Typha > 10 mm'. A sample whose particles were retained by the 4 mm sieve: 'Typha > 4 mm' and following the same logic the samples 'Typha > 2 mm'; 'Typha > 1 mm'; 'Typha > 500 μm'; 'Typha > 250 μm'; and 'Typha low pass', corresponding to particles passing under the 250 μm threshold.

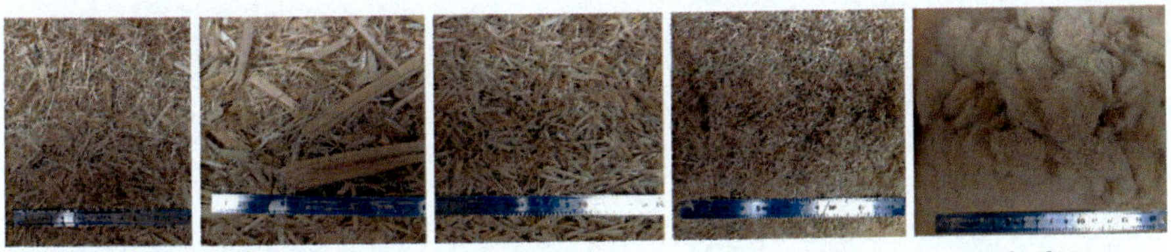

Raw Typha *Typha > 10mm* *Typha > 2mm* *Typha > 500μm* *Typha low pass*

Fig. 1. Typha particles under investigation (selection)

Stems. In this paper, two types of stems are used and characterized. The first ones are called "flat stems". As can be seen in Fig. 2(a), these elements are parts of stems and are characterized by a flat base and a domed top. The second ones are called "cylindrical stems" because they correspond to complete circular sections of stems, as shown in Fig. 2(b). It is assumed that larger the diameter is, the more the sample is in the lower part of the stems and the smaller the diameter is, the more the sample belongs to the upper part of the stem.

(a) (b)

Fig. 2. The two types of Typha stems (a) Flat stems and (b) Cylindrical stems.

2.2 Characterization Methods

Particles and Stems Size Distributions. The particle size analysis of Typha particles (defibrated particles) was carried out by image analysis on a 9.40 g sample, i.e. 2508 particles, distributed without contact points, on 8 black A4 sheets. Photographs were taken with a digital camera mounted on a tripod, with a resolution of 130 px/mm. The images were then analyzed using ImageJ software, and only particles with a surface area greater than or equal to 1 mm^2 were retained for evaluation. In the software, after an initial image processing (black and white conversion, segmentation), each particle is identified by an ellipse. The characteristics of this particle can thus be determined: its surface, length, width, and even its slenderness (length/width ratio). The data were analyzed using a surface weighted analysis. The statistical distributions were compared to normal and lognormal distributions, whose parameters (mean and standard deviation) were calculated.

For the stems, width were measured directly on a representative batch of 200 stems.

Bulk Densities. Bulk particle density measurements were carried out on 10 trials according to the RILEM protocol [16], with a container volume close to 12 L and a mass of particles of about 350 g. Two density values were evaluated, a minimum value of the "loose" mixture corresponding to the value according to the RILEM protocol and a value of the mixture after settling, called "packed".

To determine the bulk density of the stems, 50 mm long samples were cut from cylindrical stems. Then, the samples were divided into three categories:

- With lumen from the lower part of the stem;
- Without lumen from the lower part of the stem;
- Without lumen from the upper part of the stem.

It should be noted that none of the samples from the upper part of a stem had a lumen.

Open Porosity. The measurement protocol used is based on the works of Leclaire et al. [17]. It enables to calculate the open porosity (directly accessible by air) of materials by determining directly their skeleton volume. Porosity measurements were carried out on both raw Typha particles, defibrated particles, and the particle size fractions obtained by mechanical sieving in order to assess whether an effect of particle size was visible. The measurements were repeated 4 times on samples of the order of 2 to 3g in mass.

For Typha stems, both skeleton volume and porosity are determined for the 3 categories of stems defined in the previous paragraph.

Airflow Resitivity. The measurement protocol of the airflow resistivity complies with standard ISO 9053-1 [18]. For a sample of known section and thickness, the device enables to apply an air flow on one side and to measure the pressure difference existing between both sides. The resistivity of loose Typha particles stacks was evaluated for the different particle samples, and as a function of their density. The measurements were carried out on circular samples of 10 cm diameter, and with a height of 5cm (except for the two finest sizes, the volume being insufficient), and were repeated 3 times in each configuration.

For stems, two configurations are considered in this study. The first one corresponds to a stack of stems parallel to the direction of the flow and the second one corresponds to a stack of stems perpendicular to the flow direction, as shown in Fig. 3.

Fig. 3. The two stack-configurations of Typha stems: (a) parallel to the flow (b) perpendicular to the flow.

Acoustical Properties. The sound absorption coefficient was obtained by measurements made with a Kundt tube at normal incidence according to the standard ISO 10534-2 [19]. A B&K type 4106 device with sections of 40 mm and 100 mm has been used. The acoustic characterization of the Typha particle stacks was carried out with a circular cross-section and a diameter of 10 cm. A thickness of 5cm was chosen for the characterizations, except when the volume of particles did not allow it (case of the two finest sizes). Measurements were carried out for all particle sizes and bulk densities. In each configuration, the measurements were repeated 3 times in order to assess the variability of the results.

For stems, as for resistivity characterization, both parallel and perpendicular configurations are studied. For the parallel configuration, it was decided to carry out tests for stack thicknesses of 50, 70 and 90 mm. For each thickness, measurements were carried out for density values of 70, 80 and 90 kg.m^{-3}, and even 100 kg.m^{-3} for the 50 mm thickness. In order to evaluate the repeatability and fidelity of the tests, for each configuration of the thickness-density pair, the tests were repeated 5 times. The first three measurements were carried out without changing the arrangement of the stems in the sample holder, and the last two measurements were carried out using the same stems but mixing them between each measurement. For practical reasons, a 40 mm diameter

Kundt tube sample holder was used. For the perpendicular configuration, it was chosen to carry out tests for stack thicknesses of 20, 50 and 100 mm. However, in contrast to the previous tests carried out for the parallel stem configuration, it was not possible to vary the density values for a given thickness. In order to assess the fidelity of the tests, for each thickness, the tests were repeated 3 times by changing the arrangement of the stems in the sample holder. For practical reasons, a 10cm diameter Kundt tube sample holder was used.

2.3 Modelling Approach

The method used to model the acoustic performance of Typha aggregates and stems is based on two well known semi-phenomenological approaches based on the pore geometry. The JCA (Johnson - Champoux - Allard) model, based on the successive works of [20] and [21], is based on the use of 5 input parameters: the porosity, the resistivity, the tortuosity, the viscous and thermal characteristic length, has been used to model stems. The so-called JCAL model (Johnson - Champoux - Allard - Lafarge) is a variant of the JCA model based on the work of [22] integrating a 6th parameter: the thermal permeability and has been used to describe particles properties.

The porosity and resistivity values used are those characterized experimentally and presented in the paper. The other four parameters, which are related to viscous and thermal effects within the pore network of the materials, are determined indirectly from measurements made with the Kundt tube as described by [23, 24].

3 Results

3.1 Physical Characteristics

Particles Size Distributions. The results show large particle sizes compared to other bio-based materials such as hemp shiv, with lengths and widths of up to 100 and 11 mm respectively. The distributions are fairly true to lognormal distributions, which is very often characteristic of the way plant particles are produced. The characteristics of the distributions are given as an indication for the three analyses carried out, with a surface weighting allowing to limit the bias linked to the thresholding of small particles, and gives an average length and width of about 25 and 3 mm respectively.

A final piece of information provided by the image analysis is the average surface mass of the particles (ratio between the mass of the particles and their cumulative surface area). The latter is 140 g.m^{-2}, and is therefore in the low range of values observed for hemp shiv ([117; 328 g.m^{-2}] in [13]) (Fig. 4 and Table 1).

Stem Size. In an indicative representation of a batch of 200 stems, 188 are of the "flat" type and 12 are of the "cylindrical" type.

The diameter of the cylindrical stems is shown in Table 2. "Cylindrical" stems are used in this paper to characterize bulk densities, skeleton densities and porosities of the Typha stems.

The width of the "flat" stems is shown in Table 3. "Flat" stems are used in this paper to characterize sound absorption properties and resistivities of the Typha stems.

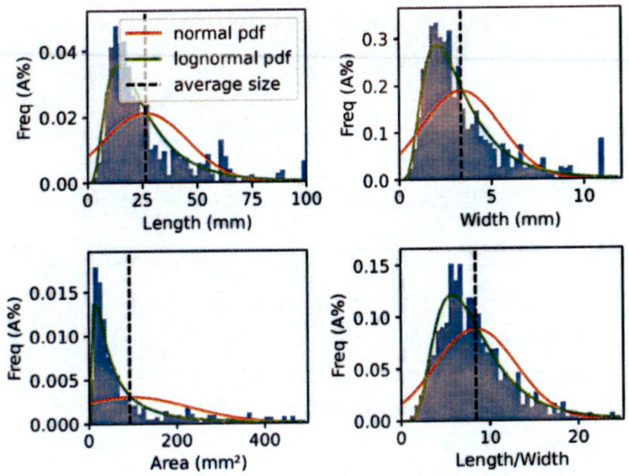

Fig. 4. Raw Typha particle size distribution - Surface weighted analysis of length, width, area and slenderness

Table 1. Raw Typha particle size distribution statistical parameters

Surface weight	*Normal law*	*Lognormal law*		
	mean	std	mean	std
Length (mm)	26,3	19,0	21,2	1,9
Width (mm)	3,4	2,1	2,8	1,8
Surface (mm^2)	92,6	134,3	47,3	3,1
Length/Width	8,4	4,5	7,5	1,6

Table 2. Diameters for "cylindrical" stems

Number of stems	Mean diameter (mm)	Min diameter (mm)	Max diameter (mm)
12	11,6	3,6	20,1

Table 3. Width for "flat" stems

Number of stems	Mean diameter (mm)	Min diameter (mm)	Max diameter (mm)
188	9,1	3,7	18,0

Bulk Densities. The values obtained, and presented in Table 4 range between 30 and 60 kg.m^{-3} and are very low compared to those for hemp, wood and flax aggregates. One reason for this is the high fiber content, which acts like a spring in the mixture and prevents the particles from being compacted (Table 4).

Table 4. Raw Typha mass density

	Loose density	Packed density
Mean	33,8 kg.m^{-3}	51,5 kg.m^{-3}
Std	4,8 kg.m^{-3}	8,1 kg.m^{-3}

For the samples from the lower part of the stems, whether a lumen is present or not, the results are relatively homogeneous with an average bulk density of between 211 and 222 kg.m^{-3}. For the samples from the upper part of the stems, the density is close to the previous values but shows a much higher standard deviation that is difficult to explain at this level of analysis (Table 5).

Table 5. Typha stems bulk densities

Sample type	Diameter values (mm)	Bulk density (kg.m^{-3}) Mean Standard dev	
		Mean	Standard dev.
With lumen (stem lower part)	[9,5–1,20]	211	21
Without lumen (stem lower part)	[10,5–19,9]	222	28
Without lumen (stem upper part)	[3,6–8,4]	202	97

Open Porosity. The results of the measurements reported in Table 6 show values between approximately 650 and 1100 kg.m^{-3}, which is within the range of the skeletal densities observed on biobased materials, while being largely lower than the densities of the plant components (cellulose, lignin, pectin, etc.) which are rather of the order of 1450 kg.m^{-3} [25]. This position shows that part of the porosity is not accessible with the porosimeter used (air volume comparison). This can be explained in part by the presence of very small pores that could only be accessed with Helium pycnometry, and/or after grinding the particles.

These values allow in fine to deduce directly the air porosity ϕ of the materials from the apparent density $\boldsymbol{\rho_a}$ of the particles by the relation $\phi = 1 - \frac{\rho_a}{\rho_s}$. Thus, for raw Typha, the values found in Table 6 lead to porosities of 95% ("loose") and 93% ("packed").

Concerning stems, the samples located in the upper part of the stems have overall higher skeletal densities than the samples located in the lower part, as shown in the Table 7. There was no significant effect of the presence of a lumen. Some values are even close to the densities of the constituent elements of plants. This may suggest that the internal structuring is not quite the same, with certainly smaller and less accessible pores for the lower part of the stems. For the lower part of the stems, the values obtained are consistent with those of the particle size greater than 2 mm presented in Table 6.

Table 6. Skeleton densities evaluated on the different Typha particle samples

Material	Frame density ($kg.m^{-3}$)	
	Mean	Std
Defibred Typha	675	27
Raw Typha	711	97
Typha > 10 mm	655	29
Typha > 4 mm	641	16
Typha > 2 mm	868	65
Typha > 1 mm	979	70
Typha > 500μm	1042	63
Typha > 250μm	972	47
Low pass Typha	1096	61

Table 7. Skeleton densities evaluated on the different Typha stem samples

Sample type	Frame density ($kg.m^{-3}$)		Porosity (%)	
	Mean	Std	Mean	Std
With lumen (lower part)	757	20	72,4	1,6
Without lumen (lower part)	705	54	69,5	2,3
Without lumen (upper part)	1086	360	82,4	5,1

Airflow Resistivity. The results obtained are summarized in Fig. 5.

We observe increasing flow resistance values as a function of the density, with values ranging from 1000 to more than 15000 $N.m^{-4}s$. The stacks made up of the smallest particles are also the most resistive. These results can be explained by the fact that the denser stacks, or those made up of finer particles, are characterized by smaller pore sizes, thus slowing down the airflow.

Furthermore, a very limited effect of the fibers in the resistivity value can be observed, the main effect visible here being the effect of these fibers on the bulk density of the stack. The different samples also show a relatively similar behavior, except for the two finest grain sizes for which the resistivity values increase at least by a factor of 2.

For stem stacks, the results are very homogeneous as shown in Table 8, and even though the parallel configuration shows very slightly higher values than the perpendicular configuration, the overall values remain relatively low and are, by way of comparison, below the values shown by the majority of insulating wools such as hemp, flax, kenaf, etc. [26].

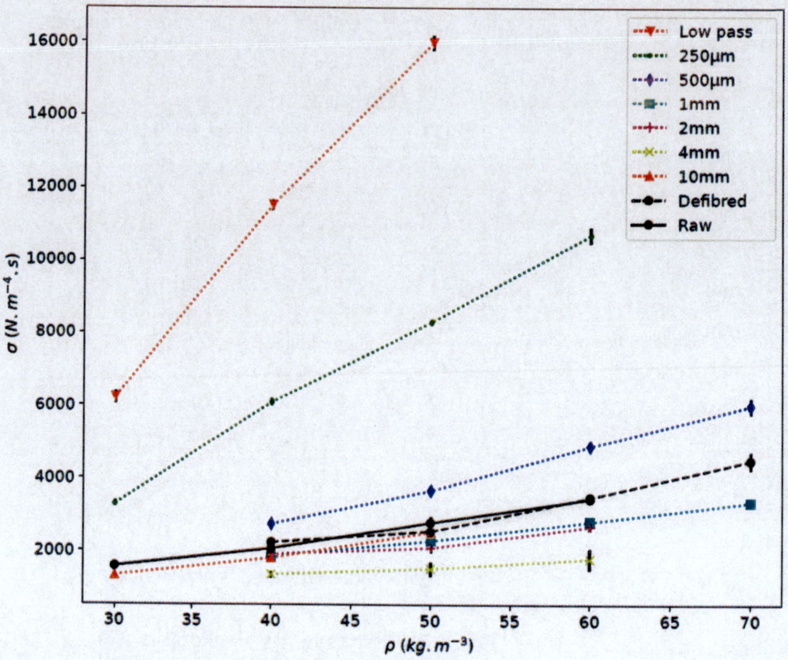

Fig. 5. Resistivity of Typha particle stacks as a function of particle size and bulk density

Table 8. Resistivity of Typha stems for both perpendicular and parallel stack configurations

Configuration stack	Thickness (mm)	Resistivity ($N.m^{-4}.s^{-1}$)
Perpendicular	20	302
Parallel	50	523
Perpendicular	53	388
Parallel	70	776
Parallel	90	672
Perpendicular	95	378

3.2 Acoustical Properties

Loose Particles. Results are presented to show the effect of density and grain size of the stacks on their acoustic properties.

In Fig. 6(a), it can be seen that the increase in bulk density is accompanied by an increase in the acoustic absorption and attenuation of the Typha particle stacks. While the sound attenuation index remains very modest, reaching a mere 2dB in this case, the sound absorption is more interesting, with a maximum level at the peak ranging from 40 to 80%. These performance levels and behaviors are characteristic of granular stacks. For other particles [13], we see the appearance of an optimal density in terms of absorption, but the latter does not seem to be achieved in the case of Typha.

The effect of particle size is also very visible Fig. 6(b) with a slight increase in attenuation for the finest particles (an increase linked to that of resistivity) and with performance levels that remain in the same ranges as before. In terms of sound absorption, the results show an increase in the coefficient for fine particles up to an optimum particle size (Typha > 500 μm) below which absorption falls back. As the thickness is different for the finer particle sizes, it is very likely that this relapse is at least partly related to the associated decrease in thickness.

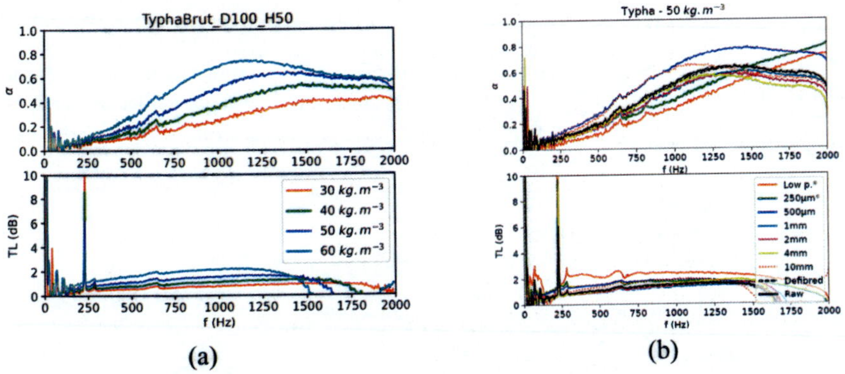

Fig. 6. Effect of density on the sound absorption and transmission loss of raw loose Typha particles, for a thickness of 50 mm (a) and effect of particle size distribution on the sound absorption and transmission loss of loose Typha particles for a given bulk density of 50 kg.m^{-3} and a thickness of 50 mm (* in the legend indicates a limited thickness) (b)

Parallel Configuration of Stems. For a fixed density, the increase in thickness will contribute to a low frequency shift of the absorption peaks, as shown by Fig. 7(a). However, as thicknesses increase, this effect tends to be moderated. Indeed, the difference observed between thicknesses of 70 and 90 mm remains relatively moderate (of the order of about 50 Hz), whereas the difference between thicknesses of 50 and 70 mm is more significant (of the order of about 400 Hz). Furthermore, for a thickness of 90 mm, the absorption level of the first peak appears to increase slightly from 0.95 to 1.

For a given thickness, the increase in the density of the samples has the effect of shifting the absorption peaks towards the lower frequencies, as shown in Fig. 7(b).

Perpendicular Configuration of Stems. For this configuration, only the evolution of the thickness could be studied. As with the parallel configuration, increasing the thickness of the samples significantly lowers the frequency of the absorption peaks, as shown in Fig. 8.

3.3 Modelling of the Behavior

Loose Particles. Modelling was carried out on all configurations. All the results are not presented here, but a typical example of comparison between the measurements and the

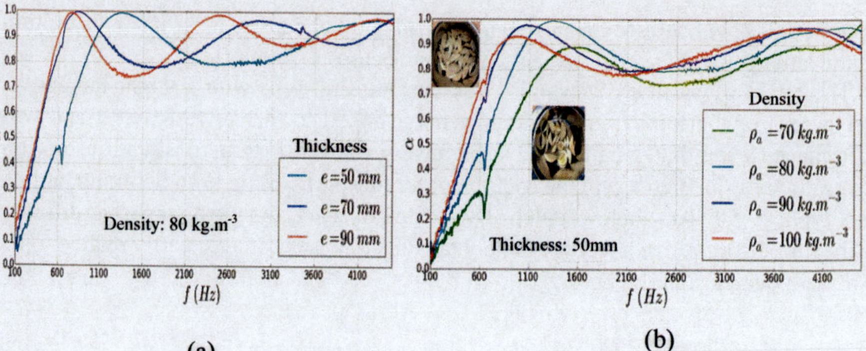

Fig. 7. Effect of thickness on the sound absorption of parallel Typha stem stacks, for a given bulk density of 80 kg.m^{-3} (a) and effect of density on the sound absorption of perpendicular Typha stem stacks for a given thickness of 50 mm (b)

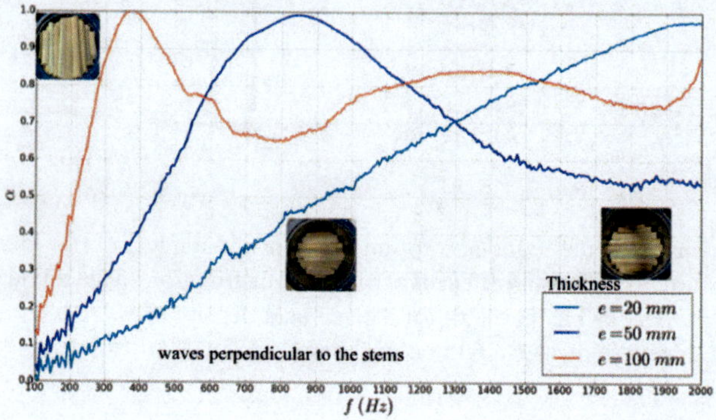

Fig. 8. Effect of thickness on the sound absorption of perpendicular Typha stem stacks

model is given in Fig. 9 for raw Typha with a bulk density of 40 kg.m^{-3}. In this figure, we can see a very good agreement between model and measurement, with differences of less than 5% in absorption and less than 1 dB in attenuation (the noisy zone at low frequencies corresponds to the limit of validity linked to the geometric configuration of the Kundt tube). It can be seen that this approach, based on the hypothesis of a single porosity scale, is perfectly satisfactory. Indeed, it was observed on the real part of the incompressibility modulus that measurement remains in the range of the theoretical asymptotic values evaluated from the open porosity.

On this point, Typha particles therefore show a different behavior from oak, flax shiv or particles from certain wood species, for which double porosity effects have been observed [13, 14, 20].

Stems. The JCA modelling method was used to simulate the behavior of Typha stems. Modelling was carried out for both parallel and perpendicular configurations.

As shown in Fig. 10, there is very good agreement between model and measurement.

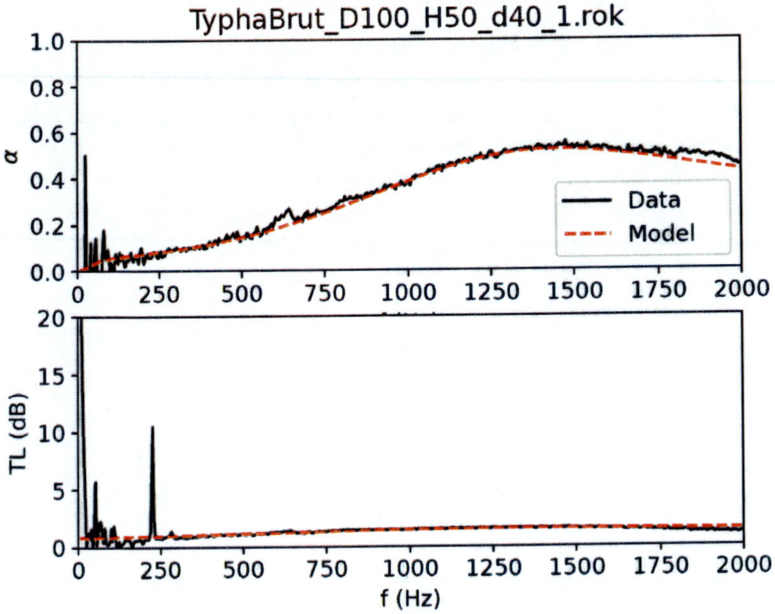

Fig. 9. Sound absorption modelling of the Raw Typha particle stack for a density of 40 kg.m^{-3}

It can be seen that for a thickness of 50 mm and a density of 70 kg.m^{-3}, the perpendicular configuration has the advantage of a first absorption peak at a lower frequency of the order of 650 Hz, whereas it is close to 1600 Hz for the parallel configuration. Moreover, the value of the acoustic absorption coefficient of this first peak is close to 1 for the perpendicular configuration whereas it is close to 0.9 for the parallel configuration. On the other hand, the perpendicular configuration will show a slightly more selective character with a narrower peak and a more marked decrease in sound absorption than the parallel configuration.

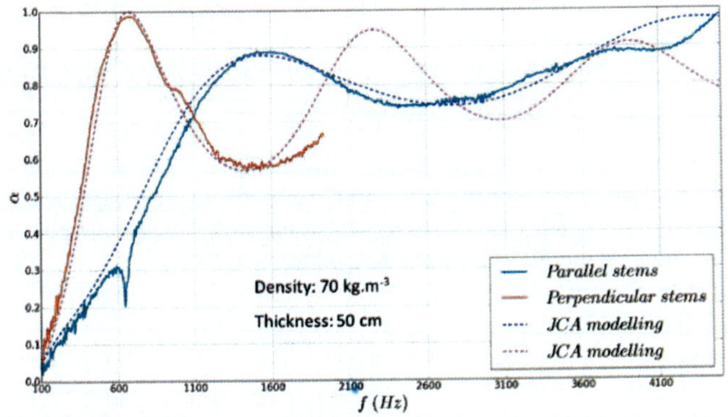

Fig. 10. Sound absorption modelling of both parallel and perpendicular configuration stem stacks for a density of 70 kg.m^{-3} and a thickness of 50 mm

4 Conclusion

This paper presents acoustical investigations on Typha stems and particles realized in the context of TyCCAO project. Results show promising results in term of sound absorption level, and evidences of the effects of density, particle size and orientation on this performance. It was also shown that the observed behavior can be modelled under a simple porosity approach for these materials. Further investigations will now aim at understanding the evolution of the acoustical parameters as a function of the configuration.

References

1. Varis, O., Fraboulet-Jussila, S.: water resources development in the lower senegal river basin: conflicting interests, environmental concerns and policy options. Int. J. Water Resour. Dev. **18**(2) (2002)
2. Mietton, M., Dumas, D., Hamerlynck, O., Kane, A., Coly, A.: Water management in the Senegal River Delta: a continuing uncertainty. Hydrol. Earth Syst. Sci. Discussions Eur. Geosci. Union **4**(6), 4297–4323 (2007)
3. Kane, I., Akpo, L.E.: Effects of repetitive cuttings of the stem on the growth of *Typha australis*. Int. J. Inf. Res. Rev. **4**(1), 3545–3548 (2017)
4. Kotschoubey, N.: Rehabilitation of the ecological functions of the senegal river delta. Bus. Public Adm. Stud. **11**(1), 25–45 (2017)
5. Castellanet, C., et al.: Etude pour l'évaluation des besoins pour le contrôle du Typha dans le delta du fleuve au Sénégal et en Mauritanie. Rapport d'état des lieux et bilan des méthodes de lutte contre le Typha. OMVS, GRET. Rapport 114 p. (2019)
6. Caro, R., de Frutos, H., Kitwana, A., Shen, A.: Typha charcoal in senegal: changing a national threat into durable wealth. Massachusetts Institute of Technology, Technical report, (15.915), p. 31 (2011)
7. Lumia, G.: Bio-based insulation materials: an opportunity for the renovation of European residential building stock. Evaluation of carbon uptake benefits through a dynamic life cycle assessment (DLCA), Ph.D. thesis (2017)
8. Pittau, F., Krause, F., Lumia, G., Habert, G.: Fast-growing bio-based materials as an opportunity for storing carbon in exterior walls. Build. Environ. **129**, 117–129 (2018)
9. Oldham, D., Egan, C., Cookson, R.: Sustainable acoustic absorbers from the biomass. Appl. Acoust. **72**(6), 350–363 (2011)
10. Asdrubali, F., Schiavoni, S., Horoshenkov, K.: A review of sustainable materials for acoustic applications. Build. Acoust. **19**(4), 283–312 (2012)
11. Glé, P.: Acoustics of building materials based on plant fibers and particles: tools for characterization, modelling and optimisation, INSA de Lyon, Ph.D. thesis (2013)
12. Arenas, J.P., Asdrubali, F.: Eco-materials with noise reduction properties. In: Martínez, L.M.T., Kharissova, O.V., Kharisov, B.I. (eds.) Handbook of Ecomaterials, pp. 1–26. Springer, Cham (2018). https://doi.org/10.1007/978-3-319-48281-1_137-1
13. Glé, P., Gourdon, E., Arnaud, L.: Modelling of the acoustical properties of hemp particles. Cons. Build. Mat. (2012)
14. Chabriac, P.A., Gourdon, E., Glé, P., Fabbri, A. Lenormand, H.: Agricultural by-products for building insulation: acoustical characterization and modeling to predict micro-structural parameters. Cons. Build. Mat. **112** (2016)
15. Blinet, T., Falwisanner, T., Glé, P., Guigou-Carter, C., Kadri, E.: Propriétés acoustiques des matériaux biosourcés - Rapport final étude CSTB-Cerema-DHUP (2018)

16. Amziane, S., Collet, F., Lawrence, M., Magniont, C., Picandet, V., et al.: Recommendation of the RILEM TC 236-BBM: characterisation testing of hemp shiv to determine the initial water content, water absorption, dry density, particle size distribution and thermal conductivity. Mater. Struct. **50**(3) (2017)
17. Leclaire, P., Umnova, O., Horoshenkov, K.V., Maillet, L.: Porosity measurement by comparison of air volumes. Rev. Sci. Instrum. **74**(3), 1366–1370 (2003)
18. ISO 9053-1. Acoustics - Determination of airflow resistance - Part 1: Static method (2018)
19. NF EN ISO 10534-2. Determination of sound absorption factor and impedance of impedance tubes - Part 2: Transfer function method (2003)
20. Johnson, D.L., Koplik, J., Dashen, R.: Theory of dynamic permeability and tortuosity in fluid-saturated porous media. Fluid Mech. **176**, 379–402 (1987)
21. Champoux, Y., Allard, J.F.: Dynamic tortuosity and bulk modulus in air-saturated porous media. Appl. Phys. **70**, 1975–1979 (1991)
22. Lafarge, D., Lemarinier, P., Allard, J.F., Tarnow, V.: Dynamic compressibility of air in porous structures at audible frequencies. J. Acoust. Soc. Am. **102**(4), 1995–2006 (1997)
23. Panneton, R., Olny, X.: Acoustical determination of the parameters governing viscous dissipation in porous media. J. Acoust. Soc. Am. **119**, 2027–2040 (2006)
24. Olny, X., Panneton, R.: Acoustical determination of the parameters governing thermal dissipation in porous media. J. Acoust. Soc. Am. **123**, 814–824 (2008)
25. Glé, P., et al.: Densities of hemp shiv for building: from multiscale characterisation to application. Ind. Crops Prod. **164** (2021)
26. Piégay, C., Glé, P., Gourdon, E., Gourlay, E., Marceau, S.: Acoustical model of vegetal wools including two types of fibers. Appl. Acoust. **129**, 36–46 (2018)

'Smartifying' Construction for Circular and Zero-Carbon Biobased Buildings (SmartBioC)

Hector F. Archila[1(✉)], Rebecca Lashley[1], Jessica Lamond[1], Abhinesh Prabhakaran[1], Ashleigh Msipo[2], and Edwin Zea Escamilla[3]

[1] University of the West of England, Bristol BS16 1QY, UK
hector.archila@uwe.ac.uk
[2] Amphibia BASE Ltd., Chester CH3 5JQ, UK
[3] ETH Zürich, 8093 Zürich, Switzerland

Abstract. SmartBioC focuses on creating a user-friendly digital tool that allows users to select and specify biobased materials and building components for a modular housing unit based on a set of performance indicators including carbon footprint, thermal performance, cost, social value, health and wellbeing. The ultimate aim of the tool is to speed up the uptake of circular biobased materials to provide zero-carbon, healthy and socially and economically viable solutions for the construction industry. Smart construction integrating the use of digital technologies and modern methods of construction (MMC) has the potential to improve the affordability, efficiency, and sustainability of new and refurbished buildings. Aligned to a circular economic model, opportunities exist to optimise the use, reuse, and disposal of biobased materials within the expanding MMC housing market, thereby improving carbon sequestration and mitigating climate change. Together with industry partners, SmartBioC's research team is using the UK's Design Council's Beyond Net Zero Framework of exploring, reframing, creating and catalysing to facilitate a collaborative and iterative process where the end-user is at the centre and determines the final outcome. SmartBioC 'explores' data obtained on biobased materials and 'reframes' it for alignment with indicators relatable for end-users. A BIM Object library of MMC premanufactured components with alternative biobased material configurations 'created' using Autodesk Revit is then exported into gaming platform Unity. Rapid prototyping and testing in Unity allow distribution of a web-based tool (html) for user feedback and development purposes (catalysing). User-friendly digital tools like SmartBioC give end-users, designers, decision-makers and specifiers, the ability to select circular biobased materials whilst visualising their design in 3D, along with information about the environmental, economic and social impacts of their choices. The adoption by the construction industry of zero-carbon, healthy and socially and economically viable biobased building solutions is imperative if the catastrophic impacts of predicted global warming are to be averted.

Keywords: Biobased Construction · Serious Games · Nature-Based Solutions · Smart construction · Digital Technologies · Circular Economy · Life Cycle Assessment · Modern Methods of Construction

S. Amziane et al. (Eds.): ICBBM 2023, RILEM Bookseries 45, pp. 926–936, 2023.
https://doi.org/10.1007/978-3-031-33465-8_71

1 Introduction

Arguments based on sustainability exist in favour of utilising digitisation in construction globally and locally (HM Government, 2020; Thelen, Zijlstra and Zandbergen, 2021). In the UK, the Ministry of Housing, Communities & Local Government (MHCLG, 2019) and the Construction Leadership Council (HM Government, 2018) emphasise the use of digital technologies and industrialised manufacturing as the key to unlocking current competitiveness and sustainability issues in the industry. Digital technologies can accelerate the transition of the construction industry, as outlined by the UK Government's Construction Strategy 2025, into a 50% less greenhouse gasses polluting industry, with a 50% increase on output speed and 33% cheaper (DBIS, 2013). Conversely, decarbonisation and climate change mitigation targets for the global construction industry, can significantly benefit from the mainstream use of biobased building materials within a circular economy approach. The UK Green Building Council (UKGBC) 2030's sector ambition statement sets biobased solutions in all building and infrastructure as a priority for reducing both GHG emissions and resource depletion caused by the industry (UKGBC, 2019). Therefore, the selection and/or development and implementation of digital technologies that can, swiftly enable the uptake of circular biobased materials in the construction industry becomes paramount.

Biobased building materials - whether plant or animal based -, are organic and renewable alternatives to man-made (technical materials such as steel and plastic) that sequester carbon during growth and that, depending on their transformation process – can result on a zero or negative carbon balance. Biobased materials such as natural fibres, wool and bio-adhesives can also offer a series of health, wellbeing and environmental benefits. However, at present, biobased materials (a.k.a. biogenic) and sustainable construction and/or green building technologies face several technical, commercial and social barriers, which hinder their rapid and cost-effective uptake in construction applications. Chan et al., (2017) appropriately group the following factors as interconnected barriers to the adoption of 'green building' technologies: 1) Stakeholders' reluctance to change; 2) Limited knowledge, information and awareness; 3) Higher cost compared to conventional technologies; 4) Market and financial constraints; and 5) Technological risk and obstacles. These five broad categories encompass similar barriers outlined by several authors surveying stakeholders across the global built-environment and reviewing the literature in the topic. Darko et al., (2017), Wong and Voon (2020) survey construction experts in the US and Malaysia, whilst Li et al., (2018) and Gan et al., (2015) include owners, end-users and other organisations in China and discuss influential factors in sustainable construction's decisions. Systematic and bibliometric reviews by Darko and Chan (2017) and Det Udomsap and Hallinger (2020), respectively, provide an overview of these barriers across the literature.

The five factors identified by Chan et al., (2017) depict interconnected and at times 'wicked issues' (Goel, 2019) surrounding sustainable construction which are experienced across the supply chain and that respond to broader global constraints. For instance, in terms of cost (2), market and financial aspects (4), biobased building materials are subjected to an 'economy of scale's' focus in the current global economic model. This model understandably prioritises manufactured and financial value (*e.g.*, highest output at the lowest cost) over the natural, human and social value of activities and output of the construction industry and beyond (*e.g.,* knowledge, training, biodiversity gain, human health and local impact). This long-established 'linear economy' model influences a 'take-use-dispose' approach, instigates monoculture and biodiversity loss, and creates social and economic inequalities worldwide. A systemic shift in the construction industry is therefore needed to address the 'wicked issues' facing sustainable construction. Campbell, Hairstans and Jones (2020) argue that a shift to the 'five capitals' model by Porritt (2012) including natural, human and social value, not only manufactured and financial value, can increase sustainability in modern methods of construction (MMC) *i.e.* offsite manufacturing (OSM). Particularly, in terms of fostering natural capital in construction, the authors highlight the current role of timber along with the future potential of other new biobased materials such as hemp, bamboo, straw and mycelium. a) Natural capital refers to the environmental sustainability and includes all resources (e.g., renewable and non-renewable) that are required for any building. b) Human capital measures all benefits to individuals including their physical and mental health and wellbeing, knowledge, skills, motivation, recreation, etc. c) Social capital: indicates the building's contribution to society, communities, family, businesses, culture and to government institutions, schools and networks, etc. d) Manufactured capital evaluates the goods, facilities, infrastructure & technologies required in construction to 'build buildings'. Finally, e) Financial capital measures the financial & economic value of the building choices.

The Alliance for Sustainable Building Products (ASBP) confirms the recent growth in the use of biobased materials in construction, as well as their painstakingly slow adoption (House of Commons EAC, 2022). Therefore, wholistic, systemic and easy to deploy and use solutions are imperative to generate the 'shift' in the construction industry. Innovative solutions to the built environment challenges that integrate new digital technologies could encourage the uptake of biobased materials in the construction industry within a circular economy framework (rather than a linear-economy model). Initiatives in the field, targeted at overcoming such challenges include the EU funded H2020 Building as Material Banks (BAMB) project, which is developing a *'Materials Passports Platform'* and a web-based Circular Building Assessment (CBA) tool, both aimed at maintaining the value of building materials throughout the building's life from design, construction, and management through to operations, refurbishing and dismantling (Honic, Kovacic and Rechberger, 2019). These types of initiatives are complex and require gathering and analysis of sizeable amounts of material-based data along with the integration of different concepts and metrics to capture the natural, human, social, manufacturing and financial impacts of biobased building products. Transparent information and user-friendliness of digital platforms that appeal to a new generation of digitally savvy building users and stakeholders become fundamental to tackle barriers for the adoption of biobased building materials. In particular, regarding the 'reluctance to change' (1) and the perceived limited knowledge, information and awareness of sustainable building alternatives (2). Kempeneer, Peeters and Compernolle (2021) remark that twenty-first century buildings users are better informed, globally connected and seek a series of environmental and social values (capitals) in the products and services they consume.

Therefore, data science and digital engineering technologies can play a significant role on the delivery transparent and comprehensive information systems in the current climate emergency. SmartBioC explores how digital imaging technologies, data science, building information modelling (BIM) and serious games tools can accelerate the uptake of circular biobased materials to provide zero-carbon and socially and economically viable solutions for the construction industry. These tools are explored as a way of encouraging decision makers (*e.g.*, architects, contractors and developers) and end-users to adopt biobased building materials. Alongside the integration of material information databases (*e.g.*, data on LCA, thermal performance and strength), the visualisation aspect and the interactivity of a digital platform are key to provide a biobased building decision tool that is fit for the future. This 'smart' platform can help users and the industry make informed decisions on biobased building material choices with the ability to increase human, social and natural capital, whilst maintaining manufactured and financial capital. This paper presents the conceptual framework and the methods adopted for 'smartifying' construction for circular and zero-carbon biobased buildings (SmartBioC). Early concept modelling using Autodesk Revit and the workflow developed are also presented.

2 Integrating Digital Technologies for Smart Biobased Construction (SmartBioC)

The development of the digital tool(s) at the core of SmartBioC is aligned to the Beyond Net Zero's systematic design approach by the Design Council (2018) which supports action towards the achievement of sustainability goals and climate commitments. As such, SmartBioC's design process comprises four stages: exploring, reframing, creating and catalysing, which entail the mining of data (explore) that is 'decodified' (reframe) through a smart visualisation (create) strategy enabling user interaction (catalyse).

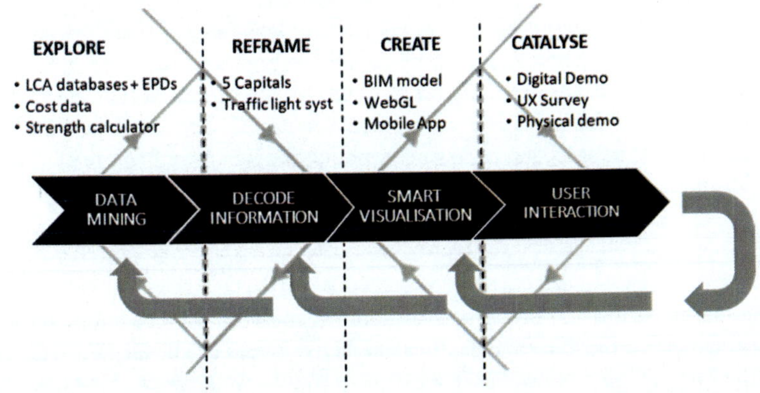

Fig. 1. SmartBioC's methodology using Beyond Net Zero's approach (Design Council, 2018)

Figure 1 illustrates these four stages of the framework applied to the project and their continuous iteration, whilst Fig. 2 illustrates the workflow generated with the digital interfaces for the development of the SmartBioC tool(s).

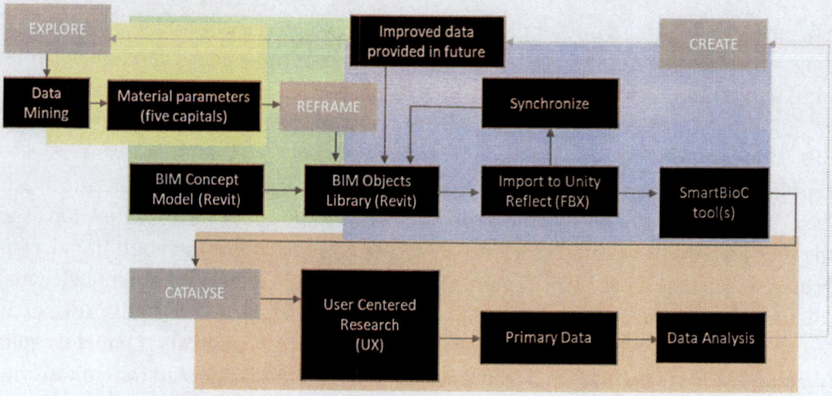

Fig. 2. SmartBioC's workflow with digital interfaces.

2.1 EXPLORE - Data Mining

Initial data parameters on physical and mechanical properties, thermal and VOC emissions, along with information on price and CO_2 have been sourced for a series of biobased materials directly from suppliers and retailers in the UK (e.g., estimated price) and from Environmental Product Declarations (EPDs). Some of this data is presented in Table 1 for structural timber products such as CLT, LVL and Glulam, timber-bamboo hybrids (TimBam) and for insulation materials made from straw, hemp and agricultural fibre (ECOboard).

2.2 REFRAME – Decode Information

The Five Capitals Framework by Porritt (2012) has been used for indexing the data acquired from biobased materials into indicators that can potentially ease their interpretation by users. Overall, these capitals simplify the wealth of data on sustainability and encompass all the UN Sustainable Development Goals (SDGs). Data on the biobased material parameters, metrics and assessment methodologies have been related to these five capitals as shown in Table 2.

2.3 CREATE – Smart Visualisation

Following the gathering and processing of data for individual biobased materials and their indexing against the five capitals, the next step consists of visually showcasing the information for user interaction. This stage follows a two-fold approach with a series of software interfaces. Firstly, BIM Object libraries of three main biobased building components (floors, walls and roofs) are 'created' for a 3D model building using Autodesk Revit (Fig. 3).

The three biobased building components are conceived for design for manufacture and assembly (DfMA) and for future disassembly, and contain the parameters and metrics defined in the previous stage. Figure 4 displays the floorplan of the model building unit with the wall build-up, which includes the main structural support using TimBam panels

Table 1. Biobased material parameters

Product	Density kg/m^3	Strength N/mm^2	MOE ‖ (E_0) N/mm^2	λ W/mK	U value	CO_2 ($kgCO_2e$)	VOC emissions	Est. Price £
TimBam5[r] Amphibia BASE	581	77.5 (B)	21,740	0.13	0.295	0.19/kg–109.5/m^3	Formaldehyde free PUR	£ 825[r] per m^3
CLT KLH	480	24 (B)	12,000	0.12	0.375[f]	0.25/kg[s]–120 /m^3	VOC free & formaldehyde free PUR (EN 15425)	£ 715[o] per m^3
Glulam (GL24h) Buckland Timber	420	24 (B)	11,500	0.13	0.295[h]	0.28/kg[s]–134.4 /m^3	Formaldehyde emission class E1 in accordance with EN 14080:2013	£ 2,400[p] per m^3
Beam BauBuche GL75 Pollmeier	730	75 (B)	16,800	0.17	0.566[i]	[1,468 /m^3] biogenic only	VOC free & formaldehyde free PUR (EN 15425)	£ 3,725[q] per m^3
ECOBoard Softboard ECOBOARD Int b.v	489	7.8 (B)	1,041.90	0.065	1.626[j]	0.98 /kg	No formaldehyde or other VOCs (MDI polyurethane)	-
Hemp Blocks The Hemp Block Company	330	0.9 (C)	-	0.07	0.22	100/m^3	No VOC reported	£10.77*/block
Straw Bale Panels EcoCocon	110[c]	-	-	0.645	0.123[l] 0.145[m]	2.48/m^2	No VOC reported	-

Key nomenclature: B (bending); C (compression); ‖ (parallel to grain); ⊥ (perpendicular); λ: Lambda (thermal conductivity); Rs-F (Resistance to Fire); R-F (Reaction to fire); ETA (European Technical Approval); REI marking identifies the fire-resistance rating of a structure; * rough pre-pandemic commercial prices.

Other details: a) 28 days; b) 3 months; c) compressed straw; d) fresh mortar; e) dry hardened mortar; f) = 0.320 mm/ 0.12 = 2.667 = 1/2.667; g) = 0.440 mm/0.13 = 3.385 = 1/3.385; h) = 0.440 mm/0.13 = 3.385 = 1/3.385; i) = 0.3 mm/0.17 = 1.765 = 1/1.765; j) = 0.040 mm/0.065 = 0.615 = 1/0.615; l) with clay and wood fibreboard; m) without; n) = 0.100 mm/0.30 = 0.333 = 1/0.333; o) calculated from 210 mm floor price = £150/m^2; p) Douglas Fir; q) calculated from supplier in Canada; r) calculated values for 2-ply Timber and 3-ply Bamboo laminate; s) revised biogenic factors by ICE database in IStructE carbon guide

(65.75 mm), straw bale insulation panels (140mm), agricultural fibre panels (18 mm) and internal lining plasterboard made from lime and food crop by-product (12.5 mm).

Secondly, the BIM Object libraries are exported into Unity to generate the bulk of the SmartBioC tool. The Unity Reflect plugin allows streaming of data from Revit into Unity. Gaming platform Unity is then used for 'creating' a WebGL 'Builder' and a mobile App 'Viewer' which are purposedly chosen to enable 'easy' distribution and user interaction. In effect, users with access to a computer web-browser can select different building components (BIM Objects) and 'build' their own one-storey design in 3D using the WebGL builder (alike IKEA's online design tool for rooms or kitchens). The 'builder' application displays the five capitals weightings for each building component available using a 'spider' radar chart with one capital per axis (Fig. 5). The users are

Table 2. Metrics and assessment methodologies for capital parameters

Capitals	Metrics	Assessment Methodologies
Natural	kWh/m^2/y and kgCO$_2$e/year	CIBSE TM 54 Evaluating Operational Energy Use of Buildings at Design Stage (2013) *Update 2022*
		Passivhaus PPHP
	kWh/m^2/y and kgCO$_2$e/floor area m^2	RICS Whole Life Carbon Assessment for Built Environment, 2017
	m^3/person/year	England and Wales building regulations water calculator
	increase in new flora or fauna species on site	BREEAM 2018 bio-diversity credits
		Urban Green Factor, London Plan
Human	m^2 per person	BCO, DfE, HQM
	Response time	Usable Buildings Trust
	CO2, CO, NOX, PM25, PM10, Mould, VOC	CIBSE TM40, WELL v2
	Degrees centigrade	CIBSE
	Reverberation time and Noise Rating NR appropriate to use	
	Open window within 7 m	BREEAM, WELL
	Biophilia - contact to views, places, plants, natural materials	
		Good Homes Alliance overheating guidance
Social	kgCO$_2$ e per km per person per annum	BREEAM 2018 Transport Credits
	Various metrics	WELL Building Standard v2 Preconditions, 2019
		RIBA Social Value Toolkit, 2019
Financial	£/m^2, m^3	CMS Global Consistency in Presenting Construction and Other Life Cycle Costs, 2019
Manufactured	Density (kg/m^3); Strength & MOE (N/mm^2); MC; Service class; Fire resistance; etc	Performance: Physical, mechanical, durability, fire
	λ (W/Mk); U value	Thermal performance
	Circularity	Adaptability, reusability, adaptability
	kgCO$_2$ e per km	Transport

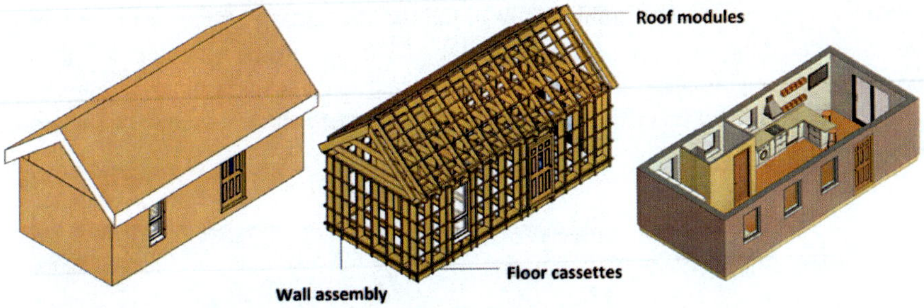

Fig. 3. 3D model building views with components

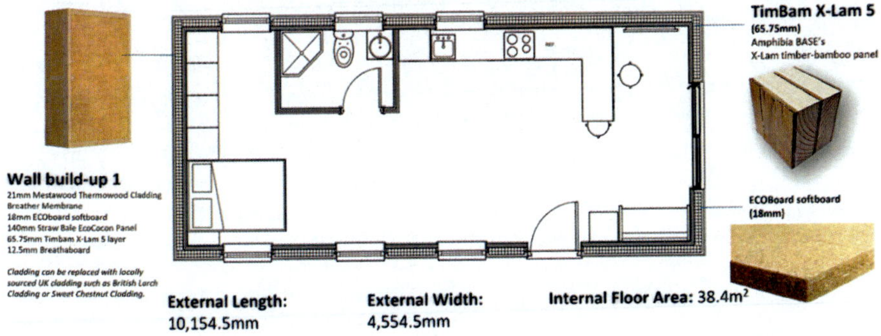

Fig. 4. Floorplan of the model building unit with details of the wall build-up

then, free to choose their combination of floor, roof and wall assemblies based on the natural, human, social, manufactured and financial impact weightings (i.e., five capitals) presented to them. The target of this 'serious game' application is for the players (users) to achieve the highest score in the five 'capitals' (green zone), which measure the overall sustainability of their building.

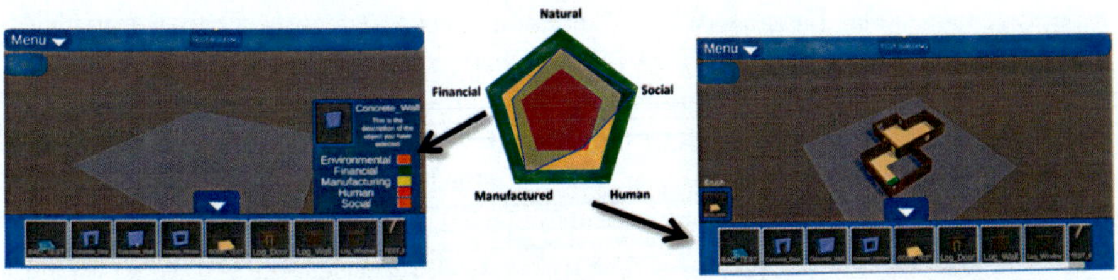

Fig. 5. Screenshots of the WebGL builder with the menu of building components available (objects) and the five capitals weighting per object.

Alongside the builder, the accompanying mobile app developed provides an Augmented Reality (AR) 'Viewer' which allows the user to navigate, visualise and interact with the individual components (objects) and the constructions by them created (using the builder). The aim of the combined use of the Builder and the Viewer is to help users better understand the overarching impacts of their building choices and ultimately enable them to make use of biobased building solutions which could have positive impacts on the environment, their health and wellbeing, the communities they live in and the economy (Fig. 6).

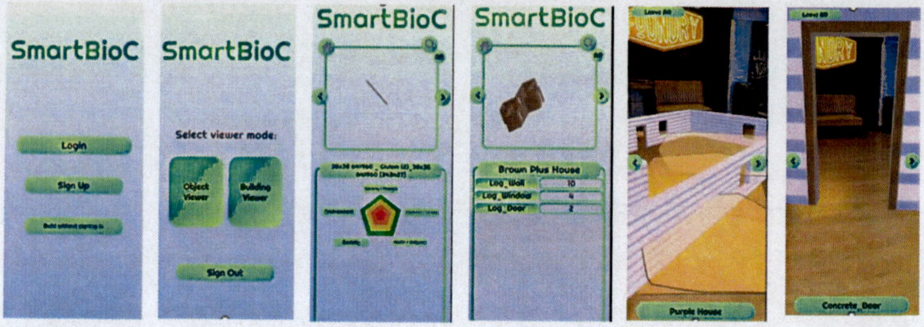

Fig. 6. Screenshots of the mobile App ‘Viewer’ and the ‘user journey’. The images from left to right display: 1) logging in into the App; 2) choosing the object or ‘building’ viewer; 3) visualising the radar chart for the object or building and their attributes (4); 5) placing of the object/building onto a surface; and 6) interacting in AR with the object/building.

2.4 CATALYSE – User Interaction

To date the WebGL ‘Builder’ and the mobile App ‘Viewer’ are still under development and the ‘catalysing’ phase involving user interaction has not formally started. In this phase ‘user-experience’ data will be collected for improving the SmartBioC tool.

3 Conclusions

Digital technologies can offer a low-investment solution to tackle recurrent technical, commercial and social barriers, which hinder the rapid and cost-effective uptake of biobased and climate-neutral solutions in construction. Researchers and entrepreneurs developing biobased building solutions are frequently confronted with these barriers. Smart tools like SmartBioC can both help rapidly simulate building alternatives and capture user preferences, whilst educating users on the benefits of adopting zero-carbon, healthy and socially and economically viable biobased building solutions. SmartBioC is an exploratory study into the effectiveness of this ‘fairly’ low investment digital tools to enable mainstream adoption of biobased building solutions. This digital tool(s) can provide a testbed for biobased solutions with potential users to validate their business case and foster research, development and investment. It is however important to highlight that highly specialised, and nowadays in high demand IT skills, is paramount, along with the analysis of vast amounts of information relating the’sustainability’ of building solutions. For instance, there are dozens of LCA databases and somehow conflicting and/or complex methods of assessment which hinder understanding by end-users and construction professionals. The SmartBioC project observes the need for a shifting of value in the built environment, from a concentrated focus on financial and manufactured 'capitals' to a balanced focus on the environmental, social and human capitals encompassing any building. User testing of the digital tools will assess their potential to unravel the positive impacts that balanced biobased building solutions can have on the environment, human health and wellbeing, society and the economy.

Acknowledgements. The authors are grateful to UWE's Vice-Chancellor's Early Career Researcher (VC ECR). Development Awards for the financial support provided to the principal investigator Dr Hector Archila. Amphibia BASE Ltd is grateful to the UK government's Department for Work and Pensions for funding provided through the Kickstart Scheme for two young placement posts contributing to the research.

References

Campbell, A., Hairstans, R., Jones, G.: Refocusing modern methods of construction on the climate emergency: a "five capitals" model for action. Struct. Eng. 22–28 (2020). https://www.istructe.org/IStructE/media/Public/TSE-Archive/2020/Refocusing-modern-methods-of-construction-on-the-climate-emergency-a-5-capitals-model.pdf. Accessed 9 Aug 2022

Chan, A.P.C., Darko, A., Ameyaw, E.E., Owusu-Manu, D.: Barriers affecting the adoption of green building technologies. J. Manag. Eng. **33**(3) (2017). https://doi.org/10.1061/(ASCE)ME.1943-5479.0000507

Darko, A., Chan, A.P.C., Ameyaw, E.E., He, B., Olanipekun, A.O.: Examining issues influencing green building technologies adoption: the United States green building experts' perspectives. Energy Build. **144**, 320–332 (2017). https://doi.org/10.1016/J.ENBUILD.2017.03.060

Darko, A., Chan, A.P.C.: Review of barriers to green building adoption. Sustain. Dev. **25**(3), 167–179 (2017). https://doi.org/10.1002/SD.1651

DBIS, D. for B. I. & S.: Construction 2025. London (2013)

Design Council: Beyond Net Zero. A Systemic Design Approach. London (2018)

Det Udomsap, A., Hallinger, P.: A bibliometric review of research on sustainable construction, 1994–2018. J. Clean. Prod. **254** (2020). https://doi.org/10.1016/J.JCLEPRO.2020.12007

Gan, X., Zuo, J., Ye, K., Skitmore, M., Xiong, B.: Why sustainable construction? Why not? An owner's perspective. Habit. Int. **47**, 61–68 (2015). https://doi.org/10.1016/J.HABITATINT.2015.01.005

Goel, A.: Sustainability in construction and built environment: a "wicked problem"? Smart Sustain. Built Environ. **8**(1), 2–15 (2019). https://doi.org/10.1108/SASBE-06-2018-0030/FULL/PDF

HM Government: Industrial Strategy: Construction Sector Deal (2018). https://www.constructionleadershipcouncil.co.uk/wp-content/uploads/2018/07/construction-sector-deal-print-ready.pdf. Accessed 31 Mar 2023

HM Government: The Construction Playbook Government Guidance on sourcing and contracting public works projects and programmes (2020)

Honic, M., Kovacic, I., Rechberger, H.: Concept for a BIM-based material passport for buildings. In: He, Y.-H., et al. (eds.) IOP Conference Series: Earth and Environmental Science Concept for a BIM-Based Material Passport for Buildings You May Also Like Machine-Learning Dessins d'enfants: Explorations via Modular and Seiberg-Witten Curves Assessment of Malaysian E-Passport PKI. IOP Publishing Ltd., Brussels (2019). https://doi.org/10.1088/1755-1315/225/1/012073

House of Commons EAC, E. A. C.: Building to net zero: costing carbon in construction. First Report of Session 2022–23. London (2022). https://committees.parliament.uk/publications/22427/documents/165446/default/. Accessed 31 May 2022

Kempeneer, S., Peeters, M., Compernolle, T.: Bringing the user back in the building: an analysis of ESG in real estate and a behavioral framework to guide future research. Sustainability **13**(6), 3239 (2021). https://doi.org/10.3390/su13063239

Li, H., Zhang, X., Thomas Ng, S., Skitmore, M.: Quantifying stakeholder influence in decision/evaluations relating to sustainable construction in China – a Delphi approach. J. Clean. Prod. **173**, 160–170 (2018). https://doi.org/10.1016/J.JCLEPRO.2017.04.151

Ministry of Housing, Communities and Local Government (MHCLG): MODERN METHODS OF CONSTRUCTION - Introducing the definition framework (2019). https://www.cast-consultancy.com/wp-content/uploads/2019/03/MMC-I-Pad-base_GOVUK-FINAL_SECURE.pdf. Accessed 7 Feb 2023

Porritt, J.: Capitalism as if the world matters. Capitalism as if the World Matters, pp. 1–360 (2012). https://doi.org/10.4324/9781849770675/CAPITALISM-WORLD-MATTERS-JONATHON-PORRITT

Thelen, D., Zijlstra, R., Zandbergen, J.: Digitalization of the built environment: Towards a more sustainable construction sector - World Business Council for Sustainable Development. Geneva, Switzerland (2021). https://www.wbcsd.org/Programs/Cities-and-Mobility/Sustainable-Cities/Transforming-the-Built-Environment/Digitalization/Resources/Digitalization-of-the-built-environment-Towards-a-more-sustainable-construction-sector. Accessed 31 Mar 2023

UKGBC, UK Green Building Council: Climate resilience and embracing nature : An ambition for the built environment. London, UK (2019). https://www.ukgbc.org/ukgbc-work/climate-resilience-and-embracing-nature-an-ambition-for-the-built-environment/

Wong, S., Voon, Y.: Barriers affecting the adoption of green building technologies, architects' perspectives. In: ICCREM 2020. American Society of Civil Engineers (ASCE) (2020)

Structural Analysis of Cross-Laminated Guadua-Bamboo (G-XLam) Panels Using Design Methods for CLT

Hector F. Archila[1,2(✉)], Edwin Zea-Escamilla[3], and Kent Harries[4]

[1] University of the West of England, Bristol, UK
hector.archila@uwe.ac.uk
[2] Amphibia Group SAS, Bogotá, Colombia
[3] ETH Zürich, Zürich, Switzerland
[4] University of Pittsburgh, Pittsburgh, USA

Abstract. Densified lamellas of bamboo *Guadua angustifolia* Kunth (Guadua) were arranged orthogonally and glued together to form three- and five-layer cross laminated (XLam) panels with symmetrical compositions, which are refereed as G-XLam3 & G-XLam5, respectively. This XLam composition is typical of plywood and cross-laminated timber (CLT) panels, hence analytical design methods developed for these systems have been applied to analyse the structural performance of G-XLam panels. For structural analysis G-XLam panels were considered as multi-layered systems composed of contiguous lamellas with orthotropic axes orientated at 0° and 90°. Timber engineering methods normalized by international standards were used in this study for the prediction of the mechanical properties of G-XLam panels. Bending stiffness properties of these panels were predicted using the transformed cross section method to find the modulus of elasticity of beams from the moduli of elasticity of the individual layers. As the elastic properties of the laminate system depend on the elastic properties of the individual layers, the elastic parameters for small clear specimens obtained in previous studies were used for the numerical analysis of G-XLam panels. Furthermore, compression stiffness and shear stiffness of the G-XLam panels were also calculated. Compared to test results and finite elements (FE) models undertaken prior to this study, analytical values provided a reasonably accurate prediction of the elastic properties of G-XLam3 and G-XLam5 panels. Differences in the results are assumed to be caused by manufacture flaws and thickness variation within the individual lamellas. Overall, this study showcases the potential of using timber engineering methods for the structural analysis of novel engineered bamboo products as well as their potential for structural applications. This design method also enables the design and analysis of novel composite laminated structures combining the high density and high axial stiffness of densified bamboo lamellas with low grade softwood or low-density core materials.

Keywords: bamboo · timber engineering · cross-laminated timber · engineered bamboo products · plywood

S. Amziane et al. (Eds.): ICBBM 2023, RILEM Bookseries 45, pp. 937–945, 2023.
https://doi.org/10.1007/978-3-031-33465-8_72

1 Introduction

The structural response under load of cross-laminated Guadua panels (G-XLam) depends on the mechanical properties of the material used to form the layers of the panel and the arrangement of layers to form the panel structure. In timber engineering, several methods have been used to analyse the mechanical response of multi-layered materials such as plywood and CLT (Bodig and Jayne, 1982; Ceccotti, 2008; BSI, 2010, 2011; Gagnon and Pirvu, 2011; BSI, 2014). The structural design of CLT is dictated by European standards (TRADA, 2009; BSI, 2010, 2011, 2014). Two of the calculation methods used for the determination of CLT design values for floor plates and wall panels in Eurocode 5 (BSI, 2014) are the theory of mechanically jointed beams and the composite theory method (TRADA, 2009). The former, also referred to as the Gamma method, considers layers acting only in the load direction connected by imaginary fasteners. The composite theory or 'k-method' based on plywood analysis theory, assumes that the longitudinal modulus of elasticity E_0 is thirty times the transverse modulus E_{90}. Both theories neglect shear deformation for span to width ratios ≥ 30.

As an alternative, BS EN 14272 (BSI, 2011) derives the mechanical properties of plywood panels in bending, compression and tension from the independent properties of rigidly connected layers and their transformed cross section. Thus, composite theory of laminated systems and equivalent section area are used for the prediction of the stiffness and strength values. This is in agreement with the method given by (Bodig and Jayne, 1982) for the analysis of multi-layered orthotropic wood laminates. A similar approach, known as the shear analogy, was developed by Kreuzinger (Blass et al. 1995) for calculating the stiffness and strength values of solid panels with cross-laminated layers in bending including those with span to width ratios less than 30. The BS EN 14272 procedure is widely accepted for the analysis of cross-laminated timber systems (Blass and Fellmoser, 2004; Ceccotti, 2008; Gagnon and Pirvu, 2011; Okabe et al., 2013) and is used in this study for the prediction of the elastic properties of G-XLam panels in bending, compression and shear.

2 Analytical Design Methods for G-XLam Panels

The mechanical properties of G-XLam panels are dependent on the direction of applied stress, the arrangement of the individual layers that form the panel and the specific mechanical properties of the constituent material. A thermoset epoxy resin was used and assumption of a perfectly rigid connection between layers was made; thus, the effect of adhesive is neglected for the analysis of the elastic properties of G-XLam panels. Usually the lay-up of CLT and plywood is symmetrical, therefore, calculation of their mechanical properties is based on a mechanically symmetrical composition of alternated cross layers.

2.1 Determination of Modulus of Elasticity of the Panels in Bending (Ep_m)

For the determination of the G-XLam panel modulus of elasticity in bending (Ep_m), the gross section stiffness, $Ep_m \cdot I_0$ is equated to the sum of the second moments of area of the individual lamella stiffnesses. Ep_m is found (BS EN 14272) using the notation provided in Fig. 1, with the location of the neutral axis denoted $\overline{y}$, a panel has n lamellas; the origin is at top surface and all equations consider unit width (i.e., $b = 1$).

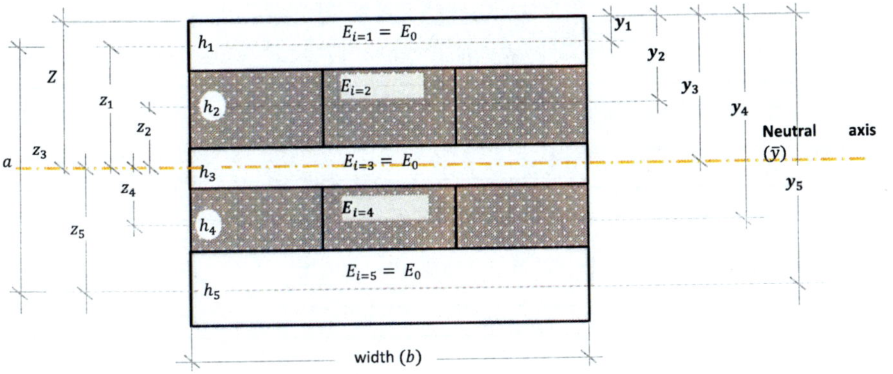

Fig. 1. Typical cross section of a five-layer cross-laminated panel using the shear analogy method.

The location of the centroid of each lamella is:

$$y_i = \sum_{j=1}^{i-1} (h_j) + h_i/2$$

The location of the neutral axis of the transformed cross section is given as:

$$\overline{y} = \frac{\sum_{i=1}^{n} E_i h_i y_i}{\sum_{i=1}^{n} E_i h_i}$$

For the typical case of a symmetric section having an odd number of alternating lamellas:

$$\overline{y} = \frac{\sum_{i=1}^{n} h_i}{2}$$

The distance from the neutral axis to the centroid of each lamella is:

$$z_i = y_i - \overline{y}$$

The gross moment of inertia of the panel section is:

$$I_0 = \left(\frac{1}{12}\right) \sum_{i=1}^{n} h_1^3$$

The gross section flexural stiffness, $Ep_m I_0$ is equated to the sum of the second moments of area of the individual lamella stiffnesses. The equivalent gross section modulus, Ep_m, is:

$$Ep_m = \frac{\left(\frac{1}{12}\right) \sum_{i=1}^{n} E_i h_i^3 + \sum_{i=1}^{n} E_i h_i z_i^2}{I_0}$$

The elastic section moduli are:

$$S_0 = \frac{\left(\frac{1}{12}\right)\sum_{i=1}^{n} E_i h_i^3 + \sum_{i=1}^{n} E_i h_i z_i^2}{Ep_m \bar{y}}$$

$$S_n = \frac{\left(\frac{1}{12}\right)\sum_{i=1}^{n} E_i h_i^3 + \sum_{i=1}^{n} E_i h_i z_i^2}{Ep_m \left(\sum_{i=1}^{n} h_i - \bar{y}\right)}$$

For the typical case of a symmetric section having an odd number of alternating lamellas:

$$S_0 = S_n = S = \frac{\left(\frac{1}{12}\right)\sum_{i=1}^{n} E_i h_i^3 + \sum_{i=1}^{n} E_i h_i z_i^2}{Ep_m \bar{y}} = \frac{2I_0}{\sum_{i=1}^{n} h_i}$$

The moment capacity of the panel is therefore:

$$M = fr_n S_n \leq fr_0 S_0$$

This assumes that failure of the extreme tension and/or compression fibres of the section constitute the moment limit state. For example, we can determine modulus of rupture from a permitted strain and assume that the outermost lamella is longitudinally oriented:

$$f_{rT} = E_0 \varepsilon_T \textit{ and } f_{rC} = E_0 \varepsilon_C$$

BS EN 338 (BSI, 2016) and BS EN 14272 (BSI, 2011) assume that the modulus of elasticity in bending of the longitudinal layers ($E_{0,m}$) is 30 times greater than the moduli of elasticity in bending of the cross layers ($E_{90,m,t,c}$) for softwood and 15 times greater for hardwood.

Other gross section material properties are similarly obtained using transformed section properties.

2.2 Determination of the Panel Shear Modulus

In timber engineering, the shear modulus parallel to the grain (G_0) for softwood CLT panels is generally considered to range between 1/12 and 1/20 of the longitudinal modulus of elasticity of the panel (E_0) and typically assumed to be 1/16 of E_0. . Furthermore, the shear modulus perpendicular to grain known also as the rolling shear modulus (G_R) is assumed to be 1/10 of G_0 (Gagnon and Pirvu, 2011) (Fig. 2).

The gross section shear modulus is:

$$G_{12} = \frac{\sum_{i=1}^{n-1} G_{12i} \cdot h_i}{\sum_{i=1}^{n-1} h_i} \textit{ and } G_{21} = \frac{\sum_{i=1}^{n-1} G_{21i} \cdot h_i}{\sum_{i=1}^{n-1} h_i}$$

where $_{12}$ is in the plane of the section shown and $_{21}$ is out of plane.

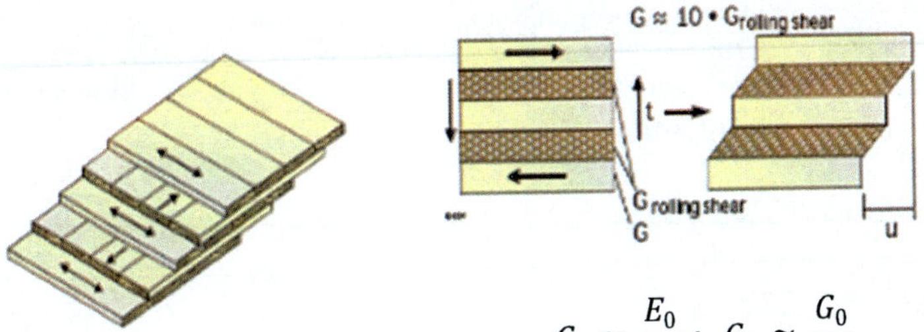

$$G_0 \approx \frac{E_0}{16};\ G_R \approx \frac{G_0}{10}$$

Fig. 2. Shear modulus parallel (G_0) and perpendicular to grain (G_R) of a cross laminated panel. Image by (Gagnon and Pirvu, 2011).

2.3 Determination of the Modulus of Elasticity in Compression and Tension of Cross-Laminated Panels

The MOE of cross laminated panels in compression and tension ($Ep_{t,c}$) can be calculated according to BS EN 14272 (BSI, 2011). The gross section in-plane tension and compression moduli are:

$$Ep_t = \frac{\sum_{i=1}^{n-1} E_{t,i} \cdot h_i}{\sum_{i=1}^{n-1} h_i} \text{ and } Ep_c = \frac{\sum_{i=1}^{n-1} E_{c,i} \cdot h_i}{\sum_{i=1}^{n-1} h_i}$$

$Ep_{t,c}$ is equal to the summation of the elastic properties of the compound layers in the panel and assumption of one unit of width is also made in this equation.

2.4 Prediction of the Elastic Properties of G-XLam Panels Using Stiffness Values from Their Individual Constituting Layers

Calculation of the elastic properties of three- and five-layer cross-laminated Guadua (G-XLam3 & G-XLam5) panels in compression (Ep_c), shear (G_{12}) and bending (Ep_m) was undertaken using the equations above. The moduli of elasticity in bending parallel ($Ep_{m,0}$) and perpendicular ($Ep_{m,90}$) to the direction of outer skins of the panels (considered as the main direction) are calculated. An exemplary calculation of $Ep_{m,0}$ for a G-XLam3 and G-XLam5 panel can be seen in Fig. 3 and Fig. 4, respectively.

Furthermore, the moduli of elasticity in compression along the longitudinal ($Ep_{c,0}$) and transverse ($Ep_{c,90}$) axes, and the in-plane shear modulus (G_{12}) are calculated. Values obtained previously for E_0, E_{90} and G_o (G_{12}) through testing of small clear specimens (Archila et al., 2014; Archila, Ansell and Walker, 2014) were used as the characteristic properties of the individual layers in the three and five layers cross laminated panels (G-XLam3 and G-XLam5).

For the calculations, mostly the notation system from BS EN 14272 (BSI, 2011) has been adopted and related to the methodology herein presented, which were used for calculating the elastic properties of the G-XLam panels as explained below:

$A_i = h_i \cdot b_i \cdot E_i$; where $b = 1$;

$I_i = A_i \cdot \frac{h_i^2}{12} = \sum_{i=1}^{n} E_i \cdot \frac{h_i^3}{12}$ and $J_i = A_i \cdot C_i = \sum_{i=1}^{n} E_i \cdot \frac{h_i^3}{12}$ in (kN.mm^2);

$C_i = z_i^2$ in mm^2; $Ep_m \cdot \mathrm{I} = Ep_m \cdot \frac{h_i^3}{12} = 12 \cdot \frac{\sum_{i=1}^{n} I_i + J_i}{\sum_{i=1}^{n} h_i^3}$; where $\mathrm{I} = \frac{b \cdot h_i^3}{12}$.
and, for the panel shear modulus (G_{12}): $N_i = h_i \cdot G_{vi}$; where G_{vi} refers to the shear modulus of the individual layers.

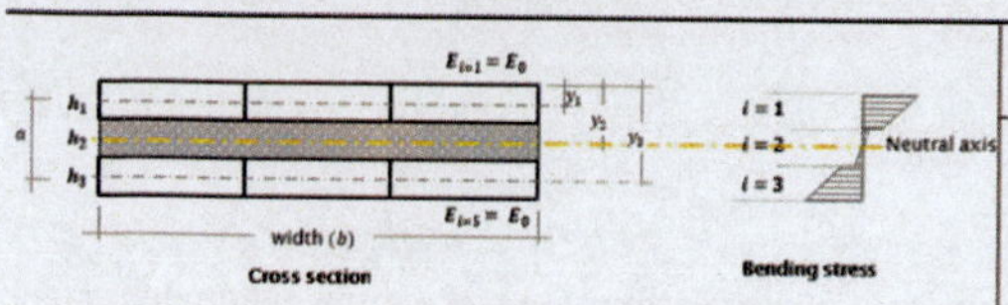

Type of panel: G-XLam5 2 L + 1 T layers THM modified Guadua Calculation including cross layers	$Ep_{m,0}$ - G-XLam3 (2L + 1T)
$Ep_m = \frac{12 \cdot \sum_{i=1}^{n} E_i \cdot h_i \cdot z_i^2 + \sum_{i=1}^{n} E_i \cdot h_i^3}{(\sum_{i=1}^{n} h_i)^3}$	$E_0 = 21.74\ kN/mm^2$
	$E_{90} = 1.02\ kN/mm^2$
	$G_0 = 1.32\ kN/mm^2$

G-XLam3		thickness	Property	A_i	T_i	Y_i	B_i	z_i	C_i	I_i	J_i
		h_i	E_i	h_i * b_i * E_i	t_{i-1}+T_{i-1}	T_i + t_i/2	A_i * Y_i	|Z_{ax} - Y_i|	z_i^2	A_i * h_i^2/12	A_i * C_i
Layers	Grain	(mm)	(kN/mm²)	(kN)	(mm)	(mm)	(kN*mm)	(mm)	(mm)²	(kN.mm²)	(kN.mm²)
h_1	L, (0°)	5.70	21.74	123.87	0.00	2.85	352.85	5.70	32.46	335.05	4,020.58
h_2	T, (90°)	5.70	1.02	5.81	5.70	8.55	49.66	0.00	0.00	15.72	0.00
h_3	L, (0°)	5.70	21.74	123.87	11.39	14.24	1,764.27	5.70	32.46	335.05	4,020.58
$\sum_{i=1}^{n}$ =		17.09		253.55			2,166.79			685.81	8,041.16

$z_{ax} = \frac{\sum_{i=1}^{n} B_i}{\sum_{i=1}^{n} A_i}$	8.55 mm (z_{ax} = 17.09/2)	I = 415.95 mm⁴	$Ep_m \cdot I$ = 8,726.97 kN.mm²	Ep_m = 20.97 kN/mm²

Fig. 3. Calculation of the bending modulus of elasticity ($Ep_{m,0}$) of a G-XLam3 panel with main fibre-direction along its length (E_0) based on the elastic properties of all its compounding layers ($2L$, $E_0 + 1T$, E_{90}) and on a unit of width.

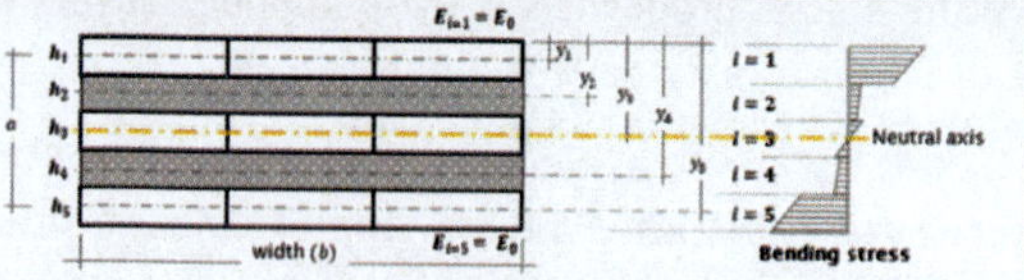

Type of panel: G-XLam5 3 (L) + 2 (T) layers THM modified Guadua Calculation including cross layers	$Ep_{m,0}$ - G-XLam5 (3L + 2T)
$Ep_m = \frac{12 \cdot \sum_{i=1}^{n} E_i \cdot h_i \cdot z_i^2 + \sum_{i=1}^{n} E_i \cdot h_i^3}{(\sum_{i=1}^{n} h_i)^3}$	$E_0 = 21.74\ kN/mm^2$
	$E_{90} = 1.02\ kN/mm^2$
	$G_0 = 1.32\ kN/mm^2$

G-XLam5		thickness	Property	A_i	T_i	Y_i	B_i	z_i	C_i	I_i	J_i
		h_i	E_i	h_i * b_i * E_i	t_{i-1}+T_{i-1}	T_i + t_i/2	A_i * Y_i	|Z_{ax} - Y_i|	z_i^2	A_i * h_i^2/12	A_i * C_i
Layers	Grain	(mm)	(kN/mm²)	(kN)	(mm)	(mm)	(kN*mm)	(mm)	(mm)²	(kN.mm²)	(kN.mm²)
h_1	L, (0°)	5.25	21.74	114.15	0.00	2.63	299.63	10.50	110.25	262.18	12,584.54
h_2	T, (90°)	5.25	1.02	5.36	5.25	7.88	42.17	5.25	27.56	12.30	147.60
h_3	L, (0°)	5.25	21.74	114.15	10.50	13.13	1,498.16	0.00	0.00	262.18	0.00
h_4	T, (90°)	5.25	1.02	5.36	15.75	18.38	98.40	5.25	27.56	12.30	147.60
h_5	L, (0°)	5.25	21.74	114.15	21.00	23.63	2,696.69	10.50	110.25	262.18	12,584.54
$\sum_{i=1}^{n}$ =		26.25		353.15			4635.05			811.13	25,464.28

$z_{ax} = \frac{\sum_{i=1}^{n} B_i}{\sum_{i=1}^{n} A_i}$	13.13 mm (z_{ax} = 26.25/2)	I = 1,507.48 mm⁴	$Ep_m \cdot I$ = 26,275.41 kN.mm²	Ep_m = 17.43 kN/mm²

Fig. 4. Calculation of the bending modulus of elasticity ($Ep_{m,0}$) of a G-XLam5 panel with main fibre-direction along its length (E_0) based on the elastic properties of all its compounding layers ($2L$, $E_0 + 1T$, E_{90}) and on a unit of width.

Bending strength of the panels (fp_m) was calculated as the sum of the strengths of the individual layers and using results of MOR_b obtained from small clear specimens testing in (Archila et al., 2014; Archila, Ansell and Walker, 2014). For this calculation, cross layers (E_{90}) were not taken into account and a symmetrical composition of the cross laminated (G-XLam) panel was assumed.

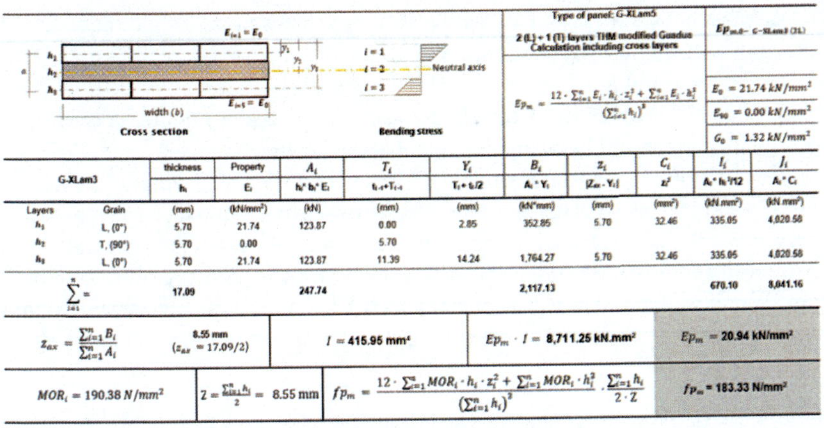

Fig. 5. Calculation of the bending modulus of elasticity (Ep_m) of a G-XLam5 panel with main fibre-direction along its length (E_0) based on the elastic properties of these layers ($2L$, E_0) and not taking into account cross layers ($1T$, E_{90}).

Finally, z_{ax} (kN.mm^2) in symmetrical panel compositions is $T/2$ (half the total thickness of the panel); z_i is the distance between the centre point of each layer and the central axis and Y_i is the distance between the external face and the central axis of each layer in (mm), as illustrated in Figs. 3 and 5.

3 Results and Discussion

A summary of all the predicted elastic values for G-XLam panels in bending (Ep_m), compression (Ep_c) and shear (G_v) obtained are presented in (Table 1).

Table 1. Predicted elastic values for G-XLam panels in bending, compression and shear.

Property	Main direction of layers	Symbol	G-XLam 3		G-XLam 5	
			2L, E_0+1T, E_{90}	2L, E_0	3L, E_0+2T, E_{90}	2L, E_0
Bending	Longitudinal (L)	$Ep_{m,0}$	**20.97 GPa**	20.94 GPa	**17.43 GPa**	17.22 GPa
	Transverse (T)	$Ep_{m,90}$	1.79 GPa	0.98 GPa	5.33 GPa	0.81 GPa
Bending strength	Longitudinal (L)	fp_m	–	183.33 MPa	–	150.78 MPa
Compression	Longitudinal (L)	$Ep_{c,0}$	**14.83 GPa**	14.49 GPa	13.45 GPa	13.05 GPa
	Transverse (T)	$Ep_{c,90}$	7.93 GPa	0.68 GPa	9.31 GPa	0.61 GPa
Shear	–	G_{12}	**0.92 GPa**	0.88 GPa	0.84 GPa	0.79 GPa

L = main load direction (longitudinal) & T = 90° to the main load direction (transverse)

As can be seen, no-significant contribution to the stiffness of the panels is given by the cross layers (those transversally orientated or at 90° to the main direction of the panel). A very small influence of the rolling shear (G_R) on the mean panel modulus of rigidity (G_v) is also observed. Overall, G_v calculations including G_R presented an improvement in axial stiffness between 5% and 6%, while stiffness values in the longitudinal direction of G-XLam panels in bending are about 35% higher than those in compression. Similar behaviour for structural and non-structural wood panels is reported by BSI (2008) and Marcroft (2012). This reduction in the elastic properties in compression can be explained by the effects of shear and buckling during in-plane loading.

Bending strength (fp_m) values predicted for G-XLam3 and G-XLam5 panels were 183.33 MPa and 150.78 MPa, respectively. These results overestimate the actual capacity of the panel, as the *MOR* values of the individual layers used for this calculation were obtained from three-point bending tests. As showcased in Table 2, elastic values in bending, compression and shear calculated for G-XLam panels are generally higher than the mean values of conventional structural and non-structural wood products.

Table 2. Mean elastic values of structural and non-structural wood panels (all values in GPa).

Property (GPa)	Main layer direction (external)	Symbol	Multilay. Solid wood panels [a] (20–30 mm)	Birch plywood [b] (8.5–25 mm)	CLT 3	CLT 5	CLT M1 BSP (C24) crossplan	G-XLam 3 2L + 1T	G-XLam 5 3L + 2T
Bending modulus	0° (L)	Ep_m	8.2	9.7	–	–	11.6	**20.97**	**17.43**
	90° (T)	$Ep_{m,90}$	0.55	6.1	–	–	0.37	1.79	5.33
Compression modulus	0° (L)	$Ep_{c,0}$	3.5	9	7.42	6.74	–	**14.83**	13.45
	90° (T)	$Ep_{c,90}$	2.9	7.9	3.91	4.62	–	7.93	9.31
Shear modulus	–	G_{12}	0.47		–	–	0.65	**0.92**	0.84

[a] (BSI, 2008); [b] (Marcroft, 2012); [c] (Mayr-Melnhof Kaufmann Group, 2009)

4 Conclusions

Given the mechanical properties of the individual components (layers) of the panelised system G-XLam, its bulk mechanical behaviour could be predicted through different mathematical expressions that consider the variation of number of layers, orientation, thickness and mechanical properties of the individual layers. These formulae are based on analytical design methods for plywood and CLT that are accepted by international standards, including European and Canadian standards.

Stiffness values obtained using this design method provide design values for structural design with an adequate match with test results using physical (contact) measurement techniques performed by (Archila et al., 2017). Currently, multiple laminations with engineered bamboo that achieve comparable sections to structural timber products result in overengineered products, which are costly and have a higher carbon footprint. This lack of competitiveness has been a barrier to the mainstream use of engineered bamboo products in structural applications. Analytical methods like the one developed in this research can aid the design of hybridised engineered bamboo products mixed with weaker and cheaper lamellas that achieve the same performance of structural timber products such as CLT, whilst maintaining an effective section, a competitive cost and a low carbon footprint.

Acknowledgement. The authors are grateful to MINCIENCIAS, for the financial support provided to Amphibia Group SAS (Colombia) for the project *'G-XLam Hybrid: Sistemas laminados híbridos de Guadua para construcción prefabricada'* part of the postdoctoral programme (Convocatoria No. 784-2017 Estancias Postdoctorales 2017).

References

Archila, H.F., Brandon, D., Ansell, M.P., Walker, P., Ormondroyd, G.A.: Evaluation of the mechanical properties of cross laminated bamboo panels by digital image correlation and finite element modelling. In: Salenikovich, A.J. (ed.) WCTE 2014, World Conference on Timber Engineering, Quebec, QC: WCTE 2014, p. 8 (2014). ISBN 978-086488560-9

Archila, H.F., Rhead, A., Ansell, M.P., Walker, P., Lizarazo-Marriaga, J.: Elastic response of cross-laminated engineered bamboo panels subjected to in-plane loading. In: ICE Proceedings: Construction Materials, pp. 1–12 (2017). https://doi.org/10.1680/jcoma.16.00080

Archila, H.F., Ansell, M.P., Walker, P.: Elastic properties of thermo-hydro-mechanically modified bamboo (Guadua angustifolia Kunth) Measured in Tension, Key Engineering Materials, CH, 600, pp. 111–120 (2014). https://doi.org/10.4028/www.scientific.net/KEM.600.111

Blass, H.J.: Timber engineering. - 1 : Basis of design, material properties structural components and joints. Almere, The Netherlands: Centrum Hout (1995)

Blass, H.J., Fellmoser, D.P.: Design of solid wood panels with cross layers. In: 8th World Conference on Timber Engineering, vol. 14, pp. 1001–1006 (2004)

Bodig, J., Jayne, B.A.: Mechanics of Wood and Wood Composites. Reprint edn. Van Nostrand Reinhold Company, New York (1982)

BSI: BS EN 12369-3:2008, Wood-based panels—Characteristic values for structural design. London, UK (2008)

BSI (2016) BS EN 338: 2016 Structural timber—Strength classes. London, UK

BSI: BS EN 408:2010, Timber structures—Structural timber and glued laminated timber—Determination of some physical and mechanical properties. London, UK (2010)

BSI: BS EN 14272:2011, Plywood—Calculation method for some mechanical properties. London, UK (2011)

BSI: BS EN 1995-1-1:2004 +A2:2014, Eurocode 5: Design of timber structures—Part1-1: General -Common rules and rules for buildings. London, UK (2014)

Ceccotti, A.: New technologies for construction of medium-rise buildings in seismic regions: the XLAM case. Struct. Eng. Int. **18**(2), 156–165 (2008). https://doi.org/10.2749/101686608784218680

Gagnon, S., Pirvu, C.: CLT Handbook - cross laminated timber. Canadian. Edited by Sylvain Gagnon and Ciprian Pirvu. FPInnovations, Québec (2011)

Marcroft, J.: Concise Eurocodes: Design of Timber Structures. BS EN 1995-1-1: Eurocode 5. London, UK (2012)

Mayr-Melnhof Kaufmann Group: Manual Cross-laminated timber panels M1 BSP cross plan. Gaishorn am See, Austria (2009)

Okabe, M., Yasumura, M., Kobayashi, K., Fujita, K.: Prediction of bending stiffness and moment carrying capacity of sugi cross-laminated timber. J. Wood Sci. **60**(1), 49–58 (2013). https://doi.org/10.1007/s10086-013-1377-8

TRADA: TRADA Technology Guidance Document 10 (GD10) Cross-laminated Timber (Eurocode 5) Design Guide for Project Feasibility. High Wycombe, UK (2009)

Environmental Performance and Life Cycle Assessment of Bio-Based Building Materials and Constructions

Methodology for Calculating the Environmental Impacts of Different Classes of Sawn Timber Strength

André Dias[1(✉)], Alfredo Dias[1], José Silvestre[2], and Jorge de Brito[2]

[1] ISISE, Department of Civil Engineering, University of Coimbra, R. Luis Reis dos Santos 290, 3030-790 Coimbra, Portugal
adias@serq.pt

[2] CERIS, Department of Civil Engineering, Architecture and Georesources of Instituto Superior, Lisbon, Portugal

Abstract. In the European Union, when sawn timber has a structural use, it must be classified in strength classes according to the standard EN 14081-1. Strength classes classify each sawn timber element according to its strength, stiffness and density properties. The environmental impacts of sawn timber for structural use have been calculated in several life cycle assessment studies, however, these studies use 1 m^3 of sawn timber as functional unit, without specifying the strength class. This means that the environmental impacts calculated for one cubic meter of sawn timber do not take into account its structural use. The structural use of sawn timber products should be taken into account when calculating environmental impacts because in the process of design a timber-based structure, the structural class is essential to determine the volume of timber required. Thus, the main goal of this study is to develop and present a methodology to quantify the environmental impacts of various strength classes by using the results of mechanical grading for a sample of sawn timber of a given wood species from a specific region of origin. This methodology consists of two steps: i) data collection (identification and collection of existing data for combinations of wood species and regions of origin – allocation yields), and ii) allocation of environmental impacts to the various strength classes. In this study, the methodology is applied to a case study of Maritime pine wood from Portugal.

Keywords: Life Cycle Assessment · Sawn timber · Strength Classes · Allocation

1 Introduction

Sawnwood (SW), apart from roundwood, is the most rudimentary structural wood product. The most common fabrication process starts with cutting roundwood logs, followed by debarking, crosscutting, sawing, drying, planning and grading (and, if required, finger jointing) [1]. The performance requirements of structural solid timber are defined by EN 14081-1 [2].

S. Amziane et al. (Eds.): ICBBM 2023, RILEM Bookseries 45, pp. 949–958, 2023.
https://doi.org/10.1007/978-3-031-33465-8_73

The main disadvantages of structural SW are the limitation of dimensions (tree size and machinery limitations) and the high variability of properties (due to defects). The main advantage of SW is the less energy, equipment required during manufacturing, and low the environmental impacts when compared with other materials (for example, steel and concrete).

Life Cycle Assessment (LCA) is a methodology regulated by the following ISO standards: (ISO 14040 [3], and ISO 14044 [4]) that intends to quantify the environmental impacts of products and services. In construction sector, the environmental impacts of products may be calculated through the (EN 15804 + A2 [5], and EN 16485 [6]) standards. The last is specific for sawnwood products.

LCA methodology has been used by academia and industry to quantify and compare the environmental impacts of products and solutions used in the construction sector [7–11]. In these studies, SW have been highlighted due to their environmental characteristics. One of the main advantages of using wood-based products is related to the trees' ability to "capture" carbon dioxide that exists in the atmosphere and retain the carbon in the wood products.

Dias et al. [12] reviewed 14 EPDs that quantified the environmental impacts of SW. The authors identified that 4 EPDs indicated the strength class of SW (C24), 3 given a range of mechanical properties and none of the others referred the strength classes or mechanical properties of products. This study proposed also a methodology that estimates the strength classes of sawnwood products assessed in EPDs, allowing the engineers, architects and LCA expects to compare the environmental impacts of products with different strength classes. This methodology was based on visual grading methodology (EN 14081-1 [2]) and depends on the representativity of the studies assessed. The methodology presented for predicting the strength classes may have some limitations since it depends on the approved and available data for visual classification, and for some cases it may not represent all the strength classes of the products.

Dias et al. [13] quantified and compared the environmental impacts of various structural solutions for a structurally equivalent functional unit. The equivalent functional unit was defined based on the structural requirements of European standards for a residential floor. The structural solutions varied the wood-based products and their strength classes: SW (C18, C24, C35, D24, D30, D35), glulam (GL24, GL28, GL32), LVL and I-Joist. The authors noted that, for the majority of environmental categories, LVL had the highest impacts followed by I-Joists. In the other hand, SW had the lowest impacts followed by glulam. For SW and glulam, the increase of strength classes decreased the environmental impacts of structural solutions. Although this study considered different strength classes, it did not calculate the environmental impacts of each strength class. In this study, it was assumed that the structural class only influenced the volume of the structural solutions.

As shown by previous studies, the attribution of environmental impacts to different strength classes of SW will allow the environmental impacts of classified SW to be quantified more accurately. This classification can be made based on the results of the mechanical grading of wood of a given combination of wood species and region of origin. In the literature review, no studies that quantify and/or allocate the environmental impacts of different strength classes of SW mechanically graded were found. Thus, this study aims to develop and give a method to quantify the environmental impacts of SW from a specific strength class that has been mechanically determined.

2 Methodology for Estimation of Allocation Percentages of Each Strength Class

2.1 Allocation

Allocation is defined as "partitioning the input or output flows of a process or a product system between the product system under study and one or more other product systems" [4]. Allocation procedures shall be applied when the LCA deals with systems involving multiple products and recycling systems. The sum of inputs and outputs of allocated products shall be equal to the inputs and outputs of the unit process before allocation.

Analysing the system of mechanical grading of SW, the input is one cubic meter of unclassified SW, and the output is the classified SW (and rejected) of the various strength classes. However, since wood of a certain strength class may perform the functions of a lower strength class, the distinction between types is not clear and may take on different strength classes depending on the intended use.

The methodology for allocating environmental impacts to SW strength classes presented in this study does not represent the same allocation presented by the LCA standards [3–5]. Still, the objective of both processes is the same: to quantify environmental impacts of different coproducts of the same system.

2.2 Description of Methodology

The methodology proposed in this study begins with the quantification of the percentage of wood elements of each strength class, quantified through published mechanical grading reports and/or academic studies. This method identifies the combinations of strength classes of a SW sample and calculates the percentage of wood elements that can be assigned to each specific strength class. The percentage calculated is used to allocate the environmental impacts of one cubic meter of unclassified SW to one cubic meter of classified SW.

Firstly, studies and/or reports that performed destructive tests or mechanical grading for the determination of strength classes of wood samples shall be found in literature. This information must be collected for the wood species and geographical representativeness of the products that the study aims to compare.

The data that must be identified in each study and/or report are: i) the combination of strength classes considered in the mechanical grading procedure and ii) the percentage of wood pieces allocated to each strength class. The percentage of each strength class is collected to allocate the inputs and outputs within the system boundary to each strength class.

Collection of Data – Identification of Combinations of Data
This step consists in the collection of data regarding the datasets to be used in environmental comparisons. For each combination of wood species and country of origin, this

information can be found in reports and/or studies that performed destructive and/or non-destructive tests for the determination of bending or tension properties and/or strength classes of SW. Usually, studies and reports developed for mechanical grading perform these tests and, when published, they provide information that can be considered on this methodology.

The collection of data for each combination of wood species and country of origin shall identify:

Combinations of strength classes reported,

For each combination of strength classes, the percentage and/or amount of wood pieces that belonged to each strength class.

Allocation of Strength Classes – Based on Tests Results
The values of percentage and/or the number of specimens of each strength class are used to allocate the environmental impacts per cubic meter of wood to the strength classes determined. The allocation shall model the amount of wood pieces that corresponds to each strength class in a wood sample.

When various combinations of strength classes are identified, the combination that considers the highest number of strength classes shall be used. In the case of various combinations with the same number of strength classes, the combination that includes strength classes with the lowest difference of strength and stiffness properties between them (for example, the use of a C18/C24/C35 combination instead of a C18/C24/C40 combination) must be considered.

The allocation procedure is applied by considering a grading stage at the end of production stages which considers the complete wood sample as input and the amount of each strength class graded as output. At the end of this methodology, the environmental impacts of a wood sample can be allocated to each strength class of SW.

3 Maritime Pine Strength Classes

Maritime pine (*Pinus Pinaster Ait.*) is a medium-size softwood tree (Fig. 1a) native to the western Mediterranean basin. The tree reaches between 20–30 m of height [14]. In Europe, the Maritime pine occupies the Iberian Peninsula, the South and West of France, the West of Italy and Western Mediterranean isles (Fig. 1b). In Portugal, the Maritime pine was widely used for boatbuilding during the 15th century and as construction product, and it also has been used during the last centuries for dune stabilization and to prevent the erosion of the coast [14]. The wood of Maritime pine can also be used as construction material, furniture, poles and posts.

According to EN 1912 [15], the range of Maritime pine structural classes is between C18 and C24. Machado et al. [16] identify the possibility of also considering Maritime pine wood as C30. The structural classes identified by LNEC [17], Machado et al. [16] and EN 1912 [15] are given in Table 1.

DEC-UC & SerQ [18] graded 866 wood pieces into three combinations of strength classes. The mean values of bending moment, modulus of elasticity and density of this sample are shown in Table 2. The strength classes considered in the assessment and their yield are shown in Table 3.

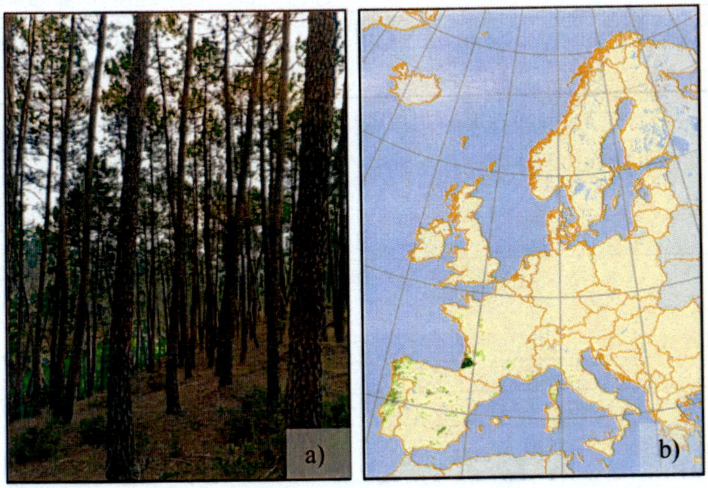

Fig. 1. a) Maritime pine forest; b) distribution map of Maritime pine [14]

Table 1. Strength classes, country of origin and publishing country of Maritime pine SW

Strength Class	Country of Origin
C30	Portugal
C24	Spain
	France
	Portugal
C18	Spain
	France
	Portugal
	Spain

Table 2. Mean values of bending strength, modulus of elasticity and density of Maritime pine SW from Portugal

	Bending moment	Modulus of elasticity	Density (12% of MC)	Number of specimens
Units	N/mm^2	N/mm^2	kg/m^3	p
Maritime pine	54.1	12900	597	866

Table 3. Mechanical grading combinations, strength classes and yield of Maritime pine SW

Combination	Strength classes	Yield [%]
C40/C24/C18/Rejected	C40	74.0
	C24	14.9
	C18	5.6
	Rejected	5.5
C35/C24/C18/Rejected	C35	81.9
	C24	6.9
	C18	6.6
	Rejected	4.6
C24/Rejected	C24	97.6
	Rejected	2.4

4 Estimation of Allocation Percentages of SW Strength Classes

The collection of data considers the mechanical grading results found on the literature results shown in Sect. 3 for Maritime pine from Portugal. The combination of strength classes used in the allocation of strength classes is shown in Table 4.

Table 4. Yield of Maritime pine strength classes used for mechanical grading allocation.

Wood specie	Combinations assumed	Strength class	Yield [%]
Maritime pine	C35/C24/C18/Rejected	C35	81.9
		C24	6.9
		C18	6.6
		Rejected	4.6
		Total	100.0

The proposed allocation procedure modelled a process for each strength class, assuming that the output was the amount of wood from a sample that was able to be used as a specific strength class. The percentage of wood able to be used as a specific strength class (allocation yield) was calculated by dividing the amount of the wood sample (for example, 1 m^3) per summed yield of wood able to be used as a specific strength class. The summed yield of wood able to be used as a specific strength class was calculated by summing the yield of wood classified with a higher strength class than the strength class that is being calculated. The allocation yield of rejected wood was considered 100.0% because all the other wood can be used as ungraded wood. The representation of this allocation procedure is shown in Fig. 2.

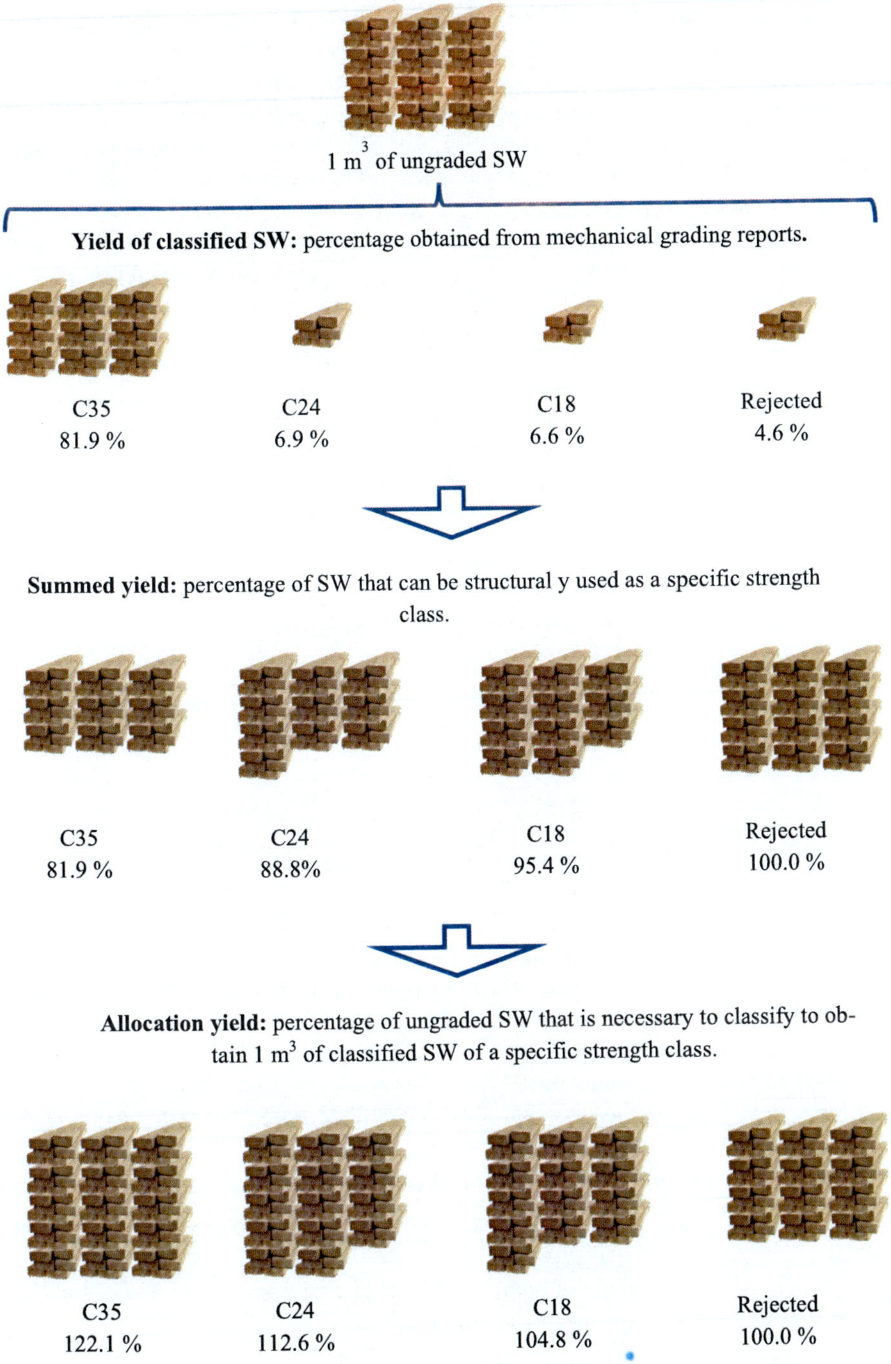

Fig. 2. Description of methodology for determination of allocation yields

For example, to calculate the allocation yield of C24 strength class of Maritime pine, firstly the summed yield of C24 strength class shall be calculated by summing the yield of C35 and C24 strength classes (81.9% + 6.9% = 88.8%). Thus, to calculate the allocation yield used for environmental allocation procedures of class C24, the total volume of wood sample graded (considered 1.0 m^3) shall be divided by the summed yield (1 m^3/88.8% = 112.6%). This percentage represents the cubic meters of wood that it is necessary to saw to obtain one cubic meter of wood of this structural class. In this case, to obtain 1 m^3 of C24 it is necessary to saw 1.126 m^3 of Maritime pine wood. Finally, to obtain the environmental impacts of C24 strength class, the environmental impacts per cubic meter of wood shall be multiplied by the allocation yield of C24 strength class. The values of Maritime pine grading method are shown in Table 5.

Table 5. Yield of strength classes used for allocation of environmental impacts.

Strength class	Yield [%]	Summed yield [%]	Allocation yield [%]
C35	81.9	81.9	122.1
C24	6.9	88.8	112.6
C18	6.6	95.4	104.8
Rejected	4.6	100.0	100.0
Total	100.0	-	-

5 Conclusions

This study presented a methodology to allocate environmental impacts of SW to SW of different strength classes. The methodology is based on the LCA and the mechanical grading procedures, and is divided in two main phases: i) collection of data – identification of combinations of data, and ii) allocation of strength classes – based on tests results. The first intends to collect and select data related with combinations of strength classes used by reports of mechanical graded SW. The latest, use the data collected to determine the allocation yield of various strength classes identified in the previous phase. There are three types of yields that must be calculated:

- Yield of classified SW: percentage obtained from mechanical grading reports.
- Summed yield: percentage of SW that can be structural y used as a specific strength class.
- Allocation yield: percentage of ungraded SW that is necessary to classify to obtain 1 m^3 of classified SW of a specific strength class.

This methodology was applied in this study to determine the allocation yield for the following combination of maritime-pine strength classes: Rejected/C18/C24/C35. The allocation yields are: 122.1%, 112.6%, 104.8% and 100% for C35, C24, C18 and rejected, respectively, which means that to obtain the environmental impacts of a specific strength classes, the environmental impacts of unclassified maritime pine SW must be multiplied by those percentages.

In the future, the authors would like to apply this method to the calculation of environmental impacts of structurally applied Maritime pine SW. Additionally, a methodology for environmental and structural optimization of timber structures could be developed.

References

1. Moore, J.R., Cown, D.J.: 2 - Processing of wood for wood composites. In: Ansell, M.P.B.T.-W.C. (ed.) Woodhead Publishing, pp. 27–45 (2015). ISBN 978-1-78242-454-3
2. CEN EN 14081-1: Timber Structures - Strength Graded Structural Timber with Rectangular Cross Section. European Commission Standard (2016)
3. ISO ISO 14040: Environmental Management – Life Cycle Assessment—Principles and Framework. International Organization for Standardization (2006)
4. ISO ISO 14044: Environmental Management - Life Cycle Assessment - Requirements and Guidelines. International Organization for Standardization, p. 1 (2006)
5. CEN EN 15804+A2:2019 - Sustainability of Construction Works—Environmental Product Declarations—Core Rules for the Product Category of Construction Products (2019)
6. CEN EN 16485:2014 - Round and Sawn Timber, Environmental Product Declarations, Product Category Rules for Wood and Wood-Based Products for Use in Construction, p. 27 (2014)
7. Dodoo, A., Gustavsson, L., Sathre, R.: Lifecycle primary energy analysis of low-energy timber building systems for multi-storey residential buildings. Energy Build. **81**, 84–97 (2014). https://doi.org/10.1016/j.enbuild.2014.06.003
8. Pajchrowski, G., Noskowiak, A., Lewandowska, A., Strykowski, W.: Wood as a building material in the light of environmental assessment of full life cycle of four buildings. Constr. Build. Mater. **52**, 428–436 (2014). https://doi.org/10.1016/j.conbuildmat.2013.11.066
9. Røyne, F., Peñaloza, D., Sandin, G., Berlin, J., Svanström, M.: Climate impact assessment in life cycle assessments of forest products: implications of method choice for results and decision-making. J. Clean. Prod. **116**, 90–99 (2016). https://doi.org/10.1016/j.jclepro.2016.01.009
10. Robertson, A.B., Lam, F.C.F., Cole, R.J.: A comparative cradle-to-gate life cycle assessment of mid-rise office building construction alternatives: laminated timber or reinforced concrete. Buildings **2**, 245–270 (2012). https://doi.org/10.3390/buildings2030245
11. Skullestad, J.L., Bohne, R.A., Lohne, J.: High-rise timber buildings as a climate change mitigation measure – a comparative LCA of structural system alternatives. Energy Proc. **96**, 112–123 (2016). https://doi.org/10.1016/j.egypro.2016.09.112
12. Dias, A.M.A., Dias, A.M.P.G., Silvestre, J.D., de Brito, J.: Comparison of the environmental and structural performance of solid and glued laminated timber products based on EPDs. Structures **26**, 128–138 (2020). https://doi.org/10.1016/j.istruc.2020.04.015
13. Dias, A.M.A., Dias, A.M.P.G., Silvestre, J.D., de Brito, J.: Comparison of environmental impacts of structurally equivalent wood-based floors. Eur. J. Wood Wood Prod. **79**, 1607–1622 (2021). https://doi.org/10.1007/s00107-021-01742-x
14. Viñas, R.A., Caudullo, G., Oliveira, S., de Rigo, D.: Pinus Pinaster in Europe: distribution, habitat, usage and threats. Eur. Atlas For. Tree Species **2016**, 128–129 (2016)
15. CEN EN 1912 - Structural Timber - Strength Classes - Assignment of Visual Grades and Species, pp. 1–16 (2012)

16. Machado, J.S., Lourenço, P.B., Palma, P.: Assessment of the structural properties of timber members in situ – a probabilistic approach, pp. 1–12 (2011)
17. LNEC Ficha M2: Pinho Bravo Para Estruturas. Série Madeira Para Construção; Lisboa (1997)
18. DEC-UC: SerQ Derivation of Settings for the Handheld MTG Grading Device 920 for Maritime Pine Grown in Portugal (Restricted Access). Rep. RT-FCTUC-SERQ-1401 (2014)

Life Cycle Assessment of OSB Panels Produced with Alternative Raw Materials

Estefani Sugahara[1,2](✉), Andre Dias[2,3], Edson Botelho[1], Cristiane Campos[1], and Alfredo Dias[2,3]

[1] UNESP, São Paulo State University, Guaratinguetá 12516-410, Brazil
estefani.sugahara@unesp.br
[2] University of Coimbra, 3030-788 Coimbra, Portugal
[3] SerQ, Innovation and Competence Forest Centre, 6100-711 Sertã, Portugal

Abstract. To produce sustainable materials, it is essential to start by quantifying their environmental impacts. In this context, Life Cycle Assessment (LCA) methodology is an efficient tool to assess the environmental performance and impacts of products and/or services. Due to the global increasing market for OSB (Oriented strand board) panels and the concern to produce composites materials in an environmentally sustainable way, and based on the knowledge gaps and existing demands, the overall objective of this research was to quantify the environmental impacts of OSB panels produced with Eucalyptus ssp. (reforested wood) and castor oil-based bicomponent polyurethane adhesive (adhesive derived from vegetal oil). The environmental impacts were calculated using the methodology given by EN 15804+A2 standard. This work identified the system boundaries, environmental impacts, and environmental hotspots in OSB production. The results of this new type of panel were compared with the environmental impacts of traditional panels given by state-of-process available in international databases. This study also includes recommendations for the industry on how to reduce the environmental impacts of this new type of panel.

Keywords: Environmental Impacts · Life Cycle Assessment · Construction industry · OSB panels

1 Introduction

The Oriented Strand Board (OSB) is a multi-layered structural panel made from wood strands, flaked from roundwood and bonded together with a synthetic resin adhesive with typical thickness in a range of 6 to 40 mm. OSB varies in color from a light straw color to a medium brown depending on the wood species, resin system and pressing conditions. Waterproof and boil proof resin binders are combined with the strands to improve internal strength, rigidity and moisture resistance [1].

In this type of board, the strands of wood are glued in an aligned way and the different layers are orientated perpendicular between them [2]. The resins usually used in the OSB production are phenol formaldehyde, melamine formaldehyde or methylene diphenyl

S. Amziane et al. (Eds.): ICBBM 2023, RILEM Bookseries 45, pp. 959–972, 2023.
https://doi.org/10.1007/978-3-031-33465-8_74

diisocyanate. The panels density depends on the wood species and the manufacturing process itself, however, the typical density ranges between 600 and 680 kg/m^3 [3].

This type of panel is one of the most used wood products in construction and its relevance is increased by the fact that the demand for wood-based products has recently increased worldwide [4] since these have been progressively more used in the construction industry, because of the contemporary trend of selecting materials with low environmental impact [5].

Because of the growing concerns about quality, price and the social and environmental impacts of processes and products, added to the possibility of shortage supply of specific wood species due to high production volumes, the research for alternative raw-materials, as different wood species and adhesives potentially less harmful and that have renewable origin, became an essential matter [6, 7].

The studies that evaluate the performance of wood panels found in literature review is based on the measurement of its physical and mechanical properties [8] but to produce more sustainable building materials, it is essential to quantify their environmental impacts. In this way, Life Cycle Assessment (LCA) emerged as a tool that can quantify and compare the environmental impacts of different materials and construction solutions [9–11].

Wood-based panels are usually known as a potentially carbon-neutral material, however, there are several other crucial questions associated with the life cycle of wood-based panels. Moreover, wood-based panels generally have a long service life and a multifunctional production chain. Thus, due to the methodological challenges related to the wood-based panel supply chain, it is essential for decision makers, companies and LCA researchers to understand the different methodologies to treat these issues and how important are those to the LCA results [12].

In this context, several researches assessed the environmental aspects related to wood-based products destined for different uses in civil construction such as roundwood [13–15], commercial wood-based panels [3, 12, 16, 17], wood-based floors [18–20], and glued laminated timber [18, 21, 22]. Nevertheless, the differences among LCA research of wood-based products are mostly related to the definition of the system boundaries as well as to the life cycle inventory data quality and selection of environmental impact categories and methods for the life cycle impact assessment [3].

Thus, since the global increasing market for OSB and the concern to produce composite materials in an environmentally sustainable way, and based on the knowledge gaps and existing demands, this paper aimed to quantify the potential environmental impacts related to the experimental OSB production with alternative raw materials by using LCA methodology in a cradle-to-gate perspective.

In this way, this study applied the LCA to quantify the environmental impacts of OSB panels produced with *Eucalyptus ssp.* and castor oil-based bicomponent polyurethane adhesive, that is an adhesive derived from vegetal oil, biodegradable, non-pollutant and well known as a renewable product [4] that is used in alternative to formaldehyde-based adhesives, which emit formaldehyde in the pressing process [5]. This work identified the system boundaries, environmental impacts, and environmental hotspots in OSB production by using the methodology given by EN 15804+A2 standard [23].

That is a current and extremely important topic, presenting great relevance for the scientific, technological, economic and innovation development, justified by the lack of research related to experimental panels, produced with reforestation wood and without the use of products that have potential toxicity such as chemical preservatives of wood and formaldehyde-based adhesives [6].

2 Methods

2.1 Case of Study

In this study, LCA model was applied using the software SimaPro (PhD with Share & Collect, release 9.1.0.7) to implement the LCI worldwide datasets of environmental impacts such as Ecoinvent 3, ELCD, and following the procedures of the ISO 14040 [24], ISO 14044 [25] and EN 15804+A2 [23] standards. In this way, the four stages of an LCA study (goal and scope definition, inventory, impact assessment and interpretation) were described.

2.2 Goal and Scope Definition

This work aims to quantify the environmental impacts of experimental OSB panels produced with *Eucalyptus ssp.* (reforested wood with density of 520 kg/m^3 obtained from the southeast region of Brazil) and castor oil-based bicomponent polyurethane adhesive (adhesive derived from vegetal oil) and compare the results with the environmental impacts of traditional panels. The data were obtained through a survey of manufacturers following the ISO14040 standard or replacing some specific parameters of products and processes available in the software dataset.

The scope of this study was from cradle-to-gate, which means that it covers the raw materials extraction and production, and production of OSB panels excluding the construction, product use and end of life phases.

The intended audience for this work includes worldwide OSB and raw materials researchers, industry, decision-makers, and other LCA practitioners. The adopted functional unit was 1 m^3 of OSB, considering this new type of panel produced with a nominal density of 780 kg/m^3 without covering, and following the existent European production processes available in the Ecoinvent database.

The assessment of the environmental impacts associated with the OSB production system was conducted considering characterization factors reported by EN 15804+A2 Method V1.00/EF 3 Normalization and weighting set [23].

The environmental impact categories considered were those given by the most recent version of EN 15804+A2 standard [23], which are: Climate change (CC), Ozone depletion (OD), Ionising radiation (IR), Photochemical ozone formation (POF), Particulate matter (PM), Human toxicity (non-cancer, HTNC), Human toxicity (cancer, HTC), Acidification (AC), Eutrophication (freshwater, EF), Eutrophication (marine, EM), Eutrophication (terrestrial, ET), Ecotoxicity (freshwater, EFW), Land use (LU), Water use (WU), Resource use (fossils, RUF), Resource use (minerals and metals, RUMM).

2.3 Description of the Product and System Under Assessment

The system boundaries of the OSB life cycle under study (cradle-to-gate) is shown in Fig. 1.

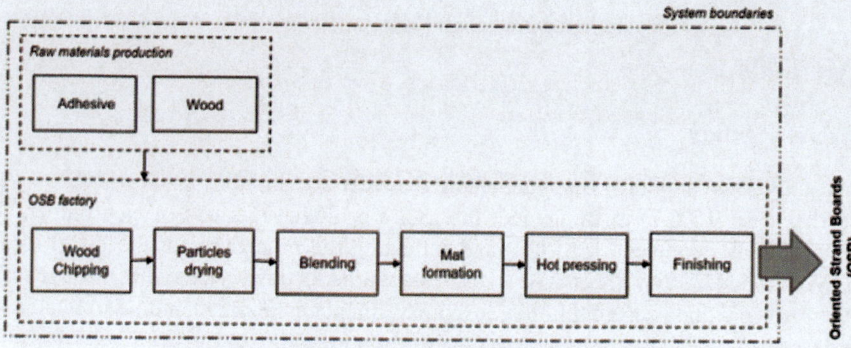

Fig. 1. System boundaries of the OSB panels.

The type of board assessed in this work is intended for structural uses compatible with the physical and mechanical requirements of the European standard EN 300 [2]. According to the inventory of the Ecoinvent 3 database of OSB production used in this study as a model for assessment and comparison (Oriented strand board production RER, Process identifier IMOADATA000043921900016, Energy values: 1, Technology level: 3), the OSB production consists of: production of strands, drying and placing of strands into a rotating drum and blended with a fine mist of wax and resin. The coated strands are then transported to the forming line where a mat of flakes is laid on a continuous belt conveyor with a cut-off saw. The mat is pressed under high temperature and pressure and the OSB is formed. Finally, in the finishing area the product is trimmed and cut to produce the finished OSB board [1].

2.4 Life Cycle Inventory

The inventory data was modeled from templates obtained from Ecoinvent Database available in Simapro Software, listed in the Table 1. The modifications made to the original files used in this study, obtained through a survey of manufacturers following the ISO 14040 standard or replacing some specific parameters of products and processes available in the software dataset, are listed in Table 2.

Table 1. Summary of descriptors of materials and processes from Ecoinvent 3 inventory used in this study.

Activity name	ISIC rev.4 Ecoinvent	CPC classification	Reference product	Abbreviation
Oriented strand board production RER (Europe)	1621: Manufacture of veneer sheets and wood-based panels	31432: Oriented strand board (OSB)	Oriented strand board	OSB_S
Hardwood forestry, eucalyptus ssp., sustainable forest management RoW (Rest-of-World)	0220: Logging	031: Wood in the rough	Roundwood, eucalyptus ssp. from sustainable forest management, under bark	HW_S
Market for roundwood, eucalyptus ssp. from sustainable forest management, under bark GLO (Global)	0220: Logging	031: Wood in the rough	Roundwood, eucalyptus ssp. from sustainable forest management, under bark	MW_S
Palm oil mill operation RoW (Rest-of-World)	1040: Manufacture of vegetable and animal oils and fats	21651: Palm oil, crude	Palm oil, crude	OMO_S
Palm oil refinery operation GLO (Global)	1040: Manufacture of vegetable and animal oils and fats	21652: Palm oil, refined	Palm oil, refined	ORO_S
Market for palm oil, crude GLO (Global)	1040: Manufacture of vegetable and animal oils and fats	21651: Palm oil, crude	Palm oil, crude	MOC_S
Phenolic resin production RER (Europe)	2013: Manufacture of plastics and synthetic rubber in primary forms	34790: Other plastics in primary forms; ion exchangers	Phenolic resin	A_S

Table 2. Summary of modifications made to the original files used in this study.

New file	Original file	Description of actions taken
OSB_S1	OSB_S	The inputs related to the values of "Electricity from grid times share of total production in specific country" were added and the total value was gathered in a single input (Activity link: market group for electricity, medium voltage - RER)
HW_S	HW_N	The quantity of carbon dioxide sequestrated was replaced according to the density of the wood used in the experimental OSB (520 kg/m^3 at 12% of moisture content)
MW_S	MW_N	The input HW_S was replaced by HW_N
OMO_S	OMO_N	The input "Palm fruit bunch {RoW} | market for palm fruit bunch | Cut-off, U" replaced by "Castor bean {GLO} | market for castor bean | Cut-off, U"
ORO_S	ORO_N	The input "Palm oil, crude {GLO} | market for | Cut-off, U" replaced by "Castor oil, crude {GLO} | market for | Cut-off, U_PhD_Est"
MOC_S	MOC_N	The input "Palm oil, crude {MY} | palm oil mill operation | Cut-off, U" and "Palm oil, crude {RoW} | palm oil mill operation | Cut-off, U" replaced by "Castor oil, crude {RoW} | Castor oil mill operation | Cut-off, U_PhD_Est"
A_S	A_N	For the polyurethane adhesive derived from castor oil, the input "Chemical factory, organics {RER} | construction | Cut-off, U" was maintained and the other inputs and outputs were all excluded and replaced by the new ones provided by the manufacturer, along with their respective values
OSB_N	OSB_S1	OSB_S was used as a template. However, for the experimental OSB panel the wood was replaced by HW_N, and the adhesive was replaced by "A_N". Was considered for the OSB production a nominal density of 780 kg/m^3 for the panels, a density of 520 kg/m^3 at 12% of moisture content for the wood and 10% of adhesive based on the dry mass of wood strands. The inputs chemical additives for wood and paraffin were excluded as well as the outputs related to the glue drying process as Acetaldehyde, Formaldehyde, Methanol and NMVOC (non-methane volatile organic compounds, unspecified origin)

Thus, Table 3 presents the Global inventory data per functional unit corresponding to the OSB standard production (OSB_S1) and the experimental OSB panels (OSB_N).

Table 3. Global inventory data per functional unit.

	Unit	OSB_ S1	OSB_N
INPUTS FROM ENVIRONMENT			
Water	m^3	0.29	0.29
INPUTS FROM TECHNOSPHERE **Materials/fuels**			
Chemical, organic	kg	5.21	0.00
Dust collector, electrostatic precipitator	p	4.73	4.73
Lubricating oil	kg	0.06	0.06
Methylene diphenyl diisocyanate	kg	18.30	0.00
Polyurethane adhesive derived from castor oil	kg	0.00	70.86
Paraffin	kg	10.94	0.00
Wood under bark	m^3	1.91	1.36
Tap water	kg	203.88	203.88
Wooden board factory	p	3.79	3.79
Electricity/heat			
Diesel, burned in building machine	MJ	33.52	33.52
Electricity, medium voltage	kWh	116.25	116.25
Furnace, wood chips	p	2.36	2.36
Furnace, wood chips, with silo	p	9.99	9.99
Heat, district or industrial, natural gas	MJ	344.66	344.66
Heat, district or industrial, other than natural gas	MJ	2.49	2.49
OUTPUTS **Emissions to air**			
Acetaldehyde	kg	0.00056	0.00
Ammonia	kg	0.0062	0.0062
Arsenic	kg	3.57E-6	3.57E-6
Benzene	kg	0.0033	0.0033
Benzene, ethyl-	kg	0.00010	0.00010
Benzene, hexachloro-	kg	2.58E-11	2.58E-11
Benzo(a)pyrene	kg	1.79E-6	1.79E-6
Bromine	kg	0.00022	0.00022

(continued)

Table 3. (*continued*)

	Unit	OSB_ S1	OSB_N
Cadmium	kg	2.50E-6	2.50E-6
Calcium	kg	0.021	0.021
Carbon dioxide, biogenic	kg	364.85	364.85
Carbon monoxide, biogenic	kg	0.72	0.72
Chlorine	kg	0.00064	0.00064
Chromium	kg	1.42E-5	1.42E-5
Chromium VI	kg	1.43E-7	1.43E-7
Copper	kg	7.87E-5	7.87E-5
Dinitrogen monoxide	kg	0.0082	0.0082
Dioxin, 2,3,7,8 Tetrachlorodibenzo-p-	kg	1.07E-10	1.07E-10
Fluorine	kg	0.00018	0.00018
Formaldehyde	kg	0.02	0.02
Hydrocarbons, aliphatic, alkanes, unspecified	kg	0.0033	0.0033
Hydrocarbons, aliphatic, unsaturated	kg	0.011	0.011
Lead	kg	8.94E-5	8.94E-5
Magnesium	kg	0.0013	0.0013
Manganese	kg	0.00060	0.00060
Mercury	kg	1.07E-6	1.07E-6
Methane, biogenic	kg	0.0054	0.0054
Methanol	kg	0.018	0.018
m-Xylene	kg	0.00043	0.00043
Nickel	kg	2.15E-5	2.15E-5
Nitrogen oxides	kg	0.43	0.43
Non-methane volatile organic compounds, unspecified	kg	0.33	0.00
PAH, polycyclic aromatic hydrocarbons	kg	3.97E-5	3.97E-5
Particulates, < 2.5 um	kg	0.0089	0.0089
Particulates, > 10 um	kg	0.099	0.099
Phenol, pentachloro-	kg	2.90E-8	2.90E-8
Phosphorus	kg	0.0011	0.0011
Potassium	kg	0.084	0.084
Sodium	kg	0.0047	0.0047
Sulfur dioxide	kg	0.0090	0.0090

(*continued*)

Table 3. *(continued)*

	Unit	OSB_ S1	OSB_N
Toluene	kg	0.0011	0.0011
Zinc	kg	0.0011	0.0011
Waste to treatment			
Biowaste	Kg	0.26	0.26
Wood ash mixture, pure	kg	1,704	1,704

3 Results and Discussion

This section presents the life cycle impact assessment (LCIA) and interpretation phases. The environmental impacts obtained for each impact category per 1 m^3 of OSB_ S1 and OSB_N are presented in Table 4.

Table 4. Potential environmental impacts per 1 m^3 of OSB_ S1 and OSB_N

Impact category		Unit	OSB_S1	OSB_N
Climate change	CC	kg CO2 eq	−974.58	−262.40
Ozone depletion	OD	kg CFC11 eq	4.02E-05	1.64E-05
Ionising radiation	IR	kBq U-235 eq	16.84	11.24
Photochemical ozone formation	POF	kg NMVOC eq	1.87	1.26
Particulate matter	PM	disease inc	1.23E-05	2.10E-05
Human toxicity, non-cancer	HTNC	CTUh	2.02E-05	2.54E-05
Human toxicity, cancer	HTC	CTUh	2.54E-06	3.96E-07
Acidification	A	mol H + eq	1.63	2.23
Eutrophication, freshwater	EF	kg P eq	0.015	0.015
Eutrophication, marine	EM	kg N eq	0.51	0.84
Eutrophication, terrestrial	ET	mol N eq	4.74	9.31
Ecotoxicity, freshwater	EFW	CTUe	12316.53	5599.08
Land use	LU	Pt	103089.37	24012.47
Water use	WU	m3 depriv	124.69	52.43
Resource use, fossils	RUF	MJ	4748.44	2024.94
Resource use, minerals and metals	RUMM	kg Sb eq	0.00035	0.0011

Comparing OSB_N and OSB_S1, it is noted that OSB_S1 has 9 categories with major environmental impact, while OSB_N has 8 categories. OSB_N performed better in OD, IR, POF, HTC, EF, EWF, LU, WU, RFU. OSB_S1 performed better in CC, PM, HTCN, A, EF, EM, ET, RUMM.

The assessment of the environmental impacts (relative contributions to each impact category from the main contributing factors involved in the OSB cradle-to-gate life cycle) associated with the OSB_N (Fig. 2) were conducted considering characterization factors reported by EN 15804+A2 Method V1.00/EF 3 Normalization and weighting set [27].

A detailed analysis of the impacts of each category of OSB_N panels is presented in Fig. 2.

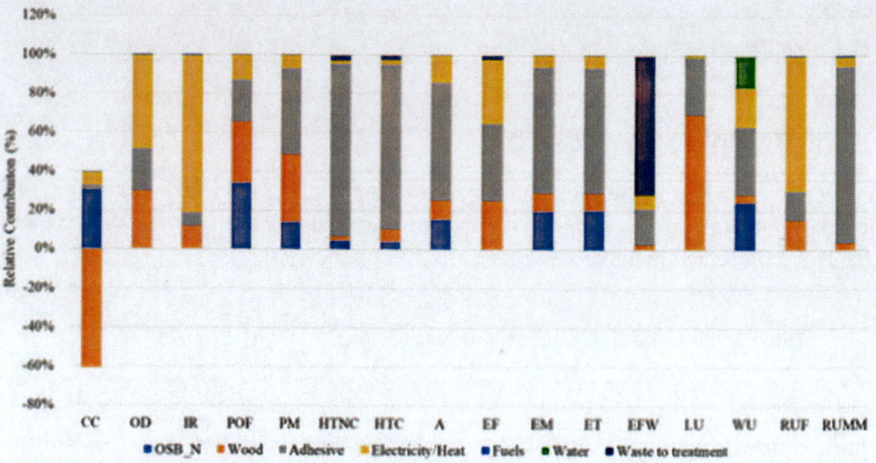

Fig. 2. Assessment of the environmental impacts involved in the OSB_N cradle-to-gate life cycle.

Climate change potential (CC): In this work, for experimental OSB_N, "Production stage" was responsible for 30% of the total CO_2 emissions produced throughout the product system. On the other hand, the Wood has a contribution of -61%, removing carbon dioxide from the atmosphere. For standard OSB_S1 the Production presented 17% of the total CO_2 emissions, and Wood has a contribution of −73%.

Ozone depletion potential (OD): In this category, for OSB_N the Electricity/Heat have the main contribution with 48% (where 24% is relative to electricity, 19% is relative to heat and 4% is relative to diesel,) representing almost the half of the total, being followed by Wood (30%) and Adhesive (22%). In OSB_S1 Adhesive represented the major contribution (64%), followed by Electricity/Heat (20%) and Wood (16%).

Ionising radiation (IR): In OSB_N the Electricity/Heat contributes with 81% of the total, the Wood with 12% and Adhesive with 7%, while for in OSB_S1, Electricity/Heat represents 54%, Adhesive 35%, and Wood 11%.

Photochemical ozone formation (POF): In this category for OSB_N the major values were found in Production (35%), followed by Wood (31%), Adhesive (22%) and Electricity/Heat (13%). For OSB_S1 were found 42% for Production, 32% for Adhesive, 18% for Wood and 8% for Electricity/Heat.

Particulate matter (PM): The main contribution for OSB_N was related to Adhesive (44%), followed by Wood (35%), Production (14%) and Electricity/Heat (7%). For OSB_S1 Adhesive (48%) also presented the main contribution, followed by Production (24%), Wood (16%) and Electricity/Heat (12%).

For Human toxicity (non-cancer, HTNC) of OSB_N the Adhesive represents 88% of the contribution, the Production 5%, Waste to treatment 3%, Wood 2% and Electricity/heat 2%. For Human toxicity (cancer, HTC), of OSB_N the Adhesive also presents the major contribution (84%), followed by Wood (6%), Production (4%), Waste to treatment (2%) and Electricity/heat (2%).

For HTNC of OSB_S1 Adhesive presented 85%, Production 6%, Waste to treatment 4%, Wood 3% and Electricity/heat 2%. For HTC of OSB_S1 Adhesive presented 78%, Production 20% and Wood 2%.

Acidification (AC): For OSB_N Adhesive was 61% of the contribution, the production 15%, Electricity/heat 14% and Wood 10%. For OSB_S1 Adhesive was 50% of the contribution, the production 21%, Electricity/heat 19% and Wood 10%.

In EF category (Eutrophication potential, fraction of nutrients reaching freshwater end compartment) for OSB_N the major contributions were found in Adhesive (40%), followed by Electricity/heat (33%), Wood (25%) and Waste to treatment (2%). For EM (Eutrophication potential, fraction of nutrients reaching marine end compartment), Adhesive presented 65%, followed by Production with 20%, Wood 9% and Electricity/heat 6%. For ET (Eutrophication potential, Accumulated Exceedance) were found 64% for Adhesive, 21% for Production, 9% for Wood and 6% for Electricity/heat.

For OSB_S1 in EF category, the main contribution was found in Wood (34%) followed by Electricity/heat (33%), Adhesive (31%) and Waste to treatment (2%). For EM, Adhesive presented 44%, Production 33%, Wood 13% and Electricity/heat 10%. For ET were found 40% for Production, 33% for Adhesive, 14% for Wood and 13% for Electricity/heat.

Ecotoxicity (freshwater, EFW): For OSB_N Waste to treatment represents the major contribution in this category (71%), followed by Adhesive (19%), Electricity/heat (7%) and Wood (3%). For OSB_S1 Adhesive presented 62%, Waste to treatment 32%, Electricity/heat 3% and Wood 2%.

Land use (LU): In this category, for OSB_N Wood presented major contribution (69%), followed by Adhesive (30%) and Electricity/heat (1%). For OSB_S1 Wood presented 100%.

Water use (WU): In OSB_N Adhesive presented contribution of 35%, Wood 24%, Electricity/heat 20%, Water 17% and Wood 4%. In OSB_S1 Adhesive showed 72%, Wood 10%, Electricity/heat 8%, Water 7% and Wood 3%.

For RUF (abiotic depletion potential for fossil resources) in OSB_N Electricity/heat presented major contribution (69%), followed by Wood (16%) and Adhesive (15%). For RUMM (abiotic depletion potential for non-fossil resources - minerals and metals) in OSB_N Adhesive presented major contribution (90%), followed by Electricity/heat (5%) and Wood (4%).

For RUF in OSB_S1 Adhesive presented the main contribution (62%), Electricity/heat 29% and Wood 9%. For RUMM in OSB_S1 Adhesive presented major contribution (74%), followed by Electricity/heat (17%), Wood (4%) and Fuels (2%).

For OSB_N assessment it can be observed that, Adhesive has major contribution in 9 of 16 categories (PM, HTCN, HTC, A, EF, EM, ET, WU and RUMM), Electricity/heat in 3, Production in 2, Wood and Waste to treatment 1 each.

In the assessment of the environmental impacts involved in the OSB_S1, it is verified that the Adhesive it has also the major quantity of categories with main contribution, surpassing the amount found for OSB_N reaching 11 of 16 categories (OD, POF, PM, HTCN, HTC, A, EM, EFW, WU, RUF and RUMM). Production and Wood reached 2 each, and Electricity/heat only 1.

In this way, it is concluded that in this assessment the adhesive has an important significance in the environmental impact categories analyzed. Changes in the resin type could have a significant impact on individual impacts or energy consumption [26]. The substitution of the adhesive for alternatives with lower impacts could result in important improvements in the environmental profile, however, could change the technical properties of the wood panels [16].

Nevertheless, the changes in the physical and mechanical properties of wood panels produced with alternative adhesives, as the castor oil-based bicomponent polyurethane adhesive, has shown potential for use in the wood-based panels production. Those panels meet the basic requirements determined by the normative instruments or showing up compatible with those presented by commercial ones [4, 5, 7] or even presenting improvements compared to these [27].

However, to efficiently improve the overall environmental performance of OSB_S1 panels compared with the OSB_N, it must be improved the aspects related with the Production and Electricity/heat categories.

4 Conclusions

This study performed a cradle-to-gate LCA of OSB panels produced with *Eucalyptus ssp.* and castor oil-based bicomponent polyurethane adhesive. This study identified the system boundaries, environmental impacts, and environmental hotspots in OSB production. The potential environmental impacts were assessed in terms of sixteen impact categories: CC, OD, IR, POF, PM, HTCN, HTC, AC, EF, EM, ET, EFW, LU, WU, RUF, RUMM.

Were identified that in the OSB production the adhesive has a strong influence in the environmental impacts in most categories. So, for the overall impacts evaluated it was concluded that the adhesive is the main environmental hotspot of this evaluation. In this way it is crucial that this raw material be carefully chosen and be researched alternatives for more ecologically appropriate adhesives in order to preserve the environment and human health. However, special attention also must be given to make changes in order to improve the impacts associated to Production and Electricity/heat categories.

It is considered that the type of wood and adhesive used to produce the experimental OSB_N panels (*Eucalyptus ssp.* and castor oil-based adhesive) can be considered good and more ecological alternatives to produce OSB panels, since the categories Adhesive and Wood performed better in OSB_N if compared to OSB_S1.

References

1. Ecoinvent 3.6 Oriented Strand Board Production RER - Process; Dataset documentation generated by: Frank Werner, Zurich, Switzerland (2019)

2. BSI, E.S. BS EN 300:2006 - Oriented Strand Boards (OSB). Definitions, Classification and Specifications, vol. 3, p. 24 (2006)
3. Ferro, F.S., Silva, D.A.L., Rocco Lahr, F.A., Argenton, M., González-García, S.: Environmental aspects of oriented strand boards production. A Brazilian case study. J. Clean. Prod. **183**, 710–719 (2018). https://doi.org/10.1016/j.jclepro.2018.02.174
4. Barbirato, G.H.A., Fiorelli, J., Mejiaa, J., Sarasini, F., Tirillò, J., Ferrante, L.: Quasi-static and dynamic response of oriented strand boards based on Balsa wood waste. Compos. Struct. **219**, 83–89 (2019). https://doi.org/10.1016/j.compstruct.2019.03.062
5. Pozzer, T., Gauss, C., Ament Barbirato, G.H., Fiorelli, J.: Trapezoidal core sandwich panel produced with sugarcane Bagasse. Constr. Build. Mater. **264**, 120718 (2020). https://doi.org/10.1016/j.conbuildmat.2020.120718
6. Sugahara, E.S., et al.: Thermal treatment of OSB panels: a systematic and bibliometric overview. Eur. J. Wood Wood Prod. (2022). https://doi.org/10.1007/s00107-022-01849-9
7. Sugahara, E., et al.: Study of the influence of heat treatment on OSB panels produced with eucalyptus wood in different layer compositions, pp. 1–18 (2022)
8. Garrido, N., Martins, J., Pereira, J., Carvalho, L.M., Mendes, J., Costa, C.: Effect of recycled wood and alternative species on particleboard machining. In: 1st International Conference on Processing Technologies for the Forest and Bio-based Products Industries (2010)
9. Werner, F., Richter, K.: Wooden building products in comparative LCA: a literature review. Int. J. Life Cycle Assess. **12**, 470–479 (2007). https://doi.org/10.1065/lca2007.04.317
10. Islam, H., Jollands, M., Setunge, S., Haque, N., Bhuiyan, M.A.: Life cycle assessment and life cycle cost implications for roofing and floor designs in residential buildings. Energy Build. **104**, 250–263 (2015). https://doi.org/10.1016/j.enbuild.2015.07.017
11. John, S., Nebel, B., Perez, N., Buchanan, A.: Environmental impacts of multi-storey buildings using different construction materials. Res. Rep. **2009**, 204 (2008)
12. Garcia, R., Freire, F.: Carbon footprint of particleboard: a comparison between ISO/TS 14067, GHG protocol, PAS 2050 and climate declaration. J. Clean. Prod. **66**, 199–209 (2014). https://doi.org/10.1016/j.jclepro.2013.11.073
13. Dias, A.M., Machado, J.S., Dias, A.M.P.G., Silvestre, J.D., de Brito, J.: Influence of the wood species, forest management practice and allocation method on the environmental impacts of roundwood and biomass. Forests **13**, 1357 (2022). https://doi.org/10.3390/f13091357
14. Dias, A.C., Arroja, L.: Environmental impacts of eucalypt and maritime pine wood production in Portugal. J. Clean. Prod. **37**, 368–376 (2012). https://doi.org/10.1016/j.jclepro.2012.07.056
15. Kuka, E., Cirule, D., Andersone, I., Miklasevics, Z., Andersons, B.: Life cycle inventory for currently produced pine roundwood. J. Clean. Prod. **235**, 613–625 (2019). https://doi.org/10.1016/j.jclepro.2019.07.004
16. Silva, D.A.L., Lahr, F.A.R., Garcia, R.P., Freire, F.M.C.S., Ometto, A.R.: Life cycle assessment of medium density particleboard (MDP) produced in Brazil. Int. J. Life Cycle Assess. **18**, 1404–1411 (2013). https://doi.org/10.1007/s11367-013-0583-3
17. Benetto, E., Becker, M., Welfring, J.: Life cycle assessment of oriented strand boards (OSB): from process innovation to ecodesign. Environ. Sci. Technol. **43**, 6003–6009 (2009). https://doi.org/10.1021/es900707u
18. Dias, A.M.A., Dias, A.M.P.G., Silvestre, J.D., de Brito, J.: Comparison of environmental impacts of structurally equivalent wood-based floors. Eur. J. Wood Wood Prod. **79**(6), 1607–1622 (2021). https://doi.org/10.1007/s00107-021-01742-x
19. Dias, A.M.A., Santos, P.G.G., Dias, A.M.P.G., Silvestre, J.D., de Brito, J.: Life cycle assessment of a preservative treated wooden deck. Wood Mater. Sci. Eng. (2021). https://doi.org/10.1080/17480272.2021.1897673
20. Dossche, C., Boel, V., de Corte, W.: Comparative material-based life cycle analysis of structural beam-floor systems. J. Clean. Prod. **194**, 327–341 (2018). https://doi.org/10.1016/j.jclepro.2018.05.062

21. Dias, A.M.A., Dias, A.M.P.G., Silvestre, J.D., de Brito, J.: Comparison of the environmental and structural performance of solid and glued laminated timber products based on EPDs. Structures **26**, 128–138 (2020). https://doi.org/10.1016/j.istruc.2020.04.015
22. Hill, C.A.S., Dibdiakova, J.: The environmental impact of wood compared to other building materials. Int. Wood Prod. J. **7**, 215–219 (2016). https://doi.org/10.1080/20426445.2016.1190166
23. CEN, E.C. for S. EN 15804:2012+A2 - Sustainability of Construction Works - Environmental Product Declarations - Core Rules for the Product Category of Construction Products, Belgium, p. 72 (2019)
24. ISO, I.O. for S. ISO 14040:2006 - Environmental Management - Life Cycle Assessment - Principles and Framework, Switzerland, p. 28 (2006)
25. ISO, I.O. for S. ISO 14044:2006(E) - Environmental Management - Life Cycle Assessment - Requirements and Guidelines, Switzerland, p. 54 (2006)
26. Puettmann, M., Kaestner, D., Taylor, A.: Module E Life Cycle Assessment of Oriented Strandboard (OSB) Production (2016)
27. Sugahara, E.S., et al.: High-density particleboard made from agro-industrial waste and different adhesives. Bioresources **14**, 5162–5170 (2019). https://doi.org/10.15376/biores.14.3.5162-5170

Generation of Electric Energy from Gasifying Rice Husk and Utilization of Its By-Products to Cementitious Materials

Yuma Shimamoto[1](✉) and Tetsuya Suzuki[2]

[1] Tokyo University of Agriculture and Technology, Tokyo 134-0006, Japan
simamoto@go.tuat.ac.jp

[2] Niigata University, Niigata 950-2181, Japan

Abstract. In recent years, a system has been developed to generate electricity energy by rice husk gasification and reuse the energy as a power source for agricultural facilities, but the use of "rice husk ash" discharged by gasification is limited. The authors have been developing structural materials using "rice husk ash". The reason for focusing on rice husks and rice husk ash is that they can contribute to resource reuse, energy generation, and improvement of mechanical properties by pozzolanic reaction. In term of the reduction of environmental burdens, this paper analyzes both the carbon dioxide emissions of producing rice husk using a rice husk biomass power plant operated by a cogeneration system as a case study, and the effectiveness in the reduction of carbon dioxide generated in the production of the material by adding rice husk ash. As a result, the cogeneration system creates twice as large energy as the input and utilizing rice husks decrease the amount of carbon dioxide emissions by 47 tons in a year. Replacing the solidifier with rice husk ash reduces carbon dioxide emissions in production of the material. In particular, cement materials have the lowest carbon dioxide emissions. Compared to MgO materials, carbon dioxide emissions in cement material are about 1/3. Therefore, the utilization of rice husk and rice husk ash for structural material lighten the environmental burden.

Keywords: Rice husk · Carbon dioxide emission · Structural material

1 Introduction

In recent years, materials utilizing industrial wastes such as fly ash and blast furnace slag as cement substitutes have been developed for the sustainable development of society [1]. Now, sustainable agriculture is needed to solve environmental and economic problems. Using waste agricultural materials is one of the most effective approaches. Paddy fields provide a particularly large quantity of rice husks. A total of 1.5 hundred million tons of rice husk were wasted worldwide in 2014 [2]. In Japan, unutilized agricultural by-products amount to 12 million tons in 2021[3].

Some studies have proposed a rice husk biomass power plant as a rice husk utilization method [4]. However, rice husk ash, which is the powder left after burning rice husks, has

S. Amziane et al. (Eds.): ICBBM 2023, RILEM Bookseries 45, pp. 973–981, 2023.
https://doi.org/10.1007/978-3-031-33465-8_75

not been utilized. The authors have been developing structural materials using "rice husk ash", which is incinerated ash from rice husks, an agricultural by-product. The reason for focusing on rice husks and rice husk ash is that they can con-tribute to resource reuse, energy generation, and improvement of mechanical properties.

Previous studies have shown that rice husk ash improved the durability of concrete through the pozzolanic reaction [5]. This reaction is a chemical reaction between cement and silica. Rice husk ash contains a lot of silica. Our research has proven that rice husk ash is a factor that contributes to the improvement of the durability of cement improved soil [6]. By the authors, development of evaluation method for physical properties of ductile materials, such as cement improved soil include rice husk ash, by using acoustic emission method and related elastic wave methods [7, 8].

However, the environmental friendliness of rice husk ash has been studied less often than its durability. To realize a practical structural material made using rice husk, it is necessary to obtain a stable supply of rice husk ash which imposes a low environmental burden. The process can be explained from at least two phases: rice husk ash production and the production of the material by mixing in the rice husk ash.

In term of the reduction of environmental burdens, this paper analyzes both the carbon dioxide emissions of producing rice husk using a rice husk biomass power plant operated by a cogeneration system as a case study, and the effectiveness in the reduction of carbon dioxide generated in the production of the material by adding rice husk ash. In addition, to clarify the relationship between solidification materials and carbon dioxide emissions in the production of structural materials, we calculate the amount of carbon dioxide emitted during the manufacture of structural materials when three solidification materials (CaO, MgO, cement) are used.

2 Material and Methods

2.1 Rice Husk Biomass Power Plant

In this study, an analytical investigation was conducted on the rice husk biomass power plant actually operated in Tainai City, Niigata Prefecture, Japan. The energy balance was estimated according to input and output of energy from September 21th to November 9th, 2014 (operating hours: 360 h). The input of energy was electricity that powered a gas furnace and pelletized rice husks (65 kWh) and the use of auxiliary fuel (3,000 L). The output of energy was electricity (35,562 kWh) and heat (168 GJ). The input and output of energy were converted to crude oil equivalent and carbon dioxide emissions [9, 10]. The analytical system and appearance of the cogeneration plant are shown in Figs. 1 and 2.

2.2 Method of Estimating CO_2 Emissions by Production of the Cement Improved Soil with Rice Husk Ash

This paper compared the carbon dioxide emissions by producing the improved soil mixed with different kinds of solidification material (MgO, CaO, and Portland cement). Carbon dioxide emissions for each improved soil and concrete are estimated based on carbon

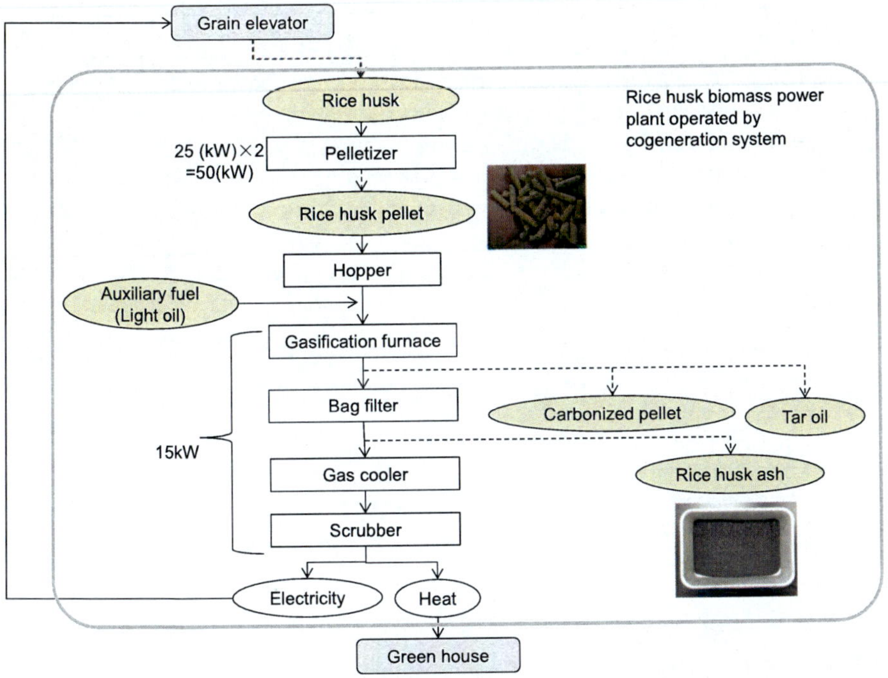

Fig. 1. Diagram of the rice husk biomass power plant operated by the cogeneration system.

Fig. 2. Appearance of gas generation system.

dioxide emissions intensity and each specified mix proportion (Table 1). The interval of rice husk ash replacement ratio is set at 10% as default. In some cases, the interval is set at 15%. This is because concrete has shown a good performance with 10–25% of the cement replaced fly ash which is a pozzolanic material. Specified mix proportion of concrete is set as follow: The maximum size of the coarse aggregate: 25 mm, Slump flow: 10 cm, Air: 5%, Water-cement ratio: 47%, Sand percentage: 35.7%.

Carbon dioxide emissions intensity of MgO is 3.24 kg-CO_2/kg, that of CaO is 1.12 kg-CO_2/kg, that of Portland cement is 0.882 kg-CO_2/kg, and that of aggregates are 0.00565 kg-CO_2/kg [11]. Carbon dioxide intensity are shown in Table 1. In addition, compressive strength test was conducted in the cement improved soil mixed with rice husk ash. There are eight experiment cases (Table 2). Cylinder specimen of 50 mm in diameter and 100 mm in height were made (curing age: 28 days). Based on the compressive test, the carbon dioxide emission per compressive strength is calculated.

Rietveld analysis of X-ray diffraction patterns (XRD/Rietveld). Rietveld analysis is iteratively comparing the experimental pattern with a pattern simulated based on the presumed amounts, crystal parameters, and equipment parameters of a mixture of known phases [12]. In this study, XRD was carried out in θ/2θ scans. The beam source was a long-line focus copper anode. The running conditions were: 40 kV, 30 mA, step-scan mode: 2θ range 5–70°, 2θ step 0.02° with 3 s/step.

Table 1. Carbon dioxide intensity

Classified item	Carbon dioxide intensity (kg-CO_2/kg)
Aggregates	0.00565
MgO	3.24
CaO	1.12
Cement	0.882

Table 2. Specified mix proportion of the improved soil mixed with rice husk ash

Rice husk ash replacement ratio	Soil (kg/m^3)	Solidification material (kg/m^3)	Rice husk ash (kg/m^3)	Water (kg/m^3)
0%	1,783	125	0	166
10%	1,783	112	12	166
25%	1,783	94	31	166
40%	1,783	75	50	166
50%	1,783	62	62	166
60%	1,783	50	75	166
70%	1,783	37	88	166
80%	1,783	25	100	166

3 Result and Discussion

3.1 Environmental Burdens Imposed by the Rice Husk Biomass Power Plant

Figure 3 shows the energy balance in the rice husk biomass power plant. This figure suggests the utilization of rice husk must lighten the environmental burden. The energy produced by using rice husks accounted for about 68% of all energy input. The rice husk biomass power plant produced 15 kl of crude oil equivalent during an approximately 2-month surveillance period. According to an interview survey, the number of annual operating times was 270 days. These results show the power plant generates 50 kl of crude oil equivalent annually. Utilizing rice husks decreased the amount of carbon dioxide emissions by 14 tons. Therefore, rice husks decreased the amount of carbon dioxide emissions by 47 tons in a year.

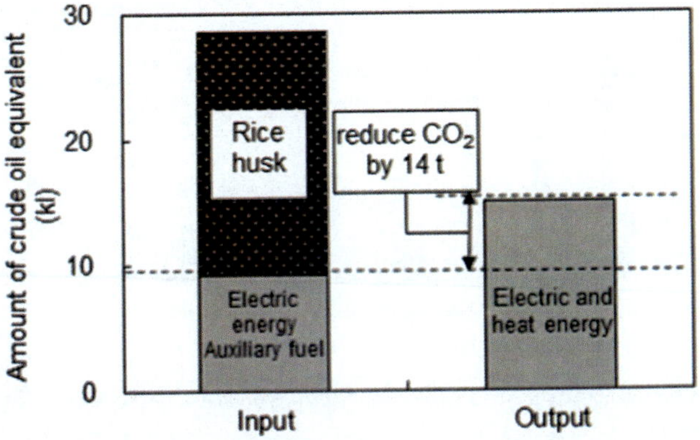

Fig. 3. Energy balance in the rice husk biomass power plant during the surveillance period.

3.2 Environmental Burdens Imposed by Producing Improved Soil Mixed with Rice Husk Ash

Figure 4 shows the carbon dioxide emitted by producing improved soil mixed with different kinds of solidification material. Mixing cement improved soil with rice husk ash decreased the carbon dioxide emissions. The amount of the carbon dioxide emitted by producing MgO improved soil was two to three times larger (200–300 kg-CO_2) than the amount of carbon dioxide emissions in CaO improved soil or cement improved soil. MgO improved soil with rice husk ash replacement ratio less than 25% emitted the same amount of carbon dioxide as concrete. This occurred because MgO is obtained from seawater whose carbon dioxide emissions intensity of MgO is large. These results suggest CaO improved soil or cement improved soil is more appropriate than MgO improved soil as an environmentally friendly material.

Figure 5 shows the compressive strength of the cement improved soil mixed with rice husk ash (curing age: 28 days). Replacing cement with rice husk reduced the compressive strength. The result could have been caused by the decreased cement and the inhibition of the hydration reaction. Compressive strength 0.3 N/mm^2 or less is required for greenery material. The soil with more than 50% of the cement replaced rice husk ash passed the target value. Carbon dioxide emissions per compressive strength are shown in Figs. 5 and 6. The carbon dioxide emissions per compressive strength was largest in the case of the soil with 50% of the cement replaced with rice husk ash. Carbon dioxide emissions per compressive strength in soil with 60 to 70% of the cement replaced with rice husk ash were 30 kg-CO_2/MPa lower than in the case of soil with 50% of the cement replaced with rice husk ash. Our previous research cleared plant roots developed by replacing more than 60% of the cement with rice husk ash [6]. Therefore, the appropriate replacement ratio of rice husk ash is 60 to 70% to develop an agro-infrastructural material with low carbon dioxide emissions and high plant life friendliness.

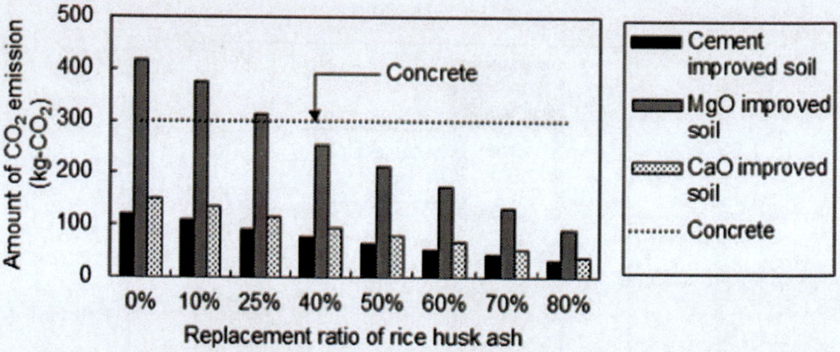

Fig. 4. Carbon dioxide emitted by producing improved soil mixed with different kinds of solidification materials.

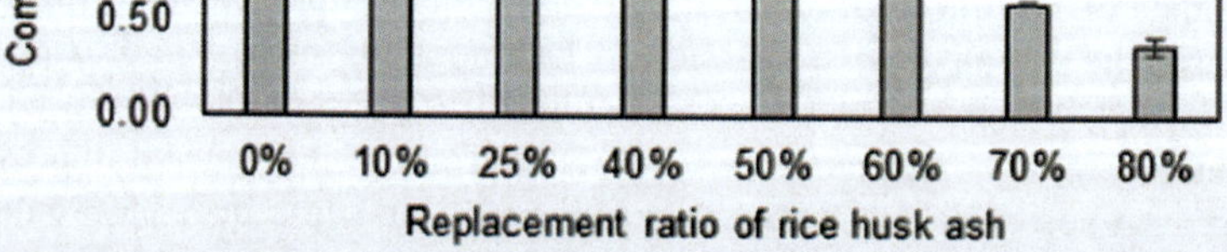

Fig. 5. Compressive strength of the cement improved soil mixed with rice husk ash (curing age: 28 days).

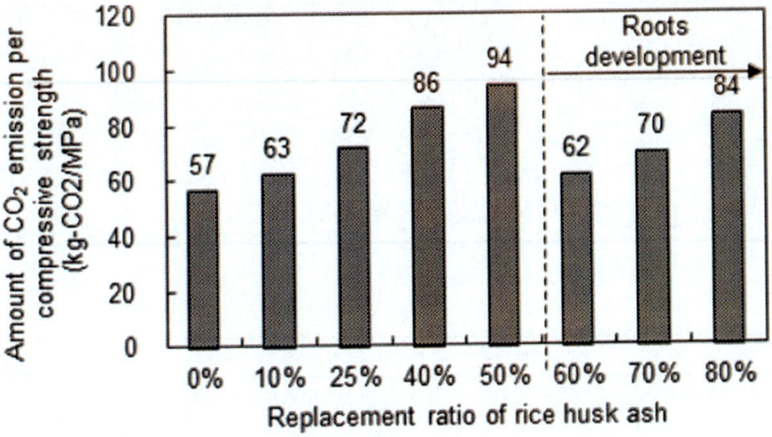

Fig. 6. Carbon dioxide emissions per compressive strength.

3.3 Phase Composition of Rice Husk Ash

The quality of rice husk ash was examined by XRD Rietveld analysis. The X-ray diffraction result of rice husk ash is shown in Fig. 7. The results of Rietveld analysis of the anhydrous phases are given in Table 3. The rice husk ash in the biomass power plant had a smaller amount of amorphous silica than that in an electric furnace. This is probably a result of the higher temperature of the power plant: the electric furnace was set up at 500 °C and the power plant between 500 to 1000 °C. Previous research clarified that amorphous silica was produced at 500 to 600 °C [13, 14]. Pozzolanic reaction in rice husk ash produced in a power plant must be less active.

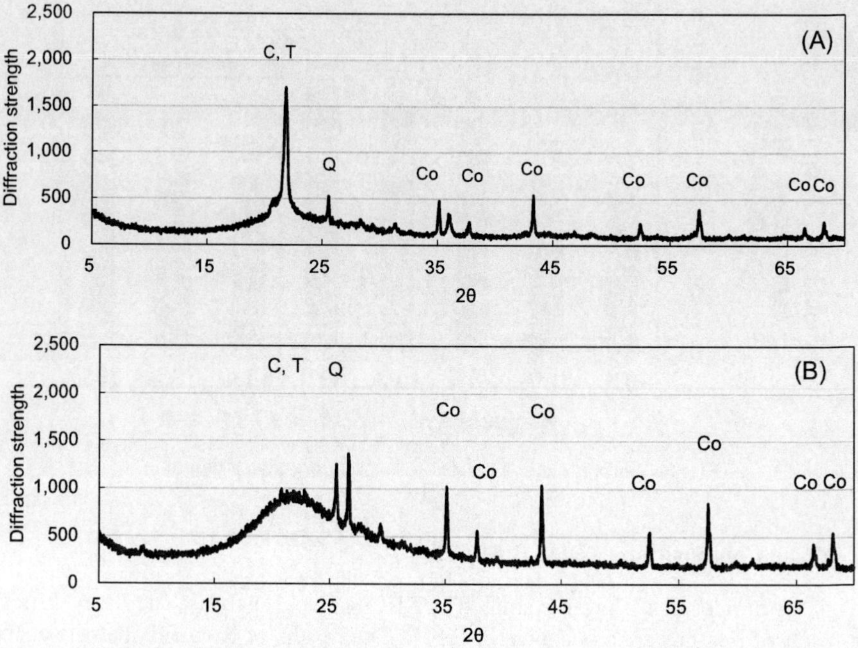

Fig. 7. X-ray diffraction result of rice husk ash. (A) rice husk ash from biomass power plant, (B) rice husk ash fired in an electric furnace at 500 °C. *C: Cristobalite, T: Tridymite, Q: Quartz, Co: Corundum (internal standard material)

Table 3. Phase composition of rice husk ash by XRD/Rietveld analysis

Method	Amorphous silica (%)	Quart (%)	Cristobalite (%)	Tridymite (%)
Rice husk biomass power plant	61.73	0.23	32.83	5.21
Electric furnace	97.83	2.17	0.00	0.00

4 Conclusion

This research presents durable structural materials with plant life friendliness: agro-infrastructural material mixed with rice husk ash. In this paper, the environmental burden of producing rice husk ash was estimated for the case of a rice husk biomass power plant. The utilization of rice husk and rice husk ash lightened the environmental burden. In the future, by constructing a recycling system for rice husks, an agricultural by-product, this research will contribute to the realization of a low-carbon society through the sustainable supply of energy in agricultural regions and the construction of agricultural facilities.

References

1. Nayak, D.K., Abhilash, P.P., Kumar, R.S.R., Kumar, V.: Fly ash for sustainable construction: a review of fly ash concrete and its beneficial use case studies. Clean. Mater. **6**, 100143 (2022)
2. Food and Agriculture Organization of United Nations. http://faostat.fao.org. Accessed 8 Nov 2014
3. Ministry of Agriculture, Forestry and Fishers. https://www.maff.go.jp/j/shokusan/biomass/attach/pdf/index-36.pdf. Accessed 12 Jan 2023
4. Ramchandra, P.: Potential applications of rice husk ash waste from rice husk biomass power plant. Renew. Sustain. Energy Rev. **53**, 1469–1485 (2016)
5. Mehta, P.K.: Properties of blended cements made from rice husk ash. J. Am. Concr. Inst. **74**(9), 440–442 (1977)
6. Shimamoto, Y., Suzuki, T., Morii, T.: Ability of cement improved soil mixed with rice husk ash as greening material. J. Jpn. Soc. Revegetat. Technol. **43**(1), 127–132 (2017). (in Japanese)
7. Shimamoto, Y., Suzuki, T.: Damage evaluation of heavily cracked concrete by initial AE energy parameter. Adv. Exp. Mech. **5**, 122–127 (2020)
8. Shimamoto, Y., Tayfur, S., Alver, N., Suzuki, T.: Identifying effective AE parameters for damage evaluation of concrete in headwork: a combined cluster and random forest analysis of acoustic emission data. J. Paddy Water Environ. (2022). https://doi.org/10.1007/s10333-022-00910-w
9. Ministry of Economy: Trade and Industry, Japan. http://law.e-gov.go.jp/htmldata/S54. Accessed 8 Nov 2014
10. Ministry of the Environment, Japan. http://ghg-santeikohyo.env.go.jp/. Accessed 8 Nov 2014
11. Japan Environmental Management Association for Industry, CFP Database ver. 4.01. http://www.cms-cfp-japan.jp/. Accessed 8 Nov 2014
12. Rietveld, H.: A profile refinement method for nuclear and magnetic structure. J. Appl. Crystallogr. **2**(2), 65–71 (1969)
13. Nair, D.G., Fraaij, A., Klaassen, A.K., Kentgens, P.M.: A structural investigation relating to the pozzolanic activity of rice husk ashes. Cem. Concr. **38**, 861–869 (2008)
14. Savant, N.K., Snyder, G.H., Datnoff, L.E.: Silicon management and sustainable rice production. Adv. Agron. **58**, 151–199 (1997)

Comparative Life Cycle Assessment of Two "Vegetarian Architecture" Pavilions

Redina Mazelli[1](✉), Arthur Bohn[1], Edwin Zea Escamilla[2], Guillame Habert[2], and Andrea Bocco[1]

[1] DIST, Politecnico di Torino, Turin, Italy
redina.mazelli@polito.it
[2] Chair for Sustainable Construction, ETHz, Zürich, Switzerland

Abstract. The goal to achieve carbon neutrality in the built environment by 2040 becomes more challenging every day. Moreover, the challenge goes beyond energy efficiency and embodied energy, and includes the quality of life of buildings' occupants. Under this framework it becomes clear that a radical change is needed in the current construction practices. In this sense, 'Vegetarian Architecture' advocates a closer relationship with construction materials and an increased use of regionally produced and locally available materials and building techniques.

The paper showcases two 'Vegetarian Architecture' pavilions built in Northern Italy: one with a load-bearing straw bale structure and the other with a light-earth infilled timber frame. We detail the local material selection and transportation processes, as well as provide bills of quantities. The comparative Life Cycle Assessment considers production, construction, and operation stages following EN15978, with Ecoinvent 3.8 database and OpenLCA software. IPCC2021 evaluates carbon footprint, and EN 16449 assesses stored biogenic CO_2. The contribution analysis found that as material production approaches carbon zero, transportation becomes more important. However, due to the low carbon footprint of the two pavilions studied, transportation distances could be increased without a significant impact. Additionally, the use of fast-growing, locally sourced, bio-based materials like straw and reed allows the pavilions to store significant amounts of biogenic CO_2, making them carbon negative. Vegetarian Architecture not only promotes alternative materials but also facilitates the development of local value chains and inter-sectoral synergies.

Keywords: LCA · Real scale pavilion · Biobased materials

1 Introduction

The urgent need to reduce carbon emissions and achieve carbon neutrality in the built environment by 2040 has become a global challenge, requiring significant changes in the construction industry [1]. The adoption of sustainable practices and the use of low-carbon materials are key to achieving this goal [2]. Moreover, beyond the challenge of energy efficiency and carbon neutrality, the quality of life of building occupants also needs to be taken into consideration. Thus, a radical change in the construction practices and materials is necessary.

S. Amziane et al. (Eds.): ICBBM 2023, RILEM Bookseries 45, pp. 982–999, 2023.
https://doi.org/10.1007/978-3-031-33465-8_76

The concept of 'Vegetarian Architecture' advocates for a closer relationship with construction materials and increased use of regionally produced and locally available materials and building techniques. Furthermore, Vegetarian Architecture advocates for the use of low-carbon, locally sourced, and renewable materials. It is an innovative approach that showcases buildings based on simple technology, frugality, good craftsmanship, and careful choice of natural materials. A key concept in Vegetarian Architecture approach is the importance of choosing materials that are readily available in the local area, such as straw, reed, timber, and earth-based materials. By using these materials, the carbon footprint of construction can be significantly reduced, and local economies can be supported. In addition to promoting low carbon construction practices, Vegetarian Architecture also focuses on community development initiatives. These initiatives can include landscape restoration projects and community revitalisation programs that create more sustainable and liveable environments.

In this paper we present a comparative Life Cycle Assessment of two 'Vegetarian Architecture' pavilions constructed in the city of Grugliasco, Northern Italy. The first pavilion, light-earth infilled timber frame, and the second, load bearing straw bale, are constructed using local materials, and their transportation to the construction site was carefully considered. Our objective is to show the environmental savings potential of this type of construction and to quantify both carbon footprint and biogenic CO_2 storage of the studied pavilions.

1.1 Life Cycle Assessment (LCA)

Life Cycle Assessment (LCA) is the most established methodology for the calculation of whole life environmental impact [3]. LCA proposes a relation input-output between human activities and the natural environment [3]. An LCA accounts for all the energy, materials, processes and land-use inputs required to produce a unit of a certain product. LCA takes a holistic approach and is built up on existing methodologies like material and energy flow analysis. LCA is described in the ISO norm series 14040 [4] and contains four methodological steps: (i) definition of goal and scope; (ii) development of life cycle inventories, (iii) environmental impact assessment and (iv) interpretation of results. LCA has been used extensively for buildings and construction materials [5] and serves as the base methodology for environmental declaration schemes like Environmental Product Declarations (EPDs) [6] and Product Environmental Footprint (PEF) [7]. Beyond providing information on the environmental hotspots of products and services, LCA produces results aimed to support decision making processes. These can be aimed to optimise production processes and/or to compare alternatives.

1.2 Case Studies

The case studies are two prototypes currently under construction as part of the research project "Experimental Modules of Vegetarian Architecture" developed at DIST (Politecnico di Torino).

The project aims at the construction and monitoring of real scale pavilions following the principles of 'Vegetarian Architecture' – a theoretical stance based on principles learnt from agriculture and nutrition [8], which advocates:

- building with natural, renewable, locally available construction materials, free of toxic chemicals so that they can be absorbed by the environment, and as little processed as possible [9–15];
- minimising the import of energy intensive, high-tech components;
- giving preference to labour-intensive, small-scale production processes, simple constructional technologies, and passive solar design.

The research is expected to lead to a deeper scientific understanding of the construction techniques employed and in particular of the use of bio-based materials through comparative analysis of the field performance of the pavilions, including environmental impact, hygrothermal behaviour, non-toxicity, and durability. It is also expected to highlight data on the construction market in northern Italy, its gaps and its potential to create new synergies aiming at lowering the carbon emissions of the sector.

The chosen building technologies derive from the international panorama of non-conventional constructions that focus particularly on bio-based materials, and more specifically on a meticulous research on ecologically oriented buildings that served as case studies for previous research [8, 16, 17]. Locally available materials have been prioritised. This has brought to light some shortcomings of the construction sector in the region: local and little processed materials are hardly available and challenging to find in a market dominated by conventional, highly processed materials.

The construction activity is carried out through training workshops with the help of local craftsmen, in an attempt to incorporate research with hands-on teaching initiatives. The construction site is located about 9 km west of Turin. Its area is approximately 1700 m^2, of which 450 m^2 are available for building (Fig. 1). There are no obstructions except a shed in the northern edge. Three pavilions have been designed, namely in

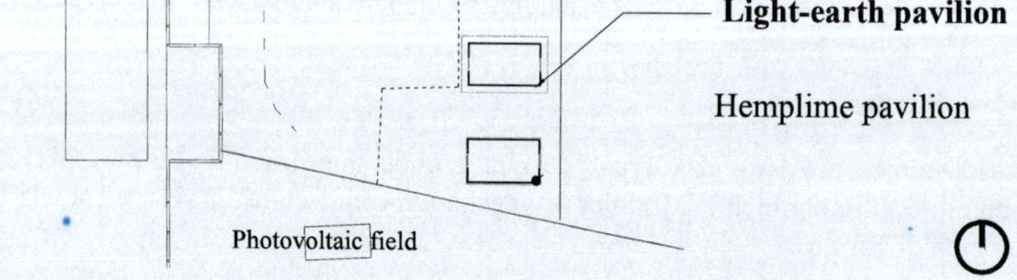

Fig. 1. Construction site plan, scale 1:1000

load bearing strawbale, light-earth infilled timber frame, and hempcrete. They are to be positioned in front of each other in the aforementioned order (north to south), with their longer axis running east-west; on the shorter façades openings will be present. Each pavilion will have a gross internal area (GIA) of approximately 30 m^2. The roofs will have different geometries depending on their building technology, which affects the shape of the volume. No building services are planned except electricity.

Only the first two pavilions are discussed in this paper (Fig. 2), as the third is still under the technical feasibility study [18].

Fig. 2. Sketches of the two case studies (left to right: light-earth pavilion and load bearing strawbale pavilion)

The light-earth infilled timber frame is inspired by the attempts to revive the traditional wattle and daub techniques [19–21] following in particular the studies and the work of the German architect Franz Volhard [22]. The gabion foundation extends 50 cm above ground; the galvanised steel cages are filled with 50–100 mm aggregate, recycled from construction demolition waste at a facility 17 km from the building site. The timber framed structure consists of untreated, local larch wood. The ventilated floor is insulated with 30 cm of reed mats and is enclosed by larch boards. The walls are composed of 15 cm of light earth mixture, whose density varies from 300 to 1000 kg/m^3 depending on the wall's exposure, plus 18 cm of reed mat insulation on the inside. The walls are lime putty and *cocciopesto* rendered on the outer face, and clay plastered on the inside. The build-up of the single-pitch roof echoes that of the wall, i.e., 15 cm light earth mixture plus 5 cm of reed mat insulation on the inside, enclosed by larch ceiling boards. The roofing consists of larch shingles on timber battens that create an 8-cm ventilation gap. The two 4,6 m^2 French doors on the east and west façades are triple glazed; the frame is made of bamboo.

The load-bearing strawbale pavilion is designed based on careful research on strawbale buildings [23–27] and taking as reference the work of the Swiss architect Werner Schmidt [28–30] who was consulted for the project. The above-ground foundation structure is composed of twelve blocks of Luserna stone (a gneiss from approx. 50 km SW of Turin, in the Alps) upon which lays a grid of dry-joint chestnut beams. Two timber boxed slabs filled with organic wheat straw bales constitute the ground floor and the roof: the latter is slightly larger so as to protrude 80 cm all around the building line, and bears a turf layer upon a ventilation gap, obtained by placing 15 cm chestnut joists on the slab. Walls are load-bearing rice straw bales pinned by hazelnut sticks; they are rendered with hydraulic natural lime, while their internal face is clay plastered on a jute fibre mesh. The internal flooring is a lime and *cocciopesto* (crushed bricks or tiles) mix. Two triple glazed French doors of 5 m^2 each are present on the east and west façades; local wood is used for the frame.

Figure 3 presents the sections of the pavilions.

From top to bottom:

Light-earth pavilion.

1_1 cm larch shingles; 4 × 4 cm larch transversal battens; 4x4 cm larch longitudinal battens; 1 cm larch planking; 8 × 20 cm larch beams infilled with larch sticks enrolled with 15 cm light-earth mixture plus 5 cm reed mats; 1.5 cm larch planking.

2_3.7 cm earth plaster; 3 − 5 + 5 + 5 cm reed mats; 8 × 24 cm larch stud infilled with 5 cm light-earth framed by 3 × 3 cm horizontal larch laths; 3.7 cm earth/*cocciopesto*/putty/sand lime render.

3_2.5 cm larch planking; 5 × 15 cm larch bearers infilled with 5 cm reed mats; 5 × 15 cm larch joists infilled with reed mats; 2.5 cm larch planking; 5cm larch anchor planking; 50 cm gabions infilled with recycled aggregate.

Strawbale pavilion.

1_10 cm turf; geotextile membrane; 5 cm expanded clay; EPDM waterproof membrane; 9 × 15 cm larch beams; 2.5 cm larch planking; 10 × 33 cm fir beams infilled with 33 cm low-density wheat straw bales (70 kg/m^3); 2.5 cm fir planking.

2_3.7 cm earth plaster on jute net; 45 cm rice straw bales pinned with hazel sticks; 3.7 cm NHL/earth/sand render.

3_1 + 4 cm NHL/*cocciopesto*/Roman cement flooring; 2 cm fir boards; 10 × 33 cm fir beams infilled with 33 cm low-density wheat straw bales; 2 cm larch boards; 20 × 20 cm chestnut beams; 30 cm Luserna stone blocks.

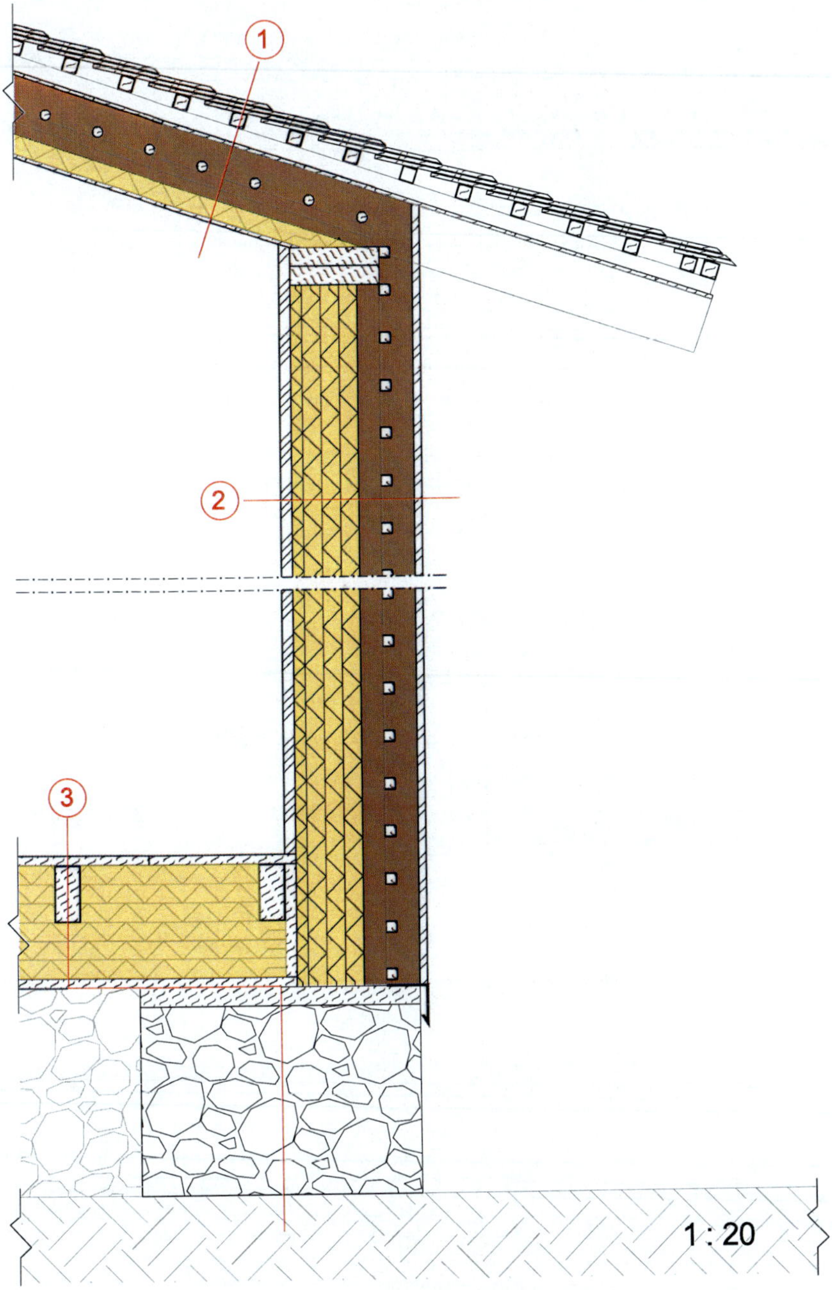

Fig. 3. Sections of the two pavilions, scale 1:20

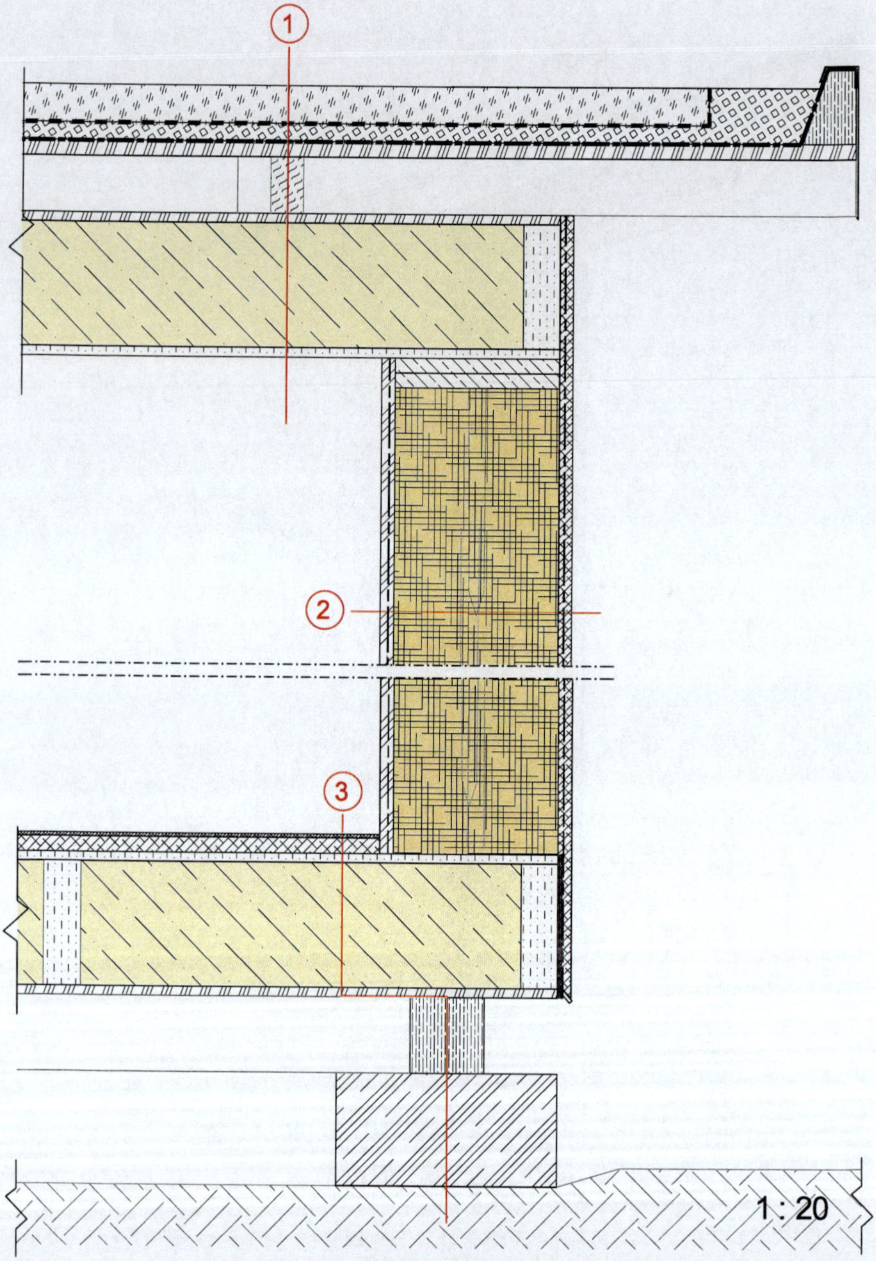

Fig. 3. (*continued*)

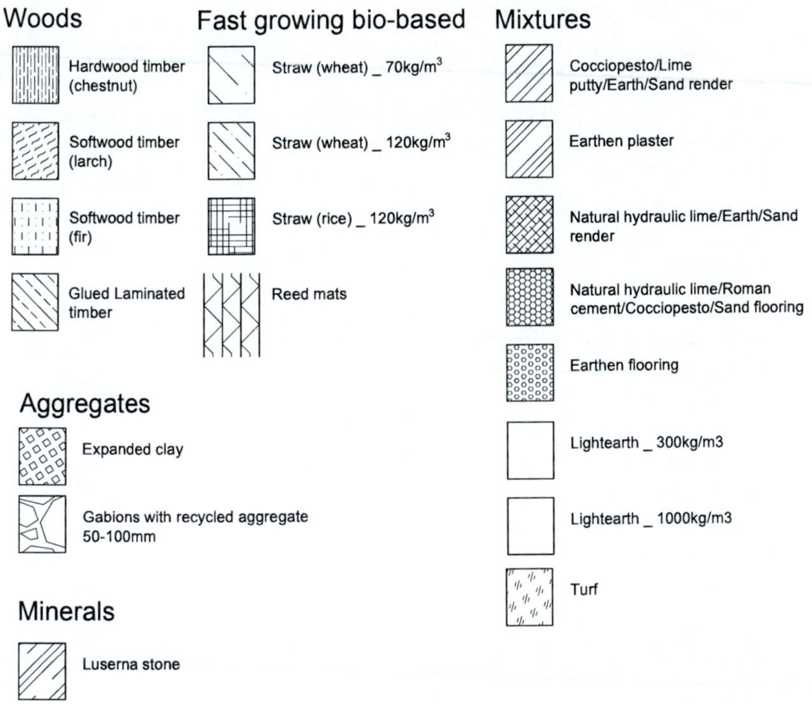

Fig. 3. *(continued)*

2 Data and Methods

2.1 Methods

We will now present the methodological approach used for the comparative life cycle assessment of the studied cases. Moreover, we will describe the method used for the calculation of the biogenic carbon content of the bio-based materials of each pavilion. For the development of the LCA framework of the pavilions we followed the ISO 14040 norm [4]. In this section we present the first three steps while the final step will be presented in Sect. 3. The goal was to assess the life cycle impacts of the two 'Vegetarian Architecture' pavilions in Grugliasco, Northern Italy, that were described in Sect. 1.2. To do so we considered the life cycle stages A1 to A5 as described in the norm EN15978 [31]. These cover the production of the construction materials, their transport from factory to construction site and the erection of the buildings. The life cycle inventories (LCIs) were developed using the bills of quantities and transport distances presented in Sect. 2.2. We used the software OpenLCA 1.11 [32] in combination with the database Ecoinvent 3.8 [33] to develop the LCIs. In this process, we selected datasets representative for the Italian context; when those datasets were not available, we used datasets representing European averages (RER). To calculate the Life Cycle Impact Assessment we selected the evaluation method IPCC2021 [34] which has been implemented for the Ecoinvent database in the openLCA software. For our comparative study we used the category

Global Warming Potential 100a (GWP_{100}). This evaluation method produces results in $kgCO_2eq$, which represent the global warming potential of all greenhouse gases associated to the production of the studied pavilions. To account the biogenic CO_2 content of the pavilions, we followed the methodology described in the norm EN 16449 [35]. For this calculation, specific carbon and water content were defined for each biomass source. In order to better understand the result, a reference model was developed to represent a pavilion utilising conventional construction materials – reinforced concrete and brick.

2.2 Data

The life cycle inventories of the pavilions are presented in Table 1.

Table 1. Life cycle inventories of the pavilions

Material	Light-earth pavilion	Strawbale pavilion
	[kg]	[kg]
softwood timber	8940,0	6571,0
hardwood timber		1450,9
glued laminated timber		310,9
wood fiber		28,1
straw (wheat)	1089,9	1840,0
straw (rice)		3511,2
reed	4320,3	46,1
bamboo	46,8	
jute		39,4
stone		3460,3
earth	29609,8	5160,0
natural hydraulic lime		579,3
lime putty	626,3	
sand	5917,1	8278,8
gravel	1277,9	
recycled aggregate	32500,0	
cocciopesto	1036,1	1514,8
expanded clay		1187,1
turf		1014,7
glass	355,0	340,6
EPDM		126,1

(*continued*)

Table 1. (*continued*)

Material	Light-earth pavilion	Strawbale pavilion
	[kg]	[kg]
fleece		7,9
stainless steel	11,5	
galvanized steel	116,6	

Table 2 presents and overviews the actual transport distances form production site to construction site for all construction materials used in the pavilions.

Table 2. Transport distances from provider to construction site

Material	Distance (one way)	Weight (light-earth pavilion)	Weight (strawbale pavilion)
	[km]	[kg]	[kg]
straw (wheat)[a]	61	324	1840
timber (strawbale pavilion)	84		8265
jute mesh	237		39
reed mats	512		46
rice straw bales[b]	85		3511
sand 0–2 mm	71	1279	1136
sand 0–1 mm	100	350	187
cocciopesto 0–10 mm	261		1212
cocciopesto 0–1 mm	111	1036	552
Luserna stone blocks	57		3460
turf	3		2202
clay	43	29610	5160
NHL5	446		511
sand 0–6 mm	9	4914	6706
root barrier waterproof membrane	315		135
doors	32		451

(*continued*)

Table 2. (*continued*)

Material	Distance (one way)	Weight (light-earth pavilion)	Weight (strawbale pavilion)
	[km]	[kg]	[kg]
recycled aggregate 50/100	17	32500	
wire mesh for gabions	10	117	
timber (solidwood, larch)	84	7920	
wheat straw bales (lightearth infill)	69	766	
reed mat panels	512	4320	
doors	32	402	

[a]Two trips were necessary to transport the material due to the volume: the bales were placed on pallets – easier for storage but occupying more space in the truck (had the bales been placed one by one in the truck, one trip would have been enough)
[b]Two trips were necessary to transport the material due to the volume: the bales were placed in the box one by one, reaching the full volume capacity of the truck

3 Results and Discussion

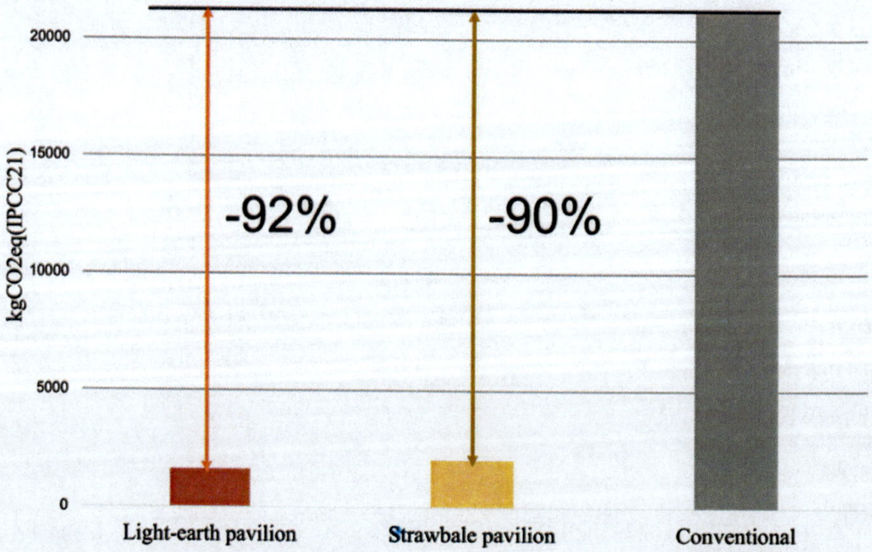

Fig. 4. Carbon footprint and environmental savings potential

The results of the comparative LCA are presented in Fig. 4. From this figure, we can observe that the light earth pavilion withholds the smaller carbon footprint (around 1.6 tCO_2eq) and that the straw bale pavilion's footprint is slightly larger (around 2 tCO_2eq). Nevertheless, both pavilions present a significant environmental saving, 92 and 90% respectively, when compared to the conventional construction reference. This reference represents a "what if" scenario portraying the construction of a fictitious pavilion with same dimensions and energy performance as the Vegetarian Architecture pavilions. It was calculated using the parametric approach developed by Eleftheriou et al. [36], which allowed to carry out the main structural checks for foundation and superstructure, and therefore quantify the material requirements for the foundation, the vertical load bearing structure, and the roof. The insulation material was assumed as XPS.

To better understand these results, a contribution analysis was performed (Fig. 5). From Fig. 5, we can observe that the main contributors to the environmental impact are energy-intensive products like glass and expanded clay. Products and materials that require higher energy inputs in their production inherently contribute more to emissions than low processed materials. In both cases transport (phase A4) makes a relatively significant contribution. This highlights, first, the importance of including these values in LCAs and, second, that the use of materials with a small carbon footprint allows to increase transport distances. Indeed, despite the efforts to choose locally available materials, the providers are located up 512 km from the building site for both pavilions (Fig. 6). For the light-earth pavilion, 75% of material in weight comes from the range of 16–64 km away, while for the strawbale pavilion 68% of the materials is transported from a 32–128 km range. In the first pavilion, for approximately 90% of the weight the provider corresponds to the manufacturer and the raw materials are locally sourced; in

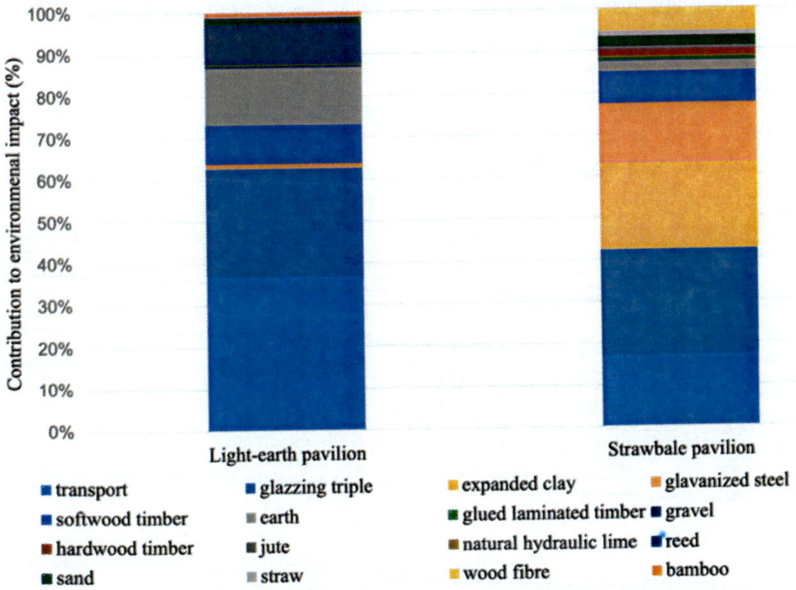

Fig. 5. Contribution Analysis

the second one, 64% of the weight. The remaining materials are either imported or of mixed origin.

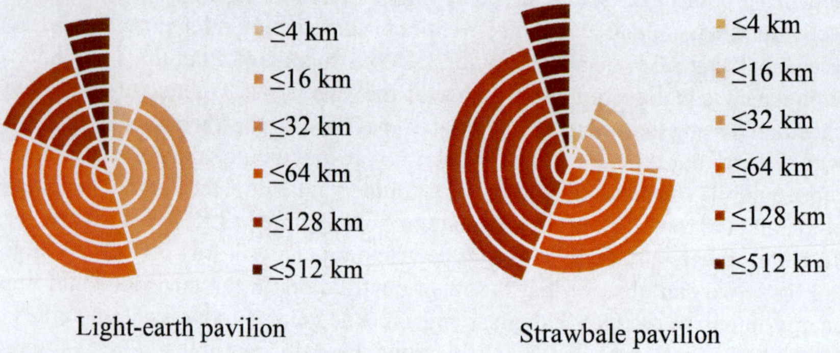

Fig. 6. Concentric charts representing the transport distances from the manufacturer to the construction site and the corresponding quantities of material used in the pavilions.

While in relative terms the source and provision of materials is considerably 'more local' in the case of the light-earth pavilion, this does not correspond to the transport contribution in the overall impact: on the contrary, this impact is approximately twice compared to the strawbale pavilion. This is due to a substantial difference in the weight of the two pavilions (Fig. 7), despite maintaining similar values of gross internal areas and volumes. It should however be noted that this is not reflected in the contribution of the materials impact, showing that the conventional 'lighter is better' assumption can be misleading [37]. The difference in weight between the two pavilions derives from different building technique choices especially for the foundations and the perimeter walls, as shown in Fig. 7. This corresponds to the use of two high-density materials: CDW recycled aggregate for the foundations, and high-density light-earth mixes for the walls, both of which are low impact materials.

The next step was to calculate the biogenic CO_2 stored in the bio-based materials of each pavilion. The results of this calculation are presented in Fig. 8. From this figure we can observe that the bio-based materials used in each pavilion store more than 20 t of biogenic CO_2.

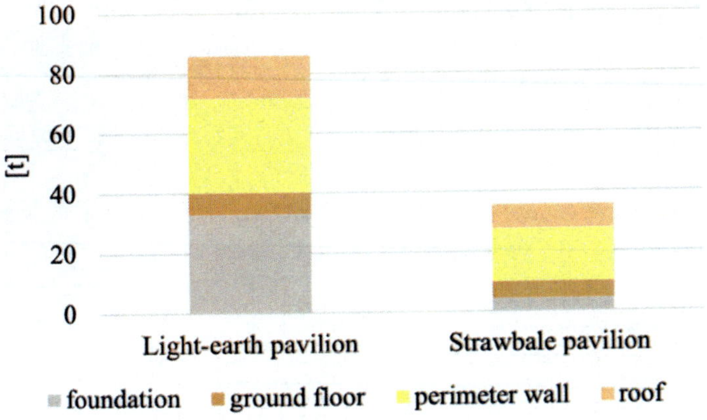

Fig. 7. Weight per building component

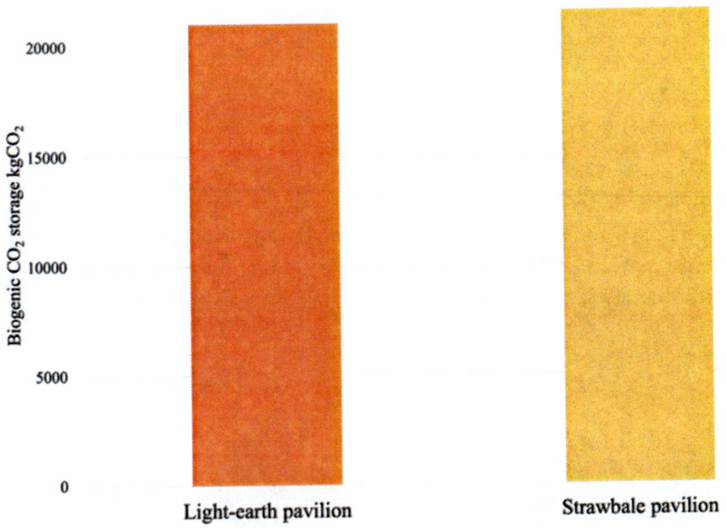

Fig. 8. Biogenic CO_2 storage

The IPCC method assigns an impact factor of -1 $kgCO_2eq$ to for the capture process of CO_2 into biomass of forestry and agriculture. By applying this factor to the results of the biogenic storage we were able to convert these values to the same units ($kgCO_2eq$) used for the carbon footprint. By adding the two results we obtained the CO_2 balance of the pavilions (Fig. 9). From Fig. 9 we can observe that both pavilions withhold a negative CO_2 balance. This means that they not only compensate for the emissions incurred but also that each pavilion stores more than ten times the CO_2 emitted for the production of their materials, their transport and the construction.

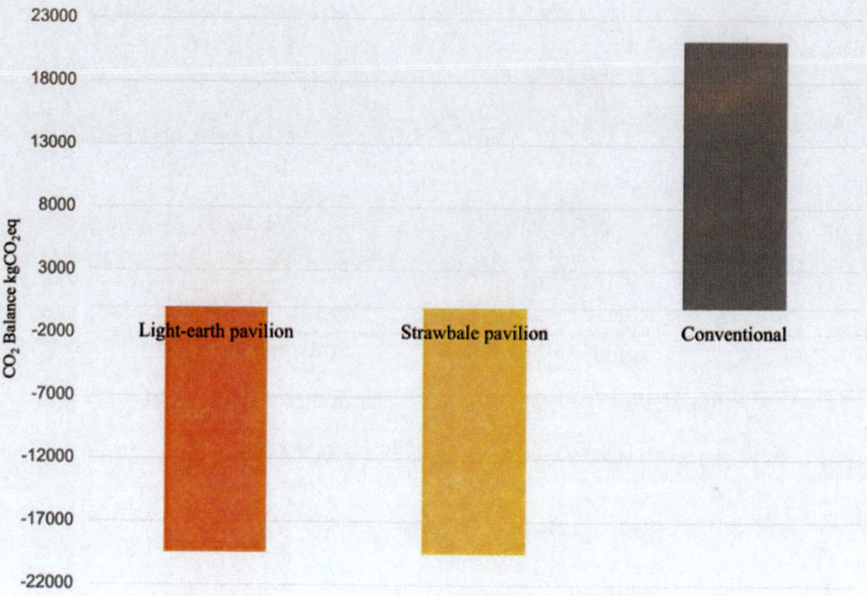

Fig. 9. CO_2 balance

In terms of material choices, this translates into approximately the same weight of bio-based materials used in each pavilion. In relative terms, it corresponds to 17% of the weight in the case of the light-earth pavilion and 39% for the strawbale pavilion. In both cases, approximately 40% of these materials are fast-growing; the rest is timber (Fig. 10).

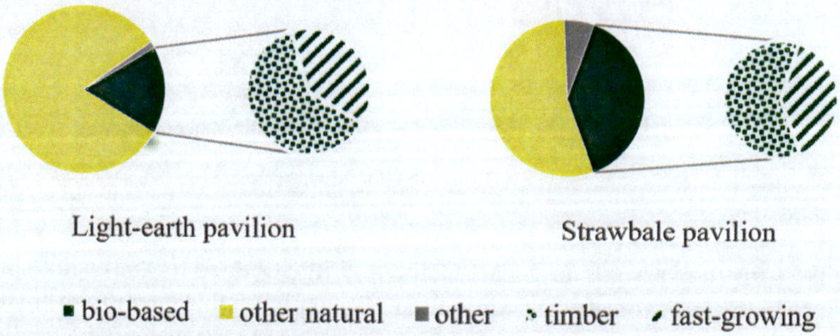

Fig. 10. Building weight contribution by material category (The building materials used were aggregated into three macro-categories: bio-based, other natural and other. The materials falling under the 'other natural' category have mineral origin and have undergone little processing. All materials that have undergone energy-intensive processing and chemical reactions for their transformation are considered as 'other'. The recycled aggregate and *cocciopesto* are classified as 'other natural' since the secondary raw materials, despite being themselves highly processed (fired bricks and tiles, concretes, asphalts etc.), undergo only physical processes (crushing and sieving).)

4 Conclusions

In this paper, we studied the life cycle of two 'Vegetarian Architecture' pavilions for the life cycle stage module A, cradle to handover. The results showed that the use of Vegetarian Architecture principles, and of low processed, local materials allowed for the production of near zero carbon buildings. The environmental saving potential compared with conventional construction was calculated at around 90%. Moreover, the calculations of the stored biogenic CO_2 showed that the studied pavilions can store around 20 tCO_2eq each. This amount is around ten times the CO_2 emitted by the production of construction materials and their transport to the construction site.

It should be noted that the pavilions have an experimental character, and they will be used exclusively for monitoring and for gathering research data. In a scenario of upscaling these building technologies, it should be considered that further components – such as building services – would be needed, leading to an inevitable increase of the overall impact. However, the objective of further research is to check the hygrothermal performance of the pavilions, which is expected to minimise, if not eliminate, the need for heating and cooling. Of course, the passive energy performance could be improved e.g. by optimising the position of the openings, which were not placed so to maximise solar irradiation due to building site constraints.

Another limitation of the experiment is the small GIA of the pavilions, which does not correspond to the current size of residential buildings in developed countries. On the other hand, the results of this analysis reinforce the validity of the principle that a careful choice of materials and building techniques is not enough to reach true sustainability: the floor area per capita should be reduced, or, to put it simply, we should live in smaller houses [38].

It is also important to underline that the achievement of so low an environmental impact of the two pavilions, is, in part, due to the design choices for the foundations. Previous research showed that the foundations and below-grade insulation appear to strongly influence the overall impact, even in buildings otherwise made of low impact materials [16, 39]. Indeed, the foundations of the analysed pavilions are made of either very little transformed material, or of recycled material, zeroing the amount of concrete used and drastically limiting that of steel. However, it should be noted that the pavilions are one storey structures, erected on firm ground and located in a non-seismic area. Further research is needed to investigate the constraints of the building technologies employed and their conformity with building and seismic regulations, as well as their application in taller buildings.

We can conclude that the studied pavilions present a radical solution to the challenge of achieving a net-zero built environment. Moreover, 'Vegetarian Architecture' with-holds a significant regenerative potential, contributing to healing the environment with each new building.

References

1. UNFCCC: United Nations - PARIS AGREEMENT (2018)

2. Salzer, C., Wallbaum, H., Alipon, M., Lopez, L.F.: Determining material suitability for low-rise housing in the Philippines: physical and mechanical properties of the bamboo species Bambusa blumeana. BioResources **13**(1), 346–369 (2018)
3. Hellweg, S., i Canals, L.M.: Emerging approaches, challenges and opportunities in life cycle assessment. Science **344**(6188), 1109–1113 (2014)
4. ISO, ISO 14040: environmental management - life cycle assessment - principles and framework. International Organization for Standardization, Geneva, Switzerland (2007)
5. Bribián, I.Z., Usón, A.A., Scarpellini, S.: Life cycle assessment in buildings: state-of-the-art and simplified LCA methodology as a complement for building certification. Build. Environ. **44**, 2510–2520 (2009)
6. EU-Standards: EN 15804 - Product Category Rules (PCR) for EPD development in the construction sector. EU (2015)
7. E. JRC: Product Environmental Footprint (PEF) Guide. European Commission, Joint Research Centre, Ispra, Italy, Ares(2012)873782-17/07/2012 (2012)
8. Bocco Guarneri, A.: Vegetarian Architecture: Case Studies on Building and Nature. Jovis Verlag, Berlin (2020)
9. Wolley, T.: Natural Building. A Guide to Materials and Techniques. The Crowood Press, Ramsbury (2016)
10. Wolley, T.: Low Impact Building. Housing Using Renewable Materials. Wiley-Blackwell, Chichester (2013)
11. Berge, B.: The Ecology of Building Materials, 2nd edn. Architectural Press, London (2009)
12. Ghavami, K. (ed.): Non-conventional materials and technologies for sustainable engineering (Proceedings of 14th NOCMAT, 24–27 March 2013, João Pessoa). Trans Tech Publications, Dürnten (2014)
13. Walker, P., Ghavami, K., Paine, K., Heath, A., Lawrence, M., Fodde, E. (eds.): 11th International Conference on Non-conventional Materials and Technologies - NOCMAT 2009 Proceedings. University of Bath, Bath (2009)
14. Harries, K., Sharma, B. (eds.): Nonconventional and Vernacular Construction Materials. Woodhead Publishing, Sawston (2016)
15. Harries, K., Papadopoulos, C. (eds.): Proceedings of 18th International Conference on Non-conventional Materials and Technologies (NOCMAT 2022). Zenodo (2022). https://doi.org/10.5281/zenodo.6998159
16. Bocco, A., Bocci, M.: Reflections on the environmental impact of 'Vegetarian' buildings, and on the reliability of databases. Constr. Technol. Archit. **1**, 395–404 (2022). https://doi.org/10.4028/www.scientific.net/CTA.1.395
17. Mazelli, R., Bocci, M., Bohn, A., Zea Escamilla, E., Habert, G., Bocco, A.: Fair play: why reliable data for low-tech construction and non-conventional materials are needed. In: Conference Proceedings CONF.ITECH 2022, Rome (2022, in print)
18. Bohn, A., Mazelli, R., Bocco, A.: Load bearing Hempcrete? A preliminary research. In: Proceedings of 18th International Conference on Non-conventional Materials and Technologies (NOCMAT 2022), June 2022 (2022). https://doi.org/10.5281/zenodo.6631769
19. Volhard, F.: Light Earth Building: A Handbook for Building with Wood and Earth (2016)
20. Marcom, A.: Construire en terre-paille. Terre Vivante (2011)
21. Le projet CobBauge. https://www.cobbauge.eu/en/. Accessed 14 Feb 2023
22. Schauer+Volhard. https://www.schauer-volhard.de/. Accessed 14 Feb 2023
23. Jones, B.: Building with Straw Bales: A Practical Manual for Self-builders and Architects. Green Books, Thrissur (2015)
24. Minke, G., Krick, B.: Straw Bale Construction Manual: Design and Technology of a Sustainable Architecture. Birkhäuser, Basel (2020)
25. Magwood, C., Mack, P.: More Straw Bale Building: A Complete Guide to Designing and Building with Straw. New Society Publishers, Gabriola (2005)

26. Floissac, L.: La construction en paille. Principes fondamentaux, Techniques de mises en oeuvre, Exemples de réalisations. Terre Vivante (2012)
27. Walker, P., Thomson, A., Maskell, D.: Straw bale construction. In: Harries, K.A., Sharma, B. (eds.) Nonconventional and Vernacular Construction Materials (Second Edition), 2nd edn, pp. 189–216. Woodhead Publishing (2020). https://doi.org/10.1016/B978-0-08-102704-2.00009-3
28. Bocco Guarneri, A.: WERNER SCHMIDT architekt: Ecology Craft Invention/Ökologie Handwerk Erfindung. Ambra Verlag (2013)
29. Bocco Guarneri, A.: Architect Werner Schmidt's straw-bale construction. In: KEM, vol. 600, pp. 727–738, (2014). https://doi.org/10.4028/www.scientific.net/KEM.600.727
30. Atelier SCHMIDT GmbH—Atelier Werner Schmidt—Architektur—Truns. https://www.atelierschmidt.ch/. Accessed 14 Feb 2023
31. European Standards: CSN EN 15978 Sustainability of construction works - Assessment of environmental performance of buildings - Calculation method. European Standards (2011)
32. GD: OpenLCA The open source life cycle and sustainability assessment Software (2017)
33. Ecoinvent Association: EcoInvent Database. Ecoinvent Association (2020)
34. Edenhofer, O., Pichs-Madruga, R., Sokona, Y., Farahani, E., Kadner, S., Seyboth, K.: IPCC, 2014: Climate Change 2014: Mitigation of Climate Change. Contribution of Working Group III to the Fifth Assessment Report of the Intergovernmental Panel on Climate Change. Cambridge, United Kingdom & New York, NY, USA (2014)
35. EN-16449: Wood and wood-based products—Calculation of the biogenic carbon content of wood and conversion to carbon dioxide. European Committee for Standardization Brussels, Belgium (2014)
36. Eleftheriou, E., Lopez Muñoz, L.F., Habert, G., Zea Escamilla, E.: Parametric approach to simplified life cycle assessment of social housing projects. Sustainability **14**(12), 7409 (2022)
37. Ruby, I., Ruby, A. (eds.): The Materials Book, 1st edn. Ruby Press, Berlin (2020)
38. Vale, R.J.D., Vale, B.: Time to Eat the Dog: The Real Guide to Sustainable Living. Thames & Hudson, London (2009)
39. Hoxha, E., Habert, G., Lasvaux, S., Chevalier, J., Le Roy, R.: Influence of construction material uncertainties on residential building LCA reliability. J. Clean. Prod. **144**, 33–47 (2016). https://doi.org/10.1016/j.jclepro.2016.12.068

Whole Life Carbon (WLC) Assessment of Building Envelope with Bamboo Bio-Concrete Simulated in Tropical and Subtropical Climates

Arthur Ferreira de Araujo[1(✉)], Mariana Sanches de Proença Franco[1], Carolina Goulart Bezerra[1], Lucas Rosse Caldas[1], Nicole Pagan Hasparyk[2], and Romildo Dias Toledo Filho[1]

[1] Programa de Engenharia Civil (PEC/COPPE), Universidade Federal do Rio de Janeiro (UFRJ), Rio de Janeiro, RJ, Brazil
arthur.araujo@coc.ufrj.br
[2] Furnas Centrais Elétricas Eletrobrás, Rio de Janeiro, Brazil

Abstract. Two main challenges of the construction and building sector are improving energy efficiency and reducing embodied environmental impacts. Low carbon products incorporating bio-based and waste materials are essential to reduce Greenhouse Gas (GHG) emissions from buildings' whole life-cycle. In this context, bio-concretes can be considered innovative and promising solutions that reduce buildings' embodied and operational carbon. However, studies are still needed to better understand how beneficial these materials are when compared to conventional ones and when the Whole Life Carbon (WLC) approach is used. This study aims firstly to quantify bio-concretes production-related GHG emissions by a cradle-to-gate Life Cycle Assessment (LCA). Secondly, bio-concretes application in an office building's internal and external walls and roof is assessed and compared with conventional envelope materials by a cradle-to-grave LCA, where the operational energy use stage is estimated by thermal-energy simulation. Two cities of different Brazilian climatic zones were chosen, Rio de Janeiro (RJ state) and Curitiba (Paraná state), with tropical monsoon (Am) and subtropical highland (Cfb) climates, respectively, according to Köppen-Geiger Climate Classification. Simulations were performed in the DesignBuilder software, in which buildings' annual energy consumption was estimated for room equipment and cooling. BBC production results show negative carbon footprint. Building envelope WLC results indicate that BBC use has the potential to reduce emissions, mainly due to embodied emissions reduction, in comparison with conventional materials. However, design-related parameter variation can considerably influence these results. Finally, this research contributes to the construction sector to achieve low carbon targets and transition towards a circular economy by presenting the evaluation of an innovative and promising material.

Keywords: Life Cycle Assessment (LCA) · Thermal-energy simulation · GHG emissions · Office building · Bio-concretes

S. Amziane et al. (Eds.): ICBBM 2023, RILEM Bookseries 45, pp. 1000–1011, 2023.
https://doi.org/10.1007/978-3-031-33465-8_77

1 Introduction

The construction industry emits pollutants and generates waste on a global scale, significantly damaging the environment [1, 2]. In addition, this sector greatly contributes to natural resource consumption [3].

Regarding greenhouse gas (GHG) emissions, the cement industry is an important source of CO_2 emission into the environment with a share of about 5–7% of the total, and according to an estimate, each ton of cement produced emits the same amount of CO_2 [2] Moreover, buildings were found to be responsible for 37% (27% due to direct and 10% due to indirect emissions) of global energy-related CO_2 emissions in 2020 [4].

The International Panel on Climate Change (IPCC) has investigated different global warming scenarios and "related emission reduction pathways" [5], and in total 80% of the global current carbon emissions would have to be saved by 2040 to ensure a relatively safe transition to a zero-carbon society [6]. In this background, the reduction of environmental impacts throughout the building's life cycle is, therefore, an important target. Since this includes buildings' operational and embodied emissions, the building envelope plays a key role when thermal performance is considered. With the increasing development of low-energy buildings and energy supply systems with low climate impact, the additional reduction potential is mainly found in mitigation production stages, and therefore the focus has been shifting to emissions related to material input [7]. The life cycle assessment (LCA) is a widely used and accepted methodology for assessing the environmental performance of buildings [8].

The LCA is a recommended process for estimating GHG emissions associated with buildings. In addition, it is considered perhaps the most appropriate methodology to quantify the potential environmental impacts of concretes and other construction materials, including new concrete mixtures, as it is a scientifically acknowledged methodology and allows systemic assessment [3, 9–11].

In the context, alternative solutions out of local resources, such as earth, bio-based and reused materials like forest products and waste materials from especially fast-growing plants (straw fibers, hemp, flax, bamboo, etc.) can positively contribute to climate regeneration through the capacity of restoring the carbon from the atmosphere into the land via photosynthesis [12]. Thus, bio-concretes, which partially or totally substitute mineral aggregates for a bio-based material, can be considered innovative and promising solutions that reduce buildings' embodied and operational carbon [13].

In this scenario, Building Information Modeling (BIM) has been seen as an important collaborator and facilitator in the thermal-energy evaluation of buildings and can be integrated with LCA [14]. The digital model provides an overview of the complete project for having the possibility of containing information that supports the construction, manufacture, and supply of necessary inputs. In this way, the design realization also enables to quickly evaluate a wide range of solutions, providing accurate data and minimizing the risk of future mistakes and waste generation [15]. Researchers and practitioners suggest an integration of LCA and BIM during early-stage decision-making in the design process to potentialize the reduction of negative environmental impacts of buildings throughout their life cycle [16].

This study aims firstly to quantify bamboo bio-concrete (BBC) production-related GHG emissions by a cradle-to-gate LCA. Secondly, the BBC application in an office building's internal and external walls and roof is assessed and compared with conventional materials by a Whole Life Carbon (WLC) assessment, where the operational energy use input is estimated by thermal-energy simulation, in a BIM environment, of a three-story office building.

2 Methodology

The WLC assessment was used in this study. In this approach, the LCA methodology is employed, with a focus on the quantification and evaluation of embodied and operational GHG emissions (in $kgCO_2$-eq./functional unit). The LCA is divided into four phases: (a) Definition of goal and scope; (b) Life Cycle Inventory (LCI) analysis; (c) Life Cycle Impact Assessment (LCIA); and (d) Interpretation. The methodology is divided into two parts: (1) GHG emissions of the BBC production; (2) GHG emissions of building envelope with BBC considering a cradle-to-grave scope.

2.1 GHG Emissions of Bamboo Bio-Concrete (BBC) Production

In this stage, the goal was to estimate the GHG emissions of BBC production. Emissions in $kgCO_2$-eq per m^3 of BBC were quantified considering a cradle-to-gate scope: production and extraction of raw materials (A1), transportation (A2), and BBC production (A3), according to EN 15804 + A2 [17]. For the LCI, Ecoinvent v. 3.8 and data from literature were used, considering that most of them came from the Brazilian context, Table 1 details datasets inventory used for BBC production. In the LCIA the EN 15804 + A2 method was used considering the Global Warming Potential (GWP) impact category, where considered fractions were: fossil (GWP_{Fossil}) and Land Use and Land Use Change (GWP_{LULUC}). The biogenic carbon (GPW_{Bio}) for the bamboo particles case was quantified according to the Guest et al. [18] method. A sensitivity analysis was performed for transportation and biogenic carbon, assuming optimistic, intermediate, and pessimist scenarios. More details can be found in Araujo et al. (2022) [19], where very similar approaches and data regarding LCA biogenic carbon were adopted. SimaPro v. 9.4 software was used.

Table 1. Bamboo bio-concrete datasets inventory.

Material/Process	Dataset/Reference
Cement	Cement, limestone 6–10% {BR} I cement production, limestone 6–10% (Ecoinvent v3.8)
Fly ash	Modeled according to Chen *et al.* (2010) [20] (drying and storage processes)
Rice husk ash	Modeled according to Chen *et al.* (2010) [20] (drying and storage processes)
Additive $CaCl_2$	Calcium chloride {RoW} I soda production, solvay process (Ecoinvent v3.8)
Bamboo	Bamboo culm {BR} I bamboo forestry, sustainable forest management (Ecoinvent v3.8)
Alcaline treatment $Ca(OH)_2$	Lime, hydrated, packed {RoW} I production (Ecoinvent v3.8)
Water	Tap water {BR} I market for tap water (Ecoinvent v3.8)
Electricity (for mixing)	Machining processes only of 'Concrete, 25MPa {BR}I concrete production, 25MPa, ready-mix, with cement limestone 6–10%' (Ecoinvent v3.8)
Electricity (for biomass processing)	Electricity, medium voltage {BR-South-eastern grid} I market for electricity, medium voltage (Ecoinvent v3.8)
Transport	Transport, freight, lorry 16–32 metric ton, euro3 {BR} I market for transport, freight, lorry 16–32 metric ton, EURO3 (Ecoinvent v3.8)

2.2 GHG Emissions of Building Envelope with Bamboo Bio-Concrete (BBC)

In this part, the GHG emissions in $kgCO_2$-eq. per m^2 of an office building considering 50 years of service life were considered, assuming a cradle-to-grave scope: materials production (A1–A3), transportation to the construction site (A4), replacements (B4), operational energy (B6), transportation to landfill (C2) and final disposal (C4). In the LCI the Ecoinvent v. 3.8 and data from literature were used and the same LCIA method employed in the first part is used. SimaPro v. 9.4 software was also used in this part.

- For the operational energy (B6) a thermal-energy simulation, in a BIM environment, of a three-story office building (Fig. 1) was performed. The building is modeled in Autodesk Revit and the thermal-energy simulation in DesignBuilder with EnergyPlus simulation machine. The inputs used in the simulation are detailed in Table 2. Two cities with different climates are evaluated: Curitiba (Cfb - Subtropical highland climate) and Rio de Janeiro (Am - Tropical monsoon climate), according to the Köppen-Geiger climate classification. These variations modeled were called scenarios: C9-F, C9-B, C14-F, C14-B, B9-F, B9-B, B14-F, and B14-B. The first letter stands for the type of wall simulated (C – concrete block; B – bamboo bio-concrete), followed by a number that represents the thickness of the wall in cm (9 and 14). The

Table 2. Properties of building elements and materials

Material		e^* (m)	λ^{**} [21, 22] W/(m.K)	ρ^{***} [21, 22] kg/m^3	c^{****} [21, 22] J/(kg.K)	U^{*****}W/(m^2.K)
Walls						
B9	Bio-concrete	0.09	0.38	1118	1200	1.59
	Cement mortar	0.025 (x2)	1.15	1900	1000	
B14	Bio-concrete	0.14	0.38	1118	1200	2.02
	Cement mortar	0.025 (x2)	1.15	1900	1000	
C9	Concrete block	0.09	1.75	2300	1000	2.68
	Cement mortar	0.025 (x2)	1.15	1900	1000	
C14	Concrete block	0.14	1.75	2300	1000	2.78
	Cement mortar	0.025 (x2)	1.15	1900	1000	
Flooring						
Inside flooring	Ceramic tile	0.01	0.80	1700	850	2.99
	Cement mortar	0.05	1.15	1900	1000	
	Concrete slab	0.12	1.75	2300	1000	
Outside flooring	Cement mortar	0.05	1.15	1900	1000	3.11
	Concrete slab	0.12	1.75	2300	1000	
Roofing						
F	Fiber-cement roof tile	0.006	0.95	2200	840	6.84
B	Bio-concrete roof tile	0.03	0.38	1118	1200	4.57

*e – thickness (m)
**λ – thermal conductivity (W/mK)
***ρ – apparent density (kg/m^3)
****c – specific heat (J/kgK)
*****U – thermal transmittance (W/m^{2}K), calculated according the NBR 15220 [22]

last letter indicates the type of roofing (F – fiber-cement; B – bamboo bio-concrete). Considering both cities, the total of scenarios modeled was 16. In summary:

- C9-F: 9 cm thick concrete block wall with fiber-cement roof tile
- C9-B: 9 cm thick concrete block wall with bio-concrete roof tile
- C14-F: 14 cm thick concrete block wall with fiber-cement roof tile
- C14-F: 14 cm thick concrete block wall with bio-concrete roof tile
- B9-F: 9 cm thick bio-concrete wall with fiber-cement roof tile
- B9-B: 9 cm thick bio-concrete wall with bio-concrete roof tile
- B14-F: 14 cm thick bio-concrete wall with fiber-cement roof tile
- B14-B: 14 cm thick bio-concrete wall with bio-concrete roof tile

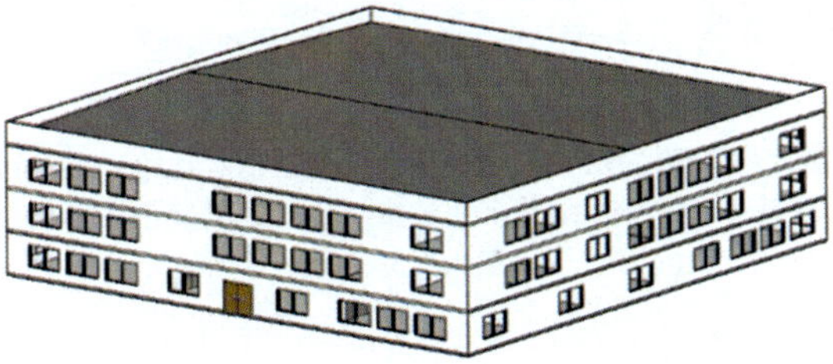

Fig. 1. Three-story office building design evaluated in this study.

3 Results and Discussion

3.1 GHG Emissions of Bamboo Bio-Concrete (BBC) Production

Figure 2a shows the GHG emissions of BBC production in $kgCO_2$-eq./m^3 where the error bar represents a standard deviation of $\pm$43.1 $kgCO_2$-eq./m^3 due to transport and biogenic carbon sensitivity analysis. BBC production average estimated emissions is -113.1 $kgCO_2$-eq./m^3. Such a negative carbon footprint is given mainly due to the significant biomass content (50% by volume) and also due to the partial replacement of cement by SCM, in this case, fly ash (30%) and rice husk ash (25%). In comparison, a BBC (50% of bamboo by volume) with only cement as binder material emits around 133 $kgCO_2$-eq./m^3 [23], whereas for a BBC (45% of bamboo by volume), with binders' proportion of 30:30:40 (cement: metakaolin: fly ash) and with 20% of binders' matrix replacement by processed earth, called earth-based BBC, emissions go down to -142 $kgCO_2$-eq./m^3 [20].

Figure 2b shows materials and processes contribution shares on GHG emissions, where cement is the most impactful input representing 77.1% of positive emissions followed by transport (12.8%), CaCl2 additive, and electricity (both with 2.9%). These findings are in accordance with the previous literature [20, 23]. Biogenic carbon accounts for -435.6 kgCO2-eq./m^3 on average for the optimistic, intermediate, and pessimist scenarios. The biogenic carbon that is removed from the atmosphere via photosynthesis

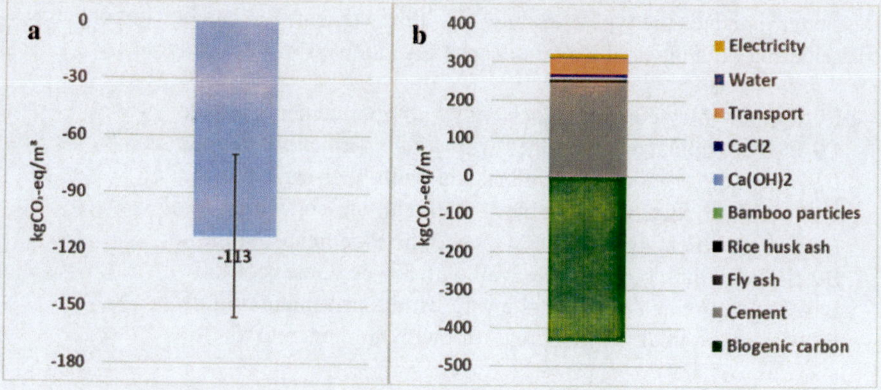

Fig. 2. **a.** GHG emissions of BBC production (error bar represents a standard deviation of transport and biogenic carbon sensitivity analysis); **b.** Materials and processes contribution shares.

is stored in the biomass and mineralized the bio-concrete production, thus it is accounted as negative emission. The by-product ashes integrated into the cementitious matrix have low embodied emissions since they require little processing.

3.2 GHG Emissions of Building Envelope with Bamboo Bio-Concrete (BBC)

This section provides LCA results in kgCO_2-eq/(m^2 of building over 50 years) from both embodied and operational carbon, where operational carbon was thermal-energetically simulated for the cities of Rio de Janeiro and Curitiba, and where BBC application is compared with conventional envelope solutions as presented in Fig. 3. For Rio de Janeiro, 1310.7 kgCO_2-eq/(m^2 of building over 50 years) was found for an average of all scenarios, where embodied carbon accounted for 22.7% of emissions, and operational carbon 77.3%. For Curitiba, a similar pattern was found, with an average for all scenarios of 1122.4 kgCO_2-eq/(m^2 of building over 50 years), being 26.5% of emissions attributed to embodied carbon, and 73.5% to operational carbon.

This emission pattern was expected for a Brazilian office building over 50 years. On the other hand, it was anticipated more significant variations between the studied solutions, in the sense of BBC having better thermal properties than conventional materials. Such small differences are attributed to the little difference between scenarios found in operational carbon simulation, as presented at the end of this section. With that said, for both cities, the most and least impactful scenarios were, C9-F (9 cm thick concrete block wall with BBC roof tile) and B14-B respectively (14 cm thick BBC wall with BBC roof tile).

Embodied Emissions

Taking a closer look at life cycle embodied emissions, as expected, production (A1–A3) and replacements (B4) are the most impactful stages being responsible on average for 48.9% and 46.4%, respectively, as presented in Fig. 4. In contrast with operational stage, greater differences between scenarios with BBC and conventional solutions can

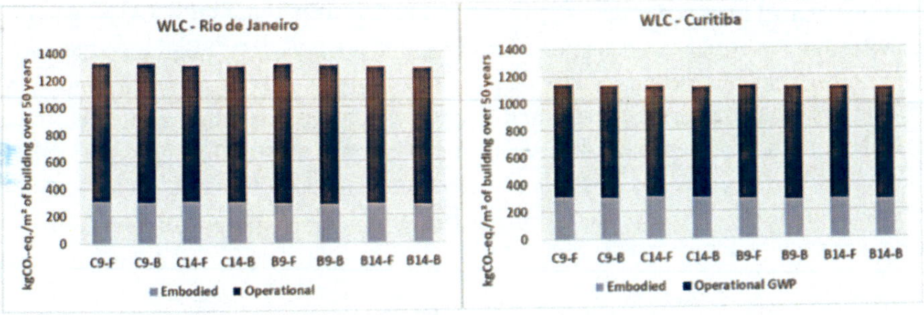

Fig. 3. Embodied and operational emissions' contribution shares.

be identified. The most and least impactful scenarios were C14-F with 313.8 $kgCO_2$-eq/(m^2 of building over 50 years) and B14-B with 280.6 $kgCO_2$-eq/(m^2 of building over 50 years), accounting for 10.6% of reduction.

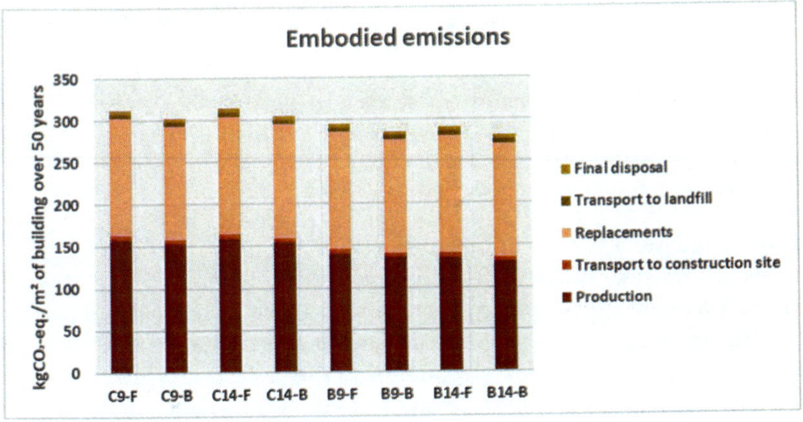

Fig. 4. Life cycle stages' contribution shares on embodied emissions.

Regarding the production stage (A1–A3), Fig. 5 presents the building envelope systems (Fig. 5a) and materials and processes (Fig. 5b) contribution shares, where the flooring system considers not only the ground floor but also other stories' horizontal compartmentalization, which includes reinforced concrete slabs with a cement mortar screed and ceramic tile cladding; and where the walls system considers not only the façades but also the internal vertical compartmentalization with the substrate of each scenario with cement mortar and painting, besides windows and doors. In Fig. 5a, since the flooring system did not vary from one scenario to another: it has the same emission value for all of them, with 66.8 $kgCO_2$-eq/(m^2 of building over 50 years), whereas walls and roofing systems varied according to the solutions studied. Overall, BBC solutions emit less than conventional ones, as expected. This can be evidenced by the comparison of C14-F and B14-B, with a reduction in production emissions of 18.2%.

As presented in Fig. 5b, on average the most impactful materials were as follows: cement mortar (40.2%) used on screed (~3/5) and walls (~2/5); reinforced concrete slab (32.9%), including concrete (19.7%) and reinforcing steel (13.2%); and aluminum (22.1%), used on windows (~9/10) and doors (~1/10). Besides that, BBC walls and roof tile solutions present negative emissions. The emissions reduction due to BBC solutions is more noticeable when this material is applied on the walls due to greater material mass than on the roof tiles (see C9-B, C14-B, B9-F, B9-B, B14-F, and B14-B below-zero values shown in Fig. 5b).

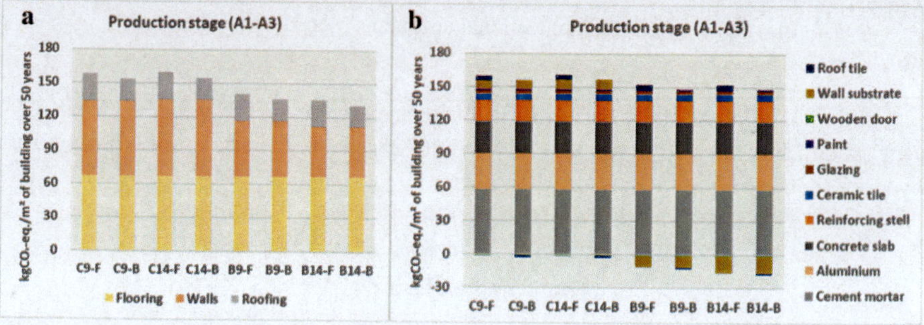

Fig. 5. **a**. Systems contribution shares on production stage; **b.** Materials contribution shares on production stage.

Operational Emissions

This subsection provides the thermal-energy simulation results in kWh/year in Fig. 6, which considers room (computers, equipment, process, etc.) and cooling-related electricity consumption for the cities of Rio de Janeiro and Curitiba. On average difference, Curitiba consumes 18.6% less electricity than Rio de Janeiro. Whereas differences between roof tile solutions were <0.1% and between walls solutions were <1.3% for all scenarios and can be considered negligible. Such small differences can be attributed to the great contribution of room-related electricity consumption on the thermal performance of an office building typology. For the wall solutions, it is important to notice that a concrete block wall is being compared with a BBC massive wall, which influences results in favor of the block, which has air gaps favoring the thermal properties of the element.

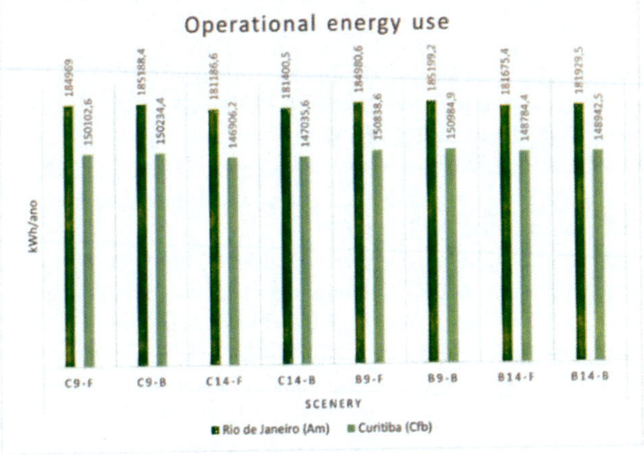

Fig. 6. Annual operational energy use.

4 Conclusions

This study quantified the bamboo bio-concrete (BBC) production-related GHG emissions by a cradle-to-gate LCA. After this, the BBC application in an office building's internal and external walls and roof was assessed and compared with conventional materials by a Whole Life Carbon (WLC) Assessment for two cities of different climates.

BBC production average estimated emissions is -113.1 $kgCO_2$-eq./m^3. A such negative value is attributed mainly due to the accounting of biogenic carbon related to biomass (bamboo particles) content in the mixture (50% by volume). Besides that, the adoption of SCM (fly ash and rice husk ash) in replacement of cement also plays an important role, since cement is the main emitter in BBC production (77% of positive emissions).

Over 50 years, on average for the WLC results, operational energy represented about 75% of emissions whereas embodied emissions, about 25%. Small differences were found between scenarios for the same city. For Rio de Janeiro, the average WLC emission is 1310.7 $kgCO_2$-eq/(m^2 of building over 50 years) and for Curitiba, 1122.4 $kgCO_2$-eq/(m^2 of building over 50 years).

Life cycle embodied emissions came mainly from production (49%) and replacements' stages (46%). In the production stage, greater differences between BBC and conventional solutions can be identified, reaching -18% when comparing a scenario with only conventional solutions (C14-F) with BBC on walls and roofing systems (B14-B). Regarding operational emissions, BBC and conventional solutions present negligible differences, pointing out that both solutions can be used for this building typology. For Curitiba (subtropical highland - Cfb) electricity consumption was about 19% lower than for Rio de Janeiro (tropical monsoon - Am).

For future work, it is recommended to evaluate different climates, building typologies and features (e.g., window-to-wall - WWR ratio, type of glass, shading devices, solar orientation, etc.), and also other types of waste materials in bio-concrete production.

References

1. Leoto, R., Lizarralde, G.: Challenges in evaluating strategies for reducing a building's environmental impact through Integrated Design. Build. Environ. **155**(March), 34–46 (2019). https://doi.org/10.1016/j.buildenv.2019.03.041
2. Ahmad, M.R., Chen, B.: Influence of type of binder and size of plant aggregate on the hygrothermal properties of bio-concrete. Constr. Build. Mater. **251**, 118981 (2020). https://doi.org/10.1016/j.conbuildmat.2020.118981
3. Cabeza, L.F., Rincón, L., Vilariño, V., Pérez, G., Castell, A.: Life cycle assessment (LCA) and life cycle energy analysis (LCEA) of buildings and the building sector: a review. Renew. Sustain. Energy Rev. (2014). https://doi.org/10.1016/j.rser.2013.08.037
4. United Nations Environment Programme UNEP: 2021 Global Status Report for Buildings and Construction: Towards a Zero-emission, Efficient and Resilient Buildings and Construction Sector. Nairobi (2021)
5. IPCC: Global warming of 1.5 °C. An IPCC Special Report on the impacts of global warming of 1.5 °C (Masson-Delmotte, V., et al. (eds.)) (2018)
6. Pittau, F., et al.: Carbon storage project Technical report: methodology for biogenic carbon accounting and carbonation in LCA of buildings and construction products (2022). https://doi.org/10.3929/ethz-b-000554234
7. Weißenberger, M., Jensch, W., Lang, W.: The convergence of life cycle assessment and nearly zero-energy buildings: the case of Germany. Energy Build. **76**, 551–557 (2014). https://doi.org/10.1016/j.enbuild.2014.03.028
8. Zimmermann, R.K., Bruhn, S., Birgisdóttir, H.: BIM-based life cycle assessment of buildings—An investigation of industry practice and needs. Sustain. (Switz.) **13**(10), 5455 (2021). https://doi.org/10.3390/su13105455
9. Gettu, R., Pillai, R.G., Santhanam, M., Basavaraj, A.S., Rathnarajan, S., Dhanya, B.S.: Sustainability-based decision support framework for choosing concrete mixture proportions. Mater. Struct. **51**(6), 1–16 (2018). https://doi.org/10.1617/s11527-018-1291-z
10. Hoxha, E., et al.: Biogenic carbon in buildings: a critical overview of LCA methods. Build. Cities **1**, 504–524 (2020). https://doi.org/10.5334/bc.46
11. Caldas, L.R., da Gloria, M.Y.R., Pittau, F., Andreola, V.M., Habert, G., Toledo Filho, R.D.: Environmental impact assessment of wood bio-concretes: evaluation of the influence of different supplementary cementitious materials. Constr. Build. Mater. **268**, 121146 (2021). https://doi.org/10.1016/j.conbuildmat.2020.121146
12. Pittau, F., Krause, F., Lumia, G., Habert, G.: Fast-growing bio-based materials as an opportunity for storing carbon in exterior walls. Build. Environ. **129**, 117–129 (2018). https://doi.org/10.1016/j.buildenv.2017.12.006
13. Siqueira, T.P.L., Sá, T.S.W., Caldas, L.R., Hasparyk, N.P., Toledo Filho, R.D.: Bio-concretes and GHG emissions reduction: what the scientific literature says?. In: IOP Conference Series: Earth and Environmental Science, vol. 1101, no. 2, p. 022026 (2022). https://doi.org/10.1088/1755-1315/1101/2/022026
14. Anton, L., Díaz, J.: Integration of LCA and BIM for sustainable construction. World Acad. Sci. Eng. Technol. Int. J. Civ. Environ. Eng. **7**(5), 1356–1360 (2014)
15. Franco, M.S.P.F., Araujo, A.F., Caldas, L.R., Hasparyk, N.P., Toledo Filho, R.D.: Avaliação Do Desempenho Termoenergético De Edificações De Escritórios Públicas Com Diferentes Soluções De Parede Externa De Bioconcreto De Bambu (BCB). CONGRESSO LUSO-BRASILEIRO DEMATERIAIS DE CONSTRUÇÃO SUSTENTÁVEIS. Salvador, Brasil (2022)
16. Azizoglu, B., Seyis, S.: Analyzing the benefits and challenges of building information modelling and life cycle assessment integration. In: Ofluoglu, S., Ozener, O.O., Isikdag, U. (eds.)

EBF 2019. CCIS, vol. 1188, pp. 161–169. Springer, Cham (2020). https://doi.org/10.1007/978-3-030-42852-5_13
17. CEN, European Committee for Standardization: 15804:2012+A2:2019. Sustainability of Construction Works - Environmental Product Declarations - Core Rules for the Product Category of Construction Products. CEN, Belguim, Brussels (2019)
18. Guest, G., Cherubini, F., Strømman, A.H.: Global warming potential of carbon dioxide emissions from biomass stored in the anthroposphere and used for bioenergy at end of life. J. Ind. Ecol. **17**, 20–30 (2013). https://doi.org/10.1111/j.1530-9290.2012.00507.x
19. Araujo A., Da Silva N., Sá T., Caldas L., Toledo Filho, R.: Potential of earth-based bamboo bio-concrete in the search for circular and net-zero carbon solutions to construction industry. IOP Conf. Ser.: Earth Environ. Sci. **1122**, 012043 (2022)
20. Chen, C., Habert, G., Bouzidi, J.A., Ventura, A.: LCA allocation procedure used as an incitative method for waste recycling: an application to mineral additions in concrete. Resour. Conserv. Recycl. **54**(12), 1231–1240 (2010). https://doi.org/10.1016/j.resconrec.2010.04.001
21. Andreola, V. M.: Durabilidade ao Intermperismo Natural e ao Ataque Biológico de Bioconcretos de Bambu. Tese de Doutorado. PEC-COPPE-UFRJ, p. 183 (2021)
22. ABNT: Associação Brasileira de Normas Técnicas, NBR 15220 - Desempenho Térmico de Edificações (2005)
23. Caldas, L., Da Glória, M.Y.R., Santos, D., Andreola, V.M., Pepe, M., Toledo Filho, R.: Carbon footprint of bamboo particles, rice husk and wood shavings-cement composites. Acad. J. Civ. Eng. **35**(2) 499–506 (2017)

Comparative Life Cycle Assessment of Rammed Earth Stabilized with Different Biopolymers

Liudmila Lavrik[1], Alessia Emanuela Losini[2](✉), Paola Gallo Stampino[1], Marco Caruso[3], Anne-Cecile Grillet[2], Monika Woloszyn[2], and Giovanni Dotelli[1]

[1] Department of Chemistry, Materials and Chemical Engineering "Giulio Natta", Politecnico di Milano, 20133 Milan, Italy
{paola.gallo,giovanni.dotelli}@polimi.it
[2] LOCIE, CNRS, Universite Savoie Mont Blanc, 73000 Chambery, France
{alessia.losini,anne-cecile.grillet,monika.woloszyn}@univ-smb.fr
[3] Material Testing Laboratory, Politecnico di Milano, 20133 Milan, Italy
marco.caruso@polimi.it

Abstract. Rammed Earth (RE) is a traditional technique based on the compaction of local soil, avoiding long-distance transportation and carbon emissions processes typical of concrete production. Nevertheless, RE presents limited mechanical performances and low resistance against water. Cement is usually added to improve RE properties to comply with the construction standards. Consequently, the environmental impacts increase and the recyclability of the soil at the end of life is lost. In the present study, some biopolymers were considered for RE stabilization: lignin sulfonate, tannins, and sheep wool fibers. These additives were found to be effective to improve the unconfined compressive strength of the RE in a previous study by Losini et al. [1]. In this study, the analysis focuses on the environmental impacts of the production phase of a stabilized rammed earth (RE) wall via Life Cycle Assessment (LCA).

Through a comparative LCA from cradle to gate, the environmental impacts generated by the construction phase of the RE wall stabilized with three natural additives, is performed on a functional unit equal to $1m^2$ wall with a thickness of 30 cm. The unstabilized RE is then compared with the stabilized materials, both in terms of improvements of compressive strength and LCA environmental indicators.

As a result, lignin and sheep wool fibers stabilized walls show almost negligible differences in terms of environmental impact while improving the mechanical properties up to +38% and +6% respectively. Despite the relatively high enhancement of the mechanical strength by +13%, the tannins stabilizer performs around +95% rise in all environmental indicators compared to the unstabilized material.

Keywords: sustainable building materials · rammed earth · bio-stabilizers · recycled materials · Life Cycle Assessment · circular economy · environmental impact

S. Amziane et al. (Eds.): ICBBM 2023, RILEM Bookseries 45, pp. 1012–1032, 2023.
https://doi.org/10.1007/978-3-031-33465-8_78

1 Introduction

The need for sustainable development necessarily involves buildings, as large consumers of non-renewable resources, in particular, energy ones, and large producers of waste. According to the Intergovernmental Panel on Climate Change (IPCC) 6th assessment report in 2022, buildings accounted for 31% (128.8 EJ in 2019) of primary energy produced worldwide and 34% of energy-related CO_2 emissions [2]. In addition, the waste generation due to demolition at the end of life accounts in Europe for about 35% of the total waste generation [3]. Nowadays, considering the shortage of natural resources, coupled with the growing demand for raw materials and the emerging issues generated by construction and demolition wastes, the investigation of new more sustainable materials together with the utilization of wastes and possible recycling after the usage phase has become a principal field of study in civil engineering.

From a wider perspective, it may be claimed that the designing phase of a building itself plays an essential role: the selection and use of sustainable low-impacting materials and a forward-looking strategy for the utilization of the wastes at the end-of-life of the structures can generate tangible economic and environmental benefits and externalities for the community.

The use of sustainable material as raw earth in buildings boasts a millennial tradition, and even nowadays on our planet, about thirty percent of the world's population lives in houses made of earth [4]. In particular, Rammed Earth (RE) is a traditional technique based on the compaction of soil layer by layer into the formwork, where clay is the only binder and soil, and gravels play a structural role. The soil can be provided when excavating the foundation, avoiding long-distance transportation and carbon emissions processes typical of concrete production. Hence, RE construction offers the potential of reusing nonhazardous wastes from excavations, which represents around 7–10 million tons per year in France [5]. RE walls can also contribute to the passive regulation of indoor climate, thanks to its hygroscopic properties given by clay particles, able to store and realize water during condensation and evaporation processes [6, 7]. Moreover, due to the lack of soil can be reused at the end of life, answering to the logic of the circular economy.

Despite the numerous environmental advantages, RE presents a low mechanical strength and a non-negligible sensitivity to water, along with a large variability of material properties due to soil spatial variability [8]. In some cases, cement and lime are usually added to improve RE properties to comply with the construction standards. However, this practice may reduce the hygroscopicity of the material and its recyclability. The ancient practice to use biopolymers, as wastes or by-products with low economic value may be a valid alternative to chemical stabilizers [9]. Recent studies show the large availability of biopolymers and fibers which can be added to earth materials to improve their properties [10, 11]. This practice may highly contribute to the goals of sustainability and the boost of the circular economy, aiding the conservation of resources through recycling and decreasing the consumption of raw materials (Fig. 1).

In recent years, interest in such easily available and eco-sustainable materials have been increased, since the need for sustainable development necessarily involves the building and construction sector, as a large consumer of non-renewable resources, in particular, energy ones, and as a large producer of waste. Nevertheless, the existing

Fig. 1. Rammed earth walls (Festival Autour du pisè, la Batie-Divisin, Rhone-Alpes, France).

expertise and awareness of RE material properties and design procedures are still considerably less than other civil engineering materials like steel, concrete, and timber. In particular, the environmental performance of construction materials is a crucial aspect of the environmental sustainability of the construction sector. For this aim Life Cycle Assessment (LCA) is generally used, a scientific methodology implemented with the aim to create the selection of more environmentally sustainable materials and reduce the overall environmental impact of buildings.

The present research focuses on the LCA analysis of the environmental impacts of the production phase of a RE wall stabilized by three biopolymers (lignin sulfonate, tannins and wool), which were found to be effective to improve the unconfined compressive strength of the RE in a previous study by Losini et al. [1]. The unstabilized RE is then compared with the stabilized materials, both in terms of improvements in compressive strength and LCA environmental indicators.

2 Materials and Methods

2.1 Soil and Wall Composition

In the present research, as a case study, it was considered to use the local soils excavated directly at the construction site. The soil named ABS used for the experimental part of the present study was provided by the company Minerali Industriali S.r.l., it was excavated in Italy in the cave near Lozzolo (Piemonte region, Italy). To obtain an appropriate grain size distribution for RE, the soil was mixed with sand. The sand was washed and transported from the riverbed of Piave river located in Trevis, Italy, provided by the collaborating company Dal Zotto S.r.l. (Veneto region, Italy). The mixture obtained by blending ABS soil (25%) and sand (75%) will be named MIX. More details on the soil composition are reported in a previous article by Losini et al., where the material is experimentally tested [1].

MIX will be called the unstabilized RE, and the environmental impacts of walls stabilized with three different biopolymers, i.e., lignin sulfonate, tannins and wool sheep fibers are compared.

2.2 Additives

The additives were selected on a larger base of biopolymers for earth stabilization, presented by Losini et al. [11] in a review, including 50 independent studies. Three out of five biopolymers initially tested showed good results to improve the compressive strength of RE, as reported by a previous article Losini et al. [1]. The production characteristics of the additives were investigated with the collaboration of the same suppliers which participated in the study by Losini et al. [1].

Lignin Sulfonate. Lignin sulfonate is a lignin-based by-product of the paper and pulp production industry, obtained from sulfite pulping process. Considering the worldwide annual production of lignin in pulp mills is around 50 million tons [12], the main challenge is to manage and dispose of this by-product avoiding the increase of environmental impacts by sending it to recycling or reuse. In the last years, the major part of the lignin-based by-products were straightforward released into the rivers as black liquor or burned [13]. This kind of waste management method not only pollutes the environment but also causes severe waste of natural resources. Given that, to improve the disposal phase of its life cycle, lignin sulfonate can be used as an earth stabilizer.

Over recent years, lignin sulfonate was recognized as a promising soil stabilizer for cohesive and non-cohesive soils since it has the ability to enhance the mechanical strength of clay [13]. According to the previous study presented by Losini et al. [1] RE was stabilized with 1% wt of lignin sulfonate, which showed a reduction of the specific surface area, the possible formation of aggregates and an improvement of compressive strength by 38% in the unconfined compressive strength. Currently, lignosulfonates are applied mostly as concrete additives, dust control on roads, drilling muds, animal feed binders, agricultural dispersants and as potential applications as green fuels (through gasification, pyrolysis, or hydro-liquefaction) [14].

The information on lignin sulfonate used in this study was provided by Burgo Group S.r.l. (Mosaico S.r.l., Vicenza, Italy). The additive was produced in the factory of Tolmezzo located in the province of Udine in Italy. It appears as a brown powder (Fig. 2a) completely soluble in water (>10,000 mg/L) with a typical coffee smell, and a commercial value of around 0.55 EUR/kg.

Tannins. Tannins are the most widespread phenolic compounds in nature after lignin [15]. They are generally obtained by means of hot water extraction method from wood and bark. Tannins are used as natural stabilization aid for beer production or as natural antioxidants to enhance the properties of wine but is also used as retaining of chrome-tanned leathers, antibacterial agents, antifungal, anticorrosion and as a base agent for natural glue production. RE stabilized with 1% wt tannins showed a reduction of the specific surface area, possible formation of aggregates and an increment of compressive strength by 13%, according to a previous study of Losini et al. [1]. Tannins also seem to connect grains and lock particles due to phenolic elements polymerizing and adhering to soil particles [1, 16]. They are also able to form chemical bonds with active sites on mineral surfaces leading to a change of macroscopic behavior [15, 17].

In the present study, the powder of tannins produced by Silvateam S.p.a. was used (a factory in Bergamo located in the province of Lombardy in Italy). Tannins can be obtained from chestnut wood (Castanea sativa) from Italian production, but the quality used by Losini et al. [1] in previous experimental study was extracted from the quebracho tree, a typical plant of Argentina. It appears as a brown powder completely miscible in water. Since the color of tannins is dark brown by adding it in the soil-sand mixture the appearance of the material can be changed and give it a very dark brown color (Fig. 2b). Thus, it depends on the purpose of the application and the preferences because this can be or cannot be desirable. It appears as a brown powder, and it is completely miscible in water. This biopolymer presents a commercial value of around 2–3 EUR/kg.

Wool. 75% of European wool production is categorized as waste from the main function of sheep farming for milk or meat production or other textile industries since sheep need to be sheared for health reasons [18, 19]. It has extremely low economic value or sometimes even negative since the costs of transportation and disposal are required. The keratin present inside the material can improve mechanical properties and resistance to chemical attack, but protective treatments are necessary in order to make it not attackable by insects and parasites [20]. Sheep wool fibers can be used as reinforcement in cement mortars or for the production of insulating panels with excellent hygrothermal and acoustic properties [21–24]. A precedent study by Losini et al. reports that sheep wool fibers added as 0.25% wt to RE improve the compressive strength of RE by 6.10%, connecting the grains by locking them into the fibers network, thanks to the rough surface of the fibers which helps the adhesion with clay particles [1].

(a) (b) (c)

Fig. 2. a) lignin sulfonate powder, b) tannins powder, c) sheep wool fibers.

The greasy wool (Fig. 2c) considered in the study is produced by The Wool Company consortium, from Miagliano Wool Collection and Enhancement Center (Biella, Piemonte region, Italy), where it is provided by the European suppliers, since for them the material does not have any market value and is considered as a waste. These wool fibers have a variable length, from a few millimeters to 10 cm.

Walls Composition for RE. Four types of wall assembly were considered in the study: the unstabilized RE (MIX), the walls stabilized with lignin sulfonate and tannins as 1% of dry matter, respectively abbreviated as LIG and TAN, and RE walls stabilized with 0.25% of wool sheep fibers, WOOL. In Table 1 are reported the main information about the RE wall composition, with the relative compressive strength tested in a previous study by Losini et al. [1]. The previous study recently published reports all the details about the compressive strength test, therefore, here is reported just a resume of the results obtained in Table 1. Although in is article all studied materials' characteristics are not mentioned, more detailed tests on a microstructural characterization and hydrothermal properties were tested and published in the work by Losini et al. [1, 25].

Table 1. The Unconfined compressive strength (UCS) values and its improvement percentages with respect to the unstabilized RE mixture [1].

Mixtures	Biopolymers (%)	UCS (MPa)	Improvement UCS (%)
MIX	-	2.51 ± 0.17	-
LIG	1% lignin sulfonate	3.47 ± 0.60	38.27%
TAN	1% tannins	2.85 ± 0.26	13.26%
WOOL	0.25% sheep wool fibers	2.67 ± 0.17	6.10%

3 Life Cycle Assessment – Environmental Impact

3.1 Goal and Scope

The goal of LCA study is to compare the environmental impacts generated by the construction of RE walls stabilized with three different natural additives. The comparative assessment of the environmental performances is done with respect to MIX.

The study is motivated by a lack of references in academic literature regarding the environmental sustainability of rammed earth materials stabilized with considered natural additives. Indeed, the theory that earth-based materials are environmentally sustainable is largely assessed merely with the use of qualitative methodologies rather than quantitative ones.

The comparative environmental impact assessment of the RE wall construction process is enclosed within the LCA boundaries, which allows identifying the most environmentally sustainable wall production, considering at the same moment the compressive strength of the studied walls.

3.2 Functional Unit and System Boundaries

According to ISO 14040 standard, the functional unit of the analysis is defined as 1 m^2 RE wall 30-cm thick "at the gate", i.e., at the end of the wall manufacturing process on the construction site, that is assumed to be in Biella city (Lombardy region, Italy). The thickness of the wall was defined as the minimum one allowed by international guidelines for RE construction and the current thickness of RE construction in Europe [26, 27]. The choice of functional unit is a logical consequence of the primary function of the system on which the LCA is carried out, namely the production of the final RE wall.

The reference flow for one functional unit of the stabilized RE walls is shown in Table 2 together with a summary of earth-based preparation mixture compositions

Table 2. Analyzed RE walls compositions and corresponding reference flows.

Functional Unit: 1 m^2 wall 30 cm thick			
Wall type	Composition	Reference Flow (kg)	Density (g/cm^3)
MIX	- Soil, - Sand, - Water	885	2.084
LIG	- Soil, - Sand, - Water - Lignin sulfonate	893	2.086
TAN	- Soil - Sand - Water - Tannins	893	2.057
WOOL	- Soil - Sand - Water - Wool	898	2.069

The following process units are considered in the system boundary (Fig. 3):

- Raw materials extraction (excavation of soil and sand)
- Natural additive production (manufacturing phase)
- Transportation of natural additives and sand to the construction site
- RE wall production process

The system boundary of the study can be considered as "cradle-to-gate" approach. Since the main goal of the analysis is to compare the environmental impacts of differently stabilized RE walls with respect to the MIX. The impacts considered as independent from the choice of the mixtures' components and are the same for all analyzed RE walls are excluded from the study (e.g., energy expenditure for erecting the wall formwork).

3.3 Data Sources

Due to the lack of primary data regarding emissions of materials' production processes used in the study, the Ecoinvent 3.5 database was used. The locations of raw material excavation sites and production industries were verified with the suppliers and distances were calculated by the means of Google Maps.

3.4 Life Cycle Impact Assessment

The study has been performed according to the state of art of the LCA methodology, in accordance with the following standards and guidelines:

- EN ISO 14040: 2006 Environmental management – Life cycle assessment – Principles and framework
- EN ISO 14044: 2006 Environmental management – Life cycle assessment – Requirements and guidelines
- EN 15804:2012 + A1:2013 Sustainability of construction works - Environmental product declarations - Core rules for the product category of construction products.

The assessment of the potential environmental impacts of the products and components under study has been performed with SimaPro® 9.0, and the use of the implemented databases (Ecoinvent 3.5 database). Results presented and discussed in the study are calculated according to the Life Cycle Impact Assessment (LCIA) method recommended by EN 15804:2012 + A1:2013. The environmental impact categories addressed by such method are the following:

- abiotic resource depletion potential for elements (ADP);
- abiotic resource depletion potential of fossil fuels (ADPf);
- acidification potential of land and water (AP);
- eutrophication potential (EP);
- global warming potential over 100 years (GWP);
- depletion potential of the stratospheric ozone layer (ODP);
- formation potential of tropospheric ozone photochemical oxidants (POCP).

4 Life Cycle Inventory

4.1 Wall Production Process

Soil and sand are excavated from the quarries (locations mentioned in Sect. 2.1) and after processing are stored at the construction site. Once the stabilizing additive arrives at the production company gate, it is imported at the storage in the required quantity. From the storage place, the components of RE wall are sent to a blending machine, where they are mixed with water. The energy consumption of the blending machine is assumed to be 0.776 kWh per cycle. With one cycle it produces 0.6 m^3 of earth-based mixture [28]. After mixing, according to the rammed earth technique considered in this study, the material is poured into formwork and is rammed with a backfill tamper. A tamper is an air-powered tool, driven by compressed air supplied by an air compressor. According to the on-site earth compaction recommendations [29] in order to compact

1 m^2 wall with a thickness of 30 cm it is needed 20 min of pneumatic rammer work (32 cfm according to the datasheet of average backfill tamper), that means it requires 18.13 m3 of compressed air. During the production cycle, there is no waste produced since all the residues are used as input materials and all packaging is sent to be refilled. The production process of stabilized RE wall is summarized in Fig. 3.

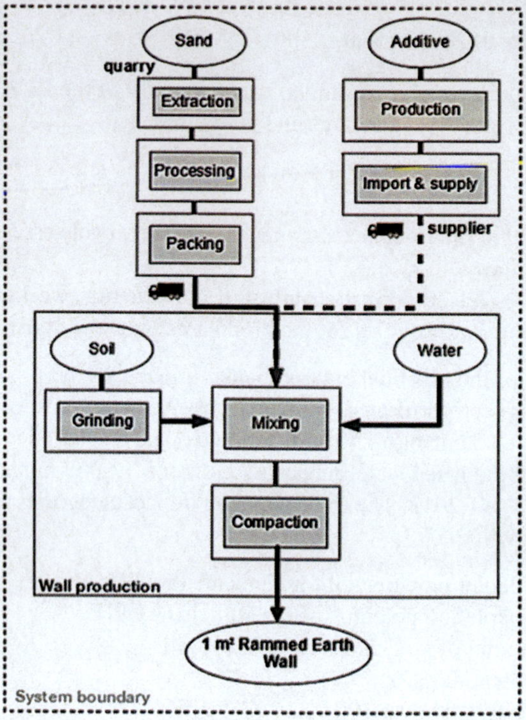

Fig. 3. Process scheme of RE wall production.

4.2 Lignin Sulfonate Production

At the first stage of lignin recovery, the precipitation process includes an injection of liquid CO_2 in the black liquor with the purpose of decreasing its pH level. Lignin precipitates partially and a small quantity of H_2S is also released and further neutralized in the non-condensing gas scrubber that uses NaOH in excess to assume stoichiometric conditions [30].

In the subsequent steps, since the precipitated lignin is wet, it requires to be filtered, washed, and dried. The filtrate (lignin depleted black liquor) from the filtration step is recycled in the multi-effect evaporators. At the washing step of lignin, the water and H_2SO_4 are inputs of the process. Hence, the diluted washing effluent contains excess H_2SO_4 and further is neutralized at the wastewater treatment plant, by rising the lime mud rate, in its turn rising $CaCO_3$ purchase [30]. The main steps of lignin recovery can be seen in Fig. 4.

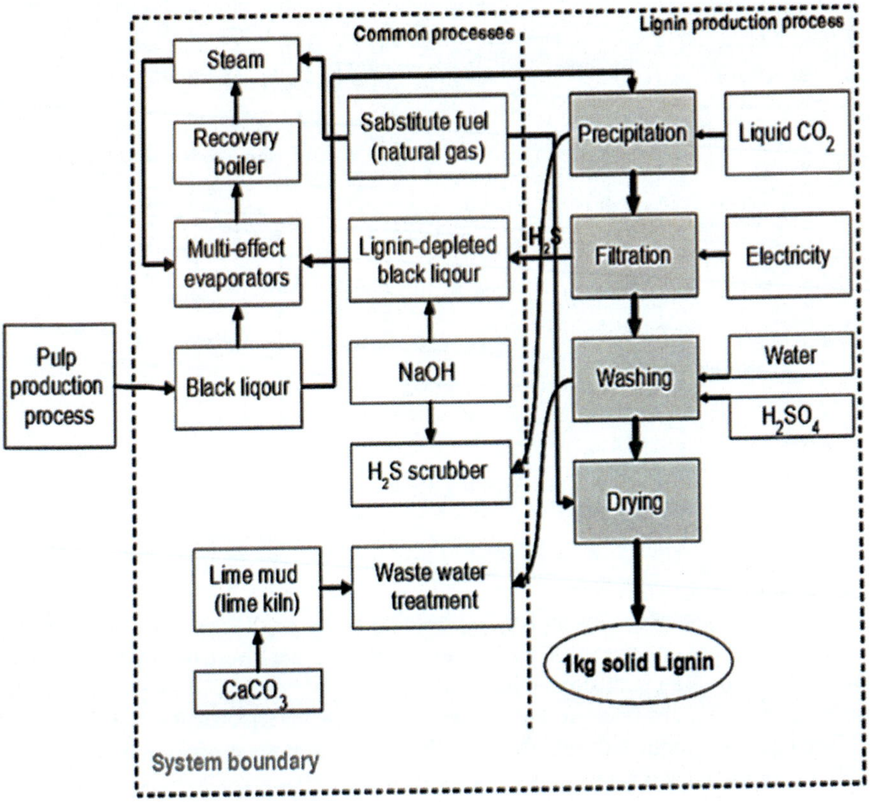

Fig. 4. Process scheme of 1 kg solid lignin sulfonate production, inspired by [30].

4.3 Tannins Production

As one of the possible tannin extraction scenarios at the material supplier plant was considered a hot water extraction method. The environmental impact is analyzed by combining one step of hot water extraction with evaporation and spray drying. For the extraction process with hot water, a milled bark powder is blended with diatomaceous earth and filled into an ASE extraction cell. Tap water is added and heated to 90 °C within five minutes. After 20 min at constant temperature and pressure (10 MPa), the extract is poured into the collection vessel. The resulting liquid extracts are concentrated first by an evaporator to 30% (w/w), and then by spray drying into the solid phase [31]. The tannins extraction process can be seen in Fig. 5.

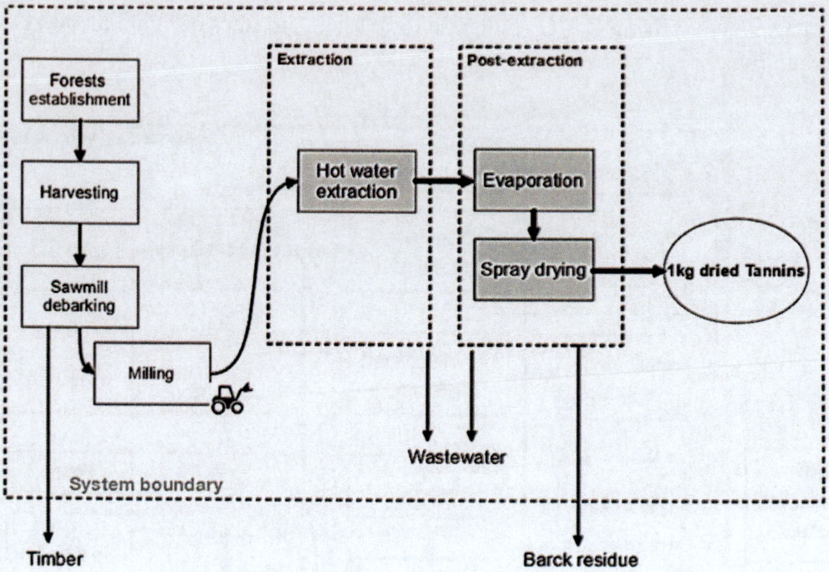

Fig. 5. Process scheme of 1 kg dried tannins production, inspired by [31].

4.4 Wool Treatment

The greasy wool comes from suppliers all over Europe for whom it does not have any market value and is considered a waste. Hence, the allocation of upstream processes between the high value industries' outputs and the low or negative wool values are performed on an economic allocation basis. The sheep wool material is allocated at zero and the more valuable products of industries 100% allocated to the upstream interventions.

Following the cut-off in attributional approach, the only burden assigned to the Wool Collection Centre may be the treatment process, including scouring and rinsing (Fig. 6). Wool in most cases comes dirty and contaminated with natural substances that must be removed before processing by means of the scouring process. Thus, the raw wool passes through four bowls until at the end it is rinsed in another tank of four bowls containing clean water [32]. Then it is dried to about 20% moisture content.

4.5 Transportation

It is essential to account for transportation, that is responsible for an extremely huge part of the cumulative energy demand. For example, kraft mills that produce lignin sulfonate are frequently distant from consumption centers or construction sites, that they tend to rely on trucking. The simplification considered for this study was to assume that all material suppliers are in Italy, as well as the construction site. Packaging transportation is also included.

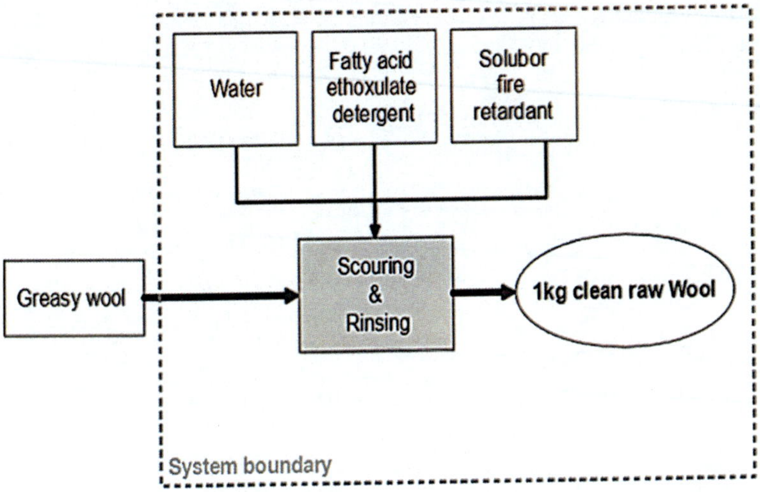

Fig. 6. Process scheme of 1 kg clean raw wool production, inspired by [32].

4.6 Packaging

It is assumed that the natural additive materials and sand come out packed in bags with a capacity of 1 ton and made of polypropylene (160 gr/m^2 density and dimension 55.6 × 54 × 20 cm), further it is transported by trucks (16–32 metric tons, EURO 4) to the construction site where the soil is excavated and dried directly without packaging.

5 Results - Environmental Impact

Since the wall production phase and MIX components are the same for all studied RE walls, only the choice of additives affects significantly the overall environmental impact of the studied RE walls. The results of the comparative LCA analysis of RE walls are reported in Table 3 and Fig. 7, where the environmental impact categories comparatively represent impacts related to each stabilization wall type.

According to the results, TAN has the highest environmental impact in all analyzed categories (increase by 93% using tannins compared to MIX), while WOOL environmental performance is almost equal to MIX (increase by 7%). LIG has an impact increase in all categories on average by 57% respect to MIX.

Table 3. LCA results of RE walls production for the seven environmental categories (EN 15804 (Calculated according to the EN 15804:2012 + A1:2013 method (FU: 1 m^2 of a 300-mm thick RE wall).)).

Impact category	AP	EP	GWP	ODP	POCP	ADP	ADPf
Unit	kg SO2eq	kg PO4-eq	kg CO2 eq	kg CFC-11 eq	kg C2H4 eq	kg Sbeq	MJ
MIX	0.204	0.037	45.40	7.15E-06	0.0082	1.15E-04	753
LIG	0.458	0.083	113.0	1.81E-05	0.0184	2.72E-04	1780
TAN	6.040	0.450	1360	1.24E-04	0,256	4.77E-04	17200
WOOL	0.220	0.040	49.30	7.81E-06	0.0088	1.24E-04	812

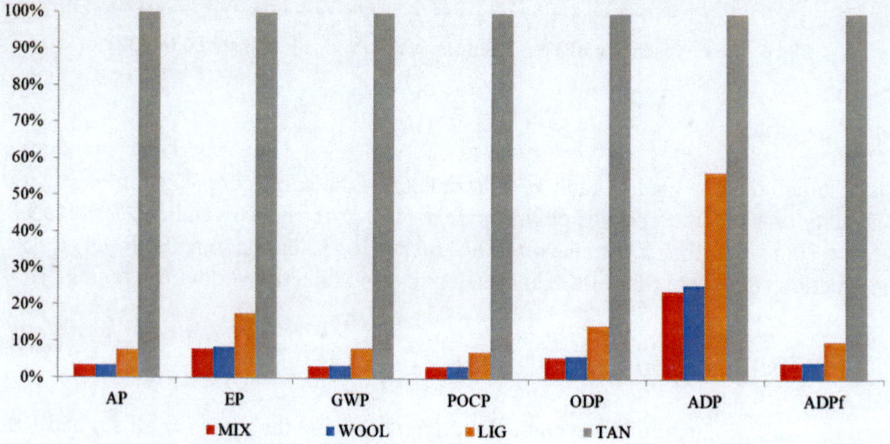

Fig. 7. Comparative LCA results for four wall types (here, the impact of tannins is assumed to be 100% in all categories).

5.1 Rammed Earth Wall

The LCA analysis of MIX showed that its environmental impact is mainly due to the transport and production of the mixture components. Namely, the supply and excavation of sand from the quarries. Figure 8 shows that in all impact categories, mainly the transport and less sand production generate the main impact flow (approximately 98% contribution).

5.2 Lignin Sulfonate

The results of LCA study in Fig. 9. LCA results of 1m2 LIG production. Shows that for most of the impact categories (ADPf, POCP, GWP, and AP) the impact of LIG has been increased in comparison to the MIX. However, the highest increase of LIG impact can be noticed in ADP category due to the use of chemicals in the production phase.

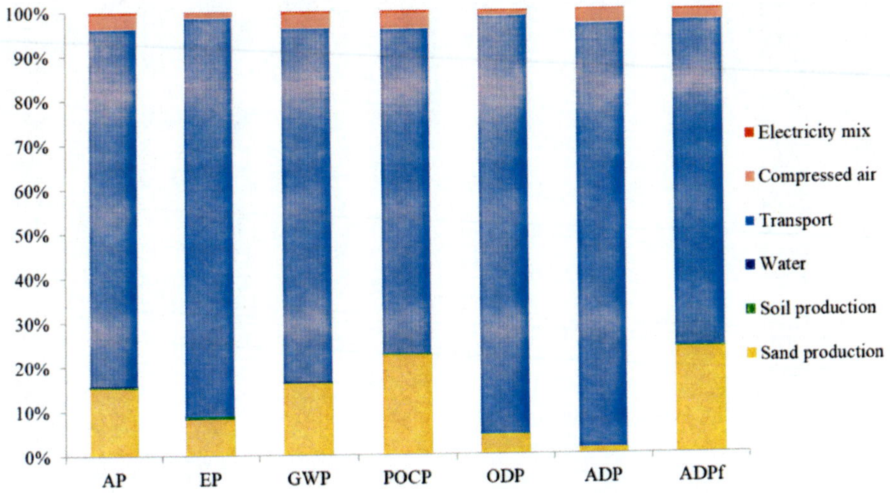

Fig. 8. LCA results of 1m^2 MIX production.

In Fig. 9 the main processes contributing to the environmental impact of LIG are the transport and production of lignin sulfonate. The trend is the same as for MIX, in all impact categories, the transport of additive generates the main impact flow, resulting in the main reason as well for increased impact in ADPf abiotic depletion (fossil fuel).

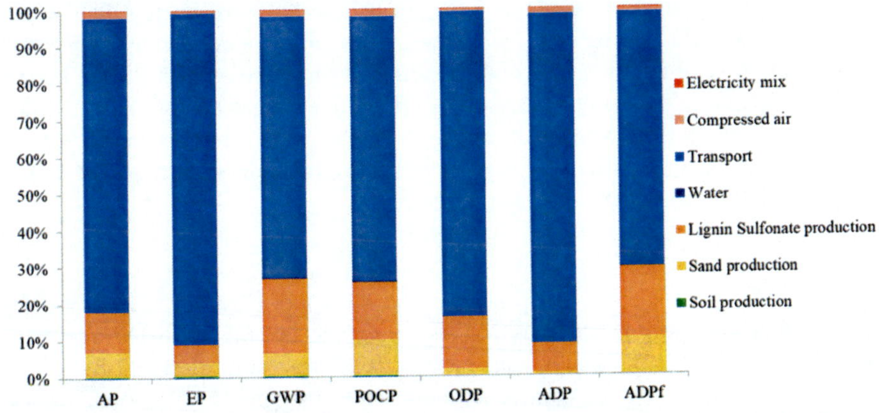

Fig. 9. LCA results of 1 m^2 LIG production.

The production phase accounts for about 14% of total impacts in almost all categories due to the use of chemicals and steam production (based on natural gas). The production of liquid CO_2, H_2SO_4 and NaOH represent more than 50% of the overall impacts in GWP, ADP, ODP, POCP categories, mainly because of electricity consumption (CO_2 and NaOH) and process SO_2 emissions (H2SO4) (Fig. 10).

In order to decrease the environmental performance of LIG production, potential improvements can be applied in order to reduce the emissions from transport, namely,

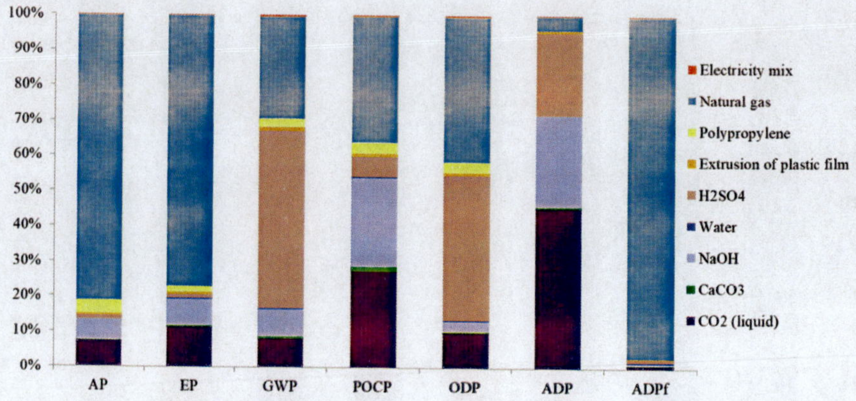

Fig. 10. LCA results of 1 kg of lignin sulfonate production.

the location of a supplier company providing the additive should be located closer to the construction site.

5.3 Tannins

As it is seen in Fig. 11, TAN has the main impact contributors which are the tannins production (around 99–95% in all impact categories) and the transportation of additives to the construction site (around 5-1% in all impact categories). At the stage of tannins production, evaporation and extraction are indicated as the most energy demanding process since most of the water contained in the extracts must be evaporated. Moreover, the high energy consumption impact of the production stage is mostly due to the Italian residual electricity mix, where the main electricity source is natural gas (63.61%) [33]. The key to improve environmental performance seems to have a more sustainable source of heat for water evaporation, like the waste of bark residue may be burned and the gained energy used within the system for evaporation [31].

5.4 Wool

The environmental performance of WOOL is almost the same as for MIX. The main activity accounted for in the analysis and differentiated the results of the assessment is the treatment phase of the wool, namely scouring, which includes the wool washing and cleaning with a big amount of warm water needed to be heated up before the treatment by means of natural gas, that can be observed in Figs. 12 and 13.

5.5 Functional Unit as 1 MPa of Compressive Strength

Functional unit is a benchmark of the functional performance gained as dimensionless ratios of selected characteristics of compared products. For the study of earth-based construction material using the unit of volume may not always be the most suitable option since commonly it is not the primary function of the construction materials.

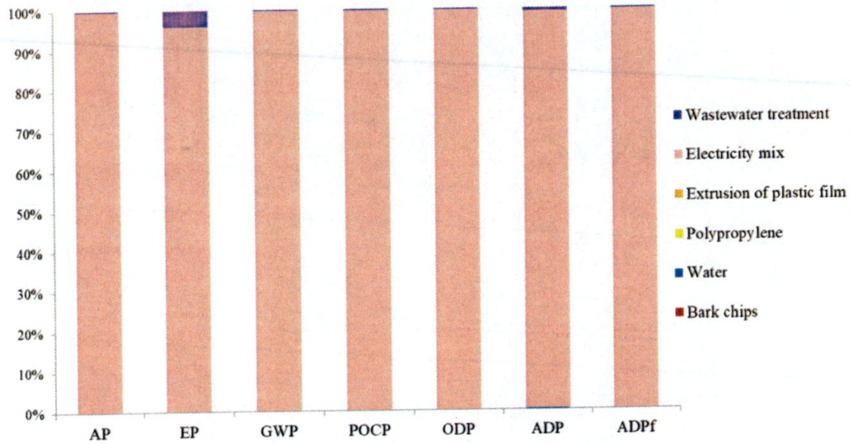

Fig. 11. LCA results of 1 m^2 TAN production.

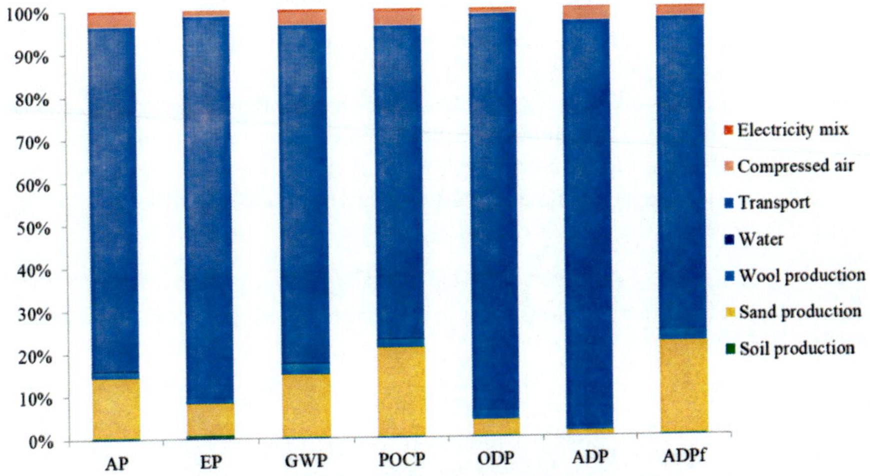

Fig. 12. LCA results of 1 m^2 WOOL production.

Instead, the compressive strength of the wall construction material is frequently seen as a functional requirement. The function of RE is the function of the wall made of it, and it depends on the wall structure in which earth mixture is applied [34]. Hence, the comparison of the environmental impacts of different RE walls is conducted on the base of volume (as it is represented above) and per unit of 1 MPa compressive strength.

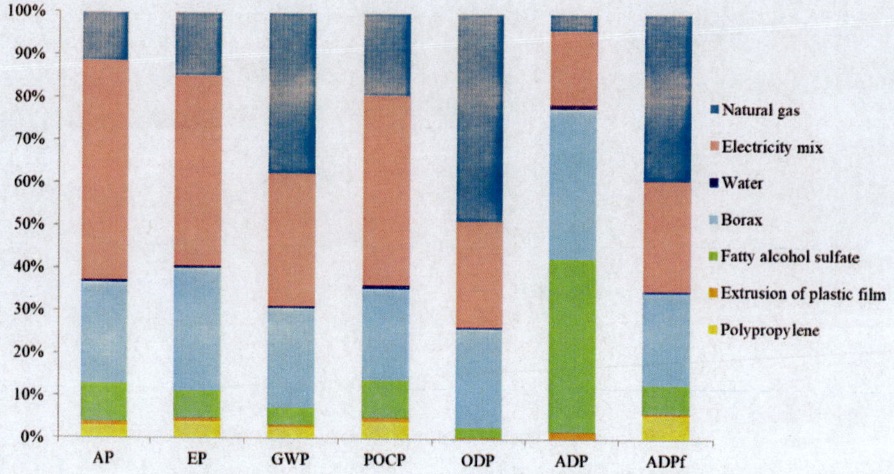

Fig. 13. LCA results of 1 kg of WOOL production.

Table 4 shows the LCA results of environment impact assessment per functional unit as 1 MPa of each analyzed walls, while Fig. 14 gives more representative view of the impact in each category on the graph. So, it can be clearly noticed that in both chosen functional units, the main highlighted tendency remains. That is, the tannins stabilized wall consistently has the highest impact values in all categories, whereas the wool and lignin sulfonate stabilizers maintain slightly higher impacts per 1 MPa compressive strength with respect to the unstabilized RE.

Table 4. LCA results for the 7 environmental impact categories per unit compressive strength (1 MPa).

Impact category	UCS	AP	EP	GWP	ODP	POCP	ADP	ADPf
Unit	MPa	kg SO2 eq/1MPa	kg PO4-eq/1MPa	kg CO2 eq/1MPa	kg CFC-11 eq/1MPa	kg C2H4 eq/1MPa	kg Sb eq/1MPa	MJ/1MPa
MIX	2,51 ± 0.17	81.4	14.9	18.1	2.85	3.25	45.7	300
LIG	3.47 ± 0.60	132	23.8	32.7	5.21	5.31	78.3	512
TAN	2.85 ± 0.26	2120	158	477	43.5	89.7	167	6040
WOOL	2.67 ± 0.17	82.4	15.0	18.5	2.92	3.28	46.4	304

6 Discussion

A comparative analysis of LIG, TAN, WOOL in relation to MIX has been conducted, with a focus on the environmental performance and consideration of the material's mechanical strength.

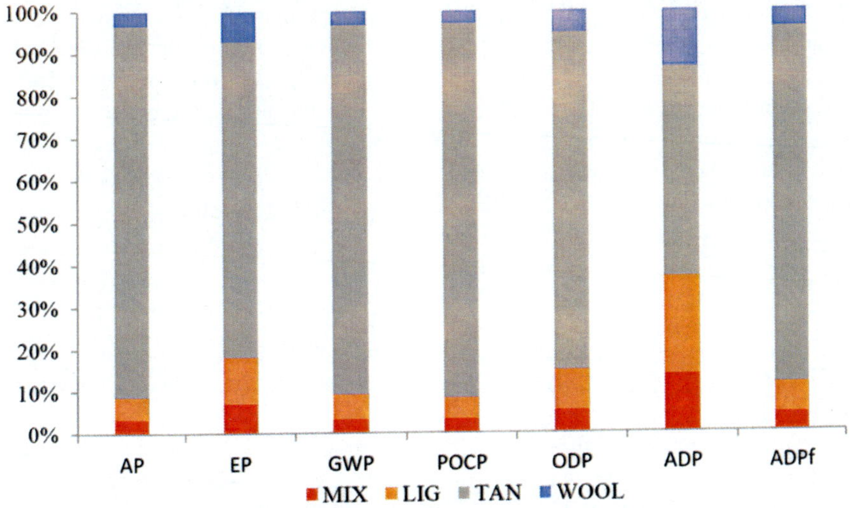

Fig. 14. LCA results for impact categories per unit compressive strength (1 MPa).

The LCA analysis demonstrated that the choice of the stabilizer has a relevant impact on the overall environmental performance. For MIX, most of the impact is attributed to the transportation of sand and its excavation. These impacts are the same and always included in the impacts of other the analyzed walls.

WOOL has the lowest environmental impact with respect to other stabilized walls. Its environmental impact is very close to MIX. However, WOOL has a relative increase in compressive strength of about 6%. Hence, wool as an additive in RE wall can be considered as reinforcement material for some projects that require higher compressive strength performance and enhancement of the material elasticity.

In the case of LIG, the maximum increase is achieved regarding the compressive strength of the material, that is about 38.3% respect to MIX. The overall environmental impact rises respect to MIX and consists of about 5–10% in almost all analyzed categories except the abiotic depletion category (increase by about 33%) due to a high amount of chemicals usage during the production phases of lignin. Lignin sulfonate as an additive in RE wall can be chosen for constructions requiring higher compressive strength performances while at the same time preserving their environmental sustainability.

Despite the relatively high enhancement of the compressive strength by 13%, TAN performs around 95% rise in all environmental categories compared to MIX. This result is attributed to a high energy demand during the evaporation and extraction phases of tannins production. Thus, in this research, based on the environmental assessment, the tannins as additive in RE wall may be dismissed as not environmentally sustainable construction material in case of local soil stabilization.

However, it appears paramount how a rigorous analysis of the real feasibility of any of these investigated cases should be performed. For instance, the choice of a stabilizer may also be considered in terms of the location of the factories producing by-products or waste.

By varying the percentage of RE wall composition, it may affect the gained balance, for example, by changing the sand content, or by reducing the transportation distances. Similarly, sourcing sand from a closer quarry, or having the production of additives closer, might improve overall RE wall performance in all impact categories.

Clearly, the present research has no claim to be exhaustive in terms of the whole set of assessments, both economic and technical, which any company should specifically consider sourcing by-products or waste, and ultimately attain its own goals.

7 Conclusions

In conclusion of the research, it can be noted that two biopolymers have been identified as the most optimal stabilizers for rammed earth construction in terms of both compressive strength and environmental sustainability.

The utilization of lignin sulfonate as a stabilizer of RE results in a significant enhancement of mechanical strength, with an increase of 38.3%. However, this is accompanied by an increase in environmental impacts, with a rise of 10% due to the production phase of the additive. On the other hand, utilizing wool as a stabilizer in RE results in a moderate enhancement of compressive strength (about +6%) and the environmental impact remains comparable to that of unstabilized rammed earth. Tannins provide a discrete improvement in the mechanical strength (+13%), but all the impact in the environmental categories rise around 95%, due to the high energy demand during the production phase. It is worth noting that the transportation distance is a significant determinant of the environmental performance of all RE walls, particularly in relation to the distance between the production site of biopolymers and the construction site.

Therefore, lignin sulfonate and wool stabilizers may be suggested as reinforcement of RE construction materials for projects requiring higher compressive strength, having a reduced environmental impact, while the use of tannins is less recommended due to the strong increase of environmental impact related to its production.

For future investigations, a cradle-to-gate LCA may give a complete overview of the use of the biopolymers and their impact on the whole life cycle of the building, including the effects of the energy consumption and hygrothermal performance of RE.

Acknowledgment. The authors would like to thank Dr.Ing. Claudio Tonin of CNR of Biella for kindly providing the wool fibers from Biella the Wool Company, Minerali Industriali S.r.l., Dal Zotto S.r.l., Silvateam S.p.a, and The Wool Biella Company for the collaboration to suppling the materials used for the study.

References

1. Losini, A.E., et al.: Mechanical and microstructural characterization of rammed earth stabilized with five biopolymers. Materials **15**, 3136 (2022). https://doi.org/10.3390/ma15093136
2. IEA: World Energy Outlook 2022. International Energy Agency (2022). https://www.iea.org/Reports/World-Energy-Outlook-2022/Executive-Summary

3. Villoria Sáez, P., Osmani, M.: A diagnosis of construction and demolition waste generation and recovery practice in the European Union. J. Clean. Prod. **241**, 118400 (2019). https://doi.org/10.1016/j.jclepro.2019.118400
4. Minke, G.: Building with earth, Design and Technology of a Sustainable Architecture. Birkhäuser – Publishers for Architecture (2012)
5. Hale, S.E., et al.: The reuse of excavated soils from construction and demolition projects: limitations and possibilities. Sustainability **13**, 1–15 (2021). https://doi.org/10.3390/su13116083
6. Cascione, V., Maskell, D., Shea, A., Walker, P.: A review of moisture buffering capacity: from laboratory testing to full-scale measurement. Constr. Build. Mater. **200**, 333–343 (2019). https://doi.org/10.1016/j.conbuildmat.2018.12.094
7. Giuffrida, G., Caponetto, R., Nocera, F.: Hygrothermal properties of raw earth materials: a literature review. Sustainability **11**, 5342 (2019). https://doi.org/10.3390/su11195342
8. Redige sous Direction Collectif des Associaiton et Federation Terre Crue, Guide des bonnes pratiques de la constuction en terre crue, complet, edition 13 décembre 2018 (n.d.). https://www.rehabilitation-bati-ancien.fr/espace-documentaire/guide-des-bonnes-pratiques-la-construction-en-terre-crue. Accessed 18 Apr 2022
9. Vissac, A., Bourgès, A., Gandreau, D., Anger, R., Fontaine, L.: Argiles & Biopolymères, les stabilisants naturels pour la construction en terre. CRAterre éditions, Villefontaine, France (2017)
10. Laborel-Préneron, A., Aubert, J.E., Magniont, C., Tribout, C., Bertron, A.: Plant aggregates and fibers in earth construction materials: a review. Constr. Build. Mater. **111**, 719–734 (2016). https://doi.org/10.1016/j.conbuildmat.2016.02.119
11. Losini, A.E., Grillet, A.C., Bellotto, M., Woloszyn, M., Dotelli, G.: Natural additives and biopolymers for raw earth construction stabilization – a review. Constr. Build. Mater. **304**, 124507 (2021). https://doi.org/10.1016/j.conbuildmat.2021.124507
12. Xiao, B., Sun, X.F., Sun, R.: Chemical modification of lignins with succinic anhydride in aqueous systems. Polym. Degrad. Stab. **71**, 223–231 (2001). https://doi.org/10.1016/S0141-3910(00)00133-6
13. Tingle, J.S., Santoni, R.L.: Stabilization of clay soils with nontraditional additives. Transp. Res. Rec. **II**(1819), 72–84 (2003). https://doi.org/10.3141/1819b-10
14. Holladay, J.E., White, J.F., Bozell, J.J., Johnson, D.: Top Value-Added Chemicals from Biomass Volume II—Results of Screening for Potential Candidates from Biorefinery Lignin. Prepared for the U.S. Department of Energy under Contract DE-AC05-76RL01830, Pacific Northwest Natl. Lab. Natl. Renew. Energy Lab. II (2007). https://www.pnnl.gov/main/publications/external/technical_reports/PNNL-16983.pdf
15. Kaal, J., Nierop, K.G.J., Verstraten, J.M.: Retention of tannic acid and condensed tannin by Fe-oxide-coated quartz sand. J. Colloid Interface Sci. **287**, 72–79 (2005). https://doi.org/10.1016/j.jcis.2005.01.104
16. Arkema, R.E.: Tannin in earthen architecture: a case-study on the use of tannin as an amendment for adobe preservation. J. Chem. Inf. Model. **53**, 1689–1699 (2013)
17. Kam, S., et al.: Permeability to water of sintered clay ceramics. Appl. Clay Sci. **46**, 351–357 (2009)
18. Fantilli, A.P., Sicardi, S., Dotti, F.: The use of wool as fiber-reinforcement in cement-based mortar. Constr. Build. Mater. **139**, 562–569 (2017). https://doi.org/10.1016/j.conbuildmat.2016.10.096
19. Petek, B., Marinšek Logar, R.: Management of waste sheep wool as valuable organic substrate in European Union countries. J. Mater. Cycles Waste Manag. **23**(1), 44–54 (2020). https://doi.org/10.1007/s10163-020-01121-3
20. Alessandro, F., Laura, M.: L'isolamento ecoefficiente, Guid. All'uso Dei Mater. Nat. Milano, Ed. Ambient (2006)

21. Aymerich, F., Fenu, L., Meloni, P.: Effect of reinforcing wool fibres on fracture and energy absorption properties of an earthen material. Constr. Build. Mater. **27**, 66–72 (2012). https://doi.org/10.1016/j.conbuildmat.2011.08.008
22. Statuto, D., Sica, C., Picuno, P.: Experimental development of clay bricks reinforced with agricultural by-products. In: Actual Tasks on Agricultural Engineering (2018)
23. Parlato, M.C.M., Porto, S.M.C.: Organized framework of main possible applications of sheep wool fibers in building components. Sustainability **12**, 761 (2020). https://doi.org/10.3390/su12030761
24. Galán-Marín, C., Rivera-Gómez, C., Petric, J.: Clay-based composite stabilized with natural polymer and fibre. Constr. Build. Mater. **24**, 1462–1468 (2010). https://doi.org/10.1016/j.conbuildmat.2010.01.008
25. Losini, A.E., Grillet, A., Vo, L., Dotelli, G., Woloszyn, M.: Biopolymers impact on hygrothermal properties of rammed earth : from material to building scale. Bulging Environ. **232**, 110087 (2023). https://doi.org/10.1016/j.buildenv.2023.110087
26. Maniatidis, V., Walker, P.: A review of rammed earth construction (2003). http://staff.bath.ac.uk/abspw/rammedearth/review.pdf
27. El-Nabouch, R., Bui, Q.B., Plé, O., Perrotin, P.: Characterizing the shear parameters of rammed earth material by using a full-scale direct shear box. Constr. Build. Mater. **171**, 414–420 (2018). https://doi.org/10.1016/j.conbuildmat.2018.03.142
28. Arrigoni Marocco, A.: How sustainable are natural construction materials? Stabilised rammed earth, hempcrete and other strategies to reduce the life-cycle environmental impact of buildings, Doctoral T, Italy, Milan (2017)
29. Burroughs, S.: Recommendations for the selection, stabilization, and compaction of soil for rammed earth wall construction. J. Green Build. **5**, 101–114 (2010). https://doi.org/10.3992/jgb.5.1.101
30. Bernier, E., Lavigne, C., Robidoux, P.Y.: Life cycle assessment of kraft lignin for polymer applications. Int. J. Life Cycle Assess. **18**, 520–528 (2013). https://doi.org/10.1007/s11367-012-0503-y
31. Ding, T., Bianchi, S., Ganne-Chédeville, C., Kilpeläinen, P., Haapala, A., Räty, T.: Life cycle assessment of tannin extraction from spruce bark. IForest **10**, 807–814 (2017). https://doi.org/10.3832/ifor2342-010
32. Moreira Cardoso, A.A.: Life Cycle Assessment of Two Textile Products Wool and Cotton, Porto University, vol. 123 (2013). https://repositorio-aberto.up.pt/bitstream/10216/85778/2/26984.pdf
33. AIB: European Residual Mixes 2011, Association of Issuing Bodies (2011)
34. Panesar, D.K., Francis, J.: Influence of limestone and slag on the pore structure of cement paste based on mercury intrusion porosimetry and water vapour sorption measurements. Constr. Build. Mater. **52**, 52–58 (2014). https://doi.org/10.1016/j.conbuildmat.2013.11.022

Author Index

S. Amziane et al. (Eds.): ICBBM 2023, RILEM Bookseries 45, pp. 1033–1036, 2023.
https://doi.org/10.1007/978-3-031-33465-8